Advances in Intelligent Systems and Computing

Volume 1301

AF864193

The series "Advances in Intelligent Systems and Computing" contains publications on theory, applications, and design methods of Intelligent Systems and Intelligent Computing. Virtually all disciplines such as engineering, natural sciences, computer and information science, ICT, economics, business, e-commerce, environment, healthcare, life science are covered. The list of topics spans all the areas of modern intelligent systems and computing such as: computational intelligence, soft computing including neural networks, fuzzy systems, evolutionary computing and the fusion of these paradigms, social intelligence, ambient intelligence, computational neuroscience, artificial life, virtual worlds and society, cognitive science and systems, Perception and Vision, DNA and immune based systems, self-organizing and adaptive systems, e-Learning and teaching, human-centered and human-centric computing, recommender systems, intelligent control, robotics and mechatronics including human-machine teaming, knowledge-based paradigms, learning paradigms, machine ethics, intelligent data analysis, knowledge management, intelligent agents, intelligent decision making and support, intelligent network security, trust management, interactive entertainment, Web intelligence and multimedia.

The publications within "Advances in Intelligent Systems and Computing" are primarily proceedings of important conferences, symposia and congresses. They cover significant recent developments in the field, both of a foundational and applicable character. An important characteristic feature of the series is the short publication time and world-wide distribution. This permits a rapid and broad dissemination of research results.

Indexed by SCOPUS, DBLP, EI Compendex, INSPEC, WTI Frankfurt eG, zbMATH, Japanese Science and Technology Agency (JST), SCImago.

All books published in the series are submitted for consideration in Web of Science.

More information about this series at http://www.springer.com/series/11156

Tofigh Allahviranloo · Soheil Salahshour ·
Nafiz Arica
Editors

Progress in Intelligent Decision Science

Proceeding of IDS 2020

Springer

Editors
Tofigh Allahviranloo
Faculty of Engineering and Natural Sciences
Bahcesehir University
Istanbul, Turkey

Soheil Salahshour
Faculty of Engineering and Natural Sciences
Bahçeşehir University
Istanbul, Turkey

Nafiz Arica
Faculty of Engineering and Natural Sciences
Bahcesehir University
Istanbul, Turkey

ISSN 2194-5357 ISSN 2194-5365 (electronic)
Advances in Intelligent Systems and Computing
ISBN 978-3-030-66500-5 ISBN 978-3-030-66501-2 (eBook)
https://doi.org/10.1007/978-3-030-66501-2

This Springer imprint is published by the registered company Springer Nature Switzerland AG
The registered company address is: Gewerbestrasse 11, 6330 Cham, Switzerland

Contents

Intelligent Control Systems

Artificial Intelligence and Deep Learning

Data Mining and Knowledge Discovery

Identification of Sport News in Turkish Tweets Using Deep Learning Architectures

Mansur Alp Toçoğlu[1](✉) and Aytuğ Onan[2]

[1] Faculty of Technology, Department of Software Engineering, Manisa Celal Bayar University, 45400 Manisa, Turkey
mansur.tocoglu@cbu.edu.tr

[2] Faculty of Engineering and Architecture, Department of Computer Engineering, İzmir Katip Çelebi University, 35620 İzmir, Turkey
aytug.onan@ikc.edu.tr

Abstract. In social media and microblogging platforms, it is very popular to share news about sport activities from all around the world. This makes it important to extract information about sports out of text crawled from such platforms. In the scope of this paper, a binary classification is presented to identify tweets containing any information of sports in Turkish. To do so, firstly a dataset, composed of two categories (namely, sport and non-sport), is collected from eleven different Twitter news accounts. Afterwards, a preprocess phase takes place to remove the punctuation marks, the extra spaces, and the numeric characters. In the classification phase, accuracy values of four deep-learning architectures (namely, convolutional neural network, recurrent neural network, gated recurrent unit, and long short-term memory) are calculated to show the classification performances of each architecture. At last, the deep learning classification accuracy values are compared to the most commonly used supervised learning algorithms (namely, Naïve Bayes algorithm, Support Vector Machines, Random Forest, Dense Artificial Neural Network and Decision Tree).

Keywords: Sport topic · Deep learning · Machine learning

1 Introduction

As the information and communication technologies are advancing in recent years, large amount of user-generated information becomes available on the Web. This innovative development provides prominent sources for humankind to share their opinions about every aspect of their life by using social media and microblogging platforms [1]. Twitter is one of the most popular microblogging site, where users can easily share and access many actual topics about any kind of information, such as politics [2], sports [3], etc. [4].

In today's world sport has an important part in humans' lives in many forms such as football, volleyball, rugby, etc. Some people do the sport activities professionally and some do as an amateur for an entertainment activity. Besides doing sports, people are

T. Allahviranloo et al. (Eds.): IDS 2020, AISC 1301, pp. 3–13, 2021.
https://doi.org/10.1007/978-3-030-66501-2_1

interested in being up to date about sport activity news from foremost broadcasts. This has been usually done with the purchase of newspapers. However, with the advancement of technology, habits of reading newspapers evolved to a stage, where actual data is obtained by using online newspapers, social media and microblogging platforms. Twitter is one of the most frequently used platform providing information about sport activities from well-known sport broadcast accounts (namely, Fanatik, BeinSport, TRTSpor). This continues to increase day by day as the usage of social media and microblogging platforms take part in humans' daily lives more intensely. In this regard, the automatic identification of sport topics in news can be viewed as a challenging task in natural language processing.

In the literature, there are studies focusing on different type of topics of tweets. However, the identification of sport content in news is not a common research area. In this regard, we present a supervised machine learning classification approach to identify sport activities in Turkish news tweets. To the best of our knowledge, this is the first study about classifying sport activities in Turkish news using supervised machine learning based approach. To do so, firstly we collected tweets of 11 Twitter broadcast accounts in two categories (namely sport and non-sport). In the next step, we examined the classification performances of different stemming approaches (first five characters, Snowball Stemmer), n-gram models (unigram, bigram, trigram), weighting schemes (term frequency-based representation and term frequency-inverse document frequency based representation), traditional machine learning algorithms (Naïve Bayes algorithm, Support Vector Machines, Random Forest, Dense Artificial Neural Network and Decision Tree) and deep learning architectures (Convolutional Neural Network, Recurrent Neural Network, Gated Recurrent Unit, and Long Short-Term Memory).

The rest of this paper is structured as follows: In Sect. 2, studies about classification of tweets in the literature are presented. In Sect. 3, the methodology of the study (namely, dataset collection process, feature engineering and classification algorithms) has been introduced. In Sect. 4, experimental procedure and the empirical results of study have been presented. Finally, Sect. 5 concludes the paper and provides a projection for further studies on this topic.

2 Related Work

In literature, there are studies based on classification of tweets. Sriram et al. [5] focused on classifying tweets to predefined generic classes namely news, events, opinions, deals, and private messages by using author information and domain specific features extraction. In the study [6], the authors introduced a Wikipedia-based classification technique to classify tweets by mapping them onto most similar Wikipedia pages and used the distances between pages to compute semantic distances between tweets. They examined that the proposed technique yielded more accurate results than the alternative techniques. In another study [7], Yerva et al. presented various techniques to classify tweets whether they are related to a company or not by using company profiles. Khan et al. [8] proposed a framework to classify tweets in two categories which are political and non-political. To do so, they collected around 20,000 tweets and demonstrated the applicability of Naïve Bayes algorithm. They used precision efficacy criterion for evaluation of the results. Duwairi and Qarqaz [9] used supervised machine learning technique for sentiment analysis in Arabic tweets. To do so, they collected 10,000 tweets and 500 Facebook reviews in various domains, such as, news and sport. They preprocessed the raw data by removing empty tweets, duplicated tweets, and reviews which have only emoticons in their content. To validate the sentiment of the collected dataset, a number of volunteers are asked to label each document for categories namely positive, negative, neutral, or other. In the study [10], the authors, focused on classifying real time Twitter trending topics into 18 general categories such as sports, politics, technology, etc. to aid users to look at smaller subset of topics in case of searching for any information in Twitter. In the classification phase, they examined two classification methods which are bag-of-word text classification and network based classification by using traditional machine learning algorithms such as Support Vector Machines, ZeroR, etc. classifiers.

3 Methodology

This section presents the dataset collection, feature extraction process and the classification algorithms utilized in the experimental analysis.

3.1 Database Collection and Preprocessing

We utilized Tweepy [11], which is an easy-to-use Python library for accessing the Twitter API, to fetch tweets of news. Table 1 shows the category and broadcast distribution of the news of the dataset used in the empirical analysis. The dataset is composed of two categories (namely, sport and non-sport) from eleven different Twitter Turkish well-known news broadcast accounts. Table 2 shows the descriptive information of the raw dataset. There is a total of 30,597 tweets, where 15,393 of them are labelled as sport news and the rest 15,204 tweets labeled as non-sport news. The total number of vocabularies of the dataset is 428,906 and the number of the total number of unique vocabularies computed as 42,520.

Table 1. Category and broadcast distribution of the news of the dataset used in the analysis.

Sport news		Non-sport news	
Broadcast account	# of news	Broadcast account	# of news
BeinSport	3,166	CNN	2,410
Fanatik	3,058	HaberTurk	2,725
NTVSpor	3,064	Hurriyet	2,147
SportX	2,963	Milliyet	2,597
TRTSpor	3,142	Sozcu	2,583
		TRT	2,742
Total	15,393	Total	15,204

In the preprocess stage, we preprocessed the collected dataset to make it ready for the classification phase of the study. In the first step, we removed all punctuation marks,

Table 2. Descriptive information of the raw dataset used in the analysis.

Category	# of documents	Total vocabulary	Total unique vocabulary
Sport	15,393	242,770	22,190
Non-sport	15,204	186,136	28,625
Total	30,597	428,906	42,520

numeric characters, and extra spaces. Then, we stemmed each term by using a pure Python stemming library Snowball-stemmer (SS) [12] and fixed prefix stemming (FPS) [13]. FPS simply gets the first n characters of a term, trims out the rest. We decided to choose n as five to represent the first five characters (F5). We chose F5, instead of F4 or F7, since it has been shown that it has optimum performance in terms of effectiveness measures among others [13]. At last, we removed stopwords from the relevant dataset by using the Turkish language stopword list provided in Python natural language toolkit (NLTK).

3.2 Feature Extraction Schemes

N-gram modelling is a popular feature representation scheme for language modelling and natural language processing tasks. An n-gram is a contiguous sequence of n items from a given instance of text document. In this scheme, items may be phonemes, syllables, letters, words or characters. In natural language processing tasks, word-based n-grams and character n-grams have been widely utilized. N-gram of size 1 has been referred as unigram, N-gram of size 2 has been referred as bigram and N-gram of size 3 has been referred as trigram. In the empirical analysis, we utilized n-gram models of unigram, bigram and trigram.

In the vector space model (VSM), we have considered two different weighting schemes to represent news documents, namely, term frequency-based representation (TF) and term frequency-inverse document frequency based representation (TF-IDF).

3.3 Classification Algorithms

In the classification stage, we achieved classification results from both traditional supervised machine learning algorithms (namely, Naïve Bayes algorithm, Support Vector Machines, Random Forest, Decision Tree and Dense Artificial Neural Networks) and well-known deep learning architectures (namely, Convolutional Neural Network, Recurrent Neural Network, Long Short-Term Memory Networks and Gated Recurrent Units).

Naïve Bayes algorithm (NB) is a probabilistic classification algorithm based on Bayes' theorem. It has a simple structure due to the assumption of conditional independence. Despite its simple structure, it can be effectively utilized in a wide range of applications, including text mining and web mining [14].

Support Vector Machines (SVM) are supervised learning algorithms that can be utilized to solve classification and regression problems. They can be applied effectively to classify both linear and non-linear data [15]. SVM build a hyperplane in a higher dimensional space to solve classification or regression problem. The hyperplane aims to make a good separation by achieving the largest distance to the nearest training data points of classes (known as functional margin).

Decision Tree (DT) [16] is supervised learning algorithm which creates a tree using the concept of information entropy in the training phase. The tree is formed by two major parts which are nodes and branches. In addition, there are two kind of nodes which are internal and leaf. The internal nodes represent attributes and the leaf nodes stands for classification labels.

Random Forest (RF) is a supervised learning algorithm, combining bagging algorithm and random method of subspace [17]. Decision trees are used as the base learning algorithm in this algorithm. Each tree is constructed based on training data bootstrap samples. A random selection of features is used to provide the variety among the base learners. In response, the model can yield promising learning models on datasets with noisy or irrelevant data.

Artificial Neural Network (ANN) [18] is a supervised machine learning algorithm which is inspired by neurons in brain. It is composed of densely interconnected group of artificial neurons. Each neuron receives a signal and processes it then sends the signal to another neuron connected to it. Neural network consists of three layers which are input, hidden and output layers. Each layer can have more than one neurons.

Convolutional Neural Network (CNN) is type deep neural networks which is widely used in image and video recognition, recommender systems and natural language processing. CNN architecture composes of layers which are embedding, convolution, pooling, flattening and fully connected artificial neural network [19].

Recurrent Neural Network (RNN) is a type of feedforward artificial neural network which can handle variable-length sequence inputs [20]. Unlike traditional feedforward neural networks, RNN uses feedback loops to process sequences in order to maintain memory over time. In the traditional RNN algorithm, recurrent units have very simple structures that have no memory units and additional gates. There is only a simple multiplication of inputs and previous outputs, which is passed through the corresponding activation function. RNN is applicable to tasks of unsegmented, connected handwriting recognition or speech recognition.

Long Short Term Memory (LSTM) is an artificial neural network. Unlike simple RNN, an LSTM recurrent unit contains gates, which are used to maintain memory for long periods of time [21].

Gated Recurrent Units (GRU) is a type of Recurrent Neural Network which has a gating mechanism [22]. GRU is similar to LSTM as they both have forget gates. However, as GRU has less parameters than LSTM as it does not have an output gate.

4 Experimental Procedure and Results

In this section, experimental procedure and experimental results have been presented.

4.1 Experimental Procedure

In the empirical analysis, 10-fold cross validation has been utilized in all cases. In this scheme, the original dataset is randomly divided into ten mutually exclusive folds. Training and testing process are repeated ten times and each part is tested and trained ten times. The results reported in this section are the average results for 10-folds. In the empirical analysis, two different feature extraction methods (namely, TF based representation and TF-IDF weighting scheme) have been taken into consideration. In addition, the performances of four different deep learning architectures (CNN, RNN, LSTM, GRU) are compared to each other and to traditional machine learning algorithms. Unless otherwise stated, we stemmed each term using SS.

4.2 Experimental Results

Figure 1 shows the comparison of SVM classification accuracy values of four different feature sizes and two different schemes (namely, TF and TF-IDF) obtained in the empirical analysis. We achieved the highest predictive performance as 0.968 in terms of accuracy value by setting VSM with 2,000 features and TF-IDF based representation. The calculated accuracy values have an intention of decline as the feature sizes decrease.

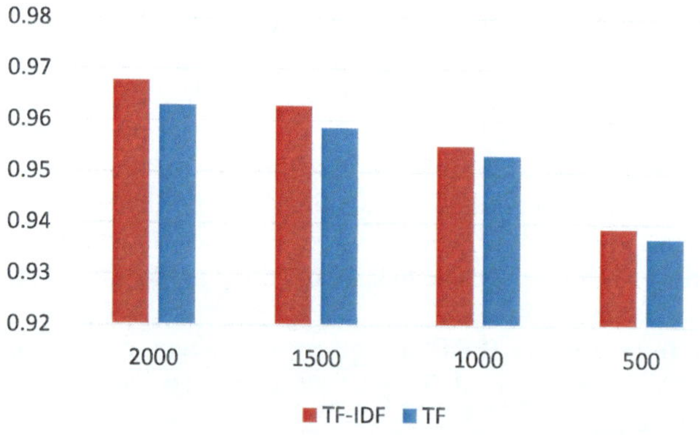

Fig. 1. Comparison of SVM accuracy results of different feature sizes and two different schemes

Table 3 shows the comparison of SVM classifier accuracy values obtained by using different N-gram models (namely unigram, bigram and trigram) and different feature sizes. Beside using only one n-gram model at a time, we also focused on achieving classification results of joint n-grams such as, using unigrams, bigrams and trigrams at the same time which is represented as (uni-bi-tri) grams in the empirical analysis. As it can be observed from the results listed in Table 3, we obtained the highest performance as 0.967 by using single unigram representation with 2,000 feature size and the lowest performance as 0.675 by using single trigram representation with 500 feature size. Generally, single unigram representations provided the highest accuracy results among the other models. The second highest performances are achieved by all joint n-gram models.

Table 3. SVM classifier accuracy results obtained by compared representation of N-gram models and feature sizes.

Feature Size	N-grams representation	Accuracy value
2,000	Unigrams	**0.967709032**
2,000	(uni-bi) grams	0.964343019
2,000	Bigrams	0.91508976
2,000	(uni-bi-tri) grams	0.964146801
2,000	Trigrams	0.750106618
1,500	Unigrams	0.96277406
1,500	(uni-bi) grams	0.960518944
1,500	Bigrams	0.907572693
1,500	(uni-bi-tri) grams	0.960911101
1,500	Trigrams	0.736902337
1,000	Unigrams	0.955060969
1,000	(uni-bi) grams	0.953622975
1,000	Bigrams	0.895675923
1,000	(uni-bi-tri) grams	0.953361142
1,000	Trigrams	0.714154375
500	Unigrams	0.93898102
500	(uni-bi) grams	0.935255167
500	Bigrams	0.861097554
500	(uni-bi-tri) grams	0.935091405
500	Trigrams	0.675066727

Figure 2 shows the comparison of accuracy values obtained by setting feature size of VSM as 2,000 features and using different stemming approaches (namely, F5, SS and raw), two different weighting schemes (namely, TF and TF-IDF), traditional machine learning algorithms (namely, NB, SVM, RF, ANN, DT) and unigram feature representation. In overall results, the highest predictive performances are obtained by using SS stemming approach, TF-IDF based representation and classifiers namely SVM and ANN, whereas the lowest predictive performances are obtained by using the raw dataset, TF based representation and classifiers namely DT and NB. We achieved the highest predictive performance as 0.968 in terms of accuracy by using stemming approach SS, TF-IDF based representation and SVM as the supervised classifier. On the other hand, we obtained the lowest predictive performance as 0.933 in terms of accuracy by using the raw dataset, TF based representation and DT as the supervised classifier.

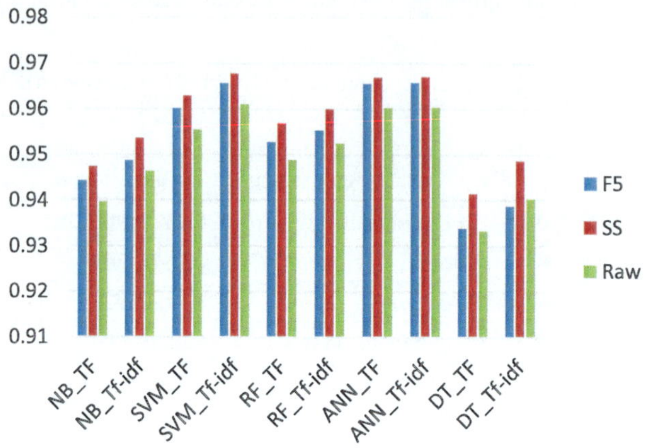

Fig. 2. Comparison of accuracy values of different stemming approaches, two different schemes and traditional machine learning algorithms.

Figure 3 presents the comparison of accuracy values of different stemming approaches (namely, F5, SS and raw) and four different deep learning architectures (namely, CNN, RNN, LSTM and GRU). In all architectures, we set vector size and dimension values as 100. In addition, in all cases we used embedding layer, which does not require stopword elimination. In general, among the stemming approaches, SS provided the highest results whereas the raw dataset, without applying any stemming approach, produced the lowest results. On the other hand, CNN deep learning architecture achieved to obtain the highest predictive performances among the architectures in most cases.

Table 4 shows the comparison of accuracy values of four different deep learning architectures in terms of different vector sizes and dimensions in 50 to 200 value ranges. In overall terms, all architectures provided similar results by using different vector size and dimension values. However, CNN architecture generally achieved slightly higher performances among others.

Figure 4 shows the calculated accuracy values of the traditional machine learning algorithms and four different deep learning architectures in comparison. In general, deep learning architectures provided higher results compared to traditional machine learning classifiers. Among all classifiers, CNN obtained the highest classification results in terms of accuracy value as 0.97, whereas DT algorithm provided the lowest accuracy value as 0.949.

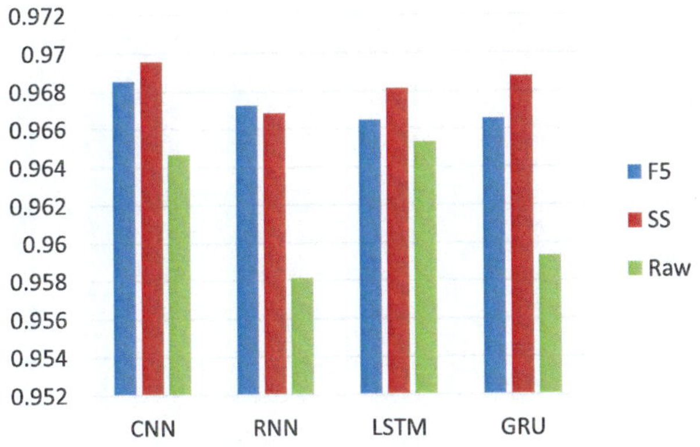

Fig. 3. Comparison of accuracy values of different stemming approaches and four different deep learning architectures.

Table 4. Comparison of accuracy values of four different deep learning architectures in terms of different vector sizes and dimensions.

Vector size	Dimension	CNN	RNN	LSTM	GRU
50	50	0.9708	0.9677	0.9702	0.9677
100	50	0.9707	0.9675	0.9694	0.9667
150	50	**0.9711**	0.9663	0.9695	0.9666
200	50	0.9708	0.9647	0.9697	0.9666
50	100	0.9699	0.9664	0.9684	0.9673
100	100	0.9696	0.9668	0.9681	0.9688
150	100	0.9698	0.9675	0.9671	0.9680
200	100	0.9697	0.9682	0.9664	0.9681
50	150	0.9690	0.9656	0.9680	0.9680
100	150	0.9687	0.9661	0.9663	0.9673
150	150	0.9688	0.9693	0.9690	0.9672
200	150	0.9687	0.9711	0.9674	0.9677
50	200	0.9694	0.9668	0.9708	0.9661
100	200	0.9695	0.9699	0.9688	0.9664
150	200	0.9690	0.9709	0.9694	0.9664
200	200	0.9688	0.9668	0.9684	0.9679

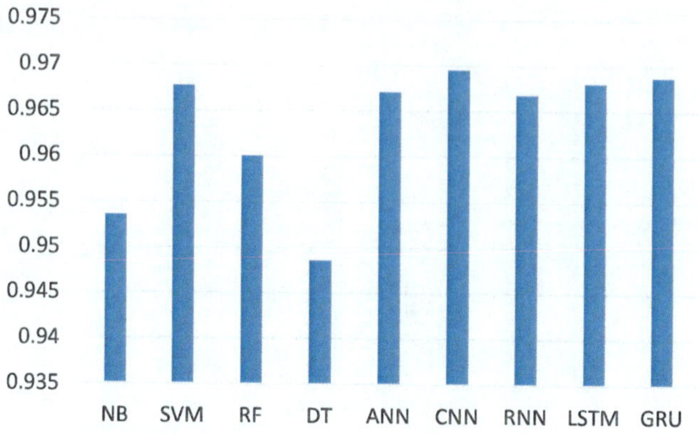

Fig. 4. Comparison of accuracy values of traditional machine learning algorithms and four different deep learning architectures.

5 Conclusion

In this study, we focused on identifying information of sports in Turkish tweets by using binary classification methods. To do so, first we collected a dataset composed of 30,597 tweets from 11 different broadcast accounts containing information of sports and non-sports activities. After the data creation stage, we implemented empirical analysis to compare performances of different stemming approaches (F5, SS, raw), n-gram models (unigram, bigram, trigram), weighting schemes (TF and TF-IDF), traditional machine learning algorithms (NB, SVM, RF, DT, ANN) and deep learning architectures (CNN, RNN, LSTM, GRU). The results indicate that TF-IDF based representation yields better predictive performance for sport identification, compared to the TF based representation. In addition, dataset stemmed by using SS stemming approach provided higher accuracy values in all cases. Generally, single unigram representations provided the highest accuracy results among the other n-gram models. On the other hand, when the results are examined between classifiers, generally deep learning architectures provided higher performances compared to traditional machine learning algorithms.

For future work, we might extend the size of the dataset. In addition, different embedding schemes such as word2vec, fastText, gloVe, and doc2vec can be used to examine their performances.

References

1. Pak, A., Paroubek, P.: Twitter as a corpus for sentiment analysis and opinion mining. In: Proceedings of 7th International Conference on Language Resources and Evaluation, Malta, pp. 1320–1326 (2010)
2. Lin, Y.R., Margolin, D., Keegan, B., Lazer, D.: Voices of victory: a computational focus group framework for tracking opinion shift in real time. In: Proceedings of the 22nd International Conference on World Wide Web, Rio de Janeiro, pp. 737–748. ACM (2013)

3. Nichols, J., Mahmud, J., Drews, C.: Summarizing sporting events using twitter. In: Proceedings of the 17th ACM International Conference on Intelligent User Interfaces, Lisbon, pp. 189–198. ACM (2012)
4. Becker, H., Iter, D., Naaman, M., Gravano, L.: Identifying content for planned events across social media sites. In: Proceedings of the 5th ACM International Conference on Web Search and Data Mining, New York, pp. 533–542. ACM (2012)
5. Sriram, B., Fuhry, D., Demir, E., Ferhatosmanoglu, H., Demirbas, M.: Short text classification in twitter to improve information filtering. In: Proceeding of the 33rd International ACM SIGIR Conference on Research and Development in Information Retrieval, Geneva, pp. 841–842. ACM (2010)
6. Genc, Y., Sakamoto, Y., Nickerson J.V.: Discovering context: classifying tweets through a semantic transform based on Wikipedia. In: Proceedings of the International Conference on Foundations of Augmented Cognition, Orlando, FL, pp. 484–492. Springer (2011)
7. Yerva, S.R., Miklos, Z., Aberer, K.: What have fruits to do with technology? The case of orange, blackberry and apple. In: Proceedings of the International Conference on Web Intelligence, Mining and Semantics, Sogndal, Norway, pp. 1–10. ACM (2011)
8. Khan, I., Naqvi, S.K., Alam, M., Rizvi, S.N.A.: An efficient framework for real-time tweet classification. Int. J. Inf. Technol. **9**(2), 215–221 (2017). https://doi.org/10.1007/s41870-017-0015-x
9. Duwairi, R.M., Qarqaz, I.: Arabic sentiment analysis using supervised classification. In: Proceedings of the International Conference on Future Internet of Things and Cloud, Barcelona, pp. 579–583. IEEE (2014)
10. Lee, K., Palsetia, D., Narayanan, R., Patwary, M.A., Agrawal, A., Choudhary, A.: Twitter trending topic classification. In: Proceedings of the 11th IEEE International Conference on Data Mining Workshops, Vancouver, BC, pp. 251–258. IEEE (2011)
11. Roesslein, J.: Tweepy Documentation. http://docs.tweepy.org/en/latest/. Accessed 14 Apr 2020
12. Porter, M.F.: Snowball: a language for stemming algorithms. http://snowball.tartarus.org/texts/introduction.html. Accessed 14 Feb 2020
13. Can, F., Kocberber, S., Balcik, E., Kaynak, C., Ocalan, H.C., Vursavas, O.M.: Information retrieval on Turkish texts. J. Am. Soc. Inform. Sci. Technol. **59**(3), 407–421 (2008)
14. Zhang, H.: The optimality of Naive Bayes. In: Proceedings of the 17th International Florida Artificial Intelligence Research Society Conference, Florida, pp. 562–567. AAAI Press (2004)
15. Cortes, C., Vapnik, V.: Support-vector networks. Mach. Learn. **20**(3), 273–297 (1995)
16. Quinlan, J.R.: Induction of decision trees. Mach. Learn. **1**, 81–106 (1986)
17. Breiman, L.: Random forests. Mach. Learn. **45**, 5–32 (2001)
18. Caudill, M., Butler, C.: Understanding Neural Networks: Computer Explorations. MIT Press, Cambridge (1992)
19. Krizhevsky, A., Sutskever, I., Hinton, G.E.: ImageNet classification with deep convolutional neural networks. In: Proceedings 25th International Conference on Neural Information Processing Systems, Nevada, pp. 1097–1105. Curran Associates, Inc. (2012)
20. Hochreiter, S., Schmidhuber, J.: Long short-term memory. Neural Comput. **9**(8), 1735–1780 (1997)
21. Sailunaz, K., Alhajj, R.: Emotion and sentiment analysis from Twitter text. J. Comput. Sci. **36**, 101003 (2019)
22. Cho, K., Merrienboer, B.V., Gulcehre, C., Bahdanau, D., Bougares, F., Schwenk, H., Bengio, Y.: Learning phrase representations using RNN encoder-decoder for statistical machine translation. In: Proceedings of Conference on Empirical Methods in Natural Language Processing. arXiv preprint arXiv:1406.1078, Qatar (2014)

Real-Time News Grouping: Detecting the Same-Content News on Turkish News Stream

Yavuz Kömeçoğlu[1], Başak Buluz Kömeçoğlu[2](✉), and Burcu Yılmaz[2]

[1] Kodiks Bilisim, Istanbul, Turkey
komecoglu.yavuz@gmail.com
[2] Gebze Technical University, Kocaeli, Turkey
{bbuluz,byilmaz}@gtu.edu.tr

Abstract. The increasing number of news sources makes analysis of the news difficult and increase the need for automated systems. This paper presents a system that clusters news with similar content in real time. The system uses the Apache Solr database to capture texts from the news source and its MoreLikeThis (MLT) search component to extract the 5 most similar news from thousands of previously recorded news. The new news will be included in the cluster with the most similar of the 5 news obtained by pre-filtering. Therefore, the main problem sought in this study is finding the news that most closely resembles the new news. For this, a 2-step approach has been proposed. The majority of the news published in different sources in the media is created by the same or broadening/shortening of the same text. For this reason, the 'citation rate' was calculated primarily among news pairs. If there are news pairs that exceed the citation threshold, the pair with the highest citation rate is included in the same cluster. Otherwise, the numerical representations of the texts at different levels were used in order to determine the similarity semantically. The results of the study show that the proposed 2-stage approach reduces the sensitivity of embeddings at different levels to text lengths. Thus, it achieved up to 7.6% improvement compared to clustering approach only with embeddings. The system proposed in these study has a structure that can be used in real life applications in terms of real-time clustering with a high F-score rate of over 90%.

Keywords: Real-time news grouping · B-Cubed · Text clustering · Natural language processing

1 Introduction

Natural language processing is a field of research area in computer science, where products reflecting natural language are taken as input and processed in computerized environment. It is possible to list many different problems that are included in this research area and researchers have been developing algorithmic approaches for the solution of these problems for many years. When the studies in this research area are examined, it is

T. Allahviranloo et al. (Eds.): IDS 2020, AISC 1301, pp. 14–25, 2021.
https://doi.org/10.1007/978-3-030-66501-2_2

seen that the classification problems have been studied comprehensively. However, less attention was paid to clustering problems that need to be solved are especially encountered in real life problems. Text clustering is unlike the determination of texts belonging to a certain class, it is the task of identifying groups by combining texts with similar properties but not having a label value.

As a natural result of the widespread use of the Internet, the number and variety of news sources continue to increase exponentially. Processes such as finding what is meaningful for the user and highlighting the important ones among the growing data heap such as avalanche are a great task for media monitoring agencies (MMA). Dozens of editors are employed in MMAs for tasks that require attention and require a lot of manpower, and they are expected to review all the news one by one. However, the fact that the data flows continuously and that the news sources that provide the data flow are increasing day by day, this causes the problem to come to a level that cannot be solved with the employment of the editor. It is precisely at this point that designing automated systems that can group similar news on streaming data and thus increasing productivity while reducing workload is engaged as a solution for MMAs. The main objective of this study is to design a real-time system capable of clustering text-based news with the same content quickly and accurately. The novel contributions of the paper are as follows:

1. *Real-time news clustering system:* We propose a real-time system, which cluster the news that comes instantly from different news sources, with high speed and accuracy and can be easily integrated into real life applications.
2. *Pre-trained word and document vectors:* We created the Turkish News Corpus, which includes 975,809 news text, 17,470,827 sentences and 1,673,884,673 words (the number of unique word is 422,346) and trained word2vec and doc2vec models with this corpus. The pre-trained words and document embeddings obtained as a result of this training will be explicitly shared for use in other studies.
3. *Speed/performance comparison of different levels of text representations in the clustering task:* Representations at the level of word, sentence and document were compared in terms of success and speed for the clustering problem on Turkish texts.
4. *Creating Turkish Test text clustering dataset:* In order to measure the performance of the system, a Grouped Turkish News Texts Test Set consisting of 40,000 news with 5223 groups was created.
5. *Determination of the sensitivity of text representations to text length:* By comparing the measured performances of different levels of text representations using or without the 'citation rate' approach, their sensitivity to text lengths was determined.

This paper is organized as follows. The first section gives a brief overview. The second section includes the related works in the literature. The preprocessing step and the proposed approaches are explained in the third section. Datasets, experiments, and evaluation metric are given in the fourth section. The results of the experiments are evaluated in the fifth section. The article is concluded and future studies are listed in the sixth section. Thanks to the media monitoring agency that provided the data in the last part.

2 Related Works

The Topic Detection and Tracking (TDT) task aims to help search and edit texts that are news from different media platforms and other sources. While achieving this goal, the news with the same content should be clustered as 'stories' as a sub-task [1]. When the studies focusing on the solution of the mentioned sub-task are examined, it is seen that the way of determining the news clusters in a data collection is followed. We divided the related proposed methods in the literature into several groups. These approaches are; vector space models, k-means variations, generative algorithms, dimensionality reduction based methods and phrased based methods [2, 3]. Vector space models were originally developed for the problem of automatic indexing, but they are also successful methods for clustering where clusters are distributed homogeneously and the number of clusters is known. On the other hand, it is known that hierarchical and sectional algorithms always play a dominant role in solving clustering problems. Among these algorithms, k-means and its extensions are frequently used [4–7, 23]. However, the decrease in their effectiveness especially on a large corpus puts these algorithms in a disadvantageous position. The biggest problem of generative algorithms is their sensitivity to outliers and this makes it difficult to work on data that is not homogeneously distributed. Gaussian Model [8], Expectation Maximization [9, 10] and von Mises-Fisher model [11, 12] can be listed for the most known of these generative algorithms that take the number of clusters as input. Clustering algorithms that do not need to take the number of clusters as inputs and work very successfully if the vector model can be expressed as bipartite graph are called "spectral clustering" [13, 14]. Dimension reduction is a preprocessing step that improves clustering performance as well as it can be used for clustering on its own. Text clustering applications of methods such as Principal Component Analysis [15] and Nonnegative Matrix Factorization [16–18], which are frequently used in image processing applications, also draw attention. In addition, Soft Spectral Coclustering [19] and Lingo [20] are examples of other methods that can be clustered with the dimension reduction approach. Unlike other clustering algorithms, clustering methods based on word sequences are referred to as Phrased-based generally have lower cluster performance compared to other methods [21, 22].

Almost all of the case studies of text clustering approaches, which are divided into basic categories, are applications that work on collections of pre-recorded news texts. As in this study, the number of systems running in real time is very low. One of the first of the clustering algorithms working on the flowing data in the literature was published in 2002 and it suggests applying the k-means clustering algorithm to each piece by dividing the data flow into pieces [27]. Another approach proposed as an alternative to this method was presented by Brain B. and others in the following year, and the most noticeable difference in this approach is the use of the 'sliding window' structure [28]. In the study conducted by Linghui G., Jianping Z. and Shiyong Z. in 2011, text aggregation was performed using adaptive text attribute selection on flowing data [25]. In 2018, the BEstream method, which can detect clusters in single-pass flowing data by elliptical function, was introduced by Niwan W., Saranya M. and Chidchanok L. [26]. In another study presented by Miranda S., Zinotins A. and others in the same year, a system that could work with news texts flowing simultaneously in German, Spanish and English languages is proposed [24].

3 Methodology

3.1 Preprocessing

Preprocessing can include many different methods depending on the nature of the problem being studied and the potential of the data source to generate noisy data. The news texts used in the study include the news gathered from the internet news sources and the texts obtained from the printed publications by the optical character recognition (OCR) system. Especially in the texts obtained by scanning from OCR, distorted characters and typographical errors are frequently seen. Correction of these errors instantly, manually or automatically slows down the operation of the system. For this reason, a preprocessing step for the correction of the texts has not been carried out, and despite these misspellings, different approaches mentioned in Sect. 3.2 have been exhibited in order to perform the clustering task successfully. The texts, which are considered as a whole, are reduced to the level of words primarily by tokenization. Then all characters are converted to lowercase and the texts are standardized. The texts are free of all numerical characters and other symbols. Cleaning the stop words is other important preprocessing step and it was also used in this study. The list of stop words cleared in this study contains 438 different words obtained from the Turkish News Texts Corpus.

3.2 Our Approach

In the task of grouping news with the same content, four different approaches have been adopted. The details of these approaches are described below.

Word Vector Based Clustering: Within the scope of the study, the Turkish News Texts Corpus has been created in order to vectorize the words. After the pre-processing steps were applied on this large corpus created without leaving the news domain, the Word2Vec [29] model, which is frequently used in the literature, was trained. Thus, a vector for 422,346 different words in the corpus was created. The preparation of word vectors by training beforehand was used to rapidly vectorize the impressions of news texts to be grouped instantly. The way to take the average of 100-dimensional vectors of the words in the news texts created with the Word2Vec model was followed and the embedding of the entire text was obtained. The cosine similarity was calculated between the vectors of each new news that was instantly included in the stream and the vectors of the 5 most similar news texts recorded in the last 3 days obtained with the help of the MLT query. For the similarity rate, the minimum 90% similarity value was determined as the threshold. The new news is included in the set containing the news text with the highest similarity score (provided that the similarity rate is min. 90%). If none of the news compared exceeds this threshold, the new news is considered to be the first element of a new cluster in which none of the other news is included, and a new cluster id is assigned.

Sentence Vector Based Clustering: Multilingual sentence encoding model based on Transformer and CNN model architectures, which was proposed in 2018 by Google Researchers Yinfei Yang, Daniel Cer and others, was used to create sentence representations. This model has been trained in 16 languages including Turkish simultaneously using a deep learning technique called "Multi-task Dual-Encoder Model" [30]. Thus, sentences of different languages are matched in vector space using the same model. With this important feature, it is possible to examine the semantic similarities or differences between the sentences from different languages supported by the model without any intermediate operation. Unlike the previous method, document embeddings created by sentence vectors instead of word vectors are also clustered with the same threshold value, depending on the similarity of cosine similarity. Pre-trained sentence vectors were used directly from Tensorflow Hub.

Document Vector Based Clustering: Fixed-length feature vector required by many machine learning methods used in different tasks as input was commonly obtained by the Bag-of-words method in applications. Bag-of-words is open to development in terms of the fact that it ignores the semantic structure of the words and does not care about the order of the word. Word2Vec, which is used within the scope of this study and enables great developments in the field of natural language processing, does not lose the relationships and affinities between the words due to the conditional probability principle and semantic similarity. In this respect, it leads to the creation of word representations that can express the semantic integrity to a large extent. Following the Word2Vec method, in 2014, 'Distributed Representations of Sentences and Documents' was introduced by the same team [31]. This time, different from the creation of individual representations of the words that make up a text, it provides to obtain the fixed-size representation vectors of documents of different sizes directly. Only paragraph tokens have been added, with almost no changes to the Word2Vec structure. In this clustering approach, 100-dimensional document vectors created by Doc2Vec. The cluster id likely to belong to the news text determined again based on the cosine similarity metric and 90% threshold value such as word and sentence level clustering approaches.

Two-Stage Clustering: When the news texts flowing continuously from different platform are examined, it is seen that many of them were created by shortening/expanding the original news texts. The representation of new news texts created by shortening and expanding the news has been achieved in previous approaches. However, considering the sensitivity of the methods according to the text size, the similarity with the original text cannot be made properly. In order to reduce this sensitivity, a two-stage approach has been applied shown in Fig. 1. From this point of view, when comparing the news text to be grouped and the news previously included in the system, if a citation rate is determined between each other and this rate is above a certain threshold value, this information will be quite meaningful in determining similarity. For the 'citation rate' approach, let's get the 5 most similar news among the news published in the last 3 days determined by the new news (s1, s2, s3, s4, s5). A short text was determined by measuring the number of words for each the number of words for each element in the news set shown as A = {n, s1, s2, s3, s4, s5}, respectively. One-to-one match ratio of the unique words contained in the long text in the text pair to be compared with the unique words contained in the short text was determined as the "citation rate". The thresholdvalue for citation rate was determined as 80%, and news texts with the highest citation rate were considered to be in the same group. In this second stage, if the no news pair reaches 80% threshold, the similarity of the vectors obtained with the different level text representation methods in terms of cosine similarity is calculated and this time grouped according to 90% threshold value.

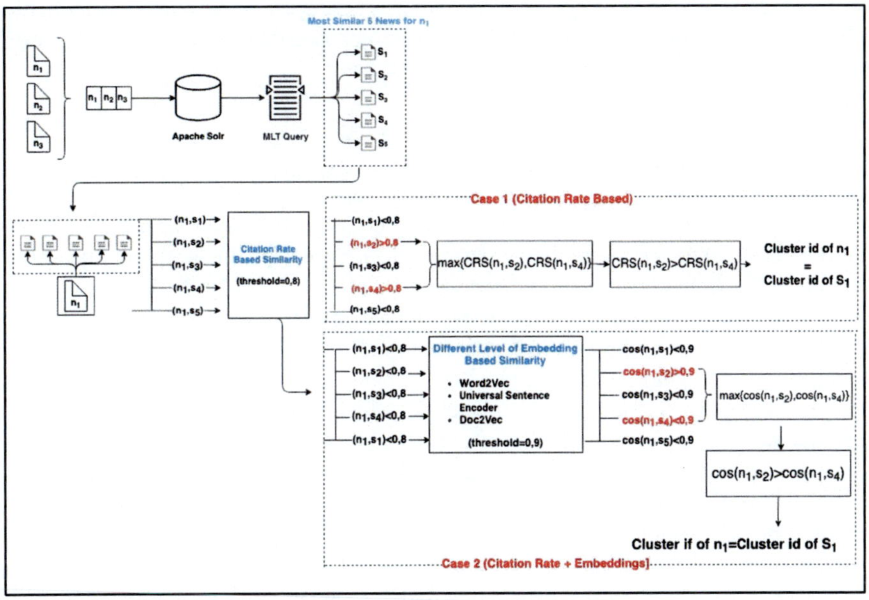

Fig. 1. Architecture of the two-stage similarity approach.

4 Experiments

The test dataset prepared for using within the scope of the study, the Turkish News Texts Corpus used for the training of text representations and the evaluation criteria used in the measurement of performance are discussed in detail in this section.

Turkish News Texts Corpus: In order to give text-based data as input to machine learning methods, numerical representations of these texts are used as expected. The corpus used in the training of the models used in the creation of the representations is quite high and in the same domain. The representations are richer in terms of meaning. Within the scope of this study, 3 different levels of representation, namely word, sentence and document, were created. The Word2Vec used for word embeds and the Doc2Vec models used in the creation of document embeds are trained on the Turkish News Texts Corpus, which consists of 975,809 news texts provided by the Interpress MMA. There are 17,470,827 sentences and 422,346 unique words in the corpus, with a total number of 1,673,884,673 words. For the word embeddings and document embeddings, the corpus was preprocessed before these models were trained.

Grouped Turkish News Texts Test Set: The test dataset was used to measure the performance of grouping news with the same content in real time streaming and was created with 40,000 news texts provided by the Interpress MMA. The news in the test dataset were grouped primarily through two-step grouping proposed in the study. Afterwards, news groups were examined by editors and incorrect groupings were corrected manually. The test dataset, consisting of 5223 different groups and with a total of 40,000 news, was used to simulate a real-time system.

4.1 Evaluation Metrics: B-Cubed

The clustering task has a structure that has complex objectives such as keeping objects closer to each other according to their structural similarities, while keeping objects that are not similar from each other as far away as possible. At this point, besides the internal metrics that take into account the closeness of the elements that need to be in the cluster and the distance from the other clusters, external metrics are generally used, which compare with the gold standard created by humans. In measuring the performance of the system proposed in the study, B-Cubed [33], which is stated to be the most successful external metric to be used in measuring cluster performance [32], was used.

The most important feature that distinguishes B-Cubed from other evaluation metrics is that it can meet all 4 constraints on evaluation metrics for clustering tasks. These constraints are; cluster homogeneity, cluster completeness, rag bag and clusters size versus quantity. B-Cubed metrics parse the evaluation process by predicting precision and recall for each item in the distribution. Item precision refers to how many items in the same set belong to their category. On the other hand, a recall of an item indicates how many items in the item's own category are in the same set shown in Fig. 2.

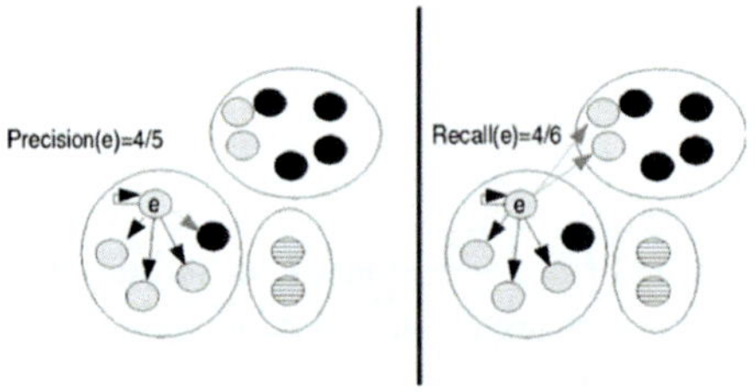

Fig. 2. Example of computing the B-Cubed precision and recall for one item [32].

5 Results and Analysis

The news grouping problem discussed in this study is designed to work on streaming data for real life applications. The proposed system in this study has a two-part structure. In the first part, the news that flows instantly from the news source is recorded in the Apache Solr database, and then 5 news items that have been recorded in the last three days with the help of the MLT query are extracted from the database. In the second part, many different approaches have been tried to perform clustering process quickly and reliably. For clustering, semantic similarity is taken into consideration. For this reason, 3 different level methods in obtaining document representations such as word, sentence and document. The reference news and the 5 most similar news were passed through various pre-processing steps and then document vectors were obtained. By calculating the cosine similarity between the embedding of the reference news and each of the 5 similar news, the similarity of the text pairs was determined. The threshold for the similarity rate is 90%. If the pairs of text have a similarity ratio over 90%, it is likely to be in the same cluster. If this threshold value is exceeded for more than one text, the reference text is included in the same set as the other text with the highest rate. If no 90% similarity ratio can be obtained between any pair of compared text, a new set id for the reference text is given.

It is noteworthy that the texts in different news sources are often exactly the same, or created by expanding or shortening the text in the original news source. For this reason, it was predicted that the rate of one-to-one citation in the system design process would play an important role in comparing the similarity of news content, and the system design was changed. A 2-stage clustering approach was applied. The one-to-one citation rate was obtained by calculating the exact match ratio of the unique words in the shorter news and the longer news in the text pair. Numerical representations of documents were obtained by embedding, if no text pair whose similarity was compared based on the citation rate could exceed the threshold. Then, similarity rates of vector pairs were determined by cosine similarity. 90% threshold value was applied for the similarity rate as previously done. Table 1 shows the precision, recall and f-score values calculated based on the B-Cubed evaluation metric using word, sentence and document embeddings. Determining that the pair has the same content directly when a news pair has a higher citation rate than the threshold value was a factor that increased the system's success in the clustering task. This positive effect is seen in Table 2. One of the most important elements for a real-time system is speed. For this reason, working speeds are as important as the success of the proposed methods. When Table 1 and Table 2 are examined, the first thing to notice is that the 2-stage clustering approach causes a negligible change.

Table 1. The precision, recall and f-score values calculated based on the B-Cubed.

Level of embedding	Similarity	Precision	Recall	F-score	Speed (sn)
Word	Cosine	0.7999	0.9176	0.8547	0.8
Sentence	Cosine	0.9818	0.7885	**0.8746**	0.72
Document	Cosine	0.9996	0.7259	0.8410	**0.59**

Table 2. The precision, recall and f-score values calculated based on the B-cubed for 2-stage clustering.

Level of embedding	Similarity	Precision	Recall	F-score	Speed (sn)
Word + C	Cosine	0.8042	0.9550	0.8731	0.76
Sentence + C	Cosine	0.9833	0.7929	0.8779	0.68
Document + C	Cosine	0.9952	0.8508	**0.9174**	**0.62**

The findings obtained as a result of these methods are listed below:

- The "Multi-task Dual-Encoder Model" performed a more successful clustering performance in terms of F-score compared to other methods in the clustering approach, which was achieved only by obtaining word/sentence/document representations, regardless of the citation rate. However, when evaluated in terms of speed, Doc2Vec was the fastest working method.
- In a 2-step clustering approach, a significant 7.6% F-score increase was observed in the clustering performance achieved by the Doc2Vec method. The F-score of 91.74%, the highest clustering performance was obtained.
- Among the methods used to create word, sentence and document embeds, the greatest increase in performance through citation rate was observed in Doc2Vec. This increase also indicates that Doc2Vec method is sensitive to text lengths.
- 2-stage clustering method has been found to cause a negligible time increase or a positive effect in terms of time. For this reason, it has been observed that using the citation rate for the system designed in real time doesn't cause loss in terms of speed, but it is very useful in terms of clustering performance.
- When the F-score values are compared, Word2Vec has the worst performance in terms of both speed and performance among the proposed methods. This situation did not change for 2-step clustering performances.
- As a result of all tests, it is determined that the most suitable approach for the system proposed in the study in terms of speed/performance is to perform the 2-step clustering method with the help of Doc2Vec document representations.

6 Conclusion and Future Work

A real-time system has been proposed that accurately determines which cluster belongs to the content of news texts from different news sources with an F-score achievement of over 90%. In this system, which is predicted to provide significant benefits compared to manual processes in terms of performance and speed, different levels of text representation methods such as words, sentences and documents were used and time/performance measurement was compared. In order to prevent the sensitivity of text representation methods against the length of the text and thus increase the clustering performance, the 'citation rate' approach is also integrated into the system architecture. With this improvement, a 2-stage clustering method was presented and an F-score increase of up to 7.6% was observed. In real-time systems, the shortest possible processing time and fast return is one of the most important issues. For this purpose, it is planned to carry out studies to increase the speed as much as possible in future studies. Also, after grouping the same news, it is aimed to reduce the manual processes as much as possible by focusing on the categorization of the grouped news texts and automatic summaries (Table 3).

Table 3. The news samples determined to belong to the same cluster by the system proposed in the study.

Cluster: 1	
News 1	Türkiye'de de korona virüsünün 47 kişide pozitif çıkmasıyla birlikte Erzurum'da her gün binlerce insanın ziyaret ettiği tarihi mekanlar boş kaldı. Sağlık Bakanı Fahrettin Kocanın 'evden çıkmayın' uyarıları sonrası Erzurumlular önlem amaçlı evlerinden çıkamazken, birçok turistik bölge de ıssız kaldı
News 2	Koronavirüs endişesi sebebiyle Erzurum'da virüsten korunmak için maske, eldiven, kolonya gibi ürünlerle önlem alan vatandaşlar gerekmediği sürece dışarıya çıkmamaya özen gösteriyor. Sağlık Bakanı Fahrettin Kocanın 'evden çıkmayın' uyarıları sonrası Erzurum'un en önemli simgelerinden birisi olan ve her gün binlerce turistin ziyaret ettiği Çifte Minareli Medrese ve Üç Kümbetler adeta sessizliğe gömüldü
News 3	Korona virüsünün dünyada hızla yayılmasının ardından Türkiye'de de önlemler üst seviyeye çıktı. Erzurumlular önlem amaçlı evlerinden çıkamazken, birçok turistik bölge de ıssız kaldı.Erzurum'un en önemli simgelerinden birisi olan Çifte Minareli Medrese ve Üç Kümbetler adeta sessizliğe gömüldü. Camilerde cemaat halinde ibadet yapmanın askıya alınması üzerine Tarihi Ulu Cami'nin kapısına ise 'Yaygın hastalık sonlanıncaya kadar camimizde ibadetler askıya alınmıştır. Lütfen namazlarınızı evinizde kılınız' notu asıldı
Cluster: 2	
News 1	Galatasaray ile Kasımpaşa, Süper Lig'de yarın 21. kez karşı karşıya gelecek.İki ekip arasında geride kalan 30 lig maçından 19'unu Galatasaray kazandı, 6 müsabaka berabere sonuçlandı, Kasımpaşa 5 kez galip geldi.Galatasaray'ın 62 golüne, Kasımpaşa 29 golle yanıt verdi
News 2	Galatasaray, Süper Lig'in 21. hafta maçında yarın Kasımpaşa ile deplasmanda karşı karşıya gelecek. Recep Tayyip Erdoğan Stadı'nda saat 19.00'da başlayacak maçı hakem Mete Kalkavan yönetecek. İki takım arasında sezonun ilk yarısında maçı Galatasaray, 1-0'lık skorla kazanmıştı
News 3	Kasımpaşa'nın rakibi Galatasaray. Lacivert-beyazlılar, Süper Lig'in 4. haftasında yarın deplasmanda Galatasaray'la karşılaşacak. Lige kötü bir başlangıç yapan ve 3 haftada sadece 1 puan alabilen lacivert-beyazlılar, Türk Telekom Stadı'nda sezonun ilk galibiyetini almak için mücadele edecek

Acknowledgment. We would like to thank to Interpress Media Monitoring Agency, which provides news data for Turkish News Texts Corpus and Grouped Turkish News Texts Test Set

References

1. Staykovski, T., Barrón-Cedeno, A., Da San Martino, G., Nakov, P.: Dense vs. sparse representations for news stream clustering. In: Text2Story@ ECIR, pp. 47–52 (2019)
2. Andrews, N.O., Fox, E.A.: Recent developments in document clustering. Technical report, Virginia Tech. (2007)
3. Blokh, I., Alexandrov, V.: News clustering based on similarity analysis. Proc. Comput. Sci. **122**, 715–719 (2017)
4. Dangre, N., Bodke, A., Date, A., Rungta, S., Pathak, S.S.: System for Marathi news clustering. Proc. Comput. Sci. **92**, 18–22 (2016)
5. Bisandu, D.B., Prasad, R., Liman, M.M.: Clustering news articles using efficient similarity measure and N-grams. Int. J. Knowl. Eng. Data Mining **5**(4), 333–348 (2018)
6. Lwin, M.T., Aye, M.M.: A modified hierarchical agglomerative approach for efficient document clustering system. Am. Sci. Res. J. Eng. Technol. Sci. (ASRJETS) **29**(1), 228–238 (2017)
7. Bouras, C., Tsogkas, V.: Assisting cluster coherency via n-grams and clustering as a tool to deal with the new user problem. Int. J. Mach. Learn. Cybernet. **7**(2), 171–184 (2014)
8. Liu, X., Gong, Y., Xu, W., Zhu, S.: Document clustering with cluster refinement and model selection capabilities. In: Proceedings of the 25th Annual International ACM SIGIR Conference on Research and Development in Information Retrieval, pp. 191–198 (2002)
9. Neal, R., Hinton, G.: A view of the em algorithm that justifies incremental, sparse, and other variants. In: Jordan, M.I. (ed.) Learning in Graphical Models. Kluwer (1998)
10. Liu, X., Gong, Y., Xu, W., Zhu, S.: Document clustering with cluster refinement and model selection capabilities. In: SIGIR 2002: Proceedings of the 25th Annual International ACM SIGIR Conference on Research and Development in Information Retrieval, pp. 191–198. ACM Press (2002)
11. Banerjee, A., Ghosh, J.: Frequency sensitive competitive learning for clustering on high-dimensional hyperspheres. In: IEEE International Joint Conference on Neural Networks, Honolulu, Hawaii, pp. 1590–1595 (2002)
12. Banerjee, A., Dhillon, I.S., Ghosh, J., Sra, S.: Clustering on the unit hypersphere using von Mises-Fisher distributions. J. Mach. Learn. Res. **6**, 1345–1382 (2005)
13. Ravi, K., Santosh, V., Adrian, V.: On clusterings: good, bad and spectral. J. ACM **51**(3), 497–515 (2004)
14. Kummamuru, K., Dhawale, A., Krishnapuram, R.: Fuzzy co-clustering of documents and keywords. In: FUZZ 2003: 12th IEEE International Conference on Fuzzy Systems, pp. 772–777 (2003)
15. Dell, Z., Yisheng, D.: Semantic, hierarchical, online clustering of web search results. In: 6th Asia Pacific Web Conference (APWEB), Hangzhou, China, pp. 69–78 (2004)
16. Wei, X., Xin, L., Yihong G.: Document clustering based on non-negative matrix factorization. In: SIGIR 2003: Proceedings of the 26th Annual International ACM SIGIR Conference on Research and Development in Information Retrieval, New York, NY, USA, pp. 267–273 (2003)
17. Chris, D., Xioafeng, H., Horst D.S.: On the equivalence of nonnegative matrix factorization and spectral clustering. In: Proceedings of SIAM International Conference on Data Mining, pp. 606–610 (2005)
18. Chris, D., Tao, L., Wei, P.: NMF and PLSI: equivalence and a hybrid algorithm. In: SIGIR 2006: Proceedings of the 29th Annual International ACM SIGIR Conference on Research and Development in Information Retrieval, pp. 641–642, New York, NY, USA (2006)
19. Derek, G., Padraig, C.: Producing accurate interpretable clusters from high dimensional data. In: 9th European Conference on Principles and Practice of Knowledge Discovery in Databases, vol. 3721, pp. 486–494. University of Dublin, Trinity College, Dublin (2005)

20. Stanislaw, O., Jerzy, S., Dawid, W.: Lingo: search results clustering algorithm based on singular value decomposition. In: Klopotek, M.A., Wierzchon, S.T., Trojanowski, K. (eds.) Intelligent Information Systems, Advances in Soft Computing, pp. 359–368. Springer (2004)
21. Sven, E., Benno, S., Martin, P.: The suffix tree document model revisited. In: Tochtermann, K., Maurer, H. (eds.) Proceedings of the 5th International Conference on Knowledge Management (I-KNOW 2005), Graz, Austria, pp. 596–603. Know-Center (2005). Journal of Universal Computer Science
22. Hammouda, K.M., Kamel, M.S.: Efficient phrase-based document indexing for web document clustering. IEEE Trans. Knowl. Data Eng. **16**(10), 1279–1296 (2004)
23. Llewellyn, C., Grover, C., Oberlander, J.: Summarizing newspaper comments. In: Eighth International AAAI Conference on Weblogs and Social Media (2014)
24. Miranda, S., Znotiņš, A., Cohen, S.B., Barzdins, G.: Multilingual clustering of streaming news. arXiv preprint arXiv:1809.00540 (2018)
25. Gong, L., Zeng, J., Zhang, S.: Text stream clustering algorithm based on adaptive feature selection. Expert Syst. Appl. **38**(3), 1393–1399 (2011)
26. Wattanakitrungroj, N., Maneeroj, S., Lursinsap, C.: BEstream: batch capturing with elliptic function for one-pass data stream clustering. Data Knowl. Eng. **117**, 53–70 (2018)
27. O'callaghan, L., Mishra, N., Meyerson, A., Guha, S., Motwani, R.: Streaming-data algorithms for high-quality clustering. In: Proceedings 18th International Conference on Data Engineering, pp. 685–694. IEEE (2002)
28. Babcock, B., Datar, M., Motwani, R., O'Callaghan, L.: Maintaining variance and k-medians over data stream windows. In: Proceedings of the Twenty-Second ACM SIGMOD-SIGACT-SIGART Symposium on Principles of Database Systems, pp. 234–243 (2003)
29. Mikolov, T., Sutskever, I., Chen, K., Corrado, G.S., Dean, J.: Distributed representations of words and phrases and their compositionality. In: Advances in Neural Information Processing Systems, pp. 3111–3119 (2013)
30. Chidambaram, M., Yang, Y., Cer, D., Yuan, S., Sung, Y.H., Strope, B., Kurzweil, R.: Learning cross-lingual sentence representations via a multi-task dual-encoder model. arXiv preprint arXiv:1810.12836 (2018)
31. Le, Q., Mikolov, T.: Distributed representations of sentences and documents. In: International Conference on Machine Learning, pp. 1188–1196 (2014)
32. Amigó, E., Gonzalo, J., Artiles, J., Verdejo, F.: A comparison of extrinsic clustering evaluation metrics based on formal constraints. Inf. Retrieval **12**(4), 461–486 (2009)
33. Bagga, A., Baldwin, B.: Entity-based cross-document coreferencing using the vector space model. In: Proceedings of the 17th International Conference on Computational Linguistics, vol. 1, pp. 79–85. Association for Computational Linguistics (1998)

Statistical Analysis of Behavioral Intention Towards Private Umbilical Cord Blood Banking

Sonia Ghayem[1(✉)] and Reza Foudazi[2]

[1] Islamic Azad University - Center Tehran Branch, Tehran, Iran
soniaghayem@gmail.com
[2] Islamic Azad University - South Tehran Branch, Tehran, Iran
rezaf1989@gmail.com

Abstract. In this paper we propose a conceptual framework to identify the key dimensions affecting behavioral intention to bank umbilical cord blood in Iran. We examine the impact of awareness, reference group, usability, disease history, and price on perceived risk and behavioral intention to use umbilical cord blood banking service. To evaluate the proposed model of umbilical cord blood banking behavioral intention and to test our hypotheses, we apply field exploratory research. We use a five-point Likert scale to form a questionnaire to collect the data. The model is estimated with a sample of 242 Royan cord blood bank customers in Tehran. We use Pearson correlation and structural equation modeling to analyze the structural relationships between research variables, perceived risk, and behavioral intention. This research gives novelty on the determinants of behavioral intention in private umbilical cord blood banking, which adds value to literature and future managerial practices. Results show that usability is the primary determinant in cord blood banking.

Keywords: Statistical analysis · Umbilical cord blood banking · Perceived risk · Behavioral intention · Stem cells

1 Introduction

Riding on a trend that the numbers of cord blood banks are growing in the last decade (Lim and Phan 2014), it is envisaged that Iran is looking for a new era of promoting this new technology. In view of the fact that Iran is in the top 10 of countries in the world that produce, culture, and freeze human embryonic stem cells (Shaiegan 2010), it is necessary to undergo research that may assist in explaining the factors influencing behavioral intention in cord blood banking and subsequently hasten the process of commercialization.

The cloning of Dolly the sheep and the multilineage differentiation of human embryonic stem cells have led to improvements in regenerative medicine (Wilmut *et al.* 2007). Umbilical cord blood (UCB) is one of the most absorbing bio-objects in medical applications (Brand *et al.* 2008). Privately banked UCB might offer treatment for regenerative medicine, Parkinson, or diabetes (Goodarzi *et al.* 2015). The use of UCB stem cells has

T. Allahviranloo et al. (Eds.): IDS 2020, AISC 1301, pp. 26–41, 2021.
https://doi.org/10.1007/978-3-030-66501-2_3

no concern with an embryo's destruction. Instead, it is effective in research and therapeutic procedures, which do not involve any moral and ethical concerns (Yang *et al.* 2003). Umbilical cord blood banks store umbilical cord blood stem cells of the newborn for the private use of the family. These stem cells will be used in regenerative medicine. Regenerative medicine suggests the process of restoring or reengineering the structure and function of damaged human cells, tissues, and organs to establish normal function (Armson *et al.* 2015). With the exceptions of countries such as Italy and Spain, the privately run cord blood banks are now operating in almost all parts of the world (Sullivan 2008). One reason for worldwide private cord blood banking is the interest of parents in giving their children biological insurance in circumstances when they develop different health problems in the future (Martin *et al.* 2008). In countries such as Iran, the broad perceptions and ethical concepts of stem cells being unproblematic can open the doors of growth and develop a sense of belief in the donors as well as the common individuals so that they can trust cord blood banks in a better way.

Although several studies revolve around UCB banking points of argument (Gluckman 2009; Fasouliotis and Schenker 2000; Kline and Bertolone 1998) remarkably, no research has examined the key contributing factors in private cord blood banking in the marketing-oriented literature review. Seeing this gap in previous studies, we study the factors motivating people to stock UCB.

Optimization methods use sophisticated analyzes to predict the outcome (see, for example, Mohamadi and Bahrini (2020); Ghassemi (2019); Peykani et al. (2019); Ghassemi et al. (2017); Esmaeili et al. (2015)) and may fail to solve the issue, as simulating a real problem such as in the UCB banking is not easy. Rather, in this research, we used a conceptual statistical model of UCB banking behavioral intention, and tested hypotheses which are created from the model and applied field exploratory research.

The research draws on the concept of perceived risk to propose a conceptual model to address the hypothesized relationships. We propose a conceptual framework to point out whether UCB awareness, reference group, usability, disease history, and price may affect the perceived risk and accordingly behavioral intention to bank UCB stem cells for personal use. The variables were selected based on an extensive review of the literature (Yang *et al.* 2003). The contribution of this study would be a valuable one because the sample population of the study is already composed of private cord blood bank customers, whereas most of the past research studies that have offered insights into this issue revolve around pregnant women at hospitals (Fox *et al.* 2007; Yang *et al.* 2003; Katz *et al.* 2011).

2 Theoretical Background and Hypotheses

In the profit-oriented industry of private UCB banking, patients are considered as consumers, and consumer behavior entails considerable risks and requires prudent decision making because of the uncertainty of the outcome (Fanoodi et al. 2019). The theory of perceived risk was first introduced by Bauer (1960) in the field of consumer behavior. Since 1960 (Cox 1967), this theory was considerably applied in many studies, and the impact of perceived risk in consumer behavior has been affirmed globally. In the case of UCB banking offering state-of-the-art technology, perceived risk was anticipated to play a role. The higher the risk perceived; the fewer consumers are inclined to pay for UCB banking.

2.1 UCB Awareness

Awareness indicates the level of consumers' understanding, recognition, and recall (Aaker 1991). UCB awareness is the level of experience and knowledge of UCB banking and its therapeutic applications (Yang et al. 2003). It is essential to remark that several studies have highlighted the effect of awareness on behavioral intention to use a service (Peykani et al. 2020; Yang et al. 2003; Homburg et al. 2010; Dowling and Stealin 1994). According to Katz et al. (2011), about 92% of all pregnant women had a lack of knowledge toward UCB banking and showed their eagerness to be informed on UCB applications (80% in the UK to 98% in Italy). The findings of Fox et al. (2007), Yang et al. (2003), and Bhundari et al. (2017) also have indicated the noticeably poor understanding of pregnant women about the applications and the use of UCB stem cells in regenerative medicine. Besides, several studies also reported a lack of knowledge among healthcare professionals, which hinders UCB banking (Moustafa and Younes 2015, Hatzistilli et al. 2014). Many studies have tested and obtained support for the knowledge and awareness impact on customers' perceived risk and, consequently, on behavioral intention (see Sharp (1995) and Keller (1998), for example).

Drawing from the literature, when the level of UCB banking awareness is upgraded, we can establish the following hypothesis because high awareness reduces consumer risks:

H_1: UCB banking awareness has a negative effect on perceived risk.

2.2 Reference Group

Based on Park and Lessig (1977), a reference group is a group where the individuals hold their attitudes in conformity to reduce the uncertainty about the outcome. Fishbein and Ajzen (1975) believe that an individual's intention to purchase a product or use a service is affected by the community and the people around them. Numerous studies demonstrated that family members and friends' reference groups could influence behavioral intentions (Turner 1991; Whittler and Spira 2002). As an example of brand selection, see the studies in Moutinho (1987); Bearden and Etzel (1982); Moschis (1976); Park and Lessig (1977). Barkhordari et al. (2019) and Venkatesh and Davis (2000) have proven the influence of social factors on willingness to use new technology.

A high-risk level imbues the UCB banking decision-making process because the benefit is not received at the time of investment. The results of Katz et al. (2011) reveal that the majority of pregnant women (91%) believe that fathers need to be involved in the decision related to bank UCB stem cells (87% in the United Kingdom to 93% in Spain). This shows that fathers remain the basis in the decision-making process, and they should be enlightened regarding the innovative service of UCB banking. Thus, we can establish the following hypothesis:

H_2: A reference group has a negative effect on UCB banking perceived risk.

2.3 Usability

Usability is the degree a person believes that using a particular system would enhance their performance. The UCB banking context is the effectiveness and satisfaction level of users who accept bank UCB stem cells for personal use and result in a favorable outcome. The technology acceptance model is widely used by scholars to explore technology adoption and usage (Koufaris 2002; Moon and Kim 2001). Fox et al. (2007) demonstrated that the women bank their child's UCB stem cells as biological insurance (83%). However, UCB stem cells' private storage remains controversial; the therapeutic use of UCB stem cells for transplantation is rising (Armson et al. 2015). According to Fox et al. (2007), private UCB banking's feasibility is reduced because of the high degree of uncertainty about the banked UCB stem cells' future benefits. The probability that a child will ever use their banked cord blood ranges from 1:1000 to 1:200000, so people should not take the risk (Fisk et al. 2005).

H_3: The usability of UCB has a negative effect on perceived risk.

2.4 Price

Cord blood banking is a procedure that requires fee payment in two phases; enrolment, collection, and storage are the initial charges that cover the first one-year, and the second is the annual storage fee. The initial amount varies from one bank to another, depending on the storage period, from $1000 to $2000. After the first year, storage fees range from $100 to $150 (Yang et al. 2003; Martin et al. 2008). The results of a study in France, Germany, Italy, Spain, and the UK regarding pregnant women and their understanding and feeling of cord blood stem cells and UCB banking showed that due to the costs, more than half of them would not store cord blood in private banks (Katz et al. (2011)). There was a relatively positive response to the view between the buying intentions and the cost of operation being low.

H_4: Price has a positive effect on UCB banking perceived risk.

2.5 Disease History

In the healthcare industry, a family disease history consists of information about disorders from which the patient's direct blood relatives have suffered. When a healthy child with an ill brother is born, and his umbilical cord is preserved, the stem cells are beneficial. It is an accepted fact that UCB storage in families with high risk makes UCB banking cost-effective. Fox et al. (2007) revealed that of his research population, only 5% had a history of a family disease, whereas only 2% had a child who had the disease. Fisk et al. (2005) also found out that a family with no family disease history does not need banked cord blood. Different researches have different views on UCB banking and the disease history influence. Generally speaking, if a family has a history of some diseases such as aplastic anemia, thalassemia, leukemia, and other similar disorders, the need for stem cell transplantation could be higher than the rest of the population. Therefore, there is increased use of UCB banking among the people with a disease, which has reduced the UCB banking perceived risk. The perception of UCB banking not meeting the consumers' expectations among the high-risk families will be reduced (Fisk et al. 2005). Thus, we hypothesize the following:

H_5: Disease history has a negative effect on UCB banking perceived risk.

2.6 Perceived Risk

The amount and nature of risk perceived by consumers in considering a purchase is the definition of perceived risk in business. Researchers such as Stone and Gronhaug (1993) define perceived risk in terms of the uncertainty and results, whereas others (Roselius 1971; Bettman 1973; Dowling and Staelin 1994) consider it as the decision-making importance and any likelihood of negative results that come with the decision. The doubt in technology increases the risks associated with cord blood banking and its small chances of being put to use. According to Littler and Melanthiou (2006), there is a high risk among consumers during the early stages of introducing innovation. The consumers' high-risk perception should be reduced by taking practical actions to attract extensive expert attention and hence advance the UCB banking service development. This will, therefore, be of great practical importance. The consumers' adoption and willingness to UCB banking are highly dependent on their perceived risk, whereas its usability is the leading cause of the consumers' perceived risk (Fisk et al. 2005). Therefore, improving the UCB usability and its perceived risk will positively support the adoption of UCB banking.

2.7 Behavioral Intention

A consumer's willingness to buy or use a service in the future is called behavioral intention (Fishbein and Ajzen 1975). This is emphasized continuously in the literature that if the perceived risk is high, there will be a lower proneness to use a service (Lin 2008; Mitchell 1999; Dowling and Staelin 1994). In the theory of consumers' perceived risk, it is shown that when the consumers were deciding on buying a product, their level of perception would eventually affect their decisions (Taylor 1974). Also, different risk levels would create various changes with additional services and persons. As a result, perceived risk had considerable influence on the consumers' behavioral intention. Through the above discussion, we can establish the following hypothesis:

H_6: There is a negative relationship between perceived risk and intention to bank UCB.

Through the inference of the above literature and hypotheses, we can propose the conceptual model of this research as Fig. 1.

We map marketing concerns about private cord blood banking from those who are already banking UCB stem cells of their children in Royan cord blood bank. What is motivating people to bank UCB stem cells for personal use, and what is their effect on perceived risk? By arraying these research questions, we identify the plausible conceptual model. In some respects, this big picture is simplistic, but it is usefully showing the different factors affecting perceived risk in UCB banking.

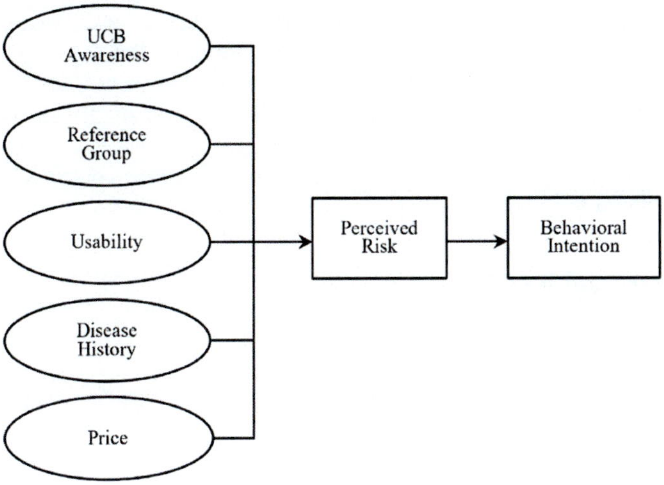

Fig. 1. A conceptual model.

The usefulness of the proposed model lies in its ability to provide a broad view of perceived risk in UCB banking for personal use and to highlight the factors affecting perceived risk to take into account in the development of private cord blood banks.

3 Research Methodology

The study aims to determine causal relationships between variables. Based on the purpose of the study and the data collection method, the research is descriptive and correlational and specifically used SEM to analyze the structural relationship between research variables. In the proposed conceptual model, UCB awareness, reference group, usability, disease history, and price are exogenous latent variables, where perceived risk and behavioral intention are endogenous latent variables. Despite this, UCB awareness, reference group, usability, disease history, and price can be described as independent variables where the perceived risk is an intervening variable, and behavioral intention is the dependent variable.

The survey population consisted of people who already decided to store UCB in Tehran's Royan cord blood bank. The Royan Institute for Reproductive Biomedicine has been chosen as one of the well-versed and leading stem cell institutions in Asia. It has so far been able to store more than 35000 stem cell samples in its UCB bank. The random sampling is employed. The random selection of the sample units is based on the different segments of a population, which has high information, and research characteristics of interest (e.g., Bosch and Wildner 2003; Guarte and Barrios 2006).

Of 450 questionnaires administered, 316 returned in Royan. We discarded 74 incomplete questionnaires, and in the end, we produced 242 valid questionnaires. The valid return rate was around 54%. Each variable is evaluated and measured with 5-point Likert-type scales with anchors of 1 to indicate "strongly disagree" and 5 to indicate "strongly agree." To check how the drivers of UCB banking affect perceived risk and behavioral intention as well as attitudes toward private UCB banking, the questionnaires consisted of two parts: the first part included demographic questions (sex, age, education, and annual income level); the second part focused on knowledge about UCB and examined the preference of participants regarding UCB private banking and the reasons for their choice. We adopted all of the measures from Tanaka (2004) and researches conducted by Yang et al. (2003). We made necessary modifications to address the particular context of the cord blood banking in Iran.

4 Analysis and Results

4.1 Confirmatory Factor Analysis (CFA)

In order to assess reliability, we pretested 40 questionnaires. The Cronbach α reliability coefficient values are shown in Table 1. Only questions 4 and 7 were filtered out.

Table 1. Reliability of each measurement variable.

Variables	Number of measurement variables	Cronbach α	Excluded questions	Cronbach α after excluding
Awareness	3	0.77	Quest. 4	0.78
Price	2	0.63	Quest. 7	0.72
Reference group	3	0.72		
Usability	5	0.73		
Disease history	3	0.81		
Perceived risk	3	0.76		
Behavioral intention	3	0.75		

Since all Cronbach α values were greater than 0.7, all 7 constructs were deemed acceptable (Nunnally 1978). To test the validity, we used factor analysis, which is a statistical technique frequently used in social studies. Exploratory factor analysis of the exogenous and endogenous variables revealed the values of Kaiser-Meyer-Olkin (KMO) 0.84 and 0.82, respectively, which shows the adequacy of sample size for research variables. Furthermore, the significant coefficient of the Bartlett test was zero (less than 0.50). To check the validity of the measurement model, we used the confirmatory factor analysis. The results revealed that all factor loadings were greater than 0.3, which indicates a significant level of convergent validity.

4.2 Investigating the Measurement Model Variables

Before getting into the research hypotheses and conceptual model, it is necessary to ensure the validity of the measured variables. Thus, we measure the exogenous and endogenous latent variables in the following models of research discussed by first-order CFA (see Table 2) The CFA results suggest that our measurement model and variables provide an excellent fit to the data.

Table 2. CFA of the measurement model.

Construct	Measure	Mean	Std dev	Loading factor
UCB awareness	1. UCB banking is essential for the family	4.63	0.54	0.76
	2. I think others should be encouraged to store UCB	4.64	0.55	0.80
	3. I agree that UCB banking should be popularized	4.60	0.60	0.65
Price	4. The operation fee is reasonable	3.52	0.83	0.60
	5. Annual storage fee is reasonable	3.84	0.85	0.94
Reference group	6. Family's and friends' advice to store UCB has great importance	4.54	0.60	0.64
	7. Doctor's recommendation to store UCB has great importance	4.66	0.52	0.75
	8. "Media" suggestion to store UCB has great importance	4.79	0.42	0.70
Usability	9. UCB is viable biological insurance	4.49	0.59	0.64
	10. Own cord blood is thought to be safer than others	4.49	0.67	0.55
	11. The overall chance of using banked UCB for me is high compared with conventional treatments	3.80	0.79	0.55
	12. UCB stem cells can be used to treat many diseases and abnormalities my baby might encounter in his/her life	4.59	0.58	0.61
	13. UCB is useful for treatment	4.47	0.62	0.73
Disease history	14. Hematopoietic and nonhematopoietic diseases in the family have influenced my decision in UCB banking	4.12	0.94	0.73
	15. The anticipation of feeling guilty if the child needs cord blood, and it has not been stored	4.38	0.88	0.81

(continued)

Table 2. (*continued*)

Construct	Measure	Mean	Std dev	Loading factor
	16. I don't have a family history of the disease, but I am affluent enough for this service	4.23	0.85	0.77
Perceived risk	17. Cord blood banking may improve the health of my family	4.14	0.80	0.64
	18. I would participate in cord blood banking only if my relatives or friends do so	3.75	0.75	0.75
	19. I would participate in cord blood banking if cord blood banking technology were mature enough	4.21	0.82	0.77
Behavioral intention	20. I have a good image of cord blood banking	4.14	0.78	0.69
	21. I have specific plans to participate in cord blood banking	4.13	0.74	0.64
	22. I have the intention to use cord blood banking technology	4.26	0.76	0.80

4.3 Correlation Analysis of the Research Variables

For testing the relationship between variables, we used Pearson correlation analysis, where the null hypothesis and hypothesis 1 are as follows:

H_0: The correlation coefficient is zero (no significant relationship between two variables).
H_1: The correlation coefficient is not zero (relationship between two variables).

As the results of Table 3 show, among the independent variables, usability has a significant relationship with perceived risk, and perceived risk has a significant relationship with behavioral intention.

4.4 Evaluation of Structural Research Model (Path Analysis)

The null hypothesis and hypothesis 1 for verification or rejection are as follows:

H_0: There is no significant relationship between two variables.
H_1: There is a significant relationship between two variables.

Table 3. Pearson correlation coefficient analysis

		UCB awareness	Price	Reference group	Usability	Disease history	Perceived risk	Behavioral intention
Awareness	Pearson correlation	1	.25**	.55**	.46**	.13*	-.09	.02
	sig. (2-tailed)	---	.00	.00	.00	.04	.19	.80
Price	Pearson correlation	.25**	1	.20**	.08	-.01	.01	-.03
	sig. (2-tailed)	.00	---	.00	.23	.89	.92	.63
Reference group	Pearson correlation	.55**	.20**	1	.44**	.19**	-.01	.05
	sig. (2-tailed)	.00	.00	---	.00	.00	.88	.41
Usability	Pearson correlation	.46**	.08	.44**	1	.25**	-.24**	.02
	sig. (2-tailed)	.00	.23	.00	---	.00	.00	.76
Disease history	Pearson correlation	.13*	-.01	.19**	.25**	1	-.11	.05
	sig. (2-tailed)	.04	.89	.00	.00	---	.09	.46
Perceived risk	Pearson correlation	-.09	.01	-.01	-.24**	-.11	1	.23**
	sig. (2-tailed)	.19	.92	.88	.00	.09	---	.00
Behavioral intention	Pearson correlation	.02	-.03	.05	.02	.05	.23**	1
	sig. (2-tailed)	.79	.63	.41	.76	.46	.00	---

Sample size for all the analysis is N=242
* Correlation is significant at the 0.05 level (2-tailed).
** Correlation is significant at the 0.01 level (2-tailed).

If the significance value of the test in the regression analysis is greater than 1.96, the null hypothesis is rejected, and hypothesis 1 is verified and vice versa. Table 4 shows the verification or rejection among research variables in summary.

Table 4. Structural equation modeling: causal relationship analysis

Research variables	Effect	Significance	Results
The effect of UCB awareness on perceived risk	0.02	0.13	Not supported
The effect of price on perceived risk	0.02	0.22	Not supported
The effect of the reference group on perceived risk	0.23	1.54	Not supported
The effect of usability on perceived risk	0.41	2.76	Supported
The effect of disease history on perceived risk	0.04	0.48	Not supported
The effect of perceived risk on behavioral intention	0.30	3.34	Supported

In the test of the research hypotheses by structural equation model, the results demonstrate an acceptable model fit: $\chi^2 = 251.74$, DF = 193, RMSEA = 0.036, NFI = 0.90, NNFI = 0.96, PNFI = 0.75, CFI = 0.97, IFI = 0.97, RFI = 0.88, GFI = 0.91, and AGFI = 0.89. Fit statistics are all within the suggested threshold (Browne and Cudeck 1992; Hu and Bentler 1999). Based on the above statistical data, all SEM indicators were acceptable. These results reported in Table 5 confirmed that the conceptual model of perceived risk is an acceptable and appropriate theoretical model for examining the determinants of cord blood banking.

Table 5. Measurement item properties

Research variables	Significance	*T*-value	Mean	Result
UCB awareness	.00	53.87	4.62	Very high
Price	.00	14.17	3.68	High
Reference group	.00	62.04	4.66	Very high
Usability	.00	47.03	4.37	Very high
Disease history	.00	25.58	4.24	Very high
Perceived risk	.00	24.79	4.03	Very high
Behavioral intention	.00	29.86	4.18	Very high

5 Discussion and Conclusions

5.1 Summary of the Results

The focus of this research is getting a clearer understanding of private UCB banking perceived risk determinants in Iran. We have examined the key factors that may impact consumers' perceived risk and the behavioral intention in private cord blood banking.

The results show that 71% of the population have a university degree and 26% have a high school or a diploma degree. These findings recommend that there is a relationship between educational level and UCB banking (Katz *et al.* 2011; Yang *et al.* 2003).

About 71% of the customers are from low-income-level families (less than $750 per month), 22% of customers are from middle-income-level families (an income between $750 and $1500) per month, and 7% have an income above $1500 per month. The results reveal that family income is not a determinant for private cord blood banking in Iran. In other words, the price of private cord blood banking was affordable so that even the families with lower income levels could take part. The finding of this research supports the study of Katz *et al.* (2011), which highlights that the choice of UCB banking is independent of income.

The empirical result of this research revealed the following significance. We classified them based on our hypotheses.

1. Hypothesis 1 of this research: UCB awareness did not reveal a negative influence on the perceived risk. Several studies in the past revolved around knowledge and attitude of pregnant women about UCB banking such as in Canada and Turkey (Fernandez et al. 2003; Dinc and Sahin 2009) and it has been found that women have a lack of knowledge about UCB banking (Huang 2003). The sample population of the past studies was taken from pregnant women in hospitals; however, exceptionally, this research is focused on Royan cord blood bank customers who are already aware of UCB banking. The results show that people who have the intention to bank UCB in Royan are confidently aware of UCB applications. Another possible reason is that UCB banking awareness is not the primary source of perceived risk in Iran, and other factors such as usability, culture, and reputation are critical. Besides, the demographic results of this study revealed that the majority of the respondents were educated, which could have been the reason for a high level of knowledge and UCB awareness.

2. Hypothesis 2 of this research: reference group did not reveal a negative influence on the perceived risk in UCB banking. This can be explained by mentioning the point that UCB banking is a socially accepted behavior in Iran, which has some of the most liberal laws on stem cell research in the world, and both the government and the society show a supportive attitude toward UCB banking for personal use. Although random sampling is used and past studies heavily focused on women's attitude (Fernandez *et al.* 2003; Kharaboyan *et al.* 2007; Dinc and Sahin 2009), surprisingly, 85% of the respondents were men. A reason to explain this result is that in many developing countries such as Iran, men are the moneymakers, and they have the final word in decision making for the family. The role of fathers in the decision-making process in UCB banking has been studied by Katz *et al.* (2011), which supports our findings. The result is the key indicator that to commercialize cord blood banking, not only pregnant women but also fathers should be involved in getting access to appropriate information about UCB banking because they involve an important part of responsibility in deciding whether to bank their child's cord blood.
3. Hypothesis 3 of this research: the empirical findings of this research revealed a significant and negative influence of usability on UCB banking perceived risk. Besides, in the past, the studies reported by several researchers exhibit noticeable uncertainty about UCB banking usability for personal use (Fisk *et al.* 2005). Private cord blood banking has been criticized by giving unreal hopes to future parents to save UCB stem cells with very little chance of getting used in the future (Brown 2007; Fox *et al.* 2007). In line with the literature, the negative influence of usability on perceived risk is verified in this research.
4. Hypothesis 4 of this research: price did not reveal a positive influence on UCB banking perceived risk. According to literature, the price was a barrier to entry of people to private cord blood banks (Katz *et al.* 2011), but in Iran, price is not a determinant for UCB banking. The main reason was the low cost of private UCB banking in Iran, which is roughly $500 for the operation fee and $40 for the annual fee; however, in developed countries such as the United States, private cord blood banking cost ranges from $1000 to $2000 for operation fee and approximately $100 to $200 for an annual fee. Accordingly, we can find out that in a low-cost private cord blood bank, price is not a determining factor of the respondent's choice.
5. Hypothesis 5 of this research: disease history did not reveal a negative influence on UCB banking perceived risk, which responded to the fact that having a disease history in a family would not promote the feasibility of UCB banking for personal use in Iran. This can be explained by pointing out that the price people pay to bank their child's UCB stem cells in Iran compared with that of developed countries is almost half. For that reason, for a family to decide on banking their child's UCB stem cells, having a disease history will not play a role. People in Iran are affluent to privately save their child's UCB stem cells, which is highly commercialized as a "once-in-a-lifetime opportunity" and "biological insurance" that may save their child's life in the future.
6. Hypothesis 6 of this research: perceived risk revealed significant and negative influence on intention to bank UCB. This is because private UCB banking is a high-risk service that is not a necessity for families. This result is in line with several studies,

such as the theory of perceived risk suggested by Taylor (1974) and Lin (2008), which indicates the significant effect of perceived risk on the intention to purchase a product or use a service.

5.2 Implications for Managers

Whether and why consumers will adopt a new medical service should be well understood as a vital insight for managers concerned with marketing innovations. To alleviate perceived risk, in particular, cord blood banks must be directly addressed. The result indicates new paths for future explorations and provides insights to administrators of UCB banking and can be incorporated into government regulatory policy. There is increasing literature focusing on the economization of human biological life, and this case study will make a significant contribution to this. The usability of UCB stem cells is the leading cause of consumers' perceived risk. Therefore, to promote the adoption of UCB banking, service must start from improving the usability and reducing the perceived risk.

5.3 Limitations and Future Research

This research inevitably has limitations. The sample population of this research was taken from Royan cord blood bank customers, and the respondents already had knowledge and awareness about UCB banking for personal use. This paper creates an avenue for future studies, especially in the gray areas identified. The cross-regional samples in future studies may move on to examine whether the customers in developed countries perceive UCB banking differently from those in Iran and other developing countries. The storage of UCB should not be treated as a commercial activity and instead as an intention to advance the development of stem cell research for a safe and healthy life. As we tried to keep the analysis as simple as possible, more advanced statistical techniques such as robust estimation of multivariate processes, which is not sensitive to outlier data, as applied in Ebadi et al. (2011), can also be implemented.

References

Aaker, D.: Managing Brand Equity, pp. 19–32. Free Press, New York (1991)

Armson, B.A., Allan, D.S., Casper, R.F.: Umbilical cord blood: counseling, collection, and banking. J. Obstet. Gynaecol. Can. **37**(9), 832–844 (2015)

Barkhordari, A., Malmir, B., Malakoutikhah, M.: An analysis of individual and social factors affecting occupational accidents. Saf. Health Work **10**(2), 205–212 (2019)

Bauer, R.A.: Consumer behavior as risk taking. In: Hancock, R.S. (ed.) Dynamic Marketing for a Changing World, pp. 389–398. American Marketing Association, Chicago (1960)

Bearden, W.O., Etzel, M.J.: Reference group influence on product and brand purchase decisions. J. Consum. Res. **9**, 183–194 (1982)

Bettman, J.R.: Perceived risk and its components: a model and empirical test. J. Mark. Res. **10**, 184–190 (1973)

Bhandari, R., Lindly, A., Bhatla, D., Babic, A., Mueckle, K., Rao, R., Brooks, P., et al.: Awareness of cord blood collection and the impact on banking. Pediatr. Blood Cancer **64**(7), e26412 (2017)

Bosch, V., Wildner, R.: Optimum allocation of stratified random samples designed for multiple mean estimates and multiple observed variables. Commun. Stat. Theory Methods **32**(10), 1897–1909 (2003)

Brand, A., Rebulla, P., Engelfriet, C.P., Reesink, H.W., Beguin, Y., Baudoux, E., Kögler, G., Ebrahimi, M., Grazzini, G., Costa, A.N., Bosi, A., Sacchi, N., Lombardini, L., Pupella, S., Lecchi, L., Garcidueñas, E.D., van Beckhoven, J.M., de Wit, H.J., Fibbe, W.E., Zhiburt, E.B., Bart, T., Beksaç, M., Navarrete, C., Regan, F.: Cord blood banking. Vox Sang. **95**(4), 335–348 (2008)

Brown, N.: Shifting tenses—from regimes of truth to regimes of hope? Configurations **13**(3), 331–355 (2007)

Browne, M.W., Cudeck, R.: Alternative ways of assessing model fit. Sociol. Methods Res. **21**(2), 230–258 (1992)

Cox, D.F.: Risk Taking and Information Handling in Consumer Behavior. Harvard University Press, Boston (1967)

Dinc, H., Sahin, N.H.: Pregnant women's knowledge and attitudes about stem cells and cord blood banking. Int. Nurs. Rev. **56**(2), 250–256 (2009)

Dowling, G., Staelin, R.: A model of perceived risk and intended risk-handling activity. J. Consum. Res. **21**(6), 119–134 (1994)

Ebadi, M., Shahriari, H., Abdollahzadeh, M.R., Bahrini, A.: Robust estimation of parameters in multivariate processes. In: 2011 IEEE International Conference on Quality and Reliability, pp. 585–589. IEEE, September 2011

Esmaeili, M., Bahrini, A., Shayanrad, S.: Using game theory approach to interpret stable policies for Iran's oil and gas common resources conflicts with Iraq and Qatar. J. Ind. Eng. Int. **11**(4), 543–554 (2015). https://doi.org/10.1007/s40092-015-0120-z

Fanoodi, B., Malmir, B., Jahantigh, F.F.: Reducing demand uncertainty in the platelet supply chain through artificial neural networks and ARIMA models. Comput. Biol. Med. **113**, 103415 (2019)

Fasouliotis, S., Schenker, J.: Human umbilical cord blood banking and transplantation: a state of the art. Eur. J. Obstet. Gynecol. Reprod. Biol. **90**(1), 13–25 (2000)

Fernandez, C.V., Gordon, K., Van den Hof, M., Taweel, S., Baylis, F.: Knowledge and attitudes of pregnant women with regard to collection, testing and banking of cord blood stem cells. Can. Med. Assoc. J. **168**(6), 695–698 (2003)

Fishbein, M., Ajzen, I.: Belief, Attitude, Intention and Behavior: An Introduction to Theory and Research. Addison-Wesley, Reading (1975)

Fisk, N.M., Roberts, I.A., Markwald, R., Mironov, V.: Can routine commercial cord blood banking be scientifically and ethically justified? PLoS Med. **2**(2), 44 (2005)

Fox, N.S., Stevens, C., Ciubotariu, R., Rubinstein, P., McCullough, L.B., Chervenak, F.A.: Umbilical cord blood collection: do patients really understand? J. Perinat. Med. **35**, 314 (2007)

Ghassemi, A., Hu, M., Zhou, Z.: Robust planning decision model for an integrated water system. J. Water Resour. Plan. Manag. **143**(5), 05017002 (2017)

Ghassemi, A.: System of Systems Approach to Develop an Energy-Water Nexus Model Under Uncertainty (Doctoral dissertation) (2019)

Gluckman, E.: History of cord blood transplantation. Bone Marrow Transplant. **44**(10), 621–626 (2009)

Goodarzi, P., Aghayan, H.R., Larijani, B., Soleimani, M., Dehpour, A.R., Sahebjam, M., Ghaderi, F., Arjmand, B.: Stem cell-based approach for the treatment of Parkinson's disease. Med. J. Islamic Republic Iran **29**, 168 (2015)

Guarte, J.M., Barrios, E.B.: Estimation under purposive sampling. Commun. Stat. Simulation Comput. **35**(2), 277–284 (2006)

Hatzistilli, H., Zissimopoulou, O., Galanis, P., Siskou, O., Prezerakos, P., Zissimopoulos, A., Kaitelidou, D.: Health professionals' knowledge and attitude towards the umbilical cord blood donation in Greece. Hippokratia **18**(2), 110 (2014)

Homburg, C., Martin, K., Jens, S.: Brand awareness in business markets: when is it related to firm performance? Int. J. Res. Mark. **27**(3), 201–212 (2010)
Hu, L.T., Bentler, P.M.: Cutoff criteria for fit indexes in covariance structure analysis: conventional criteria versus new alternatives. Struct. Equ. Model.: Multi. J. **6**(1), 1–55 (1999)
Huang, C.C.: A Cognition and Attitude Study for Umbilical Cord Blood Banking: An Example of Working Employees in the Taiwan Hsinchu Science-Based Industrial Park. National Chiao Tung University, Hsin-Chu (2003)
Katz, G., Mills, A., Garcia, J., Hooper, K., McGuckin, C., Platz, A., Rebulla, P., Salvaterra, E., Schmidt, A.H., Torrabadella, M.: Banking cord blood stem cells: attitude and knowledge of pregnant women in five european countries. Transfusion **51**(3), 578–586 (2011)
Keller, K.L.: Strategic Brand Management: Building, Measuring, and Managing Brand Equity, pp. 45–56. Prentice-Hall, New York (1998)
Kharaboyan, L., Knoppers, B.M., Avard, D., Nisker, J.: Knowledge and attitudes of pregnant women with regard to collection, testing and banking of cord blood stem cells. Can. Med. Assoc. J. **168**, 695–698 (2007)
Kline, R.M., Bertolone, S.J.: Umbilical cord blood transplantation: providing a donor for everyone needing a bone marrow transplant. South. Med. J. **91**(9), 821–828 (1998)
Koufaris, M.: Applying the technology acceptance model and flow theory to online consumer behavior. Inf. Syst. Res. **13**(2), 205–223 (2002)
Lim, I.J., Phan, T.T.: Epithelial and mesenchymal stem cells from the umbilical cord lining membrane. Cell Transplant. **23**(4–5), 497–503 (2014)
Lin, W.-B.: Investigation on the model of consumers' perceived risk – integrated viewpoint. Expert Syst. Appl. **34**(2), 977–988 (2008)
Littler, D., Melanthiou, D.: Consumer perceptions of risk and uncertainty and the implications for behaviour towards innovative retail services: the case of internet banking. J. Retail. Consum. Serv. **13**(6), 431–443 (2006)
Martin, P., Nik, B., Andrew, T.: Capitalizing hope: the commercial development of umbilical cord blood stem cell banking. New Genet. Soc. **27**(2), 127–143 (2008)
Mitchell, V.: Consumer perceived risk: conceptualizations and models. Eur. J. Mark. **33**(1/2), 163–195 (1999)
Mohamadi, M., Bahrini, A.: A Nash–Stackelberg equilibrium model for internet and network service providers in the demand market—a scenario-based approach. Wireless Netw. **26**(1), 449–461 (2020). https://doi.org/10.1007/s11276-019-02155-8
Moon, J.W., Kim, Y.G.: Extending the TAM for a world-wide-web context. Inf. Manag. **38**(2), 217–230 (2001)
Moschis, G.P.: Social comparison and informal group influence. J. Mark. Res. **13**, 237–244 (1976)
Moustafa, M.F., Youness, E.M.: Nurses' knowledge about umbilical cord blood banking and it's barriers. IOSR J. Nurs. Health Sci. **4**(2), 44–53 (2015)
Moutinho, L.: Consumer behaviour in tourism. Eur. J. Mark. **21**(10), 5–44 (1987)
Nunnally, J.: Psychometric Theory. McGraw-Hill, New York, NY (1978)
Park, C., Lessig, V.: Students and housewives: differences in susceptibility to reference group influence. J. Consum. Res. **4**(2), 102–110 (1977)
Peykani, P., Mohammadi, E., Emrouznejad, A., Pishvaee, M.S., Rostamy-Malkhalifeh, M.: Fuzzy data envelopment analysis: an adjustable approach. Expert Syst. Appl. **136**, 439–452 (2019)
Peykani, P., Mohammadi, E., Saen, R.F., Sadjadi, S.J., Rostamy-Malkhalifeh, M.: Data envelopment analysis and robust optimization: a review. Expert Syst. e12534 (2020)
Roselius, R.: Consumer rankings of risk reduction methods. J. Mark. **35**(1), 55–61 (1971)
Shaiegan, M.: Stem cell therapy and research status in Iran: at a glance. Iran. J. Blood Cancer **2**(3), 99–100 (2010)
Sharp, B.: Brand equity and market-based assets of professional services firms. J. Prof. Serv. Mark. **13**(1), 3–13 (1995)

Stone, R., Gronhaug, K.: Perceived risk: further considerations for the marketing discipline. Eur. J. Mark. **27**(3), 39–50 (1993)
Sullivan, M.J.: Banking on cord blood stem cells. Nat. Rev. Cancer **8**(7), 555–563 (2008)
Tanaka, Y.: Major psychological factors affecting acceptance of gene-recombination technology. Risk Anal. **24**(6), 1575–1583 (2004)
Taylor, J.W.: The role of risk in consumer behavior. J. Mark. **38**, 54–60 (1974)
Turner, J.C.: Social Influence. Open University Press, Buckingham (1991)
Venkatesh, V., Davis, F.D.: A theoretical extension of the technology acceptance model: four longitudinal field studies. Manag. Sci. **46**(2), 186–204 (2000)
Whittler, T.E., Spira, J.S.: Model's race: a peripheral cue in advertising messages. J. Consum. Psychol. **12**(4), 291–301 (2002)
Wilmut, I., Schnieke, A.E., McWhir, J., Kind, A.J., Campbell, K.H.: Viable offspring derived from fetal and adult mammalian cells. Cloning Stem Cells **9**(1), 3–7 (2007)
Yang, M., Tsai-Chi, R.K., Roseanne, M.J.: The marketing strategies analysis for the umbilical cord blood banking service. Int. J. Health Care Qual. Assur. **16**(6), 293–299 (2003)

Improved Weighted Random Forest for Classification Problems

Mohsen Shahhosseini(✉) and Guiping Hu

Department of Industrial and Manufacturing Systems Engineering, Iowa State University, Ames, IA, USA
mohsen@iastate.edu

Abstract. Several studies have shown that combining machine learning models in an appropriate way will introduce improvements in the individual predictions made by the base models. The key in making well-performing ensemble model is in the diversity of the base models. Of the most common solutions for introducing diversity into the decision trees are bagging and random forest. Bagging enhances the diversity by sampling with replacement and generating many training data sets, while random forest adds selecting random number of features as well. This has made random forest a winning candidate for many machine learning applications. However, assuming equal weights for all base decision trees does not seem reasonable as the randomization of sampling and input feature selection may lead to different levels of decision-making abilities across base decision trees. Therefore, we propose several algorithms that intend to modify the weighting strategy of regular random forest and consequently make better predictions. The designed weighting frameworks include optimal weighted random forest based on accuracy, optimal weighted random forest based on area under the curve (AUC), performance-based weighted random forest, and several stacking-based weighted random forest models. The numerical results show that the proposed models are able to introduce significant improvements compared to regular random forest.

Keywords: Ensemble · Weighted random forest · Optimization · Stacking

1 Introduction

Ensemble learning is defined as a method to combine predictions of multiple machine learning models in order to design a "committee" of decision makers. Intuitively, a committee of decision makers should have lower overall prediction error. This has been investigated in many studies and the results show that improvements can be made on the machine learning models by combining them in an appropriate way [1]. There are several methods to combine base learners including linear/product combination of the learners [2], bagging [3], random forests [4], boosting [5] etc.

It should be noted that including diverse decision makers in the "committee" of base models is necessary to make better judgments, as there will be no improvement from a set of identical models. This is identified as "diversity" of the base learners. It has been proved that well-performing ensemble models demonstrate diversity in base

T. Allahviranloo et al. (Eds.): IDS 2020, AISC 1301, pp. 42–56, 2021.
https://doi.org/10.1007/978-3-030-66501-2_4

learners as well as achieve superior performance individually. There have been a handful of methods to introduce diversity in the ensemble models including but not limited to bagging, random forests, or boosting methods.

In the bagging method proposed in [3], N samples with replacement of the training data are considered and N training data sets are generated. A learning algorithm, traditionally a decision tree, is built on each of the N training data sets and the final prediction is formed by the simple average or voting over the class label. In essence, bagging tries to incorporate diversity in the ensemble model by creating random differences between the input data sets. On the other hand, random forest learning algorithm [4] adds another source of diversity to the bagging procedure. In the random forest method, in addition to sampling with replacement for creating N training data sets, a random number of features are chosen each time to reduce the amount of correlation between created trees. The final result is again simple average or voting over all the predictions made by the constructed trees of the forest (see Fig. 1).

Although random forest has shown remarkable performance and is used widely in many applications, but it seems that some improvements in the way the base learners are combined can be made to make the random forest predictions even better. Naturally, assuming equal weights for all base learners, i.e. using a simple average for the final predictions, does not seem reasonable. The reason lies behind the randomization of sampling and input feature selection. Specifically, this randomization cannot guarantee that all built trees have the same decision-making abilities [6]. Therefore, having a weighting procedure to weight the trees based on their performance appears reasonable.

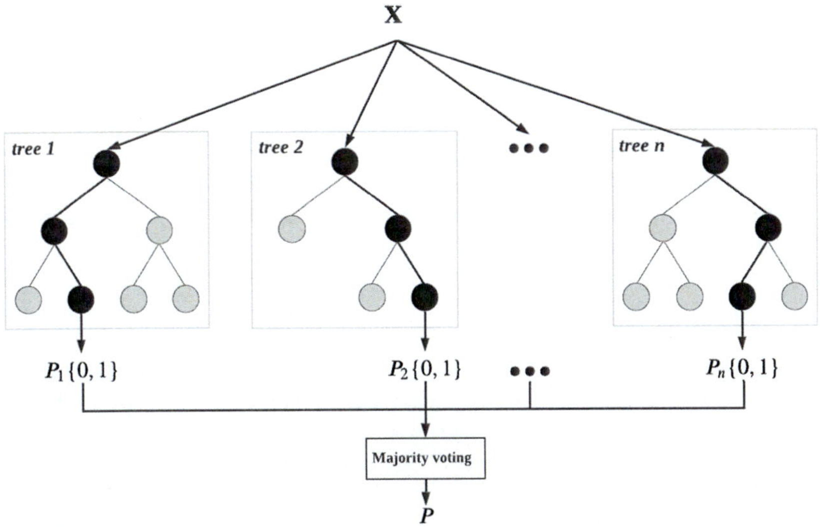

Fig. 1. Random forest classifier uses majority voting of the predictions made by randomly created decision trees to make the final predictions

2 Background

Several studies have proposed variations of weighted random forest and demonstrated modest improvements in their numerical results compared to traditional random forest predictions [7]. For instance, [8] examined some adjustments in the random forest algorithm to boost its performance. The author modified the evaluation measure and used a combination of five attribute evaluation measures instead of Gini index. In addition, this paper identified instances that are most similar to the target instance and proposed a weighted random forest that used the voting margin of the similar instances as the weights. The numerical results on several data sets showed that the proposed approach is able to improve accuracy and AUC in classification problems.

[6] proposed tree weighted random forest (TWRF) method for classifying high dimensional noisy data. They claimed that since random forest gets affected by noisy data and is prone to make wrong decisions, a new approach for weighting the trees according to their classification ability can be regarded as a remedy. They used out-of-bag (OOB) subset of the training data to evaluate the trees of the forest and eventually used OOB accuracies as the tree weights. The results demonstrated superiority of TWRF compared to regular random forest.

In addition, [9] takes the random forest weighting approaches one step ahead by introducing two weight vectors: a weight vector of classifiers and a weight vector of instances. Specifically, they assigned higher weights to hard-to-classify instances, while assigning performance-based weights to the base tree classifiers. Investigating the method on 28 data sets, it was revealed that the proposed weighted voting method outperforms regular random forest with majority voting, significantly.

In another study, [10] proposed a probabilistic weighted scheme for combining trees of the forest considering four combination methods of majority vote, weighted majority vote, recall combiner and the naïve Bayes combiner. Experimental results with 73 data sets demonstrated that the naïve Bayes combiner is slightly better than other candidates, especially for large balanced data sets, while weighted majority vote is a better weighting method for small unbalanced data sets.

Furthermore, there have been several studies that made use of weighted random forest variations in different applications. [11] proposed a weighted random forest model to identify genetic variants associated with complex disease in high-dimensional data. [12] used performance based weighted random forest model to design an automated trading system that improves the profitability and stability of trading seasonality events. An application of weighted random forest in credit card fraud detection was studied in [13]. Other application areas of weighted random forest models include speech language therapy [14], ischemic stroke lesion segmentation [15], survival analysis [16] and etc.

In this study, we propose optimization, performance, and stacking based weighting mechanisms that try to combine the trees of the forest in a more appropriate manner. The prerequisite for integrating the merits of model combinations is having well-diverse input models. To this end, we try to keep the trees of the forest shallow. This prevents having near identical, i.e. not diverse, trees as input features. The designed models include optimal weighted random forest based on accuracy, optimal weighted random forest based on area under the curve (AUC), performance-based weighted random forest, and several stacking-based weighted random forest models.

3 Materials and Methods

As stated, to achieve superior performance for the ensemble model, the initial base learners should be diverse. In other words, the correlation between the base models should be low. Therefore, the assumption of the designed ensemble random forest models is that the trees of the forest should be constructed shallow, i.e. they don't acquire large depths. This results in the base decision trees having an acceptable amount of difference from each other. The designed improved weighted random forest models are explained below.

3.1 Optimal Weighted Random Forest Based on Accuracy

The inspiration to design an optimal weighted random forest comes from the optimization model proposed in [17], which minimized mean squared error (MSE) of a linear combination of several base regressors. Here we propose an optimization model to minimize prediction accuracy of a weighted random forest ensemble model for binary classification, in which the weights are the decision variables. The out-of-bag predictions generated by k-fold cross-validation are considered as emulators of unseen test observations and used as inputs to the optimization problem. The mathematical model is as follows.

$$\begin{gathered} Max\, accuracy\left(Y, \left\lfloor \sum\nolimits_{j=1}^{k} w_j\hat{Y}_j + 0.5 \right\rfloor\right) \\ s.t. \\ \sum\nolimits_{j=1}^{k} w_j = 1, \\ w_j \geq 0,\ \forall j = 1, \ldots, k. \end{gathered} \tag{1}$$

where $w_j's$ are the weights corresponding to decision tree j ($j = 1, \ldots, k$), Y is the vector of actual response values, and $\hat{Y}_j$ is the out-of-bag prediction of decision tree j. *accuracy*() measures the proportion of correct predictions (both true positives and true negatives) among the total number of cases examined. Also, $\left\lfloor \sum\nolimits_{j=1}^{k} w_j\hat{Y}_j + 0.5 \right\rfloor$ calculates the nearest integer among class labels (0 and 1) for the ensemble model (see Fig. 2).

3.2 Optimal Weighted Random Forest Based on Area Under the Curve (AUC)

Area under the ROC curve (AUC) is a measure used mainly for comparing different classifiers. ROC is a popular curve that plots the trade-off between true positive and false positive rates at different classification thresholds. The area under this graph, i.e. AUC, is useful for comparing binary classifiers as it takes into account all possible thresholds. In addition, the accuracy has an intrinsic drawback of reporting very high accuracy when classifying highly imbalanced data sets [18]. The following optimization model intends to find the optimal weights of combining trees of a random forest model by optimizing the ensemble's AUC.

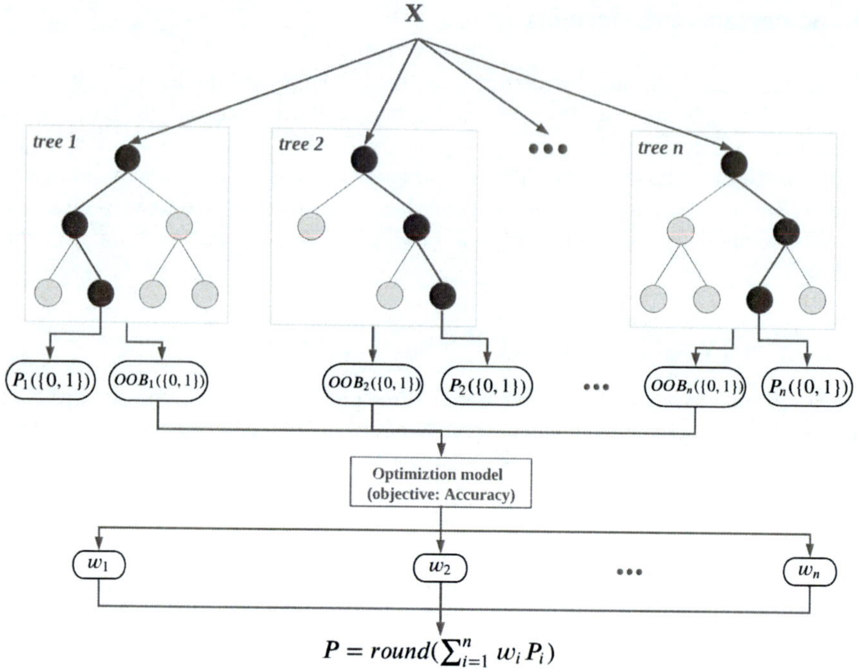

Fig. 2. Optimal weighted random forest classifier uses out-of-bag (OOB) binary predictions of the randomly created decision trees to optimize prediction accuracy

$$
\begin{gathered}
Max\,AUC\left(Y, \sum_{j=1}^{k} w_j \hat{P}_j\right) \\
s.t. \\
\sum_{j=1}^{k} w_j = 1, \\
w_j \geq 0, \forall j = 1, \ldots, k.
\end{gathered} \tag{2}
$$

The $\hat{P}_j$ in the above formulation refers to the out-of-bag probability vector of each base classifier, and $AUC()$ calculates the area under the ROC curve for the built ensemble. Figure 3 depicts the proposed framework to create an optimal weighted random forest using out-of-bag probabilities of true class.

3.3 Performance-Based Weighted Random Forest Model

We design a performance-based weighted random forest models using out-of-bag accuracy. Each classifier's out-of-bag performance is measured with accuracy and the normalized accuracy of each classifier is used as its weight in forming the ensemble. The procedure is demonstrated in Fig. 4.

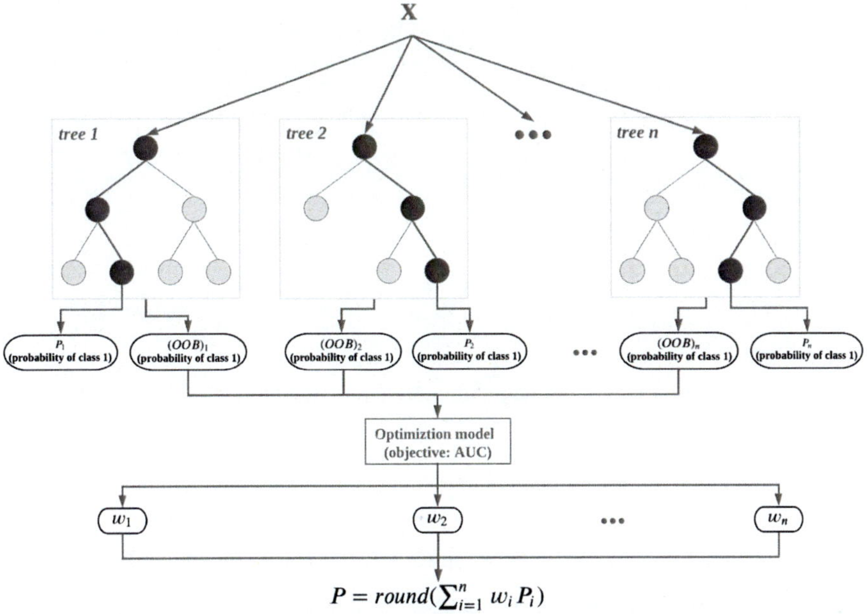

Fig. 3. Optimal weighted random forest classifier uses out-of-bag (OOB) probability predictions of true class made by randomly created decision trees to optimize AUC

3.4 Stacking-Based Random Forest Models

Stacking is a well-known ensemble learning procedure which combines multiple base learners by performing at least another level of learning task. It uses out-of-bag predictions of the base learners, and the actual response values of the training data to form the independent and dependent variables of the 2nd-level learning task, respectively [19]. Here, we use the following steps to make use of the out-of-bag predictions made by trees of the forest and train another machine learning model on top of them to end up with an improved random forest (see Fig. 5).

1. Learn a random forest model using the training data
2. Obtain decision trees of the forest and generate out-of-bag predictions for each of them by using k-fold cross-validation.
3. Create a new data set with out-of-bag predictions as the input variables and actual response values of data points in the training set as the response variable.
4. Learn a second-level machine learning model on the created data set and make predictions for unseen test observations.

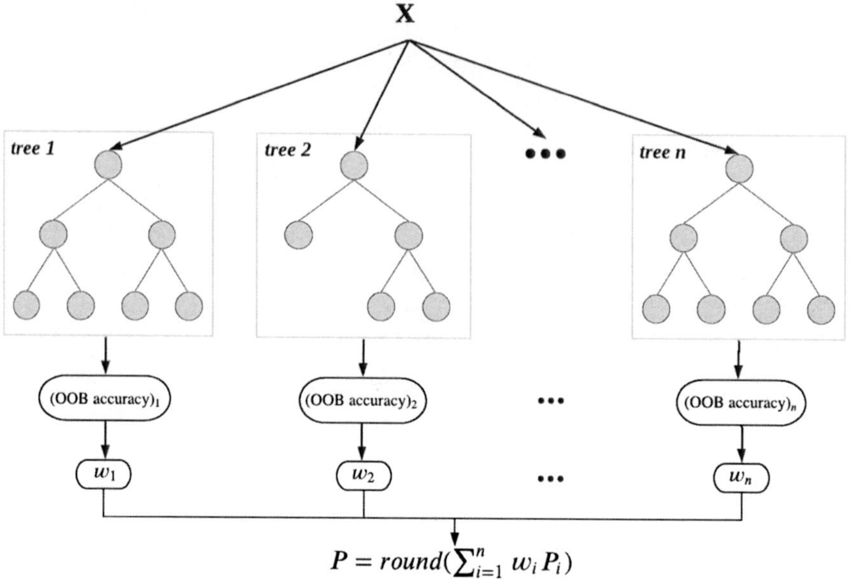

Fig. 4. Performance-based weighted random forest classifier uses accuracy of out-of-bag (OOB) predictions as the ensemble weights

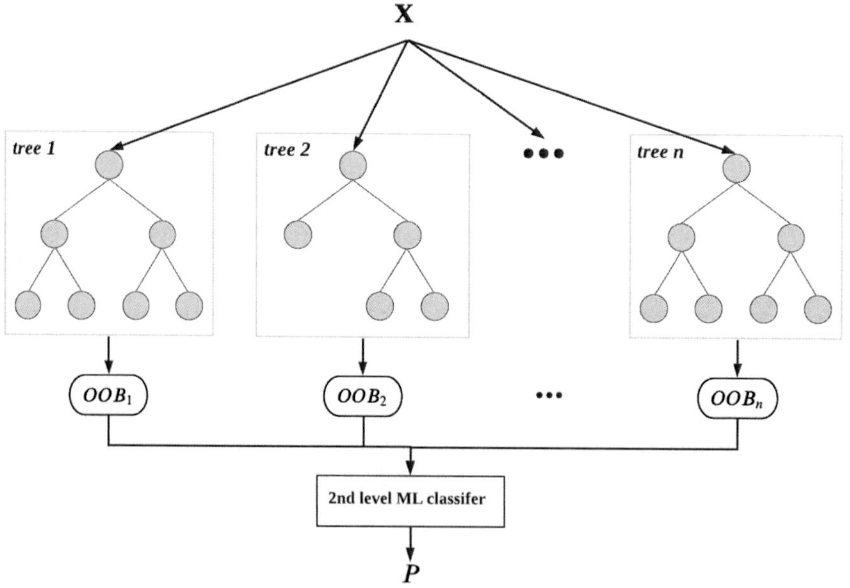

Fig. 5. Stacking-based weighted random forest classifier trains a second level machine learning model on the out-of-bag (OOB) predictions made by each randomly created decision tree

We have selected three machine learning models as the second level classifier which include random forest, logistic regression, and K-nearest neighbors. In addition, two scenarios are considered for each 2nd level classifier: using binary classifications of out-of-bag predictions or using out-of-bag predictions of the probability an observation belongs to the majority class. In the second case instead of having binary predictions as input variables, the probability of the true class (class 1) is used.

4 Results and Discussion

To examine the performance of proposed improved weighted random forest classifiers, 25 public binary classification data sets from UCI machine learning repository [20] were used. Minimal pre-processing tasks including treating missing values, one hot encoding etc. were performed on each data set to make them ready for training classification models. To evaluate the true performance of built models, 25% of each data set is reserved as the test set and the remaining 75% is used to build and optimize ensemble models.

For training the regular random forest and subsequently n randomly created decision trees, the number of trees (n) is set at 100 trees and for keeping the trees shallow and hence not correlated with each other, the maximum depth of the trees is set at the half of the common choice for the maximum depth of random forest trees, i.e. square root of the number of features ($\sqrt{p}/2$). Out-of-bag predictions of the decision trees are generated using 5-fold cross-validation. For solving the optimization problems, Sequential Least Squares Programming algorithm (SLSQP) from Python's SciPy optimization library was used [21]. Finally, the entire process is repeated 50 times to avoid biased results due to the randomness in partitioning the data set into training and test sets, and in forming decision trees. The reported numerical results are the average of 50 runs for each data set.

Table 1 represents the details of 25 example data sets obtained from UCI machine learning repository including the data set size, number of features and the proportion of class labels. As it can be seen in this table, the chosen data sets cover data sets with different levels of size and class label proportions.

The complete experimental results of all designed ensemble models applied on the example data sets are shown in Table 2. Stacking-based ensembles have created in two ways: using binary OOB predictions and using OOB probability predictions of the true class label. As the results suggest in 22 out of 25 considered data sets at least one of the designed models provide improvements over regular random forest (first column of the table). In addition, it appears that the stacking-based random forest with a second random forest model as the second level classifier when having binary OOB binary predictions outperforms other models more often (10 data sets out).

Figure 6 intends to better compare the performance of proposed improved random forest classifiers with regular random forest by depicting average accuracy of designed models over all data sets. Figure 6 (a) presents a chart comparing average accuracy scores of all models with regular random forest classifier. Based on this figure it is evident that RF stacked RF using binary OOB predictions outperforms all designed classifiers as well as regular random forest. Figure 6 (b) compares the average improvements each

Table 1. Details of example data sets downloaded from UCI machine learning repository

#	Data set	Size	Features	Class 0 (%)	Class 1 (%)
1	Acute inflammations [22]	120	6	42%	58%
2	Adult	48842	14	24%	76%
3	APS failure at Scania trucks	60000	171	2%	98%
4	Audit [23]	777	18	39%	61%
5	Australian credit approval	690	14	44%	56%
6	Autism screening adult	704	21	30%	70%
7	Bank marketing [24]	45211	17	12%	88%
8	Breast cancer (Wisconsin)	569	31	37%	63%
9	Breast cancer (Yugoslavia)	286	9	29%	71%
10	Census-income (KDD)	99762	40	6%	94%
11	Cervical cancer (risk factors) [25]	858	36	9%	91%
13	Climate model simulation crashes	540	18	9%	91%
12	Cylinder bands	512	39	42%	58%
14	Default of credit card clients [26]	30000	24	22%	78%
15	Divorce predictors [27]	170	54	49%	51%
16	Drug consumption (quantified) [28]	1885	32	33%	67%
17	EEG eye state	14980	15	45%	55%
18	Electrical grid stability simulated	10000	14	36%	64%
19	Extension of Z-Alizadeh Sani [29]	303	59	29%	71%
20	Hepatitis	155	19	20%	80%
21	Horse colic	368	27	36%	64%
22	HTRU2 [30]	17898	9	9%	91%
23	Indian liver patient dataset (ILPD)	583	10	28%	72%
24	Internet advertisements	3279	1558	14%	86%
25	Ionosphere	351	34	36%	64%

model could provide over regular random forest classifier. This figure suggests that four of the designed models improve the prediction accuracy of regular random forest with RF stacked RF (binary) as the best-performing model. This classifier could improve regular random forest predictions by 0.5%.

Table 2. Experimental results of designed improved random forest classifiers compared to regular random forest classifier. The best-performing classifier for each data set is shown in bold. The last row shows the average accuracy of all models considering all data sets.

#	RF[1]	Optimal WRF (Acc.)[2]	Optimal WRF (AUC)[3]	RF stacked RF (binary)[4]	RF stacked RF (prob.)[5]	Log. stacked RF (binary)[6]	Log. stacked RF (prob.)[7]	KNN stacked RF (binary)[8]	KNN stacked RF (prob.)[9]	Performance based WRF[10]
1	91.00%	98.20%	98.20%	**100.00%**	97.00%	**100.00%**	90.27%	**100.00%**	**100.00%**	97.73%
2	83.75%	82.12%	78.67%	**85.18%**	82.73%	84.67%	83.27%	84.28%	84.75%	82.12%
3	99.21%	99.21%	99.20%	**99.24%**	99.21%	99.21%	99.20%	99.23%	99.20%	99.21%
4	99.69%	99.69%	99.67%	99.94%	**99.95%**	**99.95%**	**99.95%**	**99.95%**	**99.95%**	99.70%
5	85.86%	85.69%	74.42%	84.88%	83.87%	**86.24%**	84.84%	85.11%	85.57%	85.99%
6	99.92%	99.62%	99.62%	**100.00%**	**100.00%**	**100.00%**	**100.00%**	99.91%	99.93%	99.62%
7	88.92%	89.25%	88.43%	**90.04%**	85.44%	90.02%	89.25%	89.74%	89.78%	89.25%
8	94.69%	94.73%	94.22%	94.77%	93.93%	94.71%	94.76%	**94.83%**	94.69%	94.76%
9	74.89%	**75.43%**	73.00%	72.89%	62.49%	72.54%	72.43%	73.26%	73.77%	75.31%
10	94.79%	94.72%	94.60%	95.39%	95.00%	**95.53%**	95.33%	95.06%	95.19%	94.71%
11	**90.49%**	90.47%	90.31%	89.51%	25.13%	90.14%	90.12%	90.22%	90.08%	90.47%
12	91.04%	91.04%	90.79%	90.39%	91.07%	91.32%	91.01%	90.89%	**91.41%**	91.04%
13	72.01%	69.90%	66.67%	**73.67%**	70.06%	69.78%	69.01%	71.42%	71.93%	69.82%
14	80.25%	80.69%	79.07%	81.21%	81.39%	**82.14%**	80.76%	79.69%	79.82%	80.69%
15	97.26%	97.26%	97.26%	97.02%	97.12%	97.35%	**97.40%**	97.26%	97.26%	97.26%
16	70.19%	71.18%	69.28%	69.76%	69.36%	71.22%	70.17%	68.48%	70.32%	**71.29%**

(continued)

Table 2. *(continued)*

#	RF[1]	Optimal WRF (Acc.)[2]	Optimal WRF (AUC)[3]	RF stacked RF (binary)[4]	RF stacked RF (prob.)[5]	Log. stacked RF (binary)[6]	Log. stacked RF (prob.)[7]	KNN stacked RF (binary)[8]	KNN stacked RF (prob.)[9]	Performance based WRF[10]
17	66.29%	64.23%	59.89%	**73.97%**	64.50%	70.70%	66.22%	71.46%	67.68%	64.31%
18	99.97%	96.10%	96.23%	99.97%	99.97%	**99.98%**	99.97%	**99.98%**	**99.98%**	98.52%
19	82.55%	83.47%	77.24%	**84.79%**	81.71%	80.11%	80.29%	83.61%	83.84%	83.76%
20	**83.33%**	83.11%	81.50%	79.56%	79.83%	80.33%	80.11%	80.56%	79.83%	83.00%
21	85.47%	85.23%	77.36%	84.83%	37.09%	82.53%	37.33%	**85.63%**	37.09%	85.25%
22	97.07%	96.78%	95.66%	**97.77%**	97.72%	**97.77%**	97.61%	97.63%	97.72%	96.85%
23	71.74%	71.75%	70.99%	68.10%	52.06%	69.52%	28.15%	68.54%	51.06%	**71.77%**
24	96.88%	96.93%	96.79%	**97.50%**	97.35%	97.26%	97.27%	97.27%	97.37%	96.93%
25	**92.86%**	92.77%	89.66%	92.61%	91.59%	90.32%	90.09%	92.30%	92.00%	92.75%
Ave.	**87.61%**	**87.58%**	**85.55%**	**88.12%**	**81.42%**	**87.73%**	**83.39%**	**87.85%**	**85.21%**	**87.68%**

1 Regular random forest classifier
2 Optimal weighted random forest based on accuracy
3 Optimal weighted random forest based on AUC
4 Stacking-based random forest with random forest as the 2nd level classifier using binary OOB predictions
5 Stacking-based random forest with random forest as the 2nd level classifier using probability OOB predictions
6 Stacking-based random forest with logistic regression as the 2nd level classifier using binary OOB predictions
7 Stacking-based random forest with logistic regression as the 2nd level classifier using probability OOB predictions
8 Stacking-based random forest with KNN as the 2nd level classifier using binary OOB predictions
9 Stacking-based random forest with KNN as the 2nd level classifier using probability OOB predictions
10 Performance-based weighted random forest

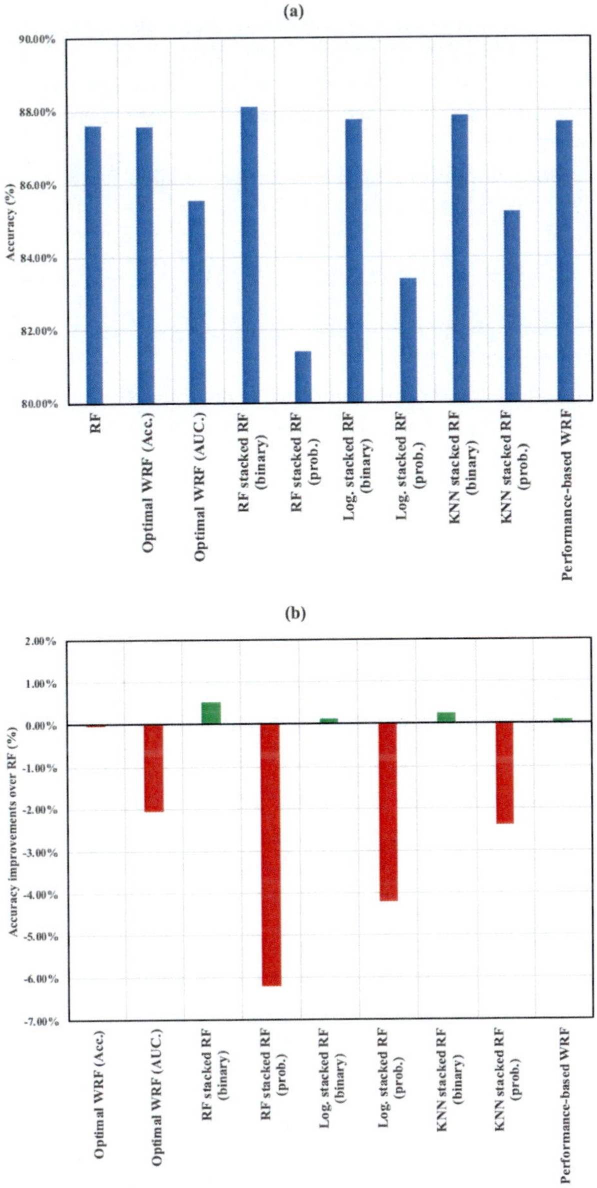

Fig. 6. Comparing proposed improved random forest classifiers with regular random forest

5 Conclusion

This study was an attempt to improve well-performing machine learning model, random forest, as a classifier. To this end, several models based on ensemble learning were designed here. The proposed models include optimal weighted random forest using out-of-bag accuracy and AUC, stacking-based random forest, and performance-based random forest. The experimental results of applying the designed models on 25 public data sets demonstrated that four of the designed models could provide some improvements over the regular random forest classifier. From them, a stacking-based random forest model which trains a 2nd level random forest on inner randomly created decision trees outperformed all designed models. The possible future research directions following this study could be as follows.

- Using other optimization tools to find a better optimal solution for the weights ([31, 32, 33, 34, 35]);
- Adding instance-weighting to the developed framework to assign higher weights to hard-to-classify instances ([8, 9]);
- Developing a similar framework to improve random forest regressor; and
- Applying similar concept on other fields or research such as data envelopment analysis (DEA) ([36, 37, 38, 39])
- Combining bagged and boosted trees to improve both bias and variance of the predictions.

References

1. Brown, G.: Ensemble learning. In: Sammut, C., Webb, G.I. (eds.) Encyclopedia of Machine Learning and Data Mining, pp. 393–402. Springer, US, Boston, MA (2017)
2. Kuncheva, L.I., Classifiers, C.P.: Methods and Algorithms. John Wiley & Sons, New York, NY (2004)
3. Breiman, L.: Bagging predictors. Mach. Learn. **24**(2), 123–140 (1996)
4. Breiman, L.: Random forests. Mach. Learn. **45**(1), 5–32 (2001)
5. Freund, Y., Schapire, R.E.: Experiments with a new boosting algorithm. Paper presented at the ICML (1996)
6. Li, H.B., Wang, W., Ding, H.W., Dong, J.: Trees weighting random forest method for classifying high-dimensional noisy data. In: Paper presented at the 2010 IEEE 7th International Conference on E-Business Engineering, 10–12 November (2010)
7. Pham, H., Olafsson, S.: Bagged ensembles with tunable parameters. Comput. Intell. **35**(1), 184–203 (2019)
8. Robnik-Šikonja, M.: Improving random forests. In: Paper presented at the Machine Learning: ECML 2004, Berlin, Heidelberg (2004)
9. Kim, H., Kim, H., Moon, H., Ahn, H.: A weight-adjusted voting algorithm for ensembles of classifiers. J. Korean Stat. Soc. **40**, 437–449 (2011)
10. Kuncheva, L.I., Rodríguez, J.J.: A weighted voting framework for classifiers ensembles. Knowl. Inf. Syst. **38**(2), 259–275 (2014)
11. Winham, S.J., Freimuth, R.R., Biernacka, J.M.: A weighted random forests approach to improve predictive performance. Stat. Anal. Data Min. ASA Data Sci. J. **6**(6), 496–505 (2013)
12. Booth, A., Gerding, E., McGroarty, F.: Automated trading with performance weighted random forests and seasonality. Expert Syst. Appl. **41**(8), 3651–3661 (2014)
13. Xuan, S., Liu, G., Li, Z.: Refined weighted random forest and its application to credit card fraud detection. In: Paper presented at the Computational Data and Social Networks, Cham

14. Byeon, H., Cha, S., Lim, K.: Exploring factors associated with voucher program for speech language therapy for the preschoolers of parents with communication disorder using weighted random forests, p. 10. Int. J. Adv. Comput. Sci., Appl (2019)
15. Sunil Babu, M., Vijayalakshmi, V.: An effective approach for sub-acute ischemic stroke lesion segmentation by adopting meta-heuristics feature selection technique along with hybrid naive bayes and sample-weighted random forest classification. Sens. Imaging **20**(1), 7 (2019)
16. Utkin, L.V., Konstantinov, A.V., Chukanov, V.S., Kots, M.V., Ryabinin, M.A., Meldo, A.A.: A weighted random survival forest. Knowl.-Based Syst. **177**, 136–144 (2019)
17. Shahhosseini, M., Hu, G., Pham, H.: Optimizing Ensemble Weights and Hyperparameters of Machine Learning Models for Regression Problems. arXiv preprint arXiv:1908.05287(2019)
18. James, G., Witten, D., Hastie, T., Tibshirani, R.: An Introduction to Statistical Learning, vol. 112. Springer, New York (2013)
19. Wolpert, D.H.: Stacked generalization. Neural Networks **5**(2), 241–259 (1992)
20. Dua, D., Graff, C.: UCI Machine Learning Repository Irvine, CA: University of California, School of Information and Computer Science. [http://archive.ics.uci.edu/ml]
21. Jones, E., Oliphant, T., Peterson, P.: SciPy: Open source scientific tools for Python (2001)
22. Czerniak, J., Zarzycki, H.: Application of rough sets in the presumptive diagnosis of urinary system diseases. In: Artificial Intelligence and Security in Computing Systems, pp. 41–51. Springer, Boston, MA (2003)
23. Hooda, N., Bawa, S., Rana, P.S.: Fraudulent firm classification: a case study of an external audit. Appl. Artif. Intell. **32**(1), 48–64 (2018)
24. Moro, S., Cortez, P., Rita, P.: A data-driven approach to predict the success of bank telemarketing. Decis. Support Syst. **62**, 22–31 (2014)
25. Fernandes, K., Cardoso, J. S., Fernandes, J.: Transfer learning with partial observability applied to cervical cancer screening. In: Iberian Conference on Pattern Recognition and Image Analysis, pp. 243–250. Springer, Cham, June 2017
26. Yeh, I.C., Lien, C.H.: The comparisons of data mining techniques for the predictive accuracy of probability of default of credit card clients. Expert Syst. Appl. **36**(2), 2473–2480 (2009)
27. Yöntem, M.K., Adem, K., İlhan, T., Kılıçarslan, S.: Divorce prediction using correlation based feature selection and artificial neural networks. Nevşehir Hacı Bektaş Veli Üniversitesi SBE Dergisi **9**(1), 259–273 (2019)
28. Fehrman, E., Muhammad, A.K., Mirkes, E.M., Egan, V., Gorban, A.N.: The five factor model of personality and evaluation of drug consumption risk. In: Data Science, pp. 231–242. Springer, Cham (2017)
29. Alizadehsani, R., Habibi, J., Hosseini, M.J., Mashayekhi, H., Boghrati, R., Ghandeharioun, A., Sani, Z.A.: A data mining approach for diagnosis of coronary artery disease. Comput. Methods Programs Biomed. **111**(1), 52–61 (2013)
30. Lyon, R.J., Stappers, B.W., Cooper, S., Brooke, J.M., Knowles, J.D.: Fifty years of pulsar candidate selection: from simple filters to a new principled real-time classification approach. Mon. Not. R. Astron. Soc. **459**(1), 1104–1123 (2016)
31. Donate, J.P., Cortez, P., SáNchez, G.G., De Miguel, A.S.: Time series forecasting using a weighted cross-validation evolutionary artificial neural network ensemble. Neurocomputing **109**, 27–32 (2013)
32. Peykani, P., Mohammadi, E., Saen, R.F., Sadjadi, S.J., Rostamy-Malkhalifeh, M.: Data envelopment analysis and robust optimization: a review. Expert Syst. e12534 (2020)
33. Cielen, A., Peeters, L., Vanhoof, K.: Bankruptcy prediction using a data envelopment analysis. Eur. J. Oper. Res. **154**(2), 526–532 (2004)
34. Peykani, P., Mohammadi, E., Emrouznejad, A., Pishvaee, M.S., Rostamy-Malkhalifeh, M.: Fuzzy data envelopment analysis: an adjustable approach. Expert Syst. Appl. **136**, 439–452 (2019)

35. Peykani, P., Mohammadi, E., Pishvaee, M.S., Rostamy-Malkhalifeh, M., Jabbarzadeh, A.: A novel fuzzy data envelopment analysis based on robust possibilistic programming: possibility, necessity and credibility-based approaches. RAIRO-Oper. Res. **52**(4–5), 1445–1463 (2018)
36. Zheng, Z., Padmanabhan, B.: Constructing ensembles from data envelopment analysis. INFORMS J. Comput. **19**(4), 486–496 (2007)
37. Peykani, P., Mohammadi, E.: Window network data envelopment analysis: an application to investment companies. Int. J. Ind. Math. **12**(1), 89–99 (2020)
38. Hong, H.K., Ha, S.H., Shin, C.K., Park, S.C., Kim, S.H.: Evaluating the efficiency of system integration projects using data envelopment analysis (DEA) and machine learning. Expert Syst. Appl. **16**(3), 283–296 (1999)
39. Peykani, P., Mohammadi, E., Seyed Esmaeili, F.S.: Stock evaluation under mixed uncertainties using robust DEA model. J. Qual. Eng. Prod. Optim. **4**(1), 73–84 (2019)

Ontology-Based Elements Evaluation for Emergency Response Apps: A Delphi Technique Approach

Mohd Khairul Azmi Hassan[1(✉)] and Norshita Mat Nayan[2]

[1] Kulliyyah of Information and Communication Technology, International Islamic University Malaysia, Gombak, Malaysia
mkazmi@iium.edu.my

[2] Institute of IR4.0, National University of Malaysia, Bangi, Malaysia
norshitaivi@ukm.edu.my

Abstract. *Objectives.* This article explains the findings of the suitability and completeness of our Communication and Tracking Ontology (CTO) by evaluating the input using Delphi methodology on emergency response experts with more than five years of ground-based expertise, experience in assisting rescue operations, and participation in the disaster control centre.

Methods. The study was carried out in three rounds by getting and analysing the opinions from experts. The research question was identified and circulated to all participant. SPSS application software was then used to analyse the reports from experts by evaluating the min, median, as well as interquartile. Therefore, all objects were analysed, and those objects with a high value were selected.

Results. The results in this article were from three rounds of discussion with experts. Towards the end of this article, we showed a statistically significant result from all rounds interviews. We first, carried out with fifteen (15) main components in CTO based on literature reviews and later we dropped four (4) out of fifteen (15) components based on experts' recommendation during the first-round session. We reconciled our CTO to eleven (11) components at the second round of session and the outcome was that the quartile deviation value was $\leq$0.5 and all the experts firmly agreed on the elements of CTO at final round assessment.

Conclusions. Based on the result, it shows the CTO ontology is suitable and effective to apply in mobile apps system to store knowledge of disaster and information of emergency response on the device. Moreover, the apps that will develop can utilise CTO as a reference and guideline especially to those who interested in this area.

Keywords: Delphi technique · Mobile apps · Communication · Emergency response · Ontology · Evaluation · Disaster relief · Natural disaster

T. Allahviranloo et al. (Eds.): IDS 2020, AISC 1301, pp. 57–70, 2021.
https://doi.org/10.1007/978-3-030-66501-2_5

1 Overview

Earthquake is unpredictable, that is the only tragedy that can't be expected where and when it will happen. This will end in serious incidents, deaths and lost peoples. For instance, in Sichuan (2008) about 69,000 people were killed, 374,000 wounded and around 4.8 million homeless. While in 2010 Haiti disaster caused more than 200,000 deaths, 300,000 were injured and leaving more than 1 million people homeless [1]. Another major catastrophe was the Pacific Ocean tragedy and the subsequent earthquake in Japan (2011), which cost over $300 billion to their government and causing unimaginable damage to the Japanese nation, their environment and the world's economy.

This study will discuss the assessment of CTOs that may be utilised for connecting survivors, first responders and families through smartphone devices and disaster response apps. We are interested in using ontologies as our core technology as it is a concise way to provide a hierarchical overview of a diverse domain area, including several specific actors with various skills and tasks. On the basis of these principles, a number of relevant existing and stable ontologies have been merged, Inc. Disaster Management, AEMET Weather Ontology [2], Friend of a Friend (FOAF) and Management of a Crisis (MOAC) [3] ontology. The current CTO will promote contact between various parties and help monitor citizens using cell phones as a device. In this post, we will provide a step-by-step assessment of CTO.

2 Background and Related Work

Over the last 20 years, several types of incidents have existed across the world, such as epidemics, eruptions, explosions, flooding, forest fires, hailstorms, heat waves, hurricanes, ice storms, landslides, maelstroms, mudslides, thunderstorms, tornadoes, tsunamis, typhoons, volcanoes and wildfires [4–9]. Flooding was the worst natural catastrophe, 43% ahead of every weather incident reported by the UN Disaster Risk Management Office [10]. While another disaster centre such as Epidemiology of Disasters Centre Research recorded at least 3,062 natural floods occurred from 1995 to 2015.

Following the devastating flood in Japan in July 2018 [11], in which 200 people perished, studies were taken into consideration the possibility of supplying public warning pagers to citizens residing in high-risk flood zones to consider the victim's life situation. Affordability (low cost) devices are the main motivation. Nonetheless, these apps only provide minimal features, i.e. transmitting warnings and getting a simple response (e.g., figuring out whether the person is still alive).

In comparison, storms were the second most commonly experienced occurrence, at 28%, while earthquakes were the third at 8%. Following Fig. 1 indicate natural disasters which frequently occurs between 1995 and 2015.

Between them, the earthquake is one of the worst-understood events that can happen without any alert sign and no one can know exactly, even geologists, when or where it will occur [12]. Earthquakes, for instance, are caused by a sudden release of stress along faults in the crust. Due to the continuous motion of tectonic plates, tension rises in the rock on both ends. When the pressure exceeds the rock's strength, therefore it will be

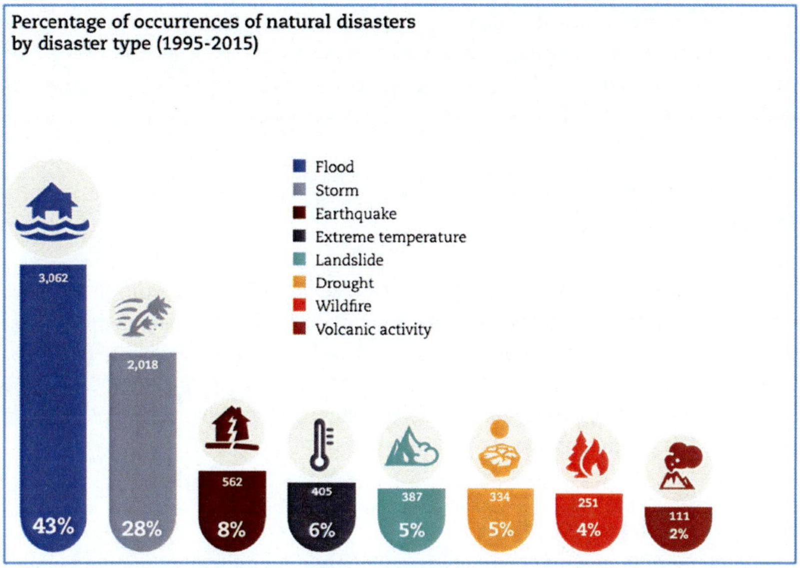

Fig. 1. Percentage of occurrences of natural disasters by disaster type (adapted from [10])

released. The seismic energy released propagates like waves across the earth along its ground surface, and these waves induce the trembling that we feel.

The key reason for implementing an emergency management programme in mobile applications is to provide better contact and assist coordination between people, families, rescuers and organisations who are directly or indirectly engaged in the earthquake disaster, thereby ensuring speedier relief to people as possible [13]. As a result, the number of deaths, the delay in access to basic needs and the longer recovery time has risen [14].

Early on, a variety of experiments have demonstrated that the field of emergency management can benefit in particular from Semantic Web technologies. Ontology has been used and some of the apps have been developed to locate missing persons, such as PeopleFinder, which has been developed from the experience of Katrina hurricane and the SEA-EAT blog that was launched in 2014 during the Asian tsunami. This has been built into an information-sharing programme for missing people, requests for assistance and news alerts.

There are many categories of disasters in the planet, such as Flood, Earthquake, Fire, Forest Fire, Hailstorm, Hurricane, Heat Wave, Landslides, Ice Storm, Mudslide, Maelstrom, Tornado, Thunderstorm, Typhoon, Tsunami, Wildfire and Volcanic. [4–7, 9, 15]. Based on initial analysis, the existing ontology including Management of a Crisis (MOAC) [3], AEMET Weather [2], Friend of Friend (FOAF) and Disaster Management ontology could not be applied to communication and tracking people in earthquake disaster. The research was thus carried out using Delphi Technique to assess the completeness and will be addressed in the following section.

Many sources have indicated that many of the civilians who had to be evacuated had trouble locating the closest evacuation centres that the government and the companies had set up for them, thereby obtaining the requisite assistance in a timely manner. In this reason, the way in which information on emergency centres for these people is presented is a significant issue in the future.

3 Problem Statement

Until now, most of the work is done to establish and offer a definition of ontology information to government and Non-Profit Organization (NGO) to help people in disaster relief operation. Depending on the literature reports, there is no appropriate ontology that crowdsourcing can use to make them a part of volunteering in the situation of disaster relief. The current ontologies of earthquake disaster relief service are still insufficient to be applied by the public volunteer.

4 Research Objective

The key goals of this research are:

1. To Identify CTO's elements to use in serving and storing crisis and emergency information.
2. To recognise the completeness of CTO Classes to monitor the emergency response activity.

5 Methodology

5.1 Three Round Delphi Technique

Many of the other methods are relevant in the field of research. However, in order to solve a problem in expert group, the Delphi technique is the best way to collect constructive criticism from experts. In this study, the Delphi method was used as the main approach of the survey, interviews and observation involved to determine the appropriateness of our CTO development. The survey technique was used to collect information in this research. In addition, quantitative and qualitative methods are also used. The methods are used by gather expert opinion and judgment to accumulate data and obtain feedback which will then obtain reliable results [16–19]. Delphi Technique was used to investigate the suitability of CTO.

Generally, the Delphi method is a practice to achieve the most accurate agreement of opinion from a panel of experts by a sequence of extensive questionnaires interspersed with monitored reviews. Most scholars have proposed a few conditions under which the analysis is suitable for the usage of the Delphi method, such as: (i) the issue may not lend itself to specific empirical techniques but may profit from contextual decisions on a collective basis; (ii) the persons required to contribute to the evaluation of a large or complex issue do not have a background of adequate communication and can reproduce it with respect to experience or expertise; (iii) more people are required that can engage successfully in a face-to-face exchange; (iv) time and cost make regular community sessions unfeasible; and (v) the variability of the groups must be preserved in order to assure the consistency of the findings. Delphi is a method used to collect expert opinions.

According to [20–22], some steps have been introduced to describe the meanings. In the first position, analysis problems need to be defined through knowing the issue statements. The questionnaires would then be built on the basis of the research question identified. Delphi relies on new ideas and suggestions [23–25]. There are some benefits to Delphi, as stated by [26–28] such as producing new ideas and advice while there are knowledge gaps in specific fields. However, Delphi needs a high cost [29–31] and systematic method to be introduced.

5.2 Steps Involved in Delphi

Delphi is a technique for searching a new idea from expert groups for majority views or among knowledgeable communities in the specific area in research field. In order to accomplish the main point of the article; to determine the suitability of CTO to be used in the representation and storing of disaster and emergency knowledge; and also to define the completeness of CTO classes, Delphi techniques have been used to validate the model and concepts by receiving expert ideas. Table 1 shows the process of verification and validation of the CTO involved in the Delphi technique.

Based on the description given in Table 1, diagram in Fig. 2 description exemplifies the essential measures of the Delphi technique [32]:

Delphi is a technique in which researchers seek to establish a hypothesis or a simple description of possible inventions [33]. The key purpose of this approach is to obtain an appreciation of the theories of experts on particular topics discussed in this paper on the nature of ontology in the context of the emergency response method for disaster management system model. Outstandingly, Delphi is a powerful and usable tool for dealing with such particular issues [34]. Steps in Delphi:

1. Problem identification

   - Identify the issue by evaluating the documents including the literature studies, the original report and other policy records in stakeholder papers.

2. Experts Classification

   - Identify experts by evaluating their competence in a specific field, professional experience with crisis recovery and emergency response operations.
   - Classify the experts by measuring their expertise in a devoted area, working experience in disaster management and emergency response operation.
   - All experts were verified by searching in the area of ontology, communication in disaster management.
   - Snowball method was used to identify experts.

Table 1. Delphi process for CTO validation and verification

Method	Round 1	Round 2	Round 3
Expert selection in Delphi technique: 1. Experts were consulted to through the questionnaires that contained elements in CTO raised through literature review 2. Ten (10) expert was selected and invited to join this validation and verification process, but only four (4) of them agreed and involved until end of the process 3. The process of identifying the experts by searching them in their area of their research in the website that related to emergency response and communication areas and also in ontology fields	Data collection as below: 1. Expert invitation: Invite respondents to involve in the research. Expert will be asked to give a feedback about the developed ontology, (communication and tracking ontology - CTO) 2. Expert demographic: Some information about respondents based on a few characteristics like work experiments, position, age, working years and others	1. Experts will share their opinions on competencies of the model. Their will give a rank for the related elements given 2. Experts will use their expertise in their fields to give a feedback by categorising, ranking, adding and/or deleting, in whatever ways they see appropriate	1. Experts will be asked again about their opinion based on result/feedback analysis in Round 2 2. Experts will be asked about the important of CTO elements 3. Experts will be asked to identify the interrelationship between the competencies in the CTO 4. The process will stop while get the similar result/feedback among them

3. Instrument Development

- Develop the instruments based on literature review and previous research.
- The questionnaires were designed with an initial of the classification of communication in emergency response.
- Each item must reflect to the construct selection.
- Construct and items departed from previous literature review and previous survey.
- Experts were asked to give their feedback expressing their level of agreement with the elements of CTO.
- Likert scale of 5 points was used to analyze the degree of expert's agreement in CTO.

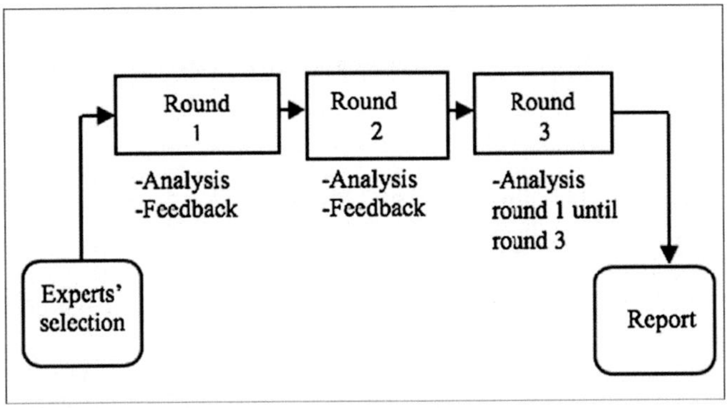

Fig. 2. Steps in Delphi technique

- The experts were also invited to propose additional elements or to remove the elements in CTO based on their opinion.

4. Result Analysis

- Data for each round of Delphi process need to analyse.
- Analyse the results by using statistical tools - SPSS.
- Experts were asked to response in Likert scale 5 points on the questionnaires about their agreement of all the items.
- The consensus value for each round will be identify and measured and based on that value the next step of Delphi process will be plan.

5. Redistribute instruments

- Questionnaire for survey has been developed and distributed after analysis have been carried out. Analysis from the previous distributions would be used to create a new collection of instructions/questions for the next cycle of Delphi.
- Expert received a set of new questionnaires for each round of Delphi.
- The questionnaires were modified based on response from expert for each level of Delphi.
- New questionnaires with a change suggested in the previous level of Delphi will send again to experts until the process done which is get same idea/opinion from all experts.

6. Result finding

- Analysis from expert response will continuously repeat until all the experts agree with the results.
- Final round was presented to experts.
- The CTO model containing all the amendment suggested by experts in all round of Delphi.

- Must have one agreement from expert surveyed.
- The model was validated and various elements were adjusted to the dimensions that closely match their characteristics.

Also, as we agree, Delphi methodology is a tool used to collect expert opinions and feedback on recent results, and also to collect their thoughts about anything in the future or something without answers [35]. Apparently unclear is the concept of an expert or procedure for assessing an expert. There are in reality still disputes regarding the word. According to [36], the term expertise may be applied to anyone who has knowledge and experience in specific fields.

Another concern that has been discussed is the number of experts involved. In the study as [37] indicated, that the number of Delphi panels is approximately 8–12 people, while [38] claimed that there are 7–12 experts, and [39] suggested that there are approximately 10–15 experts. However, according to [40, 41], the number of experts depends on the area to be researched and the need for it and the ability of the expert to engage before the conclusion of the research is the most important [42, 43].

6 Result and Finding Discussion

Several scholars have investigated and tested the methods used in the Delphi methodology [43].

Figure 3 illustrates the Delphi processes used in this analysis for this study.

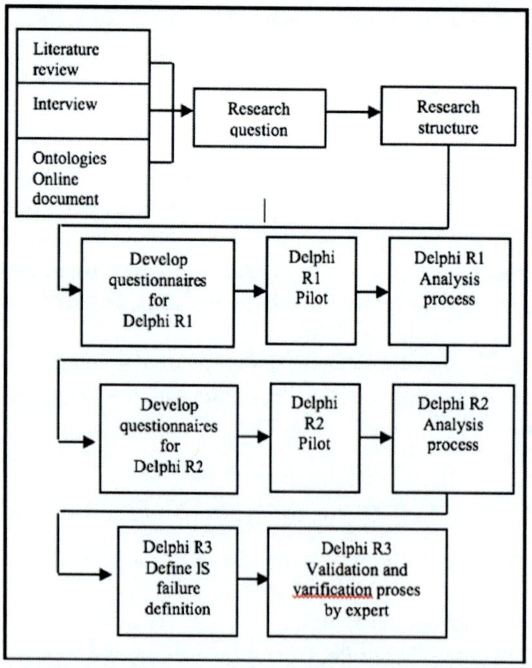

Fig. 3. Delphi process involved in this research

Interviews, comprehensive analyses of literature as well as electronic ontology documentation were performed and carried out until the Delphi methodology was used. These procedures are important for the development of a pilot and an internal review to improve the research issue to be carried out. Therefore, analysis problems and the study methodology were developed. The study questions established concentrate on the initial analysis. Several researchers and professionals familiar in emergency management have been chosen to verify the pilot project issues that arose.

The questionnaires were distributed to 10 chosen experts. Experts have been chosen on the basis of the parameters listed above. Nevertheless, only four researchers have committed to be active in both testing processes. Throughout the first phase of Delphi round 1, all aspects of the CTO were decided on the basis of the analysis of the systematic literature review. On this basis, the ontology was established and submitted to experts for the evaluation and validation process. All experts agreed with the current CTO. Nonetheless, four of the 15 elements had to be revised to have it better specified on the basis of the earthquake circumstance and expert feedback. Table 2 presents the output of result Delphi Round 1 in Delphi step.

Table 2. Item analysis for Delphi round 1

ITEM	Q3	Q1	QUARTILE DEVIATION
ITEM01	5	4	0.5
ITEM02	5	3	1
ITEM03	5	4	0.5
ITEM04	5	3.1	1
ITEM05	5	4	0.5
ITEM06	3	2.75	0.1
ITEM07	4	3	0.5
ITEM08	3.25	1.75	0.5
ITEM09	4.25	3.75	0.25
ITEM10	4.25	3.75	0.25
ITEM11	5	4	0.5
ITEM12	5	3	1
ITEM13	4	3	0.5
ITEM14	5	4	0.5
ITEM15	5	3	1

Quartile Deviation Value

0.1 - ≤ 0.5 = 11 item
0.6 - ≤ 1.0 = 0 item
≥ 1.0 = 4 items low and no consensus

Based on Table 2, the consequence review for the first round of Delphi reveals that the quartile deviation value for all elements is between 0.1 and 1.0. The significance of the quartile deviation indicates out of the 15 items of ontology, only 4 items are to be removed out as they do not have a solid expert agreement because the importance is small because there is no consensus. Experts have recommended to withdraw four of the categories from the CTO as they agree that it is unimportant to retain the details in the knowledge base. According to experts, information of the tragedy will not only be preserved but must also be tailored to the emergency response situation. Therefore, Delphi round two continued to reassess the ontology again on the basis of the elements proposed by them.

In round two, 11 items were submitted to the experts to be assessed. Based on the study, the quartile deviation value of all elements is less than 1. This indicates that all the experts have agreed on all the elements. Any of the 11 elements in the litigation was classified as ≤0.5, which shows that all the experts strongly approved on certain elements. No new item has been introduced by the experts during this phase. Quartile deviation analysis use SPSS as presented in Table 3.

Table 3. Item analysis for Delphi round 2

ITEM	Q3	Q1	QUARTILE DEVIATION
ITEM01	5	4	0.5
ITEM02	5	4	0.5
ITEM03	5	4	0.5
ITEM04	3	2.75	0.1
ITEM05	4	3	0.5
ITEM06	3.25	1.75	0.5
ITEM07	4.25	3.75	0.25
ITEM08	4.25	3.75	0.25
ITEM09	5	4	0.5
ITEM10	4	3	0.5
ITEM11	5	4	0.5

Quartile Deviation Value
0.0 – ≤ 0.5 = 11 items
0.7 – ≤ 1.0 = 0 item

The questionnaires were sent again to the experts in the 3rd round of Delphi to demonstrate the findings of the 2nd round of Delphi. The expert standard of consensus improved across all measurements and exceeded 100% of the appropriate standards. Feedback results from experts in the second phase of Delphi was evaluated. Based on the study, the importance of an expert agreement is seen in Table 4 below. The value determined on the basis of the quartile deviation. All of the 11 factors in dispute were classified as $\leq$0.5, which indicates that all the experts completely agreed with these elements in the CTO. In this phase, no new items were added by the experts and 11 elements were included in the final categories in Communication and Tracking Ontology (CTO).

Table 4. Item analysis Delphi round 3

ITEM	Value of expert agreement
ITEM01	0.4
ITEM02	0.5
ITEM03	0.4
ITEM04	0.1
ITEM05	0.4
ITEM06	0.5
ITEM07	0.3
ITEM08	0.3
ITEM09	0.3
ITEM10	0.4
ITEM11	0.5

Value of Agreement

0.0 - ≤ 0.5 = 11 item
0.8 - ≤ 1.0 = 0 item

Based on the assessed importance value analysed in Delphi Round 3, a final result of has been created as in Fig. 4 below:

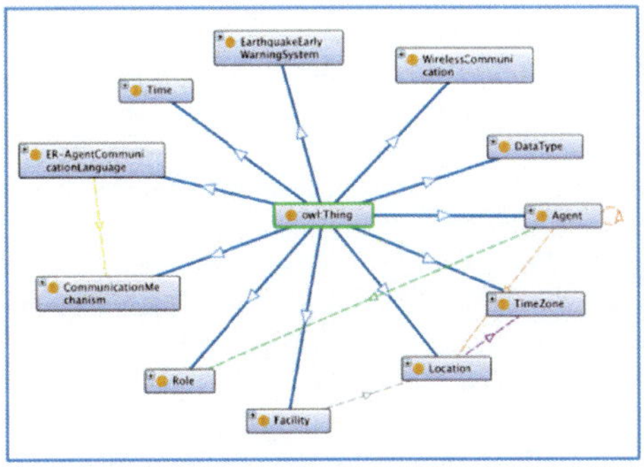

Fig. 4. Main classes domain of CTO

7 Conclusion

Based on this study by using the Delphi technique, the CTO developed in this research is suitable for mobile application system to be used and store the knowledge of disaster and emergency response on the device. After the first round of verification process, fifteen of classes has been verified and four of them need to change and removed for second round process. The research was conducted with the intention to become guidance and evaluation in order to measure the suitability of CTO classes in the emergency response area. In general, we believe the findings of this research can be a guideline to those who involve in this area especially to the system developers in order to develop a user-friendly mobile app that can be used in emergency response to decrease the number of fatalities and missing people after the disaster. According to research done before, research findings can be obtained from individual perspectives or group of people that call as an expert, are capable to handle identified problems because they have ability and skill to predict the future.

References

1. Yates, D., Paquette, S.: Emergency knowledge management and social media technologies: a case study of the 2010 Haitian earthquake. Int. J. Inf. Manage. **31**, 6–13 (2011). https://doi.org/10.1016/j.ijinfomgt.2010.10.001
2. Atemezing, G., Corcho, O., Garijo, D., Mora, J., Poveda, M., Rozas, P., Vila-Suero, D., Villazón-Terrazas, B.: Transforming meteorological data into linked data. Semant. Web. **1**, 1–5 (2013)
3. Ortmann, J., Limbu, M., Wang, D., Kauppinen, T.: Crowdsourcing linked open data for disaster management. In: Framework, pp. 11–12 (2011)
4. Malizia, A., Onorati, T., Diaz, P., Aedo, I.: SEMA4A: an ontology for emergency notification systems accessibility. Expert Syst. Appl. **37**, 3380–3391 (2010). https://doi.org/10.1016/j.eswa.2009.10.010
5. Yan, L.: A survey on communication networks in emergency warning systems. Sci. Comput. **100** (2011)
6. Sakaki, T., Okazaki, M., Matsuo, Y.: Earthquake shakes Twitter users: real-time event detection by social sensors. WWW'2010 Proc. 19th Int. Conf. World wide web. 851 (2010). https://doi.org/10.1145/1772690.1772777
7. Chou, C.H., Zahedi, F.M., Zhao, H.: Ontology for developing web sites for natural disaster management: methodology and implementation. IEEE Trans. Syst. Man Cybern. Part A Syst. Hum. **41**, 50–62 (2011)
8. Associated Press & NORCb Center: Communication During Disaster Response and Recovery. 1–5 (2013)
9. Sharmeen, Z., Martinez-Enriquez, A.M., Aslam, M., Syed, A.Z., Waheed, T.: Multi agent system based interface for natural disaster. Lecture Notes Computer Science, vol. 8610, pp. 299–310 (2014). https://doi.org/10.1007/978-3-319-09912-5_25
10. UNISDR, CRED: The human cost of weather-related disasters 1995–2015 (2015)
11. Ogura, J., Berlinger, J.: Japan floods: death toll rises to 200 as UN offers assistance. https://edition.cnn.com/2018/07/10/asia/japan-floods-intl/index.html
12. Planet Science: Can we predict earthquakes? http://www.planet-science.com/categories/over-11s/natural-world/2011/03/can-we-predict-earthquakes.aspx

13. Hassan, M.K.A., Chen-Burger, Y.-H.: ER-agent communication languages and protocol for large-scale emergency responses. In: Jezic, G., Chen-Burger, Y.-H.J., Howlett, R.J., Jain, L.C., Vlacic, L., Šperka, R. (eds.) Agents and Multi-Agent Systems: Technologies and Applications 2018, pp. 134–143. Springer International Publishing, Cham (2019)
14. Ratnam, K.R., Karunaratne, D.D.: Application of ontologies in disaster management. Int. Conf. Adv. ICT Emerg. Reg. (2008)
15. Liu, S., Shaw, D., Brewster, C.: Ontologies for crisis management: a review of state of the art in ontology design and usability. ISCRAM 2013–10th Int. Conf. Inf. Syst. Cris. Response Manag, pp. 349–359 (2013)
16. Maria and Esperanza.: Approach to the integration on qualitative and quantitative method in software engineering. Philisophiocal Foundations on Information Systems Engineering (2006)
17. Dalkey, N., Helmer, D.: An experimental application of the Delphi method to the use of experts. Manage. Sci. **9**, 458–467 (1963)
18. Ludwig, B.: Predicting the future: have you considered using the Delphi methodology? J. Extension **35**(5), 233–239 (1997)
19. Stewart, J.: Is the Delphi techniques a qualitative method. Med. Educ. **35**, 922–923 (2001)
20. Powell, C.: The Delphi techniques: myths and realities. Methodological Issues Nurs. Res. **41**(4), 376–382 (2003)
21. Cantrill, J.A., SIbbald, B., Beutow, S.: The Delphi and niminal group technique in health service research. Int. J. Pharm. Pract. **4**, 67–74 (1966)
22. Helmer, O.: Analysis of The Future: The Delphi Method. Prentice Hall, Upper Saddle River; NJ (1968)
23. Steward, J., O'Halloran, C., Harrigan, P., Spencer, J.A.: Identifying appropriate tasks for the preregistration year: modified Delphi technique. Br. Med. J. **319**(7704), 224–729 (1999)
24. Meyer, J.H.: Rethinking The Outlook of Colleges Whose Roots Have Been in Agriculture. University of California, Davis, CA (1995)
25. Miller, G.: The development of the indicators for sustainable tourism: result of Delphi survey of tourism researcher. Tourism Manage. **22**(4), 351–362 (2001)
26. Jones, C.G.: A Delphi evaluation of agreement between organizations (1975)
27. Anderson, D.H., Schneider, I.E.: Using the Delphi process to identify significant recreation research-based innovations. J. Park Recreation Adm. **11**(1), 25–36 (1993)
28. Bogdan, R.C., Biklen, S.K.: Qualitative Research for Education: An Introduction to Theory and Methods, 2nd edn. Allyn and Bacon, Needham Heights, MA (1992)
29. Martino, J.P.: Technological Forecasting for Decision Making. North-Holland, New York (1983)
30. Ludwig, B.G.: Use of the Delphi technique to assist in determining future. In: Proceedings of the 12th European Seminar of Extension Education. Thessaloniki, Greece (1995)
31. Wicklein, R.C.: Identifying critical issues and problems in technology education using a modified-Delphi technique. J. Technol. Educ. **5**(1), 54–71 (1993)
32. Wiersma, W., Jurs, S.G.: Research Methods in Education, an Introduction. Pearson, Boston (2005)
33. Dalkey, N.C., Helmer, O.: An experimental application of the Delphi method to the use of experts. Manage. Sci. **9**(3), 458–467 (1963)
34. Anderson, D.H., Schneider, I.E.: Using the Delphi process to identify significant recreation research-based innovations. J. Park Recreation Adm. **11**(1), 25–36 (1993)
35. Yousuf, M.I.: Using experts' opinions through Delphi technique. Pract. Assess. Res. Eval. A Peer-reviewed Electron. J. **12** (2007). ISSN 1531-7714
36. Cantrill, J.A., Sibbald, B., Buetow, S.: The delphi and nominal group techniques in health service research. Int. J. Pharm. Pract. **4**(2), 67–74 (1996)

37. Cavalli-Sforza, V., Ortolano, L.: Delphi forecasts of land-use-transportation interactions. J. Transp. Eng. **110**(3), 324–339 (1984)
38. Phillips, R.: New applications of the Delphi technique. Ann. Pfeiffer Company **2**, 191–196 (2000)
39. Turoff, M.: The design of policy Delphi. Technol. Forecast. Soc. Change **2–2**, 149–171 (1970)
40. Mullen, P.M.: Delphi: myths and reality. J. Health Organ. Manage. **17–1**, 37–52 (2003)
41. Adler, M., Ziglio, E.: Gazing into the Oracle: The Delphi Method and its Application to Social Policy and Public Health. Jessica Kingsley Publishers, London (1996)
42. Delbeq, A., Van de Ven, A., Gustafson, D.H.: Group Techniques for Program Planning: A Guide to Nominal Group and Delphi Processes. Scott, Foresman and Company, Glenview, USA (1975)
43. Norshita, M.N., Halimah, B.Z., Tengku Mohammad, T.S.: Public user assessment of Malaysia's e-government applications. World Acad. Sci. Eng. Technol. **43**, 813–817 (2010)

A Comparative Study on Visualization Technique for Home Network

Erman Hamid[1](✉), Mei Choo Ang[2], and Azizah Jaafar[2]

[1] Faculty of Communication and Information Technology, Universiti Teknikal Malaysia Melaka, Hang Tuah Jaya, Melaka, Malaysia
erman@utem.edu.my

[2] Institute of IR 4.0, Universiti Kebangsaan Malaysia, Bangi, Selangor, Malaysia

Abstract. There has been a number of researches carried out on Visualization Technique that applied into Computer Networking including Enterprise Network and Home Network. Many researchers have stated that the way data being displayed in Home Network application is the most important aspects in understanding network data issues concerning Home Network. This paper reviewed the existing research related to Visualization, Home Network and Network Management; intended to (1) identify the visualization techniques that being used in existing Network Management Tools and (2) recommend potential future work that could be conducted in order to archive desirable goals of Home Networking.

Keywords: Home Network · Human computer interaction · Visualization

1 Introduction

Home Network is a collection of resource including hardware peripherals and computers interconnected [1] by networking media uses to share resources and information [2] at home [3]. It is also referred as a Computer Network [4] that widely used not only at home, but also in various activities including business and office administration [5–8].

At home, network serves as a platform that interconnect digital devices such as personal computers, printers, mobile computing devices – and also human as a user [7]. Home Network must be properly managed to ensure stability, accountability and security [2, 7, 9]. Good network management is depends on how network data from network management tools is visualized and understand by network user [10]. The more user understand data that visualized, the more it will help in managing the network [11].

Home Networks are vastly expanding, and have become popular nowadays [6]. The activities is not only related to e-mailing and information browsing, but it includes every part of household's activities including electronic shopping, real-time communication, entertainment activities and internet banking. Home Network is growing rapidly and it looks like one computer is no longer enough to be shared in a Home Network. A Home Network with multiple computers (including smart phones) interconnected means there are a greater need on network management.

T. Allahviranloo et al. (Eds.): IDS 2020, AISC 1301, pp. 71–85, 2021.
https://doi.org/10.1007/978-3-030-66501-2_6

The **effectiveness** of home networking depends upon its level of usability. It is fully depended on how much it is used and understands by the user – and this is where network management take place [12]. An optimum home networking requires a full user understanding and efficient network management in terms of performance and security. Good network management are always came from good user interpretation when monitoring network management tools [13]. In easy words – what is being displayed must be clearly understand by user. The tools should have the good ability in visualizing network data to ensure the information carries could be understood by user.

Visualization refers to the methods of delivering information using 'holistic understanding' [14]. visualization elements enable users to get a comprehensive and clear understandings about what is being shown [15]. Visual element with helpful environment make users feel supported when dealing with interfaces, and influence user to achieve their objectives. Visualization is different with 'Graphics'. Graphics focuses on how to represent data on a computer screen while the visualization emphasis on how to deliver data in interactive way and easy for user understandings. Visualization serves to help users identify the most important information of data that being displayed [16]. Effective visualization able to avoid users being diverted from the real meaning of an interface. It help users explore and understand data better [17]. It is also able to stimulate the understanding that ultimately allows users to translate their understanding when communicating.

A big picture or complex idea of network data can be delivered easily using the right **Visualization Technique**. The effectiveness of home networking depends upon its level of Usability that came from the most helpful visualization approach. It is fully depended on how much the visual of the network tools understands by the user [18]. An optimum home networking requires a full user understanding, and efficient network management in terms of performance and security [19]. Nowadays, there are too many network management tools and the existing visualization technique applies are more suitable to general Computer Network – not Home Network. This emerges the needs to studies the proper visualization way for network management tools that focus to the needs of Home Network itself.

2 Methodology

This study employed **Content Analysis** technique; which means the data from the previous research were analyzed and summarized scientifically [20]. It is a series of activities in analyzing and summarizing the scientific data that produced by previous research [21]. The 'data' referring to the form of spoken, written or visual of pictures, words, symbols, meanings, ideas, diagrams, or any message that can be analyzed [22]. Figure 1 show the steps taken in extracting the problem and solutions in home networking.

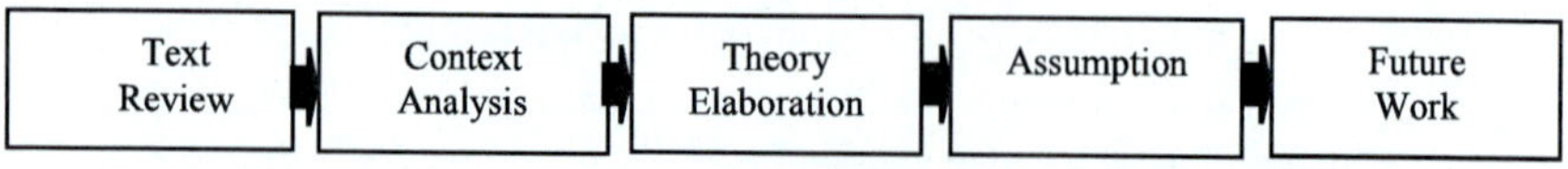

Fig. 1. Steps taken in finding the visualization technique for home network.

In this paper, data gathered, reviewed and analyzed from twenty major selected papers and thirty more additional related references. It started with the review on research that being carried regarding network visualization. Analysis began by constructing a corpus of papers on network visualization by searching the term of "Visualization", "Network Visualization", and "Visualization Technique in Network/Home Network". Each paper that selected was examined, and evaluated to see either it is suit to the topics discussed. It was later reviewed to fulfill the good visualization criteria as concluded by [17, 19, 23–26] including: (1) Understanding, (2) Interpreting, (3) Experiencing, (4) Describing and (5) Remembering. The steps taken in evaluating the visualization criteria displayed in Fig. 2 below.

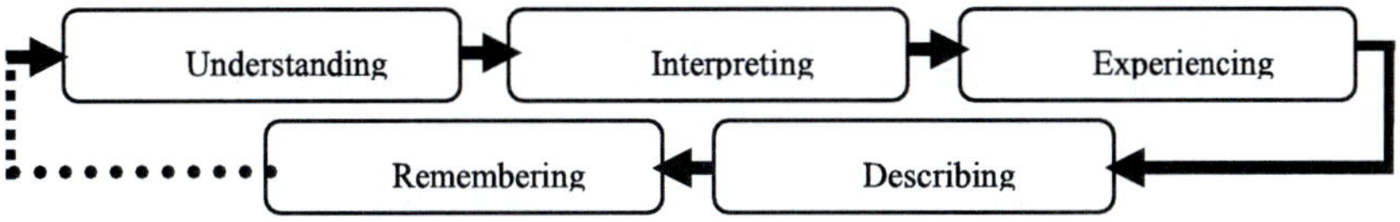

Fig. 2. Steps in evaluating visualization criteria

The study focuses on the research paper related to network problem and visualization technique proposed in computer network. From the 20 major papers selected, 16 of them are focusing on enterprise network, leaving only 4 papers focusing on 'Home Network'. This shows that there are too many research focusing on 'big network' like enterprise network that shown the needs to do more research on Visualization Technique for 'small network' like Home Network itself. Figure 2 shown 'Types of Network' that being select in this study.

Literacy studies on visualization techniques have been made by highlighting the research of various researchers. Data collection methods of home networking research includes ethnographic observation methods, questionnaires, interviews/field observations, content analysis, and heuristic analysis [1, 2, 6, 7, 9, 26–30].

Basically, the data collection techniques used by the various studies referenced are observation, interviews, and questionnaires [27, 28]. Jeonghwa Yang and Edwards [26], Poole and Chetty [1], Jeonghwa Yang and Edwards [26] and Bly [6] began with interviews, while Crabtree [2] began with the questionnaire. All studies were conducted on the environment of 15–30 study participants, dealing with various ages and computer literacy background. All these researchers further their data collection with the field observation method, which is done on the user home, on the street, and in the office.

Crabtree [2], Jeonghwa Yang and Edwards [26], R. E. Grinter [7], Poole and Chetty [1], Shehan and Edwards [13], and Bly [6] in their respective research conducted in the US, UK, and Europe have involved more than 95 participants accumulated with quantities between 15 to 30 participants used in each study. The small number of participants is due to the study platform that involves private premises where the Home Network is located as a private network, thus complicating the larger number of study samples [2, 13]. Refer to Fig. 3 to see the number of participants for each study for all five major surveys.

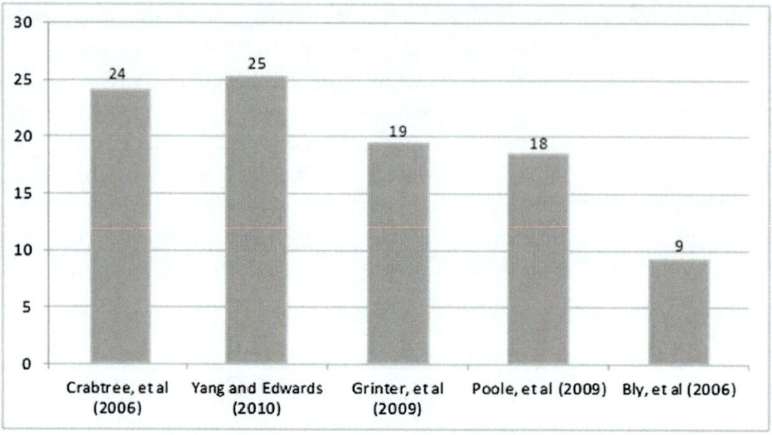

Fig. 3. Number of participants

Studies show that users use network management tools to manage their respective networks. The patterns are involving related activities connecting their networks to internet service providers (ISPs), connecting computers to networks, and solutions to network problems. Most of them prefer to use tools available in their respective operating systems (e.g. TCP/IP rules in Windows operating systems) and tools inserted by vendors (Internet Service Providers) while operating their networks [28]. This strongly reflects the need for a good visualization technique on the network management tool interface.

This fact explains that Home Network management tools with good visualization techniques are needed to assist users in using their respective networks [1–3, 5–9, 13, 26–29, 31–34]. In other words, the usability problem can be solved with the network management tool itself if it is presented with appropriate visualization techniques.

We than developed a set of rubric question as a guide to analyze each paper. For each of that, we produced summaries to describe the conceptual models for each of the questions: (1) Network visualization problem that being catered, (2) Visualization technique that being implemented, and (3) The contribution to Home Networking. For each of that, we overcome a summary, and produce a conceptual model that shown as an answer for each of question. Based on the work of previous researchers, basic guidelines were recommended to develop the visualization element of the interface for Home Network Management Tools in this paper.

3 Discussion

The discussion below is divided into four sections; (1) Why visualization? (2) Technique that being proposed and their (3) Contributions. Based on the analysis conducted on previous studies, several important criteria are integrated and concluded into the solution of the problem. As a result, Theory Model of proper visualization technique for home networking user are proposed.

3.1 Why Visualization?

The first question examined in this study is why visualization? It focuses on why visualization is chosen as a method to gain usability of network usage. Why it is used as an element to enhance user understandings in observing network.

According to Liao [10], networks are becoming increasingly complex, and the need to establish a good visualization for network data becomes a very big challenge. The ability to match and visualize the dynamic network data is not easy. It involves big dynamic data and must be simplifies in a small screen – to be finally understand by the user. This makes the correct visualization techniques are necessary to reflect the complex network data.

Shannon [11] focused on a very dynamic network data and try to overcome the problem on how to visualize it in easy understanding way. The network data is complex, and appeared in screen in fast tempo. It affects the quantity of data that could be observed and understand by user. Some data might be captured, but some might be lost. For expert, it should be no problem in understanding it, but when it comes to normal user – the very dynamic network data are too difficult to be understand. The simple but sufficient visualizations technique might help.

Xydas [35] in their study titled 3D Graph Visualization of Web Normal and Malicious Traffic, focus on the problems of network data visualization on web platform. It caused by the emerging of internet and too many of network data that displayed through web based application. There is a great need in terms of monitoring network activity especially in knowing the status of access to the organization's website. There are also important needs in viewing the network traffic of the company local area network in order to ensure efficient networking and network security. The network data are in large-scale, and the only way to understand are to represent data with the right visualization techniques.

Shannon [36] in their research sees network as a huge and big in activity, and classified it as complex. Visualizing the complex network is quite a challenging task and usually deals with a lot of effort, especially in producing the right visual data in order to fulfill the user understandings. The abstract that derived from the complex environment should be simplifies to reduces complexity.

Reduce Complexity. Network data are too dynamic and displayed in the form that are hard to be understand. Showtime [36], Graphemes [11] and Polyplane [37] focused on a dynamic network data and displayed them in the better form such as 2D, 3D and 'time to time' animated graph.

Simplify Understandings. ENAVis by Liao et al. (2010), Showtime by Shannon et al. (2010b), and 3D Visualization Prototype System by Xydas et al. (2006) developed due to the needs to simplify the visualization of network data because of its complexity. It is classified as a big challenge because it is not an easy work to visualize the dynamic network data into a form that readable and understand by users.

Improves Navigation. Castelli [27], Wilson [38] and Samuel [39] explained that good visualization has the ability to guide users to use home network management tools efficiently, while at the same time stimulating the usability of the home network. It means users will not be lost with the assist of a good visualization interface.

Provide Communication. Xu [32], Guimaraes [40] and Moallem [28] agrees with the fact that precise visualization techniques provide good communication between home networks, network management tools, and home network users themselves. There is a clear relationship between the accuracy of Visualization techniques used in stimulating Usability.

Provide Helps. Dimara and Bezerianos [41], Boy [42] and Knowles [43, 44] refers to the ability of visualization appearance on the user interface in replacing the human function as an assistant to the user in using the home network. Appropriate visualization techniques are able to serve as a good helper in brightening user problems in the network.

3.2 The Techniques

The interface is an area that focuses on the interaction between humans and computer systems, intermediaries and the basic processes that produce such interactions [45–49]. It can be in the form of a visual display, or an audio bounce that conveys the meaning of both the sender and the receiver of information [50].

Visual interface is communicative, in a form that is able to enable both parties to communicate to achieve communication objectives [51–53]. In the context of network management (including Home Networks), the visual interface is able to be an intermediary element, that enables users to understand the network management elements [54].

The visual interface must consist of a mixture of principles, practices, and people, related to how an idea or objective of delivery, is conveyed to the recipient through a particular approach [46]. Kotze [55] define visual interface including (i) method of representing the meaning to be conveyed; (ii) assisted delivery of intermediary tools, and follow-up of interventions; and (iii) the method of sending/receiving information that is the exchange material.

This definition is equivalent to the meaning of Interface by Karam [56, 57] which classifies it into four specific functions namely application domain, technology support, system response, and style/signal gesture. The application domain involves scope or field covered by Interface theory in conveying a purpose; assisted by support tools based on existing technology. The functionality of this Interface is guided by system response elements with characteristic features and motion signals that allow interaction to occur between the system and humans.

There are many ways to present data and information. It all depends on the type of data and information to be conveyed, the category of data and information recipients, and the approach to be taken in conveying the data and information [58]. In order to ensure that the data presented is meaningful, it needs to be generated effectively, interestingly and creatively, aimed at connecting users and systems in a beneficial way [59].

Visualization display must be entirely in line with the needs of the user, in a more appropriate sense - it should be developed according to the needs of the user [60]. All interactions between humans (user) and technology (systems) are based on the senses. It is because human beings are provided with the ability to receive input and interpret data into information [58]. The variety of types of data and information, causing it to need to be adapted to the type of target users.

Graph techniques are used in structured mathematical representations, describing an implicit meaning, in the form of axes, nodes, and dots connected to form corners, arcs, or lines garisan [11, 61]. It is popularly used as an interface visualization in many fields including representing data in a network [62].

Direct manipulation technique, is a method of visualization display in the interface that involves the continuous representation of objects of interest on the screen, with a fast, easy-to-understand display, in the form of an object display that collects user participation in the interface that allows users to manipulate visualization objects to achieve needs [63, 64].

Concept mapping techniques are commonly used to clarify things in the appearance of a screen that is able to translate linked meanings [40, 65]. Mapping techniques The popular concept used is a network tool, representing a large network structure, which contains detailed descriptions of the network data involved such as router-switch connections, up to the detailed data structure used in the network itself [27].

The user adaptation technique is an interface visualization technique that is able to adapt to the characteristics of the user himself [66]. It is as if the interface was developed to suit the level of a user, driving the user to use the system without difficulty [67]. It is also called user-based interface design, in which the features of the Interface displayed seem to be adjusted in line with the capabilities and desires of the user [15].

Embedded visualization technique is a technique that provides flexibility, incorporating several tasks into a visual display. This technique is able to reduce the number of objects that need to be displayed on the screen, without reducing the functionality of the Interface [68]. Embedded visualization provides flexibility, reduces the use of space on the screen, and even users are free to choose the screen display they want to use according to needs [69].

The augmentation visualization technique is a feature of a visual display that mimics the real situation [70]. It seems to be 'alive', similar to the real world but represented on a computer screen in the form of visualization objects [71]. It is able to match what users want to see in a real environment, yet demands the same use of high-scale hardware and software [72].

Social computing techniques, on the other hand, lead to creating or creating social connections to the appearance of software and technology [73]. In the context of visualization appearance, social computing requires the social needs of users to be taken into account in the development of visualization objects, leading to consumer levels, user habits in everyday practice, and suitability in their respective platforms [74].

The heat map visualization technique is a presentation method visualization of detailed data is displayed in a color-based matrix [75]. Network data uses this technique extensively in representing large-scale data simultaneously [40].

The tree mapping technique is a visualization approach describing data as a set of nested rectangles [76].Each branch of a tree is given a rectangle, which is then joined by a smaller rectangle to represent the meaning of the sub-branch. Each sub-branch then has a branch, which is advertised [77, 78].

Hyperbolic tree technique is a method of information visualization based on geometric graphs [78, 79]. It is widely used to describe large-scale network data, with representatives of the central nodes connected to all the nodes around it, easily depicting computers connected within a network [80].

Problem-solving environment techniques are techniques that only emphasize the final impact, and the extent to which the goal of a visualization is achieved [67, 81]. In this technique, the assisted user is presented with the problem in visual form and developed to solve the problem using visualization elements as well, to finally achieve the targeted goal [65].

Highlights of visualization techniques in the Interface show that the problem-solving environment techniques are suitable as a core technique in the direction of the study because of its principles that are more about how to help the problems of network users. However, other techniques in the science of visualization design can be sub-techniques in helping to develop problem solving environment techniques. That is, the problems of users in managing the Home Network can be addressed by the 'problem solving' feature represented in the objects of visualization, which is responsible for leading users in solving everything that happens in their own Home Network.

3.3 The Contributions

Visualization is basically translated as a method of describing abstract data into a form that is easier to understand [41]. It should be able to represent and retain users, as well as achieve the basic objectives of the relationship between users and systems [42].

Effective visualization should be able to manipulate the breadth of the human mind, thus enabling a holistic understanding to be gained, based on the experience of the user exploring the visual display of the interface. It directly gives the user a true picture of what the Intermediate Interface is trying to convey [82, 83].

Visualization display is done to meet the most basic need - to convey information [61]. It is a primitive form of communication that has been defined for a long time - from 3000 BC [84]. Objective-based understanding helps, which helps stimulate the usability of the elements that convey the visuals of this interaction, visualization leans on 2 general goals; (i) Exploration, and (ii) Explanation [85]. It covers how an Interface is presented with the ability to give users browsing experience, and at the same time provide users with an explanation of each activity and response that users feel throughout using the visualization element. It is as if the user is being assisted or is having an assistant when alone using the system with the visual help of the Interface [82].

Appearance and representation of visual elements on the Interface in assisting users when using this system develop the basic features of the principle of visualization. It is a visual display of the system that provides a holistic understanding to the user, based on the browsing experience, adapting the actual image of the representative trying to represent, and appearing as a natural helper to the user during use.

In line with the need for the role of assistant to the user in the use of home network management tools, this study will focus on highlighting the role of natural assistant from the basic visualization principle as the focus of the study. It will be further explored to see how the role of this natural assistant can be translated into the visual display of the

Interface that will be developed. The basic principles of visualization can be summarized in graphic info as presented in Fig. 4 below.

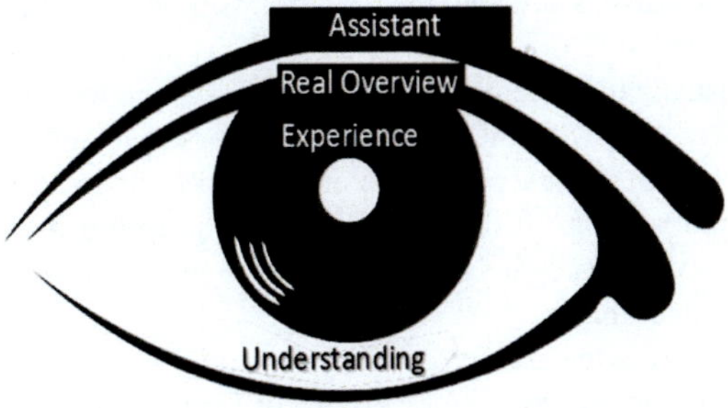

Fig. 4. Visualization principle

4 Conclusion

There are various mechanisms used to display meaning through the elements of visualization, but they are only effective based on certain principles and purposes [86]. Another field, another locust, that can be inferred by the visualization method which is how it should be highlighted to enable the purpose of consumption to be achieved.

A good visualization display does not necessarily guarantee that all users who interact with the interface achieve what they are trying to get [28]. The ability of user to perceive the meaning of a visualization element featured, does not necessarily guarantee similarity among other users [87]. It clearly demands a critical view and a specific focus on the relationship between users, interface and visualization design techniques to be aligned in a single line of understanding, so that the targeted objectives, can be swallowed well. They are mutually supportive, and have a profound effect on each other, and can be clearly explained on the basis of knowledge [42].

Measurement of the ability to achieve an objective, or more simply called the goal of using a tool or system, or environment, is more accurately referred to as Usability [88]. It is a term referred to as a measure of ability, or the ability to use a tool (or system, or environment) to achieve a specific goal. It is a simple phrase that symbolizes the degree of effectiveness of one's ability to use something. For example, when user A, user B and user C use a system, the percentage of understanding and success of each user's use of the tool is certainly not equal. It is possible that user C is more skilled and understands what they are facing, compared to user A, while user C may continue to dissolve in uncertainty when using the system. This is called Usability. It is interpreted as a measure of the extent to which a user achieves objectives when using a system or environment.

How to achieve usability? In the context of using an interface for a network management tool, it involves the role of the tool itself, usability objectives, and how users interact with the interface visual in enabling usability [41]. It is based on a process, starting from the user does not know, to the user knows - after using and interacting with the intermediate elements, then supporting the achievement of a usability.

It is based on the appearance of the interface element (in the context of this study - visualization) which conveys data into the form of information, which is then interpreted by the user into knowledge (knowledge), to be finally finalized as wisdom. The data presented by the visualization element will become very meaningful after going through the process of interacting with the user, which allows it to unfold into information. Habits, derived from repeated use, will be interpreted as knowledge (knowledge), which reaches its peak into wisdom. It defines the level of usability itself from the level of information, to the level of knowledge (knowledge) to be preferably the level of wisdom as the highest level in usability.

Figure 5 adapted from the article "An Overview of Understanding" by N. Sherdroff in the book R.S. Wurman entitled "An Overview of Understanding", clearly reflects the relationship between the elements of Visualization, Users and Usability; which contains levels of usability from the raw data level, to information, to knowledge (knowledge) and ends with wisdom [89]. The increase in context, with a focus on features that lead to the needs of the User, will change the degree of interpretation of the data, into understandable information, leading to the manipulation of knowledge (knowledge), which is ultimately translated as wisdom. Wisdom will enable users to use the data, information and knowledge (knowledge) received from the process towards this Usability to be practiced in future use, both through the tools being used, as well as the external environment associated with the tools.

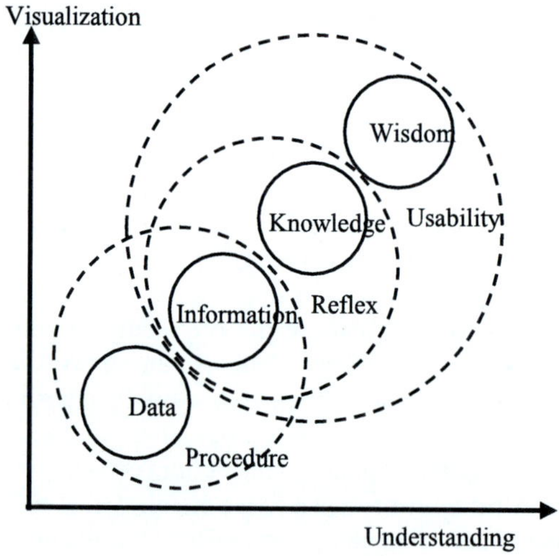

Fig. 5. Relation between procedure, visualization interaction and usability

The relation between procedure, visualization interaction and usability shown in Fig. 5 will be the basic of guideline in developing the visualization element of the interface for Home Network Management Tools.

Acknowledgement. Corresponding author (Erman Hamid) gratefully acknowledges concession and study leaves granted by the Universiti Teknikal Malaysia Melaka (UTeM) for his PhD. study. Acknowledgement to University Kebangsaan Malaysia (UKM) for the contribution of expertise and infrastructure, and to the Ministry of Higher Education (MOHE) Malaysia for the financial support of the programme.

References

1. Poole, E.S., Chetty, M.: More than meets the eye: transforming the user experience of home network management. In: Proceedings of the 7th ACM Conference on Designing Interactive Systems. ACM, pp. 455–464 (2008)
2. Crabtree, A., Mortier, R., Rodden, T., Tolmie, P.: Unremarkable networking: the home network as a part of everyday life. In: Proceedings of the Designing Interactive Systems Conference. ACM, pp. 554–563 (2012)
3. Yang, J., Edwards, W.K., Haslem, D.: Eden: supporting home network management through interactive visual tools. In: Proceedings of the 23nd Annual ACM Symposium on User Interface Software and Technology. ACM, pp. 109–118 (2010)
4. Comer, D., Ralph, D.: Computer Networks and Internets. Lecture Notes (1997)
5. Grinter, R.E., Edwards, W.K., Newman, M.W., Ducheneaut, N.: the work to make a home network work. In: Ecscw 2005, vol. 200, pp. 469–488 (2005). https://doi.org/10.1007/1-4020-4023-7_24
6. Bly, S., Schilit, B., McDonald, D.W.: Broken expectations in the digital home. In: CHI'2006 Extended Abstracts on Human Factors in Computing Systems. ACM, pp. 568–573 (2006)
7. Grinter, R.E., Greenhalgh, C., Steve Benford, W., Edwards, K., Chetty, M., Poole, E.S., Sung, J.Y., et al.: The ins and outs of home networking. ACM Trans. Comput.-Hum. Inter. **16**, 1–28 (2009). https://doi.org/10.1145/1534903.1534905
8. Poole, E.S., Chetty, M., Morgan, T.: Computer help at home: methods and motivations for informal technical support. In: Proceedings of the SIGCHI Conference on Human Factors in Computing Systems. ACM, pp. 739–748 (2009)
9. Elmore, B., Hamilton, S., Ivaturi, S.: Designing software for consumers to easily set up a secure home network. In: CHI'2007 Extended Abstracts on Human Factors in Computing Systems. ACM, pp. 1735–1740 (2007)
10. Liao, Q., Striegel, A., Chawla, N.: Visualizing graph dynamics and similarity for enterprise network security and management. In: Proceedings of the Seventh International Symposium on Visualization for Cyber Security, pp. 34–45 (2010)
11. Shannon, R., Quigley, A., Nixon, P.: Graphemes: self-organizing shape-based clustered structures for network visualisations. In: CHI'2010 Extended Abstracts on Human Factors in Computing Systems. ACM, pp. 4195–4200 (2010)
12. Erman, H., Azizah, J., Mei Choo, A.: A comparative study on human-computer interactiob's influence to home network management. In: 2nd International Conference on Interactive Digital Media (ICIDM 2013), pp. 218–225 (2013)
13. Shehan, E., Edwards, W.K.: Home networking and HCI: what hath god wrought? In: Proceedings of the SIGCHI Conference on Human Factors in Computing Systems. ACM (2007)
14. Le Grand, B., Soto, M.: Information management–topic maps visualization. XML Europe, pp. 1–11 (2000)
15. van Wijk, J.J.: Views on visualization. IEEE Trans. Vis. Comput. Graph. **12**, 421–432 (2006). https://doi.org/10.1109/TVCG.2006.80

16. Löwgren, J.: Visualization and HCI: a brief survey (1996)
17. Ten, D.W.H., Manickam, S., Ramadass, S., Al Bazar, H.A.: Study on advanced visualization tools in network monitoring platform. In: 2009 Third UK Sim European Symposium on Computer Modeling and Simulation. IEEE, pp. 445–449 (2009). https://doi.org/10.1109/EMS.2009.24
18. Herman, I.: Graph visualization and navigation in information visualization: a survey. IEEE Trans. Vis. Comput. Graph. **6**, 24–43 (2000)
19. Ball, R., Fink, G.A., North, C.: Home-centric visualization of network traffic for security administration. In: Proceedings of the 2004 ACM Workshop on Visualization and Data Mining for Computer Security. ACM, pp. 55–64 (2004)
20. Berger, A.A.: Media Research Technique (1998)
21. Neuendorf, K.A.: A flowchart for the typical process of content analysis research. In: The Content Analysis Guidebook, p. 320 (2002)
22. Neuman, W.L.: Social Research Methods: Qualitative and Quantitative Approaches. Relevance of social research, vol. 8 (2003)
23. Ahmad, F., Sumner, T.: Improving conceptual learning through customized knowledge visualization. In: 2010 Third International Conference on Knowledge Discovery and Data Mining, pp. 407–410. IEEE (2010). https://doi.org/10.1109/WKDD.2010.68
24. Nuutinen, J.A., Sutinen, E.: Visualization of the learning process using concept mapping. In: Proceedings of the 3rd IEEE International Conference on Advanced Learning Technologies, 2003, pp. 348–349. IEEE (2003)
25. Fulcher, J.A.: Visualization as a key element in learning. In: Proceedings of IEEE Region 10 International Conference on Electrical and Electronic Technology, pp. 462–472. IEEE (2001)
26. Yang, J., Edwards, W.K.: A study on network management tools of householders. In: Poceedings of the 2010 ACM SIGCOMM workshop on Home networks, vol. 82, pp. 1–5. ACM (2010)
27. Castelli, N., Ogonowski, C., Jakobi, T., Stein, M., Stevens, G., Wulf, V.: What happened in my home ?: an end-user development approach for smart home data visualization. In: 2017 CHI Conference on Human Factors in Computing Systems, pp. 853–866 (2017)
28. Moallem, A.: Why should home networking be complicated? In: Advances in Usability Evaluation: Part II. CRC Press, pp. 4169–4178 (2013)
29. Edwards, W.K., Grinter, R.E., Mahajan, R., Wetherall, D.: Advancing the state of home networking. Commun. ACM **54**, 62 (2011). https://doi.org/10.1145/1953122.1953143
30. Edwards, W.K., Newman, M.W., Shehan Poole, E., Arbor, A.: The infrastructure problem in HCI. In: Proceedings of the SIGCHI Conference on Human Factors in Computing Systems, pp. 423–432 (2010)
31. Rana, A.I., Jennings, B.: Semantic aware processing of user defined inference rules to manage home networks. J. Netw. Comput. Appl. **79**, 68–87 (2017). https://doi.org/10.1016/j.jnca.2016.11.020
32. Xu, K., Wang, X., Wei, W., Song, H., Mao, B.: Toward software defined smart home. IEEE Commun. Mag. **54**, 116–122 (2016). https://doi.org/10.1109/MCOM.2016.7470945
33. Chetty, M., Feamster, N.: Refactoring network infrastructure to improve manageability: a case study of home networking. ACM SIGCOMM Comput. Commun. Rev. **42**, 54–61 (2012)
34. Mortier, R., Rodden, T., Lodge, T., McAuley, D., Rotsos, C., Moore, A.W., Koliousis, A., Sventek, J.: Control and understanding: owning your home network. In: 2012 Fourth International Conference on Communication Systems and Networks (COMSNETS 2012), pp. 1–10. IEEE (2012). https://doi.org/10.1109/COMSNETS.2012.6151322
35. Xydas, I., Miaoulis, G., Bonnefoi, P.-F., Plemenos, D., Ghazanfarpour, D.: 3D graph visualisation of web normal and malicious traffic. In: Tenth International Conference on Information Visualisation (IV'2006), pp. 621–629. IEEE (2006). https://doi.org/10.1109/IV.2006.2

36. Shannon, R., Quigley, A., Nixon, P.: Showtime: increasing viewer understanding of dynamic network visualisations. In: Proceedings of the International Conference on Advanced visual Interfaces, pp. 377–380 (2010)
37. Hong, S.H., Murtagh, T.: Visualisation of large and complex networks using polyplane. In: Graph Drawing, pp. 471–481 (2005)
38. Wilson, C., Hargreaves, T., Hauxwell-Baldwin, R.: Benefits and risks of smart home technologies. Energy Policy **103**, 72–83 (2017). https://doi.org/10.1016/j.enpol.2016.12.047
39. Samuel, S., Issac, S.: A review of connectivity challenges in IoT-smart home. In: 2016 3rd MEC International Conference on Big Data and Smart City, ICBDSC 2016, pp. 364–367 (2016). https://doi.org/10.1109/ICBDSC.2016.7460395
40. Guimaraes, V.T., Dal Sasso Freitas, C.M., Ramin Sadre, L.M., Granville, L.Z.: A survey on information visualization for network and service management. IEEE Commun. Surv. Tutorials **18**, 285–323 (2015). https://doi.org/10.1109/COMST.2015.2450538
41. Dimara, E., Bezerianos, A.: The attraction effect in information visualization. IEEE Trans Vis. Comput. Graph. **23**, 471–480 (2016). https://doi.org/10.1109/TVCG.2016.2598594
42. Boy, J., Rensink, R.A., Bertini, E., Fekete, J.D.: A principled way of assessing visualization literacy. IEEE Trans. Vis. Comput. Graph. **20**, 1963–1972 (2014). https://doi.org/10.1109/TVCG.2014.2346984
43. Knowles, B., Blair, L., Coulton, P., Lochrie, M.: Rethinking plan A for sustainable HCI. In: Proceedings of the 32nd Annual ACM Conference on Human Factors in Computing Systems - CHI'2014, pp. 3593–3596 (2014). https://doi.org/10.1145/2556288.2557311
44. Knowles, B., Blair, L., Hazas, M., Walker, S.: Exploring sustainability research in computing: where we are and where we go next. In: Proceedings of the 2013 ACM International Joint Conference on Pervasive And Ubiquitous Computing, pp. 305–314. ACM (2013)
45. Shneiderman, B., Plaisant, C.: Designing the User Interface: Strategies for Effective Human-Computer Interaction. Introduction, vol. 2 (2005). https://doi.org/10.1016/0166-3615(93)90066-A
46. Dix, A.: Human-computer interaction, foundations and new paradigms. J. Vis. Lang. Comput. Elsevier, 1–13 (2016). https://doi.org/10.1016/j.jvlc.2016.04.001
47. Obrenović, Ž.: Rethinking HCI education: teaching interactive computing concepts based on the experiential learning paradigm. Interactions **19**, 66–70 (2012). https://doi.org/10.1145/2168931.2168945
48. Posard, M.N., Rinderknecht, R.G.: Do people like working with computers more than human beings? Comput. Hum. Behav. **51**, 232–238. Elsevier Ltd. (2015). https://doi.org/10.1016/j.chb.2015.04.057
49. Håkansson, M., Sengers, P.: Beyond being green : simple living families and ICT. beyond being green: simple living families and ICT. In: Proceedings of the SIGCHI Conference on Human Factors in Computing, pp. 2725–2734. ACM (2013)
50. Viller, S., Worthy, P., Bodén, M., Weigel, J., Fitzpatrick, G., Rodden, T., Matthews, B.: IoT: designing for human values. In: Proceedings of the 2016 ACM Conference Companion Publication on Designing Interactive Systems, pp. 61–64 (2016). https://doi.org/10.1145/2908805.2913019
51. Gonzalez, C.S., Ordonez, C.A.C.: Special section on human-computer interaction and education. IEEE Revista Iberoamericana de Tecnologias del Aprendizaje **11**, 10–11 (2016). https://doi.org/10.1109/RITA.2016.2518439
52. Harvey, J., Golightly, D., Smith, A.: HCI as a means to prosociality in the economy. In: Proceedings of the 32nd Annual ACM Conference on Human Factors in Computing Systems-CHI'2014, pp. 2955–2964 (2014). https://doi.org/10.1145/2556288.2557367
53. Hourcade, J.P., Bullock-Rest, N.E.: HCI for peace: a call for constructive action. In: Chi 2011, pp. 443–452 (2011). https://doi.org/10.1145/1978942.1979005

54. Saizmaa, T., Kim, H.-C.: A holistic understanding of HCI perspectives on smart home. In: 2008 Fourth International Conference on Networked Computing and Advanced Information Management 2, pp. 59–65. IEEE (2008). https://doi.org/10.1109/NCM.2008.141
55. Kotzé, P., Wong, W., Jorge, J., Dix, A., Silva, P.A. ed. Creativity and HCI: from experience to design in education. In: IFIP – The International Federation for Information Processing, vol. 289. Boston, MA. Springer US (2009). https://doi.org/10.1007/978-0-387-89022-7
56. Karam, M.: A taxonomy of gestures in human computer interactions. Technical Report ECSTR-IAM05-009, Electronics and Computer Science, pp. 1–45 (2005)
57. Gillan, D.J., Cooke, N.J.: Methods of cognitive analysis for HCI. In: Conference companion on Human factors in computing systems-CHI'1995, pp. 349–350 (1995). https://doi.org/10.1145/223355.223716
58. Luu, L.: Five Principles for Effective Data Visualizations, pp. 1–9 (2017)
59. Jeffery, C., Al-Gharaibeh, J.: Writing virtual environments for software visualization. Writing Virtual Environments for Software Visualization, pp. 1–155 (2015). https://doi.org/10.1007/978-1-4614-1755-2
60. Thomas, J.J., Cook, K.A.: Visualization viewpoints: a visual analytics agenda. IEEE Comput. Graph. Appl. **26**, 10–13 (2006). https://doi.org/10.1109/MCG.2006.5
61. Shneiderman, B., Ben, S.: The Golden Rules Of User Interface Design. In: Designing the User Interface, pp. 1–28 (1997)
62. Myers, B., Hudson, S.E., Pausch, R.: Past, present, and future of user interface software tools. ACM Trans. Comput.-Hum. Inter. **7**, 3–28 (2000). https://doi.org/10.1145/344949.344959
63. Kwon, B.C., Javed, W., Elmqvist, N., Yi, J.S.: Direct manipulation through surrogate objects. In: Proceedings of the International Conference on Human Factors in Computing Systems (CHI'2011), pp. 627–636 (2011). https://doi.org/10.1145/1978942.1979033
64. Allen, M., Mclachlan, P.: NAV - Network Analysis Visualization. University of British Columbia, Online, Columbia (2009)
65. Barksdale, J.T., McCrickard, D.S.: Concept mapping in agile usability. In: Proceedings of the 28th of the International Conference Extended Abstracts on Human Factors in Computing Systems - CHI EA'2010, p. 4691 (2010). https://doi.org/10.1145/1753846.1754212
66. Abduljalil, S., Kang, D.K.: Analysis of human factors in software application design for effective user experience. In: 2011 13th International Conference on Advanced Communication Technology (ICACT), pp. 1446–1451 IEEE (2011)
67. Smith, C.L., Kantor, P.B.: User Adaptation: Good Results from Poor Systems. Sigir, pp. 147–154 (2008). https://doi.org/10.1145/1390334.1390362
68. Harrer, A., Göhnert, T.: Integrated representations and small data: towards contextualized and embedded analytics tools for learners. In: Proceedings of the 5th International Conference on Learning Analytics and Knowledge - LAK'2015, pp. 406–407 (2015). https://doi.org/10.1145/2723576.2723665
69. Saito, K., Kouzou Ohara, M.K., Motoda, H.: Graph embedding on spheres and its application to visualization of information diffusion data. In: Proceedings of the 21st International Conference Companion on World Wide Web - WWW'2012 Companion, pp. 1137 (2012). https://doi.org/10.1145/2187980.2188253
70. Rautek, Peter and Media Informatics. Augmented Visualization Gábor Sörös. Human-Computer Interaction (2010)
71. Mann, S.: Phenomenal augmented reality: advancing technology for the future of humanity. IEEE Consumer Electron. Mag. **4**, 92–97 (2015). https://doi.org/10.1109/MCE.2015.2463312
72. Billinghurst, M., Clark, A., Lee, G.: A survey of augmented reality. Foundations and Trends® in Human–Computer Interaction. **8**, 73–272 (2015). https://doi.org/10.1561/1100000049

73. Hsiao, I-H., Guerra, J., Parra, D., Bakalov, F., König-ries, B., Brusilovsky, P.: Comparative social visualization for personalized e-learning. In: Proceedings of the International Working Conference on Advanced Visual Interfaces-AVI'2012, p. 303 (2012). https://doi.org/10.1145/2254556.2254614
74. Karahalios, K.G., Viégas, F.B.: Social visualization. In: CHI'2006 Extended Abstracts on Human Factors in Computing Systems-CHI EA'06, 1667 (2006). https://doi.org/10.1145/1125451.1125758
75. Kumar, C., Poppinga, B., Haeuser, D., Heuten, W., Boll, S.: Assessing end-user interaction for multi-criteria local search with heatmap and icon-based visualizations. In: Proceedings of the 1st ACM SIGSPATIAL International Workshop on MapInteraction, pp. 13–19 (2013). https://doi.org/10.1145/2534931.2534937
76. Wattenberg, M.: A note on space-filling visualizations and space-filling curves. In: Proceedings - IEEE Symposium on Information Visualization, INFO VIS, pp. 181–186 (2005). https://doi.org/10.1109/INFVIS.2005.1532145
77. Bruls, M., Huizing, K., van Wijk, J.J.: Squarified Treemaps, pp. 33–42 (2000). https://doi.org/10.1007/978-3-7091-6783-0_4
78. Sathiyanarayanan, M., Burlutskiy, N.: Visualizing social networks using a treemap overlaid with a graph. Procedia Comput. Sci. **58**, 113–120 (2015). https://doi.org/10.1016/j.procs.2015.08.037
79. Li, N.: Hyperbolic Tree + Time Disc: visualizing hierarchical time-series data. In: VINCI'2014 Proceedings of the 7th International Symposium on Visual Information Communication and Interaction, p. 188 (2014). https://doi.org/10.1145/2636240.2636862
80. Clough, D., Rivera, S., Kuttel, M.: Panopticon : A Scalable Monitoring System. In: Proceedings of the 2010 Annual Research Conference of the South African Institute of Computer Scientists and Information Technologists, pp. 39–47. ACM (2010)
81. Villalon, J.J., Calvo, R.A.: Concept map mining: a definition and a framework for its evaluation. In: Proceedings-2008 IEEE/WIC/ACM International Conference on Web Intelligence and Intelligent Agent Technology - Workshops, WI-IAT Workshops 2008, pp. 357–360 (2008). https://doi.org/10.1109/WIIAT.2008.387
82. Cao, N., Cui, W.: Introduction to Text Visualization (2016). https://doi.org/10.2991/978-94-6239-186-4
83. Osburn, D.B.: Executive summary. In: National Visualization and Analytics Center-US Department of Homeland Security, pp. 1–4 (2010). https://doi.org/10.1053/j.ajkd.2006.03.010
84. Taylor, T.: Principles of Data Visualization - What We See in a Visual. Fusion Charts (2014)
85. Shneiderman, B.: Designing for fun. Interactions **11**, 48 (2004). https://doi.org/10.1145/1015530.1015552
86. Kriglstein, S., Wallner, G., Pohl, M.: A user study of different gameplay visualizations. In: Proceedings of the 32nd Annual ACM Conference on Human Factors in Computing Systems - CHI'2014, pp. 361–370 (2014). https://doi.org/10.1145/2556288.2557317
87. Shneiderman, B.: The eyes have it: a task by data type taxonomy for information visualizations. In: Proceedings 1996 IEEE Symposium on Visual Languages, pp. 336–343 (1996). https://doi.org/10.1109/VL.1996.545307
88. Bode, S., Riebisch, M.: Usability-focused architectural design for graphical user interface components. In: 2008 International Conference on Computational Intelligence for Modelling Control & Automation, pp. 1246–1251. IEEE (2008). https://doi.org/10.1109/CIMCA.2008.120
89. Wurman, R.S.: Information Anxiety 2, vol. 2 (2000)

A Sequential Decomposition Approach and Integrating the Concepts for Super-Networks in the Activity-Based Models

Mahdieh Allahviranloo(✉)

The City College of New York, 160 Convent Avenue, New York, NY 10031, USA
mallahviranloo@ccny.cuny.edu

Abstract. In this paper, a sequential decomposition method to model the travel behavior of individuals in a mixed transportation network (transit and roads) is proposed. A mixed binary model is developed assuming the mode choice of the individuals depend on the accessibility level for different modes and parking fees of the vehicles at different locations, where the parking fees vary depending on the demand level and time of day. The efficiency of the proposed model is tested using simulated data. It is shown that by slicing the activity patterns and parallelizing some segments of the algorithm, the computation time increases linearly as the population size increases.

Keywords: Super-networks · Activity-based models · Transportation systems

1 Introduction

Modeling mobility behavior in transportation super-networks, sometimes called as a network of networks, is a complex problem [1]. Most transportation problems are classified as NP-hard problems and the computation time scales up polynomial as the network size gets larger. The complexity is higher when we are dealing with multiple modes of transportation systems [2]. In this paper, we present a method to model the modal transfer. The proposed model is a decomposition-based approach, which will result in improving the computation efficiency of the algorithm. It uses the concepts of activity-based travel demand models to infer travel behavior [3–7]. The term 'activity' in these models refers to the participation of the population in any out of home action during the day, such as going to work, or reading a book in a café, visiting a friend in a park, etc. [3]. In travel demand analysis, forecasting the type of activities is important since it is argued that travel is a derived demand and people travel to be able to perform their daily activity routines at different locations [5, 6, 8, 9]. Some activities (e.g. going to work or attending a lecture at school) are mandatory and they cannot be replaced, relocated, or transferred to others whereas some are more flexible. Being probably the most fundamental step in the field of transportation, inferring travel demand is critical for successful executions of transportation projects. Demand models, which date back

T. Allahviranloo et al. (Eds.): IDS 2020, AISC 1301, pp. 86–98, 2021.
https://doi.org/10.1007/978-3-030-66501-2_7

to more than half a century ago, have been vastly developed and fully studied whether it is regional level models (e.g. four-step models, direct demand models) or they are individual-based models (activity-based models [3–7], tour-based models [5], simulations, etc. [5, 9, 10]). Having their roots in econometrics, these models relate the mobility behavior and travel-related choices to the demographics of the travelers and the attributes of the environments (accessibility, land use, can be examples of those attributes), and operational time-windows for activity locations [11, 12].

Realistic representation of the transportation models and capturing the true behavior of the population and their choice trends requires the use of super-networks or in other words, multimodal systems of networks [13]. A group of transportation modelers in the Netherlands has developed interesting multistate super-network models to study transition among different modes, and its impact on the duration of the activities or parking time [13–18]. Despite the existence of truly rich and novel works in the field, the integration of different modes of transportation and the development of mathematical models to study super-networks has not been fully investigated. Whether it is the complexity of these models, which holds researchers back, or the promising results achieved by advanced simulation-based models and frameworks that make simulation-based models more appealing compared to the super-network analysis, the multimodal system analysis is still truly an intriguing problem that deserves the attention of the academic community. In this paper, we will discuss a methodology where we can solve activity-based models in a super-network using a sequential approach. Kitamura proposed conditional sequential models to capture mobility behavior where each choice depends on the other choices made during the day [19, 20]. Markovian models are also appropriate models to infer the next activity and its properties of any individual based on the attributes of previous and future activities in the agenda [21–23]. In this paper, we use a stepwise model within the context of super-networks. The paper is outlined in five sections. We present the methodology and formulations in Sect. 2. Section 3 explains the model using a simple example. We present a discussion on application and scalability in Sect. 4 and finally, the paper is concluded in Sect. 5.

2 Methodology and Model Formulation

2.1 Assumptions

Before discussing the model, the set of assumptions that were made to develop the model is presented here. These assumptions are as follows:

Assumption 1: network topology for each mode of the transportation system is known. The routes, the location of stations (for non-automobile modes like transit stations), and the frequency of the service provided by each mode are inputs to the model.

Assumption 2: a detailed itinerary of the daily activity of individuals including activity sequences, activity types, the list of candidate locations to perform flexible activities, and the duration of each activity are known.

Assumption 3: activities are classified as mandatory and flexible activities. Flexible activities can be performed at different candidate locations and the duration and start time for flexible activities are flexible as well.

Assumption 4: if the person is traveling by a personal vehicle (car, personal bike), the transfer from the personal mode to the public mode, depends on the availability of the parking garage/station for the mode.

At any time during the day, the person can decide to leave the vehicle (or bike, scooter) parked at the parking location, and take transit to the next destinations if there is a nearby transit station. The decision depends on the overall cost/gain from the traveler perspective.

Assumption 5: we use dummy links to connect networks associated with different transportation modes. Travel times on these links if the connecting nodes are nearby are considered to be zero and the transition between the personal travel mode and the transit can occur if there is a transit station right above the node. This assumption is made for our case example, however, it can be easily adjusted for the real world networks without changing the overall solution and modeling approach.

2.2 Preliminaries

Activity patterns of the individuals are sequences of activities, each represented by $Y = \langle y_1, y_2, \ldots, y_n \rangle$, where n is the total number of activities on the agenda of each individual. Each activity $i, i \in \{1, \ldots, n\}$, is identified by some attributes such as the type of activity, start time, location, and the duration of the activity.

Figure 1 illustrates the trip trajectories that the individual has during the day.

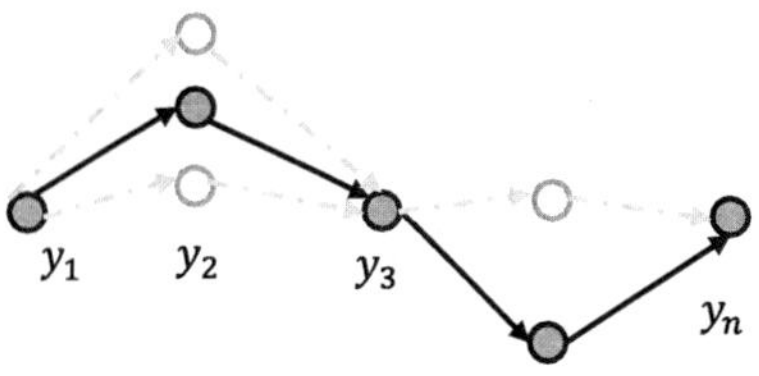

Fig. 1. Illustration of activity sequence

In this trajectory, activity 2 (represented by y_2), can be carried out in three different locations. An example of this type of activity can be grocery shopping which the traveler has multiple options to chose from. Activity 3 needs to be carried out in a designated place. The links connecting the activities represent trips and depending on the utility, access to the transit station or other modes of transportation, and the parking fees, the drivers may decide to (a) drive to all the destinations, or (2) to park in some intermediate nodes and then come back later during the day to pick up the vehicle, or (c) use alternative modes for all the trips. Travel cost per each segment of the trip depends on the distance between the nodes. Most transit systems have zone-based pricing mechanisms, therefore, we assume the decision to choose the next activity location and whether to drive to the next location or to park in the current location and take the transit will be a sequential decision. In fact, the decision at state 't' depends on the trip attributes at the state '$t + 1$' and state '$t + 2$'. The reason for selecting only two proceeding activities is the embedded uncertainty in travel time for future activities and variable parking fees. Figure 2 illustrates the decision procedure. At state 't', the traveler will only consider the proceeding two states. Assuming the person has traveled to state 't' with a car and if s/he decides to park the vehicle at the current location, a dummy node, depicted by the red circle, will be added. The functions of the dummy nodes are different from the other nodes. In dummy nodes, the transfers between different modes are allowed whereas in the other nodes the mode change is not allowed. At any state, once the vehicle (or any other personal transportation model) is used it should be dropped off at the next destination. Immediately after parking the vehicle, a new dummy node will be added to the network and the settings of the model will be updated.

Consider a person drives to the location 'a' and spends 30 min at this location. At any time after finishing the activity at location 'a', the person can pick up the car. It can be right after finishing the activity 'a' or it can be anytime later during the day after finishing any other activities. The decision on when to pick up the vehicle depends on the parking cost for the vehicle, travel cost with the alternative modes, and the schedule of the current and the next activity.

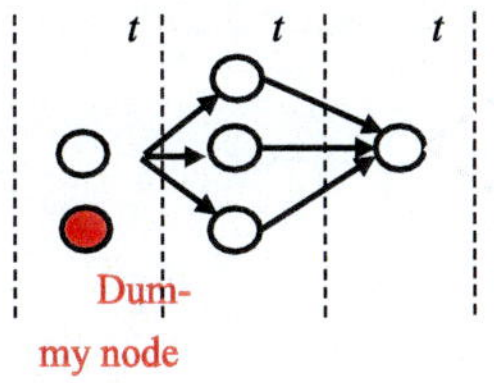

Fig. 2. Sequential decision making

2.3 Model Components

For the formulation of this model, we define the following sets, decision variables, and the parameters. In these equations 't' represents the state at each activity sequence.

Sets. n_t: set of nodes at state t, n_t^v: is the node that the personal vehicle (or bike or any other personal mode of transportation) is located at time t. N_t: set of all nodes at time t, where $N_t = \{n_t, n_{t+1}, n_{t+2}, n_t^v\}$. h: is the node associated with the home location of the individual. We set home location apart from the other nodes because people can return home at any time during the day, however, this condition does not apply for other nodes. V: represents the set of different modes of transportation, and V_p: represents the set of personal transportation modes.

Decision Variables. $x_{i,j}^k$ is a binary variable, takes the value of 1 if a trip is made from node i to node j by mode k. Y_i^t: is a binary variable used to identify if the personal transportation mode will be parked at node i at time state t. T_i^k: is a continuous variable representing arrival time to node i by mode k. TL_i represents the latest arrival time to any location regardless of the mode taken. We define this variable since the total number of trips to the place that the personal vehicle is parked can be more than once. P_i^t is the total parking fee paid for parking the personal vehicle in location i at time t.

Parameters. $C_{i,j}^k$: is the trip cost (or time) to travel from node i to node j by mode k. tt^k: is the travel time matrix by mode k. p_i^t represents the parking fees for mode k. It is assumed that the parking fees vary based on the demand level at different times of the day.

We will incorporate the variable parking fee strategy by slicing the time of the day to 24 h. Before departing from their location at the time 't', individuals decide based on the price of parking and travel time/cost trade-off for their next activities in the itinerary, at the beginning of the day. However, as time passes, the parking fees will be updated according to the demand level for parking, Fig. 3.

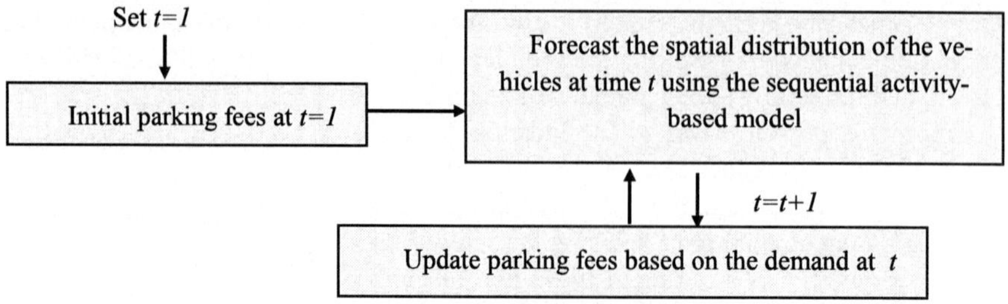

Fig. 3. Time-variant parking fee

2.4 Model Formulation

Depending on the state that the driver is making the decision, three types of models are used. M is used for the intermediate states of the day; M1 is applied for the first trip of the day, @state $t = 1$; And M2 is used for the last trip of the day, @state $t = n$. First, the equations for the M model are presented followed by the relevant modifications for M1 and M2.

Formulation of M. The objective function of M:

$$Min\ Z(x, Y, T, TL, P) = \left\{ \sum\nolimits_{i,j \in N_t, k \in V} \left(tt_{ij}^k + cost_{ij}^k \right) \times x_{ij}^k + \sum\nolimits_{j \in \{n_{t+1}, n_t^v\}} P_i^t \right\} \quad (1)$$

Subject to the following set of constraints:

$$\sum\nolimits_{i \in N_t, i \neq n_t,} x_{n_t,i}^k = 1, k\ is\ mode\ used\ in't-1'step \quad (2)$$

$$\sum\nolimits_{i \in N_t, i \neq n_t, k \in V} x_{i,n_t}^k = 0, \quad (3)$$

$$\sum\nolimits_{i \in n_{t+1}} \sum\nolimits_{j \in \{n_t, n_t^v\}, i \neq j, k \in V} x_{j,i}^k = 1, \quad (4)$$

$$\sum\nolimits_{i \in n_{t+2}} \sum\nolimits_{j \in \{n_{t+1}, n_t^v\}, i \neq j, k \in V} x_{j,i}^k = 1, \quad (5)$$

$$\sum\nolimits_{j \in N_t, j \neq, n_t^v, k \in V} x_{n_t^v,j}^k = \sum\nolimits_{j \in N_t, j \neq, n_t^v, k \in V} x_{j,n_t^v}^k, \quad (6)$$

$$\sum\nolimits_{j \in \{n_{t+2}, n_t^v\}} x_{ij}^k = \sum\nolimits_{j \in \{n_t, n_t^v\}} x_{ji}^k, \forall i \in \{n_{t+1}\}, k \in V, \quad (7)$$

$$T_i^k + tt_{ij}^k + S_i \leq T_j^k,\ if\ x_{ij}^k = 1,\ \forall i, j \in \{n_t, n_{t+1}, n_t^v\} \quad (8)$$

$$a_i \leq T_i^k \leq b_i\ \forall i \in N_t k \in V \quad (9)$$

$$T_i^k \leq TL_i, \forall i \in N_t \tag{10}$$

$$x_{n_t,j}^k \leq M.Y_j^t\,, \forall j \in \{n_{t+1}, n_t^v\}, k \in V_p \tag{11}$$

$$\sum\nolimits_{i \in \{n_{t+1}, n_t^v\}} Y_i^t = 1, \tag{12}$$

$$P_i^t = (TL_i - t) \times p_i^t, \text{ if } Y_i^t = 1, \forall i = n_t^v \tag{13}$$

$$P_i^t = S_i \times p_i^t, \text{ if } Y_i^t = 1, \forall i \in n_{t+1} \tag{14}$$

$$x_{ij}^k \in \{0, 1\}, Y_j^t \in \{0, 1\}, TL_i \geq 0, T_i^k \geq 0, P_i^t \geq 0. \tag{15}$$

The objective function of the model M is represented by Eq. 1. It minimizes the total travel time, travel cost, and the total parking fee for the vehicle owners.

Constraint 2 ensures that the person will depart from the current node to pursue the next activity in the agenda, using the same mode of transportation that he has traveled to node n_t. Keep in mind that we have separated the nodes that the vehicle is parked by n_t^v. Constraint 3, guarantees that the current node cannot be the destination for any trip. Meaning that the current activity location can only generate trips because the person has already arrived at the current location and since this is a binary model, the trips cannot be destined to the already visited locations. Constraint 4, ensures that a trip should be made either from the current location, 'n_t', of the person or from the location of the parked personal vehicle, 'n_t^v' to the location at 't + 1'. Constraint set 5, ensures that a trip should be made from either the intermediate node, 't + 1', or the location of the parked personal vehicle to the activity located at the state of 't + 2'. Constraint 6, allows to model to switch between the modes in the parking location. This is the only node that this transition between the modes is allowed. Constraint 7, conserves flow in the network, and it also ensures that the same mode is used for entering and exiting trips from the intermediate nodes. Constraints 8, is set to compute the arrival time to each location which cannot be earlier than the latest arrival time to the previous location plus the time spent on the previous location and the travel time between the locations of the previous and current activities. Constraint 9 is the time windows constraint, meaning arrival time to each activity location should be within a designated operating hour, e.g., some workplaces have fixed operating hours. Constraint 10, computes the latest arrival time to each location, TL_i. Constraints 11 and 12, are added to find the parking location of the personal mode. Given that the vehicle is parked at node n_t^v, the driver can still leave the vehicle there or drive it to the next activity location which is going to be a node in the set n_{t+1}. Constraint 13 and constraint 14 compute the total parking cost to park the vehicle whether in its current parking location or it is the location of the next following activity.

Modification to M1. In the *M1* model, n_t and n_t^v, both refer to the nodes associated with the home location, $n_t = n_t^v = h$, and the person will start the initial trip from home. This model is the same as model M except we exclude Eq. (2) from the model.

Modification to M2. In this model, which represents the last segment of the daily activity chain, n_{t+2} refers to the home location. If the person has parked the personal vehicle at any location other than home, the last return home trip should be made by this mode, making sure that the personal vehicles (bike, car, scooter, etc.) be returned to the home. To accommodate these conditions, the following changes are made to the *M* model:

- Equation 2 will not be used in the model.
- Personal vehicles should be picked up from the parked location, which is handled by adding the following constraint to the model:

$$\sum\nolimits_{i\in n_t^v} x_{ij}^k = 1 \text{ , } k : personal\ mode, j \in N_t \tag{16}$$

3 Case Study Experiment

In this section, we discuss the results of the application of the model on a very small sample set to illustrate the type of data that the model would require.

We tested the model on a simulated sample set. The sample consists of 100 individuals. Three modes of transportation including personal vehicles, bus, and subway are defined for this example. Figure 4 illustrates the sample network used for this study and Fig. 5 illustrates the concept of dummy links used to connect transit and street networks. In these figures, the lines in gray illustrate the street network, the green lines represent bus service lines and the red lines illustrate the subway network. Node IDs are listed on them and the large circles show the transfer nodes where the transition from one mode to another is feasible.

Assuming the equal length of 3 miles for all the links, the travel time on the road network is computed using the shortest path method, at the average speed of 20 mph. The travel time with transit (including access, transfer and wait time) is assumed to be twice as the travel time with personal car. The monetary value of the time is assumed to be \$10/h. Subsidized transit fare is assumed to be \$0.25 per mile of transit distance. Fuel cost for the personal vehicle is computed based on an average of 30 mpg fuel consumption at the price of \$3/gallon. We assume a variable parking fee at each node, where the base fee is a function of the time of day and the location.

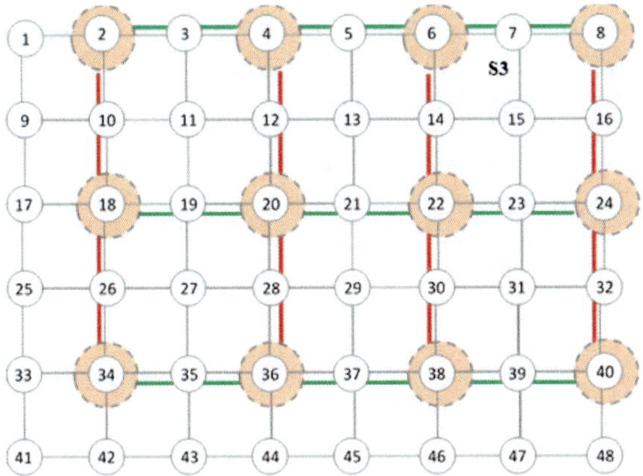

Fig. 4. Illustration of super-network designed for the case study

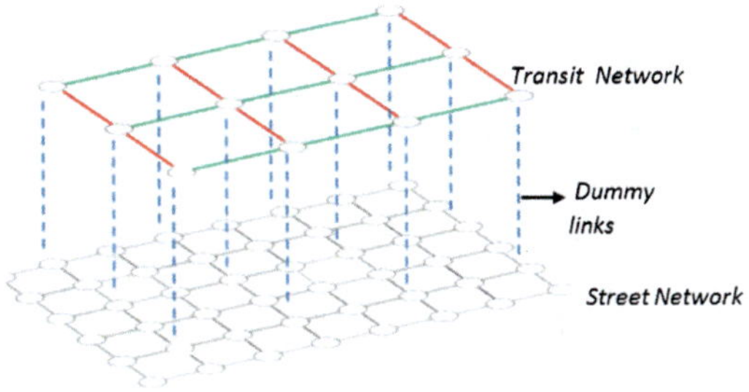

Fig. 5. Dummy link connecting different networks of transportation systems

Table 1 presents the agenda including the location, schedule, and the type of different activities that individuals plan to pursue during the day, for three individuals randomly selected from the population. According to this table, person 1 has 3 activities on her agenda. Her home is located at node 14 and she plans to conduct the first activity at node 15 and should arrive there anytime between 06:00 to 06:30. She will spend about 1.5 h on node 15. Her second activity will be carried out at node 23 with a duration of 9 h, and she can conduct her last activity at either of nodes 5, 13, or 21. The results of the model for the three individuals are presented in Table 1. Person 1, will use a car for the first two activities but then after finishing activity 2 at node 23, the person will return home at 06:00 PM and use transit to go to node 22. The model predicts the optimum routing, scheduling, and mode choices. According to the results of the model, person 2 and 3, both will use transit for their daily trips.

We tested the impacts of the value of time and parking fees on the percentage of transit usage in the sample population. Assuming a stepwise function for the base-charging fee in the parking locations, the algorithm starts predicting activity patterns; however, as time passes the parking fees will be adjusted according to the demand and the number of visits at each location.

Table 1. Sample activity itinerary of individuals in The test study

Individual	Itinerary and the activity patterns (VOT = \$10/h. Non-peak hour rate was set to \$3/h, peak hour rate was \$7/h, peak hour is [7:00–09:00] and [17:00–19:00].)
1	{ Home: Node 14 Act1: node15 @ [06:00–06:30], ~1.5 h, Act2: node23 @ [9:00–10:00], ~9 h Act3: nodes {5,13,22} @ [20:00–21:00], ~1 h} Results: 14 → 15 (@06:10, by car) → 23 (@9:00, by car) → 14 (@18:00, by car) → 22 (@20:00 by transit) → 14 (@21:40)
2	{Home: Node 2 Act1: node 11 @ [08:00–09:00], ~7 h, Act2: nodes {17,19} @ [16:00–17:00], ~1 h Act 3: node 3 @ [20:00–21:00], ~1 h} Results: 2 → 11 (@08:00 by transit) → 19 (@ 16:40, transit) → 3 (@20:00, transit) → 2
3	{Home: Node 39 Act1: node 40 @[05:00–06:00], ~1 h, Act2: nodes 37 @[09:00–10:00], ~8 h Act 3: node 21 @[19:00–20:00], ~1 h} Results: 06:00, transit) → 37 (@ 09:00, transit) → 21 (@19:00, transit) → 39 (@21:10, transit)

Figure 6(a), illustrates the percentage of transit trips as a function of the value of time (VOT). The hourly parking fees are kept in the range of \$3 to \$5, and the peak hours are assumed to be [06:00 to 09:00] and [16:00–19:00]. As seen in the graph, when VOT increases the share of transit reduces, and for VOT higher than \$36/h, due to longer travel time for the transit system, all the trips will be shifted to personal vehicles and the share of transit reduces to 0. According to this figure, for some ranges in the VOT, the share is constant (e.g. the share of transit is 0.4 when VOT is in the range of 10\$/h to 14\$/h). Figure 6(b) illustrates the change in transit share as the average parking fee changes in the network. For higher values of the parking fee, the share of transit reaches its maximum which is %78, since some nodes are not accessible by transit and some travelers have no choice other than driving. In this network, 78% is the upper bound for the transit share.

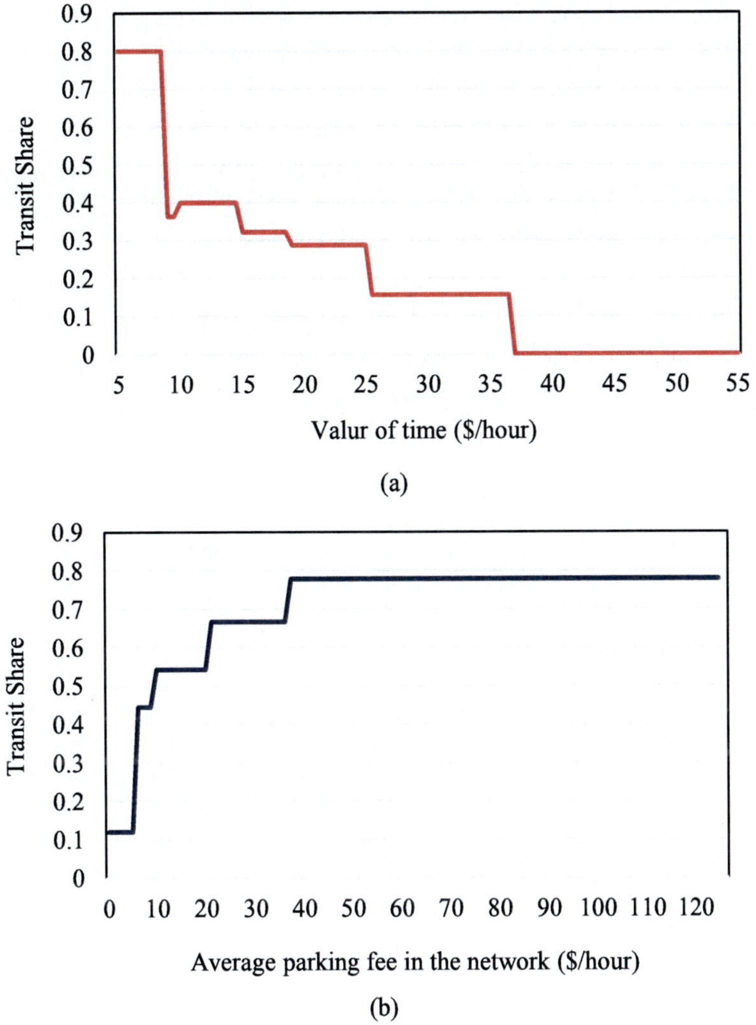

Fig. 6. Illustration of super-network designed for the case study

4 A Discussion on the Model

4.1 Scalability

The simulation was coded in Matlab 2019a and the optimization models were executed using AMPL. The computation time for daily pattern generation of 100 individuals using the developed model, in a single-core computer without parallel computation, was −3 s. The computation profile generated for the code shows ~80% of the run time is related to the segments of the code that can be parallelized. The remaining 20% of the run time is mostly the time needed to setting the code path and calling the intermediate functions of the code, which their computation time does not depend on the problem size. Table 2 partitions the code based on parallel computation capabilities.

Table 2. Parallel Sections of the model

Parts of code that cannot be executed in parallel	Parts of code that can be executed in parallel
Data processing and preparation	Updating the activity pattern of each individual at time '*t*'
Generation of the input data for each time step '*t*'	Running AMPL optimization code for each individual at time '*t*'
Collecting the results of population mobility behavior at time step 't' to update the parking fees for time '*t* + *1*'	

We set the sections that can be executed in parallel and evaluated the model performance through the simulation for a population size of 10^5 and 10^6, using 24 cores. The execution time for each scenario was 3.5 h and 5.6 h respectively.

4.2 Further Applications

The development of models to solve problems related to super-networks and smooth integration of different modes of transportation is important. Considering a growing interest in the use of shared mobility systems, such as bike-sharing, scooter sharing, ride-sharing, etc., the transition from one mode to another mode is growing. Flexibility and affordability of these shared systems are the main two contributing factors in their market growth, which have changed the mobility behavior in the population. People who may rely on one mode of transportation now can seamlessly transfer from one to another, without wasting time in transfer between the modes. Therefore, in the transportation field, the demand for the development of efficient and scalable models is rising and due to its simplicity, the work presented here can be easily adapted to different algorithms and models to estimate the travel behavior of the population with a fine resolution.

5 Conclusions

The emergence of new modes of transportation provides a smooth transition between different modes, demanding modeling multimodal systems. Due to differences in the attributes of transportation systems, integration between the modes is not an easy task and often is solved through heuristic methods. For instance, transit systems have fixed routes, with more regulated schedules and service frequencies and limited capacity. The price of transit fare and the travel time with the transit system will be different from the personal vehicle. For the case of shared bike systems, the station locations and the availability of bikes at each station are important factors affecting their usage and utilization.

In this paper, we develop a simple, yet innovative, and efficient method to predict the mobility behavior of the population in super networks. In the proposed model, users will have the choice of using transit or their vehicle (or any other personal mode). For the personal modes, the travelers have to leave the vehicles in the parking locations whereas other modes do not have this constraint. Itinerary of the travelers including the set of activities, their schedule, the locations are provided in the system. Also if the activity has time and location flexibility, this information is also provided in the system.

We infer the sequence of activities using a sequential pattern segmentation method. The decision to travel with personal vehicle or public transportation depends on the travel time and travel cost using each mode and the cost of parking for personal vehicles. If a trip is made by vehicle, at the park location, a dummy node will be added to the network where it will serve as a transfer node, and the person can park the vehicle in the parking and transfer to another mode. However, at the end of the day, the parked vehicle should be dropped off at the home location. Travelers decide whether to use transit or not and where to execute their flexible activities based on the other activities during the day. Parking prices at different locations follow a dynamic model where they depend on the demand level and time of day.

The model was tested on a small sample consisted of 100 individuals and also the scalability of the model for larger datasets were tested. Using parallel coding, the computation time for the model increases linearly with the size of the population. Although the work presented here is simple in its formulation, it provides a new approach to break down large transportation problems and solve the activity generation problem in a multimodal system efficiently.

References

1. Gao, J., Buldyrev, S.V., Stanley, H.E., Havlin, S.: Networks formed from interdependent networks. Nat. Phys. (2011)
2. Liao, F., Arentze, T., Timmermans, H.J.: Super-network approach for multi-modal and multi-activity travel planning. Transportation Research Record, pp. 38–46 (2010)
3. Ben-Akiva, M., McFadden, D., Train, K., Walker, J., Bhat, C., Bierlaire, M., Bolduc, D., Brownstone, D., Bunch, D.S., Daly, A., De Palma, A., Gopinath, D., Karlstorm, A., Munizaga, M.A.: Hybrid choice models : progress and challenges. Massachusetts Inst. Technol. Mark. Lett. **13**(3), 163–175 (2000)
4. Bowman, J.L.: Activity-based disaggregate travel demand model system with activity schedules. Transp. Res. Part A **35**, 1–28 (2000)
5. RDC Activity-Based Modeling System For Travel Demand Forecasting. Washington D.C. (1995)
6. Recker, W.W., McNally, M.G., Root, G.S.: A model of complex travel behavior: Part I—Theoretical development. Transp. Res. Part A **20**, 307–318 (1986)
7. Root, G.S., Recker, W.W.: Toward a dynamic model of individual activity pattern formulation. Recent Adv. Travel Demand Anal. pp. 371–382 (1981)
8. Supernak, J.: Temporal utility profiles of activities and travel: uncertainty and decision making. Transp. Res. Part B Methodol. **26**, 61–76 (1992)
9. Bhat, C.R., Guo, J.Y., Srinivasan, S., Sivakumar, A.: A comprehensive econometric micro-simulator for daily activity-travel patterns (CEMDAP). Transp. Res. Rec. **1894**, 57–66 (2004)

10. Becker, H.D., Guggisberg Bicudo, B., Axhausen, K.W.: A MATSim 2015 scenario for Basel, Switzerland. In: 18th Swiss Transport Research Conference, Ascona (2018)
11. Recker, W.W.: The household activity pattern problem: general formulation and solution. Transp. Res. Part B: Methodol. **29**, 61–77 (1995)
12. Recker, W.W.: A bridge between travel demand modeling and activity-based travel analysis. Transp. Res. Part B Methodol. **35**, 481–506 (2001)
13. Carlier, K., Fiorenzo Catalano, S., Lindvel, C., Bovy, P.: A supernetwork approach towards multimodal travel modelling. In: Proceedings of the 82nd Annual Meeting of the Transportation Research Board, Washington, D.C. (2003)
14. Liao, F., Arentze, T., Timmermans, H.: Multi-state super-networks: recent progress and prospects. J. Traffic Transp. Eng. **1**(1), 13–27 (2014)
15. Fu, X., Lam, W.H.K.: A network equilibrium approach for modelling activity-travel pattern scheduling problems in multi-modal transit networks with uncertainty. Transportation **41**(1), 37–55 (2014)
16. Feixiong, L.: Modeling duration choice in space-time multi-state supernetworks for individual activity-travel scheduling. Transp. Res. Part C **69**, 16–35 (2016)
17. Feixiong Liao, L., Arentze, T., Timmermans, H.: Incorporating space–time constraints and activity-travel time profiles in a multi-state supernetwork approach to individual activity-travel scheduling. Transp. Res. Part B: Methodol. **55**, 41–58 (2013)
18. Feixiong Liao, L., Arentze, T., Timmermans, H.: Multi-state supernetworks: recent progress and prospects. J. Traffic Transp. Eng. **1.1**, 13–27 (2014)
19. Kitamura, R., Chen, C., Pendyala, R., Narayanan, R.: Micro-simulation of daily activity-travel patterns for travel. Transportation **27**, 25–51 (2000)
20. Kitamura, R., Fujii, S.: Time-use data, analysis and modeling: toward the next generation of transportation planning methodologies. Transp. Policy **4**, 225–235 (1997)
21. Allahviranloo, M., Recker, W.: Daily activity pattern recognition by using support vector machines with multiple classes. Transp. Res. Part B Methodol. **58**, 16–43 (2013)
22. Allahviranloo, M., Chastanet, L., Rehmann, J.: Mobility knowledge discovery to generate activity pattern trajectories. In: Proceeding of IEEE 20th International Conference on Intelligent Transportation Systems, Yokohama, Japan (2017)
23. Allahviranloo, M., Recker, W.W.: Mining activity pattern trajectories and allocating activities in the network. Transportation (2015)

Non-rigid Joint Segmentation and Registration Using Variational Approach for Multi-modal Images

Mazlinda Ibrahim[1(✉)], Lavdie Rada[2], Adela Ademaj[3], and Ke Chen[4]

[1] Centre for Defence Foundation Studies, National Defence University of Malaysia, 57000 Kuala Lumpur, Malaysia
mazlinda@upnm.edu.my

[2] Biomedical Engineering Department, Bahcesehir University, Besiktas, Istanbul, Turkey
lavdie.rada@eng.bay.edu.tr

[3] Biomedical Engineering Department, Heidelberg University, Heidelberg, Germany
oh288@uni-heidelberg.de

[4] Centre for Mathematical Imaging Techniques and Department of Mathematical Sciences, The University of Liverpool, Liverpool L69 7ZL, UK
K.Chen@liv.ac.uk

Abstract. Segmentation and registration are two common problems in medical imaging in particular and computer vision in general. These two problems have been studied substantially in the past two decades and often required as simultaneous tasks. Combination of these tasks in a single framework has proven to yield better results in terms of accuracy. In this paper, a model for the joint segmentation and registration based on the variational approach is presented for non-rigid multi-modal. The model is based on an active contour without edges for the segmentation task, normalized gradient fields, and the linear curvature model for image registration. The discrete functional of the new joint model is optimized using coarse to fine registration. The model is evaluated on synthetic and medical images and compared with the existing model. The proposed model is comparable based on the two evaluation criteria used.

Keywords: Segmentation · Non-parametric image registration · Regularization · Variational model · Optimization · Multi-modality images

1 Introduction

Image segmentation and registration are closely related to image processing tasks in a wide range of problems in medical imaging and pattern recognition. Image registration aims to find an optimal geometric transformation so that the appearance of the template image is similar in some sense to the reference image. Meanwhile, in image segmentation, images are partition into several meaningful parts

T. Allahviranloo et al. (Eds.): IDS 2020, AISC 1301, pp. 99–112, 2021.
https://doi.org/10.1007/978-3-030-66501-2_8

with distinguishable characteristics to identify the relevant structure. Normally, these two tasks are tackled independently. But, the effectiveness of the tasks can be improved significantly by coupling these two tasks [3,4,14,16].

In the variational approach, active contour model without edges is used to sub-divide the images into foreground and background for the segmentation task. For the variational image registration, the transformation is modeled as an optimal physical orientation process preserving one-to-one deformation between two given images using different transformation modeling such as elastic, non-elastic, and diffusion via the regularization functional. The joint registration and segmentation task competes the simultaneous registration followed segmentation. Commonly, in the joint frameworks the segmentation functional is combined with registration functional to form the final joint functional. Then, the functional is solved using optimization techniques.

Many examples were such a joint task will be useful are found in medical imaging since medical images are acquired from different imaging machines such as magnetic resonance imaging (MRI), computed tomography (CT) and X-ray. Different imaging machines produce different intensity values for the same objects or features. At the same time, the images are different due to the patients' motion and diseases progression. Thus, image registration and segmentation are required to further analyze and combine information from these images. Radiologist needs a fully automatic tool to understand the images due to the time consuming for the manual delineation of the objects of interest. A simple example, such as a lung segmentation for MRI for the assessment of the pulmonary perfusion, the amount of time needed by a specialist in a field to accomplish the task varies from 30 min to one hour per patient. This leads to a huge time consuming and automatic segmentation is required. To cope with such task, Kohlman and his colleagues [13] introduced a morphological operation based segmentation. This work focused only on the segmentation task without considering other image modalities taken for the same patient in the different time.

The combined task of image segmentation and registration can be divided into two groups, first, joint image segmentation and registration (JSR) framework, second, simultaneous registration and segmentation (SRS). Some authors used the wording *joint* instead of *simultaneous* and vice versa although it was not used *per se* as defined below. The JSR models use either the segmentation of the reference or the template images as the input in the models. As such [7,14,16] and [20]. The segmentation and registration tasks are treated within each iteration step. The first group is also called as *regmentation* and it is different from atlas based registration. In the atlas based registration, the atlas acts as the reference image and the template image is registered with the atlas. The resulting transformation from the registration is then applied to the atlas which was labeled beforehand. Then, the process produced the segmentation of the template image.

The SRS models are two steps process: registration and segmentation. These steps are iteratively repeated until reaching the stopping criteria. The iterative updates to registration and segmentation are typically performed in alternating steps. Thus SRS also can be regarded as alternating registration and segmentation. As an example the models in [2,9,18] and [22]. In addition, the SRS models do not require the segmentation of the reference or the template images in the initialization step.

In 2011, Guyader and Vese [14] presented a JSR model where the features in the template and reference images are implicitly modeled with level set functions. The model used the non-linear elastic regularization to control the transformation. The segmentation process is based on the active contour without edges and the numerical method used is based on the finite difference and Augmented Lagrangian methods. Later, in 2014, Ozere and Guyader [16] improved the previous distance measure in [14] where the new similarity measure is based on weighted total variation (TV). These two models ([14] and [16]) are only for mono-modal images where the intensity values of the images are comparable. To the best of our knowledge, there is no extension of their work for the multi-modal images.

A different approach for the JSR model was presented by Hodneland et al. in [10] where they used a time normalized, continuous Mahalanobis distance to a training set and applied the shape information. The training set is utilized to compute the training masks and guided the actual segmentation problem (so-called supervised segmentation). The similarity measure is the normalized gradient fields (NGF) and the model was applied to 4D dynamic contrast enhanced DCE-MRI of the kidneys from ten healthy patients. The regularization term in the proposed model is the first order linear elastic model. The drawback of this model is the requirement of a training set and shape information.

In the JSR work introduced by Swierczynski et al. [20] advantages of the joint framework has been shown in comparison with sequential segmentation and registration performed separately. However, their work is only valid for mono-modal images. They also pointed out several advantages of the JSR models. First, the model will eliminate the problem of initializing surface based segmentation method. Second, the joint tasks will incorporate prior knowledge about the segmentation of the images into registration. Third, the separate tasks of segmentation and registration are more prone to image noise and algorithm initialization.

In this paper, a novel joint segmentation and registration based on variational approach is proposed for multi-modal images. The multi-modal images are images that have been acquired through different sensors or imaging machines. The proposed model is an extension for the works in [1] and [12]. In [12], the model is only suitable for the mono-modal JSR. When the images have been acquired through different imaging machine, the functional in [12] is no longer valid. Thus we have to modify the functional using normalized gradient fields (NGF). The authors in [1] used the mutual information (MI) based similarity measure for the joint tasks. They compared the proposed model with MI and NGF. But, these two models used for the comparison in [1] are not the JSR models.

Our proposed model improves the existing model in [1] based on the test images and allows large and smooth transformation fields. The model obtained higher values of the performance metric and dice coefficient than the model in [1] for the particular images. However, there are set of images which showed that the model in [1] is superior to the proposed model.

The contributions of this work can be shortly summarized as follows:

- A joint variational framework for image segmentation and registration based on active contour without edges which was solved using level set, normalized gradient fields (NGF) and the linear curvature is proposed for multi-modal images.
- Evaluation of the framework on a set of synthetic and medical images.

The outline of this paper is as follows. In Sect. 2 the mathematical background of the joint models is introduced. In Sect. 3 the existing model in [1] which is based on MI is reviewed. In Sect. 4 model based on NGF and the algorithm is introduced. The numerical results are in Sect. 5. Finally, in Sect. 6 the paper is concluded.

2 Background

Some notation and mathematical background related to JSR are introduced. Let T denote the template image and R the reference image, $R, T : \Omega \subset \mathbb{R}^2 \to \mathbb{R}$, given as compactly support functions and denote by $\boldsymbol{\varphi} = \boldsymbol{\varphi}(\boldsymbol{x}) : \Omega \to \mathbb{R}^2$, the unknown transformation aiming for $T(\boldsymbol{\varphi}(\boldsymbol{x})) \approx R(\boldsymbol{x})$ with $\boldsymbol{x} = (x_1, x_2)$. In the non-parametric (variational approach) image registration, the transformation is written as $\boldsymbol{\varphi}(\boldsymbol{x}) = \boldsymbol{x} + \boldsymbol{u}(\boldsymbol{x})$, with $\boldsymbol{u}(\boldsymbol{x})$ the displacement vector field defined as $u(\boldsymbol{x}) : \Omega \to \mathbb{R}^n$, ($n = 2$ or 3). This transformation enables us to focus on the unknown displacement vector $\boldsymbol{u}(\boldsymbol{x}) = (u_1(\boldsymbol{x}), u_2(\boldsymbol{x}))$. Here $\boldsymbol{u}(\boldsymbol{x})$ is searched over admissible functions in the set $\mathcal{U}$, a linear subspace of a Hilbert space with Euclidean scalar product.

The segmentation of the template image is represented by the zero level line $\phi_0 : \Omega \to \mathbb{R}$ to represent target contour Γ given as

$$\begin{cases} \Gamma = \partial\Omega_1 = \{(x_1, x_2) \in \Omega | \phi_0(x_1, x_2) = 0\}, \\ \text{inside}(\Gamma) = \Omega_1 = \{(x_1, x_2) \in \Omega | \phi_0(x_1, x_2) > 0\}, \\ \text{outside}(\Gamma) = \Omega_2 = \{(x_1, x_2) \in \Omega | \phi_0(x_1, x_2) < 0\}. \end{cases} \tag{1}$$

The JSR model based on active contour without edges is given by the following functional

$$\begin{aligned} \min_{c_1, c_2, \boldsymbol{u}(\boldsymbol{x})} \mathcal{J}(c_1, c_2, \boldsymbol{u}(\boldsymbol{x})) = \lambda_1 & \int_\Omega |R(\boldsymbol{x}) - c_1|^2 H_\epsilon(\phi_0(\boldsymbol{x} + \boldsymbol{u}(\boldsymbol{x}))) \, \mathrm{d}\boldsymbol{x} \\ & + \lambda_2 \int_\Omega |R(\boldsymbol{x}) - c_2|^2 (1 - H_\epsilon(\phi_0(\boldsymbol{x} + \boldsymbol{u}(\boldsymbol{x})))) \, \mathrm{d}\boldsymbol{x} \\ & + \mathcal{D}(T, R, \phi_0(\boldsymbol{x}), \boldsymbol{u}(\boldsymbol{x})) + \alpha \mathcal{S}^{\mathrm{LC}}(\boldsymbol{u}) \end{aligned} \tag{2}$$

where

$$\mathcal{S}^{\mathrm{LC}}(\boldsymbol{u}) = \int_\Omega (\Delta u_1)^2 + (\Delta u_2)^2 \, \mathrm{d}\boldsymbol{x}. \tag{3}$$

The functional in (3) is the linear curvature model for image registration [8]. The constants c_1 and c_2 in (2) are the average intensities inside and outside the curve Γ in the reference image. The curve Γ is represented by the zero level line as in Eq. (1) and H_ϵ is a regularized Heaviside function

$$H_\epsilon(z) = \frac{1}{2}\left(1 + \frac{2}{\pi}\arctan\frac{z}{\epsilon}\right)$$

with its corresponding Delta function

$$\delta_\epsilon(z) = \frac{\mathrm{d}H_\epsilon(z)}{\mathrm{d}z} = \frac{\epsilon}{\pi(\epsilon^2 + z^2)}.$$

In [12], the authors used $\mathcal{D}(T, R, \phi_0(\boldsymbol{x}), \boldsymbol{u}(\boldsymbol{x})) = \mathcal{D}^{\text{SSDH}}(T, R, \phi_0(\boldsymbol{x}), \boldsymbol{u}(\boldsymbol{x}))$ where

$$\mathcal{D}^{\text{SSDH}}(T, R, \phi_0(\boldsymbol{x}), \boldsymbol{u}(\boldsymbol{x})) = \lambda_3 \int_\Omega (T((\boldsymbol{x} + \boldsymbol{u}(\boldsymbol{x})) - R(\boldsymbol{x})))^2 H_\epsilon(\phi_0(\boldsymbol{x} + \boldsymbol{u}(\boldsymbol{x})))\,\mathrm{d}\boldsymbol{x} \tag{4}$$

for mono-modal JSR. $\mathcal{D}^{\text{SSDH}}$ is the modified version of the sum of the squared difference (SSD)

$$\mathcal{D}^{\text{SSD}}(T, R, \boldsymbol{u}(\boldsymbol{x})) = \frac{1}{2}\int_\Omega (T(\boldsymbol{x} + \boldsymbol{u}(\boldsymbol{x})) - R(\boldsymbol{x}))^2\,\mathrm{d}\boldsymbol{x} \tag{5}$$

The functional in Eq. (4) is weighted by the parameter λ_3 and the term $H_\epsilon(\phi_0(\boldsymbol{x} + \boldsymbol{u}(\boldsymbol{x})))$. By adopting a level set formulation, $\phi_0(\boldsymbol{x})$ as in Chan-Vese [6], the Eq. (2) is minimize with respect to c_1 and c_2 to obtain:

$$\begin{aligned} c_1 &= \frac{\int R(\boldsymbol{x}) H_\epsilon(\phi_0(\boldsymbol{x} + \boldsymbol{u}(\boldsymbol{x})))\,\mathrm{d}\boldsymbol{x}}{\int H_\epsilon(\phi_0(\boldsymbol{x} + \boldsymbol{u}(\boldsymbol{x})))\,\mathrm{d}\boldsymbol{x}}, \\ c_2 &= \frac{\int R(\boldsymbol{x})(1 - H_\epsilon(\phi_0(\boldsymbol{x} + \boldsymbol{u}(\boldsymbol{x}))))\,\mathrm{d}\boldsymbol{x}}{\int 1 - H_\epsilon(\phi_0(\boldsymbol{x} + \boldsymbol{u}(\boldsymbol{x})))\,\mathrm{d}\boldsymbol{x}}. \end{aligned} \tag{6}$$

There are two approaches for solving the functional in Eq. (2): optimize then discretize vs discretize then optimize. In the first approach the Euler-Lagrange equations are discretized to find the minimizer. Meanwhile, in the second approach, the functional is discretize using finite difference method and the Quasi Newton method is applied to the discrete functional in the optimization process. Here, the closed form solutions in Eq. (6) is used to update the constants c_1 and c_2 based on the first approach and proceed with the second approach to find the deformation (transformation) field $\boldsymbol{u}(\boldsymbol{x})$. Next, we review the existing similarity measures for multi-modal images: MI and NGF.

3 Mutual Information (MI)

Mutual information based similarity measures for image registration measures the joint distribution of the intensities values in the reference and template images. If these two images are independent, then there is a relation between the intensities values in the images so their mutual information is equal to zero. If the intensities values of the reference image are one to one mapping with the intensities values of the template image, then the mutual information is at the maximum value. The authors in [21] highlighted that the MI based measures fail when the images contain features or objects with different intensities in the template images have similar intensities in the reference images. In addition, the measure also having vital issues such as the optimal size of the bins for the intensities values and inadequate of the spatial information. There exist variants of the MI such as normalized mutual information (NMI) and conditional mutual information (CMI). In [5], the authors highlighted the disadvantages and advantages of several variants of the MI. As such, the NMI does not take into account the spatial and geometric information. The MI is given by

$$\mathcal{D}^{\text{MI}}(T, R, \boldsymbol{u}(\boldsymbol{x})) = \int_{\Omega} \rho_{[T]} \log \rho_{[T]} \; \mathrm{d}t + \int_{\Omega} \rho_{[R]} \log \rho_{[R]} \; \mathrm{d}r - \int_{\Omega^2} \rho_{[T,R]} \log \rho_{[T,R]} \; \mathrm{d}(t,r) \tag{7}$$

where $\rho_{[T]}$, $\rho_{[T]}$, and $\rho_{[T,R]}$ are the marginal and joint densities distribution of the template and reference images.

In [1], the MI term is modified with $TH_\epsilon(\phi_0(\boldsymbol{x} + \boldsymbol{u}(\boldsymbol{x})))$ and $RH_\epsilon(\phi_0(\boldsymbol{x} + \boldsymbol{u}(\boldsymbol{x})))$ where

$$\begin{aligned}\mathcal{D}^{\text{MIH}}(T, R, \phi_0(\boldsymbol{x}), \boldsymbol{u}(\boldsymbol{x})) = \lambda_3 \Bigg[& \int_{\Omega} \rho_{[TH_\epsilon(\phi_0(\boldsymbol{x}+\boldsymbol{u}(\boldsymbol{x})))]} \log \rho_{[TH_\epsilon(\phi_0(\boldsymbol{x}+\boldsymbol{u}(\boldsymbol{x})))]} \; \mathrm{d}t \\ & + \int_{\Omega} \rho_{[RH_\epsilon(\phi_0(\boldsymbol{x}+\boldsymbol{u}(\boldsymbol{x})))]} \log \rho_{[RH_\epsilon(\phi_0(\boldsymbol{x}+\boldsymbol{u}(\boldsymbol{x})))]} \; \mathrm{d}r \\ & - \int_{\Omega^2} \rho_{[TH_\epsilon(\phi_0(\boldsymbol{x}+\boldsymbol{u}(\boldsymbol{x}))), RH_\epsilon(\phi_0(\boldsymbol{x}+\boldsymbol{u}(\boldsymbol{x})))]} \\ & \log \rho_{[TH_\epsilon(\phi_0(\boldsymbol{x}+\boldsymbol{u}(\boldsymbol{x}))), RH_\epsilon(\phi_0(\boldsymbol{x}+\boldsymbol{u}(\boldsymbol{x})))]} \; \mathrm{d}(t,r) \Bigg]\end{aligned} \tag{8}$$

Thus, the final functional for the JSR based on MI is given by

$$\begin{aligned}\min_{c_1, c_2, \boldsymbol{u}(\boldsymbol{x})} \mathcal{J}(c_1, c_2, \boldsymbol{u}(\boldsymbol{x})) = & \lambda_1 \int_{\Omega} |R(\boldsymbol{x}) - c_1|^2 H_\epsilon(\phi_0(\boldsymbol{x} + \boldsymbol{u}(\boldsymbol{x}))) \, \mathrm{d}\boldsymbol{x} \\ & + \lambda_2 \int_{\Omega} |R(\boldsymbol{x}) - c_2|^2 (1 - H_\epsilon(\phi_0(\boldsymbol{x} + \boldsymbol{u}(\boldsymbol{x})))) \, \mathrm{d}\boldsymbol{x} \\ & + D^{\text{MIH}}(T, R, \phi_0(\boldsymbol{x}), \boldsymbol{u}(\boldsymbol{x})) + \alpha \mathcal{S}^{\text{LC}}(\boldsymbol{u})\end{aligned} \tag{9}$$

and the model is denoted by Model JSR-MIH in the next section. According to [1], the term D^{MIH} is evaluated using TH_ϵ and RH_ϵ for ease of the computation which is just an approximation to the functional.

4 Normalized Gradient Fields (NGF)

The NGF distance measure [15] has been successfully proven to be well suited for multi-modal image registration. This distance measure assumes that the intensity changes are preserved across imaging machines. The changes in the intensities values captured the edges in the images via computation of the image gradients: ∇T and ∇R. The NGF evaluates the angle between the image gradients. When T and R are aligned, the NGF has a lower value with minimum value of one. The maximum value of the NGF is obtained when the angle between the gradient of images is orthogonal. The NGF is given by

$$\mathcal{D}^{\mathrm{NGF}}(T, R, \boldsymbol{u}(\boldsymbol{x})) = \int_{\Omega} 1 - (nT(\boldsymbol{x} + \boldsymbol{u}(\boldsymbol{x}))^{\mathrm{T}} nR(\boldsymbol{x}))^2 \, \mathrm{d}\boldsymbol{x} \tag{10}$$

where

$$nT(\boldsymbol{x} + \boldsymbol{u}(\boldsymbol{x})) = \frac{|\nabla T(\boldsymbol{x} + \boldsymbol{u}(\boldsymbol{x}))|}{\sqrt{|\nabla T(\boldsymbol{x} + \boldsymbol{u}(\boldsymbol{x}))|^2 + \eta^2}} \tag{11}$$

for the template image and similarly for the reference image. This measure calculates the gradient of the images. Thus, information such as shape and edges are automatically recorded when we calculate the measure. The NGF is at disadvantages when the gradients are null or very weak [21]. The edge parameter η in (11) should be tuned in the range of the gradients of the major features [10]. As such [17], used different values for the edge parameters for the reference and template images. In addition, the parameter represents modality dependent and aims to filter the noise.

The NGF in (10) is modified using $H_\epsilon(\phi_0(\boldsymbol{x} + \boldsymbol{u}(\boldsymbol{x})))$, to obtain

$$\mathcal{D}^{\mathrm{NGFH}}(T, R, \phi_0(\boldsymbol{x}), \boldsymbol{u}(\boldsymbol{x})) = \lambda_3 \int_{\Omega} 1 - (nT(\boldsymbol{x} + \boldsymbol{u}(\boldsymbol{x}))^{\mathrm{T}} nR(\boldsymbol{x}))^2 H_\epsilon(\phi_0(\boldsymbol{x} + \boldsymbol{u}(\boldsymbol{x}))) \, \mathrm{d}\boldsymbol{x} \tag{12}$$

The regularized Heaviside term in the functional $\mathcal{D}^{\mathrm{NGFH}}$ aims to control the deformation for the features in the foreground. Finally, the Model JSR-NGFH is given by

$$\begin{aligned} \min_{c_1, c_2, \boldsymbol{u}(\boldsymbol{x})} \mathcal{J}(c_1, c_2, \boldsymbol{u}(\boldsymbol{x})) = {} & \lambda_1 \int_{\Omega} |R(\boldsymbol{x}) - c_1|^2 H_\epsilon(\phi_0(\boldsymbol{x} + \boldsymbol{u}(\boldsymbol{x}))) \, \mathrm{d}\boldsymbol{x} \\ & + \lambda_2 \int_{\Omega} |R(\boldsymbol{x}) - c_2|^2 (1 - H_\epsilon(\phi_0(\boldsymbol{x} + \boldsymbol{u}(\boldsymbol{x})))) \, \mathrm{d}\boldsymbol{x} \\ & + D^{\mathrm{NGFH}}(T, R, \phi_0(\boldsymbol{x}), \boldsymbol{u}(\boldsymbol{x})) + \alpha \mathcal{S}^{\mathrm{LC}}(\boldsymbol{u}) \end{aligned} \tag{13}$$

The algorithm for the two models is given by Algorithm 4. Refer to [11] for the discrete formulation, numerical discretization using finite difference method and the LBFGS method.

Algorithm 1: Joint Image Segmentation and Registration
Input: Reference image R, template image T
Output: $T(\boldsymbol{x}+\boldsymbol{u}(\boldsymbol{x})), \phi_0(\boldsymbol{x}+\boldsymbol{u}(\boldsymbol{x}))$

1. Segment the template image to produce $\phi_0(\boldsymbol{x})$ using Chan-Vese segmentation model.
2. Coarsening T, R and $\phi_0(\boldsymbol{x})$ using multilevel to produce T^{level}, R^{level} and $\phi_0(\boldsymbol{x})^{\text{level}}$.
3. Initialize $\boldsymbol{u}(\boldsymbol{x})^{\text{minlevel}}$ as the zero matrices. Set the parameters $\lambda_1, \lambda_2, \lambda_3$ and α.
4. **For** level = minlevel : maxlevel
   (a) $\left(\boldsymbol{u}(\boldsymbol{x})^{\text{level}}, c_1, c_2\right) \leftarrow \text{LBFGS}\left(\mathcal{J}(T, R, \phi_0(\boldsymbol{x})^{\text{level}}, \boldsymbol{u}(\boldsymbol{x})^{\text{level}}, \lambda_1, \lambda_2, \lambda_3, \alpha)\right)$
   (b) interpolate $\boldsymbol{u}(\boldsymbol{x})^{\text{level}}$ and $\phi_0(\boldsymbol{x})^{\text{level}}$ to level = level + 1
5. **End**
6. Compute the transformed template image: $T(\boldsymbol{x}+\boldsymbol{u}(\boldsymbol{x}))$ and the segmentation of the reference image $\phi_0(\boldsymbol{x}+\boldsymbol{u}(\boldsymbol{x}))$

5 Numerical Experiments

In this section, the performance of the proposed model: Model JSR-NGFH is compared with Model JSR-MIH in [1]. To measure the quality of the registered images, use

$$p = |\mathcal{D}^{\text{MI}}(T(\boldsymbol{x}+\boldsymbol{u}(\boldsymbol{x})), R)| - |\mathcal{D}^{\text{MI}}(T(\boldsymbol{x}), R)| \tag{14}$$

and the dice coefficient metric

$$dc = \text{Dice}(T(\boldsymbol{x}+\boldsymbol{u}(\boldsymbol{x})), R) = \frac{2|\text{Seg}(T(\boldsymbol{x}+\boldsymbol{u}(\boldsymbol{x}))) \cap \text{Seg}(R(\boldsymbol{x}))|}{|\text{Seg}(T(\boldsymbol{x}+\boldsymbol{u}(\boldsymbol{x})))| + |\text{Seg}(R(\boldsymbol{x}))|}. \tag{15}$$

The values for p where $0 \leq p < \infty$ indicates the increase of the value of the mutual information after registration. A larger value of p shows better registration results compared to the smaller value of p. Meanwhile, for dc where $dc = 0$ means no overlap between the segmentation of the reference and the transformed template images. $dc = 1$ points out perfect the agreement between the segmentation of the reference and the transformed template images. Thus, we want large value of p and dc as close as to 1.

Three experiments are conducted for both models. In the experiments, $\lambda_1 = \lambda_2 = 1$ are used for both models and varies λ_3 and α. The approach, implementation, parameters as in [15], and $\eta = 1$ are used for the computation of the NGFH. The parameters λ_3 and α for all experiments are shown in Table 1 for both models. Several tests are run using $\lambda_3, \alpha \in \{1, 0.5, 0.25, 0.125, 0.0625, 0.03125, 0.015625, 0.01, 0.001\}$. The optimal choices of the parameters are chosen based on the registered image is very close to the reference image and the transformation is free from breaking and folding.

Table 1. Values of the parameters λ_3 and α for models JSR-MIH and JSR-NGFH.

Experiment	JSR-MIH		JSR-NGFH	
	λ_3	α	λ_3	α
1	0.001	0.125	0.125	0.125
2	0.001	0.03125	0.0625	0.125
3	1	0.001	0.03125	0.5

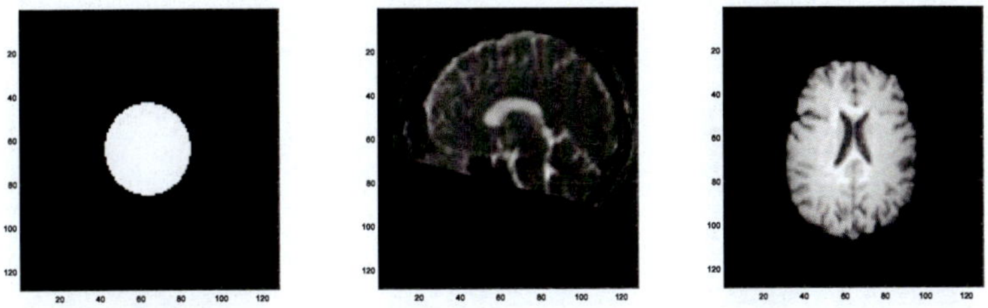

(a) R (b) R (c) R

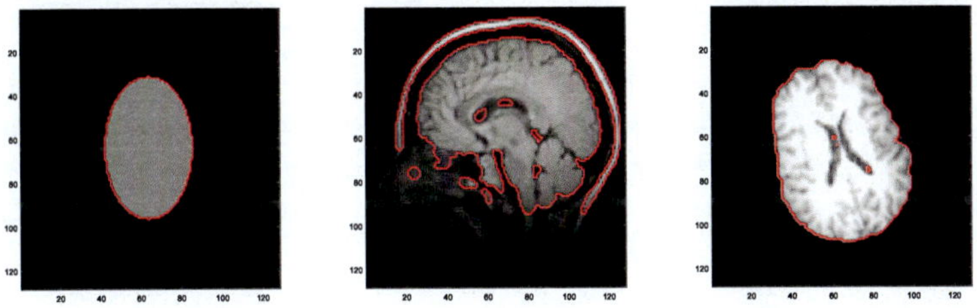

(d) T, $\phi_0(\boldsymbol{x})$ (e) T, $\phi_0(\boldsymbol{x})$ (f) T, $\phi_0(\boldsymbol{x})$

Fig. 1. Reference R, template T and the segmentation of the template images $\phi_0(\boldsymbol{x})$ (red curves) for Experiment 1 (first column), 2 (second column) and 3 (third column).

Images for Experiment 1 are inspired from [21] where the reference image R is a white circle in the middle and the template image T is a gray ellipse as shown in Fig. 1 (a) and (d) respectively. In [21], the authors showed that for this type of images MI and NGF fail to register. The results for both models are shown in Fig. 2. It is observed that both models manage to solve the problem where the gray ellipse are transformed into the circle as shown in Fig. 2 (a) and (d). However using the values of p and dc, model JSR-MIH performs better than JSR-NGFH.

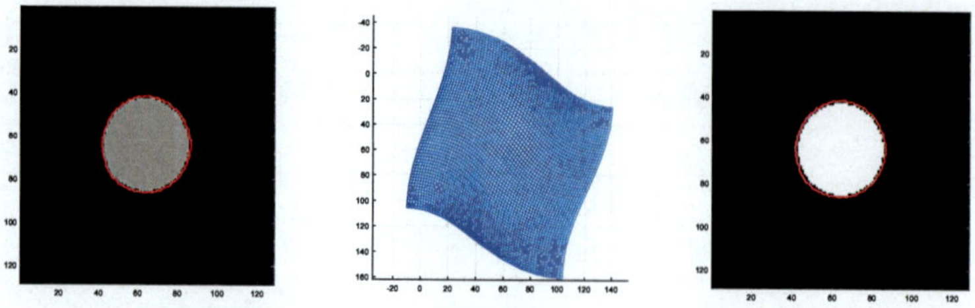

(a) $T(\boldsymbol{x}+\boldsymbol{u}(\boldsymbol{x}))$,$\phi_0(\boldsymbol{x}+\boldsymbol{u}(\boldsymbol{x}))$, $p = 0.07378$

(b) $\boldsymbol{x}+\boldsymbol{u}(\boldsymbol{x})$

(c) R,$\phi_0(\boldsymbol{x}+\boldsymbol{u}(\boldsymbol{x}))$, $dc = 0.9072$

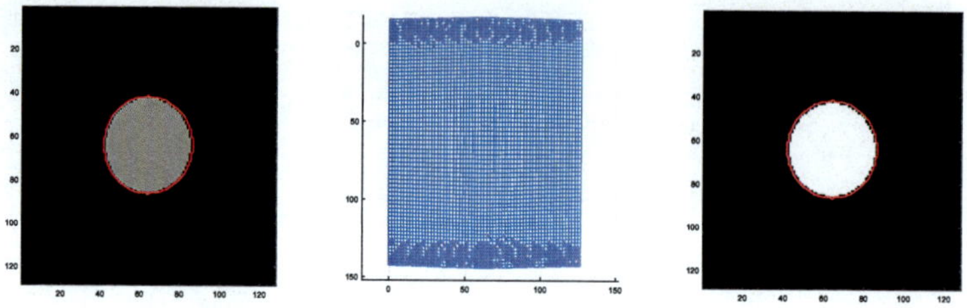

(d) $T(\boldsymbol{x}+\boldsymbol{u}(\boldsymbol{x}))$,$\phi_0(\boldsymbol{x}+\boldsymbol{u}(\boldsymbol{x}))$,$p = 0.07239$

(e) $\boldsymbol{x}+\boldsymbol{u}(\boldsymbol{x})$

(f) R,$\phi_0(\boldsymbol{x}+\boldsymbol{u}(\boldsymbol{x}))$,$dc = 0.9050$

Fig. 2. Results for Experiment 1 using Model JSR-MIH(top row) and JSR-NGFH (bottom row). First columns are the transformed template images and segmentation of the images (red curves). The second columns are the transformation fields. The third columns show the reference images with the segmentation of the transformed template image. Based on the values of p and dc model JSR-MIH is at advantages compared to the model JSR-NGFH.

The reference and template images for Experiment 2 are depicted in Fig. 1 (b) and (e) respectively. There are MRI of the human's brain where the reference image is the T2 weighted images. Meanwhile, the template image is the T1 weighted image which highlights the fat tissue during the imaging process. The results for both models are shown in Fig. 3. It is observed that both models manage to solve the problem as shown in Fig. 3 (a) and (d). However, using the values of p and dc model JSR-MIH performs better than JSR-NGFH. The models are able to solve real medical 2D images.

The reference and template images for Experiment 3 are depicted in Fig. 1 (c) and (f) respectively. There are MRI of the human's brain with low bias field. The bias field or so-called as the intensity in-homogeneity is a smooth variation of intensities across the image. This effect causes variation in the intensity of the same objects or features in the different location within the image.

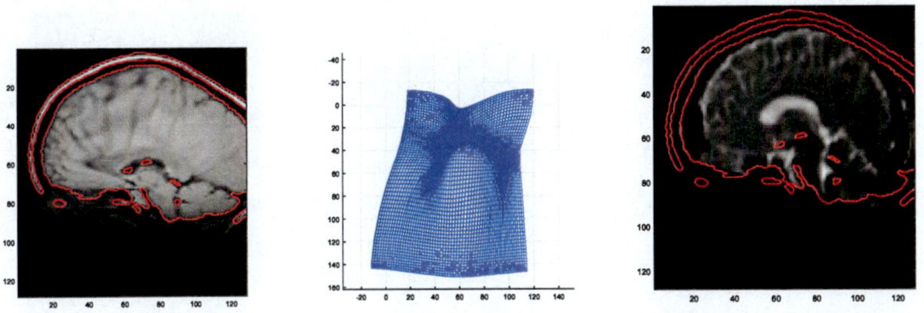

(a) $T(\boldsymbol{x}+\boldsymbol{u}(\boldsymbol{x}))$,$\phi_0(\boldsymbol{x}+\boldsymbol{u}(\boldsymbol{x}))$, $p = 0.0733$

(b) $\boldsymbol{x}+\boldsymbol{u}(\boldsymbol{x})$

(c) R,$\phi_0(\boldsymbol{x}+\boldsymbol{u}(\boldsymbol{x}))$, $dc = 0.8236$

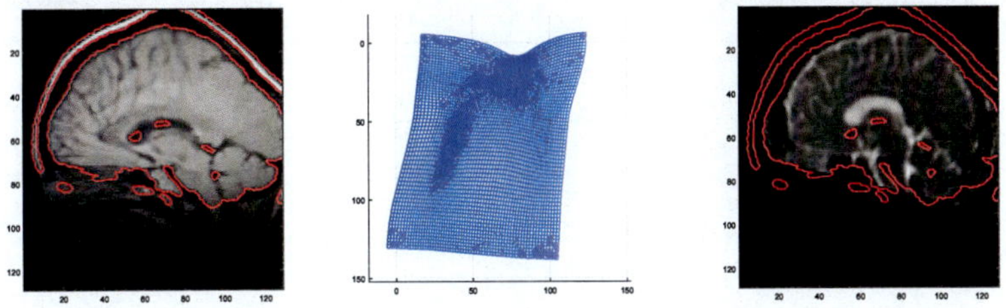

(d) $T(\boldsymbol{x}+\boldsymbol{u}(\boldsymbol{x}))$,$\phi_0(\boldsymbol{x}+\boldsymbol{u}(\boldsymbol{x}))$, $p = 0.05786$

(e) $\boldsymbol{x}+\boldsymbol{u}(\boldsymbol{x})$

(f) R,$\phi_0(\boldsymbol{x}+\boldsymbol{u}(\boldsymbol{x}))$, $dc = 0.7784$

Fig. 3. Results for Experiment 2 using Model JSR-MIH(top row) and JSR-NGFH (bottom row). First columns are the transformed template images and segmentation of the images (red curves). The second columns are the transformation fields. The third columns show the reference images with the segmentation of the transformed template image. Based on the values of p and dc model JSR-MIH is at advantages compared to the model JSR-NGFH.

According to [19], the bias field invalidates the piece wise constant property of medical images. One might argue that Experiment 3 is a mono-modality images. Thus, any multi-modality models for image registration are able to solve the mono-modal problems. The results for both models are shown in Fig. 4. Both models manage to solve the problem as shown in Fig. 4 (a) and (d). However, using the values of p and dc model JSR-NGFH performs better than JSR-MIH. The models are able to solve problems related to the low bias field and real medical 2D images.

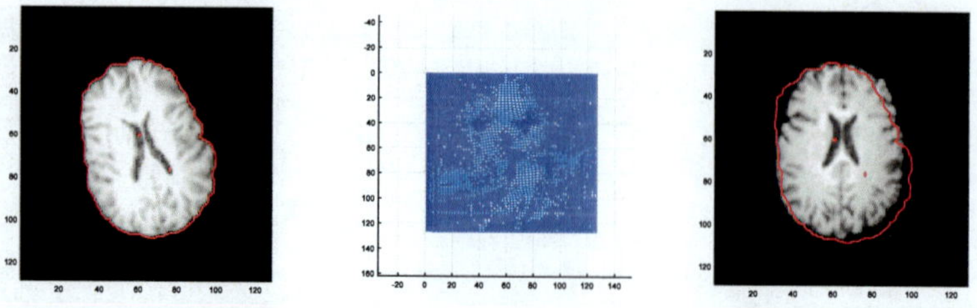

(a) $T(\boldsymbol{x}+\boldsymbol{u}(\boldsymbol{x}))$,$\phi_0(\boldsymbol{x}+\boldsymbol{u}(\boldsymbol{x}))$, $p = 0.0058$

(b) $\boldsymbol{x}+\boldsymbol{u}(\boldsymbol{x})$

(c) R,$\phi_0(\boldsymbol{x}+\boldsymbol{u}(\boldsymbol{x}))$, $dc = 0.9920$

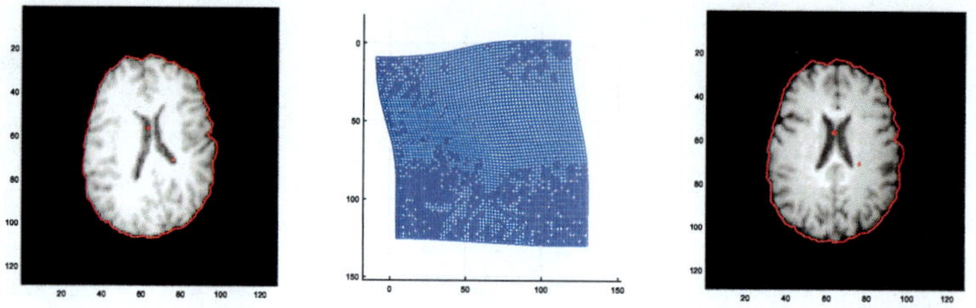

(d) $T(\boldsymbol{x}+\boldsymbol{u}(\boldsymbol{x}))$,$\phi_0(\boldsymbol{x}+\boldsymbol{u}(\boldsymbol{x}))$, $p = 0.14912$

(e) $\boldsymbol{x}+\boldsymbol{u}(\boldsymbol{x})$

(f) R,$\phi_0(\boldsymbol{x}+\boldsymbol{u}(\boldsymbol{x}))$, $dc = 0.9960$

Fig. 4. Results for Experiment 3 using Model JSR-MIH(top row) and JSR-NGFH (bottom row). First columns are the transformed template images and segmentation of the images (red curves). The second columns are the transformation fields. The third columns show the reference images with the segmentation of the transformed template image. Based on the values of p and dc model JSR-NGFH is at advantages compared to the model JSR-MIH.

6 Conclusions

This paper focused on the joint work between segmentation and deformable image registration for multi-modal images using varational approaches. The proposed model is based on the active contour model for image segmentation, normalized gradient fields and linear curvature model for image registration. The model is solved using discretize then optimize approach and multilevel techniques for fast implementation. The proposed model: JSR-NGFH is compared with the model JSR-MIH in [1]. The types of images where the two models are at advantages and disadvantages have been shown. Future works including incorporate the regridding techniques and diffeomorphic transformations.

Acknowledgment. MI would like to acknowledge funding from the Ministry of Higher Education of Malaysia (RACER/1/2019/ICT01/UPNM/1).

References

1. Ademaj, A., Rada, L., Ibrahim, M., Chen, K.: A variational joint segmentation and registration framework for multimodal images. In: Annual Conference on Medical Image Understanding and Analysis, Springer, pp. 305–316 (2019)
2. Aganj, I., Fischl, B.: Multimodal image registration through simultaneous segmentation. IEEE Sig. Process. Lett. **24**(11), 1661–1665 (2017)
3. Aviles-Rivero, A.I., Williams, G., Graves, M.J., Schonlieb, C.-B.: Compressed sensing plus motion (cs+ m): a new perspective for improving undersampled mr image reconstruction. *arXiv preprint* arXiv:1810.10828 (2018)
4. Barles, G., Cardaliaguet, P., Ley, O., Monteillet, A.: Existence of weak solutions for general nonlocal and nonlinear second-order parabolic equations. Nonlinear Anal. Theor. Method Appl. **71**(7–8), 2801–2810 (2009)
5. Borvornvitchotikarn, T., Kurutach, W.: A taxonomy of mutual information in medical image registration. In: International Conference on Systems, Signals and Image Processing (IWSSIP), pp. 1–4. IEEE (2016)
6. Chan, T.F., Vese, L.A.: Active contours without edges. IEEE Trans. Image Process. **10**(2), 266–277 (2001)
7. Debroux, N., Ozeré, S., Le Guyader, C.: A non-local topology-preserving segmentation-guided registration model. J. Math. Imaging Vis. **59**(3), 432–455 (2017)
8. Fischer, B., Modersitzki, J.: A unified approach to fast image registration and a new curvature based registration technique. Linear Algebra Appl. **380**, 107–124 (2004)
9. Gass, T., Szekely, G., Goksel, O.: Simultaneous segmentation and multiresolution nonrigid atlas registration. IEEE Trans. Image Process. **23**(7), 2931–2943 (2014)
10. Hodneland, E., Hanson, E.A., Lundervold, A., Modersitzki, J., Eikefjord, E., Munthe-Kaas, A.Z.: Segmentation-driven image registration-application to 4d dce-mri recordings of the moving kidneys. IEEE Trans. Image Process. **23**(5), 2392–2404 (2014)
11. Ibrahim, M.: Variational models and numerical algorithms for effective image registration. PhD thesis, University of Liverpool (2015)
12. Ibrahim, M., Chen, K., Rada, L.: An improved model for joint segmentation and registration based on linear curvature smoother. J. Algorithms Comput. Technol. **10**(4), 314–324 (2016)
13. Kohlmann, P., Strehlow, J., Jobst, B., Krass, S., Kuhnigk, J.-M., Anjorin, A., Sedlaczek, O., Ley, S., Kauczor, H.-U., Wielpütz, M.O.: Automatic lung segmentation method for mri-based lung perfusion studies of patients with chronic obstructive pulmonary disease. Int. J. Comput. Assist. Radiol. Surg. **10**(4), 403–417 (2015)
14. Le Guyader, C., Vese, L.A.: A combined segmentation and registration framework with a nonlinear elasticity smoother. Comput. Vis. Image Underst. **115**(12), 1689–1709 (2011)
15. Modersitzki, J.: Flexible Algorithms for Image Registration. SIAM publications (2009)
16. Ozeré, S., Le Guyader, C.: A joint segmentation-registration framework based on weighted total variation and nonlinear elasticity principles. In: IEEE International Conference on Image Processing (ICIP), pp. 3552–3556. IEEE(2014)
17. Rühaak, J., König, L., Hallmann, M., Papenberg, N., Heldmann, S., Schumacher, H., Fischer, B.: A fully parallel algorithm for multimodal image registration using normalized gradient fields. In: IEEE 10th International Symposium on Biomedical Imaging, pp. 572–575. IEEE (2013)

18. Shah, P., Gupta, M.D.: Simultaneous registration and segmentation by l1 minimization. In: International Workshop on Machine Learning in Medical Imaging, Springer, pp. 128–135 (2012)
19. Song, S., Zheng, Y., He, Y.: A review of methods for bias correction in medical images. Biomed. Eng. Rev. **1**(1) (2017)
20. Swierczynski, P., Papież, B.W., Schnabel, J.A., Macdonald, C.: A level-set approach to joint image segmentation and registration with application to ct lung imaging. Comput. Med. Imaging Graph. **65**, 58–68 (2018)
21. Theljani, A., Chen, K.: An augmented lagrangian method for solving a new variational model based on gradients similarity measures and high order regularization for multimodality registration. Inverse Problems and Imaging (2019)
22. Xue, Z., Wong, K., Wong, S.T.: Joint registration and segmentation of serial lung ct images for image-guided lung cancer diagnosis and therapy. Comput. Med. Imaging Graph. **34**(1), 55–60 (2010)

Intelligent Control Systems

A Mathematical Approach to Improve Energy-Water Nexus Reliability Using a Novel Multi-stage Adjustable Fuzzy Robust Approach

Afshin Ghassemi(✉) and Michael J. Scott

University of Illinois at Chicago, Chicago, IL 60607, USA
afshinghassemi@gmail.com, mjscott@uic.edu

Abstract. A system of a systems approach that analyzes energy and water systems simultaneously is called energy-water nexus. Neglecting the interrelationship between energy and water drives vulnerabilities whereby limits on one resource can cause constraints on the other resource. Power plant energy production directly depends on water availability, and an outage of the power systems will affect the wastewater treatment facility processes. Therefore, it is essential to integrate energy and water planning models. As mathematical energy-water nexus problems are complex, involve many uncertain parameters, and are large-scale, we proposed a novel multi-stage adjustable Fuzzy robust approach that balances the solutions' robustness against the budget-constraints. Scenario-based analysis indicates that the proposed approach generates flexible and robust decisions that avoid excessive costs compared to conservative methods.

Keywords: Energy-water Nexus · Renewable energy · Optimization under uncertainty · Fuzzy logic · Robust optimization

1 Introduction

In mathematical energy-water nexus planning, the aim is to offer a general model that generates energy decisions and water decisions while covering these two complex systems' interconnections. The energy-water nexus planning has been thoroughly studied via discussions of policies and regulations during the last fifty years [1–4]. The recent research in this area provided more accurate data for energy-water nexus planning (e.g., energy for water and water for energy) [5–9]. The last class of the studies applied mathematical approaches toward energy-water nexus problems, where limited numbers of them propose a general mathematical model or framework for energy-water nexus planning[10–15]. The growing demand for shared resources is predicted to be one of the primary subjects of conflicts in the approaching years, not only on the connected energy-water area but in many other critical areas (see [16,17]). When studying the energy

T. Allahviranloo et al. (Eds.): IDS 2020, AISC 1301, pp. 115–123, 2021.
https://doi.org/10.1007/978-3-030-66501-2_9

planning and water planning simultaneously in the general model, the uncertainty is considered in both the supply and demand sides of the energy-water nexus [18–20]. Overlooking these randomnesses in systems can undoubtedly lead to infeasible solutions. Some studies have analyzed uncertainty for at least part of the energy-water nexus planning [21–26]. Ghassemi [27]offers a robust framework for their research. One of the main methods to deal with uncertainty is using Fuzzy logic which has vast applications in optimization under uncertainty [28,29]. This research aims to extend on that method and propose a flexible, multi-stage, Fuzzy robust approach for energy-water nexus planning. The coupling of the Fuzzy method and robust optimization makes it possible to cover diverse types of uncertainties while keeping the system's flexibility.

The new, flexible, multi-stage, Fuzzy Robust method is described in detail Sect. 2. The next section describes the energy-water systems profile and provides the results of the scenario-base analysis Sect. 3. The discussion of this study and future research plans are presented in Sect. 4.

2 Optimization under Uncertainty Using Robust and Fuzzy Methods

Using the Fuzzy and robust optimization method leads to an increase in the total systems cost. Models that are flexible to generate a range of solutions are more preferred than the traditional single-solution methods. The managers would be able to generate decisions with a certain level of robustness based on their desired criteria, such as the budget constraints.

In energy-water nexus planning, dismissing the possible scenarios' details will lead to loss of information and irreversible consequences.

A new two-stage adjustable robust approach is introduced in [27] that provides a balance between solution robustness and the budget constraints. In this work, the Fuzzy decision models' concepts are integrated into that proposed model, and a new adjustable, multi-stage, robust, Fuzzy approach is proposed.

2.1 Fuzzy Logic Integration

The possibility, necessity, and credibility indexes represent the chances of occurrence of Fuzzy events with an optimistic, pessimistic, and moderate view towards planning, respectively. Here, the conversion of Fuzzy chance constraints to their equivalent crisp for a specific confidence level is provided. In this paper, $\tilde{\nu}$ and γ are Fuzzy and crisp numbers. It is assumed that the uncertain parameter is a Fuzzy number with trapezoidal distribution that is showed by $\tilde{\nu}\,(\nu_1, \nu_2, \nu_3, \nu_4)$ while $\nu_1 < \nu_2 < \nu_3 < \nu_4$.

To integrate the Fuzzy logic, a Fuzzy number with trapezoidal distribution of $\tilde{\nu}\,(\nu_1, \nu_2, \nu_3, \nu_4)$ is presented. The triple (Ψ, P(Ψ), Pos) represent a possibility space that a universe set Ψ is a non-empty set, covering all possible events and P(Ψ) be the power set of Ψ.

The credibility measure is defined as the average of the possibility and necessity measures as $Credibility\{.\} = \frac{1}{2}(Possibility\{.\} + Necessity\{.\})$.

The credibility profile measures are defined as follows:

- $Credibility\{\varnothing\} = 0$
- $Necessity\{\Psi\} = 1$
- $if \;\; A \in P(\Psi) \;\Rightarrow\; 0 \leq Credibility\{A\} \leq 1$
- $if \;\; A_i \in P(\Psi) \;\; and \;\; Sup_i(Credibility\{A_i\}) < 0.5 \Rightarrow Credibility\{\cup_i A_i\} = Sup_i(Credibility\{A_i\})$
- $if \;\; A \in P(\Psi) \Rightarrow Credibility\{A\} + Credibility\{A^C\} = 1 \quad (Self - Duality)$
- $if\, A, B \in P(\Psi)\, and\, A \subseteq B \Rightarrow Credibility\{A\} \leq Credibility\{B\}$ (*Monotonicity*)
- $if\, A, B \in P(\Psi) \Rightarrow Credibility\{A \cup B\} \leq Credibility\{A\} + Credibility\{B\}$ (*Subadditivity*)
- $Possibility\{A\} \geq Credibility\{A\} \geq Necessity\{A\}$

Based on the credibility measure, the Eqs. in 1 and 2, transforms the Fuzzy chance constraints for confidence level of α as:

$$Credibility\{\tilde{\nu} \leq \gamma\} = \begin{cases} 0, & if \;\; \nu_1 \geq \gamma; \\ \dfrac{\gamma - \nu_1}{2(\nu_2 - \nu_1)}, & if \;\; \nu_1 \leq \gamma \leq \nu_2; \\ \dfrac{1}{2}, & if \;\; \nu_2 \leq \gamma \leq \nu_3; \\ \dfrac{\gamma - 2\nu_3 + \nu_4}{2(\nu_4 - \nu_3)}, & if \;\; \nu_3 \leq \gamma \leq \nu_4; \\ 1, & if \;\; \nu_4 \leq \gamma. \end{cases} \tag{1}$$

$$Credibility\{\tilde{\nu} \geq \gamma\} = \begin{cases} 1, & if \;\; \nu_1 \geq \gamma; \\ \dfrac{2\nu_2 - \nu_1 - \gamma}{2(\nu_2 - \nu_1)}, & if \;\; \nu_1 \leq \gamma \leq \nu_2; \\ \dfrac{1}{2}, & if \;\; \nu_2 \leq \gamma \leq \nu_3; \\ \dfrac{\nu_4 - \gamma}{2(\nu_4 - \nu_3)}, & if \;\; \nu_3 \leq \gamma \leq \nu_4; \\ 0, & if \;\; \nu_4 \leq \gamma. \end{cases} \tag{2}$$

The Eqs. in 1 and 2 can be summarized as:

$$Credibility\{\tilde{\nu} \leq \gamma\} \geq \alpha \quad \Leftrightarrow \quad \begin{cases} (2 - 2\alpha)\nu_3 + (2\alpha - 1)\nu_4 \leq \gamma & if \;\; \alpha > 0.5; \\ (1 - 2\alpha)\nu_1 + 2\alpha\nu_2 \leq \gamma & if \;\; \alpha \leq 0.5. \end{cases} \tag{3}$$

$$Credibility\{\tilde{\nu} \geq \gamma\} \geq \alpha \quad \Leftrightarrow \quad \begin{cases} (2\alpha - 1)\nu_1 + (2 - 2\alpha)\nu_2 \geq \gamma & if \quad \alpha > 0.5; \\ 2\alpha\nu_3 + (1 - 2\alpha)\nu_4 \geq \gamma & if \quad \alpha \leq 0.5. \end{cases} \tag{4}$$

2.2 The Multi-stage Fuzzy Robust Approach Architecture

In energy-water nexus planning, the system could face different scenarios, and each of these scenarios has a designated probability of happening that a Fuzzy number is assigned to represent that. In the multi-stage Fuzzy robust approach, three types of decisions should be processed. The first two are the Fuzzy weights' discrete decisions and then the Fuzzy numbers in the uncertain model. The third decision is related to parameters with continues ranged uncertainty that is calculated by the robust optimization. The third decision is related to parameters with continues ranged uncertainty, which is calculated by the robust optimization.

In the robust approach, the first step calculates the total system cost based on the uncertain model. The next step is to convert the weights based on one of the mentioned Fuzzy methods (necessity, possibility, or credibility) to nominal values and then normalize the new weights by dividing the weight for each scenario to the sum of the weights for all possible scenarios. The last step is to find the sum-product of each scenario's weights multiplied by the corresponding total system cost. Figure 1 illustrates the multi-stage Fuzzy robust approach for energy-water system robust optimization.

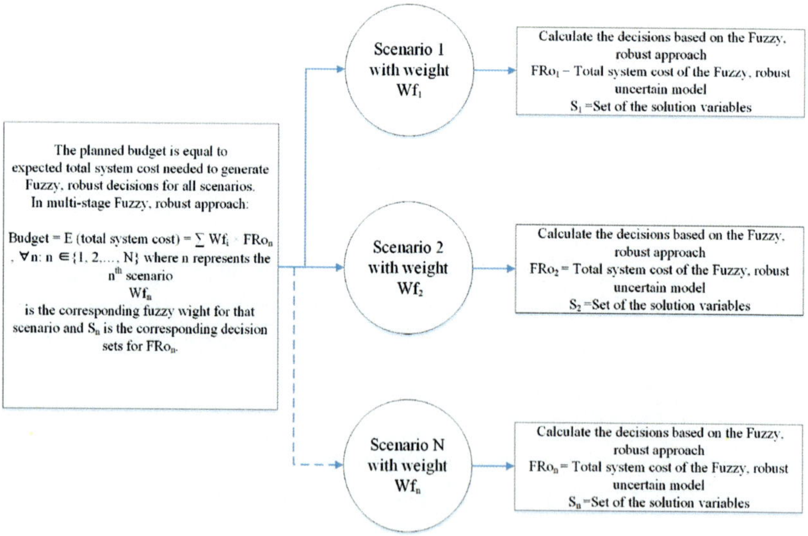

Fig. 1. The architecture of the multi-stage adjustable Fuzzy robust approach to find the expected total system cost

3 Scenario Analysis

In this section, first, the parameters and the features of the energy-water nexus model are explained. Then, using a large-size example, the proposed model's performance is examined, and the results are investigated.

3.1 The Energy-Water System Profile

The uncertain energy-water nexus model and the parameters such as power generator information and scenario parameters are the same as the author's previous work [27] with two main changes. The weights are considered to be represented by trapezoidal Fuzzy numbers. The efficiency parameters are modeled by both robust and Fuzzy methods. The planning horizon is 10×24, and the model generates online hourly decisions. The final model is a Fuzzy robust, multi-period, linear mixed-integer optimization problem. The energy-water nexus model covers water and wastewater extraction, transmission, purification, and all the associated costs. The energy part covers thermal power systems, such as power transmission systems, power generators, and batteries. The objective function includes two categories of penalty costs for unmet demand and untreated wastewater to reduce unwanted results.

3.2 Numerical Analysis

In order to demonstrate the benefits of the new multi-stage Fuzzy robust planning method, the expected total system cost is calculated on two levels of $\alpha = 0$ and $\alpha = 0.5$ (α works as a control parameter for the level of conservativity in Fuzzy methods) using the credibility Fuzzy method. The robust levels are calculated for $\Gamma = 0$, $\Gamma = 0.25$, $\Gamma = 0.5$, $\Gamma = 0.75$, $\Gamma = 1$ (the parameter Γ is a control of how conservative a robust method is), that balances the decision robustness against the total system cost. The larger the Γ or α the more lavish budget is needed to prepare for the risks.

In the first numerical example (see Fig. 2), different levels of Γ (where the $\alpha = 0$) can be picked based on the budget restrictions. In the second experiment (see Fig. 3) the $\alpha = 0.5$ and the parameterΓ is used to control the consevativity levels.

The multi-stage Fuzzy robust method incurs a lower total cost than all the other methods besides the deterministic model, which has inadequate dependability for real-world situations and can show infeasible decisions and water or power outages. If all the uncertain data had been added simultaneously to the model, the total system cost would have increased unreasonably, and the outcomes would have become impractical.

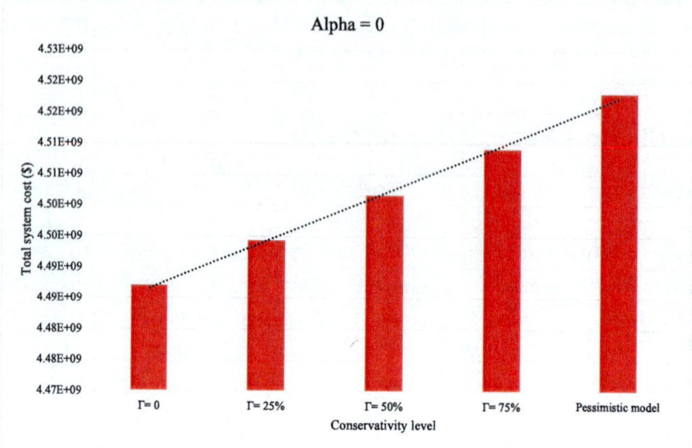

Fig. 2. Multi-stage adjustable Fuzzy robust approach to find the expected total system cost for $\alpha = 0$

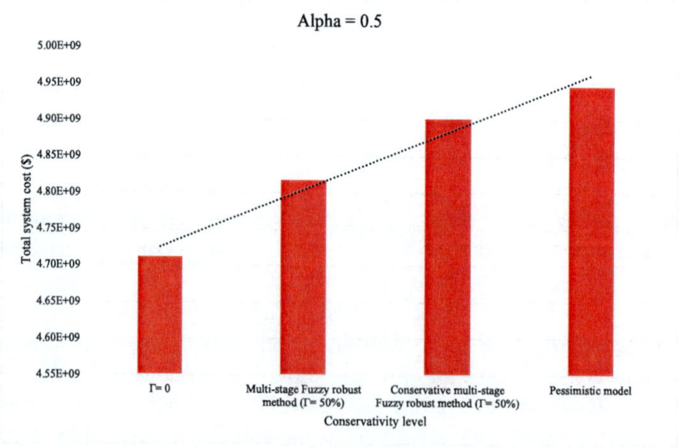

Fig. 3. Multi-stage adjustable Fuzzy robust approach to find the expected total system cost for $\alpha = 0.5$

4 Discussion

One of the main challenges of energy-water nexus planning is to find an efficient way to deal with the uncertainties of the parameters. There are different types of randomness in energy-water nexus models that each requires a specific method to address. Fuzzy logic and robust optimizations are two popular methods to solve this issue. In this work, a new multi-stage adjustable Fuzzy robust approach is proposed to generate flexible, affordable, and efficient, reliable decisions without unreasonable cost-load on the total system cost.

The authors' proposed a multi-stage adjustable Fuzzy robust decision approach built on the authors' previous work that has lots of benefits; (1) yields flexible decisions that can be adjusted based on the budget constraints, (2) operates great with ranged uncertainty using robust optimization, (3) considers policy-makers and managers views to apply mathematically to the deterministic model using the Fuzzy logic, and (4) brings more cost-effective decisions to the system, in comparison with traditional conservative models. For future studies, the goal is to integrate the concept of Z-number [30] with the currently Fuzzy robust developed model.

References

1. Carrillo, A.M.R., Frei, C.: Water: a key resource in energy production. Energy Policy **37**(11), 4303–4312 (2009)
2. Mackres, E.: Addressing the energy-water nexus: a blueprint for action and policy agenda. Proc. Water Environ. Fed. **2011**(6), 1242–1244 (2011)
3. Park, L., Croyle, K.: California's water-energy nexus: pathways to implementation. GEI Consultants, Inc. (2012)
4. Tarroja, B., AghaKouchak, A., Sobhani, R., Feldman, D., Jiang, S., Samuelsen, S.: Evaluating options for balancing the water-electricity nexus in California: part 1 - securing water availability. Sci. Total Environ. **497**, 697–710 (2014)
5. Wolff, R.B.: Energy down the drain, the hidden costs of California's water supply. Technical report, Natural Resources Defense Council (2004)
6. Wakeel, M., Chen, B., Hayat, T., Alsaedi, A., Ahmad, B.: Energy consumption for water use cycles in different countries: a review. Appl. Energy **178**, 868–885 (2016). http://linkinghub.elsevier.com/retrieve/pii/S0306261916308893
7. DOE: Energy demands on water resources (report to Congress on the interdependency of energy and water). Technical report, U.S. Department of Energy (2006)
8. Mielke, E., Diaz Anadon, L., Narayanamurti, V.: Water consumption of energy resource extraction, processing, and conversion. Energy Technology Innovation Policy Research Group, October 2010. http://belfercenter.ksg.harvard.edu/files/ETIP-DP-2010-15-final-4.pdf
9. Stillwell, A.S., King, C.W., Webber, M.E., Duncan, I.J., Hardberger, A.: The energy-water nexus in Texas. Ecol. Soc. **16**(1), p. 2 (2011). http://www.ecologyandsociety.org/vol16/iss1/art2/
10. Yang, J., Chen, B.: Energy-water nexus of wind power generation systems. Appl. Energy **169**, 1–13 (2016)

11. DeNooyer, T.A., Peschel, J.M., Zhang, Z., Stillwell, A.S.: Integrating water resources and power generation: the energy-water nexus in Illinois. Appl. Energy **162**, 363–371 (2016)
12. Sanders, K.T., Webber, M.E.: Evaluating the energy and CO2 emissions impacts of shifts in residential water heating in the United States. Energy **81**, 317 – 327 (2015). http://www.sciencedirect.com/science/article/pii/S0360544214014170
13. Santhosh, A., Farid, A.M., Youcef-Toumi, K.: Real-time economic dispatch for the supply side of the energy-water nexus. Appl. Energy **122**, 42–52 (2014)
14. Chen, S., Chen, B.: Urban energy-water nexus: a network perspective. Appl. Energy **184**, 905–914 (2016)
15. Wang, S., Liu, Y., Chen, B.: Multiregional input–output and ecological network analyses for regional energy–water nexus within China. Appl. Energy **227**, 353 – 364 (2018). http://www.sciencedirect.com/science/article/pii/S030626191731694X, Transformative Innovations for a Sustainable Future – Part III
16. Esmaeili, M., Bahrini, A., Shayanrad, S.: Using game theory approach to interpret stable policies for Iran's oil and gas common resources conflicts with Iraq and Qatar. J. Ind. Eng. Int. **11**, 543–554 (2015)
17. Mohamadi, M., Bahrini, A.: A nash-stackelberg equilibrium model for internet and network service providers in the demand market-a scenario-based approach. Wireless Netw. **26**(1), 449–461 (2020)
18. Hassanzadeh, E., Elshorbagy, A., Wheater, H., Gober, P., Nazemi, A.: Integrating supply uncertainties from stochastic modeling into integrated water resource management: Case study of the Saskatchewan River basin. J. Water Resour. Plann. Manage. **142**(2), 5015006, August 2015. https://doi.org/10.1061/(ASCE)WR.1943-5452.0000581
19. Arandia, E., Ba, A., Eck, B., McKenna, S.: Tailoring seasonal time series models to forecast short-term water demand. J. Water Resour. Plann. Manage. **142**(3), 4015067, December 2015. https://doi.org/10.1061/(ASCE)WR.1943-5452.0000591
20. Ju, L., Tan, Z., Yuan, J., Tan, Q., Li, H., Dong, F.: A bi-level stochastic scheduling optimization model for a virtual power plant connected to a wind-photovoltaic-energy storage system considering the uncertainty and demand response. Appl. Energy **171**, 184–199 (2016). http://www.sciencedirect.com/science/article/pii/S0306261916303312
21. Ghassemi, A., Hu, M., Zhou, Z.: Robust planning decision model for an integrated water system. J. Water Resou. Plann. Manage. **143**, 1–12 (2017)
22. Babayan, A., Kapelan, Z., Savic, D., Walters, G.: Least-cost design of water distribution networks under demand uncertainty. J. Water Resour. Plann. Manage. **131**(5), 375–382 (2005)
23. Weini, Z.: Water network design and management via stochastic programming. Ph.D. thesis, University of Arizona (2013)
24. Parisio, A., Rikos, E., Tzamalis, G., Glielmo, L.: Use of model predictive control for experimental microgrid optimization. Appl. Energy **115**, 37 – 46 (2014). http://www.sciencedirect.com/science/article/pii/S0306261913008477
25. Coelho, V.N., Coelho, I.M., Coelho, B.N., Reis, A.J., Enayatifar, R., Souza, M.J., Guimarães, F.G.: A self-adaptive evolutionary fuzzy model for load forecasting problems on smart grid environment. Appl. Energy **169**, 567 – 584 (2016). http://www.sciencedirect.com/science/article/pii/S0306261916301684
26. Suganthi, L., Iniyan, S., Samuel, A.A.: Applications of fuzzy logic in renewable energy systems – a review. Renew. Sustain. Energy Rev. **48**, 585 – 607 (2015). http://www.sciencedirect.com/science/article/pii/S136403211500307X

27. Ghassemi, A.: System of Systems Approach to Develop an Energy-Water Nexus Model Under Uncertainty. Ph.D. thesis (2019)
28. Peykani, P., Mohammadi, E., Pishvaee, M., Rostamy-Malkhalifeh, M., Jabbarzadeh, A.: A novel fuzzy data envelopment analysis based on robust possibilistic programming: Possibility, necessity and credibility-based approaches. RAIRO - Oper. Res. **52**, 1445 – 1463 (12 2018)
29. Peykani, P., Mohammadi, E., Emrouznejad, A., Pishvaee, M.S., Rostamy-Malkhalifeh, M.: Fuzzy data envelopment analysis: an adjustable approach. Expert Syst. Appl. **136**, 439 – 452 (2019). http://www.sciencedirect.com/science/article/pii/S0957417419304397
30. Zadeh, L.A.: A note on z-numbers. Inf. Sci. **181**(14), 2923–2932 (2011). http://www.sciencedirect.com/science/article/pii/S0020025511001046

A New Combination Method for Fuzzy Optimal Control

Sahar Askari[1] and Saeid Abbasbandy[2(✉)]

[1] Department of Applied Mathematics, Hamedan Branch, Islamic Azad University, 65138 Hamedan, Iran
[2] Department of Mathematics, Faculty of Science, Imam Khomeini International University, 34194-288, Qazvin, Iran
abbasbandy@sci.ikiu.ac.ir

Abstract. In this paper, we present a new hybrid method for solving fuzzy optimal control problems (FOCP). This hybrid method consists of a polynomial and an improved multi-layer perceptron (IMLP) network neural network. An improved neural network is a two-layer neural network. The first layer consists of inputs, weights, and six non-linear sigmoid transfer functions per α-cut and every training point. The second layer, which is the same output layer, includes the weights of the output layer and the neural network outputs, and six linear transfer functions per α-cut and each educational point. Artificial Neural Network training is based on the optimization technique on the target function. This objective function is the error function, and is equal to the sum of the squared errors that are based on the Pontryagin minimization principle.

Keywords: Fuzzy optimal control problems · IMLP networks · Pontryagin minimum principle · BFGs method

1 Introduction

The most important factor of modern control is monitoring control theory. Optimal control deals with the problem of finding a control law for a given system such that a certain optimality criterion is achieved. A control problem includes a cost functional that is a function of state and control variables. An optimal control is a set of differential equations describing the paths of the control variables that minimize the cost function. The study of optimal control greatly attracted the attention of many mathematicians because of the necessity of strict expression form in optimal control theory. With the more use of methods and results on mathematics and computer science, optimal control theory has greatly achieved development, and been applied to many fields such as production engineering, programming, economy and management. The optimal control can be derived using Pontryagin's maximum principle (a necessary condition also known as Pontryagin's minimum principle or simply Pontryagin's Principle), [2] or by solving the Hamilton–Jacobi–Bellman equation (a sufficient condition). Many studies have been conducted on optimal control, including one that can be mentioned [1, 3, 5, 9, 10, 12,

T. Allahviranloo et al. (Eds.): IDS 2020, AISC 1301, pp. 124–140, 2021.
https://doi.org/10.1007/978-3-030-66501-2_10

14, 19, 22, 25, 29]. The fuzzy optimal control problem is described as a problem of satisfaction of an objective function and transversality conditions in a certain degree instead of the crisp optimal control problem. Which is often, using PMP [23], or using solving the Hamilton–Jacobi–Bellman (HJB) partial differential equation (PDE) [15, 19], the optimization control problems (OCP) calculated. Effati and Pakdaman in 2012 used neural networks to approximate the answer to OCP [6]. The concept of the fuzzy set [29] has recently expanded in various research fields such as optimization, differential equations and OCP. Although the notion of fuzzy sets is widely spread to various control optimization problems, establishing necessary optimality conditions for fuzzy optimal control problems is seldom available in literatures [16, 17, 25, 26, 28].

In this paper, we present a new method based on Semi-Taylor by using FMLP network to solve FOCP. The present article consists of following sections: In Sect. 2, the basic and required concepts are expressed. In Sect. 3, the estimate of control optimization problems has been stated using Combination Method. In Sect. 4, The numerical examples are presented. In Sect. 5, the conclusion is presented and finally, the references are listed.

2 Definitions and Required Theorems

In this section we introduce notations, definitions, some preliminary notions, and required theorems, which is used to propose the model.

Definition 2.1. The fuzzy number $\tilde{a} : \mathbb{R} \to [0, 1]$ is a mapping with the properties:

(i) $\tilde{a}$ is normal, i.e. there exists $x_0 \in \mathbb{R}$, such that $\tilde{a}(x_0) = 1$.
(ii) $\tilde{a}$ is fuzzy convex, i.e.

$$\forall x, y \in \mathbb{R} \; and \; \lambda \in [0, 1], \tilde{a}(\lambda x + (1 - \lambda)y) \geq min\{\tilde{a}(x), \tilde{a}(y)\}.$$

(iii) $\tilde{a}$ is upper semi continuous.
(iv) $cl\{s \in \mathbb{R} | \tilde{a}(s)\rangle 0\}$, is compact in $x \in \mathbb{R}$. For $0 < \alpha \leq 1$, denote $[\tilde{a}]^{\alpha} = \{x \in \mathbb{R} | \tilde{a}(x) \geq \alpha\}$, $[\tilde{a}]_{\alpha}$ is the α-level set

If $\tilde{a} : \mathbb{R} \to [0, 1]$, then $\tilde{a}$ is fuzzy convex, so $[\tilde{a}]^{\alpha}$ is closed and bounded in R, i.e.$[\tilde{a}]^{\alpha} = [\underline{a}(\alpha), \overline{a}(\alpha)]$, where $\underline{a}(\alpha) = inf\{x \in \mathbb{R} : \tilde{a}(x) \geq \alpha\} > -\infty$ and $\overline{a}(\alpha) = sup\{x \in \mathbb{R} : \tilde{a}(x) \geq \alpha\} < \infty$. For more detail see [4].

Definition 2.2. If A and B be fuzzy numbers with $[A]^{\alpha} = [\underline{a}(\alpha), \overline{a}(\alpha)]$, $[B]^{\alpha} = [\underline{b}(\alpha), \overline{b}(\alpha)]$ and $\alpha \in [0, 1]$, then fuzzy operation between them are defined as follows [31].

$$[A + B]^{\alpha} = [\underline{a}(\alpha) + \underline{b}(\alpha), \overline{a}(\alpha) + \overline{b}(\alpha)],$$
$$[kA]^{\alpha} = [k\underline{a}(\alpha), k\overline{a}(\alpha)], \qquad k \geq 0,$$
$$[kA]^{\alpha} = [k\overline{a}(\alpha), k\underline{a}(\alpha)], \qquad k < 0.$$

So, the fuzzy number $\tilde{a}$ is triangular if $\underline{a}(1) = \overline{a}(1), \underline{a}(\alpha) = \underline{a}(1) - (1-\alpha)\big(\underline{a}(1) - \overline{a}(0)\big)$ and $\overline{a}(\alpha) = \underline{a}(1) + (1-\alpha)\big(\overline{a}(0) - \underline{a}(1)\big)$. The triangular fuzzy number $\tilde{a}$ is generally denoted by $\tilde{a} = <\underline{a}(0), \underline{a}(1), \overline{a}(0)>$.

Definition 2.3. Let $\tilde{x} : T \to F(\mathbb{R})$ is Hukuhara differentiable at $t_0 \in T \subseteq \mathbb{R}$ if for some $h_0 > 0$ the Hukuhara difference $\tilde{x}(t_0 + \Delta t) \sim_h \tilde{x}(t_0), \tilde{x}(t_0) \sim_h \tilde{x}(t_0 + \Delta t)$, exist in E for all $0 < \Delta < h_0$ and if there exist an element $\tilde{\tilde{x}}(t_0) \in F(\mathbb{R})$ such that

$$\lim_{\Delta t \to 0+} d_\infty \left(\frac{\tilde{x}(t_0 + \Delta t) \sim_h \tilde{x}(t_0)}{\Delta t}, \tilde{\tilde{x}}(t_0) \right) = 0 \tag{2.1}$$

and

$$\lim_{\Delta t \to 0+} d_\infty \left(\frac{\tilde{x}(t_0) \sim_h \tilde{x}(t_0 + \Delta t)}{\Delta t}, \tilde{\tilde{x}}(t_0) \right) = 0 \tag{2.2}$$

The fuzzy set $\tilde{\tilde{x}}(t_0)$ is called the Hukuhara derivative of $\tilde{x}$ at t_0 [2]. Recall that $U \sim_h V = W \in F(\mathbb{R})$ are defined on level sets, where $[U]^\alpha \sim_h [V]^\alpha = [W]^\alpha$ for all $0 \le \alpha \le 1$. By consideration of definition of the metric d_∞, all the level set mappings $[\tilde{x}(.)]^\alpha$ are Hukuhara differentiable at t_0 with Hukuhara derivatives $[\tilde{\tilde{x}}(t_0)]^\alpha$ for each $0 \le \alpha \le 1$, when $\tilde{x} : T \to F(\mathbb{R})$ is Hukuhara differentiable at with Hukuhara derivative $\tilde{\tilde{x}}(t_0)$.

Definition 2.4. (see [14]). (Fuzzy-valued function). The function $\tilde{f} : V \to F(\mathbb{R})$, F is called a fuzzy-valued function if for *any* $x \in V, \tilde{f}(x)$ is a fuzzy number. We also denote $\tilde{f}(x;\alpha) = \left[\underline{f}(x;\alpha), \overline{f}(x;\alpha)\right]$, where $\underline{f}(x;\alpha) = min\{\tilde{f}(x;\alpha)\}$ and $\overline{f}(x;\alpha) = max\left\{\tilde{f}(x;\alpha)\right\}$. Therefore any fuzzy-valued function $\tilde{f}$ may be understood by $\underline{f}(x;\alpha)$ and $\overline{f}(x;\alpha)$ being respectively a bounded increasing function of α and a bounded decreasing function of α for $\alpha \in [0, 1]$. Also it holds $\underline{f}(x;\alpha) \le \overline{f}(x;\alpha)$ for any $\alpha \in [0, 1]$.

Definition 2.5. We say that $\tilde{f} : V \to F(\mathbb{R})$, is continuous at $x \in V$, if both $\underline{f}(x;\alpha)$ and $\overline{f}(x;\alpha)$ are continuous functions of $x \in V$, for all $\alpha \in [0, 1]$.

Definition 2.6. Suppose that $\tilde{f} : V \to F(\mathbb{R})$, is fuzzy-valued function with $\left[\tilde{f}(x)\right]^\alpha = \left(\underline{f}(x;\alpha), \overline{f}(x;\alpha)\right)$. If the partial derivatives of $\underline{f}(x;\alpha)$ and $\overline{f}(x;\alpha)$ with respect to $x \in \mathbb{R}$ exist and the interval $\left(\underline{f}'(x;\alpha), \overline{f}'(x;\alpha)\right)$ defines the α-level set of a fuzzy number for $x \in \mathbb{R}$, $\alpha \in [0, 1]$. Then For $x \in \mathbb{R}$, $\alpha \in [0, 1]$, $\tilde{f}(x)$ is called differentiable and we write,

$$\tilde{f}'(x;\alpha) = \left(\underline{f}'(x;\alpha), \overline{f}'(x;\alpha)\right). \tag{2.3}$$

Definition 2.7. (See [8]), suppose that $\tilde{f} : V \to F(\mathbb{R})$, is integrable with respect to x, if both $\underline{f}(x;\alpha)$ and $\overline{f}(x;\alpha)$ are Lebesgue integrable functions of $x \in R$, for all $\alpha \in [0, 1]$

and $\left(\int \underline{f}(x;\alpha)dx, \int \overline{f}(x;\alpha)dx\right)$, defines the α-level set of a fuzzy number. Denoted the integral of fuzzy function $\tilde{f}$ with respect to x for $\alpha \in [0,1]$ by

$$\int [\tilde{f}(x)]^{\alpha} dx = \int \underline{f}(x;\alpha)dx, \int \overline{f}(x;\alpha)dx \tag{2.4}$$

Definition 2.8. Let $\tilde{x}(.)$ and $\tilde{x}(.)+\delta\tilde{x}(.)$ be fuzzy functions for which the fuzzy functional $\tilde{J}$ is defined. The increment of $\tilde{J}$, denoted by $\Delta\tilde{J}$, is defined as $\Delta\tilde{J} := \tilde{J}(\tilde{x}+\delta\tilde{x}) -_h \tilde{J}(\tilde{x})$, where $\delta\tilde{x}(.)$ is known as the variation of $\tilde{x}(.)$. In order to emphasize that the increment $\Delta\tilde{J}$ depends on the fuzzy functions $\tilde{x}$ and $\delta\tilde{x}$, we may denote $\Delta\tilde{J}$ by $\Delta\tilde{J}(\tilde{x},\delta\tilde{x})$.

Definition 2.9. *(Distance measure between fuzzy functions).* Consider that $\tilde{f} : S \subseteq \mathbb{R} \to \mathbb{F}$ and $\tilde{g} : S \subseteq \mathbb{R} \to \mathbb{F}$ are two fuzzy functions. The distance measure between $\tilde{f}$ and $\tilde{g}$ is defined by

$$D_{\mathbb{F}}\left(\tilde{f}(x),\tilde{g}(x)\right) = \sup_{0\leq\alpha\leq 1} H\left(\left[\tilde{f}\right]^{\alpha}(x), \left[\tilde{g}\right]^{\alpha}(x)\right)$$

$$= max\left\{ \sup_{z\in\tilde{f}(x)} d\left(z, \left[\tilde{g}\right]^{\alpha}(x)\right), \quad \sup_{y\in\tilde{g}(x)} d\left(\left[\tilde{f}\right]^{\alpha}(x), y\right)\right\}, \quad \forall x \in S,$$

where H is the well-known Hausdorff metric on the family of all nonempty compact subsets of $\mathbb{R}$, and $d(a,B) = inf_{b\in B} d(a,b)$. For notational convenience, we define

$$\left\|\tilde{f}(x)\right\|_{\mathbb{F}}^{2} = D_{\mathbb{F}}\left(\tilde{f}(x),\tilde{f}(x)\right), \quad \forall x \in S,$$

for any $\tilde{f} : S \subseteq \mathbb{R} \to \mathbb{F}$.

Definition 2.10. Let the increment of $\tilde{J}$ can be written as

$$\Delta\tilde{J}(\tilde{x},\delta\tilde{x}) := \delta\tilde{J}(\tilde{x},\delta\tilde{x}) + \eta(\tilde{x},\delta\tilde{x}).\|\delta\tilde{x}\|_F \tag{2.5}$$

where $\delta\tilde{J}$ is linear in $\delta\tilde{x}$. If for any $\varepsilon > 0$,

$$D_F\left(\eta(\tilde{x},\delta\tilde{x}),\tilde{0}\right) < \varepsilon, \quad as \quad \|\delta\tilde{x}\|_F \to 0. \tag{2.6}$$

Then, we say that $\tilde{J}$ is differentiable on $\tilde{x}$.

Definition 2.11. A fuzzy functional $\tilde{J}$ with domain $\tilde{C}[t_0,t_f]$, the class of all fuzzy continuous functions on $\tilde{C}[t_0,t_f]$, has a fuzzy relative minimizer $\tilde{x}^* = \tilde{x}^*(t)$, if the increment of $\tilde{J}$ is fuzzy non-negative, that is, $\tilde{J}(\tilde{x}) \geq \tilde{J}(\tilde{x}^*)$, for all fuzzy functions $\tilde{x}$ in $\tilde{C}[t_0,t_f]$. Notice that the above-mentioned inequality holds if and only if $\underline{J}(\tilde{x};\alpha) \geq \underline{J}(\tilde{x}^*;\alpha)$ and $\overline{J}(\tilde{x},\alpha) \geq \overline{J}(\tilde{x}^*;\alpha)$, for all $\alpha \in [0,1]$.

Theorem 2.12. (Fuzzy fundamental theorem) Suppose that $\tilde{x}, \delta\tilde{x} \in \tilde{\mathbb{C}}\left[t_0, t_f\right]$ are fuzzy functions of $t \in \left[t_0, t_f\right]$ and $\tilde{J}(\tilde{x})$ is differentiable fuzzy functional of $\tilde{x}$. If $\tilde{x}^*$ is a fuzzy minimizer of $\tilde{J}$, then the variation of $\tilde{J}$ regardless of any boundary conditions must vanish on $\tilde{x}^*$, that is,

$$\delta\tilde{J}(\tilde{x}, \delta\tilde{x}) = \tilde{0}, \tag{2.7}$$

for all admissible $\delta\tilde{x}$ having the property $\tilde{x}, \delta\tilde{x} \in \tilde{C}\left[t_0, t_f\right]$.

Proof. See [7]

Now we consider the fuzzy initial value problem

$$\tilde{\dot{x}}(\mathrm{t}) = \tilde{f}(t, \tilde{x}(t)), \tilde{x}(0) = 0. \tag{2.8}$$

Where $\tilde{f} : [0, T] \times F(\mathbb{R}) \to F(\mathbb{R})$ is obtained by Zadeh's extension principle from a continuous function

$\tilde{f} : [0, T] \times \mathbb{R} \to \mathbb{R}$ Note that $\tilde{f}$ is continuous because f is continuous (see [27]), and if $f : \mathbb{R} \times \mathbb{R} \to \mathbb{R}$ is a continuous function then according to Zadeh's extension principle one can extend f to $\tilde{f} : \mathcal{F}(\mathbb{R}) \times \mathcal{F}(\mathbb{R}) \to \mathcal{F}(\mathbb{R})$ by the equation $\tilde{f}(\tilde{u}, \tilde{v})(z) = \underset{z=f(s,t)}{\mathrm{Sup}} \min(\tilde{\mu}(s), \tilde{v}(t))$ It is well known that $\left[\tilde{f}(\tilde{\mu}, \tilde{v})\right]^{\alpha} = \tilde{f}\left([\tilde{\mu}]^{\alpha}, [\tilde{v}]^{\alpha}\right), \alpha \in [0, 1], \tilde{\mu} \in F(\mathbb{R}), \tilde{v} \in F(\mathbb{R})$ we have $\left[\tilde{f}(t, \tilde{x})\right]^{\alpha} = \tilde{f}\left(t, [\tilde{x}]^{\alpha}\right)$ where $f(t, A) = \{f(t, a) | a \in A\}$ Associated with (2.15) we can consider the following crisp differential equation

$$\dot{x}(\mathrm{t}) = f(t, x(t)), x(0) = x_0 \tag{2.9}$$

where $\dot{x}(\mathrm{t})$ is the derivative of a crisp function $x : [0, T] \to \mathbb{R}$.

Theorem 2.13. (see [9]) let $\tilde{x} \in \mathcal{F}(\mathbb{R})$. Suppose that f is a continuous function and for each $x_0 \in \mathbb{R}$ there exists a unique solution $x(., x_0)$ for (2.16) and that $x(t, .)$ is continuous in $\mathbb{R}$ for each $t \in [0, T]$. Then:

If $\tilde{f}$ is non-decreasing with respect to the second argument, then the fuzzy solution of (2.11) and the solution of (2.12).

Definition 2.14. (Multi-layer perceptron Network (MLP) and it training)

MLP is a class of feedforward artificial neural network. An MLP consists of at least three layers of nodes. Except for the input nodes, each node is a neuron that uses a nonlinear activation function. MLP utilizes a supervised learning technique called backpropagation for training. Its multiple layers and non-linear activation distinguish MLP from a linear perceptron. It can distinguish data that is not linearly separable. Multi-layer networks use a variety of learning techniques, the most popular being back-propagation (BP), which is based on The error correction learning rule. So, to calculate sensitivities for the different layers of neurons in the MLP network the Derivative of conversion neurons functions is required. So functions used that have derivative. One of these functions is the sigmoid function defined. it is A real, bounded, and derivative function. The sigmoid function has a positive derivative, and has the following general relationship

$$\sigma(n) = \frac{1}{1 - e^{-n}}. \tag{2.10}$$

Theorem 2.15. (The World approximation Builder)

The multi-layer perceptron network with one hidden layer with a sigmoid function in the middle layer and linear transformation functions in output layer are able to approximate All functions in any degree of the integral of the square (see [12]).

Definition 2.16. (BFGS method)

For compute the value of some predefined error-function the output values are compared with the correct answer. which it by various techniques, the error is then fed back through the network. Using this information, the algorithm adjusts the weights of each connection in order to reduce the value of the error function by some small amount. After repeating this process for a sufficiently large number of training cycles, the network will usually converge to some state where the error of the calculations is small. In this case, one would say that the network has learned a certain target function. To adjust weights properly, must minimize unconstrained optimization problem. For that purpose, minimization techniques such as the steepest descent method and the conjugate gradient or Quasi-Newton methods can be employed. The Newton method is one of the important algorithms in nonlinear optimization. The main disadvantage of the Newton method is that it is necessary to evaluate the second derivative matrix (Hessian matrix). Quasi-Newton methods were originally proposed by Davidon in 1959 and were later developed by Fletcher and Powell (1963). The most fundamental idea in quasi-Newton methods is the requirement to calculate an approximation of the Hessian matrix. Here the Quasi-Newton BFGS (Broyden Fletcher Goldfarb Shanno) method is used. This method is quadratically (see [21]).

3 Solution of FOCP Using Combination Method

In this solution, we consider the fuzzy optimal control problem which is as follows

$$\begin{aligned} &\textit{Minimize} \quad \tilde{J}(\tilde{u}) : \int_{t_0}^{t_f} \tilde{g}(\tilde{x}(t), \tilde{u}(t), t)dt \\ &\textit{subject to} \quad \tilde{\dot{x}}(t) = \tilde{h}(\tilde{x}(t), \tilde{u}(t), t), \\ &\tilde{x}(t_0) = \tilde{x}_0, \qquad \tilde{x}(t_f) = \tilde{x}_f. \end{aligned} \tag{3.1}$$

Where $\tilde{g}$ and $\tilde{h}$ assign a fuzzy number to the fuzzy point $(\tilde{x}(t), \tilde{u}(t), t) \in \mathbb{F}^2 \times \mathbb{R}$, where the fuzzy state $\tilde{x}(t)$ and the fuzzy control $\tilde{u}(t)$ are fuzzy functions of t belonging to the specified interval $[t_0, t_f]$. Assuming that the integrand $\tilde{g}$ and fuzzy function $\tilde{h}$ have continuous first and second partial derivatives with respect to all of their arguments.

Suppose $\tilde{x} = \tilde{x}(t)$ is admissible, if it satisfies the endpoints conditions and also is twice continuously differentiable with respect to $t \in [t_0, t_f]$. Also an admissible fuzzy control $\tilde{u} = \tilde{u}(t)$ is that $\tilde{u}$ is not bounded. In order to attack this problem, we adopt fuzzy Lagrange multiplier. To begin with, we formulate the fuzzy augmented functional as follows:

$$\tilde{J}_a(\tilde{u}) = \int_{t_0}^{t_f} \tilde{g}_a\left(\tilde{x}(t), \tilde{u}(t), \tilde{P}(t), \tilde{\dot{x}}(t), t\right)dt. \tag{3.2}$$

Where

$$\tilde{g}_a\left(\tilde{x}(t), \tilde{u}(t), \tilde{P}(t), \tilde{\dot{x}}(t), t\right) := \tilde{g}(\tilde{x}(t), \tilde{u}(t), t) + \tilde{P}(t)\tilde{h}(\tilde{x}(t), \tilde{u}(t), t) - \tilde{\dot{x}}(t). \tag{3.3}$$

In order to simplify the result presentations, we limit ourselves to the special case stated in the following assumption.

Remark 3.1. To simplify the variational equations, we assume that $\underline{J}_a(\tilde{u})$ or $\overline{J}_a(\tilde{u})$ is stated in terms containing only $\underline{x}(t;\alpha)$ and $\underline{\dot{x}}(t;\alpha)$ or ($\overline{x}(t;\alpha)$ and $\overline{\dot{x}}(t;\alpha)$) In this case, we may write $\underline{J}_a(\underline{u};\alpha)$ and $\overline{J}_a(\overline{u};\alpha)$ instead of $\underline{J}_a(\tilde{u};\alpha)$ and $\overline{J}_a(\tilde{u};\alpha)$, respectively.

On the extremal $\tilde{u}^*$, the variation of $\tilde{J}_a$ must be zero, that is, $\delta\tilde{J}_a(\tilde{u}^*) = \tilde{0}$. This admits for all $\alpha \in [0, 1]$,

$$\delta\underline{J}_a(\underline{u}^*;\alpha) = 0, \tag{3.4}$$

and

$$\delta\overline{J}_a(\overline{u}^*;\alpha) = 0. \tag{3.5}$$

Originally, For Relation (3.4), we have

$$\begin{aligned}\delta\underline{J}_a\left(\underline{u}^*;\alpha\right) = \int_{t_0}^{t_f}\Bigg\{\Bigg[&\frac{\partial \underline{g}_a}{\partial \underline{x}}\left(\underline{x}^*(t), \underline{u}^*(t), \underline{P}^*(t), \underline{\dot{x}}^*(t), t;\alpha\right)\\ &-\frac{d}{dt}\left(\frac{\partial \underline{g}_a}{\partial \underline{\dot{x}}}\underline{x}^*(t), \underline{u}^*(t), \underline{P}^*(t), \underline{\dot{x}}^*(t), t;\alpha\right)\Bigg]\delta\underline{x}\\ &+\left[\frac{\partial \underline{g}_a}{\partial \underline{x}}\left(\underline{x}^*(t), \underline{u}^*(t), \underline{P}^*(t), \underline{\dot{x}}^*(t), t;\alpha\right)\right]\delta\underline{u}\\ &+\left[\frac{\partial \underline{g}_a}{\partial \underline{x}}\left(\underline{x}^*(t), \underline{u}^*(t), \underline{P}^*(t), \underline{\dot{x}}^*(t), t;\alpha\right)\right]\Bigg\}\delta\underline{P}dt = 0,\end{aligned} \tag{3.6}$$

with due attention to the fuzzy augmented integrand function $\tilde{g}_a$ defined in (3.3), the latter equation can be written

$$\int_{t_0}^{t_f}\Bigg\{\Bigg[\frac{\partial \underline{g}_a}{\partial \underline{x}}\left(\underline{x}^*(t), \underline{u}^*(t), t;\alpha\right) + \underline{P}^*(t;\alpha)\left(\frac{\partial \underline{g}_a}{\partial \underline{x}}\left(\underline{x}^*(t), \underline{u}^*(t), t;\alpha\right)\right)$$

$$-\frac{d}{dt}\Big(-\underline{P}^*(t;\alpha)\Big)]\delta\underline{x}$$

$$+\left[\frac{\partial \underline{g}_a}{\partial \underline{x}}\Big(\underline{x}^*(t),\underline{u}^*(t),t;\alpha\Big)+\underline{P}^*(t;\alpha)\left(\frac{\partial \underline{g}_a}{\partial \underline{x}}\Big(\underline{x}^*(t),\underline{u}^*(t),t;\alpha\Big)\right)\right]\delta\underline{u}$$

$$+\underline{h}\Big[\underline{h}\Big(\underline{x}^*(t),\underline{u}^*(t),t;\alpha\Big)-\underline{\dot{x}}^*(t;\alpha)\Big]\delta\underline{P}\Big\}dt=0. \tag{3.7}$$

Since the constraint must be satisfied by the extremal $\tilde{u}^*$, we find that

$$\underline{\dot{x}}^*(t;\alpha)=\underline{h}\Big(\underline{x}^*(t),\underline{u}^*(t),t\Big) \tag{3.8}$$

and hence the coefficient of $\delta\underline{P}$ in (3.7) is zero. Moreover, the arbitrary fuzzy Lagrange multiplier $\underline{P}^*$ can be chosen such that the coefficient of $\delta\underline{x}$ does not appear in the above integral. So

$$\underline{P}^*(t;\alpha)=\left[\frac{\partial \underline{g}_a}{\partial \underline{x}}\Big(\underline{x}^*(t),\underline{u}^*(t),t;\alpha\Big)+\underline{P}^*(t;\alpha)\left(\frac{\partial \underline{g}_a}{\partial \underline{x}}\Big(\underline{x}^*(t),\underline{u}^*(t),t;\alpha\Big)\right)\right]. \tag{3.9}$$

There exists nevertheless a term inside the integral (3.7) to deal with. Since the equality (3.7) must be satisfied, we get

$$\frac{\partial \underline{g}_a}{\partial \underline{x}}\Big(\underline{x}^*(t),\underline{u}^*(t),t;\alpha\Big)+\underline{P}^*(t;\alpha)\left(\frac{\partial \underline{g}_a}{\partial \underline{x}}\Big(\underline{x}^*(t),\underline{u}^*(t),t;\alpha\Big)\right)=0. \tag{3.10}$$

We can construct the well-known Hamiltonian as:

$$H(\tilde{x}(t),\tilde{u}(t),\tilde{p}(t),t):=\tilde{g}(\tilde{x}(t),\tilde{u}(t),t)+\tilde{P}(t)\tilde{h}(\tilde{x}(t),\tilde{u}(t),t), \tag{3.11}$$

where $\tilde{p}$ is the so-called fuzzy Lagrange multiplier. with the $\alpha-$ level set

$$\Big[\tilde{H}(\tilde{x}(t),\tilde{u}(t),\tilde{p}(t),t)\Big]^{\alpha}=\Big(\underline{H}\Big(\underline{x}(t),\underline{u}(t),\underline{p}(t),t;\alpha\Big),\bar{H}(\bar{x}(t),\bar{u}(t),\bar{p}(t),t;\alpha)\Big) \tag{3.12}$$

where

$$\underline{H}\Big(\underline{x}(t),\underline{u}(t),\underline{p}(t),t;\alpha\Big)=\underline{g}\Big(\underline{x}(t),\underline{u}(t),t;\alpha\Big)+\underline{P}(t)\underline{h}\Big(\underline{x}(t),\underline{u}(t),t;\alpha\Big) \tag{3.13}$$

$$\bar{H}(\bar{x}(t),\bar{u}(t),\bar{p}(t),t;\alpha)=\bar{g}(\bar{x}(t),\bar{u}(t),t;\alpha)+\bar{P}(t)\bar{h}(\bar{x}(t),\bar{u}(t),t;\alpha) \tag{3.14}$$

Using the left-hand Hamiltonian function $\underline{H}\Big(\underline{x}(t),\underline{u}(t),\underline{p}(t),t;\alpha\Big)$ defined in (3.13), the Eqs. (3.8)–(3.10) can be written more compactly as follows:

$$\underline{\dot{x}}^*(t)=\frac{\partial\underline{H}\Big(\underline{x}^*(t),\underline{u}^*(t),\underline{P}^*(t),t\Big)}{\partial\underline{P}^*} \tag{3.15}$$

$$\underline{\dot{P}}^*(t) = -\frac{\partial \underline{H}\left(\underline{x}^*(t), \underline{u}^*(t), \underline{P}^*(t), t\right)}{\partial \underline{x}^*} \tag{3.16}$$

$$0 = \frac{\partial \underline{H}\left(\underline{x}^*(t), \underline{u}^*(t), \underline{P}^*(t), t\right)}{\partial \underline{u}^*} \tag{3.17}$$

for all $\alpha \in [0, 1]$, $t \in \left[t_0, t_f\right]$. According to the above, we say

Theorem 3.2. Let $\left[\underline{x}^*\right]^\alpha = \underline{x}^*(t; \alpha)$ be an admissible fuzzy function, i.e., it is twice continuously differentiable fuzzy function. Then, in order that $\left[\underline{x}^*\right]^\alpha$ give a relative (local) minimum to the fuzzy functional $\underline{J}_a$ in FOCP, it is necessary that for all $\alpha \in [0, 1]$, $t \in \left[t_0, t_f\right]$.

Again, following the scheme of obtaining (3.15)–(3.17) and adapting it to the case under consideration involving (3.5), one may show that for all $\alpha \in [0, 1]$, $t \in \left[t_0, t_f\right]$.

$$\dot{\bar{x}}(t) = \frac{\partial \bar{H}(\bar{x}^*(t), \bar{u}^*(t), \bar{p}^*(t))}{\partial \bar{p}^*} \tag{3.18}$$

$$\dot{\bar{p}}^*(t) = -\frac{\partial \bar{H}(\bar{x}^*(t), \bar{u}^*(t), \bar{p}^*(t))}{\partial \bar{x}^*} \tag{3.19}$$

$$0 = \frac{\partial \bar{H}(\bar{x}^*(t), \bar{u}^*(t), \bar{p}^*(t))}{\partial \bar{u}^*} \tag{3.20}$$

Theorem 3.3. Let $\left[\bar{x}^*\right]^\alpha = \bar{x}^*(t; \alpha)$ be an admissible fuzzy function, i.e., it is twice continuously differentiable fuzzy function. Then, in order that $\left[\bar{x}^*\right]^\alpha$ give a relative (local) minimum to the fuzzy functional $\bar{J}_\alpha$ in (FOCP), it is necessary that for all $\alpha \in [0, 1]$, $t \in \left[t_0, t_f\right]$.

Now, using Theorem 3.2 and Theorem 3.3, let $\left[\tilde{x}^*\right]^\alpha = \tilde{x}^*(t; \alpha)$ be an admissible fuzzy function, i.e., it is twice continuously differentiable fuzzy function. Then, in order that $\left[\tilde{x}^*\right]^\alpha$ give a relative (local) minimum to the fuzzy functional $\tilde{J}_\alpha$ in (FOCP), it is necessary that for all $\alpha \in [0, 1]$, $t \in \left[t_0, t_f\right]$. Indeed, for the Hamiltonian function in (3.11) and using (3.12) necessary conditions for the binary $(\left[\underline{x}^*\right]^\alpha, \left[\bar{x}^*\right]^\alpha)$ the optimal solution of (3.1) is the existence of a costate vector function $\left(\left[\underline{u}^*\right]^\alpha, \left[\bar{u}^*\right]^\alpha\right)$ that satisfies the following differential Eqs. (3.15)–(3.20), where $\left(\left[\underline{x}^*\right]^\alpha, \left[\bar{x}^*\right]^\alpha\right)$ is the same as $\left[\tilde{x}^*\right]^\alpha$ and $\left(\left[\underline{u}^*\right]^\alpha, \left[\bar{u}^*\right]^\alpha\right)$ is the same as $\left[\tilde{u}^*\right]^\alpha$.

Now, we are trying to propose an approximation method for solving the equations arising in PMP. This proposed method is a three-equation system with initial conditions and boundary conditions in it. The proposed equation system is as follows:

$$\begin{cases} \tilde{x}_T = \tilde{V}(t, \alpha) + f_1\left(t, \tilde{N}_x(t, \tilde{\varphi}_x; \alpha)\right) \\ \tilde{p}_T = f_2(t)\tilde{N}_p\left(t, \tilde{\varphi}_p; \alpha\right) \\ \tilde{u}_T = \tilde{N}_u(t, \tilde{\varphi}_u; \alpha) \end{cases} \tag{3.21}$$

Where, $\forall i = 1, 2.$, $f_i(t)$ is an arbitrary function in terms of the variable t, so that the conditions of the issue are true. $\varphi_{.}$ containing all weights $\widetilde{N}_{.}$ of relation (3.21). We introduce three fuzzy neural networks for each equation as follows:

$$\tilde{N}_x(t, \tilde{\varphi}_x; \alpha) = \sum_{i=1}^{k} \tilde{v}_{ix}\sigma\left(\tilde{Z}_{ix}\right), \tag{3.22}$$

$$\tilde{N}_p\left(t, \tilde{\varphi}_p; \alpha\right) = \sum_{i=1}^{k} \tilde{v}_{ip}\sigma\left(\tilde{Z}_{ip}\right), \tag{3.23}$$

$$\tilde{N}_u(t, \tilde{\varphi}_u; \alpha) = \sum_{i=1}^{k} \tilde{v}_{iu}\sigma\left(\tilde{Z}_{iu}\right). \tag{3.24}$$

We also define the function $\tilde{V}(t; \alpha)$ as follows [7].

$$\tilde{V}(t, \alpha) = \left(1 + \frac{t}{1!} + \frac{t^2}{2!} + \frac{t^3}{3!}\right)\tilde{x}(t_0; \alpha) \tag{3.25}$$

Via replacing the approximate solutions into the fuzzy Hamiltonian function, we can define an approximate fuzzy Hamiltonian $\tilde{H}_T$ that is fuzzy Hamiltonian function $\tilde{H}$ where we replaced the functions $\tilde{x}$, $\tilde{p}$ and $\tilde{u}$ by their corresponding approximate format

$$\begin{aligned}&\left[\tilde{H}(\tilde{x}_T(t), \tilde{u}_T(t), \tilde{p}_T(t), t)\right]^{\alpha} \\ &= \left(\underline{H}\left(\underline{x}_T(t), \underline{u}_T(t), \underline{p}_T(t), t; \alpha\right), \bar{H}(\bar{x}_T(t), \bar{u}_T(t), \bar{p}_T(t), t; \alpha)\right)\end{aligned} \tag{3.26}$$

with the $\alpha-$ level set, we have, $\tilde{x}_T = \left(\underline{x}_T, \bar{x}_T\right), \tilde{p}_T = \left(\underline{p}_T, \bar{p}_T\right), \tilde{u}_T = \left(\underline{u}_T, \bar{u}_T\right)$ where

$$\begin{cases} \underline{x}_T = \underline{V}(t; \alpha) + f_1(t)\underline{N}_x\left(t, \underline{\varphi}_x; \alpha\right) \\ \underline{p}_T = f_2(t)\underline{N}_p\left(t, \underline{\varphi}_p; \alpha\right) \\ \underline{u}_T = \underline{N}_u\left(t, \underline{\varphi}_u; \alpha\right) \\ \bar{x}_T = \bar{V}(t, \alpha) + f_1(t)\bar{N}_x(t, \bar{\varphi}_x; \alpha) \\ \bar{p}_T = f_2(t)\bar{N}_p\left(t, \bar{\varphi}_p; \alpha\right) \\ \bar{u}_T = \bar{N}_u(t, \bar{\varphi}_u; \alpha) \end{cases} \tag{3.27}$$

Also, with the $\alpha-$ level set, we have, $\tilde{N}_x = \left(\underline{N}_x, \bar{N}_x\right), \tilde{N}_p = \left(\underline{N}_p, \bar{N}_p\right), \tilde{N}_u = \left(\underline{N}_u, \bar{N}_u\right)$, where

$$\begin{cases} \begin{cases} \underline{N}_x\left(t, \underline{\varphi}_x; \alpha\right) = \sum_{i=1}^{k} \underline{v}_{ix} \sigma\left(\underline{Z}_{ix}\right), \underline{Z}_{ix} = \underline{w}_{1ix} t + \underline{w}_{2ix} \alpha + \underline{b}_{ix} \\ \bar{N}_x(t, \bar{\varphi}_x; \alpha) = \sum_{i=1}^{k} \bar{v}_{ix} \sigma\left(\bar{Z}_{ix}\right), \bar{Z}_{ix} = \bar{w}_{1ix} t + \bar{w}_{2ix} \alpha + \bar{b}_{ix} \end{cases} \\ \begin{cases} \underline{N}_p\left(t, \underline{\varphi}_p; \alpha\right) = \sum_{i=1}^{k} \underline{v}_{ip} \sigma\left(\underline{Z}_{ip}\right), \underline{Z}_{ip} = \underline{w}_{1ip} t + \underline{w}_{2ip} \alpha + \underline{b}_{ip} \\ \bar{N}_p(t, \bar{\varphi}_p; \alpha) = \sum_{i=1}^{k} \bar{v}_{ip} \sigma\left(\bar{Z}_{ip}\right), \bar{Z}_{ip} = \bar{w}_{1ip} t + \bar{w}_{2ip} \alpha + \bar{b}_{ip} \end{cases} \\ \begin{cases} \underline{N}_u\left(t, \underline{\varphi}_u; \alpha\right) = \sum_{i=1}^{k} \underline{v}_{iu} \sigma\left(\underline{Z}_{iu}\right), \underline{Z}_{iu} = \underline{w}_{1iu} t + \underline{w}_{2iu} \alpha + \underline{b}_{iu} \\ \bar{N}_u(t, \bar{\varphi}_u; \alpha) = \sum_{i=1}^{k} \bar{v}_{ix} \sigma\left(\bar{Z}_{iu}\right), \bar{Z}_{iu} = \bar{w}_{1iu} t + \bar{w}_{2iu} \alpha + \bar{b}_{iu} \end{cases} \end{cases} \tag{3.28}$$

Where $\sigma(Z) = \frac{1}{1+e^{-Z}}$, and also, with the $\alpha-$ level set, we have, $\tilde{V}(t; \alpha) = \left(\overline{V}, \overline{V}\right)$, where

$$\underline{V}(t, \alpha) = \left(1 + \frac{t}{1!} + \frac{t^2}{2!} + \frac{t^3}{3!}\right)\underline{x}(t_0; \alpha) \tag{3.29}$$

$$\bar{V}(t, \alpha) = \left(1 + \frac{t}{1!} + \frac{t^2}{2!} + \frac{t^3}{3!}\right)\bar{x}(t_0; \alpha) \tag{3.30}$$

On the other hand the trial solutions (3.27) must satisfy conditions (3.18)–(3.20), so we replace them in (3.18)–(3.20):

$$\tilde{\dot{P}}_T(t) + \frac{\partial \tilde{H}(\tilde{x}_T(t), \tilde{u}_T(t), \tilde{p}_T(t), t)}{\partial \tilde{x}_T} = 0 \tag{3.31}$$

$$\tilde{\dot{x}}_T(t) - \frac{\partial \tilde{H}(\tilde{x}_T(t), \tilde{u}_T(t), \tilde{p}_T(t), t)}{\partial \tilde{p}_T} = 0 \tag{3.32}$$

$$\frac{\partial \tilde{H}(\tilde{x}_T(t), \tilde{u}_T(t), \tilde{p}_T(t), t)}{\partial \tilde{u}_T} = 0 \tag{3.33}$$

To solve relations (3.31)–(3.33), we define three error function corresponding to each equation as follows:

$$\tilde{E}(\varphi, t) = \left[\tilde{\dot{P}}_T(t) + \frac{\partial \tilde{H}(\tilde{x}_T(t), \tilde{u}_T(t), \tilde{p}_T(t), t)}{\partial \tilde{x}_T}\right]^2$$

$$+\left[\tilde{\dot{x}}_T(t)-\frac{\partial\tilde{H}(\tilde{x}_T(t),\tilde{u}_T(t),\tilde{p}_T(t),t)}{\partial\tilde{p}_T}\right]^2+\left[\frac{\partial\tilde{H}(\tilde{x}_T(t),\tilde{u}_T(t),\tilde{p}_T(t),t)}{\partial\tilde{u}_T}\right]^2 \tag{3.34}$$

where φ is a vector containing all weights of there MLP networks (3.28). Scilicet φ contains all weights $\tilde{v}_x, \tilde{w}_{1x}, \tilde{w}_{2x}, \tilde{v}_p, \tilde{w}_{1p}, \tilde{w}_{2p}, \tilde{v}_u, \tilde{w}_{1u}$, and $\tilde{w}_{2u}$.

With the $\alpha-$ level set, for relation (3.34) and By deriving the functions contained in relation (3.27) we, instead of solving Eqs. (3.31), (3.32) and (3.33), we discretize the interval $[t_0, t_f]$ (by m points) and solve the following unconstrained fuzzy optimization problem:

$$\begin{aligned}
&E(\varphi, t,)\\
&=min\left[\sum_{i=1}^{k}\left[\left(\frac{\partial \underline{V}}{\partial t}+f_1'(t)\underline{Net}_p+f_1(t)\frac{\partial \underline{Net}_p}{\partial t}\right)+\frac{\partial \underline{H}\left(\underline{x}_T(t),\underline{u}_T(t),\underline{p}_T(t),t;\alpha\right)}{\partial \underline{x}_T}\right]^2\right.\\
&+\sum_{i=1}^{k}\left[\left(\underline{V}'(t)+f_1'(t)\underline{Net}_x+f_1(t)\frac{\partial \underline{Net}_x}{\partial t}\right)-\frac{\partial \underline{H}\left(\underline{x}_T(t),\underline{u}_T(t),\underline{p}_T(t),t;\alpha\right)}{\partial \underline{p}_T}\right]^2\\
&+\sum_{i=1}^{k}\left[\frac{\partial \underline{H}\left(\underline{x}_T(t),\underline{u}_T(t),\underline{p}_T(t),t;\alpha\right)}{\partial \underline{u}_T}\right]^2\\
&+\sum_{i=1}^{k}\left[\left(f_2'(t)\overline{Net}_p+f_2(t)\frac{\partial \overline{Net}_p}{\partial t}\right)+\frac{\partial \bar{H}(\bar{x}_T(t),\bar{u}_T(t),\bar{p}_T(t),t;\alpha)}{\partial \bar{x}_T}\right]^2\\
&+\sum_{i=1}^{k}\left[\left(\frac{\partial \bar{V}}{\partial t}+f_1'(t)\overline{Net}_x+f_1(t)\frac{\partial \overline{Net}_x}{\partial t}\right)\right.\\
&\left.\left.-\frac{\partial \bar{H}(\bar{x}_T(t),\bar{u}_T(t),\bar{p}_T(t),t;\alpha)}{\partial \bar{p}_T}\right]^2+\sum_{i=1}^{k}\left[\frac{\partial \bar{H}(\bar{x}_T(t),\bar{u}_T(t),\bar{p}_T(t),t;\alpha)}{\partial \bar{u}_T}\right]^2\right]\to 0
\end{aligned} \tag{3.37}$$

Where

$$\frac{\partial \underline{V}}{\partial t}=\left(1+t+\frac{t^2}{2}\right)\underline{x}(t_0;\alpha), \tag{3.38}$$

$$\frac{\partial \bar{V}}{\partial t}=\left(1+t+\frac{t^2}{2}\right)\bar{x}(t_0;\alpha). \tag{3.39}$$

In this work, we minimized the relation (3.37) using the BFGS Quasi-Newton method (see [19]).

4 Numerical Examples

In this section, two examples that illustrate the efficiency, simplicity and precision of the proposed method are presented.

Example 4.1. Find the fuzzy control that minimize

$$\tilde{J}(\tilde{u}) = \int_0^1 \tilde{u}^2(t)dt, \tag{4.1}$$

subject to

$$\tilde{\dot{x}}(\tilde{u}) = \tilde{u}(t) - (0, 1, 3)\tilde{x}(t), \qquad t \in [0, 1],$$

with boundary conditions

$$\tilde{x}(0) = 1 = (1, 1, 1), \quad \tilde{x}(1) = 0 = (0, 0, 0).$$

In First step we must construct the Hamiltonian function

$$\tilde{H}(\tilde{x}(t), \tilde{u}(t), \tilde{p}(t), t) := \tilde{u}^2(t) + \tilde{p}(t)\big[\tilde{u}(t) - (0, 1, 3)\tilde{x}(t)\big] \tag{4.2}$$

and hence the α-level set of $\tilde{H}$ is characterized by

$$\underline{H}\Big(\underline{x}(t), \underline{u}(t), \underline{p}(t), t; \alpha\Big) = \underline{u}^2(t; \alpha) + \underline{p}(t, \alpha)\Big[\underline{u}(t; \alpha) - (3 - 2\alpha)\underline{x}(t; \alpha)\Big] \tag{4.3}$$

$$\bar{H}(\bar{x}(t), \bar{u}(t), \bar{p}(t), t; \alpha) = \bar{u}^2(t; \alpha) + \bar{p}(t; \alpha)[\bar{u}(t; \alpha) - (\alpha)\bar{x}(t; \alpha)] \tag{4.4}$$

According to (3.15)–(3.17), we must have:

$$\begin{cases} \dot{\underline{x}}(t; \alpha) = \underline{u}(t; \alpha) - (3 - 2\alpha)\underline{x}(t; \alpha) \\ \dot{\bar{x}}(t; \alpha) = \bar{u}(t; \alpha) - (\alpha)\bar{x}(t; \alpha) \\ \dot{\underline{p}}(t; \alpha) = (3 - 2\alpha)\underline{p}(t; \alpha) \\ \dot{\bar{p}}(t; \alpha) = (\alpha)\bar{p}(t; \alpha) \\ 0 = 2\underline{u}(t; \alpha) + \underline{p}(t; \alpha) \\ 0 = 2\bar{u}(t; \alpha) + \bar{p}(t; \alpha) \end{cases} \tag{4.5}$$

We can select the approximate solution according to the initial conditions $x(0) = 1 = (1, 1, 1)$ and $x(1) = 0 = (0, 0, 0)$, as follows:

$$\begin{cases} \tilde{x}_T = (1 - t)\Big(1 + \frac{t}{1!} + \frac{t^2}{2!} + \frac{t^3}{3!} + t\tilde{N}_x(t, \tilde{\varphi}_x; \alpha)\Big) \\ \tilde{p}_T = \tilde{N}_p\big(t, \tilde{\varphi}_p; \alpha\big) \\ \tilde{u}_T = \tilde{N}_u(t, \tilde{\varphi}_u; \alpha) \end{cases} \tag{4.6}$$

In this example, error function for two neurons units with sigmoid function in the hidden layer and for m = 50 equally spaced points inside the interval [0, 1] is trained (Fig. 1).

E(*P*) = 5.057868969038801e-04

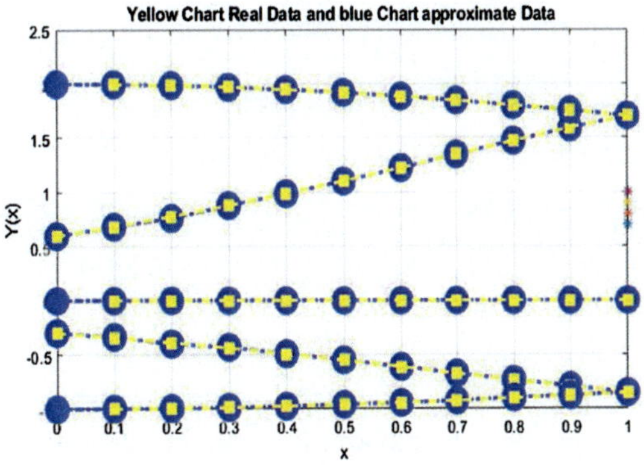

Fig. 1. The control and state functions and Lagrange multipliers for Example 1.

Example 4.2. Find the fuzzy control that minimize

$$\tilde{J}(\tilde{u}) = \int_0^1 \left[(2 - \tilde{x}(t))^2 + \tilde{u}^2(t)\right]dt, \tag{4.7}$$

subject to

$$\tilde{\dot{x}}(\tilde{u}) = -(0, 0.25, 0.5)\tilde{x}(t) + \tilde{u}(t), \qquad t \in [0, 1],$$

with boundary conditions

$$\tilde{x}(0) = 0 = (0, 0, 0), \qquad \tilde{x}(1) = 2 = (2, 2, 2).$$

In First step we must construct the Hamiltonian function

$$\tilde{H}(\tilde{x}(t), \tilde{u}(t), \tilde{p}(t), t) := \left[(2 - \tilde{x}(t))^2 + \tilde{u}^2(t)\right] + \tilde{p}(t)\left[-(0, 0.25, 0.5)\tilde{x}(t) + \tilde{u}(t)\right] \tag{4.8}$$

and hence the α-level set of $\tilde{H}$ is characterized by

$$\underline{H}\left(\underline{x}(t), \underline{u}(t), \underline{p}(t), t; \alpha\right) := \left[\left(2 - \underline{x}(t; \alpha)\right)^2 + \underline{u}^2(t; \alpha)\right] + \underline{p}(t; \alpha)\left[-(0.5 - 0.25\alpha)\underline{x}(t; \alpha) + \underline{u}(t; \alpha)\right] \tag{4.9}$$

$$\bar{H}(\bar{x}(t), \bar{u}(t), \bar{p}(t), t; \alpha) := \left[(2 - \bar{x}(t; \alpha))^2 + \bar{u}^2(t; \alpha)\right] + \bar{p}(t; \alpha)[-(0.25\alpha)\bar{x}(t; \alpha) + \bar{u}(t; \alpha)] \tag{4.10}$$

According to (3.15)–(3.17), we must have:

$$\begin{cases} \dot{\underline{x}}(t;\alpha) = -(0.5-0.25\alpha)\underline{x}(t;\alpha) + \underline{u}(t;\alpha) \\ \dot{\bar{x}}(t;\alpha) = -(0.25\alpha)\bar{x}(t;\alpha) + \bar{u}(t;\alpha) \\ \dot{\underline{p}}(t;\alpha) = -2\left(2-\underline{x}(t;\alpha)\right) + \underline{p}(t;\alpha)(-(0.5-0.25\alpha)) \\ \dot{\bar{p}}(t;\alpha) = -2(2-\bar{x}(t;\alpha)) + \bar{p}(t,\alpha)(-0.25\alpha) \\ 0 = 2\underline{u}(t;\alpha) + \underline{p}(t;\alpha) \\ 0 = 2\bar{u}(t;\alpha) + \bar{p}(t;\alpha) \end{cases} \tag{4.11}$$

We can select the approximate solution according to the initial conditions $x(0) = (0,0,0)$ and $x(1) = (2,2,2)$, as follows:

$$\begin{cases} \tilde{x}_T = \mathrm{t}\left(2 + (\mathrm{t}-1)\left(1 + \frac{t}{1!} + \frac{t^2}{2!} + \frac{t^3}{3!} + \tilde{N}_x(t,\tilde{\varphi}_x;\alpha)\right)\right) \\ \tilde{p}_T = \tilde{N}_p\left(t,\tilde{\varphi}_p;\alpha\right) \\ \tilde{u}_T = \tilde{N}_u(t,\tilde{\varphi}_u;\alpha) \end{cases} \tag{4.12}$$

In this example, error function for two neurons units with sigmoid function in the hidden layer and for m = 50 equally spaced points inside the interval [0, 1] is trained (Fig. 2).

E(*P*) = 0.043252952533444

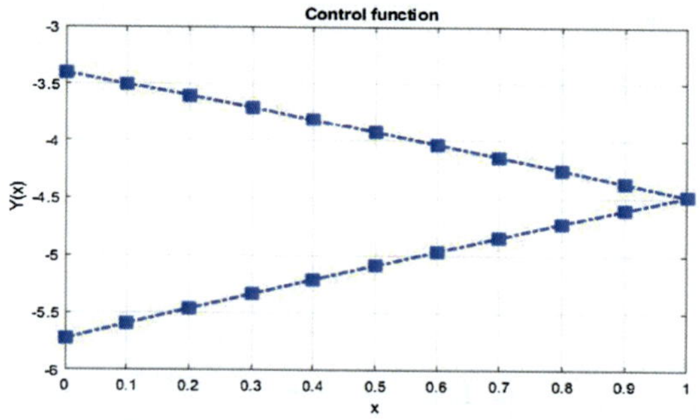

Fig. 2. Control function for Example 2.

5 Conclusion

In this article, a model based on IMLP network and semi-Taylor for an approximate solution to FOCP is proposed. For the first time, we demonstrate, the ability of improved neural networks and semi-Taylor to approximate the solutions of FOCP. This approach allows to transform a fuzzy problem to a set of crisp problems via cut-α, that can be solved with IMLP network method. The IMLP network training is based on minimizing the squared error function. One of the advantages of the proposed method is that, provides a general form of algorithms that can be designed based on the neural network to approximate the solution of FHJB equation and also for problems arising in calculus of variations, formula of each of the functions used in this method is determined by the type of problem. Another advantage of the proposed method is that, to achieve more precise solutions, we can set the number of hidden layers and training points depending on the type of problem. The main reason semi-Taylor for using neural networks was their applicability in function approximation. In some cases, increase the number of points or hidden layer of the neural network and in some cases reduce them. The proposed solution is an appropriate function for state, co-state and control functions. Using an examples, we summarized and highlighted the main feature of the article.

References

1. Berkani, S., Manseur, F., Maidi, A.: Optimal control based on the variational iteration method. Comput. Math. Appl. **64**, 604–610 (2012)
2. Bede, B., Gal, S.G.: Generalizations of the differentiability of fuzzy-number-valued functions with applications to fuzzy differential equations. Fuzzy Sets Syst. **151**, 581–599 (2005)
3. Clever, D., Lang, J., Ulbrich, S., Ziems, J.C.: Combination of an adaptive multilevel SQP method and a space-time adaptive PDAE solver for optimal control problems. Procedia Comput. Sci. **1**, 1435–1443 (2010)
4. Cheng, T., Sun, H., Qu, Z., Lewis, F.L.: Neural network solution for suboptimal control of non-holonomic chained form system. Trans. Inst. Measurement Control **31**(6), 475–494 (2009)
5. Cheng, T., Lewis, F.L.: Neural network solution for finite horizon H-infinity constrained optimal control of nonlinear systems. J. Control Theory Appl. **5**(1), 1–11 (2007)
6. Effati, S., Pakdaman, M.: Optimal control problem via neural networks. Neural Comput. Appl. **23**, 2093–2100 (2012)
7. Ezadi, S., Parandin, N., Homashi, A.G.: Numerical solution of fuzzy differential equations based on semi-taylor by using neural network. J. Basic Appl. Sci. Res. **3**(1s), 477–482 (2013)
8. Farhadinia, B.: Pontryagin's minimum principle for fuzzy optimal control problems. Iran. J. Fuzzy Syst. **11**, 27–43 (2014)
9. Farahi, M.H., Keshtegar, M., Najariyan, M.: Fuzzy optimal control of a poisoning-pest model by using -cuts. J. Eng. Res. Technol. **1**(3), 79–82 (2014)
10. Garg, D., Patterson, M., Hagera, W.W., Raoa, A.V., Bensonb, D.A., Huntington, G.T.: A unified framework for the numerical solution of optimal control problems using pseudo-spectral methods. Automatica **46**, 1843–1851 (2010)
11. Gerdts, M.: A non-smooth Newton's method for control-state constrained optimal control problems. Math. Comput. Simul. **79**, 925–936 (2008)
12. Hornick, K., Stinchcombe, M., White, H.: Multilayer feedforward networks are universal approximators. Neural Netw. **2**, 359–366 (1989)
13. Hilscher, R.S., Zeidan, V.: Hamilton-Jacobi theory over time scales and applications to linear-quadratic problems. Nonlinear Anal. **75**, 932–950 (2012)
14. Wu, H.-C.: Duality theory in fuzzy optimization problems. Fuzzy Optim. Decis. Making **3**, 345–365 (2004)
15. Krabs, W., Pickl, S.: An optimal control problem in cancer chemotherapy. Appl. Math. Comput. **217**, 1117–1124 (2010)
16. Kirk, D.E.: Optimal Control Theory—An Introduction. Dover Publications, Mineola (2004)
17. Ku, C.C., Huang, P.H., Chang, W.J.: Passive fuzzy controller design for nonlinear systems with multiplicative noises. J. Franklin Institute **347**, 732–750 (2010)

18. Kwun, Y.C., Kim, J.S., Park, M.J., Park, J.H.: Nonlocal controllability for the semi linear fuzzy integro differential equations in n-dimensional fuzzy vector space. Advances in Difference Equations (2009)
19. Liu, C., Nocedal, J.: On the limited memory BFGS method for large scale optimization. Math. Program. **45**(3), 503–528 (1989)
20. Lewis, F., Syrmos, V.L. (eds.): Optimal Control. Springer, Cham (2016)
21. Luenberger, D.G. (ed.): Linear and Nonlinear Programming, 2nd edn. Addison Wesley, Boston (2008)
22. Modares, H., Naghibi Sistani, M.B.: Solving nonlinear optimal control problems using a hybrid IPSOSQP algorithm. Eng. Appl. Artif. Intell. **24**, 476–484 (2011)
23. Najariyan, M., Farahi, M.H.: Optimal control of fuzzy linear controlled system with fuzzy initial conditions. Iran. J. Fuzzy Syst. **10**(3), 21–35 (2013)
24. Pontryagin, L.S., Boltyanskii, V.G., Gamkrelidze, R.V., Mishchenko, E.F.: The mathematical theory of optimal processes, translated from the Russian by K. N. Trirogoff; edited by L. W. Neustadt. Interscience Publishers John Wiley & Sons, Inc., New York (1962)
25. Pinch, R.: Optimal Control and the Calculus of Variations. Oxford Science Publications, Oxford University Press, New York (1993)
26. Phu, N.D., Tung, T.T.: Existence of solutions of fuzzy control differential equations. J. Sci. Technol. Dev. **10**, 5–12 (2007)
27. Roman, F., Barros, L., Bassanezi, R.: A note on the Zadeh's extensions, Fuzzy sets on Banach spaces. Inform. Sci. **144**, 227–247 (2002)
28. Ross, I.M.: A Primer on Pontryagin's Principle in Optimal Control. Collegiate Publishers (2009). ISBN 978-0-9843571-0-9
29. Nik, H.S., Effati, S., Shirazian, M.: An approximate-analytical solution for the Hamilton–Jacobi–Bellman equation via homotopy perturbation method. Appl. Math. Model. **36**(11), 5614–5623 (2012)
30. Takagi, T., Sugeno, M.: Fuzzy identification of systems and its applications to modeling and control. IEEE Trans. SMC **15**, 116–132 (1985)
31. Yang, D., Cai, K.Y.: Finite-time quantized guaranteed cost fuzzy control for continuous- time nonlinear systems. Expert Syst. Appl. (2010). https://doi.org/10.1016/j.eswa.03.024
32. Zadeh, L.A.: Fuzzy sets. Inf. Control **8**, 338–353 (1965)
33. Zimmermann, H.J.: Fuzzy Set Theory and Its Applications. Kluwer Academic, Boston (1991)

Accuracy of Non-inertial Reference Frame on Sloshing Phenomenon with Intelligent Control in Rectangular Tanks Under Oblique Excitation

Laleh Noeiaghdam(✉) and Akbar Safarzadeh

Faculty of Engineering, Mohaghegh Ardabili University, Ardabil 56199-11367, Iran
l_noeiaghdam@yahoo.com, Safarzadeh@uma.ac.ir

Abstract. In this paper, the 3D numerical modelling of the sloshing phenomenon in rectangular tanks was conducted under the oblique excitation. The effect of the aspect ratio on the maximum side wall, bed hydrodynamic forces and the maximum water run up over the side wall investigated under the intelligent controls. The non-inertial reference frame method was used to apply the dynamic excitation to the fluid body, and the volume of fluid (VOF) method was used to define the water surface variations. The experimental benchmark test case was used to verify the non-inertial reference frame method and effect of the excitation direction, frequency and amplitude were studied. Results show that the non-inertial reference method is more accurate than other method to predict the sloshing phenomenon. Using Flow-3D software, intelligent controls and big data algorithms the optimal numerical results are obtained. Then to study the effect of the geometrical parameters on the sloshing phenomenon, three rectangular tanks with different length/width ratio were considered and for various frequencies ratio were loaded under harmonic excitations. Based on these considerations, for all cases the maximum pressure force on the side wall occurs when the excitation is parallel with the side wall, and the maximum water surface occurs for the longitudinal excitation. By increasing the length/width ratio the maximum pressure force on the side wall decreases and on the other hand the maximum pressure force on the tank bottom increases.

Keywords: Rectangular tank · Sloshing · Excitation direction · Intelligent controls · Big data algorithms

1 Introduction

Water storage tanks are one of the most important civil structures, and studying their behavior during earthquakes is more important than their economic issues. When a filled tank is subjected to disturbances from external sources, the liquid inside it undergoes motion called sloshing. Topics that mostly considered in sloshing are the amplitude of vibration, fluid depth, properties of external excitation, physical properties of the fluid and the geometry of the tank. These parameters have a direct effect on the dynamic

T. Allahviranloo et al. (Eds.): IDS 2020, AISC 1301, pp. 141–152, 2021.
https://doi.org/10.1007/978-3-030-66501-2_11

stability and operation of fluid storage tanks, but the main issue in the study of sloshing is to obtain the hydrodynamic pressure distributions, forces, moments and oscillations of the liquid surface. The phenomenon of sloshing occurs in many practical applications, such as the behavior of the fluid stored in the tank of fuel carriers, the vibrations of the fluid in the fuel storage tank of the spacecraft or rocket, and the vibrational behavior of large fluid storage tanks under earthquake excitation. If the excitation frequency of the external agent is close to the normal frequency of system, because of resonance, intensive loads is applied to the reservoir structure. The investigations about the seismic behavior of reservoirs and attempts to reduce structural damage caused by sloshing have been done in many countries with high seismicity under the control system.

One of the first studies in the analytical and experimental investigation of hydrodynamic pressure in rectangular tanks that were subjected to horizontal motion was carried out by Hoskin et al. in [1]. Then analytical studies in the field of reservoir hydrodynamics were performed by Jacobsen in [2] and Housner in [3]. The results of these analyzes showed that the hydrodynamic pressure induced by seismic excitation is separated into impulsive and convective components using lumped mass approximation. The impulsive pressure is caused by the portion of liquid accelerating with the tank and the convective pressure is caused by the portion of liquid oscillating in the tank Housner in [4]. He formulated his idea for estimating the fluid response in rigid rectangular and cylindrical tanks under earthquake stimulation. Figure 1 shows the mass and spring model proposed by Housner in [4].

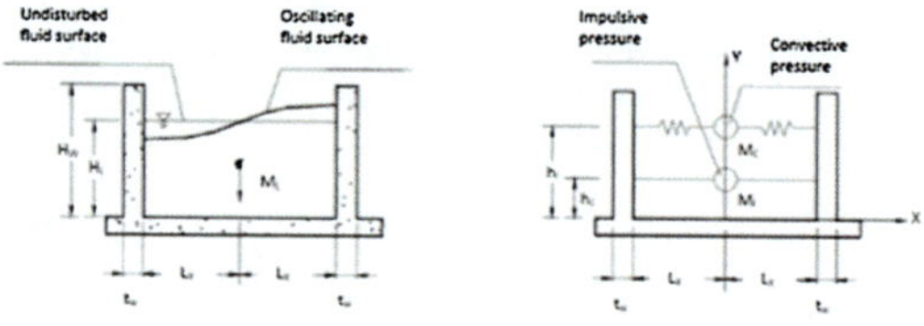

(a) Fluid motion in the tank (b) Mechanical model of liquid

Fig. 1. Housner's model

Haroun and Housner in [5] Investigated the hydrodynamic of reservoirs with flexible walls located on a rigid bed and showed the base shear and turning moment calculated in the rigid tank is much less than the flexible one. By using linear wave theory, Su et al. in [6] presented the natural vibrational modes of a rectangular reservoir as Eq. (1):

$$\omega_m = \sqrt{g\frac{m\pi}{L}\tanh\left(\frac{m\pi}{L}d\right)} \tag{1}$$

Where d is the water depth inside the tank, L is the length of the tank in the direction of excitation, g is the gravitational acceleration of the earth and m is the oscillator. Minowa in [7] investigated the effect of reservoir wall flexibility and hydrodynamic pressure on the wall and conducted its experimental studies to determine the dynamic properties of rectangular reservoirs. Yang Hsiang Chen et al. in [8] used the boundary element method to calculate nonlinear reservoir sloshing, unstable wave surfaces and base shear without assuming water depth. They concluded that the calculation of the base shear using hydrostatic pressure is correct when the force frequency is not larger than the first fundamental frequency of the reservoir. Harmonic excitation with any large amplitude can only cause a harmonic wave, and the linear model cannot accurately model the resonance phenomenon.

Gao Mimi in [9] in a study entitled Numerical modeling of liquid sloshing in rectangular tanks using the CPM method (compatible particle method) has investigated this issue. Ekeda et al. in [10] theoretically and experimentally investigated nonlinear fluid fluctuations in a square reservoir exposed to horizontal oblique. The effects of reservoir aspect ratio, fluid level and direction of excitation on steady wave types were discussed in detail. Wu Chih-Hua et al. in [11] by using the finite difference method and experimental results, the square reservoir model was subjected to harmonic excitation with different frequencies.

Most research in the past on the dynamic and static behavior of cylindrical tanks has been done by considering the simplest hypotheses and less attention has been paid to rectangular tanks. To detect the effect of excitation angle, previous studies have been performed on square tanks. In this research, we try to investigate the dynamic response of a rectangular tank under harmonic angular excitation. In addition, the effect of geometric dimension ratio on the maximum compressive force on the floor and wall of the tank and also the maximum elevation of water on the tank wall is investigated.

2 Theoretical Foundations

2.1 The Equations

The equations of unsteady fluid, including sloshing, are the same as the Navier-Stokes equations, which are based on the law of mass conservation (continuity equation) and momentum (Newton's second law).

$$\frac{\partial \rho}{\partial t} + \frac{\partial \rho u}{\partial x} + \frac{\partial \rho v}{\partial y} + \frac{\partial \rho w}{\partial z} = 0 \tag{2}$$

Equation 2 is a form of continuity equation and is applicable to all types of fluid flow, including non-steady, turbulent and compressible. In this relation ρ is the density u, v and w is the velocity component in the x, y and z direction. The equation governs the motion of the Navier-Stokes three-dimensional equations are:

$$\begin{aligned} &\frac{\delta u}{\delta t} + \frac{1}{V_F}\left\{uA_x\frac{\delta u}{\delta x} + vA_yR\frac{\delta u}{\delta y} + wA_z\frac{\delta u}{\delta z}\right\} - \xi\frac{A_y v^2}{xV_F} \\ &\quad = -\frac{1}{\rho}\frac{\delta p}{\delta x} + G_x + f_x - b_x - \frac{RSOR}{\rho V_F}u \end{aligned} \tag{3}$$

$$\begin{aligned} &\frac{\delta v}{\delta t} + \frac{1}{V_F}\left\{uA_x\frac{\delta v}{\delta x} + vA_yR\frac{\delta v}{\delta y} + wA_z\frac{\delta v}{\delta z}\right\} - \xi\frac{A_y uv}{xV_F} \\ &\quad = -\frac{1}{\rho}\frac{\delta p}{\delta y} + G_y + f_y - b_y - \frac{RSOR}{\rho V_F}v \end{aligned} \tag{4}$$

$$\frac{\delta w}{\delta t}+\frac{1}{V_F}\left\{uA_x\frac{\delta w}{\delta x}+vA_yR\frac{\delta w}{\delta y}+wA_z\frac{\delta w}{\delta z}\right\}$$
$$=-\frac{1}{\rho}\frac{\delta p}{\delta z}+G_z+f_z-b_z-\frac{RSOR}{\rho V_F}w \quad (5)$$

Where G_x, G_y and G_z are body acceleration f_x, f_y and f_z are viscous accelerations, b_x, b_y and b_z are flow losses in porous media or across porous baffle plates, and the final terms account for the injection of mass at a source represented by geometry component. For a variable dynamic viscosity, the viscos accelerations are:

$$\frac{\delta w}{\delta t}+\frac{1}{V_F}\left\{uA_x\frac{\delta w}{\delta x}+vA_yR\frac{\delta w}{\delta y}+wA_z\frac{\delta w}{\delta z}\right\}=-\frac{1}{\rho}\frac{\delta p}{\delta z}+G_z+f_z-b_z-\frac{RSOR}{\rho V_F}w \quad (6)$$

$$\rho v_F f_x=wsx-\left\{\tfrac{\delta}{\delta x}(A_x\tau_{xx})+R\tfrac{\delta}{\delta y}\left(A_y\tau_{xy}\right)+\tfrac{\delta}{\delta z}(A_z\tau_{xz})\right.$$
$$\left.+\tfrac{\xi}{x}\left(A_x\tau_{xx}-A_y\tau_{yy}\right)\right\} \quad (7)$$

$$\rho v_F f_y=wsy-\left\{\frac{\delta}{\delta x}\left(A_x\tau_{xy}\right)+R\frac{\delta}{\delta y}\left(A_y\tau_{yy}\right)+\frac{\delta}{\delta z}(A_z\tau_{yz})+\frac{\xi}{x}\left(A_x+A_y\tau_{xy}\right)\right\} \quad (8)$$

$$\rho v_F f_z=wsz-\left\{\frac{\delta}{\delta x}(A_x\tau_{xz})+R\frac{\delta}{\delta y}\left(A_y\tau_{yz}\right)+\frac{\delta}{\delta z}(A_z\tau_{zz})+\frac{\xi}{x}(A_x\tau_{xz})\right\} \quad (9)$$

Where

$$\tau_{xx}=-2\mu\left\{\frac{\delta u}{\delta x}-\frac{1}{3}\left(\frac{\delta u}{\delta x}+R\frac{\delta v}{\delta y}+\frac{\delta w}{\delta z}+\frac{\xi u}{x}\right)\right\} \quad (10)$$

$$\tau_{yy}=-2\mu\left\{R\frac{\delta v}{\delta y}+\xi\frac{u}{x}-\frac{1}{3}\left(\frac{\delta u}{\delta x}+R\frac{\delta v}{\delta y}+\frac{\delta w}{\delta z}+\frac{\xi u}{x}\right)\right\} \quad (11)$$

$$\tau_{xz}=-2\mu\left\{\frac{\delta w}{\delta z}-\frac{1}{3}\left(\frac{\delta u}{\delta x}+R\frac{\delta v}{\delta y}+\frac{\delta w}{\delta z}+\frac{\xi u}{x}\right)\right\} \quad (12)$$

$$\tau_{xy}=-\mu\left\{\frac{\delta v}{\delta x}+R\frac{\delta u}{\delta y}+\frac{\delta w}{\delta z}-\frac{\xi v}{x}\right\} \quad (13)$$

$$\tau_{xz}=-\mu\left\{\frac{\delta u}{\delta z}+\frac{\delta w}{\delta x}\right\} \quad (14)$$

$$\tau_{yz}=-\mu\left\{\frac{\delta v}{\delta z}+R\frac{\delta w}{\delta y}\right\} \quad (15)$$

2.2 Volume of Fluid Method (VOF)

Fluid configurations are defined in terms of a volume of fluid (VOF) function F(x,y,z,t). This function represents the volume of the fluid unit per unit volume and is expressed as the following equation:

$$\tfrac{\delta F}{\delta t}+\tfrac{1}{V_F}\left[\tfrac{\delta}{\delta x}(FuA_x)+R\tfrac{\delta}{\delta y}\left(FvA_y\right)+\tfrac{\delta}{\delta z}(FwA_{xz})+\xi\tfrac{FuA_x}{x}\right]$$
$$=FDIF+FSOR \quad (16)$$

$$FDIF = \frac{1}{V_F}\left[\frac{\delta}{\delta x}\left(v_f A_x \frac{\delta F}{\delta x}\right) + R\frac{\delta}{\delta y}\left(v_f A_y \frac{\delta F}{\delta y}\right) + \frac{\delta}{\delta z}\left(v_f A_z \frac{\delta F}{\delta z}\right) + \xi \frac{F v_x A_x}{x}\right] \quad (17)$$

The diffusion coefficient is defined as $V_F = \frac{C_F \mu}{\rho}$ where c_F is a constant value and its inverse referred to as a turbulent Schmidt number. The definition of F depends on the cases being modeled. Non-compressible cases include a single fluid with a free-surface or two fluids and no free-surfaces. For a fluid, F represents the volume fraction occupied by the fluid. Therefore, where F = 1, there is fluid, and where F = 0, there are bubbles and cavities. Voids are regions without fluid mass that have a uniform pressure assigned to them. Physically, they represent regions filled with a vapor or gas whose density is insignificant with respect to the fluid density. Problems with two fluids may include two incompressible fluids and one compressible fluid. F represents the volume fraction of incompressible fluid. Complement areas with a volume fraction of 1-F represent the second fluid.

2.3 Non Inertial Reference Frame Method

For applying to fluid sloshing in moving containers, for example, it is useful to modify the equations of motion to have a formulation for general non-inertial accelerations. The form of such accelerations on a fluid element at r and moving with velocity u is:

$$G = g - \left\{\dot{U} + \dot{\Omega} \times (r - R) + 2\Omega \times u + (\Omega \times (r - R))\right\} \quad (18)$$

In this equation g is a constant gravitational acceleration, Ω is rotation ratio, $\dot{\Omega}$ is deformed rotation ratio and $\dot{U}$ is the translational acceleration of point R in the mesh coordinate system with respect to an inertial reference frame.

All of big data algorithms and intelligent control algorithms are defined in Flow- 3D software. There are many mathematical relations and algorithms in this software that according to our model we can choose some of these relations and apply them.

3 Modeling

3.1 Basic Modeling

In the present study, by examining the possibilities, capabilities and limitations of the existing software, Flow-3D software was used to simulate the flow field and the finite difference method and finite volume method were used to calculate the temporal and spatial variables in motion equations. Using this software, intelligent controls and big data algorithms the optimal numerical solutions can be obtained. The finite volume method is preferable to other separation methods in solving the fluid flow field because of its hide conservation of mass property. Also, non-inertial coordinate system method has been used to analyze the flow inside the moving tank. According to the purpose of the present study, it is necessary to investigate the capability of the numerical model to simulate the phenomenon of sloshing under angular excitation relative to the sides of the tank. Therefore, Wu Chih-Hua's [11] experimental data has been used to evaluate the function of the non-inertial coordinate system method in predicting the phenomenon of sloshing on rectangular reservoir under oblique excitation. Based on the presented results, the sloshing phenomenon depends on the fluid depth, reservoir dimensions, amplitude, excitation frequency and density of fluid. Figure 2 shows the experimental model and schematic diagram of a tank connected to a vibrating table that moves at different excitation angles.

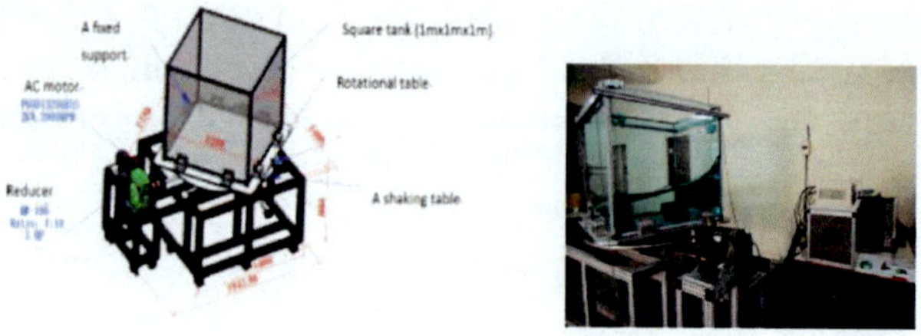

Fig. 2. Laboratory model and schematic design of a rectangular tank under angular excitation Wu et al. in [11]

In this experiment, a square tank with dimensions of 1 m, with water depth 25 cm and a gravitational acceleration 9.81 $\frac{m}{s^2}$ is considered. For meshing, non-uniform horizontal and vertical lines in the longitudinal direction and uniform in the transverse direction are used. In order to better understand the field geometry in the areas near the water surface, finer meshing is used. Due to the angular motion of the tank, the dimensions of the mesh are somehow chosen that the tank does not go out of the mesh area. The body of the tank is also considered rigid and impervious. Figure 3 shows the geometry and meshing of the model.

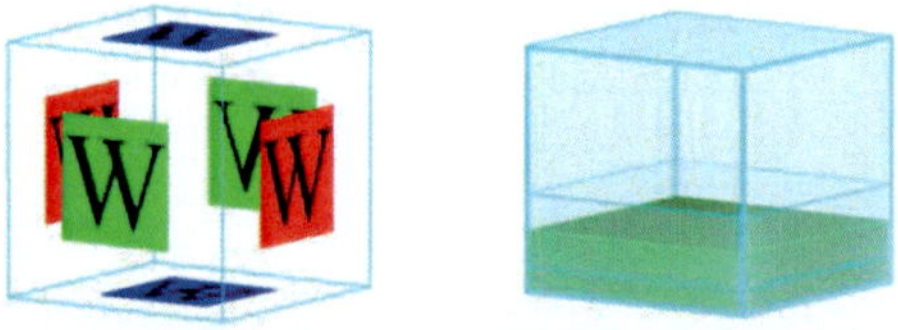

Fig. 3. Boundary conditions and networking used in rectangular reservoir modeling

For the initial conditions, we consider the fluid with VOF = 1. In this research, different points of the reservoir are used to collect hydraulic data. The properties of these points are given with names P_1 and P_2 in Fig. 4.

The time series of the reservoir oscillations along the x and y axes is entered as a predefined option from the non-inertial coordinate system of the software. The loading is done harmonically and follows Eqs. (19) and (20).

$$x(t) = A\,L \sin\,(\omega t)\,cos(\theta) \tag{19}$$

$$y(t) = A\,L \sin\,(\omega t)\,sin(\theta) \tag{20}$$

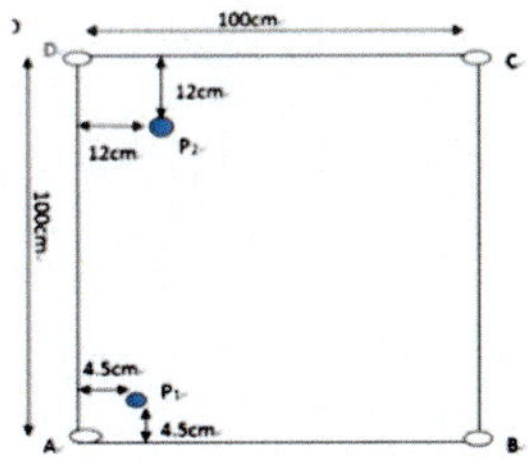

Fig. 4. Specifications of measuring points of different hydraulic parameters in the tank

In these equations, $x(t)$ is location of the tank, A is the excitation amplitude. L is the length of the tank, ω is the excitation frequency and θ is the excitation angle. The amount of excitation amplitude is 0.005 m and modeling has been done for different loading frequencies.

3.2 Accuracy of the Present Numerical Model with Experimental Data

In the present model, the tank is subjected to 30° angular excitation for 60 s with a frequency equal to 0.93 of the normal frequency of the tank and the accuracy of the results at a point P_1 that is 4.5 cm away from the left wall of the tank, shown in Fig. 5. The accuracy of the results at the point P_2 is also investigated according to Fig. 6 under angular excitation of 15° degrees for 60 s with a frequency equal to 0.97 of the natural frequency of the reservoir. The lowest natural tank frequency is obtained from Eq. (1), which is equivalent to 4.49 rad/ s.

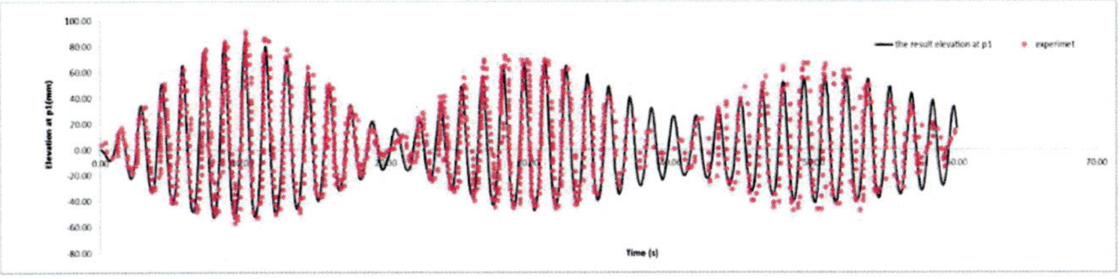

Fig. 5. Comparison of experimental and numerical results in 1 L/B = and θ = 30° and ω = 0/93ω_1, in 60 s

The results of experimental measurements are well consistent with numerical analysis. Wave oscillations in experimental results are exactly the same as numerical analysis, but there is a slight difference between their peak points, which is about 6% of the water depth.

3.3 The Effect of Geometric Dimension Ratio on the Phenomenon of Rectangular Tanks Slashing Under Angular Excitation

Following validation, the tank modeled in the laboratory work of Wu Chih-Hou et al. in [11], is stimulated with length/width ratios of 1, 1.5 and 2, at angles of 0, 15, 30 and 45°.

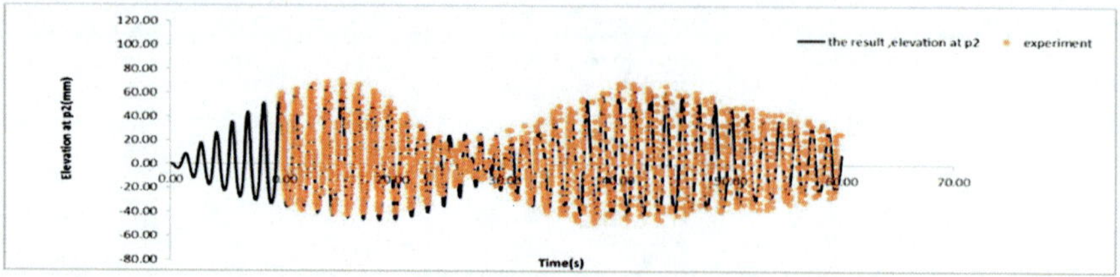

Fig. 6. Comparison of experimental and numerical results in 1 L/B = and θ = 15° and ω = 0/97ω_1, in 60 s

The maximum free surface height and maximum force on the right wall and the bottom of the reservoir are examined and the results are presented in Table 1.

Table 1. The values of maximum force on the wall and floor of the tank with different aspect ratios

θ°	F_{max} (N)	$\frac{\omega}{\omega_1}$	L/B: 1 H_{max}	L/B: 1 $\omega_1 = 4.49$	L/B: 1.5 H_{max}	L/B: 1.5 $\omega_1 = 3.13$	L/B: 2 H_{max}	L/B: 2 $\omega_1 = 2.39$
0	F_{wall}	0.93	0.311	422.2043	0.3013	415.797	0.2956	405.3561
	F_{bed}			2470.781		3687.573		4914.342
	F_{wall}	0.97	0.398	531.39933	0.35	503.528	0.325	468.9977
	F_{bed}			2524.573		3703.464		4921.403
	F_{wall}	1.03	0.458	588.7803	0.381	554.907	0.331	487.1469
	F_{bed}			2579.962		3723.184		4913.934
15	F_{wall}	0.93	0.308	416.8078	0.3	412.4324	0.2933	401.2997
	F_{bed}			2470.522		3687.085		4913.711
	F_{wall}	0.97	0.375	512.767	0.344	496.5635	0.322	463.569
	F_{bed}			2524.564		3701.074		4920.169
	F_{wall}	1.03	0.447	568.58	0.375	546.964	0.327	478.5328
	F_{bed}			2567.527		3719.77		4918.448
30	F_{wall}	0.93	0.2976	405.2724	0.294	401.1596	0.2879	390.3098
	F_{bed}			2471.804		3685.487		4911.698
	F_{wall}	0.97	0.345	488.8745	0.334	480.4116	0.3137	446.4655
	F_{bed}			2538.57		3697.403		4916.906
	F_{wall}	1.03	0.377	510.1125	0.356	519.0652	0.316	457.1758
	F_{bed}			2547.31		3710.083		4915.676
45	F_{wall}	0.93	0.2839	386.6913	0.2858	383.3414	0.28	373.6039
	F_{bed}			2472.302		3683.493		4909.293
	F_{wall}	0.97	0.326	472.9553	0.318	450.593	0.2999	419.4474
	F_{bed}			2545.006		3691.789		4912.289
	F_{wall}	1.03	0.321	453.9512	0.33	475.889	0.3	423.3296
	F_{bed}			2543.102		3699.118		4911.993

In order to study the process of force changes, Figs. 7 to 10 are presented. According to Fig. 7 in all tanks, as the length/width ratio increases, the lateral force on the wall decreases. The notable point in this figure is the force changes in the tank with $\frac{L}{B} = 1$

and at an angle of 45°. For a frequencies ratio more than 1, the amount of force on the wall is reduced. That is because of changes in the shape and pattern of the resulting waves inside the square reservoir.

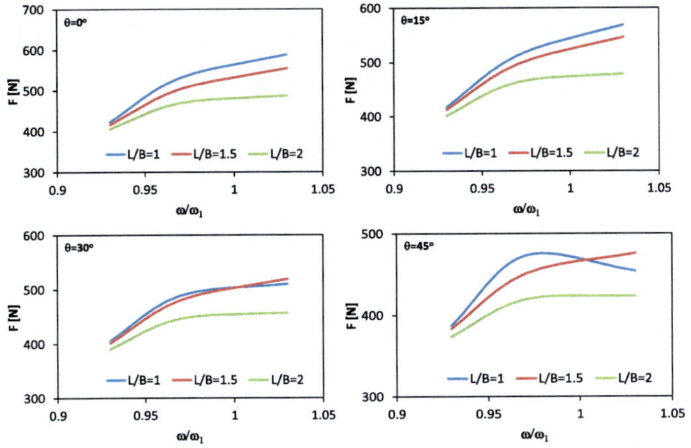

Fig. 7. Changes in the force applied to the wall of the tank with different length/width ratio

According to Fig. 8, the maximum force on the wall occurs at 0° excitation (along the longitudinal side of the tank) and by increasing the excitation angle, the maximum force on the wall for all excitation frequencies decreases. In all tanks, by increasing excitation frequency, the amount of force applied to the side wall increases.

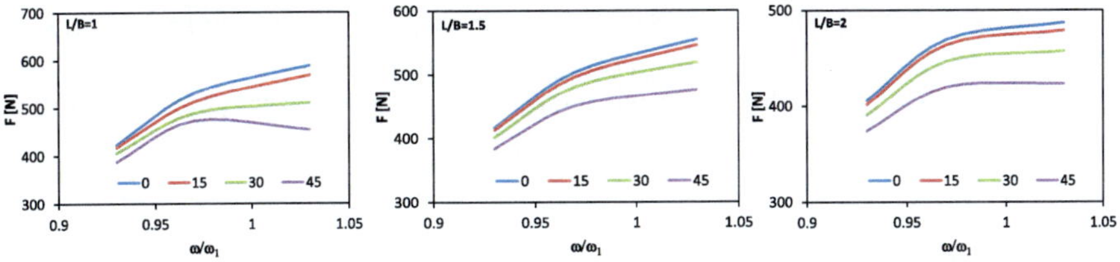

Fig. 8. Changes in the force applied to the wall of the tank under oblique excitation

Figures 9 and 10 show the force changes on the bottom of different tanks. According to Fig. 9, in the square tank, for the excitation angles of 0° and 15°, by increasing excitation frequency, the force on the floor increases linearly, but for the excitation angles of 30° and 45°, the amount of force on the bottom remains almost constant for the frequencies larger than natural frequency. In the tank with a length/width ratio of 1.5 for all excitation angles, the amount of force applied to the tank floor increases linearly. In a tank with a length/width ratio of 2, the amount of force applied to the floor does not change for the excitation angles as well as the excitation frequencies.

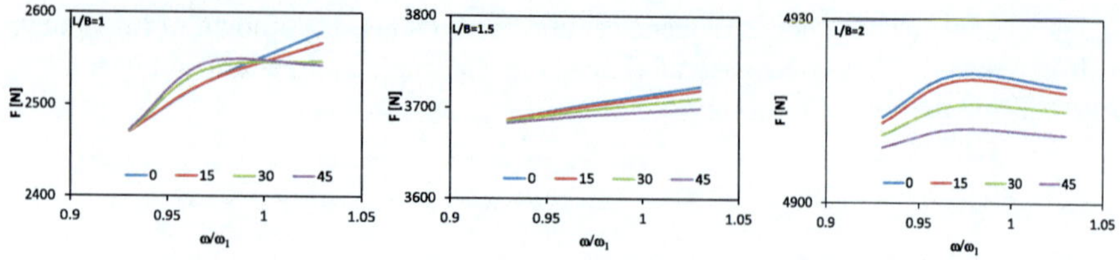

Fig. 9. Changes in the force applied to the bottom of the tank under oblique excitation

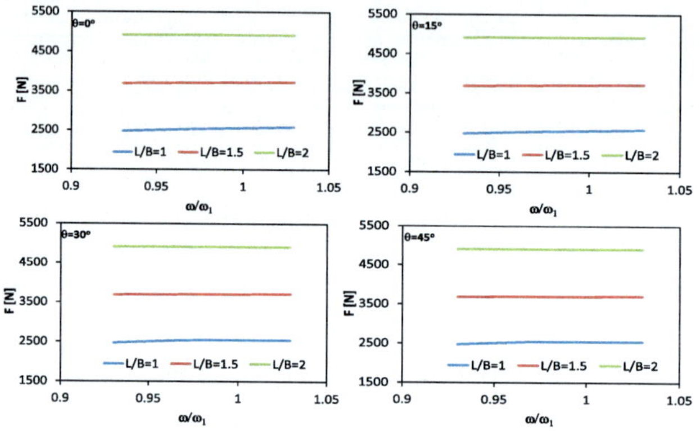

Fig. 10. Changes in the force applied to the bottom of the tank with different length/width ratio

According to Fig. 10, by increasing length/width ratio the amount of hydrodynamic force on the tank bottom for all angles of excitation increases, and changes for the excitation frequency are linear.

Figures 11 and 12 show the raise of the water surface under angular excitation in different reservoirs. In Fig. 11, by increasing the excitation angle, the maximum elevation of the water surface decreases. It means that the maximum elevation of the water surface is allocated to longitudinal excitation.

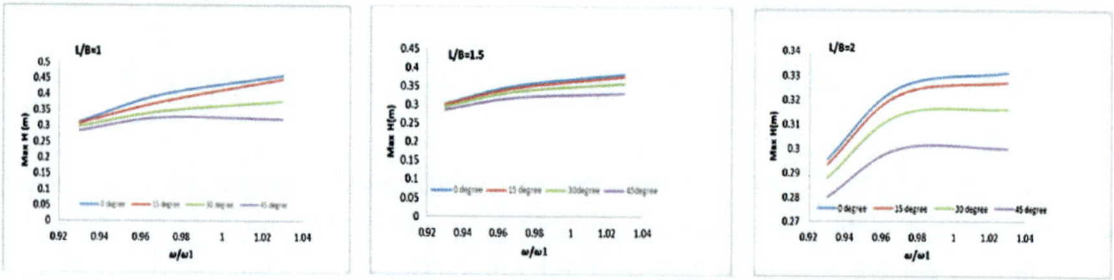

Fig. 11. Changes in the maximum elevation of water surface under oblique excitation

Figure 12 shows that, by increasing the length/width ratio, the maximum elevation of the water surface decreases. It means that the maximum elevation of the water surface is allocated to square reservoirs.

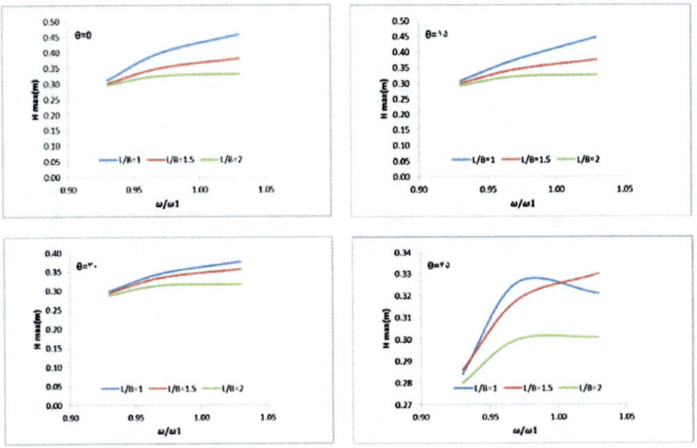

Fig. 12. Changes in the maximum elevation of water surface with different length/width ratio

4 Conclusions

In this paper, three-dimensional numerical modeling of the sloshing phenomenon in rectangular tanks under angular harmonic excitation with intelligent control in FLOW-3D software is performed using the non-inertial coordinate system method and the following results are obtained:

1- Based on the validations made with the available experimental data, the method of non-inertial coordinate system has a very good accuracy for simulating the sloshing phenomenon.
2- By increasing the excitation frequency with constant amplitude, without considering of the aspect ratio, the maximum force on the wall and floor of the tank increases.
3- The maximum force is applied to the wall for excitation in the longitudinal direction of the tank.
4- The maximum elevation of the water surface is caused by the excitation angle of zero degree and in the longitudinal direction of the reservoir.
5- As the length of the tank increases, the force on the wall decreases for all frequency ratios.
6- As the length of the tank increases, the force on the bottom of the tank increases for all frequency ratios.
7- As the length of the tank increases, the maximum elevation of the water level decreases for all frequency ratios.
8- As the excitation amplitude increases, more of the surface layers are displaced and a larger mass of the fluid approaches the tank wall. As a result, the force is applied to the wall and floor increases.

References

1. Hoskins, L.M., Jacobsen, L.S.: Water pressure in a tank caused by a simulated earthquake. Bull. Seismol. Soc. Am. **24**(1), 1–32 (1934)
2. Jacobsen, L.S.: Impulsive hydrodynamics of fluid inside a cylindrical tank and of fluid surrounding a cylindrical pier. Bull. Seismol. Soc. Am. **39**(3), 189–204 (1949)
3. Housner, G.W.: Earthquake pressures on fluid containers. Bull. Seismol. Soc. Am. **53** (1954)
4. Housner, G.W.: The dynamic behavior of water tanks. Bull. Seismol. Soc. Am. **53**(2), 381–387 (1963)
5. Haroun, M.A., Housner, G.W.: Seismic design of liquid storage tanks. J. Tech. Councils ASCE **107**(1), 191–207 (1981)
6. Subhash, S., Bhattachryya, S.K.: Finite element analysis of fluid-structure interaction effect on liquid retaining structures due to sloshing. Comput. Struct. **59**(6), 1165–1171 (1996)
7. Minowa C.: Experimental studies of seismic properties of various type water tanks. In: Proceedings of eighth WCEE, San Francisco, pp. 945–952 (1984)
8. Chen, Y.H., Hwang, W.S., Ko, C.H.: Sloshing behaviours of rectangular and cylindrical liquid tanks subjected to harmonic and seismic excitations. Earthq. Eng. Struct. Dyn. **36**(12), 1701–1717 (2007)
9. MIMI, G.: Numerical simulation of liquid sloshing in rectangular tanks using consistent particle method and experimental verification (2011)
10. Ikeda, T., et al.: Nonlinear liquid sloshing in a square tank subjected to obliquely horizontal excitation. J. Fluid Mech. **700**, 304–328 (2012)
11. Wu, C.-H., Chen, B.-F., Hung, T.-K.: Hydrodynamic forces induced by transient sloshing in a 3D rectangular tank due to oblique horizontal excitation. Comput. Math Appl. **65**(8), 1163–1186 (2013)

Convergence of Q-Value in Case of Gaussian Rewards

Konatsu Miyamoto[1,2(✉)], Masaya Suzuki[2], Yuma Kigami[3], and Kodai Satake[4]

[1] Dynamic Pricing Technology.inc, Osaka Umeda, JP, Japan
u472532f@ecs.osaka-u.ac.jp
[2] Osaka Univercity, Osaka, JP, Japan
[3] Matrix Flow.inc, Tokyo, JP, Japan
[4] Kindai Univercity, Osaka, JP, Japan
https://dynamic-pricing.tech/1

Abstract. In this paper, as a study of reinforcement learning, we converge the Q function to unbounded rewards such as Gaussian distribution. From the central limit theorem, in some real-world applications it is natural to assume that rewards follow a Gaussian distribution, but existing proofs cannot guarantee convergence of the Q-function. Furthermore, in the distribution-type reinforcement learning and Bayesian reinforcement learning that have become popular in recent years, it is better to allow the reward to have a Gaussian distribution. Therefore, in this paper, we prove the convergence of the Q-function under the condition of $E[r(s,a)^2] < \infty$, which is much more relaxed than the existing research. Finally, as a bonus, a proof of the policy gradient theorem for distributed reinforcement learning is also posted. ...

Keywords: Reinforcement learning · Stochastic approximate · Convergence theory · Mathematics · Probability

1 Introduction

In recent years, Reinforcement Learning(RL) has come into fasion. General method in ordinary Reinforcement Learning using Markov decision processes use a state action value functions [1]. Agents created by these algorithms take strategies to maximize the expected value of the cumulative reward. However, in practical use, there are many situations where it is necessary to consider not only expected values but also risks. Therefore, Distributional Reinforcement Learning(DRL) that considers the distribution of cumulative rewards has also been studied. DRL research presents a particle method of risk responsive algorithm [2]. As for similar research, there are [3,4], which is equivalent to [2] mathematically, but used the different algorithm and parametric methods [5]. [4] discusses the convergence of measures in discrete steps. Another way to practice DRL is using the Bayesian approach. In [22], it is regarded as an estimation of the uncertainty of the expected value. But in fact, the Bayesian inferece can

T. Allahviranloo et al. (Eds.): IDS 2020, AISC 1301, pp. 153–165, 2021.
https://doi.org/10.1007/978-3-030-66501-2_12

approximate the distribution of uncertain objecsion. It can perform distributed reinforcement learning. There are other existing papers on Bayesian reinforcement learning. We want to take [6,7] up this time. It is a method using Gaussian processes, and it can be said that the reward follows Gaussian distributions. [5] also supports unbounded rewards like Gaussian distributions. We want to show that the approximation of the cumulative reward distribution converges even in unbounded rewards. In this paper, we prove the convergence of the normal state action value function as a preliminary step. In addition, we perform the convergence proof for Q functions with continuous concentration domain, taking Deep Q-learning(DQN) into consideration.

1.1 Related Works

The proof history of Q-function convergence is long. For example, there are papers such as [8–10], and [11] using [10]. A paper on an unusual proof method is [12] using ordinary differential equations. For DQN, there is a study [13] summarizing the approximation error. The approximation error due to the neural network is verified there. Other research results include [14–18]. All of these studies assume that rewards are bounded. That is, there is a certain constant $R_{max} < \infty$ and

$$|r(s,a)| \leq R_{max} \; a.e. \tag{1}$$

holds. Therefore, Gaussian distributions cannot be assumed. In this paper, we prove the convergence of the Q function under condition ,

$$\forall s, a \in S \times A, E[r(s,a)^2] < \infty \tag{2}$$

which is more relaxed than (1.1), with normal distribution in mind. Finally, we prove the convergence of the Q function in the domain of continuous concentration under ideal conditions. This is a frequent concept in reinforcement learning.

2 Background

2.1 Transition Kearnel

Let two tuples $(S, \mathcal{S}), (T, \mathcal{T})$ be both measurable spaces. Transition kernel $k : S \times \mathcal{T} \to \mathbb{R}_+$ is defined to satisfy the following two conditions.

$$\cdot \, \forall \, B \in \mathcal{T}, k(\cdot, B) \; on \; S \; is \; measurable \tag{3}$$

$$\cdot \, \forall \, s \in S, k(s, \cdot) is \; measure \; on \; T \tag{4}$$

This is used in situations where s is fixed and the distribution on T is fixed.

2.2 Markov Decision Process

Assume that both the set of states S and the set of actions A are finite sets. A transition kernel p is defined on $(S \times A, 2^{S \times A}), (S \times \mathbb{R}, 2^S \otimes \mathcal{B}(\mathbb{R}))$. That is, $p(r, s|s, a)$ is a probability measure that governs the distribution of the next state $s \in S$ and immediate reward $r \in \mathbb{R}$ when an action $a \in A$ is taken in state $s \in S$ is there. The strategy $\pi : S \rightarrow \mathcal{P}(A)$ is the action probability determined from the current situation, as can be seen from the definition. The deterministic approach is that for any s, there is a a and $\pi(a|s) = 1$. A set of random variables s_t, a_t, r_t taking values in $S, A, \mathbb{R}$ is written as $(s_t, a_t, r_t)_{t=0}^{\infty}$. This stochastic process is called Markov decision process(MDP).

2.3 Optimal Measures and State Action Value Functions

Put the whole set of policies as Π. The state action value function $Q^{\pi} : S \times A \rightarrow \mathbb{R}$ for the policy π is defined as follows.

$$Q^{\pi}(s, a) := E[\sum_{t=0}^{\infty} \gamma^t R_t | s_0 = s, a_t = a(r_t, s_{t+1}), p(r_t, s_{t+1}|s_t, a_t), \pi(a_t|s_t)] \quad (5)$$

Furthermore, the state value function $V^{\pi}(s)$ is defined as follows.

$$V^{\pi}(s) := \sum_{a \in A} \pi(a|s) Q^{\pi}(s, a) \quad (6)$$

Define the optimal strategy π^* as

$$\pi^* := argmax_{\pi \in \Pi} V^{\pi}(s_0) \quad (7)$$

In addition, the state action value function Q^{π^*} for the optimal policy is called the optimum state action value function, and simply expressed as Q^*. The action that takes the maximum value for the optimal state action function is the optimal policy.

$$\pi^*(a|s) = \{ \begin{matrix} 1 \; argmax_{a \in A} Q(s, a) \\ 0 \; else \end{matrix} \quad (8)$$

holds for any s, a.

3 Update of State Action Value Function and Robbins Monro Condition

Update the Q-unction as follows

$$Q_{n+1}(s, a) = (1 - \alpha(s, a, s_t, a_t, t)) Q_t(s, a) + \alpha(s, a, s_t, a_t, t)[r_t(s_t, a_t) + \max_{b \in A} Q_t(s_{t+1}, b)] \quad (9)$$

The following sequence $\{c_t\}_{t=0}^{\infty}$ satisfies the Robbins-Monro condition.

$$\forall t, c_t \in [0,1] \tag{10}$$

$$\sum_{t=0}^{\infty} c_t = \infty \tag{11}$$

$$\sum_{t=0}^{\infty} c_t^2 < \infty \tag{12}$$

Using this, the mapping $\alpha : S \times A \times S \times A \times \mathbb{N} \to (0,1)$ is defined as follows.

$$\alpha(s,a,s_t,a_t,t) = c_t 1_{s_t=s,a_t=a} \tag{13}$$

In addition,it is assumed that this also satisfies the Robbins Monroe condition stochastically uniformly for arbitrary s, a.

$$\sum_{t=0}^{\infty} \alpha(s,a,s_t,a_t,t) = \infty \ a.e. \tag{14}$$

$$\sum_{t=0}^{\infty} \alpha(s,a,s_t,a_t,t)^2 < \infty \ a.e. \tag{15}$$

4 Proof of Q-Function Convergence for Unbounded Rewards

Consider a real-valued function $w_t(x)$ on a finite set $\mathcal{X}$.

Theorem 1. *Convergence of Q-value in case of Gaussian rewards*

$\mathcal{X}$ is finite set. Let ramdom value $r_t(x), \mathcal{X} := S \times A$. Let W be the set of functions $f : \mathcal{X} \to \mathbb{R}$ and $||f||_W$ is defined as $||f||_W := \max_{x\in\mathcal{X}} f(x)$. For any s, a, let $E[r^2(s,a)] < \infty$.

$$||Q_t - Q^*||_W \to 0 \tag{16}$$

proof. *In line with the proof of [9]. The F condition is relaxed and the statement is stronger, so it needs to be done more precisely. Consider a stochastic process of $\Delta_t(x) := Q_t(x) - Q^*(x)$. Since $Q^*(x)$ is a constant, $V(\Delta_t(x)) = V(Q_t(x))$. Putting $F_t(x) := r_t(x) + \gamma \sup_b Q_t(X(s,a),b) - Q^*(x)$, this is $\mathcal{F}_{t+1}$ measurable stochastic process. Furthermore, if we put $G_t(x) := r_t(x) + \gamma \sup_b Q_t(X(s,a),b)$, by definition $G_t - E[G_t(x)|\mathcal{F}_t] = F_t - E[F_t(x)|\mathcal{F}_t]$. The two stochastic processes $\delta_t, w_t \in W$ are taken so that $\Delta_0(x) = \delta_0(x) + w_0(x)$. Define time evolution as*

$$\delta_{t+1}(x) = (1 - a_t(x))\delta_t(x) + a_t(x)E[F_t(x)|\mathcal{F}_t] \tag{17}$$

$$w_{t+1}(x) = (1 - a_t(x))w_t(x) + a_t(x)p_t(x) \tag{18}$$

However, $p_t(x) := F_t(x) - E[F_t(x)|\mathcal{F}_t]$. *At this time,* $\Delta_t(x) = w_t(x) + \delta_t(x)$. *First, we show that* w_t *converges to 0 for* $\mathcal{X}$ *with probability 1 by using Lemma 2. By definition,* $E[p_t|\mathcal{F}_t] = 0$, *so* $\sum_t E|[p_t|\mathcal{F}_t]| = 0$ *holds. From Lemma 1 and the definition of* p_t, G_t, $E[p_t^2] \leq 4E[G_t^2]$ *holds. Putting* $L_t(\omega) := \sup_x |Q_t(x)|$, *this random variable is* $\mathcal{F}_t$ *-measurable and takes a finite value with probability 1. Since* L_0 *is a finite value, a certain constant* K_0 *can be taken so that* $E[L_0^2] \leq K_0^2 C_R$ *holds. And the following holds with probability 1.*

$$L_{t+1} \leq \max(L_t, (1-b_t)L_t + b_t(\sup_x |r_t(x)| + \gamma L_t)) \tag{19}$$

Using the above formula, the following holds

$$E[L_{t+1}^2] \leq \max(E[L_t^2], E[((1-b_t)L_t + b_t(\sup_x |r_t(x)| + \gamma L_t))^2]) \tag{20}$$

Suppose there is $K_t \in \mathbb{R}$ *that is* $E[L_t^2] \leq K_t^2 C_R$. *At this time, put* $H_t := \sup_x |r_t(x)| + \gamma L_t$

$$E[H_t^2] = E[\sup_x |r_t(x)|^2] + 2E[\sup_x |r_t(x)| \gamma L_t] + \gamma^2 E[L_t^2] \tag{21}$$

$$\leq C_R + 2\gamma\sqrt{C_R K_t^2 C_R} + K_t^2 C_R \tag{22}$$

$$= (1+\gamma K_t)^2 C_R \tag{23}$$

Then,

$$E[((1-b_t)L_t +_t (\sup_x |r_t(x)| + \gamma L_t))^2] \tag{24}$$

$$\leq (1-b_t)^2 E[L_t^2] + 2(1-b_t)b_t\sqrt{E[L_t^2]E[H_t^2]} + b_t^2\gamma^2 E[H_t^2] \tag{25}$$

$$\leq (1-b_t)^2 K_t^2 C_R + 2(1-b_t)b_t K_t(1+\gamma K_t)C_R + (1+\gamma K_t)^2 C_R \tag{26}$$

$$= ((1-b_t)K_t + b_t(1+\gamma K_t))^2 C_R \tag{27}$$

$$= (K_t + b_t(1-(1-\gamma)K_t))^2 C_R \tag{28}$$

Putting $K_{t+1} = max(K_t, K_t + b_t(1-(1-\gamma)K_t))$, $EL_{t+1}^2 \leq K_{t+1}C_R$ *can be said. Since* $K_0 \in \mathbb{R}$ *exists,* $K_t \in \mathbb{R}$ *exists for any* t, *and* $E[L_t^2] \leq K_t^2 C_R$ *can be said. It is clear from the equation that* $K_{t+1} = K_t$ *when* $K_t > \frac{1}{1-\gamma}$, *and* $K_t \leq \frac{1}{1-\gamma}$ *Then,* $K_{t+1} \leq \frac{1}{1-\gamma} + 1$ *holds. Therefore, it was shown earlier that* K_t *exists for any* t, *in addition* $K_t \leq K^* := \max(K_0, \frac{1}{1-\gamma} + 1)$ *can be also said.* $|G_t(x)| \leq |r_t(x)| + \gamma L_t$ *holds, so the following equation hold.for all* x

$$\frac{1}{4}E[p_t^2(x)] \leq E[G_t^2(x)] \tag{29}$$

$$= E[r_t(x)^2] + 2\gamma\sqrt{E[r_t(x)^2]E[L_t^2]} + E[L_t^2] \tag{30}$$

$$\leq (1+\gamma K^*)C_R \tag{31}$$

Then,

$$\sum_t E[a_t^2 p_t^2] \leq \sum_t 4b_t^2(1+\gamma K^*)C_R \tag{32}$$

$$\leq 4M(1+\gamma K^*)C_R < \infty \tag{33}$$

holds for all x. When we use Lemma2,putting

$$U_t : S = a_t(x)p_t(x) \tag{34}$$

$$T(w_t, \omega) : = (1 - a_t(x))w_n \tag{35}$$

$\sum_t E[U_t^2] < \infty$*can be said. Since* $E[U_t|\mathcal{F}_n] = 0$, $\sum_t |E[U_t|\mathcal{F}_n]| = 0$ *holds. Then, for any* $\epsilon > 0$, *set* $\alpha = \epsilon, \beta_t(x) = b_t^2(x)$ *and* $\gamma_t(x) = \epsilon(2a_t(x) - a_t^2(x))$, *then*

$$T^2(w_t, \omega) \leq \max(\alpha, (1+\beta_t)w_t^2 - \gamma_t) \tag{36}$$

$$\sum_t \gamma_t = \infty \ a.e \tag{37}$$

holds. The latter is based on Robbins Monro conditions. Therefore, $w_t(x) \to 0$ *holds for any x. Define the linear operator* $\mathcal{T} : W \to W$ *as follows: for* $q \in W$

$$\mathcal{T}q(s,a) = \int_{\mathrm{IR}} \sum_{s'} [r(s,a) + \gamma \sup_b q(s', b)] p(dr, s'|s, a) \tag{38}$$

$$= E[r(s,a) + \sup_b q(X(s,a), b)] \tag{39}$$

Q^* *is a fixed point for this operator. For any* $q_1, q_2 \in W$

$$||\mathcal{T}q_1 - \mathcal{T}q_2||_W = sup_{s,a}[|\int_{\mathrm{IR}} \sum_{s'} [r(s,a) + \gamma \sup_b q_1(s', b)] p(dr, s'|s, a) \tag{40}$$

$$- \int_{\mathrm{IR}} \sum_{s'} [r(s,a) + \gamma \sup_b q_2(s', b)] p(dr, s'|s, a)|] \tag{41}$$

$$\leq \int_{\mathrm{IR}} \sum_{s'} [\gamma | \sup_b q_1(s', b) - \sup_b q_2(s', b)|] p(dr, s'|s, a) \tag{42}$$

$$\leq \int_{\mathrm{IR}} \sum_{s'} [\gamma sup_b |q_1(s', b) - q_2(s, b)|] p(dr, s'|s, a) \tag{43}$$

$$= \gamma ||q_1 - q_2||_W \tag{44}$$

Thus $\mathcal{T}$ is a reduction operator.

$$|E[F_t(x,a)|\mathcal{F}_t]| \leq \int_{\mathbb{R}} \sum_{s'} |r(s,a) + \gamma \sup_b Q_t(s',b) - Q^*(s,a)| p(dr, s'|s,a) \quad (45)$$

$$\begin{aligned} &= |\mathcal{T}Q_t(x,a) - Q^*(s,a)| && (46)\\ &= |\mathcal{T}Q_t(x,a) - \mathcal{T}Q^*(s,a)| && (47)\\ &\leq \gamma ||\Delta_t||_W && (48)\end{aligned}$$

Then,

$$\begin{aligned} ||\delta_{t+1}|| &\leq (1 - a_t(x))||\delta_t|| + a_t(x)||\delta_t + w_t|| && (49)\\ &\leq (1 - a_t(x))||\delta_t|| + a_t(x)(||\delta_t|| + ||w_t||) && (50)\end{aligned}$$

$||w_t(x)||$ converges uniformly to 0 with a probability of 1 for any x as described above. Therefore, from Lemma 3, $||\delta_{t+1}(x)|| \to 0$ for any x. That is, for any x, $||\Delta_t(x)||_W \to 0$, which holds the main theorem assertion.

5 Theorem for SARASA

The method in Chap. 3 is called Q-learning, and the value is updated before performing the next action. On the other hand, SARASA updates the value after performing the following actions.

$$Q_{t+1}(s,a) = (1 - \alpha(s,a,s_t,a_t,t))Q_t(s,a) + \alpha(s,a,s_t,a_t,t)(r_t(s,a) + Q_t(s_{t+1},a_{t+1}))$$

a_{t+1} is often stochastically determined by softmax function or the like.

Theorem 2. *Suppose that the Q function is updated by the above SARASA method. At this time,*

$$||Q_t - Q^*||_W \to 0 \ in \ t \to \infty \quad (51)$$

proof. *Put $L'_t := max_{x,y \ in \ mathcalX} |Q_t(x) - Q_t(y)|$ It is clear from the definition that $L'_t \leq 2L_t$. Later along this follows the proof of Theorem 1.*

6 Convergence Proof for Unbounded Rewards Under Continuous Concentration

For example, in a situation such as DQN, an update for one s, a has an effect on other state actions. As a simple model to take such situations into account, we put the ripple function $f(x_1, x_2)$ defined on the compact set $\mathcal{X}^2$. This satisfies the next conditions.

$$\begin{aligned} &f(x,x) = 1 && (52)\\ &f(x,y) \ is \ continue. && (53)\end{aligned}$$

Theorem 3. *If Q^* is a continuous function, it can be used to depart from any continuous function and have the same convergence on the compact set. Let $\mathcal{X} \subset \mathbb{R}^d$ be a simple connected compact set. Let Q^*, Q_0 be a continuous function on $\mathcal{X}$. Let W be a continuous function on $\mathcal{X}$. $||f||_W := \max_{x \in \mathcal{X}} f(x)$*

$$\begin{aligned} Q_{t+1}(s,a) &= (1 - f(s,a,s_t,a_t)\alpha(s,a,s_t,a_t,t))Q_t(s,a) \\ &+ f(s,a,s_t,a_t)\alpha(s,a,s_t,a_t,t)(r_t(s,a) + \max_{b \in A} Q_t(s_{t+1},b)) \end{aligned} \tag{54}$$

At this time, $||Q_t - Q^*||_W \to 0$

proof. *Consider a finite set $K_N := \{x_1, x_2, x_3, ..., x_N\}$ on $\mathcal{X}$. Limiting Q to K converges to a correct function uniformly over K from Theorem 1. For any ϵ Since Q^* is a continuous function, the function whose value is defined on a dense set is uniquely determined. Convergence can be said.*

7 Conclusion and Future Work

As we mentioned earlier, we want to prove the convergence of the distribution. An order evaluation of the expected value should be also performed. We also want to estimate the convergence order for a specific neural network such as [13]. According to [13], as with Theorem 3, in the domain of continuous concentration, as $R_{max} := \sup r(\omega, s, a)$, using constants C_1, C_2, ξ, α

$$||Q^* - Q_n||_W \leq C_1 \cdot (\log n)^{\xi} n^{-\alpha} + C_2 R_{max} \tag{55}$$

is established. However, when r follows a normal distribution, $R_{max} = \infty$, so the upper limit of the error is infinite, and this unexpected expression has no meaning. In case of using unbounded rewards, stronger inequality proofs are needed.

A Lemmas and proofs

Lemma 1. *Consider a random variable Y and a partial σ-algebla $\mathcal{G}$. If $Z := Y - E[Y|\mathcal{G}]$, the following equation holds.*

$$E[Z^2] \leq 4E[Y^2] \tag{56}$$

We quote the important theorem.

Lemma 2. *Convergence theorem for stochastic systems[19] Consider the following stochastic process.*

$$X_{t+1} := T(X_0, ..., X_t, \omega) + U_t(\omega) \tag{57}$$

This satisfies the following equation with probability 1.

$$|T(x_1, x_2, ..., x_t, \omega)|^2 \leq max(\alpha, (1 + \beta_t(\omega))x_t^2 - \gamma_t) \tag{58}$$

However, with $\alpha > 0$*, with probability 1,* $\beta_t(\omega) < M', \sum_t \beta_t < \infty$ *holds, and with probability 1, Let* $\sum_t \gamma_t(\omega) = \infty$.

$$\sum_t E[U_t^2] < \infty \tag{59}$$

$$\sum_t E[U_t|\mathcal{F}_t] < \infty \tag{60}$$

At this time, there exists a certain $N(\omega)$*, and it holds for any* $n > N(\omega)$

$$\limsup_{t\to\infty} |X_t|^2 < \alpha \ a.e. \tag{61}$$

If β, γ are taken again for any α and the same can be said, "uniform convergence to 0" can be said that is much stronger than approximate convergence.

Lemma 3. $x_0 \in \mathbb{R}$ *is assumed to be a real number.*

$$x_{n+1} = (1 - a_n)x_n + \gamma a_n|x_n| \tag{62}$$

$\gamma \in (0, 1)$ *is a constant.At this time,* $x_n \to 0$ *holds with probability 1.*

proof. *Look at each* ω*. That is,* $\{a_n\}_{n=0}^{\infty}$ *is constant sequence that satisfies* $\sum_{n=0}^{\infty} a_n = \infty, \sum_{n=0}^{\infty} a_n^2 < \infty$*.* X_n *is nonnegative for a sufficiently large n, so it is bounded below. In addition, since* $x_n \geq x_{n+1}$ *is apparent from the equation,* $\{x_n\}_{n=1}^{\infty}$ *is a monotonically decreasing sequence. The sequence converges because it is bounded and monotonically decreasing below. Putting* $b_n := a_n - \gamma a_n$*, this satisfies* $\sum_{n=0}^{\infty} b_n = \infty, \sum_{n=0}^{\infty} b_n^2 < \infty$ *. You can say*$x_n = x_1\Pi_{i=1}^{n}(1-b_i)$*, and the convergence destination x is* $x_1\Pi_{n=1}^{\infty}(1-b_n)$*.* $c_n = \Pi_{i=1}^{n}(1-b_i)$*, the infinite product of* $n \to \infty$ *is* $\sum_{n=0}^{\infty} b_n = \infty$*, but diverges. However, since it is* $0 \leq c_n \leq 1$*,* $c_n \to 0$ *is known, and* $x_n \to 0$ *can be said.*

Lemma 4. *Let* $\epsilon > 0$.

$$x_{n+1} = (1 - a_n)x_n + \gamma a_n|x_n + \epsilon| \tag{63}$$

Then $x_n \to \epsilon\frac{\gamma}{1-\gamma}$ *holds.*

proof.

$$x_{n+1} - x_n = -a_n((1 - \gamma)x_n - \epsilon\gamma) \tag{64}$$

$$= -a_n(1 - \gamma)(x_n - \epsilon\frac{\gamma}{1 - \gamma}) \tag{65}$$

The difference from $\epsilon\frac{\gamma}{1-\gamma}$ *is reduced by* $a_n(1-\gamma)$*. If* $y_n := x_n - \epsilon\frac{\gamma}{1-\gamma}$*, by definition it is clearly* $y_{n+1} - y_n = x_{n+1} - x_n$*. Moreover,*

$$y_{n+1} - y_n = -a_n(1-\gamma)(y_n) \tag{66}$$

$$y_{n+1} = (1 - a_n(1-\gamma))y_n \tag{67}$$

After that, it is $x_n \to \epsilon\frac{\gamma}{1-\gamma}$ *because it is* $y_n \to 0$ *by the same argument in Lemma 3.*

Lemma 5. *Suppose that the sequence* $\{c_n\} \subset \mathbb{R}_+$ *converges uniformly to 0 on a set of probabilities 1. That is, for any* $\epsilon_1 > 0$*, there is a certain* $N_{\epsilon_1}(\omega)$*, and when* $n > N_{\epsilon_1}(\omega)$*,* $|c_n| < \epsilon_1$ *holds with probability 1. At this time,*

$$x_{n+1} = (1-a_n)x_n + \gamma a_n|x_n + c_n| \tag{68}$$

x_n *converges to 0.*

proof.

$$z_{N_{\epsilon_1}} = x_{N_{\epsilon_1}} \tag{69}$$

$$z_{n+1} = (1-a_n)z_n + \gamma a_n|z_n + \epsilon_1| \tag{70}$$

$|Z_n| \geq |x_n|$ *for such* $n > N$*.* $Z_n \to \epsilon_1\frac{\gamma}{1-\gamma}$ *from Lemma 4. That is, for any* $\epsilon_2 > 0$*, there is a certain* $N_{\epsilon_2} > N_{\ epsilon_1}$*, and* $n > N_{\epsilon_2}$ *for any* n*,* $z_n < \epsilon_1\frac{\gamma}{1-\gamma} + \epsilon_2$ ϵ_1, ϵ_2 *can be arbitrarily taken, so if we define a new* $\epsilon := \epsilon_1\frac{\gamma}{1-\gamma} + \epsilon_2$*, this is also* $\epsilon > 0$ *can be taken arbitrarily. Within the range of Using* $z_n > x_n$*, there is a* N_{ϵ_2} *for any* ϵ *and* $x_n < \epsilon$*for* $n > N_{\epsilon_2}$ *.*

B Strict Proof of Policy Gradient thorem and Distributionaly

We prove the famous policy gradient theorem using the Q function and its version in distributed reinforcement learning [23].

Theorem 4. *Policy Gradient thorem*

Consider the gradient of the policy value function $J(\theta) := E[Q(x, \pi_\theta(x))]$*. At this time, it is assumed that* π, Q *is implemented by a neural network, the activation function is Lipschitz continuous, and* $\nabla_\theta Q(x,a) = 0$*. Then, The following equation holds,*

$$\nabla_\theta J(\theta) = E_\rho[\nabla_\theta \pi(\theta) \nabla_a Q(s,a)|_{a=\pi(x)}] \tag{71}$$

However, ρ *is memory data in general implementation. Next, consider the case of distributed reinforcement learning. If a random variable representing the cumulative reward sum is expressed as* Z*, then* $Q(s,a) = E_n[Z(s,a)]$ *holds. Suppose* Z *is a neural network with stochastic output.*

$$Z(\omega)(s,a) = f_\omega(s,a) \tag{72}$$

Then,

$$\nabla_\theta J(\theta) = E_\rho[\nabla_\theta \pi(x) E_n[\nabla_a Z(x,a)]|_{a=\pi(x)}] \tag{73}$$

proof. *The interchangeable conditions of differentiation and Lebesgue integration are described as follows. Suppose there is a function $f(x,\omega)$ that can be Lebesgue integrable over Ω and differentiable by x. At this time, there is an integrable function $\phi(\omega)$, and x can be differentiated almost everywhere on Ω by x and $|\nabla_x f(x,\omega)_i| \leq \phi(\omega)$ holds, then $\int_\Omega f(x,\omega)d\mu(\omega)$ is differentiable by x, and holds,*

$$\nabla_x \int_\Omega f(x,\omega)d\mu(\omega) = \int_\Omega \nabla_x f(x,\omega)d\mu(\omega) \tag{74}$$

When $\mu(\Omega) < \infty$, An example of a function class that satisfies this is the "Lipschitz continuous function". Neural networks is generally combinations of linear transformations and Lipschitz continuous activation map. General activation functions such as sigmoid, ReLu, Reaky Relu, and Swish are all Lipschitz continuous functions. Moreover, if the Lipschitz constant of the function f is written as $||f||_L$, then considering two Lipschitz continuous functions f, g, $||f \circ g||_L \leq ||f||_L||g||_L$. From this, $\pi_\theta(x), Q(x,a), Q(x,\pi_\theta(x))$ are Lipschitz continuous for x, a, respectively. Although it is not Lipschitz continuous for θ, it is Lipschitz continuous for each element, and the definition and definition of ∇ allow the exchange of differentiation and integration. That is, the following holds from the differential chain rule,

$$\nabla_\theta J(\theta) = E_\rho[\nabla_\theta \pi_\theta(x) \nabla_a Q(s,a)|_{a=\pi_\theta(x)}] \tag{75}$$

Similarly, $\nabla_a E[Z(s,a)] = E[\nabla_a f_\omega(s,a)]$ and f_ω is Lipschitz continuous functions for any ω , For distribution type

$$\nabla_\theta J(\theta) = E_\rho[\nabla_\theta \pi(x) E_n[\nabla_a f_\omega(x,a)]|_{a=\pi(x)}] \tag{76}$$

$$= E_\rho[\nabla_\theta \pi(x) E_n[\nabla_a Z(x,a)]|_{a=\pi(x)}] \tag{77}$$

As described above, the policy gradient theorem is established because the policy is Lipschitz-continuous for each parameter, and is obviously not for a policy function composed of ODEnet [24], hypernet [25], or the like that reuses parameters.

C Notation

Let $(A, \mathcal{O})$ be a topological space.

- $\sigma(\mathcal{O})$:Smallest σ-algebla containing all $o \in \mathcal{O}$.
- $\mathcal{B}(A) := \sigma(\mathcal{O})$
- $\mathcal{P}(A)$:If A is a finite set, it is the set of all probability measures defined by measurable space $(A, 2^A)$, and if it is an infinite set, $(A, \mathcal{B}(A))$
- $\mathcal{F} \otimes \mathcal{G} := \sigma(\mathcal{F} \times \mathcal{G})$

References

1. Watldns, C.J.C.H.: Learning from delayed rewards. PhD Thesis, University of Cambridge, England (1989)
2. Morimura, T., Sugiyama, M., Kashima,H., Hachiya, H., Tanaka, T.: Nonparametric return distribution approximation for reinforcement learning. In: International Conference on Machine Learning (2010)
3. Bellemare, M.G., Dabney, W., Munos, R.: A distributional perspective on reinforcement learning. In: International Conference on Machine Learning, pp. 449–458 (2017)
4. Rowland, M., Bellemare, M.G., Dabney, W., Munos, R., Teh, Y.W. An analysis of categorical distributional reinforcement learning. In: Artificial Intelligence and Statistics (AISTATS) (2018)
5. Morimura, T., Sugiyama, M., Kashima, H., Hachiya, H., Tanaka, T.: Parametric return density estimation for reinforcement learning. In: Conference on Uncertainty in Artificial Intterigence (2010)
6. Azizzadenesheli, K., Brunskill, E., Anandkumar, A.: Efficient exploration through Bayesian deep q-networks. arXiv preprint arXiv:1802.04412.2018 (2018)
7. Rasmussen, C.E., Kuss, M.: Gaussian processes in reinforcement learning. In: Thrun, S., Saul, L.K., Schölkopf, B. (eds.) Advances in Neural Information Processing Systems 16, pp. 751–759. USA. MIT Press, Cambridge, MA (2004)
8. Watkins, C.J.C.H., Dayan, P.: Q-learning,Machine Learning, vol. 8, pp. 279-292 Kluwer Academic Publishers, Boston. Manufactured in The Netherlands (1992)
9. Tsitsiklis, J.N.: Asynchronous stochastic approximation and Q-learning. Mach. Learn. **16**, (1994)
10. Jakkola, T., Jordan, M.I., Singh, S.P.: On the convergence of stochastic iterative dynamic programming algorithms. Neural Comput. **6**(6), 1185–1201 (1994)
11. Melo, F.S.: Convergence of Q-learning: A simple proof, Institute Of Systems and Robotics. Tech, Rep (2001)
12. Borkar, V.S., Meyn, S.P.: The O.D.E. method for convergence of stochastic approximation and reinforcement learning. SIAM J. Control Optim. **38**(2), 447–469 (2000)
13. Zhuoran, Y., Xie, Y., Wang, Z.: A Theoretical Analysis of Deep Q-Learning, arXiv preprint arXiv:1901.00137v2.2019
14. Scherrer, B., Ghavamzadeh, M., Gabillon, V., Lesner, B., Geist, M.: Approximate modified policy iteration and its application to the game of tetris. J. Mach. Learn. Res. **16**, 1629–1676 (2015)
15. Farahmand, A., Szepesvári, C., Munos, R.: Error propagation for approximate policy and value iteration. In: Advances in Neural Information Processing Systems (2010)
16. Antos, A., Szepesvári, C., Munos, R.: Learning near-optimal policies with Bellman-residual minimization based fitted policy iteration and a single sample path. Mach. Learn. **71**, 89–129 (2008)
17. Munos, R.: Performance bounds in lp norm for approximate value iteration. SIAM J. Control Optim. **46**(2), 541–561 (2007)
18. Munos, R.: Error bounds for approximate policy iteration. In: ÍCML 2003: Proceedings of the 20th Annual International Conference on Machine Learning (2003)
19. Venter, J.: On Dvoretzky stochastic approximation theorems. Ann. Math. Stat. **37**, 1534–1544 (1966)
20. Berti, P., Crimaldi, I., Pratelli, L., Rigo, P.: Rate of convergence of predictive distributions for dependent data. Bernoulli **15**, 1351–1367 (2009)

21. Berti, P., Pratelli, L., Reigo, P.: Limit theorems for predictive sequences of random variables. Technical Report 146, Dip. EPMQ, Univ. Pavia. Available at economia (2002)
22. Ghavamzadeh, M., Mannor, S., Pineau, J., Tamar, A.: Bayesian reinforcement learning: a survey. Found. Trends Mach. Learn. **8**(5–6), 359–483 (2015)
23. Tessler, C., Tennenholtz, G., Mannor, S. :Distributional policy optimization: An alternative approach for continuous control. arXiv preprint arXiv:1905.09855,2019
24. Chen, T.Q., Rubanova, Y., Bettencourt, J., Duvenaud, D.: Neural ordinary differential equations, arXiv preprint arXiv:1806.07366 (2018)
25. Ha, D., Dai, A., Le, Q.V.: HyperNetworks. In: International Conference on Learning Representations (2017)

Check for updates

Evaluation of the Impact of Controllable Load Condition on Optimal Sizing of an Isolated Microgrid

Gulfem Er(✉), Gurkan Soykan, Ethem Canakoglu, Recep Tufan Bektas, Yaren Sahin, and Belgin Erdogan

Bahcesehir University, 34349 Istanbul, Turkey
gulfem.er@eng.bau.edu.tr

Abstract. Self-sufficient energy systems such as isolated microgrid have been used in rural areas to generate electrical energy. This study aims to investigate the impact of controllable loads on the optimal capacity configuration for an isolated microgrid formed from renewable resources and a hybrid energy storage system (HESS), which contains battery and supercapacitor. To find the optimal design of the proposed system, a mathematical model for minimizing life cycle cost (LCC) is proposed for a given level of reliability. In particular, the total life cycle cost is minimized while keeping the reliability index defined as loss of power supply probability (LPSP), within the required limits in addition to the other constraints. The simulation results for the system are analyzed by considering the different levels of LPSP. The analysis shows that the availability of controllable loads affects the life cycle cost of the isolated microgrid system positively.

Keywords: Controllable load · Hybrid energy storage · Isolated microgrid

1 Introduction

Fossil fuels are non-sustainable energy resources, and the formation of these resources takes millions of years. As the world population and gross domestic product (GDP) increase, total energy consumption also increases; therefore, concerns regarding the detrimental consequences of fossil fuel usage has pushed governments to discover renewable energy resources as an alternative. Many governments have focused on developing their renewable energy infrastructures. By 2016, 173 countries adjusted to renewable development, and 146 countries acknowledged several policies. This strategy has led to an increase of 20% in electricity production from renewable resources [1].

In the traditional grid, an energy storage system is one of the possible solutions to solve the demand-based problems. After starting to use renewable resources, it can be used to handle the problem that occurs from the intermittent

T. Allahviranloo et al. (Eds.): IDS 2020, AISC 1301, pp. 166–177, 2021.
https://doi.org/10.1007/978-3-030-66501-2_13

nature of renewable resources. Energy storage systems in terms of used technologies are classified into three types; mechanical, electrical and chemical. Also, they are categorized in terms of energy and power. Depending on this, high power devices are given as supercapacitor, superconducting magnetic energy storage, and flywheel whereas high energy devices are presented as pumped hydro, compressed air energy storage, fuel cell, and battery [2]. The combination of two or more different storage systems, called the hybrid energy storage system, provides higher power quality, higher efficiency, and less interruption.

Microgrids are mainly considered as a potential solution for either preventing the effects of major power disruption events on the grid or the lack of distribution infrastructure in rural areas. They can be viewed as a small scale power grid that has its own local generation and consumption. In terms of the usage purpose, microgrids are designed as either on-grid or off-grid system. Renewable energy sources are preferred as the main generation resources considering environmental issues on microgrid. Because of the intermittent structure of RESs, energy storage systems should be utilized as a solution to microgrid that has renewable energy sources. Especially, the hybrid energy storage system can be used to improve the performance of off-grid microgrid system.

During the design of a microgrid system, that is operated either on-grid or off-grid, cost, reliability, availability, and environment are considered in order to determine the optimal capacity. By using either one of them or the combination of at least two of them, the capacity configuration problem is defined as an optimization problem with either single objective or multi objective considering the constraints [3–10]. Because of the usage of RESs in the microgrid, the cost and reliability based objectives are more important than environment based objectives.

In the literature, several studies deal with the capacity configuration problem for the off-grid mode based microgrids that consist of distributed generation resources and HESS. In [11–15], the cost-related single objective function are used for different systems.

In this work, the impact of controllable load condition is investigated during the optimal sizing of an isolated microgrid system with wind, photovoltaic(PV), battery and supercapacitor. The optimization problem is described as minimizing the LCC with the system and component constraints. Moreover, LPSP is used as reliability constraint during the optimal capacity configuration in order to evaluate the effect of controllable load situation.

2 Microgrid System Configuration

Microgrids are low voltage smart distribution networks, including diverse distributed energy resources, which can be operated either connected to the distribution grid or as islanded if it is separated from the grid [16]. In these systems, it is necessary to model all components - which are PV panel, wind turbine, battery, and supercapacitor (SC) in this study - detailed because of the uncertainty of renewable resources. A representative configuration of an isolated microgrid with HESS is shown in Fig. 1.

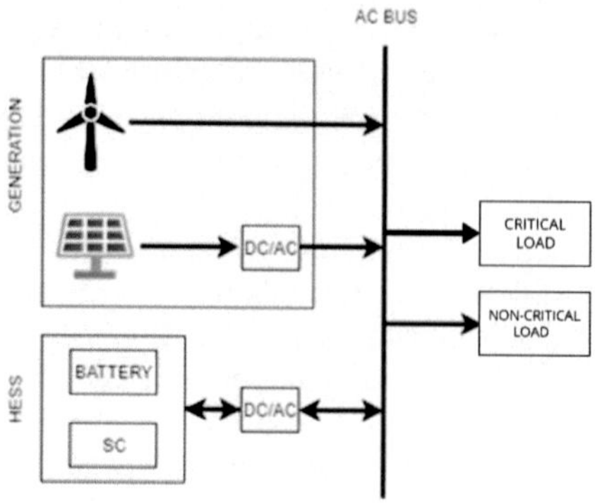

Fig. 1. System configuration

3 Modelling of System Components

3.1 Wind Modelling

The output power of a wind turbine can be defined as:

$$p_{wt}(t) = \begin{cases} 0 & v(t) \leq v_{cut-in} \text{ or } v(t) \geq v_{cut-out} \\ P_r \frac{v(t)-v_{cut-in}}{v_r - v_{cut-in}} & v_{cut-in} < v(t) < v_r \\ P_r & v_r < v(t) < v_{cut-out} \end{cases} \tag{1}$$

In Eq. (1), P_r represents rated power of the wind turbine $(Watt)$, $v(t)$ is the wind speed (m/s) at time t, v_r is the rated speed (m/s), $v_{(cut-in)}$ and $v_{(cut-out)}$ represent cut-in and cut-out wind speed respectively. Cut-in value corresponds to the wind speed that the turbine begins to rotate, whereas, cut-out value is the wind speed that the turbine must stop rotating in order to protect itself.

Since a microgrid problem is solved, it is assumed that the wind regime for all the turbines is the same. Hence, the total output power of all wind turbines at time t can be defined as:

$$P_{WT}(t) = N_{WT} \times p_{wt}(t). \tag{2}$$

3.2 Photovoltaic Modelling

The electrical power that is generated by PV array can be expressed as:

$$p_{pv}(t) = R(t) \times A_{PV} \times \eta_{PV} \tag{3}$$

where A_{PV} represents the area of each PV panel (m^2), η_{PV} is efficiency of the PV collector unit and $R(t)$ represents the incident solar irradiance of selected surface (W/m^2) for a specific time. The total output power of photovoltaic system at time t can be shown as:

$$P_{PV(t)} = N_{PV} \times p_{pv(t)} \tag{4}$$

3.3 HESS Modelling

In this paper, there are two types of storage systems, which are battery and supercapacitor. The charging situation of battery and SC is represented by state of charge (SOC) index. The dynamic energy equation for the battery is given as:

$$\left(-P_{Bat}^{+}(t)\times\frac{\Delta t}{\eta_{bat}^{+}}\right)+Tot_{Bat}(t)+(P_{Bat}^{-}(t)\times\Delta t\times\eta_{bat}^{-})=Tot_{Bat}(t+1) \quad (5)$$

where $P_{Bat}^{-}(t)$ and $P_{Bat}^{+}(t)$ show the power flow in a unit time for battery charging and discharging, Δt is time interval, η_{bat}^{+} is battery discharging efficiency, $Tot_{Bat}(t)$ and $Tot_{Bat}(t+1)$ are total energy stored in batteries at time t and $t+1$, η_{bat}^{-} is charging efficiency for the battery.

SOC of the battery can be achieved by:

$$SOC_{Bat}(t)=\frac{Tot_{Bat}(t)}{E_{Bat}\times N_{Bat}} \quad (6)$$

where E_{Bat} represents the capacity of a single battery and N_{Bat} shows the number of batteries.

Just like a battery storage system, the dynamic energy equation for the supercapacitors can be defined as:

$$\left(-P_{SC}^{+}(t)\times\frac{\Delta t}{\eta_{sc}^{+}}\right)+Tot_{SC}(t)+(P_{SC}^{-}(t)\times\Delta t\times\eta_{sc}^{-})=Tot_{SC}(t+1) \quad (7)$$

where $P_{SC}^{-}(t)$ and $P_{SC}^{+}(t)$ show the power flow in a unit time for supercapacitor charging and discharging, Δt is time interval, η_{sc}^{+} is SC discharging efficiency, $Tot_{SC}(t)$ and $Tot_{SC}(t+1)$ are the total energy stored in supercapacitors at time t and $t+1$, η_{sc}^{-} is SC charging efficiency. The SOC of SC at time t can be achieved from:

$$SOC_{SC}(t)=\frac{Tot_{SC}(t)}{E_{SC}\times N_{SC}} \quad (8)$$

where E_{SC} represents capacity of the SC, and N_{SC} represents number of the SCs that were used.

4 Evaluating the Impact of Controllable Load Condition on Optimal Sizing of Isolated Microgrid

In this paper, a microgrid system comprises different components, including PV panels, wind turbines, batteries, and supercapacitors, as shown in Fig. 1. These power sources have various effects on cost. But the configuration is also important for the reliability of the system. Therefore, the cost is defined as the objective of the problem whereas the reliability is included in the LPSP constraint. The objective and the constraint functions are explained in this section.

4.1 Objective

Objective function is defined as the minimization of life cycle cost which is the sum of the net present value of all components [17]. The total LCC can be calculated as:

$$\min \sum_{m\in\{PV,WT,Bat,SC\}} LCC_m = \frac{1}{CRF(i,n)} \sum_{m\in\{PV,WT,Bat,SC\}} AC_m \quad (9)$$

where (CC) is capital cost, $(O\&M)$ is operation and maintenance cost for each component. In this paper the LCC is calculated converting the annual cost by dividing the capital recovery factor defined as

$$CRF(i,n) = \frac{i(1+i)^n}{(1+i)^n - 1} \quad (10)$$

where the annual cost of each component is calculated using the life time of that component therefore the replacement cost is eliminated from the analysis. In Eq. 10, i shows the interest rate and n represents the planning horizon.

Cost of Wind Turbine: The annual cost of wind turbines is based on their maintenance cost, $O\&M_{WT}$, and capital cost, CC_{WT}. The annualized cost for wind turbines is

$$AC_{WT} = (CC_{WT} + O\&M_{WT}) \times N_{WT} \quad (11)$$

where N_{WT} is the number of wind turbines that will be build in the microgrid. The equations of the costs are given as:

$$CC_{WT} = CRF \times A_{WT} \times C_{WT} \quad (12)$$

$$O\&M_{WT} = A_{WT} \times C_{Mnt-WT} \quad (13)$$

where A_{WT} represents the swept area, C_{WT} and C_{Mnt-WT} show the unit cost and the annual maintenance cost for each wind turbine.

Cost of PV: Same as wind turbines, annualized cost of PV comprises of the capital cost and maintenance cost for each PV panel.

$$AC_{PV} = (CC_{PV} + O\&M_{PV}) \times N_{PV} \quad (14)$$

The capital cost is defined as:

$$CC_{PV} = CRF(i,n) \times A_{PV} \times C_{PV}, \quad (15)$$

and the maintenance cost is defined as:

$$O\&M_{PV} = A_{PV} \times C_{Mnt-PV} \quad (16)$$

where C_{PV} denotes collector cost of PV, and C_{Mnt-PV} denotes annual maintenance cost per unit area of PV panel.

Cost of Battery: Battery life cycle is taken as 5 years hence the replacement cost is also included. Annualized cost for N_{Bat} number of batteries is calculated as:

$$AC_{Bat} = (CRF_{Bat(i,5)} \times C_{Bat} + C_{Mnt-Bat}) \times N_{Bat} \tag{17}$$

C_{Bat} is the cost of unit battery and $C_{Mnt-Bat}$ is the maintenance cost of battery. The CRF for batteries is calculated for 5 years, which is equal to the lifetime of battery.

Cost of Super Capacitor: The lifetime of superconductors is equal to the lifetime of the project. Therefore the recovery factor is the same as the recovery factor for the wind turbines and PV panels. The annualized cost for SC is

$$AC_{SC} = (CRF(i,n) \times C_{SC} + C_{Mnt-SC}) \times N_{SC} \tag{18}$$

where C_{SC} is cost of unit SC, C_{Mnt-SC} is the maintenance cost of SC and N_{SC} is number of SC that are used.

4.2 Reliability Constraint

The reliability of the system is reflected by LPSP, which is the probability of the power not supplied. For a given system parameters scenario, the LPSP formulation is

$$LPSP = \frac{\sum_{t=1}^{T} DE(t)}{Tot_{demand}} \times 100. \tag{19}$$

Total demand is written as:

$$Tot_{demand} = \sum_{t=1}^{T} P_d(t) \tag{20}$$

where $DE(t)$ is the instant deficit power and Tot_{demand} is the total demand.

4.3 Other Constraints

The SOC values are limited due to the physical limitations of the battery systems between the minimum ($SOC_{Bat-min}$) and maximum ($SOC_{Bat-max}$) levels defined as:

$$SOC_{Bat-min} \leq SOC_{Bat}(t) \leq SOC_{Bat-max}. \tag{21}$$

The level of the SOC for supercapacitor is also limited just as the batteries defined as:

$$SOC_{SC-min} \leq SOC_{SC}(t) \leq SOC_{SC-max}. \tag{22}$$

The Eqs. (21) and (22) are nonlinear therefore the limits for the SOC is redefined through $Tot_{Bat}(t)$ variable as:

$$SOC_{Bat-min}E_{Bat}N_{Bat} \leq Tot_{Bat}(t) \leq SOC_{Bat-max}E_{Bat}N_{Bat} \tag{23}$$

$$SOC_{SC-min}E_{SC}N_{SC} \leq Tot_{SC}(t) \leq SOC_{SC-max}E_{SC}N_{SC}. \tag{24}$$

The total energy in storage system is limited through constraints (23), and (24) whereas the power flow is limited by constraints (25), and (26). $P_{Bat}^{-}(t)$ and $P_{Bat}^{+}(t)$, should not exceed these limits:

$$P_{Bat}^{-}(t) \leq N_{Bat} \times P_{Bat-max}^{-} \tag{25}$$

$$P_{Bat}^{+}(t) \leq N_{Bat} \times P_{Bat-max}^{+} \tag{26}$$

where $P_{Bat-max}^{-}$ and $P_{Bat-max}^{+}$ are maximum battery charging and discharging power capacities.

There is an area restriction in the microgrid system, which is ensured by the following constraint.

$$(A_{PV} \times N_{PV}) + (WT_d \times WT_{RS}) \times N_{WT} \leq A_{MG} \tag{27}$$

where WT_d represents the diameter of a turbine blade, and WT_{RS} shows the required area for a turbine. A_{MG} is the available space for the planning system. The relation between the demand/supply of the electricity and the storage system at every time period can be modeled as:

$$P_{Bat}^{+}(t) - P_{Bat}^{-}(t) + P_{SC}^{+}(t) - P_{SC}^{-}(t) = P_d(t) - P_g(t) - DE(t) + S(t) \tag{28}$$

where $DE(t)$ is the deficit power and $S(t)$ is the excessive power at time t. Demand consists of critical demand, $P_{d-c}(t)$, and noncritical demand, $P_{d-nc}(t)$. In terms of criticality, demand can be defined as:

$$P_d(t) = P_{d-c}(t) + P_{d-nc}(t) \tag{29}$$

$P_g(t)$ is the total power generation at time t which is shown as:

$$P_g(t) = P_{WT}(t) + P_{PV}(t) \tag{30}$$

There is a limit for deficit power:

$$\sum_{t=1}^{T} DE(t) \leq Tot_{demand} \times LPSP \tag{31}$$

$$DE(t) \leq P_{d-nc}(t) \tag{32}$$

5 Simulation and Results

A case study is implemented to analyze the changes in the optimal solution in accordance to required LPSP values. The hourly load data is taken from [5], wind speed and solar radiation data are taken from [12], which are shown in Fig. 2, 3 and 4 respectively. The interval time is taken as 5 min however the original values were hourly; therefore, through the interpolation method, hourly data is converted to five-minute data. All parameters used in the simulations are

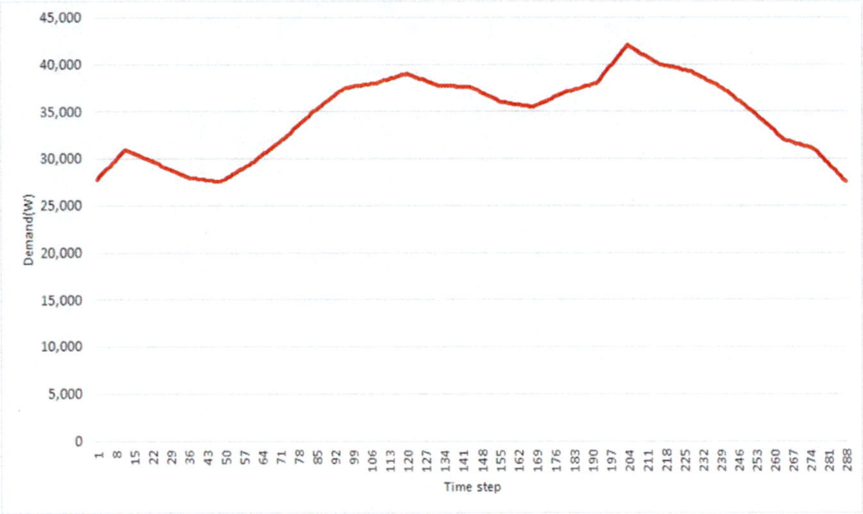

Fig. 2. Daily load

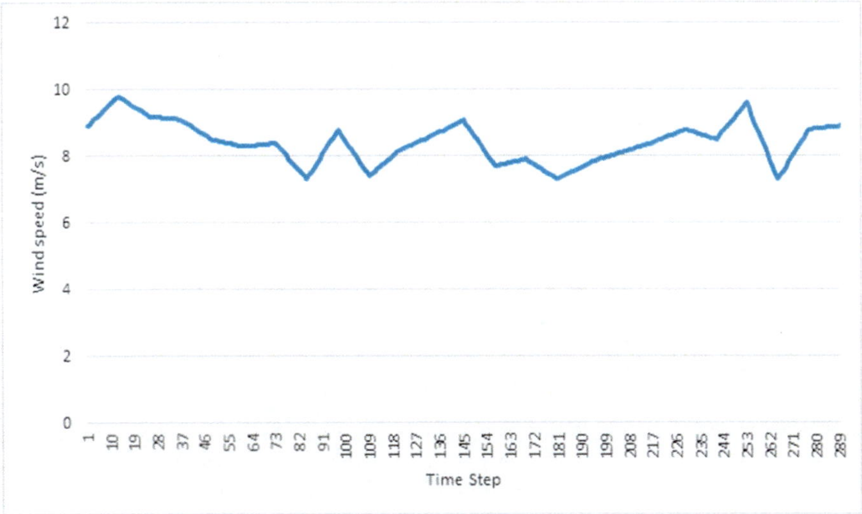

Fig. 3. Daily wind speed

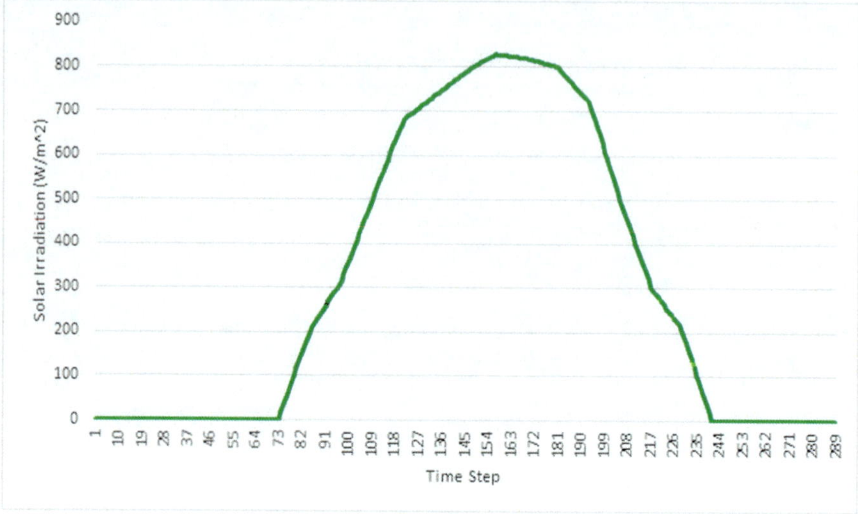

Fig. 4. Daily solar irradiance

given in Table 1 and Table 2. Critical demand is set to 10% of the total demand. Results are given in Table 3.

Table 1. Component Parameters

Type	Capital Cost	O& M Cost	Rep. Cost	Lifetime	Other Parameters
Wind Turbine	150 \$/unit	10 \$/unit-year	150 \$/unit	25 years	Capacity = 500 W v_{cut-in} = 3 m/s v_r = 11 m/s $v_{cut-out}$ = 25 m/s Swept area = 5.73 m^2 Diameter = 2.7 m^2 WT_{RS} = 25 m^2
PV	189 \$/unit	3 \$/unit-year	189 \$/unit	25 years	Capacity = 300 W Dimensions = 1956 mm×9992 mm× 40 mm Area = 1.95 m^2 η_{PV} = 15.5%
Battery	300 \$/unit	10 \$/unit-year	300 \$/unit	5 years	Capacity = 4 kWh $SOC_{Bat-max}$ = 80% $SOC_{Bat-min}$ = 20% η_{bat}^{+} = 90% η_{bat}^{-} = 80% $P_{Bat-max}^{+}$ = 6.910 kW $P_{Bat-max}^{-}$ = 4.610 kW
Supercapacitor	2500 \$/unit	98 \$/unit-year	2500 \$/unit	25 years	Capacity = 1 kWh SOC_{SC-max} = 95% SOC_{SC-min} = 5% η_{sc}^{+} = 98% η_{sc}^{-} = 98%

Table 2. Parameters for the designed system

Parameter	Value
Interest rate (Available space (m^2)	10,000
Project life time (years)	25

Table 3. Optimal results of the system

LPSP	Nwt	Npv	Nbat	Nsc	LCC($)	Installed Power(kW)	Storage Capacity(kWh)
0	92	47	11	1	159,032	60.1	45
0.02	87	55	7	1	152,271	60	29
0.04	86	52	5	1	148,008	58.6	21
0.06	83	53	4	1	143,482	57.4	17
0.08	83	47	2	1	139,333	55.6	9
0.1	80	49	1	1	135,229	54.7	5

In each solution, represented in a row in Table 3, the asset composition of renewable energy sources as well as the storage sizes, changes. Therefore, the cost of the solutions differ. The value of LPSP is set between 0–0.1. LPSP of 0 means the load will always be satisfied whereas LPSP of 0.1 means 10% of the load throughout the day cannot be satisfied, resulting with a low reliable network. As the reliability of a solution, defined by the LPSP constraint, gets less, the LCC gets smaller. It means that if 10% of load can be controlled on the customer side, the designed system supplies the necessary power to the customer with a low LCC. If the amount of controllable load decreases, LCC increases for the designed system.

6 Conclusion

In the case of an inadequate infrastructure on the distribution system in rural areas, the microgrid concept is a possible solution to supply power to local loads by using the local generation. The proposed microgrid system consists of intermittent renewable energy resources because of the environmental benefits of them and hybrid energy storage systems to prevent failure by the lack of available supplies during the off-grid operation mode of the microgrid. The impact of controllable load condition on the optimal sizing for the designed system is evaluated on this study. A case study was investigated to understand the effect of controllable loads on LCC. Simulation results have shown that when the amount of the controllable loads increase on the customer side, the life cycle cost of an optimal designed system decreases.

References

1. Wang, X., Li, N., Sun, W., Xu, S., Zhang, Z.: Quantitative analysis of distributed and centralized development of renewable energy. Global Energy Interconnection **1**, 576–584 (2018)
2. Hemmati, R., Saboori, H.: Emergence of hybrid energy storage systems in renewable energy and transport applications - a review. Renew. Sustain. Energy Rev. **65**, 11–23 (2016)
3. Hajiaghasi, S., Salemnia, A., Hamzeh, M.: Hybrid energy storage system for microgrids applications: a review. J. Energy Storage **21**, 543–570 (2019)
4. Fathima, A.H., Palanisamy, K.: Optimization in microgrids with hybrid energy systems - a review. Renew. Sustain. Energy Rev. **45**, 431–446 (2015)
5. Adefarati, T., Bansal, R.C.: Reliability, economic and environmental analysis of a microgrid system in the presence of renewable energy resources. Appl. Energy **236**, 1089–1114 (2019)
6. Abbes, D., Martinez, A., Champenois, G.: Eco-design optimisation of an autonomous hybrid wind-photovoltaic system with battery storage. IET Renew. Power Gener. **6**, 358–371 (2012)
7. Diab, A.A.Z., Sultan, H.M., Mohamed, I.S., Kuznetsov, O.N., Do, T.D.: Application of different optimization algorithms for optimal sizing of pv/wind/diesel/battery storage stand-alone hybrid microgrid. IEEE Access **7**, 119223–119245 (2019)
8. Wang, S., Li, Z., Wu, L., Shahidehpour, M., Li, Z.: New metrics for assessing the reliability and economics of microgrids in distribution system. IEEE Trans. Power Syst. **28**, 2852–2861 (2013)
9. Baharozu, E., Soykan, G.: Economic and environmental evaluation of a designed smart micro-grid system in Turkey. In: 4th International Istanbul Smart Grid Congress and Fair (ICSG), pp. 1–4. IEEE (2016)
10. Costa, P.M., Matos, A.M.: Economic analysis of microgrids including reliability aspects. In: 9th International Conference on Probabilistic Methods Applied to Power Systems, pp. 1–8. IEEE (2006)
11. Homayouini, F., Roshandel, R., Hamidi, A.A.: Sizing and performance analysis of standalone hybrid photo-voltaic/battery/hydrogen storage technology power generation systems based on the energy hub concept. Int. J. Green Energy **14**, 121–134 (2017)
12. Liu, Z., Chen, Y., Zhuo, R., Jia, H.: Energy storage capacity optimization for autonomy microgrid considering CHP and EV scheduling. Appl. Energy **210**, 1113–1125 (2018)
13. Jacob, A.S., Banerjee, R., Ghosh, P.C.: Sizing of hybrid energy storage system for a PV based microgrid through design space approach. Appl. Energy **212**, 640–653 (2018)
14. Wang, H., Wang, T., Xie, X., Ling, Z., Gao, G., Dong, X.: Optimal capacity configuration of a hybrid energy storage system for an isolated microgrid using quantum-behaved particle swarm optimization. Energies **11**, 453–467 (2018)

15. Zhang, W., Maleki, A., Rosen, M.A., Liu, J.: Optimization with a simulated annealing algorithm of a hybrid system for renewable energy including battery and hydrogen storage. Energy **163**, 191–207 (2018)
16. Logenthiran, T., Srinivasan, D., Khambadkone, A., Raj, T.S.: Optimal sizing of an islanded microgrid using evolutionary strategy. In: 2010 IEEE 11th International Conference on Probabilistic Methods Applied to Power Systems, pp. 12–17. IEEE (2010)
17. Soykan, G., Er, G., Canakoglu, E.: Evaluation of the effect of weather conditions on optimal sizing of an isolated microgrid. In: 2020 IEEE International Conference on Environment and Electrical Engineering and 2020 IEEE Industrial and Commercial Power Systems Europe (EEEIC/I CPS Europe), pp. 1–6 (2020)

Check for updates

Smart Room Temperature Controller IoT System

Erman Hamid(✉), Muhammad Akmal Zolkepli, Nazrulazhar Bahaman, Syarulnaziah Anawar, and Zakiah Ayop

Faculty of Communication and Information Technology, Universiti Teknikal Malaysia Melaka, Hang Tuah Jaya, Melaka, Malaysia
erman@utem.edu.my

Abstract. This research aims to develop and implement a type of smart room temperature controller especially for the small industries and home who are not afford to spent a lot of money to install the high technology room temperature controller system in the market. SRTCIoT is proposed with the objectives to implement with lower cost. Furthermore, it is developed to connect to internet connection and provide a real-time notification if any temperature changes in the room that exceeds the set limit. The system is based on LM35 temperature sensor, LCD, LED, DC Fan, and NodeMCU ESP8266 Microcontroller, and implemented using Arduino IDE and ESP Notify Android Application for the notification.

Keywords: Room temperature controller · LM35 · ESP notify android application

1 Introduction

Room temperature is a general term that refers to a specific temperature in an enclosed space. For human comfort, the proper room temperature is highly dependent on individual needs and other various factors. Smart Room Temperature Controller IoT System (SRTCIoT) is a digital device that replaces the common thermostats found in homes, apartments and industry. One of the drawbacks of the traditional thermostat is that it is commonly left at a single out of sheer convenience. What would be far more efficient is to have a thermostat that knows when you need it and when you don't, so that it could vary the temperature and save energy.

2 Data Collection and Analysis

The temperature control system is more like a programmable thermostat. It is able to maintain the desired temperature level regardless of changing and unexpected environmental conditions and weather [1]. The advantage of having a temperature control system from a common thermostat is that it saves energy and costs. It also maintains different temperatures automatically at different times, regardless of whether it is day or night

T. Allahviranloo et al. (Eds.): IDS 2020, AISC 1301, pp. 178–184, 2021.
https://doi.org/10.1007/978-3-030-66501-2_14

[2]. Typically, a feedback system has a control loop, including sensors, actuators, and control algorithms. It is arranged in such a way aims to arrange a variable at a certain reference value or set point [3]. It is able to increase the fuel supply to the furnace when the measured temperature decreases.

A temperature control system consists of a small programmable digital logic controller device. It's wired to a heating and/or cooling system [4]. The thermostat gauge is mounted on a special wall. It is a temperature control system that is included for the purpose of measuring temperature. Named as a set point, the system will use a heating and/or air conditioning unit (if needed) as an effector, to standardize settings for the programmed duration.

The result shows that the functionality of the current temperature control system in the existing market is not too comprehensive [1, 3–5]. Most of the temperature control systems that being used nowadays do not come with cooling systems and built-in heating features. Therefore, they need to be connected to third-party heaters and cooling systems. Coping with the difficulties that are posed by some of these old legacy heaters and cooling systems and the connectors or configuration has been a major hindrance to the use of temperature control systems. Most temperature control system has been bought by users, only to finally discover at installation point that the system cannot work efficiently with the existing heating and cooling system installations [6]. This means that the existing heating and cooling system have to be abandoned and completely uninstalled and replace with more compatible systems.

The cost of the temperature control system increases proportionally with the amount of the operational accuracy and flexibility and provided by the system [3, 4]. This will also affect the installation caused by the more sophisticated temperature control system that difficult to install. Temperature Control System have to be able to respond efficiently and can be adapted to rapid temperature swings [6]. Temperature swings are instances when the exterior temperature rises or falls rapidly from the current state with a large difference margin. The industrial age has resulted in very unstable and unpredictable temperature conditions as a result of the greenhouse effect. Modern Temperature Control System must be able to efficiently manage these unforeseen contingencies.

3 Methodology

Rapid Application Development (RAD) model is chosen in this project. It is because the suitability as it designs to produce a high-quality output in a shortest time compare to traditional lifecycle software development [4, 7, 8]. Besides that, RAD methodology can provide some output of the product in very limited time and getting feedback from the end users regarding their requirements [9, 10] (Fig.1).

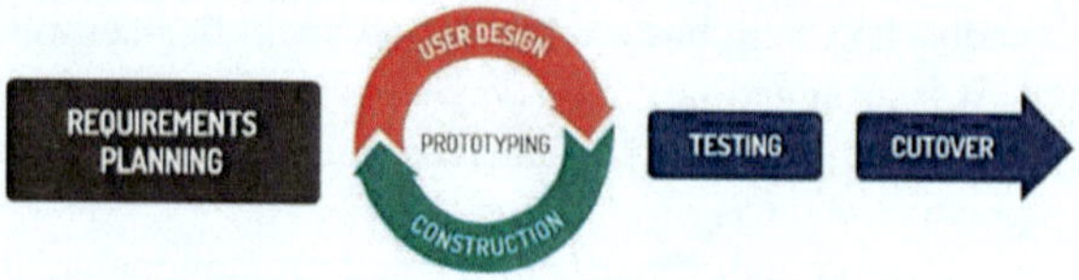

Fig. 1. Process of RAD

4 System Architecture

Arduino IDE is used to program the NodeMCU ESP8266 Microcontroller. When the temperature exceeds the desired temperature, the microcontroller needs to stimulate the LM35 sensor. It happens in the way of the LM35 sensor detecting the temperature and triggers the LED, DC Fan. It finally sends a notification to the phone. Therefore, the connection must be made correctly using the suitable components. The system architecture is shown in Fig. 2 below:

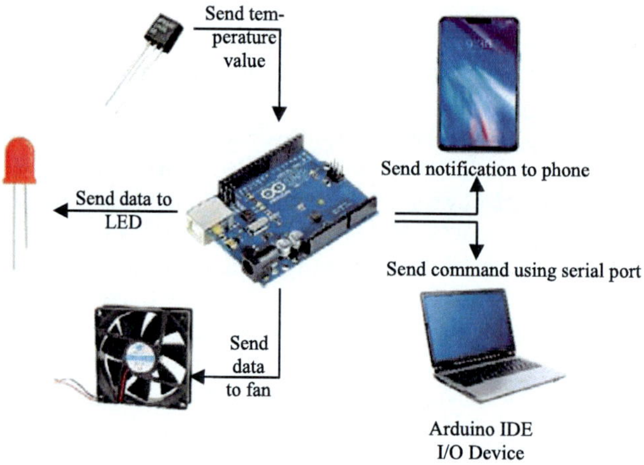

Fig. 2. System architecture

5 System Requirements

The system requirements of the system are divided into three parts which are hardware and software requirement, data requirement, and functional requirement.

5.1 Hardware and Software Requirement

Wires, LED, LCD, DC Fan, and LM35 are the essential hardware needed in the development of this project. The temperature sensor will be attached to the NodeMCU ESP8266

Table 1. Details of each pins numbers

Hardware	Wire	Pins
Led Light (Yellow)	Power	G
	Signal	D6
LM35 Temp Sensor	VCC	3V
	OUT	A0
	GND	3V
DC Fan	Yellow	3V
	Red	D5
	Black	G
LCD	GND	G
	VCC	3V
	SDA	D2
	SCL	D1

Microcontroller, and use to receive the command from the Arduino IDE application to control the hardware. Table 1 shows the details of each pin's numbers.

NodeMCU ESP8266 Microcontroller is connected to the PC via USB port. It is also including a jumper wires for connection to LCD, DC Fan, LED, and LM35 temperature sensor. All connection is based on the Table 1.

The Arduino Integrated Development Environment or Arduino Software (IDE), contains a text editor for writing code, a message area, a text console, a toolbar with buttons for common functions and a series of menus. It connects to the Arduino and Genuino hardware to upload programs and communicate with them. The functionalities of Smart Room Temperature Controller IoT System is shown in Fig. 3.

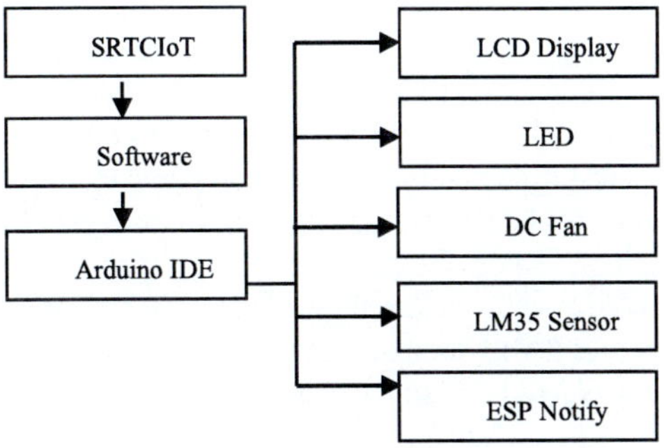

Fig. 3. Functionalities of SRTCIoT

Liquid Crystal Display (LCD) – This is required to display a message and the value of a variable. In temperature control application, it is required to display the value of the temperature dynamically.

LED – Is a two-lead semiconductor light source. In this project, LED will light up when the temperature is below the desired temperature.

DC Fan - The functionality of DC Fan in this project it will turn on when the temperature is exceeding the set limit.

LM35 Sensor – The LM35 series are precision integrated-circuit temperature sensors that attached to pin 3V, pin A0, and pin 3V at NodeMCU ESP8266 board which will detect the room temperature.

ESP Notify Android Application – Is a phone application that can be used to send notification that will be programme on Arduino IDE.

5.2 Data Requirement

The Fig. 4 below show the data flow of the system from Arduino IDE to the ESP Notify Android Application. The Arduino IDE will calculate right after the LM35 sensor receive the temperature value. The command receive by the NodeMCU ESP8266 is from Arduino IDE that send thru via serial port. If the temperature value is exceeding the set limit, ESP Notify will have sent alert or notification thru user's phone.

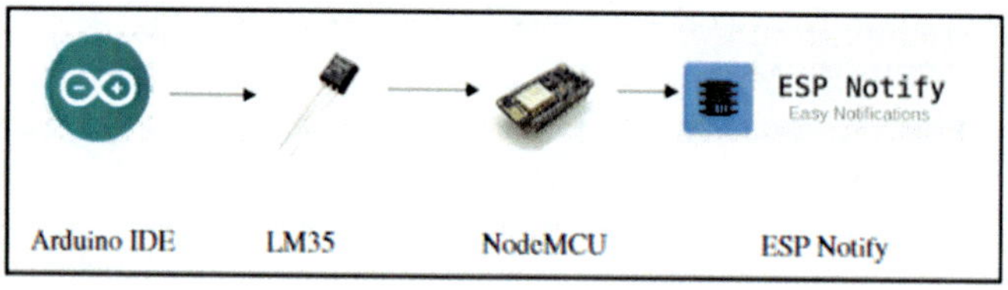

Fig. 4. Data flow

5.3 Functional Requirement

This project is divided into several blocks including, input/output block, microcontroller block, and development block. Figure 5 shows the detail of block diagram for this project.

Microcontroller – NodeMCU ESP 8266 Microcontroller will receive the command from Arduino IDE and on the same time sending command to the LM35 sensor. All code that used to receive and calculate temperature value will be coded in this microcontroller by using IDE cable.

Input/output – LM35 will acts as the temperature sensor which will detect the current room temperature. It will send the temperature value to the microcontroller and it will directly have calculated to degree Celsius.

Development – The main function will be located at the development block. The function consists of receiving temperature value, calculate to degree Celsius, and send notification if the temperature value exceeds the set limit. The NodeMCU ESP8266 will be directly connected by using USB cable. Arduino IDE software will be able to send and receive command from NodeMCU ESP8266 Microcontroller.

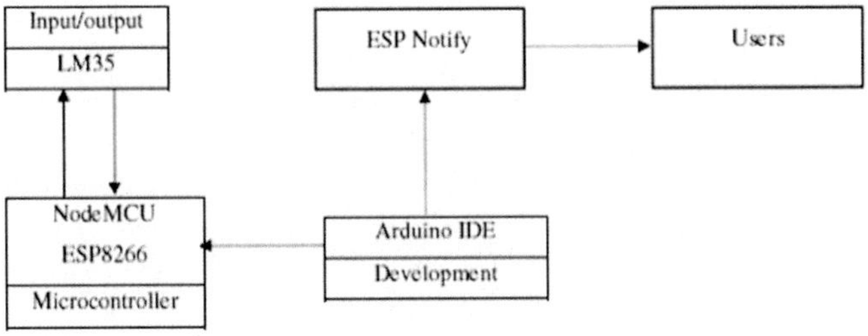

Fig. 5. Functional requirement

6 Project Limitations

This project consists of some limitations which stated as below.

Internet Connection – The internet connection must be connected to the Arduino IDE software application to ensure the notification is able to send out. The connection signal and speed are important to prevent any delay is happening.

LED – The LED light is too small and the brightness is low compared to bulb. This is because LED is more suitable to attach to NodeMCU ESP8266 Microcontroller because it small compare to bulb which is bigger than LED.

7 Result

The prototype of this system was finally successfully developed. It achieves the first objectives with the ability to detect current room temperature by using a valid temperature sensor. It also meets the next objective when it able to send notifications to the user when the temperature value exceeds the set level. Overall, the system meets 99% of its objectives. Table 2 below shows the test results for SRTCNP.

Table 2. Result

No of detection	Correct detection	Error detection	Success %
100	99	1	99

8 Conclusion

This paper aims to expand the focus of the study on room temperature control system focused on a medium level room temperature control system with minimal cost. It ultimately achieves the goal, with findings that increase the weakness of the current room temperature control system (as it exists) in the market by achieving the set objectives.

Acknowledgement. The authors would like to thank Universiti Teknikal Malaysia Melaka for the support in doing this projects.

References

1. Zulkifli, A.F.: Temperature control. Automatic Room Temperature Control with Security System (Doctoral dissertation, UMP) (2009). https://doi.org/10.1007/978-3-319-76617-1_6
2. Kumar, S., Jasuja, A.: Air quality monitoring system based on IoT using Raspberry Pi. In: Proceeding - IEEE International Conference on Computing, Communication and Automation, ICCCA 2017 (2017). https://doi.org/10.1109/CCAA.2017.8230005
3. Chandramohan, A.J., Nagarajan, R., Satheeshkumar, K., Ajithkumar, N., Gopinath, P.A., Ranjithkumar, S.: Intelligent smart home automation and security system using Arduino and Wi-fi. Int. J. Eng. Comput. Sci. (2017) https://doi.org/10.18535/ijecs/v6i3.53
4. Javale,D., Mohsin, M., Nandanwar, S., Shingate, M.: Home automation and security system using android ADK. Int. J. Electr. Commun. Comput. Technol. **3**, 382–385 (2013)
5. Rosenbaum, Primary Examiner-mark. 1980. United States Patent 19 U. S. Patent
6. Azid, S.I., Kumar, S.: Analysis and performance of a low cost SMS based home security system. International Journal of Smart Home **5**, 15–24 (2011)
7. Agarwal, R., Prasad, J., Tanniru, M., Lynch, J.: Risks of rapid application development. Commun. ACM **43**, 1-es (2002). https://doi.org/10.1145/352515.352516
8. Hamid, E., Gee, L.C., Anawar, S., Ayob, Z.: Implementation of intelligent automated gate system with QR Code - An IOT System to help gate management. Int. J. Adv. Comput. Sci. Appl. **9**, 359–363 (2018)
9. Wang, A.I., Bjørnsgård, T., Saxlund, K.: Peer2Me - rapid application framework for mobile peer-to-peer applications. In: Proceedings of the 2007 International Symposium on Collaborative Technologies and Systems, CTS, pp. 379–388 (2007). https://doi.org/10.1109/CTS.2007.4621778
10. Kriglstein, S., Wallner, G., Pohl, M.: A user study of different gameplay visualizations. In: Proceedings of the 32nd Annual ACM Conference on Human Factors in Computing Systems - CHI 2014, pp. 361–370 (2014). https://doi.org/10.1145/2556288.2557317

Solving a Class of Optimal Control Problems by Using Chebyshev Polynomials and Recurrent Neural Networks

M. J. Ebadi(✉) and Hossein Jafari

Department of Mathematics, Chabahar Maritime University, Chabahar, Iran
{ebadi,jafari}@cmu.ac.ir

Abstract. In this paper, a recurrent neural network (RNN) is used to solve a class of optimal control problems (OCPs). These problems can be transferred to a constrained optimization problem subject to linear equality constraints. To solve such problems, some terms of the first kind of Chebyshev polynomials are entered into dynamic state parametrizations instead of $x(t)$, and then the terms are substituted into the performance index and the integral is taken from it. Finally, we have a constrained optimization problem that can be solved by an RNN. An illustrative example is presented to show the effectiveness and performance of the proposed method.

Keywords: Recurrent neural networks · Optimal control problem · Chebyshev polynomials

2010 MSC: 90C30; 65K05

1 Introduction

Consider a dynamic optimization problem with the performance index:

$$J = \left\{ \int_{t_0}^{t_1} g(t, X(t), U(t))dt + q(X(t_1)) \right\}, \tag{1}$$

subject to a class of differential equations on the fixed time interval $[t_0, t_1]$:

$$U(t) = f(t, X(t), \dot{X}(t)), \tag{2}$$

with initial and boundary conditions:

$$X(t_0) = x_0, \qquad X(t_1) = x_1, \tag{3}$$

where $X(.) : [t_0, t_1] \rightarrow \mathbb{R}$ is the state variable, $U(.) : [t_0, t_1] \rightarrow \mathbb{R}$ is the control variable, and g, q and f are real-valued continuously differentiable functions. The

T. Allahviranloo et al. (Eds.): IDS 2020, AISC 1301, pp. 185–194, 2021.
https://doi.org/10.1007/978-3-030-66501-2_15

OCPs arise in a broad range of real-world applications such as finance, economics and engineering. Because the analytical solutions of OCPs can not be easily obtained, the most logical way to solve such problems is finding a numerical solution. However, many numerical approaches have been proposed to solve them. In the direct methods of solving OCPs, the optimal solution can be found by direct minimization of the performance index with the constraints, which leads to a mathematical programming problem. In these methods, the parametrization technique is implemented in three different ways: control-state parametrization, state parametrization and control parametrization. In [1], the authors converted the OCP into a nonlinear optimization problem by finding unknown polynomial coefficients in the state parametrization approach. A numerical technique to solve nonlinear constrained OCPs was proposed in [2]. In [3], the authors presented an algorithm to solve the controlled Duffing Oscillator (CDO) and OCPs in which the solution is on the base of the state parametrization considering the state variable as a linear combination of Chebyshev polynomials of the first kind with unknown coefficients. Then, extended state parametrization is given to the solution of the CDO and nonlinear OCPs. Many researchers have focused on using neural networks in different fields of science and engineering [4–10]. This paper is grounded in solving OCPs using direct methods. In our proposed approach, the state parametrization method is utilized to convert an OCP into a mathematical programming problem. It can be either convergent to the solution of the equations system or non-convergent in the case that a solution of the mathematical programming problem is not available. Therefore, we employ an RNN model which guarantees reaching the solution of the mathematical programming problem. The good thing about the present combined method is that it can be useful even with a small number of Chebyshev basis. The remainder of this paper is illustrated as follows. In Sect. 2, we introduce the preliminary details about Chebyshev polynomials. Section 3 gives some preliminary details about the proposed neural network method and explains how the proposed method is applied to solve an OCP. In Sect. 4, the combined method is applied to an optimal control problem as an example to demonstrate the advantages of our method in comparison with two existing methods. Section 5 contains the sum up of the paper.

2 Preliminaries

The aim of OCP is finding control $U(.)$, which transfers the system (2) from position $X(t_0) = x_0$ to position $X(t_1) = x_1$ within the time $(t_1 - t_0)$, such that the performance index J, which is given in (1) is minimized.
For using Chebyshev polynomials, if $t_0 \neq -1$ or $t_1 \neq 1$, we introduce the transformation:

$$t = \frac{t_1 - t_0}{2} z + \frac{t_1 + t_0}{2}. \tag{4}$$

This change of variable makes $z \in [-1, 1]$, corresponding to $t \in [t_0, t_1]$. We can obtain the OCP in Eqs. (1)-(2) using Eq. (4) as follows.

The optimal control:

$$U(z) = f(\frac{t_1 - t_0}{2} z + \frac{t_1 + t_0}{2}, X(z), \dot{X}(z)), \tag{5}$$

and its corresponding trajectory $x(z)$ with initial conditions:

$$X(-1) = x_0, \qquad X(1) = x_1, \tag{6}$$

minimizes:

$$J = \left\{ \frac{t_1 - t_0}{2} \int_{-1}^{1} g(\frac{t_1 - t_0}{2} z + \frac{t_1 + t_0}{2}, X(z), U(z)) dz + q(X(1)) \right\}. \tag{7}$$

2.1 Chebyshev Polynomials of the First Kind

In this section, we review the Chebyshev polynomials of the first kind briefly. For more details the reader can refer to [11,12,14–16].

Definition 1. *A unique polynomial of degree n which is defined as follows is Chebyshev polynomial $T_n(t)$ of the first kind:*

$$T_n(t) = \begin{cases} \cos(n \cos^{-1} t), & if\ |t| \leq 1; \\ \cosh(n \text{ arccosh } t), & if\ t \geq 1; \\ (-1)^n \cosh(n \text{ arccosh}(-t)), & if\ t \leq -1 \end{cases} \tag{8}$$

such that

$$t = \cos \theta. \tag{9}$$

Since the range of variable t is $[-1, 1]$, then the range of θ can be $[0, \pi]$. In the recurrence relations, the Chebyshev polynomials of the first kind are as follows:

$$\begin{aligned} &T_0(t) = 1 \\ &T_1(t) = t \\ &T_{n+1}(t) = 2tT_n(t) - T_{n-1}(t) \\ &T_{n+1}(t) = tT_n(t) - \sqrt{(1 - t^2\{1 - [T_n(t)]^2\})} \end{aligned} \tag{10}$$

for $n \geq 1$, in addition to

$$\begin{aligned} (t-1)[T_{2n+1}(t) - 1] &= [T_{n+1}(t) - T_n(t)]^2 \\ 2(t^2 - 1)[T_{2n}(t) - 1] &= [T_{n+1}(t) - T_{n-1}(t)]^2 \end{aligned} \tag{11}$$

According to the Eqs. (8) or (10) some Chebyshev polynomials of the first kind are:

$$\begin{aligned} T_0(t) &= 1 \\ T_1(t) &= t \\ T_2(t) &= 2t^2 - 1 \\ T_3(t) &= 4t^3 - 3t \\ T_4(t) &= 8t^4 - 8t^2 + 1 \\ T_5(t) &= 16t^5 - 20t^3 + 5t \\ T_6(t) &= 32t^6 - 48t^4 + 18t^2 - 1 \\ &. \\ &. \\ &. \end{aligned} \tag{12}$$

T_n can be found by the following ordinary and exponential generating functions:

$$\begin{aligned} \sum_{n=0}^{\infty} T_n(t)x^n &= \frac{1-tx}{1-2tx+x^2} \\ \sum_{n=0}^{\infty} T_n(t)\frac{x^n}{n!} &= \frac{1}{2}(e^{(t-\sqrt{t^2-1})x} + e^{(t+\sqrt{t^2-1})x}) \\ &= e^{tx}\cosh(x\sqrt{t^2-1}). \end{aligned} \tag{13}$$

For 2−dimensional multi pole expansion and potential theory, the relevant generating function is

$$\sum_{n=1}^{\infty} T_n(t)\frac{x^n}{n!} = \ln\frac{1}{\sqrt{1-2tx+x^2}} \tag{14}$$

Chebyshev polynomials can be defined as follows:

$$T_n(t) = \frac{n}{2}\sum_{r=0}^{[\frac{n}{2}]}\frac{(-1)^r}{n-r}\binom{n-r}{r}(2t)^{n-2r} = \sum_{m=0}^{[\frac{n}{2}]}\binom{n}{2m}(t)^{n-2m}(t^2-1)^m, \tag{15}$$

where $\binom{n}{k}$ is a binomial coefficient and $[t]$ is the floor function, or

$$T_n(t) = 2^{n-1}\prod_{k=1}^{n}\{t - cos[\frac{(2k-1)\pi}{2n}]\}, \tag{16}$$

which is a product representation of Chebyshev polynomials. These polynomials are orthogonal with respect to the weighting function $(1-t^2)^{-\frac{1}{2}}$ as follows:

$$\int_{-1}^{1}\frac{T_n(t)T_m(t)dt}{\sqrt{1-t^2}} = \begin{cases} \frac{1}{2}\pi\delta_{\mathrm{nm}} & \text{for m} \neq 0,\ \text{n} \neq 0 \\ \pi & \text{for m} = \text{n} = 0 \end{cases}, \tag{17}$$

where δ_{nm} is the Kronecker delta. Also, these polynomials satisfy the following discrete identity:

$$\sum_{k=1}^{m} T_i(t_k)T_j(t_k) = \begin{cases} \frac{1}{2}m\delta_{\mathrm{ij}} & \text{for i} \neq 0,\ \text{j} \neq 0 \\ \pi & \text{for i} = \text{j} = 0 \end{cases}, \quad (18)$$

where t_k are the m zeros of $T_m(t)$ for $k = 1, ..., m$

3 Material and Method

3.1 Recurrent Neural Network Model

In this section, we implement an RNN model which is proposed in [13] to solve the following equality constrained OP

$$\min f(x) \quad subject\ to\ h_i(x) = 0, \qquad i = 1, 2, \cdots, m, \quad x \in \mathbb{R}^n, \quad (19)$$

where the functions $f : \mathbb{R}^n \to \mathbb{R}$ and $h_i : \mathbb{R}^n \to \mathbb{R}, i = 1, 2, \cdots, m$ are considered to be continuously differentiable everywhere in $\mathbb{R}^n$ and $m < n$.

Assumption 1. The matrix $\nabla h(x)$ has full row rank.

Their proposed model is

$$\begin{aligned} \dot{x}(t) &= -(I - P)\nabla f(x(t)) - \rho \nabla h(x(t))(\nabla h(x(t))^T \nabla h(x(t)))^{-1} h(x(t)), \\ x(0) &= x_0 \in \mathbb{R}^n, \end{aligned} \quad (20)$$

where $P = \nabla h(x(t))(\nabla h(x(t))^T \nabla h(x(t)))^{-1} \nabla h(x(t))^T$ and ρ is a positive number.

Theorem 1 (LaSalle's Theorem)[13]. *Suppose that $x(t)$ is a solution of $\dot{x}(t) = \phi(x)$, $x(0) = x_0$ where $\phi : \mathbb{R}^n \to \mathbb{R}^n$ is continuous. If $[0, \omega)$ be the maximal existence interval of $x(t)$. Consider $V : \mathbb{R}^n \to \mathbb{R}$ is a Lipschitz continuous function such that $D^+V(x(t)) \leq 0$ for all $t \in [0, \omega)$. Then, $\Lambda^+ \subset E$, where $D^+V(x(t))$ is the upper right Dini derivative of V, Λ^+ is the positive limit set of the solution $x(t)$ and $E = \{x \in \mathbb{R}^n : D^+V(x) = 0\}$*

Theorem 2 *[13]. Suppose that $x(t)$ is a solution of (20) starting at an arbitrary point $x_0 \in \mathbb{R}^n$ and Λ^+ is its positive limit set. Then $\Lambda^+ \subset D$, where D is the desirable points set of problem (19).*

Proposition 1 *[13]. If $x(t)$ is a solution of (20) and is bounded then it extends to infinite time and has limit points, each of which is a desirable point of (19).*

By selecting $h(x(t)) = Ax - b$, where A is a full row rank matrix with $rank\{A\} = m < n$ and b is an m vector, the problem (19) transformed to

$$\min f(x) \quad subject\ to\ Ax - b = 0,\ x \in \mathbb{R}^n, \quad (21)$$

and the neural network (20) changes to

$$\dot{x}(t) = -(I - P)\nabla f(x(t)) - \rho A^T (AA^T)^{-1}(Ax - b), \quad x(0) = x_0 \in \mathbb{R}^n \quad (22)$$

3.2 Proposed Method

In this method, we substitute $x(t)$ by some terms of first kind of Chebyshev polynomials in the dynamic state. Next, we substitute these new $x(t)$ and $u(t)$ into the performance index and then take integral in the interval $[-1, 1]$. Afterwards, we apply the first kind of Chebyshev polynomials corresponding to the substituted terms in the dynamic state to the boundary or initial conditions of the optimal control. Now, we have a constrained optimization problem(OP) subject to linear equality constraint similar to (21). Then, we apply RNN (22) on this OP.

4 Numerical Experiments

Example 1. Consider the following OCP.

$$\begin{aligned} &J = \frac{1}{2}\int_0^1 (U(\tau)^2 + X(\tau)^2)d\tau, \qquad \tau \in [0, 1],\\ &\text{subject to: } \mathrm{U}(\tau) = \dot{\mathrm{X}}(\tau) + \mathrm{X}(\tau),\\ &\text{with boundary condition}: \quad \mathrm{X}(0) = 1. \end{aligned} \tag{23}$$

By using Pontryagin's maximum principle the exact solution is 0.1929092981 [17] and according to [3], the analytical solution is as follows:

$$\begin{aligned} J &= \frac{e^{-2\sqrt{2}}}{2}((\sqrt{2}+1)e^{4\sqrt{2}} - 1))(\frac{2\sqrt{2}-3}{-(e^{\sqrt{2}})^2 + 2\sqrt{2} - 3})^2\\ &+\frac{e^{-2\sqrt{2}}}{2}((\sqrt{2}-1)e^{2\sqrt{2}} - 1))(1 - \frac{2\sqrt{2}-3}{-(e^{\sqrt{2}})^2 + 2\sqrt{2} - 3})^2. \end{aligned} \tag{24}$$

For using Chebyshev polynomials, we should transform the bounds of the integral in performance index to $[-1, 1]$ by using $t = 2\tau - 1$. With this transformation on all equations in (23) we have

$$\begin{aligned} &J = \frac{1}{4}\int_{-1}^1 (u(t)^2 + x(t)^2)dt,\\ &\text{subject to}: \quad \mathrm{u(t)} = 2\dot{\mathrm{x}}(\mathrm{t}) + \mathrm{x(t)}, \qquad \mathrm{t} \in [-1, 1],\\ &\text{with boundary conditions}: \quad \mathrm{x}(-1) = \mathrm{x}_0 = 1. \end{aligned} \tag{25}$$

To better understanding the method, we apply 3 terms of the first kind of Chebyshev polynomials to find analytical approximation of performance index. The unknowns a_0, a_1, a_2 and a_3 have to be used. We substitute the following 3 terms of the first kind of Chebyshev polynomials and its derivative respect to t to the dynamic state

$$x(t) = a_0 + a_1 t + a_2(2t^2 - 1) + a_3(4t^3 - 3t), \quad \dot{x}(t) = a_1 + 4a_2 t - 12a_3 t^2 - 3a_3, \tag{26}$$

we have

$$u(t) = 2\dot{x}(t)+x(t) = 2a_1+a_0+8a_2t-24a_3t^2-6a_3+a_1t+a_2(2t^2-1)+a_3(4t^3-3t). \tag{27}$$

Now, we substitute (26) and (27) into J in (25) and taking integral on the interval $[-1, 1]$ we have

$$J = a_0^2+2a_0a_1-\frac{2a_0a_2}{3}+2a_0a_3+\frac{7a_1^2}{3}+2a_1a_2+\frac{18a_1a_3}{5}+\frac{167a_2^2}{15}+2a_2a_3+\frac{983a_3^2}{35}. \tag{28}$$

Then, to find the constraints we substitute (26) into the boundary conditions in (25) we have

$$a_0 - a_1 + a_2 - a_3 = 1 \tag{29}$$

Now, we have a constrained OP with objective function J in (28) and linear equality constraints (29) as follows:

$$\begin{aligned} \min\ & a_0^2 + 2a_0a_1 - \frac{2a_0a_2}{3} + 2a_0a_3 + \frac{7a_1^2}{3} + 2a_1a_2 + \frac{18a_1a_3}{5} + \frac{167a_2^2}{15} + 2a_2a_3 + \frac{983a_3^2}{35}. \\ s.t.\ & a_0 - a_1 + a_2 - a_3 = 1 \end{aligned} \tag{30}$$

After solving this problem by RNN (22) the solutions for $u(t)$ and $x(t)$ with 3 terms of the first kind of Chebyshev polynomials are as follows:

$$\begin{aligned} x_3(t) &= 0.507447 - 0.330459t + 0.133538t^2 - 0.028556t^3 \\ u_3(t) &= -0.153471 + 0.203693t + 0.304874t^2 - 0.028556t^3, \quad J = 0.1929316. \end{aligned} \tag{31}$$

We solve this problem with 5 terms of the first kind of Chebyshev polynomials and reach to 0.1929092980 which the error is about $9.9e^{-11}$. In Fig. 1, we show the approximation of X and U with different terms of Chebyshev polynomials. As it can be clearly seen in Fig. 1, the approximation of X and U with 5 terms of the first kind of Chebyshev polynomials are consistent with their analytical solutions. In Table 1, the error comparison of the proposed method and two other methods, Kafash [3] and Mehne [1] methods, with four terms of the first kind of Chebyshev polynomials is given. Table 2 shows the values of cost functional J and error of our proposed method.

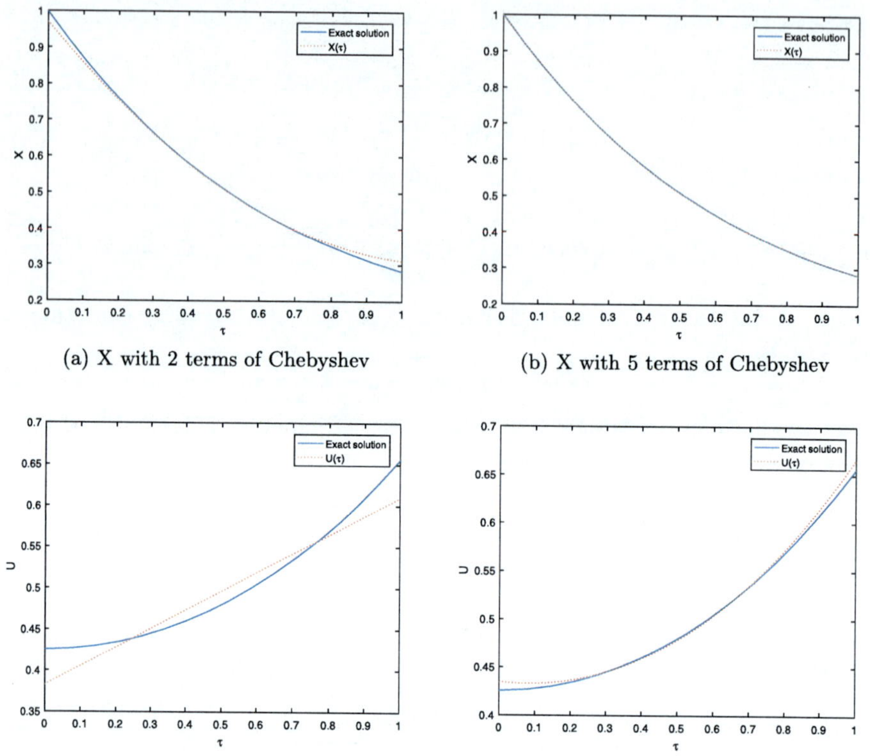

(a) X with 2 terms of Chebyshev

(b) X with 5 terms of Chebyshev

(c) U with 2 terms of Chebyshev

(d) U with 5 terms of Chebyshev

Fig. 1. Approximation of X and U with different terms of Chebyshev polynomial

Table 1. Optimal cost functional J in Example 1 with Four terms of Chebyshev polynomial

Herein	Error	Kafash method [3]	Error	Mehne method [1]	Error
0.192909447	$1.5e^{-7}$	0.192909776	$4.7e^{-7}$	0.193828723	$9.2e^{-4}$

Table 2. Optimal cost functional J in Example 1 with Four terms of Chebyshev polynomial

Number of terms	Optimal value of J (Herein)	Error
1	0.251362735	$5.0e^{-2}$
2	0.194298642	$1.3e^{-3}$
3	0.192931605	$2.2e^{-5}$
4	0.192909447	$1.5e^{-7}$

5 Conclusion

In this paper, a class of OCPs was solved using an RNN. Such problems have been transferred to a constrained optimization problem subject to linear equality constraints. In order to solve those, some terms of the first kind of Chebyshev polynomials were entered into dynamic state parametrizations instead of $x(t)$, and then the terms were substituted into the performance index and the integral was taken from it. In the end, a constrained optimization problem was obtained that has been solved by an RNN. The effectiveness and performance of the proposed method were shown in an illustrative example.

References

1. Mehne, H.H., Borzabadi, A.H.: A numerical method for solving optimal control problems using state parametrization. Numer. Algor. **42**, 165–169 (2006). https://doi.org/10.1007/s11075-006-9035-5
2. Vlassenbroeck, J.: A Chebyshev polynomial method for optimal control with state constraints. Automatica **24**(4), 499–506 (1988). https://doi.org/10.1016/0005-1098(88)90094-5
3. Kafash, B., Delavarkhalafi, A., Karbassi, S.M.: Application of Chebyshev polynomials to derive efficient algorithms for the solution of optimal control problems. Sci. Iran. D, Comput. Sci. Eng. Electr. Eng. **19**(3), 795–805 (2012). https://doi.org/10.1016/j.scient.2011.06.012
4. Ebadi, M.J., Hosseini, A., Hosseini, M.M.: A projection type steepest descent neural network for solving a class of nonsmooth optimization problems. Neurocomputing **235**, 164–181 (2017). https://doi.org/10.1016/j.neucom.2017.01.010
5. Ebadi, M.J., Hosseini, M.M., Karbassi, S.M.: An efficient one-layer recurrent neural network for solving a class of nonsmooth pseudoconvex optimization problems. J. Theoret. Appl. Inf. Technol. **96**(7), 1999–2014 (2018). http://www.jatit.org/volumes/Vol96No7/21Vol96No7.pdf
6. Ebadi, M.J., Hosseini, A., Jafari, H.: An efficient one-layer recurrent neural network for solving a class of nonsmooth optimization problems. J. New Res. Math. **6**(24), 97–110 (2020). http://journals.srbiau.ac.ir/article_15615_f34599f523793828ae53dca49834f495.pdf
7. Heydarpoor, F., Karbassi, S.M., Bidabadi, N., Ebadi, M.J.: Solving multi-objective functions for cancer treatment by using Metaheuristic Algorithms. Int. J. Combinat. Optim. Probl. Informat. **11**(3), 61–75 (2020). https://ijcopi.org/ojs/article/view/124/136
8. Farahani, H., Ebadi, M.J., Jafari, H.: Finding inverse of a fuzzy matrix using eigenvalue method. Int. J. Inn. Technol. Explor. Eng. **9**(2), 3030–3037 (2019). https://doi.org/10.35940/ijitee.B6295.129219
9. Heydarpour, F., Abbasi, E., Ebadi, M.J., Karbassi, S.M.: Solving optimal control problem of cancer treatment by artificial neural networks. Int. J. Interact. Multimed. Artif. Intell. **6**(4), 18–25 (2020). https://doi.org/10.9781/ijimai.2020.11.011
10. Radmanesh, M., Ebadi, M.J.: A local mesh-less collocation method for solving a class of time-dependent fractional integral equations: 2D fractional evolution equation. Eng. Anal. Boundary Elem. **131**, 372–381 (2020). https://doi.org/10.1016/j.enganabound.2020.01.017

11. Stoer, J., Bulirsch, R.: Introduction to Numerical Analysis, 2nd Edn., Springer-Verlag, New York, Translated by Bartels, R., Gautschi, W., Witzgall, C. (1993). https://doi.org/10.1007/978-0-387-21738-3
12. Mason, J.C., Handscomb, D.C.: Chebyshev Polynomials. A CRC Press Company, United States of America (2003)
13. Barbarosou, M.P., Maratos, N.G.: A nonfeasible gradient projection recurrent neural network for equality-constrained optimization problems. IEEE Trans. Neural Networks **19**(10), 1665–1677 (2008). https://doi.org/10.1109/TNN.2008.2000993
14. Rivlin, T.J.: Chebyshev Polynomials. Wiley, New York (1990)
15. Watkins, W., Zeitlin, J.: The Minimal Polynomial of $cos(2\pi/n)$. Amer. Math. Monthly **100**, 471–474 (1993). https://doi.org/10.2307/2324301
16. Zwillinger, D. (ed.): CRC Standard Mathematical Tables and Formulae. CRC Press, Boca Raton (1995)
17. Feldbaum, A.: Principes théoriques des systèmes asservis optimaux. Mir, Moscow (1973)

Artificial Intelligence and Deep Learning

An Overview of Fake Face Detection Approaches

Duygu Cakir(✉), Merve Aritürk, and Özge Yücel

Bahcesehir University, Istanbul, Turkey
{duygu.cakir,merve.ariturk,ozge.yucelkasap}@eng.bau.edu.tr

Abstract. Human face has been one of the most popular objects in computer vision. With the advancements in computing power over the past decades, deep face understanding systems have reached (and surpassed at some cases) human performance and hence, the problem of facial analysis has changed its route from analyzing to generating. These adversarial generation techniques have been used in generating fake news, fake political and financial statements, and even fake porn videos, leveraging the wide availability of large-scale public databases and machine learning techniques. However, there are only a few studies to detect the generated face from real. This study investigates the abundantly used fake face generation models and examines the techniques for detecting facial manipulation.

Keywords: Face synthesis · Fake face generation · Facial manipulation · Fake face detection · Generative adversarial networks

1 Introduction

Human face has been one of the most popular and well-studied non-rigid, yet complex and unresolved objects in computer vision. It has been studied by a variety of fields from psychologists to actors. It has been tried to be detected [1], tracked [2], frontalized and normalized [3], verified [4], recognized [5], and analyzed both attribute-wise [6] and expression-wise [7]. Although the computer still lacks the instinct to interpret everything at once, with the advancements in computing power over the past decades, especially deep face understanding systems have reached (and surpassed at some cases) human performance and hence, the problem of facial analysis has changed its route from analyzing to generating.

With the success of facial analysis algorithms, particularly on face identification and verification, various types of attacks are becoming large threats. These adversarial techniques are divided into two sub categories: *spoofing* and *perturbation*. Spoofing attacks are executed by physically demonstrating a face as an input to the system (by showing a picture or a silicone mask), and perturbation attacks are executed by manipulating the input face by adding/removing additional attributes.

T. Allahviranloo et al. (Eds.): IDS 2020, AISC 1301, pp. 197–211, 2021.
https://doi.org/10.1007/978-3-030-66501-2_16

The first type of facial reenactment is the 3D silicone face mask. They perform a skin-like appearance and motion to spoof the detectors [8]. However, since the scope of this paper is the digital manipulators and not the physical attackers, they won't be examined any further, but have been studied deeply in [9,10]. The rest of the adversarial techniques will be covered under four types: ***1. Facial attribute synthesis***: modifies some of the attributes of the face (such as glasses, gender, hair etc.); ***2. Facial expression synthesis***: manipulates the facial expression of a subject; ***3. Face swapping***: swaps a subject's face with another subject; ***4. Entire face synthesis***: creates a brand new face which does not exist in real life. All these adversarial attacks are used in creating fun apps like Zao[11], FaceApp[12], Reflect[13], Snapchat[14] as well as creating fake news, fake political and financial statements, and even fake porn videos with celebrity faces, leveraging the wide availability of large-scale public databases. This results in a critical question: Is computer vision ready to detect the real ones from fakes, even though they are inseparable by the human eye?

The remainder of this study will investigate different fake face generation methods, types of manipulation detection techniques, main issues in this field, followed by conclusion with some closing remarks and keypoints for future studies.

2 Fake Face Generator Methods

For the most part, a face can be manipulated in two different ways: 1. by the facial region (or by the facial attribute) and 2. by the entire face. The facial region can be manipulated by changing an attribute such as adding glasses, switching the hair color, removing a mole; or by replacing the facial expression such as putting on a smile. The entire face can be manipulated by swapping identities between two people, generating a whole new face that does not exist, or by using a physical mask, i.e., fooling the system by the image, image series, or 3D silicone mask of the input subject. The physical attacks are named as ***spoofing*** attacks, which are presented to the system by physically demonstrating a face, hence they won't be examined any further in this study, however, comprehensive state-of-the art research have been published by [15–17]. On the other hand, the digital attacks, also named as ***perturbation*** attacks, manipulate the input face by manipulating attributes or the entire face digitally. Figure 1 shows some examples of spoofing attacks as well as some real and fake examples of attribute manipulation on face images.

There are many studies that use geometry-driven [19–23], morphing [24–26] or statistical [27–29] models to modify/synthesize a facial expression in the entire face. Being one of the first studies on face swapping in an image sequence, [30] doesn't require additional hardware. All it needs is a single-camera video to warp the source to the target face using a 3D multilinear model. [31,32] also use 3D modelling techniques to map the face or the mouth region to the target faces. [33] presents real-time facial reenactment between two image sequences applying lip sync between the source and the target. Another similar approach is adopted by [34] in which they use a Recurrent Neural Network (RNN) to learn the mapping

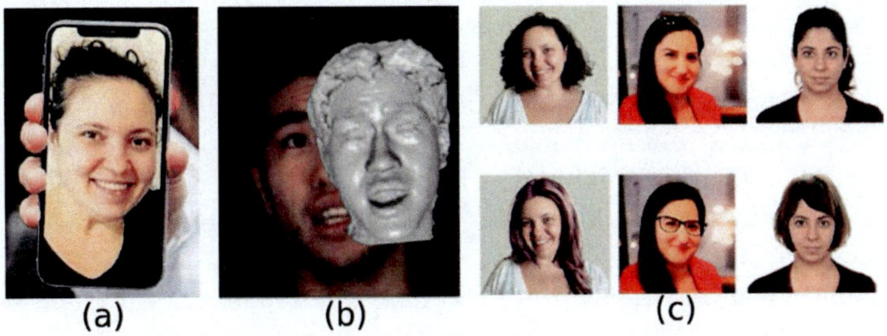

Fig. 1. (a) is a spoofing attack in which the forger presents the face through a picture (b) is a spoofing attack that can be done by printing the face as a physical mask. The image was retrieved from the BP4D dataset [18] (c) is a perturbation attack on facial attributes such as changing the hair color, glasses, or hair style.

of audio features to mouth shapes, and also apply 3D pose matching which creates more realistic results than the previous study. [35] and [36] are other recent 3D reconstruction studies that work on expression swapping.

Current manipulation and synthesis methods usually use the help of AutoEncoders [37] and deep learning techniques, especially variations of Generative Adversarial Networks (GANs) [38] to manipulate the viewpoint, resolution, attributes such as age, hair / skin / eye color of a face, or to generate a whole new one.

To defeat the challenge of training instability in a GAN, SNGAN was proposed by [39] which used spectral normalization to stabilize the training. Although the authors did not use their method in face synthesis, their work has been used in fake face generation studies [40]. Viewpoint inconsistency is also a major issue when it comes to face swapping. To this end, DR-GAN [41] was implemented to frontalize or rotate a face with an arbitrary pose to explicitly disentangle the face representation.

PGGAN (also known as ProGAN) is a milestone in entire face synthesis that is proposed by [42], which was then improved with StyleGAN [43]. PGGAN described a new training methodology by growing the generator and the discriminator progressively to speed up the training and increasing the stability to produce high quality images. Their next generation method StyleGAN improved the method by using bilinear up/downsampling operations and tuned hyperparameters. Their proposed architecture automatically learned high-level attributes and enabled automatic synthesis. Some other face synthesis algorithms are CramerGAN [44] and InterFaceGAN [45].

Facial manipulation can also be accomplished by attribute manipulation or expression swapping. One popular technique is StarGAN [46], which proposed an approach that can learn image-to-image mappings for multiple domains within a single network. Another popular method that has been used in expression swapping was the Neural Textures [47].

3 Geometry Based Detectors

Considering their impact on social media, partial and full face manipulation by GANs is a major public concern. While there are so many areas that can produce and use fake faces, the algorithms that can detect these fake faces or locate the image forgery are also quite in demand. Manipulations, specifically on face images, are becoming so advanced that they nearly leave almost no clues behind. Hence, studies that can detect such manipulations on facial synthesis started in the near future and continue to increase day by day.

Many different studies conducted in recent past are able to detect whether a face is fake or not by examining pixel by pixel data with various methods, usually without using the whole face. These methods roughly address whether the overall geometrical structure of the image is distorted or not. Properties such as texture integrity, brightness, and pixel values are expected to be similar between changed regions.

Geometry-based forgery can be done with several techniques. The first of these is by combining certain parts of two or more images into a single image. It is very difficult to be perceived by the human eye since the toning and colors are the same, but there are many studies that can determine this. One other technique is to combine specific regions from two or more images into a new image by manipulating the scales (such as expanding/shrinking, lengthening/shortening). In this process, the similarity values between neighboring pixels are calculated and resized on the new plane.

In another type of geometric image forgery, the process is performed where the data are brought close to each other by looking at certain statistical properties of the images. In this way, forgery is made on the image in a way that the human eye cannot detect, yet again some studies can detect these fake areas.

Table 1 presents most recent studies that can successfully detect fake images using several different geometry-based approaches. We considered the papers according to what each study is based on, their classifiers, datasets for real images, and fake dataset generation models. Since almost every study uses different evaluation metrics, they were not included in this table to make a fair comparison.

In their study, [55] pointed out the imperfection of upsampling methods on face manipulation techniques based on GANs. They also introduced a new approach that used a gray-scale fakeness map to get more information about the fake parts of the image. They made a comparison with several different methods and observed that they were more successful in various metrics. Another study that relied on fake locations, named "Face X-ray [56], tried to detect face swap images by locating the fake regions. They tried to use the advantages of noise and error level analysis to find these fake regions in an image which are commonly very large and relatively easy to locate. To test the model, they mostly used face swapping videos. On the other hand, when the image was completely synthetic, this approach had some lacking parts.

Matern et al. [51] proposed a method that uses some visual aspects, such as the eye region, the eye color, or missing details around the eye or the teeth to

Table 1. Geometry based detectors

Study	Based on	Classifier(s)	Dataset(s)	Fake dataset Generator model(s)
Li et al. (2018) [48]	Disparities in color components	SVM	CelebA, CelebA-HQ, LFW	DFC-VAE, DCGAN, WGAN-GP, PGGAN
Li et al. (2018) [49]	Limited resolutions, Distinctive artifacts	CNN	UADFV, DeepFakeTIMIT	VidTIMIT, Youtube
Yang et al. (2019) [50]	Locations of facial landmark points	SVM	CelebA	PGGAN
Matern et al. (2019) [51]	Segmentation, Facial landmarks	KNN, Neural Network, Logistic Regression	Unpublished	PGGAN, Glow, Youtube
He et al. (2019) [52]	Chrominance components in multi color spaces	CNN, Random Forest	CelebA	PGGAN
Nataraj et al. (2019) [53]	Co-occurance matrix on color channels	CNN	Unpublished	CycleGAN, StarGAN
Wang et al. (2019) [54]	Similarity between the morphed andthe original facial images	SVM	CelebA, CelebA-HQ	StyleGAN
Huang et al. (2020) [55]	Semantic Segmentation Map	Classifier does not affect theprediction, thus not needed	CelebA, FFHQ	STGAN, StarGAN, StyleGAN, AttGAN, IcGAN, PGGAN, StyleGAN2
Li et al. (2020) [56]	Self-supervised data generation	Neural Network	FaceForensics++	DFD, DFDC, Celeb-DF
Liu et al. (2020) [57]	Texture Statistics	CNN	CelebA, CelebA-HQ, FFHQ	StyleGAN, PGGAN, DRAGAN, DCGAN, StarGAN

detect fake images. They considered logistic regression and Multi-Layer Perceptron (MLP) as their classifiers. Their results indicated that the approach has 85.1% AUC for the MLP classifier system and with a 86.6% of accuracy, this approach can identify deepfakes. However, they used a private database to test the model, which makes it hard to compare their results with others'.

In [48], the authors used statistical differences and mismatched color profiles to separate GAN-generated images from real images. For the experiments, they used many types of generative models and conducted that whether the generative models were available or not, the proposed method successfully separated fake images from real images. Being another study working with colors, [53] proposed an approach based on both co-occurrence matrices on color channels and Convolutional Neural Network (CNN) frameworks to detect fake images. The authors performed experiments on two different GAN architectures, CycleGAN [58] and StarGAN [46] and claimed that their proposed detection system was found effective and generalizable.

Rather than working on specific regions, a detection method that relies on facial landmarks was proposed by [50]. They mentioned that even though GAN-based fake face generator algorithms can produce a real-like face with many actualist characteristics; landmark locations on the face like nose, mouth or tips of the eyes are distinguishing features that have a confirming role when it comes to verifying a fake image. Therefore, they suggested that a Support Vector Machine (SVM) can be trained by using these landmark point locations.

A promising approach designed by Li and Lyu [49] aimed to detect deepfake videos by training a CNN with synthetic and real images. The experiments that they conducted have a success rate between 84% to 99%. Even though they employed a deep approach, their aim was to detect the boundaries and their artifacts. [52] also applied an ensembling approach via CNNs using multi-color spaces as an input for manipulation detection. They used all the images available in the CelebA dataset and 10000 fake images generated by PGGAN. The authors compared their approach with the state-of-the-art methods and demonstrated that for almost every case, their approach outperformed the others by having an accuracy of 99% and above. More recently, [57] carried out a new system, called Gram-Net, for robust fake image detection. Experimental results on 800 images from different GAN datasets outperformed the previously mentioned detector [53] and ResNet-18 [59].

4 Deep Detectors

There are many applications that use Deep Neural Networks (DNNs) to detect fake faces or locate the synthesized regions. This section will provide a list of the studies that use DNNs to detect synthesized faces and some of their methods in detail. Table 2 contains the methods they combined to detect the forgery, deep network types, their training datasets, and their generated datasets - if any.

The research of [62] focused on training and blending parameters to generate low and high quality faces and publish a publicly available dataset. Another contribution of their study was to analyze the vulnerability of face recognition systems when they are exposed to fake faces. Another research that centered upon the vulnerability of detection systems [63] suggested a deep metric learning to make a finer detection. They heuristically suggested that the less the number of new images, the better the performance of detecting new types of generated images.

A deep forgery discriminator (DeepFD) was proposed by [60] attract attention to the difficulty of using a binary classifier on forgery detection. Their extended work [67] introduced a network for identifying the fake images using contrastive loss and reduced DenseNet. The pairwise information coming from a two-streamed network was used as an input for the training process, which they called the Common Fake Feature Network (CFFN). Their network improved the

Table 2. Deep detectors

Study	Based on	Classifier(s)	Dataset(s)	Fake dataset Generator model(s)
Hsu et al. (2018) [60]	DeepFD Random Forest Classifier (RFC) SVM Linear Classification (LC)	CNN	CelebA	DCGAN WGAP WGAN-GP LSGAN PGGAN
Tariq et al. (2018) [61]	ShallowNetV1 ShallowNetV2 ShallowNetV3	CNN	CelebA PGGAN Human created	Datasets are used as they are,human created images are made by Adobe Photoshop
Korshunov and Marcel (2019) [62]	VGG Facenet-based face recognition algorithms	LSTM	DeepFakeTIMIT	faceswap-GAN
Xuan et al. (2019) [63]	Multi-class classification network architecture based on deep metric learning	SRM Resnet18	CelebA	BEGAN DCGAN Glow PGGAN StyleGAN IntroVAE WGAN-GP
Fu et al. (2019) [64]	Gaussian Filter High-pass filter SGD	Dual-Channel CNN	CelebA-HQ PGGAN	Datasets are used with some pre-processing operations
Tariq et al. (2019) [65]	FakeFaceDetect Xception DenseNet	Shallow CNN	CelebA PGGAN	Datasets are used as they are
Zhuang and Hsu (2019) [66]	Common Fake Features (CFF) Learning Method	Coupled Deep Neural Network (CDNN)	CelebA	DCGAN WGAP WGAN-GP LSGAN PGGAN
Hsu et al. (2020) [67]	Common Fake Feature Network(CFFN)	CNN	CelebA	DCGAN WGAP WGAN-GP LSGAN PGGAN

performance by investigating cross-layer information and generalized the detection of deepfakes. Another pairwise study was proposed by [66], which hosted a two-step pairwise learning approach to learn common fake features over the dataset. In their model, the triplet loss was employed to find the relation between real and fake images, followed by learning the discriminative features.

Pre-processing techniques have shown to improve the quality of fake face detection systems. [64] applied a pre-processing algorithm on the images to improve detection against some widely used post-processing operations. [61] also

Table 3. Synthesized face datasets

Dataset	Model(s) Used in Generation	Source	# of Data		Data Type(s)	Resolution	Attributes	Year
			Real	Fake				
CelebA-HQ [42]	PGGAN [42]	Celeb-A [68]	-	30,000	Image	1024×1024	40	2017
DeepfakeTIMIT [69]	CycleGAN [58] and FaceNet [5]	VidTIMIT [70]	-	640	Video	64×64 and 128×128	2	2018
CelebDF [71]	swapping faces with improved DeepFake	Youtube	590	5,639	Video	256×256	3	2019
DFDC [72]	-	Crowdsourcing	1,131	4.113	Video	Multi resolution	3	2019
FaceForensics++ [73]	Face2Face, FaceSwap, DeepFakes, NeuralTextures	Youtube	1000	4000	Video	Mostly 640×480	-	2019
FFHQ [43]	StyleGAN [43]	flickr	-	70,000	Image	1024x1024	40	2019
UADFV [50]	DNN model with FakeAPP [74]	Youtube	49	49	Video	294×500	-	2019
DFFD[75]	FaceApp[12], StarGAN[46]	CelebA, FFHQ	58,703+ 1000	240,336+ 3000	Image+ Video	-	68	2019
faceSwap-GAN [62]	FaceSwap	VidTIMIT [70]	320	640	Video	64×64, 128×128, 256×256 [76]	2	2019
iFakeFaceDB [76]	StyleGAN [43], GANprintR [76]	Online + Public databases	-	87,000	Image	224×224	-	2020

employed pre-processing to detect human created (such as Adobe Photoshop) as well as GAN-based face images in an end-to-end pipeline. A year later, the same team named their method as FakeFaceDetect [65] which was claimed to be robust against arbitrarily changed or removed metadata of the original images to eliminate human-hand interventions.

5 Synthesized Face Datasets

Whether it is a research paper or a Research & Development project, everyone will agree that data is the most important part, especially if it is of good quality with proper attributes. Large datasets of images or videos of numerous kinds have to be collected to achieve successful results. In face-related studies, the most important issue of collecting large data comes with the problem of consent. It is impossible to go out and take pictures of hundreds of human beings in an hour without their permission. Therefore, it is crucial to be able to produce good quality faces tagged with many attributes. These faces should not be repetitive as well as unpersuasive both for the human eye and the computer.

Fake face generation algorithms mentioned in the previous section usually publish their own synthetic data for the public use. They collect data from the internet or use pre-published datasets to synthesize their own. In these research papers, most, if not all use the CelebA [68] dataset which contains more than 200,000 number of face images of 10,177 unique identities of celebrities, all marked with 5 landmark locations and tagged with 40 binary attributes.

Most popular datasets, up to this date, that contain fake faces have been summarized in Table 3 to facilitate the jobs of researchers who work on Computer-Generated Imagery (CGI). Details of the datasets contain how they are collected, the type and amount of data they contain, the methods they use to produce the fake data, and their attributes have been given and sorted with respect to their release dates.

6 Related Issues

Fake face generation methods, deepfakes, fake face synthesizers, adversarial perturbation attacks... Though studies name these algorithms differently, they all come to the same end, and they all keep on succeeding like a race. New types of attacks continue to be developed since, as [42] defines, *"there is a long way to true realism"*. Each time a new attack comes up, a corresponding detector is also being implemented. However, there are some issues related to this race, and the root cause of this weakness has to be found and handled to be a generalized algorithm.

In The Wild Data: Most approaches for fake face generation and detection are focused on controlled environments, using the same level of brightness, occlusion, or compression level. Methods to manipulate or detect in-the-wild data (uncontrolled scenarios) should also be considered in adversarial trickery methods and their detectors.

Robustness: The new types of deepfake images are coming forward one after another, and the existing detection methods cannot (or may not) cope well with this fast-forward technology. Robust, more generalized, and previously unseen solutions need to be developed for new face manipulation attacks.

Quality Degradation: GAN fingerprints in the generated image make it easy to perform deepfake detection, at least still for now. Since the detector works pixel-by-pixel, or by the attribute, the more compressed the data is, the less successful the detector, while the devil is in the details. Image quality degradation decreases the performance of the detection systems. The solution must be prepared for all scenarios such as compression level, noise, blur, brightness, occlusion, etc.

Adversarial Vulnerability: The vulnerability of adversarial attacks are coming to an alarming stage, if not already. As mentioned by [77], this vulnerability comes in two different ways: ***1. Believability:*** The quality of the manipulated faces are becoming more and more realistic, hence humans are easily fooled by these outputs, which is the root cause of the generator algorithms; ***2. Accessibility:*** Since the manipulation algorithms are popular, the implementors serve their software to the end-user, and the end-users can easily access these methods served through a web page or a mobile app. In 2018, FakeApp required large amounts of images belonging to the subject to make a simple manipulation, however since the public used this popular app to generate involuntary pornography, it was banned for a while due to the violation. The following year, Zao came up with a new mobile app, which only required a few selfie images, as low as 6. In late 2019, a ground-breaker algorithm was implemented by Samsung researchers, called the Living Portraits [78] which required a single image to master, so to speak, the puppet image. Although the last one still lacks the talent of fooling the human eye, it will get better at some point.

To surpass all these issues, and more, 2 major challenges have been started by NIST and Facebook. The first one by NIST (National Institute of Standards and Technology) is the Media Forensics challenge, started in 2018. The second one is the global Deepfake Detection Challenge (DFDC) which has been started in 2019 by Facebook's AI Red Team [72]. Each year they publish datasets containing manipulated faces, and the competitors are expected to gain the highest fake face detection and localization accuracy. Although these are the only challenges organized so far, there are many datasets containing fake faces that are prepared for the public use that are mentioned in the previous section.

7 Conclusion

Computer vision has had prosperous success over the past decade on synthesizing fake images and detecting those adversarial attacks. For the media forensics departments, the first thing to look for is the camera fingerprints of the image such as optical lens, compression, color filter, frame rate, etc. However, the manipulation techniques have been generating tremendously successful outputs to surpass these "physical" obstacles. The problem has been shifted from detecting the camera fingerprints to geometry based abnormalities and hidden GAN fingerprints. While the accessibility of creating deepfakes is increasing, although the geometrical abnormality is still widely visible to the human eye, it has been observed that the GAN-generated faces are detected better for the reason that GAN fingerprints make it easier to detect deepfakes.

References

1. Viola, P., Jones, M.J.: Robust real-time face detection. Int. J. Comput. Vision **57**(2), 137–154 (2004)
2. Tao, H., Huang, T.S.: Explanation-based facial motion tracking using a piecewise bezier volume deformation model. In: Proceedings. 1999 IEEE Computer Society Conference on Computer Vision and Pattern Recognition (Cat. No PR00149), vol. 1, pp. 611–617. IEEE (2019)
3. Cao, K., Rong, Y., Li, C., Tang, C., Change Loy, C.: Pose-robust face recognition via deep residual equivariant mapping. In: Proceedings of the IEEE Conference on Computer Vision and Pattern Recognition, pp. 5187–5196 (2018)
4. Taigman, Y., Yang, M., Ranzato, M., Wolf, L.: Deepface: closing the gap to human-level performance in face verification. In: Proceedings of the IEEE Conference on Computer Vision and Pattern Recognition, pp. 1701–1708 (2014)
5. Schroff, F., Kalenichenko, D., Philbin, J.: Facenet: a unified embedding for face recognition and clustering. In: Proceedings of the IEEE Conference on Computer Vision and Pattern Recognition, pp. 815–823 (2015)
6. Kumar, N., Berg, A.C., Belhumeur, N., Nayar, S.K.: Attribute and simile classifiers for face verification. In: 2009 IEEE 12th International Conference on Computer Vision, pp. 365–372. IEEE (2009)
7. Zheng, W., Tang, H., Lin, Z., Huang, T.S.: A novel approach to expression recognition from non-frontal face images. In: 2009 IEEE 12th International Conference on Computer Vision, pp. 1901–1908. IEEE (2009)
8. Manjani, I., Tariyal, S., Vatsa, M., Singh, R., Majumdar, A.: Detecting silicone mask-based presentation attack via deep dictionary learning. IEEE Trans. Inf. Forensics Secur. **12**(7), 1713–1723 (2017)
9. Hadid, A., Evans, N., Marcel, S., Fierrez, J.: Biometrics systems under spoofing attack: an evaluation methodology and lessons learned. IEEE Signal Process. Mag. **32**(5), 20–30 (2015)
10. Tolosana, R., Gomez-Barrero, M., Busch, C., Ortega-Garcia, J.: Biometric presentation attack detection: beyond the visible spectrum. IEEE Trans. Inf. Forensics Secur. **15**, 1261–1275 (2019)
11. Zao: https://zaodownload.com/. Accessed 25 Jun 2020
12. Faceapp: https://faceapp.com/app. Accessed 25 Jun 2020
13. Reflect: https://reflect.tech/. Accessed 25 Jun 2020
14. Snapchat: https://www.snapchat.com/. Accessed 25 Jun 2020
15. Jia, S., Guo, G., Xu, Z.: A survey on 3D mask presentation attack detection and countermeasures. Pattern Recogn. **98**, 107032 (2020)
16. Li, L., Correia, P.L., Hadid, A.: Face recognition under spoofing attacks: countermeasures and research directions. IET Biometrics **7**(1), 3–14 (2017)
17. Ramachandra, R., Busch, C.: Presentation attack detection methods for face recognition systems: a comprehensive survey. ACM Comput. Surv. (CSUR) **50**(1), 1–37 (2017)

18. Zhang, X., Yin, L., Cohn, J.F., Canavan, S., Reale, M., Horowitz, A., Liu, P., Girard, J.M.: Bp4d-spontaneous: a high-resolution spontaneous 3d dynamic facial expression database. Image Vision Comput. **32**(10), 692–706 (2014)
19. Pighin, F., Hecker, J., Lischinski, D., Szeliski, R., Salesin, D.H.: Synthesizing realistic facial expressions from photographs. In: ACM SIGGRAPH 2006 Courses, p. 19 (2006)
20. Nebel, J.C.: Soft tissue modelling from 3d scanned data. In: Deformable Avatars, pp. 85–97. Springer (2001)
21. Singh, K., Fiume, E.: Wires: a geometric deformation technique. In: Proceedings of the 25th Annual Conference on Computer Graphics and Interactive Techniques, pp. 405–414 (1998)
22. Zhang, Q., Liu, Z., Quo, G., Terzopoulos, D., Shum, H.-Y.: Geometry-driven photorealistic facial expression synthesis. IEEE Trans. Vis. Comput. Graph. **12**(1), 48–60 (2005)
23. Halder, S., Bhattacharjee, D., Nasipuri, M., Basu, D.K., Kundu, M.: Face synthesis (fasy) system for generation of a face image from human description, arXiv preprint arXiv:1005.4034 (2010)
24. Dai, Z.-l., Zhu, H.-Y., Zhang, S., Jia, J., Cai, L.-H.: Mpeg-4 based facial expression image morphing. J. Image Graph. **5** (2009)
25. Xudong, L.: Image morphing method for facial expression image and animation synthesis. Geomatics Inf. Sci. Wuhan Univ. **9**, 796–799 2007
26. Fu, T., Foroosh, H.: Expression morphing from distant viewpoints. In: 2004 International Conference on Image Processing, 2004, ICIP 2004, vol. 5, pp. 3519–3522. IEEE (2004)
27. Wang, X., Tang, X.: Face photo-sketch synthesis and recognition. IEEE Trans. Pattern Anal. Mach. Intell. **31**(11), 1955–1967 (2008)
28. Zhang, S., Gao, X., Wang, N., Li, J., Zhang, M.: Face sketch synthesis via sparse representation-based greedy search. IEEE Trans. Image Process. **24**(8), 2466–2477 (2015)
29. Liu, F., Zeng, D., Li, J., Zhao, Q.-J.: On 3d face reconstruction via cascaded regression in shape space. Front. Inf. Technol. Electron. Eng. **18**(12), 1978–1990 (2017)
30. Dale, K., Sunkavalli, K., Johnson, K., Vlasic, D., Matusik, W., Pfister, H.: Video face replacement. In: Proceedings of the 2011 SIGGRAPH Asia Conference, pp. 1–10 (2011)
31. Garrido, P., Valgaerts, L., Rehmsen, O., Thormahlen, T., Perez, P., Theobalt, C.: Automatic face reenactment. In: Proceedings of the IEEE Conference on Computer Vision and Pattern Recognition, pp. 4217–4224 (2014)
32. Garrido, P., Valgaerts, L., Sarmadi, H., Steiner, I., Varanasi, K., Perez, P., Theobalt, C.: Vdub: Modifying face video of actors for plausible visual alignment to a dubbed audio track. In: Computer Graphics Forum, vol. 34, no. 2, pp. 193–204. Wiley Online Library (2015)
33. Thies, J., Zollhofer, M., Stamminger, M., Theobalt, C., Nießner, M.: Face2face: real-time face capture and reenactment of rgb videos. In: Proceedings of the IEEE Conference on Computer Vision and Pattern Recognition, pp. 2387–2395 (2016)
34. Suwajanakorn, S., Seitz, S.M., Kemelmacher-Shlizerman, I.: Synthesizing obama: learning lip sync from audio. ACM Trans. Graph. (TOG) **36**(4), 1–13 (2017)
35. Liu, F., Zhu, R., Zeng, D., Zhao, Q., Liu, X.: Disentangling features in 3d face shapes for joint face reconstruction and recognition. In: Proceedings of the IEEE Conference on Computer Vision and Pattern Recognition, pp. 5216–5225 (2018)

36. Tran, L., Liu, X.: On learning 3d face morphable model from in-the-wild images. IEEE Trans. Pattern Anal. Mach. Intell. **43**, 157–151 (2019)
37. Kingma, D.P., Welling, M.: Auto-encoding variational bayes, arXiv preprint arXiv:1312.6114 (2013)
38. Goodfellow, I., Pouget-Abadie, J., Mirza, M., Xu, B., Warde-Farley, D.. Ozair, S., Courville, A., Bengio, Y.: Generative adversarial nets. In: Advances in Neural Information Processing Systems, pp. 2672–2680 (2014)
39. Miyato, T., Kataoka, T., Koyama, M., Yoshida, Y.: Spectral normalization for generative adversarial networks, arXiv preprint arXiv:1802.05957 (2018)
40. Yu, N., Davis, L.S., Fritz, M.: Attributing fake images to gans: Learning and analyzing gan fingerprints. In: Proceedings of the IEEE/CVF International Conference on Computer Vision (ICCV), pp. 7556–7566, October 2019
41. Tran, L., Yin, X., Liu, X.: Disentangled representation learning gan for pose-invariant face recognition. In: Proceedings of the IEEE Conference on Computer Vision and Pattern Recognition, pp. 1415–1424 (2017)
42. Karras, T., Aila,T., Laine, S., Lehtinen, J.: Progressive growing of gans for improved quality, stability, and variation, arXiv preprint arXiv:1710.10196 (2017)
43. Karras, T., Laine, S., Aila, T.: A style-based generator architecture for generative adversarial networks. In: Proceedings of the IEEE Conference on Computer Vision and Pattern Recognition, pp. 4401–4410 (2019)
44. Bellemare, M.G.,. Danihelka, I., Dabney, W., Mohamed, S., Lakshminarayanan, B., Hoyer, S., Munos, R.: The cramer distance as a solution to biased wasserstein gradients, arXiv preprint arXiv:1705.10743 (2017)
45. Shen, Y., Yang, C., Tang, X., Zhou, B.: Interfacegan: interpreting the disentangled face representation learned by gans, arXiv preprint arXiv:2005.09635 (2020)
46. Choi, Y., Choi, M., Kim, M., Ha, J.-W., Kim, S., Choo, J.: Stargan: unified generative adversarial networks for multi-domain image-to-image translation. In: Proceedings of the IEEE Conference on Computer Vision and Pattern Recognition, pp. 8789–8797 (2018)
47. Thies, J., Zollhöfer, M., Nießner, M.: Deferred neural rendering: image synthesis using neural textures. ACM Trans. Graph. (TOG) **38**(4), 1–12 (2019)
48. Li, H., Li, B., Tan, S., Huang, J.: Detection of deep network generated images using disparities in color components, arXiv preprint arXiv:1808.07276 (2018)
49. Li, Y., Lyu, S.: Exposing deepfake videos by detecting face warping artifacts, arXiv preprint arXiv:1811.00656 (2018)
50. Yang, X., Li, Y., Qi, H., Lyu, S.: Exposing gan-synthesized faces using landmark locations. In: Proceedings of the ACM Workshop on Information Hiding and Multimedia Security, pp. 113–118 (2019)
51. Matern, F., Riess, C., Stamminger, M.: Exploiting visual artifacts to expose deepfakes and face manipulations. In: 2019 IEEE Winter Applications of Computer Vision Workshops (WACVW), pp. 83–92. IEEE (2019)
52. He, P., Li, H., Wang, H.: Detection of fake images via the ensemble of deep representations from multi color spaces. In: 2019 IEEE International Conference on Image Processing (ICIP), pp. 2299–2303. IEEE (2019)
53. Nataraj, L., Mohammed, T.M., Manjunath, B., Chandrasekaran, S., Flenner, A., Bappy, J.H., Roy-Chowdhury, A.K.: Detecting gan generated fake images using co-occurrence matrices. Electron. Imaging **2019**(5), 1–532 (2019)
54. Wang, R., Juefei-Xu, F., Xie, X., Ma, L., Huang, Y., Liu, Y.: Amora: black-box adversarial morphing attack, arXiv preprint arXiv:1912.03829 (2019)

55. Huang, Y., Juefei-Xu, F., Wang, R.,. Xie, X., Ma, L., Li, J., Miao, W., Liu, Y., Pu, G.: Fakelocator: robust localization of gan-based face manipulations via semantic segmentation networks with bells and whistles, arXiv preprint arXiv:2001.09598 (2020)
56. Li, L., Bao, J., Zhang, T., Yang, H., Chen, D., Wen, F., Guo, B.: Face x-ray for more general face forgery detection. In: Proceedings of the IEEE/CVF Conference on Computer Vision and Pattern Recognition, pp. 5001–5010 (2020)
57. Liu, Z., Qi, X.,Torr, P.H.: Global texture enhancement for fake face detection in the wild. In: Proceedings of the IEEE/CVF Conference on Computer Vision and Pattern Recognition, pp. 8060–8069 (2020)
58. Zhu, J.-Y.,. Park, T., Isola, P., Efros, A.A.: Unpaired image-to-image translation using cycle-consistent adversarial networks. In: Proceedings of the IEEE International Conference on Computer Vision, pp. 2223–2232 (2017)
59. He, K., Zhang, X., Ren, S., Sun, J.: Deep residual learning for image recognition. In: Proceedings of the IEEE Conference on Computer Vision and Pattern Recognition, pp. 770–778 (2016)
60. Hsu, C.-C., Lee, C.-Y., Zhuang, Y.-X.: Learning to detect fake face images in the wild. In: 2018 International Symposium on Computer, Consumer and Control (IS3C), September 2018
61. Tariq, S., Lee, S., Kim, H., Shin, Y., Woo, S.S.: Detecting both machine and human created fake face images in the wild. In: Proceedings of the 2nd International Workshop on Multimedia Privacy and Security, ser. MPS 2018, pp. 81–87. Association for Computing Machinery, New York, NY, USA (2018). https://doi.org/10.1145/3267357.3267367
62. Korshunov, P., Marcel, S.: Vulnerability assessment and detection of deepfake videos. In: International Conference on Biometrics (ICB), vol. 2019, pp. 1–6 (2019)
63. Xuan, X., et al.: Scalable fine-grained generated image classification based on deep metric learning. arXiv preprint arXiv:1912.11082 (2019)
64. Fu, Y., Sun, T., Jiang, X., Xu, K., He, P.: Robust gan-face detection based on dual-channel cnn network. In: 2019 12th International Congress on Image and Signal Processing, BioMedical Engineering and Informatics (CISP-BMEI), pp. 1–5, October 2019
65. Tariq, S., Lee, S., Kim, H., Shin, Y., Woo, S.: Gan is a friend or foe?: a framework to detect various fake face images. In: Proceedings of the 34th ACM/SIGAPP Symposium on Applied Computing, pp. 1296–1303, April 2019
66. Zhuang, Y., Hsu, C.: "Detecting generated image based on a coupled network with two-step pairwise learning. In: IEEE International Conference on Image Processing (ICIP), vol. 2019, pp. 3212–3216 (2019)
67. Hsu, C.-C., Zhuang, Y.-X., Lee, C.-Y.: Deep fake image detection based on pairwise learning. Appl. Sci. **10**, p. 370, January 2020
68. Liu, Z., Luo, P., Wang, X., Tang, X.: Deep learning face attributes in the wild. In: Proceedings of the IEEE International Conference on Computer Vision, pp. 3730–3738 (2015)
69. Korshunov, P., Marcel, S.: Deepfakes: a new threat to face recognition? assessment and detection, arXiv preprint arXiv:1812.08685 (2018)
70. Sanderson, C., Lovell, B.C.: Multi-region probabilistic histograms for robust and scalable identity inference. In: International Conference on Biometrics, pp. 199–208. Springer, (2009)
71. Li, Y., Yang, X., Sun, P., Qi, H., Lyu, S.: Celeb-df: a large-scale challenging dataset for deepfake forensics. In: Proceedings of the IEEE/CVF Conference on Computer Vision and Pattern Recognition, pp. 3207–3216 (2020)

72. Dolhansky, B., Bitton, J., Pflaum, B., Lu, J., Howes, R., Wang, M., Ferrer, C.C: The deepfake detection challenge dataset, arXiv preprint arXiv:2006.07397 (2020)
73. Rossler, A., Cozzolino, D., Verdoliva, L., Riess, C., Thies, J., Nießner, M.: Faceforensics++: learning to detect manipulated facial images. In: Proceedings of the IEEE International Conference on Computer Vision, pp. 1–11 (2019)
74. Fakeapp: https://www.malavida.com/en/soft/fakeapp/. Accessed 25 Jun 2020
75. Dang, H., Liu, F., Stehouwer, J., Liu, X., Jain, A.K.: On the detection of digital face manipulation. In: Proceedings of the IEEE/CVF Conference on Computer Vision and Pattern Recognition, pp. 5781–5790 (2020)
76. Neves, J.C.,. Tolosana, R., Vera-Rodriguez, R., Lopes, V., Proenca, H.P., Fierrez, J.:Ganprintr: improved fakes and evaluation of the state of the art in face manipulation detection. IEEE J. Sel. Top. Signal Process. **14**, 1038–1048 (2020)
77. Kietzmann, J., Lee, L.W., McCarthy, I.P., Kietzmann, T.C.: Deepfakes: trick or treat? Bus. Horiz. **63**(2), 135–146 (2020)
78. Zakharov, E., Shysheya, A., Burkov, E., Lempitsky, V.: Few-shot adversarial learning of realistic neural talking head models. In: Proceedings of the IEEE International Conference on Computer Vision, pp. 9459–9468 (2019)

Audio to Video: Generating a Talking Fake Agent

Duygu Cakir(✉) and Özge Yücel Kasap

Bahcesehir University, Istanbul, Turkey
{duygu.cakir,ozge.yucelkasap}@eng.bau.edu.tr

Abstract. People have been interested in creating avatars of their own, to expose themselves in an exaggerated version or to hide themselves. From time to time, it is for the reason that they are embarrassed of their visual appearance or just because they just don't want to be seen. However, there are certain times that the other side on the network wants to see a real-looking talking person. Especially with the rise of using online services, particularly online education, it has become more important to engage the user to the system, more than ever. This paper will propose an end-to-end pipeline on the way to controlling a single fake image (a face that does not exist) by using an audio as the input, and propose a chain of methods to create a Talking Fake Agent (TFA), that is meant to "puppet" the image through a voice with no real face in front of it. As a result, it is strongly believed that it will increase the users' engagement to the system and increase their overall satisfaction.

Keywords: Face synthesis · Voice profiling · Audio to face · Lip sync · Virtual agents

1 Introduction

This year has been a huge breakdown in the 21st century, the pandemic has taught us a lot about our lifestyles, social and professional relations, family affairs, educational consistency, and most important of all, taking care of our health. It has also made us find alternative solutions effectively and act quicker, meanwhile trying to find a long-term settlement since it is not certain how long this will last and how much it will affect our lives. Even if it ends, we may never work, study, or shop the way we did before. One thing we are sure of is the advantage of an online system will never decrease anymore, the need for it will grow each day. This brings up a few questions: Is it possible to engage more users to online systems? Is it possible to work with less people on site preserving their social distances, without having to let go of employees? Is it possible to work more from home and less from the office?

Studies have shown that using an avatar (i.e. a self-representing online character) can help the user get engaged with the system and not leave too early [1–3]. Apart from engaging users to the system, an avatar also has the effect of

T. Allahviranloo et al. (Eds.): IDS 2020, AISC 1301, pp. 212–227, 2021.
https://doi.org/10.1007/978-3-030-66501-2_17

changing the behaviour of a user. If the user chooses an attractive avatar, he becomes friendlier [4]. All the research that we've come up with investigates the usage of computer-drawn avatars, but what if the avatar looks like a real person? Not the exact same person, but a real-looking version of the speaker?

A system that employs a real-looking virtual / digital assistant can be online no matter where the service is given, whether it is a contact center office or the agent's home. A company that uses this kind of online customer service can be able to lower the number of seats - not the employees - since the speaker only needs a high quality microphone and a stable internet connection. It can also be used in online education to replace the visuals of the teacher or the student. It has been observed that the *zoom fatigue* [5] students are lacking eagerness to open their webcams during a class [6], hence this makes them lose their attention and get engaged with other stuff. The evening news can be brought to the audience with respect to their choice or user habits (personalization), or a shopping assistant in the form of a desired face can help the customer buy more.

Since machine learning techniques have become easier to implement and databases have become widely accessible, there have been groundbreaking changes in the world of computer vision, especially with the implementation of Generative Adversarial Networks (GANs) [7], it has become easy to conduct GAN-based fun apps such as Snapchat [8], Reflect [9], or [10] however, it has also become easy to perform forgery attacks that create fake news [11–13] or political / financial statements [14]. There are also endless creative areas where people can use GANs, such as converting photos into famous artists' styles [15], swapping the look of animals between species [16], reconstructing colors from a B&W photo [17], etc... GANs can also be used to manipulate or synthesize a face in the forms of; 1. manipulating a facial attribute such as putting on glasses, changing the hair color, or aging [18], 2. modifying the facial expression such as putting on a smile or making the image cry [19], 3. swapping faces between two individuals [20], and 4. synthesizing an entire face, i.e., creating a face that does not exist at all, such as "create a young, bald, brown-eyed man with a mustache" [21,22].

This work presents an end-to-end pipeline to generate a non-existent talking face for the voice that is behind the scenes, named as the Talking Fake Agent (TFA). Throughout the pipeline, the voice will be decoded to its attributes, the attributes will generate a fake face, the fake face will be used to produce a long video, and finally the video will be the training set for the new incoming voice. With the inputs of the voice and the training set, the final output will be the fake face's lips that are synchronized with the voice. The rest of the paper will give the recent work on GAN-based face synthesis methods, voice profiling studies, and lip synching research followed by the proposed approach. Last but not least, the conclusion section will evaluate the issues on the proposed pipeline and give hints on further studies that can be done on this area.

2 Related Work

2.1 GAN-Based Face Synthesis

GAN models, without a doubt, are one of the most important developments in the field of deep learning introduced by Goodfellow et al. [7]. These models have the ability to produce synthetic but highly realistic images and form the basis of many adversarial attack studies in the literature [23–27]. The GAN models have two deep networks called the Generator, which is responsible for producing fake images and the Discriminator, which tries to distinguish the given real images from these fake ones. These are like opposite working players. The learning process ends with the discriminator perceiving the fake image as real.

There are many diverse application areas that GANs can be used in, such as music generation [28], text to image generation [29], image to image generation [30], and video generation [31]. GANs that are used to synthesize fake faces (*deepfakes*), achieve results that seem almost real and indistinguishable by the human eye [32–36].

As aforementioned, the main aim of GANs is to generate images from the given input image while generator and discriminator are trained to minimize their losses. Designed by [37], Progressive Growing of GANs (PGGAN) can generate high resolution photo-realistic fake faces up to 1024 $\times$ 1024 pixels. The training process of PGGAN starts with images having a low resolution (such as 4 $\times$ 4 pixels) and progressively the resolution is increased by addition of new layers to the generator and the discriminator. The Style Generative Adversarial Network (StyleGAN), was introduced by [38] and was not just capable of generating high-quality fake faces but also presented a way of controlling over the style of the generated image by changing the architecture of the generator. Along with increasing the training performance, [39] identified and fixed numerous image quality problems in StyleGAN and announced StyleGAN2 as its successor.

Face2Face was probably one of the most famous face generation techniques that had been implemented by [40], in which it transmitted one person's facial expressions to another person while retaining the identity of the target person. One feature that distinguishes Face2Face from other methods is that it performed the transfer process in real time. Another encouraging method named FaceSwap [41] took the facial region from the source video and transferred it to a target video. It was a graphics-based approach where the face region was extracted by detecting facial landmarks. A 3D head model was used as a template to compose the model using these landmarks, then the composed 3D head model was reflected back to the target image. Being one of the many other commercial apps, FaceApp [42] used distinctive GANs to modify facial attributes such as aging or inserting facial accessories.

Produced from the words "deep" as in deep learning and "fake" as in fake face, the term Deepfakes [10] became a synonym among face transfer techniques. As it was used in several areas like celebrity fake pornographic videos, fake news, scams, and also to generate financial fraud in accounting [43].

A method that is not meant to generate fake faces, CycleGAN [28] addressed the problem of the training data coming in pairs which causes a lot of effort to work with. Without explicit pairs of images, it implements image translations by using two generators and a discriminator called cycle consistency. Surely, soon after the model was proposed, it was applied to synthesize non-existent faces [44]. A technique named StarGAN [34] was implemented to translate images from one domain to the other. Unlike other GANs, the discriminator had two tasks: in addition to identifying if an image is fake or real, it also predicted the domain of the given input image. It was even used for emotion generation and recognition [45]. In NeuralTextures [46], to learn the neural texture of the target person, they used the original data from the input video. Together with photo-metric reconstruction loss, adversarial loss was also used in the training part.

Considering that different GAN models and their diverse set of application areas are increasing day by day, it would not be wrong to say that GAN-based fake face generation will increase in the same way too.

2.2 Voice Profiling

Our voice, as well as being a channel for conveying a person's deepest thoughts and feelings, is also a tool that gives a person's characteristic features. During the speech, not only words, but also the speaker's profile prints (i.e. attributes) such as identity [47], age [48], gender [49] or even the mood [50] is transmitted. Voice profiling aims to extract meaningful information from the input voice to generate a list of its attributes. If a human can hear the voice and match the face belonging to that voice [51,52], a computer also can. This section provides a summary of several studies that try to extract attributes of an audio clip to create a profile for the voice, which will be the input for the face.

In a relatively early age and gender estimation study, [53] proposed a method that used Weighted Supervised Non-Negative Matrix Factorization (WSNMF) and General Regression Neural Network (GRNN) in a hybrid architecture. [54] proposed a new cost function for the same task by fine-tuning the age and gender Deep Neural Networks (DNN) jointly to optimize network parameters. Another one [50] used DNNs to identify age and emotion; and used identity-Vectors (I-Vectors) [55] on top of a DNN for gender, language, and accent classification.

An image can be generated from textual data [56], such as "*the small bird with a pink breast*" can be reconstructed to an image as described in the text, so why not create a face from attributes such as "*a Caucasian man in his early 40s*" ? There are studies not only focusing on extracting the attributes, but also trying to reconstruct the face behind the voice. A recent research [57] used GANs with two discriminators: the first one to evaluate if the generated *image* is real-looking or not, the second one to see if the generated *identity* matches the actual identity. [58] proposed the same result with a different approach and named their framework *Speech2Face*, this time the model being VGG-Face to extract face features. One recent study [59] followed a completely different path before generating the face. Their proposed approach employed a network to define the geometry of the face and another one to use the landmark information to reconstruct the face. [60] trained their whole system with a conditional GAN using an aligned audio and video channel. Existing datasets did not have enough high quality for their proposed method, hence they also collected data from Youtube to build up their own dataset which contains high quality audio-visual signals.

Children are a whole different case when it comes to identification, recognition, and verification problems, hence their datasets are designed and collected only for their problem space. Being one of the very few, [61] presented a child emotional speech corpus *EmoChildRu* to investigate the emotional state of a child as well as his gender and age (in months). [62] is another study that automatically analyzed the emotions of children from their voice. Both studies used a simpler yet strong classifier (SVM) in their studies.

2.3 Speech-Driven Lip Synching

There are many researches to puppet the lips of a subject with an incoming audio, whether it's the subject speaking or a voice belonging to a complete stranger. They all have a few things in common though: 1. if they are mapping video to video, they need to train their dataset with a "satisfactory amount of time" -long footage of a real, existing person; 2. almost all of them change the mapping of the mouth region, not the whole face. This section will provide some recent studies which claim to have successful outputs on lip synching.

Long short-term memory (LSTM) networks [63] have become a popular neural network type (to be more specific, a Recurrent Neural Network type) on problems with given time series. Hence, it was inevitable to use LSTMs for image sequence problems. [64] employed an attention-based bidirectional LSTM to extract information from speech, and not only generated lip movements but also generated temporal transitions of the lips. In their broad spectrum method proposed in [65], the system reconstructed the 3D face (not just the mouth region) with the audio input using a DNN. While being able to represent only the neutral face is a problem by itself, they found a solution to represent emotions. [66,67] have also used audio to video conversion in their studies.

In their work, [88] used the same speaker's voice and video in training, and tried to sync lip movements with the same speaker's new-coming voice while preserving temporal consistency. Being a more dynamic and challenging study, [89] both the source and the target could be an arbitrary speaker as well as the voice that is going to be synchronized. Using a 3D face reconstruction, they factorized each image in the image sequence, then the source audio was translated into expression parameters which decoded the expression, geometry, and pose of the speaker. Finally, the mouth region landmarks were projected from the reconstructed 3D face to the target using a GAN-based network for the texture details. Their competitor [90] used the same approach, however at the last step, insead of GANs, they used a weighted median texture for the mouth texture. [91] also used videos for the source and the target, however, their DNN-based method not only copied the mouth region but also took full control over the target face

Table 1. Audio Visual Datasets

Dataset	Type		Attributes	Year
	Audio	Video		
AVOZES [68]	x	x	Age, gender, education, height, weight, native language of speaker, some parental information medical conditions and etc.	2004
Vid-TIMIT [69]	x	x	Head rotation	2004
AVICAR [70]	x	x	Gender	2004
GRID [71]	x	x	Gender, age, accent	2006
IV2 [72]	x	x	Age class, color of skin, color of eyes, color of hair, and presence of beard, mustache and glasses	2008
TED-LIUM [73]	x	x	Gender	2012
MOBIO [74]	x	x	Gender, nativeness	2012
LibriSpeech [75]	x	–	Gender	2015
VoxCeleb1 [76]	x	x	Ethnicity, accent, profession, age, gender and nationality	2017
LRW [77]	x	x	-	
LRS2 [78]	x	x	-	2017
MODALITY [79]	x	x	Gender, nativeness, age	2017
Audio Set [80]	x	x	-	2017
VoxCeleb2 [81]	x	x	Ethnicity, accent, profession and age	2018
LRS3-TED [82]	x	x	-	2018
AVSpeech [83]	x	x	-	2018
Audio-Visual TED [84]	x	x	Name, occupation, short biography	2019
Common voice [85]	x	x	Age, gender, accent	2019
BVC Voice Dataset [86]	x	–	Gender, age, ethnicity	2019
VGGSOUND [87]	x	x	-	2020

such as transferring the pose, expression, and eye motion. Other studies that used videos for both the source and target are [40, 46].

Instead of mapping the input audio to a pre-recorded video, there are also some methods which create videos from a single still image. Being one, Speech2Vid [92] took the input audio and an image which the system had never encountered before, and generated a talking video of the image using CNNs. [93] also generated a talking video from a single image, however, this time it was driven by another video, not an audio. While they can work in real-time, the pre-training takes a large corpus of talking videos to be trained. Other audio-to-video methods that used a single photo to generate videos can be found in [94–96]. In their proposed method named X2face [97], the target sequence (or the driving frame) could both be a video or an audio to generate a video to puppet the source. Having everything all at once, they chose to leave out the pose and the emotion in their output video.

As can be seen from the related work given above, there many different audio-visual conversion systems that use DNNs. To bypass severe overfitting issues, such DNNs depend on a high quality dataset. There are many audio-visual datasets, including, but not limited to, the following in Table 1, which summarizes some large and small scale state-of-the-art audio-visual corpora.

3 Proposed Method

The proposed model employs four different methods in a sequence to convert an audio to a GAN-generated video. Although the methods may vary and can be replaced with better resulting studies, the overall structure remains the same. The main steps of the general pipeline is as follows:

1. Decode the attributes of the speaker
2. Generate a GAN-based non-existent face image with respect to these attributes
3. Create 10–15 h of video for the next training step
4. Puppet the GAN-face with the new audio coming from the speaker

The first three steps are necessary and time consuming, however these are only executed once and by the time they are finished, the training phase is completed and the real-time step begins. Figure 1 shows the overall outline of the proposed pipeline. The details of each step with a recommended method for each of them is explained in the following sub-sections.

3.1 Speech to Attributes

The initial stage of the algorithm starts with converting the speaker's voice to a GAN-generated face. In order to present a realistic image, the input voice has to be correlated with the output image, i.e., a young woman's voice should not be mapped to an elderly male. Figure 2 holds the outline in which the speaker's voice and an audio-visual dataset enters the system and the output is a set of attributes that belong to the voice. The voice dataset should be a set that holds labels such as age, gender, and ethnicity. For this part of the proposed algorithm, the recommended method to convert a voice to a set of attributes is Speech2Face [58] and the suggested dataset to be used in training is VoxCeleb [76].

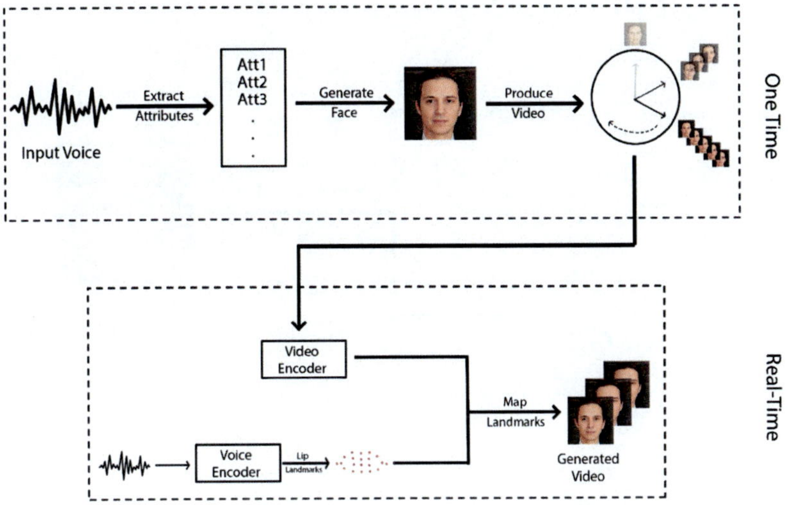

Fig. 1. The end-to-end TFA pipeline

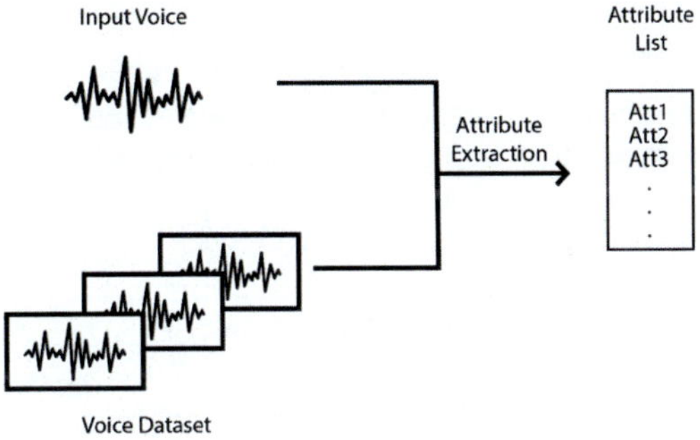

Fig. 2. Extracting attributes of the input voice

3.2 Synthesize a Face for the Voice

After retrieving a voice profile, a realistic face should be generated, which should match the input voice with a high level of believability to make the viewer engage. This part can be personalized according to the speaker's desires or the viewers habits, however it still needs to comply with certain attributes like age, gender, and ethnicity. Figure 3 shows a brief summary of this step. Since it holds real faces with the information of these attributes, the suggested dataset for this part would be CelebA [98]. A specific GAN-based method won't be recommended for this part, since there are hundreds (as of now) methods that can be used to synthesize a face with some selected attributes attached to it. The most popular methods can be found in Sect. 2.1.

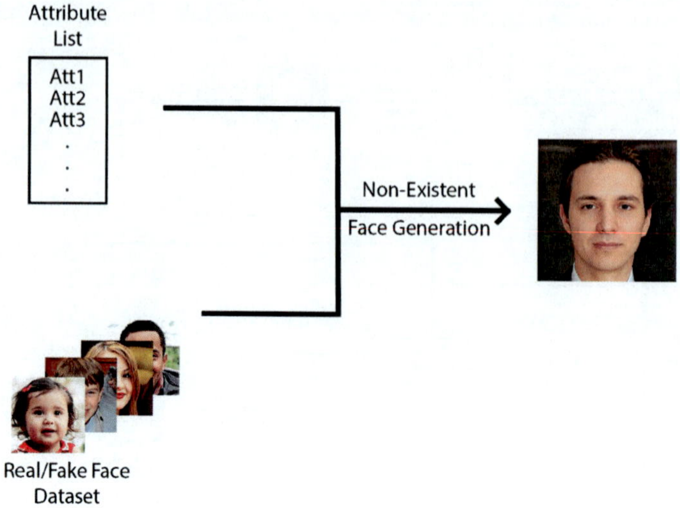

Fig. 3. Synthesizing a proper GAN-based face for the voice

3.3 Produce the Training Set

In their study, [90] have seen that the more videos they used in training, the more realistic lip movements they gained, therefore, only a single shot deepfake won't be enough to synchronize lip landmarks and the training set should be at least 10–15 h (or more). In order to get a 10-h-long video, the GAN-face should be augmented to a 10-h-long video of the speaker himself, and the system should transform the landmark information of the image sequence to the deepfake image to generate a long video to be the training set of the next step (Fig. 4). The ground-breaking researches [92, 93] which claim they can map only a few shots of source images, as few as one, is recommended for this step of the pipeline.

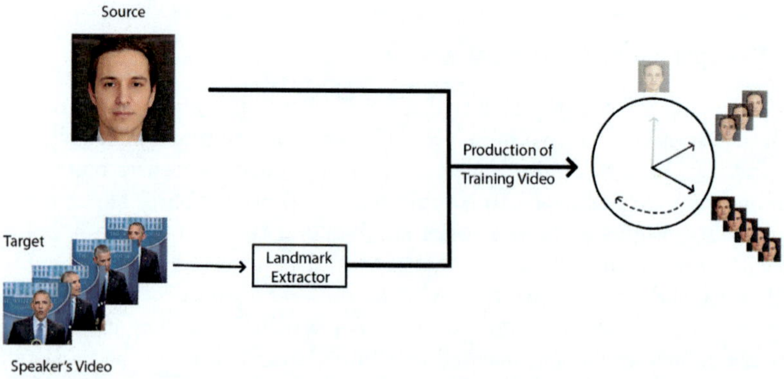

Fig. 4. Producing a series of videos from a single shot image

3.4 Generate the Virtual Agent

After generating a dataset of at least ten hours of video of the speaker, the time-consuming part for the speaker and the computer will be finished and the training video set will be ready for the real-time part. Being a recent study, [89] uses GANs for the mouth region textures and generates breath-taking results than most of the others. Therefore, the same model is recommended to be employed for the last part of this algorithm, in which the video and a new voice belonging to the speaker enters the system, the landmarks for the voice are extracted and then mapped to each frame on the videoto generate the final result: the real-looking, talking virtual agent (Fig. 5).

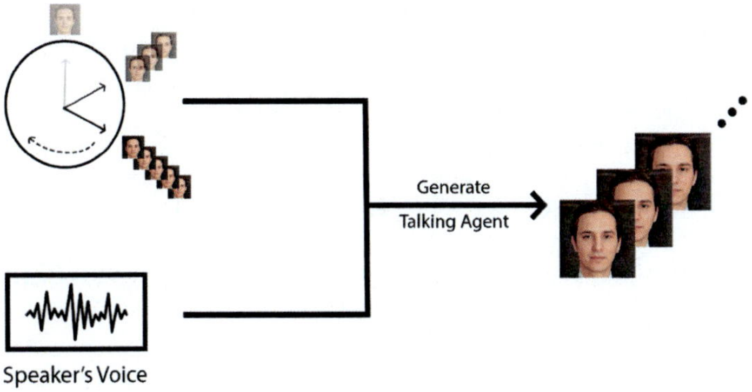

Fig. 5. Generation of the real-looking Fake Agent

4 Concluding Remarks

With the advancements in technology, especially GAN-based methods, it is becoming harder to fool the human eye with the synthesized images. These methods can be used to forge systems and fool human beings, but mostly, they are used for good, such as trying to increase the engagement with current online systems. In this paper, an end-to-end pipeline named TFA has been proposed to convert the voice of a person to a real-time video of a GAN-generated face. The limitations of TFA and further contributions are as follows:

- The TFA approach does not take emotional states into consideration. Until a decent "face-puppetting" work is implemented, emotions have to be left out. In order to leave any emotions outside, the generated video sequence for the first phase of training has to be recorded as if the speaker gives a political statement, i.e. no emotions and exaggerated expressions / mimics should be taped. Otherwise while the other side is feeling annoyed, the generated face can be expressing a happy emotion, which can destroy the whole purpose of the application.

- Even though it will only be once, constructing TFA will be time consuming and exhausting, since the initial training video sequence has to be approximately 10–15 h (maybe more) to get proper results, both machine and human-wise. The more high quality videos there are in the initial phase, the more convincing the output will be.
- For the reason that creating a real-time result will need a high computing power, an everyday user is assumed to not have the opportunity to use this type of application. However, with the increase in the power of computing in cloud technologies and the ease of their accessibility, it can be used by the single user in a very near future.

Taking the limitations into consideration, two contributions beyond TFA can be suggested for further studies; **1.** As mentioned above, this study does not investigate the emotional states on the voice, however, it would be nice to add emotional expressions into the problem. **2.** In the current pipeline of TFA, we extract the attributes of the current virtual agent from his voice and according to those attributes, generate a suitable face for it. Instead, the ability to choose the agent as desired for the user can be easily added to this method. In this way, instead of talking to a different person each time the user connects to the system, the users can manage the agent according to their liking, or, even better, automatically bring the agent that will connect them to the system with respect to their habits.

References

1. McMahan, A.: Immersion, engagement and presence. The Video Game Theory Reader **67**, 86 (2003)
2. Lee, H., Doh, Y.Y.: A study on the relationship between educational achievement and emotional engagement in a gameful interface for video lecture systems. In: International Symposium on Ubiquitous Virtual Reality. IEEE **2012**, 34–37 (2012)
3. Mahyar, N., Kim, S.-H., Kwon, B. C.: Towards a taxonomy for evaluating user engagement in information visualization. In: Workshop on Personal Visualization: Exploring Everyday Life, vol. 3, p. 2 (2015)
4. Yee, N., Bailenson, J.: The proteus effect: the effect of transformed self-representation on behavior. Human Commun. Res. **33**(3), 271–290 (2007)
5. Zoom fatigue' is real. here's why you're feeling it, and what you can do about it. Accessed 20 Aug 2020. https://news.northeastern.edu/2020/05/11/zoom-fatigue-is-real-heres-why-youre-feeling-it-and-what-you-can-do-about-it/
6. Kozar, O.: Perceptions of webcam use by experienced online teachers and learners: a seeming disconnect between research and practice. Comput. Assisted Lang. Learn. **29**(4), 779–789 (2016)
7. Goodfellow, I., Pouget-Abadie, J., Mirza, M., Xu, B., Warde-Farley, D., Ozair, S., Courville, A., Bengio, Y.: Generative adversarial nets. In: Advances in Neural Information Processing Systems 2014, pp. 2672–2680
8. "Snapchat," https://www.snapchat.com/. Accessed 25 June 2020
9. "Reflect," https://reflect.tech/. Accessed 25 June 2020
10. "Deepfakes web," https://deepfakesweb.com/. Accessed 25 June 2020

11. Botha, J., Pieterse, H..: Fake news and deepfakes: a dangerous threat for 21st century information security. In: ICCWS 2020 15th International Conference on Cyber Warfare and Security. Academic Conferences and publishing limited, 2020, p. 57 (2020)
12. Tariq, S., Lee, S., Kim, H., Shin, Y., Woo, S.S.: Gan is a friend or foe? a framework to detect various fake face images. In: Proceedings of the 34th ACM/SIGAPP Symposium on Applied Computing, pp. 1296–1303 (2019)
13. Zhou, X., Zafarani, R.: Fake news: a survey of research, detection methods, and opportunities, arXiv preprint arXiv:1812.00315 (2018)
14. Zhang, X., Ghorbani, A.A.: An overview of online fake news: characterization, detection, and discussion. Inf. Process. Manage. **57**(2), 102025 (2020)
15. Liu, H., Michelini, P.N., Zhu, D.: Artsy-gan: a style transfer system with improved quality, diversity and performance. In: 2018 24th International Conference on Pattern Recognition (ICPR). IEEE, pp. 79–84 (2018)
16. Choi, Y., Uh, Y., Yoo, J., Ha, J.-W.: Stargan v2: diverse image synthesis for multiple domains. In: Proceedings of the IEEE/CVF Conference on Computer Vision and Pattern Recognition, pp. 8188–8197 (2020)
17. Zhu, J.-Y., Park, T., Isola, P., Efros, A.A.: Unpaired image-to-image translation using cycle-consistent adversarial networks. In: 2017 IEEE International Conference on Computer Vision (ICCV) (2017)
18. Yang, H., Huang, D., Wang, Y., Jain, A.K.: Learning continuous face age progression: a pyramid of gans. In: IEEE Transactions on Pattern Analysis and Machine Intelligence (2019)
19. Wang, X., Wang, Y., Li, W.: U-net conditional gans for photo-realistic and identity-preserving facial expression synthesis. In: ACM Transactions on Multimedia Computing, Communications, and Applications (TOMM), vol. 15, no. 3s, pp. 1–23 (2019)
20. Li, L., Bao, J., Yang, H., Chen, D., Wen, F.: Faceshifter: towards high fidelity and occlusion aware face swapping," arXiv preprint arXiv:1912.13457 (2019)
21. Tolosana, R., Vera-Rodriguez, R., Fierrez, J., Morales, A., Ortega-Garcia, J.: Deepfakes and beyond: a survey of face manipulation and fake detection, arXiv preprint arXiv:2001.00179 (2020)
22. Di, X., Patel, V.M.: Face synthesis from visual attributes via sketch using conditional vaes and gans, arXiv preprint arXiv:1801.00077 (2017)
23. Radford, A., Metz, L., Chintala, S.: Unsupervised representation learning with deep convolutional generative adversarial networks, arXiv preprint arXiv:1511.06434, (2015)
24. Wang, T.-C., Liu, M.-Y., Zhu, J.-Y., Tao, A., Kautz, J., Catanzaro, B.: High-resolution image synthesis and semantic manipulation with conditional gans. In: Proceedings of the IEEE Conference on Computer Vision and Pattern Recognition, pp. 8798–8807 (2018)
25. Langr, J., Bok, V.: GANs in Action: Deep Learning with Generative Adversarial Networks. Manning Publications (2019)
26. Isola, P., Zhu, J.-Y., Zhou, T., Efros, A.A.: Image-to-image translation with conditional adversarial networks. In: Proceedings of the IEEE Conference on Computer Vision and Pattern Recognition, pp. 1125–1134 (2017)
27. Odena, A.: Semi-supervised learning with generative adversarial networks, arXiv preprint arXiv:1606.01583 (2016)
28. Zhu, J.-Y., Park, T., Isola, P., Efros, A.A.: Unpaired image-to-image translation using cycle-consistent adversarial networks. In: Proceedings of the IEEE International Conference on Computer Vision, pp. 2223–2232 (2017)

29. Zhang, H., et al.: Stackgan: text to photo-realistic image synthesis with stacked generative adversarial networks. In: Proceedings of the IEEE International Conference on Computer Vision, pp. 5907–5915 (2017)
30. Welander, P., Karlsson, S., Eklund, A.: Generative adversarial networks for image-to-image translation on multi-contrast mr images-a comparison of cyclegan and unit, arXiv preprint arXiv:1806.07777 (2018)
31. Li, H.: Statistical learning method beijing (2012)
32. Richardson, E., Sela, M., Or-El, R., Kimmel, R.: Learning detailed face reconstruction from a single image. Proceedings of the IEEE Conference on Computer Vision and Pattern Recognition, pp. 1259–1268 (2017)
33. Richardson, E., Sela, M., Kimmel, R.: 3d face reconstruction by learning from synthetic data. In: Fourth International Conference on 3D Vision (3DV). IEEE **2016**, pp. 460–469 (2016)
34. Choi, Y., Choi, M., Kim, M., Ha, J.-W., Kim, S., Choo, J.: Stargan: unified generative adversarial networks for multi-domain image-to-image translation. In: Proceedings of the IEEE Conference on Computer Vision and Pattern Recognition, pp. 8789–8797 (2008)
35. Liu, M.-Y., Tuzel, O.: Coupled generative adversarial networks. In: Advances in Neural Information Processing Systems, pp. 469–477 (2016)
36. Perarnau, G., Van De Weijer, J., Raducanu, B., Álvarez, J.M.: Invertible conditional gans for image editing, arXiv preprint arXiv:1611.06355 (2016)
37. Karras, T., Aila, T., Laine, S., Lehtinen, J.: Progressive growing of gans for improved quality, stability, and variation, arXiv preprint arXiv:1710.10196 (2017)
38. Karras, T., Laine, S., Aila, T.: A style-based generator architecture for generative adversarial networks. In: Proceedings of the IEEE Conference on Computer Vision and Pattern Recognition, pp. 4401–4410 (2019)
39. Karras, T., Laine, S., Aittala, M., Hellsten, J., Lehtinen, J., Aila, T.: Analyzing and improving the image quality of stylegan. In: Proceedings of the IEEE/CVF Conference on Computer Vision and Pattern Recognition, pp. 8110–8119 (2020)
40. Thies, J., Zollhofer, M., Stamminger, M., Theobalt, C., Nießner, M.: Face2face: real-time face capture and reenactment of RGB videos. In: Proceedings of the IEEE Conference on Computer Vision and Pattern Recognition, pp. 2387–2395 (2016)
41. Kowalski, M.: FaceSwap (2016). https://github.com/MarekKowalski/FaceSwap/
42. FaceApp (2016). https://faceapp.com/ap
43. Schreyer, M., Sattarov, T., Reimer, B., Borth, D.: Adversarial learning of deepfakes in accounting, arXiv preprint arXiv:1910.03810 (2019)
44. Lu, Y., Tai, Y.-W., Tang, C.-K.: Attribute-guided face generation using conditional cyclegan. In: Proceedings of the European Conference on Computer Vision (ECCV), pp. 282–297 (2018)
45. Banerjee, A., Kollias, D.: Emotion generation and recognition: a stargan approach, arXiv preprint arXiv:1910.11090 (2019)
46. Thies, J., Zollhöfer, M., Nießner, M.: Deferred neural rendering: image synthesis using neural textures. ACM Trans. Graph. (TOG) **38**(4), 1–12 (2019)
47. Villalba, J., Lleida, E.: Speaker verification performance degradation against spoofing and tampering attacks. FALA workshop, pp. 131–134 (2010)
48. Bahari, M.H., McLaren, M., Van Leeuwen, D., et al.: Age estimation from telephone speech using i-vectors (2012)
49. Kotti, M., Kotropoulos, C.: Gender classification in two emotional speech databases. In: 2008 19th International Conference on Pattern Recognition. IEEE, pp. 1–4 (2008)

50. Goel, N.K., Sarma, M., Kushwah, T.S., Agrawal, D.K., Iqbal, Z., Chauhan, S.: Extracting speaker's gender, accent, age and emotional state from speech. Language, vol. 68, no. 87.62, pp. 95–27 (2018)
51. Mavica, L.W., Barenholtz, E.: Matching voice and face identity from static images. J. Experimental Psychol. Human Perception Performance **39**(2), 307 (2013)
52. Smith, H.M., Dunn, A.K., Baguley, T., Stacey, P.C.: Matching novel face and voice identity using static and dynamic facial images. Attention, Perception, Psychophysics **78**(3), 868–879 (2016)
53. Bahari, M.H., Van Hamme, H.: Speaker age estimation and gender detection based on supervised non-negative matrix factorization. In: IEEE Workshop on Biometric Measurements and Systems for Security and Medical Applications (BIOMS). IEEE **2011**, pp. 1–6 (2011)
54. Qawaqneh, Z., Mallouh, A.A., Barkana, B.D.: Age and gender classification from speech and face images by jointly fine-tuned deep neural networks. Expert Syst. Appl. **85**, 76–86 (2017)
55. Dehak, N., Kenny, P.J., Dehak, R., Dumouchel, P., Ouellet, P.: Front-end factor analysis for speaker verification. IEEE Trans. Audio, Speech, Lang. Process. **19**(4), 788–798 (2010)
56. Reed, S., Akata, Z., Yan, X., Logeswaran, L., Schiele, B., Lee, H.: Generative adversarial text to image synthesis, arXiv preprint arXiv:1605.05396 (2016)
57. Wen, Y., Raj, B., Singh, R.: Face reconstruction from voice using generative adversarial networks. In: Advances in Neural Information Processing Systems, pp. 5265–5274 (2019)
58. Oh, T.-H., Dekel, T., Kim, C., Mosseri, I., Freeman, W.T., Rubinstein, M., Matusik, W.: Speech2face: learning the face behind a voice. In: Proceedings of the IEEE Conference on Computer Vision and Pattern Recognition, pp. 7539–7548 (2019)
59. Zhang, J., Liu, L., Xue, Z., Liu, Y.: Apb2face: audio-guided face reenactment with auxiliary pose and blink signals. In: ICASSP 2020 IEEE International Conference on Acoustics, Speech and Signal Processing (ICASSP), IEEE, pp. 4402–4406 (2020)
60. Duarte, A., Roldan, F., Tubau, M., Escur, J., Pascual, S., Salvador, A., Mohedano, E., McGuinness, K., Torres, J., Giro-i Nieto, X.: Wav2pix: speech-conditioned face generation using generative adversarial networks. In: ICASSP, pp. 8633–8637 (2019)
61. Kaya, H., Salah, A.A., Karpov, A., Frolova, O., Grigorev, A., Lyakso, E.: Emotion, age, and gender classification in children's speech by humans and machines. Comput. Speech Lang. **46**, 268–283 (2017)
62. Ringeval, F., Marchi, E., Grossard, C., Xavier, J., Chetouani, M., Cohen, D., Schuller, B.: Automatic analysis of typical and atypical encoding of spontaneous emotion in the voice of children (2016)
63. Hochreiter, S., Schmidhuber, J.: Long short-term memory. Neural Comput. **9**(8), 1735–1780 (1997)
64. Tian, G., Yuan, Y., Liu, Y.: Audio2face: Generating speech, face animation from single audio with attention-based bidirectional LSTM networks. In: IEEE International Conference on Multimedia & Expo Workshops (ICMEW), IEEE **2019**, 366–371 (2019)
65. Karras, T., Aila, T., Laine, S., Herva, A., Lehtinen, J.: Audio-driven facial animation by joint end-to-end learning of pose and emotion. ACM Trans. Graph. (TOG) **36**(4), 1–12 (2017)
66. Thies, J., Elgharib, M., Tewari, A., Theobalt, C., Nießner, M.: Neural voice puppetry: audio-driven facial reenactment, arXiv preprint arXiv:1912.05566 (2019)

67. Tzirakis, P., Papaioannou, A., Lattas, A., Tarasiou, M., Schuller, B., Zafeiriou, S.: Synthesising 3d facial motion from in-the-wild" speech, arXiv preprint arXiv:1904.07002 (2019)
68. Goecke, R., Millar, J.B.: The audio-video australian english speech data corpus avozes. In: Proceedings of the 8th International Conference on Spoken Language Processing INTERSPEECH, pp. 2525–2528 (2004)
69. Sanderson, C., Paliwal, K.K.: Identity verification using speech and face information. Digital Signal Process. **14**(5), 449–480 (2004)
70. Lee, B., Hasegawa-Johnson, M., Goudeseune, C., Kamdar, S., Borys, S., Liu, M., Huang, T.: Avicar: audio-visual speech corpus in a car environment. In: Eighth International Conference on Spoken Language Processing (2004)
71. Cooke, M., Barker, J., Cunningham, S., Shao, X.: An audio-visual corpus for speech perception and automatic speech recognition. J. Acoustical Soc. Am. **120**(5), 2421–2424 (2006)
72. Petrovska-Delacrétaz, D., et al.: The iv 2 multimodal biometric database (including iris, 2d, 3d, stereoscopic, and talking face data), and the iv 2-2007 evaluation campaign. In: 2008 IEEE Second International Conference on Biometrics: Theory, Applications and Systems. IEEE, pp. 1–7 (2008)
73. Rousseau, A., Deléglise, P., Esteve, Y.: Ted-lium: an automatic speech recognition dedicated corpus. In: LREC, pp. 125–129 (2002)
74. McCool, C., et al.: Bi-modal person recognition on a mobile phone: using mobile phone data. In: 2012 IEEE International Conference on Multimedia and Expo Workshops. IEEE, pp. 635–640 (2012)
75. Panayotov, V., Chen, G., Povey, D., Khudanpur, S.: Librispeech: an asr corpus based on public domain audio books. In: 2015 IEEE International Conference on Acoustics, Speech and Signal Processing (ICASSP), IEEE, pp. 5206–5210 (2015)
76. Nagrani, A., Chung, J.S., Zisserman, A.: Voxceleb: a large-scale speaker identification dataset, arXiv preprint arXiv:1706.08612 (2017)
77. Chung, J.S., Zisserman, A.: Lip reading in the wild. In: Asian Conference on Computer Vision. Springer, pp. 87–103 (2016)
78. Afouras, T., Chung, J.S., Senior, A., Vinyals, O., Zisserman, A.: Deep audio-visual speech recognition. In: IEEE Transactions on Pattern Analysis and Machine Intelligence (2018)
79. Czyzewski, A., Kostek, B., Bratoszewski, P., Kotus, J., Szykulski, M.: An audio-visual corpus for multimodal automatic speech recognition. J. Intell. Inf. Syst. **49**(2), 167–192 (2017)
80. Gemmeke, J.F., Ellis, D.P., Freedman, D., Jansen, A., Lawrence, W., Moore, R.C., Plakal, M., Ritter, M.: Audio set: an ontology and human-labeled dataset for audio events. In: 2017 IEEE International Conference on Acoustics, Speech and Signal Processing (ICASSP). IEEE, pp. 776–780 (2007)
81. Chung, J.S., Nagrani, A., Zisserman, A.: Voxceleb2: deep speaker recognition, arXiv preprint arXiv:1806.05622 (2018)
82. Afouras, T., Chung, J.S., Zisserman, A.: Lrs3-ted: a large-scale dataset for visual speech recognition, arXiv preprint arXiv:1809.00496 (2018)
83. Ephrat, A., Mosseri, I., Lang, O., Dekel, T., Wilson, K., Hassidim, A., Freeman, W.T., Rubinstein, M.: Looking to listen at the cocktail party: a speaker-independent audio-visual model for speech separation, arXiv preprint arXiv:1804.03619 (2018)

84. Chao, G.-L., Hu, C.C., Liu, B., Shen, J.P., Lane, I.: Audio-visual ted corpus: enhancing the ted-lium corpus with facial information, contextual text and object recognition. In: Adjunct Proceedings of the 2019 ACM International Joint Conference on Pervasive and Ubiquitous Computing and Proceedings of the 2019 ACM International Symposium on Wearable Computers, pp. 468–473 (2009)
85. Ardila, R., Branson, M., Davis, K., Henretty, M., Kohler, M., Meyer, J., Morais, R., Saunders, L., Tyers, F.M., Weber, G.: Common voice: a massively-multilingual speech corpus, arXiv preprint arXiv:1912.06670 (2019)
86. Iloanusi, O., Ejiogu, U., Okoye, I.-E., Ezika, I., Ezichi, S., Osuagwu, C., Ejiogu, E.: Voice recognition and gender classification in the context of native languages and lingua franca. In: 2019 6th International Conference on Soft Computing & Machine Intelligence (ISCMI). IEEE, pp. 175–179 (2019)
87. Vedaldi, A., Zisserman, A., Chen, H., Xie, W.: Vggsound: a large-scale audio-visual dataset. In: Proceedings of the International Conference on Acoustics, Speech, and Signal Processing (2020)
88. Nakashima, Y., Yasui, T., Nguyen, L., Babaguchi, N.: Speech-driven face reenactment for a video sequence. ITE Trans. Media Technol. Appl. **8**(1), 60–68 (2020)
89. Song, L., Wu, W., Qian, C., He, R., Loy, C.C.: Everybody's talkin': let me talk as you want, arXiv preprint arXiv:2001.05201 (2020)
90. Suwajanakorn, S., Seitz, S.M., Kemelmacher-Shlizerman, I.: Synthesizing obama: learning lip sync from audio. ACM Trans. Graph. (TOG) **36**(4), 1–13 (2017)
91. Kim, H., Garrido, P., Tewari, A., Xu, W., Thies, J., Niessner, M., Pérez, P., Richardt, C., Zollhöfer, M., Theobalt, C.: Deep video portraits. ACM Trans. Graph. (TOG) **37**(4), 1–14 (2018)
92. Jamaludin, A., Chung, J.S., Zisserman, A.: You said that?: synthesising talking faces from audio. Int. J. Comput. Vis. **127**(11–12), 1767–1779 (2019)
93. Zakharov, E., Shysheya, A., Burkov, E., Lempitsky, V.: Few-shot adversarial learning of realistic neural talking head models. In: Proceedings of the IEEE International Conference on Computer Vision, pp. 9459–9468 (2019)
94. Song, Y., Zhu, J., Li, D., Wang, X., Qi, H.: Talking face generation by conditional recurrent adversarial network, arXiv preprint arXiv:1804.04786 (2018)
95. Zhou, H., Liu, Y., Liu, Z., Luo, P., Wang, X.: Talking face generation by adversarially disentangled audio-visual representation. In: Proceedings of the AAAI Conference on Artificial Intelligence, vol. 33, pp. 9299–9306 (2019)
96. Vougioukas, K., Petridis, S., Pantic, M.: End-to-end speech-driven realistic facial animation with temporal gans. In: CVPR Workshops, pp. 37–40 (200)
97. Wiles, O., Sophia Koepke, A., Zisserman, A.: X2face: a network for controlling face generation using images, audio, and pose codes. In: Proceedings of the European conference on computer vision (ECCV), pp. 670–686 (2018)
98. Liu, Z., Luo, P., Wang, X., Tang, X.: Deep learning face attributes in the wild. In: Proceedings of International Conference on Computer Vision (ICCV), December 2015

Classification of Recyclable Materials Using Efficient Deep Learning Models and Benchmarking of GPU Performance

Yavuz Kömeçoğlu[1] and Merve Ayyüce Kızrak[2(✉)]

[1] Kodiks Information Technology, Machine Learning Research and Development Engineer, İstanbul, Turkey
yavuz.komecoglu@kodiks.com

[2] Bahçeşehir University, Faculty of Engineering and Natural Sciences, Department of Electrical-Electronics Engineering, İstanbul, Turkey
merveayyuce.kizrak@eng.bau.edu.tr

Abstract. One of the consequences of climate change and global warming results from excessive consumption of sources. To slow down global warming and increase energy saving, recycling, within the framework of waste management, needs to be widely implemented. Waste management and recycling is not only environmentally advantageous but also of great importance for a sustainable economy. Preferring smart systems instead of human workers is a socially important step that allows people to work in more welfare environments. Intelligent waste management approaches open up a major research area.

The main objective of the study is aimed to contribute to the efficient collection of recyclable materials from the end consumer, to optimize the collection process and to reduce the workload in the waste institution. In addition to the TrashNet dataset used in the previous classification of recyclable materials, an expanded dataset is collected, and a more advanced version is obtained. Data from three different classes, including glass, plastic, and metal waste, were collected and the current dataset was enhancement from 2527 to 6136. The new extended dataset is called TrashX. Therefore, not only the methods used in the literature have been improved, but also the convolutional neural network-based models used are tested. All results are evaluated according to performance criteria. In this research, 6 different recyclable waste classifications are made on a progressed dataset consisting of 6136 RGB images. Within the scope of this study, the largest dataset in the literature was created. For this purpose, high performance and robust models such as MobileNet, RecycleNet, and EfficientNet are offered. One of the most important factors of the study is that the performance of the models is evaluated in terms of time on different hardware. This benchmarking light on researchers to improve intelligence recycling and waste management systems.

Finally, the experiments are run to compare the performances of the methods for both TrashNet and the TrashX datasets. The experimental results demonstrate that EfficientNet-b3 efficiency 93.8% and 97.3% in terms of accuracy for Trashnet and TrashX datasets separately and thus it outperforms the many recent approaches for trash classification on both experimental datasets.

T. Allahviranloo et al. (Eds.): IDS 2020, AISC 1301, pp. 228–239, 2021.
https://doi.org/10.1007/978-3-030-66501-2_18

Keywords: Waste management · Trash classification · Recyclable materials · Artificial intelligence · Computer vision · Deep learning

1 Introduction

Nature is a highly dynamic system and in perfect balance. All populations in the ecosystem are in a constant state of change within the food-cycle. Everything exists in the process, breaks down and is reused. Increasing population and industrialization and waste management have become a more important issue. Recycling practices are one of the main elements of waste management. Recycling of reusable materials provides critical benefits to society, not only for the environment but also socio-economically. Annual solid waste in the world is expected to reach 2.2 billion tons by 2025, which will cost $ 375.5 billion in waste management [1]. There are several obstacles to efficient waste management. The government plan, budgeting, and regulations are often at the top of the obstacles. One of the obstacles that are difficult to overcome is the education of the household. Another obstacle is the resistance to belief in technology [2, 3]. Today, this resistance is decreasing, and this is a positive development.

An effective recycling system is vital for a sustainable future in rapidly changing climate conditions. Recycling can be used to recover raw resources, conserve energy, reduce greenhouse gas emissions, water pollution, reduce new storage areas, etc. Nowadays, recycling works are limited to human abilities [4, 5]. However, using technology in recycling systems in recent years provides unnecessary workload and energy saving. Intelligent systems are expected to increase efficiency for the separation of waste and reusable components. Therefore, optimization of the entire recycling process is an important development to use artificial intelligence for social good.

1.1 Related Works

The first pattern recognition approach for waste sorting has a Bayes framework structure that requires feature extraction and therefore domain expertise [6]. Convolutional neural networks (CNN) approaches, which have proven success in object recognition, and studies using different architectures contribute to the literature. In order to study recyclability with computer vision, Stanford University has classified the data using a CNN approach with a dataset called TrashNet as part of a student project [7, 8]. After that, another successful study is the publication where comparisons of another deep learning approach defined as RecycleNet in 2018 is published. In the same year, another study adopted a hybrid approach to classifying recycling materials [9]. The alternative study conducted in 2019 focuses on measuring the fine-tuning effect on existing deep learning models [10]. In another study published in the same year, a hardware structure is proposed for sorting the materials on the conveyor [11]. The most important common feature of these studies is the utilization of camera images of recyclable materials and their focus on classification with CNN-based deep learning models. Anh H. Vo et al. developed a novel ResNext-based convolutional neural network model called DNN-TC for garbage classification. They created a new data set for this study, called VN-garbage.

In this database, there are 5904 images from three different classes. As a result of the study, DNN-TC reached 94% for TrashNet dataset and 98% for VN-garbage dataset [12].

Recyclable material classification is a difficult task, the dataset and the appropriate deep-learning model must be rigorously studied to achieve reliable and successful results. It may also bring about application-specific hardware designs. Effective results are required at different temperatures, humidity and light conditions such as factory, indoor and outdoor.

2 Dataset

The main purpose of this study is to classify recyclable materials based on images only. With a smart system, a sustainable economy is aimed at a safer and robust recycling system in dumps. This work aims to get to know the common recyclable materials such as glass, paper, cardboard, plastic, metal, and garbage. For this, both the TrashNet dataset that was published explicitly was utilized and the dataset was expanded by collecting and labeling extra data. The properties of the new dataset are examined under our dataset heading.

2.1 TrashNet Dataset

The TrashNet dataset was published as part of a student project at Stanford University [8]. It contains images of the recyclable materials of six different classes. Images in this dataset have a white background. Each image has a different pose, exposure, lighting and the shape of the material is not expected to be shapely. Each image was resized to 512 × 384 pixels. The original size of this dataset is 3.5 GB in size and contains 2527 images.

2.2 Expanded Dataset: TrashX

In addition to the TrashNet dataset, data collection and labeling have been carried out again in order to robust model performance, ensure diversity and achieve effective results in real life.

We collected 3609 images from three different classes, including glass, plastic and metal waste, and increased the current data set from 2527 to 6136. Called this expanded dataset TrashX. In this research, all experiments were trained, validated and tested on the TrashX dataset. In all models used in the study, the training set is 70%, the validation set is 13% and the test set is 17%. Examples from TrashNet and the new dataset are shown in Fig. 1. The information of images for each class was presented in Table 1.

3 Models and Methods

In this study, we experimented on several different state-of-the-art deep convolutional neural network architectures as well as RecycleNet [9] model for reusable material classification. This research focuses on the sorting of recyclable materials in the mobile environment. Therefore, cost-efficient methods are tried. Experiments are run on different equipment and benchmarking is introduced.

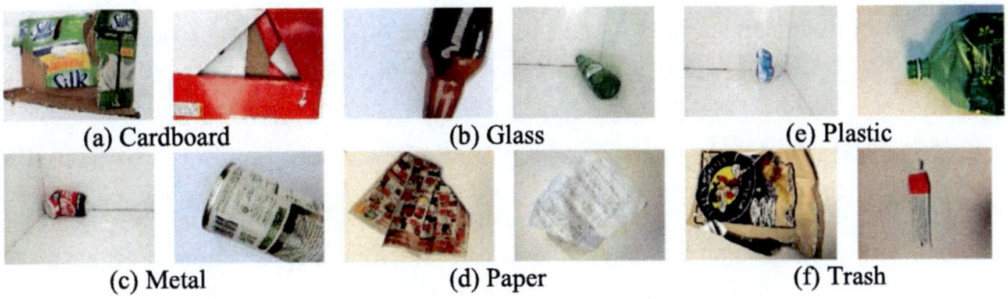

Fig. 1. Sample images of TrashNet and TrashX datasets.

Table 1. TrashNet [8] dataset and TrashX dataset information.

TrashNet Dataset				
Recyclable materials	# of Each material	Training set (70%)	Validation set (17%)	Test set (13%)
Paper	594	403	83	108
Glass	501	354	65	82
Plastic	482	347	61	74
Metal	410	286	56	68
Cardboard	403	287	46	70
Trash	137	91	17	29
Total	**2527**	**1768**	**328**	**431**
TrashX Dataset				
Recyclable materials	# of Each material	Training set (70%)	Validation set (17%)	Test st (13%)
Paper	594	412	79	103
Glass	918	643	109	166
Plastic	2468	1740	316	412
Metal	1616	1140	211	265
Cardboard	403	262	65	76
Trash	137	100	17	20
Total	**6136**	**4297**	**797**	**1042**

3.1 MobileNet

MobileNet, in 2017 published by Google researchers and especially known for its effective work on mobile devices, reduces the hyper-parameter computational cost by making use of the Inception module [13]. Due to the MobileNets, they are able to lower resource requirements for trains faster with that is one of the useful properties. Thus, it is among the most preferred deep learning models on mobile devices. The input size of MobileNet is determined as 224 × 224 as in many other convolutional neural network models.

3.2 DenseNet

Through the skip connection between the layers, it works more efficiently than the previous models. Because this approach allows the vanishing gradient problem to be avoided during backpropagation as the model is deeper. The skip connection structure of the model is shown in the figure. As DenseNet has these additional connections, the computational cost is high. However, it is one of the important deep learning approaches preferred to increase accuracy. The input size of this model is shown as 224 × 224 [14]. DenseNet is not used directly in the study, but a variation of this, RecycleNet is used.

3.3 RecycleNet

RecycleNet is a DenseNet based deep learning model that classifies the TrashNet dataset in the most efficient way, published in 2018. This model is designed to be used on mobile devices with a high accuracy rate as in DenseNet besides reducing the parameter cost as in MobileNet [13–15]. Therefore, a new skip connection structure is adopted. Another advantage of the RecycleNet model is that it is faster than DenseNet in the prediction time.

This model additionally enables as to control the intensity of various dense blocks independently, by evolving m. In Eq. (1) x_1 provides the number of layers, H_1 the function of operations, hyperparameter control of the layers of m dense. For example, when $m = 3$ is selected, the number of skip connections decreases to one in three. Due to this approach, it reduces the number of parameters in a 121-layered network from 7 million to about 3 million [9].

$$x_1 = H_1\left(\left[x_{l-1}, x_{l-m-1}, x_{l-2m-1}, \ldots, x_0\right]\right) \quad (1)$$

In this study, the RecycleNet model is also used for benchmarking and the input size of the model is given as 224 × 224. Figure 2 shows the RecycleNet model.

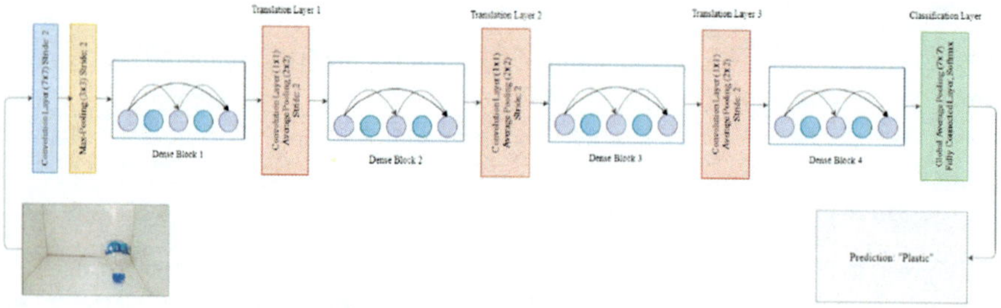

Fig. 2. RecycleNet model [9].

3.4 EfficientNet

EfficientNet is an up-to-date advanced model based on a convolutional neural network. EfficientNet B0-B7 is a group of 8 models. Although more accuracy is obtained as the convolution layers increase, it is known that in recent years, fewer parameters must be calculated in order to the models to work faster, especially on edge devices. Therefore, smaller models have started to be preferred in recent years. The model is used depthwise convolution and pointwise convolution that divides the original convolution into two stages to significantly reduce the cost of calculation, with a minimum loss of accuracy. EfficientNet-b0 model input size is 224 × 224, whereas b3 version model's input size is 300 × 300. EfficientNet, scaling on depth, width and input resolution, on the other hand, is among the most successful and fast models of the last year with its the Fig. 3 shows the scaling structure that summarizes EfficientNet [16].

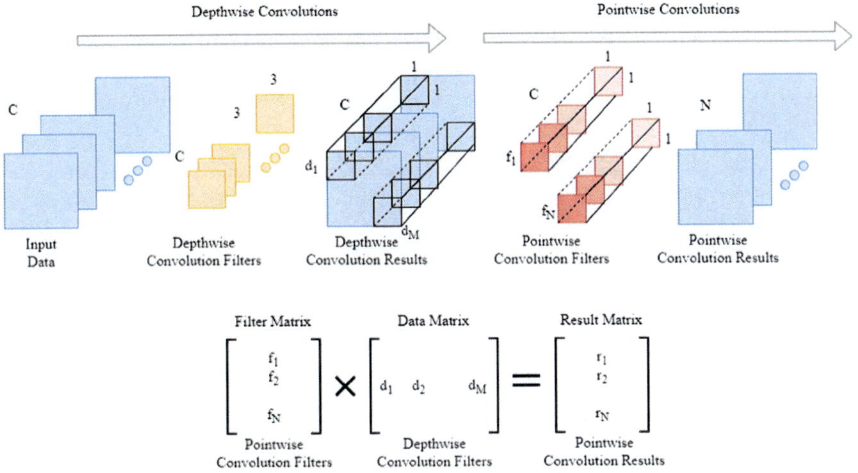

Fig. 3. A basic representation of depthwise and pointwise convolutions.

The B7 model from the EfficientNet group accomplishes the classification accuracy 84.4% top-1 and 97.1% top-5 on the ImageNet dataset. The most important property is the 8.4 times smaller model and 6.1 times faster than the model that has the closest classification accuracy as a result of its scaling [16].

4 Experiments

In this study, firstly, the importance of expanding the data set was emphasized and the data set was reproduced by collecting real data. The TrashNet data set was increased to 2.428 times and a new data set consisting of 6136 recyclable materials was created. In the selection of the model, especially suitability for edge work with high performance and low computing is taken into consideration. Besides these issues, experiments are carried out in different working environments. Today, data is produced in different formats due to portable devices. Within the scope of the study, data of waste are collected and also processed. Some of these processes are performed on physical servers, cloud environments, and edge devices. Contributing to the relevant literature by providing an end-to-end performance analysis is one of the main objectives.

Programming Language, Libraries and versions: The main experiments of this research were performed using the 3.7 version of Python as a programming language, 2.1 recent version of TensorFlow as a deep learning library.

GPU's Hardware: The chosen deep learning models are trained for both TrashNet [8] and extended datasets on three different local GPUs produced by NVIDIA. Table 2 shows the CUDA core, architecture, and memory speed information of the three GPUs.

Table 2. Property of GPUs.

GPU's	GTX 1080 Ti	GTX 2080 Ti	Titan Xp
CUDA cores	3584	4352	3840
GPU Architecture	Turing	Turing	Pascal
Memory speed	1584 MHz	1545 MHz	1582 MHz

Cloud GPUs: In 2018, Google opened the Colab environment to Python programmers for free. In this way, it allows you to develop deep learning applications by using many different frameworks such as TensorFlow, Keras, PyTorch. Specially to use this cloud service, it is sufficient to have a Google account. In this work, the results are examined using a data center accelerator with Tesla P100 GPUs, which are accessible in the Colab cloud, to evaluate performance [17].

Evaluation Metrics: Various evaluation metrics are needed to measure performance in classification problems. It is beneficial to use the most preferred metrics, especially when using tagged datasets. Based on this study, accuracy, precision, recall, and F1-score metrics were calculated to evaluate the results obtained.

5 Results and Discussion

The most important aspect of this study from other similar studies is that it creates an TrashX dataset. Experiments are carried out both in the TrashNet dataset and extended dataset with four robust deep learning models. Nesterov momentum was 0.9 and the learning rate was 0.0001 in all models selected in the study. Stochastic gradient descent is used as an optimization algorithm and batch size 32 is selected. To generalize the models, the dataset is expanded with real data. As a performance criterion, validation and test accuracy and real-time prediction performance are measured in multi-class cross-entropy loss in GPUs.

Figure 4 shows the accuracy and loss curves for the best results of the EfficientNet-b3 model. Table 3 shows the test and the validation accuracy of different CNN-based deep learning models, as well as precision, recall and F1-score metrics for testing. Moreover, Fig. 4 shows the loss and accuracy performance of validation and test sets over 120 epochs for TrashNet and the TrashX datasets. The most remarkable result is the test accuracy of the EfficientNet-b3 model in the extended dataset.

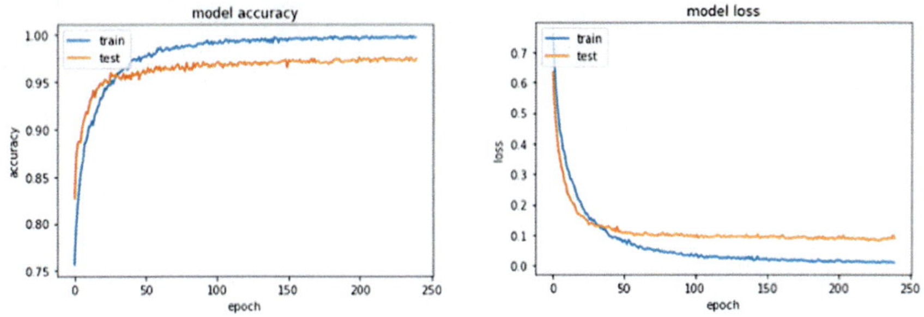

Fig. 4. The best result of EfficientNet-b3 model accuracy and loss.

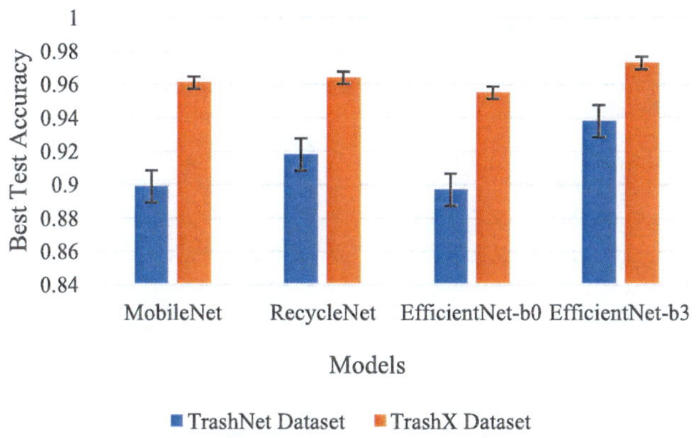

Fig. 5. Comparison model's test results of the datasets.

Besides, Table 4 presents the confusion matrices of best model experimental methods for both TrashNet and TrashX datasets. The most inaccurate predictions of all experimental methods for the TrashNet dataset are in the trash class which has different types of objects.

The other two classes, where there are many misclassifications, are glass and plastic materials. The positive effect of the TrashX dataset on the best model experiment is clearly shown in these classes.

Table 3. Comparison test results of the datasets.

Comparison results of training on TrashNet dataset						
Model	Epoch	Validation accuracy	Test accuracy	Test precision	Test recall	Test F1-Score
MobileNet	97/120	0.887	0.899	0.900	0.888	0.907
RecycleNet	56/120	0.899	0.918	0.920	0.919	0.928
EfficientNet-b0	180/240	0.887	0.897	0.880	0.870	0.870
EfficientNet-b3	240/240	0.928	**0.938**	0.940	0.940	0.940
Comparison results of training on TrashX dataset						
Model	Epoch	Validation accuracy	Test accuracy	Test precision	Test recall	Test F1-Score
MobileNet	66/120	0.953	0.961	0.950	0.950	0.950
RecycleNet	38/120	0.966	0.964	0.960	0.960	0.960
EfficientNet-b0	155/240	0.948	0.955	0.950	0.950	0.950
EfficientNet-b3	240/240	0.978	**0.973**	0.970	0.970	0.970

Table 4. Confusion matric of the best model for TrashNet and TrashX datasets.

TrashNet dataset

Actual Classes \ Predicted Classes	Cardboard	Glass	Metal	Paper	Plastic	Trash
Cardboard	58	0	0	1	0	0
Glass	0	64	4	0	2	1
Metal	0	2	52	0	1	1
Paper	1	0	1	78	0	1
Plastic	1	1	2	1	49	1
Trash	1	1	2	0	1	14

TrashX dataset

Actual Classes \ Predicted Classes	Cardboard	Glass	Metal	Paper	Plastic	Trash
Cardboard	75	0	0	1	0	0
Glass	0	159	5	0	2	0
Metal	0	1	261	0	1	2
Paper	4	0	2	95	0	2
Plastic	0	3	4	0	405	0
Trash	0	0	1	0	0	19

At this phase of the study, we measured the performance of models' training and prediction times for different GPUs. GTX 2080 Ti gives the highest performance in terms of speed for the best model EfficientNet-b3. Performance evaluation is shown in Table 5. Figure 6 shows GPU performance for TrashNet and TrashX dataset.

Table 5. Comparison of GPU's and training and prediction time performance.

Comparison of GPU's performance for the best model on TrashNet dataset			
GPUs	Training time (sec per epoch)	First prediction time (sec)	Second prediction time (sec)
GTX 1080 Ti	24	20.67	0.02
TitanXp	12	3.57	0.01
GTX 2080 Ti	8	2.86	0.01
Google Colab - Tesla P100	9	7.54	0.01
Comparison of GPU's performance for the best model on TrashX Dataset			
GPUs	Training time (sec per epoch)	First prediction time (sec)	Second prediction time (sec)
GTX 1080 Ti	232	17.79	0.02
TitanXp	201	3.74	0.02
GTX 2080 Ti	132	2.14	0.01
Google Colab - Tesla P100	141	8.56	0.02

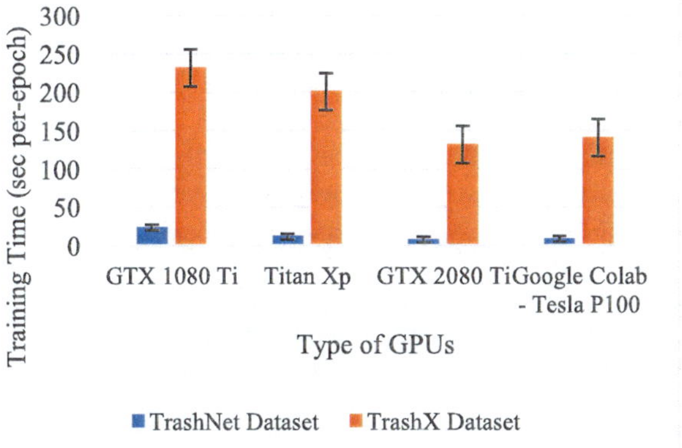

Fig. 6. Comparison of GPU's and training and prediction time performance.

Our proposed model is EfficientNet-b3 for this trash classification problem. The model obtains the highest accuracy and the smallest loss value after 240 epochs, which has proven that EfficientNet-b3 could quickly achieve the stable and generalization for the expanded dataset. The findings suggest that these approaches could also be useful for real-time edge computing systems such as intelligence recycle bins.

6 Conclusion

Using artificial intelligence to sort waste is an issue that has been on the agenda in recent years. The data to be collected from the intelligence waste recycling system to be designed depends on environmental factors such as lighting, deformities and material diversity. For instance, deformed substances do not lose their properties, but it becomes difficult to identify from the camera. In this paper, the classification of waste materials is considered as an important problem and results on solutions are clearly presented. When selected robust CNN-based models are meticulously trained, they can provide industrial solutions.

The main motivation of this study is to measure the performance of recent deep learning models on different hardware in a larger dataset. Thus, it creates a perspective for the hardware to be preferred in real-life practical applications. Among the models trained with 4297 images, the best test accuracy is achieved with the EfficientNet-b3 model with 97.3%. It is a success well above the previous studies in this field. In order to achieve, 240 epoch training was performed, and each epoch was trained for 132 s. Thus, the fastest hardware is measured as GTX 2080 Ti. As a result, a significant study has been shown to use artificial intelligence for social good and ecological awareness.

7 Future Work

There is a need to design sustainable cities by contributing to the realization of the sustainable development goals of the United Nations with a smartly designed waste management facility based on artificial intelligence. Intelligent waste management is one of the most important tools for stopping global warming is expenditure and energy saving. This is of major importance for a sustainable economy. As the studies soon, it is among the targets to design waste facilities that will be used in daily life with more sophisticated approaches. Further research on try edge device applications for waste material classification is necessary. In this way, the benefits that can be obtained by using artificial intelligence for humans and the environment worldwide can be measured.

Acknowledgments. The authors wish to acknowledge the support of the NVIDIA GPU Grant Program with the donation of the TITAN Xp GPU. The authors would also like to thank CRK Technology, which permits the data to be expanded to almost 2.5 times for this study.

References

1. Williams, P.T.: Waste Treatment and Disposal. John Wiley & Sons, Hoboken (2005)
2. Palmer, J.A.: Environmental thinking in the early years: understanding and misunderstanding of concepts related to waste management. Environ. Educ. Res. **1**(1), 35–45 (1995)

3. Amick, D.S.: Reflection on the origins of recycling: a paleolithic perspective. Lithic Technol. **39**(1), 64–69 (2014)
4. Davidson, Cummings L.: Voluntary strategies in the environmental improvement: recycling as cooptation. J. Voluntary Action Res. **6**(3–4), 153–160 (1977)
5. Alvarez-Ch´avez, C.R., Edwards, S., Moure-Eraso, R., Geiser, K.: Sustainability of bio-based plastics: general comparative analysis and recommendations for improvement. J. Clean. Product. **23**(1), 47–56 (2012)
6. Liu, C., Sharan, L., Adelson, E.H., Rosenholtz, R.: Exploring features in a Bayesian framework for material recognition. In: Conference on Computer Vision and Pattern Recognition (CVPR), pp. 239–246. IEEE (2010)
7. Yang, M., Thung, G.: Classification of Trash for Recyclability Status, arXiv preprint (2016)
8. Thung, G.: Trashnet, GitHub repository (2016)
9. Bircanoğlu, C., Atay, M., Beşer, F., Genç, O., Kızrak, M.A.: RecycleNet: intelligent waste sorting using deep neural networks. In: IEEE International Conference on INnovations in Intelligent SysTems and Applications (INISTA) (2018)
10. Özkaya, U., Seyfi, L.: Fine-tuning models comparisons on garbage classification for recyclability. In: 2nd International Symposium on Innovative Approaches in Scientific Studies (2019)
11. Seredkin, A.V., Tokarev, M.P., Plohih, A., Gobyzov, O.A., Markovich, D.M.: Development of a method of detection and classification of waste objects on a conveyor for a robotic sorting system. J. Phys. Conf. Ser. **1359** (2019)
12. Vo, A.H., Son, L.H., Vo, M.T., Le, T.: A novel framework for trash classification using deep transfer learning. IEEE Access **7**, 178631–178639 (2019)
13. Howard, A.G., Zhu, M., Chen, B., Kalenichenko, D., Wang, W., Weyand, T., Andreetto, M., Adam, H.: MobileNets: Efficient Convolutional Neural Networks for Mobile Vision Applications, arXiv preprint arXiv:1704.04861 (2017)
14. Huang, G., Liu, Z., Weinberger, K.Q., Van Der Maaten, L.: Densely connected convolutional networks. In: Proceedings of the IEEE Conference on Computer Vision and Pattern Recognition, vol. 1, no. 2, (2017)
15. Chollet, F.: Xception: Deep Learning with Depthwise Separable Convolutions, arXiv preprint (2016)
16. Tan, M., Le, Q.V.: EfficientNet: rethinking model scaling for convolutional neural networks. In: Thirty-sixth International Conference on Machine Learning (ICML) (2019)
17. Google colaboratory (2017). https://colab.research.google.com/

Cluster-Based Monitoring and Location Estimation for Crowd Counting

Merve Ayyüce Kızrak[1(✉)] and Bülent Bolat[2]

[1] Faculty of Engineering and Natural Sciences, Department of Electrical & Electronics Engineering, Bahçeşehir University, Istanbul, Turkey
merveayyuce.kizrak@eng.bau.edu.tr

[2] Faculty of Electrics-Electronics Engineering, Department of Electronics and Communications Engineering, Yıldız Technical University, Istanbul, Turkey
bbolat@yildiz.edu.tr

Abstract. Crowd management and monitoring are important research topics in order to ensure personal and community safety by making use of video images. The crowd tracking system (CMS) includes tasks such as density variation in images, irregular distribution of people and objects, overcrowding, exposure estimation. People come together for various purposes in areas designed for socializing, such as parks, stadiums, airports, hospitals, and shopping malls. Generally, these areas are monitored by closed-circuit TeleVision (CCTV). However, this type of system brings problems. The main ones are portability, flexible accessibility, limited coverage area, high power consumption. Crowd density is often examined by people for behavioral analysis or to identify suspicious people. Computer vision and deep learning approaches are maintained in order to prevent mistakes caused by human error and to make a faster evaluation.

In this study, a method for detecting event changes with cluster-based inspection in crowd images is proposed. Gaussian YOLOv3 model is used for object recognition. The proposed clustering approach is used to track changes in the number, coordinate, and direction information of the cluster. Behavior change time, location, and classification are achieved as a result of this information extractions. These two event changes are taken into consideration, especially because sudden changes in state occur in walking and running behavior. Six different video sequences in the PET2009 dataset are used for the study. Accuracy performance is achieved between 83.2% and 96.4%. The results obtained to achieve the success that can be compared with similar ones in the literature.

Keywords: Crowd counting · Crowd monitoring · Location estimation · Cluster-Based crowd analysis · Computer vision · Deep learning · Artificial intelligence

1 Introduction

Crowd is a community made up of many people coming together. If this is a group, it may mean that they are meeting for a specific purpose. Sometimes the crowd moves

T. Allahviranloo et al. (Eds.): IDS 2020, AISC 1301, pp. 240–253, 2021.
https://doi.org/10.1007/978-3-030-66501-2_19

scattered and sometimes they adopt movements with similar patterns. One of the important purposes of crowd monitoring is to detect unexpected events. Crowd monitoring is a critical issue to ensure safety in case of sudden situation changes, such as a collapse, fire, explosion, rather than ordinary behavior of people. For this, the processing of video data collected from security cameras, especially at crowded points of cities, enables safer environments by reducing latency. As the use of camera systems becomes widespread, crowd monitoring methods are also developing. These methods address major problems such as density estimation, tracking, counting, localization, scene understanding, and behavior analysis. In the literature, clustering, object detection, and regression approaches can be used together or separately for crowd counting and monitoring.

It is possible to classify crowd-related research according to the method and the way the problem is handled. Figure 1 shows a taxonomy of researches in the literature on crowd-related studies [1].

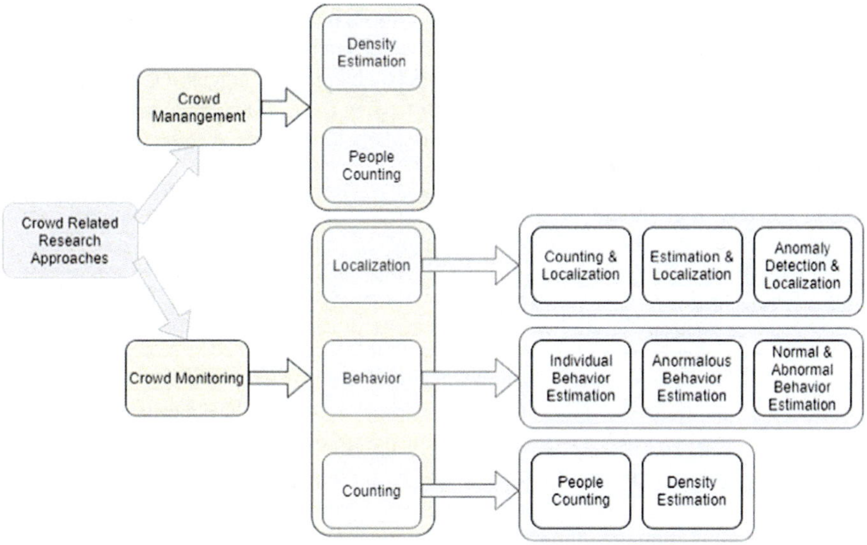

Fig. 1. Taxonomy of crowd related research approaches [1].

In this study, object detection and clustering-based approaches are used on crowd monitoring problems. Localization is also an important factor while performing the task of counting people in crowded images. YOLOv3 deep learning architecture, which is one of the up-to-date and real-time object detection approach and the KNN clustering approach is used at the stage of behavior analysis for the detection of people [2, 3]. Thus, the behavioral changes in the crowd are detected. Cluster centroid used as a presentment of groups of people on the stage. The distance between the centroid each cluster, the change and speed of the center position, and the perception of the walking and running behaviors of the crowd behavior are provided. Crowd groups of people are segmented by this approach. Each video sequence is expressed as a time series for crowd event change detection. Only the coordinate information of the detected objects is used for event detection. It does not use information such as the shape and sound of the objects in the image. In the study, experiments and comparisons with other studies in the literature are made using the PETS2009 dataset as a crowd video [4].

The organization of the study is as follows. In the second part, the recent literature review was summarized. In Sect. 3, details about available datasets published for crowd analysis were explained. There was more comprehensive information about the dataset used for this research. The 4th section of the study was about the problem statement and the proposed method. The software, hardware, and performance metrics used for the experiments of the study were explained in Sect. 5. In the 6th section, the discussion on the results and the last section, the 7th section, provides information about the benefits of research and future studies.

2 Related Works

There are different approaches in the literature for behavior change and analysis among crowd monitoring methods. Detecting rapid event changes is critical. Space-time related approaches are frequently used to capture these changes in images. Detecting motion and direction from the image is important for finding the event change and the direction of the event. Recent studies related to behavior analysis and crowd count in the literature are shown in Table 1.

Table 1. Recent literature of crowd monitoring.

Ref	Type of crowd analysis	Methods	Performance
[5]	Behavior detection	Spatio-Temporal Texture (STT)	The STT approach shows comparable results of Spatio-temporal Compositions (STC) and Inference according to Composition (IBC)
[6]	Monitoring/Behavior detection	Isometric Mapping (ISOMAP)	Feature space reduction
[7]	Behavior detection	Spatio-Temporal model	Accuracy 98%–88%
[8]	Anomaly detection	Optical flow horn schunck algorithm	Calculation of the distance between centroids
[9]	Behavior detection	Proximate median filter foreground segmentation	Lower false rate
[10]	Real-Time behavior detection	Scale-Invariant Feature Transform (SIFT)	Accuracy 95%

(*continued*)

Table 1. *(continued)*

Ref	Type of crowd analysis	Methods	Performance
[11]	Behavior detection	Holistic approach	The performance of holistic approaches yields better results
[12]	Monitoring event detection	Fixed-Width clustering & YOLO	Accuracy 80%–95.7%
[13]	Monitoring/Counting	Size estimation & counting	Accuracy 90%
[14]	Monitoring/Counting	Neural network & regression trees	The best estimation of BPNN
[15]	Anomaly detection	Optical flow method and SVM	Accuracy 87.4%
[16]	Monitoring behavior detection	EHCAF	Accuracy 99.1- FNR of 2.8%,
[17]	Evacuation behavior	Legion Evac software	Positive correlation scores
[18]	Behavior detection of violent	Deep Neural Networks & Hybrid Random Matrix (HRM)	Accuracy 90.17%–91.61%

3 Datasets

Video frames collected from different fields are shared as open datasets. Many are used to test the performance of applications for different tasks and to compare them with other studies in the literature. Datasets used for crowd-based studies are shown in Table 2.

Table 2. Crowd datasets overview [1].

Datasets	Description	#of Images	Resolution	Min	Ave	Max	Overall count	Accessibility
UCSD [19]	People counting	2000	238 × 158	11	25	46	49885	Yes
MALL [20]	People counting	2000	320 × 240	13	–	53	62325	Yes
UCF_CC_50 [21]	Density estimation	50	Variable	94	1279	4543	63974	Yes
Shanghai Tech A, B [22]	Crowd counting	482	768 × 1024	33	501	3139	241677	Yes

(continued)

Table 2. (*continued*)

Datasets	Description	#of Images	Resolution	Min	Ave	Max	Overall count	Accessibility
UCF-QNRF [23]	Crowd counting and localization	716	400 × 300 9000 × 6000	9	123	578	88488	Yes
WorldExpo'10 [24]	Cross scene crowd counting	3980	576 × 720	1	50	253	199923	Yes
CUHK [25]	Crowd behavior	1535	Variable	49	815	12865	–	Yes

The UCSD dataset consists of images taken from the streets for the human counting problem. MALL dataset is collected with a security camera in a shopping center. It is preferred for counting people indoors. UCF_CC_50 is a challenging dataset that includes various stages such as protests, stadiums, concert venues, and crowds at different densities and is used for density mapping. Shanghai Tech dataset consists of two parts named A and B. A consists of images taken randomly from the internet. Part B consists of footage collected from the metropolitan street in Shanghai. UCF-QNRF dataset is an updated one shared for crowd counting and localization. It is a difficult dataset due to the high-density change and resolution change. WorldExpo'10 is a dataset prepared for inter-scene crowd counting. However, it is insufficient to be used for accuracy assessment due to its low density. It is a dataset published for behavioral analysis obtained from airports, shopping malls, parks, and streets. In this study, experiments are carried out on the PETS2009 dataset instead of the datasets were explained.

3.1 PET2009

PET2009 is a dataset which contains different mass activities. The images in the dataset have different camera views. The dataset has 8 subsets under 3 main headings. Details and samples of the dataset are as in Table 3 and Fig. 2. Hierarchical structure of images in the dataset is shown in Fig. 3.

Table 3. Details of PETS2009 dataset [4].

S1: Person Count and Density Estimation		S2: People Tracking		S3: Flow Analysis and Event Recognition	
Set name	*Element*	*Set name*	*Element*	*Set name*	*Element*
S1.L1 walking	Medium density crowd, overcast	S2.L1 walking	Sparse crowd	S3 Multiple Flows	Dense crowd, running

(*continued*)

Table 3. (*continued*)

S1: Person Count and Density Estimation		S2: People Tracking		S3: Flow Analysis and Event Recognition	
Set name	*Element*	*Set name*	*Element*	*Set name*	*Element*
S1.L2 walking	High density crowd, overcast	S2.L2 walking	Medium density crowd	S3 Event Recognition	Dense crowd
S1.L3 running	Medium density crowd, bright sunshine and shadows	S2.L3 walking	Dense crowd		

Fig. 2. The 8 Camera Views Used in the Dataset.

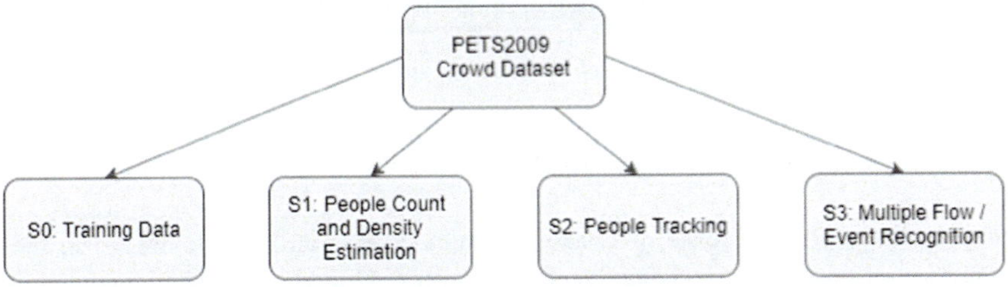

Fig. 3. Hierarchical structure of images in the dataset [4].

The part of the dataset named S3 consists of four different sequences with timestamps 14–16, 14–27, 14–31, and 14–33. S3, multiple flow analysis and event recognition sub-datasets are used for this research. These videos include running and walking as well as various crowd behaviors. The details of the dataset used are presented in Table 4. Examples of walking and running behaviors in the dataset used are shown in Fig. 4.

Table 4. Summary of video sequences prepared for behavior analysis in the PETS2009 dataset.

Behavior	Video	Frame	Sequence
Walking/Running	S3-HL, 14–16, View-001-p1	107	Seq1
Walking/Running	S3-HL, 14–16, View-001-p2	114	Seq2
Walking/Running	S3-HL, 14–33, View-001	377	Seq3
Walking/Running	S3-HL, 14–31, View-001	130	Seq5

Fig. 4. Samples of the behavior analysis sequences.

4 Problem Statement and Methods

Videos consist of a sequence of images. Various objects and people are set in these images, and these objects change position in time. Crowd monitoring analyzes the motion, speed, and behavior of groups of people within the frame. The relationship between frames is comprehended thence. Video frames are used in the input of the method used for this study. The people in the frames are detected using by the YOLOv3 (You Look Only Ones version 3) that up-to-date deep learning method. When there are people from different groups in the image, the KNN clustering algorithm is used for detection of the groups. Walking and running behaviors are predicted after these processes. The handling of the problem is shown in Fig. 5.

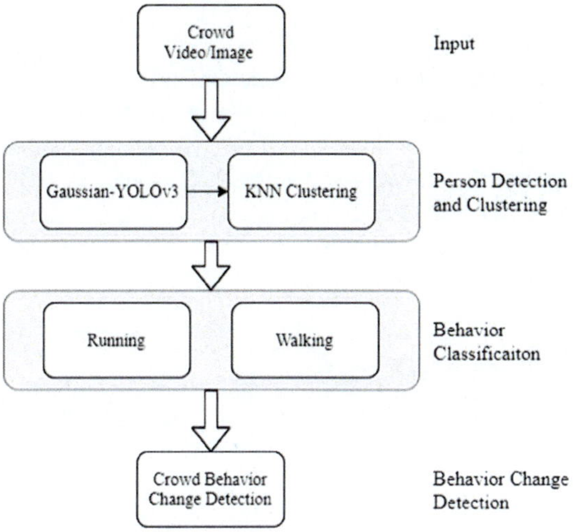

Fig. 5. Problem statement and method.

Walking and running actions can be defined as follows:

- The act of walking shows the change in the position of a group of people within the frame at an average speed.
- Similar to the act of walking, but causes a faster change between frames.

In order to determine the walking and running actions, the center of gravity velocities of each cluster detected in the image is calculated as time series. Using the velocity change curve shown in Fig. 6, behavior change times in the image are also determined.

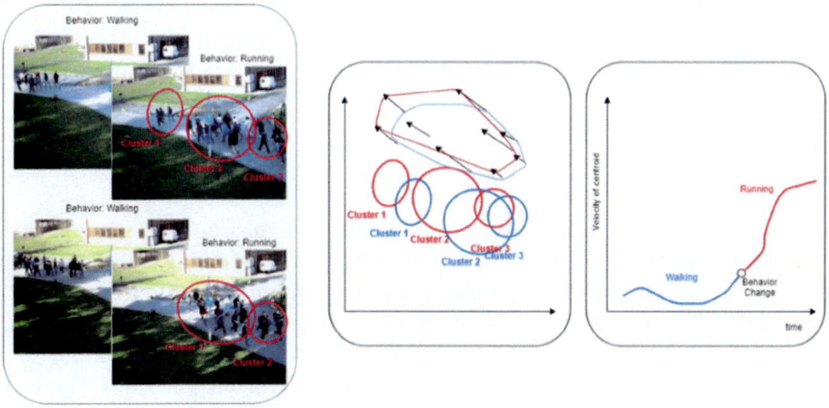

Fig. 6. Cluster-based crowd monitoring approach.

4.1 Gaussian-YOLOv3

You Look Only Ones (YOLO) deep learning architectures are designed to detect objects and their locations in the image in real-time. Feature map estimates are shown with a bounding box with specified coordinates. YOLOv3 calculates whether there is an object in the image, the type of the object is probabilistic. Gaussian YOLOv3 allows a precise improvement in object coverage of bounding boxes. Figure 7 shows YOLOv3 model architecture and feature map structure [2].

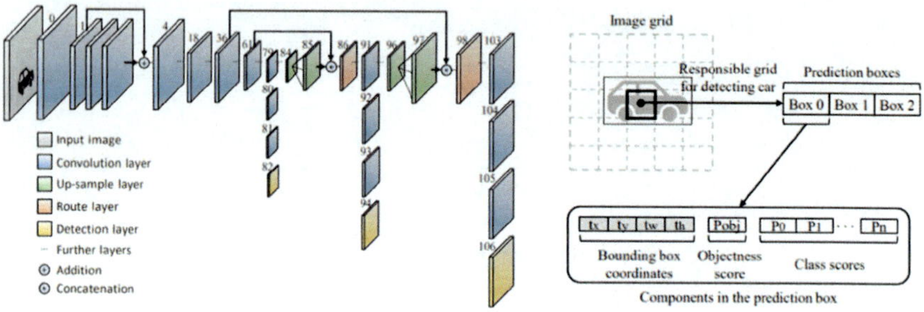

Fig. 7. YOLOv3 deep learning model architecture [2].

In the YOLOv3 model, bounding box coordinates are calculated using regression. In such a manner, it is possible to perform localization. However, there is always the problem of uncertainty. This problem can be overcome by using the Gauss model. The y output consisting of Gaussian parameters for an x test input can be expressed as follows [2]:

$$p(y|x) = N(y; \mu(x), \Sigma(x)) \tag{1}$$

where $\mu(x)$ and $\Sigma(x)$ are the mean and variance functions, separately.

Uncertainty of bounding box, each of the bounding box coordinates in the prediction feature map is modeled as the mean (μ) and variance (Σ) for the prediction as shown in Fig. 8.

4.2 KNN Clustering

The K-nearest neighbor (KNN) clustering algorithm is a robust classifier model which an old but successful approach used to classify unlabeled points using labeled points and distance information. KNN is a fast and efficient algorithm makes it preferred in real-time studies. While assigning new labels, different distance metrics such as Manhattan, Euclidean and Minkowski can be used and the algorithm calculates according to the information labeled as "k" in the number of neighbors [3]. In this study, it is proposed to detect the crowds grouping in the image and to examine their speed and direction with respect to each other (Fig. 9).

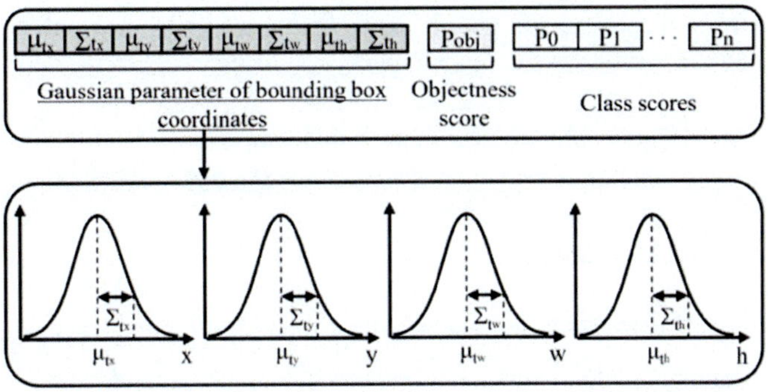

Fig. 8. Elements in the prediction box of the proposed algorithm [2].

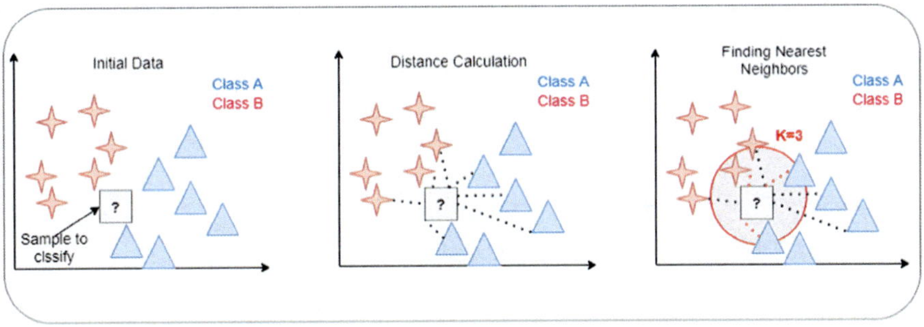

Fig. 9. Basic representation of KNN algorithm.

5 Experiments

Within the scope of the experiment, the number of people in the PETS2009 dataset, and their walking/running events are tried to be determined. Details on the preferred programming language, hardware requirement, parameter regulation, and evaluation metrics are explained.

Programming Language, Libraries and Versions: The main experiments of this research were performed using the 3.7 version of Python as a programming language, 2.1 recent version of TensorFlow as a deep learning library.

GPU's Hardware: The chosen deep learning models are trained on TitanXp local GPUs produced by NVIDIA.

Parameter Regulation: According to research, the number of clusters with a fixed width and two classes (k = 2) is chosen in order not to decrease the processing speed. The threshold setting for the Gaussian YOLOv3 is set to 0.

Evaluation Metrics: Sequences 1, 2, 3, and 5 in the dataset are used. The results obtained in these sequences were compared in terms of both the number of people in the frame and the behavior class with the ground truth values, World Coordinate System (WCS) results, and up-to-date studies in the literature. The accuracy evaluation was calculated as follows and the results are shown in Table 6.

$$1 - \text{Detected} - \text{Ground Truth} / \text{Total Number of Frame} - \text{Ground Truth} \quad (2)$$

6 Results and Discussion

Among the crowd monitoring approaches, an up-to-date literature review is included, especially on behavioral change and predicting the number of people. The performance of the PETS2009 dataset for 4 video sequences is evaluated using Gaussian YOLOv3, a special version of YOLOv3 which is a current real-time deep learning model. KNN clustering method, which is known to be efficient, is also used to identify groups in the crowd. The behavior change time calculated for sequence 1 is shown graphically in Fig. 10. x-axis frame number, y-axis is the average value of the Euclidean distance between both sets. In Fig. 11, the images that are detected as frames are given in which a group of people switched from walking to running. This refers to the 52nd frame in Fig. 10.

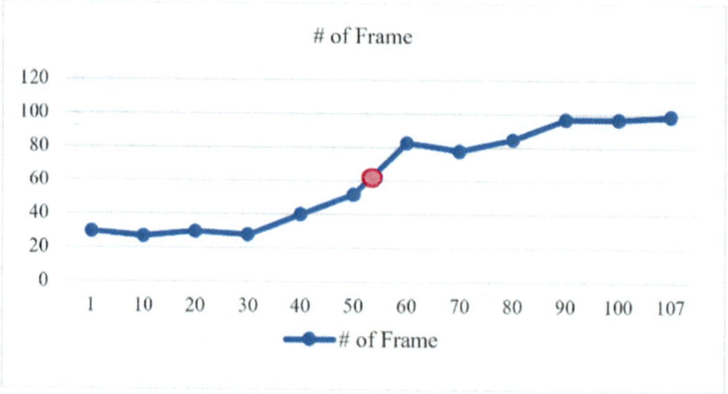

Fig. 10. Sample output time series obtained for walking/running behavior for sequence 1.

Table 5 shows the crowd counting results in each sequence obtained with the ground truth and the proposed method. Table 6 shows the accuracy of the results obtained. As a result, our cluster-based crowd monitoring approach performs well for all four video sequences compared to the method we propose with other (SOTA) methods.

Fig. 11. A group of people changing from walking behavior to running behavior.

Table 5. Results of detection: walking/running behavior.

Video Sequences	Ground Truth	Our Method
Seq1	38	39
Seq2	56	62
Seq3	335	339
Seq5	76	83

Table 6. Accuracy comparisons of the proposed method and recent approaches for 4 video sequences.

Ref	Accuracy result of Seq 1	Accuracy result of Seq 2	Accuracy result of Seq 3	Accuracy result of Seq 5
WCS	85.5%	63.8%	81.0%	87.0%
BRIASS	79.7%	50.0%	69.0%	–
Event detection [12]	95.7%	82.8%	95.2%	90.7%
Proposed method	96.4%	83.2%	95.6%	93.8%

7 Conclusion and Future Work

In this study, a new perspective on detecting changes in crowd behavior in video frames is presented by using up to date deep learning model Gaussian YOLOv3 and efficient KNN clustering approaches together. A comparable achievement was come off with studies that followed behavior classification and detection of behavior change on the same dataset. With the Gaussian YOLOv3 model, people and their locations are detected in real-time. With clustering, it was ensured that the groups with different behaviors at the time of the image were determined. Six video series under the PETS2009 dataset are preferred for this study. Accuracy performance between 83.2% and 96.4% was accomplished.

In the future, it will be studied on the recognition of behaviors such as merging and splitting in addition to walking and running with different clustering approaches in larger and more crowded data sets. Crowd monitoring researches, which is critical for the safety of society and people, are aimed to be used in the control of the COVID-19 epidemic that continues today and affects the whole world.

Acknowledgments. The authors wish to acknowledge the support of the NVIDIA GPU Grant Program with the donation of the TITAN Xp GPU.

References

1. Khan, A., Shah, J.A., Kadir, K., Albattah, W., Khan, F.: Crowd monitoring and localization using deep convolutional neural network: a review. Appl. Sci. **10**, 4781 (2020). https://doi.org/10.3390/app1014478
2. Choi, J., Chun, D., Kim,H., Lee, H.-J., GaussianYOLOv3: an accurate and fast object detector using localization uncertainty for autonomous driving. In: International Conference on Computer Vision (ICCV), (2019), https://arxiv.org/abs/1904.04620 (2020)
3. Cunningham, P., Delany, S.J.: k-Nearest Neighbour Classifiers, Technical Report UCD-CSI-2007-4 March 27 (2007)
4. Ferryman, J., Shahrokni, A.: PETS2009: dataset and challenge. In: Twelfth IEEE International Workshop on Performance Evaluation of Tracking and Surveillance (2009)
5. Wang, J. Xu, Z.: Crowd anomaly detection for automated video surveillance. In: Proceedings of the 6th International Conference on Imaging for Crime Prevention and Detection (ICDP-15), London, UK, 15–17 July (2015)
6. Rao, A.S., Gubbi, J., Palaniswami, M.: Anomalous crowd event analysis using isometric mapping. In: Advances in Signal Processing and Intelligent Recognition Systems, Springer: Berlin/Heidelberg, Germany, pp. 407–418 (2016)
7. Fradi, H., Luvison, B., Pham, Q.C.: Crowd behavior analysis using local mid-level visual descriptors. IEEE Trans. Circuits Syst. Video Technol. **27**, 589–602 (2016)
8. Palanisamy, G., Manikandan, T.: Group behaviour profiling for detection of anomaly in crowd. In: Proceedings of the International Conference on Technical Advancements in Computers and Communications (ICTACC), Melmaurvathur, India, pp. 11–15 (2017)
9. Xu, F., Rao, Y., Wang, Q.: An unsupervised abnormal crowd behavior detection algorithm. In: Proceedings of the International Conference on Security, Pattern Analysis, and Cybernetics (SPAC), Shenzhen, China,, pp. 219–223 (2017)
10. Choudhary, S., Ojha, N., Singh, V.: Real time crowd behavior detection using SIFT feature extraction technique in video sequence. In: Proceedings of the International Conference on Intelligent Computing and Control Systems (ICICCS), Madurai, India, pp. 936–940 (2017)
11. Chibloun, A., Fkihi, S.E., Mliki, H., Hammami, M., Thami, R.O.H.: Abnormal crowd behavior detection using speed and direction models. In: Proceedings of the 9th International Symposium on Signal, Image, Video and Communications (ISIVC), Rabat, Morocco, pp. 197–202 (2018)
12. Yang, M., Rashidi, L., Rao, A.S., Rajasegarar, S., Ganji, M., Palaniswami, M., Leckie, C.: Cluster-based crowd movement behavior detection. In: Proceedings of the Digital Image Computing: Techniques and Applications (DICTA), Canberra, Australia, pp. 1–8 (2018)

13. Bharti, Y., Saharan, R., Saxena, A.: Counting the number of people in crowd as a part of automatic crowd monitoring: a combined approach. In: Information and Communication Technology for Intelligent Systems; Springer: Berlin/Heidelberg, Germany, pp. 545–552 (2019)
14. Mohamed, S.A.E., Parvez, M.T.: Crowd modeling based auto activated barriers for management of pilgrims in mataf. In: Proceedings of the International Conference on Innovative Trends in Computer Engineering (ITCE), Aswan, Egypt, 19–21 Februry, pp. 260–265 (2019)
15. Yimin, D., Fudong, C., Jinping, L., Wei, C.: Abnormal behavior detection based on optical flow trajectory of human joint points. In: Proceedings of the Chinese Control and Decision Conference (CCDC), Nanchang, China, 3–5 June, pp. 653–658 (2019)
16. Sadiq, F.I., Selamat, A., Ibrahim, R., Krejcar, O.: Enhanced approach using reduced SBTFD features and modified individual behavior estimation for crowd condition prediction. Entropy **21**, 487 (2019)
17. Alginahi, Y.M., Mudassar, M., Kabir, M.N., Tayan, O.: Analyzing the crowd evacuation pattern of a large densely populated building. Arab. J. Sci. Eng. **44**, 3289–3304 (2019)
18. Gao, M., Jiang, J., Ma, L., Zhou, S., Zou, G., Pan, J., Liu, Z.: Violent crowd behavior detection using deep learning and compressive sensing. In: Proceedings of the Chinese Control and Decision Conference (CCDC), Nanchang, China, 3–5 June, pp. 5329–5333 (2019)
19. Chan, A.B., Liang, Z.-S.J., Vasconcelos, N.: Privacy preserving crowd monitoring: counting people without people models or tracking. In: Proceedings of the IEEE Conference on Computer Vision and Pattern Recognition, Anchorage, AK, USA, 23–28 June, pp. 1–7 (2008)
20. Chen, K., Gong, S., Xiang, T., Loy, C.C., Cumulative Attribute Space for Age and Crowd Density Estimation, In Proceedings of the IEEE Conference on Computer Vision and Pattern Recognition, Portland, OR, USA, 23–28 June, pp. 2467–2474 (2013)
21. Idrees, H., Saleemi, I., Seibert, C., Shah, M.: Multi-source multi-scale counting in extremely dense crowd images. In: Proceedings of the IEEE Conference on Computer Vision and Pattern Recognition, Portland, OR, USA, 23–28 June, pp. 2547–2554 (2013)
22. Zhang, Y., Zhou, D., Chen, S., Gao, S., Ma, Y.: Single-image crowd counting via multi-column convolutional neural network. In: Proceedings of the IEEE Conference on Computer Vision and Pattern Recognition, Las Vegas, NV, USA, 27–30 June (2016)
23. Idrees, H., Tayyab, M., Athrey, K., Zhang, D., Al-Maadeed, S., Rajpoot, N.: Composition loss for counting, density map estimation and localization in dense crowds. In: Proceedings of the European Conference on Computer Vision (ECCV), Munich, Germanypp. 8–14 September, pp. 532–546 (2018)
24. Zhang, C., Li, H., Wang, X., Yang, X.: Cross-scene crowd counting via deep convolutional neural networks. In: Proceedings of the IEEE Conference on Computer Vision and Pattern Recognition, Boston, MA, USA, 7–12 June (2015)
25. Shao, J., Kang, K., Loy, C.C., Wang, X.: Deeply learned attributes for crowded scene understanding. In: Proceedings of the IEEE Conference on Computer Vision and Pattern Recognition, Boston, MA, USA, 7–12 Jun, pp. 4657–4666 (2015)

Parkinson's Disease Diagnosis Based on Vertical Ground Reaction Force Signals of Gait Using Deep Neural Network

Saeideh Masoumi[1](✉), Maryam Rastgarpour[2], and Nader Jafarnia Dabanloo[1]

[1] Islamic Azad University, Science and Research Branch, Tehran, Iran
s666.masoumi@gmail.com, jafarnia@srbiau.ac.ir
[2] Islamic Azad University, Saveh Branch, Saveh, Iran
m.rastgarpour@iau-saveh.ac.ir

Abstract. Parkinson's disease (PD) is one of the most common neurodegenerative disorders globally, and sometimes it takes a long time to diagnose. Therefore, it is essential to use a smart and accurate system to quickly diagnose it in the early stages and check the disease's progress. Parkinson affects the way that people walk, and these changes can be used to diagnose it. One type of data in gait analysis is the vertical ground reaction force (VGRF), which is studied as a signal or time series. In this study, continuous wavelet transform (CWT), due to its high resolution in both frequency and time domains, analyzes the gait signals from 8 sensors under each person's foot, and two dimensional (2D) data are given as input to the deep neural network. Then, good effects on the accuracy, the deep neural network is used in data extraction and data classification fields. This research's fundamental is the convolutional neural network (CNN) because of its high feature extraction strength. In this study, the result of separating two signals feet on the binary classification accuracy of the problem is observed, too. In conclusion, it was found that the accuracy rate of the proposed method is 88.47%, which has a good effect on diagnosing people with PD from others.

Keywords: Parkinson's disease (PD) · Gait analysis · Vertical ground reaction force (VGRF) · Continuous wavelet transform (CWT) · Deep neural network (DNN) · Convolutional neural network (CNN)

1 Introduction

Parkinson's disease (PD) is a neurological disorder that mainly affects dopamine neurons in a specific part of the brain. Usually, The symptoms of this disease appear slowly over the years. The progression of symptoms often varies from person to person due to the variety of signs of PD. People with PD may experience symptoms such as restless tremors, bradykinesia, limb stiffness, and difficulty walking and balance. The cause is mostly unknown. There is no definitive cure

T. Allahviranloo et al. (Eds.): IDS 2020, AISC 1301, pp. 254–264, 2021.
https://doi.org/10.1007/978-3-030-66501-2_20

for the disease, but medications and surgery can help. Although Parkinson's itself is not fatal, the disease's complications can be severe [1]. Sometimes it can take up to a year for a PD to be diagnosed after a thorough examination of the patient's neurological history and clinical evaluation. Also, due to the lack of accuracy in objective measurements, there is a high probability of PD error detection [2]. Therefore, early and accurate diagnosis of PD is essential for better treatment and disease control. The differences in the way that healthy people and people who have Parkinson walk, gait analysis, or especially walking analysis of these patients can play an important role in diagnosing the PD.

Gait analysis is the scientific study of animal movements, especially human movement. Determining the gait characteristics, disorder, and classification is the primary goal of gait analysis to diagnose better and treat patients [3]. In gait analysis, kinetic criteria focus mainly on the forces acting between the foot and the ground and are measured in units of force [4]. The vertical reaction force of the ground is also one of the kinetic criteria. The forces between the feet and the ground are some features that gait analysis focuses on them. According to Newton's third law, there is an equal reaction for every action in the opposite direction. Due to the gravitational force of ground, there is a force contrary to human weight direction, known as the ground reaction force. This force is applied in different coordinates. If this force applied in the vertical direction is called the vertical ground reaction force (VGRF).

Different methods analyze a signal. Fourier Transform is one of these methods which decomposes a signal into frequency components but has zero resolution in the time domain. Therefore, a modified version of it, Short Time Fourier Transform (STFT), is more common; however, STFT faces the Heisenberg's Uncertainty Principle problem [5]. Wavelet Transform has a high resolution in both time and frequency domains. Wavelet Transform determines both the number of frequencies in the signal and the frequency where they occur. Wavelet Transform achieves this ability by working on different scales. It first considers the signal with a large scale or window and analyzes its characteristics. In the next step, it considers small scales or windows and obtains small signal characteristics. Wavelet Transform uses several functions called wavelet functions; each has a different scale [6]. In this study, Continuous Wavelet Transform is conducted for VGRF signals measured from 16 sensors under the right and left human's feet.

Numerous research has been done on diagnosing Parkinson's disease based on various features and classifiers on VGRF data. First, we mentioned some of these studies based on the features that are used and their methods. Some approaches have used time-domain characteristics, such as K-nearest neighbour (K-NN), classification and regression tree (CART), random forest (RF), support vector machine (SVM), K-means, and Gaussian mixture model (GMM) by Nicolas Khoury et al. [7]. This method provided an accuracy of 80–91%. Some have used frequency domain features like neural network with weighted fuzzy membership functions (NEWFM) by Sang-Hong Lee et al. [8]. They used frequency features, which obtained from Wavelet Transform and reported an accuracy of 74–77%. Besides, in some studies, they achieved frequency features from Fast

Fourier Transform (FFT), such as Enas Abdulhay et al. [9] that performed the Medium Gaussian SVM technique and the final result was accuracy 92.7%. Also, Shyam V. Perumal et al. [4] worked on frequency data presented by FFT, and they used Latent Dirichlet Allocation (LDA) as a classifier, and the outcome of this proposed method was the accuracy of 86.9%. Also, there are some methods used both time and frequency features. Two of them are RF by Koray Açıcı et al. [10] and SVM by Mahamad Nabab Alam et al. [11] that Koray Açıcı et al. used FFT. The first approach reached an accuracy of 74–98%, and the achieved result was accuracy in 85–95%. Many other classifier have been used to diagnose Parkinson's patients, too. Recently, some approaches based on deep learning, especially deep neural networks, are conducted, and they are mentioned in this section. Nancy Jane et al. [12] constructed a Q-backpropagated time-delay neural network (Q-BTDNN) classifier with considering left and right signals separating, and the best result was accuracy 94.19%. Imanne El Maachi et al. [13] developed a deep neural network (DNN) on one-dimensional VGRF signals and achieved 98.7% accuracy. They also used another type of data to classified patients into five different levels of PD severity and reached 85.3% accuracy in this part. In another work, Yi Xia et al. [14] diagnosed PD from VGRF data using a convolutional neural network (CNN) with Long Short-Term Memory (LSTM) as several different architectures, CNN-LSTM, DCNN, DALSTM, DCLSTM, DCALSTM. In addition to binary classification, patients were classified into several categories with different severities, and in binary classification using DCALSTM structure, they saw a very reasonable 99% accuracy. All of the methods were described, used VGRF data from PhysioNet online dataset [15].

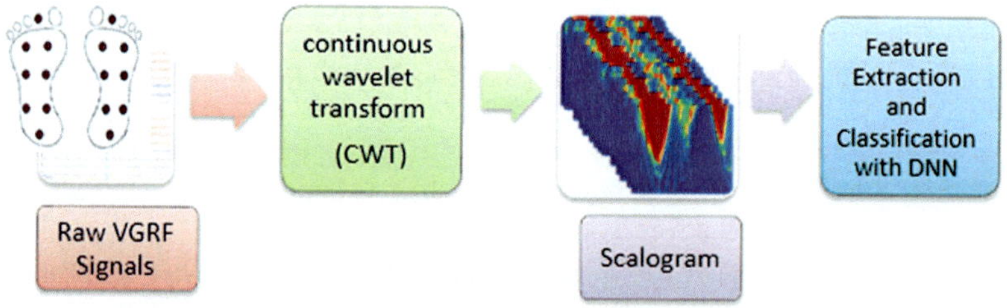

Fig. 1. The general architecture of the proposed method. 16 scalogram for 16 VGRF signals. The DNN includes CNN for feature extraction and the FC network for classification.

The proposed method is based on deep learning. In this study, our data are VGRF signals from PhysioNet online dataset [15], and we used Continuous Wavelet Transform (CWT) for preprocessing that we discuss them in Sect. 2. Both time and frequency features were received from CWT and presented like a scalogram, in other words, one dimension is time, and another one is frequency. Our method includes two architectures. For the first structure, we used one CNN to extract features, and after that, we developed it with a fully Connected (FC) network for classification. The second structure uses two CNNs, each for one foot for feature extraction, then these extracted features concatenated. So, a DNN was presented as a classifier in this structure. The entire network is called DNN. The general proposed method structure is illustrated (see Fig. 1), and in Sect. 3, the details of this DNN are explained. Eventually, the results of this research are reviewed and analyzed in Sect. 4.

2 Data and Preprocessing

2.1 Data Description

The PhysioNet [15] online dataset used in this study includes VGRF recorded from 16 sensors underneath feet at a sampling rate of 100 Hz. How these sensors are placed under the left, and the right foot is shown (see Fig. 2). This dataset is collected from 3 different sources Ga [16], Si [17], and Ju [18]. The PhysioNet dataset includes VGRF data from 93 patients with Parkinson's disease (PD) with an average age of 66.3 years and includes 66% of men with moderate disease severity and 73 control subjects (CO) with an average age of 66.3 years and includes 55% of men. The number of records in this dataset is 303 that Ju, Ga, and Si have 129, 110, and 64 records, respectively.

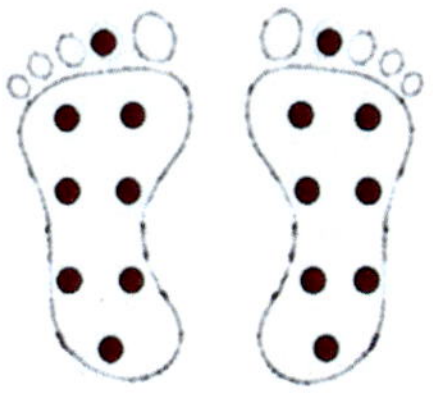

Fig. 2. The positions of 16 sensors under the left and right foot.

Each record in this dataset contains 19 columns as follows:

- Column 1, time in seconds
- Columns 2–9, VGRF from 8 sensors under the left foot
- Columns 10–17, VGRF from 8 sensors under the right foot
- Column 18, the total force under the left foot
- Column 19, the total force under the right foot

2.2 Data Equalization

Some records have different time samples. It means that each file has a table with a different number of rows. So, we select the same volume of rows. In this study, 40 first seconds (about 4000 first rows) were chosen, and the selected size was based on the minimum time in all records that exist in this dataset.

2.3 Wavelet Transform to Use both Time and Frequency Features

In this research, continuous wavelet transform (CWT) is a new solution for preprocessing with deep neural network classification and is also useful for using both time and frequency domain features compared to FFT. Original CWT is defined as (1)[6]:

$$F(a,b) = \frac{1}{\sqrt{a}} \int_{-\infty}^{+\infty} f(t)\psi^*(\frac{t-b}{a})dt \tag{1}$$

In (1), t is the time, $f(t)$ is the signal, a is the scale ($a > 0$), b is the translational value, and $\psi(t)$ is the wavelet function, and the * is the complex conjugate symbol [19]. Here, a and b in order are considered 32 and 1, the Morlet function is used as $\psi(t)$ and regarded as (2):

$$\psi(t) = e^{-\frac{t^2}{2}} cos(5t) \tag{2}$$

3 Proposed Method

3.1 Architecture of Model

CNNs like traditional deep neural networks (DNNs) are built of layers of neurons with learnable weights and biases. Each neuron receives some inputs, computes a dot product, and, finally, nonlinear function optionally. This process is iterated layer by layer until the prediction of the network is made. Traditional DNNs are essential; however, they do not have a good scale in computational resources. A CNN can be formed of different building blocks. Three primary layers for CNNs are convolutional (Conv) layer, pooling layer, and fully connected (FC) layer [19].

In this research, we use two structures in the feature extraction and classification sections. The first structure is to consider all sensors in general. The network uses a CNN to extract the feature and an FC network to classify. Inputs are given to the network in the form of tensors with size (32, 4000, 16). The hyper-parameters of this DNN are presented in Table 1. In Table 1, the convolution and the batch normalization layers are mentioned as Conv and BN in sequence. All Conv and Dense layers use the ReLU activation function, and only the sigmoid activation function is used in the output layer for binary classification. There are also stride with size (2, 2) in the two Max pooling layers.

Also, the first proposed structure diagram is depicted (see Fig. 3).

Table 1. Layer description for the proposed deep neural network.

Part	Layer No	Layer type	Number of unit	Kernel size
CNN	1	Conv 2D	4	(3,4)
	2	BN	-	-
	3	Max Pooling 2D	-	(2,2)
	4	Conv 2D	32	(3,16)
	5	BN	-	-
	6	Max Pooling 2D	-	(2,2)
	7	Flatten	-	-
	8	Dense	256	-
	9	BN	-	-
FC	10	Dense	256	-
	11	Dense	256	-
	12	Output	1	-

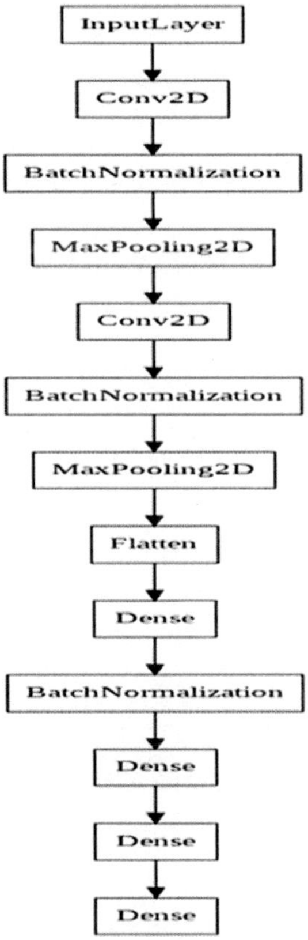

Fig. 3. The first proposed structure for the DNN.

3.2 Model Structure for the Left and Right Signals

In the second structure, the left and right foot signals are analyzed separately. Therefore, in this structure, two CNN networks are used to extract features. The inputs of this network are in the form of tensors with dimensions (32, 4000, 8). The hyper-parameters related to this deep neural network architecture are the same as the first structure (see Table 1). The second structure diagram is demonstrated (see Fig. 4).

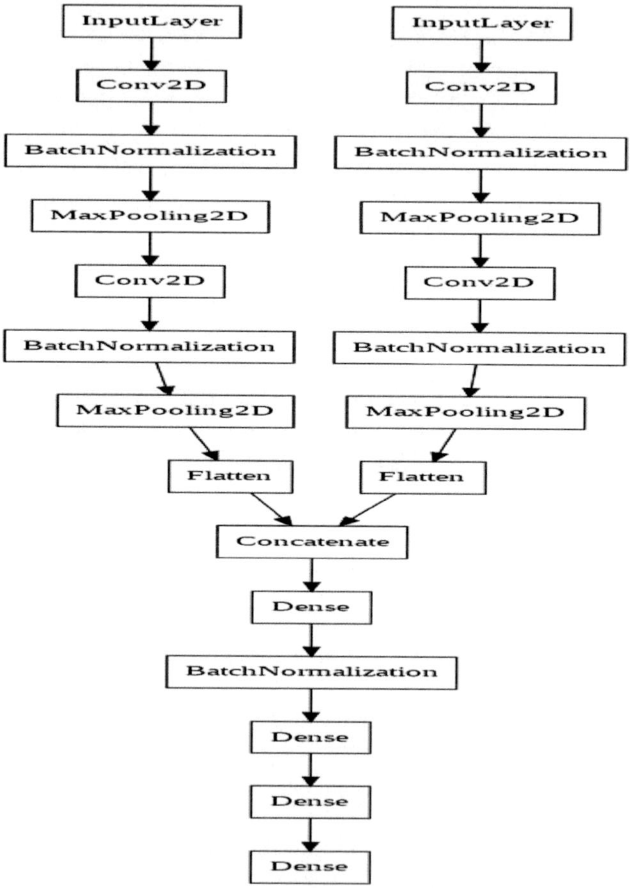

Fig. 4. The second proposed structure for the analysis of the left and right VGRF signals.

Table 2. The value or methods are used for both proposed architectures in this study.

Learning Parameter	Value or Method
Epoch	50
Learning rate	0.001
Learning metric	RMSprop
Batch size	10

3.3 Learning Parameters

In the learning part, 5fold cross-validation is used instead of the holdout method for more confidence in the result. After several experiments, the learning parameters were considered as values or methods are shown as a summary in Table 2. For example, the number of epochs was selected 10 at first, but after checking, 50 epochs was more appropriate.

3.4 Evaluation Metric

The evaluation metric of this research is accuracy. If we consider the CO group as the negative group (N) and people with PD as the positive group (P), accuracy is calculated as (3):

$$Accuracy = \frac{TP + TN}{TP + TN + FP + FN} \tag{3}$$

In (3), TP is the number of PD subjects, and TN is the number of CO subjects that are correctly classified. Also, FP is the number of CO subjects and FN number of PD subjects that are wrongly classified.

4 Results

The results obtained from both proposed structures are plotted in terms of the number of epochs (see Fig. 5).

In the first structure, which consisted of a CNN, and the data were analyzed together, the average accuracy was 83.46%. In the second structure and the presence of two CNNs to examine each foot's signals separately, the average accuracy was 88.47%. As a result, separate analyzing the features of the left and right foot VGRF has a good effect on the accuracy of the problem and has achieved higher accuracy compared to the first architecture. The results of these two DNN are presented in Table 3 in brief. Besides, the results of the two proposed structures on the dataset separated by collection source are also presented in Table 4.

As it is observed, the accuracy of dataset separation is also higher in the second structure than in the first structure, as seen in the aggregate data.

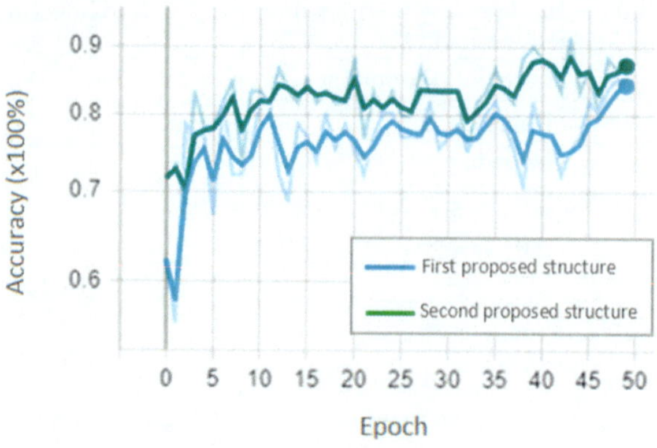

Fig. 5. The accuracy of both proposed structures. The blue line represents the accuracy of the first DNN structure, and the green line presents the accuracy of the second DNN structure.

Table 3. The result of the proposed model. The average accuracy of two DNN structure for total data.

Proposed Structure	Accuracy
The basic (first) structure of the proposed DNN	83.46%
Second DNN structure for the left and right VGRF signals	88.47%

Table 4. The average accuracy of the proposed model for Ju, Ga and Si data (separated by collection source).

Proposed structure	Ju	Ga	Si
First proposed structure	81.42%	75.45%	70.13%
Second proposed structure	84.62%	81.82%	76.92%

In addition, it can be seen that the highest and lowest accuracies are for Ju with the highest and Si with the lowest number of data in sequence. Also, in the total data, the highest accuracy has been observed. This subject indicates that as the data increase, the accuracy of the proposed method will increase.

Our proposed method is more accurate than some machine learning methods while it is not the best one. Some methods, such as DCALSTM by Yi Xia et al. [14] and DNN with 1D-Convnet by Imanne El Maachi et al. [13], are more accurate than the method presented in this study. This case could be due to the study of two structures in this research that some hyper-parameters of the deep neural network like the number of units or kernel size and learning parameters such as the number of epochs or learning rate are considered equally for both structures to pay more attention to the result of data separation of two legs. However, our proposed preprocessing and modelling have been more accurate than some of the previous methods such as NEWFM by Sang-Hong Lee et al. [8] and LDA by Shyam V. Perumal et al. [4]. Also, all these methods had different preprocessing and features for classifying PD subjects and CO subjects. For VGRF data from the PhysioNet dataset, the proposed method's performance is compared with previous studies in Table 5. These are some of the new and popular methods that worked on VGRF data for PD detection. Many other methods use different preprocessing and different features to diagnosis people with Parkinson's disease. In this study, we focused on the feature extraction and classification parts of each model for comparing.

Table 5. The comparison of the result of the proposed method with the results of other methods.

Method	Accuracy
NEWFM (Sang-Hong Lee et al. 2012) [8]	74–77%
LDA (Shyam V. Perumal et al. 2016) [4]	86.9%
Q-BTDNN (Nancy Jane et al. 2016) [12]	94.19%
RF (Koray Açıcı et al. 2017) [10]	74–98%
SVM (Mahamad Nabab Alam et al. 2017) [11]	85–95%
Medium Gaussian SVM (Enas Abdulhay et al. 2018) [9]	92.7%
K-NN, CART, K-means, GMM (Nicolas Khoury et al. 2019) [7]	80–91%
DALSTM, DCLSTM, DCALSTM (Yi Xia et al. 2019) [14]	96–99%
Deep 1D-Convnet (Imanne El Maachi et al. 2020) [13]	98.7%
Proposed DNN	83–88%

5 Conclusion

Because the diagnosis of Parkinson's disease sometimes takes a long time, the accuracy and speed of diagnosis in this disease are still medical challenges. Machine learning systems have been beneficial in this regard. Due to gait disorders in people with Parkinson's, our proposed system uses VGRF signals received from gait. In the proposed method, the wavelet transform was used for preprocessing. In the feature extraction and classification sections, we present a DNN that uses two networks, CNN and FC. We also performed this method by examining the data of each foot separately by two CNNs. Finally, the proposed method was presented with two different structures (see Fig. 3 and Fig. 4), according to the study, the results of the first and second structures were 83.46% and 88.47% accuracy respectively (see Table 3), which indicates the reasonable accuracy of the proposed DNN for the diagnosis of patients with PD. Besides, the result of the proposed method is compared to the results of other methods (see Table 5).

For future works, the study of hyper-parameters and conducting researches with using other transformations instead of CWT or different wavelet functions will be considered to improve the accuracy.

References

1. Parkinson's disease foundation. http://www.pdf.org
2. National parkinson foundation. http://www.parkinson.org
3. Akhtaruzzaman, M., Shafie, A.A., Khan, M.R.: Gait analysis: systems, technologies, and importance. J. Mech. Med. Biol. **16**(07), 1630003 (2016)
4. Perumal, S.V., Sankar, R.: Gait and tremor assessment for patients with parkinson's disease using wearable sensors. Ict Express **2**(4), 168–174 (2016)
5. Gao, R.X., Yan, R.: From fourier transform to wavelet transform: a historical perspective. In: Wavelets, Springer, pp. 17–32 (2011)
6. Dyllon, S., Xiao, P.: Wavelet transform for educational network data traffic analysis. Wavelet Theor. Appl. p. 155 (2018)
7. Khoury, N., Attal, F., Amirat, Y., Oukhellou, L., Mohammed, S.: Data-driven based approach to aid parkinson's disease diagnosis. Sensors **19**(2), 242 (2019)
8. Lee, S.H., Lim, J.S.: Parkinson's disease classification using gait characteristics and wavelet-based feature extraction. Expert Syst. Appl. **39**(8), 7338–7344 (2012)
9. Abdulhay, E., Arunkumar, N., Narasimhan, K., Vellaiappan, E., Venkatraman, V.: Gait and tremor investigation using machine learning techniques for the diagnosis of parkinson disease. Future Gener. Comput. Syst. **83**, 366–373 (2018)
10. Açıcı, K., Erdaş, Ç.B., Aşuroğlu, T., Toprak, M.K., Erdem, H., Oğul, H.: A random forest method to detect parkinson's disease via gait analysis. In: International Conference on Engineering Applications of Neural Networks, Springer, pp. 609–619 (2017)
11. Alam, M.N., Garg, A., Munia, T.T.K., Fazel-Rezai, R., Tavakolian, K.: Vertical ground reaction force marker for parkinson's disease. PloS one **12**(5), e0175951 (2017)
12. Jane, Y.N., Nehemiah, H.K., Arputharaj, K.: A q-backpropagated time delay neural network for diagnosing severity of gait disturbances in parkinson's disease. J. Biomed. Inform. **60**, 169–176 (2016)
13. El Maachi, I., Bilodeau, G.A., Bouachir, W.: Deep 1d-convnet for accurate parkinson disease detection and severity prediction from gait. Expert Syst. Appl. **143**, 113075 (2020)
14. Xia, Y., Yao, Z., Ye, Q., Cheng, N.: A dual-modal attention-enhanced deep learning network for quantification of parkinson's disease characteristics. IEEE Trans. Neural Syst. Rehabil. Eng. **28**(1), 42–51 (2019)
15. Hausdorff, J.M., et al.: Gait in parkinson's disease. http://www.physionet.org
16. Yogev, G., Giladi, N., Peretz, C., Springer, S., Simon, E.S., Hausdorff, J.M.: Dual tasking, gait rhythmicity, and parkinson's disease: which aspects of gait are attention demanding? Eur. J. Neurosci. **22**(5), 1248–1256 (2005)
17. Frenkel-Toledo, S., Giladi, N., Peretz, C., Herman, T., Gruendlinger, L., Hausdorff, J.M.: Treadmill walking as an external pacemaker to improve gait rhythm and stability in parkinson's disease. Mov. Disord. Official J. Mov. Disord. Soc. **20**(9), 1109–1114 (2005)
18. Hausdorff, J.M., Lowenthal, J., Herman, T., Gruendlinger, L., Peretz, C., Giladi, N.: Rhythmic auditory stimulation modulates gait variability in parkinson's disease. Eur. J. Neurosci. **26**(8), 2369–2375 (2007)
19. Pinaya, W.H.L., Vieira, S., Garcia-Dias, R., Mechelli, A.: Convolutional neural networks. In: Machine Learning, Elsevier, pp. 173–191 (2020)

Genetic Algorithms and Neural Networks by Data Mining

Djamila Benhaddouche(✉)

University of Science and Technology of Oran USTO-MB, Oran, Algeria
benhaddouche.djamila@gmail.com

Abstract. We propose, in our study, an approach of hybridization (Genetic Algorithms/Neural Networks) in order to apply it to solve our problems, namely "optimization of the budget estimates of the Aval activity of Sonatrach". Indeed, within the downstream activity, the budget estimates of the expenditure are done every 5 years. Those must approach reality as much as possible in theory; therefore we took as bases training the table which contains the real expenditure (entries) 3 years previous. While studying the genetic algorithms and the networks of neurons separately, the idea came to us to use the genetic algorithms for the adjustment of the synaptic weights at the time of the phase of training of our network of neurons, and to use the capacity of prediction of the network of neurons to predict our budgetary expenditure. One starts from a population of more than 100 individuals (chromosomes), each one of them represents a configuration of weight, for the network of neurons, drawn by chance between [-1,1]. For each individual, one will calculate the average error between the calculated exits and the exits wished for all the examples of our base of training. The function fitness is anything else only this average error. One selects then the individuals ready to reproduce who are those which have the weakest error. One uses the operators of crossing and change to generate a new generation. This process will be repeated to L 'obtaining of an individual whose error will be minimal. The configuration of weight represented by this individual will be to use for our prediction.

Keywords: Dated genetic mining · Algorithms · Neural networks · Learning · Optimization · And prediction

1 Introduction

The objective of our work is to carry out a hybrid system between two techniques of dated mining: genetic algorithms and networks of neurons, which would be able to solve our problems to know: the optimization of the budget estimates of the expenditure within the downstream activity of Sonatrach.

2 Approach of Hybridization

2.1 Presentation

By studying the genetic algorithms and the networks of neurons, one can note that:
• The convergence of a network of neurons depends, partly, of the initial choice of only

T. Allahviranloo et al. (Eds.): IDS 2020, AISC 1301, pp. 265–270, 2021.
https://doi.org/10.1007/978-3-030-66501-2_21

one configuration of weights drawn by chance at the beginning from the training. • The genetic algorithms work on a broad field of exploration, and that thanks to their operators (selection, crossing, change), they make it possible to diversify and make evolve their populations to optimal solutions (method of optimization). Our approach uses the genetic algorithms for the generation and the adjustment of the synaptic weights at the time of a supervised training of our network of not buckled neurons. This last will be exploited thereafter for the prediction.

2.2 Operation

One starts from a population from 100 to 999 individuals (chromosomes), each one of them represents a configuration of weight (weight = gene) for the network of neurons. At the beginning, they are drawn by chance between [–1, 1]. For each individual, one will calculate the average error between the calculated exits and the exits wished for all the examples of our base of training, and this by using his genes. The function fitness is anything else only this average error which it is necessary to minimize by using the operators of selection, crossing and change. For that one selects the individuals ready to reproduce (those which have the weakest error), and one uses the operators of crossing and change to generate a new generation. This process will be repeated until obtaining of an individual whose error will be minimal.

Algorithm. Initialization of the matrix of the chromosomes (starting population); To import the base of training; To repeat to build the network of neurons with the weights (genes) of the first chromosome; To propagate the entries of the base of training through the network of neurons; To calculate the average error obtained between the calculated exits and the awaited exits; To associate this error the chromosome (fitness); To pass to the following chromosome; Until: to pass all the chromosomes To select the chromosomes ready to reproduce; Reproduction (crossing, change) in order to pass to a new generation; Until: A number of generations reached/Obtaining of an individual whose error is minimal.

2.3 Parameters

Selection:

We used the method of selection of the caster because it makes it possible to the best individuals to be more often selected.

Crossing:

We used the method of crossing uniform. This type of crossing is adapted for the chromosomes of big sizes and allows a greater diversification of the popultion.

Change:

According to the probability of change chosen, one modifies genes of the chromsome selected in their adding a certain actual value Q in the interval.

[10–6, 10–7].

Topology of the network of neurons:

For reasons of simplicity, flexibility and facility of comprehension, we used.

Perceptron multi-layer (PMC) whose configuration (a number of under layers, a number of neurons per layer.....) will be in experiments selected at the time of the phase of training.

3 Design and Implementation

3.1 The Data Base

We took, as bases training, the data base of the entries which contains the expenditure carried out during the three (3) previous years (there remains heard that the Aval activity carried out some modifications for reasons of confidentiality).

3.2 What is What an Entry?

The principle of any countable system consists of the monetary recording of an economic act at a given time. This recording is called an entry.

In accountancy one can say that an entry is economic event which will start the recording. This event is the movement of treasury (cashing, withdrawal).

Our center of interest was fixed on the withdrawal (expenditure), it is what is called of accountancy: loads.

3.3 Description of the Data Base

Our data base contains 44925 entries representing the expenditure of the activity between January 2003 and November 2005. It is made up of 7 fields.

3.4 Environment of Programming

For reasons of flexibility, its facility of use and for its speed of input/output, the language C++ builder version 6.0 pennies Windows XP is appropriate better for the development of our application. Windows XP has a good management of the memory as well as a directed architecture event. The language C++ builder is an environment of development graphic, flexible and easy to use, it makes it possible to carry out in a very simple way the interface of the applications because it rests on a very complete whole of visual components ready with employment. It is characterized by the concept of project which allows the implementation of several modules communicating between them, and pertaining.

3.5 Experimental Step

Our work followed the course of a process dated mining:

3.5.1 Comprehension of the Problem

After several interviews with the financial ones of the Downstream of Sonatrach, we noted that the latter were based on standards and averages of consumption, calculated starting from the preceding exercises, in order to establish the budget estimates.

Let us consider that the budgetary expenditure depends on several factors, those can be only fluctuating. Therefore, the use of the standards and the averages of consumption, for the establishment of the forecasts, can sometimes be the cause of a budgetary variation. From where need for analyzing and for including/understanding the variations of the former budgetary expenditure to predict the future.

3.5.2 Identification of the Relevant Data

In order to predict the budgetary expenditure, it is natural to take as bases training, the table of the entries.

3.5.3 Cleaning of the Data Base

This stage consists with:

- To eliminate the recordings with erroneous values (Null…).
- To eliminate the data let us double.
- To eliminate the recordings with missing values.

3.5.4 Coding of the Data

The entries of our network of neurons are:

- The code of the countable account of the expenditure.
- The period (countable month).

This stage consists in coding these fields of data in order to present them in entry for the network of neurons. For certain handling, it is necessary to modify the type of certain attributes (ex: the attribute PERIOD). Codified fields: Period: coded into binary on 8 bits as follows: Field uncoded intermediate Coding final Coding to the same project. Our data base is managed under Oracle.

01/2003 1 00000001 02/2003 2 00000010 03/2003 3 00000011…………….. 11/2005 35 00100011 Code countable account: coded into binary on 7 bits sequentially. We used a real coding on the level of the genetic algorithm.

3.5.5 Discovered Model

This stage consists with:

- -To adjust the parameters of the genetic algorithm (rate of change, cuts starting population, numbers generations…..).
- To adjust the parameters of the network of neurons (choice of topology, cuts sample of the training…)

- To import the base of training.
- -To launch the training.
- To emphasize of the result of the training thanks to a graph (left calculated, left desired).

3.5.6 Evaluation and Test

This stage consists with:

To visualize the history of the former trainings.
To import the file of test.
To launch the test.
To emphasize of the result of the test thanks to a graph (left calculated, left desired).

3.5.7 Implementation and Exploitation

Choice of the year of prediction.
Seized by the rate of inflation of the dinar envisaged.
Establishment of the budget estimates.

4 Results

The establishment of the budget estimates for the year 2007, with a rate of inflation of the dinar envisaged with 2.5% gave the following results (Figs. 1 and 2):

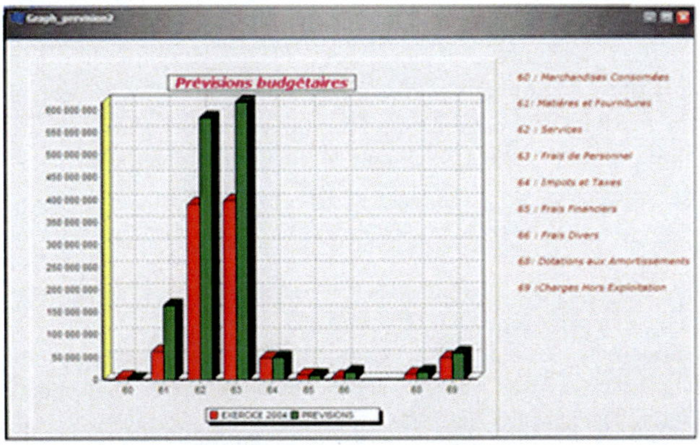

Fig. 1. Histograms representing a comparison enters the expenditure of 2004 and the forecasts of the year 2007.

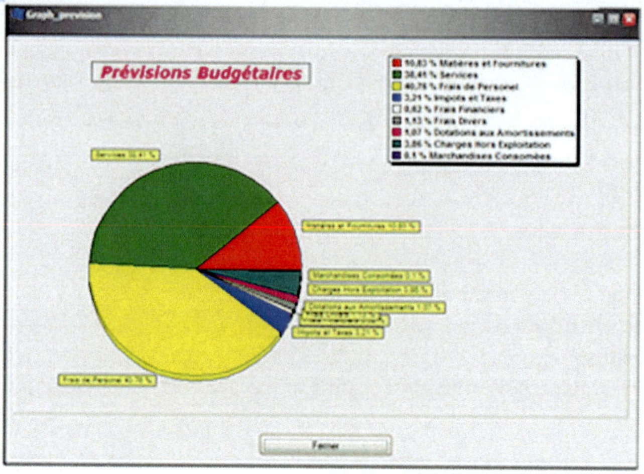

Fig. 2. Interface representing the distribution of the estimated budget.

5 Conclusion

Through this work, we propose an approach of hybridization between two techniques dated mining that we implemented and tested on the information system of the activity downstream (Sonatrach). Indeed, we proceeded to several trainings which enabled us to better observe the results of our work. We noticed that the time of training was tiny room thanks to the speed of the genetic algorithms and that the average error decreased from generation to generation. We also noted that the widening of the starting population contributed in a positive way to convergence of the system. Nevertheless, adjustment of the parameters such as: the number of under layers, the number of neurons by under layer and skews, were the disadvantage of our system. We think that our hybrid system will be more powerful if these parameters were adjusted (optimized) automatically. Therefore one proposes in prospect, the estimate and the optimization of these parameters by the genetic algorithm parallel to the adjustment of the synaptic weights.

References

1. Document of the company Sonatrach Downstream.
2. Back, B., Laitinen, T., Sere, K., Van Wezel, M.: Choosing Bankruptcy Predictors Using Discriminating Analysis, Logit Analysis, and Genetic Algorithms, September 1996
3. Gegout, C.: Initialization of the Networks of NonRecurring Neurons with real coefficients by Evolutionary Algorithms, Higher teacher training school of Lyon (2003).
4. Reisdroph, K.: The programmer, learn C++ Builder of 14 days (1998).
5. Soutou, C., Tests, O.: SQL for Oracle, December 2003.
6. Plane national accountant.

Proposing a Novel Artificial Neural Network Based Methodology for Forecasting Risk of Covid-19 Pandemic

Taha Osama Binhuraib(✉) and Gül T. Temur

Bahcesehir University, Yildiz Mah. Ciragan Cad., 34349 Besiktas/Istanbul, Turkey
tahaosamaa.binhuraib@bahcesehir.edu.tr, gul.temur@eng.bau.edu.tr

Abstract. The corona virus formally known as Covid-19 has taken the world by storm. In this article, we aim to analyze how harnessing the prowess of different computational methods –Machine Learning in particular; thus, helping policy makers take effective decisions by utilizing efficient approaches. Firstly, popularly known approaches (linear regression and logistic growth) are tried for existing data but it is figured out that accuracy rates are not satisfactory enough. Therefore, secondly, an artificial neural network (ANN) based methodology that consists of input data in different characters is proposed. It is figured out that the test results are more accurate with an R-squared score of (0.81) on an unseen data set, and handling the defined forecasting problem by using this multi-faceted data set is beneficial for policy makers, doctors and/or health managers that aim to have foresight on target groups at higher risk.

Keywords: Artificial neural network · Covid-19 · Forecasting · Pandemic

1 Introduction

Using environmental data to predict epidemics began in India in the early 1920s, amassing 30 years of records of various tropical diseases and nearly 50 years of meteorological data [1]. By concatenating the data, different risk maps were formed [2–4]. They provide governments with early warning systems and give them a valuable 2–3-month lead-time. Early and accurate predictions could prove vital in the fight against any epidemic; giving governments the time needed to develop mitigation plans and potentially save the lives of millions. In the related literature, differential equation models such as Susceptible, Infected, Recovered (SIR) model have been traditionally applied for infectious diseases [5]. In countries, where less governmental intervention was applied, the accuracy of SIR models showcased relative accuracy [6]. Besides that, researchers mostly have preferred to consider past data on number of cases or date of the cases only [7]. However, it is explicit that human behaviors, sociological events and the many unknown variables have impact on cases and their risks. Therefore, this study proposes an approach considering factors of life conditions besides traditional factors, and analyzes their relative performances in predicting the number of cases. Firstly, simple regression models are considered. Then a single layer artificial neural network (ANN) that has multi-dimensional input data is postulated.

T. Allahviranloo et al. (Eds.): IDS 2020, AISC 1301, pp. 271–281, 2021.
https://doi.org/10.1007/978-3-030-66501-2_22

In the light of these explanations, the study is designed as follows: Sect. 2 clarifies the relevant literature on forecasting the number of cases. Section 3 explains the traditional computational methods used. Section 4 describes the proposed ANN methodology. Section 5 criticizes the results. Section 6 concludes the study and offers further researches.

2 Literature Review on Forecasting Covid-19 Cases

In the literature related to mathematical epidemiology, the most widely used models for predicting the spread of a given disease are as the following [8]:

- The SIR Model
- The SIS Model
- The SIER Model

The models aforementioned are considered to be mathematical models that do not utilize machine learning algorithms and are mostly solved using ordinary differential equations.

As the nature of viruses are dependent on an array of factors, machine learning algorithms have been utilized in order to better predict spread viruses. In Table 1, the most notable algorithms used to solve different problems are shown.

Table 1. The most notable algorithms.

Authors	ML Algorithm	Outbreak
[9]	Neural Network	H1N1 flu
[10]	Random Forest	Swine fever
[11]	Bayesian Network	Dengue/Aedes
[12]	Random Forest	Influenza
[13]	Neural Network	Zika

To our knowledge, there is a lack of studies considering extra parameters besides physiological ones. In order to fill this gap, this study utilizes a comprehensive input data including day, population, population density, pollution, tourist number, number of universities and BCG (Bacillus Calmette–Guerin) vaccine regulation existence. The contribution of the study is originated from the consideration of multi-faceted factors simultaneously. Although the efficacy of infectious diseases is assumed to be based on some biological and physical reasons, factors such as the living habits of the people, the city they live in and their life culture are also effective in this process. For this reason, it would be more meaningful to evaluate such situations from a socio-technical perspective, not just a technical one. In our study, we tried to include parameters that take into account living conditions as much as we could. It contributes to the literature since it is a study that evaluates the patient's physical conditions as well as the effectiveness of living conditions.

3 Application of Traditional Computational Models

As a first step, forecasting procedure is conducted using conventional methods to test whether a more advanced method is needed. For this reason, linear and polynomial regressions, and logistic growth are chosen. The time series data was collected from Worldometers [14]. All the data used in the traditional methods are the number of cumulative cases from the first 123 in Turkey

Linear Regression: It is probably the first algorithm that comes to mind when dealing with time series data. Numerous predictions have been made using this approach, unfortunately due to the highly exponential and complex nature of viruses, the predictions made were not as accurate as intended. Linear regression simply aims to find a relation between two variables. Considering the simplicity of how this algorithm works, it is without a doubt a powerful tool that can give some useful insight to the data at hand. In Table 2, we share the results of linear regression using the Scikit learn library [15].

Table 2. Linear regression error [15].

Mean Absolute Error	14539
R-Squared	0.944

Although The R-Squared may seem relatively high, the expected positive relationship between the *"cumulative day"* variable and the number of cases accounts for a high R-squared score. Figure 1 illustrates the positive relationship between the two variables.

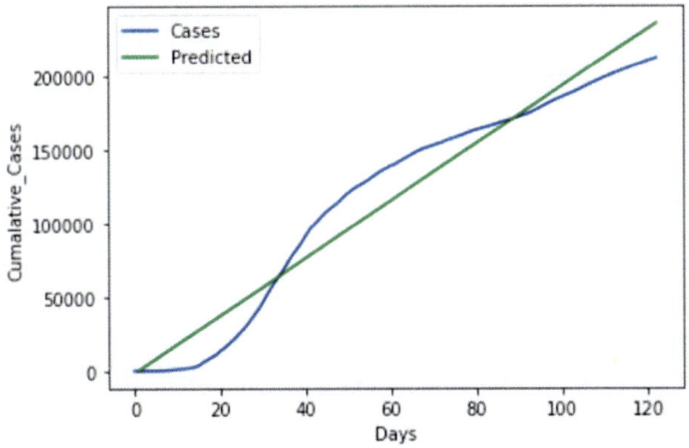

Fig. 1. Linear regression (predicted) plotted against real data.

Polynomial Regression: It models the relationship between an x independent variable and a y dependent variable as an nth degree polynomial with respect to x. The main goal of polynomial regression is to find a non-linear relationship between the independent variable x and the dependent variable y.

The general formula for an nth degree polynomial regression model is given in Eq. (1).

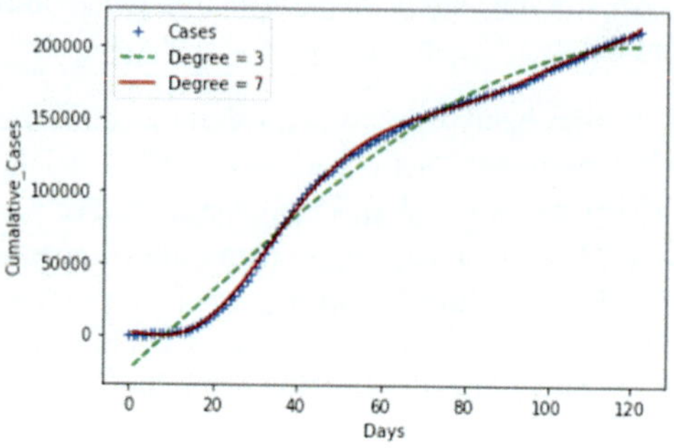

Fig. 2. Polynomial regression of 3rd & 7th order plotted against the real data

$$y = \beta_0 + \beta_1 x + \beta_2 x^2 + \beta_3 x^3 + \ldots + \beta_n x^n + \varepsilon \tag{1}$$

The Sklearn python library was used in order to optimize for the equation given above. Which aims to minimize the residual sum of squares between the observed targets in the dataset, and the targets predicted by the linear approximation, called the "residual sum of squares." The objective function is as shown in Eq. (2).

$$\|X_w - y\|_2^2 \tag{2}$$

Using the data gathered since the first case appeared in Turkey until the 123rd day. The results are given in Table 3.

Polynomial regression models can easily fit the data with higher degrees. As seen from Figs. 2 and 3 increasing the degree of the polynomial, enhanced the model's ability to fit the data.

Logistic Growth: Upon studying the nature in which viruses spread [16], it is observed that the rate increases "exponentially" until it reaches a point of maximum growth rate, which is called the inflection point. The rate then starts decreasing until the number of infected reaches the total number of susceptible people "L". Equation (3) describes the $P(t)$ which is the logistic function of t (t is the nth day after the first case).

$$P(t) = \frac{L}{1 + Ae^{-bt}} \tag{3}$$

The description above resembles that of a logistic growth model. Many predictions have been made using the logistic growth model, achieving higher accuracy rates and with lines that better fit the data. Figure 4 represents a logistic growth function.

Table 3. Polynomial regression error.

Mean Absolute Error	Value
2nd degree Mean Absolute Error	7778
3rd degree Mean Absolute Error	8198
4th degree Mean Absolute Error	4361
5th degree Mean Absolute Error	1798
6th degree Mean Absolute Error	1878
7th degree Mean Absolute Error	1189
8th degree Mean Absolute Error	921
10th degree Mean Absolute Error	3255

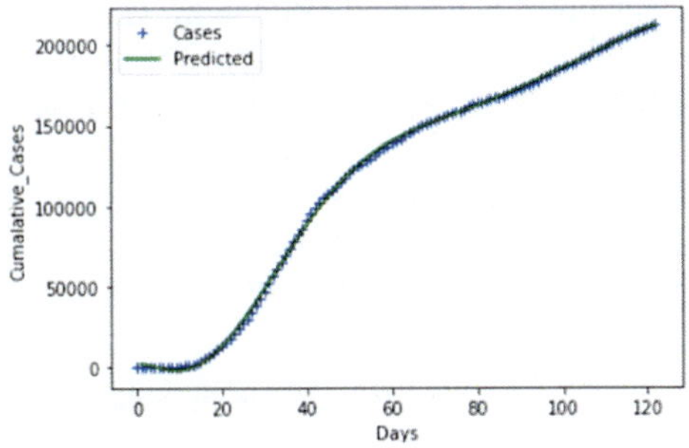

Fig. 3. Polynomial regression of 8th order plotted against the real data.

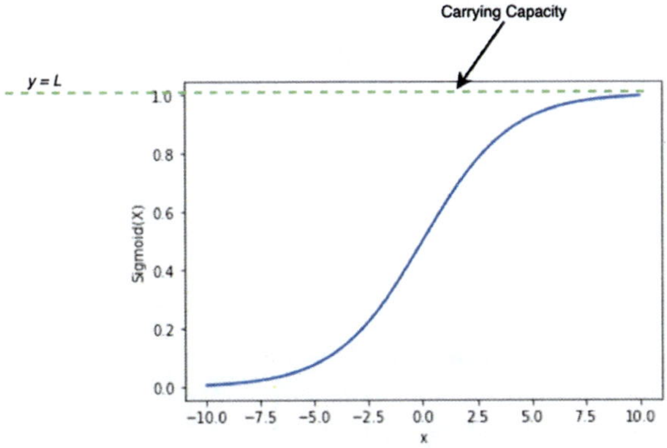

Fig. 4. Logistic growth function

The results for the above model were obtained using the SciPy scientific library which uses a Nonlinear Least Squares optimization technique, which aims to minimize

Eq. (4)

$$S = \sum_{i=1}^{m} r_i^2 \tag{4}$$

The minimum value is achieved when the gradient is equal to zero (as seen in Eq. (5)).

$$\frac{\partial S}{\partial \beta_j} = 2\sum_i \frac{r_i \partial r_i}{\partial \beta_j} = 0 \; j = (1, \ldots, n) \tag{5}$$

$$\beta = (\beta_1, \beta_2, .., \beta_n)$$

Using an iterative method is then used to optimize for the vector β parameters of size n.

As seen from Fig. 5, the number of cases started increasing exponentially after the 120th day. Due to the nature of logistic growth functions, 2nd or more waves cannot be accounted for. A simple way to overcome this would be to re-fit the curve. Keeping in mind that the model will be limited in the number of input point (days). Also considering how these models tend to overfit the data, in our case with a MAE of 9738 as seen from Table 4. such models are not suitable to be generalized and to be used for future predictions. The simplicity of these models cannot capture the complexity of an epidemic. More robust models with additional parameters are needed. Using more modern approaches, we will consider the use of an artificial neural network.

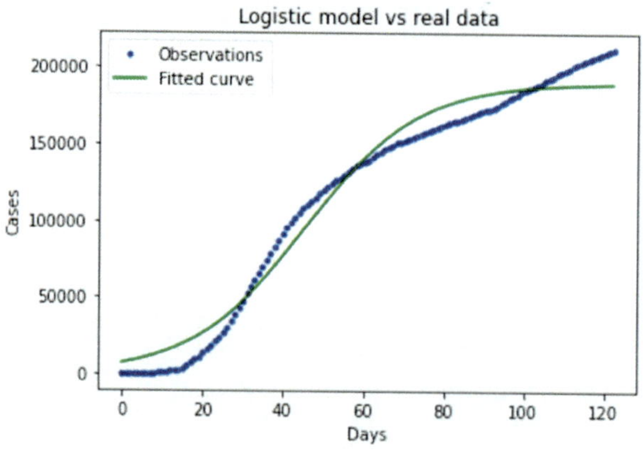

Fig. 5. Logistic growth fit against observations.

Table 4. Logistic growth error

The mean absolute error	9738.352

4 Design of a Novel Artificial Neural Network Approach

In the previous section, it is figured out that traditional methods are not strong enough to give satisfactory results. Therefore, as an alternative method, ANNs are utilized for the design of a novel approach for forecasting the number of cases. Inspired by biological neural models and highly resemblant of how animals learn by experience, ANNs aim to learn and extract patterns from large amounts of data [17]. Today, ANNs are considered to be the most powerful tool in machine learning. The main advantage in using this model is that it gives data scientists the option to be creative in determining the feature space, and using a neural network model might provide useful insight into the data and help governments determine the variables that affect the spread of the disease, giving them the opportunity to take policy decisions accordingly. Although simpler methods such as a Pearson coefficient could be used in determining variables that negatively or positively affect the number of cases. We are aiming for a more robust method that in essence could help us find hidden patterns in the data.

Considering the success of these models in prediction, we aim to develop a simple 1 hidden layer neural network in order to predict the number of Covid-19 cases.

4.1 Data Set Design

The unpredictable nature of the corona virus with its dependence on an array of variables has rendered traditional methods incapable of accurately predicting unseen data. The majority of models are most likely to overfit and thus generalizations of the trained models are not possible. This has motivated us to try and develop a neural network that will encompass different variables such as socio-economic, weather, vaccine regulations, density, and population. The features for the network is shown in Fig. 6.

PARAMETERS OF INPUT LAYER:
$x1$: *Day*
$x2$: *Population*
$x3$: *Population density*
$x4$: *Pollution*
$x5$: *Tourists*
$x6$: *Number of universities*
$x7$: *BCG vaccine regulation*

Fig. 6. The input parameters for the proposed model.

$$Activation\ function = \max(x, 0) \quad (6)$$

The Sklearn library was used in order to generate the neural network. As seen from Fig. 7, the network consists of 7 input points and a hidden layer with 3 nodes, each with a "Relu" (Eq. 6) activation function. The solver *'Adam'* which refers to a stochastic gradient-based optimizer proposed by Kingma, Diederik, and Jimmy Ba [18] was used to train the model.

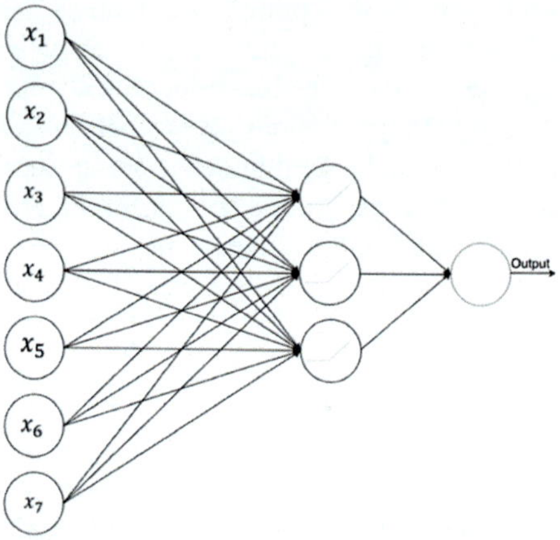

Fig. 7. Proposed neural network architecture.

The explanations for each data are given below:

- *Day:* *n*th day after the first case in given city.
- *Population:* Population of city. This input feature was normalized.
- *Population density:* Density (per km^2)
- *Pollution:* Fine particulate matter (PM2.5) is an air pollutant that can penetrate into the lung and impair lung function [19]. The data was collected from a Real-time Air Quality Index website
- *Tourists:* Number of tourist in millions per year.
- *Number of universities:* Number of universities in given city
- *BCG vaccine regulation:* 1: The country currently has universal BCG vaccination program. 2: The country used to recommend BCG vaccination for everyone, but currently does not. 3: The country never had universal BCG vaccination programs. One-hot encoding could be used in future implementations. [20].
- *Number of cases:* number of cases in given day.

Due to a lack of data, only 200 input points are used in order to determine the soundness of the method. Due to the limited space, only a part of data is shared at Table 5.

Table 5. Example of data used.

city_name	day	Population	density	pollution pm2.5	Tourist in millions/yr	No of universities	BCG vaccine	No_ofCases
New York	32	19453561	159	44	66	240	3	83506
New Jersey	1	8882190	467	30	101	47	3	4
Istanbul	27	15190000	2523	70	6	58	1	22171
London	16	8908081	4543	52	30	48	2	1588
London	17	8908081	4543	52	30	48	2	1965
London	18	8908081	4543	52	30	48	2	2189
New Jersey	6	8882190	467	30	101	47	3	23
New Jersey	7	8882190	467	30	101	47	3	29

Figure 8 illustrates how the model successfully converged after 28377 iterations with a final mean absolute error of 14626. Although this result may seem relatively high compared to the more traditional models used. A neural network's ability to combine a myriad of parameters and find patterns in data through hidden layers gives us the confidence that such a model could be generalized and is scalable. Considering that many countries exceeded 200.000 cases an absolute error that is 14626 is within the range to drive policy changes.

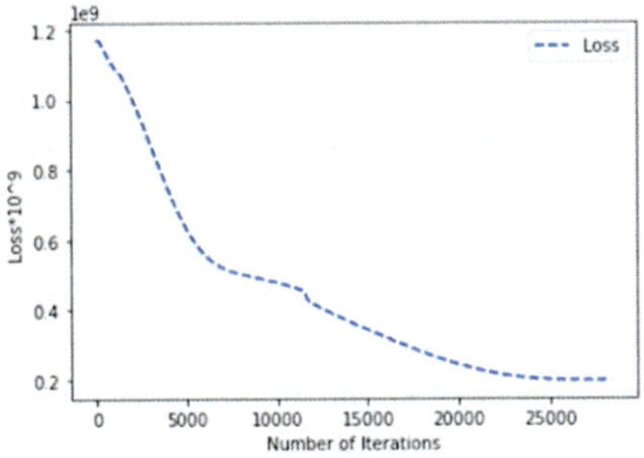

Fig. 8. The number of iterations plotted against the loss.

The Sklearn's split module was used in order to split the data into 75% training data and the remaining 25% into testing data. The mean absolute error was calculated using only the test data (data the model has not seen before).

We would like to note the different approaches taken in the process of developing this model. Using the Pytorch deep learning [21] library a single hidden layer neural network with a "*sigmoid*" activation function did not converge, the same model with a "*tanh*" activation function also did not converge.

Since the day variable is an input; this model could serve as an early detection system, alerting policy makers when a sharp increase in the number of cases is expected.

5 Discussion

From a technical stand point, the clear disadvantage in this model and the type of data provided is the low variance between the input point within the same city—the population, density, vaccine regulations and socio economic factors do not change in a time series problem. That is why it is important to collect data from as many cities as possible. We must concede that the algorithm will most likely give the highest weight to the "*day*" variable. That is why a larger feature space could offset that bias and help us determine variables that significantly play a role in the spread of the virus. Although the predictive capabilities of this model may seem unsatisfactory, it could be seen as a starting point for further research that introduces input spaces with higher dimensions for epidemiological predictions and analyses.

6 Conclusion

The Covid-19 virus has affected the whole world in a way that creates fluctuations in micro and macro dimensions. Knowing the nature and extent of these fluctuations is important for companies to develop strategies to extend their lifetimes, and for countries to have a vision to know what breakthroughs they need to take and when. In this study, based on the saturation of the relevant literature with the use of traditional methods, a new methodology with ANN, which is a more advanced technique, is proposed for forecasting the number of cases. It is obviously seen that the ANN model was able to relatively accurately predict unseen data with a mean absolute error of 14626 cases. Comparing this to how more traditional methods only try to fit a curve into the data and generally overfit. It is reasonable to assume that this model could be used by policy makers to take more affective steps against Covid-19.

In the future, the researchers who are interested in this topic can gather more sociological data and generate the model again to compare the results. For determination of the best number of layers or learning algorithms, heuristic algorithms such as Genetic Algorithm can be utilized rather than manuel trials. On the other hand, a multi-faceted generalizable early warning system tool can be developed to predict the number of cases on time.

References

1. Rogers, L.: Climate and disease incidence in India with special reference to leprosy, phthisis, pneumonia and smallpox. J. State Med. **33**, 501–510 (1925)
2. Rogers, L.: The world incidence of leprosy in relation to meteorological conditions and its bearing on the probable mode of transmission. Trans. Royal Soc. Tropical Med. Hygiene **16**, 440–464 (1923)
3. Rogers, L.: Relationship between pneumonia incidence and climate in India. Lancet **256**, 1173–1177 (1925)
4. Rogers, L.: Small-pox and climate in India: forecasting of epidemics. Med. Res. Council Reports **101**, 2–22 (1926)
5. Allen, L.J.S., Brauer, B., Driessche, P.V., Wu, J.: Mathematical Epidemiology, 45–52 (2008)
6. Ardabili, S.F., Mosavi, A., Ghamisi, P., Ferdinand, F., Varkonyi-Koczy, A. R., Reuter, U., Rabczuk, T., Atkinson, P. M.: COVID-19 Outbreak prediction with machine learning (2020)
7. Zhou, T., Liu, Q., Yang, Z., Liao, J., Yang, K., Bai, W., Lu, X., Zhang, W.: Preliminary prediction of the basic reproduction number of the Wuhan novel coronavirus 2019-nCoV. J. Evidence Based Med. **13**(1), 3–7 (2020)
8. Allen, L.J.S., Brauer, B., Driessche, P.V., Wu, J.: Mathematical Epidemiology, 19–52 (2008)

9. Koike, F., Morimoto, N.: Supervised forecasting of the range expansion of novel non-indigenous organisms: alien pest organisms and the 2009 H1N1 flu pandemic. Global Ecol. Biogeogr **27**, 991–1000 (2018)
10. Liang, R., Lu, Y., Qu, X., Su, Q., Li, C., Xia, S., Liu, Y., Zhang, Q., Cao, X., Chen, Q.: Prediction for global African swine fever outbreaks based on a combination of random forest algorithms and meteorological data. Transboundary Emer. Dis. **67**, 935–946 (2020)
11. Raja, D.B., Mallol, R., Ting, C.Y., Kamaludin, F., Ahmad, R., Ismail, S., Jayaraj, V.J., Sundram, B.M.: Artificial intelligence model as predictor for dengue outbreaks. Malays. J. Public Health Med. **19**, 103–108 (2019)
12. Tapak, L., Hamidi, O., Fathian, M., Karami, M.: Comparative evaluation of time series models for predicting influenza outbreaks: Application of influenza-like illness data from sentinel sites of healthcare centers in Iran. In: BMC Res. Notes, p. 12 (2019)
13. Akhtar, M., Kraemer, M.U.G., Gardner, L.M.: A dynamic neural network model for predicting risk of Zika in real time. BMC Med **17**(171) (2019)
14. Worldometers. https://www.worldometers.info/. Accessed 15 Aug 2020
15. Scikit. https://scikit-learn.org/stable/. Accessed 21 Aug 2020
16. Batista, M.: Estimation of the final size of the second phase of coronavirus epidemic by the logistic model. https://www.medrxiv.org/content/10.1101/2020.03.11.20024901v1.full.pdf. Accessed 15 Aug 2020
17. Graupe, D.: Principles of Artificial Neural Networks, 2nd edn., pp. 1–9. World Scientific, London (2013)
18. Kingma, D.P. ve Ba, J.: Adam: a method for stochastic optimization. In: 3rd International Conference for Learning Representations, San Diego (2015)
19. Information on Corona Virus Home Page, https://www.health.ny.gov/environmental/indoors/air/pmq_a.htm. Accessed 21 Aug 2020
20. Seger, C.: An investigation of categorical variable encoding techniques in machine learning: binary versus one-hot and feature hashing (Dissertation) (2018). http://urn.kb.se/resolve?urn=urn:nbn:se:kth:diva-237426. Accessed 12 July 2020
21. PyTorch Home Page. https://pytorch.org/. Accessed 20 Aug 2020

Improving the Prediction of Cryptocurrencies Movement Using Adversarial Training

Milad Mohammad Alemansour[1](✉) and Maryam Rastgarpour[2]

[1] Science and Research Branch, Islamic Azad University, Tehran, Iran
Milad.Alemansour@gmail.com
[2] Islamic Azad University, Saveh Branch, Saveh, Iran
M.Rastgarpour@iau-saveh.ac.ir

Abstract. Today, financial markets play a crucial role in managing order and advancing goals in communities worldwide. By increasing Internet usage by people around the world, a new concept of money (digital currency) has developed, followed by a new branch of this concept called cryptocurrency. This new concept has removed the boundaries and dominance of any state and central oversight. That is why investors have shifted focus from the traditional market to the digital market to profit in this market. This study examines and provides a machine learning strategy to predict the cryptocurrencies' movement; the purpose is to predict the price trend. The proposal presented in this study will help improve the neural network prediction model's generalization by adversarial learning. Anticipating the trend in financial markets is always based on the trading symbol's price as an input feature in neural networks and naturally changes over time. Similarly, in this study, price variables will be added for simulation, and the model trained so that they can perform better under small but deliberate fluctuations. Also, one dataset gathered to decipher popular and old currencies in digital financial markets, each of which includes five common symbols. Finally, the data set used with state-of-art neural network models by examining the different configurations and proposed models, the experiments, and results obtained were evaluated Accuracy and MCC. Finally, with a precision of %57.14 in a 24-h time step, it has the best performance among other proposed methods.

Keywords: Deep learning · Recurrent neural network · Cryptocurrency · Financial market · Time series prediction · Adversarial training

1 Introduction

Financial market forecasting in different research fields has attracted the attention of many researchers. Technical analysis focuses on engineering features to

T. Allahviranloo et al. (Eds.): IDS 2020, AISC 1301, pp. 282–293, 2021.
https://doi.org/10.1007/978-3-030-66501-2_23

select the technical indicators that have the most diagnostic power, but fundamental analysis usually involves processing unstructured data. One of the essential branches of machine learning algorithms is the neural network, which has the advantages of data-driven, numerical, and compatibility compared to conventional statistical methods. Thus, neural networks have a higher capacity to analyze noise data and find connections between data, to predict time series such as transaction price indices in financial markets. Deep learning algorithms can be trained with large volumes of nonlinear data to find more abstract nonlinear relationships between data with a deep neural network structure and several hidden layers. Compared to conventional machine learning algorithms, deep learning algorithms can solve nonlinear problems more accurately [1].

Predicting the movement of traditional financial markets such as corporate stocks has always been one of the most challenging activities due to a large amount of information processing, non-linearity of prices, and factors involved in it [2], such as the economic conditions of the countries worldwide. Some investors use technical indicators and oscillators to build charts and patterns in order to discover the price trend in this market and design a strategy accordingly, in order to have the highest profitability and return on investment ultimately. This method is called technical analysis [3], and its users also believe that price movements are repeated in a particular process. Technical analysis or volatile charts will look at the index's value and price movements in the financial market analysis (economic) as an entry point. In the last decade, with the arrival of a new type of digital currencies, the financial market has gone out of its traditional state. Due to the emergence of digital currencies, investors are always trying to maximize their profitability by trading at the right time, since the market price data of cryptocurrencies are very diverse in terms of time [4].

Moreover, typically in a nonlinear pattern, it is challenging to predict the future price of these currencies. With the increasing globalization of the economy and the evolution of information technology, analyzing market data to predict the future of cryptocurrencies is increasingly challenging [5]. It has become essential and valuable. Predicting informed information about the current situation provides the base price trend of cryptocurrencies. Therefore, the cryptocurrency price forecasting process can be used in decision making for customers in finalizing their trades. Cryptocurrency movement prediction can be addressed as a classification task. After defining the label space and features to describe a stock at a time, we can apply standard supervised learning methods such as support vector machines and neural networks to build the predictive model. Price movement prediction can be considered a classification task to find the best points to trade (buy or sell). We reviewed recent papers, and it can separate into two sections, price movement prediction, and adversarial training.

Recent works for price movement prediction mainly focused on using Neural networks in their studies. In the cryptocurrency domain, Researchers interested in Deep Learning (DL) [6] and Recurrent Neural Netowrks(RNN) such as Long Short-Term Memory (LSTM)[7]. RNNs is useful when working with high-frequency time-series data, Long term predictability, and non-linear fore-

casting [4,5]. For the stock domain, researchers have similar tastes [8–11], and it mainly fall under two categories Fundamental Analysis(FA) [12–14] , and Technical Analysis (TA) [3,15] for their datasets. Adversarial Training works in many studies for training a classification model. Mostly, these papers focus on computer visions and image classification [16]. Recently, several studies extends adversarial training to other domains, such as text classification [17] and graph nodes classification [6].
The main parts of this paper are summarized as :

- In Sect. 2, we discuss our dataset and calculating technical features for pre-processing.
- We explain our proposed methods and baselines with it in Sect. 3.
- We compare our proposed method, along with other existing method in Sect. 4.

Table 1. Dataset Indicators

Name	Symbol
Bitcoin	BTC
Bitcoin Cash	BCH
Ethereum	ETH
Litecoin	LTC
Ripple	XRP

2 Data and Preprocessing

2.1 Dataset

We collected five top trending cryptocurrency price history from Yahoo! Finance with 24 h time-steps (Table 1). Each Indicator contains 730 records (including weekends) and EOD data, and total indicators have 3650 records.

2.2 Calculate Technical Features

We use EOD data to calculate 11 temporal features (x_t^c) to describe the trend of a cryptocurrency indicator "c" at trading day "t" (Table 2). The purpose of using these features is to normalize each indicator's price and highlight the interaction of different price labels like high and close price.

Table 2. Features to describe each time-step of a cryptocurrency

Features	Calculation *e.g.*
c_high, c_low, c_open	$c\_open = \frac{open_t}{close_t} - 1$
n_adjust_close, n_close	$n\_close = \frac{close_t}{close_{t-1}} - 1$
$[5, 10, 15, 20, 25, 30]\ day$	$5day = \frac{\sum_{i=0}^{4} adj\_close_{t-i}/5}{adj\_close_t} - 1$

2.3 Classification

Each trading day must have a label to present signals of buy and sell for each record of indicators. These signals calculated from movement percent for cryptocurrency indicator "c" at time "T" (1). Price records with movement percent ≥0.55% and ≤-0.5% respectively mark as positive and negative.

$$MovementPercent(MP) = \frac{p^c_{T+1}}{p^c_T} - 1 \quad (1)$$
$$p \rightarrow Price$$

3 Proposed Method

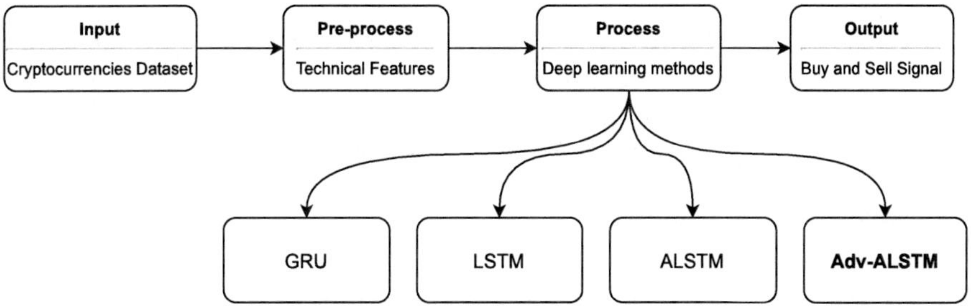

Fig. 1. The overall architecture of the proposed method

Before describing the proposed methods, we follow some rules for our problem formulation in this paper. Bold capital letters (*e.g.*, $\boldsymbol{X}$) and Bold lower letters (*e.g.*, $\boldsymbol{x}$) use to determine metrics and vectors, respectively. Also, regular lower case letters(*e.g.*, x) and Greek letters (*e.g.*, Θ) represent scalars and hyper-parameters, respectively. The symbols $tanh$ and σ are for the hyperbolic tangent and sigmoid function. The cryptocurrency movement prediction task is to learn a prediction function $\hat{y}^c = f(\boldsymbol{X}^c; \boldsymbol{\Theta})$, which maps a cryptocurrency "c" from its following features ($\boldsymbol{X}^c$) to the label space.

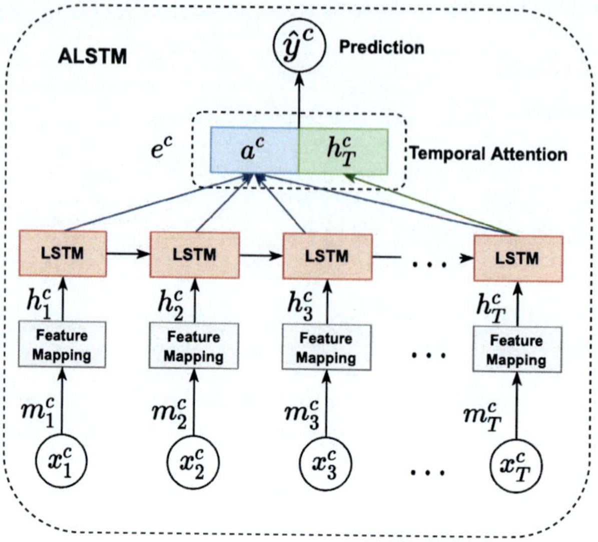

Fig. 2. Attentive LSTM flowchart

3.1 Attentive LSTM

Attentive LSTM model overall contains four layer: feature mapping layer, LSTM layer, temporal attention layer, prediction layer (see Fig. 2).

Feature Mapping Layer. Previous papers show that a more profound input gate would affect the efficiency of the LSTM model [7,18]. We followed their path to success and apply a fully connected layer to present input features into a latent representation. For each time-step, it performs as $m_t^c = tanh(\boldsymbol{W}_m x_t^c + b_m)$, which presents latent space with E dimensionality. $\boldsymbol{W}_m \in \mathbb{R}^{E \times D}$ and $b_m \in \mathbb{R}^E$ are parameters to be learned.

LSTM Layer. Capture and holding long-term dependencies is the ability of the LSTM, which has been used widely [19,20]. The general concept of LSTM is to add new input data (m_t^c) to a hidden representation of the previous time-step (h_{t-1}^c) to provide current time-step of hidden representation(h_t^c). The formulation we use is $h_t^c = LSTM(m_t^c, h_{t-1}^c)$, for detailed information of formulation can be referred to [21]. To get sequential dependencies of temporal pattern and cryptocurrency features, a LSTM model applied to input data $[m_1^c, m_2^c, \ldots, m_T^c]$ into hidden representation $[h_1^c, h_2^c, \ldots, h_T^c]$.

Temporal Attention Layer. Attention layer widely used in LSTM-based solutions to solve sequential learning problems [20,22]. The aim of using an extra layer is to compress the hidden representations at different time-steps into an overall representation. The attention mechanism provides that data at different time-steps at hidden representation might contribute to overall representation. We use an attention layer to aggregate with hidden representations as,

$$\boldsymbol{a}^c = \sum_{t=1}^{T} \alpha_t^c \boldsymbol{h}_t^c,\ \alpha_t^c = \frac{exp^{\tilde{\alpha}_t^c}}{\sum_{t=1}^{T} exp^{\tilde{\alpha}_t^c}} \\ \tilde{\alpha}_t^c = \boldsymbol{u}_a^T tanh(\boldsymbol{W}_a \boldsymbol{h}_t^c + \boldsymbol{b}_a) \tag{2}$$

where $\boldsymbol{W}_a \in \mathbb{R}^{E' \times U}$ and $\boldsymbol{u}_a \in \mathbb{R}^{E'}$ are parameters to be learned, and a^c is the aggregated representation that encodes the overall patterns in the sequence.

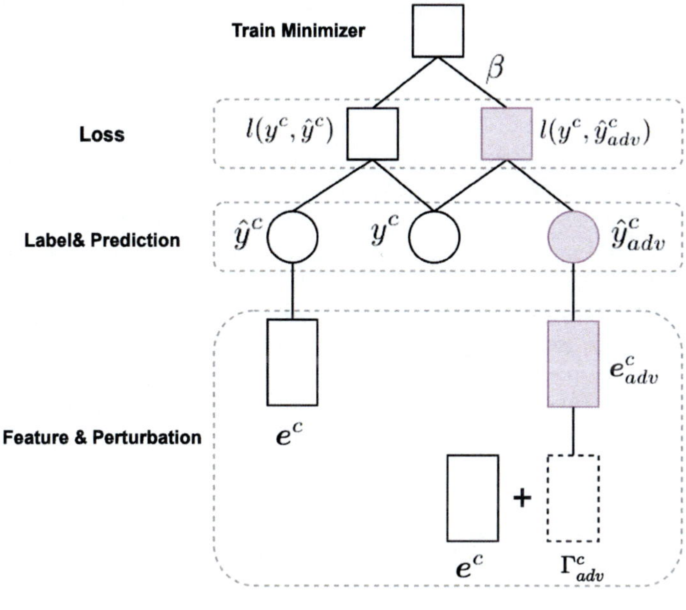

Fig. 3. Adversial Attentive LSTM flowchart

Prediction Layer. Rather than directly predicting a^c, we first examine a^c with the last hidden state h_T^c into the final latent representation of cryptocurrency indicator "c",

$$e^c = [a^{cT}, h_T^{cT}]^T \tag{3}$$

where $e^c \in \mathbb{R}^{2U}$. The intuition behind this is to highlight the most recent time-step, which assumed to be informative for the following movement [23]. We use e^c with a fully connected layer as the predictive function to estimate the classification confidence $\hat{y}^c = w_p^T e^c + b_p$.
*Note: the final prediction is $sign(\hat{y}^c)$

3.2 Adversarial Training

The normal way of training the ALSTM model is minimizing an objective function Γ (see Fig. 3) :

$$\Gamma_{normal} = \sum_{c=1}^{C} l(y^c, \hat{y}^c) + \frac{\alpha}{2}||\Theta||_F^2,$$
$$l(y^c, \hat{y}^c) = max(0, 1 - y^c\hat{y}^c). \quad (4)$$

For optimizing classification models, we use Hinge Loss [24], where researchers widely used in their papers. Also, we define a regularizer term, which is trainable parameters to prevent overfitting.

Despite the extensive usage of regular training, It is not suitable for learning the Cryptocurrency prediction model. Because regular training considers inputs as static, and it will ignore the stochastic property of these features. Note that cryptocurrency prices continuously changes with time and is affected by stochastic trading behaviors at a particular timestep and features calculated based on these prices. Regular training might lead to overfitting data for the model and lacks generalization ability. Adversarial training [16,25] trains a model with both clean and adversarial train examples (AEs) [26]. The AEs are malicious inputs generated by adding intended perturbations to features of clean samples. The perturbation, named as an adversarial perturbation (AP) is the path that leads to the most fundamental change of model prediction. Using it would be infeasible to be directly used in cryptocurrency prediction. Calculating perturbations relies on measuring the gradients regarding the input, which could take a long time (caused by the back-propagation through time-step of the LSTM layer). Besides, because the input gradients are dependent across different time-steps, there might be accidental intercommunications among the perturbations on different time-steps, which are uncontrollable. To approach explained problems, we offer to generate AEs from latent representation $\boldsymbol{e}^c$. The equation of elaborating objective function for adversarial training is:

$$\Gamma_{adv} = \sum_{c=1}^{C} l(y^c, \hat{y}^c) + \beta \sum_{c=1}^{C} l(y^c, \hat{y}^c_{adv}) + \frac{\alpha}{2}||\Theta||_F \quad (5)$$

The first term β is a hyper-parameter to balance losses for both clean and adversarial samples. The second term is an adversarial loss, where $\hat{y}^c_{adv}$ is the classification confidence of the AE of cryptocurrency c. By minimizing Γ_{adv}, the model encouraged to correctly classified clean and adversarial samples.

The following formulation generates every iteration, the latent representation of an AE($\boldsymbol{e}^c_{adv}$),

$$\boldsymbol{e}^c_{adv} = \boldsymbol{e}^c + \boldsymbol{r}^c_{adv},$$
$$\boldsymbol{r}^c_{adv} = argmax\ (l(y^c, \hat{y}^c_{adv}),\ \boldsymbol{r}^c),\ ||\boldsymbol{r}^c|| \leq \epsilon \quad (6)$$

where $\boldsymbol{e}^c$ is introduced in (3), and it is the final latent representation of cryptocurrency c. $\boldsymbol{r}^c_{adv}$ is the associated AP. "ϵ" is a hyper-parameter to control the scale of perturbation explicitly. Because it is intractable to calculate $\boldsymbol{r}^c_{adv}$ directly, we employ the fast gradient approximation method [25] as:

$$\boldsymbol{r}^c_{adv} = \epsilon \frac{\boldsymbol{g}^c}{||\boldsymbol{g}^c||}, \ \boldsymbol{g}^c = \frac{\partial l(y^c, \hat{y}^c)}{\partial \boldsymbol{e}^c}. \tag{7}$$

Calculation of perturbation is the loss function gradient regarding the latent representation $\boldsymbol{e}^c$ under an L2-norm constraint. Note that the gradient indicates the path where the loss function increases the most at the assigned point $\boldsymbol{e}^c$.In other words, it would lead to the most significant change in model prediction.

3.3 Baselines

In this paper, we used the following methods to compare results:

- **LSTM** is a neural network with two LSTM and prediction layer [15]. We can tune it with three hyper-parameters, hidden units number (U), lag size (T), and regularization term weight (λ).
- **ALSTM**, which known as the Attentive LSTM [19]. ALSTM optimized for normal training. We use similar three hyper-parameters U, T, and λ to tuned for this method.
- **GRU** is the Gated Recurrent Network [27]. Similar to LSTM and ALSTM, We use three hyper-parameters to be tuned.

3.4 Parameter Settings

We implement all models with Tensorflow and optimizing it with mini-batch Adam [28] and batch size of 1024 and an initial learning rate of 0.01. For GRU, LSTM, ALSTM, and Adv-ALSTM, we look for optimal hyper-parameters with grid search and following values,

- Number of hidden units(U): [4,8,16]
- Lag size (T): [2–5,10,15]
- Weight of regularization term(λ): [0.001,0.01,0.1,1].

In addition to previous parameters, Adv-ALSTM has more hyper-parameters to check; we use the following values:

- Balance losses (β): [0.001,0.005,0.01,0.05,0.1,0.5,1]
- Scale of perturbation controller(ϵ):[0.001,0.005,0.01, 0.05,0.1]

3.5 Evaluation Metric

Evaluation happens with two metrics, Accuracy(ACC) in range of [0,100] and Matthews Correlation Coefficient (MCC) [13] in range of [−1,1], which is shown in (8), respectively. Note that higher value indicates better performance for the model.

$$\begin{aligned} Accuracy &= \frac{TruePositives + TrueNegatives}{TotalExamples} \\ MCC &= \frac{TP \times TN - FP \times FN}{\sqrt{(TP+TN)(TP+FN)+(TN+FP)(TN+FN)}} \end{aligned} \tag{8}$$

Table 3. Performance comparison

Model	ACC	MCC	Best Time
GRU	%53.60	0.0634	10.75 s
LSTM	%54.08	0.096	15.53 s
ALSTM	%54.63	0.0840	16.46 s
Proposed Adv-LSTM	%57.14	0.137	17.95 s
RI (Proposed ADV-ALSTM)	%+2.51	+0.041	–

RI denotes the relative improvement of **Proposed Adv-ALSTM** to the other best baselines which is **ALSTM**.

Table 4. Best model hyper-parameter settings

Model	Setting
GRU	T=15, U=32, λ=0.1
LSTM	T=15, U=32, λ=0.1
ALSTM	T=15, U=8, λ=0.01
Proposed Adv-LSTM	T=15, U=16, λ=1, β=0.001, ϵ=0.01

4 Results

We use our dataset on four deep learning models and try to find the best parameter settings. We use an average of five best parameter settings output on each model for evaluation metrics.

Among the proposed models in Table 3, the Adv-ALSTM method had a higher difference in accuracy and MCC criteria than other models. The GRU and LSTM models have had similar results. However, in terms of computation time, the GRU model's performance is significantly better than expected from the other models and takes less time. Another factor to be mention is the amount of T delay, which has the highest value for all four network models.

According to the DL models chart (see Fig. 4), the ALSTM model in epochs below 100 has an outstanding accuracy compared to other models. In contrast to the model, Adv-ALSTM has better results in above 100 epochs, and it has a notable difference between it and the other three models. Besides, in the LSTM

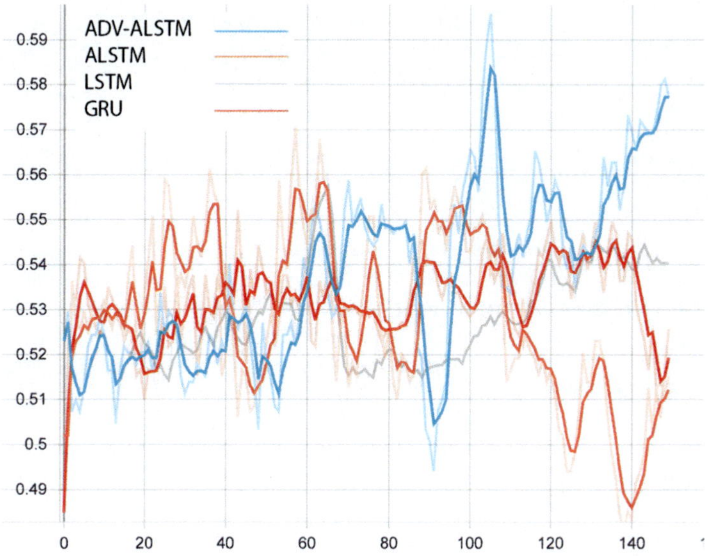

Fig. 4. Chart of evaluation results for DL methods with 0.6 smoothing in term of epochs.

model, the same ALSTM model pattern has been used but with relatively lower accuracy. Finally, the GRU model was able to fluctuate its accuracy with less fluctuation during different periods. In this model, in higher epochs, the amount of accuracy is not significantly different from its lower epochs.

5 Conclusion

In recent years, with the spread of topics related to deep learning and predicting time series movement, the attention of financial markets has shifted to using this tool to achieve higher returns and profits in this area. On the other hand, with the advent of cryptocurrencies and the fact that their use is borderless, in a situation where each country's boundaries and laws influence financial laws, it was able to gain special attention and credibility among users.

In this paper, we proposed a new time series dataset of top five trend cryptocurrencies. Also, neural network solutions for cryptocurrency movement prediction could suffer from weak generalization ability since they require to deal with the stochasticity of cryptocurrency features. We use different settings to get the best results in all four models and show an average of 5 best settings of each state-of-art model and proposed method (Table 4). Results show (Table 3) that the Adv-ALSTM model with %57.14 for ACC and 0.137 for MCC has the best cryptocurrency price movement prediction results (see Fig. 4).

In the future, we plan to apply other methods such as CNN or hybrid models along with adversarial training to improve cryptocurrencies' predict price movement.

References

1. Shintate, T., Pichl, L.: Trend prediction classification for high frequency bitcoin time series with deep learning. J. Risk Finan. Manage. **12**(1), 17 (2019)
2. Nguyen, T.H., Shirai, K., Velcin, J.: Sentiment analysis on social media for stock movement prediction. Expert Syst. Appl. **42**(24), 9603–9611 (2015)
3. Chong, E., Han, C., Park, F.C.: Deep learning networks for stock market analysis and prediction: methodology, data representations, and case studies. Expert Syst. Appl. **83**, 187–205 (2017)
4. Lahmiri, S., Bekiros, S.: Cryptocurrency forecasting with deep learning chaotic neural networks. Chaos, Solitons & Fractals **118**, 35–40 (2019)
5. Ji, S., Kim, J., Im, H.: A comparative study of bitcoin price prediction using deep learning. Mathematics **7**(10), 898 (2019)
6. Feng, F., Chen, H., He, X., Ding, J., Sun, M., Chua, T.S.: Enhancing stock movement prediction with adversarial training. arXiv preprint arXiv:1810.09936 (2018)
7. Graves, A., Mohamed, A.R., Hinton, G.: Speech recognition with deep recurrent neural networks. In: 2013 IEEE International Conference on Acoustics, Speech and Signal Processing, pp. 6645–6649. IEEE (2013)
8. Li, Z., Han, J., Song, Y.: On the forecasting of high frequency financial timeseries based on arima model improved by deep learning. J. Forecasting (2020)
9. Zhou, X., Pan, Z., Hu, G., Tang, S., Zhao, C.: Stock market prediction on high-frequency data using generative adversarial nets. Math. Prob. Eng. **2018**, 1–11 (2018). https://doi.org/10.1155/2018/4907423
10. Hiransha, M., Gopalakrishnan, E.A., Menon, V.K., Soman, K.: Nse stock market prediction using deep-learning models. Procedia Comput. Sci. **132**, 1351–1362 (2018)
11. Choi, H.K.: Stock price correlation coefficient prediction with arima-lstm hybrid model. arXiv preprint arXiv:1808.01560 (2018)
12. Hu, Z., Liu, W., Bian, J., Liu, X., Liu, T.Y.: Listening to chaotic whispers: a deep learning framework for news-oriented stock trend prediction. In: Proceedings of the Eleventh ACM International Conference on Web Search and Data Mining, pp. 261–269 (2018)
13. Xu, Y., Cohen, S.B.: Stock movement prediction from tweets and historical prices. In: Proceedings of the 56th Annual Meeting of the Association for Computational Linguistics (volume 1: Long Papers), pp. 1970–1979. Association for Computational Linguistics, Melbourne, Australia (2018) https://www.aclweb.org/anthology/P18-1183
14. Zhang, X., Zhang, Y., Wang, S., Yao, Y., Fang, B., Philip, S.Y.: Improving stock market prediction via heterogeneous information fusion. Knowledge-Based Syst. **143**, 236–247 (2018)
15. Nelson, D.M., Pereira, A.C., de Oliveira, R.A.: Stock market's price movement prediction with LSTM neural networks. In: 2017 International Joint Conference on Neural Networks (IJCNN), pp. 1419–1426. IEEE (2017)
16. Kurakin, A., Goodfellow, I., Bengio, S.: Adversarial machine learning at scale. arXiv preprint arXiv:1611.01236 (2017)

17. Miyato, T., Dai, A.M., Goodfellow, I.: Adversarial training methods for semi-supervised text classification. arXiv preprint arXiv:1605.07725 (2016)
18. Wu, L., Quan, C., Li, C., Ji, D.: Parl: Let strangers speak out what you like. In: Proceedings of the 27th ACM International Conference on Information and Knowledge Management, pp. 677–686 (2018)
19. Qin, Y., Song, D., Chen, H., Cheng, W., Jiang, G., Cottrell, G.: A dual-stage attention-based recurrent neural network for time series prediction. arXiv preprint arXiv:1704.02971 (2017)
20. Chen, L., Zhang, H., Xiao, J., Liu, W., Chang, S.F.: Zero-shot visual recognition using semantics-preserving adversarial embedding networks. In: Proceedings of the IEEE Conference on Computer Vision and Pattern Recognition, pp. 1043–1052 (2018)
21. Hochreiter, S., Schmidhuber, J.: Long short-term memory. Neural Computation **9**(8), 1735–1780 (1997)
22. Cho, K., Van Merriënboer, B., Gulcehre, C., Bahdanau, D., Bougares, F., Schwenk, H., Bengio, Y.: Learning phrase representations using rnn encoder-decoder for statistical machine translation. arXiv preprint arXiv:1406.1078 (2014)
23. Fama, E.F., French, K.R.: Size, value, and momentum in international stock returns. J. Finan. Econ. **105**(3), 457–472 (2012)
24. Rosasco, L., Vito, E.D., Caponnetto, A., Piana, M., Verri, A.: Are loss functions all the same? Neural Comput. **16**(5), 1063–1076 (2004)
25. Goodfellow, I.J., Shlens, J., Szegedy, C.: Explaining and harnessing adversarial examples. arXiv preprint arXiv:1412.6572 (2014)
26. Szegedy, C., Zaremba, W., Sutskever, I., Bruna, J., Erhan, D., Goodfellow, I., Fergus, R.: Intriguing properties of neural networks. arXiv preprint arXiv:1312.6199 (2013)
27. Cho, K., Van Merriënboer, B., Bahdanau, D., Bengio, Y.: On the properties of neural machine translation: Encoder-decoder approaches. arXiv preprint arXiv:1409.1259 (2014)
28. Kingma, D.P., Ba, J.: Adam: A method for stochastic optimization. arXiv preprint arXiv:1412.6980 (2014)

Decision Making Under Uncertainty

A Multinomial Goal Programming Model Using Markowitz Model Based on Energy Portfolio Under Uncertainty

Mohammad Hossein Karimi and Emran Mohammadi(✉)

School of Industrial Engineering, Iran University of Science and Technology, Tehran, Iran
e_mohammadi@iust.ac.ir

Abstract. Energy is a vital part of human life but by excessive energy consumption, there are more concerns on the agendas of many governments so that it is very essential to have competent energy management. Although the Markowitz model (MM) is a suitable tool for optimizing the portfolio in the financial field, it is practical in the field of energy. For this purpose, energy portfolio optimization (EPO) is a quantitative pragmatic tool according to the capacities and constraints of a country, and considering risk management, provides beneficial information for energy policymakers which helps them in decision making. In this paper, in addition to applying the Markowitz model, the issue of uncertainty is also considered for better decision making. Evaluating CO_2 emission is necessary for energy management systems, as a result, there is an objective function for calculating the amount of CO_2 emission. Also, we use a goal programming approach for optimizing objective functions. We selected Iran as the case study which has enough fossil and renewable energy resources. Finally, this model presents an optimal energy portfolio from different resources. It is very practical in energy policy.

Keywords: Energy portfolio optimization · Uncertainty · Goal programming

1 Introduction

Energy is an important factor for industrialization of the economy and is essential for economic and social development also life is not possible without the use of different energy carriers but it is an undeniable fact that the rapid growth of energy consumption in developing and newly industrialized countries has hampered the future of global energy security so that governments are looking for solutions to these problems. They need a tool that provides the right optimization of the energy resources by considering their limitations. Energy portfolio optimization (EPO) is an emerging issue that gives decision-makers the ability to create an optimal portfolio of energy consisting of all renewable and fossil resources. Many thinkers use financial models such as the Markowitz model (MM) so that to calculate the percentage of each energy resource in the portfolio we use the Markowitz mean-variance portfolio optimization theory. In addition, one of the major concerns of governments is reducing the amount of CO_2 emission in energy generation (EG) so that we proposed an objective function to reduce the amount of CO_2 emission.

T. Allahviranloo et al. (Eds.): IDS 2020, AISC 1301, pp. 297–309, 2021.
https://doi.org/10.1007/978-3-030-66501-2_24

In the real-world, energy production depends on the situation especially in renewable energies (RE). For instance, the specific wind speed is needed to be able to run the turbines to generate wind power but this situation is not common to generate a particular amount of wind power during a year. As a result, we consider our energy portfolio under uncertainty. Robust optimization (RO), is a convenient method that can be used to deal with uncertainty. In this paper, we use the Bertsimas and Sim (B&S) approach to consider uncertainty.

It is evident that we must optimize a multi-objective functions program so that we implemented a polynomial goal programming (PGP) method to solve multi-objective functions such as maximizing the expected return of the portfolio and minimizing variance and minimizing the amount of CO_2 emission of each energy resources.

In the literature, we review all articles including "Portfolio Optimization" and "Energy" statements by VOS viewer application and we find keywords that were used in Fig. 1.

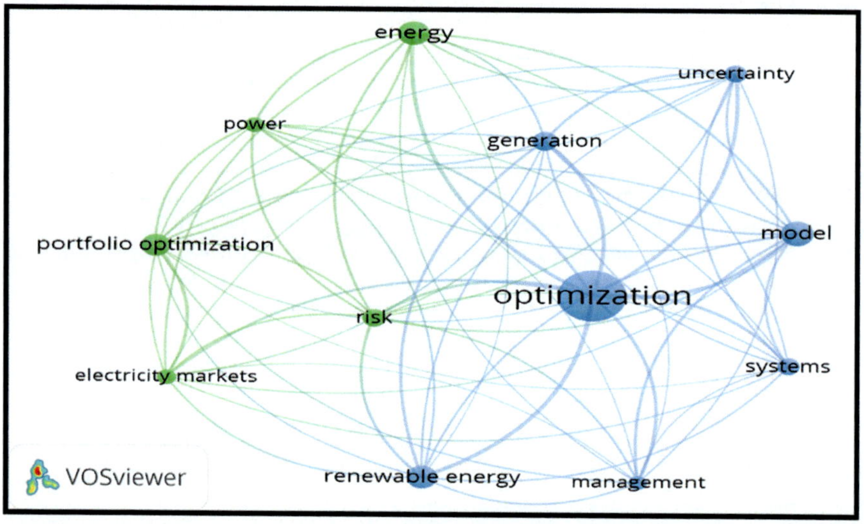

Fig. 1. The keywords that were used in articles including "Portfolio Optimization" and "Energy" statements

It is evident that uncertainty issues are one of the least used keywords so that it is necessary to consider uncertainty issues in future research in both fields of portfolio optimization and energy. Zhu and Fan [1] presented a mean-variance portfolio optimization theory based on certain parameters for China's generating portfolio and policy implications in 2010. Gökgöz and Atmaca [2] proposed Markowitz's mean-variance approach in the Turkish electricity market in 2012. Soroudi and Amraee [3] introduced a review paper about uncertainty in decision making include fuzzy arithmetic, probabilistic modeling, robust optimization, and hybrid approach in 2013. Lorcaa and Prina [4] presented a power portfolio optimization model considering conditional value at risk and stochastic evolution of local electricity prices in 2014. Mathuriaa et al. [5] generalized mean-variance portfolio theory with cost side uncertainties arising from fuel and emission market volatilities using the stochastic approach in 2015. Peykani and Roghanian [6] presented the application of robust optimization in the portfolio selection problem in 2015. Akter et al. [7] used portfolio theory to asset-based biosecurity decision analysis using the stochastic approach in Australia in 2015. Upton Jr and Snyder [8] introduced an analysis of renewable portfolio standards in the united state in the fields of energy and environmental policy based on certain data in 2017. Neto et al. [9] presented the Portfolio optimization of renewable energy assets in Brazil and stochastic models are used to generate synthetic time series for the random variables water inflow, wind speed, solar irradiance, and temperature of the photovoltaic panel in 2017. Stempien and Chan [10] proposed a mean-variance portfolio optimization model based on certain data to manage the energy resources in Singapore in 2017. Cucchiella et al. [11] introduced a portfolio analysis in the Italian electricity market whose purpose was to evaluate the profitability of investments in renewable technologies for the production of electricity in 2017. Gökgöz and Atmaca [12] used semi-variance and mean-variance portfolio optimization in emerging electricity markets in Turkey in 2017. Tavakoli et al. [13] presented an energy management scheme for optimal resilience and operational cost in commercial building microgrids using conditional value at risk and stochastic approach in 2018. Peykani et al. [14] used robust optimization to deal with uncertainty for performance assessment of stock in 2018. deLlano-Paz et al. [15] proposed a mean-variance model for power generation and pollutant emissions in the European Union based on certain parameters in 2018. Moore et al. [16] introduced a model based on the portfolio optimization for nanomaterial use in clean energy technologies with uncertainty analysis in 2018. Tan et al. [17] presented Optimization models for financing innovations in green energy technologies which the second model is a robust mixed-integer linear programming (MILP) to maximize the System readiness level (SRL) of a network of interdependent component technologies in 2019. Lan Bai et al. [18] improved the portfolio performance of renewable energy stocks using the robust portfolio (RP) approach in China which RP produced the best performance of portfolio using the data from the 10 sector stock indices in the China stock market in 2019. Peykani et al. [19] introduced an adjustable fuzzy chance-constrained network DEA approach in 2020.

In this research, we introduced a Bertsimas and Sim (B&S) approach for portfolio optimization problem by using robust optimization also the Markowitz mean-variance portfolio optimization and CO_2 emission calculator models are used. In addition, we consider four available energy resources of Iran contain Coal-fired, Wind power, Nuclear power, and Gas which consist of renewable and fossil energies as existing technologies in the energy portfolio then we implemented a polynomial goal programming (PGP) method to optimize three objective functions.

2 Nomenclatures and Background

2.1 The Nomenclatures of Paper

The nomenclatures of paper are shown in Table 1.

Table 1. These indices, parameters and variables are used in the paper

Indices	i	Energy Source i = 1,2,3,4
	j	Energy Source j = 1,2,3,4
Parameters	$\bar{R}_i$	The Return of Energy Source i which is Uncertainty Parameter
	σ_i^2	The Variance Energy Source i
	σ_i	The Standard deviation Energy Source i
	σ_j	The Standard deviation Energy Source j
	COE_i	The Amount of CO_2 Emission of Energy Source i
	ρ_{ij}	The Correlation Between i and j
	r	The Price of Robustness
Decision Variables	X_i	The Percentage of the Energy Source i in Portfolio
	X_j	The Percentage of the Energy Source j in Portfolio

2.2 Energy Portfolio Optimization and Risk Measures

Firstly, the portfolio optimization model is proposed by Markowitz [20] in 1952 which is shown below. Although it is used for the financial problem, we used it in the energy field to make an optimized portfolio of energy sources to help energy policymakers.

$$Max \sum_{i=1}^{4} X_i \bar{R}_l \tag{1}$$

$$Min \sum_{i=1}^{4} X_i X_j \sigma_i^2 + \sum_{i=1}^{4}\sum_{j=1}^{4} X_i X_j \rho_{ij} \sigma_i \sigma_j \tag{2}$$

S. t.

$$\sum_{i=1}^{4} X_i = 1 \tag{3}$$

$$X_i X_j \geq 0 \ \forall \ i, j \tag{4}$$

The portfolio objectives contain (1) and (2) which are maximizing expected return of each energy source and minimizing risk by mean-variance model respectively also in Eq. (3) summation of all percentages must be 1 and (4) represented decision variables must be positive also we considered two upper limit constraints for wind power and nuclear power because just particular regions in Iran have limited capacity to generate renewable energy so that we obtained these limitations from the website of the renewable energies organization of Iran [21] and computed maximum shares that wind power and nuclear power could be included in the energy portfolio.

$$X_2 \leq 0.0588 \tag{5}$$

$$X_3 \leq 0.2614 \tag{6}$$

It is important to note that $\bar{R}_i$ is the return of energy source i that is an uncertain parameter in the presented model in this research which we explain completely in Sect. 2.4.

2.3 Calculate the Amount of CO_2 Emission

As we mentioned above, CO_2 emission during energy generation is a problematic issue which each government faced and it is become to a global issue so that to generate the optimized energy portfolio, we consider an objective function to minimize the amount of CO_2 emission of each energy source. Finally, the presented model in this paper not only optimizes the energy portfolio but also calculates the amount of CO_2 emission of each energy source in the portfolio. An objective function which calculates the amount of CO_2 emission is (5).

$$Min \sum_{i=1}^{4} X_i COE_i \tag{7}$$

2.4 Robust Optimization

Robust optimization (RO) is an applicable method that can be used to deal with uncertain parameters in every optimization problem. Bertsimas and Sim [22] in 2004 presented one of the main robust optimization approaches that is used in our model.

It is an undeniable fact that the volume of energy generation varies during a specific period because it depends on some factors, as a result, it is hard to estimate the exact amount of energy generation so that in this model we defined a range with clear boundaries for the expected return of energy source i which takes value in $[R_i\text{-}R_i, R_i\text{+}R_i]$. R_i is a center point that is a nominal value and R_i is the perturbation of R_i. Again we mentioned $\bar{R}_i$ is the return of energy source i that is an uncertain parameter in the presented model in this research. To deal with an uncertain parameter we used Bertsimas and Sim approach so that to optimize the objective function (1), instead of (1) we consider Eqs. (8), (9), (10), and (11).

$$Max \;\; U \tag{8}$$

S. t.

$$U - \sum_{i=1}^{4} X_i R_i + Z\Gamma + \sum_{i=1}^{4} P_{1i} \leq 0 \tag{9}$$

$$Z + P_{1i} \geq \hat{R}_{1i} X_i, \quad \forall i \tag{10}$$

$$X_i, P_{1i}, Z \geq 0, \quad \forall i \tag{11}$$

It is evidence that U is considered as a parameter because $\bar{R}_i$ is an uncertain parameter in the objective function (1).

3 The Proposed Energy Portfolio Optimization Using Robust Approach

In this section, we proposed our model which contains expected return, risk measurement, and CO_2 emission calculation under uncertainty based on Bertsimas and Sim (B&S) approach. We implemented a polynomial goal programming (PGP) method which Aksaraylı and Pala [23] in 2018 presented in their paper to optimize multi-objective functions. The process of this method contains two stages. First, we solve each objective function one by one to reach the optimum values of each objective.

$$OBJ\,(1)\begin{cases} Max\ \sum_{i=1}^{4} X_i\bar{R}_l \\ S.\,t.\ \sum_{i=1}^{4} X_i = 1 \\ X_2 \le 0.0588 \\ X_3 \le 0.2614 \\ X_i \ge 0 \quad \forall i \end{cases} \tag{12}$$

We changed (12) to (13) to solve objective function because $\bar{R}_i$ is an uncertain parameter.

$$OBJ\,(1)\begin{cases} Max\ U = T^* \\ \text{S. t. } \sum_{i=1}^{4} X_i = 1 \\ X_2 \le 0.0588 \\ X_3 \le 0.2614 \\ U - \sum_{i=1}^{4} X_iR_i + Z\Gamma + \sum_{i=1}^{4} P_{1i} \le 0 \\ Z + P_{1i} \ge \hat{R}_{1i}X_i, \quad \forall i \\ X_i, P_{1i}, Z \ge 0, \quad \forall i \end{cases} \tag{13}$$

$$OBJ\,(2)\begin{cases} Min\ \sum_{i=1}^{4} X_iX_j\sigma_i^2 + \sum_{i=1}^{4}\sum_{j=1}^{4} X_iX_j\rho_{ij}\sigma_i\sigma_j = S^* \\ \text{S. t. } \sum_{i=1}^{4} X_i = 1 \\ X_2 \le 0.0588 \\ X_3 \le 0.2614 \\ X_iX_j \ge 0 \ \forall\, i, j \end{cases} \tag{14}$$

$$OBJ(3)\begin{cases} Min\ \sum_{i=1}^{4} X_i COE_i = E^* \\ \text{S. t. } \sum_{i=1}^{4} X_i = 1 \\ X_2 \leq 0.0588 \\ X_3 \leq 0.2614 \\ X_i \geq 0 \quad \forall i \end{cases} \tag{15}$$

T^*, S^* and E^* are gained from (13), (14) and (15) respectively then in the second stage we solved this following model.

$$OBJ(4)\begin{cases} Min\left(1+\left|\frac{d_1}{T^*}\right|\right)^{\lambda_1}+\left(1+\left|\frac{d_2}{S^*}\right|\right)^{\lambda_2}+\left(1+\left|\frac{d_3}{E^*}\right|\right)^{\lambda_3} \\ \text{S. t.} \\ U+d_1=T^* \\ \sum_{i=1}^{4} X_i X_j \sigma_i^2 + \sum_{i=1}^{4}\sum_{j=1}^{4} X_i X_j \rho_{ij}\sigma_i\sigma_j - d_2 = S^* \\ \sum_{i=1}^{4} X_i COE_i - d_3 = E^* \\ U - \sum_{i=1}^{4} X_i R_i + Z\Gamma + \sum_{i=1}^{4} P_{1i} \leq 0 \\ Z + P_{1i} \geq \hat{R}_{1i} X_i, \quad \forall i \\ \sum_{i=1}^{4} X_i = 1 \\ X_2 \leq 0.0588 \\ X_3 \leq 0.2614 \\ X_i, X_j, P_{1i}, Z \geq 0, \quad \forall\, i,j \\ d_i \geq 0, \quad \forall i \end{cases}$$

(16)

Goal variables d_1, d_2, and d_3 are used in order to minimize the deviations from the aspired levels. OBJ(4) is solved based on various values of λ_i to prioritize objectives and find the best portfolio with optimum shares. We used GAMS software to solve the presented model.

4 Case Study and Numerical Results

4.1 Data Description

In this section, the proposed model is presented based on real data which is obtained from the Iran power ministry website [24] and Iran renewable energies organization website [21] then we normalized data using the Euclidean norm method. In addition, we noted that R_i is the perturbation of R_i. In our proposed model R_i equals 2σ to determine the best boundaries also it is necessary to note that in this paper, Γ equals zero which means we considered the nominal solution.

4.2 Results

We solved this model by variety values of λ_i which totally we generated fifteen different scenarios to analyze the sensitivity of deviation variables, decision variables, and objective functions. Different values of λ_i are shown in Table 2 which are selected randomly. Also the values of d_i are displayed in Table 3 for each scenario then we showed the percentages of each selected energy source include Coal-fired, Wind Power, Nuclear Power, and Gas for each scenario in Table 4. Finally, the amount of each objective function consists of Expected Return, Risk Measurement, and The Amount of CO_2 Emission is demonstrated in Table 5 for each scenario.

Table 2. Different values of λ_i for each scenario

Scenario	Weights		
	λ_1	λ_2	λ_3
1	0.2	0.5	0.3
2	0.1	0.6	0.3
3	0.1	0.5	0.4
4	0.1	0.4	0.5
5	0.2	0.2	0.6
6	0.2	0.1	0.7
7	0.3	0.1	0.6
8	0.3	0.2	0.5
9	0.4	0.2	0.4
10	0.5	0.2	0.3
11	0.6	0.2	0.2
12	0.7	0.2	0.1
13	0.5	0.4	0.1
14	0.4	0.4	0.2
15	0.4	0.3	0.3

Table 3. The values of d_i for each scenario

Scenario	Deviation variables		
	d_1	d_2	d_3
1	0.034	0	0.099
2	0.034	0	0.1
3	0.035	0.001	0.096
4	0.036	0.003	0.091
5	0.042	0.025	0.064
6	0.059	0.195	0
7	0.059	0.195	0
8	0.039	0.013	0.077
9	0.037	0.006	0.086
10	0.035	0.002	0.094
11	0.034	0	0.1
12	0.032	0	0.107
13	0.032	0	0.107
14	0.033	0	0.102
15	0.035	0.001	0.097

Table 4. The percentages of each selected energy sources for each scenario

Scenario	Energy sources			
	Coal-fired	Wind power	Nuclear power	Gas
1	36%	6%	26%	32%
2	36%	6%	26%	32%
3	35%	6%	26%	33%
4	33%	6%	26%	35%
5	24%	6%	26%	44%
6	0	6%	26%	68%
7	0	6%	26%	68%
8	28%	6%	26%	40%
9	31%	6%	26%	37%
10	34%	6%	26%	34%
11	36%	6%	26%	32%
12	39%	6%	26%	29%
13	39%	6%	26%	29%
14	37%	6%	26%	31%
15	35%	6%	26%	33%

Table 5. The amount of each objective functions for each scenario

Scenario	Objective functions		
	Expected return	Risk measurement	CO_2 emission
1	0.591	0.170	0.476
2	0.591	0.170	0.477
3	0.592	0.171	0.473
4	0.593	0.173	0.468
5	0.599	0.195	0.441
6	0.616	0.365	0.337
7	0.616	0.365	0.337
8	0.596	0.183	0.414
9	0.594	0.177	0.423
10	0.592	0.173	0.431
11	0.591	0.170	0.477
12	0.589	0.170	0.484
13	0.589	0.170	0.484
14	0.590	0.170	0.479
15	0.592	0.171	0.474

4.3 Sensitivity Analysis

In this section, we analyzed results based on scenarios that different values of λ_i are considered.

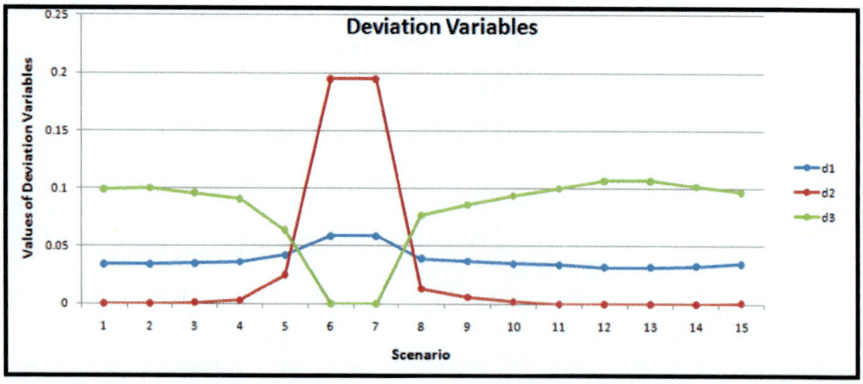

Fig. 2. The trends of d_1, d_2 and d_3 in each scenario

Figure 2 shows the trends of d_1, d_2, and d_3 in each scenario. The trends of d_2 and d_3 demonstrate that both are in conflict. In addition, in scenarios 6 and 7 when λ_3 has the most value it means that reducing the amount of CO_2 emission is a more vital issue than other factors so that in order to satisfy this issue the model considers the value of d_3 is zero. Despite this consideration, in scenarios 6 and 7 given that the value of λ_2 is 0.1, the risk is the least valuable objective. As a result, d_3 has the most value.

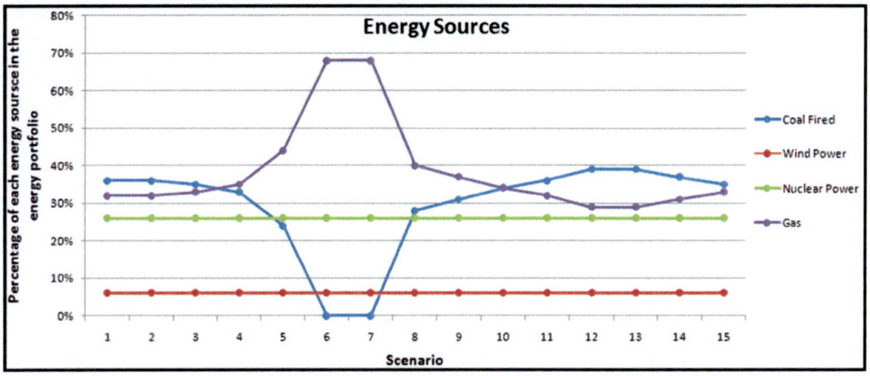

Fig. 3. The trends of energy sources in each scenario

Figure 3 shows the trends of the percentage of each energy source in the energy portfolio. The percentages of Wind Power and Nuclear Power in the portfolio are stable in each scenario. Wind Power and Nuclear Power contain 6% and 26% of the energy portfolio respectively in each scenario as a result they assert that it is necessary to use all capacities for renewable energies. It is a fact that renewable energies are more affordable than fossil energies and they are confident choice to decrease CO_2 production. The result of our presented model confirms the previous sentence. Also, the percentages of Coal-Fired and Gas have opposite relation for instance in scenario 6 and 7, λ_3 has the most value it means that reducing the amount of CO_2 emission is more important than others

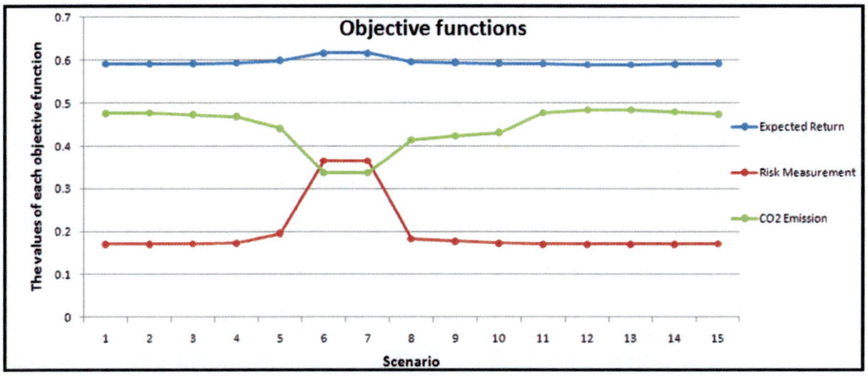

Fig. 4. The trends of each objective function value

as a result the model offers an energy portfolio which the shares of Coal-Fired and Gas are 0% and 68% respectively because Coal Fired has the most CO_2 production.

Figure 4 shows the trends of values of objective functions in each scenario. The expected return has obtained the correct values also the second and third objective functions both are in conflict, for example, the value of λ_2 is 0.1 in both scenarios 6 and 7, the value of risk increases sharply while by increasing the value of λ_2, the amount of risk is in the low levels in other scenarios. In addition, the values of λ_3 in scenarios 6 and 7 are 0.7 and 0.6 respectively so that reducing the amount of CO_2 emission is more significant than others as a result the value of CO_2 emission reaches the lowest level.

5 Conclusion and Future Directions

In this study, an energy portfolio optimization model based on the robust approach is proposed. This model contained maximizing the expected return, minimizing the level of risk, and minimizing the amount of CO_2 emission as objective functions. Also, we considered the return parameter as an uncertain parameter which Bertsimas and Sim approach is used to deal with this uncertain parameter. We implemented a polynomial goal programming method to solve this model also we considered Iran as a case study because this country has a variety of energy sources. This presented model is a practical tool for energy policymakers to create the best energy portfolio for their region considering their concerns, values, priorities, and limitations. Although Markowitz's energy portfolio optimization model mostly is used in financial problems, we used it in the energy field. For the future studies, the portfolio optimization models for dealing with uncertainty parameters could be presented based on other approaches such as fuzzy and stochastic approaches.

References

1. Zhu, L., Fan, Y.: Optimization of China's generating portfolio and policy implications based on portfolio theory. Energy **35**(3), 1391–1402 (2010)
2. Gokgoz, F., Atmacal, M. E.: Financial optimization in the turkish electricity market: Markowitz's mean-variance approach. Renew. Sustain. Energy Rev. **16**(1), 357–368 (2012)
3. Soroudi, A., Amraee, T.: Decision making under uncertainty in energy systems: State of the art. Renew. Sustain. Energy Rev. **28**, 376–384 (2013)
4. Lorca, A., Prina, J.: Power portfolio optimization considering locational electricity prices and risk management. Electric Power Syst. Res. **109**, 80–89 (2014)
5. Mathuria, P., Bhakar, R., Li, F.: GenCo's optimal power portfolio selection under emission price risk. Electric Power Syst. Res. **121**, 279–286 (2015)
6. Peykani, P., Roghanian, E.: The application of data envelopment analysis and robust optimization in portfolio selection problem. J. Oper. Res. Appl. **12**(44), 61–78 (2015)
7. Akter, S., Kompas, T., Ward, M.B.: Application of portfolio theory to asset-based biosecurity decision analysis. Ecol. Econ. **117**, 73–85 (2015)
8. Upton Jr., G.B., Snyder, B.F.: Funding renewable energy: An analysis of renewable portfolio standards. Energy Econ. **66**, 205–216 (2017)
9. Neto, D.P., Domingues, E.G., Coimbra, A.P., Almeida, A.T., Alves, A.J., Calixto, W.P.: Portfolio optimization of renewable energy assets: Hydro, wind, and photovoltaic energy in the regulated market in Brazil. Energy Econ. **64**, 238–250 (2017)

10. Stempien, J.P., Chan, S.H.: Addressing energy trilemma via the modified Markowitz Mean-Variance Portfolio Optimization theory. Appl. Energy **202**, 228–237 (2017)
11. Cucchiella, F., Gastaldi, M., Trosini, M.: Investments and cleaner energy production: A portfolio analysis in the Italian electricity market. J. Cleaner Prod. **142**(1), 121–132 (2017)
12. Gokgoz, F., Atmaca1, M. E.: Portfolio optimization under lower partial moments in emerging electricity markets: Evidence from Turkey. Renew. Sustain. Energy Rev. **67**, 437–449 (2017)
13. Tavakoli, M., Shokridehaki, F., Akorede, M.F., Marzband, M., Vechiu, I., Pouresmaeil, E.: CVaR-based energy management scheme for optimal resilience and operational cost in commercial building microgrids. Electrical Power and Energy Syst. **100**, 1–9 (2018)
14. Peykani, P., Mohammadi, E., Sadjadi, S.J., Rostamy-Malkhalifeh, M.: A robust variant of radial measure for performance assessment of stock. In: the 3th International Conference on Intelligent Decision Science, Iran (2018)
15. deLlano-Paz, F., Calvo-Silvosa, A., Antelo, S.I., Soares, I.: Power generation and pollutant emissions in the European Union: A mean-variance model. J. Cleaner Prod. **181**, 123–135 (2018)
16. Moore, E.A., Babbitt, C.W., Gaustad, G., Moore, S.T.: Portfolio optimization of nanomaterial use in clean energy technologies. Envir. Sci. Technol. **52**(7), 4440–4448 (2018)
17. Tan, R.R., Aviso, K.B., Ng, D.K.S.: Optimization models for financing innovations in green energy technologies. Renew. Sustain. Energy Rev. **113**, 109258 (2019)
18. Bai, L., Liu, Y., Wang, Q., Chen, C.: Improving portfolio performance of renewable energy stocks using robust portfolio approach: Evidence from China. Physica A, **533**, 122059 (2019)
19. Peykani, P., Mohammadi, E., Emrouznejad, A.: An Adjustable Fuzzy Chance-Constrained Network DEA Approach with Application to Ranking Investment Firms. Expert Systems with Applications (2020)
20. Markowitz, H.: Portfolio selection. J. Finance **7**(1), 77–91 (1952)
21. The renewable energies organization of Iran website, http://www.satba.gov.ir/en/home
22. Bertsimas, D., Sim, M.: The price of robustness. Oper. Res. **52**(1), 35–53 (2004)
23. Aksaraylı, M., Pala, O.: A polynomial goal programming model for portfolio optimization based on entropy and higher moments. Expert Syst. Appl. **94**, 185–192 (2018)
24. Iran power ministry website https://www.moe.gov.ir/?lang=en-us

A New Fuzzy Approach for Multi-period Portfolio Optimization Under Uncertainty

Zahra Khandan[1] and Emran Mohammadi[2(✉)]

[1] Department of Industrial Engineering, Science and Research Branch, Islamic Azad University, Tehran, Iran
z.khandan96@gmail.com
[2] Iran University of Science & Technology, Tehran, Iran
e_mohammadi@iust.ac.ir

Abstract. Portfolio optimization is one of the most important issues in the financial world, so investors are trying to make decisions that are most in line with the real world. But the uncertainty in data and parameters, and the contradiction in the investor's goals, adds to the complexity of the stock portfolio optimization problem, and the other hand because of the efficient market, it is necessary to use multi-period models that, unlike single-period models, allow the investor to review their wealth at the beginning of each period. In this paper We formulate a bi-objective mean- CVaR portfolio optimization model based on General Fuzzy theory. Then the proposed model is solved by the Epsilon constraint method. Finally, we use the data of 15 companies from different industries operating in the Iran Stock Exchange Market in 1398, we examine the validity of the model and its efficiency.

Keywords: Portfolio · Multi-period optimization · Uncertainty · General fuzzy measure · Epsilon constraint

1 Introduction

Portfolio optimization is a way to quantify different investment options so that at a certain level of risk, the return is maximized or at a certain level of return, the risk is minimized. Different goal functions and different approaches can be considered to design the optimal investment portfolio. But the important point in presenting the optimal model is to take into account the uncertainties that exist over time, legal and policy constraints and, other cases, so the design of the stock portfolio should be done in such a way that a combination of assets to achieve to the goals, consider. On the other hand, due to the dynamics and change of factors affecting the value of assets, methods should be used and policies should be considered that in certain periods to review the composition of the investment portfolio. This means that unlike single-period portfolios, the investor is allowed to reconsider his or her capital allocation method, So this model of investment is a method that does not consider only one period and updates information by considering several periods. Much research has been done on multi-period portfolio selection, most of which

T. Allahviranloo et al. (Eds.): IDS 2020, AISC 1301, pp. 310–320, 2021.
https://doi.org/10.1007/978-3-030-66501-2_25

offer single models based on probabilistic theory, regardless of market uncertainty; But since historical data contain little information about the input of the models, and on the other hand, these impact factors are evaluated with some ambiguity; Because they are usually non-statistical and because of the practical barriers to data acquisition, the nature of inaccuracies in measurement and the ambiguity in human perceptions that exist in many cases in the real world; Therefore, randomness is not the only type of uncertainty in reality, and professional knowledge should be used to make accurate predictions because in the real world we are dealing with uncertain, unknown and ambiguous data.

Following extensive fuzzy theory Zadeh [1] and using a fuzzy approach in modeling is the only practical way to deal with inaccurate, vague, and ambiguous information. Various theories have been proposed to expand the principle of fuzzy Zadeh development, including the theory of possibility Zadeh [2] fuzzy measures such as possibility measure, necessity measure, credibility.

First Markowitz's Mean-variance portfolio selection model was introduced. Markowitz initially aimed to maximize the expected returns. Markowitz then considered the expected return to be favorable and the resulting variance (return) to be undesirable, and then proceeded by assuming that the expected return was maximized for a certain level of risk [3]. Black and Litterman [4] argue that variance optimization can create extreme or non-visual weight for some portfolio assets. Mohammadi and Mohammadi [5] equipped the portfolio analysis tools with a solid technique and extended the conventional formula for the minimax measure of regret in multi-objective problems.

Peykani et al. [6] Presents a new model of robust data envelopment analysis that can be used in the presence of discrete and continuous uncertainties.

In the mentioned researches, the modeling is done in one cycle, while in the real world, the decision-maker is often faced with a series of decisions in different periods, so that each decision affects the decision of the next stage.

Tobin [7] examined that multi-period stock portfolio are equal to the assumption that the investment horizon of equal periods is equal, and that the process of random production of price changes is constant, and that successive changes in prices are reciprocal, independent, and then asset returns are independent. Li and Ng [8] considered an optimal analytical solution for the mean-variance formula in multi-year stock portfolio selection. In particular, the optimal policy of analytical investment portfolio and analytical expression of boundary and mean and efficient boundary in this paper for the multivariate formula of mean-variance of expression and an efficient algorithm to find the optimal policy of stock portfolio to maximize a desirable performance of value Expected and variance presented the ultimate wealth. Wei and Ye [9] considered a stock portfolio selection model with several rounds of a variance imposed by a bankruptcy limit in a random market. The stochastic returns of Barisk assets all depend on the random market situation that is assumed to follow the Markov chain. Then a solution scheme is created, dynamic programming is used to solve an auxiliary problem, which in turn is provided to achieve an optimal portfolio policy. The investment policy resulting from this model can help investors not only achieve optimal returns in terms of average variance but also have good risk control over bankruptcy. Califore [10] addresses a series of decision-making issues for the allocation of financial assets. The proposed model in which the optimal multi-period portfolio settings are determined with the aim of minimizing a cumulative risk on the investment horizon while meeting the limitations of portfolio diversity in each period and achieving the final expected wealth and linear control policies are used. Najafi and Moshakhian [11] in a paper called Using a combination of genetic algorithm and particle optimization designed to solve the model and since the performance of the algorithms depends on the correct choice of the parameter, the Taguchi method is used to adjust the parameter of the algorithm. Sun et al. [12] Create a minimal model for a multi-period portfolio selection problem. An analytical solution is obtained and numerical simulations show the superiority of the multi-period model over the corresponding first period as well as the market index. Hasanloo [13] has proposed a new multi-period model of stock portfolio selection with different borrowing rates and loans based on the concept of random programming. In this proposed model for evaluating the uncertainty and random nature of stock portfolio selection, it is assumed that the rate of return and the rate of borrowing and lending are independently and normally distributed and provide a genetic algorithm for solving nonlinear programming. Mohebbi and Najafi [14] propose a multi-cycle stock portfolio model considering transaction costs and the possibility of risk-free investment, and model financial market uncertainty through the integration of a scenario tree and fuzzy credit theory in this paper. The proposed model is also formulated in this research as a two-dimensional mean-VaR[1] problem using comparability, threshold, liquidity, and class constraints in order to create a more realistic model. Zhang and Li [15] presented a valid multi-purpose mean-half entropy model with an underlying risk for multi-period portfolio selection. Also, realistic constraints such as liquidity, cardinality constraints, transaction costs, and buying thresholds are considered. Liu et al. [16] Consider a multi-period stock portfolio optimization problem and, inspired by the Sharp ratio or the "reward-to-volatility" ratio, the low potential and potential volatility ratio for stock selection by measuring the effects of negative volatility. And positive in choosing the investment portfolio. Based on two new definitions, they present a fuzzy multi-period portfolio performance evaluation model. Then, using the weighted total approach, the proposed model is transformed into a single programming problem and a PSO algorithm with a feasibility constraint mechanism is designed for the solution. Gupta et al. [17] Use intuitive fuzzy to deal with the uncertainty in the market and consider the degree of membership and the degree of non-membership and the degree of doubt in their model and introduce the optimistic model and the pessimistic model and examine the results.

2 Formulation of General Fuzzy Multi-period Portfolio Optimization

In this section, a General Fuzzy multi-period portfolio problem is discussed. First, the problem description used in following section are introduced. Then, the general fuzzy return and Conditional Value-at-Risk of multi period portfolio are formulated, finally objective functions and constraint which are used in portfolio optimization model are presented.

[1] Value at Risk.

2.1 Problem Description and Notations

x_{it} : *the investment proportion of risky asset i ar period t*

x_{ft} : *the investment proportion of riskfree asset i ar period t*

$\tilde{r}_{it}$: *the fuzzy return rate of asset i ar period t*

W_t : *the holding wealth at the beginning ofperiod t*

c_{it} : *the unit transaction cost of risky asset i at period t*

$\tilde{L}_{it}$: *the fuzzy turnover rate ofasset i arperiod t*

β : *the minimum acceptable General Fuzzy confidence level for the satisfaction of chance constraint corresponding to portfolio liquidity*

α : *the minimum confidence level for risk*

z_{it} : *a binary variable indicating wheter the risky asset i contained in the portfolio*

$$z_{it} = \begin{cases} 1 & \text{if risky asset } i \text{ is contained in the portfolio} \\ 0 & \text{otherwise,} \end{cases}$$

K : *the desired number of asset held in the portfolio at period t*

l_{it} : *the minimal fraction of wealth allocated to the risky asset i at period t*

u_{it} : *the maximal fraction of wealth allocated to the risky asset i atperiod t*

$\lambda, \Delta, \Delta F$: *the optimistic − pessimistic parameter*

2.2 Objective Functions

- Terminal Wealth

In the portfolio optimization the Net expected return at period t can be calculated by:

$$E(r_{Nt}) = \sum_{i=1}^{n} E(\tilde{r}_{it}) + r_{ft}\left(1 - \sum_{i=1}^{n} x_{it}\right) - \sum_{i=1}^{n} C_{it}|x_{it} - x_{it-1}| \tag{1}$$

So, the expected wealth in the beginning of period t + 1 is determined by

$$W_{t+1} = \left[W_t\left(1 + E\left(r_{Nt}\right)\right)\right]$$
$$= \left[W_1 \prod_{t=1}^{T}\left(1 + \left(\sum_{i=1}^{n} E(\tilde{r}_{it})x_{it} + r_{ft}\left(1 - \sum_{i=1}^{n} x_{it}\right) - \sum_{i=1}^{n} c_{it}\left|x_{it} - x_{it-1}\right|\right)\right)\right] \quad (2)$$

General Fuzzy measure For defuzzification and dealing with uncertainty by Mousazadeh [18]:

$$E^{Me}[\xi] = \int_{0}^{+\infty} Me\{\xi \geq x\}\, dx - \int_{-\infty}^{0} Me\{\xi \leq x\}\, dx \quad (3)$$

- Minimizing Risk
  For minimizing Risk, we consider Conditional Value-at-Risk, So, based on General Fuzzy measure, the formula of conditional Value-at risk calculated by:

$$Min(\xi_{CVaR}(\alpha)) = GF(\xi_{CVaR}(\alpha))$$
$$= \sum_{i}^{n} \left\{ \begin{array}{ll} -a_{it} + \left(\left(a_{it} - b_{it}\right)\frac{\alpha}{\Delta F}\right) & \alpha \leq \Delta \\ -d_{it} + \left(\left(d_{it} - c_{it}\right)\frac{\alpha - \Delta F}{(1 - \Delta F)}\right) & \alpha \geq \Delta \end{array} \right\} + \left(\frac{1}{(1 - \alpha)t}\right) y(t) \quad (4)$$

2.3 Constraints

- Liquidity
  The chance constraint for General Fuzzy event that the portfolio liquidity:

$$GF\left(\sum_{i=}^{n} \tilde{L}_{it}^{s} x_{it}^{s} \geq L\right)$$
$$= \sum_{i=1}^{n} \left(\left\{ \begin{array}{l} \left(\frac{\beta_0 - \Delta}{1 - \Delta}\right) L_{a_{it}}^{s} + \left(1 - \frac{\beta_0 - \Delta}{1 - \Delta}\right) L_{b_{it}}^{s}\, if\ \beta_0 \succ \Delta \\ \left(\frac{\beta_0}{\Delta}\right) L_{c_{it}}^{s} + \left(1 - \frac{\beta_0}{\Delta}\right) L_{d_{it}}^{s}\, if\ \beta_0 \leq \Delta \end{array} \right\} \right) \quad (5)$$

- Cardinality
  The maximum number of assets which are allowed to be in portfolio:

$$\sum_{i=1}^{n} Z_{it}^{s} = K \quad (6)$$

- Threshold
  The minimum and maximum amount of investment:

$$l_{it} \leq x_{it} \leq u_{it} \quad (7)$$

Thus, the General Fuzzy multi-period portfolio optimization model will be as follows:

$$Max\left[W_1\prod_{t=1}^{T}\left(1+\left(\sum_{i=1}^{n}\left(\frac{1-\lambda}{2}(r_1+r_2)+\frac{\lambda}{2}(r_3+r_4)\right)x_{it}+r_{ft}\left(1-\sum_{i=1}^{n}x_{it}\right)-\sum_{i=1}^{n}c_{it}|x_{it}-x_{it-1}|\right)\right)\right]$$

$$Min(\xi_{CVaR}(\alpha)) =$$

$$GF(\xi_{CVaR}(\alpha)) = \sum_{i}^{n}\left\{\begin{array}{ll} -a_{it}+\left((a_{it}-b_{it})\frac{\alpha}{\Delta F}\right) & \alpha \le \Delta \\ -d_{it}+\left((d_{it}-c_{it})\frac{\alpha-\Delta F}{(1-\Delta F)}\right) & \alpha \ge \Delta \end{array}\right\}+\left(\frac{1}{(1-\alpha)t}\right)y(t)$$

$$\sum_{i=1}^{n} Z_{it}^{s} = K$$

$$l_{it} \le x_{it} \le u_{it}$$

$$Z_{it} = \begin{cases} 1 & \textit{if risky asset i contained in portfolio} \\ 0 & \textit{other wise} \end{cases}$$

$$i = 1, \ldots, n, t = 1, \ldots, T, j = 1, \ldots, J$$

3 Solution Algorithm

In this method, one of the objective functions is selected to be optimized while the other(s) are converted into additional constraints [19], leading to a solution that can be proven to always be weakly Pareto optimal

Assume the following problem:

$$\begin{aligned} & Max\ (f_1(x), f_2(x), \ldots, f_p(x)) \\ & st \\ & x \in s \end{aligned} \tag{8}$$

Where x is the vector of decision variables, $f_1(x), f_2(x), \ldots, f_p(x)$ are the p objective functions and S is the feasible region. In the epsilon constraint method, we optimize one of the objective functions using the other objective functions as constraints, in cooperating them in the constraint part of the model as shown below [20, 21]:

$$\begin{aligned} & \max f_1(x) \\ & st \\ & f_1(x) \ge e_2, \\ & \ldots \\ & f_p(x) \ge e_p, \\ & x \in S \end{aligned} \tag{9}$$

By parametrical variation in the RHS of the constrained objective (e_i) functions the efficient o the problem is obtained.

For estimating epsilon, each time we put one of the objective functions as an objective function, and the others as constraints, and we find the minimum and maximum amount of them. Then we use below formula for finding the amount of epsilon:

$$e_j = n_j + \left[\frac{t}{r-1}\right](m_j - n_j)$$
$$t = 0, 1, \ldots, r-1 \tag{10}$$

Where n_j and m_j are the minimum and maximum amount of objective function, and r is the number of points that we find between the distance of minimum and maximum amount of objective functions.

4 Empirical Study and Result

For evaluating the proposed model of multi-period portfolio optimization, we consider an empirical study in the Iran Stock Exchange market. We consider 15 companies from different industries with the trapezoidal number for return and liquidity, the result is shown in the below table (Fig. 1, Fig. 2, Table 1, Table 2, Table 3, Table 4):

Table 1. The result of multi-period portfolio optimization for different β at α = 0.05

α = 0.05			Adjustable parameter	
β = 90%	β = 80%	β = 70%		
10696.813	10708.894	10714.848	Wealth	0
0.001596	0.001260	0.001251	Risk	
11336.3178	11351.766	11354.669	Wealth	0.25
0.023	0.032	0.0277	Risk	
12033.656	12036.274	12037.270	Wealth	0.5
0.0448	0.046	0.046	Risk	
12855.780	12858.301	12859.167	Wealth	0.75
0.079	0.08	0.0798	Risk	
13776.0766	13776.0766	13815.687	Wealth	1
0.1462	0.1468	0.1326	Risk	

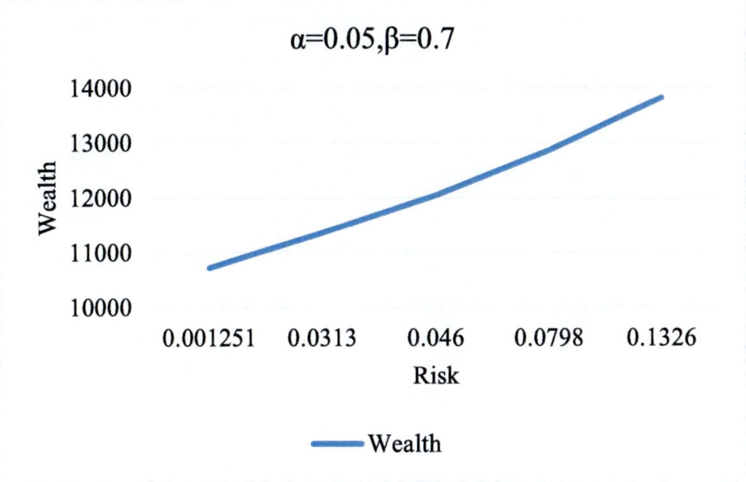

Fig. 1. The efficient frontier for $\alpha = 0.05$

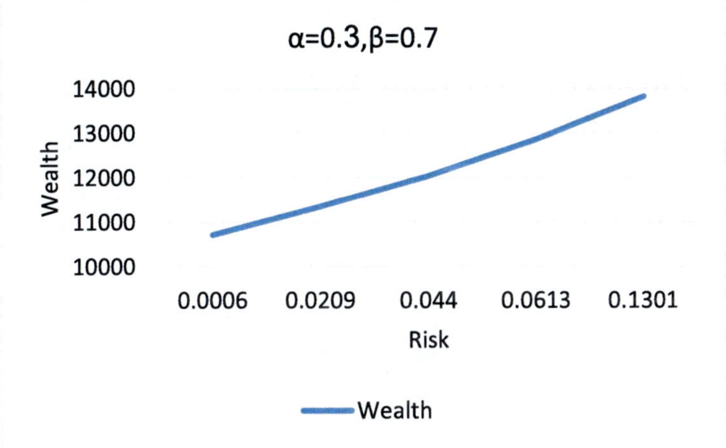

Fig. 2. The efficient frontier for $\alpha = 0.1$

Table 2. The result of multi-period portfolio optimization at $\alpha = 0.05$

$\lambda, \Delta, \Delta F$	Near zero	0.25	0.5	0.75	Near 1
wealth	10714.848	11354.669	12037.270	12859.167	13815.687
Risk	0.001251	0.0277	0.046	0.0798	0.1326

Table 3. The result of multi-period portfolio optimization for different β at $\alpha = 0.1$

$\alpha = 0.1$			Adjustable parameter	
$\beta = 90\%$	$\beta = 80\%$	$\beta = 70\%$		
10694.549	10708.818	10714.871	Wealth	0
0.0007	0.00067	0.0006	Risk	
11344.352	11351.81	11354.672	Wealth	0.25
0.015	0.014	0.0209	Risk	
1999.083	12036.274	12037.270	Wealth	0.5
0.022	0.045	0.044	Risk	
12855.757	12858.302	12859.167	Wealth	0.75
0.067	0.0678	0.0613	Risk	
138515.3514	13815.3302	13815.334	Wealth	1
0.1311	0.1337	0.1301	Risk	

Table 4. The result of multi-period portfolio optimization at $\alpha = 0.1$

$\lambda, \Delta, \Delta F$	Near zero	0.25	0.5	0.75	Near 1
wealth	10714.871	11354.672	12037.270	12859.167	13815.334
Risk	0.0006	0.0209	0.044	0.0613	0.1301

For $\beta = 0.7$, $\alpha = 0.05$, $K = 5$, $c_{it} = 0.003$, $W_t = 10000$, $l_{it} = 0$, $u_{it} = 0.5$, $r_{ft} = 0.12$, we have:

On the sixth day: if we are pessimistic, our wealth will be 10714.848. If we are optimistic the wealth will be 13815.687 and CVaR will be 13.26% or 1831.96 currency and for credibility mode we have 12037.270 wealth and CVaR will be 4.6%.

For $\beta = 0.7$, $\alpha = 0.1$, $K = 5$, $c_{it} = 0.003$, $W_t = 10000$, $l_{it} = 0$, $u_{it} = 0.5$, $r_{ft} = 0.12$, we have;

According to the above tables and graphs, as wealth increases, risk increases, which is consistent with portfolio theory, so the investor who intends to increase investment must bear more risk.

In this paper introduces a new approach to optimizing a multi-period portfolio optimization based on General Fuzzy theory. The presentation Model has made it possible for the investor to be able to apply his manner by changing the parameter to optimistic-pessimistic, and there is no need to model in credibility, necessity or possibility mode.

5 Conclusion

In this paper, we present a new approach to optimizing a multi-period portfolio under uncertainty. We formulated the bi-objective model with maximizing the expected value of wealth and minimizing the mean-conditional value at risk based on General Fuzzy theory by taking liquidity, cardinality, threshold as constraints to bring the model closer to the real world, and We used trapezoidal fuzzy numbers to calculate the turnover and return rate of assets. In the proposed model, we put an adjustable parameter that causes more flexibility on the possibility and necessity, and the investor can apply his manners in the model and change the existing conditions in the real world by changing the pessimistic optimistic parameter in the model. Then we used the Epsilon constraint method to solve the model. Finally, we considered 15 companies from different industries in the Iranian stock exchange market and optimized the multi-period stock portfolio by considering daily information.

Therefore, according to the result instead of separate modelling in the case of necessity, possibility, and credibility, by placing different numbers instead of the optimistic and pessimistic parameter, we obtained the results in the case of possibility, necessity, and validity, and learned about the model results in other cases.

As future work, we intend to consider CDaR in the objective function instead of conditional Value at Risk, we are also planning to estimate the result with using of other fuzzy numbers such as Z-number, etc., using scenario trees to deal with uncertainty, and genetic algorithms to solve models on a larger scale.

References

1. Zadeh, L.: Fuzzy Sets. *Information and Control,* pp. 338–353 (1965)
2. Zadeh, L.: Fuzzy sets as basis for theory of possibility. Fuzzy sets and Syst. **1**, 3–28 (1978)
3. Markowitz, H.: Portfolio Selection. J. Finance **7**, 77–91 (1952)
4. Black, F., Litterman, R.: Global portfolio optimization. Financial Anal. J. **48**(5), 28 (1992)
5. Mohammadi, S.E., Mohammadi, E.: Robust portfolio optimization based on minimax regret approach in Tehran stock exchange market. J. Ind. Syst. Eng. **11**, 51–62 (2018)
6. Peykani, P., Mohammadi, E., Emrouznejad, A., Pishvaee, M.S.: Fuzzy data envelopment analysis: an adjustable approach. Expert Syst. Appl. **136**, 439–452 (2019)
7. Tobin, J.: Liquidity preference as behavior toward risk. Risk Averation and Portfolio choice (1965)
8. Li, D., NG, W.: Mathematical finance. optimal dynamic portfolio selection: multiperiod mean-variance formulation. **10**, 387–406 (2000)
9. Wei, S., Ye, Z.: Multi-period optimization portfolio with bankruptcy control in stochastic market. Math. Comput. **186**, 414–425 (2007)
10. Calafiore, G.C.: Multi-period portfolio optimization with linear control policies. Automatica **44**, 2463–2473 (2008)
11. Najafi, A.A., Mushakhian, S.: Multi-stage stochastic mean–semivariance–CVaR. Appl. Math. Comput. **256**, 445–458 (2015)
12. Sun, Y., Aw, G., Teo, K.L., Zhu, Y., Wang, X.: Multi-period portfolio optimization under probabilistic risk measure. Finance Res. Lett. **18**, 60–66 (2016)
13. Hassanlou, K.:A multi period portfolio selection using chance constrained programming. Decis. Sci. Lett. pp. 221–232, (2017)
14. Mohebbi, N., Najafi, A.A.: Credibilistic multi-period optimization based on scenario tree. Phys. A **492**, 1302–1316 (2018)
15. Zhang, J., Li, Q.: Entropy, credibilistic mean-semi-entropy model for multi-period portfolio selection with background risk. vol. 21 (2019)

16. Liu, Y., Zhang, W., Gupta, P.: Multi-period portfolio performance evaluation model based on possibility theory. Transactions on Fuzzy Systems (2019)
17. Gupta, P., Mehlawat, M.K., Yadav, S., Kumar, A.: Intuitionistic fuzzy optimistic and pessimistic multi-period portfolio Intuitionistic fuzzy optimistic and pessimistic multi-period portfolio. Soft Computing (2020)
18. Mousazadeha, M., Torabia, S.A., Pishvaeeb, M. S.: Health service network design: a robust possibilistic approach. In: International Transactions in Operational Research, pp. 337–373 (2018)
19. Miettinen, K., Molina, J., González, M., Hernández-Díaz, A., Caballero, R.: Using box indices in supporting comparison in multiobjective optimization. Euro. J. Oper. Res. **197**, 17–24 (2009)
20. Cohon, J.: Multicriteria programming: brief review and application. Design optimization, pp. 163–191 (1985)
21. Chankong, V., Haimes, Y.Y.: Multiobjective decision making: theory and methodology (1983)

A New Solution for the Generalized Shortest Path Problem

Majid Darehmiraki(✉)

Department of Mathematics, Behbahan Khatam Alanbia University of Technology, Khouzestan, Iran
darehmiraki@bkatu.ac.ir

Abstract. Many real-world problems face a kind of uncertainty. One of these problems is finding the shortest path. Given that the distance between the two paths is often ambiguous. Therefore, in this article, we try to examine this problem in a hesitant fuzzy atmosphere. First, we provide a new distance to calculate the distance between two hesitant fuzzy sets, and then we try to solve it with the help of the Dijkstra algorithm. Finally, we present an example of the efficiency of our proposed method.

Keywords: Shortest path · Distance function · Hesitant fuzzy · Dijkstra algorithm

1 Introduction

The shortest path problem is a well-known problem that is used in a variety of applications, including being at the heart of networking problems. In conventional problems, the distance between nodes is known, and fuzzy numbers can be used for an unknown environment to achieve optimal results. Calculating the minimum path cost from each vertex is called the shortest individual path. Especially in the process of finding the shortest path, it is very important to find the path that has the least number of bends and gives you the most optimal result. And the cost of mapping of lengths and bends. The shortest normal path to get the path is the minimum cost from the initial node to the end, and the cost is to add the cost of the curves in the path [1, 3, 5, 9, 11, 13].

The problem of the shortest path plays an essential role in hybrid optimization due to its basic aspects and wide range of applications. Examining the shortest path is an essential part of communications, computer networks, manufacturing systems, and transportation. The weight of the path will indicate the time of transport from one end to the other, i.e. the travel time from origin to destination. By speeding up some routes, the transmission efficiency can be improved to minimize travel time between some source and terminal pairs by minimizing link weight. For quick travel destinations, there is a need to reduce travel time, reduce travel conditions, and reduce the total cost of meeting faster needs [10].

In practice, the length of arcs in a graph or network depends on various parameters (i.e., capacity, cost, demand, traffic frequency, time, etc.) and cannot be easily

T. Allahviranloo et al. (Eds.): IDS 2020, AISC 1301, pp. 321–327, 2021.
https://doi.org/10.1007/978-3-030-66501-2_26

determined. Therefore, we can use the fuzziness into a graph to handle this type of uncertain situation. In environments with uncertainty, it is best to use fuzzy concepts to represent the length of the arcs. Also, in order to provide an algorithm to solve such problems, we need a tool to rank and calculate the distance between two fuzzy sets. These types of problems are known as the fuzzy shortest path problem (FSPP), and so far various solutions have been proposed for them. Dubois and Prade [8], have extended the standard Ford–Moore–Bellman algorithm to solve the FSPP. They have applied the concept of criticality of any path. A fuzzy relationship to choose a specific path is introduced by Chanas and Kamburowski [6] and applied it in their introduced algorithm for solving FSPP. A method for solving FSPP using the concept of possibility measurement was described by Yager [14]. He considered a possibility of visiting among two nodes and determined the total possibility of any path between a starting vertex and destination vertex utilizing the T-norm. Liao et al. investigated the constrained shortest path problem, where edge weights are presented by trapezoidal fuzzy numbers.

After Zadeh first introduced fuzzy sets, various extensions were proosed for these sets, including intuitive fuzzy sets [7], interval fuzzy sets, and hesitant fuzzy sets. Hesitant fuzzy sets are used in various fields, for example, references [2, 4] can be seen. In this article, we use hesitant fuzzy sets (HFSs) to overcome the ambiguity of the problem. In this way, we display the length of the arcs in the problem with hesitant fuzzy sets, and then use a new distance function to calculate the distance between the two HFSs in order to use the Dijkstra algorithm.

2 Preliminary Concepts

This section provides an introduction to HFSs based on what will be discussed in future sections.

Definition 2.1: Consider the reference set X. An HFS is a set of values that, when used on X, it returns a subset of [0.1]. Xia and Xu [12] described HFS using the following notation:

$$H = \{x.h_H(x)|x \in X\}$$

where $h_H(x)$ is a set of multiple values within [0.1] and represents the degree of possible membership for element $x \in X$ relative to set H. It is easier to call $h_H(x)$ the hesitant fuzzy element (HFE).

Example 2.1. If $V = \{v_1, v_2, v_3\}$ is the discourse set, $p(v_1) = \{0.2, 0.4, 0.6\}$, $p(v_2) = \{0.3, 0.4\}$ and $p(v_3) = \{0.3, 0.2, 0.5, 0.6\}$ are the HFEs of $v_i (i = 1, 2, 3)$ to a set P respectively. Therefore P can be considered as a HFS, i.e.,

$$P = \{\langle v_1, \{0.2, 0.4, 0.6\}\rangle, \langle v_2, \{0.3, 0.4\}\rangle, \langle v_3, \{0.3, 0.2, 0.5, 0.6\}\rangle\}$$

From a mathematical point of view, a HFS P can be seen as a FS if there is only one element in $P(v)$. In this situation, HFSs include FSs as a special case.

In order to compare hesitant fuzzy sets in practical applications we need score functions. Basically, we are looking for functions for ranking HFEs, which results in ranking functions for HFSs as well. Then for convenience, p stands for the HFE $p(v)$ for $v \in V$, and we assume that $|p| = n$, which

$$p = \left\{p^{(1)}, p^{(2)}, \ldots, p^{(n)}\right\}$$

Ranking function $E : [0, 1]^n \to [0, 1]$ must satisfy in the following conditions

1) Monotone non-decreasing property: if $p^{\sigma(1)} \leq \hat{p}^{\sigma(1)}, p^{\sigma(2)} \leq \hat{p}^{\sigma(2)}, \ldots, p^{\sigma(n)} \leq \hat{p}^{\sigma(n)}$, then

$$E\left(p^{\sigma(1)}, p^{\sigma(2)}, \ldots, p^{\sigma(n)}\right) \leq E\left(\hat{p}^{\sigma(1)}, \hat{p}^{\sigma(2)}, \ldots, \hat{p}^{\sigma(n)}\right)$$

2) Boundary conditions property:

$$E(0, 0, \ldots, 0) = 0, E(1, 1, \ldots, 1) = 1.$$

Also, the empty HFE and the full HFE are $p = \{0\}$ and $p = \{1\}$.
Some of the most popular score functions are

a) The minimum score function:

$$E_{min}(p) = min\left\{p^{(1)}, p^{(2)}, \ldots, p^{(n)}\right\}.$$

b) The maximum score function:

$$E_{min}(p) = max\left\{p^{(1)}, p^{(2)}, \ldots, p^{(n)}\right\}.$$

c) The arithmetic-mean score function:

$$E_{AM}(p) = \frac{1}{n}\sum_{i=1}^{n} p^{(i)}.$$

Some of the operators on hesitant fuzzy elements are listed below:

a) $h_1(x) \cup h_2(x) = \bigcup_{\gamma_1 \in h_1(x).\gamma_2 \in h_2(x)} max\{\gamma_1 . \gamma_2\}$
b) $h_1(x) \cap h_2(x) = \bigcap_{\gamma_1 \in h_1(x).\gamma_2 \in h_2(x)} min\{\gamma_1 . \gamma_2\}$
c) $(h_1(x))^\lambda = \bigcup_{\gamma_1 \in h_1(x)}\{\gamma_1^\lambda\}$
d) $\lambda(h_1(x)) = \bigcup_{\gamma_1 \in h_1(x)}\{1 - (1 - \gamma_1)^\lambda\}$

So far, various distances have been proposed to calculate the distance between two hesitant fuzzy sets, some of which are mentioned below [8]:

1) Normalized Hamming distance:

$$\boldsymbol{d_{hnh}(M, N)} = \frac{1}{n}\sum_{i=1}^{n}\left[\frac{1}{l_{x_i}}\sum_{j=1}^{l_{x_i}}\left|h_M^{\sigma(j)}(x_i) - h_N^{\sigma(j)}(x_i)\right|\right]$$

2) Normalized Euclidean distance:

$$d_{hnh}(M,N) = \left[\frac{1}{n}\sum_{i=1}^{n}\left[\frac{1}{l_{x_i}}\sum_{j=1}^{l_{x_i}}\left|h_M^{\sigma(j)}(x_i) - h_N^{\sigma(j)}(x_i)\right|^2\right]\right]^{\frac{1}{2}}$$

3) Normalized Hausdorff distance:

$$d_{ghnh}(M,N) = \left[\frac{1}{n}\sum_{i=1}^{n} max_j\left|h_M^{\sigma(j)}(x_i) - h_N^{\sigma(j)}(x_i)\right|^{\lambda}\right]^{\frac{1}{\lambda}}$$

4) Normalized Euclidean-Hausdorff distance:

$$d_{ghnh}(M,N) = \left[\frac{1}{n}\sum_{i=1}^{n} max_j\left|h_M^{\sigma(j)}(x_i) - h_N^{\sigma(j)}(x_i)\right|^{2}\right]^{\frac{1}{2}}$$

5) Weighted Hausdorff distance

$$d_{ghwh}(M,N) = \left[\int_a^b w(x) max_j\left|h_M^{\sigma(j)}(x_i) - h_N^{\sigma(j)}(x_i)\right|^{\lambda} dx\right]^{\frac{1}{\lambda}}$$

3 Proposed Distance

In this section, we present a new distance function in order to calculate the distance between two HFSs. Before introducing the new distance, we define the following definition, in which the conditions necessary for a distance function are stated.

Definition: Let h_1 and h_2 are two HFSs on $X = \{x_1, \ldots, x_n\}$ and $d(h_1, h_2)$ a be a distance function between h_1 and h_2. Therefore

1) $0 \leq d(h_1, h_2) \leq 1$.
2) $d(h_1, h_2) = 0$ *if and only if* $h_1 = h_2$.
3) $d(h_2, h_1) = d(h_1, h_2)$.

The number of values in the hesitant fuzzy sets for each member may be different from the other. In order to solve this problem, we first arrange values in $h_A(x)$ for each element in ascending order then $m = max\{l(h_A(x)) : x \in A\}$ and repeat the largest value in $h_A(x)$ as long as the number of $h_A(x)$ members is equal to m. Of course, this is an optimistic trend, and in the pessimistic state, we repeat the smallest amount available.

Definition: Let M and N are two HFSs on $X = \{x_1, \ldots, x_n\}$. The distance between them is described as follows

$$d(M,N) = \frac{\sum_{i=1}^{n}\int_1^{l_{x_i}}\left|f_M^i(u) - f_N^i(u)\right|du}{\sum_{i=1}^{n}\int_1^{l_{x_i}} u\left|f_M^i(u) - f_N^i(u)\right|du}, \quad (1)$$

where $f_N^i(u)$ and $f_M^i(u)$ are membership functions of element of X.

Definition: Suppose $x \in X$ and A is a HFSs. Then

$$R(x) = \frac{\int_1^{l_x} f(x)dx}{\int_1^{l_x} xf(x)dx}, \tag{2}$$

is a ranking function where $f(x)$ is a function that interpolate values in $h_A(x)$.

It also clearly applies to the two conditions mentioned for ranking functions.

4 Hesitant Dijkstra Algorithm

Here we first calculate the length of the arcs with the help of the ranking function introduced in (2) and then use the algorithm described below.

The Dijkstra algorithm is one of the most effective algorithms for finding the shortest directional path from a certain vertex s to other vertices of a network with the length of the nonnegative edges, first introduced by Dijkstra in 1959. Suppose the graph G is a graph with nonnegative weights; We want to specify the shortest path from a certain vertex s to other vertices. This algorithm divides the graph into two sets. The set s includes the vertices for which the weight of the shortest paths is calculated. And the $v - s$ set includes vertices for which the weight of the shortest paths has not been calculated. At each step, the algorithm selects a vertex from $v - s$ for which the weight of the shortest path is the lowest, then adds that vertex to the set s, and then releases the edges attached to that vertex.

```
Dijkstra(G, w, s)
Initialize-single-source(G, s)
s → Ø
Q → V(G)
while Q ≠ Ø
        do u → Extract − min (Q)
            s → s ∪ {u}
            for each vertex v ∈ Adj[u]
                do Relax(u, v, w)
```

One of the most important applications of this algorithm is to calculate the shortest distance between two points in a city by land.

5 Numerical Results

Here is a numerical example to illustrate the effectiveness of the proposed method.

Example: Consider a network with 6 vertices and 7 edges, as described below (Fig. 1).

The length of the routes is as follows:

$$\left\{ \begin{array}{c} A = 5, 0.3, 0.7, 0.8, B = 3, 0.08, 0.2, \\ C = 7, 0.5, 0.2, G = 5, 0.5, 0.4, 0.2 \end{array} \right\},$$

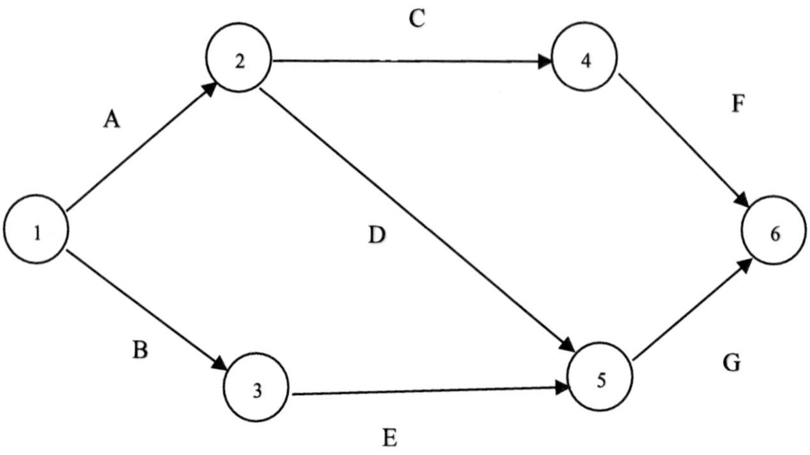

Fig. 1. The graph of the example.

$$\{D = 10, 0.03, 0.2, 0.8, E = 3, 0.1, 0.2, 0.3, F = 12, 0.5, 0.9\}.$$

In this case, the distance matrix for this graph using formula 2 is as follows:

$$\begin{bmatrix} 0 & 0.5762 & 0.3333 & 0 & 0 & 0 \\ 0 & 0 & 0 & 0.5364 & 0.5244 & 0 \\ 0 & 0 & 0 & 0 & 0.4286 & 0 \\ 0 & 0.0000 & 0 & 0 & 0 & 0.5171 \\ 0 & 0 & 0 & 0.0000 & 0 & 0.5849 \\ 0.0 & 0 & 0 & 0 & 0.0000 & 0 \end{bmatrix}$$

The following table shows the length of the shortest path from vertex (1) to the rest of the vertices:

1	2	3	4	5	6
0.0	0.5762	0.33333	1.1126	0.7619	1.3468

6 Conclusion

The shortest path problem of the problems that are always discussed in the field of transportation. In this article, we examined it in a different space. Given the widespread use of hesitant fuzzy sets, we investigated this problem in hesitant fuzzy space. Of course, this was an initial work in this field, and more advanced work can be done in this field.

It should be noted that uncertainty is not only related to the geometric distance between two nodes, and this uncertainty can also be considered in various factors such as time.

References

1. Ahuja, R.K., Mehlhorn, K., Orlin, J., Tarjan, R.E.: Faster algorithms for the shortest path problem. J. ACM (JACM) **37**(2), 213–223 (1990)
2. Akram, M., Adeel, A., Alcantud, J.C.R.: Hesitant fuzzy N-soft sets: a new model with applications in decision-making. J. Intell. Fuzzy Syst. **36**(6), 6113–6127 (2019)
3. Arsham, H.: Stability analysis for the shortest path problems. Congressus Numerantium 171–210 (1998)
4. Bisht, K., Kumar, S.: Hesitant fuzzy set based computational method for financial time series forecasting. Granular Comput. **4**(4), 655–669 (2018)
5. Broumi, S., Nagarajan, D., Bakali, A., Talea, M., Smarandache, F., Lathamaheswari, M.: The shortest path problem in interval valued trapezoidal and triangular neutrosophic environment. Complex Intell. Syst. **5**(4), 391–402 (2019)
6. Chanas, S., Kamburowski, J.: The fuzzy shortest route problem. In: Interval and fuzzy mathematics, Proceedings of Polish Symposium, Technical University of Poznan, Poznan, pp. 35–41 (1983)
7. Darehmiraki, M.: A novel parametric ranking method for intuitionistic fuzzy numbers. Iran. J. Fuzzy Syst. **16**(1), 129–143 (2019)
8. Dubois, D., Parde, H.: Fuzzy Sets and Systems: Theory and Applications, vol. 144. Academic Press, New York (1980)
9. Liao, X., JiaYi, W., Li, M.: An algorithmic approach for finding the fuzzy constrained shortest paths in a fuzzy graph. Complex Intell. Syst. 1–11 (2020)
10. Roditty, L., Zwick, U.: On dynamic shortest paths problems. In: European Symposium on Algorithms. Springer, Heidelberg (2004)
11. Xu, Z.S., Xia, M.M.: Distance and similarity measures for hesitant fuzzy sets. Inf. Sci. **181**, 2128–2138 (2011)
12. Xia, M., Xu, Z.: Hesitant fuzzy information aggregation in decision making. Inter. J. Approx. Reasoning **52**(3), 395–407 (2011)
13. Yang, C.-D., Lee, D.-T., Wong, C.-K.: On bends and lengths of rectilinear paths: a graph-theoretic approach. Int. J. Comput. Geom. Appl. **2**(01), 61–74 (1992)
14. Yager, R.R.: Paths of least resistance in possibilistic production systems. Fuzzy Sets Syst. **19**(2), 121–132 (1986)

A New Model for Ranking Z-numbers to Make Decisions with High Sensitivity

Somayeh Ezadi[1], Tofigh Allahviranloo[2(✉)], Soheil Salahshour[2], and Nermine Ahmed El Sissi[3]

[1] Department of Mathematics, Science and Research Branch, Islamic Azad University, Tehran, Iran

[2] Faculty of Engineering and Natural Sciences, Bahcesehir University, Istanbul, Turkey
tofigh.allahviranloo@eng.bau.edu.tr

[3] Faculty of Engineering and Natural Science, Bahcesehir University, Istanbul, Turkey

Abstract. In this paper, a new Multi-criteria decision making (MCDM) method based on Z-number is proposed to deal with linguistic decision making problems. The decision making process can be easily realized step by step with the arithmetic operations on Z-numbers. A few numerical examples and a practical example on MCDM is used to illustrate the efficiency of the proposed method.

Keywords: Z-numbers · Ranking of Z-numbers · Multi-criteria decision making (MCDM) method

1 Introduction

Many issues, such as optimization, data analysis, decision making, forecasting, social and economic systems, etc., are modeled by taking into account complex concepts or variables such as Z-numbers. These numbers are a combination of membership functions and probabilistic distribution functions (not statistical probability) that greatly add complexity and ambiguity to their data structure and structure. Obviously, in comparison with fuzzy numbers or fuzzy data, this type of data also involves another type of uncertainty derived from distributive functions. Z-numbers were introduced by Zade in 2011 [15]. A Z-number is a pair of fuzzy numbers to form (A, B), where A is a fuzzy subset of constraints that can have values of x, and B is a fuzzy subset of the reliability scale of component A. Therefore, in comparison with the fuzzy numbers, Z-numbers are more capable of describing human knowledge. However, in the procedure of applying Z-numbers such as decision-making, we have to face an issue, that is how to address the restriction and the reliability of Z-number [15]. Also, efforts have been made to apply Z-number in various ituations such as in computing with words (CWW) [15] and decision making problems [9, 14]. We introduced a Numerical Solution of Linear Regression Based on Z-Numbers by Improved Neural Network [8]. Also, decision making is an important process for most organizations. For example, with the development of society, a lot of harmful substances have affected human health, which leads to a high probability of human diseases. Therefore, medical diagnosis is particularly important. It

T. Allahviranloo et al. (Eds.): IDS 2020, AISC 1301, pp. 328–342, 2021.
https://doi.org/10.1007/978-3-030-66501-2_27

normally involves evaluation of prioritize of alternatives based on a given set of criteria. Usually, these criteria conflict each other and no solution can satisfy all criteria simultaneously. This problem is known as the Multiple Criteria Decision Making (MCDM) problems, and Since decision maker's opinion is usually expressed in a fairly vague manner where it cannot be measured numerically. It is common to use terms such as "poor", "excellent" and "very good "to illustrate the judgment on an alternative, closer to how human perceive on an object or situation.

On the other hand, if only to rely on the fuzzy numbers, there may exist the limitation to appropriately describing reliability of information. So using Z numbers can be very effective. In this regard, proposed an approach to convert Z-numbers into fuzzy numbers, in which the second component is defuzzified to a crisp number [10], numerous researchers proposed some useful methods to deal with the problems in the uncertain environment by applying the approach. Bakar [4] introduced a multi-layer method to rank Z-numbers, in which there are two layers, namely, Z-number conversion and fuzzy number ranking. We [6] introduced a New Multi-Layer Method for Z-number Ranking using Hyperbolic Tangent Function and Convex Combination, this procedure consists of two major parts, i) the conversion of a Z-number to a generalized fuzzy number based on the method given by [10] and ii) the ranking procedure using Hyperbolic Tangent Function and Convex Combination. Also, we [7] introduced Two new methods for ranking of Z-numbers based on sigmoid function and sign method, in which there are two layers, namely, Z-number conversion [10] and fuzzy number ranking using sigmoid function and sign function. Other methods have been introduced for ranking Z-numbers, including [3, 6, 7, 9, 10, 12, 13]. Eventually, the Z-Advanced numbers process was introduced by Allahviranloo and Ezadi [2]. In this paper, a new method for ranking Z-numbers is introduced. The rest of paper contains the following sections:

In Section two, the basic concepts are expressed. In Sect. 3, the algorithm is called the proposed method. In Sect. 4, numerical examples are presented, In Sect. 5, an applied example are presented, and finally, in Sect. 6, the conclusion is expressed.

2 Preliminaries

In this section, some definitions are brought.

Definition 2.1. A fuzzy number $\widetilde{A} = (a, b, c, d; \omega)$ is described as any fuzzy subset of the real line $\mathbb{R}$ with membership function $\mu_{\tilde{A}}$ which processes the following properties:

(a) $\mu_{\tilde{A}}$ is a continuous mapping from $\mathbb{R}$ to the closed interval $[0, \omega]$, $0 \leq \omega \leq 1$.
(b) $\mu_{\tilde{A}} = 0$, for all $x \in (-\infty, a]$.
(c) $\mu_{\tilde{A}}$, is strictly increasing on $[a, b]$.
(d) $\mu_{\tilde{A}} = \omega$, for all $x \in [b, c]$, where ω is constant and $\omega \in (0, 1]$.
(e) $\mu_{\tilde{A}}$, is strictly decreasing on $[c, d]$.
(f) $\mu_{\tilde{A}} = 0$, for all $x \in [d, \infty)$, where a, b, c, d are real numbers. We may let $a = -\infty$, $ora = b$, $orc = d$, $ord = +\infty$.

If $\omega = 1$, in (d) $\widetilde{A}$ is a normal fuzzy number, and if $0 < \omega < 1$, in (d) $\widetilde{A}$ is a non-normal fuzzy number. The image (opposite) of $\widetilde{A} = (a, b, c, d; \omega)$ can be given by $-\widetilde{A} = (-d, -c, -b, -a; \omega)$.

Definition 2.2. (Definition of Z-number).

Zadeh [15] proposed a subject called z-number along with an unknown variable X that has two components $(\widetilde{A}, \widetilde{B})$. The first component, $\widetilde{A}$, is a restriction (constraint) on the values which a real-valued uncertain variable, X, is allowed to take. The second component, $\widetilde{B}$ is a measure of reliability (certainty) of the first component. Zadeh introduced $(X, \widetilde{A}, \widetilde{B})$ as Z-valuation and showed that this amount is equal to this that X is equal to $(\widetilde{A}, \widetilde{B})$. Here Z provides information about the value of variable X. An example of this valuation of Z is as follows:

Example: (about 65 min, very sure), (about 20 min, sure). This valuation for Z according to proposal of Zadeh is observed as a restriction in X and is interpreted as follows:

$$Prob\left(x \ is \ \widetilde{A}\right) is \ \widetilde{B}$$

Actually, this means that

$$R(x) : x \ is \ \tilde{A} \rightarrow Poss(x = u) = \mu_{\tilde{A}}(u)$$

$$P(x \ is \ \tilde{A}) = \int_R \mu_{\tilde{A}}(u) P_x(u) du \ is \ \tilde{B}$$

In which μ_A is the membership function of the fuzzy set $\widetilde{A}$ and u is a value of x. $P_x(u)$ is the probability density function of x and $P(x = u)$ the probability function of x. Where we don't know the basic probability distribution, it is clear from these data that the probability distribution is itself a fuzzy number.

Note that for $Z = (\widetilde{A}, \widetilde{B})$, (where $\widetilde{A} = (a_1, a_2, a_3, a_4, \omega_{\tilde{A}})$ and $\widetilde{B} = (b_1, b_2, b_3, b_4, \omega_{\widetilde{B}})$);

(I) Convert the $\widetilde{B}$ (reliability) into crisp number by using [10]

$$\alpha = \frac{\int_X x\mu_{\widetilde{B}} dx}{\int_X \mu_{\widetilde{B}} dx} \tag{2.1}$$

Where $\int$ denotes an algebraic integration.

(II) Add the weight of the $\widetilde{B}$ to the $\widetilde{A}$ (restriction). The weighted Z-number is denoted as

$$\tilde{z}^{\alpha} = \left\{(x, \mu_{\tilde{A}^{\alpha}}(x)) | \mu_{\tilde{A}^{\alpha}} = \alpha \mu_{\tilde{A}}(x), x \in [0, 1]\right\} \tag{2.2}$$

Note that, α represents the weight of the reliability component of Z – number.

3 Algorithm of the Proposed Method to Rank the Z Numbers

Let $Z = (\widetilde{A}, \widetilde{B})$ be a Z-number such that $\widetilde{A} = (a_1, a_2, a_3, a_4, \omega_{\tilde{A}})$ and $\widetilde{B} = (b_1, b_2, b_3, b_4; \omega_{\underset{\sim}{B}})$. Clearly, $\widetilde{A}$ and $\widetilde{B}$ are trapezoidal fuzzy numbers. The membership functions $\widetilde{A}$ and $\widetilde{B}$ can also be shown below and

$$f_{\tilde{A}}(x) = \begin{cases} f^L_{\tilde{A}}(x) = \frac{x-a_1}{a_2-a_1} & \textit{if} \quad a_1 \le x \le a_2 \\ \omega_{\tilde{A}} & \textit{if} \quad a_2 \le x \le a_3 \\ f^R_{\tilde{A}}(x) = \frac{a_4-x}{a_4-a_3} & \textit{if} \quad a_3 \le x \le a_4 \\ 0 & \textit{Otherwise} \end{cases} \tag{3.1}$$

$$f_{\underset{\sim}{B}}(x) = \begin{cases} f^L_{\underset{\sim}{B}}(x) = \frac{x-b_1}{b_2-b_1} & \textit{if} \quad b_1 \le x \le b_2 \\ \omega_{\underset{\sim}{B}} & \textit{if} \quad b_2 \le x \le b_3 \\ f^R_{\underset{\sim}{B}}(x) = \frac{b_4-x}{b_4-b_3} & \textit{if} \quad b_3 \le x \le b_4 \\ 0 & \textit{Otherwise} \end{cases} \tag{3.2}$$

$f^L_{\tilde{A}}(x) : [a_1, a_4] \to [0, \omega_{\tilde{A}}]$ and $f^R_{\tilde{A}}(x) : [b_1, b_4] \to [0, \omega_{\underset{\sim}{B}}]$ are two strictly monotonic and continuous mapping from R to closed interval $[0, \omega_{\tilde{A}}]$. If $\omega_{\tilde{A}} = 1$, then $\widetilde{A}$ is normal fuzzy number; otherwise, it is a trapezoidal fuzzy number and is usually denoted by $\widetilde{A} = (a_1, a_2, a_3, a_4, \omega_{\tilde{A}})$ or $\widetilde{A} = (a_1, a_2, a_4, \omega_{\tilde{A}})$ if $\omega = 1$. In Particular, when, $a_2 = a_3$ the trapezoidal fuzzy number is reduced to a triangular fuzzy number denoted by $\widetilde{A} = (a_1, a_2, a_4, \omega_{\tilde{A}})$ or $\widetilde{A} = (a_1, a_2, a_4)$ if $\omega_{\tilde{A}} = 1$. Also, If $\omega_{\underset{\sim}{B}} = 1$, then $\widetilde{B}$ is normal fuzzy number; otherwise, it is a trapezoidal fuzzy number and is usually denoted by $\widetilde{B} = (b_1, b_2, b_3, b_4, \omega_{\underset{\sim}{B}})$ or $\widetilde{B} = (b_1, b_2, b_4, \omega_{\underset{\sim}{B}})$ if $\omega_{\underset{\sim}{B}} = 1$. In Particular, when, $b_2 = b_3$ the trapezoidal fuzzy number is reduced to a triangular fuzzy number denoted by $\widetilde{B} = (b_1, b_2, b_4, \omega_{\underset{\sim}{B}})$ or $\widetilde{B} = (b_1, b_2, b_4)$ if $\omega_{\underset{\sim}{B}} = 1$. So triangular fuzzy numbers are special cases of trapezoidal fuzzy numbers. Since $f^L_{\tilde{A}}(x), f^R_{\tilde{A}}(x), f^L_{\underset{\sim}{B}}(x)$ and $f^R_{\underset{\sim}{B}}(x)$ are Every four strictly monotonic and continuous functions, their inverse functions exist and should be continuous and strictly monotonic. Let $g^L_{\tilde{A}} : [0, \omega_{\tilde{A}}] \to [a_1, a_4]$ and $g^R_{\tilde{A}} : [0, \omega_{\tilde{A}}] \to [a_1, a_4]$, $g^L_{\underset{\sim}{B}} : [0, \omega] \to [b_1, b_4]$ and $g^R_{\underset{\sim}{B}} : [0, \omega] \to [b_1, b_4]$ be the inverse functions of $f^L_{\tilde{A}}(x), f^R_{\tilde{A}}(x), f^L_{\underset{\sim}{B}}(x)$ and $f^R_{\underset{\sim}{B}}(x)$ respectively. Then $g^L_{\tilde{A}}(y)$, $g^R_{\tilde{A}}(y)$, $g^L_{\underset{\sim}{B}}(y)$, and $g^R_{\underset{\sim}{B}}(y)$ should be integrable on the close interval $[0, \omega]$, ($\omega = \omega_{\tilde{A}}$ for $g^L_{\tilde{A}}(y)$ and $g^R_{\tilde{A}}(y)$, $\omega = \omega_{\underset{\sim}{B}}$ for $f^L_{\underset{\sim}{B}}(x)$ and $f^R_{\underset{\sim}{B}}(x)$). In other words, both $\int_a^b g^L_{\tilde{A}}(y)dy$, $\int_a^b g^R_{\tilde{A}}(y)dy$, $\int_a^b g^L_{\underset{\sim}{B}}(y)dy$, and $\int_a^b g^R_{\underset{\sim}{B}}(y)dy$ should exist. In the case of trapezoidal fuzzy number, the inverse function $g^L_{\tilde{A}}(y)$, $g^R_{\tilde{A}}(y)$, $g^L_{\underset{\sim}{B}}(y)$, and $g^R_{\underset{\sim}{B}}(y)$ can be analytically expressed as

$$g^L_{\tilde{A}}(y_{\tilde{A}}) = a_1 + (a_2 - a_1)y_{\tilde{A}}/\omega_{\tilde{A}},\ 0 \le y_{\tilde{A}} \le \omega_{\tilde{A}} \tag{3.3}$$

$$g_{\tilde{A}}^{L}(y_{\tilde{A}}) = a_4 + (a_4 - a_3)y_{\tilde{A}}/\omega_{\tilde{A}},\ 0 \le y_{\tilde{A}} \le \omega_{\tilde{A}} \tag{3.4}$$

$$g_{\tilde{B}}^{L}\left(y_{\tilde{B}}\right) = b_1 + (b_2 - b_1)y_{\tilde{B}}/\omega_{\tilde{B}},\ 0 \le y_{\tilde{B}} \le \omega_{\tilde{B}} \tag{3.5}$$

$$g_{\tilde{B}}^{L}\left(y_{\tilde{B}}\right) = b_4 + (b_4 - b_3)y_{\tilde{B}}/\omega_{\tilde{B}},\ 0 \le y_{\tilde{B}} \le \omega_{\tilde{B}} \tag{3.6}$$

Now, a method for ranking trapezoidal Z-numbers is presented, which proposes the following algorithm:

Step 1. Computing x_{0A}, x_{0B}, y_{0A} and y_{0B}.

The values of x_{0A}, x_{0B}, y_{0A} and y_{0B} are calculated using the center of gravity method. So, we have

$$x_{0\tilde{A}} = \frac{\int_{a_1}^{a_2} x(x-a_1)dx + \int_{a_2}^{a_3} x dx + \int_{a_3}^{a_4} x(a_4-x)dx}{\int_{a_1}^{a_2} (x-a_1)dx + \int_{a_2}^{a_3} dx + \int_{a_3}^{a_4} (a_4-x)dx}$$

$$= \frac{1}{3}\left[a_1 + a_2 + a_3 + a_4 - \frac{a_3a_4 - a_1a_2}{(a_3+a_4)-(a_1+a_2)}\right] \tag{3.7}$$

$$x_{0\tilde{B}} = \frac{\int_{b_1}^{b_2} x(x-b_1)dx + \int_{b_2}^{b_3} x dx + \int_{b_3}^{b_4} x(b_4-x)dx}{\int_{b_1}^{b_2} (x-b_1)dx + \int_{b_2}^{b_3} dx + \int_{b_3}^{b_4} (b_4-x)dx}$$

$$= \frac{1}{3}\left[b_1 + b_2 + b_3 + b_4 - \frac{b_3b_4 - b_1b_2}{(b_3+b_4)-(b_1+b_2)}\right] \tag{3.8}$$

$$y_{0\tilde{A}} = \frac{\int_0^{\omega_{\tilde{A}}} y_{\tilde{A}} g_{\tilde{A}}^{L}(y_{\tilde{A}})dy_{\tilde{A}} + \int_0^{\omega_{\tilde{A}}} y_{\tilde{A}} g_{\tilde{A}}^{R}(y_{\tilde{A}})dy_{\tilde{A}}}{\int_0^{\omega_{\tilde{A}}} g_{\tilde{A}}^{L}(y_{\tilde{A}})dy_{\tilde{A}} + \int_0^{\omega_{\tilde{A}}} g_{\tilde{A}}^{R}(y_{\tilde{A}})dy_{\tilde{A}}}$$

$$= \frac{\omega_{\tilde{A}}}{3}\left[1 + \frac{a_3 - a_2}{(a_3+a_4)-(a_1+a_2)}\right] \tag{3.9}$$

$$y_{0\tilde{B}} = \frac{\int_0^{\omega_{\tilde{B}}} y_{\tilde{B}} g_{\tilde{B}}^{L}\left(y_{\tilde{B}}\right)dy_{\tilde{B}} + \int_0^{\omega_{\tilde{B}}} y_{\tilde{B}} g_{\tilde{B}}^{R}\left(y_{\tilde{B}}\right)dy_{\tilde{B}}}{\int_0^{\omega_{\tilde{B}}} g_{\tilde{B}}^{L}\left(y_{\tilde{B}}\right)dy_{\tilde{B}} + \int_0^{\omega_{\tilde{B}}} g_{\tilde{B}}^{R}\left(y_{\tilde{B}}\right)dy_{\tilde{B}}}$$

$$= \frac{\omega_{\tilde{B}}}{3}\left[1 + \frac{b_3 - b_2}{(b_3+b_4)-(b_1+b_2)}\right] \tag{3.10}$$

Since non-normal triangular fuzzy numbers are special cases of normal trapezoidal fuzzy numbers with $a_2 = a_3$ ($b_2 = b_3$) formulas (3.7)–(3.10) can be simplified as

$$x_{0\tilde{A}} = \frac{1}{3}[a_1 + a_2 + a_4] \tag{3.11}$$

$$x_{0\tilde{B}} = \frac{1}{3}[b_1 + b_2 + b_4] \tag{3.12}$$

$$y_{0\tilde{A}} = \frac{\omega_{\tilde{A}}}{3} \tag{3.13}$$

$$y_{0\tilde{B}} = \frac{\omega_{\tilde{B}}}{3} \tag{3.14}$$

Obviously, for normal trapezoidal (triangular) fuzzy numbers $\omega = 1$, the formulas of $y_{0\tilde{A}}, y_{0\tilde{B}}$ in relations (3.11) and (3.14) can be simplified as

$$y_{0\tilde{A}} = \frac{1}{3}\left[1 + \frac{a_3 - a_2}{(a_3 + a_4) + (a_1 + a_2)}\right] \tag{3.15}$$

$$y_{0\tilde{B}} = \frac{1}{3}\left[1 + \frac{b_3 - b_2}{(b_3 + b_4) + (b_1 + b_2)}\right] \tag{3.16}$$

$$y_{0\tilde{A}} = \frac{1}{3} \tag{3.17}$$

$$y_{0\tilde{B}} = \frac{1}{3} \tag{3.18}$$

Step 2. Introducing the sign function.

Definition 3.1. Let R^Z stands the set of non-normal Z-numbers, ω be a constant provided that $0 < \omega \leq 1$ and $\gamma : R^Z \to \{-\omega, 0, \omega\}$ be a function that is defined as [1]:

$$\forall ZR^Z : \gamma(Z) = sign\left[\left(\int_0^{\omega_{\tilde{A}}} \left(g_{\tilde{A}}^L(y_{\tilde{A}}) + g_{\tilde{A}}^R(y_{\tilde{A}})\right)dy_{\tilde{A}}\right) + (1-\lambda)\left(\int_0^{\omega_{\tilde{B}}} \left(g_{\tilde{B}}^L\left(y_{\tilde{B}}\right) + g_{\tilde{B}}^R\left(y_{\tilde{B}}\right)\right)dy_{\tilde{B}}\right)\right]$$

According to the definition 1 and the concept of Z-numbers, we define the function of λ as follows:

$$\gamma(Z) = \begin{cases} 1 & \int_0^{\omega_{\tilde{A}}} \left(g_{\tilde{A}}^L(y_{\tilde{A}}) + g_{\tilde{A}}^R(y_{\tilde{A}})\right)dy_{\tilde{A}} + \int_0^{\omega_{\tilde{B}}} \left(g_{\tilde{B}}^L\left(y_{\tilde{B}}\right) + g_{\tilde{B}}^R\left(y_{\tilde{B}}\right)\right)dy_{\tilde{B}} > 0 \\ 0 & \int_0^{\omega_{\tilde{A}}} \left(g_{\tilde{A}}^L(y_{\tilde{A}}) + g_{\tilde{A}}^R(y_{\tilde{A}})\right)dy_{\tilde{A}} + \int_0^{\omega_{\tilde{B}}} \left(g_{\tilde{B}}^L\left(y_{\tilde{B}}\right) + g_{\tilde{B}}^R\left(y_{\tilde{B}}\right)\right)dy_{\tilde{B}} = 0 \\ -1 & \int_0^{\omega_{\tilde{A}}} \left(g_{\tilde{A}}^L(y_{\tilde{A}}) + g_{\tilde{A}}^R(y_{\tilde{A}})\right)dy_{\tilde{A}} + \int_0^{\omega_{\tilde{B}}} \left(g_{\tilde{B}}^L\left(y_{\tilde{B}}\right) + g_{\tilde{B}}^R\left(y_{\tilde{B}}\right)\right)dy_{\tilde{B}} < 0 \end{cases} \tag{3.19}$$

$$R(Z) = \sqrt{\left(\lambda x_{0\tilde{A}} + (1-\lambda)x_{0\tilde{B}}\right)^2 + \left(\lambda y_{0\tilde{A}} + (1-\lambda)y_{0\tilde{B}}\right)^2}, \ \lambda = [0, 1] \tag{3.20}$$

Then $Rank(Z) = \gamma(Z)R(Z)$.

Here, the method $Rank(Z)$ is utilized to rank Z numbers. Therefore, for any two Z numbers Z_1 and Z_2.

1. $Rank(Z1) > Rank(Z2)$ if and only if $Z_1 > Z_2$
2. $Rank(Z1) < Rank(Z2)$ if and only if $Z_1 < Z_2$.
3. $Rank(Z1) = Rank(Z2)$ if and only if $Z_1 = Z_2$.

4 Numerical Examples

Example 1. Let $\widetilde{A}$ (0.1, 0.3, 0.3, 0.5; 1), $\widetilde{B}_1 = (0.1, 0.3, 0.3, 0.5; 1)$, $\widetilde{B}_2 = (0.2, 0.3, 0.3, 0.4; 1)$. The Ranking for $Z_1 = \left(\widetilde{A}, \widetilde{B}_1\right)$ *and* $Z_2 = \left(\widetilde{A}, \widetilde{B}_2\right)$ per convex combination can be seen in Table 2. The calculated results by the existing methods can be seen in Table 1. The results of the proposed method compared with existing methods and can be seen in Table 3.

Table 1. The calculated results by the existing methods for Example 1.

Methods	Z_1	Z_2
Mohamad's method [12]	0.0774	0.0774
Bakar's method [4]	0.0288	0.0288
Kang's method [9]	0.3000	0.3000

Table 2. The calculated results by the proposed method for Example 1 ($\gamma = 1$).

λ_i	$x_{\tilde{A}}$	$x_{\tilde{B}}$	$y_{\tilde{A}}$	$y_{\tilde{B}}$	$Rank(Z_1)$	$Rank(Z_2)$
$\lambda = 0$	0.3	0.3	0.33	0.33	0.66	0.66
$\lambda = 0.5$	0.3	0.3	0.33	0.33	0.63	0.63
$\lambda = 1$	0.3	0.3	0.33	0.33	0.6	0.6

Table 3. A comparison of the proposed method with the existing methods for Example 1.

Methods	Ranking
Mohamad's method [12]	$Z_1 \approx Z_2$
Bakar's method [4]	$Z_1 \approx Z_2$
Kang's method [9]	$Z_1 \approx Z_2$
The proposed method	$\mathbf{Z_1 \approx Z_2}$

Example 2. Let $\widetilde{A}$ (0.1, 0.3, 0.3, 0.5; 1), $\widetilde{B}_1 = (0.1, 0.3, 0.3, 0.5; 1)$, $\widetilde{B}_2 = (0.1, 0.2, 0.4, 0.5; 1)$. The Ranking for $Z_1 = \left(\widetilde{A}, \widetilde{B}_1\right)$ *and* $Z_2 = \left(\widetilde{A}, \widetilde{B}_2\right)$ per convex combination can be seen in Table 4. The results of the proposed method compared with existing methods and can be seen in Table 5.

Table 4. The calculated results by the proposed method for Example 2, ($\gamma = 1$).

λ_i	$x_{\tilde{A}}$	$x_{\tilde{B}}$	$y_{\tilde{A}}$	$y_{\tilde{B}}$	$Rank(Z_1)$
$\lambda = 0$	0.3	0.3	0.33	0.33	0.66
$\lambda = 0.5$	0.3	0.3	0.33	0.33	0.63
$\lambda = 1$	0.3	0.3	0.33	0.33	0.6
λ_i	$x_{\tilde{A}}$	$x_{\tilde{B}}$	$y_{\tilde{A}}$	$y_{\tilde{B}}$	$Rank(Z_2)$
$\lambda = 0$	0.3	0.3	0.33	0.44	0.77
$\lambda = 0.5$	0.3	0.3	0.33	0.44	0.69
$\lambda = 1$	0.3	0.3	0.33	0.44	0.6

Table 5. A comparison of the proposed method with the existing methods for Example 2.

Methods	*Ranking*
Mohamad's method [12]	$Z_1 \approx Z_2$
Bakar's method [4]	$Z_1 \approx Z_2$
Kang's method [9]	$Z_1 \approx Z_2$
The proposed method ($\lambda = 0$)	$Z_1 \prec Z_2$
The proposed method ($\lambda = 0.5$)	$Z_1 \prec Z_2$
The proposed method ($\lambda = 1$)	$Z_1 \approx Z_2$

Example 3. Let $\tilde{A}_1 = (-0.03, 0, 0, 0.3)$, $\tilde{A}_2 = (0, 0.0484, 0.0968, 0.9)$, $\widetilde{B} = (0.1, 0.3, 0.3, 0.5)$. The Ranking for $Z_1 = \left(\tilde{A}_1, \widetilde{B}\right)$ *and* $Z_2 = \left(\tilde{A}_2, \widetilde{B}\right)$ per convex combination can be seen in Table 6.

Result $Z_1 \preccurlyeq Z_2$.

Table 6. The calculated results by the proposed method for Example 3, ($\gamma = 1$).

λ_i	$x_{\tilde{A}}$	$x_{\tilde{B}}$	$y_{\tilde{A}}$	$y_{\tilde{B}}$	$Rank(Z_1)$
$\lambda = 0$	0	0.3	0.33	0.33	0.66
$\lambda = 0.5$	0	0.3	0.33	0.33	0.5
$\lambda = 1$	0	0.3	0.33	0.33	0.3
λ_i	$x_{\tilde{A}}$	$x_{\tilde{B}}$	$y_{\tilde{A}}$	$y_{\tilde{B}}$	$Rank(Z_2)$
$\lambda = 0$	0.31	0.3	0.35	0.33	0.68
$\lambda = 0.5$	0.31	0.3	0.35	0.33	0.65
$\lambda = 1$	0.31	0.3	0.35	0.33	0.61

Example 4. Let $\tilde{A}_1 = (-\frac{5}{2}, -\frac{1}{2}, 0, 1)$, $\widetilde{A_2} = (-\frac{9}{4}, -\frac{1}{4}, -\frac{1}{4}, \frac{3}{4})$, $\widetilde{B} = (0.1, 0.3, 0.3, 0.5)$. The Ranking for $Z_1 = \left(\tilde{A}_1, \widetilde{B}\right)$ *and* $Z_2 = \left(\tilde{A}_2, \widetilde{B}\right)$ per convex combination can be seen in Tables 7 and 8.

Table 7. The calculated results by the proposed method for Example 4, ($\gamma = 1$).

λ_i	$x_{\tilde{A}}$	$x_{\tilde{B}}$	$y_{\tilde{A}}$	$y_{\tilde{B}}$	$Rank(Z_1)$
$\lambda = 0$	−0.56	0.3	0.37	0.33	0.7
$\lambda = 0.5$	−0.56	0.3	0.37	0.33	0.53
$\lambda = 1$	−0.56	0.3	0.37	0.33	0.26
λ_i	$x_{\tilde{A}}$	$x_{\tilde{B}}$	$y_{\tilde{A}}$	$y_{\tilde{B}}$	$Rank(Z_2)$
$\lambda = 0$	−0.58	0.3	0.33	0.33	0.6
$\lambda = 0.5$	−0.58	0.3	0.33	0.33	0.51
$\lambda = 1$	−0.58	0.3	0.33	0.33	0.28

Table 8. Result of Ranking for Example 4.

λ_i	Result
$\lambda = 0$	$Z_1 \succcurlyeq Z_2$
$\lambda = 0.5$	$Z_1 \succcurlyeq Z_2$
$\lambda = 1$	$Z_1 \preccurlyeq Z_2$

Example 5. Let $Z_1 = ((0.1, 0.3, 0.3, 0.5; 1), (0.1, 0.2, 0.4, 0.5; 1))$, $Z_2 = ((0.1, 0.2, 0.4, 0.5; 1), (0.1, 0.3, 0.3, 0.5; 1))$. The Ranking for Z_1 *and* Z_2 per convex combination can be seen in Table 9.

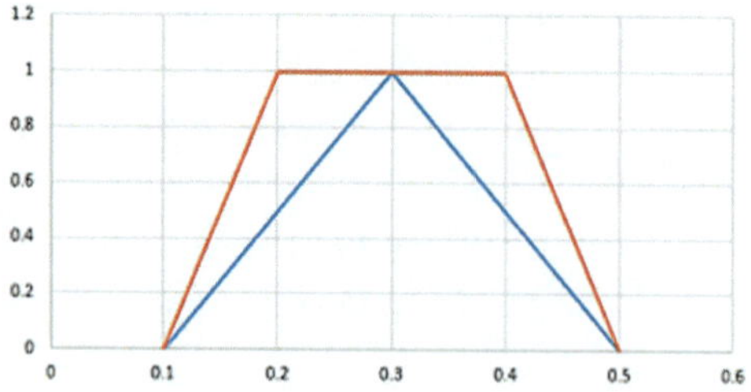

$Z_1 = ((0.1, 0.3, 0.3, 0.5; 1), (0.1, 0.2, 0.4, 0.5; 1))$

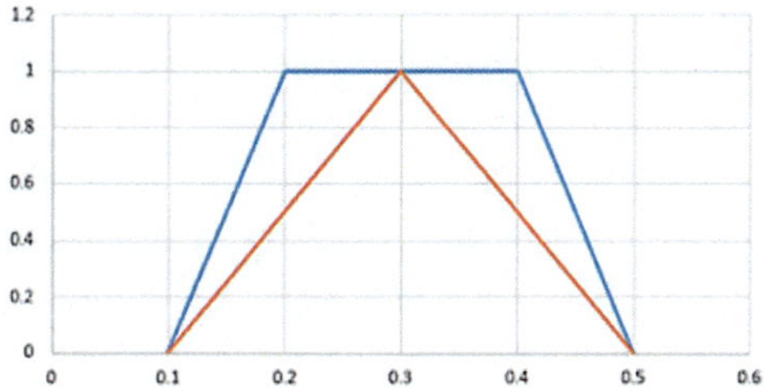

$$Z_2 = ((0.1, 0.2, 0.4, 0.5; 1), (0.1, 0.3, 0.3, 0.5; 1))$$

Table 9. The calculated results by the proposed method for Example 5, ($\gamma = 1$).

λ_i	$x_{\tilde{A}}$	$x_{\tilde{B}}$	$y_{\tilde{A}}$	$y_{\tilde{B}}$	$Rank(Z_1)$
$\lambda = 0$	0.3	0.3	0.33	0.44	0.77
$\lambda = 0.5$	0.3	0.3	0.33	0.44	0.69
$\lambda = 1$	0.3	0.3	0.33	0.44	0.6
λ_i	$x_{\tilde{A}}$	$x_{\tilde{B}}$	$y_{\tilde{A}}$	$y_{\tilde{B}}$	$Rank(Z_1)$
$\lambda = 0$	0.3	0.3	0.44	0.33	0.77
$\lambda = 0.5$	0.3	0.3	0.44	0.33	0.69
$\lambda = 1$	0.3	0.3	0.44	0.33	0.6

Result $Z_1 \approx Z_2$.

In the Example 5, it is clear that the two Z-numbers are not equal, but are the same rank with the proposed method. Also some of the methods available show that these two numbers are identical. We also looked at some ways to determine their degree of similarity, such as:

1) Mod $\left(\tilde{A}\right) = \mathrm{M}(\tilde{A}) = \frac{w*(a_2+a_3)}{4}$, for $\tilde{A} = (a_1, a_2, a_3, a_4; \omega_{\tilde{A}})$.

2) divergence $\left(\tilde{A}\right) = w(a_4 - a_1)$,

3) Left spread $\left(\tilde{A}\right) = w(a_2 - a_1)$,

4) Right spread $\left(\tilde{A}\right) = w(a_4 - a_3)$.

Their results were also identical for both Z-numbers. Therefore, the following algorithm was developed to determine their degree of similarity.

Step 1. Computing x_{0A} and y_{0B}

$$x_{0\tilde{A}} = \frac{\int_{a_1}^{a_2} x(x - a_1)dx + \int_{a_2}^{a_3} x dx + \int_{a_3}^{a_4} x(a_4 - x)dx}{\int_{a_1}^{a_2} (x - a_1)dx + \int_{a_2}^{a_3} dx + \int_{a_3}^{a_4} (a_4 - x)dx}$$

$$= \frac{1}{3}\left[a_1 + a_2 + a_3 + a_4 - \frac{a_3a_4 - a_1a_2}{(a_3 + a_4) - (a_1 + a_2)}\right] \tag{4.1}$$

$$y_{0\tilde{B}} = \frac{\int_0^{\omega_{\tilde{B}}} y_{\tilde{B}} g_{\tilde{B}}^L\left(y_{\tilde{B}}\right) dy_{\tilde{B}} + \int_0^{\omega_{\tilde{B}}} y_{\tilde{B}} g_{\tilde{B}}^R\left(y_{\tilde{B}}\right) dy_{\tilde{B}}}{\int_0^{\omega_{\tilde{B}}} g_{\tilde{B}}^L\left(y_{\tilde{B}}\right) dy_{\tilde{B}} + \int_0^{\omega_{\tilde{B}}} g_{\tilde{B}}^R\left(y_{\tilde{B}}\right) dy_{\tilde{B}}}$$

$$= \frac{\omega_{\tilde{B}}}{3}\left[1 + \frac{b_3 - b_2}{(b_3 + b_4) - (b_1 + b_2)}\right] \tag{4.2}$$

Step 2. Computing $STD\left(\tilde{A}\right)$.

First, add the weight of the $\tilde{B}$ to the $\tilde{A}$ (restriction). The weighted Z-number is denoted as [10]

$$\tilde{Z}^{\alpha} = \left\{(x, \mu_{\tilde{A}^\alpha}(x)) | \mu_{\tilde{A}^\alpha} = \alpha \mu_{\tilde{A}}(x), x \in [0, 1]\right\}$$

Where $\alpha = y_{0\tilde{B}}$. So, we have

$$\tilde{A} = (a_1, a_2, a_3, a_4, \omega_{\tilde{A}}^*) \tag{4.3}$$

Where $\omega_{\tilde{A}}^* = \omega_{\tilde{A}} * \alpha$.

The standard deviation is then calculated as follows

$$STD\left(\tilde{A}\right) = \frac{\sqrt{\sum_{i=1}^{4} (a_i - x_{0\tilde{A}})^2}}{3} \tag{4.4}$$

Step 3. Computing $Rank(Z)$

$$Rank(Z) = \frac{x_{0\tilde{A}} * \omega_{\tilde{A}}^*}{1 + STD(\tilde{A})} \tag{4.5}$$

Where γ is the function of the sign. $Rank(Z) \in [-1, 1]$, i = 1,..., n. The larger value of$Rank(Z)$, the higher the preference of Z.

Example 6. Let $Z_1 = ((0.1, 0.3, 0.3, 0.5; 1), (0.1, 0.3, 0.3, 0.5; 1))$ and $Z_2 = ((0.1, 0.3, 0.3, 0.5; 1), (0.2, 0.3, 0.3, 0.4; 1))$.

The Ranking for Z_1 *and* Z_2 per convex combination can be seen in Table 10.

Example 7. The Ranking for Z_1 *and* Z_2 in example 5 can be seen in Table 11.

$Z1 = ((0.1, 0.3, 0.3, 0.5), (0.1, 0.2, 0.4, 0.5))$ and $Z2 = ((0.1, 0.2, 0.4, 0.5), (0.1, 0.3, 0.3, 0.5)))$

Table 10. The calculated results by the proposed method for Example 6.

Z_i	$x_{\tilde{A}}$	$x_{\tilde{B}}$	$STD(\tilde{A})$	$Rank(Z)$
Z_1	0.3	0.333	0.44	0.09
Z_2	0.3	0.444	0.58	0.12
Result				$Z_1 < Z_2$

Table 11. The calculated results by the proposed method for Example 7.

Z_i	$x_{\tilde{A}}$	$x_{\tilde{B}}$	$STD(\tilde{A})$	$Rank(Z)$
Z_1	0.3	0.444	0.09	0.12
Z_2	0.3	0.333	0.1	0.09
Result				$Z_1 > Z_2$

5 Applied Example

Here is a case in point of mixture of motor vehicle for journey to demonstrate the process of the projected loom. There are three dissimilar choices, namely train, auto and bus. Take the three most important criteria (outlay, trip time, soothe) into reflection. For every automobile, according to the meticulous case, the outlay is the mainly momentous aspect, which can be described by means of linguistic variable "Very High", and the sureness of the outlay is also very sturdy, described by means of linguistic variable "Very High". Likewise, the trip time and the soothe can also be described by means of linguistic beneath the idea of Z-number. The linguistic criteria assessment of the three motor vehicles can be gained described as Table 12.

Table 12. Linguistic criterion table

	Outlay (Rupees) (V H; V H)	Trip time (min) (H; V H)	Soothe (M; V H)
Train	((9,10,12), (VH))	((70,100,120), M)	((4,5,6), H)
Auto	((20,24,25), H)	((60,70,100), V H)	((7,8,10), H)
Bus	((15,15,15), H)	((70,80,90), H)	((1,4,7), H)

According to the membership function, the linguistic erratic can be renewed to mathematical assessment, which is described as Table 13.

In addition to the next move is to simplify the fuzzy information to evade convolution of exact operations in the assessment progression. The modified decision matrix is denoted in Table 14.

Table 13. Linguistic criterion table

	Outlay (Rupees) ((0.75,1,1), (0.75,1,1))	Trip time (min) ((0.5,0.75,1), (0.75,1,1))	Soothe ((0.25,0.5,0.75), (0.75,1,1))
Train	((9,10,12), (0.75,1,1))	((70,100,120), (0.25,0.5,0.75))	((4,5,6), (0.5,0.75,1))
Auto	((20,24,25), (0.5,0.75,1))	((60,70,100), (0.75,1,1))	((7,8,10), (0.5,0.75,1))
Bus	((15,15,15), (0.5,0.75,1))	((70,80,90), (0.5,0.75,1))	((1,4,7), (0.5,0.75,1))

Table 14. Modified decision matrix with Z number

	Outlay (Rupees) ((0.75,1,1), (0.75,1,1))	Trip time (min) ((0.5,0.75,1), (0.75,1,1))	Soothe ((0.25,0.5,0.75), (0.75,1,1))
Train	((0.1,0.11,0.14), (0.75,1,1))	((0.16,0.23,0.27), (0.25,0.5,0.75))	((0.13,0.16,0.19), (0.5,0.75,1))
Auto	((0.23,0.27,0.28), (0.5,0.75,1))	((0.14,0.16,0.23), (0.75,1,1))	((0.22,0.25,0.32), (0.5,0.75,1))
Bus	((0.17,0.17,0.17), (0.5,0.75,1))	((0.16,0.18,0.2), (0.5,0.75,1))	((0.03,0.13,0.22), (0.5,0.75,1))

At last, after normalizing the weight of criteria, according to Eq. (3.20). we can get the final priority weight of the three vehicle denoted by Table 15 and Table 16.

Table 15. Decision matrix with crisp number ($\gamma = 1$).

	Outlay (Rupees) 0.85	Trip time (min) 0.82	Soothe 0.77
Train	0.711	0.672	0.694
Auto	0.708	0.720	0.709
Bus	0.696	0.697	0.690

Table 16. The priority weight of the vehicle selection ($\gamma = 1$).

	Outlay (Rupees) 0.43	Trip time (min) 0.34	Soothe 0.23	
Train	0.711	0.672	0.694	0.330
Auto	0.708	0.720	0.709	0.339
Bus	0.696	0.697	0.690	0.331

6 Conclusions

Decision making is an important process that can determine the productivity and development of organizations. Also in the field of medical science, that Human health has received so serious threat that the medical diagnosis becomes a very worthy of study. Different methods and tools have been introduced for solving decision making problems. Each of these methods tries to prioritize effective factors in decision making so that the decision maker can make the right decision. When a person's life is in the midst of a person or when an organization's decision can have a significant impact on the community's situation, it is necessary to prioritize the factors affecting a decision with greater sensitivity. In this paper we present a new two-dimensional method for ranking Z-numbers. We showed that our proposed method has a better logic in ranking Z-numbers than some of the existing ranking methods. This new procedure of decision making provides an effective way in handling decision making process and can be used as an alternative tool in solving MCDM problems.

References

1. Abbasbandy, S., Hajjari, T.: An improvement in centroid point method for ranking of fuzzy numbers. J. Sci. I. A. U (JSIAU) **20**(78/2) (2011). Winter
2. Allahviranloo, T., Ezadi, S.: Z-Advanced numbers processes. Inf. Sci. **480**, 130–143 (2019)
3. Alive, R.A., Huseynov, O.H., Serdaroglu, R.: Ranking of Z-numbers, and its application in decision making. Int. J. **15**(06), 1503-1519 (2016)
4. Bakar, A.S.A., Gegov, A.: Multi-layer decision methodology for ranking Z-numbers. Int. J. Comput. Intell. Syst. **8**, 395–406 (2015)
5. Deng, Y., Zhu, Z.F., Liu, Q.: Ranking fuzzy numbers with an area method using radius of gyration. Comput. Math. Appl. **51**, 1127–1136 (2006)
6. Ezadi, S., Allahviranloo, T.: New multi-layer method for Z-number ranking using Hyperbolic Tangent function and convex combination. Intelligent Automation Soft Computing, pp. 1–7 (2017)
7. Ezadi, S., Allahviranloo, T.: Two new methods for ranking of Z-numbers based on sigmoid function and sign method. Int. J. Intell. Syst., 1–12 (2018)
8. Ezadi, S., Allahviranloo, T.: Numerical solution of linear regression based on Z-numbers by improved neural network. IntellIgent AutomAtIon and Soft ComputIng, pp. 1–11 (2017)
9. Kang, B. Wei, D. Li, Y. and Deng, Y.J.: Decision Making Using Z-numbers under Uncertain Environment. Comput. Inf. Syst. **7**, 2807–2814 (2012)

10. Kang, B., Wei, D., Li, Y., Deng, Y.: A method of converting Z-number to classical fuzzy number. J. Inf. Comput. Sci. **3**, 703–709 (2012)
11. Kumar, A., Singh, P., Kaur, P., Kaur, A.: A new approach for ranking of L-R type generalized fuzzy umbers. Expert Syst. Appl. **38**, 10906–10910 (2011)
12. Mohamad, D., Shaharani, S.A., Kamis, N.H.: A Z-number based decision making procedure with ranking fuzzy numbers method. AIP Conf. Proc. **1635**, 160–166 (2014)
13. Yager, R.R.: On Z-valuations using Zadeh's Z-numbers. Int. J. Intell. Syst. **27**, 259–278 (2012)
14. Salari, M., Baghepour, M., Wang, J.: Int. J. Appl. Dec. Sci. **7**(1), 97–119 (2014)
15. Zadeh, L.A.: A Note on Z-numbers. Inf. Sci. **181**, 2923–2932 (2011)

An Environmentally Sustainable Closed-Loop Supply Chain Network Design Under Uncertainty: Application of Optimization

Md. Mohsin Ahmed[1], S. M. Salauddin Iqbal[1], Tazrin Jahan Priyanka[1], Mohammad Arani[2](✉), Mohsen Momenitabar[3], and Md. Mashum Billal[4]

[1] Department of Industrial and Production Engineering, Rajshahi University of Engineering and Technology (RUET), Rajshahi 6204, Bangladesh
durjoymohsin@gmail.com, salauddin11488@gmail.com, priyanka.ruet12@gmail.com

[2] Department of Systems Engineering, University of Arkansas at Little Rock, Little Rock, USA
mxarani@ualr.edu

[3] Department of Transportation, Logistics, and Finance, North Dakota State University, Fargo, ND 58105, USA
mohsen.momenitabar@ndsu.edu

[4] Department of Mechanical Engineering (Engineering Management), University of Alberta, Edmonton, Canada
mdmashum@ualberta.ca

Abstract. Newly, the rates of energy and material consumption to augment industrial production are substantially high, thus the environmentally sustainable industrial development has emerged as the main issue of either developed or developing countries. A novel approach to supply chain management is proposed to maintain economic growth along with environmentally friendly concerns for the design of the supply chain network. In this paper, a new green supply chain design approach has been suggested to maintain the financial virtue accompanying the environmental factors that required to be mitigated the negative effect of rapid industrial development on the environment. This approach has been suggested a multi-objective mathematical model minimizing the total costs and CO2 emissions for establishing an environmentally sustainable closed-loop supply chain. Two optimization methods are used namely Epsilon Constraint Method, and Genetic Algorithm Optimization Method. The results of the two mentioned methods have been compared and illustrated their effectiveness. The outcome of the analysis is approved to verify the accuracy of the proposed model to deal with financial and environmental issues concurrently.

Keywords: Closed-Loop supply chain · Environmentally sustainable · Uncertainty · CO2 emissions · NRGA-II · Epsilon constraint method

T. Allahviranloo et al. (Eds.): IDS 2020, AISC 1301, pp. 343–358, 2021.
https://doi.org/10.1007/978-3-030-66501-2_28

1 Introduction

The sustainable supply chain has been one of the most uprising topics in the current decade as the environment is polluted at an alarming rate due to the functioning of organizations and factories. Sustainable supply chain (SSC) is, thus, a management system that integrates environmental concerns with supply chain functioning such as product design, raw material sourcing, manufacturing, assembly, transportation of final goods, and controlling the disposal and recycling of products when their life has ended (Aldemir et al. 2018). The use of the green supply chain (GSC) is generally originated from two common interrelated factors the amount of carbon emission and the cost of claiming a sustainable supply chain. Environmental issues have already absorbed the attention of the developed countries; federal law and customer pressure have increased to follow SSC and considering environmentally friendly production, consequently (Niranjan et al. 2019).

For green uplift of the total supply chain network, carbon emission at every entity must be taken into consideration. Focused on the green effect on a closed-looped supply chain and attempted to abate CO_2 emissions, furthermore, motivate the customers to utilize the recyclable product. In our research, we have formulated a mathematical model to optimize both the carbon emission and the total cost. Here, multi-objective optimization has been identified using two mathematical approaches both the exact solution method and meta-heuristic approach. A closed-loop supply chain network has also been incorporated with reverse flow. This closed-loop mainly deals with the end life of products and unused raw materials that otherwise it can create more wastage and degradation of the environment. We have also considered mixed-integer linear programming for more comprehensive supply chain network evaluation and thus considering green development. (Konyalıoğlu and Zafeirakopoulos 2020; Yang et al. 2018).

In our study, the only natural environment is being considered. The natural environment is the interaction between biology, ecology, and other forces. To maintain a safe environment this interaction must be maintained and balanced. Though industrialization that increases the way of living standard, it also has a huge escapade effect on the environment; manufacturing emits a lot of wastage and greenhouse gases that affect environmental balance. However, this topic remained unnoticed up to the 20th century. Concern for the pollution developed a concern for the environment thus leading us to save ourselves and the earth. People are now more concerned than ever before as anthropogenic activity already damage the balance of ecology and the environment. As a result, several agreements have been signed on compliance with environmental objectives. Some most important protocol and conventions are:

- Montreal protocol: It was signed to save the ozone layer and signed in 1987. The Montreal Protocol stipulated that production and use of ozone-depleting chemicals would be phased out by 2000. The primary purpose was to reduce CFCs.
- Kyoto protocol: This was a consensus reached at the 1992 Earth Summit in Rio de Janeiro under the United Nations Framework Convention on Climate Change (UNFCCC). The primary goal was to maintain a small enough degree of reduction of greenhouse gas emissions in the environment to avoid harmful anthropogenic involvement with the climate system. The Protocol has been ratified by 154 countries.

2 Literature Review

The sustainable supply chain is a set of procedures performed to monitor and facilitate the environmental and economic performance indices by distributing required resources, material, goods, and allocating organizational responsibilities and tasks (Kim et al. 2011). GSC may also be used as reducing and preferably alleviating the adverse impact of the supply chain network on the natural ecosystem (Andiç et al. 2012). In our research on the green supply chain, the literature concerned about finding the sectors and subsectors where minimization of carbon could be justified and regulate the emission of carbon and the cost of the total supply chain. In recent years mostly in the 21st century, it is the most stated topics and concern has given on environment-friendly production and supply chain. So, there are several innovative studies have been done on this concern in recent times.

Green supply chain (GSC) refers to a strategic approach considering environmental indices to design the supply chain processes. This includes a group of sub-processes like product design, raw material sourcing, manufacturing, transportation, and lifecycle management. According to Wee et al. (2011), GSC is incorporating environmental problems into supply chain management, including packaging design, content procurement, and quality, production operations, distribution of the finished product to customers, and end-of-life management of greening products. GSC thus just not merely a concept of green manufacturing, instead, it is the integration of all operations of the supply chain that needs to be environment friendly. In other words, GSC can be narrowly defined as a significant new model for businesses to achieve income and market share goals by reducing their environmental risks and impacts while increasing their environmental performance. GSC may be viewed as responsible for each segment of the overall supply chain network. Some sectors are as follows:

- *Assembly:* Carbon is emitted while assembling and manufacturing a product.
- *Disassembly:* While performing disassembly after products useful life, the operation requires fossil fuels or electricity.
- *Transportation:* This produces a huge amount of carbon emission. Transportation means like truck, launch, ship, forklift, etc. use a lot of fossil fuels like diesel, petrol, coil, etc. and burning of this produces carbon and other greenhouse gases.
- *Handling:* Handling of raw materials, machinery, equipment, etc. require some electro-mechanical devices and power.
- *Remanufacturing:* In a closed-loop supply chain, after the useful life of a product is over, it needs to be remanufactured thus reducing its impact on the environment.

Che et al. (2017) examined a multi-floor facility layout problem considering bi-objective formulation. To illustrate this problem, they modeled a bi-objective mixed-integer non-linear programming model. The main objective was to minimize the total material handling costs and the total engaged room area. They have used the ε-constraint approach to evaluate the solution and thus to figure out the optimum location. Sinha and Chaturvedi (2018), investigated two important factors in production planning for green development. They studied two objectives of carbon emission and energy consumption and offered manufacturing planning through available and new process routes. They suggested a graphical method according to the perceptions of pinch analysis. They also showed that the proposed model will reduce the extra burden of CO_2 emission and energy consumption while meeting the demand.

Talaei et al. (2016), worked on reducing carbon emission in the overall supply chain network for an environmental issue. They also considered the total cost of the supply chain. They used robust fuzzy programming to investigate this issue. They considered the ε-constraint approach and numerical model to illustrate and solve this optimization problem. Nurjanni et al. (2017), investigated the traditional supply chain management and showed that with increasing levels of industrialization and globalization, the supply chain should be more robust and efficient. They focused on green development and proposed a new supply chain network that considered environmental and financial issues. They were proposing a concept of multi-objective optimization to reduce costs and carbon emissions. Three scalarization methods were used to test the optimization process, namely, the weighted sum method, and weighted Tchebycheff, and increased weighted Tchebycheff to illustrate the trade-offs between costs and environmental concerns. Cheng et al. (2018), worked on the transit system and aimed at public transit systems with optimum designs and operating plans that can mitigate both total costs and greenhouse gas (GHG) emissions. This study investigated the potential emission impact when optimizing the total costs of the transit system. They used a continuum approximation (CA) method to show the relationship of the network. They analyzed the optimal total costs and emissions and the interaction between them. This research showed a significant reduction in GHG emissions that can be achieved when total costs are reduced simultaneously. Additionally another instance of GHG emission control could be find Dehdari Ebrahimi and Momenitabar (2017). Lahnaoui et al. (2018) proposed a method to identify the minimum cost of constructing a hydrogen infrastructure using a single objective linear optimization. They only consider the minimization of cost. This work has focused on reducing both hydrogen transportation capital and operational costs. They consider costs associated with setting up inventory and compression facilities as well as the transport network. Hence, the main aim of the study was to create a cost-effective transport network using compressed gas trucks for mobility. They also showed potential hydrogen demand figures in the country, which will rise according to their model from 2030 to 2050. The research also concentrated on whether and at what flow demand for hydrogen would be delivered from the manufacturing nodes into the global distribution network. Nie et al. (2018) worked on optimizing electric power systems with cost minimization and environmental impact reduction. They used a multistage inexact-factorial fuzzy-probability programming method to illustrate their mathematical form. They also investigated quantitative analysis for individual factors and their interaction. Their policy was to analyze individual, uncertainty recognition, and interaction consideration. They showed that the uncertainties influence Qingdao's EPS long-term planning which can be organized to a cleaner and safer model exhibiting renewable energy. They showed the relation between electricity demand and associated cost (Table 1).

Table 1. Table captions should be placed above the tables.

Author	Network	Model	Objective	Solution	Sectors Under Consideration CO_2 Emission				
					PRO	HL	TRS	DA	RM
Diabat and Simchi-Levi (Diabat 2009)	F	MILP	MO	Weighted Sum	C	–	C	C	C
Franca et al. (2010)	OL	MILP	MO	6 Sigma	C	–	–	–	–
Wang et al. (2011)	OL	MILP	MO	Facility Layout	C	C	C	–	–
Abdallah et al. (2012)	OL	MILP	SO	Net. Design	C	–	C	–	–
Elhedhli and Merrick (2012)	OL	MILP	MO	Lagrange relaxation	C	C	C	–	–
Zhang and Liu (2013)	OL	NLP	SO	Game theory I, II	C	–	–	–	–
Talaei et al. (2016)	CL	MILP	MO	ε-constraint	C	–	–	–	–
Nurjanni et al. (2017)	CL	MILP	MO	Weighted Sum	C	C	C	C	C
Nie et al. (2018)	OL	MILP	MO	Fuzzy-Probability	C	C	–	–	–
Cheng et al. (2018)	CL	MILP	MO	Continuum Approximation	C	C	–	–	–
Billal and Hossain (2020)	F	MILP	MO	Stochastic Optimization	C	C	C	–	–
This Research	CL	MILP	MO	ε-constraint & GA	C	C	C	C	C

NOTE: F = Forward logistics; RL = Reverse Logistics; OL = Open Loop; CL = Closed-Loop; SC = Supply Chain; LP = Linear Programming; PNLP = Progressive Non-Linear Programming; MILP = Mixed Integer Linear Programming; NLP = Non-Linear Programming; SO = Single Objective; MO = Multi Objective; C = Considered, DA = Disassemble, RM = Remanufacturing, TRS = Transport, HL = Handling.

3 Problem Definition

The GSC problem mathematical model consists of four types of echelons, namely plants, warehouses, clients, and disassembly centers which are shown in Fig. 1. It is assumed to be one product solely. There are some substitutes for transport (i.e. roads, railways, water, etc.) within each connection link. The structure of the product is not always identical because a specified quantity of product units is allotted by such connections and different arrangements are implemented at each stage (or echelon). Three types of products form (new product, product for disposal, and product for demolition) run through the mathematical model through the supply chain network, in which the product motion description, based on product structure, is as follows:

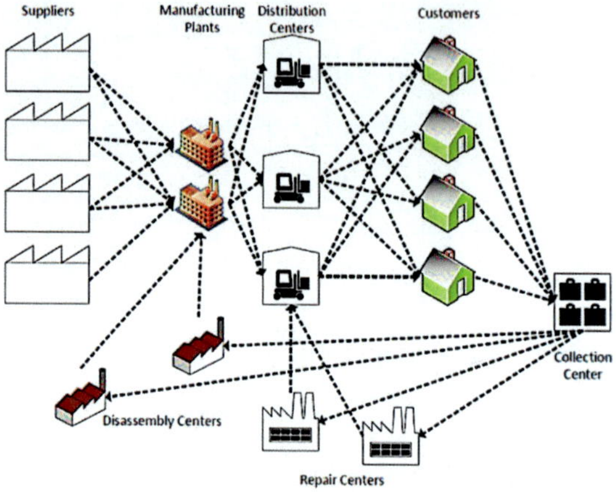

Fig. 1. Configuration of the connection between the stages of the supply chain (Alegoz et al. 2020).

- A new product, which is new or produced, would be forwarded along the network line (factory to warehouse and warehouse to a customer) to meet customers' demand.
- A product to be disposed of is collected from the customer and sent to a recycling center for disassembly.
- A product to be demolished is distributed from the disassembly center to the reproducing plant.

3.1 Assumptions

1. The demand of all customers is uncertain.
2. Each product unit manufactured at a plant is free of defects.
3. Warehouse reliability is considered.
4. In all warehouses, every client demand is always met by any factory.
5. The total product units to be disposed of that enter a disassembly center is effectively disassembled.
6. There is no restriction regarding the capacity of available transportation options.
7. It is not possible to serve customer demand from intermodal transport options (road, rail, sea, and air).

8. The distances between the network junctions used for the assessment of CO2 emissions are determined by the straight line between them. Transportation rates are used to calculate the characteristics of different types of networks (curviness', impossible link, etc.).

3.2 Notations

The following notation is used after observation of the literature and assuming practical situations (Tables 2 and 3).

Table 2. Parameters.

Parameters	Definition	Parameters	Definition
F	Potential Factories	W	Potential Warehouses
C	Customers	I	Potential disassembly Centers
TF	Transportation from factories	TW	Transportation from warehouses
TK	transportation from customers	TI	Transportation of disassembly Centers
Q_c	Customers' demand	Ta_{fw}^{tf}	Unit Cost of transport from plant to warehouse
Tb_{wc}^{tw}	Unit Costs of carriage from a warehouse to a customer	Tc_{ci}^{tk}	Unit Costs of carriage for gathering units of product that needs to be disposed of from customers to disassembly center
Td_{if}^{ti}	Unit Costs of carriage from disassembly center to the factory	L_{fw}^{tf}	Factory-to-warehouse transport rate
L_{wc}^{tw}	Warehouse-to-customer transportation rate	L_{ci}^{tk}	Transportation rate for the collection of product units to be disposed of from the customer to the disassembly center
L_{if}^{ti}	Transportation rate from disassembly center to the factory	Da_{fw}	Distance between factory and warehouse
Db_{wc}	Distance between warehouse and customer	Dc_{ci}	Distance between client and disassembly center
Dd_{if}	Distance from the disassembly center to the factory	Ra_f	Established plant fixed set-up costs
Rb_w	Established warehouse fixed set-up costs	Rd_i	Established disassembly center fixed set-up costs
Ma_f	Variable unit costs to produce a product in the factory	Mb_w	Variable unit costs to handle a product in the warehouse

(*continued*)

Table 2. *(continued)*

Parameters	Definition	Parameters	Definition
Mc_c	Unit variable costs for collecting a disposal product from customers	Md_i	Unit variable cost for disassembling a product to be disposed of in a disassembly center
Mr_f	Unit variable cost for remanufacturing a product to be demolished in the factory	Pa_f	Maximum production capacity of the factory
Pb_w	Maximum processing capacity of the warehouse	Pd_i	Disassembly Center's maximum capacity
Pr_f	Remanufacturing factory's maximum capacity	Hd	Minimum percentage of the units of the commodity to be disposed of to be obtained from consumers
Hr	Minimum percentage of the units of the product to be demolished to be shipped from the disassembly center	Ga_f	Level of CO_2 emitted for the development of one unit in a factory
Gb_w	Level of CO_2 emitted for handling one commodity in a warehouse	Gd_i	Level of CO_2 emitted to disassemble one unit of product to be disposed of in a disassembly center
Gr_f	Level of CO_2 emitted to remanufacture one unit of product to be demolished in a factory	Gta^{tf}	Level of CO_2 emitted by shipping a unit of product from a factory to a warehouse for a unit distance
Gtb^{tw}	Level of CO_2 emitted by shipping a unit of product from a warehouse to a customer for a unit distance	Gtc^{tk}	Level of CO_2 emitted by shipping to collect a unit of disposal from customer to disassembly center for a unit distance
Gtd^{ti}	Level of CO_2 emitted by shipping a unit of product to be demolished from disassembly center to factory for a unit distance		

3.3 Mathematical Formulation

The first function (1) works for minimizing the total cost. The total cost denotes the total fixed cost, the total variable cost, and the total transportation cost.

$$Min f_1 = TC = TFC + TVC + TTC \tag{1}$$

Here,

$$TFC = \sum_{f \in F} Ra_f Xa_f + \sum_{w \in W} Rb_w Xb_w + \sum_{i \in I} Rd_i Xd_i \tag{2}$$

Table 3. Decision Variables.

Variable	Definition	Variable	Definition
Xa_f	1 if the factory is opened, and 0 in otherwise	Xb_w	1 if the warehouse is opened, and 0 in otherwise
Xd_i	1 if the disassembly center is opened, and 0 in otherwise	Ya_{fw}^{tf}	Quantity of product units shipped from factory to warehouse, considering the transport options
Yb_{wc}^{tw}	Quantity of product units shipped from a warehouse to customer, considering the transport options	Yc_{ci}^{tk}	Quantity of product units to be disposed of shipped from a customer to disassembly center, considering the transport options
Yd_{if}^{ti}	Quantity of product units to be demolished shipped from a disassembly center to factory, considering the transport options	$e^{-\lambda t}$	Reliability factor: where t is time, and λ is the failure rate of a warehouse

$$TVC = \sum_{f \in F} Ma_f \sum_{w \in W} \sum_{tf \in TF} Ya_{fw}^{tf} + e^{-\lambda t} \sum_{w \in W} Mb_w \sum_{c \in C} \sum_{tw \in TW} Yb_{wc}^{tw} + \sum_{c \in C} Mc_c \sum_{i \in I} \sum_{tk \in TK} Yc_{ci}^{tk} + \sum_{i \in I} Md_i \sum_{c \in C} \sum_{tk \in TK} Yc_{ci}^{tk} + \sum_{f \in F} Mr_f \sum_{i \in I} \sum_{ti \in TI} Yd_{if}^{ti} \quad (3)$$

$$TTC = \sum_{f \in F} \sum_{w \in W} \sum_{tf \in TF} Ta_{fw}^{tf} Ya_{fw}^{tf} + e^{-\lambda t} \sum_{w \in W} \sum_{c \in C} \sum_{tw \in TW} Tb_{wc}^{tw} Yb_{wc}^{tw} + \sum_{c \in C} \sum_{i \in I} \sum_{tk \in TK} Tc_{ci}^{tk} Yc_{ci}^{tk} + \sum_{i \in I} \sum_{f \in F} \sum_{ti \in TI} Td_{if}^{ti} Yd_{if}^{ti} \quad (4)$$

The second function (5) works for minimizing the total emission of CO_2. The total emission of CO_2 denotes the emission of production, assembly, handling, disassembly, remanufacturing, and transportation.

$$Minf_2 = TE = EP + EH + ED + ER + ET \quad (5)$$

Here,

$$EP = \sum_{f \in F} Ga_f \sum_{w \in W} \sum_{tf \in TF} Ya_{fw}^{tf} \quad (6)$$

$$EA = \sum_{f \in F} Gc_f \sum_{w \in W} \sum_{tf \in TF} Ya_{fw}^{tf} \quad (7)$$

$$EH = e^{-\lambda t} \sum_{w \in W} Gb_w \sum_{c \in C} \sum_{tw \in TW} Yb_{wc}^{tw} \quad (8)$$

$$ED = \sum_{i \in I} Gd_i \sum_{c \in C} \sum_{tk \in TK} Yc_{ci}^{tk} \quad (9)$$

$$ER = \sum_{f \in F} Gr_f \sum_{i \in I} \sum_{ti \in TI} Yd_{if}^{ti} \tag{10}$$

$$ET = \sum_{tf \in TF} Gta^{tf} \sum_{f \in F} \sum_{w \in W} Ya_{fw}^{tf} Da_{fw} L_{fw}^{tf} + e^{-\lambda t} \sum_{tw \in TW} Gtb^{tw} + \sum_{w \in W} \sum_{c \in C} Yb_{wc}^{tw} Db_{wc} L_{wc}^{tw}$$
$$\sum_{tk \in TK} Gtc^{tk} \sum_{c \in C} \sum_{i \in I} Yc_{ci}^{tk} Dc_{ci} L_{ci}^{tk} + \sum_{ti \in TI} Gtd^{ti} \sum_{i \in I} \sum_{f \in F} Yd_{if}^{ti} Dd_{if} L_{if}^{ti} \tag{11}$$

The first set of constraints (12) are on the maximum capacity of the respective factory:

$$\sum_{w \in W} \sum_{tf \in TF} Ya_{fw}^{tf} \leq Pa_f Xa_f, \ \forall f \tag{12}$$

Constraints (13) on the maximum capacity of the respective warehouse:

$$\sum_{f \in F} \sum_{tf \in TF} Ya_{fw}^{tf} \leq Pb_w Xb_w, \ \forall w \tag{13}$$

Constraints (14) on the capacity of factory and warehouse:

$$\sum_{c \in C} \sum_{tw \in TW} Yb_{wc}^{tw} \leq \sum_{f \in F} \sum_{tf \in TF} Ya_{fw}^{tf}, \ \forall w \tag{14}$$

Constraints (15) and (16) on customer demand:

$$\sum_{w \in W} \sum_{tw \in TW} Yb_{wc}^{tw} \geq Q_c, \ \forall c \tag{15}$$

$$\sum_{i \in I} \sum_{tk \in TK} Yc_{ci}^{tk} \leq Q_c, \ \forall c \tag{16}$$

Constraint (17) on the maximum capacity of respective disassembly center:

$$\sum_{c \in C} \sum_{tk \in TK} Yc_{ci}^{tk} \leq Pd_i Xd_i, \ \forall i \tag{17}$$

Constraint (18) on the total number of demand of respective customer:

$$\sum_{i \in I} \sum_{tk \in TK} Yc_{ci}^{tk} \geq HdQ_c, \ \forall c \tag{18}$$

Constraint (19) on the product to be disposed of entered the respective disassembly center:

$$\sum_{f \in F} \sum_{ti \in TI} Yd_{if}^{ti} \geq Hr \sum_{c \in C} Yc_{ci}, \ \forall i \tag{19}$$

Constraint (20) on the maximum remanufacturing capacity of respective capacity:

$$\sum_{i \in I} \sum_{ti \in TI} Yd_{if}^{ti} \leq Pr_f Xa_f, \ \forall f \tag{20}$$

Non-negativity constraints (21) on decision variable:

$$Ya_{fw}^{tf}, Yb_{wc}^{tw}, Yc_{ci}^{tk}, Yd_{if}^{ti} \geq 0 \tag{21}$$

Binary number (22) explains the existence of facilities (factories, warehouses and disassembly center):

$$Xa_f, Xb_w, Xd_i \in \{0, 1\} \tag{22}$$

4 Genetic Algorithm (GA)

The genetic algorithm relies on natural selection principles and evolutionary genetics. They combine the survival of the most effective string systems. The structure allows a randomized exchange of information to create a groundbreaking algorithm for optimality searching. Genetic Algorithms have been developed by Holland (1975), a population-based probabilistic and optimization technique. They are classified as global search heuristics and are also a class of evolutionary algorithms that use techniques triggered by evolutionary biologists such as inheritance, mutation, selection, and crossover (also known as recombination). They are implemented as a computer simulation of candidate solutions to the optimization problem, which communicates a better solution. Conventionally, solutions are represented in the 0s and 1s binary strings, but other encodings are not impossible. Evolution typically starts with a population of randomly generated individuals and happens in generations. In each generation, the fitness of each person in the population is measured, several individuals are selected from the current population (based on fitness) and changed (recombined and probably mutated) to create a new population. The new population is also used for the next iteration of the algorithm. The algorithm is normally completed when either a maximum number of generations has been created or an optimum fitness level has been reached for the population. If the algorithm has been terminated due to a maximum number of generations, an optimal solution may or may not have been found. For further information, please see a closed comprehensive example by Afshar-Nadjafi and Arani (2014).

5 Epsilon Constraint Method

The ε-constraint method was suggested by Haimes Y. V. et al. (1971). This method lessens the number of objectives of multi-objective problems. Only one of the objectives is considered as main objective functions and other objective functions are turned into constraints. So, the multi-objective problems are turned into single-objective problems, then it is called ε-constraint problem. By changing ε values with a proper interval, several Pareto optimal solutions can be found. The ideal points and nadir points can be determined and eliminated then by acquiring the full Pareto front. The set of ideal points and nadir points specifies the lower and upper bounds of Pareto optimal points.

6 Case Description

The planning horizon is 1 week long. The model includes two factories (plants), two warehouses, two customers (wholesalers), and two disassembly centers. The model is a closed-loop network. Fixed costs data, variable costs data, factory capacity data, transportation costs data, and demand of customers data have been collected from the Tsiakis and Shah (2016). Distance between stages data has been used from the Google maps (2020) and the rate of CO_2 released data has also been used from the Song et al. (2014).

7 Result Analysis

In Table 2, the decision variables of the mathematical model of the GSC have been obtained with the assistance of the Genetic Algorithm (GA) optimization technique in MATLAB R2017a software. The objective functions values are: $f_1 = £\ 1476934.685$ and $f_2 = 1885905.948(\text{kg})$ (Fig. 2 and Table 4).

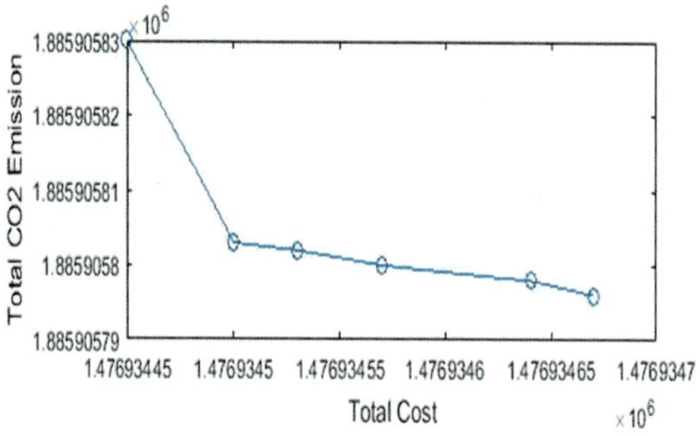

Fig. 2. Pareto Front of the Genetic Algorithm.

In this Pareto line, as the overall cost increases, the total CO2 emission decreases. But when the total cost decreases, the total emission of CO2 increases. There is a controversial relationship between them. There are some optimal points found in the Pareto front. Consequently, any company can use these optimal points for their purposes.

In Table 3, the decision variables of the mathematical model of the GSC have been obtained with the assistance of the ε-constraint technique in MATLAB R2017a software. The objective functions values are: $f_1 = £\ 1476877.1948$ and $f_2 = 1885570.948(\text{kg})$ (Fig. 3 and Table 5).

In this Pareto front, only the dominating points are used. When the total cost increases, the total CO_2 emission decreases. Otherwise, when the total cost decreases, therefore, the total CO_2 emission increases. There are some optimal points found in this Pareto front by solving this method.

Table 4. Genetic Algorithm's Output.

Variables	Values	Variables	Values	Variables	Values
Xa_1	0.719	Xa_2	0.673	Xb_1	0.805
Xb_2	1.304	Xd_1	0.724	Xd_2	0.758
Ya_{11}^1	1629.999(units)	Ya_{12}^1	1629.999(units)	Ya_{21}^2	949.999(units)
Ya_{22}^2	949.999(units)	Yb_{11}^1	1629.999(units)	Yb_{12}^1	949.999(units)
Yb_{21}^2	1629.999(units)	Yb_{22}^2	949.999(units)	Yc_{11}^1	325.999(units)
Yc_{12}^1	325.999(units)	Yc_{21}^2	142.499(units)	Yc_{22}^2	142.499(units)
Yd_{11}^1	32.599(units)	Yd_{12}^1	14.249(units)	Yd_{21}^2	16.299(units)
		Yd_{22}^2	7.125(units)		

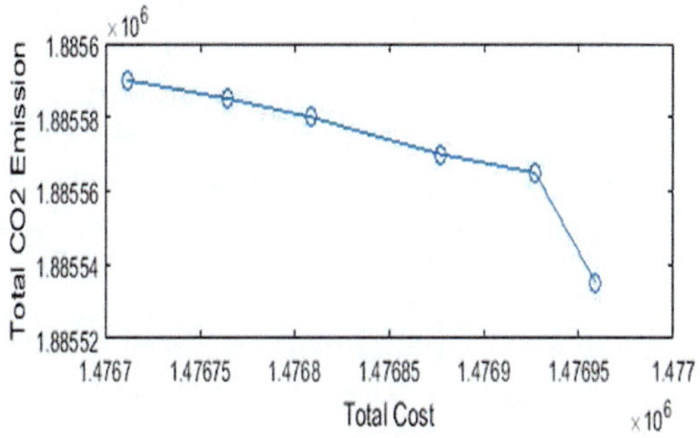

Fig. 3. Pareto Front of the ε-Constraint Method.

Table 5. ε-constraint Method Output.

Variables	Values	Variables	Values	Variables	Values
Xa_1	0.718	Xa_2	0.673	Xb_1	0.805
Xb_2	1.303	Xd_1	0.725	Xd_2	0.758
Ya_{11}^1	1629.146(units)	Ya_{12}^1	1629.142(units)	Ya_{21}^2	949.655(units)
Ya_{22}^2	949.887(units)	Yb_{11}^1	1629.119(units)	Yb_{12}^1	950.718(units)
Yb_{21}^2	1629.111(units)	Yb_{22}^2	950(units)	Yc_{11}^1	326.105(units)
Yc_{12}^1	326.115(units)	Yc_{21}^2	142.732(units)	Yc_{22}^2	142.663(units)
Yd_{11}^1	32.665(units)	Yd_{12}^1	14.333(units)	Yd_{21}^2	16.377(units)
		Yd_{22}^2	7.174(units)		

8 Conclusion and Future Research

This study is attached to the problem context. There is a controversial relationship between the two objectives of total cost and total CO_2 emission. For examining the optimal decision of the objectives, a mathematical model is formulated. The genetic algorithm has become an effective technique in solving many critical problems and has applied to practical problems. Epsilon Constraint Method is also a popular method for solving many practical problems. The mathematical model is solved by the Genetic algorithm and Epsilon Constraint Method, individually. The comparison of these two methods is one of the contributions of this study. Secondary data is used to examine to verify the mathematical formulation. For transport choices, road transport is selected as a single mode of transport for all stages of the supply chain. The cost of transport depends on the variance in the unit cost of transport and on the total amount of product units flowing. The cost of transport, therefore, has a major impact on the overall cost. The overall emission of CO_2 also depends heavily on the total distance. The distance (km) $\times$ kg is proportional to the overall CO_2 emission. Therefore, the overall emission varies with distance variability. Other factors have not had a significant impact on total CO_2 emissions. Thus, the total cost and total CO_2 emission decrease when unit transport costs and distances are reduced. The genetic algorithm and ε-constraint method have been compared. But closely the same results have been obtained by these two methods.

Currently, the GSC area increases scholars' and governments' awareness because of the recent environmental issue. A company thinks mainly the financial fact and concerns about environmental issues. So, this paper suggests a mathematical model of the GSC, and its methodology solve with a multi-objective approach. Total cost and total CO_2 emission are two main criteria to evaluate which are selected to construct a mixed-integer linear programming model. A Pareto front represents the two objectives of total cost and total emission of CO_2. One closed-loop network design is accompanied by the problem statement. The assumptions lead to a more realistic model. The proposed genetic algorithm and the ε-constraint method are intended to minimize total costs along with total CO_2 emissions. Over the last few years, researchers have proposed several models and studies for the design of the green supply chain with varying degrees of complexity. This research is an attempt to bridge between the total cost of supply and total CO_2 emissions and to find the optimum point for these two objectives.

The mathematical model of the GSC design is confined to its services. So, the model will be extended for increasing its services. Then the model is more applicable in complex situations. Further analysis can then be tackled by changing the model scenario. Other optimization techniques such as fuzzy multi-objective linear programming (FMLOP) (Momenitabar et al. 2014), Discrete Event Simulation (DES) (Arani et al. 2020) can be used to solve the formulated model. Then compare the optimization technique which leads us to draw a comprehensive comparison. For the delivery of products, different modes of transport can be used. Additionally, compare the different modes of transport and find the right one for this model. These recommendations could shed light on future research.

References

Abdallah, T., Farhat, A., Diabat, A., Kennedy, S.: Green supply chains with carbon trading and environmental sourcing: formulation and life cycle assessment. Appl. Math. Model. **36**, 4271–4285 (2012). https://doi.org/10.1016/j.apm.2011.11.056

Afshar-Nadjafi, B., Arani, M.: Multimode preemptive resource investment problem subject to due dates for activities: formulation and solution procedure. Adv. Oper. Res. **2014**, 1–10 (2014). https://doi.org/10.1155/2014/740670

Aldemir, G., Beldek, T., Celebi, D.: A Closed-Loop Sustainable Supply Chain Network Design with System Dynamics for Waste Electrical and Electronic Equipment (2018)

Alegoz, M., Kaya, O., Bayindir, Z.P.: Closing the loop in supply chains: economic and environmental effects. Comput. Ind. Eng. **142**, 106366 (2020). https://doi.org/10.1016/j.cie.2020.106366

Andiç, E., Yurt, Ö., Baltacioğlu, T.: Green supply chains: efforts and potential applications for the Turkish market. Resour. Conserv. Recycl. **58**, 50–68 (2012). https://doi.org/10.1016/j.resconrec.2011.10.008

Arani, M., Abdolmaleki, S., Liu, X.: Scenario-based simulation approach for an integrated inventory blood supply chain system. In: Winter Simulation Conference, IEEE (2020)

Billal, M.M., Hossain, M.M.: Multi-objective optimization for multi-product multi-period four echelon supply chain problems under uncertainty. J. Optim. Ind. Eng. **13**, 1–17 (2020). https://doi.org/10.22094/joie.2018.555578.1529

Che, A., Zhang, Y., Feng, J.: Bi-objective optimization for multi-floor facility layout problem with fixed inner configuration and room adjacency constraints. Comput. Ind. Eng. **105**, 265–276 (2017). https://doi.org/10.1016/j.cie.2016.12.018

Cheng, H., Mao, C., Madanat, S., Horvath, A.: Minimizing the total costs of urban transit systems can reduce greenhouse gas emissions: the case of San Francisco. Transp. Policy **66**, 40–48 (2018). https://doi.org/10.1016/j.tranpol.2018.02.009

Dehdari Ebrahimi, Z., Momenitabar, M.: Design of mathematical modeling in a green supply chain network by collection centers in the environment. Environ. Energy Econ. Res. **1**, 153–162 (2017). https://doi.org/10.22097/eeer.2017.47242

Diabat, A., Simchi-Levi, D.: A carbon-capped supply chain network problem. In: 2009 IEEE International Conference on Industrial Engineering and Engineering Management, pp. 523–527. IEEE (2009)

Elhedhli, S., Merrick, R.: Green supply chain network design to reduce carbon emissions. Transp. Res. Part D Transp. Environ. **17**, 370–379 (2012). https://doi.org/10.1016/j.trd.2012.02.002

Franca, R.B., Jones, E.C., Richards, C.N., Carlson, J.P.: Multi-objective stochastic supply chain modeling to evaluate tradeoffs between profit and quality. Int. J. Prod. Econ. **127**, 292–299 (2010). https://doi.org/10.1016/j.ijpe.2009.09.005

Haimes, Y.V., Lasdon, L.S., Wismer, D.A.: On a Bicriterion formulation of the problems of integrated system identification and system optimization. IEEE Trans. Syst. Man. Cybern. SMC-1, 296–297 (1971). https://doi.org/10.1109/TSMC.1971.4308298

Kim, J.H., Youn, S., Roh, J.J.: Green supply chain management orientation and firm performance: evidence from South Korea. Int. J. Serv. Oper. Manag. **8**, 283–304 (2011). https://doi.org/10.1504/IJSOM.2011.038973

Konyalıoğlu, A.K., Zafeirakopoulos, İ.B.: A Literature Review on Closed Loop Supply Chains (2020)

Lahnaoui, A., Wulf, C., Heinrichs, H., Dalmazzone, D.: Optimizing hydrogen transportation system for mobility by minimizing the cost of transportation via compressed gas truck in North Rhine-Westphalia. Appl. Energy **223**, 317–328 (2018). https://doi.org/10.1016/j.apenergy.2018.03.099

Momenitabar, M., Akar, N., Zaghi, D., Feili, H.R., Ghaderi, M.: Fuzzy Mathematical modeling of distribution network through location allocation model in a three-level supply chain design. J. Math. Comput. Sci. **09**, 165–174 (2014). https://doi.org/10.22436/jmcs.09.03.02

Nie, S., Huang, Z.C.C., Huang, G.H.H., Yu, L., Liu, J.: Optimization of electric power systems with cost minimization and environmental-impact mitigation under multiple uncertainties. Appl. Energy **221**, 249–267 (2018). https://doi.org/10.1016/j.apenergy.2018.03.194

Niranjan, T., Parthiban, P., Sundaram, K., Jeyaganesan, P.N.: Designing a omnichannel closed loop green supply chain network adapting preferences of rational customers. Sādhanā **44**(3), 1–10 (2019). https://doi.org/10.1007/s12046-018-1038-0

Nurjanni, K.P., Carvalho, M.S., Costa, L.: Green supply chain design: a mathematical modeling approach based on a multi-objective optimization model. Int. J. Prod. Econ. **183**, 421–432 (2017). https://doi.org/10.1016/j.ijpe.2016.08.028

Sinha, R.K., Chaturvedi, N.D.: A graphical dual objective approach for minimizing energy consumption and carbon emission in production planning. J. Clean. Prod. **171**, 312–321 (2018). https://doi.org/10.1016/j.jclepro.2017.09.272

Song, X., Fu, Y., Chen, Z., Liu, H.: Environmental impact evaluation for high-speed railway. J. Central South Univ. **21**(6), 2366–2371 (2014). https://doi.org/10.1007/s11771-014-2189-8

Talaei, M., Farhang Moghaddam, B., Pishvaee, M.S., Bozorgi-Amiri, A., Gholamnejad, S.: A robust fuzzy optimization model for carbon-efficient closed-loop supply chain network design problem: a numerical illustration in electronics industry. J. Clean. Prod. **113**, 662–673 (2016). https://doi.org/10.1016/j.jclepro.2015.10.074

Tsiakis, P., Shah, N.: Design of Multi-echelon supply chain networks under demand uncertainty design of multi-echelon supply chain networks under demand. Ind. Eng. Chem. Res. 3585–3604 (2016). https://doi.org/10.1021/ie0100030

Wang, F., Lai, X., Shi, N.: A multi-objective optimization for green supply chain network design. Decis. Support Syst. **51**, 262–269 (2011). https://doi.org/10.1016/j.dss.2010.11.020

Wee, H.M., Lee, M.C., Yu, J.C.P., Edward, W.C.: Optimal replenishment policy for a deteriorating green product: Life cycle costing analysis. Int. J. Prod. Econ. **133**, 603–611 (2011). https://doi.org/10.1016/j.ijpe.2011.05.001

Yang, L., Wang, G., Chai, Y.: Manufacturer's channel selection considering carbon emission reduction and remanufacturing. J. Syst. Sci. Syst. Eng. **27**(4), 497–518 (2018). https://doi.org/10.1007/s11518-018-5378-4

Zhang, C.T., Liu, L.P.: Research on coordination mechanism in three-level green supply chain under non-cooperative game. Appl. Math. Model. **37**, 3369–3379 (2013). https://doi.org/10.1016/j.apm.2012.08.006

A Type-2 Fuzzy TOPSIS-Based Methodology for the Analysis of Investment Decisions

Hale Gonce Kocken(✉) and Beyza Ahlatcioglu Ozkok

Department of Mathematical Engineering, Yildiz Technical University, Istanbul, Turkey
{hgonce,bahlat}@yildiz.edu.tr

Abstract. From the beginning of human history to today, all of the decisions that human beings have made in order to make their assets more efficient and valuable can be considered as investment decisions. The aim is always to maximize the return. For this purpose, researchers have developed a wide range of investment theories and decision methodologies. Multi-Attribute Decision Making techniques have found effective use for mathematical optimization to obtain both theoretical and practical results for researchers. In recent years, many extensions have been proposed that use Type-2 fuzzy sets for classical mathematical optimization techniques to address the uncertainty in nature. In this study, we used the Type-2 fuzzy TOPSIS method for the evaluation of alternative stocks in an investment decision. And a case study for the stocks selected from Borsa Istanbul (BIST) is performed to show the applicability of the proposed methodology. The results of the case study are examined, and suggestions for future studies are provided.

Keywords: Portfolio selection problem · Investment decision · Fuzzy TOPSIS · Interval Type-2 fuzzy numbers

1 Introduction

In recent decades, portfolio management and selection have taken remarkable attention in the academic and business environment since the evaluation of investment alternatives, making effective and efficient investment decisions are becoming one of the most important decision-making problems in human history. More specifically, the decision-making problem is to decide which stocks are to be chosen for investment and in what proportions they will be bought.

In the literature, there are several approaches to constructing a portfolio. Historically, the mean-variance model is the first example of a portfolio optimization problem, and it is credited to Markowitz, who presented his ideas in [1]. This model is important because mean-variance analysis provides a basis for the derivation of the equilibrium model known variously as the Capital Asset Pricing Model (CAPM), Sharpe–Lintner model, black model, and two-factor model [2, 3]. Here, we summarized some of the important mathematical models which are proposed for the portfolio selection problem. In [4], portfolio selection models are formulated by quadratic programming, based on two kinds of possibility distributions. A model using genetic algorithms for portfolio

T. Allahviranloo et al. (Eds.): IDS 2020, AISC 1301, pp. 359–367, 2021.
https://doi.org/10.1007/978-3-030-66501-2_29

selection is proposed in [5]. Two portfolio selection models based on fuzzy probabilities and possibility distributions are proposed in [6]. [7] proposed a new possibilistic programming approach based on the worst regret to the portfolio selection, considering how a model yields a distributive investment solution. [8] formulated a fuzzy goal programming with fuzzy goals and fuzzy constraints, taking into account three criteria: return, risk, and liquidity. [9] proposed a method that incorporates the grey and possibilistic regression models in formulating a novel portfolio selection model. In their paper, [10] proposed a new ranking and weighting method and applied it to the selection of stocks on the ISE. In [11], a multistage stochastic fuzzy program with soft constraints is developed in order to capture both uncertainty and imprecision, and this program is used to solve a portfolio management problem. In [12], the conventional mean-variance method is revised to determine the optimal portfolio selection under an uncertain situation. Using Sharpe's single index model in a soft framework, [13] formulated a fuzzy compromise programming problem in order to solve portfolio selection problems. [14] considered a portfolio selection problem in which the prices of the securities are treated as interval variables. In [15], two kinds of portfolio selection models based on lower and upper possibilistic means and possibilistic variance and presented an algorithm that can derive the possibilistic efficient frontier of the problem are proposed. In [3], the fuzzy AHP is combined with the portfolio selection problem. In order to measure the asymmetry of fuzzy portfolio return, a concept of skewness is defined as the third central moment in [16], and its mathematical properties are studied. In [17], they utilized the concept of interval numbers in fuzzy set theory to extend the classical mean-variance portfolio selection model into a mean-variance–skewness model with consideration of transaction cost. Based on the membership function, [18] redefined the concepts of mean and variance for fuzzy numbers, also proposed the concept of distortion, and formulated a fuzzy mean-variance-skewness portfolio selection model. The stock portfolio selection problem based on varying conservative-neutral-aggressive attitudes is studied in [19]. In their paper, the return rates of stocks are characterized by fuzzy variables. The Pareto-optimal solutions are obtained by maximizing the return and minimizing the risk subject to constraints of transaction cost and value at risk.

As can be seen in the development of literature for portfolio selection problem, as in many decision problems, fuzzy decision making is a very useful tool. After type-1 fuzzy sets were developed in 1965 by Zadeh to model the uncertainties and imprecise data, type-1 methods have been applied in many different areas, such as economics, engineering, business, etc. In portfolio selection in the fuzzy environment, a clearer evaluation of the future is aimed by taking the opinions of experts who have mastered the subject instead of using only historical data. Some of the studies that include type1 methods for solving the portfolio selection problem are [20, 21]. In 1975, Zadeh extended type 1 fuzzy sets to type 2 fuzzy sets to handle uncertainty in a better way. As a continuation of this expansion, the data of many multi-criteria decision-making problems were taken as type-2 fuzzy numbers, and new solution methods for them were discussed. The main studies include type-2 methods on multi-criteria decision-making problems are [22–26]. With this motivation, different from the literature, expert evaluations in the portfolio selection problem are taken as interval type-2 fuzzy sets, which provides more detailed information than type-1 fuzzy sets. In the light of expert evaluations, a ranking was obtained with the help of the Fuzzy TOPSIS algorithm. And, we provide a case study on BIST.

This paper is organized as follows: After the introduction section, Sect. 2 briefly reviews the interval type-2 fuzzy sets and the operations between them. In Sect. 3, the main steps of the type-2 fuzzy TOPSIS algorithm are presented. Section 4 presents the financial ratios for the evaluation of alternative stocks and a case study on BIST. Finally, conclusions are presented and point out future research in Sect. 5.

2 Interval Type-2 Fuzzy Sets

In Type 1 fuzzy sets, elements have memberships within the interval [0,1] assigned by membership functions. Type-2 fuzzy sets are introduced as an extension of Type-1 fuzzy sets ([27]). The most striking difference of this proposed expansion is that the membership functions of type-1 fuzzy sets are two-dimensional, whereas the membership functions of type-2 fuzzy sets are three-dimensional. The third dimension introduced here is the additional degrees of freedom, which allows for direct handling of uncertainty. ([23]).

In this section, brief information about interval type-2 fuzzy sets is presented.

Definition 1. A type-2 fuzzy set $\tilde{\tilde{A}}$ in the universe of discourse X can be represented by a type 2 membership function $\mu_{\tilde{\tilde{A}}}(x)$, shown as follows:

$$\tilde{\tilde{A}} = \left\{\left((x, u), \mu_{\tilde{\tilde{A}}}(x, u)\right) | \forall x \in X, | \forall u \in J_x \subseteq [0, 1], 0 \leq \mu_{\tilde{\tilde{A}}}(x, u) \leq 1\right\}$$

where $J_x \in [0, 1]$ denotes an interval [0, 1].

Definition 2. A trapezoidal interval type-2 fuzzy set is illustrated as $\tilde{\tilde{A}}_i = \left(\tilde{A}_i^U; \tilde{A}_i^L\right) = \left(\left(a_{i1}^U, a_{i2}^U, a_{i3}^U, a_{i4}^U; H_1\left(\tilde{A}_i^U\right), H_2\left(\tilde{A}_i^U\right)\right), \left(a_{i1}^L, a_{i2}^L, a_{i3}^L, a_{i4}^L; H_1\left(\tilde{A}_i^L\right), H_2\left(\tilde{A}_i^L\right)\right)\right)$ where $\tilde{A}_i^U$ and $\tilde{A}_i^L$ are type-1 fuzzy sets, $a_{i1}^U, a_{i2}^U, a_{i3}^U, a_{i4}^U, a_{i1}^L, a_{i2}^L, a_{i3}^L$ and a_{i4}^L are the reference points of the interval type-2 fuzzy set $\tilde{A}_i$, $H_j\left(\tilde{A}_i^U\right)$; denotes the membership value of the element a_{ij}^U in the upper trapezoidal membership function $\tilde{A}_i^U$; $H_j\left(\tilde{A}_i^L\right)$ denotes the membership value of the element $a_{i(j+1)}^L$ in the lower trapezoidal membership function $\tilde{A}_i^L$ where $1 \leq j \leq 2$, $H_1\left(\tilde{A}_i^U\right) \in [0, 1]$, $H_2\left(\tilde{A}_i^U\right) \in [0, 1]$,$H_1\left(\tilde{A}_i^L\right) \in [0, 1]$,$H_2\left(\tilde{A}_i^L\right) \in [0, 1]$ and $1 \leq i \leq n$ (see Fig. 1)

Let $\tilde{\tilde{A}}_1 = \left(\tilde{A}_1^U; \tilde{A}_1^L\right)$ and $\tilde{\tilde{A}}_2 = \left(\tilde{A}_2^U; \tilde{A}_2^L\right)$ be trapezoidal interval type-2 fuzzy sets.

Definition 3. The addition operation between $\tilde{\tilde{A}}_1$ and $\tilde{\tilde{A}}_2$ is defined as follows:

$$\tilde{\tilde{A}}_1 \oplus \tilde{\tilde{A}}_2 = \left(a_{11}^U + a_{21}^U, a_{12}^U + a_{22}^U, a_{13}^U + a_{23}^U, a_{14}^U + a_{24}^U; \min\left(H_1\left(\tilde{A}_1^U\right), H_1\left(\tilde{A}_2^U\right)\right), \min\left(H_2\left(\tilde{A}_1^U\right), H_2\left(\tilde{A}_2^U\right)\right)\right),$$
$$\left(a_{11}^L + a_{21}^L, a_{12}^L + a_{22}^L, a_{13}^L + a_{23}^L, a_{14}^L + a_{24}^L; \min\left(H_1\left(\tilde{A}_1^L\right), H_1\left(\tilde{A}_2^L\right)\right), \min\left(H_2\left(\tilde{A}_1^L\right), H_2\left(\tilde{A}_2^L\right)\right)\right).$$

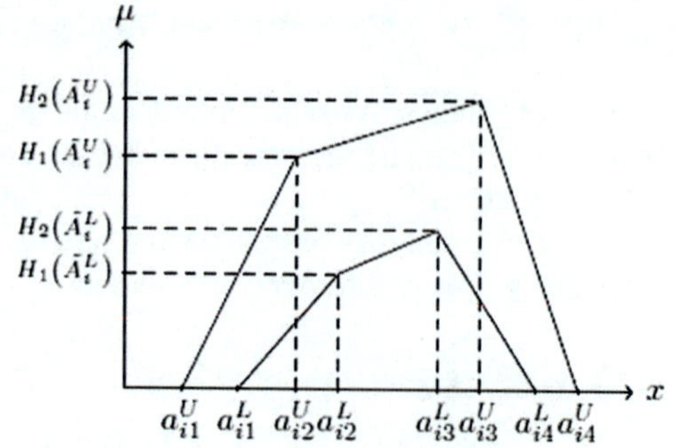

Fig. 1. The membership function of a trapezoidal interval type-2 fuzzy set.

Definition 4. The subtraction operation between $\tilde{\tilde{A}}_1$ and $\tilde{\tilde{A}}_2$ is defined as follows:

$$\tilde{\tilde{A}}_1 \Theta \tilde{\tilde{A}}_2 = \left(a_{11}^U - a_{24}^U, a_{12}^U - a_{23}^U, a_{13}^U - a_{22}^U, a_{14}^U - a_{21}^U; \min\left(H_1\left(\tilde{A}_1^U\right), H_1\left(\tilde{A}_2^U\right)\right), \min\left(H_2\left(\tilde{A}_1^U\right), H_2\left(\tilde{A}_2^U\right)\right)\right),$$
$$\left(a_{11}^L - a_{24}^L, a_{12}^L - a_{23}^L, a_{13}^L - a_{22}^L, a_{14}^L - a_{21}^L; \min\left(H_1\left(\tilde{A}_1^L\right), H_1\left(\tilde{A}_2^L\right)\right), \min\left(H_2\left(\tilde{A}_1^L\right), H_2\left(\tilde{A}_2^L\right)\right)\right).$$

Definition 5. The multiplication operation between $\tilde{\tilde{A}}_1$ and $\tilde{\tilde{A}}_2$ is defined as follows:

$$\tilde{\tilde{A}}_1 \otimes \tilde{\tilde{A}}_2 \cong \left(a_{11}^U \otimes a_{21}^U, a_{12}^U \otimes a_{22}^U, a_{13}^U \otimes a_{23}^U, a_{14}^U \otimes a_{24}^U; \min\left(H_1\left(\tilde{A}_1^U\right), H_1\left(\tilde{A}_2^U\right)\right), \min\left(H_2\left(\tilde{A}_1^U\right), H_2\left(\tilde{A}_2^U\right)\right)\right)$$
$$\left(a_{11}^L \otimes a_{21}^L, a_{12}^L \otimes a_{22}^L, a_{13}^L \otimes a_{23}^L, a_{14}^L \otimes a_{24}^L; \min\left(H_1\left(\tilde{A}_1^L\right), H_1\left(\tilde{A}_2^L\right)\right), \min\left(H_2\left(\tilde{A}_1^L\right), H_2\left(\tilde{A}_2^L\right)\right)\right).$$

Definition 6. The arithmetic operations between the trapezoidal interval type-2 fuzzy set $\tilde{\tilde{A}}_1$ and a scalar $k > 0$ can be given as:

$$k\tilde{\tilde{A}}_1 = \left(\left(ka_{11}^U, ka_{12}^U, ka_{13}^U, ka_{14}^U; H_1\left(\tilde{A}_1^U\right), H_2\left(\tilde{A}_1^U\right)\right), \left(ka_{11}^L, ka_{12}^L, ka_{13}^L, ka_{14}^L; H_1\left(\tilde{A}_1^L\right), H_2\left(\tilde{A}_1^L\right)\right)\right)$$
$$\frac{\tilde{\tilde{A}}_1}{k} = \left(\left(\frac{a_{11}^U}{k}, \frac{a_{12}^U}{k}, \frac{a_{13}^U}{k}, \frac{a_{14}^U}{k}; H_1\left(\tilde{A}_1^U\right), H_2\left(\tilde{A}_1^U\right)\right), \left(\frac{a_{11}^L}{k}, \frac{a_{12}^L}{k}, \frac{a_{13}^L}{k}, \frac{a_{14}^L}{k}; H_1\left(\tilde{A}_1^L\right), H_2\left(\tilde{A}_1^L\right)\right)\right)$$

Definition 7. The reciprocal of the trapezoidal interval type-2 fuzzy set $\tilde{\tilde{A}}_1$ is defined as follows:

$$\frac{1}{\tilde{\tilde{A}}_1} = \left(\tilde{\tilde{A}}_1\right)^{-1} = \left(\left(\frac{1}{a_{14}^U}, \frac{1}{a_{13}^U}, \frac{1}{a_{12}^U}, \frac{1}{a_{11}^U}; H_1\left(\tilde{A}_1^U\right), H_2\left(\tilde{A}_1^U\right)\right), \left(\frac{1}{a_{14}^L}, \frac{1}{a_{13}^L}, \frac{1}{a_{12}^L}, \frac{1}{a_{11}^L}; H_1\left(\tilde{A}_1^L\right), H_2\left(\tilde{A}_1^L\right)\right)\right).$$

Definition 8 [23]. The defuzzified value $DTraT\left(\tilde{\tilde{A}}_1\right)$ of a trapezoidal type-2 fuzzy set $\tilde{\tilde{A}}_1$ is given as follows:

$$DTraT\left(\tilde{A}_1\right) = \frac{\frac{\alpha}{4} + a_{11}^U + \frac{\beta}{4} + a_{11}^L}{2} \tag{1}$$

where $\alpha = \left(a_{14}^U - a_{11}^U\right) + \left(H_1\left(\tilde{A}_1^U\right)a_{12}^U - a_{11}^U\right) + \left(H_2\left(\tilde{A}_1^U\right)a_{13}^U - a_{11}^U\right)$ and $\beta = \left(a_{14}^L - a_{11}^L\right) + \left(H_1\left(\tilde{A}_1^L\right)a_{12}^L - a_{11}^L\right) + \left(H_2\left(\tilde{A}_1^L\right)a_{13}^L - a_{11}^L\right)$. After obtaining the defuzzied values, trapezoidal interval type-2 fuzzy sets can be ranked with respect to these values.

3 Type-2 Fuzzy TOPSIS

It is assumed that there are A alternatives, where $A = \{A_1, A_2, \ldots, A_n\}$ and C criteria, where $C = \{C_1, C_2, \ldots, C_m\}$. There are k DMs D_1, D_2 and D_k. The set C of criteria can be divided into two sets C^+ and C^-, where they denote set of benefit, and cost attributes, respectively, $C^+ \cap C^- = \emptyset$ and $C^+ \cup C^- = C$.

By using the scale in Table 1, the main steps of the type-2 fuzzy TOPSIS method can be given as follows [22]:

Table 1. Linguistic terms and their corresponding interval type-2 fuzzy sets [22].

Linguistic terms	Interval type-2 fuzzy set
Very low (VL)	((0, 0, 0, 0.1; 1, 1), (0, 0, 0, 0.05; 0.9, 0.9))
Low (L)	((0, 0.1, 0.1, 0.3; 1, 1), (0.05, 0.1, 0.1, 0.2; 0.9, 0.9))
Medium Low (ML)	((0.1, 0.3, 0.3, 0.5; 1, 1), (0.2, 0.3, 0.3, 0.4; 0.9, 0.9))
Medium (M)	((0.3, 0.5, 0.5, 0.7; 1, 1), (0.4, 0.5, 0.5, 0.6; 0.9, 0.9))
Medium High (MH)	((0.5, 0.7, 0.7, 0.9; 1, 1), (0.6, 0.7, 0.7, 0.8; 0.9, 0.9))
High (H)	((0.7, 0.9, 0.9, 1; 1, 1), (0.8, 0.9, 0.9, 0.95; 0.9, 0.9))
Very High (VH)	((0.9, 1, 1, 1; 1, 1), (0.95, 1, 1, 1; 0.9, 0.9))

Step 1. Considering multiple DMs, k fuzzy decision matrices $\tilde{\tilde{X}}_k$ between the criteria and the alternatives should be aggregate into the average fuzzy decision matrix $\tilde{\tilde{X}} = \left[\tilde{\tilde{x}}_{ij}\right]$ by any group decision method where $1 \leq i \leq m$ and $1 \leq j \leq n$.

Step 2. Obtain the weighting vector for each DM and aggregate into the average weighting fuzzy vector $\tilde{\tilde{W}} = \left[\tilde{\tilde{w}}_1\ \tilde{\tilde{w}}_2 \cdots \tilde{\tilde{w}}_m\right]$.

Step 3. Compute the weighted fuzzy decision matrix $\tilde{\tilde{X}}_w = \left[\tilde{\tilde{v}}_{ij}\right]$ where $\tilde{\tilde{v}}_{ij} = \tilde{\tilde{x}}_{ij} \otimes \tilde{\tilde{w}}_i$ for all $1 \leq i \leq m$ and $1 \leq j \leq n$.

Step 4. Using the ranking function *DTraT* which is given by (1), replace each entry of $\tilde{\tilde{X}}_w$ with its ranking value and construct the ranking weighted decision matrix $\tilde{\tilde{X}}_W^R = \left[DTraT\left(\tilde{\tilde{v}}_{ij}\right)\right]_{m\times n}$ where $1 \le i \le m$ and $1 \le j \le n$.

Step 5. Determine the positive ideal solution $x^+ = \left(v_1^+, v_2^+, \ldots, v_m^+\right)$ and the negative-ideal solution $x^- = \left(v_1^-, v_2^-, \ldots, v_m^-\right)$, where

$$v_i^+ = \begin{cases} \max\limits_{1\le j\le n}\left\{DTraT\left(\tilde{\tilde{v}}_{ij}\right)\right\}, & \textit{if } C_i \in C^+ \\ \min\limits_{1\le j\le n}\left\{DTraT\left(\tilde{\tilde{v}}_{ij}\right)\right\}, & \textit{if } C_i \in C^- \end{cases} \tag{2}$$

and

$$v_i^- = \begin{cases} \min\limits_{1\le j\le n}\left\{DTraT\left(\tilde{\tilde{v}}_{ij}\right)\right\}, & \textit{if } C_i \in C^+ \\ \max\limits_{1\le j\le n}\left\{DTraT\left(\tilde{\tilde{v}}_{ij}\right)\right\}, & \textit{if } C_i \in C^-. \end{cases} \tag{3}$$

Step 6. Calculate the distance $d^+\left(A_j\right)$ and $d^-\left(A_j\right)$ between each alternative A_j and the positive and negative ideal solutions x^+ and x^-, respectively:

$$d^+\left(A_j\right) = \sqrt{\sum_{i=1}^{m}\left(DTraT\left(\tilde{\tilde{v}}_{ij}\right) - v_i^+\right)^2} \tag{4}$$

$$d^-\left(A_j\right) = \sqrt{\sum_{i=1}^{m}\left(DTraT\left(\tilde{\tilde{v}}_{ij}\right) - v_i^-\right)^2} \tag{5}$$

Step 7. Calculate the relative degree of closeness $Closeness\left(A_j\right)$ of A_j with respect to the positive ideal solution x^+, shown as:

$$Closeness\left(A_j\right) = \frac{d^-\left(A_j\right)}{d^+\left(A_j\right) + d^-\left(A_j\right)}. \tag{6}$$

Step 8. Sort the values $Closeness\left(A_j\right)$ in a descending sequence where $1 \le j \le n$. The larger the value of $Closeness\left(A_j\right)$, the higher the preference of the alternative A_j, where $1 \le j \le n$.

4 A Case Study on the Portfolio Selection Problem

In this section, we will first present the financial ratios for the evaluation of alternative stocks. In order to demonstrate the applicability of our methodology, ten stocks were selected from BIST by using the past price movements. In order to evaluate our alternative stocks, we selected the following five important financial ratios:

C_1*: Market to book value ratio.*

C_2*: Net profit margin.*

C_3: *Return on capital employed (ROCE).*

C_4: *Return on assets (ROA).*

C_5: *Price-to-Earnings Ratio (P/E ratio).*

We obtained these financial ratio values from the website of finnet.com.tr on the date of 01.02.2019 for selected alternative stocks. Based on these values, a financial expert whose knowledge and experience is sufficient to evaluate the stock performance of each company for this example filled the decision matrix given in Table 2 (Step1). After this stage, we apply the steps of the interval type-2 fuzzy TOPSIS algorithm presented in Sect. 3.

Table 2. The decision matrix of the case study.

	A_1	A_2	A_3	A_4	A_5	A_6	A_7	A_8	A_9	A_{10}
C_1	VH	VH	M	L	L	M	VL	ML	ML	M
C_2	VH	MH	L	L	L	M	ML	ML	M	ML
C_3	MH	MH	VL	M	MH	ML	M	ML	VL	VH
C_4	VH	MH	VL	VL	VL	M	ML	VL	M	VL
C_5	M	ML	ML	VH	VL	L	ML	VL	M	VL

In our case study, it is assumed that all financial ratios are of equal weight that is $\tilde{\tilde{X}}_w = \tilde{\tilde{X}}$ (Step 2 and Step 3).

Using the ranking function *DTraT* given by Eq. (1), the ranking weighted decision matrix $\tilde{\tilde{X}}_W^R$ is obtained as (Step 4):

$$\begin{bmatrix} 0.9563 & 0.9563 & 0.4875 & 0.1163 & 0.1163 & 0.4875 & 0.0188 & 0.2925 & 0.2925 & 0.4875 \\ 0.9563 & 0.6825 & 0.1163 & 0.1163 & 0.1163 & 0.4875 & 0.2925 & 0.2925 & 0.4875 & 0.2925 \\ 0.6825 & 0.6825 & 0.0188 & 0.4875 & 0.6825 & 0.2925 & 0.4875 & 0.2925 & 0.0188 & 0.9563 \\ 0.9563 & 0.6825 & 0.0188 & 0.0188 & 0.0188 & 0.4875 & 0.2925 & 0.0188 & 0.4875 & 0.0188 \\ 0.4875 & 0.2925 & 0.2925 & 0.9563 & 0.0188 & 0.1163 & 0.2925 & 0.0188 & 0.4875 & 0.0188 \end{bmatrix}.$$

Considering $C_1, C_2, C_3, C_4 \in C^+$ and $C_5 \in C^-$, the positive (x^+) and negative x^- ideal solutions are determined by using Eq. (2) and Eq. (3) as $(0.9563, 0.9563, 0.9563, 0.9563, 0.0188)$ and $(0.0188, 0.1163, 0.0188, 0.0188, 0.9563)$, respectively (Step 5)

After calculating the distances between each alternative and the ideal solutions by using Eq. (4) and Eq. (5) (Step 6), the relative degree of closeness of each alternative is calculated by using Eq. (6) (Step 7), and given in Table 3.

Table 3. The closeness values.

	A_1	A_2	A_3	A_4	A_5
Closeness	0.7650	0.7436	0.3285	0.2064	0.4284
	A_6	A_7	A_8	A_9	A_{10}
Closeness	0.5252	0.3791	0.4097	0.3647	0.5332

Considering the descending sequence of the closeness values, the preferred order of the alternatives is $A_1, A_2, A_{10}, A_6, A_5, A_8, A_7, A_9, A_3, A_4$. Thus, implementing the fuzzy TOPSIS algorithm, we created a preference ranking among the candidate stocks traded on BIST.

5 Conclusion

It is clear that in the globalizing world, all decision-making problems related to finance will increase their attractiveness day by day and maintain their importance. From this point of view, the analysis of stocks traded on the stock exchange and the studies conducted to guide investors are extremely important. In this paper, we discussed the problem of portfolio selection by using the fuzzy expansion of the TOPSIS method, which is one of the multi-criteria decision-making techniques that are frequently used. We would like to expand our work at a later stage and combine it with other multi-attribute decision-making techniques to provide weight information in addition to the order of preference we have presented to the investor for stocks.

Acknowledgements. This research has been supported by Yildiz Technical University Scientific Research Projects Coordination Department. Project Number: FBA-2017-3073.

References

1. Markowitz, H.M.: Portfolio Selection/Harry Markowitz. J. Finan. **7**(1), 77–91 (1952)
2. Alexander, G.J., Sharpe, W.F., Bailey, J.V.: Fundamentals of Investments/Fundamentos de inversiones. Pearson Educación, Prentice Hall (2003)
3. Tiryaki, F., Ahlatcioglu, B.: Fuzzy portfolio selection using fuzzy analytic hierarchy process. Inf. Sci. **179**(1–2), 53–69 (2009)
4. Tanaka, H., Guo, P.: Portfolio selection based on upper and lower exponential possibility distributions. Eur. J. Oper. Res. **114**(1), 115–126 (1999)
5. Xia, Y., Liu, B., Wang, S., Lai, K.K.: A model for portfolio selection with order of expected returns. Comput. Oper. Res. **27**(5), 409–422 (2000)

6. Tanaka, H., Guo, P., Türksen, I.B.: Portfolio selection based on fuzzy probabilities and possibility distributions. Fuzzy Sets Syst. **111**(3), 387–397 (2000)
7. Inuiguchi, M., Tanino, T.: Portfolio selection under independent possibilistic information. Fuzzy Sets Syst. **115**(1), 83–92 (2000)
8. Parra, M.A., Terol, A.B., Urıa, M.R.: A fuzzy goal programming approach to portfolio selection. Eur. J. Oper. Res. **133**(2), 287–297 (2001)
9. Ong, C.S., Huang, J.J., Tzeng, G.H.: A novel hybrid model for portfolio selection. Appl. Math. Comput. **169**(2), 1195–1210 (2005)
10. Tiryaki, F., Ahlatcioglu, M.: Fuzzy stock selection using a new fuzzy ranking and weighting algorithm. Appl. Math. Comput. **170**(1), 144–157 (2005)
11. Lacagnina, V., Pecorella, A.: A stochastic soft constraints fuzzy model for a portfolio selection problem. Fuzzy Sets Syst. **157**(10), 1317–1327 (2006)
12. Huang, J.J., Tzeng, G.H., Ong, C.S.: A novel algorithm for uncertain portfolio selection. Appl. Math. Comput. **173**(1), 350–359 (2006)
13. Bilbao-Terol, A., Pérez-Gladish, B., Arenas-Parra, M., Rodríguez-Uría, M.V.: Fuzzy compromise programming for portfolio selection. Appl. Math. Comput. **173**(1), 251–264 (2006)
14. Giove, S., Funari, S., Nardelli, C.: An interval portfolio selection problem based on regret function. Eur. J. Oper. Res. **170**(1), 253–264 (2006)
15. Zhang, W.G., Wang, Y.L., Chen, Z.P., Nie, Z.K.: Possibilistic mean–variance models and efficient frontiers for portfolio selection problem. Inf. Sci. **177**(13), 2787–2801 (2007)
16. Li, X., Qin, Z., Kar, S.: Mean-variance-skewness model for portfolio selection with fuzzy returns. Eur. J. Oper. Res. **202**(1), 239–247 (2010)
17. Bhattacharyya, R., Kar, S., Majumder, D.D.: Fuzzy mean–variance–skewness portfolio selection models by interval analysis. Comput. Math Appl. **61**(1), 126–137 (2011)
18. Li, X., Guo, S., Yu, L.: Skewness of fuzzy numbers and its applications in portfolio selection. IEEE Trans. Fuzzy Syst. **23**(6), 2135–2143 (2015)
19. Zhou, X., Wang, J., Yang, X., Lev, B., Tu, Y., Wang, S.: Portfolio selection under different attitudes in fuzzy environment. Inf. Sci. **462**, 278–289 (2018)
20. Chen, C.T., Hung, W.Z., Yu, L.: A new decision-making method for stock portfolio selection based on computing with linguistic assessment. Journal of Applied Mathematics and Decision Sciences **2009**, 32 (2009)
21. Ece, O., Uludag, A.S.: Applicability of fuzzy TOPSIS method in optimal portfolio selection and an application in BIST. Int. J. Econ. Finan. **9**(10), 107–127 (2017)
22. Chen, S.M., Lee, L.W.: Fuzzy multiple attributes group decision-making based on the interval type-2 TOPSIS method. Expert Syst. Appl. **37**(4), 2790–2798 (2010)
23. Kahraman, C., Öztayşi, B., Sarı, İ.U., Turanoğlu, E.: Fuzzy analytic hierarchy process with interval type-2 fuzzy sets. Knowl. -Based Syst. **59**, 48–57 (2014)
24. Abdullah, L., Zulkifli, N.: Integration of fuzzy AHP and interval type-2 fuzzy DEMATEL: An application to human resource management. Expert Syst. Appl. **42**(9), 4397–4409 (2015)
25. Sang, X., Liu, X.: An analytical solution to the TOPSIS model with interval type-2 fuzzy sets. Soft. Comput. **20**(3), 1213–1230 (2016)
26. Celik, E., Akyuz, E.: An interval type-2 fuzzy AHP and TOPSIS methods for decision-making problems in maritime transportation engineering: the case of ship loader. Ocean Eng. **155**, 371–381 (2018)
27. Zadeh, L.A.: The concept of a linguistic variable and its application to approximate reasoning. In: Learning systems and intelligent robots, pp. 1–10. Springer, Boston, MA (1974)

An Extension of DEMATEL Under Pythagorean Fuzzy Environment

Sait Gül(✉), Çağlar Sivri, and Ozan Rıdvan Aksu

Faculty of Engineering and Natural Sciences, Management Engineering Department, Bahçeşehir University, 34353 İstanbul, Beşiktaş, Turkey
sait.gul@eng.bau.edu.tr

Abstract. The complexity around the decision maker (DM) and intrinsic vagueness in human conscious generate difficulties in expressing judgments when they should use crisp measures. Several fuzzy set extensions have been introduced by different researchers in order to ease this decision-making process. DEMATEL is a Multi-Attribute Decision Making (MADM) method aiming to reveal potential casual dependencies among the problem's attribute set by paying attention to the evaluation of decision-makers' judgments. It offers a practical way to determine the cause and effect relationship between attributes. In this study, a novel Pythagorean fuzzy set (PFS) extension of DEMATEL has been developed to unveil the interrelations among decision attributes that should be thought of in a decision problem. PFS extension of DEMATEL was carried out as multi-step approach that entire extension has been completed in 9 proposed steps. The vagueness and hesitancy of the decision makers are tried to be modelled via this newly developed fuzzy extension of DEMATEL. It has enabled more flexibility to decision-makers, especially in expressing their opinions. It was also seen that defuzzification takes place at the very end of the method. The present outputs offer application potential for real life case studies such as exposing the relationship of antivirus mask selection attributes or polar clothing selection attributes.

Keywords: Pythagorean fuzzy sets · DEMATEL · Linguistic terms · Multi-attribute decision-making

1 Introduction

Inherent dubiousness in human consciousness and complications surrounding decision-makers (DMs) give rise to obstacles in reasoning when they need to take decisions. A comprehensive study area has been created by characteristic ambiguity. Ordinary fuzzy sets were introduced by Zadeh [26] who also constructed a new formal system of mathematical logic. He proposed fuzzy sets (FS) to illustrate questionable data by means of a membership function enabling allocation of a membership degree ranging between 0 and 1 to each element in the set.

Various types of FS extensions can be found in past studies such as type-2 FS [27] as the fuzzified version of the classical FS, intuitionistic FS [3] comprising membership

T. Allahviranloo et al. (Eds.): IDS 2020, AISC 1301, pp. 368–378, 2021.
https://doi.org/10.1007/978-3-030-66501-2_30

and non-membership degrees, hesitant FS [21] for enabling a few membership values being taken by a single element in the set, Pythagorean FS (PFS) [23] as an extension of intuitionistic FS, neutrosophic set [20] containing membership, non-membership, and hesitancy degrees as freestanding constituents, spherical FS [14] as an aggregation of PFS and neutrosophic set, etc.

DEMATEL, which is used to solve complex and interrelated problems, was developed by the Battelle Memorial Institute of the Geneva Research Centre [6]. As a Multi-Attribute Decision Making (MADM) method, it purposes to unfold potential casual dependencies among the problem's attribute set by considering the assessments of DMs' judgments. It is known to be widespread as a functional means used to identify the cause and effect relationship between attributes [10].

To cope with the issue of revealing the interrelations among decision attributes that ought to be thought of in a decision problem, a novel PFS extension of DEMATEL approach has been developed within this work. In the present work, we have opted utilizing PFS in the methodology for enabling more flexibility to DMs, especially in expressing their opinions because the DMs do not have to assign membership and non-membership degrees whose sum is at most 1 [7]. Pythagorean FS is an extension of intuitionistic FS forcing DMs to assign membership and non-membership degrees by modifying the rule saying that their sum should be equal to or less than 1. In PFS, this sum can exceed 1 but their squared sum should be within the range of [0,1] [23].

The rest of the study is organized as follows: PFS concept is introduced in Sect. 2 and examples of the extensions of some MADM approaches are given in this section. Section 3 explains the details of the proposed PFS extension of DEMATEL step-by-step and an illustrative application is given in Sect. 4. The conclusions and future research agenda are summarized in Sect. 5.

2 Preliminaries of PFS and a Brief Literature Review

PFS is an extension of intuitionistic FS forcing DMs to assign membership and non-membership degrees. This sum may surpass 1 in PFS, but their squared sum should be between 0 and 1 [24].

Let a set X be a universe of discourse. A PFS P is an object having the form [24]:

$$P = \{< x, \mu_P(x), \vartheta_P(x) > |x \in X\} \tag{1}$$

where the membership and non-membership degrees are defined by $\mu_P(x) : X \to [0, 1]$ and $\vartheta_P(x) : X \to [0, 1]$, respectively. For every $x \in X$, the following rule applies:

$$0 \leq \mu_P^2 + \vartheta_P^2 \leq 1 \tag{2}$$

For any PFS, the hesitancy degree of x to P is $\pi_A(x) = \sqrt{(1 - \mu_A^2(x) - v_A^2(x)}$.

Assume that $P_1 = P(\mu_1, \vartheta_1)$ and $P_2 = P(\mu_2, \vartheta_2)$ are two Pythagorean fuzzy numbers (PFNs) and $\lambda > 0$. It is possible to present the operations determined on PFNs as follows [29].

$$P_1 \oplus P_2 = P\left(\sqrt{\mu_1^2 + \mu_2^2 - \mu_1^2\mu_2^2}, \vartheta_1\vartheta_2\right) \tag{3}$$

$$P_1 \otimes P_2 = P\left(\mu_1\mu_2, \sqrt{\vartheta_1^2 + \vartheta_2^2 - \vartheta_1^2\vartheta_2^2}\right) \tag{4}$$

$$\lambda P_1 = P\left(\sqrt{1 - (1 - \mu_1^2)^\lambda}, \vartheta_1^\lambda\right) \tag{5}$$

$$P_1^\lambda = P\left(\mu_1^\lambda, \sqrt{1 - (1 - \vartheta_1^2)^\lambda}\right) \tag{6}$$

We ought to utilize a Pythagorean fuzzy aggregated weighted averaging (PFAWA) operator for aggregating DMs' assessments in a decision model [28]. Let $P_i = P_i(\mu_i, \vartheta_i)$, $i = 1,\ldots,n$ be a compilation of PFNs and $w = (w_1, \ldots, w_n)$ be the weight vector of P_i.

$$PFAWA(P_1, \ldots, P_n) = \left[\sqrt{1 - \prod_{i=1}^{n} (1 - \mu_i^2)^{w_i}}, \prod_{i=1}^{n} \vartheta_i^{w_i}\right] \tag{7}$$

In literature there are many PFS extensions of MADM approaches. Gul et al. [8] used a PFS extension of the Analytic Hierarchy Process (AHP) with the aim of risk assessment of routes in oil transportation while Ozdemir et al. [16] used a similar methodology of AHP in an application of measuring sustainable campus services. In terms of outranking field of MADM, Chen [4] proposed the extension of ELECTRE and applied it in selecting the best financial policy while Chen [5] developed a PFS extension of PROMETHEE and present its applicability in obtaining the best bridge construction method. One of the most applied PFS integrations is made with TOPSIS. For example, Akram et al. [2] used a PFS-extended TOPSIS in location selection problem; Rani et al. [19] developed a novel type of PFS-based TOPSIS and applied it in an example of sustainable recycling partner selection problem; Ho et al. [12] analyzed stroke rehabilitation treatment alternatives with a PFS-TOPSIS method. Similarly, PFS extensions of VIKOR [11, 18], WASPAS [13], MABAC [1], MOORA [17] etc. can be found in the extensive literature on this topic.

Numerous types of fuzzy sets utilized by DEMATEL extensions are available in a great number in past studies. For example, while an intuitionistic fuzzy DEMATEL for risk analysis of gob coal spontaneous combustion in coal seam was proposed by Zhou et al. [30], the assessment of e-commerce websites was carried out via neutrosophic version of DEMATEL which was introduced by Liang et al. [15]. Spherical fuzzy version of DEMATEL is newly proposed by Gül [9] and the applicability is shown on an illustrative example. Also, the robustness of the proposition is presented by comparing the results of different types of DEMATEL extensions. As far as we could possibly know, there is just one study proposing a PFS-based DEMATEL [25], however an early defuzzification operation were executed through their work. This obstacle is attempted to be settled by keeping the process Pythagorean fuzzy until the end of the process and delaying the defuzzification operation as far as possible.

3 Proposed PFS Extension of DEMATEL

Considering MADM perspective, the decision problems are dealt with by thinking about the related attribute to acquire the most fitting option among a set of alternatives.

Revealing the interdependencies among attributes were not drawn interest by numerous approaches. DEMATEL and cognitive maps are the most cited methods that can deal with identifying these potential influences [22]. This work offers development of the PFS extension of DEMATEL as a contribution to the literature. In the proposed extension, there are 9 steps:

Step 1. m decision makers (DMs) who are known as the specialists for the decision problem and n attributes having a critical effect on the decision are available.
Step 2. The DMs are asked with respect to their inclinations or feelings on the likely impacts among attributes. After tailoring the scale introduced by Yazdi et al. [25], a PFS linguistic term set (Table 1) was obtained enabling DMs to remark their judgments. $d_{ij}^{e} = \left(\mu_{ij}^{e}, v_{ij}^{e}\right)$ where $i,j = 1,\ldots,n$ and $e = 1,\ldots,m$ exemplifies e^{th} DM's assessment about the influence of attribute i on attribute j. Pythagorean fuzzy matrix $D^{e} = \left\langle d_{ij}^{e}\right\rangle_{nxn} = \left\langle \mu_{ij}^{e}, v_{ij}^{e}\right\rangle_{nxn}$ demonstrates the direct influence assessment matrix of the eth DM.

Table 1. DEMATEL linguistic terms and the related PFNs.

Definition	*Abb.*	μ	v
Strong	S	0.90	0.10
Moderate	M	0.70	0.30
Weak	W	0.30	0.70
No influence	NI	0.10	0.90

Step 3. Proficiency level of DMs are symbolized by their weights (ω_e, $e = 1,\ldots,m$). Various approaches are available to this aim of revealing these weights.
Step 4. By means of PFAWA operator (Eq. 8), the DMs' direct influence evaluation matrices (D^{e}) are aggregated. Aggregated Direct Influence Matrix, D^{agg} in short, obtained as a result of this process (Eq. 9).

$$D^{agg} = PFAWA_{\omega}\left(D^{1}, D^{2}, \ldots, D^{m}\right) = \omega_1 D^{1} + \omega_2 D^{2} + \cdots + \omega_m D^{m}$$
$$= \left\langle \mu_{ij}^{agg}, v_{ij}^{agg}\right\rangle = \left\{\sqrt{1 - \Pi_{e=1}^{m}\left(1 - \left(\mu_{ij}^{e}\right)^{2}\right)^{\omega_{\theta}}}, \Pi_{e=1}^{m}\left(v_{ij}^{e}\right)^{\omega_{\theta}}\right\} \quad (8)$$

$$D^{agg} = \begin{bmatrix} 0 & \left\langle \mu_{12}^{agg}, U_{12}^{agg}\right\rangle & \cdots & \left\langle \mu_{1n}^{agg}, v_{1n}^{agg}\right\rangle \\ \left\langle \mu_{21}^{agg}, U_{21}^{agg}\right\rangle & 0 & \cdots & \left\langle \mu_{2n}^{agg}, v_{2n}^{agg}\right\rangle \\ \vdots & \vdots & \ddots & \vdots \\ \left\langle \mu_{n1}^{agg}, U_{n1}^{agg}\right\rangle & \left\langle \mu_{n2}^{agg}, U_{n2}^{agg}\right\rangle & \cdots & 0 \end{bmatrix} \quad (9)$$

Step 5. Existence of two independent elements in each comparison creates an opportunity of dividing the Aggregated Direct Influence Matrix (D^{agg}) into two sub-matrices. It is possible to execute matrix operations on these matrices corresponding to the membership and non-membership degrees. The initial direct influence matrix might be created via normalization of each one and associating them again in the same matrix if required. The normalization process is performed as follows.

$$X^{\mu} = sD^{\mu} \quad \text{where} \quad s = \min\left[\frac{1}{\max\limits_{i} \sum_{j=1}^{n} |\mu_{ij}|}, \frac{1}{\max\limits_{i} \sum_{i=1}^{n} |\mu_{ij}|}\right] \tag{10}$$

$$X^{v} = sD^{v} \quad \text{where} \quad s = \min\left[\frac{1}{\max\limits_{i} \sum_{j=1}^{n} |v_{ij}|}, \frac{1}{\max\limits_{i} \sum_{i=1}^{n} |v_{ij}|}\right] \tag{11}$$

The output of the aggregation step can be seen below in matrix form.

$$X^{\mu} = \begin{bmatrix} 0 & \mu_{12} \cdots & \mu_{1n} \\ \mu_{21} & 0 \cdots & \mu_{21} \\ \vdots & \vdots \ddots & \vdots \\ \mu_{n1} & \mu_{n2} \cdots & 0 \end{bmatrix}; \; X^{\upsilon} = \begin{bmatrix} 0 & \upsilon_{12} \cdots & \upsilon_{1n} \\ \upsilon_{21} & 0 \cdots & \upsilon_{21} \\ \vdots & \vdots \ddots & \vdots \\ \upsilon_{n1} & \upsilon_{n2} \cdots & 0 \end{bmatrix} \tag{12}$$

Step 6. The total influence matrix of T is obtained by the initial direct influence matrix of X. The sub-matrices of T are aggregated as shown in Eq. (15), after the conversion of sub-matrices of X.

$$T^{\mu} = X^{\mu} + (T^{\mu})' = X^{\mu}(I - X^{\mu})^{-1} = \begin{bmatrix} \mu_{11}^{T} & \cdots & \mu_{1n}^{T} \\ \vdots & \ddots & \vdots \\ \mu_{n1}^{T} & \cdots & \mu_{nn}^{T} \end{bmatrix} \tag{13}$$

$$T^{v} = X^{v} + (T^{v})' = X^{v}(I - X^{v})^{-1} = \begin{bmatrix} v_{11}^{T} & \cdots & v_{1n}^{T} \\ \vdots & \ddots & \vdots \\ v_{n1}^{T} & \cdots & v_{nn}^{T} \end{bmatrix} \tag{14}$$

$$T = \begin{bmatrix} \langle\mu_{11}^{T}, v_{11}^{T}\rangle & \langle\mu_{12}^{T}, v_{12}^{T}\rangle & \cdots & \langle\mu_{1n}^{T}, v_{1n}^{T}\rangle \\ \langle\mu_{21}^{T}, v_{21}^{T}\rangle & \langle\mu_{22}^{T}, v_{22}^{T}\rangle & \cdots & \langle\mu_{2n}^{T}, v_{2n}^{T}\rangle \\ \vdots & \vdots & \ddots & \vdots \\ \langle\mu_{n1}^{T}, v_{n1}^{T}\rangle & \langle\mu_{n2}^{T}, v_{n2}^{T}\rangle & \cdots & \mu_{nn}^{T}, v_{nn}^{T}\rangle \end{bmatrix} \tag{15}$$

Step 7. Row (r_i) and column (c_j) sums can be calculated by Eq. (16).

$$r_i = \sum_{i=1}^{n} \langle\mu_{ij}^{T}, v_{ij}^{T}\rangle; \quad c_j = \sum_{j=1}^{n} \langle\mu_{ij}^{T}, v_{ij}^{T}\rangle / \tag{16}$$

e.g. $\langle\mu_{11}^{T}, v_{11}^{T}\rangle \oplus \langle\mu_{12}^{T}, v_{12}^{T}\rangle = \left\{\sqrt{(\mu_{11}^{T})^{2} + (\mu_{12}^{T})^{2} - (\mu_{11}^{T})^{2}(\mu_{12}^{T})^{2}}, (v_{11}^{T})(v_{12}^{T})\right\}$

The values of general strength (r_i) and general weakness (c_j) of each attribute can be calculated at the end of this step and be specified as PFNs. The PFNs ought to be defuzzified in order to determine Prominence and Relation values for each column and row sum.

$$Score = \frac{1 + \mu - v + \pi(\varphi - 0.5)}{2} \tag{17}$$

where $\varphi \in \{0, 1\}$ (risk-preference coefficient) is the importance of hesitancy degree [25].
Step 8. "$r_i + c_i$" and "$r_i - c_i$" are called Prominence of attribute i and Relation of attribute i, respectively. A positive sign of the Relation value (when $r_i - c_i > 0$) indicates that attribute i nfluences other attributes, and it should be in the "cause" group; a negative sign of the Relation value (when $r_i - c_i < 0$) indicates that attribute i is influenced by others, and it should be in the "effect" group.
Step 9. Casual dependencies and influences among attributes are expressed in the casual diagram in which Prominence and Relation values are represented in horizontal and vertical axis, respectively. PFNs are defuzzied via Eq. (17) and defuzzied total influence matrix *(T)* are filtered by using a threshold (α) which is appointed by DMs. After these operations, significant influences among attributes can be remarked and expressed on casual diagram. The name of this new figure in which the significant influences are drawn one by one is Network Relation Map (NRM).

4 An Illustrative Example

An example of building contractor selection problem is solved via PFS-based DEMATEL methodology in order to present the potential influences among the attributes. A real-life case was used for this scenario and hypothetical data were generated as DMs' evaluations.

The main attributes of the problem and DMs are stated in Step 1. Three DMs out of experts of the field evaluated the influences among attributes which are Cost (C_1), Duration (C_2), Flexibility (C_3), Quality (C_4), and Uncertainty (C_5). Linguistic evaluations are presented in Step 2. Linguistic terms given in Table 2 are converted to PFNs $d_{ij}^e = \left(\mu_{ij}^e, v_{ij}^e\right)$ for $i, j = 1, \ldots 5$ and $e = 1, 2, 3$.

Table 2. DMs' evaluations on influences between attribute pairs

	C_1	C_2	C_3	C_4	C_5
C_1	0	S; S; M	M; S; S	S; W; M	M; M; W
C_2	M; M; M	0	M; M; M	M; NI; M	S; S; NI
C_3	M; M; M	M; S; M	0	M; M; NI	S; M; NI
C_4	S; S; NI	M; NI; S	W; M; S	0	M; M; M
C_5	M; M; S	S; NI; NI	M; M; NI	S; M; M	0

The weights of DMs are discussed and determined as $\omega_1 = 0.3784$, $\omega_2 = 0.3514$, $\omega_3 = 0.2703$. Aggregated direct influence matrix, D^{agg}, which is shown in Table 3 is obtained by aggregating DM's individualistic direct influence evaluation matrices (D^e) via Eq. (8). For illustration purposes, the aggregated linguistic evaluation for the pair of C_3 and C_5 is given below. The individualistic evaluations are S (0.9, 0.1), M (0.7, 0.3), NI (0.1,0.9).

Table 3. Aggregated direct influence matrix

	C_1		C_2		C_3		C_4		C_5	
C_1	0.0000	0.0000	0.8671	0.1346	0.8508	0.1515	0.7549	0.2666	0.6353	0.3772
C_2	0.7000	0.3000	0.0000	0.0000	0.7000	0.3000	0.5968	0.4413	0.8386	0.1811
C_3	0.7000	0.3000	0.7997	0.2039	0.0000	0.0000	0.6244	0.4037	0.7616	0.2664
C_4	0.8386	0.1811	0.7120	0.3279	0.7168	0.3072	0.0000	0.0000	0.7000	0.3000
C_5	0.7807	0.2229	0.6855	0.3919	0.6244	0.4037	0.8056	0.1980	0.0000	0.0000

$$\mu_{35}^{agg} = \sqrt{1 - \prod_{e=1}^{3} \left(1 - \left(\mu_{35}^{e}\right)^2\right)^{\omega_e}}$$

$$= \sqrt{1 - \left(1 - (0.9)^2\right)^{0.3784} * \left(1 - (0.7)^2\right)^{0.3514} * \left(1 - (0.1)^2\right)^{0.2703}} = 0.7616$$

$$\upsilon_{35}^{agg} = \prod_{e=1}^{3} \left(\upsilon_{35}^{e}\right)^{\omega_e} = (0.1)^{0.3784} * (0.3)^{0.3514} * (0.9)^{0.2703} = 0.2664$$

After obtaining two sub-matrices from D^{agg}, DEMATEL calculations which are given in Eqs. (10–15) are performed and in Step 7, the total influence matrix of T, which is shown in Table 4 is obtained.

Table 4. Total influence matrix

	C_1		C_2		C_3		C_4		C_5	
C_1	0.4468	0.3334	0.4739	0.3743	0.4748	0.3737	0.4728	0.3840	0.4688	0.4178
C_2	0.4381	0.4878	0.4145	0.4183	0.4385	0.4796	0.4367	0.4985	0.4423	0.4646
C_3	0.4439	0.4738	0.4459	0.4548	0.4203	0.4048	0.4431	0.4771	0.4460	0.4670
C_4	0.4584	0.4422	0.4545	0.4695	0.4558	0.4605	0.4313	0.3970	0.4549	0.4604
C_5	0.4486	0.4805	0.4453	0.5076	0.4448	0.5043	0.4509	0.4677	0.4228	0.4237

Step 8 calculates Prominence and Relation values via Eq. (16). After that, for any further evaluations that is necessary in methodology, PFNs are converted into crisp numbers via Eq. (17). The Prominence and Relation values for varying φ coefficients are given in Table 5. Considering different coefficients would not change the Cause and Effect groups, as it can be seen from the table. Cause group involves Cost (C_1) and

Quality (C_4) and Effect group involves Duration (C_2), Flexibility (C_3), and Uncertainty (C_5). It means that Duration, Flexibility and Uncertainty will be affected indirectly by any upgrade on Cost and Quality. Cost and Quality should be focused on in terms of investments and improvement plans. The presentation of NRM (Fig. 1) is included by Step 9.

Table 5. Prominence and Relation values for varying φ parameters

φ	0.5		0.6		0.7	
	$r-c$	$r+c$	$r-c$	$r+c$	$r-c$	$r+c$
C_1	0.0154	1.8190	0.0138	1.8746	0.0121	1.9302
C_2	−0.0104	1.7928	−0.0094	1.8510	−0.0083	1.9093
C_3	−0.0055	1.7978	−0.0049	1.8556	−0.0043	1.9134
C_4	0.0019	1.8050	0.0016	1.8619	0.0012	1.9188
C_5	−0.0060	1.7964	−0.0056	1.8540	−0.0053	1.9115
φ	0.8		0.9		1	
	$r-c$	$r+c$	$r-c$	$r+c$	$r-c$	$r+c$
C_1	0.0105	1.9858	0.0089	2.0414	0.0073	2.0969
C_2	−0.0073	1.9675	−0.0062	2.0257	−0.0052	2.0840
C_3	−0.0037	1.9712	−0.0031	2.0289	−0.0025	2.0867
C_4	0.0009	1.9757	0.0006	2.0326	0.0003	2.0894
C_5	−0.0049	1.9691	−0.0045	2.0267	−0.0042	2.0842

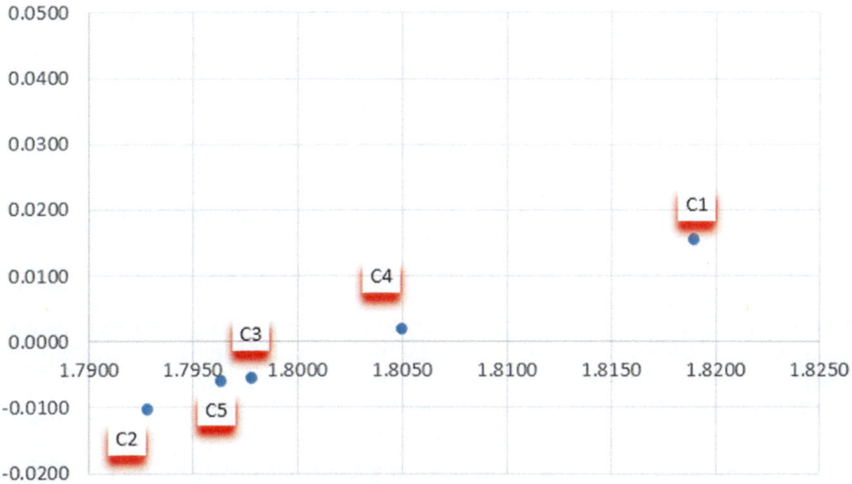

Fig. 1. Casual diagram of illustrative example.

5 Conclusions and Future Research Agenda

The PFS version of DEMATEL, which is among top cited MADM methods is the contribution of this study. The study includes introduction of PFS-based mathematical formulations. DMs became able to indicate their judgements by stating the membership and non-membership degrees as independent parameters under Pythagorean fuzzy uncertainty environment with the help of defuzzification of DEMATEL. The study also considered the hesitancy of DMs indirectly. An illustrative example of building contractor evaluation problem was given in order to show the applicability of the proposed method. In order to select the best alternative among all, relations between the attributes are appointed and final output of NRM is generated. Applying other versions of DEMATEL and comparing the results is a potential study for a further research in order to check the robustness of the proposed method which can also be applied in real-life MADM problems. Examples of such problems would be exposing the relationship of antivirus mask selection attributes or polar clothing selection attributes.

References

1. Aghamohagheghi, M., Hashemi, S.M., Tavakkoli-Moghaddam, R.: Soft computing-based new interval-valued pythagorean triangular fuzzy multi-criteria group assessment method without aggregation: application to a transport projects appraisal. Int. J. Eng. **32**(5), 737–746 (2019)
2. Akram, M., Dudek, W.A., Ilyas, F.: Group decision-making based on pythagorean fuzzy TOPSIS method. Int. J. Intell. Syst. **34**(7), 1455–1475 (2019). https://doi.org/10.1002/int.22103
3. Atanassov, K.T.: Intuitionistic fuzzy sets. Fuzzy Sets Syst. **20**, 87–96 (1986). https://doi.org/10.1016/S0165-0114(86)80034-3
4. Chen, T.Y.: An outranking approach using a risk attitudinal assignment model involving Pythagorean fuzzy information and its application to financial decision making. Appl. Soft Comput. **71**, 460–487 (2018). https://doi.org/10.1016/j.asoc.2018.06.036
5. Chen, T.Y.: A novel promethee-based outranking approach for multiple criteria decision analysis with pythagorean fuzzy information. IEEE Access **6**, 54495–54506 (2018). https://doi.org/10.1109/ACCESS.2018.2869137
6. Gabus, A., Fontela, E.: World problems, an invitation to further thought within the framework of DEMATEL. Battelle Geneva Research Centre, Geneva, Switzerland (1972)
7. Gul, M.: Application of Pythagorean fuzzy AHP and VIKOR methods in occupational health and safety risk assessment: the case of a gun and rifle barrel external surface oxidation and colouring unit. Int. J. Occup. Saf. Ergon. (2018). https://doi.org/10.1080/10803548.2018.1492251
8. Gul, M., Guneri, A.F., Nasirli, S.M.: A fuzzy-based model for risk assessment of routes in oil transportation. Int. J. Environ. Sci. Technol. **16**(8), 4671–4686 (2018). https://doi.org/10.1007/s13762-018-2078-z
9. Gül, S.: Spherical fuzzy extension of DEMATEL (SF-DEMATEL). Int. J. Intell. Syst. **35**(9), 1329–1353 (2020). https://doi.org/10.1002/int.22255
10. Gül, S., Kabak, Ö., Topcu, Y.I.: Coordinating the ISM Code and OHSAS procedures to improve the occupational health and safety at sea. J. Multi-Criteria Decis. Anal. Early View, 1–18 (2020). https://doi.org/10.1002/mcda.1710

11. He, T.T., Wei, G., Lu, J., Wei, C., Lin, R.: Pythagorean 2-Tuple linguistic VIKOR method for evaluating human factors in construction project management. Mathematics **7**(12), 1149 (2019). https://doi.org/10.3390/math7121149
12. Ho, L.-H., Lin, Y.-L., Chen, Ting-Yu.: A Pearson-like correlation-based TOPSIS method with interval-valued Pythagorean fuzzy uncertainty and its application to multiple criteria decision analysis of stroke rehabilitation treatments. Neural Comput. Appl. **32**(12), 8265–8295 (2019). https://doi.org/10.1007/s00521-019-04304-8
13. Ilbahar, E., Cebi, S., Kahraman, C.: Assessment of renewable energy alternatives with pythagorean fuzzy WASPAS method: a case study of Turkey. In: Kahraman, C., Cebi, S., Cevik Onar, S., Oztaysi, B., Tolga, A.C., Sari, I.U. (eds.) INFUS 2019. AISC, vol. 1029, pp. 888–895. Springer, Cham (2020). https://doi.org/10.1007/978-3-030-23756-1_106
14. Kutlu Gündoğdu, F., Kahraman, C.: Spherical fuzzy sets and spherical fuzzy TOPSIS method. J. Intell. Fuzzy Syst. **36**(1), 337–352 (2019). https://doi.org/10.3233/JIFS-181401
15. Liang, R., Wang, J., Zhang, H.: Evaluation of e-commerce websites: An integrated approach under a single-valued trapezoidal neutrosophic environment. Knowl.-Based Syst. **135**, 44–59 (2017). https://doi.org/10.1016/j.knosys.2017.08.002
16. Ozdemir, Y., Kayapinar Kaya, S., Turhan, E.: A scale to measure sustainable campus services in higher education: "sustainable service quality". J. Clean. Prod. **245**, 118839 (2020). https://doi.org/10.1016/j.jclepro.2019.118839
17. Perez-Dominguez, L., Rodriguez-Picon, L.A., Alvardo-Iniesta, A., Crus, D.L., Xu, Z.: MOORA under pythagorean fuzzy set for multiple criteria decision making. Complex. Manuf. Processes Syst. 2602376 (2018). https://doi.org/10.1155/2018/2602376
18. Rani, P., Mishra, A.R., Pardasani, K.R., Mardani, A., Liao, H., Streimikiene, D.: A novel VIKOR approach based on entropy and divergence measures of Pythagorean fuzzy sets to evaluate renewable energy technologies in India. J. Clean. Prod. **238**, 117936 (2019). https://doi.org/10.1016/j.jclepro.2019.117936
19. Rani, P., Mishra, A.R., Rezaei, G., Liao, H., Mardani, A.: Extended pythagorean fuzzy TOPSIS method based on similarity measure for sustainable recycling partner selection. Int. J. Fuzzy Syst. **22**(2), 735–747 (2019). https://doi.org/10.1007/s40815-019-00689-9
20. Smarandache, F.: A unifying field in logics neutrosophy: neutrosophic probability, set and logic. American Research Press, Rehoboth (1999)
21. Torra, V., Narukawa, Y.: On hesitant fuzzy sets and decision. In: IEEE International Conference on Fuzzy Systems on Proceedings, Jeju Island, South Korea, August 20–24, pp. 1378–1382 (2009). https://doi.org/10.1109/fuzzy.2009.5276884
22. Wang, W.C., Lin, Y.H., Lin, C.L., Chung, C.H., Lee, M.T.: DEMATEL-based model to improve the performance in a matrix organization. Expert Syst. Appl. **39**, 4978–4986 (2012). https://doi.org/10.1016/j.eswa.2011.10.016
23. Yager, R.R.: Pythagorean Membership Grades in Multicriteria Decision Making. IEEE Trans. Fuzzy Syst. **22**(4), 958–965 (2014). https://doi.org/10.1109/TFUZZ.2013.2278989
24. Yager, R.R.: Properties and applications of pythagorean fuzzy sets. In: Angelov, P., Sotirov, S. (eds.) Imprecision and Uncertainty in Information Representation and Processing. SFSC, vol. 332, pp. 119–136. Springer, Cham (2016). https://doi.org/10.1007/978-3-319-26302-1_9
25. Yazdi, M., Nedjati, A., Zarei, E., Abbassi, R.: A novel extension of DEMATEL approach for probabilistic safety analysis in process systems. Saf. Sci. **121**, 119–136 (2020). https://doi.org/10.1016/j.ssci.2019.09.006
26. Zadeh, L.A.: Fuzzy sets. Inf. Control **8**, 338–353 (1965). https://doi.org/10.1016/S0019-9958(65)90241-X
27. Zadeh, L.A.: The concept of a linguistic variable and its application to approximate reasoning – I. Inf. Sci. **8**(3), 199–249 (1975). https://doi.org/10.1016/0020-0255(75)90036-5

28. Zhang, X.: A novel approach based on similarity measure for Pythagorean fuzzy multiple criteria group decision making. Int. J. Intell. Syst. **31**, 593–611 (2016). https://doi.org/10.1002/int.21796
29. Zhang, X., Xu, Z.: Extension of TOPSIS to multiple criteria decision making with Pythagorean fuzzy sets. Int. J. Intell. Syst. **29**(12), 1061–1078 (2014). https://doi.org/10.1002/int.21676
30. Zhou, L., Dai, G., Qin, R., Tang, M., Qiu, J.: Risk analysis of gob coal spontaneous combustion in methane-rich, combustion-prone coal seam based on intuitionistic fuzzy DEMATEL. J. Fail. Anal. Prev. **18**(4), 975–987 (2018). https://doi.org/10.1007/s11668-018-0492-7

ANN Based Holistic Performance Valuation and Mutual Fund Selection: Evidence from Turkey

Arif Tanış[1] and Yaman Ömer Erzurumlu[2(✉)]

[1] Institute of Natural Sciences, Istanbul, Turkey
arif.tanis@bau.edu.tr
[2] Department of Engineering Management, Faculty of Engineering and Natural Sciences, Istanbul, Turkey
yamanomer.erzurumlu@eng.bau.edu.tr

Abstract. This study attempts to conduct a quantitative analysis of the mutual fund performance using ANN-Machine learning based fund scoring methodology. We generate fund scores using 62 performance measures for all equity mutual funds and fixed income mutual funds in Turkish mutual fund market and test the performance of the ANN-Machine learning generated scores using alternative simulation approaches.

The results indicate that the more holistic and the least subjective the performance measurement approach is, the better it performs for equity fund selection. The portfolio blending all performance measure approaches for fund selection performs better than other approaches. AHP based approach, which arguably the next most holistic and the least subjective in nature comes the second. The results are robust to alternative asset al.location. However, findings do not hold for fixed income mutual fund portfolios. The best performing fixed income mutual fund portfolio is based on return based performance measures alone.

From a practical perspective, our results imply that in an era which incorporating big data in performance analysis using methods like the ANN is possible, equity mutual fund performance measurement should not exclude any performance measurement approach. Omitting certain aspects of performance or subjective selection of performance measures could lead to sub-optimal results. Results also indicate that one size does not fit all. Incorporating a large size of data using artificial neural networks alone could not be the ultimate performance measurement method for all types of mutual funds.

Keywords: Mutual funds · Performance · Fund selection · MCDA · Artificial neural network · AHP

T. Allahviranloo et al. (Eds.): IDS 2020, AISC 1301, pp. 379–398, 2021.
https://doi.org/10.1007/978-3-030-66501-2_31

1 Introduction

Financial markets gain depth and stability as the number of participants and total assets under management increase. However, fund selection is still a significant obstacle for potential and existing participants. Participants seek the highest performance for their investments. However, performance could be evaluated from several perspectives: return, risk-based, manager performance, benchmark-based, etc. The variety of these perspectives further complicates the selection process, especially under limited financial literacy. The selection remains mainly based on a few performance criteria in which the investor is familiar or subjective combination of some selected performance measures capturing a limited number of aspects of performance. Although the methodologies are not revealed, the existing performance indices use limited number of selected performance measures. With the limitation of the performance measures comes the possibility of incomplete performance evaluation.

On the other hand development of artificial neural network (ANN) and Multi Criteria Decision Analysis (MCDA) methodologies enables us to analyze large numbers of performance evaluation methods to compare the mutual funds. Such evaluation of mutual fund performance provides a holistic view of performance. Literature, in general, indicates that such a holistic view is essential. Cogneau and Hübner (2009) present a detailed summary of a large set of these methods. These methods with different approaches capture various aspects of mutual fund performance. Most methods were developed to complete and contribute to the preceding ones, and each alternative view of approach is an integral part of the fund evaluation picture. All approaches have to be taken into consideration for a rather tenable deficiency in mutual fund performance evaluation and comparison.

This study attempts to analyze fund performance using a broad set of performance measures with an artificial neural network approach in Turkish equity and fixed income mutual fund universe over 3.5 years period. As of December 2019, Turkish mutual fund market consists of 459 mutual funds and has a total market capitalization of 97 billion TRY (15 billion USD). During our study period, there are 66 equity funds and 72 fixed income funds. (TEFAS 2019).

Our study incorporates seven fund performance measurement approaches in an ANN-based fund scoring and simulated portfolios. In doing so, it examines the contribution of such holistic approach based fund selection to portfolio return. In an era in which ratings are extensively used by the investors and funds attract funds based on these ratings (Del Guercio and Tkac 2008), the outcome of the underlying methodology becomes more critical (Müller and Weber 2014). Our results indicate that ANN and MCDA methodologies based selection leads to better portfolio performance for equity mutual funds than single approach models, a group of the most favorite criterion-based method, random selection, or AHP based fund selection. However, the same strategy does not lead to superior results for fixed income mutual funds.

The rest of the study proceeds as follows. Section 2 summarizes the relevant literature, development of fund performance measurement methods, and use of

ANN. Section 3 summarizes the data and methodology. Section 4 presents and discusses the results for equity and fixed income mutual funds. Section 5 concludes.

2 Background

The performance valuation measures used in this study are summarized by Ippolito (1993), Grinblatt and Titman (1992), Cesari and Panetta (2002) and specifically for hedge funds by Eling and Schuhmacher (2007). Fund performance measurement has been extensively studied over forty years with mixed results (Babalos et al. 2008; Matallin et al. 2014). Here, we summarize the performance measures in 7 underlying approaches; return, risk, manager performance, benchmark comparison, absolute risk level, resistance to extreme market conditions and relative performance under different market conditions.

First generation of performance measurement techniques depend on diversification (Sharpe 1966; Treynor 1965) and abnormal return based regression analyses. These mean variance (Markowitz 1952) and CAPM based performance measures were later criticized by Roll (1977) and Roll (1978). Most of the following studies that belong to the same body of literature evaluate fund performance based on a single factor and attempt to blend a limited number of selected measures. As an extension, multivariate models with additional factors were developed such as Fama and French (1966) and Carhart (1997). However, as Caporin et al. (2014) points out, such approaches would be incomplete due to evaluating performance based on a limited number of dimensions. Moreover, they would be sensitive to forecast errors (Kourtis 2016) or reveal incomplete results when they are based on single factor measures like risk (Lisi and Caporin 2012) or Sharpe ratio (Bacon 2009). Although, there are measures and models attempting to reduce the impact of these shortfalls such as variations of Sharpe ratio by Liu and Li (2017), squared Sharpe of Kourtis (2016), modified Sharpe ratio of Ardia and Boudt (2015) and fuzzy Sharpe ratio of Nguyen et al. (2015) or Sortino ratio based genetic algorithm model by Chen et al. (2014), mainly two points remain unclear. One, findings are inconclusive on defining the most effective underlying approach (Chen and Huang 2011). Two, results largely indicate that using a single (i.e. only risk based) or a few underlying approach for selecting performance measures leads to biased results (Caporin et al. 2014).

The second body of studies is based on mean-variance. They argue that the primary focus of the investor is the relationship between risk and return. However, an extensive literature document that several quantitative and qualitative factors impact the individuals' decision-making process. A risk-based approach alone could not be sufficient to explain the performance. Even, well-accepted rating systems are found to be incomplete due to this shortfall (Lisi and Caporin 2012). Furthermore, if we accept that the risk-based approach is superior to other approaches, the question of which method measures the risk in the most accurate way remains unanswered (Babalos et al. 2012).

Third generation of performance measures were developed as alternative or complement to the preceding parametric and rather mechanical performance measures. This group of performance measurement techniques attempt to quantify the fund manager generated value and compare the fund performance with the second best investment opportunity (Berk and Binsbergen 2012). The main goal of this family of performance measures is to distinguish managers that show skill and deliver persistent high returns. Whether fund managers generate positive alpha has been long debated since Jensen (1968). Since the early works of Treynor (1965) and Sharpe (1966) to Gruber (1996), Carhart (1997) and Fama and French (1996), literature suggests that actively managed funds deliver better performance than passive managed funds but struggle to generate positive return net of fees.

In order to examine the manager contribution in further detail, the fourth approach of performance measures were developed. This group focused on timing and selection ability of the manager. Treynor and Mazuy (1966), Jensen (1968) and Henriksson and Merton (1981) models pioneered this body of performance measurement literature. Major question of these studies was whether these fund managers were rewarded by the market. Findings were rather against such rewarding mechanism. Managers were getting some degree of share from economic surplus even if they do not display these skills (Berk and Binsbergen 2012). Recent bootstrap technique (Kosowski et al. 2006; Cheng and Yan 2017; Blake et al. 2017) and Bayesian approach based (Baks et al. 2001; Pastor and Stambaugh 2002) studies documented that some managers deliver positive alpha. From investor perspective it is vital to determine these managers. Nevertheless, selection of funds purely based on manager contribution would suffer from shortfall of single approach based selection (Barras et al. 2010).

Fifth group of analyses stands on the search for persistence of positive performance. Grinblatt and Titman (1992), Hendricks et al. (1993), Goetzmann and Ibbotson (1994) and Cremers and Ptajisto (2009) document that positive performance shows persistence. Jensen (1968), Malkiel (1995) and Carhart (1997) for US funds, Babalos et al. (2012) for Greek funds reach opposite results. Persistency was found to be existing rather for short term (Grinblatt and Titman 1992; Hendricks et al. 1993; Elton et al. 1996; Goetzmann and Ibbotson, 1994; Carhart, 1997; Droms and Walker 2006; Ferreira et al. 2012). Gruber (1996), Carhart (1997), Zheng (1999) and Bollen and Busse (2001) argued that observed persistency is a result of luck rather than skill.

Final body of performance measure studies focus on the timing ability of the manager. This body of research began with Henriksson and Merton (1981) and grew to be an alternative approach for evaluating fund performance (Ferson and Schadt 1996; Ferson and Warther 1996; Christopherson et al. 1998). These methods allow time varying expected return and risk. Studies document that when certain timing principles are followed, fund returns stay above passive portfolios (Bollen and Busse 2001) and deliver additional value (Hammami and Oueslati 2017).

Studies on Turkish mutual fund market are predominantly remained on comparison of active and passive management of mutual funds with different data sets and time periods. These studies document that active funds do not beat passive funds (Gürsoy and Erzurumlu 2001; Saim 2002; Canbaş and Kandir 2002; Gökçen and Yalçın 2015), managers do not show significant timing skill (Karacabey 2002; Korkmaz and Uygurtürk 2008) and fund performance do not display persistency (Canbaş and Kandir 2002). Studies indicate that fund selection in Turkish mutual fund market based on a limited performance criteria would lead to sub-optimal results. Gürsoy and Erzurumlu (2001) document that different performance measurement lead to different ranking of funds. Kurun et al. (2008) and Teker, Karakum and Tav (2008) suggest that risk based performance measurement lead to better results than return based performance measurement. Alptekin and Şıklar (2009) document that The Technique for Order of Preference by Similarity to Ideal Solution (TOPSIS) based multiple criteria decision making improve the selection process significantly. The only ANN based study is Akel and Karacameydan's (2012) work which they adopt the methodology of Chiang et al. (1996) using macroeconomic parameters as the selection criteria.

Hence, fund performance measurement literature indicates that using selected approaches would lead to sub-optimal results. A holistic view of combining these approaches is vital.

2.1 Artificial Neural Network

Our ANN method is a multi-layer and back propagation technique. Multi-layer nature of ANN is based on Minsky and Papert's (1969) study. Back propagation method is later developed by Werbos (1988) in 1975. It laid the ground for Wang et al.'s (2010) FAN and Anish and Majhi's (2016) FFLAN. Godefroid et al. (2017) review the studies between 1993–2013 and document that ANN based techniques could lead to better forecasting results than traditional time series techniques. Zhang et al. (2018) and Zhang et al. (1999) document that ANN could improve empirical analyses. These two studies provide guide way for our study as well.

ANN has been long a modeling and forecasting method of financial indicators (Yao and Liu 1996; Dutta et al. 2006; Panda and Narasimhan, 2007; Tseng et al. 2008; Xu et al. 2009; Qian and Rasheed, 2010). However, the studies predominantly focus on FX and stock markets. Chiang et al. (1996) is one example of application ANN to mutual fund market using macro-economic variables. They document that ANN leads to relatively better results than linear and non-linear regression models. Regarding mutual funds studies largely focus on net asset value (NAV hereafter) prediction. Ray and Vina (2004) and Sahoo and Hathy (2007) applies ANN to Indian fund market NAV. Wang and Huang (2010) document that ANN predicts NAV better than traditional models. ANN leads to better predictions for Turkish stock market as well (Diler 2003; Benli 2005; Altay and Satman 2005; Karaçor and Alptekin 2006; Avcı 2007; Altan 2008). Details of the method is summarized in the next section of Methodology.

3 Data and Methodology

The study covers all equity and bond funds that were active longer than one fiscal year during the study period. Hence, the data set is survivorship bias-free. Equity data is obtained from Rasyonet, and bond data is obtained from

Riskactive data vendors. The analysis period for equity funds is 30.06.2014 and 29.10.2017, for bond funds 02.01.2015 and 29.12.2017. We excluded mutual funds that were terminated and merged with other funds within a year of inception. The final data set consists of 66 equity funds and 79 bond funds.

Performance measurement indicators that belong to 8 groups of underlying approaches were calculated quarterly for equity funds and bond funds separately. We obtained quarterly values of 62 performance measures for equity funds and 41 performance measures for bond funds. Table 1 lists the performance measures included in the ANN analysis. Final data set consist of 868 quarterly performance measurement value per equity mutual funds (57.288 in total) and 574 quarterly performance measurement value per bond mutual funds (45.346 in total). There are two reasons for calculating performance measures quarterly. First, companies report financial statements quarterly. Second, Turkish mutual funds largely reduce redemption fee to 0% after three months of holding period. Considering the well documented negative return of active mutual funds net of fees, we simulate the portfolios on a quarterly base.

The 8 classes that represent different underlying approaches for performance measurements are:

- Return: Return for a specific period.
- Risk: Risk adjusted performance measures.
- Absolute risk level performance: Risk based parametric performance measures.
- Portfolio manager performance: Performance measures that evaluate how well manager performs with respect to general market trend.
- General risk benchmark comparison: Regression based, beta or tracking error like parametric performance measures.
- Performance criteria for resistance to extreme market conditions: Performance under sharp market swings.
- Relative performance criteria as per different market conditions: Techniques that compare the performance of the fund under different market conditions.
- Unclassified

Practitioners often pick a limited number of criteria they are familiar with and generate scoring indices. The main argument behind this approach is that a certain number of performance measures could capture the big picture, and it is more practical for application. In order to investigate whether limited selection of the performance measures is optimal we generated ANN based fund scores using one most common performance measure from each approach. Additionally, in order to test whether fund scores based on expert opinion leads to optimal results, we generated ANN based fund scores using 19 performance measures selected using Analytic Hierarchy Process (AHP hereafter). Our objective is to compare the outcome of ANN methodology with a limited set and a reduced set of expert-selected performance measures. The AHP process follows the selection of Akçay (2015) where he conducts a survey with 20 professional portfolio managers from 10 fund management companies. Participants rank performance

Table 1. Performance measurements

	Equity	Bond		Equity	Bond
Return performance criterion			**Portfolio manager performance criterion**		
Active return*	X	X	Annualized jensen alpha*	X	X
Quarterly return	X	X	Fama's measure*	X	X
Cumulative return	X	X	Fama's beta	X	X
Risk based performance criterion			Active premium	X	
Sharpe ratio*	X	X	Information ratio v1*	X	X
Treynor ratio*	X	X	Appraisal ratio	X	X
Sortino ratio*	X	X	**General risk benchmark comparison**		
Sterling ratio	X	X	Tracking error v1*	X	
M2 modigliani ratio	X	X	Tracking error v2	X	
M2 sortino ratio	X	X	R Square	X	
Adjusted sharpe ratio	X		Correlation	X	
T2 measure	X		EWMA correlation*	X	X
V2 measure	X		Active premium	X	
Pain ratio	X		CAPM beta*	X	X
Absolute risk level performance criterion			**Performance criteria for resistance to extreme market conditions**		
Standard deviation	X	X	Maximum drawdown*	X	X
Annualized volatility	X		Calmar ratio	X	X
Upside volatility	X	X	Omega ratio*	X	X
Annualized upside volatility	X		Burke ratio	X	X
Annualized downside volatility*	X		Pain index	X	X
Downside volatility	X	X	Ulcer Index	X	X
Variance	X	X	Martin Ratio	X	
Skewness	X	X	**Relative performance criteria as per different market conditions**		
Kurtosis	X		Up capture ratio*	X	X
Value at risk (VaR)	X		Down capture ratio	X	X
Cornish fisher VaR*	X		Up number ratio	X	
Expected shortfall*	X		Down number ratio	X	
Tracking error residual risk	X	X	Up percentage ratio*	X	X
Total risk	X	X	Down percentage ratio	X	X
Systemic risk	X	X	**Unclassified performance criterion**		
Unsystemic risk	X	X	Positive return days	X	X
VaR (EWMA)	X		Negative return days	X	X
EWMA volatility	X	X	EWMA covariance	X	X
			Number of observation	X	
			Positive gain percentage	X	

measures within each group in Table 1. Participants select 19 performance measures out of 77 as the most used and familiar with. These 19 AHP selected performance measures are denoted with an asterisk (*) in Table 1.

Each performance measure is normalized between −1 and 1 per quarter. A mutual fund is attained the 1 if the value is the highest for a specific performance measure in a particular quarter and −1 if it is the lowest. Hence, we obtain a (66 × 14 by 62) matrix for equity funds and (79 × 14 by 41) normalized performance score matrix for bond funds. The ANN was fed with normalized quarterly performance measurement values.

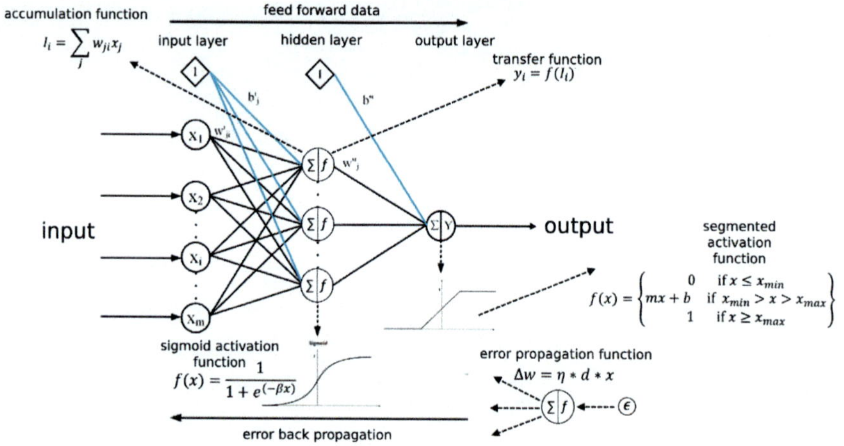

Fig. 1. Artificial neural network

As presented in Fig. 1, a typical ANN consists of three stages. An input layer with a specific number of nodes compiles input through a sigmoid activation function, one or multiple hidden layers with user defined number of k nodes in the input layer in which the input is grouped through an accumulation function with specific weights and a third output layer with a single score. The process has an iterative nature where the weights are updated through a "rprop+" (Resilient Backpropagation Algorithm with Weight Backtracking) algorithm. These weights are represented by the black arrows in Fig. 1. Blue arrows represent the intercept value of the activation function at a specific node. ANN stops iteration, and the training phase concludes once the model reaches the maximum allowed error threshold.

Our ANN model has one hidden layer for each performance measurement approach family and nodes half the number of hidden layers. SSE (sum of squared error) is used as an error threshold for halting the calculation. "rprop+" (Resilient Backpropagation Algorithm with Weight Backtracking) algorithm is used for training. Exponential smoothed accept/reject activation function is applied to output neurons. Objective function is convergence of next quarter's normalized return value by using this quarter's performance scores to capture the predictive power of performance score's of the performance. Training and testing were repeated for cross-checking with alternative lengths of data. The normalized scores were labeled under five groups. Funds were grouped as "worst" for a score between 0-0.2, "bad" for a score of 0.2–0.4, "average" for a score of 0.4–0.6, "good" for a score of 0.6–0.8 and "best" for a score of 0.8–1. The model fit threshold is set as 0.9, and the number of hidden nodes is set at half of input nodes which rounded up to a constant value.

To track and compare the cumulative return score based investment following all criteria-based and alternative approaches, we simulated portfolios consist of 5 equally distributed funds, and the allocation is revised on a quarterly base. The portfolio is reallocated to the newly determined top five funds at the beginning of the next quarter, and funds are replaced if necessary. The simulations are conducted based on the following fund score calculation methodologies; (i) following the artificial network method using all criteria and all (66) available performance measures, (ii) for each group of performance measure separately, (iii) for AHP selected 19 performance criteria. For further robustness check, we also simulated (iv) the portfolio including one most popular performance measure from each criterion (7 in total) attempting to mimic the practitioners approach (The selected performance measures are three months return, Sharpe, Jensen alpha, correlation, standard deviation, maximum drawdown, up capture ratio, positive return percentage) (v) randomly selected funds (repeated eight times, average and average plus-minus 3 standard deviations are taken as high and low scores), (vi) randomly selected funds with a buy-and-hold strategy. Portfolio simulation is repeated with alternative allocations to the top 5 mutual funds (i.e. 34%, 27%, 20%, 13%, 6%). We also simulated the portfolio return calculations with the five lowest, and the five average scored mutual funds to examine whether the ANN scores distinguish performance between the funds. Finally, the performance of the simulated portfolio is tracked over time.

4 Descriptive Statistics and Empirical Analysis

This section has two parts. First part presents the descriptive statistics for equity and fixed income mutual funds. Second part presents and discusses the portfolio simulation analyses of equity and fixed income mutual funds separately.

4.1 Descriptive Statistics of Equity Mutual Funds

Figure 2 presents the fund scores obtained from ANN. The results indicate that the scores are distributed close to normal with a negative skew. This structure is a typical financial data where the gain is unlimited, and the loss is limited with 100%. The majority of the equity funds obtain a score smaller than the average, normally distributed. The minimum score is 0.034; first quartile is 0.349, the median score is 0.459, the average score is 0.475, the third quartile score is 0.634, and the maximum score is 0.891.

Next, we obtained the ANN scores, excluding one group of performance measure at a time. We conducted a t-test of difference of scores at 5% significance. The objective is to get a preliminary idea of the relative contribution of a group of performance measures, whether excluding it leads to a significant change in scores. Rejection of the null hypothesis indicates that scores significantly change as we exclude a specific measurement group. However, results at Table 2 are inconclusive. Scores alter significantly if absolute risk, manager based, extreme market condition, and AHP based performance measure groups are excluded.

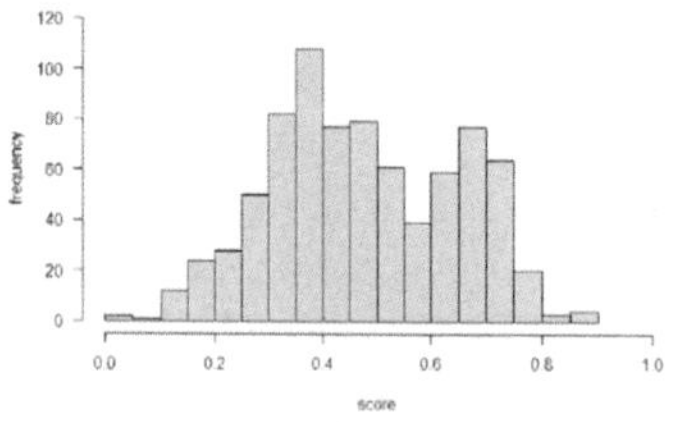

(a) Histogram

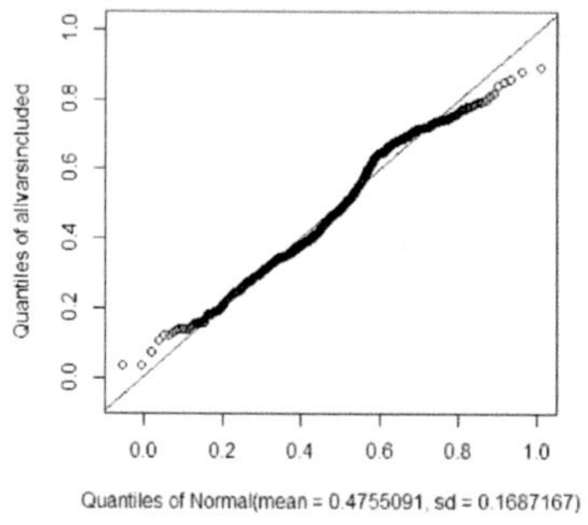

(b) QQ-Plot (NormDist)

Fig. 2. Histogram and QQ plots of all criteria based equity fund scores

Table 2. Equity funds two sample t-test comparison results of one group excluded ANN scores $\alpha = 0.05$

Sample name	df	t-value	p-value	Null hypothesis
All vs (All except AbsRisk)	788	2.815	0.004	Rejected
All vs (All except PortMan)	788	3.095	0.002	Rejected
All vs (All except GenBench)	788	0.761	0.446	Can not be rejected
All vs (All except Unclass)	788	0.662	0.507	Can not be rejected
All vs (All except Risk)	788	0.611	0.540	Can not be rejected
All vs (All except Return)	788	−0.075	0.939	Can not be rejected
All vs (All except ExtMarket)	788	3.351	0.000	Rejected
All vs (All except DiffMarket)	788	−0.108	0.913	Can not be rejected
All vs (All except AHP based)	788	2.398	0.016	Rejected

4.2 Descriptive Statistics of Fixed Income Mutual Funds

Fixed income funds naturally have different performance dynamics than equity funds. Some performance measures become ineffective due to these differences. We excluded the performance measures which do not quantitatively change the ANN results in any direction. The final list of performance measures that remain contains 41 performance measures (See Table 1). Results in Fig. 3 indicates that ANN scores obtained with 41 performance measures for fixed income funds have a non-normal distribution and skewed to the right. Moreover, Table 3 indicates that the exclusion of any group of performance measure family alone does not significantly change the ANN scores.

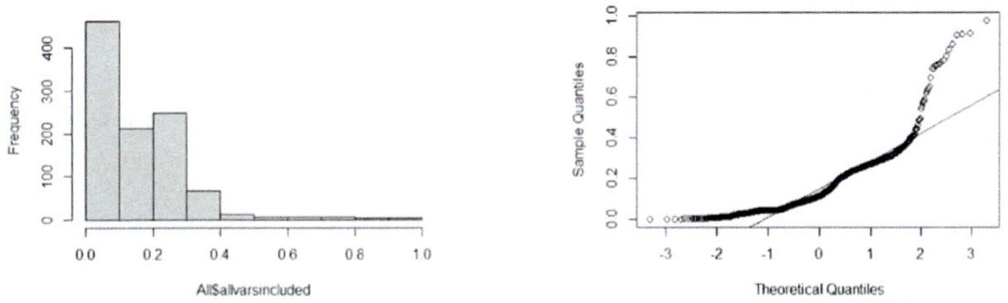

(a) Histogram (b) QQ-Plot (NormDist)

Fig. 3. Histogram and QQ plots of all criteria based fixed income fund scores

Table 3. Fixed income funds two sample t-test comparison results of one group excluded ANN scores $\alpha = 0.05$

Sample name	df	t-value	p-value	Null hypothesis
All vs (All except AbsRisk)	1026	0.108	0.913	Can not be rejected
All vs (All except PortMan)	1026	0.252	0.800	Can not be rejected
All vs (All except GenBench)	1026	0.303	0.761	Can not be rejected
All vs (All except Unclass)	1026	0.299	0.764	Can not be rejected
All vs (All except Risk)	1026	0.630	0.528	Can not be rejected
All vs (All except Return)	1026	0.433	0.664	Can not be rejected
All vs (All except ExtMarket)	1026	0.654	0.512	Can not be rejected
All vs (All except DiffMarket)	1026	0.077	0.937	Can not be rejected

4.3 Empirical Analyses

This section summarizes the return performance of equity and fixed income fund portfolio simulations. Each portfolio contains five equally distributed funds with the highest score at the beginning of a quarter. Simulations were repeated using the fund scores based on a single group of performance measure. To capture the impact of using performance measures from multiple approaches, we simulated the portfolios using an AHP selected 19 performance measures. We believe this methodology is sounder than the subjective selection of specific performance measures. We also simulate portfolios with randomly selected funds and one criterion from each approach. Finally, results are compared to the ANN-based scores using all performance measures.

4.3.1 Equity Mutual Funds

First, we conduct a preliminary test to examine whether the ANN-based scoring could distinguish the funds appropriately. We tracked the portfolios' performance, which invests in 5 funds with the highest, median, and the lowest scores. Results in Fig. 4 indicate that the scores successfully distinguish between the high (labeled max), medium (labeled med), and low (labeled min) performing funds. Alternative portfolio distributions (labeled alt) confirm the relationship. Hence, we could safely say ANN-based scoring distinguishes between the equity funds.

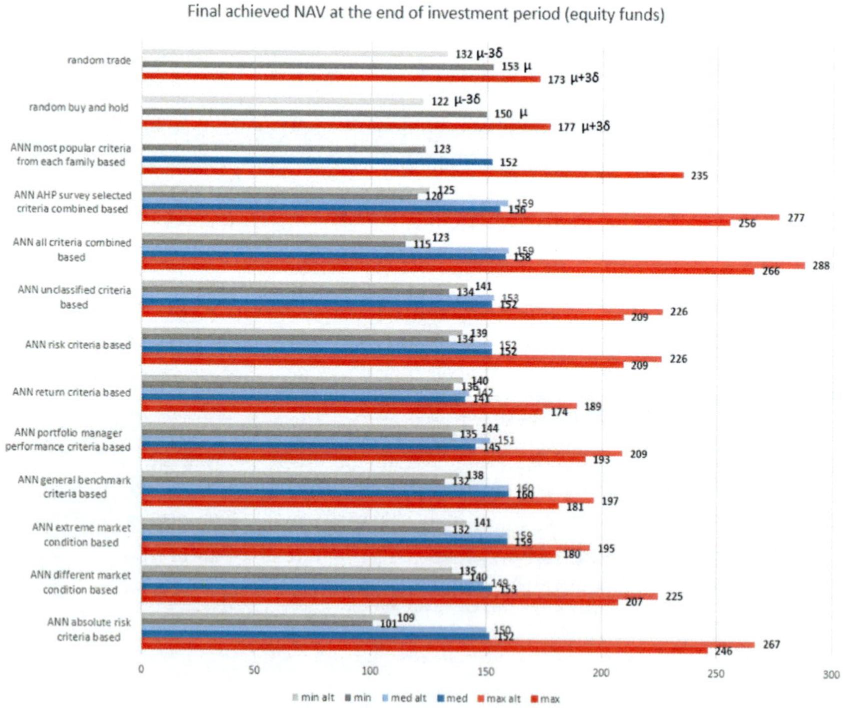

Fig. 4. Final size of equity fund portfolios invested in max, median and low scores of specific group of performance measures

Next, we report the simulated portfolio performances. Figure 5 displays the gradual track of each portfolio. The portfolio that consists of funds selected according to all criteria ANN approach based scores reaches the largest portfolio size. AHP method based portfolio in which 19 performance measures are incorporated is the second. Risk-based portfolio is the third and portfolio of 7 most common performance measures is the fourth. Other single group measures follow these four portfolios. We believe the results emphasize the importance of a holistic view of performance valuation. ANN-based portfolio incorporating all performance measures provides the most holistic view within the portfolios, AHP is the second least subjective and the second most holistic method. We repeated the tests with alternative weights of allocation, and the results qualitatively remained the same. The ranks of the first four portfolios did not change. Figure 6 documents the return of 11 portfolios per quarter. The portfolios, regardless of the underlying methodology, tend to move together. Hence, we could rule out that the portfolio with all criteria combined or AHP based portfolio do not perform better due to a small number of good quarters, neither the other portfolios underperform due to one or two bad quarters.

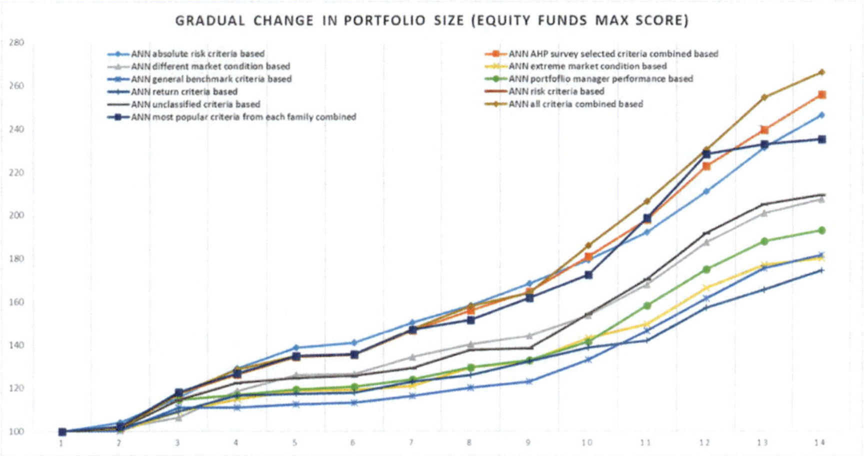

Fig. 5. Gradual change of equity fund portfolio size by specific group of performance measures

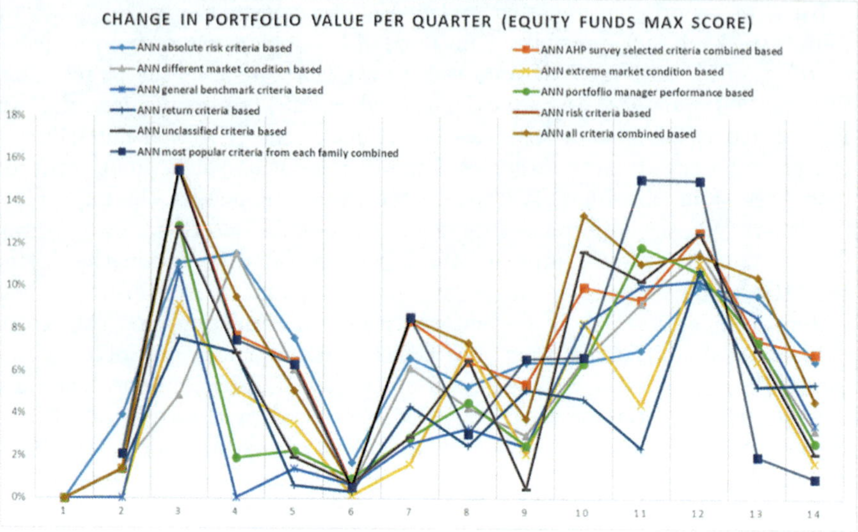

Fig. 6. Change in equity fund portfolio value per quarter

4.3.2 Fixed Income Mutual Funds

We applied the ANN-based scoring to fixed income mutual funds to examine whether the approach is fund type specific. We examined all fixed income mutual funds with complete data between January 2015–December 2017. The final data set consists of 79 of the 87 fixed income mutual funds. Excluded funds were either terminated or merged with other mutual funds within the inception year. We repeated the same scenarios and simulations for fixed income mutual funds. However, as explained above, the number of performance measures is reduced to 41. The performance criteria is kept if it contributes to the ANN procedure.

The results in Fig. 7 indicate that the ANN-based scores successfully distinguish between the high medium and low ranking funds. The results remain qualitatively the same when the portfolio is allocated alternatively. However, as Fig. 8 indicates, the rank of the portfolio performance types are different from the equity funds. Return based performance measures ANN scores reveal the highest return. All criteria based scores come the fourth. Although the portfolios tend to trend together as Fig. 9 indicates, including a larger number of performance measures to achieve a holistic view does not work for fixed income securities mutual funds as it does for equity mutual funds.

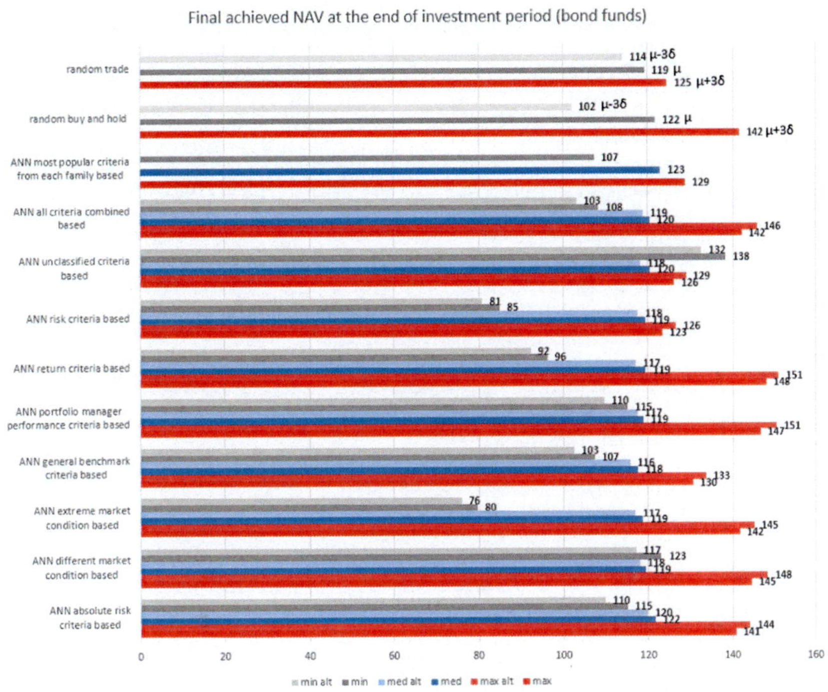

Fig. 7. Final size of fixed income fund portfolios invested in max, median and low scores of specific group of performance measures

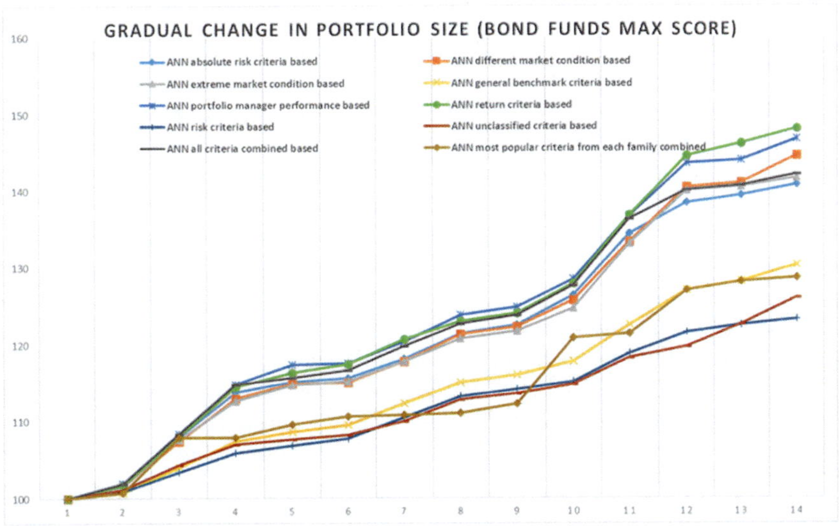

Fig. 8. Gradual change of fixed income fund portfolio size by specific group of performance measures

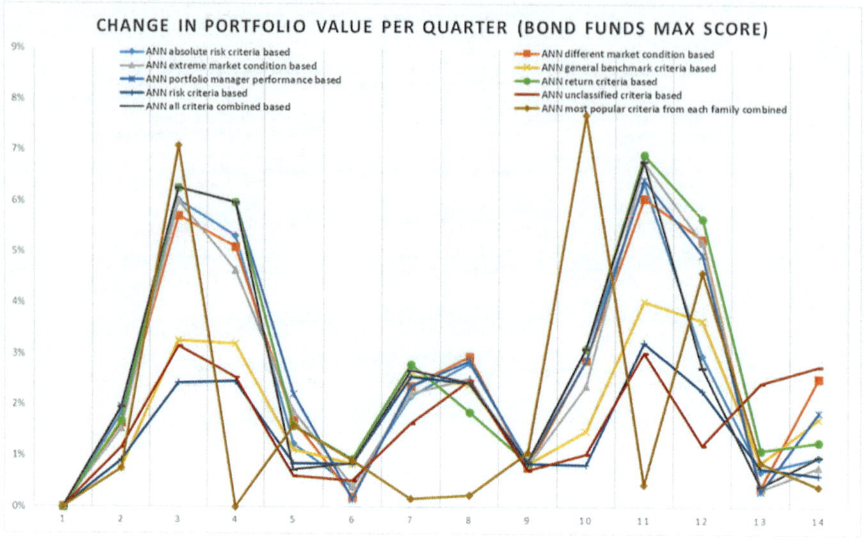

Fig. 9. Change in fixed income fund portfolio value per quarter

5 Conclusion

Artificial Neural Network is a powerful tool that enables us to conduct analyses with larger sets of input. Mutual fund performance measurements is an example of that. The scope of the measurement criterion posits whether incorporating different performance measurement families improves fund performance valuation and enables the investors to pick better performing funds. Decades of research provide us eight broad families of performance measures, each representing an alternative approach to fund performance valuation. Since each represents another aspect of performance, incorporating all in performance valuation carries the potential benefit of the holistic approach that captures all these aspects simultaneously.

This study attempts to conduct a purely quantitative analysis of the fund performance using ANN-based methodology. We generate ANN-based fund scores using 62 performance measures for all equity mutual funds that are active between June 2014 and October 2017 and 41 performance measures for all fixed income mutual funds that are active between January 2015 and December 2017 in the Turkish mutual fund market. We simulate 11 portfolios using all performance measures, AHP selected 19 performance measures, seven performance measurement family alone, one performance measure from each family, and randomly selected portfolio. We repeat the scoring on a quarterly base and update the portfolios throughout the study period.

The results indicate that the holistic nature of the performance measurement approach could be significant for equity funds. The portfolio using all performance measures for scoring performs better than other approaches, AHP based approach, which arguably the second holistic view in nature comes the second. The results are robust to alternative allocation.

However, findings do not hold for fixed income mutual fund portfolios. The best performing fixed income mutual fund portfolio is the one that is formed based on return based performance measures alone. This is understandable considering that most of these funds' investors expect to earn the underlying interest rates and portfoliio returns closely track the market interest rates. From a practical perspective, our results imply that in an era which incorporating big data in performance analysis is possible, equity mutual fund performance measurement should not exclude any performance measurement approach. Omitting certain aspects of performance or subjective limitation of the use of performance measures could lead to sub-optimal results. Results also indicate that one size does not fit all. Incorporating a large size of data using artificial neural networks alone could not be the ultimate performance measurement method for all types of mutual funds.

References

Akçay, M.B.: A Model Proposal For Performance Measurement and Performance Rating for Mutual Funds, thesis no. 410487 (2015). www.tez.yok.gov.tr. Accessed 4 Jan 2021

Akel, V., Karacameydan, F.: Yatirim Fonlari Net Varlik Degerlerinin Yapay Sinir Ağlari Yöntemiyle Tahmin Edilmesi, vol. 12, no. 2, p. 13 (2012)

Alptekin, N., Şiklar, E.: Türk hisse senedi emeklilik yatirim fonlarinin çok kriterli performans değerlendirmesi:TOPSIS metodu. In: Dumlupinar Üniversitesi Sosyal Bilimler Dergisi, vol. 25 (2009)

Altan, S.: Döviz Kuru Öngörü Performansi İçin Alternatif Bir Yaklasim: YSA, vol. 10, no. 2, pp. 1141–160 (2008)

Altay, E., Hakan Satman, M.: Stock market forecasting: artificial neural network and linear regression comparison in an emerging market. J. Finan. Manage. Anal. **18**(2), 18 (2005)

Anish, C.M., Majhi, B.: Hybrid nonlinear adaptive scheme for stock market prediction using feedback FLANN and factor analysis. J. Korean Stat. Soc. **45**(1), 64–76 (2016)

Ardia, D., Boudt, K.: Testing equality of modified Sharpe ratios. Finan. Res. Lett. **13**, 97–104 (2015)

Avci, E.: Forecasting Daily and Sessional Returns of the ISE-100 Index with Neural Networks Models, vol. 8, no. 2, pp. 128–142 (2007)

Babalos, V., Caporale, G.M., et al.: Testing for persistence in mutual fund performance and the ex-post verification problem: evidence from the Greek market. Eur. J. Finan. **14**(8), 735–753 (2008)

Babalos, V., Philippas, N., et al.: Mutual funds performance appraisal using stochastic multicriteria acceptability analysis. Appl. Math. Comput. **218**(9), 5693–5703 (2012)

Bacon, C.: How sharp is the Sharpe-ratio?-Risk-adjusted Performance Measures (2009). www.statpro.com. Accessed 4 Jan 2021

Baks, K.P., Metrick, A., Wachter, J.A.: Should investors avoid all actively managed mutual funds? a study in Bayesian performance evaluation. J. Finan. **56**(1), 45–85 (2001)

Barras, L., Scaillet, O., Wermers, R.: False discoveries in mutual fund performance: measuring luck in estimated alphas. J. Finan. **65**(1), 179–216 (2010)
Benli, Y.K.: Bankalarda mali başarisizliğin öngörülmesi lojistik regresyon ve yapay sinir aği karşilaştirmasi". In: Gazi Üniversitesi Endüstriyel Sanatlar Eğitim Fakültesi Dergisi, vol. 16, pp. 31–46 (2005)
Berk, J.B., van Binsbergen, J.H.: Measuring managerial skill in the mutual fund industry. In: National Bureau of Economic Research (2012)
Blake, D.P., et al.: New evidence on mutual fund performance: a comparison of alternative bootstrap methods. J. Finan. Quant. Anal. **52**(3), 1279–1299 (2017)
Bollen, N.P.B., Busse, J.A.: On the timing ability of mutual fund managers. J. Finan. **56**(3), 1075–1094 (2001)
Canbaş, S., Kandir, Y.: Türkiye'deki Yatirim Fonlarinin Performans Değerlendirmesi. In: Iktisat Isletme Ve Finans, vol. 17, no. 201, pp. 13–19 (2002)
Caporin, M., et al.: A survey on the four families of performance measures. J. Econ. Surv. **28**(5), 917–942 (2014)
Carhart, M.: On persistence in mutual fund performance. J. Finan. **52**(1), 57–82 (1997)
Cesari, R., Panetta, F.: The performance of Italian equity funds. J. Bank. Finan. **26**(1), 99–126 (2002)
Chen, A.N.K., Wang, S.-Y., Yu, P.-L.: Evaluating multi-criteria ratings of financial investment options. Int. Rev. Econ. Finan. **31**, 46–58 (2014)
Chen, C.R., Huang, Y.: Mutual fund governance and performance: a quantile regression analysis of Morningstar's Stewardship Grade. Corpor. Gover. Int. Rev. **19**(4), 311–333 (2011)
Cheng, T., Yan, C.: Evaluating the size of the bootstrap method for fund performance evaluation. Econ. Lett. **156**, 36–41 (2017)
Chiang, W.-C., Urban, T.L., Baldridge, G.W.: A neural network approach to mutual fund net asset value forecasting. Omega **24**(2), 205–215 (1996)
Christopherson, J.A., Ferson, W.E., Glassman, D.A.: Conditional measures of performance and persistence for pension funds. Res. Finan. **16**, 1–46 (1998)
Cogneau, P., Hübner, G.: The 101 ways to measure portfolio performance. In: Available at SSRN 1326076 (2009)
Cremers, K.J.M., Petajisto, A.: How active is your fund manager? a new measure that predicts performance. Rev. Finan. Stud. **22**(9), 3329–3365 (2009)
Del Guercio, D., Tkac, P.A.: Star power: The effect of morningstar ratings on mutual fund flow. J. Finan. Quant. Anal. **43**(4), 907–936 (2008)
Diler, A.İ.: İMKB Ulusal-100 Endeksinin Yönünün Yapay Sinir Ağlari Hata Geriye Yayma Yöntemi ile Tahmin Edilmesi, vol. 25, no. 26, pp. 65–81 (2003)
Droms, W.G., Walker, D.A.: Performance persistence of fixed income mutual funds. J. Econ. Finan. **30**(3), 347–355 (2006)
Dutta, G., et al.: Artificial neural network models for forecasting stock price index in the Bombay stock exchange. J. Emerg. Mark. Finan. **5**(3), 283–295 (2006)
Eling, M., Schuhmacher, F.: Does the choice of performance measure influence the evaluation of hedge funds. J. Bank. Finan. **31**, 2632–2647 (2007). https://doi.org/10.1016/j.jbankfin.2006.09.015
Elton, E.J., Gruber, M.J., Blake, C.R.: The persistence of risk-adjusted mutual fund performance. J. Bus. 133–157 (1996)
Fama, E.F., French, K.R.: Multifactor explanations of asset pricing anomalies. J. Finan. **51**(1), 55–84 (1996)
Ferreira, M.A., et al.: The flow-performance relationship around the world. J. Bank. Finan. **36**(6), 1759–1780 (2012)

Ferson, W.E., Schadt, R.W.: Measuring fund strategy and performance in changing economic conditions. J. Finan. **51**(2), 425–461 (1996)
Ferson, W.E., Warther, V.A.: Evaluating fund performance in a dynamic market. Finan. Anal. J. **52**(6), 20–28 (1996)
Godefroid, P., Peleg, H., Singh, R.: Learn&fuzz: Machine learning for input fuzzing. In: Proceedings of the 32nd IEEE/ACM International Conference on Automated Software Engineering, pp. 50–59 . IEEE Press (2017)
Goetzmann, W.N., Ibbotson, R.G.: Do winners repeat? J. Portf. Manage. **20**(2), 9 (1994)
Gökçen, U., Yalçin, A.: The case against active pension funds: evidence from the Turkish private pension system. Emerg. Mark. Rev. **23**, 46–67 (2015)
Grinblatt, M., Titman, S.: The persistence of mutual fund performance. J. Finan. **47**(5), 1977–1984 (1992)
Gruber, M.J.: Another puzzle: the growth in actively managed mutual funds. J. Finan. **51**(3), 783–810 (1996)
Gürsoy, C.T., Erzurumlu, Y.Ö.: Evaluation of portfolio performance of Turkish investment funds. Doğus Üniversitesi Dergisi **2**(2), 43–58 (2001)
Hammami, Y., Oueslati, A.: Measuring skill in the Islamic mutual fund industry: evidence from GCC countries. J. Int. Finan. Mark. Inst. Money **49**, 15–31 (2017)
Hendricks, D., Patel, J., Zeckhauser, R.: Hot hands in mutual funds: short-run persistence of relative performance, 1974–1988. J. Finan. **48**(1), 93–130 (1993)
Henriksson, R.D., Merton, R.C.: On market timing and investment performance. II. Statistical procedures for evaluating forecasting skills. J. Bus. **54**(4), 513–533 (1981)
Ippolito, R.A.: On studies of mutual fund performance, 1962–1991. Finan. Anal. J. **49**(1), 42–50 (1993)
Jensen, M.C.: The performance of mutual funds in the period 1945–1964. J. Finan. **23**(2), 389–416 (1968)
Karacabey, A.A.: A Tipi Hisse Senedi Fonlarinin Performanslarinin Değerlendirilmesi, vol. 2, pp. 84–99 (2002)
Karaçor, Z., Alptekin, V.: Finansal Krizlerin Önceden Tahmin Yoluyla Değerlendirilmesi: Türkiye Örneği, vol. 13, no. 2, pp. 237–256 (2006)
Korkmaz, T., Uygurtürk, H.: Türkiye'deki Emeklilik Fonlari ile Yatirim Fonlarinin Performans Karşilaştirmasi ve Fon Yöneticilerinin Zamanlama Yetenekleri. In: Kocaeli Üniversitesi Sosyal Bilimler Enstitüsü Dergisi, vol. 15, pp. 114–147 (2008)
Kosowski, R., et al.: Can mutual fund 'stars' really pick stocks? new evidence from a bootstrap analysis. J. Finan. **61**(6), 2551–2595 (2006)
Kourtis, A.: The sharpe ratio of estimated efficient portfolios. Finan. Res. Lett. **17**, 72–78 (2016)
Kurun, E., et al.: Yatirim Fonlarinin Performans Analizinde Kullanilan Risk Bazli Ölçüm Teknikleri ve Türk Yatirim Fonlari Üzerinde Bir Uygulama. In: Küresel Dalgalanmalar ve Finans Sektörüne Yansimalari Bildiriler Kitabi. Istanbul, Marmara Üniversitesi Bankacilik ve Sigortacilik Yüksekokulu (2008)
Lisi, F., Caporin, M.: On the role of risk in the Morningstar rating for mutual funds. Quant. Finan. **12**(10), 1477–1486 (2012)
Liu, Y., Li, Y.-N.: A parametric sharpe ratio optimization approach for fuzzy portfolio selection problem. In: Mathematical Problems in Engineering 2017 (2017)
Malkiel, B.G.: Returns from investing in equity mutual funds 1971 to 1991. J. Finan. **50**(2), 549–572 (1995)
Markowitz, H.: Portfolio selection. J. Finan. **7**(1), 77–91 (1952)
Minsky, M., Papert, S.: An introduction to computational geometry. In: Cambridge tiass., HIT (1969)

Müller, S., Weber, M.: Evaluating the rating of stiftung warentest: how good are mutual fund ratings and can they be improved? Eur. Finan. Manage. **20**(2), 207–235 (2014)
Nguyen, T.T., et al.: Fuzzy portfolio allocation models through a new risk measure and fuzzy sharpe ratio. IEEE Trans. Fuzzy Syst. **23**(3), 656–676 (2015)
Panda, C., Narasimhan, V.: Forecasting exchange rate better with artificial neural network. J. Pol. Model. **29**(2), 227–236 (2007)
Pastor, L., Stambaugh, R.F.: Mutual fund performance and seemingly unrelated assets. J. Finan. Econ. **63**(3), 315–349 (2002)
Qian, B., Rasheed, K.: Foreign exchange market prediction with multiple classifiers. J. Forecast. **29**(3), 271–284 (2010)
Ray, P., Vina, V.: Neural network models for forecasting mutual fund net asset value. In: 8th Capital Markets Conference, Indian Institute of Capital Markets Paper (2004)
Roll, R.: A critique of the asset pricing theory's tests part I: on past and potential testability of the theory. J. Finan. Econ. **4**(2), 129–17 (1977)
Roll, R.: Ambiguity when performance is measured by the securities market line. J. Finan. **33**(4), 1051–1069 (1978)
Sahoo, P., Hathy, P.: Prediction of Mutual Funds: Use of Neural Network Technique–The Icfai Journal of Applied Finance (2007)
Saim, K.: Türkiye'deki yatirim fonlarinin performanslarinin değerlendirilmesi. IMKB Yayinlari, Istanbul (2002)
Sharpe, W.F.: Mutual fund performance. J. Bus. **39**(1), 119–138 (1966)
TEFAS. Türkiye Elektronik Fon Alim Satim Platformu (2019). https://www.tefas.gov.tr/fonkarsilastirma.aspx. Accessed 13 Sep 2019
Teker, S., Karakurum, E., Tav, O.: Yatirim fonlarinin risk odakli performans degerlemesi (2008)
Treynor, J.L.: How to rate management of investment funds. Harvard Bus. Rev. **43**(1), 63–75 (1965)
Treynor, J.L., Mazuy, K.: Can mutual funds outguess the market? Harvard Bus. Rev. **44** (1966)
Tseng, C.-H., et al.: Artificial neural network model of the hybrid EGARCH volatility of the Taiwan stock index option prices. Physica A Stat. Mech. Appl. **387**(13), 3192–3200 (2008)
Wang, G., et al.: A new approach to intrusion detection using Artificial Neural Networks and fuzzy clustering. Expert Syst. Appl. **37**(9), 6225–6232 (2010)
Wang, K., Huang, S.: Using fast adaptive neural network classifier for mutual fund performance evaluation. Expert Syst. Appl. **37**(8), 6007–6011 (2010). https://doi.org/10.1016/j.eswa.2010.02.003
Werbos, P.J.: Generalization of backpropagation with application to a recurrent gas market model. Neural Networks **1**(4), 339–356 (1988)
Xu, L., et al.: Neural mechanism of intertemporal choice: from discounting future gains to future losses. Brain Res. **1261**, 65–74 (2009)
Yao, X., Liu, Y.: Ensemble structure of evolutionary artificial neural networks. In: Proceedings of IEEE International Conference on Evolutionary Computation, pp. 659–664. IEEE (1996)
Zhang, G., et al.: Artificial neural networks in bankruptcy prediction: General framework and cross-validation analysis. Eur. J. Oper. Res. **116**(1), 16–32 (1999)
Zhang, L., Priestley, J., Ni, X.: Comparison of Bankruptcy Prediction Models with Public Records and Firmographics, pp. 97–109 (2018). https://doi.org/10.5121/csit.2018.80309
Zheng, L.: Is money smart? a study of mutual fund investors' fund selection ability. J. Finan. **54**(3), 901–933 (1999)

Fair Allocation of Fixed Costs in Data Envelopment Analysis

Farhad Hosseinzadeh Lotfi[1], Mohsen Rostamy-Malkhalifeh[1], Mohamad Reza Mozaffari[2], Mohammad Hasan Behzadi[1], and Marzieh Ghasemi[1](✉)

[1] Science and Research Branch, Islamic Azad University, Tehran, Iran
Marziehghasemi2000@gmail.com
[2] Islamic Azad University, Shiraz Branch, Shiraz, Iran

Abstract. Data Envelopment Analysis (DEA) technique is a mathematical programming to evaluate the performance of a set of homogeneous decision making units (DMUs) with multiple inputs and outputs. One of DEA's applications is to allocate the fixed resources and to target among DMUs in a fair way. In this manuscript, a novel and practical model is developed to allocate the fixed costs. In the proposed method, each DMU first determines both the minimum and maximum values that can participate in the payment of the fixed costs so that the efficiency of itself and other units after the allocation of fixed costs is equal to one of the efficiency before the allocation of the fixed costs. Then, a convex combination of the minimum and maximum values is considered as the amount of the allocated cost to each unit. Ultimately, both the efficiency and applicability of the proposed method are investigated using a numerical example.

Keywords: Data Envelopment Analysis (DEA) · Fixed costs · Fair allocation

1 Introduction

Data Envelopment Analysis (DEA) is firstly developed by Charnes et al. [1] in 1978, by proposing the CCR model. Banker et al. [2] extended this method by proposing the BCC model. DEA is a so useful method to evaluate the performance and to obtain the patterns of different DMUs with multiple inputs and output. One of the important applications of DEA is to allocate the fixed costs or fixed resources, and target setting among DMUs in a fair way.

In today's competitive world, the subsets of an organization require to corporate with each other to reduce their costs. The used costs in common infrastructure are called the fixed costs.

Cook and Kress [3] are the first researchers who provide the fixed costs allocation idea in DEA in a fair way according to two following principles, invariance and pareto-minimality input. Jahanshahloo et al. [4], with a simple example, demonstrated that the pareto-minimality input condition does not satisfy for the cost allocation of Cook and Cress [3]. Cook and Zhu [5] extended the Cook and Cress [3] approach from the input-oriented nature to the output-oriented nature, based on Cook and Kress's invariance, and

T. Allahviranloo et al. (Eds.): IDS 2020, AISC 1301, pp. 399–405, 2021.
https://doi.org/10.1007/978-3-030-66501-2_32

then they proposed the cost-effective allocations (but not optimal). Lin [6], with the same assumptions and considering the cost allocation as a new input, found that if some of the specific constraints are added to the Cook and Zhu [5] approach, they would probably have no solutions, and their approach is improved by setting fixed objectives to achieve the cost allocation. Li et al. [7] provided another way to achieve a unique allocation. Du [8] utilized the concept of cross-efficiency in DEA to demonstrate the issue of resource allocation and fixed costs. It is shown in this paper, the proposed method is always available and ensures that all DMUs are efficient after being allocation a fixed charge as an additional input.

Beasley [9] proposed a non-linear cost allocation approach based on DEA, by maximizing the average efficiency of all DMUs and by adding additional constraints to obtain a unique cost allocation. Amirteimoori and Kordrostami [10] showed that the Beasley's method is infeasible in many examples, and then they proposed a method based on DEA for allocating the fixed costs to DMUs, allocating fixed input, and target setting. Jahanshahloo et al. [11] indicated that the methodology that Amirteimoori and Kordrostami [10] proved is infeasible, is always feasible, and the invariance principle is not necessarily established in their method. Moreover, they described two methods for the fair allocation of the fixed costs, according to the invariance and the common set of weights. Furthermore, Li et al. [12] developed a model for fair fixed resources allocation and target setting, by utilizing the invariance and common set of weights principles.

Sadeghi et al. [13] proposed two novel methods for centralized resource allocation. The first method involved maximizing the outputs and eliminating the input inefficiencies to the extent possible with the aim of achieving strong efficiency in all DMUs, while the second method optimized the total cost and revenue functions in order to reach maximum efficiency. It was assumed that all units were able to adjust their inputs and outputs in the production possibility set under study. Zhou et al. [14] used a two-stage model for resource allocation based on the assumptions of zero-sum gain and fixed-sum gain. In this respect, they divided the inputs into the two categories of discretionary and non-discretionary inputs and developed two resource allocation models based on environmental efficiency. Shi et al. [15] presented a fixed cost allocation model for two-stage systems using the game theory and DEA, in which each DMU can maximize its relative efficiency after the cost allocation. Hosseinzadeh Lotfi et al. [16] used a CSW model to allocate the budgets of Iranian natural gas companies. Moreover, Khalili et al. [17] conducted a study on resource allocation and target setting via DEA in the natural gas supply chain in Iran.

The allocation models mentioned above can be carried out in a variety of applications, focusing on different key points. Decision makers can choose either a cost allocation or fixed resource model with regard to a particular application context. In these mentioned methods, the comments of the Decision Making Units (DMUs) have not been considered on how much they are interested in participating to the fixed costs. Here, it is allowed to each unit to specify both the minimum and maximum values that can participate in the payment of costs, so that the efficiency of itself and other units after the allocation of fixed costs is equal to one of the efficiency before the allocation of the fixed costs. Afterwards, a convex combination of the maximum and minimum values is considered as the amount of the cost allocated to each unit. This paper is organized as follows:

The section two provides a brief background on DEA. Section three introduces our proposed method. Section four presents a numerical example to illustrate the application of the proposed method. Finally, section five concludes the manuscript and then suggests the future research.

2 Background

In this section, we describe the multiplier CCR model. Suppose $DMU_j(j = 1, \ldots, n)$ is a set of homogeneous DMUs that consume m inputs, $x_{ij}(i = 1, \ldots, m)$, to produce s outputs, $y_{rj}(r = 1, \ldots, s)$. If DMU_o is the unit under evaluation, its efficiency is measured using the following model, known as the multiplier CCR model (Charnes et al., 1978).

$$\begin{aligned} &\theta_o^* = \max \sum_{r=1}^{s} u_r y_{ro} \\ &s.t: \sum_{i=1}^{m} v_i x_{io} = 1 \\ &\sum_{r=1}^{s} u_r y_{rj} - \sum_{i=1}^{m} v_i x_{ij} \le 0 \; lj = 1, \ldots, n \\ &u_r, v_i \ge 0 \qquad r = 1, \ldots, s \;\; i = 1, \ldots, m \end{aligned} \tag{1}$$

u_r and v_i indicate the relative importance of the output and input vectors, respectively.

2.1 Definition

DMU_o is efficient if and only if $\theta_o^* = 1$.

3 Fixed Costs Allocation

Today's, in many organizations such as industry, banking, academia, etc., there is a general fixed cost that the organizations intend to allocate this fixed cost to each of its subsets in an appropriate way. Assume that $DMU_j(j = 1, \ldots, n)$ produces the s outputs $y_{rj}(r = 1, \ldots, s)$ by consuming the m inputs $x_{ij}(i = 1, \ldots, m)$. In addition, suppose an organization has a fixed costs F and intends to distribute this fixed costs among its subsidiaries equitably. Also, assume that $f_j(j = 1, \ldots, n)$ is variable corresponding to the fixed costs allocated to DMU_j. It is obvious that:

$$\sum_{j=1}^{n} f_j = F, \tag{2}$$

suppose θ_j^* is the efficiency score of DMU_j before the cost allocation, which is measured using the *CCR* multiplicative model. To do so, the fixed cost is considered as an additional input for simplicity with a weight of one. In this manuscript, a novel model is developed so that each DMU contains both maximum and minimum values that can participate in the payment of fixed costs, so that the efficiency of itself and other units after the allocation of fixed costs is equal to one of the efficiency before the allocation of the fixed costs.

$$\frac{\sum_{r=1}^{s} u_r y_{rj}}{\sum_{i=1}^{m} v_i x_{ij} + f_j} = \theta_j^* \quad j = 1, \ldots, n$$

We have $\theta_j^* > 0$ and

$$f_j = \frac{1}{\theta_j^*} \sum_{r=1}^{s} u_r y_{rj} - \sum_{i=1}^{m} v_i x_{ij} \quad j = 1, \ldots, n$$

Based on these explanations, the general model can be formulated as follows:

$$\begin{aligned} & Min\ or\ Max\, f_o \\ & \quad s.t: \\ & \quad f_j = \tfrac{1}{\theta_j^*} \sum_{r=1}^{s} u_r y_{rj} - \sum_{i=1}^{m} v_i x_{ij} \qquad j = 1, \ldots, n \\ & \quad \sum_{j=1}^{n} f_j = F \qquad j = 1, \ldots, n \\ & \quad f_j \geq 0, u_r, v_i \geq 0 \quad j = 1, \ldots, n \quad r = 1, \ldots, s \quad i = 1, \ldots, m \end{aligned} \tag{3}$$

In other words, the values f_j^{min} and f_j^{max} for all units are calculated using model 3. In this way, f_j is measured for each unit as a convex combination of maximum and minimum values as follows.

$$\begin{aligned} & f_j = \mu_j f_j^{min} + (1 - \mu_j) f_j^{max} \qquad 0 \leq \mu_j \leq 1 \quad j = 1, \ldots, n \\ & \text{By setting } \mu_1 = \mu_2 = \cdots = \mu_n \\ & f_j = \mu f_j^{min} + (1 - \mu) f_j^{max} \qquad 0 \leq \mu \leq 1 \quad j = 1, \ldots, n \\ & \sum_{j=1}^{n} f_j = F \rightarrow \sum_{j=1}^{n} \mu f_j^{min} + (1 - \mu) f_j^{max} = F \rightarrow \mu \sum_{j=1}^{n} f_j^{min} + (1 - \mu) \sum_{j=1}^{n} f_j^{max} = F \\ & \rightarrow \mu = \frac{F - \sum_{j=1}^{n} f_j^{max}}{\sum_{j=1}^{n} f_j^{min} - \sum_{j=1}^{n} f_j^{max}} \end{aligned} \tag{4}$$

Now, the f_j values are obtained for all units by calculating the value of μ by the above equation and using Eq. (4).

4 Numerical Example

In this matter, the data used in [3] are exploited to demonstrate the application of the proposed model in the allocation of fixed costs. Consider the 12 DMUs with three inputs and two outputs reported in Table 1. Suppose there is a fixed cost of F = 100 that should be allocated among them (12 DMUs).

Table 1. A simple example from Cook and Kress (1999).

DMU	x_1	x_2	x_3	y_1	y_2
1	350	39	9	67	751
2	298	26	8	73	611
3	422	31	7	75	584
4	281	16	9	70	665
5	301	16	6	75	445
6	360	29	17	83	1070
7	540	18	10	72	457
8	276	33	5	78	590
9	323	25	5	75	1074
10	444	64	6	74	1072
11	323	25	5	25	350
12	444	64	6	10	1199

By implementing the model 2, the efficiency of each DMU is calculated before allocating a fixed cost, as listed in the second column of Table 2. Regarding the model 3, each unit offers both the minimum and maximum values that can participate in the payment of costs. The achieved results of the implementation of the model 3 using GAMS software are provided in the third column of Table 2.

Table 2. Result for fixed cost allocation.

DMU	θ	$\left[f_j^{Min}, f_j^{Max}\right]$	f_j
1	0.757	[4.9881,10.612]	7.306
2	0.92	[2.935,11.569]	6.494
3	0.747	[4.854,15.588]	5.157
4	1.000	[2.645,12.550]	6.728
5	1.000	[2.236, 13.515]	6.886
6	0.961	[0,19.474]	8.027
7	0.860	[0, 12.546]	5.171
8	1.000	[0, 12.971]	5.347
9	1.000	[7.467, 31.730]	17.467
10	0.832	[0, 17.780]	7.329
11	0.333	[7.475, 30.608]	17.01
12	1.000	[0,17.172]	7.078

For instance, the minimum and maximum values that DMU9 offers 7.467 and 31.730, respectively to participate in the payment of costs. The value of cost per unit is calculated using Eq. (4) using Excel software, which is located in the fourth column of the table. Here, the fixed cost to be paid by DMU9 is 17,467, which is within the offered range. On the other hand, the maximum value offered for participation in the minimum and maximum payments is related to unit 9, which is an efficient unit. In the third column of this Table, where the allocated costs to each unit are visible, the lowest cost is allocated to unit 3, which is an inefficient unit. It is worthwhile to mention that the highest payment is for unit 9, which is an efficient unit.

5 Conclusion

The fixed cost allocation is a critical issue in the DEA. This paper developed a DEA-based approach to the fair allocation of fixed costs among a set of DMUs. As the fixed cost was the cost imposed by an organization on its subsets, this paper proposed a model that allowed each unit to contain both the minimum and maximum value that it could participate in the payment of costs, so that it could determine the efficiency of itself and other units after allocation, which was equal to the efficiency before allocation. Afterwards, the organization considered a convex combination of the offered by each unit, represented as the fixed cost value allocated to each unit. In this model, units were allowed to set a minimum value of zero as the suggested value for the minimum participation. Ultimately, for future research, a lower bound might be set for the fixed cost allocation so that all units could offer a minimum non-zero value for the minimum participation. This approach was demonstrated using the numerical example provided in Cook and Cross [3].

References

1. Charnes, A., Cooper, W.W., Rhodes, E.: Measuring the efficiency of decision making units. Eur. J. Oper. Res. **2**, 429–444 (1978)
2. Banker, R.D., Charnes, A., Cooper, W.W.: Some models for estimating technical and scale inefficiencies in data envelopment analysis. Manage. Sci. **30**, 1078–1092 (1984)
3. Cook, W.D., Kress, M.: Characterizing an equitable allocation of shared costs: a DEA approach. Eur. J. Oper. Res. **119**, 652–661 (1999)
4. Jahanshahloo, G.R., Hosseinzadeh Lotfi, F., Shoja, N., Sanei, M.: An alternative approach for equitable allocation of shared costs by using DEA. Appl. Math. Comput. **153**, 267–274 (2004)
5. Cook, W.D., Zhu, J.: Allocation of shared costs among decision making units: a DEA approach. Comput. Oper. Res. **32**, 2171–2178 (2005)
6. Lin, R.: Allocating fixed costs or resources and setting targets via data envelopment analysis. Appl. Math. Comput. **217**, 6349–6358 (2011)
7. Li, Y., Yang, F., Liang, L., Hua, Z.: Allocating the fixed cost as a complement of other cost inputs: a DEA approach. Eur. J. Oper. Res. **197**, 389–401 (2009)
8. Du, J., Cook, W.D., Liang, L., Zhu, J.: Fixed cost and resource allocation based on DEA cross-efficiency. Eur. J. Oper. Res. **235**, 206–214 (2014)
9. Beasley, J.E.: Allocating fixed costs and resources via data envelopment analysis. Eur. J. Oper. Res. **147**, 198–216 (2003)

10. Amirteimoori, A., Kordrostami, S.: Allocating fixed costs and target setting: a dea-based approach. Appl. Math. Comput. **171**, 136–151 (2005)
11. Reza Jahanshahloo, G., Sadeghi, J., Khodabakhshi, M.: Proposing a method for fixed cost allocation using DEA based on the efficiency invariance and common set of weights principles. Math. Meth. Oper. Res. **85**, 4982–4997 (2017)
12. Li, F., Song, J., Dolgui, A., Liang, L.: Using common weights and efficiency invariance principles for resource allocation and target setting. Int. J. Prod. Res. **55**, 4982–4997 (2017)
13. Sadeghi, J., Dehnokhalaji, A.: A comprehensive method for the centralized resource allocation in DEA. Comput. Ind. Eng. **127**, 344–352 (2019)
14. Zhou, Z., Wu, Q., Zhang, J.: A two-stage DEA model for resource allocation in industrial pollution treatment and its application in China. Expert Syst. Appl. **228**, 29–39 (2019)
15. Shi, S., Wang, P., An, Q.: Fixed cost allocation for two-stage systems with cooperative relationship using data envelopment analysis. Comput. Ind. Eng. **145** (2020)
16. Hosseinzadeh Lotfi, F., Ghazi, A.: Assessment and budget allocation of Iranian natural gas distribution company- A CSW DEA based model. Socio Econ. Plann. Sci. **66**, 112–118 (2019)
17. Khalili-Damghani, K., Sharahi, S.: Fuzzy type-II De-Novo programming for resource allocation and target setting in network data envelopment analysis: a natural gas supply chain. Expert Syst. Appl. **117**, 312–329 (2019)

Fuzzy Flower Pollination Algorithm with Chaos for Global Optimization

İlker Gölcük[1](✉) and Fehmi Burcin Ozsoydan[2]

[1] Department of Industrial Engineering, İzmir Bakırçay University, 35665 İzmir, Turkey
ilker.golcuk@bakircay.edu.tr

[2] Department of Industrial Engineering, Dokuz Eylül University, 35397 İzmir, Turkey
burcin.ozsoydan@deu.edu.tr

Abstract. This paper is concerned with development of a hybrid flower pollination algorithm by combining fuzzy logic and chaos theory. Flower pollination algorithm is one of the recently developed algorithms which mimics the pollination process of the flowers observed in nature. Flower pollination algorithm has two main procedures which are global and local pollination, respectively. These two mechanisms are guided by a switching probability which is a user-defined parameter. Setting a proper value for switching probability is a challenging issue. In this paper, fuzzy inference system is used to control switching probability by considering generation number and population diversity. Furthermore, chaotic numbers are utilized to enhance search capability of the algorithm. All of the proposed modifications are tested on unconstrained function minimization problems and non-parametric statistical tests are used to validate improvements over canonical flower pollination algorithm.

Keywords: Flower Pollination Algorithm · Fuzzy inference system · Chaotic numbers · Global optimization

1 Introduction

Optimization is defined as a process of finding the best combination of variables to minimize or maximize an objective function. Optimization problems are ubiquitous in the real-world, and many state-of-the-art methods have been proposed to deal with different problem characteristics. Traditional optimization routines are inadequate to solve real-world optimization problems in a reasonable time with acceptable accuracy. Therefore, nature-inspired algorithms are widely used for modeling and solving real-world problems. Flower Pollination Algorithm (FPA) [1] is one of the bio-inspired optimization algorithms, which is inspired by self- and cross-pollination of flowers.

FPA has been used to solve a wide array of optimization problems in the literature. Benkercha, Moulahoum, and Kabache [2] combined artificial neural networks and FPA for the training of fuzzy logic controller. Gölcük, Baykasoğlu, and Özsoydan [3] employed FPA for cluster analysis. Dahi, Mezioud, and Draa [4] used binary FPA for solving antenna positioning problems. Fuzzy logic was also integrated into the canonical FPA algorithm in the literature. Carreon, Valdez, and Castillo [5] used fuzzy FPA to

T. Allahviranloo et al. (Eds.): IDS 2020, AISC 1301, pp. 406–414, 2021.
https://doi.org/10.1007/978-3-030-66501-2_33

solve control problems. In the study, switching probability and ε (change in the error) parameters are tuned by fuzzy logic. Valenzuela, Valdez, and Melin [6] used fuzzy logic for controlling the switching probability of FPA for solving real-valued function optimization.

The present study contributes to the literature by combining chaotic numbers and fuzzy logic for improving the performance of canonical FPA. Furthermore, fuzzy logic and chaotic enhanced FPA updates the switching probability by considering both the number of iterations and the population diversity within the fuzzy inference system.

The remainder of the paper is as follows: Sect. 2 gives the basics of FPA. Section 3 presents the proposed FPA algorithm. Section 4 is dedicated to experimental results. Concluding remarks are given in Sect. 5.

2 Flower Pollination Algorithm

In this section, the fundamental components of the proposed FPA are introduced.

The Flower Pollination Algorithm (FPA) was proposed by Yang [1]. The main inspiration of the FPA is the pollination process carried out by flowers in nature. There are two types of pollinations, namely self-pollination, and cross-pollination. Self-pollination states that the same flower or plant pollinates the flowers. This situation is observed in nature when the flower encompasses both the male and female gametes. On the other hand, cross-pollination comes to mean that pollen grains are moved by insects, birds, or other animals, and the pollination happens with the help of abiotic or biotic agents.

Flowering plants obey the following four rules:

Rule 1: Cross-pollination is modeled as a process of global search so that the pollen carrying pollinators obey Lévy flights.

Rule 2: Self-pollination represents local search in the optimization domain.

Rule 3: Pollen carrying insects may develop flower constancy that is represented as a probability proportional to the similarity of two flowers involved.

Rule 4: the interplay between local and global search is controlled by a switching probability. The switching probability is represented by $p \in [0, 1][0, 1]$.

The rule 1 and rule 3 can be mathematically represented as:

$$\boldsymbol{x}_i^{t+1} = \boldsymbol{x}_i^t + \gamma L(\lambda)\left(\boldsymbol{g}^* - \boldsymbol{x}_i^t\right) \tag{1}$$

where $\boldsymbol{x}_i^t$ is the solution vector of the i-th individual at iteration t, γ is a scaling factor, and $\boldsymbol{g}^*$ is the incumbent solution.

The step-size is adopted from Levy flight, which implies strength of the pollination. The Levy distribution is given by:

$$L \sim \frac{\lambda\Gamma(\lambda)\sin(\pi/2)}{\pi}\frac{1}{s^{1+\lambda}}, \quad (s \gg s_0 > 0) \tag{2}$$

where $\Gamma(\lambda)$ represents standard gamma function. To mimic the Levy distribution, Mantegna algorithms is employed where the two Gaussian distributions U and V are calculated by the following:

$$s = \frac{U}{|V|^{1/\lambda}}, \quad U \sim N\left(0, \sigma^2\right), \quad V \sim N(0, 1) \tag{3}$$

where the standard deviation of the Gaussian normal distribution is calculated as:

$$\sigma^2 = \left[\frac{\Gamma(1+\lambda)}{\lambda\Gamma\left(\frac{1+\lambda}{2}\right)} \cdot \frac{\sin(\pi\lambda/2)}{2^{(\lambda-1)/2}} \right]^{1-\lambda} \tag{4}$$

The local pollination is calculated as:

$$\boldsymbol{x}_i^{t+1} = \boldsymbol{x}_i^t + \in \left(\boldsymbol{x}_j^t - \boldsymbol{x}_k^t \right) \tag{5}$$

where $\in$ is a uniform random number.

The switch probability p is used to control global and local optimization phases, respectively. If a uniform random number is smaller or equal to p, then the algorithm undergoes global optimization through Levy flight. Otherwise, the algorithm performs a local search.

3 Proposed Fuzzy Flower Pollination Algorithm

The proposed fuzzy FPA makes use of chaotic numbers instead of random numbers in all phases of the canonical FPA. Additionally, the Mamdani-type fuzzy inference system is used to adjust the switching probability dynamically.

3.1 Chaotic Maps

In this study, the most widely used chaotic maps are used to improve the search performance of the FPA algorithm. The employed chaotic maps are circle map, iterative map, logistic map, sine map, and singer map. Furthermore, the uniform distribution is used to analyze the effect of the chaotic maps so that the uniform random numbers are used as the sixth generator within the proposed algorithm. Mathematical formulations of the chaotic maps are given in Table 1. The generated number numbers by using chaotic maps and the uniform random number are depicted in Fig. 1.

3.2 Fuzzy Rule-Based System

A fuzzy rule-based system is used to tune the switching probability parameter of the FPA algorithm. The inputs of the fuzzy rule base are iterations and diversity. The iteration and diversity measures are calculated as a percentage in order to define corresponding membership functions more accurately. The iteration is calculated as:

$$iteration = \frac{current\ iteration}{total\ number\ of\ iterations} \tag{6}$$

Table 1. Chaotic maps and their mathematical representations

Chaotic Map	Formulation
Circle map [7]	$x_{i+1} = x_i + b - (a - 2\pi)\sin(2\pi x_i) \bmod (1)$ $a = 0.5$ and $b = 0.2$
Iterative map [8]	$x_{i+1} = \sin\left(\frac{a\pi}{x_i}\right)$ $a = 0.7$
Logistic map [9]	$x_{i+1} = ax_i(1 - x_i)$ $a = 0.7$
Sine map [10]	$x_{i+1} = \frac{a}{4}\sin(\pi x_i)$ $a = 4$
Singer map [11]	$x_{i+1} = \mu\left(7.86x_i - 23.31x_i^2 + 28.75x_i^3 - 13.302875x_i^4\right)$ $a = 4$

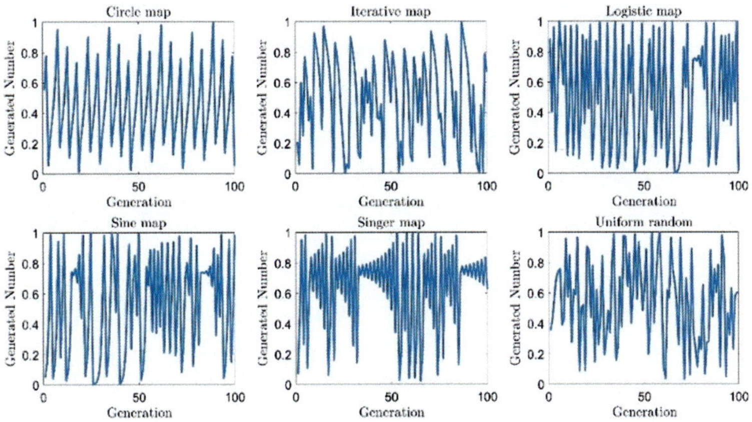

Fig. 1. Generated random numbers within fuzzy FPA algorithm

Additionally, population diversity is calculated as:

$$dist_i(t) = \sqrt{\sum_{j=1}^{nDim}\left(x_{i,j}(t) - x_{best,j}(t)\right)^2} \tag{7}$$

$$diversity(t) = \begin{cases} \frac{\frac{1}{nPop}\sum_{i=1}^{nPop} dist_i(t) - \min\{dist_i(t)\}}{\max\{dist_i(t)\} - \min\{dist_i(t)\}}, & \max\{dist_i(t)\} \neq \min\{dist_i(t)\} \\ 0, & otherwise \end{cases} \quad (8)$$

where $dist_i(t)$ is the distance between i-th individual and the incumbent solution. Based on the iteration and diversity measures, fuzzy membership functions, and fuzzy rules are generated, as given in Fig. 2 and 3, respectively.

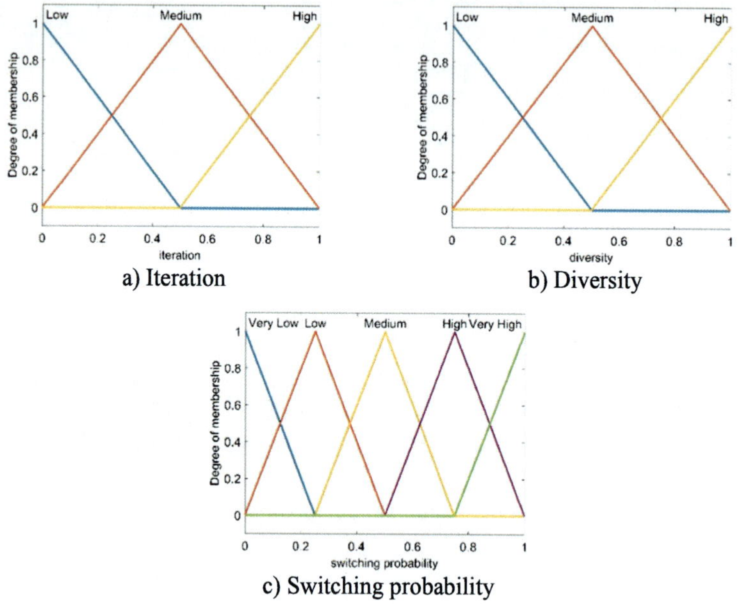

Fig. 2. Membership functions of the fuzzy rule-based system

4 Experimental Study

In this section, seven real-valued function optimization problems are used to test the performance of the proposed methods. These functions are Sphere (F1), Schwefel 2.22 (F2), Schwefel 1.2 (F3), Schwefel 2.21 (F4), Rosenbrock (F5), Step (F6), and Quartic (F7) functions, respectively. All functions are optimized for 30 dimensions. Population size is taken as 20, the maximum number of iterations is set to 1000 with 30 replications. The test functions are given in Table 2.

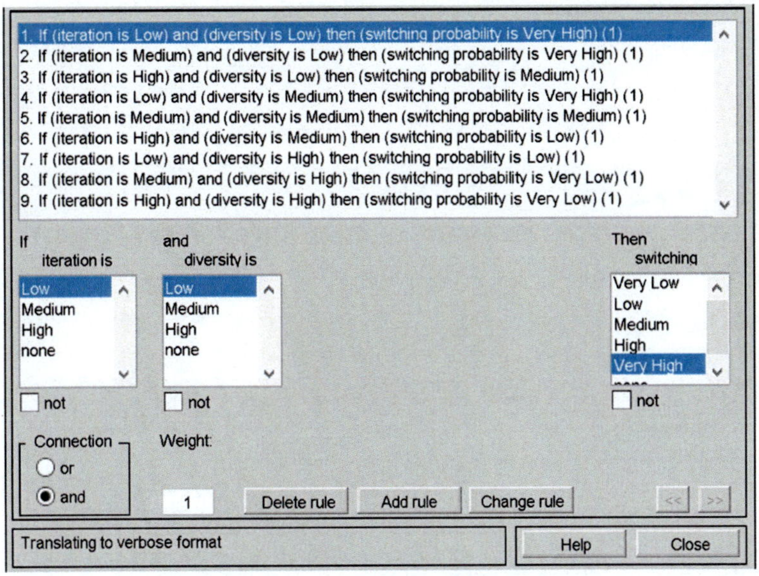

Fig. 3. Rule list of the fuzzy flower pollination

Table 2. Test functions

Function	Range
$f_1(x) = \sum_{i=1}^{n} x_i^2$	[−100, 100]
$f_2(x) = \sum_{i=1}^{n} \lvert x_i \rvert + \prod_{i=1}^{n} \lvert x_i \rvert$	[−10, 10]
$f_3(x) = \sum_{i=1}^{n} \left(\sum_{j-1}^{i} x_j\right)^2$	[−100, 100]
$f_4(x) = \max_i\{\lvert x_i \rvert, 1 \le i \le n\}$	[−100, 100]
$f_5(x) = \sum_{i=1}^{n-1} \left[100(x_{i+1} - x_i^2)^2 + (x_i - 1)^2\right]$	[−30, 30]
$f_6(x) = \sum_{i=1}^{n} ([x_i + 0.5])^2$	[−100, 100]
$f_7(x) = \sum_{i=1}^{n} ix_i^4 + random(0, 1)$	[−1.28, 1.28]

The convergence graphs are visualized in Fig. 4.

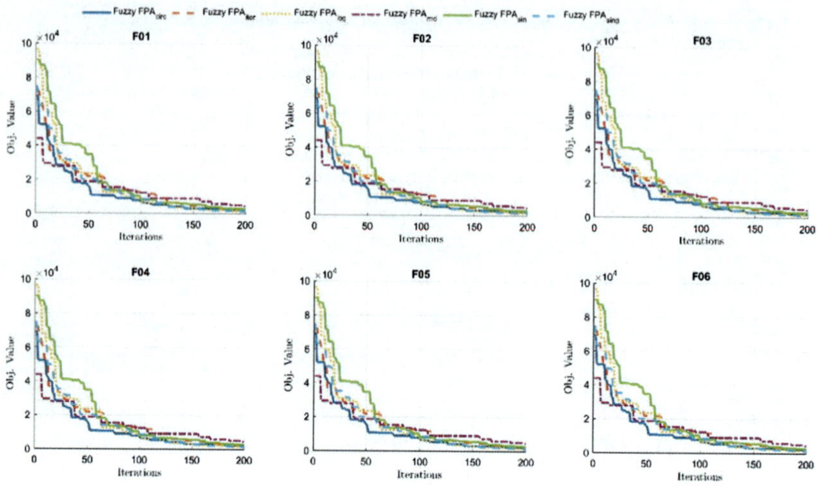

Fig. 4. Convergence graphs

Statistical test results are given in Table 3.

Table 3. Statistical test results of the FPA variants

Prob	Perf	Algorithms					
		FPA_circ	FPA_iter	FPA_log	FPA_rnd	FPA_sin	FPA_sing
F01	Best	8.1392E−65	2.1085E−66	6.1709E−71	3.0026E−01	7.4955E−70	**5.4569E−73**
	Mean	7.6000E−63	1.5882E−65	4.5022E−70	1.1068E+00	2.4096E−68	**5.5353E−72**
	Std_Dev	9.3559E−63	1.7674E−65	7.5692E−70	7.3419E−01	3.0749E−68	**6.0723E−72**
	Worst	2.3457E−62	4.2676E−65	1.7987E−69	2.8676E+00	7.8131E−68	**1.5576E−71**
F02	Best	6.9977E−41	7.0474E−42	6.4454E−42	3.3674E−01	9.4242E−42	**7.5337E−44**
	Mean	2.5938E−40	1.7974E−41	1.0715E−41	5.2025E−01	2.9995E−41	**3.7743E−43**
	Std_Dev	1.7513E−40	1.3600E−41	5.0363E−42	1.4170E−01	2.4385E−41	**2.6757E−43**
	Worst	4.3340E−40	3.4452E−41	1.8891E−41	7.6799E−01	7.1215E−41	**7.7266E−43**
F03	Best	1.9965E+00	2.3877E−03	1.6649E−03	5.4884E+03	5.7637E−03	**7.3653E−04**
	Mean	4.7938E+00	1.5613E−02	6.4591E−03	9.4992E+03	1.3820E−02	**2.6330E−03**
	Std_Dev	2.5104E+00	1.9106E−02	4.5429E−03	2.0374E+03	1.2139E−02	**2.6565E−03**
	Worst	7.6669E+00	4.8934E−02	1.1930E−02	1.1583E+04	3.5329E−02	**7.2845E−03**
F04	Best	2.6786E−04	1.2675E−05	**1.9633E−06**	6.5107E+00	2.2008E−06	4.9940E−06
	Mean	7.3337E−04	2.4304E−05	**4.4989E−06**	1.0574E+01	5.1672E−06	7.5266E−06
	Std_Dev	9.4663E−04	1.6438E−05	**1.6675E−06**	3.3431E+00	3.2505E−06	2.3658E−06
	Worst	2.4259E−03	5.2897E−05	**6.4257E−06**	1.8108E+01	1.0070E−05	1.0753E−05
F05	Best	1.9560E+01	4.8743E+00	9.3757E+00	1.1533E+02	1.2049E+01	**3.4550E−03**
	Mean	2.1877E+01	2.3065E+01	1.5309E+01	3.1808E+02	1.6816E+01	**6.5604E+00**
	Std_Dev	**2.1985E+00**	2.7006E+01	5.8758E+00	1.9469E+02	4.9104E+00	1.0792E+01
	Worst	**2.4265E+01**	7.0868E+01	2.4744E+01	5.9354E+02	2.4522E+01	2.4997E+01
F06	Best	**0.0000E+00**	**0.0000E+00**	**0.0000E+00**	2.5750E−01	**0.0000E+00**	**0.0000E+00**
	Mean	**0.0000E+00**	**0.0000E+00**	6.1630E−34	7.9893E−01	**0.0000E+00**	**0.0000E+00**
	Std_Dev	**0.0000E+00**	**0.0000E+00**	1.3781E−33	7.4213E−01	**0.0000E+00**	**0.0000E+00**
	Worst	**0.0000E+00**	**0.0000E+00**	3.0815E−33	2.8006E+00	**0.0000E+00**	**0.0000E+00**

To test the efficiency of the proposed FPA, non-parametric statistical tests are used. The Friedman test and the post-hoc analysis are given in Tables 4 and 5, respectively.

Table 4. Average ranks and obtained *p*-values of (Friedman Tests)

Algorithms	Overall avg. Ranks
FPA_rnd	6.00
FPA_circ	3.71
FPA_sin	3.54
FPA_iter	2.83
FPA_log	2.67
FPA_sing	2.25
p-values (*adj. Ties*)	0.000
decision on H0	reject *H0*

Table 5. Results of the post-hoc analysis

Significance	comparisons	z-score	unadjusted	Nemenyi	Holm
0.05	FPA_sing vs FPA_rnd	4.91	0 (+)	0 (+)	0 (+)
	FPA_log vs FPA_rnd	4.364	0 (+)	0 (+)	0 (+)
	FPA_iter vs FPA_rnd	4.146	0 (+)	0.001 (+)	0 (+)
	FPA_sin vs FPA_rnd	3.219	0.001 (+)	0.019 (+)	0.015 (+)
	FPA_circ vs FPA_rnd	3	0.003 (+)	0.04 (+)	0.03 (+)
0.10	FPA_sing vs FPA_rnd	4.91	0 (+)	0 (+)	0 (+)
	FPA_log vs FPA_rnd	4.364	0 (+)	0 (+)	0 (+)
	FPA_iter vs FPA_rnd	4.146	0 (+)	0.001 (+)	0 (+)
	FPA_sin vs FPA_rnd	3.219	0.001 (+)	0.019 (+)	0.015 (+)
	FPA_circ vs FPA_rnd	3	0.003 (+)	0.04 (+)	0.03 (+)

5 Conclusions

In this study, chaotic maps and fuzzy inference system are integrated into the FPA for real-valued function optimization problems. Circle map, iterative map, logistic map, sine map, and singer map are employed. Furthermore, iteration and population diversity are used to tune switching probability parameter of the FPA by using fuzzy inference system. Non-parametric statistical analysis shows that FPA_sing yields the best results. Also, all the proposed FPA variants outperform canonical FPA.

In the future, the proposed algorithm can be used to solve other benchmark suites. Furthermore, constrained and multi-objective problems can be used to test the performance of the proposed algorithms. Last but not least, the proposed algorithm can be tested in dynamic environments.

References

1. Yang, X.S.: Flower pollination algorithm for global optimization. In: Lecture Notes in Computer Science (including subseries Lecture Notes in Artificial Intelligence and Lecture Notes in Bioinformatics). LNCS, vol. 7445, pp. 240–249 (2012)
2. Benkercha, R., Moulahoum, S., Kabache, N.: Combination of artificial neural network and flower pollination algorithm to model fuzzy logic MPPT controller for photovoltaic systems. In: 2017 18th International Symposium on Electromagnetic Fields in Mechatronics, Electrical and Electronic Engineering, ISEF 2017 (2017)
3. Gölcük, İ., Baykasoğlu, A., Özsoydan, F.B.: A flower pollination algorithm for improving cluster analysis. In: International Conference on Applied Mathematics in Engineering (ICAME 2018), Balıkesir, Turkey, p. 186 (2018)
4. Dahi, Z.A.E.M., Mezioud, C., Draa, A.: On the efficiency of the binary flower pollination algorithm: application on the antenna positioning problem. Appl. Soft Comput. **47**, 395–414 (2016)
5. Carreon, H., Valdez, F., Castillo, O.: Fuzzy flower pollination algorithm to solve control problems. Stud. Comput. Intell. **827**, 119–154 (2020)
6. Valenzuela, L., Valdez, F., Melin, P.: Flower pollination algorithm with fuzzy approach for solving optimization problems. Stud. Comput. Intell. **667**, 357–369 (2017)
7. Hilborn, R.C.: Chaos and Nonlinear Dynamics: An Introduction for Scientists and Engineers. Oxford Univ. Press, Oxford (2006)
8. May, R.M.: Simple mathematical models with very complicated dynamics. Nature **261**, 459–467 (1976)
9. Zhenyu, G., Bo, C., Min, Y., Binggang, C.: Self-Adaptive Chaos Differential Evolution, pp. 972–975. Springer, Heidelberg (2006)
10. Devaney, R.: An introduction to chaotic dynamical systems. CRC Press, Boca Raton (2018)
11. Peitgen, H.-O., Jürgens, H., Saupe, D.: Chaos and Fractals: New Frontiers of Science. Springer (2006)

It Is Important to Take All Available Information into Account When Making a Decision: Case of the Two Envelopes Problem

Laxman Bokati, Olga Kosheleva, and Vladik Kreinovich(✉)

University of Texas at El Paso, 79968 El Paso, TX, USA
lbokati@miners.utep.edu, {olgak,vladik}@utep.edu

Abstract. In situations when we know the probabilities of all possible consequences, traditional decision theory recommends selecting the action that maximizes expected utility. In many practical situations, however, we only have partial information about the corresponding probabilities. In this case, for different possible probability distributions, we get different values of expected utility. In general, possible values of expected utility form an interval. One way to approach this situation is to use the optimism-pessimism approach proposed by Nobelist Leo Hurwicz. Another approach is to select one of the possible probability distributions – e.g., the one that has the largest possible entropy. Both approaches have led to many good practical applications. Usually, we get reasonable conclusions even when we ignore some of the available information – e.g., because this information is too vague to be easily formalized. In this paper, we show, on the example of the two envelopes problem, that ignoring available information can lead to counter-intuitive recommendations.

Keywords: Decision making under uncertainty · Hurwicz optimism-pessimism approach · Maximum entropy approach · Two envelopes problem

1 Decision Making Under Imprecise Probabilities: A Brief Reminder

Decision Theory: A Brief Reminder. Traditional decision theory (see, e.g., [1,6,8,10,12]) is based on the notion of *utility*. The natural-language word "utility" has many meanings, but in decision theory, it has a very precise – and somewhat unusual – meaning.

T. Allahviranloo et al. (Eds.): IDS 2020, AISC 1301, pp. 415–424, 2021.
https://doi.org/10.1007/978-3-030-66501-2_34

To describe this meaning, we need to select two alternatives:

- a very bad alternative A_- which is worse that anything that we will actually encounter, and
- a very good alternative A_+ which is better than anything that anything that we will actually encounter.

Then, for each number p from the interval $[0, 1]$, we can form a lottery $L(p)$ in which:

- we get A_+ with probability p and
- we get A_- with the remaining probability $1 - p$.

For $p = 0$, the lottery $L(p)$ coincides with the very bad alternative A_- and is, thus, worse that any actual alternative A; we will denote this by $L(0) = A_- < A$. For $p = 1$, the lottery $L(p)$ coincides with the very good alternative A_+ and is, thus, better than any actual alternative A: $A < A_+$.

We can ask the user to compare the alternative A with the lotteries $L(p)$ corresponding to different probabilities p. We assume that for every two alternatives A and B, the user always decides:

- whether A is better (i.e., $B < A$)
- or whether B is better (i.e., $A < B$),
- or whether A and B are of the same quality to this user; we will denote this by
$$A \sim B.$$

We also assume that the user's decisions are consistent – i.e.:

- that the preference relation $<$ is transitive and
- that $p < p'$ implies $L(p) < L(p')$.

One can see that under these assumptions, there is a threshold value p_0 such that:

- for all $p < p_0$, we have $L(p) < A$, and
- for all $p > p_0$, we have $A < L(p)$.

This threshold value is called the *utility* of the alternative A. The utility is usually denoted by $u(A)$.

By definition of the utility, for every small value $\varepsilon > 0$, we have

$$L(u(A) - \varepsilon) < A < L(u(A) + \varepsilon).$$

For very small ε, lotteries with probabilities $u(A)$ and $u(A) \pm \varepsilon$ are practically indistinguishable, so we can interpret the above double inequality as saying that the alternative A is equivalent to the lottery $L(u(A))$. We will denote this equivalence by

$$A \equiv L(u(A)).$$

Suppose now that for some action a, we have consequences $A_1, \ldots, A_n$ with probabilities $p_1, \ldots, p_n$. This means that the action a is equivalent to a lottery in which we get each A_i with probability p_i. Each alternative A_i, in its turn, is equivalent to the lottery $L(u(A_i))$ in which we get A_+ with probability $u(A_i)$ and A_- with the remaining probability $1-u(A_i)$. Thus, the action a is equivalent to a two-stage lottery, in which:

- first, we select A_i with probability p_i, and
- then, depending on which alternative we selected on the first stage, we select A_+ with probability $u(A_i)$ and A_- with probability $1 - u(A_i)$.

As a result of this two-stage lottery, we get either A_+ or A_-, and the probability of getting A_+ is equal to $p = \sum\limits_{i=1}^{n} p_i \cdot u(A_i)$. So, the action a is equivalent to the lottery $L(p)$. By definition of utility, this means that the utility $u(a)$ of the action a is equal to this probability p, i.e., that

$$u(a) = \sum_{i=1}^{n} p_i \cdot u(A_i).$$

By definition of utility, we select the action with the largest possible value of utility. The right-hand side of the above formula is the expected value of the utility, so this means that a rational person should select the alternative with the largest possible value of expected utility.

Utility is Defined Modulo a Linear Transformation. Numerical values of utility depend on the selection of the very bad and very good alternatives A_- and A_+. What if we select a different pair A'_- and A'_+?

For example, if $A_- < A'_- < A'_+ < A_+$, then for every alternative A, its utility $u'(A)$ in relation to A'_- and A'_+ means that A is equivalent to a lottery $L'(u'(A))$ in which we get A'_+ with probability $u'(A_+)$ and A'_- with the remaining probability $1 - u'(A)$. Here:

- the alternative A'_- is equivalent to a lottery $u(A'_-)$ in which we get A_+ with probability $u(A'_-)$ and A_- with the remaining probability $1 - u(A'_-)$;
- similarly, A'_+ is equivalent to a lottery $u(A'_+)$ in which we get A_+ with probability $u(A'_+)$ and A_- with the remaining probability $1 - u(A'_+)$.

Thus, the original alternative A is equivalent to a two-stage lottery in which:

- we first select A'_- or A'_+, and
- then, depending on what we selected on the first stage, select A_+ or A_-.

As a result of this two-stage lottery, we get either A_+ or A_-, and the probability p of selecting A_+ is equal to

$$p = u'(A) \cdot u(A'_+) + (1 - u'(A)) \cdot u(A'_-) = u(A'_-) + u'(A) \cdot (u(A_+) - u(A_-)).$$

By definition of utility, this probability p is the utility $u(A)$ of the alternative A in terms of A_- and A_+. Thus,

$$u(A) = u(A'_-) + u'(A) \cdot (u(A_+) - u(A_-)).$$

In other words, the utility is defined modulo a linear transformation.

Comment. This is similar to measuring quantities like time or temperature, where the numerical value depends on the selection of the starting point and of the measuring unit.

What if We Only Have Partial Information About Probabilities: Two Approaches. The formula for the expected utility assumes that we know the probability of each alternative. In many practical situations, we only have partial information about the probabilities. In this case, for different probability distributions $P = (p_1, \ldots, p_n)$, we have different values of the expected utility. If two probability distributions P and P' are possible, then we can also consider as possible the case when we have P with probability β and P' with probability $1 - \beta$. In this case, the expected utility is equal to the convex combination of expected utilities corresponding to P and P'. So, the set of possible values of expected utility is closed under convex combinations and is, thus, an interval

$$[\underline{u}(a), \overline{u}(a)].$$

How can we make a decision under this interval uncertainty? There are two approaches to solving this problem.

The first approach is based on the fact that we need to compare the action a, in particular, with lotteries $L(p)$. Thus, we need to assign, to this interval, a corresponding utility $u(a)$ depending on the bounds $u(a) = f(\underline{u}(a), \overline{u}(a))$ for some function $f(x, y)$. Utility is defined modulo a linear transformation $u(a) \to k \cdot u(a) + \ell$, for some $k > 0$ and ℓ. It is therefore reasonable to require that the function $f(x, y)$ not depend on the selection of the alternatives A_- and A_+, i.e., that it remains the same for all $k > 0$ and ℓ:

- if $z = f(x, y)$,
- then $z' = f(x', y')$ for $z' = k \cdot a + \ell$, $x' = k \cdot x + \ell$, and $y' = k \cdot y + \ell$.

In particular, every interval $[\underline{a}, \overline{a}]$ can be obtained from the interval $[0, 1]$ if we take $k = \overline{u} - \underline{u}$ and $\ell = \underline{u}$. Thus, if we denote $\alpha \stackrel{\text{def}}{=} f(0, 1)$, then the above invariance implies that

$$f(\underline{u}, \overline{u}) = \alpha \cdot (\overline{u} - \underline{u}) + \underline{u} = \alpha \cdot \overline{u} + (1 - \alpha) \cdot \underline{u}.$$

So, when we make a decision, we should select the action a for which the value

$$\alpha \cdot \overline{u} + (1 - \alpha) \cdot \underline{u}$$

is the largest possible.

This idea was first proposed by a Nobelist Leo Hurwicz. It is known as *Hurwicz optimism-pessimism approach*; see, e.g., [2,6,8]. The reason for this name is as follows:

- When $\alpha = 1$, this means that the decision maker only takes into account the best possible situation, with utility $\overline{u}$; this is what we usually mean by extreme optimism.
- Vice versa, when $\alpha = 0$, this means that the decision maker only takes into account the worst possible situation, with utility $\underline{u}$; this is what we usually mean by extreme pessimism.

Of course, in real life, these two behaviors do not make sense:

- extreme pessimism would mean not going into the street at all since there is a change of getting hit by a car, and
- extreme optimism would mean crossing the street on red light in heavy traffic.

Realistic decision making corresponds to values α between 0 and 1. For example, it is often recommended to select $\alpha = 0.5$.

An alternative approach is that instead of considering all possible probability distributions, we should select the most reasonable one. Some of the possible distributions correspond to larger uncertainty, some to smaller uncertainty. We do not want to pretend that we have less uncertainty, so it is reasonable to select the distribution with the largest possible value of uncertainty. A natural measure of uncertainty of a probability distribution is its entropy – the average number of binary questions that we need to ask to determine the actual value; see, e.g., [3,11]. Thus, out of all possible distributions, we select the one with the largest possible entropy. This approach is known as the *Maximum Entropy* approach; see, e.g., [3].

In particular, when we have a natural symmetry – e.g., we have no reason to believe that some alternatives are more probable than others – the resulting distribution is also invariant with respect to the same symmetry. So, if the class of distributions is invariant with respect to permutations, the maximum entropy distribution is also invariant with respect to all permutations – and thus, assigns equal probability to all alternatives. This is known as Laplace Indeterminacy Principle.

This invariance idea can be applied to the case when all we know is that the value x of some physical quantity is positive, but we have no information about different probabilities. Which probability density function (pdf) $\rho(x)$ should we choose? The numerical value of x depends on the choice of a measuring unit: if we select a unit which is λ times smaller, then all numerical values x are replaced by new values $x' = \lambda \cdot x$. For example, 1.7 m becomes $100 \cdot 1.7 = 170$ cm. Under this transformation, the original probability density function $\rho(x)$ takes, in the new unit, the form

$$\frac{1}{\lambda} \cdot \rho\left(\frac{x'}{\lambda}\right).$$

We want to have the pdf that does not depend on the choice of the measuring unit, so we should have

$$\frac{1}{\lambda} \cdot \rho\left(\frac{x'}{\lambda}\right) = \rho(x'),$$

or, equivalently,

$$\rho\left(\frac{x'}{\lambda}\right) = \lambda \cdot \rho(x').$$

For $x' = 1$ and $\lambda = 1/a$, we get $\rho(a) = \text{const} \cdot \dfrac{1}{a}$.

It should be noted that, strictly speaking, this is *not* a probability density function, since here, we have $\int \rho(a)\, da = \infty \neq 1$. This is known as an *improper* probability distribution.

What We Do in this Paper. Both Hurwicz and Maximum Entropy approaches have been used in many practical problems. They usually lead to intuitively acceptable results, even when we ignore some available information – e.g., when this information is too vague to be easily formalized.

In this paper, we show, on the example of the two envelopes problems, that if we ignore this information, both approaches lead to counter-intuitive results. Thus, when making a decision under uncertainty, it is important to take into account all available information.

2 Two Envelopes Problem: Description and Analysis

Formulation of the Problem. The two envelopes problem is as follows; see, e.g., [9] and references therein. Someone places some amount of money in one envelope, and a double that amount in another envelope. We do not know which envelope contains a smaller amount and which contains a larger amount.

We can pick one envelope and check how much money x it has. Then, we can make a decision:

- we can either keep this amount
- or we can select the second envelope instead.

The question is:

- should we keep the original amount or
- should we select the other envelope instead?

How Realistic Is this Situation. The above situation is over-simplified, but similar situations occur in real life.

For example, suppose that there are two competing countries producing certain military equipment. A smaller country would like to buy from one of them. For political reasons, it cannot negotiate simultaneously with both, so it starts serious negotiations with one of the countries. After negotiations, it comes up with the expected cost x of the purchase. This cost is usually rather significant. Now, this smaller country has a choice:

– it can go with this contract, or
– it can try its luck by negotiating with the competing country – taking into account that if these negotiations do not lead to cheaper prices, there will be no going back to the original negotiated contract: e.g., the first weapons-producing country will have already committed its factories to produce weapons for someone else.

What should it do?

More peaceful examples can be found: e.g., when forming political alliances inside a country or within countries, when planning mergers, etc.

At First Glance, this is Somewhat Paradoxical. Let us go back to the original two envelopes problem. From the common sense viewpoint, since our selection was random, it does not make sense to switch.

On the other hand, intuitively, we do not know these the second envelope has $2x$ or $x/2$. Thus, in line with the Laplace Indeterminacy Principle, we assume that these two amounts have equal probability 0.5. But then, the expected gain is

$$0.5 \cdot 2x + 0.5 \cdot (x/2) = 1.25x.$$

Since $1.25x > x$, it looks like it is always reasonable to switch – which contradicts to common sense.

Comment. This argument is based on the assumption that utility is proportional to money. In practice, it is proportional to square root of money (see, e.g., [4,5,7]), but the paradox remains if we take $\sqrt{2} \cdot x$ and $x/\sqrt{2}$ as the utilities of the two alternatives.

This is Not Really a Paradox. A detailed analysis shows that, in reality, there is no paradox. Indeed, suppose that the original values come with the probability density $\rho(x)$. Then, the double amounts come with probability density $\frac{1}{2} \cdot \rho\left(\frac{x}{2}\right)$. Then, if we find the value x:

– the condition probability that this is the original amount of money – and thus, that the remaining envelope will contain $2x$ – is proportional to $\rho(x)$, and
– the conditional probability that this is the double amount of money – and thus, that the remaining envelope contains $x/2$ – is proportional to

$$\frac{1}{2} \cdot \rho\left(\frac{x}{2}\right).$$

Since these two probabilities should add up to 1, we conclude that they should be equal to

$$\frac{\rho(x)}{\rho(x) + \frac{1}{2} \cdot \rho\left(\frac{x}{2}\right)}$$

and to

$$\frac{\frac{1}{2} \cdot \rho\left(\frac{x}{2}\right)}{\rho(x) + \frac{1}{2} \cdot \rho\left(\frac{x}{2}\right)}.$$

Thus, the expected value of the gain that we get when we switch is equal to

$$\frac{\rho(x)}{\rho(x) + \frac{1}{2} \cdot \rho\left(\frac{x}{2}\right)} \cdot 2x + \frac{\frac{1}{2} \cdot \rho\left(\frac{x}{2}\right)}{\rho(x) + \frac{1}{2} \cdot \rho\left(\frac{x}{2}\right)} \cdot \frac{x}{2}.$$

So:

- If this amount is smaller that x, we should not switch.
- If this amount is larger than x, we should switch.
- If this amount is equal to x, it does not matter whether we switch or not.

If we know the probability distribution, then, depending of the amount x, we should switch or not switch. For example:

- If we know that all the values x are smaller than some amount x_0, then, if we get an amount $x > x_0$, we know that this is the larger amount, so switching does not make sense.
- Similarly, if we know that all original amounts are larger than or equal to some minimal amount m, and we have a value $x < 2m$, we know that this cannot be the double amount, so we should switch.

The above formula explains why there is no paradox: the only time when the probability of each option is exactly 1/2 is when $\rho(x) = \frac{1}{2} \cdot \rho\left(\frac{x}{2}\right)$ for all x, and since this can be repeated not just for doubling money, but also for λ times larger amount, we should have $\rho(x) = \frac{1}{\lambda} \cdot \rho\left(\frac{x}{\lambda}\right)$ for all x and λ – and we already know that this leads to an "improper" – not real – probability distribution.

Idealized Formulation: What the Two Approaches Recommend. Suppose now that we do not have any information about the probability distributions. What should the above two approaches recommend?

Let us first consider the Hurwicz approach. We do not know the pdf, so we should consider all possible probability density functions.

- The worst-possible case is when $\frac{1}{2} \cdot \rho\left(\frac{x}{2}\right) = 0$, then after switching, we get $x/2$ with probability 1.
- The best-possible case is when $\rho(x) = 0$, then after switching, we get $2x$ with probability 1.

So, e.g., for $\alpha = 1/2$, the Hurwicz combination is equal to

$$0.5 \cdot 2x + 0.5 \cdot (x/2) = 1.25x > x,$$

so we arrive at the counter-intuitive conclusion that we should always switch.

So maybe we should use smaller value of Hurwicz's α? This will not help, since, as we have mentioned, we could have $\lambda \cdot x$ in the second envelope. In this case, the expected value after switching is $\alpha \cdot \lambda \cdot x + (1 - \alpha) \cdot (x/\lambda)$. For any $\alpha > 0$, for a sufficiently large λ, we get a counter-intuitive conclusion that we should switch. The only case when this conclusion is not possible is the case $\alpha = 0$ which, as we have mentioned, is not realistic at all. So, for this problem, Hurwicz approach leads to a counter-intuitive behavior.

Let us now consider the maximum entropy approach. Since all we know is that the amount is positive, we should use the corresponding improper distribution for which $\rho(x) = \frac{1}{2} \cdot \rho\left(\frac{x}{2}\right)$, so we also get $1.25x$. In other words, in this case, the Maximum Entropy approach also leads to a counter-intuitive conclusion.

So What Should We Do? Why did we get the counter-intuitive result? Because our description of the problem is not realistic. Do we really believe that an envelope can contain any amount of money? Realistically, an envelope cannot contain more than several thousand dollars – a million will not fit into an envelope. If we impose an upper limit x_0 on the original amount of money, then the conclusions change – and become more intuitive:

- For the Hurwicz approach with $\alpha = 0.5$, we still recommend switching when $x \le x_0$, but now we do not recommend switching when $x > x_0$, since in this case, $\rho(x) > 0$.
- The Maximum Entropy approach leads to the uniform distribution

  $$\rho(x) = \text{const on the interval } [0, x_0] \text{ and } \rho(x) = 0 \text{ for } x > x_0.$$

  For this distribution, the above formula also recommends switching when $x \le x_0$ and not switching when $x > x_0$.

This common recommendation seems to be in perfect accordance with common sense.

Conclusion. When making decision under uncertainty, it is important to take into account *all* available information, even seemingly useless one – otherwise, if we ignore this information, we may end up with counter-intuitive recommendations.

Acknowledgments. This work was supported in part by the National Science Foundation grants 1623190 (A Model of Change for Preparing a New Generation for Professional Practice in Computer Science) and HRD-1242122 (Cyber-ShARE Center of Excellence).

References

1. Fishburn, P.C.: Utility Theory for Decision Making. Wiley, New York (1969)
2. Hurwicz, L.: Optimality Criteria for Decision Making Under Ignorance. Cowles Commission Discussion Paper, Statistics, No. 370 (1951)
3. Jaynes, E.T., Bretthorst, G.L.: Probability Theory: The Logic of Science. Cambridge University Press, Cambridge (2003)
4. Kahneman, D.: Thinking, Fast and Slow, Farrar, Straus, and Giroux, New York (2011)
5. Kosheleva, O., Afravi, M., Kreinovich, V.: Why utility non-linearly depends on money: a commonsense explanation. In: Proceedings of the 4th International Conference on Mathematical and Computer Modeling, Omsk, Russia, 11 November 2016, pp. 13–18 (2016)
6. Kreinovich, V.: Decision making under interval uncertainty (and beyond). In: Guo, P., Pedrycz, W. (eds.) Human-Centric Decision-Making Models for Social Sciences, pp. 163–193. Springer, Heidelberg (2014)
7. Lorkowski, J., Kreinovich, V.: Granularity helps explain seemingly irrational features of human decision making. In: Pedrycz, W., Chen, S.-M. (eds.) Granular Computing and Decision-Making: Interactive and Iterative Approaches, pp. 1–31. Springer, Cham (2015)
8. Luce, R.D., Raiffa, R.: Games and Decisions: Introduction and Critical Survey. Dover, New York (1989)
9. McDonnell, M.D., Abott, D.: Randomized switching in the two-envelope problem. Proc. Roy. Soc. A **465**(2111), 3309–3322 (2009)
10. Nguyen, H.T., Kosheleva, O., Kreinovich, V.: Decision making beyond Arrow's 'impossibility theorem', with the analysis of effects of collusion and mutual attraction. Int. J. Intell. Syst. **24**(1), 27–47 (2009)
11. Nguyen, H.T., Kreinovich, V., Wu, B., Xiang, G.: Computing Statistics under Interval and Fuzzy Uncertainty. Springer, Heidelberg (2012)
12. Raiffa, H.: Decision Analysis. McGraw-Hill, Columbus (1997)

m-Polar Fuzzy Soft Graphs in Group Decision Making: A Combining Method by Aggregation Functions

Azadeh Zahedi Khameneh[1(✉)] and Adem Kilicman[1,2]

[1] Institute for Mathematical Research, Universiti Putra Malaysia, UPM, 43400, Serdang Selangor, Malaysia
{zk.azadeh,akilic}@upm.edu.my
[2] Department of Mathematics, Faculty of Science, Universiti Putra Malaysia, UPM, 43400 Serdang, Selangor, Malaysia

Abstract. Recently, study on graph theory has been fulfilled in a three-dimension space, known as fuzzy soft graph, characterized by the set of functions $([0,1]^U)^P$. This new tool is successfully applied for dealing with uncertainties arisen from multi-parameter data. However, the present format needs to be expanded into the space $((\underbrace{[0,1] \times \cdots \times [0,1]}_{m-times})^U)^P$ in case of multi-polar or multi-index parametric information. This paper introduces the concept of m-polar fuzzy soft graph. Vertex neighborhood degrees matrices, consequently regularity and irregularity for m-polar fuzzy soft graphs are considered. To overcome the consensus problem in group decision-making with m-polar fuzzy soft graphs, merging several m-polar fuzzy soft graphs into a unique one by using aggregation functions is investigated. The sufficient condition for preservation of m-polar fuzzy soft regularity under aggregation process is studied. Finally, an algorithm is designed to illustrate the application of aggregated m-polar fuzzy soft graph in group decision-making.

Keywords: m-polar fuzzy soft graph · Vertex open neighborhood matrix · Regular m-polar fuzzy soft graph · Aggregating function · Group decision making

1 Introduction

Fuzzy graph theory, discussed for the first time by Rosenfeld [1] based on Zadeh's fuzzy relations [2] and then developed by Bhattacharya [3], is one of the important mathematical tools to model objects having interconnection in a network where some aspects of them and/or some pairwise relations between them are uncertain or fuzzy. In traffic light control [4–7], image processing and coloring problem [8–10], decision making problems [11–14], selection problem [15] and many other fields the applications of fuzzy graphs can be found. After introducing the concept of fuzzy graph, more theoretical aspects have been added

T. Allahviranloo et al. (Eds.): IDS 2020, AISC 1301, pp. 425–455, 2021.
https://doi.org/10.1007/978-3-030-66501-2_35

to this new tool. One of the first attempts was carried out by Mordeson and Peng [16] to develop basic operations on fuzzy graphs. They also defined the concept of fuzzy graph complement for the first time. However, later, Sunitha and Vijaya kumar [17] modified their definition. In 1994, Zhang [18] initiated the concept of bipolar fuzzy sets whose membership degrees are in $[-1,1]$ instead of $[0,1]$ to model situations where objects can be considered to have a certain property and its counter. This framework is successfully applied when both positive and negative sides of information are given. Bipolar fuzzy sets were then extended by Chen et al. [19] to introduce the idea of m-polar fuzzy sets (or multi polar fuzzy sets) where instead of $[-1,1]$ the membership values belong to $[0,1]^m$, that is $[0,1] \times \cdots \times [0,1]$ m-times. As a result, the framework of fuzzy graphs was expanded to bipolar fuzzy sets by Akram [20] and m-polar fuzzy sets by Chen et al. [19]. Later, Yang et al. [21] showed some errors in Akram's paper and introduced the new notion of generalized bipolar fuzzy graph. The concepts of regularity and irregularity for bipolar fuzzy graphs were discussed in [22,23]. However, recently, Poulik and Ghorai [24] showed some results in [23] contain flaws and modified them. Rashmanlou et al. [25] defined two new product operations on bipolar fuzzy graphs. Ghorai and Pal [26] investigated regular and totally regular concepts for product bipolar fuzzy graphs. They also studied some operations and properties of m-polar fuzzy graphs in [27,28]. Akram and Shahzadi [29] used the concept of α-level sets in fuzzy set theory to characterize level graphs of m-polar fuzzy graphs. Regularity of m-polar fuzzy graphs was discussed by Akram and Adeel [30]. In [31], Singh developed a method to represent m-polar fuzzy graphs by lattices.

Limitations of fuzzy graph and its various extensions for visualization the parametric uncertain relations motivated researchers to develop the concept of fuzzy soft graphs on the basis of Molodtsov's soft sets [32]. In 2015, Mohinta and Samanta [33] introduced for the first time the notion of fuzzy soft graph. However, later research on this topic were led by Akram and his team. Akram and Nawaz [34–36] studied soft graphs and fuzzy soft graphs, Akram and Zafar [36] considered fuzzy soft trees, and Shahzadi and Akram [37–39] extended the framework of fuzzy soft graphs to intuitionistic fuzzy soft graphs. This concept was then developed to Pythagorean fuzzy soft graph in [40,41]. Bipolar fuzzy soft graph and its applications in decision making were also discussed by Akram et al. [42,43].

As can be seen from the literature described above, fuzzy graph theory has been extended in different frameworks. However, moving towards multi-polarity in decision sciences highlights introducing more hybrid concepts on fuzzy soft graph theory. Therefore, this paper presents the concept of m-polar fuzzy soft graph where $m \geq 2$. This issue will be more challenged when the multi-polar data comes from K different agents or decision makers. In general case, the only fuzzy soft operations to obtain the collective graph from different K fuzzy soft graphs are fuzzy soft union and fuzzy soft intersection, characterized respectively by Min and Max operators. While to model real-world situations other operators rather than Min or Max might better be used. We extend these two operators to any arbitrary aggregation function to visualize the idea of whole group, that is not necessarily the meet or join, by a unique m-polar fuzzy soft graph so-called aggregated one. From these points of view, the study makes some contributions to the literature.

The main goals of this research are to develop the concept of m-polar fuzzy soft graph and combine them by using aggregation functions. To achieve these objectives, the rest of this paper is organized as the following. In Sect. 2, we bring the definitions and properties of fuzzy graph and its extensions. Section 3 is dedicated to develop m-polar fuzzy soft graphs. We start with m-polar fuzzy soft set, discussed in [44], and continue by introducing m-polar fuzzy soft graph. The new concepts of size and order matrices for m-polar fuzzy soft graph, union and intersection of m-polar fuzzy soft graphs, vertex open and closed neighborhood degrees matrices, consequently regular and irregular m-polar fuzzy soft graphs are investigated. Corresponding to each parameter $p \in P$, the underlying fuzzy and crisp graphs are generated and connections between them are presented. Moreover, by using the concept of level sets in fuzzy set theory, level soft graph of an m-polar fuzzy soft graph is studied. Then, in Sect. 4, we propose a new method to combine several m-polar fuzzy soft graphs by using aggregation functions, so-called aggregated m-polar fuzzy soft graph. The sufficient condition for preservation of m-polar fuzzy soft regularity under aggregation process is studied. Finally, in Sect. 5, a multi-phase algorithm is designed to discuss the application of proposed method.

2 Fuzzy Graph Theory and Its Extensions

This section reviews some basic concepts in fuzzy graph theory and its extensions, including m-polar fuzzy graphs and fuzzy soft graphs.

2.1 Fuzzy Graphs and m-Polar Fuzzy Graphs

In graph theory, a pair $G^* = (V, E)$ presents the concept of graph where V is a vertex set and E is a binary relation on V, so-called the edge set. The notation uv is usually used to show the pair (u, v) and say u and v are neighbors or adjacent. A graph $H^* = (W, F)$ is called a subgraph of $G^* = (V, E)$ if $W \subseteq V$ and $F \subseteq E$.

Since in real-life situations some uncertainties may be put on vertices and/or edges, graph theory was extended into the concept of fuzzy graph.

In the following definitions let V be a non-empty set of objects, called vertices, and $\mathbb{I} = [0, 1]$.

Definition 1 *([1]). A 3-tuple $G = (V, \phi, \psi)$ where $\phi : V \to [0, 1]$ and $\psi : V \times V \to [0, 1]$ are fuzzy sets on V and V^2, respectively, such that $\psi(uv) \leq \min\{\phi(u), \phi(v)\}$ for all $u, v \in V$ is called a fuzzy graph (FG in brief) where ψ is a symmetric fuzzy relation on V, i.e. $\psi(uv) = \psi(vu)$.*

The fuzzy graph $H = (V, \sigma, \mu)$ is a fuzzy subgraph of $G = (V, \phi, \psi)$ if for all $u, v \in V$: $\sigma(v) \leq \phi(v)$ and $\mu(uv) \leq \psi(uv)$. Based on the concept of α-level set in Zadeh's fuzzy set theory [45], here also we can define α-level graph of $G = (V, \phi, \psi)$ for any $\alpha \in (0, 1]$ that gives us a crisp graph $G_\alpha = (V_\alpha, E_\alpha)$ where $V_\alpha = \{v \in V : \phi(v) \geq \alpha\}$ and $E_\alpha = \{uv \in V^2 : \psi(uv) \geq \alpha\}$. The underlying crisp graph of G is defined as a crisp graph $G_0 = (V_0, E_0)$ where $V_0 = \{v \in V : \phi(v) > 0\}$ and $E_0 = \{(u, v) \in V^2 : \psi(uv) > 0\}$, so clearly, $G_0 = \bigcup_{\alpha \in (0,1]} G_\alpha$. For any given thresholds $\alpha, \beta \in (0, 1]$ if $\alpha \leq \beta$, then $G_\beta = (V_\beta, E_\beta)$ is a subgraph of $G_\alpha = (V_\alpha, E_\alpha)$. Moreover, if $H = (V, \sigma, \mu)$ is a fuzzy subgraph of $G = (V, \phi, \psi)$, then H_α is a subgraph of G_α for any threshold value $\alpha \in (0, 1]$. If $\psi(uv) = \min\{\phi(u), \phi(v)\}$ for all $uv \in E_0$, then $G = (V, \phi, \psi)$ is called strong fuzzy graph. G is called complete fuzzy graph if for all $u, v \in V$ we have $\psi(uv) = \min\{\phi(u), \phi(v)\}$.

After extending the concept of fuzzy set into the m-polar fuzzy set, that is presented by a mapping $f : U \to [0, 1]^m$ where $m \in \mathbb{N}$ and $(0, \ldots, 0)$ and $(1, \ldots, 1)$ are the least and the greatest elements, m-polar fuzzy graph is defined as below.

Definition 2 *([19, 29]). An m-polar fuzzy graph (shortly m-PFG) is defined as a 3-tuple $G = (V, \phi, \psi)$ where $\phi : V \to [0, 1]^m$ (an m-polar fuzzy set on V) and $\psi : V \times V \to [0, 1]^m$ (an m-polar fuzzy set on V^2) such that $\pi_s \circ \psi(uv) \leq \min\{\pi_s \circ \phi(u), \pi_s \circ \phi(v)\}$ for all $u, v \in V$ and $s = 1, 2, \cdots, m$.*

An m-polar fuzzy graph $H = (V, \sigma, \mu)$ is a subgraph of m-polar fuzzy graph $G = (V, \phi, \psi)$ if for all $u, v \in V$ and $s = 1, 2, \cdots, m$: $\pi_s \circ \sigma(v) \leq \pi_s \circ \phi(v)$ and $\pi_s \circ \mu(uv) \leq \pi_s \circ \psi(uv)$. The underlying crisp graph $G_0 = (V_0, E_0)$ of the m-polar fuzzy graph $G = (V, \phi, \psi)$ is defined by vertex set $V_0 = \{v \in V : \pi_s \circ \phi(v) > 0; 1 \leq s \leq m\}$ and edge set $E_0 = \{(u, v) \in V^2 : \pi_s \circ \psi(uv) > 0; 1 \leq s \leq m\}$. The m-polar fuzzy graph $G = (V, \phi, \psi)$ is called strong if $\pi_s \circ \psi(uv) = \min\{\pi_s \circ \phi(u), \pi_s \circ \phi(v)\}$ for all $uv \in E_0$. It is called complete m-PFG if the equation is held for all $u, v \in V$.

For any vertex u of the m-polar fuzzy graph G the neighborhood set is defined as

$$N_G(u) = \{v \in V : 0 < \pi_s \circ \psi(uv) \leq \min\left(\pi_s \circ \phi(u), \pi_s \circ \phi(v)\right); 1 \leq s \leq m\}$$

The open neighborhood degree of vertex u in G is computed as the following vector

$$d_G(u) = \Big(\sum_{v \in N_G(u)} \pi_1 \circ \phi(v), \cdots, \sum_{v \in N_G(u)} \pi_m \circ \phi(v)\Big)$$

and its closed neighborhood degree is defined by $d_G[u] = d_G(u) + \phi(u)$. The m-polar fuzzy graph $G = (V, \phi, \psi)$ is called $\mathbf{t}$-regular if open neighborhood degrees of all vertices of G is the same vector $\mathbf{t} = (t_1, \cdots, t_m)$. It is called irregular m-polar fuzzy graph if there exists a vertex whose neighbors have different degrees.

Operations on m-PFG are defined by extending the corresponding operations in fuzzy graph theory. If $G_1 = (V_1, \phi_1, \psi_1)$ and $G_2 = (V_2, \phi_2, \psi_2)$ are two m-polar fuzzy graphs. Then, for instance, the union of G_1 and G_2 is an m-PFG

$G_1 \cup G_2 = (V_1 \cup V_2, \phi_1 \vee \phi_2, \psi_1 \vee \psi_2)$ where the mappings $\phi_1 \vee \phi_2 : V_1 \cup V_2 \to [0,1]^m$ and $\psi_1 \vee \psi_2 : V_1^2 \cup V_2^2 \to [0,1]^m$ are respectively defined by $\pi_s \circ (\phi_1 \vee \phi_2)(v) = \max\{\pi_s \circ \phi_1(v), \pi_s \circ \phi_2(v)\}$ and $\pi_s \circ (\psi_1 \vee \psi_2)(uv) = \max\{\pi_s \circ \psi_1(uv), \pi_s \circ \psi_2(uv)\}$ for all $s = 1, 2, \cdots, m$.

2.2 Fuzzy Soft Graphs

In 1999, the Russian mathematician Molodtsov [32] proposed the first applications of a new tool, so-called soft sets. Later, in 2001, Maji et al. [46] used this concept to develop a parameterized extension of fuzzy sets, that is called fuzzy soft set, by combining the concepts of fuzzy set and soft set where the latter is in fact a set-valued mapping $f : P \to 2^V$ such that P is a set of parameters and 2^V shows the power set of V.

In the sequence, let V and P be two non-empty sets of objects and parameters, respectively.

Definition 3 *([46]). A fuzzy soft set (FSS in brief) over V and P, denoted by f_P or (f, P), is a mapping $f : P \to [0,1]^V$. Thus for any $p \in P$, $f(p)$ is a fuzzy subset of V with membership function $f_p : V \to [0,1]$. The null FSS, denoted by $\tilde{\mathbf{0}}$, and the absolute FSS, denoted by $\tilde{\mathbf{1}}$, are defined by the membership functions $\tilde{\mathbf{0}}(p)(v) = 0$ and $\tilde{\mathbf{1}}(p)(v) = 1$ for all $v \in V$ and $p \in P$, respectively.*

Using Definitions 1 and 3, the concept of fuzzy soft graph was proposed as below.

Definition 4 *([33]). A fuzzy soft graph (FSG in brief) on V is a 4-tuple $\tilde{G} = \langle P, (V, \phi, \psi)\rangle$ where (ϕ, P) is a FSS over V and (ψ, P) is a FSS over V^2 such that for each $p \in P$, the $\tilde{G}(p) = (V, \phi(p), \psi(p))$ is a FG on V, that implies $\psi(p)(uv) \leq \min\{\phi(p)(u), \phi(p)(v)\}$.*

A fuzzy soft graph $\tilde{H} = \langle A, (V, \sigma, \mu)\rangle$ is a subgraph of $\tilde{G} = \langle P, (V, \phi, \psi)\rangle$ if $A \subseteq P$ and for each $a \in A$, the FG $\tilde{H}(a) = (V, \sigma(a), \mu(a))$ is a fuzzy subgraph of $\tilde{G}(a) = (V, \phi(a), \psi(a))$. The underlying crisp graph of the FSG $\tilde{G}$ is denoted by $\tilde{G}_0 = (V_0, E_0)$ where $V_0 = \{v \in V : \phi(p)(v) > 0; \forall p \in P\}$ and edge set $E_0 = \{uv \in V^2 : \psi(p)(uv) > 0; \forall p \in P\}$. $\tilde{G} = \langle P, (V, \phi, \psi)\rangle$ is called strong FSG if $\psi(p)(uv) = \min\{\phi(p)(u), \phi(p)(v)\}$ for all $uv \in E_0$ and all $p \in P$. And it is called complete FSG if $\psi(p)(uv) = \min\{\phi(p)(u), \phi(p)(v)\}$ for all $u, v \in V$ and all $p \in P$.

Take two fuzzy soft graphs $\tilde{G}_1 = \langle P_1, (V_1, \phi_1, \psi_1)\rangle$ and $\tilde{G}_2 = \langle P_2, (V_2, \phi_2, \psi_2)\rangle$. The union of them is defined as a FSG $\tilde{G}_1 \cup \tilde{G}_2 = \langle P_1 \cup P_2, (V_1 \cup V_2, \phi_1 \vee \phi_2, \psi_1 \vee \psi_2)\rangle$ where the mappings $\phi_1 \vee \phi_2 : P_1 \cup P_2 \to [0,1]^{V_1 \cup V_2}$ and $\psi_1 \vee \psi_2 : P_1 \cup P_2 \to [0,1]^{V_1^2 \cup V_2^2}$ are respectively defined by

$$(\phi_1 \vee \phi_2)(p)(v) = \max\{\phi_1(p)(v), \phi_2(p)(v)\}$$

and

$$(\psi_1 \vee \psi_2)(p)(v) = \max\{\psi_1(p)(v), \psi_2(p)(v)\}$$

The intersection of these two fuzzy soft graphs is defined as a FSG $\tilde{G}_1 \cap \tilde{G}_2 = \langle P_1 \cap P_2, (V_1 \cap V_2, \phi_1 \wedge \phi_2, \psi_1 \wedge \psi_2)\rangle$ where the mappings $\phi_1 \wedge \phi_2 : P_1 \cap P_2 \to [0,1]^{V_1 \cap V_2}$ and $\psi_1 \wedge \psi_2 : P_1 \cap P_2 \to [0,1]^{V_1^2 \cap V_2^2}$ are respectively defined by

$$(\phi_1 \wedge \phi_2)(p)(v) = \min\{\phi_1(p)(v), \phi_2(p)(v)\}$$

and

$$(\psi_1 \wedge \psi_2)(p)(v) = \min\{\psi_1(p)(v), \psi_2(p)(v)\}$$

The order $O(\tilde{G})$ and size $S(\tilde{G})$ of the FSG $\tilde{G} = \langle P, (V, \phi, \psi)\rangle$ are respectively defined by

$$O(\tilde{G}) = \sum_{p \in P}(\sum_{v \in V} \phi(p)(v))$$

and

$$S(\tilde{G}) = \sum_{p \in P}(\sum_{uv \in V^2} \psi(p)(uv))$$

The FSG $\tilde{G}$ is called t-regular if for each $p \in P$, $\tilde{G}(p) = (V, \phi(p), \psi(p))$ is a regular fuzzy graph of degree t. It is called irregular if for each $p \in P$, $\tilde{G}(p) = (V, \phi(p), \psi(p))$ is an irregular fuzzy graph.

3 m-Polar Fuzzy Soft Graphs

In order to model data with multi-polar or multi-feature information, the notion of m-polar fuzzy soft set was developed by Khameneh and Kilicman [44] as below.

Before continuing, let U and P show the sets of alternatives and parameters, respectively, throughout the rest of this paper.

Definition 5 *([44]). The pair (f, P) where f is the mapping $f : P \to ([0,1]^m)^U$ is defined as an m-polar fuzzy soft set (shortly m-PFSS) over U.*

Note that, the m-PFSS (f, P) can be also presented by $f(p)(u) = (\pi_1 \circ f(p)(u), \pi_2 \circ f(p)(u), \ldots, \pi_m \circ f(p)(u))$ where $\pi_s \circ f(p) : U \to \mathbb{I}$ and $\pi_s : \mathbb{I}^m \to \mathbb{I}$ is the sth projection mapping.

An m-PFSS (f, P) is called a null m-PFSS, shown by $\tilde{\mathbf{0}}$, or an absolute m-PFSS, shown by $\tilde{\mathbf{1}}$, if for any $p \in P$ and $1 \leq s \leq m$ we have $\pi_s \circ f(p)(u) = 0$ and $\pi_s \circ f(p)(u) = 1$ for all $u \in U$, respectively.

Put $S = \{1, 2, \cdots, m\}$. If I is an indexing set and $\{(f_i, P_i)\}_{i \in I}$ is a family of m-PFSSs over the common universe U and parameter sets P_i, then for any $u \in U$ and $i, j \in I$:

1. $(f_i, P_i) \tilde{\leq} (f_j, P_j)$, if $P_i \subseteq P_j$ and $\pi_s \circ f_i(p)(u) \leq \pi_s \circ f_j(p)(u)$ for all $p \in P_i$ and $s \in S$.
2. $(\tilde{\bigvee}_{i \in I} f_i)(p)(u) = \sup_{i \in I}\{f_i(p)(u)\} = (\sup_i\{\pi_1 \circ f_i(p)(u)\}, \ldots, \sup_i\{\pi_m \circ f_i(p)(u)\})$ for all $p \in \bigcup_{i \in I} P_i$.

3. $(\widetilde{\bigwedge}_{i\in I} f_i)(p)(u) = \inf_{i\in I}\{f_i(p)(u)\} = (\inf_i\{\pi_1 \circ f_i(p)(u)\}, \ldots, \inf_i\{\pi_m \circ f_i(p)(u)\})$, for all $p \in \bigcap_{i\in I} P_i$.

where the operations $\widetilde{\bigvee}$ (fuzzy soft join) and $\widetilde{\bigwedge}$ (fuzzy soft meet) are defined by "sup" and "inf", respectively.

Definition 6. *Let $\Lambda = (f, P)$ be an m-PFSS over the set U. If the m-tuple $\mathbf{a} = (a_1, \ldots, a_m) \in [0,1]^m$ is a given threshold vector on membership degrees, then the* **a***-level soft set of Λ is a soft set $L(\Lambda; \mathbf{a}) = (f_a, P)$ defined by*

$$L(f(p); \mathbf{a}) = f_a(p) = \{u \in U : \pi_s \circ f(p)(u) \geq a_s; 1 \leq s \leq m\} \quad (1)$$

Note that, for each $s \in S = \{1, 2, \cdots, m\}$, the a_s-level soft sets (f_{a_s}, P) can be also defined by $f_{a_s}(p) = \{u \in U : \pi_s \circ f(p)(u) \geq a_s\}$ that implies $f_a(p) = \bigcap_{s=1}^m f_{a_s}(p)$.

Definition 7. *An m-polar fuzzy soft relation $(\tilde{R}, P)$, m-PFSR in brief, from U to V is an m-polar fuzzy soft set over $U \times V$ characterized by the membership mapping $\tilde{R} : P \to \mathbb{I}^{m^{U\times V}}$.*

Clearly, for each $p \in P$ the mapping $\tilde{R}(p) : U \times V \to \mathbb{I}^m$ is defined as an m-polar fuzzy relation on $U \times V$.

Definition 8. *Let $U = \{u_1, \ldots, u_N\}$ and $P = \{p_1, \ldots, p_M\}$ be the finite sets of alternatives and parameters, respectively. An m-polar fuzzy soft graph (shortly m-PFSG) is a 4-tuple $\tilde{G} = \langle P, (U, f, \tilde{R})\rangle$ where (f, P) is an m-PFSS on U (say m-polar fuzzy soft vertices) and $(\tilde{R}, P)$ is an m-PFSR on U (say m-polar fuzzy soft edges) such that for each $p_y \in P$, $1 \leq y \leq M$, $\tilde{G}(p_y) = (U, f(p_y), \tilde{R}(p_y))$ is an m-polar fuzzy graph on U that is*

$$\pi_s \circ \tilde{R}(p_y)(u_i u_j) \leq \min\{\pi_s \circ f(p_y)(u_i), \pi_s \circ f(p_y)(u_j)\}\colon i, j = 1, 2, \ldots, N \quad (2)$$

for each $s \in S$. The value $\pi_s \circ f(p_y)(u_i)$ shows the sth membership degree of the vertex u_i for parameter p_y and $\pi_s \circ \tilde{R}(p_y)(u_i u_j)$ shows the sth membership degree of the edge $u_i u_j$ for parameter p_y.

Remark 1. Note that, in the m-PFSG $\tilde{G} = \langle P, (U, f, \tilde{R})\rangle$ the m-PFSR $(\tilde{R}, P)$ is symmetric on U, that means $\pi_s \circ \tilde{R}(p_y)(u_i u_j) = \pi_s \circ \tilde{R}(p_y)(u_j u_i)$ for all $s \in S$ and $p_y \in P$. Otherwise, $\tilde{G} = \langle P, (U, f, \tilde{R})\rangle$ is called an m-PFS digraph.

Example 1. Let us suppose the m-PFSR $\tilde{R} : P \to \mathbb{I}^{m^{U\times U}}$ is defined on U by given utility values $\pi_s \circ f(p_y)(u_i)$ for each $s \in S$, $p_y \in P$ and $u_i \in U$ as below:

$$\pi_s \circ \tilde{R}(p_y)(u_i u_j) = \pi_s \circ f(p_y)(u_i) \cdot \pi_s \circ f(p_y)(u_j)$$

Then clearly, $\tilde{G} = \langle P, (U, f, \tilde{R})\rangle$ is an m-PFSG on U since it is satisfied in inequality (2) of Definition 8.

Now, let $U = \{u_1, u_2, u_3\}$, $P = \{p_1, p_2\}$ and $m = 2$. Suppose that $f : P \to \mathbb{I}^{2^U}$ describes the utility values of objects u_i (for $1 \leq i \leq 3$) with respect to parameters p_1 and p_2 by 2-polar data as below:

$$f(p_1) = \{\langle u_1, (0.3, 0.4)\rangle, \langle u_2, (0.5, 0.2)\rangle, \langle u_3, (0.7, 0.5)\rangle\}$$
$$f(p_2) = \{\langle u_1, (0.2, 0.6)\rangle, \langle u_2, (0.7, 0.9)\rangle, \langle u_3, (0.6, 0.5)\rangle\}$$

Then $\tilde{R} : P \to \mathbb{I}^{2^{U\times U}}$ is computed as

$$\tilde{R}(p_1) = \{\langle u_1u_2, (0.15, 0.08)\rangle, \langle u_1u_3, (0.21, 0.2)\rangle, \langle u_2u_3, (0.35, 0.1)\rangle\}$$
$$\tilde{R}(p_2) = \{\langle u_1u_2, (0.14, 0.54)\rangle, \langle u_1u_3, (0.12, 0.3)\rangle, \langle u_2u_3, (0.42, 0.45)\rangle\}$$

which shows $\tilde{G} = \langle P, (U, f, \tilde{R})\rangle$ is a 2-polar fuzzy soft graph on U where corresponding to each parameters p_1 and p_2 we have two different 2-polar fuzzy graphs $\tilde{G}(p_1) = (U, f(p_1), \tilde{R}(p_1))$ and $\tilde{G}(p_2) = (U, f(p_2), \tilde{R}(p_2))$ as shown in Fig. 1.

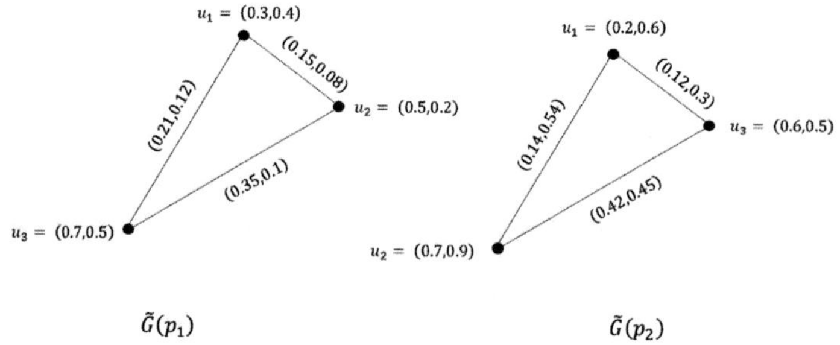

Fig. 1. 2-polar fuzzy graphs $\tilde{G}(p_1)$ and $\tilde{G}(p_2)$ of 2-PFSG $\tilde{G}$

Definition 9. *An m-polar fuzzy soft graph $\tilde{O} = \langle D, (U, h, \tilde{Q})\rangle$ is called m-PFS subgraph of $\tilde{G} = \langle P, (U, f, \tilde{R})\rangle$, shown by $\tilde{O}\tilde{\subseteq}\tilde{G}$, if $D \subseteq P$, $(h, D)\tilde{\leq}(f, P)$ and $(\tilde{Q}, D)\tilde{\leq}(\tilde{R}, P)$. Note that $\tilde{G}$ is known as m-PFS supergraph of $\tilde{O}$.*

Example 2. Consider the 2-PFSG $\tilde{G}$ given in Example 1. The 2-PFSG $\tilde{O} = \langle P, (U, h, \tilde{Q})\rangle$ given by:

$$h(p_1) = \{\langle u_1, (0.2, 0.2)\rangle, \langle u_2, (0.1, 0.2)\rangle, \langle u_3, (0.7, 0.5)\rangle\}$$
$$h(p_2) = \{\langle u_1, (0.2, 0.3)\rangle, \langle u_2, (0.1, 0.4)\rangle, \langle u_3, (0.4, 0.5)\rangle\}$$

and

$$\tilde{Q}(p_1) = \{\langle u_1u_2, (0, 0.03)\rangle, \langle u_1u_3, (0.14, 0.2)\rangle, \langle u_2u_3, (0.05, 0.1)\rangle\}$$
$$\tilde{Q}(p_2) = \{\langle u_1u_2, (0.02, 0.1)\rangle, \langle u_1u_3, (0.05, 0.3)\rangle, \langle u_2u_3, (0.1, 0.35)\rangle\}$$

is a subgraph of $\tilde{G}$.

The following proposition is the immediate result of Definitions 8 and 9.

Proposition 1. *Let* $\tilde{O} = \langle D, (U, h, \tilde{Q})\rangle$ *and* $\tilde{G} = \langle P, (U, f, \tilde{R})\rangle$ *be two* m*-PFSGs on* U *such that* $D \subseteq P$*. Then* $\tilde{O}$ *is an* m*-PFS subgraph of* $\tilde{G}$ *iff* $\tilde{O}(p)$ *is an* m*-polar fuzzy subgraph of* $\tilde{G}(p)$ *for each* $p \in D \subseteq P$*.*

Example 3. According to Examples 1 and 2, the 2-polar fuzzy graphs $\tilde{O}(p_1)$ and $\tilde{O}(p_2)$, shown in Fig. 2, are respectively the subgraphs of $\tilde{G}(p_1)$ and $\tilde{G}(p_2)$.

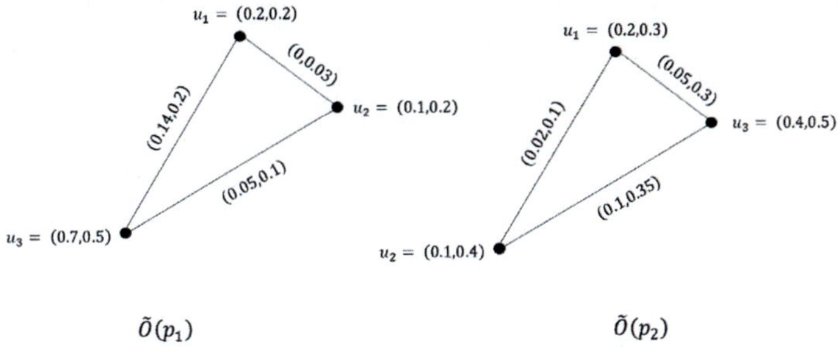

Fig. 2. 2-polar fuzzy soft subgraphs $\tilde{O}$ of $\tilde{G}$

Definition 10. *Let* $\tilde{G} = \langle P, (U, f, \tilde{R})\rangle$ *be an* m*-PFSG on* U*. The order of* $\tilde{G}$*, shown by* $O(\tilde{G})$*, and the size of* $\tilde{G}$*, shown by* $S(\tilde{G})$*, are defined as the following* $M \times m$ *matrices:*

$$O(\tilde{G}) = \begin{bmatrix} \sum_{i=1}^{N} \pi_1 \circ f(p_1)(u_i) & \cdots & \sum_{i=1}^{N} \pi_m \circ f(p_1)(u_i) \\ \vdots & \ddots & \vdots \\ \sum_{i=1}^{N} \pi_1 \circ f(p_M)(u_i) & \cdots & \sum_{i=1}^{N} \pi_m \circ f(p_M)(u_i) \end{bmatrix}$$

$$S(\tilde{G}) = \begin{bmatrix} \sum_{i,j=1}^{N} \pi_1 \circ \tilde{R}(p_1)(u_i u_j) & \cdots & \sum_{i,j=1}^{N} \pi_m \circ \tilde{R}(p_1)(u_i u_j) \\ \vdots & \ddots & \vdots \\ \sum_{i,j=1}^{N} \pi_1 \circ \tilde{R}(p_M)(u_i u_j) & \cdots & \sum_{i,j=1}^{N} \pi_m \circ \tilde{R}(p_M)(u_i u_j) \end{bmatrix}$$

Thus, corresponding to any parameter $p_y \in P$, each row of $O(\tilde{G})$ and $S(\tilde{G})$ respectively show the order and size of the m-polar fuzzy graph $\tilde{G}(p_y)$.

Note that, if $\tilde{O} = \langle P, (U, h, \tilde{Q})\rangle$ is an m-PFS subgraph of $\tilde{G} = \langle P, (U, f, \tilde{R})\rangle$, then $O(\tilde{O}) \le O(\tilde{G})$ and $S(\tilde{O}) \le S(\tilde{G})$.

Example 4. Reconsider Examples 1 and 2. The order and size matrices of the 2-PFSGs $\tilde{G}$ and $\tilde{O}$ are as below:

$$O(\tilde{G}) = \begin{bmatrix} 1.5 & 1.1 \\ 1.5 & 2 \end{bmatrix}_{2\times 2}, S(\tilde{G}) = \begin{bmatrix} 0.71 & 0.38 \\ 0.68 & 1.29 \end{bmatrix}_{2\times 2}$$

$$O(\tilde{O}) = \begin{bmatrix} 1 & 0.9 \\ 0.7 & 1.2 \end{bmatrix}_{2\times 2}, S(\tilde{O}) = \begin{bmatrix} 0.19 & 0.33 \\ 0.17 & 0.75 \end{bmatrix}_{2\times 2}$$

The following theorem shows the relationship between an m-PFSG $\tilde{G}$ and its corresponding p_y^s-underlying crisp graphs. The proof is straightforward, thus omitted.

Theorem 1. *The $\tilde{G} = \langle P, (U, f, \tilde{R})\rangle$ is an m-PFSG on U iff for any $p_y \in P$ (where $1 \leq y \leq M$) and $1 \leq s \leq m$, $G^s(p_y) = (F^s(p_y), R^s(p_y)) = (supp(\pi_s \circ f(p_y)), supp(\pi_s \circ \tilde{R}(p_y)))$ defined by*

$$F^s(p_y) = \{u_i \in U : 0 < \pi_s \circ f(p_y)(u_i)\} \tag{3}$$

$$R^s(p_y) = \Big\{u_i u_j \in U \times U : 0 < \pi_s \circ \tilde{R}(p_y)(u_i u_j) \leq \min\{\pi_s \circ f(p_y)(u_i), \pi_s \circ f(p_y)(u_j)\}\Big\} \tag{4}$$

is a crisp graph called p_y^s-underlying crisp graph of $\tilde{G}$.

Corollary 1. *Let $\tilde{G} = \langle P, (U, f, \tilde{R})\rangle$ be an m-PFSG on U. Then, for any $p_y \in P$ we can define the crisp graph $G(p_y) = (F(p_y), R(p_y))$, called p_y-underlying crisp graph of $\tilde{G}$, as below*

$$F(p_y) = supp(f(p_y)) = \{u_i \in U : f(p_y)(u_i) > \tilde{\mathbf{0}}\} \tag{5}$$

$$R(p_y) = supp(\tilde{R}(p_y)) = \Big\{u_i u_j \in U \times U : \tilde{\mathbf{0}} < \tilde{R}(p_y)(u_i u_j) \leq \inf\{f(p_y)(u_i), f(p_y)(u_j)\}\Big\} \tag{6}$$

Clearly $G(p_y) = \bigcap_{s=1}^m G^s(p_y)$. If $\tilde{O} = \langle D, (U, h, \tilde{Q})\rangle$ is an m-PFS subgraph of $\tilde{G}$, then the p_y-underlying crisp graph $O(p_y) = (H(p_y), Q(p_y))$ of $\tilde{O}$ is a subgraph of $G(p_y) = (F(p_y), R(p_y))$.

Example 5. Reconsider Example 1. Clearly, $G(p_1) = G(p_2)$ where $F(p_1) = F(p_2) = \{u_1, u_2, u_3\}$ and $R(p_1) = R(p_2) = \{u_1u_2, u_1u_3, u_2u_3\}$ are, respectively, p_1-underlying crisp graph and p_2-underlying crisp graph of $\tilde{G}$, as shown in Fig. 3. Moreover, for any $s = 1, 2$ four crisp graphs $G^1(p_1) = G^2(p_1) = G^1(p_2) = G^2(p_2)$ can be defined by vertices $F^s(p_1) = F^s(p_2) = \{u_1, u_2, u_3\}$ and edges $R^s(p_1) = R^s(p_2) = \{u_1u_2, u_1u_3, u_2u_3\}$.

Example 6. As another case, reconsider Example 2.

It is easy to check that $H(p_1) = H(p_2) = \{u_1, u_2, u_3\}$, $Q(p_1) = \{u_1u_3, u_2u_3\}$, and $Q(p_2) = \{u_1u_2, u_1u_3, u_2u_3\}$ are respectively determined the p_1-underlying crisp graph and p_2-underlying crisp graph of $\tilde{O}$, as shown in Fig. 4. Clearly, for each $i = 1, 2$, $O(p_i)$ is a subgraph of $G(p_i)$.

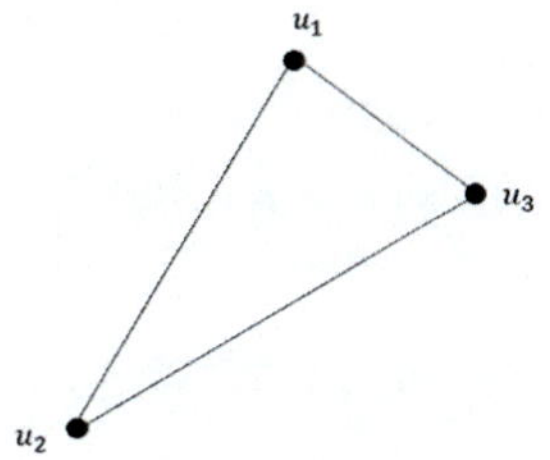

Fig. 3. Underlying crisp graph $G(p_1) = G(p_2)$ of $\tilde{G}$

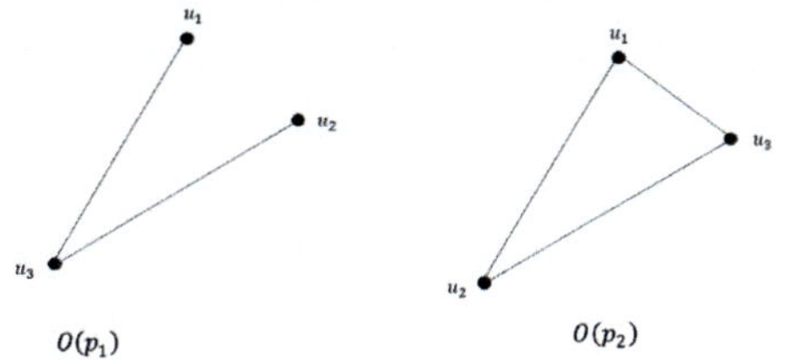

Fig. 4. Underlying crisp graphs $O(p_1)$ and $O(p_2)$ of $\tilde{O}$

Definition 11. *An m-PFSG $\tilde{G}$ is called strong if for all $p_y \in P$ and $u_i u_j \in R(p_y)$ we have*

$$\pi_s \circ \tilde{R}(p_y)(u_i u_j) = \min\{\pi_s \circ f(p_y)(u_i), \pi_s \circ f(p_y)(u_j)\}$$

for all $1 \leq s \leq m$.
The m-PFSG $\tilde{G}$ is called complete if for all $p_y \in P$ and $u_i, u_j \in F(p_y)$ we have

$$\pi_s \circ \tilde{R}(p_y)(u_i u_j) = \min\{\pi_s \circ f(p_y)(u_i), \pi_s \circ f(p_y)(u_j)\}$$

for all $1 \leq s \leq m$.

According to Definitions 8 and 11; and Theorem 1, we have the following result.

Proposition 2. *The m-PFSG $\tilde{G} = \langle P, (U, f, \tilde{R})\rangle$ is strong (complete) iff for all $p_y \in P$, the m-polar FG $\tilde{G}(p_y) = (U, f(p_y), \tilde{R}(p_y))$ is strong (complete). Moreover, if $\tilde{G}$ is complete then for any $p_y \in P$, $G^s(p_y) = (F^s(p_y), R^s(p_y))$ and $G(p_y) = (F(p_y), R(p_y))$ are also complete crisp graphs.*

Definition 12. *Let $\{\tilde{G}_k = \langle P_k, (U, f_k, \tilde{R}_k)\rangle : k = 1, \cdots, K\}$ be a set of several m-PFSGs on U.*

- *The union of $\tilde{G}_k$s is an m-PFSG $\tilde{\bigcup}_{k=1}^K \tilde{G}_k = \langle \bigcup_{k=1}^K P_k, (U, \bigvee_{k=1}^K f_k, \bigvee_{k=1}^K \tilde{R}_k)\rangle$ such that $\bigvee_{k=1}^K f_k : \bigcup_{k=1}^K P_k \to \mathbb{I}^{mU}$ and $\bigvee_{k=1}^K \tilde{R}_k : \bigcup_{k=1}^K P_k \to \mathbb{I}^{mU \times U}$ are* *defined by*

$$\pi_s \circ (\bigvee_{k=1}^{K} f_k)(p)(u) = \max_k\{\pi_s \circ f_k(p)(u) : p \in \cup_{k=1}^{K} P_k, u \in U\}$$

$$\pi_s \circ (\bigvee_{k=1}^{K} \tilde{R}_k)(p)(uv) = \max_k\{\pi_s \circ \tilde{R}_k(p)(uv) : p \in \cup_{k=1}^{K} P_k, uv \in U \times U\}$$

where $1 \leq s \leq m$, $p \in \bigcup_{k=1}^{K} P_k$ *and* $u \in U$.

- *The intersection of* $\tilde{G}_k s$ *is an m-PFSG* $\tilde{\bigcap}_{k=1}^{K}\tilde{G}_k = \langle \bigcap_{k=1}^{K} P_k, (U, \bigwedge_{k=1}^{K} f_k, \bigwedge_{k=1}^{K} \tilde{R}_k)\rangle$ *such that* $\bigcap_{k=1}^{K} P_k \neq \emptyset$ *and* $\bigwedge_{k=1}^{K} f_k : \bigcap_{k=1}^{K} P_k \to \mathbb{I}^{mU}$ *and* $\bigwedge_{k=1}^{K} \tilde{R}_k : \bigcap_{k=1}^{K} P_k \to \mathbb{I}^{mU\times U}$ *are defined by*

$$\pi_s \circ (\bigwedge_{k=1}^{K} f_k)(p)(u) = \min_k\{\pi_s \circ f_k(p)(u) : p \in \cup_{k=1}^{K} P_k, u \in U\}$$

$$\pi_s \circ (\bigwedge_{k=1}^{K} \tilde{R}_k)(p)(uv) = \min_k\{\pi_s \circ \tilde{R}_k(p)(uv) : p \in \cup_{k=1}^{K} P_k, uv \in U \times U\}$$

where $1 \leq s \leq m$, $p \in \bigcup_{k=1}^{K} P_k$ *and* $u, v \in U$.

Note that for any $1 \leq s \leq m$ *if there exists some* $1 \leq k \leq K$ *such that* $p \notin P_k$, *then we put* $\pi_s \circ f_k(p)(u) = 0$ *and* $\pi_s \circ \tilde{R}_k(p)(uv) = 0$.

Example 7. Let $U = \{u_1, u_2, u_3\}$ and $P = \{p_1, p_2\}$ are given. The 3-PFSGs $\tilde{G}_1\tilde{\cup}\tilde{G}_2$, given in Figs. 5 and 7, and the 3-PFSGs $\tilde{G}_1\tilde{\cap}\tilde{G}_2$, shown in Figs. 6 and 8, are respectively the union and intersection of two 3-PFSGs $\tilde{G}_1$ and $\tilde{G}_2$ based on different parameters p_1, p_2.

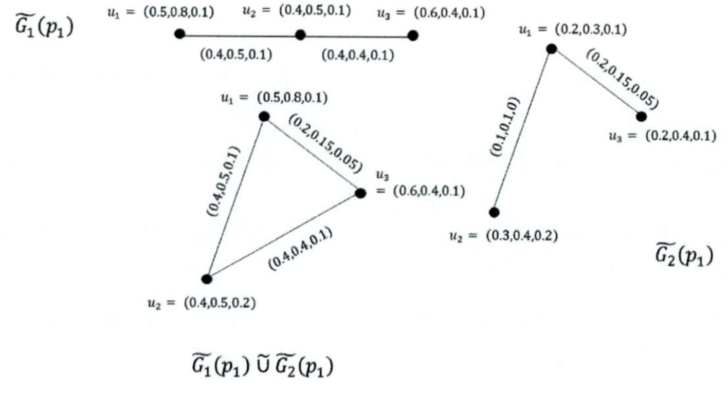

Fig. 5. Union of 3-polar fuzzy soft graphs $\tilde{G}_1, \tilde{G}_2$ corresponding to parameter p_1

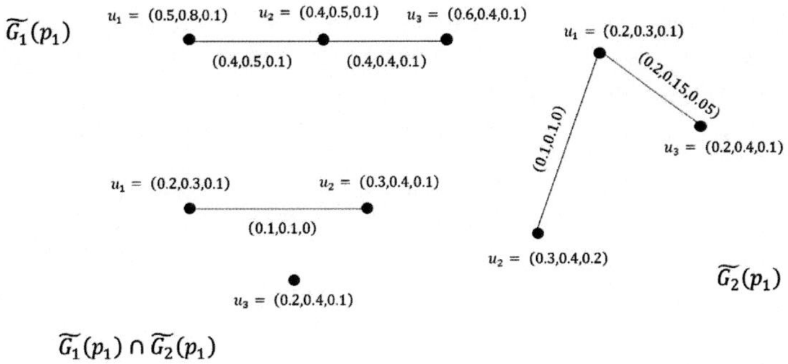

Fig. 6. Intersection of 3-polar fuzzy soft graphs $\tilde{G}_1, \tilde{G}_2$ corresponding to parameter p_1

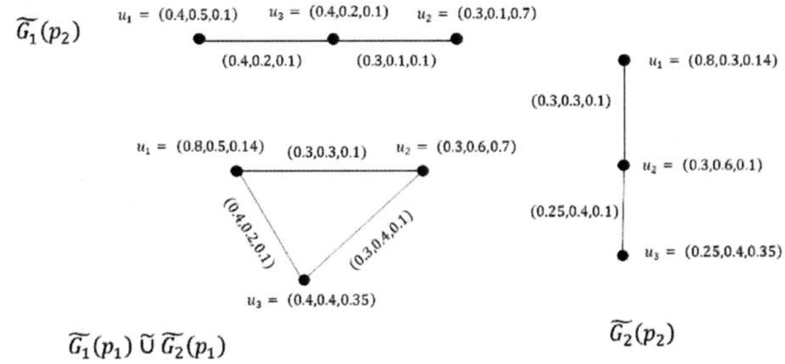

Fig. 7. Union of 3-polar fuzzy soft graphs $\tilde{G}_1, \tilde{G}_2$ corresponding to parameter p_2

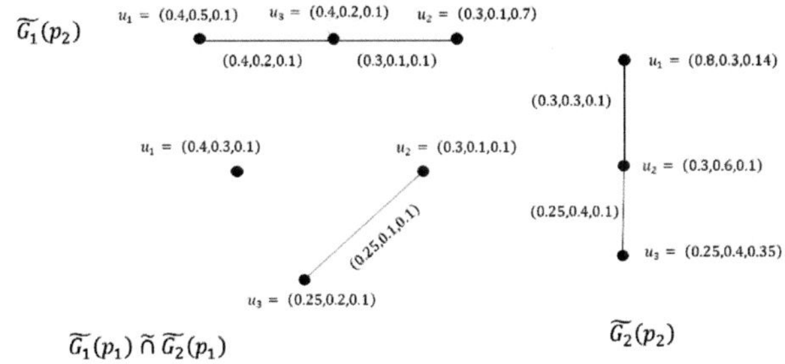

Fig. 8. Intersection of 3-polar fuzzy soft graphs $\tilde{G}_1, \tilde{G}_2$ corresponding to parameter p_2

Remark 2. Let $\{\tilde{G}_k = \langle P_k, (U, f_k, \tilde{R}_k)\rangle : k = 1, \cdots, K\}$ be a set of several m-PFSGs on U. If $\bigcap_{k=1}^{K} P_k \neq \emptyset$, then corresponding to any $p \in \bigcap_{k=1}^{K} P_k$ the m-polar fuzzy graphs $(\tilde{\bigcup}_{k=1}^{K} \tilde{G}_k)(p) = \big(U, \bigvee_{k=1}^{K} f_k(p), \bigvee_{k=1}^{K} \tilde{R}_k(p)\big)$ and $(\tilde{\bigcap}_{k=1}^{K} \tilde{G}_k)(p) = \big(U, \bigwedge_{k=1}^{K} f_k(p), \bigwedge_{k=1}^{K} \tilde{R}_k(p)\big)$ are respectively equal to the union and intersection of m-polar fuzzy graphs $\tilde{G}_k(p) = \big(U, f_k(p), \tilde{R}_k(p)\big)$, where for each $1 \leq k \leq K$ $\tilde{G}_k(p) = (U, f(p), \tilde{R}(p))$ is the m-polar fuzzy graph of $\tilde{G}_k$ corresponding to the parameter p.

In decision making problems, each decision maker has a level of satisfying that is measured based on different criteria. Objects that meet this satisfaction level are usually preferred to others.

According to Definition 6, this issue can be formulated by using the concept of level graph of an m-PFSG $\tilde{G} = \langle P, (U, f, \tilde{R})\rangle$ as introduced in the following proposition.

Proposition 3. *Let* $\tilde{G} = \langle P, (U, f, \tilde{R})\rangle$ *be an* m*-PFSG on* U*. If* $\mathbf{a} = (a_1, \ldots, a_m) \in [0,1]^m$ *is a given satisfaction threshold vector on membership degrees, then* $\tilde{G}_a = \langle P, (U, f_a, \tilde{R}_a)\rangle$ *is a soft graph called* **a***-level soft graph of* $\tilde{G}$ *where for any* $p_y \in P$*:*

$$f_a(p_y) = \{u_i \in U : (\pi_1 \circ f(p_y)(u_i), \ldots, \pi_m \circ f(p_y)(u_i)) \geq (a_1, \ldots, a_m)\} \tag{7}$$

$$\tilde{R}_a(p_y) = \{u_iu_j \in U \times U : (\pi_1 \circ \tilde{R}(p_y)(u_iu_j), \ldots, \pi_m \circ \tilde{R}(p_y)(u_iu_j)) \geq (a_1, \ldots, a_m)\} \tag{8}$$

Corollary 2. *Let* $\tilde{G} = \langle P, (U, f, \tilde{R})\rangle$ *be an* m*-PFSG on* U *and* $\mathbf{a} = (a_1, \ldots, a_m) \in [0,1]^m$ *is a given threshold vector. Then, for any* $p_y \in P$*,* $\tilde{G}_a(p_y) = (f_a(p_y), \tilde{R}_a(p_y))$ *is a subgraph of* p_y*-underlying crisp graph* $G(p_y) = (F(p_y), R(p_y))$*, discussed in Corollary 1.*

Example 8. Let $\tilde{G} = \langle P, (U, f, \tilde{R})\rangle$ be a 3-PFSG on $U = \{u_1, u_2, u_3, u_4\}$ and $P = \{p_1, p_2\}$ such that the mapping $f : P \to \mathbb{I}^{3^U}$ is given by

$$f(p_1) = \{\langle u_1, (0.5, 0.6, 0.3)\rangle, \langle u_2, (0.7, 0.7, 0.4)\rangle, \langle u_3, (0.9, 0.6, 0.6)\rangle, \langle u_3, (0.8, 0.6, 0.5)\rangle\}$$
$$f(p_2) = \{\langle u_1, (0.4, 0.45, 0.6)\rangle, \langle u_2, (0.5, 0.4, 0.7)\rangle, \langle u_3, (0.6, 0.73, 0.53)\rangle, \langle u_3, (0.65, 0.7, 0.6)\rangle\}$$

and $\tilde{R} : P \to \mathbb{I}^{3^{U\times U}}$ is defined as

$$\tilde{R}(p_1) = \{\langle u_1u_2, (0.5, 0.5, 0.3)\rangle, \langle u_1u_3, (0.5, 0.6, 0.3)\rangle, \langle u_1u_4, (0.5, 0.6, 0.2)\rangle,$$
$$\langle u_2u_3, (0.7, 0.6, 0.3)\rangle, \langle u_2u_4, (0.6, 0.6, 0.4)\rangle, \langle u_3u_4, (0.8, 0.5, 0.5)\rangle\}$$

$$\tilde{R}(p_2) = \{\langle u_1u_2, (0.4, 0.3, 0.5)\rangle, \langle u_1u_3, (0.3, 0.4, 0.5)\rangle, \langle u_1u_4, (0.4, 0.4, 0.55)\rangle,$$
$$\langle u_2u_3, (0.5, 0.4, 0.53)\rangle, \langle u_2u_4, (0.5, 0.4, 0.5)\rangle, \langle u_3u_4, (0.6, 0.7, 0.5)\rangle\}.$$

(see Fig. 9). Take $\mathbf{a} = (0.6, 0.4, 0.3) \in \mathbb{I}^3$. The $(0.6, 0.4, 0.3)$-level soft graph of $\tilde{G}$ is the graph $\tilde{G}_{(0.6,0.4,0.3)}(p_y) = \langle P, (f_{(0.6,0.4,0.3)}, \tilde{R}_{(0.6,0.4,0.3)})\rangle$ where

$$f_{(0.6,0.4,0.3)}(p_1) = \{u_2, u_3, u_4\}$$
$$f_{(0.6,0.4,0.3)}(p_2) = \{u_3, u_4\}$$

and

$$\tilde{R}_{(0.6,0.4,0.3)}(p_1) = \{u_2u_3, u_2u_4, u_3u_4\}$$
$$\tilde{R}_{(0.6,0.4,0.3)}(p_2) = \{u_3u_4\}$$

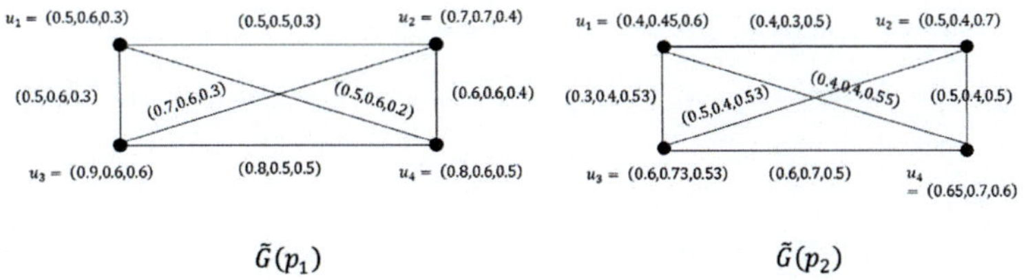

Fig. 9. 3-polar fuzzy graphs $\tilde{G}(p_1)$ and $\tilde{G}(p_2)$ of 3-polar fuzzy soft graph $\tilde{G}$

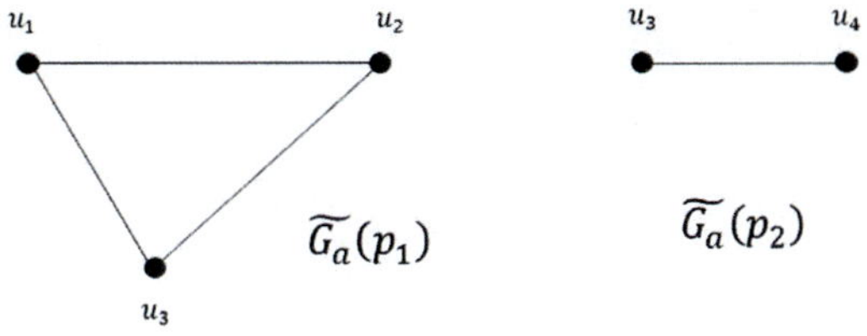

Fig. 10. Level graphs $\tilde{G}_a(p_1)$ and $\tilde{G}_a(p_2)$ for $\mathbf{a} = (0.6, 0.4, 0.3)$

Fig. 11. p_1-underlying crisp graph $G(p_1)$ and p_2-underlying crisp graph $G(p_2)$

The two graphs $\tilde{G}_{(0.6,0.4,0.3)}(p_1)$ and $\tilde{G}_{(0.6,0.4,0.3)}(p_2)$ are, respectively, subgraphs of p_1-underlying crisp graph $G(p_1) = (F(p_1), R(p_1))$ and p_2-underlying crisp graph $G(p_2) = (F(p_2), R(p_2))$ where $F(p_1) = F(p_2) = \{u_1, u_2, u_3, u_4\}$ and $R(p_1) = R(p_2) = \{u_1u_2, u_1u_3, u_1u_4, u_2u_3, u_2u_4, u_3u_4\}$ (see Figs. 10 and 11).

Theorem 2. *$\tilde{G} = \langle P, (U, f, \tilde{R})\rangle$ is an m-PFSG on U iff for all $p_y \in P$ and any $\mathbf{a} \in [0,1]^m$, $\tilde{G}_a(p_y) = (f_a(p_y), \tilde{R}_a(p_y))$ is a crisp graph on U.*

Proof. Let $\tilde{G} = \langle P, (U, f, \tilde{R})\rangle$ be an m-PFSG on U. By Proposition 3, $\tilde{G}_a(p_y) = (f_a(p_y), \tilde{R}_a(p_y))$ is a crisp graph on U. Conversely, let for any $\mathbf{a} \in [0,1]^m$ and $p_y \in P$, $\tilde{G}_a(p_y) = (f_a(p_y), \tilde{R}_a(p_y))$ is a crisp graph on U. Take $u_iu_j \in U \times U$ and let for each $s \in S$, $\pi_s \circ \tilde{R}(p_y)(u_iu_j) = b_s \in [0,1]$. Put $\mathbf{b} = (b_1, \cdots, b_s, \cdots, b_m)$. Thus $u_iu_j \in \tilde{R}_b(p_y)$ and consequently $u_i, u_j \in f_b(p_y)$ since $\tilde{G}_b(p_y)$ is a simple crisp graph; equivalently, $\pi_s \circ f(p_y)(u_i) \geq b_s$ and $\pi_s \circ f(p_y)(u_j) \geq b_s$. Thus

$$b_s = \pi_s \circ \tilde{R}(p_y)(u_i, u_j) \leq \min\{\pi_s \circ f(p_y)(u_i), \pi_s \circ f(p_y)(u_j)\}$$

This shows $\tilde{G} = \langle P, (U, f, \tilde{R})\rangle$ is an m-polar fuzzy soft graph on U.

The following results can be obtained easily.

Proposition 4. *Let $\tilde{G} = \langle P, (U, f, \tilde{R})\rangle$ be an m-PFSG on U and* $\mathbf{a}, \mathbf{b} \in [0, 1]^m$. *Then* $\tilde{G}_a \tilde{\subseteq} \tilde{G}_b$ *if* $\mathbf{a} \geq \mathbf{b}$.

Proposition 5. *Let $\tilde{G} = \langle P, (U, f, \tilde{R})\rangle$ and $\tilde{O} = \langle D, (U, h, \tilde{Q})\rangle$ be two m-PFSGs on U. For any $\mathbf{a} \in [0, 1]^m$, the soft graph $\tilde{G}_a$ is a subgraph of $\tilde{O}_a$ if $\tilde{G}\tilde{\subseteq}\tilde{Q}$.*

Theorem 3. *Let $\{\tilde{G}_k = \langle P_k, (U, f_k, \tilde{R}_k)\rangle : k = 1, \cdots, K\}$ be a set of several m-PFSGs on U. Then for any $\mathbf{a} \in [0, 1]^m$ the* **a***-level soft graph of $\tilde{\bigcup}_{k=1}^{K} \tilde{G}_k$ is the union of $(\tilde{G}_k)_a$ and the* **a***-level soft graph of $\tilde{\bigcap}_{k=1}^{K} \tilde{G}_k$ is the intersection of $(\tilde{G}_k)_a$.*

In the following subsections, two important classes of m-PFSG, so-called regular m-PFSG and irregular m-PFSG, are discussed.

3.1 Regular m-Polar Fuzzy Soft Graphs

This section is started by definition of vertex degree in an m-PFSG and then continued by studying the concept of regular m-PFSG.

Definition 13. *Let $p_y \in P$ be a given parameter. The open p_y-neighborhood degree of vertex $u_i \in U$ in m-PFSG $\tilde{G} = \langle P, (U, f, \tilde{R})\rangle$ is defined as an m-tuple*

$$\begin{aligned} d_{\tilde{G}}(p_y)(u_i) &= (d^1_{\tilde{G}}(p_y)(u_i), \ldots, d^m_{\tilde{G}}(p_y)(u_i)) \\ &= \Big(\sum_{j \neq i; u_j \in N^1_{\tilde{G}}(p_y)(u_i)} \pi_1 \circ f(p_y)(u_j), \ldots, \sum_{j \neq i; u_j \in N^m_{\tilde{G}}(p_y)(u_i)} \pi_m \circ f(p_y)(u_j) \Big) \end{aligned}$$

where for any $s = 1, \cdots, m$ the set

$$N^s_{\tilde{G}}(p_y)(u_i) = \{u_j \in F^s(p_y) : u_i u_j \in R^s(p_y), j \neq i\}$$

is the p^s_y-neighborhood of vertex u_i in $\tilde{G}$. Note that, the $N_{\tilde{G}}(p_y)(u_i) = \bigcap_{s=1}^{m} N^s_{\tilde{G}}(p_y)(u_i)$ and $N_{\tilde{G}}(u_i) = \bigcap_{y=1}^{M} \bigcap_{s=1}^{m} N^s_{\tilde{G}}(p_y)(u_i)$ are called, respectively, the p_y-neighborhood of u_i and neighborhood of u_i in $\tilde{G}$.

An open neighborhood degrees matrix of vertex u_i is defined by the following $M \times m$ matrix

$$D_{\tilde{G}}(u_i) = \begin{bmatrix} d^1_{\tilde{G}}(p_1)(u_i) & \cdots & d^m_{\tilde{G}}(p_1)(u_i) \\ \vdots & \ddots & \vdots \\ d^1_{\tilde{G}}(p_M)(u_i) & \cdots & d^m_{\tilde{G}}(p_M)(u_i) \end{bmatrix}$$

The closed p_y-neighborhood degree of u_i in the m-PFSG $\tilde{G} = \langle P, (U, f, \tilde{R})\rangle$ is defined by

$$\begin{aligned} d_{\tilde{G}}(p_y)[u_i] &= d_{\tilde{G}}(p_y)(u_i) + f(p_y)(u_i) \\ &= (d^1_{\tilde{G}}(p_y)(u_i) + f^1(p_y)(u_i), \ldots, d^m_{\tilde{G}}(p_y)(u_i) + f^m(p_y)(u_i)) \end{aligned}$$

We can also define a closed neighborhood degrees matrix of vertex u_i by the following $M \times m$ matrix

$$D_{\tilde{G}}[u_i] = \begin{bmatrix} d^1_{\tilde{G}}(p_1)[u_i] & \cdots & d^m_{\tilde{G}}(p_1)[u_i] \\ \vdots & \ddots & \vdots \\ d^1_{\tilde{G}}(p_M)[u_i] & \cdots & d^m_{\tilde{G}}(p_M)[u_i] \end{bmatrix}$$

Remark 3. Let $\tilde{G} = \langle P, (U, f, \tilde{R}) \rangle$ be an m-PFSG on U and $p_y \in P$. Then, clearly

- $d_{\tilde{G}}(p_y)(u_i) = d_{\tilde{G}(p_y)}(u_i)$, that means the open p_y-neighborhood degree of vertex u_i in m-polar fuzzy soft graph $\tilde{G}$ equals to open neighborhood degree of u_i in m-polar fuzzy graph $\tilde{G}(p_y)$. Accordingly $d_{\tilde{G}}(p_y)[u_i] = d_{\tilde{G}(p_y)}[u_i]$.
- $N^s_{\tilde{G}}(p_y)(u_i) = N_{G^s(p)}(u_i)$ means that $N^s_{\tilde{G}}(p_y)(u_i)$ is in fact the neighborhood set for vertex u_i in crisp graph $G^s(p_y)$.

Example 9. Consider the 2-PFSG $\tilde{G}$ given in Example 1. The open neighborhood degrees matrices $D_{\tilde{G}}(u_i)$ of vertices u_i (for $i = 1, 2, 3$) are as below

$$D_{\tilde{G}}(u_1) = \begin{bmatrix} 1.2 & 0.7 \\ 1.3 & 1.4 \end{bmatrix}_{2\times2}, D_{\tilde{G}}(u_2) = \begin{bmatrix} 1 & 0.9 \\ 0.8 & 1.1 \end{bmatrix}_{2\times2},$$
$$D_{\tilde{G}}(u_3) = \begin{bmatrix} 0.8 & 0.6 \\ 0.9 & 1.5 \end{bmatrix}_{2\times2}$$

where $N^s_{\tilde{G}}(p_y)(u_1) = \{u_2, u_3\}$, $N^s_{\tilde{G}}(p_y)(u_2) = \{u_1, u_3\}$ and $N^s_{\tilde{G}}(p_y)(u_3) = \{u_1, u_2\}$ for all $p_1, p_2 \in P$ and $s = 1, 2$. Furthermore, $N_{\tilde{G}}(u_1) = \{u_2, u_3\}$, $N_{\tilde{G}}(u_2) = \{u_1, u_3\}$ and $N_{\tilde{G}}(u_3) = \{u_1, u_2\}$.
The closed neighborhood degrees matrices $D_{\tilde{G}}[u_i]$ of vertices u_i (for $i = 1, 2, 3$) are shown as below

$$D_{\tilde{G}}[u_1] = \begin{bmatrix} 1.5 & 1.1 \\ 1.5 & 2 \end{bmatrix}_{2\times2}, D_{\tilde{G}}[u_2] = \begin{bmatrix} 1.5 & 1.1 \\ 1.5 & 2 \end{bmatrix}_{2\times2},$$
$$D_{\tilde{G}}[u_3] = \begin{bmatrix} 1.5 & 1.1 \\ 1.5 & 2 \end{bmatrix}_{2\times2}$$

Definition 14. *If the open p_y-neighborhood degree $d_{\tilde{G}}(p_y)(u_i)$ for all vertices of $\tilde{G}$ is the same vector $\mathbf{t} = (t_1, \cdots, t_m)$, then $\tilde{G}$ is called a p_y-regular m-PFSG of degree* $\mathbf{t}$. *In other words, the m-PFSG $\tilde{G}$ is p_y-regular if $\tilde{G}(p_y)$ is a regular m-PFG. Moreover, if the open neighborhood degrees matrix $D_{\tilde{G}}(u_i)$ for all vertices of $\tilde{G}$ is the same, then $\tilde{G}$ is called a regular m-polar fuzzy soft graph.*

$\tilde{G}$ is called a totally p_y-regular m-PFSG if for all vertices of $\tilde{G}$ the closed p_y-neighborhood degree $d_{\tilde{G}}(p_y)[u_i]$ is the same. Similarly, if the closed neighborhood degrees matrix $D_{\tilde{G}}[u_i]$ for all vertices of $\tilde{G}$ is the same, then $\tilde{G}$ is called a totally regular m-PFSG.

Example 10. From Example 9, it is easy to check that the 2-PFSG given in Example 1 is totally regular, but not regular 2-PFSG.

Corollary 3. *$\tilde{G} = \langle P, (U, f, \tilde{R})\rangle$ is a regular m-PFSG iff for all $p_y \in P$, $\tilde{G}$ is a p_y-regular m-PFSG. $\tilde{G} = \langle P, (U, f, \tilde{R})\rangle$ is a totally regular m-PFSG iff for all $p_y \in P$, $\tilde{G}$ is a totally p_y-regular m-PFSG.*

Remark 4. It is clear that the m-PFSG $\tilde{G} = \langle P, (U, f, \tilde{R})\rangle$ is p_y-regular iff $\tilde{G}(p_y) = (U, f(p_y), \tilde{R}(p_y))$ is a regular m-PFG.

Based on Definitions 13 and 14, the following results can be obtained easily.

Proposition 6. *Let $\tilde{G} = \langle P, (U, f, \tilde{R})\rangle$ be a p_y-regular m-PFSG and $f(p_y) : U \to \mathbb{I}^m$ be a constant function. Then $\tilde{G}$ is a totally p_y-regular m-PFSG.*

Proposition 7. *Let $\tilde{G} = \langle P, (U, f, \tilde{R})\rangle$ be a totally p_y-regular m-PFSG and $f(p_y) : U \to \mathbb{I}^m$ be a constant function. Then $\tilde{G}$ is a p_y-regular m-PFSG.*

Theorem 4. *If $\tilde{G} = \langle P, (U, f, \tilde{R})\rangle$ is regular and totally regular m-PFSG on U. Then for any $p_y \in P$, the mapping $f(p_y) : U \to \mathbb{I}^m$ is constant.*

Proof. Let $\tilde{G} = \langle P, (U, f, \tilde{R})\rangle$ be regular and totally regular m-PFSG on U. Then for all $u_i \in U$, the open neighborhood degrees matrices $D_{\tilde{G}}(u_i) = L$ where L is an $M \times m$ matrix with fixed entries $[l_{ys}]_{M\times m}$ such that $l_{ys} = d^s_{\tilde{G}}(p_y)(u_i)$ and the closed neighborhood degrees matrices $D_{\tilde{G}}[u_i] = T$ where T is a fixed $M \times m$ matrix with entries $[t_{ys}]_{M\times m}$ such that $t_{ys} = d^s_{\tilde{G}}(p_y)[u_i]$ for each $s \in S$. Hence $\pi_s \circ f(p_y)(u_i) = t_{ys} - l_{ys}$. Therefore, for any $u \in U$: $f(p_y)(u_i) = (\pi_1 \circ f(p_y)(u_i), \cdots, \pi_m \circ f(p_y)(u_i)) = (t_{y1} - l_{y1}, \cdots, t_{ym} - l_{ym})$. That means $f(p_y)$ is constant.

Corollary 4. *Consider the m-PFSG $\tilde{G} = \langle P, (U, f, \tilde{R})\rangle$ on U. The function $f(p_y) : U \to \mathbb{I}^m$ is constant iff the following conditions are equivalent:*

1. *$\tilde{G}$ is a regular m-PFSG.*
2. *$\tilde{G}$ is a totally regular m-PFSG.*

Theorem 5. *If $\tilde{G} = \langle P, (U, f, \tilde{R})\rangle$ is p_y-regular m-PFSG and totally p_y-regular m-PFSG. Then for all $s \in S$, the graphs $G^s(p_y) = (F^s(p_y), R^s(p_y))$ and $G(p_y) = (F(p_y), R(p_y))$ are regular.*

Proof. Consider the parameter $p_y \in P$ that the m-PFSG $\tilde{G} = \langle P, (U, f, \tilde{R})\rangle$ is p_y-regular and totally p_y-regular. Now let for some $1 \leq s \leq m$, the crisp graph $G^s(p_y) = (F^s(p_y), R^s(p_y))$ is not regular. This means that there exist two different vertices $u_i, u_j \in G^s(p_y)$ such that $\mid N_{G^s(p_y)}(u_i) \mid \neq \mid N_{G^s(p_y)}(u_j) \mid$ or equivalently, by Remark 3 we have $\mid N^s_{\tilde{G}}(p_y)(u_i) \mid \neq \mid N^s_{\tilde{G}}(p_y)(u_j) \mid$. On the other hand by Theorem 4, $f(p_y)$ is a constant function for all $u_1, \cdots, u_N \in U$, meaning that for any $s \in S$: $\pi_s \circ f(p_y)(u_1) = \cdots = \pi_s \circ f(p_y)(u_N) = l_s$ where $l_s \in [0,1]$. Since $\tilde{G}$ is p_y-regular m-PFSG then

$$\sum_{k \neq i; u_k \in N^s_{\tilde{G}}(p_y)(u_i)} \pi_s \circ f(p_y)(u_k) = \sum_{k \neq j; u_k \in N^s_{\tilde{G}}(p_y)(u_j)} \pi_s \circ f(p_y)(u_k)$$

or equivalently

$$\mid N^s_{\tilde{G}}(p_y)(u_i) \mid \cdot l_s = \mid N^s_{\tilde{G}}(p_y)(u_j) \mid \cdot l_s$$

which is a contradiction. This completes the proof.

3.2 Irregular m-Polar Fuzzy Soft Graphs

Definition 15. *If the m-PFSG $\tilde{G}$ has a vertex whose p_y-adjacent vertices have different open p_y-neighborhood degrees, then $\tilde{G}$ is called a p_y-irregular m-PFSG. Additionally, if there exists a vertex whose neighbors have different open neighborhood degrees matrices $D_{\tilde{G}}(u_i)$, then $\tilde{G}$ is called an irregular m-PFSG.*

If the m-PFSG $\tilde{G}$ has a vertex whose p_y-adjacent vertices have different closed p_y-neighborhood degrees $D_{\tilde{G}}[u_i]$, then $\tilde{G}$ is called a totally p_y-irregular m-PFSG. Moreover, $\tilde{G}$ is called total irregular m-PFSG if there exits a vertex whose neighbors have distinct closed neighborhood degrees matrices.

Example 11. Reconsider the 2-polar fuzzy soft graph $\tilde{G}$ given in Example 1. By using the open neighborhood degrees matrices $D_{\tilde{G}}(u_i)$ of vertices u_i (for $i = 1, 2, 3$) computed in Example 9, it is easy to see that $\tilde{G}$ is an irregular 2-polar fuzzy soft graph.

Corollary 5. *$\tilde{G} = \langle P, (U, f, \tilde{R}) \rangle$ is an irregular m-PFSG iff for all $p_y \in P$, $\tilde{G}$ is a p_y-irregular m-PFSG. Moreover, $\tilde{G} = \langle P, (U, f, \tilde{R}) \rangle$ is a totally irregular m-PFSG iff for all $p_y \in P$, $\tilde{G}$ is a totally p_y-irregular m-PFSG.*

Similarly to Proposition 5, we have the following result.

Proposition 8. *$\tilde{G} = \langle P, (U, f, \tilde{R}) \rangle$ is p_y-irregular m-PFSG iff the m-polar fuzzy graph $\tilde{G}(p_y) = (U, f(p_y), \tilde{R}(p_y))$ is an irregular. If $\tilde{G}$ is p_y-irregular m-PFSG then $G^s(p_y) = (F^s(p_y), R^s(p_y))$ and $G(p_y) = (F(p_y), R(p_y))$ are irregular crisp graphs.*

According to Definitions 13 and 15, the following results can be obtained easily.

Proposition 9. *If $\tilde{G} = \langle P, (U, f, \tilde{R}) \rangle$ is a p_y-irregular m-PFSG and $f(p_y) : U \to \mathbb{I}^m$ is constant. Then $\tilde{G}$ is a totally p_y-irregular m-PFSG.*

Proposition 10. *Let $\tilde{G} = \langle P, (U, f, \tilde{R}) \rangle$ be a totally p_y-irregular m-PFSG and $f(p_y) : U \to \mathbb{I}^m$ be a constant function. Then $\tilde{G}$ is a p_y-irregular m-PFSG.*

Corollary 6. *Consider the m-PFSG $\tilde{G} = \langle P, (U, f, \tilde{R}) \rangle$ on U. If for any $p_y \in P$, the mapping $f(p_y) : U \to \mathbb{I}^m$ is constant then the following conditions are equivalent:*

1. *$\tilde{G}$ is an irregular m-PFSG.*
2. *$\tilde{G}$ is a totally irregular m-PFSG.*

4 Combining m-Polar Fuzzy Soft Graphs by Aggregation Functions

Aggregation functions are in fact a mathematical solution to the consensus problem in any discipline where combining a list of data into a unique one is needed. The methods based on aggregation operators are extremely used to solve decision making problems with fuzzy information [47–49]. In general, an aggregation function (operator) is defined as a non-decreasing n-array real function $A^{(n)} : [0,1]^n \rightarrow [0,1]$ satisfying

A1. $A(x) = x$, for $n = 1$ and any $x \in [0,1]$;
A2. $A^{(n)}(x_1, \cdots, x_n) \leq A^{(n)}(y_1, \cdots, y_n)$ if $(x_1, \cdots, x_n) \leq (y_1, \cdots, y_n)$;
A3. $A^{(n)}(0, 0, \cdots, 0) = 0$ and $A^{(n)}(1, 1, \cdots, 1) = 1$.

The function $A : \bigcup_{n\in\mathbb{N}}[0,1]^n \rightarrow [0,1]$ whose restriction over $\mathbb{I}^n = [0,1]^n$ is $A|_{\mathbb{I}^n} := A^{(n)}$ is called extended aggregation function, however, if there is no confusion, we can simply use the notation A instead of $A^{(n)}$. Many mathematical functions such as mean values, median, minimum operator and maximum operator can be seen as aggregation functions. The (extended) aggregation function A is called *conjunctive, disjunctive, and idempotent (average)* whenever for all $(x_1, \cdots, x_n) \in \mathbb{I}^n$, and $n \in \mathbb{N}$, we have: $A(x_1, \cdots, x_n) \leq Min(x_1, \cdots, x_n)$, $A(x_1, \cdots, x_n) \geq Max(x_1, \cdots, x_n)$, and $Min(x_1, \cdots, x_n) \leq A(x_1, \cdots, x_n) \leq Max(x_1, \cdots, x_n)$, respectively.

In the following, a method to combine several m-polar fuzzy soft graphs by using aggregation functions is presented.

Theorem 6. *Let* $\{\tilde{G}_k = \langle P, (U, f_k, \tilde{R}_k)\rangle : k = 1, \cdots, K\}$ *be a set of several m-PFSGs on* U. *If* $A : \bigcup_{n\in\mathbb{N}}[0,1]^n \rightarrow [0,1]$ *is an extended aggregation function then* $\tilde{G}_A = \langle P, (U, A(f_k)_{k=1}^K, A(\tilde{R}_k)_{k=1}^K)\rangle$ *is also an m-polar fuzzy soft graph on* U *where*

$$A(f_k)_{k=1}^K(p_y)(u_i) = \Big(A\big(\pi_1 \circ f_1(p_y)(u_i), \cdots, \pi_1 \circ f_K(p_y)(u_i)\big), \cdots, A\big(\pi_m \circ f_1(p_y)(u_i), \cdots, \pi_m \circ f_K(p_y)(u_i)\big)\Big) \quad (9)$$

$$A(\tilde{R}_k)_{k=1}^K(p_y)(u_iu_j) = \Big(A\big(\pi_1 \circ \tilde{R}_1(p_y)(u_iu_j), \cdots, \pi_1 \circ \tilde{R}_K(p_y)(u_iu_j)\big), \cdots, A\big(\pi_m \circ \tilde{R}_1(p_y)(u_iu_j), \cdots, \pi_m \circ \tilde{R}_K(p_y)(u_iu_j)\big)\Big) \quad (10)$$

for $1 \leq y \leq M$ *and* $1 \leq i, j \leq N$.

Proof. Let $\{\tilde{G}_k = \langle P, (U, f_k, \tilde{R}_k)\rangle : k = 1, \cdots, K\}$ be a set of K m-PFSG on U and $A : \bigcup_{n\in\mathbb{N}}[0,1]^n \to [0,1]$ be the aggregation function. Construct the m-PFS structure $\tilde{G}_A = \langle P, (U, A(f_k)_{k=1}^K, A(\tilde{R}_k)_{k=1}^K)\rangle$ as defined in Eq. (9) and Eq. (10). For any $s \in S$, $p_y \in P$ and $u_i, u_j \in U$ we have

$$\begin{aligned} A\big(\pi_s \circ \tilde{R}_1(p_y)(u_iu_j), \cdots, \pi_s \circ \tilde{R}_K(p_y)(u_iu_j)\big) &\leq A\Big(\min\{\pi_1 \circ f_1(p_y)(u_i),\\ &\pi_s \circ f_1(p_y)(u_j)\}, \cdots, \min\{\pi_s \circ f_K(p_y)(u_i), \pi_s \circ f_K(p_y)(u_j)\}\Big)\\ &\leq \min\Big(A\big(\pi_s \circ f_1(p_y)(u_i), \cdots, \pi_s \circ f_K(p_y)(u_i)\big),\\ &A\big(\pi_s \circ f_1(p_y)(u_j), \cdots, \pi_s \circ f_K(p_y)(u_j)\big)\Big) \end{aligned}$$

Since the operator min dominates any aggregation function.

Thus, $A(\tilde{R}_k)_{k=1}^K(p_y)(u_iu_j) \leq \min\{A(f_k)_{k=1}^K(p_y)(u_i), A(f_k)_{k=1}^K(p_y)(u_j)\}$. This completes the proof.

Note that, if $\{\tilde{G}_k = \langle P, (U, f_k, \tilde{R}_k)\rangle : k = 1, \cdots, K\}$ is a set of m-PFSGs on U, then the m-PFSG $\tilde{G}_A = \langle P, (U, A(f_k)_{k=1}^K, A(\tilde{R}_k)_{k=1}^K)\rangle$ defined in Theorem 6 is called the aggregated m-polar fuzzy soft graph on U.

Remark 5. Let $\{\tilde{G}_k = \langle P, (U, f_k, \tilde{R}_k)\rangle : k = 1, \cdots, K\}$ be a set of m-PFSGs on U and $\tilde{G}_A = \langle P, (U, A(f_k)_{k=1}^K, A(\tilde{R}_k)_{k=1}^K)\rangle$ be the aggregated m-PFSG on U by aggregation function A.

- If $A := Max$, then $\tilde{G}_A = \bigcup_{k=1}^K \tilde{G}_k$. Moreover, if A is a disjunctive aggregation function, then $\tilde{G}_A$ is a supergraph of $\bigcup_{k=1}^K \tilde{G}_k$.
- If $A := Min$, then $\tilde{G}_A = \bigcap_{k=1}^K \tilde{G}_k$. If A is a conjunctive aggregation function, then $\tilde{G}_A$ is a subgraph of $\bigcap_{k=1}^K \tilde{G}_k$.
- If A is an average aggregation function, then $\bigcap_{k=1}^K \tilde{G}_k \leq \tilde{G}_A \leq \bigcup_{k=1}^K \tilde{G}_k$.

According to the monotonicity property of any aggregation function, the following results can be obtained easily from Theorem 6.

Proposition 11. *Let $\{\tilde{G}_k = \langle P, (U, f_k, \tilde{R}_k)\rangle : k = 1, \cdots, K\}$ be a set of m-PFSGs on U and $A, B : \bigcup_{n\in\mathbb{N}}[0,1]^n \to [0,1]$ be two extended aggregation functions such that $B \leq A$. Then, $\tilde{G}_B \lesssim \tilde{G}_A$ where*

$$\tilde{G}_B = \langle P, (U, B(f_k)_{k=1}^K, B(\tilde{R}_k)_{k=1}^K)\rangle$$

and

$$\tilde{G}_A = \langle P, (U, A(f_k)_{k=1}^K, A(\tilde{R}_k)_{k=1}^K)\rangle$$

Proposition 12. *Let* $\{\tilde{G}_k = \langle P, (U, f_k, \tilde{R}_k)\rangle : k = 1, \cdots, K\}$ *and* $\{\tilde{O}_k = \langle P, (U, g_k, \tilde{Q}_k)\rangle : k = 1, \cdots, K\}$ *be two families of m-PFSGs on U and* $A : \bigcup_{n\in\mathbb{N}}[0,1]^n \to [0,1]$ *be an extended aggregation function. If* $\tilde{G}_k \tilde{\leq} \tilde{O}_k$ *for* $k = 1, \cdots, K$*, then* $\tilde{G}_A \tilde{\leq} \tilde{O}_A$ *where*

$$\tilde{G}_A = \langle P, (U, A(f_k)_{k=1}^K, A(\tilde{R}_k)_{k=1}^K)\rangle$$

and

$$\tilde{O}_A = \langle P, (U, A(g_k)_{k=1}^K, A(\tilde{Q}_k)_{k=1}^K)\rangle$$

Proposition 13. *Let* $\{\tilde{G}_k = \langle P, (U, f_k, \tilde{R}_k)\rangle : k = 1, \cdots, K\}$ *be a family of m-PFSGs on U and* $A : \bigcup_{n\in\mathbb{N}}[0,1]^n \to [0,1]$ *be an extended aggregation function. Let for each* $p_y \in P$ *and* $1 \leq s \leq m$*, there exists* $1 \leq k, k' \leq K$ *such that*

1. $\pi_s \circ f_k(p_y)(u_i) = 0$ *and* $\pi_s \circ \tilde{R}_{k'}(p_y)(u_i u_j) = 0$ *for some* $u_i, u_j \in U$*. If* $A \leq Min$ *is a conjunctive aggregation operator then the vertices and edges of aggregated m-PFSG* $\tilde{G}_A = \langle P, (U, A(f_k)_{k=1}^K, A(\tilde{R}_k)_{k=1}^K)\rangle$ *are null m-polar fuzzy soft sets.*
2. $\pi_s \circ f_k(p_y)(u_i) = 1$ *and* $\pi_s \circ \tilde{R}_{k'}(p_y)(u_i u_j) = 1$ *for some* $u_i, u_j \in U$*. If* $A \geq Max$ *is a disjunctive aggregation operator then the vertices and edges of aggregated m-PFSG* $\tilde{G}_A = \langle P, (U, A(f_k)_{k=1}^K, A(\tilde{R}_k)_{k=1}^K)\rangle$ *are absolute m-PFSS resulting in complete crisp graph.*

Lemma 1. *Let* $\{\tilde{G}_k = \langle P, (U, f_k, \tilde{R}_k)\rangle : k = 1, \cdots, K\}$ *be a set of m-PFSGs on U and* $A \neq A_w$ *be an aggregation function. For any* $p_y \in P$ *and* $u_i \in U$*, if* $A \leq Min$ *then* $N_{\tilde{G}_A}(p_y)(u_i) = \bigcap_{k=1}^K N_{\tilde{G}_k}(p_y)(u_i)$ *and if* $A > Min$ *we have* $N_{\tilde{G}_A}(p_y)(u_i) = \bigcup_{k=1}^K N_{\tilde{G}_k}(p_y)(u_i)$*.*

Proof. Let $\{\tilde{G}_k = \langle P, (U, f_k, \tilde{R}_k)\rangle : k = 1, \cdots, K\}$ be a family of m-PFSGs on U. Suppose that $A : \bigcup_{n\in\mathbb{N}}[0,1]^n \to [0,1]$ be a conjunctive aggregation function such that $A \neq A_w$. Put $\tilde{G}_A = \langle P, (U, A(f_k)_{k=1}^K, A(\tilde{R}_k)_{k=1}^K)\rangle$ as the aggregated m-PFSG. For any $s \in S$, $p_y \in P$, $u_i \in U$ and $u_j \in N_{\tilde{G}_A}(p_y)(u_i)$, we have $0 < A\big(\pi_s \circ f_1(p_y)(u_j), \cdots, \pi_s \circ f_K(p_y)(u_j)\big)$ and

$$\begin{aligned} &0 < A\big(\pi_s \circ \tilde{R}_1(p_y)(u_i u_j), \cdots, \pi_s \circ \tilde{R}_K(p_y)(u_i u_j)\big) \\ &\leq \min\{A\big(\pi_s \circ f_1(p_y)(u_i), \cdots, \pi_s \circ f_K(p_y)(u_i)\big), A\big(\pi_s \circ f_1(p_y)(u_j), \cdots, \pi_s \circ f_K(p_y)(u_j)\big)\} \\ &\leq \min\Big(\min\{\pi_s \circ f_1(p_y)(u_i), \pi_s \circ f_1(p_y)(u_j)\}, \cdots, \min\{\pi_s \circ f_K(p_y)(u_i), \pi_s \circ f_K(p_y)(u_j)\}\Big) \end{aligned}$$

Clearly, the latter implies that for all $1 \leq k \leq K$ and $s \in S$ we have $0 < \min\{\pi_s \circ f_k(p_y)(u_i), \pi_s \circ f_k(p_y)(u_j)\}$ or equivalently $0 < \pi_s \circ f_k(p_y)(u_i)$ and $0 < \pi_s \circ f_k(p_y)(u_j)$. This means that $u_i, u_j \in \tilde{G}_k$ consequently $0 < \pi_s \circ \tilde{R}_k(p_y)(u_i u_j) \leq \min\{\pi_s \circ f_k(p_y)(u_i), \pi_s \circ f_k(p_y)(u_j)\}$ for all $1 \leq k \leq K$ since $\tilde{G}_k$ is an m-PFSG and A. This implies that u_j is an adjacent vertex to u_i in all m-PFSGs $\tilde{G}_k$, i.e. $u_j \in N_{\tilde{G}_k}(p_y)(u_i)$ for all $1 \leq k \leq K$ resulting in $u_j \in \bigcap_{k=1}^K N_{\tilde{G}_k}(p_y)(u_i)$. Conversely, let $u_j \in \bigcap_{k=1}^K N_{\tilde{G}_k}(p_y)(u_i)$. Thus $u_j \in N_{\tilde{G}_k}(p_y)(u_i)$ for all $1 \leq k \leq K$

that implies $0 < \pi_s \circ \tilde{R}_k(p_y)(u_iu_j) \leq \min\{\pi_s \circ f_k(p_y)(u_i), \pi_s \circ f_k(p_y)(u_j)\}$ for all $1 \leq k \leq K$ and $s \in S$. Then we have

$$\begin{aligned}\pi_s \circ A(\tilde{R})_{k=1}^K(p_y)(u_iu_j) &= A\big(\pi_s \circ \tilde{R}_1(p_y)(u_iu_j), \cdots, \pi_s \circ \tilde{R}_K(p_y)(u_iu_j)\big)\\ &\leq A\Big(\min\{\pi_s \circ f_1(p_y)(u_i), \pi_s \circ f_1(p_y)(u_j)\}, \cdots, \min\{\pi_s \circ f_K(p_y)(u_i), \pi_s \circ f_K(p_y)(u_j)\}\Big)\\ &\leq \min\Big(A\big(\pi_s \circ f_1(p_y)(u_i), \cdots, \pi_s \circ f_K(p_y)(u_i)\big), A\big(\pi_s \circ f_1(p_y)(u_j), \cdots, \pi_s \circ f_K(p_y)(u_j)\big)\Big)\\ &= \min\big(A(f_k)_{k=1}^K(p_y)(u_i), A(f_k)_{k=1}^K(p_y)(u_j)\big)\end{aligned}$$

for all $s \in S$. This implies that $u_j \in N_{\tilde{G}_A}(p_y)(u_i)$.
The similar method can be used to show $N_{\tilde{G}_A}(p_y)(u_i) = \bigcup_{k=1}^K N_{\tilde{G}_k}(p_y)(u_i)$ in the case of $A > Min$.

Theorem 7. *Let $\{\tilde{G}_k = \langle P, (U, f_k, \tilde{R}_k)\rangle : k = 1, \cdots, K\}$ be a set of p_y-regular m-PFSGs on U. Suppose that corresponding to parameter $p_y \in P$ the membership functions $f_k(p_y) : U \to [0,1]^m$ are constant for all $k = 1, 2, \cdots, K$. If $A \neq A_w$ is an aggregation function, then the aggregated m-PFSG*

$$\tilde{G}_A = \langle P, (U, A(f_k)_{k=1}^K, A(\tilde{R}_k)_{k=1}^K)\rangle$$

is p_y-regular.

Proof. Suppose that $\{\tilde{G}_k = \langle P, (U, f_k, \tilde{R}_k)\rangle : k = 1, \cdots, K\}$ be a set of m-PFSGs on U. Consider the fixed parameter $p_y \in P$ such that for each $1 \leq k \leq K$, the functions $f_k(p_y) : U \to [0,1]^m$ are constant, i.e. $f_k(p_y)(u_i) = (a_1^k, \cdots, a_m^k)$ for all $u_i \in U$.

Now suppose that $\tilde{G}_A$ is not p_y-regular m-polar fuzzy soft graph on U. Then there exists two vertices u, v in $\tilde{G}_A$ having different p_y-neighborhood degrees $d_{\tilde{G}_A}(p_y)(u)$ and $d_{\tilde{G}_A}(p_y)(v)$. Since the membership functions $f_k(p_y)$ are constant then for each $s \in S$ we have

$$\begin{aligned}\pi_s \circ A\big(f_1, \cdots, f_K\big)(p_y)(u_i) &= A\big(\pi_s \circ f_1(p_y)(u_i), \cdots, \pi_s \circ f_K(p_y)(u_i)\big)\\ &= A(a_s^1, \cdots, a_s^K) = a_s\end{aligned}$$

This implies that the vertex degrees of all points in $\tilde{G}_A$ are the same, therefore the cardinality of adjacent vertices of u and v must be different, i.e. $\mid N_{\tilde{G}_A}(p_y)(u) \mid \neq \mid N_{\tilde{G}_A}(p_y)(v) \mid$. Without loose of generality put $\mid N_{\tilde{G}_A}(p_y)(v) \mid = \mid N_{\tilde{G}_A}(p_y)(u) \mid + t$. Take $v' \in N_{\tilde{G}_A}(p_y)(v)$ such that $v' \notin N_{\tilde{G}_A}(p_y)(u)$.

If $A_w \neq A \leq Min$ is a conjunctive aggregation function. Then, using Lemma 1, there exists $1 \leq k \leq K$ such that $v' \notin N_{\tilde{G}_k}(p_y)(u)$ but $v' \in N_{\tilde{G}_k}(p_y)(v)$ for all $1 \leq k \leq K$. This implies $\tilde{G}_k$ is not p_y-regular which is a contradiction. If $A > Min$, by using Lemma 1, there exits $1 \leq k \leq K$ such that $v' \in N_{\tilde{G}_k}(p_y)(v)$ but $v' \notin N_{\tilde{G}_k}(p_y)(u)$ for all $1 \leq k \leq K$, which is a contradiction with p_y-regularity of $\tilde{G}_k$. This completes the proof.

5 Application

m-polar fuzzy soft graphs are widely applied in multi-agent decision making problems and expert systems, social networking problems, information sciences, computer sciences and engineering with multi polar parametric data. In this section, an example is presented to show the application of aggregated m-polar fuzzy soft graph in a group decision making problem.

5.1 m-Polar Fuzzy Soft Graphs to Model Product Manufacturing

Suppose that $U = \{u_1, u_2, u_3, u_4, u_5\}$ is the list of managers of a company. Let this group wants to decide to manufacture a product among c_1, c_2, c_3 and c_4 to sell in market. The below three parameters are necessary to consider before starting the manufacturing process:

1. Market Demand;
2. Manufacturing Cost;
3. Manufacturing Time.

We list these three points as $P = \{p_1 = \text{Demand}, p_2 = \text{Cost}, p_3 = \text{Time}\}$. In reality, these parameters are usually measured based on some m-polar information. For instance, demand can be defined by answering the following questions: Is the product applicable? Is the product attractive to people? The cost factor can be computed based on raw material cost and designer/engineer employees salaries. And time may be measured according to the following points: Is the process of raw material preparation fast? Is the product manufacturing process time consuming? Thus, by answering these questions each parameter p_i, $i = 1, 2, 3$, can be presented as a 2-polar data. The graphical model of product manufacturing preferences of each expert u_1, u_2, u_3, u_4 and u_5 can be presented by a 2-polar fuzzy soft graph where $C = \{c_1, c_2, c_3, c_4\}$ is the vertex set and the 2-polar membership value of each vertex means the degree of demand, cost, and time consuming of that product. The edge between two products shows the same applicability and attraction; using common raw materials and designer/engineer employees; and having same time consuming for raw material preparation and manufacturing.

For instance, let for any $k = 1, 2, 3, 4, 5$ the expert u_k presents his/her preference of each products c_i, $i = 1, 2, 3, 4$, based on parameters p_1, p_2 and p_3 as the vertices membership degrees of the 2-polar fuzzy soft graph $\tilde{G}_k = \langle P, (C, f_k, \tilde{R}_k)\rangle$ by the following matrix:

$$f_k = \begin{bmatrix} \begin{bmatrix} fk1_{11} & fk1_{12} \\ fk1_{21} & fk1_{22} \\ fk1_{31} & fk1_{32} \end{bmatrix}_{3\times 2} & \begin{bmatrix} fk2_{11} & fk2_{12} \\ fk2_{21} & fk2_{22} \\ fk2_{31} & fk2_{32} \end{bmatrix}_{3\times 2} & \begin{bmatrix} fk3_{11} & fk3_{12} \\ fk3_{21} & fk3_{22} \\ fk3_{31} & fk3_{32} \end{bmatrix}_{3\times 2} & \begin{bmatrix} fk4_{11} & fk4_{12} \\ fk4_{21} & fk4_{22} \\ fk4_{31} & fk4_{32} \end{bmatrix}_{3\times 2} \end{bmatrix}_{1\times 4}$$

such that $fki_{ys} = \pi_s \circ f_k(p_y)(c_i)$ and the membership value of edges in $\tilde{G}_k$ is presented as below matrix:

$$\tilde{R}_k = \begin{bmatrix} \begin{bmatrix} 0 & 0 \\ 0 & 0 \\ 0 & 0 \end{bmatrix}_{3\times 2} & \begin{bmatrix} \tilde{R}k12_{11} & \tilde{R}k12_{12} \\ \tilde{R}k12_{21} & \tilde{R}k12_{22} \\ \tilde{R}k12_{31} & \tilde{R}k12_{32} \end{bmatrix}_{3\times 2} & \begin{bmatrix} \tilde{R}k13_{11} & \tilde{R}k13_{12} \\ \tilde{R}k13_{21} & \tilde{R}k13_{22} \\ \tilde{R}k13_{31} & \tilde{R}k13_{32} \end{bmatrix}_{3\times 2} & \begin{bmatrix} \tilde{R}k14_{11} & \tilde{R}k14_{12} \\ \tilde{R}k14_{21} & \tilde{R}k14_{22} \\ \tilde{R}k14_{31} & \tilde{R}k14_{32} \end{bmatrix}_{3\times 2} \\ \begin{bmatrix} \tilde{R}k21_{11} & \tilde{R}k21_{12} \\ \tilde{R}k21_{21} & \tilde{R}k21_{22} \\ \tilde{R}k21_{31} & \tilde{R}k21_{32} \end{bmatrix}_{3\times 2} & \begin{bmatrix} 0 & 0 \\ 0 & 0 \\ 0 & 0 \end{bmatrix}_{3\times 2} & \begin{bmatrix} \tilde{R}k23_{11} & \tilde{R}k23_{12} \\ \tilde{R}k23_{21} & \tilde{R}k23_{22} \\ \tilde{R}k23_{31} & \tilde{R}k23_{32} \end{bmatrix}_{3\times 2} & \begin{bmatrix} \tilde{R}k24_{11} & \tilde{R}k24_{12} \\ \tilde{R}k24_{21} & \tilde{R}k24_{22} \\ \tilde{R}k24_{31} & \tilde{R}k24_{32} \end{bmatrix}_{3\times 2} \\ \begin{bmatrix} \tilde{R}k31_{11} & \tilde{R}k31_{12} \\ \tilde{R}k31_{21} & \tilde{R}k31_{22} \\ \tilde{R}k31_{31} & \tilde{R}k31_{32} \end{bmatrix}_{3\times 2} & \begin{bmatrix} \tilde{R}k32_{11} & \tilde{R}k32_{12} \\ \tilde{R}k32_{21} & \tilde{R}k32_{22} \\ \tilde{R}k32_{31} & \tilde{R}k32_{32} \end{bmatrix}_{3\times 2} & \begin{bmatrix} 0 & 0 \\ 0 & 0 \\ 0 & 0 \end{bmatrix}_{3\times 2} & \begin{bmatrix} \tilde{R}k34_{11} & \tilde{R}k34_{12} \\ \tilde{R}k34_{21} & \tilde{R}k34_{22} \\ \tilde{R}k34_{31} & \tilde{R}k34_{32} \end{bmatrix}_{3\times 2} \\ \begin{bmatrix} \tilde{R}k41_{11} & \tilde{R}k41_{12} \\ \tilde{R}k41_{21} & \tilde{R}k41_{22} \\ \tilde{R}k41_{31} & \tilde{R}k41_{32} \end{bmatrix}_{3\times 2} & \begin{bmatrix} \tilde{R}k42_{11} & \tilde{R}k42_{12} \\ \tilde{R}k42_{21} & \tilde{R}k42_{22} \\ \tilde{R}k42_{31} & \tilde{R}k42_{32} \end{bmatrix}_{3\times 2} & \begin{bmatrix} \tilde{R}k43_{11} & \tilde{R}k43_{12} \\ \tilde{R}k43_{21} & \tilde{R}k43_{22} \\ \tilde{R}k43_{31} & \tilde{R}k43_{32} \end{bmatrix}_{3\times 2} & \begin{bmatrix} 0 & 0 \\ 0 & 0 \\ 0 & 0 \end{bmatrix}_{3\times 2} \end{bmatrix}_{4\times 4}$$

where $\tilde{R}kij_{ys} = \pi_s \circ \tilde{R}_k(p_y)(c_i c_j)$ and

$$\pi_s \circ \tilde{R}_k(p_y)(c_i c_j) \leq \min\{\pi_s \circ f_k(p_y)(c_i), \pi_s \circ f_k(p_y)(c_j)\}$$

for $i, j = 1, 2, 3, 4$; $y = 1, 2, 3$; $s = 1, 2$; and $k = 1, 2, 3, 4, 5$. Note that here, each entry is presented as a 3×2 matrix that gives the information of underlying fuzzy graphs corresponding to different parameters. To visualize the collective idea of whole group, the aggregated graph of these five 2-polar fuzzy soft graphs can be computed by applying an aggregation function A as $\tilde{G}_A = \langle P, (C, A(f_1, \cdots, f_5), A(\tilde{R}_1, \cdots, \tilde{R}_5))\rangle$. To this end, Algorithm 1 is designed.

Algorithm 1

Finding collective m-PFSG in MAGDM problems

Input : m-polar fuzzy soft graphs $\tilde{G}_k = \langle P, (C, f_k, \tilde{R}_k)\rangle$ where $\mid C \mid = N$ and $\mid P \mid = M$.
Output: Aggregated m-polar fuzzy soft graph $\tilde{G}_A = \langle P, (C, A(f_k), A(\tilde{R}_k))\rangle$ by aggregation function A.

begin

Step 1. Enter
$N \in \mathbb{N}$: the number of alternatives.
$M \in \mathbb{N}$: the number of parameters.
$m \in \mathbb{N}$: the number of different features of the parameters.
$K \in \mathbb{N}$: the number of decision makers.
$\mathbf{a} \in [0, 1]^m$: satisfaction level.
Function A : aggregation operator.
Matrices $f_k = \left[([fki_{ys}]_{M\times m})_{1i}\right]_{1\times N}$ where for each $k = 1, \cdots, K$, $y = 1, \cdots, M$, $s = 1, \cdots, m$ and $i = 1, \cdots, N$ we have $fki_{ys} = \pi_s \circ f_k(p_y)(c_i)$.
Matrices $\tilde{R}_k = \left[([\tilde{R}kij_{ys}]_{M\times m})_{ij}\right]_{N\times N}$ where for each $k = 1, \cdots, K$, $y = 1, \cdots, M$, $s = 1, \cdots, m$ and $i, j = 1, \cdots, N$ we have $\tilde{R}kij_{ys} = \pi_s \circ \tilde{R}_k(p_y)(c_i c_j)$.

Step 2. **if** $A \leq Min$ **then**
  for $k = 1, 2, \ldots, K$ **do**
    for $i, j = 1, 2, \ldots, N$ **do**
      for $y = 1, 2, \ldots, M$ **do**
        for $s = 1, 2, \ldots, m$ **do**
          If $\pi_s \circ f_k(p_y)(c_i) = 0$ and $\pi_s \circ \tilde{R}_k(p_y)(c_i c_j) = 0$,
          Then change operator A
        end
      end
    end
  end
else
  Go to Step 3.
end

Step 3. **if** $A \geq Max$ **then**
  for $k = 1, 2, \ldots, K$ **do**
    for $i, j = 1, 2, \ldots, N$ **do**
      for $y = 1, 2, \ldots, M$ **do**
        for $s = 1, 2, \ldots, m$ **do**
          If $\pi_s \circ f_k(p_y)(c_i) = 1$ and $\pi_s \circ \tilde{R}_k(p_y)(c_i c_j) = 1$,
          Then change operator A
        end
      end
    end
  end
else
  Go to Step 4.
end

Step 4. Aggregate the m-PFSG $\tilde{G}_k$ by using the aggregation function A and based on Eq. (9) and Eq. (10) to find the collective matrices $A(f)$ and $A(\tilde{R})$.
Step 5. Compute $\mathbf{a}$-level soft graphs by using Eq. (7) and Eq. (8).
Step 6. Draw the aggregated graph.

end

Now let $A := AM$ be the mean value operator.

According to Algorithm 1, since $Min \leq AM \leq Max$, the aggregated 2-polar fuzzy soft graph $\tilde{G}_{AM} = \langle P, (C, AM(f_1, \cdots, f_5), AM(\tilde{R}_1, \cdots, \tilde{R}_5))\rangle$ can be presented by the following matrices (Step 4):

$$AM(f) = \left[([(AM(f1i_{ys}, \cdots, f5i_{ys}))_{ys}]_{3\times 2})_{1i}\right]_{1\times 4}$$
$$= \left[\begin{bmatrix} \frac{1}{5}\sum_{k=1}^{5} fk1_{11} & \frac{1}{5}\sum_{k=1}^{5} fk1_{12} \\ \frac{1}{5}\sum_{k=1}^{5} fk1_{21} & \frac{1}{5}\sum_{k=1}^{5} fk1_{22} \\ \frac{1}{5}\sum_{k=1}^{5} fk1_{31} & \frac{1}{5}\sum_{k=1}^{5} fk1_{32} \end{bmatrix}_{3\times 2} \cdots \begin{bmatrix} \frac{1}{5}\sum_{k=1}^{5} fk4_{11} & \frac{1}{5}\sum_{k=1}^{5} fk4_{12} \\ \frac{1}{5}\sum_{k=1}^{5} fk4_{21} & \frac{1}{5}\sum_{k=1}^{5} fk4_{22} \\ \frac{1}{5}\sum_{k=1}^{5} fk4_{31} & \frac{1}{5}\sum_{k=1}^{5} fk4_{32} \end{bmatrix}_{3\times 2}\right]_{1\times 4}$$

and

$$AM(\tilde{R}) = \left[([(AM(\tilde{R}1ij_{ys}, \cdots, \tilde{R}5ij_{ys}))_{ys}]_{3\times 2})_{ij}\right]_{4\times 4}$$
$$= \begin{bmatrix} \begin{bmatrix} 0 & 0 \\ 0 & 0 \\ 0 & 0 \end{bmatrix}_{3\times 2} & \cdots & \begin{bmatrix} \frac{1}{5}\sum_{k=1}^{5} \tilde{R}k14_{11} & \frac{1}{5}\sum_{k=1}^{5} \tilde{R}k14_{12} \\ \frac{1}{5}\sum_{k=1}^{5} \tilde{R}k14_{21} & \frac{1}{5}\sum_{k=1}^{5} \tilde{R}k14_{22} \\ \frac{1}{5}\sum_{k=1}^{5} \tilde{R}k14_{31} & \frac{1}{5}\sum_{k=1}^{5} \tilde{R}k14_{32} \end{bmatrix}_{3\times 2} \\ \vdots & \ddots & \vdots \\ \begin{bmatrix} \frac{1}{5}\sum_{k=1}^{5} \tilde{R}k41_{11} & \frac{1}{5}\sum_{k=1}^{5} \tilde{R}k41_{12} \\ \frac{1}{5}\sum_{k=1}^{5} \tilde{R}k41_{21} & \frac{1}{5}\sum_{k=1}^{5} \tilde{R}k41_{22} \\ \frac{1}{5}\sum_{k=1}^{5} \tilde{R}k41_{31} & \frac{1}{5}\sum_{k=1}^{5} \tilde{R}k41_{32} \end{bmatrix}_{3\times 2} & \cdots & \begin{bmatrix} 0 & 0 \\ 0 & 0 \\ 0 & 0 \end{bmatrix}_{3\times 2} \end{bmatrix}_{4\times 4}$$

for $i, j = 1, 2, 3, 4$; $y = 1, 2, 3$; $s = 1, 2$; and $k = 1, 2, 3, 4, 5$.

If threshold vector $\mathbf{a} = (a_1, a_2) \in [0, 1] \times [0, 1]$ is given, then **a**-level soft graphs of $\tilde{G}_{AM}$ can be obtained by Eq. (7) and Eq. (8), where corresponding to each parameter p_y, $y = 1, 2, 3$; we have two crisp simple graphs:

- $(\tilde{G}_{AM})_{a_1}(p_y)$ with the vertices $V_{a_1}(p_y) = \{c_i \in C : \frac{1}{5}\sum_{k=1}^{5} fki_{y1} \geq a_1\}$ and edge set $E_{a_1}(p_y) = \{c_ic_j \in C : \frac{1}{5}\sum_{k=1}^{5} \tilde{R}kij_{y1} \geq a_1\}$;
- $(\tilde{G}_{AM})_{a_2}(p_y)$ with the vertices and edges sets $V_{a_2}(p_y) = \{c_i \in C : \frac{1}{5}\sum_{k=1}^{5} fki_{y2} \geq a_2\}$ and $E_{a_2}(p_y) = \{c_ic_j \in C : \frac{1}{5}\sum_{k=1}^{5} \tilde{R}kij_{y2} \geq a_2\}$

that can be drawn easily. In fact, any of these crisp graphs visualizes the collective decision process of product manufacturing based on decision parameter p_y and at the satisfaction level a_s for $s = 1, 2$.

6 Conclusion

To visualize the uncertain relationships between some objects, arising from parametric data, in literature fuzzy graphs were combined with soft set theory to introduce the concept of fuzzy soft graph (FSG). However, FSG is failed to visualize the multi-source data in a fuzzy soft environment. To overcome this problem, the present paper introduces the new approach so-called m-polar fuzzy soft graph (m-PFSG), that gives more flexible and compatible model with the reality in comparison with standard FSG. Clearly in the case of $m = 1$, the

m-PFSG is reduced to FSG. The union and intersection of m-polar fuzzy soft graphs are formulated and underlying fuzzy and crisp graphs of an m-polar fuzzy soft graph are studied. Due to this fact that in m-PFSG theory each vertex and edge come in the form of $M \times m$ dimension, where M and m respectively refer to as the numbers of parameters and their different status, the order and size of an m-polar fuzzy soft graph as well as the vertices neighborhood degrees are proposed as some $M \times m$ matrices. Accordingly, two main classes of graphs known as regular and irregular m-polar fuzzy soft graphs are discussed. Finally, the combination of several m-PFSGs by the aggregation functions are studied. Due to this fact that many consensus situations cannot be modeled when Min or Max operators are used, especially when meet or join of different ideas are not in demand, aggregation functions can be applied to provide a new approach to get a visualized collective m-PFSG from several m-PFSGs. This new framework proposes an algorithm to solve multi-agent decision making problems. The designed algorithm is illustrated by an example for solving the selection phase in a product manufacturing process with 5 inputs and one output.

Naturally, this research can be extended for m-PFSG with complete and incomplete databases and information, m-polar intuitionistic fuzzy soft graphs and m-polar interval-valued fuzzy soft graphs.

Acknowledgements. This study was supported by the Fundamental Research Grant Scheme, No. FRGS/1/2018/STG06/ UPM/01/3 and No. FRGS/1/2019/STG06/ UPM/02/6, awarded by the Malaysia Ministry of Higher Education (MOHE).

References

1. Rosenfeld, A.: Fuzzy graphs. In: Fuzzy sets and their Applications to Cognitive and Decision Processes, Proceedings of the USA–Japan Seminar on Fuzzy Sets and their Applications, Berkeley, California, 1–4 July 1974, pp. 77–95. Elsevier (1975). https://doi.org/10.1016/B978-0-12-775260-0.50008-6
2. Zadeh, L.A.: Similarity relations and fuzzy orderings. Inf. Sci. **3**(2), 177–200 (1971). https://doi.org/10.1016/S0020-0255(71)80005-1
3. Bhattacharya, P.: Some remarks on fuzzy graphs. Pattern Recogn. Lett. **6**(5), 297–302 (1987). https://doi.org/10.1016/0167-8655(87)90012-2
4. Kulkarni, G.H., Waingankar, P.G.: Fuzzy logic based traffic light controller. In: International Conference on Industrial and Information Systems, Penadeniya, Sri Lanka, 9–11 August 2007, pp. 107–110. IEEE (2007). https://doi.org/10.1109/ICIINFS.2007.4579157
5. Karakuzu, C., Demirci, O.: Fuzzy logic based smart traffic light simulator design and hardware implementation. Appl. Soft Comput. **10**(1), 66–73 (2010). https://doi.org/10.1016/j.asoc.2009.06.002
6. Rahman, S.M., Ratrout, N.T.: Review of the fuzzy logic based approach in traffic signal control: prospects in Saudi Arabia. J. Transp. Syst. Eng. Inf. Technol. **9**(5), 58–70 (2009). https://doi.org/10.1016/S1570-6672(08)60080-X
7. Mehan, S.: Introduction of traffic light controller with fuzzy control system. Int. J. Electron. Commun. Technol. **2**(3), 119–122 (2011)

8. Thakur, G.K., Priya, B., Mishra, R.K.: An efficient coloring algorithm for time detraction of sign image segmentation based on fuzzy graph theory. J. Appl. Secur. Res. **14**(2), 210–226 (2019). https://doi.org/10.1080/19361610.2019.1581880
9. Dey, A., Pal, A.: Fuzzy graph coloring technique to classify the accidental zone of a traffic control. Ann. Pure Appl. Math. **3**(2), 169–178 (2013)
10. Gómez, D., Montero, J., Yáñez, J.: A coloring fuzzy graph approach for image classification. Inf. Sci. **176**(24), 3645–3657 (2006). https://doi.org/10.1016/j.ins.2006.01.006
11. Akram, M., Habib, A., Ilyas, F., Mohsan Dar, J.: Specific types of Pythagorean fuzzy graphs and application to decision-making. Math. Comput. Appl. **23**(3), 42 (2018). https://doi.org/10.3390/mca23030042
12. Akram, M., Habib, A., Davvaz, B.: Direct sum of n pythagorean fuzzy graphs with application to group decision-making. J. Multiple-Valued Logic Soft Comput. **33**(1–2), 75–115 (2019)
13. Akram, M.: Decision making method based on spherical fuzzy graphs. In: Kahraman, C., Kutlu Gündoğdu, F. (eds.) Decision Making with Spherical Fuzzy Sets. Studies in Fuzziness and Soft Computing, vol. 392 pp. 153–197. Springer, Cham (2021). https://doi.org/10.1007/978-3-030-45461-6_7
14. Naz, S., Ashraf, S., Karaaslan, F.: Energy of a bipolar fuzzy graph and its application in decision making. Ital. J. Pure Appl. Math. **40**, 339–352 (2018)
15. Guleria, A., Bajaj, R.K.: T-spherical fuzzy graphs: operations and applications in various selection processes. Arab. J. Sci. Eng. **45**(3), 2177–2193 (2020). https://doi.org/10.1007/s13369-019-04107-y
16. Mordeson, J.N., Peng, C.S.: Operations on fuzzy graphs. Inf. Sci. **79**(3–4), 159–170 (1994). https://doi.org/10.1016/0020-0255(94)90116-3
17. Sunitha, M.S., Vijayakumar, A.: Complement of a fuzzy graph. Indian J. Pure Appl. Math. **33**(9), 1451–1464 (2002)
18. Zhang, W.R.: Bipolar fuzzy sets and relations: a computational framework for cognitive modeling and multiagent decision analysis. In: NAFIPS/IFIS/NASA'94. Proceedings of the First International Joint Conference of The North American Fuzzy Information Processing Society Biannual Conference. The Industrial Fuzzy Control and Intelligent, San Antonio, USA, 18–21 December 1994, pp. 305–309. IEEE (1994). https://doi.org/10.1109/IJCF.1994.375115
19. Chen, J., Li, S., Ma, S., Wang, X.: m-polar fuzzy sets: an extension of bipolar fuzzy sets. Sci. World J. (2014). https://doi.org/10.1155/2014/416530
20. Akram, M.: Bipolar fuzzy graphs. Inf. Sci. **181**(24), 5548–5564 (2011). https://doi.org/10.1016/j.ins.2011.07.037
21. Yang, H.L., Li, S.G., Yang, W.H., Lu, Y.: Notes on Bipolar fuzzy graphs. Inf. Sci. **242**, 113–121 (2013). https://doi.org/10.1016/j.ins.2013.03.049
22. Akram, M., Dudek, W.A.: Regular bipolar fuzzy graphs. Neural Comput. Appl. **21**(1), 197–205 (2012). https://doi.org/10.1007/s00521-011-0772-6
23. Akram, M.: Bipolar fuzzy graphs with applications. Knowl.-Based Syst. **39**, 1–8 (2013). https://doi.org/10.1016/j.knosys.2012.08.022
24. Poulik, S., Ghorai, G.: Note on Bipolar fuzzy graphs with applications. Knowl.-Based Syst. **192**, 105315 (2020). https://doi.org/10.1016/j.knosys.2019.105315
25. Rashmanlou, H., Samanta, S., Pal, M., Borzooei, R.A.: Product of bipolar fuzzy graphs and their degree. Int. J. Gen. Syst. **45**(1), 1–14 (2016). https://doi.org/10.1080/03081079.2015.1072521
26. Ghorai, G., Pal, M.: Certain types of product bipolar fuzzy graphs. Int. J. Appl. Comput. Math. **3**(2), 605–619 (2017). https://doi.org/10.1007/s40819-015-0112-0

27. Ghorai, G., Pal, M.: On some operations and density of m-polar fuzzy graphs. Pac. Sci. Rev. Nat. Sci. Eng. **17**(1), 14–22 (2015). https://doi.org/10.1016/j.psra.2015.12.001
28. Ghorai, G., Pal, M.: Some properties of m-polar fuzzy graphs. Pac. Sci. Rev. Nat. Sci. Eng. **18**(1), 38–46 (2016). https://doi.org/10.1016/j.psra.2016.06.004
29. Akram, M., Shahzadi, G.: Certain characterization of m-polar fuzzy graphs by level graphs. Punjab Univ. J. Math. **49**, 1–12 (2017)
30. Akram, M., Adeel, A.: m-Polar fuzzy graphs and m-polar fuzzy line graphs. J. Discrete Math. Sci. Cryptogr. **20**(8), 1597–1617 (2017). https://doi.org/10.1080/09720529.2015.1117221
31. Singh, P.K.: m-polar fuzzy graph representation of concept lattice. Eng. Appl. Artif. Intell. **67**, 52–62 (2018). https://doi.org/10.1016/j.engappai.2017.09.011
32. Molodtsov, D.: Soft set theory-first results. Comput. Math. Appl. **37**(4–5), 19–31 (1999). https://doi.org/10.1016/S0898-1221(99)00056-5
33. Mohinta, S., Samanta, T.K.: An introduction to fuzzy soft graph. Math. Moravica **19**(2), 35–48 (2015)
34. Akram, M., Nawaz, S.: Operations on soft graphs. Fuzzy Inf. Eng. **7**(4), 423–449 (2015). https://doi.org/10.1016/j.fiae.2015.11.003
35. Akram, M., Nawaz, S.: On fuzzy soft graphs. Ital. J. Pure Appl. Math. **34**, 497–514 (2015)
36. Akram, M., Nawaz, S.: Fuzzy soft graphs with applications. J. Intell. Fuzzy Syst. **30**(6), 3619–3632 (2016). https://doi.org/10.3233/IFS-162107
37. Shahzadi, S., Akram, M.: Edge regular intuitionistic fuzzy soft graphs. J. Intell. Fuzzy Syst. **31**(3), 1881–1895 (2016). https://doi.org/10.3233/JIFS-16120
38. Shahzadi, S., Akram, M.: Intuitionistic fuzzy soft graphs with applications. J. Appl. Math. Comput. **55**(1–2), 369–392 (2017). https://doi.org/10.1007/s12190-016-1041-8
39. Shahzadi, S., Akram, M.: Graphs in an intuitionistic fuzzy soft environment. Axioms **7**(2), 20 (2018). https://doi.org/10.3390/axioms7020020
40. Shahzadi, G., Akram, M.: Hypergraphs based on pythagorean fuzzy soft model. Math. Comput. Appl. **24**(4), 100 (2019). https://doi.org/10.3390/mca24040100
41. Shahzadi, G., Akram, M., Davvaz, B.: Pythagorean fuzzy soft graphs with applications. J. Intell. Fuzzy Syst. **38**(4), 4977–4991 (2020). https://doi.org/10.3233/JIFS-191610
42. Akram, M., Feng, F., Borumand Saeid, A., Leoreanu-Fotea, V.: A new multiple criteria decision-making method based on bipolar fuzzy soft graphs. Iran. J. Fuzzy Syst. **15**(4), 73–92 (2018)
43. Alghamdi, M.A., Alshehri, N.O., Akram, M.: Multi-criteria decision-making methods in bipolar fuzzy environment. Int. J. Fuzzy Syst. **20**(6), 2057–2064 (2018). https://doi.org/10.1007/s40815-018-0499-y
44. Zahedi Khameneh, A., Kılıçman, A.: m-polar fuzzy soft weighted aggregation operators and their applications in group decision-making. Symmetry **10**(11), 636 (2018). https://doi.org/10.3390/sym10110636
45. Zadeh, L.A.: Fuzzy sets. Inf. Control **8**(3), 338–353 (1965). https://doi.org/10.1016/S0019-9958(65)90241-X
46. Maji, P.K., Biswas, R., Roy, A.R.: Fuzzy soft sets. J. Fuzzy Math. **9**, 589–602 (2001)
47. Qiyas, M., Abdullah, S., Ashraf, S., Aslam, M.: Utilizing linguistic picture fuzzy aggregation operators for multiple-attribute decision-making problems. Int. J. Fuzzy Syst. **22**(1), 310–320 (2020). https://doi.org/10.1007/s40815-019-00726-7

48. Ma, X., Wu, P., Zhou, L., Chen, H., Zheng, T., Ge, J.: Approaches based on interval type-2 fuzzy aggregation operators for multiple attribute group decision making. Int. J. Fuzzy Syst. **18**(4), 697–715 (2016). https://doi.org/10.1007/s40815-015-0092-6
49. Yuan, R., Tang, J., Meng, F.: Linguistic intuitionistic fuzzy group decision making based on aggregation operators. Int. J. Fuzzy Syst. **21**(2), 407–420 (2019). https://doi.org/10.1007/s40815-018-0582-4

Multi-objective Genetic Algorithm and Interpolation Based Nonlinear Control Model

Dursun Ekmekci[1](✉) and Shahnaz N. Shahbazova[2]

[1] Karabuk University, 78050 Karabuk, Turkey
dekmekci@karabuk.edu.tr
[2] Azerbaijan Technical University, Baku, Azerbaijan
shahbazova@gmail.com

Abstract. One of the most important issues for nonlinear fuzzy control is to provide a balance between interpretability and accuracy. In this context, one of the most popular techniques used to find optimal values for control parameters is metaheuristic methods. In this study, optimal parameter values are tried to be determined by using multi-objective evolutionary algorithms for a multi-output nonlinear control system. The proposed method is used to seek out the optimal parameters to increase the performance of the nano-fluid mixture. The model used dataset obtained with different parameter values.

Keywords: Nonlinear fuzzy control · Fuzzy Rule-Based Systems · Multi-objective optimization · Proportional Integral Derivative · Fuzzy genetic algorithm · Non-dominated Sorting Genetic Algorithm

1 Introduction

A typical control problem must comply with various criteria according to the different behaviors of the controlled system. The provision of these constraints strongly depends on the structure and parameters of the controller. In this context, the controller parameters should be set to values that meet the requirements for system stability. Unfortunately, this solution cannot usually be provided with a mathematical formulation. The absence of a systematic approach to assigning optimal values for many parameters of the controller is a big obstacle to obtain a successful control system. Adaptive or self-tuning methods are used to improve the performance of the controller by adjusting its parameters to the desired values. For self-tuning control, Evolutionary Algorithms (EAs) are generally preferred. In many studies given in [1], EAs have been applied successfully to obtain controller parameters or controller structure, or sometimes both.

It is a difficult problem to design any control model for non-linear or imprecisely systems. The intelligent systems proposed for such complex problems produce solutions based on information from control experts or system responses. As one of the methods commonly used in the solution of these problems, Fuzzy Logic Control (FLC) [2] can be integrated with EAs effectively. However, one of the most important issues for control

T. Allahviranloo et al. (Eds.): IDS 2020, AISC 1301, pp. 456–470, 2021.
https://doi.org/10.1007/978-3-030-66501-2_36

models such as Fuzzy Rule-Based Systems (FRBSs) is the problem of finding the right trade-off between interpretability and accuracy [3]. The interpretability is the ability to clearly express the system behavior, and accuracy is the ability to represent this system faithfully [4]. The right balance between the accuracy and interpretability of FRBSs provides not only accurate but also interpretable language models. Accuracy and interpretability are usually conflicting, and in order to optimize these objectives, the Multi-Objective Evolutionary Algorithms (MOEAs) [5] are used [6]. Trebi-Ollennu and White calculated the relative importance of objective functions for multi-objective fuzzy genetic algorithm (GA) optimization using a new membership weighting strategy [7]. Also, a neuro-fuzzy logic controller has been developed, in which all parameters can be tuned simultaneously by EAs [8].

MOEAs also commonly used for multi-objective Proportional Integral (PI) [9] and multi-objective Proportional Integral Derivative (PID) control [10, 11]. PID controller is one of the most popular control systems, which is commonly preferred due to its easiness of implementation and remarkable efficiency. However, tuning PID parameters is very difficult. In [12], the classical methods used for PID tuning in the literature were compared with EAs, and more successful results were obtained with EAs.

EAs have been successfully applied to inverted pendulum control, which is used for the design, testing, and comparison of many different control systems [13, 14]. In control fields, MOAEs are also used for critical control, predictive control [15], optimal control [16, 17], robust control [18], system identification [19], multi-objective eigen-structure assignment [20] and multi-objective fault detection.

When literature studies that use MOEAs in the control systems are examined, it is seen that MOEAs can be used for various fields of control systems. MOEAs are usually used to optimize the parameter values of the controller. In this study, MOEAs is used for the multi-output Tsukamoto fuzzy inference system. Flexible membership functions are created for input and output variables using the Multi-Objective Genetic Algorithm (MOGA) and the interpolation methods. The model investigates the optimal parameter values and optimal rules for the control system using the data set obtained from the measurements.

2 Background

In this section, the multi-objective optimization concept and as the methods MOGA and interpolation used in the proposed control model, are explained in detail. In the study, Non-dominated Sorting Genetic Algorithm II (NSGA II) is used as one of the MOGA variants.

2.1 Multi-objective Optimization

In engineering fields, many optimization problems are posed as nonlinear programming problems usually involving contradictory objectives. A solution that reaches the optimal value for one of the objectives may be considered invalid for another objective. In order to solve the Multi-Objective Optimization Problem (MOOP) where such two or more objective functions should be optimized at the same time, MOEAs are used. In other words, MOOP is the process of systematic and simultaneous optimization of the collection of objective functions [21]. MOOP (for minimization case) can be defined as follows.

$$Min f(x) : \{f_1(x), f_2(x), \ldots, f_p(x)\} \tag{1}$$

$$Subject\ to\ x \in \Omega \tag{2}$$

where Ω is the decision (variable) space and R^m, which consists of m objectives, is the objective space ($\Omega \in R^m$), $x = (x_1, x_2, \ldots, x_m)^T$ are the decision variables and $f_p : R^m \to R(p \geq 2)$ are the objectives.

Without loss of generality, this study adopts the following basic concepts of multi-objective optimization.

Pareto dominance: Feasible solutions $x \prec y$ if and only if $f_i(x) \leq f_i(y)$ ($\forall i = 1, 2, \ldots, p$) and $f_i(x) < f_i(y)$ ($\exists i \in \{1, 2, \ldots, p\}$)

External archive: It is an array that saves the nondominated solutions found so far. As illustrated in Fig. 1, when the solution process is completed, there are several solutions (referred to as Pareto-optimal solutions), which can be developed by deteriorating each other, is obtained.

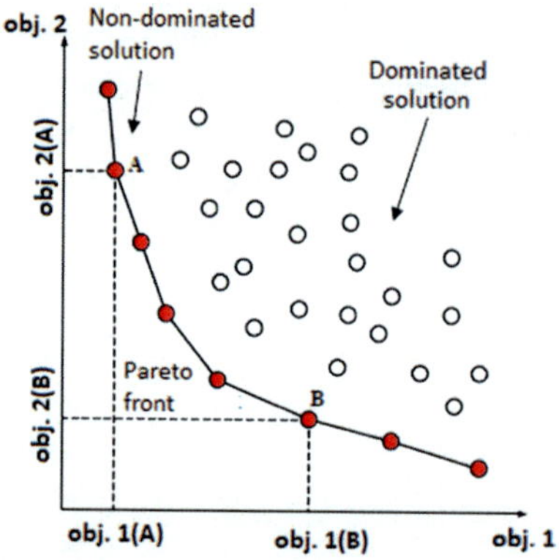

Fig. 1. Illustration of Pareto-front

All the non-dominated solutions constitute the Pareto-optimal set. Therefore, the main goal of a MOEA is to find a well-balanced approximation to the Pareto-optimal set [22].

2.2 Non-dominated Sorting Genetic Algorithm (NSGA)

GA [23], which has been applied successfully to the single-objective optimization problems, was developed for MOOP solutions, and different MOGA versions were introduced. The Non-dominated Sorting Genetic Algorithm (NSGA) [24] was one of the first such MOGAs. As the fast and elitist MOGA, the NSGA-II [25] is a heavily revised version of NSGA, developed to overcome its shortcomings. NSGA II has an elitist approach that preserves diversity and highlights non-dominant solutions [26]. In addition to the steps of the GA, NSGA-II applies dominance grading and stacking distance calculations.

NSGA-II consists of three basic components: fast non-dominated sorting approach, crowding distance assignment, and main loop.

Fast Non-dominated Sorting Approach. Solutions are classified in different dominance sets, based on their degree of dominance. The algorithm in Fig. 2 is used for this process.

Algorithm 1 Fast-non-dominated-sort

```
for all p ∈ P do
  S_p = ∅;
  n_p = 0;
  for all q ∈ P do
    if p ≺ q then
      S_p = S_p ∪ {q};
    else if q ≺ p then
      n_p = n_p + 1;
    end if
  end for
  if n_p = 0 then
    p_rank = 1;
    F_1 = F_1 ∪ {p};
  end if
end for
i=1;
while F_i ≠ ∅ do
  Q = ∅;
  for all p ∈ F_i do
    for all q ∈ S_p do
      n_q = n_q -1;
      if n_q =0 then
        q_rank = i+1;
        Q = Q ∪ {q};
      end if
    end for
  end for
  i=i+1;
  F_i = Q;
end while
```

Fig. 2. NSGA-II Algorithm [25]

By comparing the solutions in the cluster for each objective, the number of solutions dominated by the solution, and the solutions dominating this solution is recorded. If the number of solutions dominating the solution is 0, the solution is added to the first set of non-dominated solutions. Each solution of q in the S_p set of each p solution with $n_p = 0$ is visited and the n_p value is reduced by 1. If the n_p (dominance number) for solution q is 0, solution q is added to the set Q. The solutions in the Q set constitute the second set of non-dominated solutions. This process is applied to all elements of the Q set, creating the next set of non-dominated solutions. Each cluster consists of non-dominant solutions. These processes continue until all solutions are assigned to the clusters, and all clusters are determined.

Crowding Distance Assignment. Each non-dominated cluster is sorted according to the crowding distance. This ranking is determined by the density of the solutions around a particular solution. For each objective function, solutions with the minimum and maximum value are assigned infinite distance values. Thus, the extreme values are selected. For all other solutions in between, a distance value is calculated for all objective functions. The total crowding distance value is the sum of the individual distances calculated for each objective.

Main Loop. First, a random parent population (P_0) is evaluated as initial solutions. The population is sorted by their degree of non-domination. Each solution is assigned a value (1 being the best level) based on the degree of non-domination to be suppressed. This value is determined according to the fitness function. The offspring population (Q_0) of size N is generated by using the GA operators that normal binary tournament selection, recombination, and mutation.

R_t refers to a $2*N$ population consisting of a combination of parent and offspring populations. This population is ranked according to their degree of non-domination. F_1 refers to the best set of non-dominated. If $F_1 < N$, all solutions in F_1 are selected to create the next parent population ($P_t + 1$). The remaining elements of $P_t + 1$ are selected from the next non-dominant cluster based on their degree of non-domination. This procedure continues until there are no more clusters. The NSGA-II procedure is also shown in Fig. 3.

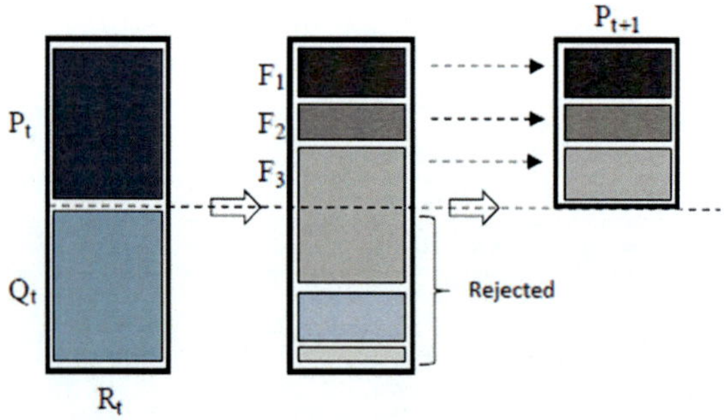

Fig. 3. NSGA-II Procedure

The total number of solutions in all clusters from F_1 to F_l will be greater than the population size. To select N solutions for the new population, the solutions in the final non-dominated order are sorted from maximum to minimum according to the population distance. These solutions are included in the population, starting with the solution that has the highest crowding distance, to fill the gaps in the population. Then, $Q_t + 1$ size of N population is derived by applying selection, crossover, and mutation processes in the new $P_t +$ 1 size of N population. Then, the $R_t + 1$ population of $2 * N$ size is evaluated by combining $P_t + 1$ with the parent population, and the processes described above are applied.

2.3 Interpolations

Generally, to determine the relations between any two variables, measurements are made. Curve fitting techniques are applied to describe the obtained discrete data points with a single function. Thus, the output values of the points other than the points used can be estimated with minor errors. Within the scope of the study, functions are created for the input and output values in the dataset. For the curved fitting, Spline, Partial Cubic Hermite Interpolated Polynomial (Pchip) [27] and Modified Akima piecewise cubic Hermite interpolation (Makima) [28] techniques are used. The curves created by these methods for randomly selected points are illustrated in Fig. 4.

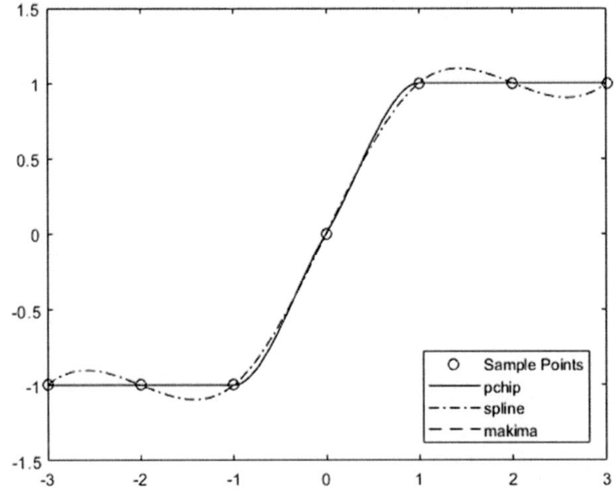

Fig. 4. Curve fitting example with Spline, Makima and PCHIP methods

3 Proposed Method

In the proposed method, optimal parameter values for the multi-objective fuzzy control model are assigned using NSGA II and interpolation techniques. Each output is treated as an objective. While determining the membership functions (MFs), for input variables 5 random points, and for output variables 3 random points are selected. Using curve fitting methods, curves involving these points are determined as MFs. In this way, a more flexible approach is adopted instead of standard MFs. In the proposed model, the Tsukamoto inference method was used.

The outputs produced by the proposed model are compared with the targets in the data set. Thus, the total error is calculated for each objective. The NSGA II method is used to select random points, and it aims to minimize the error. Curves involving the selected points are created by Spline, Makima, and Pchip techniques.

In the proposed method, the operation of the relevant rules in a control problem consisting of 2 inputs and 2 outputs is illustrated in Fig. 5. The input variables consist of three MFs, and the output variable consists of one MF. 5 points are randomly selected for the MFs of the variables, and MFs are created with the curve fitting method. For the example in Fig. 5, the relevant rules are set up as follows.

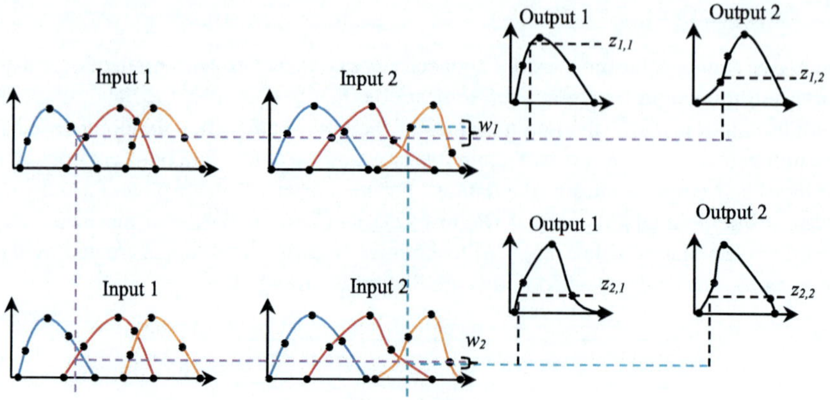

Fig. 5. In the proposed method, the operation of two sample rules

If *x*, input(1).MF(1) and *y*, Input(2).MF(3) Then $z_{1,1}$ Output(1) and $z_{1,2}$ Output(2)
If *x*, input(1).MF(2) and *y*, Input(2).MF(2) Then $z_{2,1}$ Output(1) and $z_{2,2}$ Output(2)

In the case in Fig. 5, the result is calculated as follows for objective 1.

$$Z = \left(w_1^* z_{1,1} + w_2^* z_{2,1}\right)/(w_1 + w_2) \tag{3}$$

For a problem that has *n* objectives, in the data set consisting of *m* samples, the results obtained for the inputs are compared with the targets, and the total error is calculated as follows. In the optimization process, the aim is to minimize the total error for each objective.

$$Error_i = \frac{\sum_{j=1}^{m} |Z_j - t_j|}{m} \quad i \in n \tag{4}$$

4 Experimental Study

In this section, the proposed method is applied to solve the problem described in [29]. It was run on the data set in [30] where the details of the study were shared. In the study, optimal parameter values for the control parameters of the fuzzy model were investigated using optimization techniques.

4.1 Selected Control Problem

In [29], in non-air-cooled internal combustion engines, it is aimed to prevent the water used for heat exchange from freezing and to increase the performance of the nano-fluid mixture by determining the optimum operating parameters of the engine. The authors investigated the use of dispersed aluminum oxide (Al_2O_3) and titanium dioxide (TiO_2) nanoparticles in 50:50 volumetric ratios in distilled water and ethylene glycol to enhance thermophysical properties. The nanofluid properties were measured the temperature range between 25 and 70 °C, and in the volume fractions of (0.05 and 0.3 vol.%). In the control model, the volume fractions of Al_2O_3 and TiO_2, and temperature were used as input variables. The effect of these three operating parameters on the mixtures density, viscosity, and thermal conductivity is studied.

4.2 Applying Proposed Method to the Problem

The fuzzy control model described in the previous section was run on the selected problem. In this context, the optimal values for the control parameters were defined using the data set shared in [30]. 70% of the total 55 samples in the dataset was used for training of the model and 30% for testing. (38 samples for training and 17 samples for testing) Output parameters are considered as objectives, and the problem is solved with the MOOP approach. The model designed for optimization processes is presented in Fig. 6.

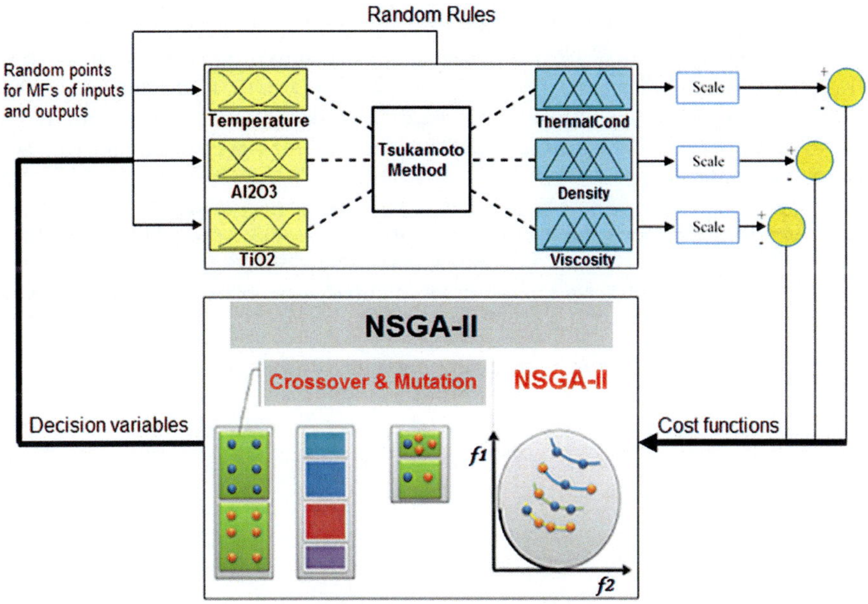

Fig. 6. Proposed model for nonlinear control

Firstly, the values in the data set are normalized. Then, 3 MFs consisting of 5 points are randomly selected in the range [0, 1] are defined for the input variables. Rules are defined randomly, considering all the objectives, and 3 points are randomly selected for the MFs of the output variables. The parameter values used for optimization are presented in Table 1.

Table 1. The NSGA II parameters.

Parameter	Value
Population size	50
Mutation	0.01
Crossover	1.0
Maximum iterations	200
Pareto front population fraction	0.35

The method was run 30 times independently with each curve fitting method (Spline, Pchip, Makima).

4.3 Results

In this section, the multi-objective optimization results obtained by each curve fitting method are presented in graphs. Since the goal is minimization for each objective in the problem, the "ideal point" is the result where the total error is the smallest. In this context, control structures formed according to the result at the ideal point are shared. Finally, the control model constructed is compared with the targets of training data and testing data in the data set.

Optimization Results. Using the Spline, Pchip and Makima interpolation techniques, the minimum erroneous solutions obtained for the objectives of the problem are presented in Fig. 7, 8 and 9. In each search, 18 different solutions were obtained for the objectives.

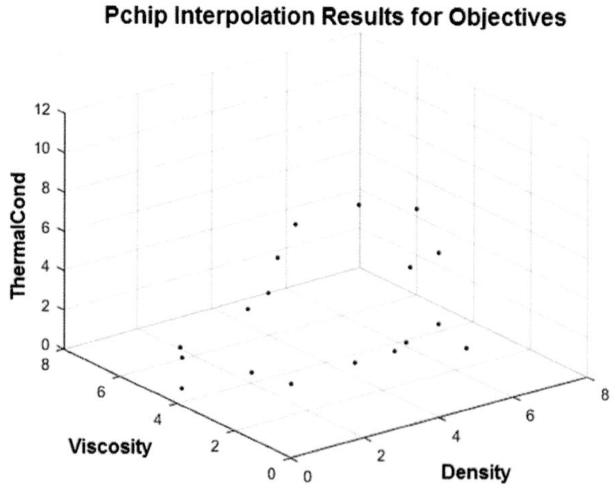

Fig. 7. Optimization results obtained by Pchip technique

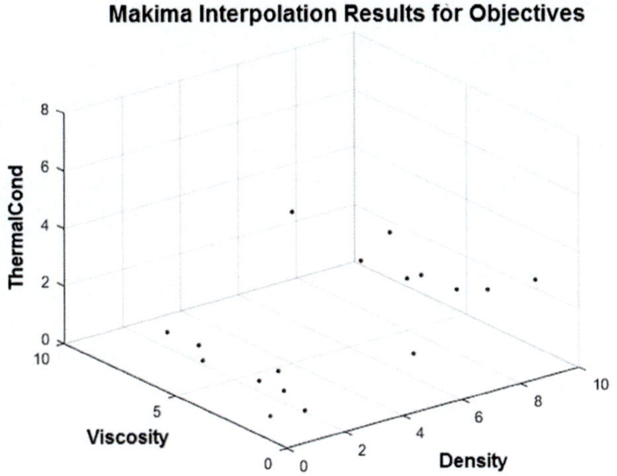

Fig. 8. Optimization results obtained by Makima technique

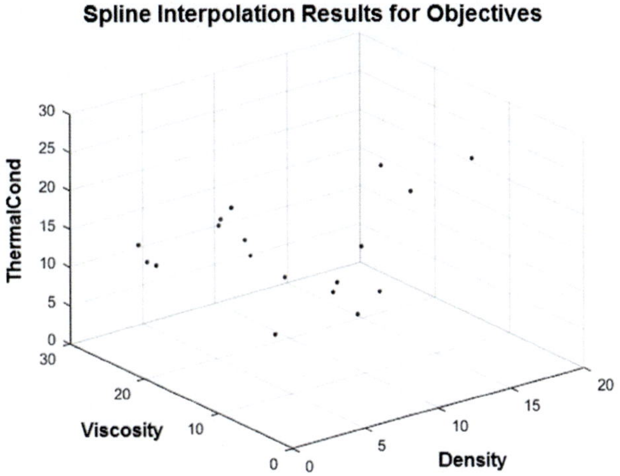

Fig. 9. Optimization results obtained by Spline technique

Control Structures. When the ideal points obtained with the methods are compared, it is seen that the Makima method achieved the most successful solution. In other words, minimum errors for output values were obtained with Makima. Membership functions created for the inputs and outputs of the fuzzy control model in this ideal solution are presented in Fig. 10.

Comparisons. The control model constructed was carried out for 38 samples determined for training and 17 samples selected for testing in the data set. The comparison for training samples is presented in Table 2, and the comparison for test samples is presented in Table 3.

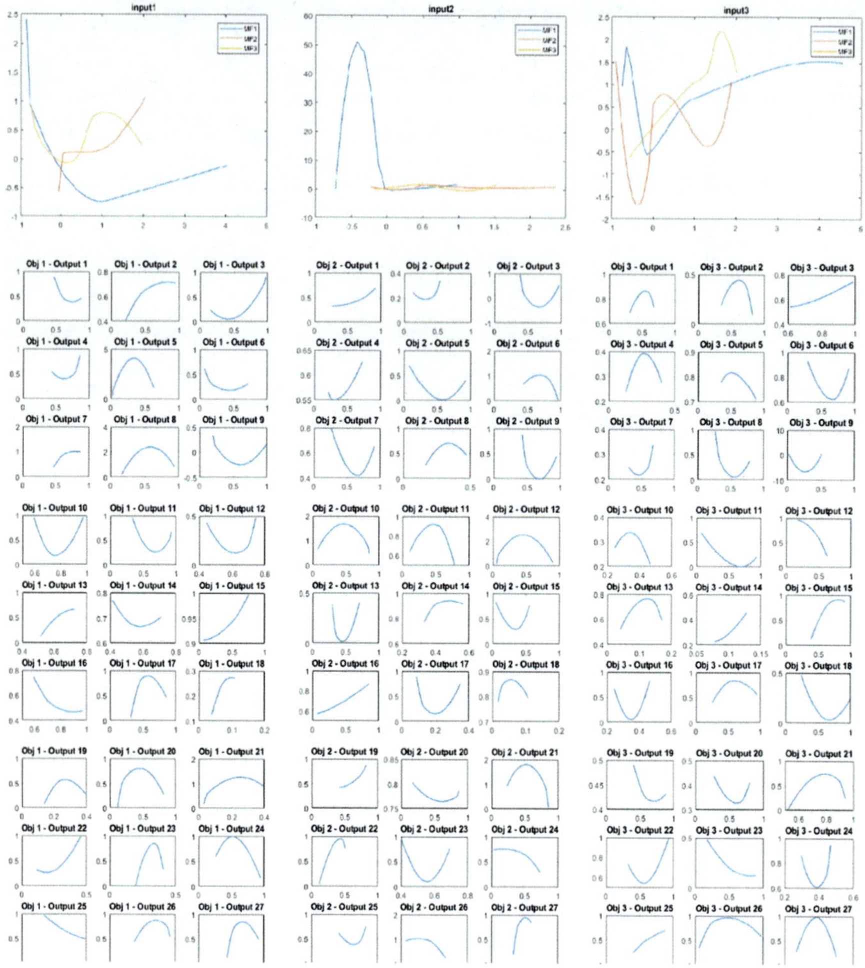

Fig. 10. Nonlinear control model of Makima ideal solution

When the results shown in Table 2 is analyzed, the maximum error rates for the outputs are approximately 0.22%, 0.3% and 0.16%, respectively. In 38 examples, the total error level of the model are 1.05%, 2.69% and 1.82%, respectively.

When the comparison results given in Table 3 are analyzed, the maximum error rates for the outputs are approximately 0.03%, 0.24%, and 0.16%, respectively. In 17 samples, the total error levels of the model are 0.25%, 1.08%, and 0.71%, respectively. The total error for the three outputs is about 2.04%.

Table 2. Comparisons with training samples.

Train	Volume fraction	Volume fraction	g.cm-3	g.cm-3	m2/s	m2/s	W.m-1.K-1	W.m-1.K-1
T(°C)	Al2O3 (%)	TiO2 (%)	Density Exp (Target)	Density Exp (Model)	Viscosity Exp (Target)	Viscosity Exp (Model)	ThermalCond Exp (Target)	ThermalCond Exp (Model)
20	0	0	1.0625	1.07117	3.707	3.84524	0.3801	0.41345
20	0	0.05	1.09098	1.10934	4.885	4.74562	0.4008	0.41903
20	0	0.3	1.22417	1.22842	7.132	7.34540	0.4391	0.48354
20	0.3	0	1.18861	1.20275	6.67	6.93563	0.4689	0.47240
25	0	0	1.05979	1.09424	3.18	3.56356	0.3823	0.40036
25	0	0.3	1.2186	1.21945	5.865	6.00345	0.4513	0.47346
25	0.05	0	1.08438	1.09132	4.12	4.53650	0.4187	0.46246
30	0	0	1.05696	1.09648	2.716	2.85463	0.3855	0.40012
30	0	0.05	1.08443	1.07934	3.587	3.93530	0.4166	0.48451
30	0	0.3	1.2131	1.19964	4.944	5.04562	0.4582	0.48713
30	0.05	0	1.08112	1.09115	3.38	3.72420	0.4211	0.48735
35	0	0.05	1.08108	1.09535	3.121	3.12645	0.4245	0.44625
35	0	0.3	1.20776	1.19975	4.354	4.62452	0.4659	0.46314
35	0.05	0	1.07781	1.08193	2.874	2.74256	0.4306	0.49135
35	0.3	0	1.17459	1.18283	4.118	4.35756	0.5101	0.49713
40	0	0.3	1.20265	1.20396	3.99	4.76356	0.4716	0.42513
40	0.05	0	1.07446	1.07345	2.489	2.24265	0.4390	0.43145
40	0.3	0	1.17017	1.17345	3.565	3.54635	0.5110	0.49462
45	0	0.3	1.19759	1.20435	3.561	3.73535	0.4756	0.43646
45	0.05	0	1.07105	1.09046	2.217	2.34523	0.4402	0.45253
45	0.3	0	1.16572	1.19462	3.111	3.25456	0.5203	0.49962
50	0	0.3	1.19268	1.21065	3.189	3.09256	0.4784	0.42345
50	0.05	0	1.0676	1.10573	1.897	1.75456	0.4470	0.43992
50	0.3	0	1.16138	1.15935	2.735	2.69945	0.5270	0.49983
55	0	0	1.04158	1.05635	1.465	1.50635	0.3977	0.40014
55	0	0.05	1.06721	1.0563	1.953	1.83456	0.4424	0.45001
55	0.3	0	1.15668	1.14563	2.426	2.53425	0.5286	0.50013
60	0	0	1.03826	1.14520	1.329	1.54265	0.4003	0.40134
60	0	0.05	1.06362	1.09452	1.662	1.74520	0.4480	0.45235

(*continued*)

Table 2. (*continued*)

Train	Volume fraction	Volume fraction	g.cm-3	g.cm-3	m2/s	m2/s	W.m-1.K-1	W.m-1.K-1
60	0.3	0	1.15233	1.13753	2.16	2.09462	0.5409	0.53012
65	0	0	1.03484	1.13241	1.211	1.35420	0.4058	0.40952
65	0	0.05	1.05999	1.29452	1.443	1.52365	0.4571	0.46943
65	0.05	0	1.05691	1.13420	1.315	1.62459	0.4728	0.46923
65	0.3	0	1.14812	1.23541	1.936	1.80942	0.5438	0.52935
70	0	0	1.03182	1.04520	1.194	1.14325	0.4113	0.42454
70	0	0.3	1.17261	1.09475	1.9097	1.88435	0.5051	0.49946
70	0.05	0	1.05291	1.04793	1.105	1.20536	0.4822	0.49023
70	0.3	0	1.14414	1.19546	1.728	1.20134	0.5552	0.55001

Table 3. Comparisons with test samples.

Train	Volume fraction	Volume fraction	g.cm-3	g.cm-3	m2/s	m2/s	W.m-1.K-1	W.m-1.K-1
T(°C)	Al2O3 (%)	TiO2 (%)	Density Exp (Target)	Density Exp (Model)	Viscosity Exp (Target)	Viscosity Exp (Model)	ThermalCond Exp (Target)	ThermalCond Exp (Model)
20	0.05	0	1.08755	1.09856	4.66	4.52256	0.4089	0.475350
25	0	0.05	1.08775	1.10909	4.236	4.26562	0.412	0.426545
25	0.3	0	1.18389	1.20193	5.594	5.63134	0.4837	0.424562
30	0.3	0	1.1792	1.21845	4.731	4.81345	0.4931	0.484624
35	0	0	1.05404	1.08533	2.361	2.63513	0.38785	0.400512
40	0	0	1.05103	1.06563	2.073	2.16242	0.3902	0.408145
40	0	0.05	1.07768	1.09542	2.655	2.59235	0.4314	0.462412
45	0	0	1.04794	1.03452	1.832	1.95634	0.39265	0.402413
45	0	0.05	1.07424	1.03695	2.402	2.52342	0.4373	0.434645
50	0	0	1.04477	1.05463	1.629	1.84624	0.3951	0.404524
50	0	0.05	1.07075	1.07454	2.109	2.62542	0.4404	0.445462
55	0	0.3	1.18746	1.19454	2.838	2.63543	0.4823	0.492451
55	0.05	0	1.06409	1.07455	1.709	1.82452	0.4536	0.441345
60	0	0.3	1.18265	1.19346	2.538	2.58234	0.4883	0.494350
60	0.05	0	1.06053	1.06934	1.47	1.57354	0.4599	0.460134
65	0	0.3	1.178	1.18035	2.287	2.37230	0.4952	0.454563
70	0	0.05	1.05601	1.07453	1.289	1.19454	0.4678	0.473513

5 Conclusion

In this study, a model including multi-objective optimization and interpolation methods is introduced for nonlinear control problems. Fuzzy inference systems generally use standard MFs for inputs, while the proposed model creates more flexible MFs for both inputs and outputs. The proposed method has been applied to determine the optimum operating parameters to increase the performance of the nano-fluid mixture. The model was designed using the data set obtained by measurements. Compared to the test data, the results produced by the model have been observed to achieve results that increase the thermal conductivity while minimizing density and viscosity.

References

1. Fleming, P., Purshouse, R.: Evolutionary algorithms in control systems engineering: a survey. Control Eng. Pract. **10**, 1223–1241 (2002). https://doi.org/10.1016/S0967-0661(02)00081-3
2. Zadeh, L.A.: Fuzzy sets. Inf. Control **8**, 338–353 (1965)
3. Gacto, M.J., Alcalá, R., Herrera, F.: A multi-objective evolutionary algorithm for tuning fuzzy rule-based systems with measures for preserving interpretability. In: 2009 International Fuzzy Systems Association World Congress and 2009 European Society for Fuzzy Logic and Technology Conference, IFSA-EUSFLAT 2009 – Proceedings, pp 1146–1151 (2009)
4. Alcalá, R., Gacto, M.J., Herrera, F., Alcalá-Fdez, J.: A multi-objective genetic algorithm for tuning and rule selection to obtain accurate and compact linguistic fuzzy rule-based systems. Int. J. Uncertainty Fuzz. Knowl. Based Syst. **15**, 539–557 (2007). https://doi.org/10.1142/S0218488507004868
5. Deb, K.: Multi-Objective Optimization Using Evolutionary Algorithms. Wiley, Hoboken (2001)
6. Dwivedi, P.K., Tripathi, S.P.: A survey on the design of fuzzy classifiers using multi-objective evolutionary algorithms. Int. J. Sci. Res. **4**, 1103–1107 (2015)
7. Trebi-Ollennu, A.: Multiobjective fuzzy genetic algorithm optimization approach to nonlinear control system design. In: UKACC International Conference on Control. Control 1996, pp 460–466. IEE (1996)
8. Seng, T.L., Bin Khalid, M., Yusof, R.: Tuning of a neuro-fuzzy controller by genetic algorithm. IEEE Trans. Syst. Man Cybern. Part B **29**, 226–236 (1999). https://doi.org/10.1109/3477.752795
9. Zuo, W.: Multivariable adaptive control for a space station using genetic algorithms. IEE Proc Control Theor. Appl. **142**, 81–87 (1995). https://doi.org/10.1049/ip-cta:19951734
10. Priyanka, T.L., Narayanan, K.R.S., Jayanthi, T., Murty, S.A.V.S.: Simulation and tuning of PID controllers using evolutionary algorithms. Int. J. Inf. Technol. Comput. Sci. **4**, 50–57 (2012). https://doi.org/10.5815/ijitcs.2012.11.07
11. Lalwani, S., Singhal, S., Kumar, R., Gupta, N.: A comprehensive survey: applications of multi-objective particle swarm optimization (MOPSO) algorithm. Trans. Comb. ISSN **2**, 39–101 (2013)
12. Bagis, A., Senberber, H.: ABC algorithm based PID controller design for higher order oscillatory systems. Elektron ir Elektrotechnika **23**, 3–9 (2017). https://doi.org/10.5755/j01.eie.23.6.19688
13. Ata, B., Coban, R.: Artificial bee colony algorithm based linear quadratic optimal controller design for a nonlinear inverted pendulum. Int. J. Intell. Syst. Appl. Eng. **3**, 1–6 (2015)

14. Tarczewski, T.: Application of artificial bee colony algorithm to auto-tuning of linear-quadratic regulator for PMSM position control. Przegląd Elektrotechniczny **1**, 59–64 (2016). https://doi.org/10.15199/48.2016.06.11
15. Sahed, O.A., Kara, K., Benyoucef, A.: Artificial bee colony-based predictive control for non-linear systems. Trans. Inst. Meas. Control **37**, 780–792 (2015). https://doi.org/10.1177/0142331214546796
16. Robandi, I., Nishimori, K., Nishimura, R., Ishihara, N.: Optimal feedback control design using genetic algorithm in multimachine power system. Int. J. Electr. Power Energy Syst. **23**, 263–271 (2001). https://doi.org/10.1016/S0142-0615(00)00062-4
17. Omran, A., El-Bayiumi, G., Bayoumi, M., Kassem, A.: Genetic algorithm based optimal control for a 6-dof non redundant stewart manipulator. Int. J. **2**, 73–79 (2008)
18. Chipperfield, A.J., Dakev, N.V., Fleming, P.J., Whidborne, J.F.: Multiobjective robust control using evolutionary algorithms. In: Proceedings of the IEEE International Conference on Industrial Technology (ICIT 1996), pp. 269–273. IEEE (1996)
19. Fonseca, C.M.: Multiobjective optimal controller design with genetic algorithms. In: International Conference on Control 1994, pp 745–749. IEE (1994)
20. Patton, R.J., Liu, G.P.: Robust control design via eigenstructure assignment, genetic algorithms and gradient-based optimisation. IEE Proc Control Theory Appl. **141**, 202–208 (1994). https://doi.org/10.1049/ip-cta:19941093
21. Marler, R.T., Arora, J.S.: Survey of multi-objective optimization methods for engineering. Struct. Multidiscip. Optim. **26**, 369–395 (2004). https://doi.org/10.1007/s00158-003-0368-6
22. Costa, L., Oliveira, P.: An elitist genetic algorithm for multiobjective optimization. In: Metaheuristics: Computer Decision-Making, pp 217–236. Kluwer Academic Publishers B. V. (2003)
23. Holland, J.H.: Genetic algorithms and adaptation. Adapt. Control Ill-Defined Syst. **16**, 317–333 (1984)
24. Srinivas, N., Deb, K.: Muiltiobjective optimization using nondominated sorting in genetic algorithms. Evol. Comput. **2**, 221–248 (1994). https://doi.org/10.1162/evco.1994.2.3.221
25. Deb, K., Pratap, A., Agarwal, S., Meyarivan, T.: A fast and elitist multiobjective genetic algorithm: NSGA-II. IEEE Trans. Evol. Comput. **6**, 182–197 (2002). https://doi.org/10.1109/4235.996017
26. Deb, K., Deb, K.: Multi-objective Optimization. In: Burke, E.K., Kendall, G. (eds.) Search Methodologies, pp 403–449. Springer, Boston (2014)
27. Fritsch, F.N., Carlson, R.E.: Monotone piecewise cubic interpolation. Soc. Ind. Appl. Math. **17**, 238–246 (1980)
28. Akima, H.: A new method of interpolation and smooth curve fitting based on local procedures. J. ACM **17**, 589–602 (1970). https://doi.org/10.1145/321607.321609
29. Said, Z., Abdelkareem, M.A., Rezk, H., Nassef, A.M.: Fuzzy modeling and optimization for experimental thermophysical properties of water and ethylene glycol mixture for Al2O3 and TiO2 based nanofluids. Powder Technol. **353**, 345–358 (2019). https://doi.org/10.1016/j.powtec.2019.05.036
30. Said, Z., Abdelkareem, M.A., Rezk, H., Nassef, A.M.: Dataset on fuzzy logic based-modelling and optimization of thermophysical properties of nanofluid mixture. Data Br. **26**, 104547 (2019). https://doi.org/10.1016/j.dib.2019.104547
31. Shahbazova, Sh.N.: Application of fuzzy sets for control of student knowledge. Appl. Comput. Math. Int. J. **10**, N 1 195–208 (2011). Special Issue on Fuzzy Set Theory and Applications

Prioritization of the Criteria Influencing Organizational Capital by Using Hesitant Fuzzy AHP

Demet Haylaz, Pınar Tuğçe Çertmek, İlayda Nur Derya, Selen Kurtuluş, Görkem Enhoş, Faik Tunç Bozbura, Ahmet Beşkese, and Gül T. Temur(✉)

Bahcesehir University, Yildiz Mah. Ciragan Cad., 34349 Besiktas/Istanbul, Turkey
{demet.haylaz,pinartugce.certmek,ilaydanur.derya,selen.kurtulus, gorkem.enhos}@bahcesehir.edu.tr, {tunc.bozbura,ahmet.beskese, gul.temur}@eng.bau.edu.tr

Abstract. In today's world, the joint impact of globalization and knowledge economy has been intensifying incessantly and consequently, many new concepts have been produced. In this sense, organizational capital is one of the prominent concepts that attracts the interests of both academicians and practitioners. Organizational capital is a component of the intellectual capital that is the collection of all assets that create the artistic ability of the organizations' potentials. In order to regulate and manage such a very potent force, the businesses should know how to measure it. This study aims to provide a methodology for the prioritization of the criteria influencing organizational capital under unstable conditions. In order to achieve this goal, a novel fuzzy based multi criteria decision making (MCDM) tool named hesitant fuzzy analytic hierarchy process (HFAHP) is used. The proposed model consists of three main attributes; knowledge management, organizational structure and management influence. The sub-attributes of knowledge management involve knowledge creation, knowledge acquisition, knowledge sharing, knowledge utilization and store knowledge. The sub-attributes of organizational structure consist of span of control, departmentalization and the number of hierarchical levels. Similarly the sub-attributes of management influence comprise rules and governance, corporate culture and strategic leadership. To outline the relative priority of every attribute, preferences of specialists were collected by administering a questionnaire which was based on pair-wise comparisons of the criteria respectively. The results of the study indicated that "Knowledge Management" had the highest importance, but the weight of "Management Influence" was very close to it. The effect of "Organizational Structure" on organizational capital was very low. Among sub criteria, "Strategic Leadership" was found as the most vital attribute which influences organizational capital.

Keywords: Hesitant fuzzy analytic hierarchy process · Multi criteria decision making · Organizational capital

T. Allahviranloo et al. (Eds.): IDS 2020, AISC 1301, pp. 471–482, 2021.
https://doi.org/10.1007/978-3-030-66501-2_37

1 Introduction

Organizational capital is one of the sub components of intellectual capital. The concept of intellectual capital was first used by John Kenneth Galbraith in 1969. Galbraith considered intellectual capital not only as a result of human intelligence but also as an accumulation of intellectual activities [1]. Intellectual capital expresses intangible assets of inherently knowledge-involving businesses and, thanks to the added value it creates and provides them with sustainable competitive advantage.

As a sub component of intellectual capital (as seen in Fig. 1), organizational capital refers to a combination of all elements which are owned by organizations, from institutionalized information technologies, databases, records and various forms of documentation, management philosophy, organizational culture, financial relations to patents [2, 3]. According to [4]; organizational capital is defined as everything which workers leave behind when they are out of work and go home. This capital includes the philosophy and system of the organization which will increase the ability of doing business. On the other hand, it is the organizational quality that encourages the implementation of knowledge on the basis of continuity of business activities. It also expresses cultural structure that encourages learning and sharing [5].

The organizational capital that shapes all the activities of an organization is generally entails the following concepts: management philosophy, organizational culture, management process, information systems, network systems, financial relations, intellectual property-related assets (patents, copyrights etc.) [6].

In this study, a novel fuzzy approach entitled Hesitant Fuzzy Sets AHP is proposed for ranking the criteria affecting organizational capital. Recognizing this goal, the study is designed as follows: Sect. 2 establishes the main and the sub criteria, Sect. 3 explains the application and its results, and last section concludes the study by discussing the results and giving comments for further research.

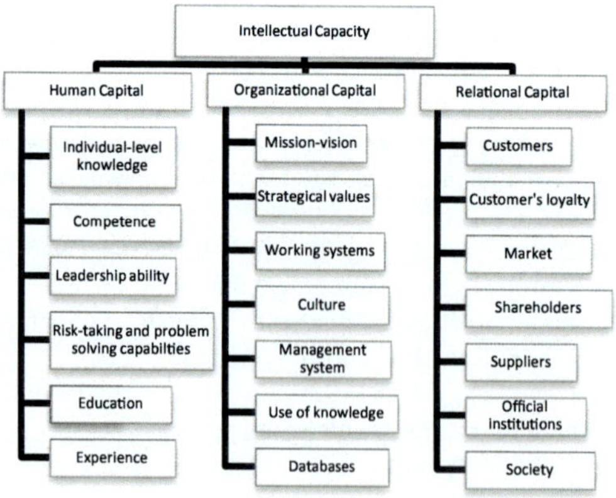

Fig. 1. Components of intellectual capital [7]

2 Criteria Influencing Organizational Capital

A review of the related literature [8–12], suggests that the criteria set that influences organizational capital may be categorized into three groups: (1) knowledge management, (2) organizational structure, and (3) management influence. In the following sub section, these criteria are clarified in detail.

Knowledge Management: One of the most important elements of organizational capital is knowledge management that can be governed by planning and organizing the usage of information technology. Information technology is the investment of an enterprise in computer and communication technology [13]. This includes hardware, software, telecommunications, data collection and demonstration tools, all data stored electronically. Leading firms, which are a strategic resource for information technologies, make significant investments in information technologies in line with the strategies they have devised, and therefore their returns are low in the short term.

Organizational Structure: An organizational structure is a system that outlines how certain activities are directed in order to achieve the goals of an organization. These activities can include rules, positions, and responsibilities. An organizational structure can (1) outline how certain activities are conducted to achieve the goals of an organization, (2) describe each employee's job and how it conforms to the whole system, and (3) give almost every employee receiving a high level of personal agency.

Management Influence: This criterion addresses the cultural and strategic viewpoints [12] and rules affecting managerial operations in the organization. Behaviors, values, beliefs, policies, strategic plans, leadership style and the adaptation process to all of these issues are the main headings of the management influence. The sub criteria of all main criteria are shown in hierarchical form as seen in Fig. 2.

3 The Proposed Methodology

The main problem of a decision making process is to select the best option from a set of options which are evaluated according to conflicting criteria. MCDM deals with such kind of situations when there are multiple criteria or objectives that have quantitative and qualitative characteristics. In order to determine the importance of criteria for MCDM problems, pairwise comparison method is found to be an efficient one that is mostly used in one of the famous techniques named Analytic Hierarchy Process (AHP) [14].

However in real life situations' choices are based on not only objective but also subjective measurements. Due to the fuzzy nature of decision process, decision makers generally can't give exact judgements. To overcome this uncertainty, fuzzy sets that were introduced by [15] are integrated into traditional approaches. These methods basically use a range of values for the corresponding linguistic term instead of a single crisp value for each term so that decision makers can decide easily and more precisely [16].

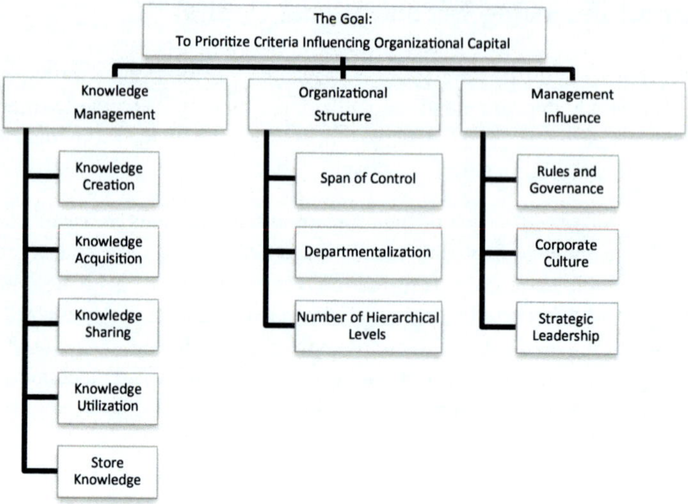

Fig. 2. The hierarchical structure of criteria influencing organizational capital

Hesitant Fuzzy Analytic Hierarchy Process: In this study, hesitant fuzzy AHP (HFAHP) methodology is chosen to determine the importance of weights of selected criteria which influence organizational capital. This method gives decision makers the advantage of choosing not just a single linguistic term from the scale but two as a between value so that when they feel uncertain about the level of importance of one criterion over another one, they have the chance of determining an interval of two terms which they are in between. Also with the help of this method, the blurriness of human cognition will mostly be eliminated. The steps of HFAHP which are adapted from [17] are given as follows.

Step 1: As a first step, pairwise comparison matrices for each hierarchical level are built. Thereafter experts are asked to evaluate the importance level of each criterion over another one within the same hierarchical level.

Step 2: According to each linguistic term obtained from the experts' evaluations, these terms are transformed into triangular fuzzy numbers with the help of the scale given in Table 1.

Step 3: Before proceeding to further calculations, for each pairwise comparison matrix, the consistency should be checked. In order to be able to calculate the consistency, first the triangular fuzzy numbers are defuzzified by the graded mean integration approach [18]. Let $A = (l, m, u)$ be a triangular fuzzy number and in order to convert this number into a crisp number with the help of graded mean integration approach Eq. (1) is used;

$$A = \frac{1 + 4\mathrm{m} + \mathrm{u}}{6} \tag{1}$$

After all triangular numbers are defuzzified and the consistency calculation steps are applied properly for each matrix, if the result shows that the matrix is inconsistent, the

Table 1. Pairwise comparison scale used in Hesitant Fuzzy AHP (adapted from [17])

Linguistic Term	Abbreviation	Triangular fuzzy number
Absolutely High Importance	AHI	(7, 9, 9)
Very High Importance	VHI	(5, 7, 9)
Essentially High Importance	ESHI	(3, 5, 7)
Weakly High Importance	WHI	(1, 3, 5)
Equally High Importance	EHI	(1, 1, 3)
Exactly Equal	EE	(1, 1, 1)
Equally Low Importance	ELI	(0.33, 1, 1)
Weakly Low Importance	WLI	(0.2, 0.33, 1)
Essentially Low Importance	ESLI	(0.14, 0.2, 0.33)
Very Low Importance	VLI	(0.11, 0.14, 0.2)
Absolutely Low Importance	ALI	(0.11, 0.11, 0.14)

results should be discussed with experts again and they should be asked to recheck their evaluations.

Step 4: If the evaluations of different experts are not close to each other for the same criterion comparisons at the same level, then the experts must be asked to recheck their evaluations.

Step 5: After renewal of the experts' evaluations and after the consistency calculations, if the results of each matrix show that it is consistent, then for each level experts' evaluations are aggregated by fuzzy envelope approach, which is proposed by [19]. According to Table 1, the linguistic terms are sorted from the lowest s_0 to the highest s_g, the lowest referring to the term "Absolutely Low Importance" and the highest to the term "Absolutely High Importance". All these terms are ranked from 0 to 10, where "Exacty Equal" has the number 5 which is right in the middle of this scale. Let's assume that the expert evaluations vary between s_i and s_j, then it can be written that $s_o < s_i < s_j < s_g$. In order to find the parameters a, b, c and d of the trapezoidal fuzzy membership function A = (a, b, c, d), Eq. (2), Eq. (3) and Eq. (4) are used;

$$\begin{aligned} a &= \min\{a_L^i, a_M^i, a_M^{i+1}, \ldots, a_M^j, a_R^j\} = a_L^i \\ d &= \max\{a_L^i, a_M^i, a_M^{i+1}, \ldots, a_M^j, a_R^j\} = a_R^j \end{aligned} \tag{2}$$

$$b = \begin{cases} a_m^i, \; if\ i+1=j \\ OWA_{w^2}\left(a_m^i, \ldots\ldots a_m^{\frac{i+j}{2}}\right), \; if\ i+j\ is\ even \\ OWA_{w^2}\left(a_m^i, \ldots\ldots a_m^{\frac{i+j-1}{2}}\right), \; if\ i+j\ is\ odd \end{cases} \tag{3}$$

$$c = \begin{cases} a_m^{j+1}, \ if\ i+1=j \\ OWA_{w^1}\left(a_m^i, a_m^{j-1}, ...a_m^{\frac{i+j}{2}}\right), \ if\ i+j\ is\ even \\ OWA_{w^1}\left(a_m^i, a_m^{j-1}, ...a_m^{\frac{i+j+1}{2}}\right), \ if\ i+j\ is\ odd \end{cases} \tag{4}$$

In hesitant fuzzy sets, weighted averaging operator (OWA) needs a weight vector for its calculations. For trapezoidal fuzzy numbers it is required to have two types of weights which are defined as the first and second types of weights and these weight vectors uses α parameter in the unit interval [0, 1] [20].

Each component of the first kind of weights $W^1 = (w_1^1, w_2^1, ..., w_n^1)$ is obtained as shown in Eq. (5).

$$w_1^1 = \alpha_2, w_2^1 = \alpha_2(1-\alpha_2), ...w_n^1 = \alpha_2(1-\alpha_2)^{n-2} \tag{5}$$

Each component of the second kind of weights $W^2 = (w_1^2, w_2^2, ..., w_n^2)$ is obtained by using Eq. (6).

$$w_1^2 = \alpha_1^{n-1}, w_2^2 = (1-\alpha_1)\alpha_1^{n-2}, ..., w_n^2 = 1-\alpha_1 \tag{6}$$

For the parameters, α_1 and α_2 the following formulas (Eq. (7) and Eq. (8)) are used;

$$\alpha_1 = \frac{g-(j-i)}{g-1} \tag{7}$$

$$\alpha_2 = \frac{(j-i)-1}{g-1} \tag{8}$$

Here, g refers to the number of terms in the scale which is shown in Table 1; j is the rank of the highest and i is the rank of the lowest evaluation value.

Step 6: After OWA operations all the trapezoidal numbers are entered into each matrix, the collaborative pairwise comparison matrix C is built by $\tilde{c}_{ij} = (\tilde{c}_{ij_1}, \tilde{c}_{ij_{m1}}, \tilde{c}_{ij_{m2}}, \tilde{c}_{ij_u})$. To calculate the reciprocal values of this matrix, the following equation (Eq. (9)) is used;

$$\tilde{c}_{ji} = \left(\frac{1}{\tilde{c}_{ij_u}}, \frac{1}{\tilde{c}_{ij_{m2}}}, \frac{1}{\tilde{c}_{ij_{m1}}}, \frac{1}{\tilde{c}_{ij_1}}\right) \tag{9}$$

Step 7: After constructing the collaborative pairwise comparison matrix entirely, the fuzzy geometric mean of each row ($\tilde{r}_i$) in this matrix is calculated by using Eq. (10).

$$\tilde{r}_i = (\tilde{c}_{i1} \otimes \tilde{c}_{i2} \ldots \otimes \tilde{c}_{in})^{1/n} \tag{10}$$

Step 8: For each criterion the fuzzy weight ($\tilde{w}_i$) is calculated by using Eq. (11).

$$\tilde{w}_i = \tilde{r}_i \otimes (\tilde{r}_1 \oplus \tilde{r}_2 \ldots \oplus \tilde{r}_n)^{-1} \tag{11}$$

Here, the value of $\tilde{r}_1 \oplus \tilde{r}_2 \ldots \oplus \tilde{r}_n$ is assumed to be 9 which is the maximum parameter of the "Absolutely High Importance" term in the pairwise comparison scale in Table 1 and the aim of this assumption is to decrease the deviation in the weights.

Step 9: The steps 1 to 8 are repeated again for each sub-criteria according to this study's model and the criterion weights for them are obtained.

Step 10: For the purpose of determining the ranking of factors i, trapezoidal fuzzy numbers are defuzzified according to Eq. (12).

$$D_i = \frac{w_{i_l} + w_{i_{m1}} + w_{i_{m2}} + w_{i_u}}{6} \tag{12}$$

After this formula all the trapezoidal fuzzy numbers are transformed into crisp values.

Step 11: Thereafter the criterion weights are defuzzified and they are normalized by dividing each element of the matrix by the column total.

Step 12: Finally, in order to find the global values for the prioritization of factors, each element of the normalized criterion weight vector of sub-criteria is multiplied by its corresponding criteria's weight vector. In the end, the factors are ranked according to the results of the multiplication.

4 An Application

In order to assess the relative importance of each entity of each criterion and to find a ranking among them, three experts were asked to fill a questionnaire based on pairwise comparisons of all criteria at each hierarchical level. After obtaining each expert's judgments, the pairwise comparison matrices for each level were constructed reflecting the experts' judgments. Then these terms were transformed into triangular fuzzy numbers according to the scale in Table 1. In order to calculate the consistency ratio, all triangular numbers were first defuzzified. After each matrix was checked for the consistency ratio, if the result confirmed its consistency, then the matrices were combined and further calculations of hesitant fuzzy AHP were applied. Since it's seen that all the CR results were below 0.1, it was concluded that all these matrices were consistent.

After checking the CR values, experts' evaluations for each level are combined by fuzzy envelope approach as explained at Step 5. In Table 2, the fuzzy envelope of three experts for the main criteria is shown.

Table 2. Hesitant fuzzy envelope of the evaluations of main criteria

	Knowledge Management	Organizational Structure	Management Influence
Knowledge Management	EE	Between VHI and WHI	Between EHI and ELI
Organizational Structure		EE	Between WLI and ESLI
Management Influence			EE

In order to find the trapezoidal fuzzy sets associated with the fuzzy envelopes, the OWA operator was used. Table 3 shows the trapezoidal fuzzy sets obtained by the OWA operator applied to Table 2.

Table 3. The trapezoidal fuzzy sets of main criteria obtained by the OWA operator

	Knowledge management	Organizational structure	Management influence
Knowledge Management	(1, 1, 1, 1)	(1, 4.78, 5.22, 9)	(0.33, 1, 1, 3)
Organizational Structure	(0.11, 0.19, 0.21, 1)	(1, 1, 1, 1)	(0.14, 0.2, 0.33, 1)
Management Influence	(0.33, 1, 1, 3)	(1, 3, 5, 7)	(1, 1, 1, 1)

Here as an example of the calculations, the associated fuzzy envelope for Knowledge Management over Organizational Structure was chosen. The detailed calculations of the value (1, 4.78, 5.22, 9) in Table 3 were given as follows: The fuzzy envelope for Knowledge Management over Organizational Structure was positioned between "Very High Importance" and "Weakly High Importance". According to the scale given in Table 1, the triangular fuzzy number for "Very High Importance" was (5, 7, 9) and for "Weakly High Importance" was (1, 3, 5). Since j is the rank of the highest evaluation value and i is the rank of the lowest evaluation value, it could be concluded that $i = 7$ and $j = 9$.The number of terms in the evaluation scale determined the value of g which was equal to 10.The next thing to do was to calculate α values by the following;

$$\alpha_1 = \frac{g - (j - i)}{g - 1} = \frac{10 - (9 - 7)}{10 - 1} = \frac{8}{9} = 0.889$$

$$\alpha_2 = \frac{(j - i) - 1}{g - 1} = \frac{(9 - 7) - 1}{10 - 1} = \frac{1}{9} = 0.111$$

In order to ascertain the a and d values of a trapezoidal fuzzy set which was defined as (a, b, c, d) the Eq. (2) was used. The values of a and d represented the min and max values of $\{a^i{}_L, a^i{}_M, a^{i+1}{}_M, \ldots\ldots, a^j{}_M, a^j{}_R\} = \{1, 3, 5, 7, 9\}$, respectively. Thus $a = 1$ and $d = 9$. In order to calculate parameter b the Eq. (3) was utilized. For the calculation of parameter c Eq. (4) was employed. Since $i + j =$ even;

$$b = OWA_{w^2}\left(a_m^i, \ldots\ldots, a_m^{\frac{i+j}{2}}\right) = 0.889 * 5 + 0.111 * 3 = 4.778$$

$$c = 2a_m^{(i+j)/2} - b = (2*5) - 4.78 = 5.22$$

Therefore the trapezoidal fuzzy set associated with the fuzzy envelope for Knowledge Management over Organizational Structure was found as (1, 4.78, 5.22, 9).

Next, the geometric mean of these values was calculated for each row by using the Eq. (10). As an example the geometric mean calculation for the first row is shown below;

$$a_g = (1 * 1 * 0.33)^{1/3} = 0.691$$
$$b_g = (1 * 4.78 * 1)^{1/3} = 1.684$$
$$c_g = (1 * 5.22 * 1)^{1/3} = 1.734$$
$$d_g = (1 * 9 * 3)^{1/3} = 3$$

Thus, the geometric mean for the first row was found as (0.691, 1.684, 1.734, 3). Next thing to do was to normalize geometric mean of all values in the trapezoidal fuzzy set by using the Eq. (11).

$$a_w = \frac{0.691}{9} = 0.077$$
$$b_w = \frac{1.684}{9} = 0.187$$
$$c_w = \frac{1.734}{9} = 0.192$$
$$d_w = \frac{3}{9} = 0.333$$

Also the trapezoidal fuzzy weights of the other factors were similarly calculated and all the results are given in Table 4.

Table 4. Trapezoidal fuzzy criteria weights of main criteria

Knowledge Management	(0.08, 0.19, 0.19, 0.33)
Organizational Structure	(0.03, 0.04, 0.05, 0.11)
Management Influence	(0.08, 0.16, 0.19, 0.31)

In order to find the weights for the remaining part of the hierarchy, the same steps were applied. After figuring out all trapezoidal fuzzy criteria weights for sub criteria, in order to obtain a prioritization among the criteria, first all the weights were defuzzified into crisp values according to the Eq. (12) and normalized. All the normalized criterion weights are shown in Tables 5, 6, 7 and 8. As an example of defuzzification calculation, the first row of Table 4 was chosen and the calculation was shown below;

$D_1 = (0.08 + 2*0.19 + 2*0.19 + 0.33)/6 = 0.195.$

After the normalization, the normalized weights of sub-criteria were multiplied by the related normalized criterion weights in order to obtain the global values. Table 9 addresses them in order.

Table 5. Normalized weights of main criteria

Knowledge Management	0.454
Organizational Structure	0.124
Management Influence	0.422

Table 6. Normalized weights of knowledge management sub criteria

Knowledge Creation	0.171
Knowledge Acquisiton	0.133
Knowledge Sharing	0.204
Knowledge Utilization	0.437
Store Knowledge	0.055

Table 7. Normalized weights of organizational structure sub criteria

Span of Control	0.27
Departmentalization	0.119
Number of Hierarchical Levels	0.611

Table 8. Normalized weights of management influence sub criteria

Rules and Governance	0.11
Corporate Culture	0.299
Strategic Leadership	0.591

According to the obtained results, for the main criteria, it is seen that the most important main criteria is Knowledge Management with its normalized weight of 45.4%, second important main criteria is Management Influence with a value of 42.2% and the last important one is Organizational Structure with 12.4%. Here it can be concluded that Knowledge Management and Management Influence have nearly the same importance level over the organizational capital but the Organizational Structure has a very low influence over organizational capital. Also, the most crucial sub-criterion for organizational capital is the "Strategic Leadership".

Table 9. Global weights of all criteria

Strategic leadership	0.249
Knowledge utilization	0.198
Culture	0.126
Knowledge sharing	0.093
Knowledge creation	0.078
Number of hierarchical levels	0.076
Knowledge acquisition	0.06
Rules and governance	0.046
Span of control	0.033
Store knowledge	0.025
Departmentalization	0.015

5 Conclusion

In today's competitive environment, competitiveness in an organizational level should be the most important agenda of enterprises since their assets depend directly on this fact. The existence of a large number of competitors competing in global markets forces businesses to make more compelling decisions and take actions more carefully. As a result of the research and studies in determining the market value of a company, it is found that this value is not only determined by the company's physical assets but also the intellectual assets play an important role, which is defined as intellectual capital. The importance of this capital is increasing day by day and this abstract concept has brought new terms which can also be defined as a resource of the companies.

Organizational capital is one of the three important components of intellectual capital. Since this capital is an intangible source of a company, it is not possible to determine its value directly. However determining the factors that affect organizational capital and prioritizing them according to their degrees will lead to the identification of the factor that deserves more importance in order to increase this capital's value and maximize organizational performance.

The main contribution of this study is the development of a framework for defining the importance ratios of predetermined factors which influence organizational capital. After defining these factors with the help of gathered information about organizational capital and its scope, the method called as Hesitant Fuzzy Analytic Hierarchy Process (HFAHP) is applied. The result of this study shows that Knowledge Management and Management Influence have nearly the same effect on organizational capital. However Organizational Structure appears to have a weak influence on organizational capital. Also as seen in evaluation of results section the most crucial factor for organizational capital turns out to be the Strategic Leadership.

In the future, researchers can apply other evolved fuzzy sets to the same topic and the results can be compared. On the other hand, in order to see how the changes on evaluations affect the results, sensitivity analysis can be conducted.

References

1. Pena, I.: Intellectual capital and business start-up success. J. Intellectual Capital **180**(3) (2002)
2. Çıkrıkcı, M., Daştan, A.: Entellektüel sermayenin temel finansal tablolar aracılığıyla sunulması. Bankacılık Dergisi **43**, 18–32 (2002)
3. Şafaklı, O.K.: Entellektüel sermayenin insani boyutu üzerine KKTC şube bankalarına yönelik bir uygulama. Endüstri ilişkileri ve insan Kaynakları Dergisi **8**(1), 79–88 (2006)
4. Bontis, N.: Intellectual capital and business performance in Malaysian industries. J.ournal Intellectual Capital 1(1) (2000)
5. Ertuğrul, M.: Bilgi çağında işletmelerin yeni kaynağı: entellektüel sermaye. Active Bankacılık ve Finans Dergisi 15 (2000)
6. Kanıbir, H.: Yeni bir rekabet gücü kaynağı olarak entellektüel sermaye ve organizasyonel performansa yansımaları. Havacılık ve Uzay Teknolojileri Dergisi **1**(3), 77–85 (1999)
7. Bozbura, F. T., Beşkese, A.: Prioritization of organizational capital measurement indicators using fuzzy AHP. In: Proceedings of the 11th International Fuzzy Systems Association World Congress, pp. 1756–1760. IFSA, Beijing (2005)
8. Fernandez, E., Montes, J.M., Vazquez, C.J.: Typology and strategic analysis of intangible resources: aA resource-based approach. Technovation **20**, 81–92 (2000)
9. Brynjolfsson, E., Hitt, L., Yang, S.: Intangible assets: computers and organizational capital. In: Brookings Papers on Economic Activity 137- 181 (2002).
10. Bozbura, F.T.: Measurement and application of intellectual capital in Turkey. Learn. Organ. Int. J. **11**(4–5), 357–367 (2004)
11. Wong, K.Y., Aspinwall, E.: Development of a knowledge management initiative and system: a case study. Expert Syst. Appl. **30**, 633–641 (2006)
12. Leitão, J., Franco, M.: Individual entrepreneurship capacity and small and medium enterprises (SME) performance: a human and organizational capital approach. Afr. J. Bus. Manage. **5**(15), 6350–6365 (2011)
13. Weil, B.: The implications of information technology infrastructure for business process redesign. 170 (1999)
14. Saaty, T.L.: The Analytic Hierarchy Process: Planning, Priority Setting, Resources Allocation. McGraw-Hill, London (1980)
15. Zadeh, L.A.: Fuzzy sets. Inf. Control **8**, 338–353 (1965)
16. Buckley, J.: Fuzzy hierarchical analysis. Fuzzy Set. System **17**, 233–247 (1985)
17. Öztayşi, B., Çevik Onar, S., Boltürk, E., Kahraman, C.: Hesitant fuzzy analytic hierarchy process. IEEE Int Conf Fuzzy, 1–7 (2015).
18. Hsieh, C.H., Chen, S.H.: A model and algorithm of fuzzy product positioning. Inf. Sci. **121**, 61–82 (1999)
19. Rodriguez, R.M., Liu, H.: A fuzzy envelope for hesitant fuzzy linguistic term set and ıts application to multicriteria decision making. Inf. Sci. **258**, 220–238 (2014)
20. Filev, D., Yager, R.: On the issue of obtaining OWA operator weights. Fuzzy Sets Syst. **94**, 157–169 (1998)

Radial Models for Classifying Flexible Measures in Two-Stage Network DEA

Seyede Nasrin Hosseini Monfared[1(✉)], Farhad Hosseinzadeh Lotfi[1], Mohammad Reza Mozaffari[2], and Mohsen Rostamy Malkhalifeh[1]

[1] Department of Mathematics, Science and Research Branch, Islamic Azad University, Tehran, Iran
nasrinhom@gmail.com, farhad@hosseinzadeh.ir, mohsen_rostamy@yahoo.com

[2] Department of Mathematics, Shiraz Branch, Islamic Azad University, Shiraz, Iran
Mozaffari854@yahoo.com

Abstract. In conventional DEA models it is assumed that each measure has to be identified whether it is an input or output. Nevertheless, in some cases flexible measures can exist which can play the role of either an input or output. In all of the previous studies did not point to classify flexible measures in two-stage network DEA. In this paper we propose new radial models are based on envelopment form of CCR such as non-oriented models under variable returns to scale, input- and output-oriented models under constant returns to scale in basic two-stage network structure. Our models can select the status of each flexible measure as an input or output and determine the relative efficiency and an efficient projection unit. The input- or output-oriented model under CRS in basic two-stage network structure may produce different efficiency scores and it can be expected that a flexible measure is selected as an input variable in one model but an output variable in another. On the other hand, the status of one flexible measure in two CRS and VRS environments may have different results. The difference between the results of the two models clearly shows the need for a model that considers VRS in the presence of flexible measure. Therefore, the main focus of this study was to propose a new non-oriented model and aggregate model that can determine a unique status for flexible measures under VRS in basic two-stage network structure. Numerical examples are used to illustrate the procedures.

Keywords: Flexible measures · Data envelopment analysis · Basic two-stage network DEA

1 Introduction

Data envelopment analysis (DEA) is an effective technique for measuring the relative efficiency of a set of homogeneous DMUs which utilize the same inputs to produce the same outputs. In conventional DEA applications, given a set of available measures, it is assumed that the status of each measure is clearly stated as an input or an output variable in the production process prior to using DEA. However, in some situations, a

T. Allahviranloo et al. (Eds.): IDS 2020, AISC 1301, pp. 483–500, 2021.
https://doi.org/10.1007/978-3-030-66501-2_38

performance measure can play input role for some DMUs and output role for others. Cook and Zhu [5] considered flexible variables and proposed a mixed integer linear programming (MILP) problem for classifying these variables. Also, they introduced the aggregate model from the perspective of the manager of the collection of DMUs. Toloo [15] claimed in practice their individual method may produce incorrect efficiency scores due to a computational problem as a result of introducing a large positive number to the model and introduces a revised model that does not need such a large positive number. Amirteimoori and Emrouznejad [1] proposed a mixed integer linear programming model by focusing on the impact of flexible measures on the definition of the production set and the assessment of technical efficiency. Toloo [16] considered alternative solutions for classifying input and output in data envelopment analysis. Kordrostami and Noveiri [10] proposed a model to evaluate the efficiency of decision-making units where flexible and negative data exist. Amirteimoori et al. [2] suggested a flexible slack-based measure of efficiency to maximize the performance. The performance measurement in the presence of interval and flexible factors have also been addressed by Kordrostami and Noveiri [9]. Tohidi and Matroud [8] proposed a new non-oriented model that not only selected the status of each flexible measure as an input or output but also determined returns to scale status. Toloo, et al. [17] proposed a non-radial directional distance method to classify flexible factors, including an application of the banking industry. Recently, Kordrostami et al. [11] proposed the integer-valued FSBM approach for evaluating the relative efficiency of DMUs where flexible and integer measures are present.

In recent years, a number of DEA studies have focused on DMUs with internal network structures. Among a wide variety of internal structures studied, one basic and popular internal structure is called a two-stage network structure. In order to identify the frontier projections in Kao and Hwang [8], Chen et al. [3] proposed to solve an additional equivalent envelopment model to determine optimal values for the intermediate measures. Chen et al. [4] showed that the envelopment model should be used for determining the frontier projection while the multiplier model should be used for finding the divisional efficiency. Kao [7] pointed out that the projection point identified by the envelopment model is not the one used by the multiplier model to measure the division efficiencies for general two-stage systems. Based on the primal-dual relationship of the two models, the envelopment model is reformulated to be able to obtain the projection point and measure the division efficiencies at the same time. Lim and Zhu [12, 13] showed that theoretically the divisional efficiency scores and frontier projections can be calculated for the multiplicative and additive models by using the primal-dual correspondence of linear programming under basic two-stage network structure. Guo et al. [6] developed envelopment network model in parametric linear form to determine frontier projection and to find divisional efficiency.

Flexible data is encountered in many real world situations in two-stage network structures. For example, in a conventional study of efficiency of bank branch operations, inputs to the first stage are resources such as various staff types, and outputs to the second stage are the standard counter transactions such as deposits and withdrawals. In order to explain the possible application of our proposed model, assume a factor such as the number of worthwhile customers that can be considered as a flexible measure to the first stage to evaluate bank branch operations. Cook and Zhu [5] declared "from one prospective, such a measuring plays the role of proxy for future investment, hence can reasonably be classified as an output. On the other hand, it can legitimately be considered as an environmental input that aids the branch in generating its existing investment portfolio". In this case, a factor such as 'deposits' can be assumed as a flexible measure to the second stage. According to Cook and Zhu [5], it is a source of revenue for the branch and can thus also be regarded as an output. It should be considered here that arguments have been made claiming that staff time expended in processing customers who are making deposits or opening deposit accounts could also be used advantageously to sell more profitable products to customers at the same time; this factor can be supposed as an input.

The presentation of models with flexible measures in two-stage network structures is essential and beneficial because there are many situations in the real world. In all of the previous studies did not point to classify flexible measures in two-stage network DEA. So we propose new models in which each one of the flexible variables is treated as either input or output to maximize the relative efficiency of the DMU under evaluation in two-stage network structures. These new models are based on envelopment form of CCR such as non-oriented models under variable returns to scale, input and output oriented models under constant returns to scale in two-stage network structures to select the status of each flexible measure as an input or output. Our models can determine the relative efficiency and an efficient projection unit. The input- or output-oriented model under CRS in basic two-stage network structure may produce different efficiency scores and it can be expected that a flexible measure is selected as an input variable in one model but an output variable in another. The status of one flexible measure in two CRS and VRS environments may have different results. In other words, a flexible measure is considered as input in CRS environment, whereas it is selected as output in VRS environment. The difference between the results of the two models clearly shows the need for a model that considers VRS in the presence of flexible measure. Therefore, the main focus of this study was to propose a new non-oriented model and aggregate model that can determine a unique status for flexible measures under VRS in basic two-stage network structure.

The paper is organized as follows. Section 2 provides the previous studies on flexible measures and two-stage network DEA in brief. Section 3 proposes new models such as non-oriented MILP problem under variable returns to scale, input- and output-oriented MILP under constant returns to scale to evaluate the relative efficiency of DMU_o and to classify flexible measures in basic two-stage network structure. In Sect. 4, numerical examples are used for clarification. Finally, conclusions are given in Sect. 5.

2 Previous Studies

In this section, we review the previous studies on flexible measures and two-stage network data envelopment analysis in brief.

2.1 Flexible Measures

Assume the purpose is to evaluate the efficiency of n decision making units $(DMU_j; j = 1, \ldots, n)$ in which there are m inputs $x_{ij}, i = 1, \ldots, m$, s outputs $y_{rj}, r = 1, \ldots, s$ and L flexible measures $w_{lj}, l = 1, \ldots, L$.

Tohidi and Matroud [8] proposed a non-oriented mix integer fractional programming model that selected the status of each flexible measure as an input or output under variable returns to scale.

$$\begin{aligned}
&\text{Min} \quad \alpha/\beta \\
&s.t. \quad \sum_{j=1}^{n} \lambda_j x_{ij} \le \alpha x_{io}, \qquad i = 1, \ldots, m \\
&\sum_{j=1}^{n} \lambda_j y_{rj} \ge \beta y_{ro}, \qquad r = 1, \ldots, s \\
&\sum_{j=1}^{n} \lambda_j w_{lj} \le \alpha w_{lo} + M d_l, \qquad l = 1, \ldots, L \\
&-\sum_{j=1}^{n} \lambda_j w_{lj} \le -\beta w_{lo} + M(1 - d_l), \qquad l = 1, \ldots, L \\
&\sum_{j=1}^{n} \lambda_j = 1, \\
&d_l \in \{0, 1\} \qquad l = 1, \ldots, L, \\
&\lambda_j \ge 0 \qquad j = 1, \ldots, n, \\
&\alpha, \beta \ge 0.
\end{aligned} \tag{1}$$

Where M is a large positive number. For each l, a binary variable $d_l \in \{0, 1\}$ is introduced to identify that factor l is an input, $d_l = 0$, or an output $d_l = 1$.

2.2 Two-Stage Network DEA Model

We assume the purpose is to evaluate the efficiency of n decision making units $(DMU_j; j = 1, \ldots, n)$ in which there are m inputs to the first stage, x_{ij}, $(i = 1, \ldots, m)$ and K outputs from the first stage, z_{kj}, $(k = 1, \ldots, K)$. These K outputs then become the inputs to the second stage and are called intermediate measures. The outputs from the second stage are y_{rj}, (r = 1, ..., s) as shown in Fig. 1.

The input-oriented two-stage network DEA for measuring the efficiency of DMU_o is given as follows:

$$
\begin{aligned}
& Min \quad \theta \\
& s.t. \quad \sum_{j=1}^{n} \lambda_j x_{ij} \le \theta x_{io}, \qquad i = 1, \ldots, m \\
& \sum_{j=1}^{n} \lambda_j z_{kj} \ge \sum_{j=1}^{n} \mu_j z_{kj}, \qquad k = 1, \ldots, K \\
& \sum_{j=1}^{n} \mu_j y_{rj} \ge y_{ro}, \qquad r = 1, \ldots, s \\
& \lambda_j \ge 0, \mu_j \ge 0, \qquad j = 1, \ldots, n
\end{aligned} \tag{2}
$$

Lim and Zhu [13] determined a frontier projection for DMU_o as $(X\lambda^*, \tilde{z}_o, Y\mu^*)$ where $\tilde{z}_o$ is any choice such that $Z\mu^* \le \tilde{z}_o \le Z\lambda^*$. An easy choice for $\tilde{z}_o$ would be $Z\lambda^*$ or $Z\mu^*$.

3 Proposed Models for Evaluating the Relative Efficiency and Determining the Status of Flexible Measures in Basic Two-Stage Network DEA

We assume the purpose is to evaluate the efficiency of n decision making units $(DMU_j; j = 1, \ldots, n)$ in which there are m inputs to the first stage, x_{ij}, $(i = 1, \ldots, m)$ and K outputs from the first stage, z_{kj}, $(k = 1, \ldots, K)$. These K outputs then become the inputs to the second stage and are called intermediate measures. The outputs from the second stage are y_{rj}, $(r = 1, \ldots, s)$.

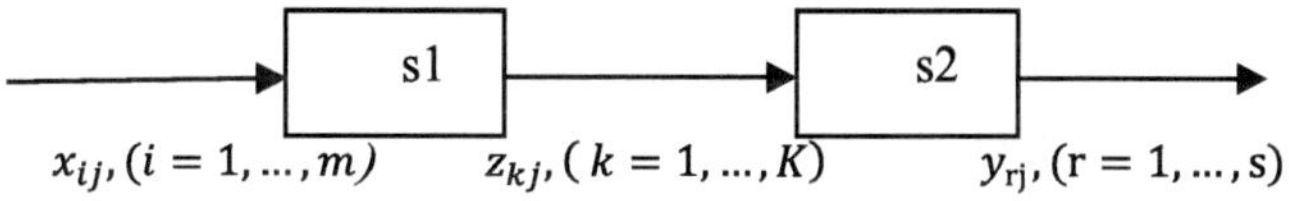

Fig. 1. Basic two-stage network structure

Suppose also that there exist L_1 flexible measures to the first stage $w^1_{l_1 j}$, $(l_1 = 1, \ldots, L_1)$ and L_2 flexible measures to the second stage $w^2_{l_2 j}$, $(l_2 = 1, \ldots, L_2)$ whose input/output status are not known, some DMUs may use these measures as inputs and other DMUs may use them as outputs.

3.1 Non-oriented Model Under Variable Returns to Scale in Basic Two-Stage Network Structure for Classifying Flexible Measures

We propose the following non-oriented fractional mix integer linear programming under variable returns to scale in basic two-stage network structure to evaluate the relative efficiency of DMU_o and to classify flexible measures.

$$
\begin{aligned}
\eta^* &= Min \quad \alpha/\beta \\
& s.t. \quad \sum_{j=1}^{n} \lambda_j^1 x_{ij} \le \alpha x_{io}, && i = 1, \ldots, m \\
& \sum_{j=1}^{n} \lambda_j^1 w_{l_1 j}^1 \le \alpha w_{l_1 o}^1 + M d_{l_1}^1, && l_1 = 1, \ldots, L_1 \\
& \sum_{j=1}^{n} \lambda_j^1 d_{l_1}^1 w_{l_1 j}^1 \ge \sum_{j=1}^{n} \lambda_j^2 d_{l_1}^1 w_{l_1 j}^1, && l_1 = 1, \ldots, L_1 \\
& \sum_{j=1}^{n} \lambda_j^2 y_{rj} \ge \beta y_{ro}, && r = 1, \ldots, s \\
& \sum_{j=1}^{n} \lambda_j^2 w_{l_2 j}^2 \le \alpha w_{l_2 o}^2 + M d_{l_2}^2, && l_2 = 1, \ldots, L_2 \\
& \sum_{j=1}^{n} \lambda_j^2 w_{l_2 j}^2 \ge \beta w_{l_2 o}^2 - M\left(1 - d_{l_2}^2\right), && l_2 = 1, \ldots, L_2 \\
& \sum_{j=1}^{n} \lambda_j^1 z_{kj} \ge \sum_{j=1}^{n} \lambda_j^2 z_{kj}, && k = 1, \ldots, K \\
& \sum_{j=1}^{n} \lambda_j^1 = 1, \sum_{j=1}^{n} \lambda_j^2 = 1 \\
& d_{l_1}^1 \in \{0, 1\}, d_{l_2}^2 \in \{0, 1\}, && \forall l_1, l_2 \\
& \lambda_j^1 \ge 0, \quad \lambda_j^2 \ge 0, && j = 1, \ldots, n \\
& \alpha \le 1, \quad \beta \ge 1,
\end{aligned} \tag{3}
$$

Where M is a large positive number. For each l_1 and l_2, binary variables $d_{l_1}^1 \in \{0, 1\}$ and $d_{l_2}^2 \in \{0, 1\}$ are introduced to identify that factor l_1 and l_2 are inputs to the first and second stage, $d_{l_1}^1 = 0$ and $d_{l_2}^2 = 0$ or outputs to the first and second stage, $d_{l_1}^1 = 1$ and $d_{l_2}^2 = 1$.

The conversion guarantees that only one of the two constraints for flexible measures to the first stage can be active at any one time. If $d_{l_1}^1 = 0$, then the constraint $\sum_{j=1}^{n} \lambda_j^1 w_{l_1 j}^1 \le \alpha w_{l_1 o}^1 + M d_{l_1}^1$ is active and the constraint $\sum_{j=1}^{n} \lambda_j^1 d_{l_1}^1 w_{l_1 j}^1 \ge \sum_{j=1}^{n} \lambda_j^2 d_{l_1}^1 w_{l_1 j}^1$ is redundant; otherwise the first constraint is redundant and the last one is active. In other words, measure l_1 will end up either as an input, $d_{l_1}^1 = 0$, or an output, $d_{l_1}^1 = 1$. Also only one of the two constraints for flexible measures to the second stage can be active at any one time. If $d_{l_2}^2 = 0$, then the constraint $\sum_{j=1}^{n} \lambda_j^2 w_{l_2 j}^2 \le \alpha w_{l_2 o}^2 + M d_{l_2}^2$ is active and the constraint $\sum_{j=1}^{n} \lambda_j^2 w_{l_2 j}^2 \ge \beta w_{l_2 o}^2 - M\left(1 - d_{l_2}^2\right)$ is redundant; otherwise the first constraint is redundant and the last one is active. In other words, measure l_2 will end up either as an input, $d_{l_2}^2 = 0$, or an output, $d_{l_2}^2 = 1$.

We use the following constraints in model (3):

$$
\sum_{j=1}^{n} \lambda_j^1 z_{kj} \ge \sum_{j=1}^{n} \lambda_j^2 z_{kj}, \qquad k = 1, \ldots, K
$$

$$\sum_{j=1}^{n} \lambda_j^1 d_{l_1}^1 w_{l_1j}^1 \geq \sum_{j=1}^{n} \lambda_j^2 d_{l_1}^1 w_{l_1j}^1, \qquad l_1 = 1, \ldots, L_1$$

If $d_{l_1}^{1*} = 1$ then the flexible measure l_1 is as an output to the first stage and this output then becomes the input to the second stage and is called intermediate measure. These constraints ensure that the amount produced of each intermediate product is no less than the consumed amount, under the assumption that there exist no supplies and demands of intermediate products from outside of the evaluated system. These network constraints hold for any network structures in which the intermediate outputs are consumed and some of them can be retained (or wasted) within the DMU.

Definition 1. DMU_o is efficient if and only if $\alpha^* = 1, \beta^* = 1$.

Theorem 1. Model (3) is feasible.

Proof. Let $d_{l_1}^1 = 0, \forall l_1 = 1, \ldots, L_1,\ d_{l_2}^2 = 1, \forall l_2 = 1, \ldots, L_2, \alpha = 1, \beta = 1$ and $\lambda_o^1 = 1, \lambda_o^2 = 1,\ \lambda_j^1 = 0, \lambda_j^2 = 0\ j = 1, .., n, j \neq o$ is a feasible solution of model (3).

Theorem 2. Model (3) has units-invariant

$$\text{If we let} \quad \begin{aligned} & x_{ij} \to \pi_i x_{ij},\ \pi_i > 0,\ i = 1, \ldots, m,\ j = 1, \ldots, n \\ & w_{l_1j}^1 \to \bar{\pi}_{l_1} w_{l_1j}^1,\ \bar{\pi}_{l_1} > 0,\ \ l_1 = 1, \ldots, L_1,\ \ j = 1, \ldots, n \\ & y_{rj} \to \gamma_r y_{rj},\ \gamma_r > 0,\ \ r = 1, \ldots, s,\ j = 1, \ldots, n \\ & w_{l_2j}^2 \to \bar{\gamma}_{l_2} w_{l_2j}^2, \bar{\gamma}_{l_2} > 0,\ \ l_2 = 1, \ldots, L_2,\ \ j = 1, \ldots, n \\ & z_{kj} \to \sigma_k z_{kj},\ \ \sigma_k > 0,\ \ k = 1, \ldots, K,\ \ j = 1, \ldots, n. \end{aligned}$$

Then the efficiency level does not alter.

Proof. We rewrite model (3) as follows:

$$\begin{aligned} \eta^* = \ & Min\ \ \alpha/\beta \\ & s.t. \sum_{j=1}^{n} \lambda_j^1 \left(\pi_i x_{ij}\right) \leq \alpha(\pi_i x_{io}), i = 1, \ldots, m \\ \text{If } d_{l_1}^1 = 0 \text{ then} \quad & \sum_{j=1}^{n} \lambda_j^1 (\bar{\pi}_{l_1} w_{l_1j}^1) \leq \alpha\left(\bar{\pi}_{l_1} w_{l_1o}^1\right), l_1 = 1, \ldots, L_1 \\ \text{If } d_{l_1}^1 = 1 \text{ then} \quad & \sum_{j=1}^{n} \lambda_j^1 (\bar{\pi}_{l_1} w_{l_1j}^1) \geq \sum_{j=1}^{n} \lambda_j^2 (\bar{\pi}_{l_1} w_{l_1j}^1),\ l_1 = 1, \ldots, L_1 \\ & \sum_{j=1}^{n} \lambda_j^2 (\gamma_r y_{rj}) \geq \beta(\gamma_r y_{ro}), \qquad r = 1, \ldots, s \\ \text{If } d_{l_2}^2 = 0 \text{ then} \quad & \sum_{j=1}^{n} \lambda_j^2 \left(\bar{\gamma}_{l_2} w_{l_2j}^2\right) \leq \alpha\left(\bar{\gamma}_{l_2} w_{l_2o}^2\right), l_2 = 1, \ldots, L_2 \end{aligned}$$

$$\text{If } d_{l_2}^2 = 1 \text{ then} \quad \begin{aligned} &\sum_{j=1}^{n} \lambda_j^2\left(\bar{\gamma}_{l_2} w_{l_2 j}^2\right) \geq \beta\left(\bar{\gamma}_{l_2} w_{l_2 o}^2\right), \; l_2 = 1, \ldots, L_2 \\ &\sum_{j=1}^{n} \lambda_j^1 (\sigma_k z_{kj}) \geq \sum_{j=1}^{n} \lambda_j^2 (\sigma_k z_{kj}), \quad k = 1, \ldots, K \\ &\sum_{j=1}^{n} \lambda_j^1 = 1, \sum_{j=1}^{n} \lambda_j^2 = 1 \\ &\lambda_j^1 \geq 0, \lambda_j^2 \geq 0, \quad j = 1, \ldots, n \\ &\alpha \leq 1, \beta \geq 1, \end{aligned}$$

Hence we remove $\pi_i, \bar{\pi}_{l_1}, \gamma_r, \bar{\gamma}_{l_2}, \sigma_k$ from constraints so we have the same model (3).

We suggest the following aggregate fractional linear programming in term of the manager of the collection of DMUs:

$$\begin{aligned} \eta^* &= Min \;\; \alpha/\beta \\ s.t. \;\; &\sum_{j=1}^{n} \lambda_j^1 x_{ij} \leq \alpha \sum_{j=1}^{n} x_{ij}, && i = 1, \ldots, m \\ &\sum_{j=1}^{n} \lambda_j^1 w_{l_1 j}^1 \leq \alpha \sum_{j=1}^{n} w_{l_1 j}^1 + M d_{l_1}^1, && l_1 = 1, \ldots, L_1 \\ &\sum_{j=1}^{n} \lambda_j^1 d_{l_1}^1 w_{l_1 j}^1 \geq \sum_{j=1}^{n} \lambda_j^2 d_{l_1}^1 w_{l_1 j}^1, && l_1 = 1, \ldots, L_1 \\ &\sum_{j=1}^{n} \lambda_j^2 y_{rj} \geq \beta \sum_{j=1}^{n} y_{rj}, && r = 1, \ldots, s \\ &\sum_{j=1}^{n} \lambda_j^2 w_{l_2 j}^2 \leq \alpha \sum_{j=1}^{n} w_{l_2 j}^2 + M d_{l_2}^2, && l_2 = 1, \ldots, L_2 \\ &\sum_{j=1}^{n} \lambda_j^2 w_{l_2 j}^2 \geq \beta \sum_{j=1}^{n} w_{l_2 j}^2 - M\left(1 - d_{l_2}^2\right), && l_2 = 1, \ldots, L_2 \\ &\sum_{j=1}^{n} \lambda_j^1 z_{kj} \geq \sum_{j=1}^{n} \lambda_j^2 z_{kj}, && k = 1, \ldots, K \\ &\sum_{j=1}^{n} \lambda_j^1 = 1, \quad \sum_{j=1}^{n} \lambda_j^2 = 1 \\ &d_{l_1}^1 \in \{0, 1\}, d_{l_2}^2 \in \{0, 1\}, && \forall l_1, l_2 \\ &\lambda_j^1 \geq 0, \lambda_j^2 \geq 0, && j = 1, \ldots, n \\ &\alpha \leq 1, \beta \geq 1 \end{aligned} \tag{4}$$

3.2 Input-Oriented Model Under Constant Returns to Scale in Basic Two-Stage Network Structure for Classifying Flexible Measures

By the following transformation, model (3) is transformed into the input-oriented following LP under constant returns to scale in basic two-stage network structure to evaluate the relative efficiency of DMU_o and to classify flexible measures.

$$\bar{\lambda}_j^1 = \frac{\lambda_j^1}{\beta}, \bar{\lambda}_j^2 = \frac{\lambda_j^2}{\beta}\ j = 1, \ldots, n, \bar{M} = \frac{M}{\beta}, \theta = \alpha/\beta.$$

$$\begin{aligned}
\theta^* &= Min\ \theta \\
&s.t.\ \sum_{j=1}^{n} \bar{\lambda}_j^1 x_{ij} \le \theta x_{io}, && i = 1, \ldots, m \\
&\sum_{j=1}^{n} \bar{\lambda}_j^1 w_{l_1 j}^1 \le \theta w_{l_1 o}^1 + \bar{M}\bar{d}_{l_1}^1, && l_1 = 1, \ldots, L_1 \\
&\sum_{j=1}^{n} \bar{\lambda}_j^1 d_{l_1}^1 w_{l_1 j}^1 \ge \sum_{j=1}^{n} \bar{\lambda}_j^2 \bar{d}_{l_1}^1 w_{l_1 j}^1, && l_1 = 1, \ldots, L_1 \\
&\sum_{j=1}^{n} \bar{\lambda}_j^2 y_{rj} \ge y_{ro}, && r = 1, \ldots, s \\
&\sum_{j=1}^{n} \bar{\lambda}_j^2 w_{l_2 j}^2 \le \theta w_{l_2 o}^2 + \bar{M}\bar{d}_{l_2}^2, && l_2 = 1, \ldots, L_2 \qquad (5) \\
&\sum_{j=1}^{n} \bar{\lambda}_j^2 w_{l_2 j}^2 \ge w_{l_2 o}^2 - \bar{M}\left(1 - \bar{d}_{l_2}^2\right), && l_2 = 1, \ldots, L_2 \\
&\sum_{j=1}^{n} \bar{\lambda}_j^1 z_{kj} \ge \sum_{j=1}^{n} \bar{\lambda}_j^2 z_{kj}, && k = 1, \ldots, K \\
&\bar{d}_{l_1}^1 \in \{0, 1\}, \bar{d}_{l_2}^2 \in \{0, 1\}, && \forall l_1, l_2 \\
&\bar{\lambda}_j^1 \ge 0, \bar{\lambda}_j^2 \ge 0, && j = 1, \ldots, n
\end{aligned}$$

Where $\bar{M}$ is a large positive number. For each l_1 and l_2, binary variables $\bar{d}_{l_1}^1 \in \{0, 1\}$ and $\bar{d}_{l_2}^2 \in \{0, 1\}$ are introduced to identify that factor l_1 and l_2 are inputs to the first and second stage, $\bar{d}_{l_1}^1 = 0$ and $\bar{d}_{l_2}^2 = 0$ or outputs to the first and second stage, $\bar{d}_{l_1}^1 = 1$ and $\bar{d}_{l_2}^2 = 1$.

It is easy to verify that model (5) has unit invariance property.

Definition 2. DMU_o is efficient if and only if $\theta^* = 1$.

Theorem 3. In model (5) it holds $0 < \theta^* \le 1$.

Proof. Let $\bar{d}_{l_1}^1 = 0, \forall l_1 = 1, \ldots, L_1, \bar{d}_{l_2}^2 = 1, \forall l_2 = 1, \ldots, L_2, \bar{\lambda}_o^1 = 1, \bar{\lambda}_o^2 = 1, \theta = 1$ and $\bar{\lambda}_j^1 = 0, \bar{\lambda}_j^2 = 0, j = 1, .., n, j \ne o$ is a feasible solution of model (5). As the objective function is a minimizing one, hence we have $\theta^* \le 1$. Now, we prove that $\theta^* \ne 0$. If we suppose the contrary of this term, then $\theta^* = 0$ and the first constraint is considered unequal. Then considering that for each j we have $\bar{\lambda}_j^1 \ge 0$, $\bar{\lambda}_j^2 \ge 0$ and, on the other hand we have for, $x_{ij} \ge 0, \forall i \in I$. Hence, based on the mentioned assumption for the first constraint in model (5), we have $\sum_{j=1}^{n} \bar{\lambda}_j^1 x_{tj} \le 0$. This equation could not exist. θ^* also could not be negative because in this case, in the first constraint on the right side of the unequal, there's a negative number and on the left side for total finite number, there's a positive number. On the other hand $x_o \ne 0, x_o \ge 0$ then there is one t that $1 < t < m$ and $x_{to} > 0$. We prove $\theta > 0$, hence we suppose the contrary of this term, then $\theta < 0$ and in the t-th constraint of inputs we have $\sum_{j=1}^{n} \lambda_j^1 x_{tj} \leq \theta x_{to}$, so on the right side of the unequal there is negative number and on the left side of the unequal there is positive number. This leads to a contradiction then we have $0 < \theta^* \leq 1$.

Definition 3. Suppose an optimal solution for model (5) is $\left(\theta^*, \bar{\lambda}^{1*}, \bar{\lambda}^{2*}, \bar{d}^{1*}, \bar{d}^{2*}\right)$ then, we determine a frontier projection for DMU_o as

$$\left(\sum_{j=1}^{n} \bar{\lambda}_j^{1*} x_{ij}, \sum_{j=1}^{n} \bar{\lambda}_j^{1*} w_{l_1 j}^1, \hat{z}_o, \sum_{j=1}^{n} \bar{\lambda}_j^{2*} y_{rj}, \sum_{j=1}^{n} \bar{\lambda}_j^{2*} w_{l_2 j}^2\right) \forall i, r, l_1, l_2$$

Where $\hat{z}_o$ is any choice such that

$$\sum_{j=1}^{n} \bar{\lambda}_j^{2*} z_{kj} \leq \hat{z}_o \leq \sum_{j=1}^{n} \bar{\lambda}_j^{1*} z_{kj}, \; k = 1, \ldots, K$$

An easy choice for $\hat{z}_o$ would be

$$\hat{z}_o = \sum_{j=1}^{n} \bar{\lambda}_j^{1*} z_{kj}, \; k = 1, \ldots, K \text{ or}$$
$$\hat{z}_o = \sum_{j=1}^{n} \bar{\lambda}_j^{2*} z_{kj}, \quad k = 1, \ldots, K.$$

Theorem 4. The following frontier projection is efficient.

Proof. Suppose an optimal solution for model (5) is $(\theta^*, \bar{\lambda}^{1*}, \bar{\lambda}^{2*}, \bar{d}^{1*}, \bar{d}^{2*}, s^{-*}, s^{+*}, s^{1*}, s^{2*}, s^{3*})$ then we have:

$$\sum_{j=1}^{n} \bar{\lambda}_j^{1*} x_{ij} = \hat{x}_{io} = \theta^* x_{io} - s_i^{-*}; \qquad i = 1, \ldots, m$$

$$\text{If } \bar{d}_{l_1}^{1*} = 0 \text{ then} \quad \sum_{j=1}^{n} \bar{\lambda}_j^{1*} w_{l_1 j}^1 = \hat{w}_{l_1 o}^1 = \theta^* w_{l_1 o}^1 - s_{l_1}^{1*}; \; l_1 = 1, \ldots, L_1$$

$$\text{If } \bar{d}_{l_1}^{1*} = 1 \text{ then} \quad \sum_{j=1}^{n} \bar{\lambda}_j^{2*} w_{l_1 j}^1 \leq \hat{w}_{l_1 o}^1 \leq \sum_{j=1}^{n} \bar{\lambda}_j^{1*} w_{l_1 j}^1; \; l_1 = 1, \ldots, L_1$$

An easy choice for $\hat{w}_{l_1 o}^1$ would be

$$\hat{w}_{l_1 o}^1 = \sum_{j=1}^{n} \bar{\lambda}_j^{1*} w_{l_1 j}^1 \; or \; \hat{w}_{l_1 o}^1 = \sum_{j=1}^{n} \bar{\lambda}_j^{2*} w_{l_1 j}^1; \; l_1 = 1, \ldots, L_1$$

$$\sum_{j=1}^{n} \bar{\lambda}_j^{2*} y_{rj} = \hat{y}_{ro} = y_{ro} + s_r^{+*}; \; r = 1, \ldots, s$$

$$\text{If } \bar{d}_{l_2}^{2*} = 0 \text{ then} \quad \sum_{j=1}^{n} \bar{\lambda}_j^{2*} w_{l_2 j}^2 = \hat{w}_{l_2 o}^2 = \theta^* w_{l_2 o}^2 - s_{l_1}^{2*}; \; l_2 = 1, \ldots, L_2$$

$$\text{If } \bar{d}_{l_2}^{2*} = 1 \text{ then} \quad \sum_{j=1}^{n} \bar{\lambda}_j^{2*} w_{l_2 j}^2 = \hat{w}_{l_2 o}^2 = w_{l_2 o}^2 + s_{l_2}^{3*}; \; l_2 = 1, \ldots, L_2$$

$$\sum_{j=1}^{n} \bar{\lambda}_j^{2*} z_{kj} \leq \hat{z}_o \leq \sum_{j=1}^{n} \bar{\lambda}_j^{1*} z_{kj}, \quad k = 1, \ldots, K$$

We solve model (5) for the evaluation of point $A = \left(\hat{x}_{io}, \hat{w}_{l_1 o}^1, \hat{y}_{ro}, \hat{w}_{l_2 o}^2, \hat{z}_o\right)$.

We achieve the optimal solution $\left(\hat{\theta}, \hat{\lambda}^1, \hat{\lambda}^2, \hat{d}^1, \hat{d}^2, \hat{s}^-, \hat{s}^+, \hat{s}^1, \hat{s}^2, \hat{s}^3\right)$, it is to show that $\hat{\theta} = 1$ and slack variables are zero. By using the definition of frontier projection, we have:

$$\sum_{j=1}^{n} \hat{\lambda}_j^1 x_{ij} = \hat{\theta}\theta^* x_{io} - \hat{\theta} s_i^{-*} - \hat{s}_i^-, i = 1, \ldots, m$$

$$\text{If } \hat{d}_{l_1}^1 = 0 \text{ then} \quad \sum_{j=1}^{n} \hat{\lambda}_j^1 w_{l_1 j}^1 = \hat{\theta}\theta^* w_{l_1 o}^1 - \hat{\theta} s_{l_1}^{1*} - \hat{s}_{l_1}^1, l_1 = 1, \ldots, L_1$$

$$\text{If } \hat{d}_{l_1}^1 = 1 \text{ then} \quad \sum_{j=1}^{n} \hat{\lambda}_j^2 w_{l_1 j}^1 \le \hat{w}_{l_1 o}^1 \le \sum_{j=1}^{n} \hat{\lambda}_j^1 w_{l_1 j}^1; l_1 = 1, \ldots, L_1$$

$$\sum_{j=1}^{n} \hat{\lambda}_j^2 y_{rj} = y_{ro} + s_r^{+*} + \hat{s}_r^+, r = 1, \ldots, s$$

$$\text{If } \hat{d}_{l_2}^2 = 0 \text{ then} \quad \sum_{j=1}^{n} \hat{\lambda}_j^2 w_{l_2 j}^2 = \hat{\theta}\theta^* w_{l_2 o}^2 - \hat{\theta} s_{l_2}^{2*} - \hat{s}_{l_2}^2, l_2 = 1, \ldots, L_2$$

$$\text{If } \hat{d}_{l_2}^2 = 1 \text{ then} \quad \sum_{j=1}^{n} \hat{\lambda}_j^2 w_{l_2 j}^2 = w_{l_2 o}^2 + s_{l_2}^{3*} + \hat{s}_{l_2}^3, l_2 = 1, \ldots, L_2$$

$$\sum\nolimits_{j=1}^{n} \hat{\lambda}_j^2 z_{kj} \le \hat{z}_o \le \sum\nolimits_{j=1}^{n} \hat{\lambda}_j^1 z_{kj}, \ k = 1, \ldots, K$$

Now we rewrite the above formula:

$$\sum_{j=1}^{n} \hat{\lambda}_j^1 x_{ij} = \tilde{\theta} x_{io} - \tilde{s}_i^-, \quad i = 1, \ldots, m$$

$$\text{If } \hat{d}_{l_1}^1 = 0 \text{ then} \quad \sum_{j=1}^{n} \hat{\lambda}_j^1 w_{l_1 j}^1 = \tilde{\theta} w_{l_1 o}^1 - \tilde{s}_{l_1}^1, l_1 = 1, \ldots, L_1$$

$$\text{If } \hat{d}_{l_1}^1 = 1 \text{ then} \quad \sum_{j=1}^{n} \hat{\lambda}_j^2 w_{l_1 j}^1 \le \hat{w}_{l_1 o}^1 \le \sum_{j=1}^{n} \hat{\lambda}_j^1 w_{l_1 j}^1; l_1 = 1, \ldots, L_1$$

$$\sum_{j=1}^{n} \hat{\lambda}_j^2 y_{rj} = y_{ro} + \tilde{s}_r^+, r = 1, \ldots, s$$

$$\text{If } \hat{d}_{l_2}^2 = 0 \text{ then} \quad \sum_{j=1}^{n} \hat{\lambda}_j^2 w_{l_2 j}^2 = \tilde{\theta} w_{l_2 o}^2 - \tilde{s}_{l_2}^2, l_2 = 1, \ldots, L_2$$

$$\text{If } \hat{d}_{l_2}^2 = 1 \text{ then} \quad \sum_{j=1}^{n} \hat{\lambda}_j^2 w_{l_2 j}^2 = w_{l_2 o}^2 + \tilde{s}_{l_2}^3, l_2 = 1, \ldots, L_2$$

$$\sum\nolimits_{j=1}^{n} \hat{\lambda}_j^2 z_{kj} \le \hat{z}_o \le \sum\nolimits_{j=1}^{n} \hat{\lambda}_j^1 z_{kj}, \ k = 1, \ldots, K$$

Where $\tilde{\theta} = \hat{\theta}\theta^*$, $\tilde{s}_i^- = \hat{\theta}s_i^{-*} + \hat{s}_i^-$, $\tilde{s}_{l_1}^1 = \hat{\theta}s_{l_1}^{1*} + \hat{s}_{l_1}^1$, $\tilde{s}_r^+ = s_r^{+*} + \hat{s}_r^+$, $\tilde{s}_{l_1}^2 = \hat{\theta}s_{l_2}^{2*} + \hat{s}_{l_2}^2$, $\tilde{s}_{l_2}^3 = s_{l_2}^{3*} + \hat{s}_{l_2}^3$.

If $\hat{\theta} < 1$ then $\hat{\theta}\theta^* < \theta^*$ so $\tilde{\theta} < \theta^*$ so this is a contradiction. Therefore $\hat{\theta} = 1$.

On the other hand, we have:

$$\begin{aligned} e\tilde{s}^- + e\tilde{s}^1 + e\tilde{s}^+ + e\tilde{s}^2 + e\tilde{s}^3 &= es^{-*} + e\hat{s}^- + es^{1*} + e\hat{s}^1 + es^{+*} \\ &\quad + e\hat{s}^+ + es^{2*} + e\hat{s}^2 + es^{3*} + e\hat{s}^3 \\ &\le es^{-*} + es^{1*} + es^{+*} + es^{2*} + es^{3*}. \end{aligned}$$

Since $es^{-*} + es^{1*} + es^{+*} + es^{2*} + es^{3*}$ is maximal, we must have $e\hat{s}^- + e\hat{s}^1 + e\hat{s}^+ + e\hat{s}^2 + e\hat{s}^3 = 0$.

So $e\hat{s}^- = 0$, $e\hat{s}^1 = 0$, $e\hat{s}^+ = 0$, $e\hat{s}^2 = 0$, $e\hat{s}^3 = 0$ and $\hat{\theta} = 1$. Hence the frontier projection is efficient.

3.3 Output-Oriented Model Under Constant Returns to Scale in Basic Two-Stage Network Structure for Classifying Flexible Measures

We introduce the following MILP model which is an output-oriented version of model (5) under constant returns to scale in basic two-stage network structure to evaluate the relative efficiency of DMU_o and to classify flexible measures.

$$\begin{aligned}
&Max \ \ \varphi \\
&s.t. \sum_{j=1}^{n} \mu_j^1 x_{ij} \le x_{io}, && i = 1, \ldots, m \\
&\sum_{j=1}^{n} \mu_j^1 w_{l_1 j}^1 \le w_{l_1 o}^1 + \widehat{M}\, d_{l_1}^{1'}, && l_1 = 1, \ldots, L_1 \\
&\sum_{j=1}^{n} \mu_j^1 d_{l_1}^{1'} w_{l_1 j}^1 \ge \sum_{j=1}^{n} \mu_j^2 d_{l_1}^{1'} w_{l_1 j}^1, && l_1 = 1, \ldots, L_1 \\
&\sum_{j=1}^{n} \mu_j^2 y_{rj} \ge \varphi y_{ro}, && r = 1, \ldots, s \\
&\sum_{j=1}^{n} \mu_j^2 w_{l_2 j}^2 \le w_{l_2 o}^2 + \widehat{M}\, d_{l_2}^{2'}, && l_2 = 1, \ldots, L_2 \\
&\sum_{j=1}^{n} \mu_j^2 w_{l_2 j}^2 \ge \varphi w_{l_2 o}^2 - \widehat{M}\left(1 - d_{l_2}^{2'}\right), && l_2 = 1, \ldots, L_2 \\
&\sum_{j=1}^{n} \mu_j^1 z_{kj} \ge \sum_{j=1}^{n} \mu_j^2 z_{kj}, && k = 1, \ldots, K \\
&d_{l_1}^{1'} \in \{0, 1\}, d_{l_2}^{2'} \in \{0, 1\}, && \forall l_1, l_2
\end{aligned}$$

$$\mu_j^1 \geq 0,\ \mu_j^2 \geq 0, \qquad j = 1, \ldots, n \tag{6}$$

Where $\widehat{M}$ is a large positive number. For each l_1 and l_2, binary variables $d_{l_1}^{1'} \in \{0, 1\}$ and $d_{l_2}^{2'} \in \{0, 1\}$ are introduced to identify that factor l_1 and l_2 are inputs to the first and second stage, $d_{l_1}^{1'} = 0$ and $d_{l_2}^{2'} = 0$ or outputs to the first and second stage, $d_{l_1}^{1'} = 1$ and $d_{l_2}^{2'} = 1$.It is easy to verify that model (6) has unit invariance property.

Definition 4. In model (6) it holds $\varphi^* \geq 1$ and DMU_o is efficient if and only if $\varphi^* = 1$.

Theorem 5. Let $(\theta^*, \bar{\lambda}^{1*}, \bar{\lambda}^{2*}, \bar{d}^{1*}, \bar{d}^{2*})$be an optimal solution of model (4), then an optimal solution of model (6) is obtained from $\left(\varphi^* = \frac{1}{\theta^*}, \mu^{1*} = \frac{\bar{\lambda}^{1*}}{\theta^*}, \mu^{2*} = \frac{\bar{\lambda}^{2*}}{\theta^*}, d^{1*'} = \bar{d}^{1*}, d^{2*'} = \bar{d}^{2*}\right)$.

Proof. As $\widehat{M}$ is a sufficient large positive, if we select $\widehat{M} > \frac{\overline{M}}{\theta^*}$ then clearly $(\varphi^*, \mu^{1*}, \mu^{2*}, d^{1*'}, d^{2*'})$ is a feasible solution of model (6). Suppose on the contrary that $\left(\bar{\varphi}, \bar{\mu}^{1*}, \bar{\mu}^{2*}, \bar{d}^{1*'}, \bar{d}^{2*'}\right)$ be the optimal solution of model (6) with $\bar{\varphi} > \varphi^*$ Similarly, $\left(\bar{\theta} = \frac{1}{\bar{\varphi}}, \hat{\lambda}^1 = \frac{\bar{\mu}^{1*}}{\bar{\varphi}}, \hat{\lambda}^2 = \frac{\bar{\mu}^{2*}}{\bar{\varphi}}, \bar{d}^{1*'}, \bar{d}^{2*'}\right)$ is a feasible solution of model (5) with $\bar{\theta} < \theta^*$, which violates optimality of θ^*. This leads to a contradiction and hence $\left(\varphi^*, \mu^{1*}, \mu^{2*}, d^{1*'}, d^{2*'}\right)$ is an optimal solution of model (6).

4 Numerical Examples

In this section, we apply non-oriented model under VRS and input- and output-oriented model under CRS for classifying flexible measures in basic two-stage network DEA to the data sets shown in Table 1. These data are random and there are 30 DMUs, three inputs to the first stage (X1, X2, X3), two intermediate measures (Z1, Z2), two final outputs from the second stage (Y1, Y2), one flexible measure to the first stage w^1 and one flexible measure to the second stage w^2.

Table 1. Data in basic two-stage network with flexible measures

DMU	X1	X2	X3	Z1	Z2	w^1	w^2	Y1	Y2
1	71.2	6.3	46.6	61.4	109	34.2	64.3	38.4	36.1
2	21.3	12.6	19.4	93.5	94.4	31.7	40	5.4	22.5

(continued)

Table 1. (*continued*)

DMU	X1	X2	X3	Z1	Z2	w^1	w^2	Y1	Y2
3	65.8	6.6	30.1	134.4	57.3	23.1	42.8	12.3	29.8
4	71.9	12.5	24.2	53	94.8	63.6	35	6.7	36
5	78.4	15.9	19.5	132	40.8	52.8	19.5	19.1	44
6	30.4	15	43.3	57.3	100.1	60.4	17.8	7.8	14.1
7	30.2	11.1	34.5	148	66.6	51.6	43.8	16.9	23.7
8	11.4	25	33.5	122	69.2	36.6	74.8	20.1	18.5
9	62.7	15.1	29.2	146.7	102.8	73.6	55.8	17.6	13.6
10	75.9	11.9	47.4	84.8	141.4	18.9	16.3	13.9	20.1
11	87.8	19.4	50.9	109.8	113.6	23.2	45.5	11.1	18.3
12	17.1	10.8	15.5	98.3	66.4	66.2	11.1	17.8	34.6
13	48.9	8.9	50.8	33.1	65.6	57.8	64.2	31.4	45.7
14	12.4	8.1	51	57.1	98.8	73.6	40.5	19.4	38.7
15	55.9	14.5	13.4	96.4	88.1	20.8	27.6	24.4	44.6
16	46.4	10.3	24.8	70.8	27.1	75.7	56.6	12.7	43.5
17	51.9	44	37.8	39.9	64.1	75.9	23.7	8.8	10.8
18	43.2	8.3	51.6	123.2	117	40.6	79.6	91.2	13.2
19	10.6	14.1	43.8	158	107.5	57.2	62.9	16.2	12.3
20	59.7	11.7	22.8	89.9	131.9	35	33	11.7	36.9
21	6.4	12.2	23.7	142.3	110.7	69.8	47.9	10.2	19.5
22	70.6	18.3	23.8	93.6	123.1	51.6	4.5	22.3	29
23	67.8	5.2	46.3	73.7	124.4	50.3	25.2	70.7	12.2
24	23.5	13.7	47.5	72.9	111.9	34.6	65.5	46.9	43.6
25	33.8	16	27.6	77.1	74.8	70.5	54.7	29.2	34
26	81.9	19.4	57.7	131	64.5	64.7	13.5	61.1	21.2
27	38	12.3	57.6	94.4	98.6	24.1	61.8	14.3	48.6
28	74.4	7.6	40	141	88.5	23.6	18.9	18.6	13.5
29	92.2	12.4	45.8	94.5	54.8	23.5	57.9	15.5	11.9
30	73.3	14.3	54.8	91.4	53.3	19.2	62.3	4.2	34.7

The results obtained from the non-oriented model (3) under VRS are presented in Table 2. As seen in Table 2, the DMU 3, 12, 14, 15, 18, 21 and 23 are efficient, and the other DMUs are non-efficient. In this model, 24 DMUs select the flexible measure to the first stage as an output measure and 6 DMUs select it as an input measure and 19 DMUs select the flexible measure to the second stage as an output measure and 11 DMUs select it as an input measure. Therefore, our proposed model selects the flexible measure to the first and second stage as an output measure. In addition, we apply our aggregate model (4) for this data set. The optimal $d_{l_1}^{1*} = 1$ and $d_{l_2}^{2*} = 1$ for the flexible measure to the first and second stage with the optimal objective function value 0.4236, are obtained. This indicates that from an aggregate efficiency perspective, the flexible measure is treated as an output to the first and second stage exactly and uniquely. The achieved status of the flexible measure by the aggregate model (4) is the same as that obtained by the model (3) from the perspective individual DMUs as expected.

Table 2. The results obtained from the non-oriented model (3) and (4) under VRS, the input-oriented model (5) under CRS and the output-oriented model (6) under CRS

DMU	η^*	d^{1*}	d^{2*}	θ^*	$\bar{d}^{1*}$	$\bar{d}^{2*}$	φ^*	$\acute{d}^{1*}$	$\acute{d}^{2*}$
1	0.9068	1	0	0.7007	1	0	1.5944	0	1
2	0.8465	1	0	0.376	1	1	2.9134	0	1
3	1	0	1	0.5397	1	1	2.6053	0	1
4	0.7841	1	1	0.4563	1	1	3.195	0	1
5	0.7583	1	1	0.6295	1	1	2.4734	0	1
6	0.6417	1	1	0.1109	0	1	9.0172	0	1
7	0.845	1	0	0.2946	0	1	3.3948	0	1
8	0.6773	1	0	0.3783	0	0	2.6436	0	0
9	0.6537	1	0	0.2225	1	0	4.4938	1	0
10	0.6422	1	1	0.2224	1	1	4.4971	1	1
11	0.4564	1	0	0.1924	1	0	5.1965	1	0
12	1	0	1	0.4557	0	1	2.1945	0	1
13	0.8216	0	1	0.4552	0	1	2.1967	0	1
14	1	0	1	0.4794	0	1	2.0861	0	1
15	1	0	1	0.5582	0	1	1.7915	0	1
16	0.8752	1	0	0.422	0	1	2.3697	0	1
17	0.4045	1	1	0.0957	0	1	10.4518	0	1
18	1	1	1	1	0	1	1	0	1
19	0.7765	1	0	0.2049	1	0	4.7756	0	0
20	0.8357	1	1	0.3619	0	1	2.7629	0	1
21	1	0	0	0.2894	1	0	3.4551	1	0
22	0.6327	1	1	0.2826	0	1	3.5388	0	1
23	1	1	1	1	0	1	1	0	1
24	0.7674	1	1	0.7349	0	1	1.3608	0	1
25	0.6462	1	1	0.4618	1	1	2.1655	1	1

(continued)

Table 2. (*continued*)

DMU	η^*	d^{1*}	d^{2*}	θ^*	$\bar{d}^{1*}$	$\bar{d}^{2*}$	φ^*	d'^{1*}	d'^{2*}
26	0.4756	1	1	0.3741	1	1	2.6734	1	1
27	0.6969	1	1	0.4978	1	1	2.0089	1	1
28	0.8862	1	1	0.2797	1	1	3.5757	1	1
29	0.6341	1	0	0.1965	1	0	5.0878	1	0
30	0.5623	1	0	0.3586	1	1	2.7883	1	1
Mean	0.4236	1	1						

The results of the input-oriented model (5) under CRS are shown in Table 2. As seen in Table 2, the DMU 18 and 23 are efficient, and the other DMUs are non-efficient. In this model, 16 DMUs select the flexible measure to the first stage as an output measure and 14 DMUs select it as an input measure and 23 DMUs select the flexible measure to the second stage as an output measure and 7 DMUs select it as an input measure. Therefore, the input-oriented model selects the flexible measure to the first and second stage as an output measure.

The results of the output-oriented model (6) under CRS are shown in Table 2. As seen in Table 2, the DMU 18 and 23 are efficient, and the other DMUs are non-efficient. In this model, 10 DMUs select the flexible measure to the first stage as an output measure and 20 DMUs select it as an input measure and 24 DMUs select the flexible measure to the second stage as an output measure and 6 DMUs select it as an input measure. Therefore, the output-oriented model selects the flexible measure to the first stage as an input measure and to the second stage as an output measure.

The input- or output-oriented model under CRS in basic and general two-stage network structures may produce different efficiency scores and it can be expected that a flexible measure is selected as an input variable in one model but an output variable in another. Therefore, the results from the input- and output-oriented models are different for DMU 1, 2, 3, 4, 5, 19. Because in the input-oriented model DMU 1, 2, 3, 4, 5, 19 select the flexible measure to the first stage as an output measure but, in the output-oriented model select the flexible measure to the first stage as an input measure. Also, the input-oriented model selects the flexible measure to the first and second stage as an output measure but, the output-oriented model selects the flexible measure to the first stage as an input measure and to the second stage as an output measure.

5 Conclusion

A conventional DEA analysis requires that each variable be specified as an input or an output before measuring efficiency. However, in real-world applications there are some measures whose status is flexible. The presentation of models with flexible measure in two-stage network structures is essential and beneficial because there are many situations in the real world. So here we propose new models in which each one of the flexible variables is treated as either input or output to maximize the relative efficiency of the DMU under evaluation in two-stage network structures. Our new models are based on envelopment form of CCR such as non-oriented models under variable returns to scale, input and output oriented models under constant returns to scale in two-stage network structure to select the status of each flexible measure as an input or output. Our models can determine the relative efficiency and an efficient projection unit. The input- or output-oriented model under CRS in basic two-stage network structure may produce different efficiency scores and it can be expected that a flexible measure is selected as an input variable in one model but an output variable in another. The status of one flexible measure in two CRS and VRS environments may have different results. In other words, a flexible measure is considered as input in CRS environment, whereas it is selected as output in VRS environment. The difference between the results of the two models clearly shows the need for a model that considers VRS in the presence of flexible measure. Therefore, the main focus of this study was to propose a new non-oriented model and aggregate model that can determine a unique status for flexible measures under VRS in basic two-stage network structure.

We believe that our proposed models can be modified into multi-stage models and our proposed models can be developed for fuzzy flexible measures and interval flexible measures in two-stage network structure. Future researchers can extend these improvements.

References

1. Amirteimoori, A., Emrouznejad, A.: Flexible measures in production process: a DEA-based approach. RAIRO Oper. Res. **45**(1), 63–74 (2011)
2. Amirteimoori, A., Emrouznejad, A., Khoshandam, L.: Classifying flexible measures in data envelopment analysis: a slack-based measure. Measurement **46**(10), 4100–4107 (2013)
3. Chen, Y., Cook, W.D., Zhu, J.: Deriving the DEA frontier for two-stage processes. Euro. J. Oper. Res. **202**(1), 138–142 (2010)
4. Chen, Y., Cook, W.D., Kao, C., Zhu, J.: Network DEA pitfalls: divisional efficiency and frontier projection under general network structures. Euro. J. Oper. Res. **226**(3), 507–515 (2013)
5. Cook, W.D., Zhu, J.: Classifying inputs and outputs in DEA. Euro. J. Oper. Res. **180**(2), 692–699 (2007)
6. Guo, C., Zhang, J., Zhang, L.: Two-stage additive network DEA: duality, frontier projection and divisional efficiency. Expert Syst. Appl. **157**, 113478 (2020)
7. Kao, C.: Efficiency measurement and frontier projection identification for general two stage systems in data envelopment analysis. Euro. J. Oper. Res. **261**(2), 679–689 (2017)
8. Kao, C., Hwang, S.N.: Efficiency decomposition in two-stage data envelopment analysis: an application to non-life insurance companies in Taiwan. Euro. J. Oper. Res. **185**(1), 418–429 (2008)
9. Kordrostami, S., Noveiri, M.J.S.: Evaluating the performance and classifying the interval data in data envelopment analysis. Int. J. Manag. Sci. Eng. Manag. **9**, 243–248 (2014)
10. Kordrostami, S., Noveiri, M.J.S.: Evaluating the efficiency of decision making units in the presence of flexible and negative data. Indian J. Sci. Technol. **5**(12), 3776–3782 (2012)
11. Kordrostami, S., Amirteimoori, A., Noveiri, M.J.S.: Inputs and outputs classification in integer-valued data envelopment analysis. Measurement **139**, 317–325 (2019)

12. Lim, S., Zhu, J.: A note on two-stage network DEA model: frontier projection and duality. Euro. J. Oper. Res. **248**(1), 342–346 (2016)
13. Lim, S., Zhu, J.: Primal-dual correspondence and frontier projections in two-stage network DEA models. Omega **83**, 236–248 (2019)
14. Tohidi, G., Matroud, F.: A new non-oriented model for classifying flexible measures in DEA. J. Oper. Res. Soc. **68**, 1019–1029 (2017)
15. Toloo, M.: On classifying inputs and outputs in DEA: a revised model. Euro. J. Oper. Res. **198**(1), 358–360 (2009)
16. Toloo, M.: Alternative solutions for classifying inputs and outputs in data envelopment analysis. Comput. Math Appl. **63**(6), 1104–1110 (2012)
17. Toloo, M., Allahyar, M., Hančlová, J.: A non-radial directional distance method on classifying inputs and outputs in DEA: application to banking industry. Expert Syst. Appl. **92**, 495–506 (2018)

Ranking of Decision-Making Units Using the Strongest Cloud Page in Data Envelopment Analysis with MPSS

Arezoo Taheri Kamran and Mohsen Rostami Mal Khalifa(✉)

Department of Mathematics, Science and Research Branch, Islamic Azad University, Tehran, Iran
mohsen_rostamy@yahoo.com

Abstract. Thus, according to the goal of the decision maker, the units are classified into separate categories. It is important to note that this goal must be clearly stated; Because if that classification changes, it may also change. The purpose of this paper is to rank decision-making units according to the distance to the strongest hyper-page with the highest number of MPSS (units with the highest productivity scale).

Keywords: Data envelopment analysis · Ranking · Decision making units

1 Introduction

The history of data envelopment analysis goes back to the idea put forward by Farrell [1]. A few years after the presentation of the paper, Farrell [1] Charans et al. [2]. Formulated the first data envelopment analysis model, which the CCR model and another model called the BCC model were performed to evaluate performance.

In data envelopment analysis, the most important goal is to determine the performance size of the units so that efficient and inefficient units can be distinguished. For this purpose, a boundary called the efficiency boundary that covers all data is considered and the efficiency of the units is determined by comparing them with such a boundary.

A set of decision-making units (DMUs) In fact, any unit that has the power and authority to decide on how and the size of the combination of inputs and outputs is called the decision-making unit, and homogeneous decision-making units are units that act similarly, That is, by receiving the same inputs, they produce similar outputs (Congruent) That means all of them m Input and s They have one type of output, Is.

In this technique, a parameter called performance size is used to evaluate the performance of decision-making units.

The efficiency of a unit is a function of various factors such as the number of units, the amount of input and output of units, the number of inputs and outputs of units; the type of production technology and the model used.

In data envelopment analysis, we typically try to improve the performance of the unit under evaluation by reducing inputs and increasing outputs.

T. Allahviranloo et al. (Eds.): IDS 2020, AISC 1301, pp. 501–508, 2021.
https://doi.org/10.1007/978-3-030-66501-2_39

DEA in fact, generalizing Farrell's work in inventing the first method is non-parametric. Farrell, using the inputs and outputs of decision-making units and the principles governing them, introduced a set called the possibility of production set and introduced a part of its boundary as a production function. This boundary is also called the efficient boundary, and the decision-making units located on this boundary are evaluated as efficient.

Since the DEA Is a technique for evaluating the relative efficiency of decision-making units, with at least one unit on the border and the rest below it.

From the perspective of data envelopment analysis in 1978, introduced by Charles and Rhodes [3] Completely different methods for evaluating performance Dmu Are expressed by complex effective processes.

Data envelopment analysis provides opportunities to study units with multiple inputs and outputs.

The method of data envelopment analysis is based on algebra and its ability is more due to the use of linear programming. Linear programming enables data envelopment analysis to use linear programming problem solving methods and dualistic theorems, thus determining the source and amount of inefficiency for each input and output.

DEA also many opportunities for collaboration between analysts and decision-makers can make it.

This cooperation can be in order to select the input and output of the units under evaluation and how to operate and model the efficient border.

Good performance means that Dmu They use inputs efficiently and deliver good outputs.

Data envelopment analysis includes techniques and methods for evaluating and measuring the productivity of decision-making units.

Because DEA is a technique for evaluating the relative efficiency of decision-making units, at least one of the units is on the border and the rest are below it.

The term strongest super plan in the term means the super plan that the sum of the distance of all strong DMUs with that super plan is the shortest possible distance.

The order of the strongest cloud page [4] The term means, finds the cloud page that to DMUs that the cloud of the screen, their total distance with the strongest DMUs has the least distance.

2 Methods Provided

In this section we consider that we have n to DMUs that use m inputs to generate S outputs. Finds the closest pattern for DMUs in a hypothetical Jg group on the Paratoo T function spacing procedure that is known to all of them [3]. In this model, it presents patterns for all DMUs in a hypothetical group simultaneously. This means that G times can be solved in this way to find these patterns for all DMUs. The idea of this model is as follows:

That is, it guarantees that $j \in J_g$, (X_j, Y_j) belongs to T.

The common coefficient of the backup super plane T can take all non-negative values (u, v). In fact $(\overline{X}_j, \overline{Y}_j)$ will be a combination of $S \in E$, Dmu_k, all on the same surface of the Parato performance line.

This is because the common backup super plane belongs to T where $-\acute{v}X + \acute{u}Y = 0$,, whose coefficient is non-zero. Patterns that are globally closer to each other can be considered Get efficient DMUs set by another Jg.

In this article, we want to estimate the model with variable scale returns. Suppose the production technology of this model does not have the principle of constant scale returns.

$$Min \sum_{j \in Jg} \left\| (X_J, Y_j) - (\widetilde{X}_J, \widetilde{Y}_j) \right\|_1^w$$

$$s.t:$$

$$\sum\nolimits_{K \in E} \lambda_{KJ} X_k = \widetilde{X}_J \qquad j \in J_g \tag{1}$$

$$\sum\nolimits_{K \in E} \lambda_{KJ} Y_k = \widetilde{Y}_J \qquad j \in J_g \tag{2}$$

$$-\acute{v}X_k + \acute{u}Y_k + u_0 + d_k = 0 \qquad k \in E \tag{3}$$

$$\overline{X}v \geq 1_m \tag{4}$$

$$\overline{Y}u \geq 1_s \tag{5}$$

$$d_k \leq M(1 - b_k) \qquad k \in E \tag{6}$$

$$\sum\nolimits_{j \in Jg} \lambda_{KJ} \leq M(1 - b_k) \qquad k \in E \tag{7}$$

$$d_k \geq 0.b_k \in \{0.1\} k \in E$$

$$\lambda_{KJ} \geq 0,\ \widetilde{X}_J \geq 0_m,\ \widetilde{Y}_j \geq 0_s \qquad \forall k, j$$

$$u_0 \quad \text{free}$$

In fact, in this model we find a superhighway that despite the efficiency on a variable scale, the sum of the distance of all DMUs with that super high is the lowest possible value. The cloud page does not pass through the source and does not have the maximum amount of productivity. In the next section, we present a new model in which the DMU selects the image as close to MPSS as possible.

3 Ranking of Decision-Making Units Using the Strongest Superpower in DEA with MPSS

In this section, we are looking to select the hyperplane to which the sum of Dmu distances is the smallest possible, so that the Dmu of the image on it has the highest productivity or is as close as possible to MPSS. For this purpose, in the objective function of this model | u_0 | We make min.

$$Min \sum_{j \in Jg} \| (X_J, Y_j) - (\widetilde{X}_J, \widetilde{Y}_j) \|_1^w + |u_0|$$

$$s.t:$$

$$\sum_{K \in E} \lambda_{KJ} X_k = \widetilde{X}_J \qquad j \in J_g \tag{8}$$

$$\sum_{K \in E} \lambda_{KJ} Y_k = \widetilde{Y}_J \qquad j \in J_g \tag{9}$$

$$-\acute{v} X_k + \acute{u} Y_k + u_0 + d_k = 0 \qquad k \in E \tag{10}$$

$$\overline{X} v \geq 1_m \tag{11}$$

$$\overline{Y} u \geq 1_s \tag{12}$$

$$d_k \leq M(1 - b_k) \qquad k \in E \tag{13}$$

$$\sum_{j \in Jg} \lambda_{KJ} \leq M(1 - b_k) \qquad k \in E \tag{14}$$

$$d_k \geq 0.b_k \in \{0.1\} k \in E$$

$$\lambda_{KJ} \geq 0, \ \widetilde{X}_J \geq 0_m, \ \widetilde{Y}_j \geq 0_s \qquad \forall k, j$$

$$u_0 \text{ free}$$

In the above model | u_0 | Makes this model nonlinear.
In the next section, we will find a way to linearize this model.

$$Min \sum_{j \in Jg} \| (X_J, Y_j) - (\widetilde{X}_J, \widetilde{Y}_j) \|_1^w + u_{10} + u_{20}$$

$$s.t:$$

$$\sum_{K \in E} \lambda_{KJ} X_k = \widetilde{X}_J \qquad j \in J_g \tag{15}$$

$$\sum_{K \in E} \lambda_{KJ} Y_k = \widetilde{Y}_J \qquad j \in J_g \tag{16}$$

$$-\acute{v} X_k + \acute{u} Y_k + u_0 + d_k = 0 \qquad k \in E \tag{17}$$

$$\overline{X}v \geq 1_m \tag{18}$$

$$\overline{Y}u \geq 1_s \tag{19}$$

$$d_k \leq M(1 - b_k) \qquad k \in E \tag{20}$$

$$\sum\nolimits_{j \in Jg} \lambda_{KJ} \leq M(1 - b_k) \qquad k \in E \tag{21}$$

$$u_0 = u_{10} + u_{20} \tag{22}$$

$$d_k \geq 0.b_k \in \{0.1\} \qquad k \in E$$

$$\lambda_{KJ} \geq 0,\ \widetilde{X}_J \geq 0_m,\ \widetilde{Y}_j \geq 0_s \qquad \forall k, j$$

$$u_{10} \geq 0,\ u_{20} \geq 0$$

4 Numerical Examples

The insurance industry is one of the sectors in the economy that has important tasks. Looking at the increase in international capital and … Paying attention to the circulation of economic wheels shows the fact that all of them are related and rely on insurance guarantees.

In this section, we will review 20 insurance companies active in the stock exchange, whose names are listed in Table 1.

In order to evaluate these 20 companies, we considered 4 input indicators and one output index as follows:

Input Index:

1. Capital
2. wage cost
3. General cost
4. Administrative costs

Output Index:
Net income
According to the following data, we can get the value of Z for 20 to Dmus.

Here, according to the inputs that we have considered for each of the 20 insurance companies, it shows the amount of productivity scale of 20 insurance companies with the realization of demand for companies.

In this article, we reviewed 20 active insurances terse in the stock market, Table 2 shows the highest productivity compared to demand realization (Table 3).

Table 1. Study insurance company

Company name	Row	Company name	Row
Alborz insurance	11	Dana insurance	1
Saman insurance	12	Asia insurance	2
Sarmad insurance	13	Parisian insurance	3
Novin insurance	14	Razy insurance	4
Day insurance	15	Hafez insurance	5
Melat insurance	16	Mihan insurance	6
Sina insurance	17	Arman insurance	7
Iranian insurance	18	Pasargad insurance	8
Kar Afaren insurance	19	Kosar insurance	9
Amin insurance	20	Taowon insurance	10

Table 2. Data

	1	2	3	4
Dmu1	0.24	0.35	0.53	0.78
Dmu2	0.37	0.81	0.14	0.69
Dmu3	0.64	0.7	0.97	0.38
Dmu4	0.91	0.5	0.61	0.33
Dmu5	0.34	0.52	0.3	0.68
Dmu6	0.25	0.51	0.32	0.64
Dmu7	0.93	0.28	0.98	0.67
Dmu8	0.78	0.84	0.1	0.73
Dmu9	0.6	0.91	0.95	0.28
Dmu10	0.67	0.4	0.53	0.39
Dmu11	0.88	0.83	0.41	0.29

(continued)

Table 2. (*continued*)

	1	2	3	4
Dmu12	0.5	0.9	0.21	0.26
Dmu13	0.84	0.37	0.67	0.21
Dmu14	0.14	0.6	0.31	0.5
Dmu15	0.14	0.3	0.16	0.36
Dmu16	0.15	0.41	0.9	0.66
Dmu17	0.53	0.42	0.92	0.25
Dmu18	0.24	0.61	0.39	0.82
Dmu19	0.29	0.41	0.55	0.19
Dmu20	0.29	0.93	0.32	0.39

Table 3. Performance size

DMU_S	Z
1	1.196
2	1.004
3	0.400
4	1.912
5	1.486
6	2.624
7	2.798
8	2.314
9	1.476
10	1.476
11	1.132
12	1.986
13	2.616
14	2.024
15	1.170
16	1.544
17	1.070
18	1.560
19	0.952
20	0.960

5 Conclusion

It should be noted that in addition to performance appraisal, another application of DEA is the ranking of decision-making units. Thus, according to the goal of the decision maker, the units are classified into separate categories.

The group benchmarking designed in this article is an intermediary approach between pure DEA and benchmarking for the entire set of DMUs. This approach is based on the Paratoo DEA performance line break method, which determines the entire set of DMUs and is homogeneous (That is, it is comparable). In this case, the small size of DMUs in the group is not a problem, but the lack of appropriate criteria causes weakness in this article. The purpose of this dissertation is to rank decision-making units according to MPSS (Units with the largest productivity scale) that this method eliminates high weakness.

References

1. Farrel, M.J.: The measurement of productive efficiency. J. Royal Stat. Soc. Series A(General) **120**(3), 253–290(1957)
2. Charnes, A., Cooper, W.W., Thrall, R.M.: A structure for ciassifying and characterzing efficiency and inefficiency in data envelopment analysis. J. Product. Anal. **2**(3), 197–237 (1991)
3. Charnes, A., Cooper, W.W., Rhodes, E.: Measuring the efficiency of decision making units. Euro. J. Oper. Res. **2**(6), 429–444 (1978)
4. Jahanshahloo, G., Farhad Hosseinzadeh, L., Hashem Nikoo, M.: Data analysis and its applications. Tehran. Islamic Azad University. Research Sciences Unit (1387)

Selecting the Best Transmitter in Wireless Device-to-Device Communications Using a Fuzzy Decision-Making Method

Alireza Pourmoslemi[1], Siavash Rajabi[2](✉), and Mehdi Salimi[3]

[1] Department of Mathematics, Payame Noor University, P.O.BOX 19395-3697, Tehran, Iran
a_pourmoslemy@pnu.ac.ir

[2] Department of Electrical Engineering, Hamedan University of Technology, 65155 Hamedan, Iran
siavash.rajabi@hut.ac.ir

[3] Center for Dynamics, Faculty of Mathematics, Technische Universität Dresden, 01062 Dresden, Germany
mehdi.salimi@tu-dresden.de

Abstract. In this paper, based on two significant network parameters, received signal to interference plus noise ratio (SINR), and transmitters battery level, a fuzzy method is proposed for pairing in a device to device (D2D) communications scenario. The proposed solution shows that in spite of the common assumptions, it is possible to have an optimal pairing solution from farther nodes instead of connecting to the nearest node in D2D communications. In other words, employing an effective type of fuzzy-based symmetrical multi-criteria decision-making procedure, we will pair D2D communications. In fact, in this study, each transmitter is seen as a fuzzy node and the decision-making function chooses the transmitter with the highest fuzzy degree.

Keywords: D2D communications · Fuzzy decision-making · Signal to interference plus noise ratio · Battery level

1 Introduction and Preliminaries

The rapid growth of the number of Internet users along with the new data devour applications, encourage the designers of the next generation of mobile networks to introduce new services to overcome these new needs. Device to Device (D2D) communications is one of these new services. D2D communications enables two nearby mobile nodes to communicate directly, without the need to use the infrastructure network routs [15]. This newly introduced short-range service increases the resource allocation efficiency in cellular networks, decreases the power consumption of mobile nodes, offloading the cellular network traffic,

T. Allahviranloo et al. (Eds.): IDS 2020, AISC 1301, pp. 509–520, 2021.
https://doi.org/10.1007/978-3-030-66501-2_40

and the possibility of public safety networks. The mentioned advantages enable the network infrastructure to support both more active nodes and new applications [18,20,26].

In the context of the shared network resources, D2D communications consists of three categories; underlay in-band, overlay in-band, and out-band communications [2]. In underlay in-band, the cellular network resources are shared between cellular and D2D users [27] while in overlay in-band the network resources of cellular and D2D users are orthogonal. It means that the cellular network resources are divided for cellular and D2D users and therefore there is no interference between these two groups. Finally, in out-band communications, the D2D nodes use the unlicensed spectrum to make a new link [41]. In this paper, we assume that D2D users have no interference with cellular users, therefore this scenario can be classified in overlay in-band or out-band groups.

In the literature, most of the works are on the problem of optimal resource allocation and interference management of cellular and D2D links. These papers can be divided to power allocation methods [3,12,23], distance-based mechanisms [33,45], mode selection algorithms [19,39], joint mode selection and power control methods [4,5], signaling mechanisms [11], interference cancellation and beam-forming methods [31,32,38], electromagnetic time reversal (EMTR) resource allocation [35], and graph-based resource management schemes [17,46]. One of the main assumptions of these problems and their associated solutions is that two nodes have chosen each other to establish a D2D connection and now the problem is how to find the optimal cellular resources to being shared and further how to manage their powers to have minimum interferences on co-channel cellular or D2D links. Another assumption is that the D2D node chooses the node with minimum distance as the best pair [25]. However, there are a lot of reasons like battery level and security consideration that the nearest node may not be the best option for pairing in a D2D communication link.

There are limited works in the literature that consider pairing as finding the proper pair for a D2D user [3,14,24,29,42]. The authors of [24] use a dynamic graph optimization framework to show the effect of D2D multi-hop communication on cellular systems. This paper considers the spatial and temporal activity of relays and D2D receivers in its system model. Pairing in [24] is considered as connection of a D2D user to another relay or evolved node base station (eNB) while its physical position changes over time. In [14] a centralized discovery scheme based on an adaptive random access procedure is proposed. The applied random access procedure prevent underutilization of cellular network resources. Joint peer discovery and resource allocation problem with the objective of maximizing the network sum-rate is considered in [42]. Their proposed solution is affected by social and physical layer information. The authors of [29] studied caching based D2D pairing and resource allocation. They proposed some approximation algorithms to solve the main problem, in centralized and distributed manner. In [3], a graph-based message passing algorithm is proposed for joint pairing, resource allocation, and power control of D2D users in underlay in-band communications.

Furthermore, the connection to farther nodes for a D2D link is only considered in [34]. This article addresses the issue of connecting to distant nodes and using stochastic geometry tools formulates the coverage probability and average rate of a typical D2D link as closed form expressions. In [34], the authors just consider the situation of connecting to farther nodes in a D2D communication scenario however in this manuscript we are trying to propose a new mechanism which takes into account two important system-level parameters (i.e. signal to interference plus noise ratio (SINR) and battery level) for decision making in D2D communications.

1.1 Fuzzy Mathematics

Since 1965 when L. A Zadeh presented his significant paper [44], fuzzy logic and fuzzy mathematics have had broad applications in computer science and engineering. The idea of Zadeh was to present a generalization of the classical notion of sets. It was to use numbers in the interval $[0, 1]$ instead of 0 and 1, as the membership degrees of elements.

Definition 1 [13]. *Let X be a collection of objects denoted by x and let $\mu : X \to [0, 1]$ be a function. Then the set of all ordered pairs $A = \{(x, \mu(x)) : x \in X\}$ is called a fuzzy set, and the function μ is called membership function which denotes the membership degree of elements in X.*

Note that for $\mu : X \to \{0, 1\}$, we have a classical (crisp) set. It means that only two degrees, 0 or 1 are assigned to each element of X. For example, in the context of crisp sets, the number 11 is not a member of the set $\{0, 1, 2, \ldots, 10\}$ and the number 5 is in the set. On the other hand, one can model the notion of *real numbers near* 0, by a fuzzy set as follows:

$$A := \{(x, \mu(x)) : x \in \mathbb{R}\}, \tag{1}$$

where $\mu(x) = \frac{1}{1+x^2}$. Let us take a rapid historical look at applications of fuzzy mathematics in science and engineering. The first application was introduced, in 1966, by Bellman, Kalaba, and Zadeh [7]. They suggested an application of fuzzy sets to pattern classification. Four years later, in 1970, an application of fuzzy sets was described by Zadeh and Bellman to solve a decision-making problem [8]. The linguistic approach to pattern classification and clustering by fuzzy logic studied in the 1970s by Zadeh [43], Bezdek [9], and Dunn [16]. The application of fuzzy mathematics was grown rapidly, in the next years. Another significant application area of fuzzy sets appeared in the 1970s, was fuzzy controllers. The first fuzzy controller was designed and built by Mamdani and Allisian [28]. Although there were a few applications of fuzzy mathematics investigated through the 1970s and 1980s, the three approaches have received the most attention: pattern classification, decision making, and control. After the success of two projects in Denmark (Controlling a cement kiln, 1980) and Japan (Controlling a subway system, 1985), the application of fuzzy mathematics in control issues prevailed. In fact, the Mamdani type of fuzzy controllers is continued by researchers and nowadays it is popular due to its simplicity and efficiency. Neuro-fuzzy systems and Fuzzy genetics algorithms were significant approaches that developed in the 1990s [22]. Recently, various fields of science and engineering, including Machine learning, Pattern recognition, Wireless sensor networks, logic programming and etc, have benefited from fuzzy mathematics. For more details see [1,6,10].

In this article, we look at D2D mobile networks from a fuzzy perspective. Our scenario is solving the decision-making problem that uses fuzzy mathematics to select the best node to receive information. In other words, we consider each of the transmitters near a typical receiver as a fuzzy node. First, we consider a set of transmitters together with a membership function which assigns to each transmitter a number in $[0,1]$. Then, a decision-making function chooses the best transmitter with the highest fuzzy degree. All of the mentioned functions will define in the next section.

Now, we need some preliminaries to define such functions. We start with the definition of triangular norms and conorms which play an important role in the theory of fuzzy sets and we will use them in the definition of our membership function.

Definition 2 [21]. *A triangular norm (briefly t-norm) is a binary function* $T : [0,1] \times [0,1] \to [0,1]$ *which is associative, commutative, non-decreasing in each place and* $T(a,1) = a$*, for all* $a \in [0,1]$ *[21]. That is*

1. $T(x,y) = T(y,x)$ *(Commutativity),*
2. $T(x,T(y,z)) = T(T(x,y),z)$ *(Associativity),*
3. $T(x,y) \leq T(x,z)$ *whenever* $y \leq z$ *(Monotonicity),*
4. $T(x,1) = x$ *(Boundary condition).*

Example 1 [21]. Four basic t-norms, T_M, T_P, T_L and T_D, That are called Minimum, Product, Lukasiewicz t-norm, and Drastic product respectively, will be defined on $[0,1]^2$ as follows:

1. $T_M(x,y) = \min(x,y)$,
2. $T_P(x,y) = x.y$,
3. $T_L(x,y) = \max(x+y-1,0)$,
4. and the fourth basic t-norm, Drastic product, is written as follows

$$T_D(x,y) = \begin{cases} 0 & ;(x,y) \in [0,1)^2, \\ \min(x,y) & ;Otherwise. \end{cases}$$

In 1961, Schweizer and Sklar introduced dual operations of t-norms as t-conorms [36]. Indeed, a t-conorm T^*, is a continuous binary operation on $[0,1]$ which is related to the continuous t-norm T through

$$T^*(x,y) = 1 - T(1-x, 1-y).$$

Another, independent axiomatic definition of t-conorms can be written as follows

Definition 3 [21]. *A triangular conorm (briefly t-conorm) is a binary operation $S : [0,1]\times[0,1] \to [0,1]$, which for all $x, y, z \in [0,1]$, satisfies Commutativity, Associativity and Monotonicity, and*

$$(4) \quad S(x,0) = x \quad for \quad all \quad x \in [0,1].$$

Example 2. Here are some examples of t-conorms, correspond to the four basic t-norms:

1. $S_M(x,y) = \max(x,y)$, (Maximum t-conorm),
2. $S_P(x,y) = x + y - x.y$, (Probabilistic sum),
3. $S_L(x,y) = \min(x+y,1)$, (Lukasiewicz t-conorm),
4. and the fourth basic t-conorm, Drastic sum, is written as follows

$$S_D(x,y) = \begin{cases} 1 & ;(x,y) \in (0,1]^2, \\ \max(x,y) & ;Otherwise. \end{cases}$$

We need to know the definition of a particular t-conorm, called Einstein sum, due to using it in our membership function.

Example 3 [21]. Let $\tau : [0,1] \times [0,1] \to [0,1]$ be a function with

$$\tau(x,y) = \frac{x+y}{1+xy}, \tag{2}$$

for all $x, y \in [0,1]$. Then τ is a t-conorm called Einstein sum.

It should be noted that we tested several t-norms and t-conorms for defining of membership function we needed and the Einstein sum was the best for our purpose. Of course, other t-norms and t-conorms can be considered for future researches. For more details about applications of this t-conorm in fuzzy systems see [30,37,40], and [47]. Now, we are ready to explain our system model based on fuzzy nodes.

2 System Model

We consider a D2D network that consists of two types of users, transmitting and receiving D2D users. Transmitting D2D users are assumed to be generated from a Poisson point process (PPP) Φ_t with density λ_t. On the other hand, the receiving D2D nodes are assumed to be generated from another independent PPP Φ_r with density λ_r. We investigate a typical D2D link in which its receiving node is placed at the origin. We assume a radius $\mathrm{R_{max}}$ around each receiver and the potential transmitter for the mentioned receiver are located in the circle with receiver center and radius $\mathrm{R_{max}}$. In other words the nodes with greater distance than $\mathrm{R_{max}}$ are not allowed to connect to each other in a D2D link. With respect to the defined density of transmitting and receiving D2D users and radius $\mathrm{R_{max}}$, there are several distant-arranged transmitters denoted by n_i around a typical receiver. We assume that all of the potential D2D links are sharing a single channel and therefore have interference with each other.

In the first step, a signaling process is exchanged between each receiver and its potential proximity transmitters. Every potential D2D transmitter has two network parameters in our system model that are sent to the typical receiver in this step. The first parameter is measured instantaneous SINR that is named as S and is defined as

$$S = \frac{H}{I + N_0},$$

where H is the main signal received power at the receiver and I is the sum of received power of co-channel D2D links and finally N_0 is the noise power. We assume the channel inversion power control at the receiver of each D2D link, therefore for a D2D link with distance d, the average transmitted power is equal to $P = d^{\alpha}$ where α is the path loss exponent. The complementary cumulative distribution function (CCDF) of the parameter S is derived as a closed-form expression in [34] and our analysis is based on these statistics. The parameter S is normalized with respect to the maximum measured SINR at the receiver. Another parameter is the normalized remained battery level denoted by E. Both values of S and E are normalized in $[0, 1]$. The feasible values of S and E are illustrated in Fig. 1. We define two thresholds for SINR and the battery level of each node that are named as S_0 and E_0, respectively. These thresholds are defined by the network protocol and guarantee the quality of service (QoS) of the D2D link. We divide the $[0, 1] \times [0, 1]$ square into four areas shown in Fig. 1. In Area B we have the nodes with acceptable battery level but poor SINR coverage. On the other hand in Area D the situation is reversed, i.e. the nodes have acceptable SINR coverage but low battery level. In Area A both SINR coverage and battery level are in a suitable condition. In Areas C the SINR coverage and battery level are both below the defined thresholds.

According to the above explanation, we consider each D2D transmitter as a fuzzy node denoted by (n_i, μ_i), where n_i is the ith arranged node based on distance and μ_i is its fuzzy membership function. Regarding to the battery level and SINR of each potential transmitter around a typical D2D receiver, the fuzzy membership function is formulated as follows

$$\mu_i(E, S) = \begin{cases} S & ; E_0 \leq E \leq 1,\ 0 \leq S < S_0, \\ E & ; 0 \leq E < E_0,\ S_0 \leq S \leq 1, \\ \tau(E, S)(S^2 + E^2) & ; 0 < E < E_0,\ 0 < S < S_0, \\ \tau(E, S) & ; E_0 \leq E \leq 1,\ S_0 \leq S \leq 1. \end{cases} \quad (3)$$

where $\tau(E, S) = \frac{E+S}{1+ES}$. The fuzzy membership function is defined in such a way that assigns the greater value to the nodes in Area A. This means that the potential nodes with acceptable SINR and battery level will be assigned with a greater fuzzy degree and it is in accordance with the engineering considerations. As it was mentioned, in Area B the nodes have acceptable battery level and their fuzzy degrees are based on their SINR, as well as in Area D the nodes have good SINR values and therefore, their fuzzy degrees are based on their battery level.

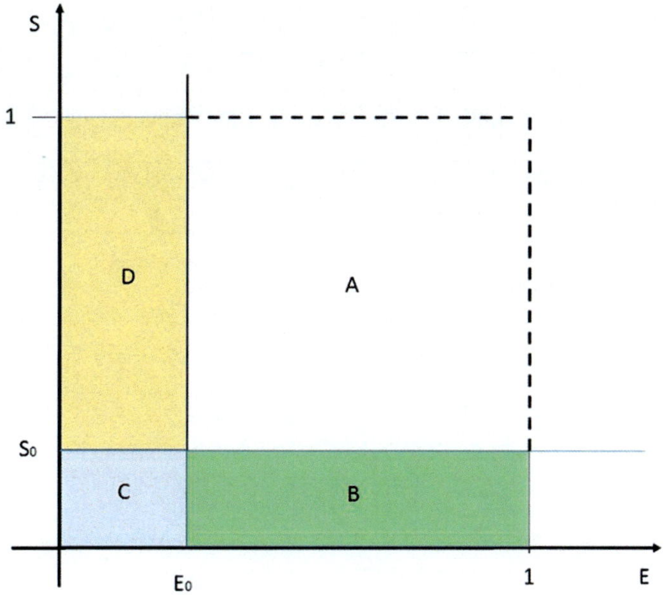

Fig. 1. The feasible values of normalized SINR and battery level and predefined thresholds.

Therefore, the node with maximum membership function will be chosen as the best D2D transmitter which is suitable for pairing with. The decision-making function on the fuzzy set $\mathcal{M} = \{(n_1, \mu_1), \ldots, (n_K, \mu_K\}$ is defined as

$$D(\mathcal{M}) = \{n_i : \mu_i = \max\{\mu_k : 1 \leq k \leq K\}. \tag{4}$$

3 Simulation Results

In this section, the effects of the two network parameters (i.e. normalized SINR and battery level) are considered. In such a way that the decision-making of a typical D2D receiver for choosing the best transmitter is investigated. The main parameters that are used in this section are summarized in Table 1.

The four randomly chosen iterations based on the normalized SINR, battery level, and fuzzy degree are shown in Table 2, and corresponding to each iteration a figure is illustrated in Fig. 2. In the first iteration, the normalized SINR, battery level, and fuzzy degree of the five distance-ordered nodes $(n_1, \ldots, n_5)$ are computed based on the simulation parameters in Table 1 and fuzzy membership function (3). According to the decision-making function (4) the node n_3 has been chosen as the best transmitter for the mentioned receiver. This choice makes sense with respect to the engineering considerations. With respect to Fig. 2, the node n_4 is located in area C. In this area both normalized SINR and battery level (0.0794 and 0.1225, respectively) are in a poor condition, therefore it is not

Table 1. Simulation parameters

Parameter	Value
R_{max} (D2D search radius)	200 m
λ_r (density of receiving D2D users)	$10(\pi 500^2)^{-1}$ m^{-2}
λ_t (density of transmitting D2D users)	$50(\pi 500^2)^{-1}$ m^{-2}
α (path loss exponent)	3.5
S_0 (SINR threshold)	0.3
E_0 (Battery level threshold)	0.25

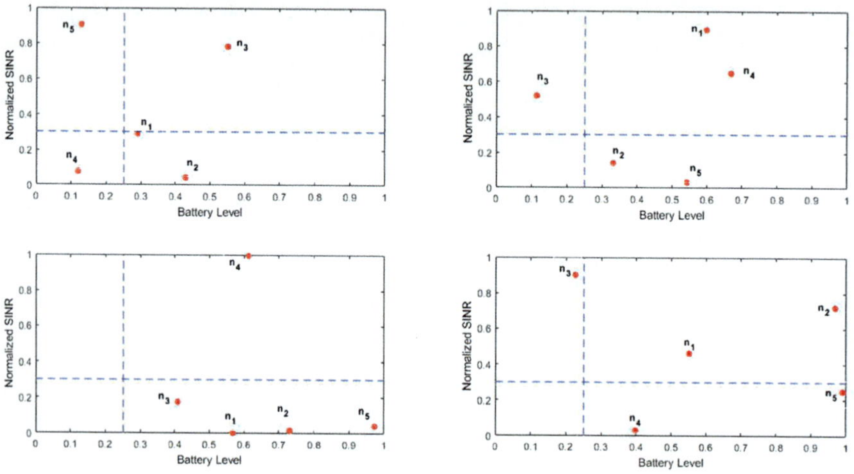

Fig. 2. The graphical view of normalized SINR and battery level for four randomly selected iterations.

acceptable to be chosen as a suitable candidate, and accordingly, a small fuzzy degree is assigned to it by the function (3). The nodes n_1 and n_2 are located in area B and they have an acceptable battery level but their SINRs are under the threshold. It means that they are assigned with lower fuzzy degrees (0.2888 and 0.0416, respectively). The node n_5 is located in Area D with acceptable normalized SINR but poor battery level (0.9048 and 0.1304, respectively) and similar to node n_4 a small fuzzy degree will be computed for it (0.1304). The node n_3 is in a proper circumstance, that is both of its parameters are acceptable (0.7787 and 0.5503, respectively) and as a result it is selected as the best choice. The iterations I_2, I_3, I_4 can be interpreted as above. According to the results illustrated in the Table 2, the decision making function has chosen the nodes with considerable values for both battery and SINR.

In the related works in D2D communications the nearest node is the best choice for pairing [23], however our proposed method shows that the best choice can be dependent on the other network parameters such as battery level and

Table 2. The results of proposed decision making solution in different iterations

Iteration number	UE index	SINR	Battery	Fuzzy degree	Decision
I_1	n_1	0.2888	0.2908	0.2888	n_3
	n_2	0.0416	0.4305	0.0416	
	n_3	0.7787	0.5503	0.9303	
	n_4	0.0794	0.1225	0.0043	
	n_5	0.9048	0.1304	0.1304	
I_2	n_1	0.8893	0.5983	0.9710	n_1
	n_2	0.1419	0.3344	0.1419	
	n_3	0.5168	0.1141	0.1141	
	n_4	0.6461	0.6680	0.9179	
	n_5	0.0335	0.5431	0.0335	
I_3	n_1	0.0014	0.5684	0.0014	n_4
	n_2	0.0169	0.7324	0.0169	
	n_3	0.1768	0.4100	0.1768	
	n_4	0.9952	0.6137	0.9988	
	n_5	0.0407	0.9720	0.0407	
I_4	n_1	0.4604	0.5524	0.8075	n_2
	n_2	0.7185	0.9683	0.9947	
	n_3	0.8982	0.2272	0.2272	
	n_4	0.0279	0.4001	0.0279	
	n_5	0.2479	0.9908	0.2479	

SINR. In this paper we added the battery level as another important parameter that affects the quality of service of a communication link. Not only battery level but also any other significant parameter can be added to our fuzzy degree function.

Moreover, the results are highly dependent to the thresholds, S_0 and E_0 and with the change in these values the fuzzy degrees and consequently the best choice will change. It is another important specification of our proposed function. This shows that the method is flexible and applicable according to the network conditions and service that are used. For example when we have the matter of transferring of high volume files, it is better to increase the E_0 while in the applications that require high speed, the threshold S_0 should be increased. This model is also flexible in the context of adding another important network parameters.

4 Conclusions

In this study, we were supposed to have a D2D network of transmitters and receivers, and the main issue was to select the best transmitter to receive the information. For this purpose, we considered the set of transmitters as a fuzzy set in which, each member was assigned a fuzzy membership degree based on two important parameters, SINR and battery level. In this way, a decision-making function selected the fuzzy node with the highest membership degree. We also simulated the results to see that, unlike conventional networks, a closer transmitter may not be selected.

References

1. Ahvar, E., Pourmoslemi, A., Piran, M.J.: FEAR: a fuzzy-based energy-aware routing protocol for wireless sensor networks. arXiv preprint arXiv:1108.2777 (2011)
2. Asadi, A., Wang, Q., Mancuso, V.: A survey on device-to-device communication in cellular networks. IEEE Commun. Surv. Tutor. **16**, 1801–1819 (2014)
3. Asvadi, R., Rajabi, S., Ghorashi, S.A., Martini, M., et al.: Power allocation for D2D communications using max-min message-passing algorithm. IEEE Trans. Veh. Technol. **69**(8), 8443–8458 (2020)
4. Awan, A.Y., Ali, M., Naeem, M., Qamar, F., Sial, M.N.: Joint network admission control, mode assignment, and power allocation in energy harvesting aided D2D communication. IEEE Trans. Ind. Inf. **16**, 1914–1923 (2019)
5. Azam, M., Ahmad, M., Naeem, M., Iqbal, M., Khwaja, A.S., Anpalagan, A., Qaisar, S.: Joint admission control, mode selection, and power allocation in D2D communication systems. IEEE Trans. Veh. Technol. **65**, 7322–7333 (2015)
6. Bělohlávek, R., Dauben, J.W., Klir, G.J.: Fuzzy Logic and Mathematics: A Historical Perspective. Oxford University Press, Oxford (2017)
7. Bellman, R., Kalaba, R., Zadeh, L.: Abstraction and pattern classification. J. Math. Anal. Appl. **13**(1), 1–7 (1966)
8. Bellman, R.E., Zadeh, L.A.: Decision-making in a fuzzy environment. Manag. Sci. **17**(4), B-141 (1970)
9. Bezdek, J.C.: Fuzzy mathematics in pattern classification. Ph.D. dissertation, Applied Mathematics, Cornell University (1973)
10. Castro, J.L., Castro-Schez, J.J., Zurita, J.M.: Use of a fuzzy machine learning technique in the knowledge acquisition process. Fuzzy Sets Syst. **123**(3), 307–320 (2001)
11. Chen, Y.: An efficient data exchange and detection scheme for two-way relay based D2D communications. In: 2016 IEEE 83rd Vehicular Technology Conference (VTC Spring), pp. 1–5. IEEE (2016)
12. Cheng, W., Zhang, X., Zhang, H.: Optimal power allocation with statistical QoS provisioning for D2D and cellular communications over underlaying wireless networks. IEEE J. Sel. Areas Commun. **34**, 151–162 (2015)
13. Cho, Y.J., Rassias, T.M., Saadati, R.: Fuzzy Operator Theory in Mathematical Analysis. Springer, Heidelberg (2018)
14. Choi, K.W., Han, Z.: Device-to-device discovery for proximity-based service in LTE-advanced system. IEEE J. Sel. Areas Commun. **33**, 55–66 (2014)
15. Doppler, K., Rinne, M., Wijting, C., Ribeiro, C.B., Hugl, K.: Device-to-device communication as an 192 underlay to LTE-advanced networks. IEEE Commun. Mag. **47**, 42–49 (2009)
16. Dunn, J.C.: A fuzzy relative of the ISODATA process and its use in detecting compact well-separated clusters. J. Cybern. **3**(3), 32–57 (1973)

17. Feng, D., Lu, L., Yuan-Wu, Y., Li, G.Y., Feng, G., Li, S.: Device-to-device communications underlaying cellular networks. IEEE Trans. Commun. **61**, 3541–3551 (2013)
18. Gandotra, P., Jha, R.K.: Device-to-device communication in cellular networks: a survey. J. Netw. Comput. Appl. **71**, 99–117 (2016)
19. Gupta, R., Ajith Kumar, P.R., Kalyanasundaram, S.: Transmission mode selection and resource allocation for D2D unicast communications. In: IEEE 84th Vehicular Technology Conference (VTC-Fall), pp. 1–6. IEEE (2016)
20. Jameel, F., Hamid, Z., Jabeen, F., Zeadally, S., Javed, M.A.: A survey of device-to-device communications: research issues and challenges. IEEE Commun. Surv. Tutor. **20**, 2133–2168 (2018)
21. Klement, E.P., Mesiar, R., Pap, E.: Triangular Norms, vol. 8. Springer, Heidelberg (2013)
22. Kosko, B.: Neural networks and fuzzy systems: a dynamical systems approach to machine intelligence, no. QA76. 76. E95 K86 (1992)
23. Lee, N., Lin, X., Andrews, J.G., Heath, R.W.: Power control for D2D underlaid cellular networks: modeling, algorithms, and analysis. IEEE J. Sel. Areas Commun. **33**, 1–13 (2014)
24. Li, Y., Song, C., Jin, D., Chen, S.: A dynamic graph optimization framework for multihop device-to-device communication underlaying cellular networks. IEEE Wirel. Commun. **21**, 52–61 (2014)
25. Lin, X., Andrews, J.G., Ghosh, A.: Spectrum sharing for device-to-device communication in cellular networks. IEEE Trans. Wirel. Commun. **13**, 6727–6740 (2014)
26. Liu, J., Kato, N., Ma, J., Kadowaki, N.: Device-to-device communication in LTE-advanced networks: a survey. IEEE Commun. Surv. Tutor. **17**, 1923–1940 (2014)
27. Mach, P., Becvar, Z., Vanek, T.: In-band device-to-device communication in OFDMA cellular networks: a survey and challenges. IEEE Commun. Surv. Tutor. **17**(4), 1885–1922 (2015)
28. Mandani, E.H., Assilian, S.: An experiment in linguistic synthesis with a fuzzy logic controller. Int. J. Man-Mach. Stud. **7**, 1 (1975)
29. Miao, L., Bai, B., Chen, W.: 4-DMWM approach for caching based optimal D2D pairing and channel allocation: centralized and distributed algorithm design. IEEE Access **4**, 9213–9224 (2016)
30. Munir, M., Kalsoom, H., Ullah, K., Mahmood, T., Chu, Y.-M.: T-spherical fuzzy Einstein hybrid aggregation operators and their applications in multi-attribute decision making problems. Symmetry **12**(3), 365 (2020)
31. Ni, Y., Jin, S., Xu, W., Wang, Y., Matthaiou, M., Zhu, H.: Beamforming and interference cancellation for D2D communication underlaying cellular networks. IEEE Trans. Commun. **64**, 832–846 (2015)
32. Ni, Y., Zhao, J., Wang, Y., Zhu, H.: Beamforming and interference cancellation for D2D communication assisted by two-way decode-and-forward relay node. China Commun. **15**, 100–111 (2018)
33. Ni, M., Zheng, L., Tong, F., Pan, J., Cai, L.: A geometrical-based throughput bound analysis for device-to-device communications in cellular networks. IEEE J. Sel. Areas Commun. **33**, 100–110 (2014)
34. Rajabi, S., Ghorashi, S., Shah-Mansouri, V.: Impact of connecting to the nth nearest node in dedicated device-to-device communications. Electron. Lett. **54**, 535–537 (2018)
35. Rajabi, S., Ghorashi, S.A., Shah-Mansouri, V., Karami, H.: Device-to-device communications using EMTR technique. IET Sig. Proc. **12**, 320–326 (2017)

36. Schweizer, B.: Associative funcrions and statistical triangle inequalities. Puble. Math. Debrecen **8**, 169–186 (1961)
37. Solanki, A., Kumar, E.: A novel technique for rule advancement in fuzzy expert system using einstein sum. 2–03 (2013)
38. Tanbourgi, R., Jäkel, H., Jondral, F.K.: Cooperative interference cancellation using device-to-device communications. IEEE Commun. Mag. **52**, 118–124 (2014)
39. Tang, H., Ding, Z.: Mixed mode transmission and resource allocation for D2D communication. IEEE Trans. Wirel. Commun. **15**, 162–175 (2015)
40. Wei, G., Zhao, X.: Induced hesitant interval-valued fuzzy Einstein aggregation operators and their application to multiple attribute decision making. J. Intell. Fuzzy Syst. **24**(4), 789–803 (2013)
41. Wu, Y., Guo, W., Yuan, H., Li, L., Wang, S., Chu, X., Zhang, J.: Device-to-device meets LTE-unlicensed. IEEE Commun. Mag. **54**, 154–159 (2016)
42. Xu, C., Gao, C., Zhou, Z., Chang, Z., Jia, Y.: Social network-based content delivery in device-to-device underlay cellular networks using matching theory. IEEE Access **5**, 924–937 (2016)
43. Zadeh, L.A.: Fuzzy sets and their application to pattern classification and clustering analysis. In: Classification and Clustering, pp. 251–299. Academic Press (1977)
44. Zadeh, L.A.: Fuzzy sets. Inf. Control **8**(3), 338–353 (1965)
45. Zhang, Z., Hu, R.Q., Qian, Y.: D2D communication underlay in uplink cellular networks with distance based power control. In: 2016 IEEE International Conference on Communications (ICC), pp. 1–6. IEEE (2016)
46. Zhang, R., Cheng, X., Yang, L., Jiao, B.: Interference-aware graph based resource sharing for device-to-device communications underlaying cellular networks. In: 2013 IEEE Wireless Communications and Networking Conference (WCNC), pp. 140–145. IEEE (2013)
47. Zhao, X., Wei, G.: Some intuitionistic fuzzy Einstein hybrid aggregation operators and their application to multiple attribute decision making. Knowl.-Based Syst. **37**, 472–479 (2013)

Solving the System Optimum Static Traffic Assignment Problem with Single Origin Destination Pair in Fuzzy Environment

Gizem Temelcan[1(✉)], Hale Gonce Kocken[2], and Inci Albayrak[2]

[1] Department of Computer Engineering, Beykoz University, Istanbul, Turkey
temelcan.gizem@gmail.com

[2] Department of Mathematical Engineering, Yildiz Technical University, Istanbul, Turkey
{hgonce,ibayrak}@yildiz.edu.tr

Abstract. This paper focuses on the system optimum traffic assignment problem (SOTAP) which aims to minimize total system travel time on a network between specified origin and destination pair. In order to represent a real-life problem dealing with imprecise data; by using fuzzy theory, the system optimum fuzzy traffic assignment problem (SOFTAP) is modeled. As a fully fuzzy quadratic programming problem (QPP), SOFTAP has linear fuzzy constraints and a nonlinear objective function expressed by sum of multiplications of each fuzzy link flow and corresponding fuzzy link travel time. Similar to the literature, BPR function is used as the link travel time function. To determine fuzzy link capacity and free flow fuzzy link travel time, average speed and vehicle length are taken as triangular fuzzy numbers and path lengths and number of lanes are taken as crisp numbers. Due to the imprecision of parameters, the travel demand of the network is also constructed as a triangular fuzzy number. Finally, SOFTAP is solved by a ranking function methodology. A numerical experiment is illustrated for the awareness of solution process.

Keywords: System optimum traffic assignment problem · Fully fuzzy programming problem · BPR link travel time function · Triangular fuzzy number

1 Introduction

Traffic assignment problem (TAP) is defined as a distribution of vehicles on a network comprising a set of nodes and a set of directed links connecting these nodes. In the assignment process, users want to eliminate the congestion effects on the network while traversing from their origin node to destination node, and thus, researchers have studied on finding a network equilibrium state of the problem. An equilibrium state is found for the existing traffic flow on the network, however, the state is changed over time and generally forced to a disequilibrium state. Since users do not want to waste their time in the congested traffic, they tend to change their routes to find any alternative route to reach their destination node, and therefore a new equilibrium state will be observed.

T. Allahviranloo et al. (Eds.): IDS 2020, AISC 1301, pp. 521–530, 2021.
https://doi.org/10.1007/978-3-030-66501-2_41

The idea of traffic assignment was presented in the beginning of 1920's. Beckmann et al. [1] formulated the first mathematical model of TAP in 1956. In literature, TAP is modeled as user equilibrium (UE) or system optimum (SO). Each model aims to minimize travel time on the network while satisfying flow conservation and non-negativity constraints. In UE model, Wardrop's first principle is adopted, that is, no users can improve their travel time by unilaterally changing routes [2]. Whereas SO model is based on Wardrop's second principle that users minimize the total system travel time by cooperating with each other. Actually, policymakers and planners choose SO model in order to minimize the total system travel time, and the chosen model gives rise to both time and cost savings. In emergency cases, ambulances, fire trucks, etc. need to reduce their travel times, while individuals can experience redundantly high travel time on the network. It will be an optimum solution if the case affects positively to the total system travel time [3]. To minimize the total system travel time, all link travel times on the network are used. Link travel time can be defined as a linear or nonlinear [3, 4] function in terms of flow, density [5] or volume-capacity ratio in the literature.

In order to represent real-life problem and deal with ambiguous, vague or imprecise information, fuzzy theory can be used. Many researchers have studied to apply fuzzy theory into TAPs. Chang and Chen [6] formulated a link-based fuzzy UE path choice model. Binetti and De Mitri [7] examined a route choice model and used fuzzy numbers to represent travel costs. Ridwan [8] introduced the first application of fuzzy individual choice in traffic assignment that is a route choice model based on the fuzzy preference. Liu and Kao [9] presented a network flow problem by using fuzzy arc lengths. Ramazani et al. [10] proposed a traffic assignment algorithm and used fuzzy theory to define drivers' perceived travel time affecting route choices. Miralinaghi et al. [11] introduced a new TAP based on fuzzy equilibrium condition, and for relaxing the UE assumptions, they used fuzzy and probability theories.

In this paper, over a sample network, by using the same parameters in [12] and using the BPR link travel time function, SOTAP problem is considered. In the proposed model, there is an interaction only between vehicle-vehicle and vehicle-roadway, i.e. uninterrupted flow is considered. Macroscopic traffic flow model examining the traffic flow from a global perspective is determined, and static models is assumed in which the traffic demand is constant and equal to outflow on the network at a particular time. In the study [12], linear fuzzy link travel time function is used, and to construct the linear function, data containing travel demand, path length, number of lanes, average speed and vehicle length are collected and used as parameters of SOTAP. In this paper, the same parameters are used whereas BPR link travel time function is determined instead. Objective function of SOTAP is in nonlinear form and constructed as a summation of multiplying each link flow and corresponding link travel time function while constraints are linear, satisfy flow conservation and guarantee non-negativity. Because the data collected include some errors or have uncertainty arising from lack of information, parameters which are travel demand, average speed and vehicle length are considered triangular fuzzy numbers. However, due to structure of other parameters that are path lengths and number of lanes are not fuzzified. Thus, the fuzzy model of SOTAP, i.e. SOFTAP, is introduced by taking the variables and the parameters as fuzzy. SOFTAP is modelled as a fully fuzzy QPP, and the fuzzy problem is solved by the modified method given in [12].

The paper is organized as follows: some definitions about fuzzy numbers are given in Sect. 2. In Sect. 3, the mathematical model of SOTAP and the construction of BPR fuzzy link travel time function are presented, respectively. Section 4 includes the solution process and a numerical experiment is illustrated in Sect. 5.

2 Preliminaries

Definition 1 [13] **.** X is a universal set and a fuzzy number $\tilde{A}$ is a convex normal fuzzy subset of X with the membership function $\mu_{\tilde{A}}(x)$ satisfying the following conditions:

- There is at least an element $r \in X$ such that $\mu_{\tilde{A}}(r) = 1$ where r is called as core of $\tilde{A}$,
- $\mu_{\tilde{A}}(x)$ is a nondecreasing on $(-\infty, r)$ and nonincreasing on (r, ∞),
- $\mu_{\tilde{A}}(x)$ is a piecewise continuous,
- $\int_{-\infty}^{\infty} \mu_{\tilde{A}}(x)dx < \infty$.

Definition 2. A fuzzy number $\tilde{A} = (a_1, a_2, a_3)$ is called a triangular fuzzy number if its membership function is given by

$$\mu_{\tilde{A}}(x) = \begin{cases} \frac{x-a_1}{a_2-a_1}, & a_1 \leq x < a_2 \\ \frac{x-a_3}{a_2-a_3}, & a_2 \leq x \leq a_3 \\ 0, & \text{otherwise} \end{cases}$$

where $a_1 \leq a_2 \leq a_3$ and $a_1, a_2, a_3 \in \mathbb{R}$.

Definition 3. A triangular fuzzy number $\tilde{A} = (a_1, a_2, a_3)$ is non-negative if $a_1 \geq 0$.

Definition 4. Two triangular fuzzy numbers $\tilde{A} = (a_1, a_2, a_3)$ and $\tilde{B} = (b_1, b_2, b_3)$ are equal iff $a_1 = b_1$, $a_2 = b_2$ and $a_3 = b_3$.

Definition 5 [14] **.** Let X, Y be universal sets $X, Y \subseteq \mathbb{R}$, $\tilde{A}$ and $\tilde{B}$ be fuzzy sets on X and Y, respectively, and f be a function. Then, the function f satisfying the condition $\mu_{\tilde{A}}(x) \leq \mu_{\tilde{B}}(f(x))$ is called a function with constraints on fuzzy domain $\tilde{A}$ and fuzzy range $\tilde{B}$.

Definition 6. A ranking function R is a mapping from the set of triangular fuzzy numbers to the set of real numbers where a natural order exists. Let $\tilde{A} = (a_1, a_2, a_3)$ be a triangular fuzzy number, then $R\left(\tilde{A}\right) = \frac{a_1+2a_2+a_3}{4}$.

Definition 7. Let $\tilde{A}$ and $\tilde{B}$ be arbitrary triangular fuzzy numbers. Ordering these triangular fuzzy numbers is given as follows:

- $\tilde{A} \overset{f}{<} \tilde{B}$ iff $R\left(\tilde{A}\right) < R\left(\tilde{B}\right)$,

- $\tilde{A} \overset{f}{>} \tilde{B}$ iff $R(\tilde{A}) > R(\tilde{B})$,
- $\tilde{A} \equiv \tilde{B}$ iff $R(\tilde{A}) = R(\tilde{B})$

where the symbols „ $\overset{f}{<}$ „and „ $\overset{f}{>}$ „represent the fuzzy order relations.

Definition 8. Basic arithmetic operations on triangular fuzzy numbers are given as follows: Let $\tilde{A} = (a_1, a_2, a_3)$ and $\tilde{B} = (b_1, b_2, b_3)$ are two arbitrary triangular fuzzy numbers, then

- $\tilde{A} + \tilde{B} = (a_1 + b_1, a_2 + b_2, a_3 + b_3)$,
- $-\tilde{A} = (-a_3, -a_2, -a_1)$,
- $\tilde{A} - \tilde{B} = (a_1 - b_3, a_2 - b_2, a_3 - b_1)$,
- Multiplication of two triangular fuzzy numbers based on extension principle is given by the following equation $\tilde{A}\tilde{B} = (a, b, c)$ where $b = a_2b_2$, $a = min\{a_1b_1, a_1b_3, a_3b_1, a_3b_3\}$ and $c = max\{a_1b_1, a_1b_3, a_3b_1, a_3b_3\}$.
- Multiplication of a triangular fuzzy number by a scalar λ is given as $\lambda\tilde{A} = (\lambda a_1, \lambda a_2, \lambda a_3)$ for $\lambda > 0$, $\lambda\tilde{A} = (\lambda a_3, \lambda a_2, \lambda a_1)$ for $\lambda < 0$.
- Division of two triangular fuzzy numbers based on extension principle is given as follows: if $0 \notin \tilde{B}$, $\frac{\tilde{A}}{\tilde{B}} = \tilde{A}\left(\frac{1}{\tilde{B}}\right) = (a, b, c)$ where $b = \frac{a_2}{b_2}$, $a = min\left\{\frac{a_1}{b_1}, \frac{a_1}{b_3}, \frac{a_3}{b_1}, \frac{a_3}{b_3}\right\}$ and $c = max\left\{\frac{a_1}{b_1}, \frac{a_1}{b_3}, \frac{a_3}{b_1}, \frac{a_3}{b_3}\right\}$.

Definition 9. Suppose that $\tilde{A}_1, \tilde{A}_2, \ldots, \tilde{A}_n$ are n triangular fuzzy numbers, then $R\left(\sum_{i=1}^{n} \tilde{A}_i\right) = \sum_{i=1}^{n} R\left(\tilde{A}_i\right)$.

3 Static System Optimum Traffic Assignment Problem in Fuzzy Environment

The following notations are defined:

Index Sets:

$N = (V, I)$: Network where V is a set of nodes and I is a set of directed links connecting the nodes
OD: A pair of origin and destination nodes, $(o, d) \in OD$, $o, d \in V$
I: Set of directed links on the network N, $i \in I$
K: Set of paths comprised of links, $k \in K$

Parameters:

F^{od}: Traffic demand between $(o, d) \in OD$

$$\delta_{i,k}^{od} = \begin{cases} 1, \text{ if } \textit{link } i \text{ is } a \text{ part of } \textit{path } k \text{ connecting } (o, d) \text{ pair} \\ 0, \text{ otherwise} \end{cases}$$

Decision Variables:

x_i: Flow on link i
t_i: Travel time on link i
f_k^{od}: Flow on path k connecting $(o, d) \in OD$
c_k^{od}: Travel time on path k connecting $(o, d) \in OD$

SOTAP aims to minimize the total system travel time on a specified network between (o, d) pair and finds link flows x_i while satisfying flow conservation constraints. Mathematical model of SOTAP is presented from (1) to (5).

$$\text{Min} \sum\nolimits_i x_i t_i(x_i) \tag{1}$$

s.t.

$$\sum\nolimits_k f_k^{od} = F^{od}, (o, d) \in OD \tag{2}$$

$$x_i = \sum\nolimits_k f_k^{od} \delta_{i,k}^{od}, (o, d) \in OD, \forall i \in I \tag{3}$$

$$c_k^{od} = \sum\nolimits_i t_i \delta_{i,k}^{od}, (o, d) \in OD, \forall k \in K \tag{4}$$

$$f_k^{od} \geq 0, (o, d) \in OD, \forall k \in K. \tag{5}$$

Here, flow conservation constraint (2) states equality between inflow and outflow. Constraint (3) defines flow on link i by using path flows. In constraint (4), path travel times are defined by using link travel times. Constraints (3) and (4) are known as path-link incidence relationships. Non-negativity constraint (5) guarantees a reliable solution space.

It is important to note that the link travel time function is assumed to be positive and increasing. To determine BPR link travel time function, some parameters such as path length, average speed can be collected by taking data. Because these data can contain some errors or have uncertainty arising from lack of information, imprecise parameters which are average speed, vehicle length and travel demand are taken as triangular fuzzy numbers. Therefore, considering a network consisting of links, and length of a link i L_i km, high speed is taken as triangular fuzzy number such that $\tilde{u}_1 = (u_{11}, u_{12}, u_{13})$ km/h in the low traffic case, whereas low speed is taken as triangular fuzzy number such that $\tilde{u}_2 = (u_{21}, u_{22}, u_{23})$ km/h in the high traffic case. In this paper, left bound of $\tilde{u}_1$ is considered to be equal to right bound of $\tilde{u}_2$, i.e. $u_{23} = u_{11}$. According to the data, parameters which are free flow fuzzy link travel time $\tilde{T}_i^1$ and fuzzy link capacity $\widetilde{cap}_i$ are determined as given in [12]. Thus, BPR link fuzzy travel time function for link i is defined as

$$\tilde{t}_i(\tilde{x}_i) = \tilde{T}_i^1 \left(1 + \alpha \left(\frac{\tilde{x}_i}{\widetilde{cap}_i}\right)^\beta\right) \tag{6}$$

where each variable $\tilde{x}_i = (x_{i1}, x_{i2}, x_{i3})$ is a fuzzy link flow, α and β are link-dependent parameters. Generally, $\alpha = 0.15$ and $\beta = 4.0$ are considered in the literature, for instance, in [5, 15–17]. By constructing BPR link fuzzy travel time, and taking fuzzy variables, the SOTAP (1)–(5) is modelled as SOFTAP. The aim of SOFTAP is to minimize the objective function which is the summation of multiplication of each fuzzy link flow $\tilde{x}_i$ and BPR link fuzzy travel time function $\tilde{t}_i(\tilde{x}_i)$. In the SOFTAP, each $\tilde{x}_i$ is non-negative triangular fuzzy variables, $\tilde{F}^{od}$ is a non-negative triangular fuzzy parameter, and $\delta_{i,k}^{od}$ is a binary parameter. Solution of SOFTAP presents the flow and travel time on each link in fuzzy version.

4 Solution Method

In this section, the modified method given in [12] is applied to solve SOFTAP. In the study [12], SOFTAP is modelled by considering linear fuzzy link travel time function, and the fuzzy model is solved by using a solution method for fully fuzzy QPPs. In this paper, link travel time function is considered as BPR function, which is nonlinear, and constructed as fuzzy. With this point of view, it is aimed to find a solution of SOFTAP.

Definition 10. If $\tilde{x}_i^*$ satisfies all constraints of SOFTAP, $\tilde{X}^* = \left[\tilde{x}_i^*\right]_{n\times 1}$ is said to be a fuzzy feasible solution.

Definition 11. A fuzzy feasible solution $\tilde{X}^* = \left[\tilde{x}_i^*\right]_{n\times 1}$ is said to be a fuzzy optimal solution if it satisfies all constraints of SOFTAP and if there exist any nonnegative $\tilde{X} = \left[\tilde{x}_i\right]_{n\times 1}$ such that $R\left(\sum_i \tilde{x}_i^* \tilde{t}_i(\tilde{x}_i^*)\right) \le R\left(\sum_i \tilde{x}_i \tilde{t}_i(\tilde{x}_i)\right)$.

SOFTAP has only equality constraints and its objective function is nonlinear. Ranking function is applied to nonlinear fuzzy objective function of SOFTAP and then a crisp nonlinear objective function is obtained. The crisp QPP is solved by using GAMS Software and the optimal solution is found. According to this, the solution algorithm of SOFTAP is given as follows.

Step 1. Load SOFTAP (1)–(5).
Step 2. Because parameter $\tilde{F}^{od}$ and variables $\tilde{x}_i$ are non-negative triangular fuzzy numbers, reformulate the constraints of SOFTAP by using fuzzy arithmetic operations presented in Definition 8.
Step 3. Convert SOFTAP into a crisp QPP by using the definition of fuzzy equality in the constraints, applying ranking function to the fuzzy objective function such that

$$\begin{aligned} R\left(\sum\nolimits_i \tilde{x}_i \tilde{t}_i(\tilde{x}_i)\right) &= R\left(\sum\nolimits_i (x_{i1}t_{i1}(x_{i1}), x_{i2}t_{i2}(x_{i2}), x_{i3}t_{i3}(x_{i3}))\right) \\ &= \sum\nolimits_i 0.25x_{i1}t_{i1}(x_{i1}) + 0.5x_{i2}t_{i2}(x_{i2}) + 0.25x_{i3}t_{i3}(x_{i3}) \end{aligned} \quad (7)$$

Step 4. Find the optimal solution x_{i1}, x_{i2} and x_{i3} of the crisp QPP, substitute them in $\tilde{x}_i = (x_{i1}, x_{i2}, x_{i3})$ for determining the fuzzy optimal solution, and calculate the fuzzy optimal value by substituting $\tilde{x}_i$ in the objective function of SOFTAP.

Fuzzy optimal solution of fuzzy flow can be interpreted that left and right bounds of the fuzzy flow refer to minimum and maximum flows, respectively, while core of fuzzy flow refers to possible flow. Similarly, left and right bounds of a fuzzy optimal travel time correspond to free flow and maximum travel time whereas core of fuzzy travel time refers to possible travel time traversing the path.

5 Numerical Experiment

In this section, a numerical experiment is illustrated to present the solution procedure of SOFTAP. In this paper, origin and destination points of the network are considered as Node A and Node B, respectively and intermediate nodes are Node C and Node D. Paths are constructed such that Path 1 including Link 1 and Link 5; Path 2 including Link 2, Link 3 and Link 5; and Path 3 including Link 2 and Link 4, respectively. Network representation of the experiment and data are given in Fig. 1 and Table 1, respectively.

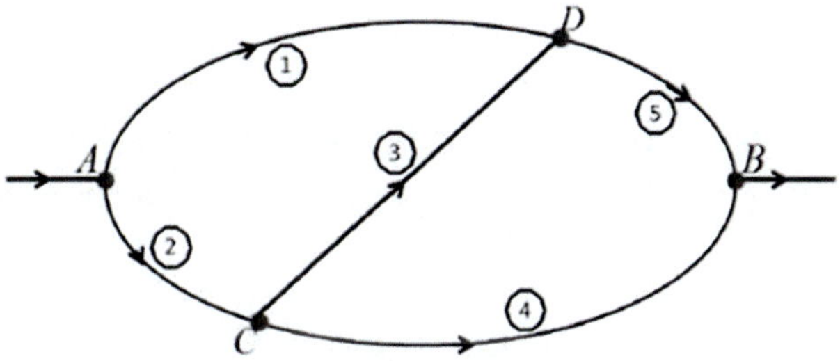

Fig. 1. The sampled network.

Table 1. Data of the sampled network.

Link	1	2	3	4	5
Node	(A, D)	(A, C)	(C, D)	(C, B)	(D, B)
L_i (km)	23.3	15.3	5.0	19.3	22.7
Number of lane	2	3	2	3	4

In this paper, length of a vehicle is taken as $\tilde{l} = (5, 6, 7)$ meters. In the low traffic case, fuzzy high speed is considered as $\tilde{u}_1 = (70, 90, 100)$ km/h, and in the high traffic case, fuzzy low speed is considered as $\tilde{u}_2 = (50, 60, 70)$ km/h. The BPR link fuzzy travel time functions are formulated as given (6). Here, one of the BPR link fuzzy travel time function, such as Link 1, is constructed as follows. The length of Link 1 is given 23.3 km in Table 1. Free flow fuzzy travel time and fuzzy capacity are determined as $\tilde{T}_1^1 = (13.98, 15.53, 19.97)$ min and $\widetilde{cap}_1 = (39.68, 55.56, 77.78)$ veh/min, respectively, by using the fuzzy high and low speeds. Thus, the BPR fuzzy travel time of Link1 is formed as $\tilde{t}_1(\tilde{x}_1) = (13.98, 15.53, 19.97)\left(1 + 0.15\left(\frac{\tilde{x}_1}{(39.68, 55.56, 77.78)}\right)^4\right)$. All BPR link fuzzy travel time functions are constructed by using the same way.

By considering fuzzy inflow as $\tilde{F} = (100, 125, 150)$, the mathematical model of the numerical experiment is presented as follows:

$$Min \sum\nolimits_{i=1}^{5} \tilde{x}_i \tilde{t}_i \tag{8}$$

s.t.

$$\sum\nolimits_{k=1}^{3} \tilde{f}_k = (100, 125, 150) \tag{9}$$

$$\tilde{x}_i = \sum\nolimits_{k=1}^{3} \tilde{f}_k \delta_{i,k}, \, i \in I \tag{10}$$

$$\tilde{c}_k = \sum\nolimits_{i=1}^{5} \tilde{t}_i \delta_{i,k}, \, k \in K \tag{11}$$

$$\tilde{f}_k \geq (0, 0, 0), k \in K. \tag{12}$$

The modified method given in Sect. 4 is applied and a fuzzy optimal solution of the experiment is found. Fuzzy path flows, fuzzy marginal path travel times and their ranking function values are given in Table 2.

Table 2. Fuzzy optimal solution.

Path	Fuzzy path flow	Fuzzy marginal path travel time	Ranking function value of marginal travel time
1	(19.16, 44.16, 55.93)	(27.65, 35.91, 112.15)	52.90
2	(8.85, 8.85, 22.08)	(27.4, 36.03, 112.15)	52.90
3	(71.99, 71.99, 71.99)	(23.61, 35.22, 117.57)	52.90

Main principle of SOTAP is that if there is flow on a path, it means the path is used. Also, all used paths have equal marginal path travel times and their marginal path travel time is less than that of unused paths. However, SOFTAP has a different principle from SOTAP. In SOFTAP, if all components of the fuzzy path flow are different than zero, it means that the considered path is used. It is seen from Table 2 that; all paths are used because all the components of fuzzy path flows are different than zero. Also, the ranking function value of all fuzzy marginal path travel times are equal, which is a desired situation. Similarly, if all components of fuzzy path flow are zero, this means that the related path is not used, and ranking function value of fuzzy marginal travel time of these paths will be greater than that of used paths.

In traffic assignment, the solution obtained by applying a method or an algorithm is desired to be closer to real solution. The measure of convergence is generally tested by a relative gap which is used in [18–20]. Relative gap is defined as

$$RGAP = \frac{\sum_{(o,d)} \sum_k f_k \left(c_k - c_{min}^{(o,d)}\right)}{\sum_{(o,d)} \sum_k f_k c_{min}^{(o,d)}} \tag{13}$$

where f_k is path flow, c_k is marginal path travel time and c_{min}^{od} is the minimum marginal path travel time, which is the marginal travel time of used paths. Relative gap takes a value between zero and one. The solution obtained is closer to the real solution whether the relative gap value is close to zero. In this paper, to show the convergence of solution of SOFTAP, ranking function is applied to all fuzzy variables in the formulation of RGAP, that is

$$RGAP = \frac{\sum_{(o,d)} \sum_k R\left(\tilde{f}_k\right)\left(R(\tilde{c}_k) - R(\tilde{c}_{min}^{od})\right)}{\sum_{(o,d)} \sum_k R(\tilde{f}_k) R\left(\tilde{c}_{min}^{od}\right)}. \tag{14}$$

The relative gap defined in (14) is used to calculate the measure of convergence of SOFTAP solution for the numerical experiment. It is seen that; the relative gap value is found 0. It shows that the solution obtained from the modified method is closer to the real solution.

6 Conclusion

In this paper, we focused on the model SOTAP in more realistic way by considering of uncertainty of data and used fuzzy numbers to construct SOFTAP. Parameters of the problem are taken as triangular fuzzy numbers. BPR function commonly used in the literature is fuzzified. Thus, BPR link fuzzy travel time function is determined. We used the method modified in [12], solved SOFTAP by converting fully fuzzy QPP into a crisp QPP, and found fuzzy optimal solution. Left and right bounds of the fuzzy optimal solution of path flow refer to minimum and maximum flow on the related path, respectively, while the core refers to the possible flow on the path. Fuzzy optimal path flow is classified as used and unused relative to whether there is flow in all the components or not. Also, the convergency of the fuzzy solution to the real solution is presented by using ranking function and the relative gap formulation, and it is seen that the value of relative gap is determined closer to zero when all the components of fuzzy flow is different than zero.

References

1. Beckmann, M., McGuire, C.B., Winsten, C.B.: Studies in the Economics of Transportation. Oxford University Press, London (1956)
2. Sheffi, Y.: Urban Transportation Networks: Equilibrium Analysis with Mathematical Programming Methods. Pretince Hall, Englewood Cliffs (1985)

3. LeBlanc, L.J., Morlok, E.K., Pierskalla, W.P.: An efficient approach to solving the road network equilibrium traffic assignment problem. Transp. Res. **9**(5), 309–318 (1975)
4. Carey, M., Ge, Y.E.: Comparing whole-link travel time models. Transp. Res. Part B Methodol. **7**(10), 905–926 (2003)
5. Kachroo, P., Sastry, S.: Traffic assignment using a density-based travel-time function for intelligent transportation systems. IEEE Trans. Intell. Transp. Syst. **17**(5), 1438–1447 (2016)
6. Chang, M.S., Chen, H.K.: A fuzzy user-optimal route choice problem using a link-based fuzzy variational inequality formulation. Fuzzy Sets Syst. **114**(2), 339–345 (2000)
7. Binetti, M., De Mitri, M.: Traffic assignment model with fuzzy travel cost. In: Proceedings of the 13th Mini-EURO Conference on Uncertainty in Transportation, Bari, Italy, pp. 805–812 (2002)
8. Ridwan, M.: Fuzzy preference based traffic assignment problem. Transp. Res. Part C Emerg. Technol. **12**(3), 209–233 (2004)
9. Liu, S.T., Kao, C.: Network flow problems with fuzzy arc lengths. IEEE Trans. Syst. Man Cybern. Part B (Cybern.) **34**(1), 765–769 (2004)
10. Ramazani, H., Shafahi, Y., Seyedabrishami, S.E.: A fuzzy traffic assignment algorithm based on driver perceived travel time of network links. Sci. Iranica **18**(2), 190–197 (2011)
11. Miralinaghi, M., Shafahi, Y., Anbarani, R.S.: A fuzzy network assignment model based on user equilibrium condition. Sci. Iranica. Trans. A Civ. Eng. **22**(6), 2012–2023 (2015)
12. Temelcan, G., Gonce Kocken, H., Albayrak, I.: System optimum fuzzy traffic assignment problem. Promet-Traffic Transp. **31**(6), 611–620 (2019)
13. Yang, R., Wang, Z., Heng, P.A., Leung, K.S.: Fuzzy numbers and fuzzification of the choquet integral. Fuzzy Sets Syst. **153**(1), 95–113 (2005)
14. Lee, K.H.: First Course on Fuzzy Theory and Applications, vol. 27. Springer, Heidelberg (2005)
15. He, X., Guo, X., Liu, H.X.: A link-based day-to-day traffic assignment model. Transp. Res. Part B Methodol. **44**(4), 597–608 (2010)
16. Lu, Z., Meng, Q., Gomes, G.: Estimating link travel time functions for heterogeneous traffic flows on freeways. J. Adv. Transp. **50**(8), 1683–1698 (2016)
17. Mtoi, E.T., Moses, R.: Calibration and evaluation of link congestion functions: applying intrinsic sensitivity of link speed as a practical consideration to heterogeneous facility types within urban network. J. Transp. Technol. **4**(2), 141–149 (2014)
18. Inoue, S.I., Maruyama, T.: Computational experience on advanced algorithms for user equilibrium traffic assignment problem and its convergence error. Procedia-Soc. Behav. Sci. **43**, 445–456 (2012)
19. Tajtehranifard, H., Bhaskar, A., Nassir, N., Haque, M.M., Chung, E.: A path marginal cost approximation algorithm for system optimal quasi-dynamic traffic assignment. Transp. Res. Part C Emerg. Technol. **88**, 91–106 (2018)
20. Boyce, D., Ralevic-Dekic, B., Bar-Gera, H.: Convergence of traffic assignments: how much is enough? J. Transp. Eng. **130**(1), 49–55 (2004)

Selection of Contract Type in Construction Projects Using Spherical AHP Method

Alper Camci(✉), Ömer Çimen, and Sait Gül

Faculty of Engineering and Natural Sciences, Management Engineering Department, Bahçeşehir University, Beşiktaş, 34353 İstanbul, Turkey
alper.camci@eng.bau.edu.tr

Abstract. Construction is a vital industry that provides solutions to many human needs. Throughout its life, a construction project is subject to uncertainties that can affect the outcomes. Because of these uncertainties and the diversity of the disciplines, trades and skills needed in construction projects, contracts are indispensable in modern construction management. The type of contract chosen determines the roles, responsibilities and liabilities of contractors and owners against the uncertainties of projects. In this study, the criteria affecting the choice of contracts and their importance are evaluated, the most common contract types used in construction projects are examined and a multi-criteria decision-making tool to select the appropriate contract is proposed. Within this study, six most common construction contracts are examined, and Firm Fixed Price Contract is selected as the most suitable type of contract after performing analytic hierarchy process. The model can contribute to the prevention of undesired results such as waste of time, inefficient project management, various damages, and contractual disputes, all of which may result from improper contract type usage. The main technical contributions of this study can be summarized as follows: (i) a spherical fuzzy version of AHP is chosen in this study to represent the hesitancy of the decision-makers more comprehensively, (ii) we developed spherical fuzzy correspondences for the intermediate level linguistic terms (2, 4, 6, 8), and (iii) a novel consistency ratio equation which is appropriate for spherical fuzzy environment is proposed in this study.

Keywords: Contract type · Construction projects · Spherical fuzzy sets · AHP

1 Introduction

Construction is a very large and diversified industry that delivers solutions ranging from basic human needs for shelter to complex engineering structures and accomplish these by converting various resources into physical infrastructure and superstructure assets. Construction work encompasses the new construction, alterations, repair and demolition of buildings and civil engineering works with the involvement of a large group of stakeholders including contractors, subcontractors, material producers, suppliers, consultants, owners, financiers and official approval authority organizations [1]. The “employer” and “contractor” are the two main stakeholders that will be mentioned frequently within the

T. Allahviranloo et al. (Eds.): IDS 2020, AISC 1301, pp. 531–547, 2021.
https://doi.org/10.1007/978-3-030-66501-2_42

scope of the study. Here, the term employer is defined as the party who plays roles of the owner and the financier of the project and makes the decisions to launch the project. The contractor is primarily responsible for the construction activities and provides the engineering, labor, materials and machinery required for the construction work. A modern construction project is a carefully choreographed sequence of steps involving many different trades, skills and knowledge areas. In order to coordinate all these diverse professional skills, there needs to be a coordinating figure who has knowledge of integrating these activities. One approach can be that the employer coordinates all the activities in a project, but usually employers do not have the competence or the resources for the task. The second and the universally accepted solution is having a contractor undertake the construction work. Contractors can employ their own tradesmen or use subcontractors. Because of this almost inevitable employer-contractor relationship, contracts are needs in order to establish the scope of work to be performed, to define the roles and responsibilities of employers and contractors, and the payment mechanisms for the jobs completed during the project's life. As the constructions projects take a long time to finish and subject to natural, legal, economic and social uncertainties and complexities, one of the most critical decisions in construction projects, the selection of the contract type. The type of the contract contributes to the success of the project and minimizes the potential conflicts when executed properly.

The aim of this study is to develop a multi-criteria decision making tool to select the appropriate construction contract type. Since the types of contracts in construction projects are decided by the employers [2] all expert opinions, criteria evaluations and weightings in this study will be made from the employer perspective. Therefore, this study aims to contribute the literature by developing an effective contract type selection method to be utilized in construction projects for employers. In this context, the main subject of the study is the selection of the contract selection criteria, evaluation of these criteria and the selection of the most appropriate construction contract type using a spherical fuzzy AHP (analytic hierarchy process) method. We chose to use a spherical fuzzy version of AHP because it can comprehensively deal with the hesitancy hidden in the preferences of decision-makers.

2 Construction Contracts

A contract can be defined as a written or oral agreement with a legal sanction for a service, sale or lease [3]. The construction contracts define the relationship structure among the key project stakeholders. A typical construction contract includes a number of annexes, including the text of the construction agreement, project documentation, general terms, and specifications, various technical and legal annexes [4]. The construction contract requires an agreement on the construction of a physical structure for a price [5]. In this study, five of the most common contract types used in Turkey plus one contract type not common in Turkey but being in use in North America as shown in Table 1.

Table 1. Construction contract alternatives considered in this study.

Contract type	Short description
Unit Price Contract (UPC)	The total price of the contract is determined by multiplying each manufacturing unit price by the quantity specified for that manufacturing
Firm Fixed Price Contract (FFPC)	Payments shall be made on a fixed amount agreed by the parties, regardless of the cost of the contractor to carry out the work, and in no case shall price adjustments be made
Fixed Price Contract with Price Adjustment (FPCPA)	It is used if the work within the scope of the contract has years. The price adjustment clause protects producers and contractors from the negative effects of inflation
Cost Plus Profit Percentage Contract (CPPPC)	The contractor is paid a fixed percentage of profit over the cost incurred under the project
Cost Plus Incentive Premium Contract (CPIPC)	The incentive premium is paid to the Contractor in case the cost incurred within the scope of the project and the planned cost and time targets are realized
Time and Material Contract (TMC)	Workmanship and equipment fees are paid at the agreed timetable. The material fee used in the project is paid over the actual cost

3 Construction Contract Selection Attributes

In the literature, there are numerous studies regarding the selection of the construction contractors using multi-criteria decision-making tools. Some of these studies are; selection of the best value contractor in road construction projects with the help of Analytic Network Process (ANP) [6], optimum contractor selection in the construction sector with fuzzy logic method [7], contractor selection using fuzzy-TOPSIS method [8], use of reconciliation tools and big data in contractor selection of construction projects [9], tender process in public institutions and the decision-making model for contractor selection [10], a contractor selection support system aimed at creating a competitive bidding environment [11]. However, the work on selecting the right construction contract is limited to Antoniou's research [12] on the selection criteria used in the selection of construction projects' contract type. Antoniou [12] focused on the selection criteria for the types of contracts used in large road construction projects, but no conclusions regarding the suitability of the types of contracts were made. One of the contributions of this study is to propose a structured, hierarchical selection criteria based on the main issues of a project. The hierarchical structure is based on the uncertainty areas (cost, time, quality, and structural) and flexibility.

One of the contributions of this study is to propose a structured, hierarchical selection criteria based on the main issues of a project. The hierarchical structure is based on the uncertainty areas (cost, time, quality, and structural) and flexibility. These main criteria and their sub-criteria are given in Table 2.

Table 2. Construction contract selection criteria and sub criteria.

Main criteria	Sub criteria	Description
Cost	Value for Money (VM)	Increasing efficiency in obtaining the resulting value
	Competitive Pricing (CP)	Contributing to competitive pricing
	Cost Risk (CR)	Transfer the cost increase risk to the Contractor
	Project Monitoring Cost (PM)	Having minimum project monitoring costs
Time	Duration of Work (DW)	Motivate the Contractor to accelerate the duration of the work
	Criticality of Schedule (CS)	Contribute to capturing critical business deadlines
	Start-up Speed (SS)	Shortening the start time of actual construction
Quality	Construction Quality (CQ)	Ensuring construction quality to the extent that the employer is satisfied
	Operational Life Quality (OQ)	Satisfactory performance in post-construction life
Structural uncertainty	Process Uncertainty (PU)	The best way to manage the project against uncertain construction methods
	Scope Uncertainty (SU)	Optimal management of the project in case of uncertain project scope
	Resource Uncertainty (RU)	Minimal impact of the project from uncertain source quality and quantity
	Dispute and Conflicts (DC)	Minimum contractual disputes
Flexibility	Compliance with Changes (CWC)	Ability to comply with the change requests
	Ease of Contract Implementation (ECI)	Contract execution is simple and understandable
	Progressive Construction (PC)	Ability to perform design and production in parallel
	Economic Flexibility (EF)	Ability to adapt to changing economic environments.

4 Spherical Fuzzy Sets

Spherical fuzzy set (SFS) [14] is introduced as a generalization and synthesis of Pythagorean fuzzy sets (PFS) [15] and neutrosophic sets (NS) [13, 16] with the aim of providing a larger preference domain for decision-makers (DMs) and allowing them to present their hesitancy about an attribute or alternative [17]. In SFSs, while the squared sum of membership, non-membership and hesitancy degrees should be between 0 and 1, each degree should be defined between 0 and 1, independently [18]. Thus, it can be emphasized that the advantage of SFSs is that they are able to collect the positive features of other fuzzy set extensions into a unique theory.

Some definitions and basic operations are summarized as follows [14, 18–20]:

Definition 1.1. An SFS $\tilde{A}_S$ of the universe of discourse X is given by

$$\tilde{A}_S = \{< x, \mu_{\tilde{A}_S}(x), \upsilon_{\tilde{A}_S}(x), \pi_{\tilde{A}_S}(x) | x \in X >\} \tag{1}$$

where $\mu_{\tilde{A}_S}(x) : X \to [0, 1]$, $\upsilon_{\tilde{A}_S}(x) : X \to [0, 1]$, $\pi_{\tilde{A}_S}(x) : X \to [0, 1]$
and

$$0 \le \mu^2_{\tilde{A}_S}(x) + \upsilon^2_{\tilde{A}_S}(x) + \pi^2_{\tilde{A}_S}(x) \le 1, \forall x \in X \tag{2}$$

For each x, the $\mu_{\tilde{A}_S}(x)$, $\upsilon_{\tilde{A}_S}(x)$, and $\pi_{\tilde{A}_S}(x)$ are the membership degree, non-membership degree, and hesitancy degree of x to $\tilde{A}_S$, respectively. By considering the relations between SFS and PFS, Kutlu Gündoğdu and Kahraman [14] developed some numerical operations for SFSs as follows.

Definition 1.2. Let X_1 and X_2 be two universes and let $\tilde{A}_S = \left(\mu_{\tilde{A}_S}, \upsilon_{\tilde{A}_S}, \pi_{\tilde{A}_S}\right)$ and $\tilde{B}_S = \left(\mu_{\tilde{B}_S}, \upsilon_{\tilde{B}_S}, \pi_{\tilde{B}_S}\right)$ be two SFSs from the universe of discourse X_1 and X_2. Basic operators are defined as follows.

Union

$$\tilde{A}_S \cup \tilde{B}_S = \left\{\max\left\{\mu_{\tilde{A}_S}, \mu_{\tilde{B}_S}\right\}, \min\left\{\upsilon_{\tilde{A}_S}, \upsilon_{\tilde{B}_S}\right\}, \min\{\left(1 - ((\max\left\{\mu_{\tilde{A}_S}, \mu_{\tilde{B}_S}\right\})^2 + (\min\left\{\upsilon_{\tilde{A}_S}, \upsilon_{\tilde{B}_S}\right\})^2\right))^{\frac{1}{2}}, \max\left\{\pi_{\tilde{A}_S}, \pi_{\tilde{B}_S}\right\}\right\}\} \tag{3}$$

Intersection

$$\tilde{A}_S \cap \tilde{B}_S = \left\{\min\left\{\mu_{\tilde{A}_S}, \mu_{\tilde{B}_S}\right\}, \max\left\{\mu_{\tilde{A}_S}, \mu_{\tilde{B}_S}\right\}, \max\{\left(1 - ((\min\left\{\mu_{\tilde{A}_S}, \mu_{\tilde{B}_S}\right\})^2 + (\max\left\{\upsilon_{\tilde{A}_S}, \upsilon_{\tilde{B}_S}\right\})^2\right))^{\frac{1}{2}}, \min\left\{\pi_{\tilde{A}_S}, \pi_{\tilde{B}_S}\right\}\right\}\} \tag{4}$$

Addition

$$\tilde{A}_S \oplus \tilde{B}_S = \left\{\left(\mu^2_{\tilde{A}_S} + \mu^2_{\tilde{B}_S} - \mu^2_{\tilde{A}_S}\mu^2_{\tilde{B}_S}\right)^{\frac{1}{2}}, \upsilon_{\tilde{A}_S}\upsilon_{\tilde{B}_S}, \left(\left(1 - \mu^2_{\tilde{B}_S}\right)\pi^2_{\tilde{A}_S} + \left(1 - \mu^2_{\tilde{A}_S}\right)\pi^2_{\tilde{B}_S} - \pi^2_{\tilde{A}_S}\pi^2_{\tilde{B}_S}\right)^{\frac{1}{2}}\right\} \tag{5}$$

Multiplication

$$\tilde{A}_S \otimes \tilde{B}_S = \left\{\mu_{\tilde{A}_S}\mu_{\tilde{B}_S}, \left(\upsilon_{\tilde{A}_S}^2 + \upsilon_{\tilde{B}_S}^2 - \upsilon_{\tilde{A}_S}^2\upsilon_{\tilde{B}_S}^2\right)^{\frac{1}{2}}, \left(\left(1-\upsilon_{\tilde{B}_S}^2\right)\pi_{\tilde{A}_S}^2 + \left(1-\upsilon_{\tilde{A}_S}^2\right)\pi_{\tilde{B}_S}^2 - \pi_{\tilde{A}_S}^2\pi_{\tilde{B}_S}^2\right)^{\frac{1}{2}}\right\} \tag{6}$$

Multiplication by a scalar (λ > 0)

$$\lambda * \tilde{A}_S = \{\left(1-\left(1-\mu_{\tilde{A}_S}^2\right)^{\lambda}\right)^{\frac{1}{2}}, \upsilon_{\tilde{A}_S}^{\lambda}, (\left(1-\mu_{\tilde{A}_S}^2\right)^{\lambda} - \left(1-\mu_{\tilde{A}_S}^2 - \pi_{\tilde{A}_S}^2\right)^{\lambda})^{\frac{zx1}{2}}\} \tag{7}$$

Power of $\tilde{A}_S$ (λ > 0)

$$\tilde{A}_S^{\lambda} = \{\mu_{\tilde{A}_S}^{\lambda}, \left(1-\left(1-\upsilon_{\tilde{A}_S}^2\right)^{\lambda}\right)^{\frac{1}{2}}, (\left(1-\upsilon_{\tilde{A}_S}^2\right)^{\lambda} - \left(1-\upsilon_{\tilde{A}_S}^2 - \pi_{\tilde{A}_S}^2\right)^{\lambda})^{\frac{1}{2}}\} \tag{8}$$

Definition 1.3. For aggregation purposes, spherical weighted arithmetic and geometric means (SWAM and SWGM) are defined as follows. The weight vector is given as $w = (w_1, w_2, \ldots, w_n)$ where $w_i \in [0, 1]$ and $\sum_{i=1}^{n} w_i = 1$.

SWAM

$$SWAM_w\left(\tilde{A}_{S1}, \tilde{A}_{S2}, \ldots, \tilde{A}_{Sn}\right) = w_1\tilde{A}_{S1} + w_2\tilde{A}_{S2} + \cdots + w_n\tilde{A}_{Sn}$$

$$= \left\{\left[1-\prod_{i=1}^{n}\left(1-\mu_{\tilde{A}_{Si}}^2\right)^{w_i}\right]^{\frac{1}{2}}, \prod_{i=1}^{n}\upsilon_{\tilde{A}_{Si}}^{w_i}, \left[\prod_{i=1}^{n}\left(1-\mu_{\tilde{A}_{Si}}^2\right)^{w_i} - \prod_{i=1}^{n}\left(1-\mu_{\tilde{A}_{Si}}^2 - \pi_{\tilde{A}_{Si}}^2\right)^{w_i}\right]^{\frac{1}{2}}\right\} \tag{9}$$

SWGM

$$SWGM_w\left(\tilde{A}_{S1}, \tilde{A}_{S2}, \ldots, \tilde{A}_{Sn}\right) = \tilde{A}_{S1}^{w_1} + \tilde{A}_{S1}^{w_2} + \ldots + \tilde{A}_{Sn}^{w_n}$$

$$= \left\{\prod_{i=1}^{n}\mu_{\tilde{A}_{Si}}^{w_i}, \left[1-\prod_{i=1}^{n}\left(1-\upsilon_{\tilde{A}_{Si}}^2\right)^{w_i}\right]^{\frac{1}{2}}, \left[\prod_{i=1}^{n}\left(1-v_{\tilde{A}_{Si}}^2\right)^{w_i} - \prod_{i=1}^{n}\left(1-v_{\tilde{A}_{Si}}^2 - \pi_{\tilde{A}_{Si}}^2\right)^{w_i}\right]^{\frac{1}{2}}\right\} \tag{10}$$

5 Spherical Fuzzy AHP

Analytical hierarchy process (AHP) which is among the most famous multi-criteria decision-making approaches, is developed by Saaty [21]. AHP also provides an effective decision-making tool for group decision-making processes. Assigning a numerical value for each criteria-based comparison supports DMs in terms of developing an efficient evaluation process and reaching a decision, since the consensus nature of group decision making makes AHP a beneficial tool that can increase the consistency and reliability of DMs' judgments [22].

In this study, AHP is extended with a spherical fuzzy set understanding in order to handle vagueness and ambiguity underlying in decision processes of DMs. In literature, there are many AHP extensions such as fuzzy AHP with T2-FSs [23, 24], intuitionistic fuzzy AHP [25, 26], hesitant fuzzy AHP [27, 28], Pythagorean fuzzy AHP [29, 30], neutrosophic AHP [31, 32]. SFS is relatively new concept in the literature and there is few numbers of extensions on decision-making approaches. In terms of AHP, Kutlu Gündoğdu and Kahraman [19, 33] developed spherical fuzzy AHP (SF-AHP), which is used by this study.

Step 1. Construction of the decision hierarchy

A hierarchal representation of the decision problem is built in this step. A hierarchy consists of the goal of the decision at the top level (Level 0), the main criteria after the goal (Level 1), the sub-criteria and the sub-sub-criteria of the related main criteria or sub-criteria respectively (Level 2, 3, ...), and finally the alternatives at the bottom level. The relations between goal, criteria, and alternatives are represented by arrows in a hierarchy. Main criteria are symbolized by the set of $C = \{C_1, \ldots, C_n\}$ which are shown in Level 1. If there are sub-criteria of some main criteria, they will be shown by related additional indices, such as $C_{11}, C_{12}, \ldots C_{nk}$. At the bottom level, a discrete set of possible alternatives $A = \{A_1, \ldots, A_m\}$ are depicted. While the criteria set is determined as an output of literature survey, expertise of DMs, and real case applications, the alternatives are mostly case-specific and unique for distinct decision problems.

Step 2. Determination of SF linguistic set

In AHP-based applications, the experts make pairwise comparisons among the elements at each level. These comparisons show the evaluations/preferences/judgments of the DMs regarding elements in the hierarchy. To compare elements, DMs use linguistic terms. Saaty proposed a 9-scale linguistic term set in which 1 represents equally importance of the compared elements and 9 shows perfect importance of an element against the other [21]. In literature there are various types of linguistic sets that are applicable for different cases. The linguistic term set which is suggested for SF-based comparisons is given in Table 3. In the original methodology, Kutlu Gündoğdu and Kahraman [19] developed the linguistic scales corresponding to 1, 3, 5, 7, 9 and their reciprocals by accepting them as SI (score index) which is defined in Eq. (11) and reciprocal type in Eq. (12). Using the same formulas, we propose SF-based linguistic terms for the intermediate SI values (2, 4, 6, 8 and their reciprocals) as a contribution to the literature.

$$SI = \sqrt{\left|100 * \left[\left(\mu_{\tilde{A}_S} - \pi_{\tilde{A}_S}\right)^2 - \left(\upsilon_{\tilde{A}_S} - \pi_{\tilde{A}_S}\right)^2\right]\right|} \tag{11}$$

$$\frac{1}{SI} = \frac{1}{\sqrt{\left|100 * \left[\left(\mu_{\tilde{A}_S} - \pi_{\tilde{A}_S}\right)^2 - \left(\upsilon_{\tilde{A}_S} - \pi_{\tilde{A}_S}\right)^2\right]\right|}} \tag{12}$$

Table 3. Linguistic terms for SF-AHP for the intermediate score index (SI) values

Linguistic terms for main SI values (Kutlu Gündoğdu et al. 2019a)			*Linguistic terms for intermediate SI values*	
SI	*Meaning*	(μ, υ, π)	*SI*	(μ, υ, π)
9	Absolutely more importance	(0.9, 0.1, 0)	8	(0.85, 0.15, 0.05)
7	Very high importance	(0.8, 0.2, 0.1)		
			6	(0.75, 0.25, 0.15)
5	High importance	(0.7, 0.3, 0.2)		
			4	(0.65, 0.35, 0.25)
3	Slightly more importance	(0.6, 0.4, 0.3)		
			2	(0.55, 0.45, 0.35)
1	Equally importance	(0.5, 0.4, 0.4)		
			1/2	(0.45, 0.55, 0.35)
1/3	Slightly low importance	(0.4, 0.6, 0.3)		
			1/4	(0.35, 0.65, 0.25)
1/5	Low importance	(0.3, 0.7, 0.2)		
			1/6	(0.25, 0.75, 0.15)
1/7	Very low importance	(0.2, 0.8, 0.1)		
			1/8	(0.15, 0.85, 0.05)
1/9	Absolutely low importance	(0.1, 0.9, 0)		

Step 3. Collecting the comparison preferences from the expert group

The DMs constitute the pairwise comparisons using SF linguistic terms given in Table 3 and build the evaluation matrices. In a group decision making environment, the traditional AHP takes the geometric averages of each comparison judgment to construct the aggregated comparison matrices. Similarly, the aggregation in SF-AHP can be made by performing SWAM operator. The SF-AHP include this aggregation process by considering the DMs' weights $w = (w_1, w_2, \ldots, w_n)$ representing their expertise and knowledge level. Rather than aggregating the individualistic comparison matrices, group consensus approach can be performed.

Step 4. Controlling the consistency ratio of each comparison

The comparisons are based on individual preferences of DMs. The inconsistency of a comparison matrix should be checked to make a more representative and consistent decision. Saaty proposed an eigenvector-based consistency measurement which is called *CR* (consistency ratio) [21]. Abdullah and Najib stated that π hesitancy degree of the aggregated three-dimensional fuzzy set based comparison matrix can be used while calculating inconsistency degree of each individualistic comparison matrix [34].

Classical definition of *CR* is given $CR = \frac{(\lambda_{max}-n)/(n-1)}{RI}$ where n is the size of matrix and *RI* is taken from Saaty's Random Index table. In this formula, $(\lambda_{max} - n)$ is assumed to be the average value of the hesitancy degrees π_i in Abdullah and Najib (2016). To make it convenient to SF-AHP, the formula is modified as $(\lambda_{max} - n) = \sqrt{\sum_i \pi^2_{\tilde{A}_{Si}}/n}$. As a result, we can give modified *CR* formula for SF-AHP in Eq. (13) as a contribution to the literature.

$$CR = \frac{\left(\sqrt{\sum_i \pi^2_{\tilde{A}_{Si}}/n}\right)/(n-1)}{RI} \tag{13}$$

If CR is less than 0.10, it can be assumed that the comparison consists of tolerable inconsistency.

Step 5. Calculation of the SF local weights of criteria and alternative

This step consists of the determination of the local importance values related to criteria, sub-criteria with respect to the related criteria and alternatives with respect to each criterion. For each comparison matrix, SWAM operator which is given in Eq. (9) is performed in order to compute the SF weights. In this formula, it is assumed that $w_1 = w_2 = \cdots = w_n = 1/n$ [19].

Step 6. Obtaining the SF global weights

After calculation of the local weights, global weights are obtained by following the hierarchical sequence. The local SF weights which are determined at each level are aggregated with the aim of ranking the alternatives. The sequence of the computations is begun from top level (main criteria) and completed with bottom level (alternatives). For the computation of global weights, there are 2 proposed approaches: the first one considers an early defuzzification of the local weights while the second continues with SF values. In order to keep the process convenient to spherical fuzzy logic, we prefer following the second one. In this case, SF global importance values are obtained by using Eq. (14).

$$\prod\nolimits_{j=1}^{n} \tilde{A}_{S_{ij}} = \tilde{A}_{S_{i1}} \otimes \tilde{A}_{S_{i2}} \otimes \ldots \otimes \tilde{A}_{S_{in}} \tag{14}$$

The final SF-AHP score ($\tilde{F}$), for each alternative is computed by performing the fuzzy arithmetic addition over each SF global importance values as given in Eq. (15).

$$\sum\nolimits_{j=1}^{n} \tilde{A}_{S_{ij}} = \tilde{A}_{S_{i1}} \oplus \tilde{A}_{S_{i2}} \oplus \ldots \oplus \tilde{A}_{S_{in}} \tag{15}$$

Step 7. Computing defuzzified values of each alternative's SF weight

As a result of the previous step, the global AHP scores are still shown as SF numbers. They are defuzzified in this step for ranking. Defuzzification of the SF numbers are based on score function given in Eq. (16).

$$S\left(\tilde{w}_j^S\right) = \sqrt{\left|100 * \left[\left(3\mu_{\tilde{A}_S} - \frac{\pi_{\tilde{A}_S}}{2}\right)^2 - \left(\frac{\upsilon_{\tilde{A}_S}}{2} - \pi_{\tilde{A}_S}\right)^2\right]\right|} \quad (16)$$

Step 8. Determination of the alternatives' priorities

As the conclusion of the whole process, SF-AHP gives a preference order of alternatives according to defuzzified final scores ($S\left(\tilde{w}_j^S\right)$). These values are ranked in descending order and called priorities. Here, the highest priority specifies the best alternative and the lowest one shows the worst alternative.

6 Construction Contract Selection Through SF-AHP

The aim of this study is to propose an MCDM approach for contractor selection problem of construction firms. This model uses a SF-AHP approach in order to show the applicability of SF sets in industrial real cases and the modelling strength regarding the representation of vagueness and imprecision.

In Step 1, the construction contract selection criteria set is depicted in Fig. 1. This hierarchical structure is the base of the SF-AHP method since the aggregation in the final step is made by considering these relationships. In this real life application, the hierarchy is formed in 4 levels: level 0 represents the goal of the problem: "Selection of the best contract type"; level 1 and 2 show the main criteria and sub-criteria sets, respectively; and finally level 3 depicts 6 prespecified alternatives.

Step 2 includes the determination of the linguistic terms that are used by the DMs while expressing their comparison judgments. In this application, the joint comparison matrices were collected from a DM group which was constituted from 5 experts working in construction projects. Their expertise originated from their experience and involvement in developing construction contracts during most of their careers. So, Step 3 covers the collection of the comparison results of the group so that they provide a consensual input.

The consistency of each matrix is computed and then, checked against inconsistency rule forcing CR be lower than or equal to 0.10 [21]. Equation (13) is used in Step 4 for this purpose.

Step 5 involves the calculation of SF local weights of criteria, sub-criteria, and alternatives level by level. Firstly, the main criteria (level 1) are compared pairwise with respect to the goal of selecting the best contract type (level 0). Then, 5 comparison matrices are built and analyzed in order to find the SF local weights of sub-criteria (level 2) with respect to the related main criteria (level 1). Finally, the six alternatives (level 3) are compared pairwise with respect to sub-criteria (level 2) so that 18 different matrices

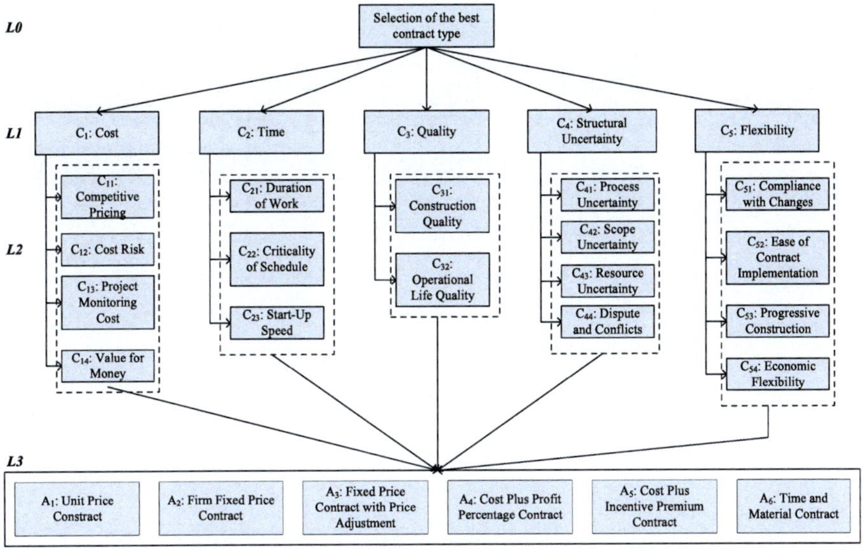

Fig. 1. Hierarchy of the contractor selection problem.

are evaluated: 4 sub-criteria of Cost, 3 sub-criteria of Duration, 2 sub-criteria of Quality, 5 sub-criteria of Uncertainty, and 4 sub-criteria of Flexibility.

As a depiction, the calculation of the local weights of main criteria are given as follows. Table 4 shows the linguistic term assignments of the DM group. The intermediate linguistic terms corresponding to 2,4,6,8 are depicted by a label of "between …, …", e.g. 4 is between 3 and 5, and depicted as "between SMI, HI". Corresponding SF numbers are given in Table 5.

Table 4. Comparison matrix of main criteria

	C_1	C_2	C_3	C_4	C_5
C_1	EI	SMI	HI	VHI	Between VHI, AMI
C_2	SLI	EI	HI	VHI	Between VHI, AMI
C_3	LI	LI	EI	SMI	SMI
C_4	VLI	VLI	SLI	EI	Between SMI, HI
C_5	Between ALI, VLI	Between ALI, VLI	SLI	Between LI, SLI	EI

In order to calculate the three parameters of SF local weights of each main criterion, SWAM operator given in Eq. (9) is run. The SF local weight of C_1 (Cost):

$$\mu = \left[1 - \prod_{i=1}^{5}\left(1 - \mu_{\tilde{A}_{Si}}^{2}\right)^{w_i}\right]^{\frac{1}{2}}$$

Table 5. SF comparison matrix of main criteria

	C_1			C_2			C_3			C_4			C_5		
C_1	0.5	0.4	0.4	0.6	0.4	0.3	0.7	0.3	0.2	0.8	0.2	0.1	0.85	0.15	0.05
C_2	0.4	0.6	0.3	0.5	0.4	0.4	0.7	0.3	0.2	0.8	0.2	0.1	0.85	0.15	0.05
C_3	0.3	0.7	0.2	0.3	0.7	0.2	0.5	0.4	0.4	0.6	0.4	0.3	0.6	0.4	0.3
C_4	0.2	0.8	0.1	0.2	0.8	0.1	0.4	0.6	0.3	0.5	0.4	0.4	0.65	0.35	0.25
C_5	0.15	0.85	0.05	0.15	0.85	0.05	0.4	0.6	0.3	0.35	0.65	0.25	0.5	0.4	0.4

$$= \left[1 - \left(1 - 0.5^2\right)^{0.2} * \left(1 - 0.6^2\right)^{0.2} * \left(1 - 0.7^2\right)^{0.2} * \left(1 - 0.8^2\right)^{0.2} * \left(1 - 0.85^2\right)^{0.2}\right]^{1/2} = 0.7238$$

$$\upsilon = \prod_{i=1}^{5} \upsilon_{\tilde{A}_{Si}}^{w_i} = 0.4^{0.2} * 0.4^{0.2} * 0.3^{0.2} * 0.2^{0.2} * 0.15^{0.2} = 0.2702$$

$$\pi = \left[\prod_{i=1}^{5}\left(1 - \mu_{\tilde{A}_{Si}}^2\right)^{w_i} - \prod_{i=1}^{5}\left(1 - \mu_{\tilde{A}_{Si}}^2 - \pi_{\tilde{A}_{Si}}^2\right)^{w_i}\right]^{\frac{1}{2}}$$

$$= \left[\left(1 - 0.5^2\right)^{0.2} * \left(1 - 0.6^2\right)^{0.2} * \left(1 - 0.7^2\right)^{0.2} * \left(1 - 0.8^2\right)^{0.2} * \left(1 - 0.85^2\right)^{0.2} - \left(1 - 0.5^2 - 0.4^2\right)^{0.2} * \left(1 - 0.6^2 - 0.3^2\right)^{0.2} * \left(1 - 0.7^2 - 0.2^2\right)^{0.2} * \left(1 - 0.8^2 - 0.1^2\right)^{0.2} * \left(1 - 0.85^2 - 0.05^2\right)^{0.2}\right]^{1/2}$$

$$= 0.2149$$

After the analysis of the whole pairwise comparison matrices, the SF local weights are calculated accordingly as given in Table 6.

Table 6. SF local weights of all items

Main criteria	*SF local weights*	*Sub-criteria*	*SF local weights*
C_1	(0.7238, 0.2702, 0.2149)	C_{11}	(0.5125, 0.4450, 0.3300)
		C_{12}	(0.7103, 0.2783, 0.2178)
		C_{13}	(0.3503, 0.6285, 0.2706)
		C_{14}	(0.5540, 0.4162, 0.3001)
C_2	(0.7052, 0.2930, 0.2128)	C_{21}	(0.7605, 0.2289, 0.1929)
		C_{22}	(0.4529, 0.5241, 0.3305)

(*continued*)

Table 6. (*continued*)

Main criteria	*SF local weights*	*Sub-criteria*	*SF local weights*
		C_{23}	(0.4047, 0.5718, 0.3266)
C_3	(0.4894, 0.5004, 0.3007)	C_{31}	(0.4018, 0.5477, 0.3222)
		C_{32}	(0.6536, 0.3162, 0.2793)
C_4	(0.4431, 0.5573, 0.2736)	C_{41}	(0.5526, 0.4281, 0.3364)
		C_{42}	(0.5925, 0.3869, 0.3137)
		C_{43}	(0.4303,0.5412,0.3353)
		C_{44}	(0.4604, 0.5147, 0.3364)
C_5	(0.3462, 0.6462, 0.2700)	C_{51}	(0.5654, 0.4120, 0.3382)
		C_{52}	(0.4521, 0.5191, 0.3543)
		C_{53}	(0.5084, 0.4427, 0.3549)
		C_{54}	(0.4923, 0.4559, 0.3677)

In Step 6, the SF local weights are aggregated by performing the multiplication operation given in Eq. (14) and these aggregated weights are called global SF weights. By multiplying the local weight of a main criterion and the local weights of its sub-criteria one by one, the global weight of each sub-criterion is computed. For illustration, the global SF weight of C_{11} (Competitive Price):

$$\tilde{A}_{S_1} \otimes \tilde{A}_{S_{11}} = \left\{ \mu_{\tilde{A}_{S_1}} \mu_{\tilde{A}_{S_{11}}}, \left(v^2_{\tilde{A}_{S_1}} + v^2_{\tilde{A}_{S_{11}}} - v^2_{\tilde{A}_{S_1}} v^2_{\tilde{A}_{S_{11}}} \right)^{\frac{1}{2}} \left(\left(1 - v^2_{\tilde{A}_{S_{11}}}\right) \pi^2_{\tilde{A}_{S_1}} + \left(1 - v^2_{\tilde{A}_{S_1}}\right) \pi^2_{\tilde{A}_{S_{11}}} - \pi^2_{\tilde{A}_{S_1}} \pi^2_{\tilde{A}_{S_{11}}} \right)^{\frac{1}{2}} \right\}$$

$$\mu_{\tilde{A}_{S_1}} \mu_{\tilde{A}_{S_{11}}} = 0.7238 * 0.5125 = 0.3710$$

$$\left(v^2_{\tilde{A}_{S_1}} + v^2_{\tilde{A}_{S_{11}}} - v^2_{\tilde{A}_{S_1}} v^2_{\tilde{A}_{S_{11}}} \right)^{\frac{1}{2}} = \left(0.2702^2 + 0.4455^2 - 0.2702^2 * 0.4455^2\right)^{1/2} = 0.5065$$

$$\left(\left(1 - v^2_{\tilde{A}_{S_{11}}}\right) \pi^2_{\tilde{A}_{S_1}} + \left(1 - v^2_{\tilde{A}_{S_1}}\right) \pi^2_{\tilde{A}_{S_{11}}} - \pi^2_{\tilde{A}_{S_1}} \pi^2_{\tilde{A}_{S_{11}}} \right)^{\frac{1}{2}} = \left(\left(1 - 0.4455^2\right) * 0.2149^2 + \left(1 - 0.2702^2\right) * 0.3300^2 - 0.2149^2 * 0.3300^2 \right)^{1/2} = 0.3647$$

After the global weights of all the sub-criteria are calculated, the global weights of each alternative with respect to each sub-criterion are determined by using Eq. (14). The total aggregated SF global weights of alternatives are then calculated by performing the addition formula given in Eq. (15) and indicated in Table 7.

In order to rank the alternatives according to their importance, Step 7 defuzzifies the aggregated SF local weights into score values (Eq. 16). In Step 8, the alternatives are ranked in descending order and these orders are called priorities. Score values and

Table 7. Total aggregated SF global weights of alternatives

Alternatives	*Aggregated SF local weight*	*Score value*	*Priority*
A_1	(0.6122, 0.0045, 0.7387)	12.6913	5
A_2	(0.7065, 0.0031, 0.6547)	16.6890	1
A_3	(0.6484, 0.0037, 0.7119)	14.2164	2
A_4	(0.6303, 0.0025, 0.7348)	13.3540	4
A_5	(0.6480, 0.0021, 0.7222)	14.0913	3
A_6	(0.5776, 0.0053, 0.7641)	11.1586	6

priorities are given in Table 7. For illustrating the defuzzification operation, an example is given as follows.

$$S\left(\tilde{w}_1^S\right) = S(0.6122, 0.0045, 0.7387)$$

$$= \sqrt{\left|100 * \left[\left(3 * 0.6122 - \frac{0.7387}{2}\right)^2 - \left(\frac{0.0045}{2} - 0.7387\right)^2\right]\right|}$$

$$= 12.6913$$

The priority order of alternatives is $A_2 \succ A_3 \succ A_5 \succ A_4 \succ A_1 \succ A_6$ or "Firm Fixed Price Contract ≻ Fixed Price Contract with Adjustment ≻ Cost Plus Incentive Premium Contract ≻ Cost Plus Profit Percentage Contract ≻ Unit Price Contract ≻ Time and Material Contract". Accordingly, Firm Fixed Price Contract is found as the best alternative for this case.

Additionally, the weights of main criteria or sub-criteria can be determined by performing the score value function if required. As an example, the score value of the first main criteria (Cost: A_1) can be computed as given below.

$$S(0.7238, 0.2702, 0.2149)$$

$$= \sqrt{\left|100 * \left[\left(3 * 0.7238 - \frac{0.2149}{2}\right)^2 - \left(\frac{0.2702}{2} - 0.2149\right)^2\right]\right|}$$

$$= 20.6246$$

The score values of the remaining main criteria are as follows: 20.0811 for Time, 13.1683 for Quality, 11.9242 for Structural Uncertainty, and 9.0195 for Flexibility. So, the priority order of the main criteria is "Cost ≻ Time ≻ Quality ≻ Structural Uncertainty ≻ Flexibility". Sub-criteria may be ordered in the same fashion for gathering additional information.

7 Conclusions

Contracts are the means by which the employers and contractors determine their rights and obligations enforceable by law. The type of construction contracts are often determined by the employers or owners of the projects before inviting contractors for bidding. Construction is an increasingly complex and uncertain industry where different stakeholders have different priorities and perceptions where the fuzzy multi-criteria decision making can be applied. In this study a spherical fuzzy AHP model is developed using 5 primary and 17 secondary selection criteria based on literature review and expert opinions with the aim of selecting the most suitable contract type that can be used in any given construction project. Among the six different contract types examined, the Firm Fixed Price Contract (FFPC) stands out as the most appropriate type of contract according to experts participated in this study. The second most appropriate type of contract was the Fixed Price Contract with Price Adjustment (FPCPA). Coming last, the Time and Material Contract (TMC) proved to be the least used type of contract in the construction sector. Among the main criteria, Cost and Time are the first and second most important ones with Cost having a slight edge over Time. However, the other three have significantly high score values that imply that all five criteria play important roles in selection.

In real life, priorities of criteria and alternatives are less distinct than what crisp multi-criteria decision tools suggest. As shown in this study, SF-AHP methodology provides decision-makers a less distinct set of scores of criteria and alternatives, which would enable the decision-makers to select different alternatives when there are subtle changes in the decision environment, rather than certain criteria and alternatives dominating the decisions. In this study, we contribute to the literature through three propositions:

i. SF version of AHP is chosen to represent the DMs' hesitancy more extensively and this is the first application in the literature of construction management.
ii. In the original SF-AHP, there are only SF numbers correspondences for main evaluation terms. In this study, the intermediate linguistic terms (2,4,6,8) and their SF numbers correspondences are defined for the first time in the literature.
iii. A novel consistency ratio equation is developed for SF-AHP.

In future researches, the technical aspects of SF-AHP should be analyzed in new real-life applications. For example, a SF-ANP (Analytic Network Process) can be developed and used since AHP does not consider the relations among selection criteria. The proposed intermediate linguistic terms and novel consistency ratio should be checked in other applications and the robustness can be analyzed through comparative analysis using different multi-criteria decision-making methods.

References

1. Hillebrandt, P.M.: Economic Theory and the Construction Industry. 3rd ed. Macmillan Press Ltd., Houndmills, Basingstoke, Hampshire (2000)
2. El-Reedy, M.A.: Construction Management for Industrial Projects: A Modular Guide for Project Managers. Wiley, Hoboken (2011)

3. Oxford Dictionaries: Definition of contract in English. https://en.oxforddictionaries.com/definition/contract. Accessed 03 Apr 2018
4. Piotrowski, C.M.: Professional Practice for Interior Designers, 5th edn. Wiley, Hoboken (2013)
5. Charrett, D.: The Application of Contracts in Engineering and Construction Projects, 1st edn. Informa Law from Routledge, London (2018)
6. Hasnain, M., Thaheem, M.J., Ullah, F.: Best value contractor selection in road construction projects: ANP-based decision support system. Int. J. Civ. Eng. **16**(6), 695–714 (2017)
7. Krishna Rao, M.V., Kumar, V.S., Kumar, P.R.: Optimal contractor selection in construction industry: the fuzzy way. J. Inst. Eng. India Ser. A **99**(1), 67–78 (2018)
8. Bintoro, I., Malani, R., Rihartanto: Modelling of contractor selection using fuzzy-TOPSIS. In: 5th International Conference on Electrical, Electronics and Information Engineering, Malang, Indonesia, 6–8 October 2018 (2018)
9. Taylan, O., Kabli, M.R., Porcel, C., Viedma, E.H.: Contractor selection for construction projects using consensus tools and big data. Int. J. Fuzzy Syst. **20**(4), 1267–1281 (2018)
10. Sabuncuoğlu, A.O., Görener, A.: Decision making through the fuzzy TOPSIS method: contractor selection in construction projects. Int. J. Res. Bus. Soc. Sci. **5**(2), 71–82 (2016)
11. Semaan, N., Salem, M.: A deterministic contractor selection decision support system for competitive bidding. Eng. Constr. Archit. Manag. **24**(1), 61–77 (2017)
12. Antoniou, F., Aretoulis, G.N., Konstantinidis, D., Kalfakakou, G.P.: Selection criteria used for the choice of contract type for major highway construction projects. Procedia Soc. Behav. Sci. **48**, 3508–3517 (2012)
13. Liang, R., Wang, J., Zhang, H.: Evaluation of e-commerce websites: an integrated approach under a single-valued trapezoidal neutrosophic environment. Knowl.-Based Syst. **135**, 44–59 (2017)
14. Kutlu Gündoğdu, F., Kahraman, C.: Spherical fuzzy sets and spherical fuzzy TOPSIS method. J. Intell. Fuzzy Syst. **36**(1), 337–352 (2019)
15. Yager, R.R.: Pythagorean fuzzy subsets. In: Proceedings of Joint IFSA World Congress and NAFIPS Annual Meeting, Edmonton, Canada, 24–28 June 2013, pp. 57–61 (2013)
16. Smarandache, F.: A Unifying Field in Logics Neutrosophy: Neutrosophic Probability, Set and Logic. American Research Press, Rehoboth (1999)
17. Kutlu Gündoğdu, F.: Principals of spherical fuzzy sets. In: Kahraman, C., Cebi, S., Cevik Onar, S., Oztaysi, B., Tolga, A., Sari, I. (eds.) Intelligent and Fuzzy Techniques in Big Data Analytics and Decision Making (INFUS 2019). Advances in Intelligent Systems and Computing, İstanbul, Turkey, 23–25 July 2020, vol. 1029, pp. 15–23 (2020)
18. Kutlu Gündoğdu, F., Kahraman, C.: A novel fuzzy TOPSIS method using emerging interval-valued spherical fuzzy sets. Eng. Appl. Artif. Intell. **85**, 307–323 (2019)
19. Kutlu Gündoğdu, F., Kahraman, C.: A novel spherical fuzzy analytic hierarchy process and its renewable energy application. Soft Comput. **24**, 4607–4621 (2020)
20. Kutlu Gündoğdu, F., Kahraman, C.: Extension of WASPAS with spherical fuzzy sets. Informatica **30**(2), 269–292 (2019)
21. Saaty, T.L.: The Analytic Hierarchy Process. RWS Publications, Pittsburgh (1980)
22. Saaty, T.L.: How to make a decision: the analytic hierarchy process. Eur. J. Oper. Res. **48**(1), 9–26 (1990)
23. Soner, O., Celik, E., Akyuz, E.: Application of AHP and VIKOR methods under interval type 2 fuzzy environment in maritime transportation. Ocean Eng. **129**, 107–116 (2017)
24. Wu, Y., Xu, C., Ke, Y., Tao, Y., Li, X.: Portfolio optimization of renewable energy projects under type-2 fuzzy environment with sustainability perspective. Comput. Ind. Eng. **133**, 69–82 (2019)

25. Ouyang, X., Guo, M.: Intuitionistic fuzzy analytical hierarchical processes for selecting the paradigms of mangroves in municipal wastewater treatment. Chemosphere **197**, 634–642 (2018)
26. Nirmala, G., Uthra, G.: AHP based on triangular intuitionistic fuzzy number and its application to supplier selection problem. Mater. Today Proc. **16**, 987–993 (2019)
27. Özkan, B., Özceylan, E., Kabak, M., Dağdeviren, M.: Evaluating the websites of academic departments through SEO criteria: a hesitant fuzzy linguistic MCDM approach. Artif. Intell. Rev. **53**, 875–905 (2020)
28. Acar, C., Beskese, A., Temur, G.T.: Sustainability analysis of different hydrogen production options using hesitant fuzzy AHP. Int. J. Hydrogen Energy **43**(39), 18059–18076 (2018)
29. Ilbahar, E., Karaşan, A., Cebi, S., Kahraman, C.: A novel approach to risk assessment for occupational health and safety using Pythagorean fuzzy AHP & fuzzy inference system. Saf. Sci. **103**, 124–136 (2018)
30. Gul, M.: Application of Pythagorean fuzzy AHP and VIKOR methods in occupational health and safety risk assessment: the case of a gun and rifle barrel external surface oxidation and colouring unit. Int. J. Occup. Saf. Ergon. (2018). https://doi.org/10.1080/10803548.2018.1492251
31. Abdel-Basset, M., Mohamed, M., Sangaiah, A.K.: Neutrosophic AHP-Delphi Group decision making model based on trapezoidal neutrosophic numbers. J. Ambient Intell. Human. Comput. **9**, 1427–1443 (2018)
32. Tey, D.J.Y., Gan, Y.F., Selvachandran, G., Quek, S.G., Smarandache, F., Son, L.H., Abdel-Basset, M., Long, H.V.: A novel neutrosophic data analytic hierarchy process for multi-criteria decision making method: a case study in Kuala Lumpur stock exchange. IEEE Access **7**, 53687–53697 (2019)
33. Kutlu Gündoğdu, F., Kahraman, C.: Spherical fuzzy analytic hierarchy process (AHP) and its application to industrial robot selection. In: Kahraman, C., Cebi, S., Cevik Onar, S., Oztaysi, B., Tolga, A., Sari, I. (eds.) Intelligent and Fuzzy Techniques in Big Data Analytics and Decision Making (INFUS 2019). Advances in Intelligent Systems and Computing, İstanbul, Turkey, 23–25 July 2020, vol. 1029, pp. 988–996 (2020)
34. Abdullah, L., Najib, L.: Sustainable energy planning decision using the intuitionistic fuzzy analytic hierarchy process: choosing energy technology in Malaysia. Int. J. Sustain. Energy **35**, 360–377 (2016)

The Design of a Novel Torque Wrench Based on TRIZ Decision Procedures

Kok Weng Ng[1], Mei Choo Ang[2(✉)], and Ismail Mostafa Mohamed Mahdy[1]

[1] Department of Mechanical, Materials and Manufacturing, Faculty of Science and Engineering, University of Nottingham Malaysia, Jalan Broga, Semenyih, Malaysia
[2] Institute of IR 4.0, Universiti Kebangsaan Malaysia, Bangi, Malaysia
amc@ukm.edu.my

Abstract. As rapid industrial growth continues, the need for accuracy and efficiency becomes ever more critical. This has led to an increasing demand for design innovations and the continuous improvement process of products have become inevitably critical. A torque wrench is an instrument developed in the 1930′s with the purpose of tightening fasteners, nuts and bolts with specific torque. Although a torque wrench with measurement and additional features such as adjustable grip features were introduced later to the market, the demands for a torque wrench with additional features were not high. The basic torque wrench without any measurement and grip adjustment features still dominate the market even though it is critical to apply the precise torque using a precise grip of the wrench when tightening fasteners, bolts and nuts for different joints to avoid loose or jammed joints. Therefore, this research work is about applying decision procedures from systematic innovation techniques such as the theory of inventive problem solving (TRIZ) to innovate and improve the design of a torque wrench (which is already a mature product and has not undergone much improvement for many years).

Keywords: Torque wrench · Design · TRIZ · Decision procedures

1 Introduction

Common hand tools are increasingly critical equipment to technicians and to the public. These common hand tools are found in any household, ranging from wrenches, screws, screwdrivers to even power tools such as drills and saws. These hand tools were once considered to be innovative instruments. The engineering of such instruments has helped to save production and assembly time in the manufacturing industry, and this is of utmost importance. Among these tools, it is common to find wrenches in any household's toolbox. All sorts of different types of wrenches with different sizes and grips were created to serve certain purposes.

The type of wrenches that this research work will be focusing on are the torque wrenches. A torque wrench is a wrench that is designed to tighten fasteners, nuts, and bolts to a pre-determined torque value, and this is critical to avoid damage caused by over-tightening or joints coming apart due to under-tightening. It is very critical to determine

T. Allahviranloo et al. (Eds.): IDS 2020, AISC 1301, pp. 548–559, 2021.
https://doi.org/10.1007/978-3-030-66501-2_43

the exact torque applied in tightening a fastener particularly in industries that deal with precious fuels and natural gas that flow in pipelines such as oil or gas companies where all fasteners must be tightened with the exact same torque to avoid any leakage from their connections. In addition to that, there are also industries such as the automotive industry where the tightening torque required for certain fasteners (in a motor block) is very specific. So, a torque wrench should have a mechanism that enables the measuring of the torque being applied to a fastener. There are torque wrenches with such measurement features or capabilities, but these types of torque wrenches are not in high demand. The common and widely demanded torque wrench are without a measuring feature and are made to tighten to a specific torque.

With different requirements of torque for different fasteners, the user needs to use the appropriate torque wrench that can deliver the right torque to tighten a fastener. This is not only an inconvenience, but it also costs more and is prone to human errors such as using the wrong torque wrench. In addition to that, the scenario of different torque wrenches for different joints also leads to the emergence of many torque wrench models.

This research work will not just apply TRIZ decision procedures to design an alternative model of torque wrench that can tighten fasteners and joints with different torque requirements and sizes of fasteners but will also explore and explain the history of this instrument, its variations, and their differences in mechanisms as well as the issues related to these torque wrenches. Subsequently, the possibility of combining more than one wrench mechanism is also explored to improve the accuracy and their range of application.

2 First Torque Wrenches

The first torque wrench as shown in Fig. 1 was patented by John H. Sharp [1]. The patent was filed on May 18th, 1931 [1]. The torque wrench by Sharp can be improved by linking a scale to the ring location of a normal wrench and by applying a force perpendicular to the handle. The weight of the force will be indicated by the scale in kg and using Eq. (1), this force can be converted to torque.

$$Torque = Weight * g * perpendicular\ distance \quad (1)$$

The scale would record the weight with g representing the acceleration of a free-falling body and the perpendicular distance would be easily measured from the pivot.

For Sharp's wrench, a dial assembled to a housing is used to indicate the torque. This housing consists of a clock arm (an indicator) which is attached to a pinion gear. This pinion gear is connected to another interim gear. The interim gear is joined to a gear-like part tied to a specific section that contains a shank, the arms and the jaw.

Sharp's wrench is held at the handle. As the person tightens the nut, the jaw of the wrench that holds the nut and the section where the arms apply a specific force will turn and this specific force will also be applied to the springs. One of the ingenuities of this design is that the forces applied by the arm from the torque will always have the same perpendicular distance pivot which falls on the centre line of the nut and the handle.

The deflection/compression and the spring's reaction force have a linear relationship as shown in Eq. (2).

$$Force = spring\ stiffnes * (L_{\circ} - L) \tag{2}$$

The magnitude of the force employed by either the arm or the spring will always be known. The torque applied on the nut can be determined using Eq. (1). The pointer can point to the correct torque by adjusting and calibrating the interim gear, arm and dial accordingly.

In addition to that, springs can be rapidly adjusted or easily changed by the screw cork-like part to apply a different range of torques. This design demonstrates the simplicity of design concept; however, Sharp's wrench is far from perfect from a practical and effectiveness perspective. Sharp mulled over using an adjustable jaw portion like the monkey wrenches. However, he did not include the adjustable jaw portion in his original design as the accuracy of the adjustable tool tend to be less. Today, Sharp's wrench would not be practical as it is cumbersome and would need a substantial amount of maintenance as the springs could lose its elasticity, although the wrench was considered good back then.

One of the early inventors of the torque wrench was Bahr [2] who developed a ratcheting torque wrench while he was an employee of the New York City Water Department. It has been claimed that the torque wrench was invented by him to prevent over-tightening of fasteners on the water mains. Bahr and his partner, George Pfefferle, invented a ratchet torque wrench, which was considered to be a revolutionary instrument back then [2]. The design of Bahr's ratchet torque wrench as shown in Fig. 1 is a lot more sophisticated. Higher sophistication leads to higher cost. The reason for higher cost is due to the difficulty of machining the parts and the increased number of parts. Moreover, based on the number of parts in Bahr's ratcheting torque wrench, and the complexity of the design, the ratcheting torque wrench back then would not be reliable. For example, the wrench would likely malfunction if it was dropped on the floor after a few times. In addition to that, the assembly and disassembly of this ratcheting torque wrench would consume more time.

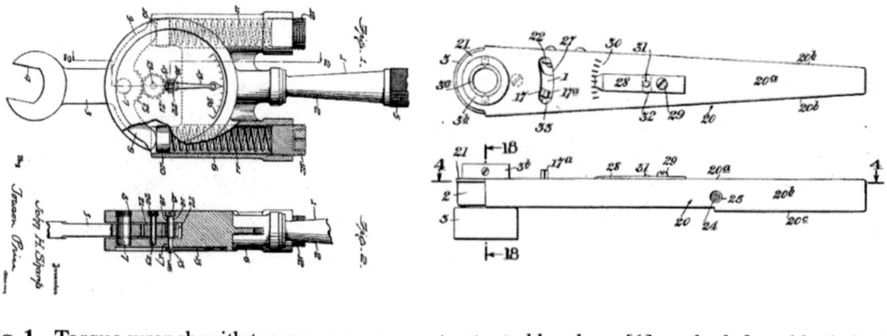

Fig. 1. Torque wrench with torque measurement patented by sharp [1] on the left and by bahr [2] on the right.

The wrench design by Bahr is superior when compared to Sharp's wrench. If we analyse the problem on creating the torque measuring wrench, we would find that the key requirement of a torque wrench is to tighten fasteners and bolts uniformly. For some engineers, it was exasperating to achieve this requirement for equal distribution of fastener stress. The amount of force required to be applied to each fastener was difficult for any worker to estimate accurately. Bahr's design is better because it provides a way to tighten a fastener to the required torque as well as prevent the over-tightening problem. The cost of manufacturing and maintenance is lower due to its design simplicity and the lower number of parts, therefore making it more profitable.

3 Present Day Torque Wrenches.

Today, torque wrenches are indispensable tools in any toolbox in a company or a personal garage. Currently, fasteners need to be accurately tightened in most machinery and this has led to the emergence of many new types of torque wrenches. Some wrenches just indicate the current applied torque, whilst for others, you could pre-set the needed torque and it would set off the alarm once the pre-set required torque was reached.

An adjustable wrench has a bar with meshed teeth mounted to the bottom of the moving side of the hub. This bar meshes with the thread in another cylinder which would be used to adjust the size of the clamp. This is the most basic design that can be found in anyone's garage or home. It is used not only to tighten fasteners, nuts, and bolts, but also to tighten round objects like pipes. With the movable side/bracket, it can close and grip any object from at least 2 sides and such a design is the most common for an adjustable clamp to hold a nut. However, this design is not ideal for torque measuring equipment because it has some drawback. First, such a design would be in contact with a maximum of 4 sides out of the 6, resulting in most of the force being applied to only two sides, which may cause some damage to the nut such as wearing out or rounding of its edges during tightening. The second problem of this design is the pivot location. As the one movable side either closes or opens for a different sized nut, the location of the pivot of the nut moves slightly off the main centre axis. This is because the jaws only move in one axis since only one jaw moves. In a normal tightening operation with a normal wrench, this would not be a problem. However, in a torque wrench, this design would create a small error as the perpendicular distance from the line of action of force would vary slightly. The next few sections will explore on TRIZ and how it can be applied in decision procedures approach to solve design problems, leading to the designing of a novel torque wrench.

4 Theory of Inventive Problem Solving (TRIZ)

TRIZ is an established methodology that has been applied by designers to solve various engineering and design problems [3–10]. TRIZ has been applied by many enterprises including Samsung, Intel, Apple and many more. The TRIZ engineering contradiction matrix is a methodology that is developed on the basis that in most engineering design problems, any improvement made on a design feature will usually cause one or more of other design features to worsen [11]. The matrix consists of 39 improving and worsening features; designers need to choose the appropriate contradictions related to their design problems and based on the matching contradicting features, several inventive principles will be recommended as a solution for the design problem. The TRIZ contradiction matrix has 39 improving and worsening features with a total of 40 inventive principles.

5 TRIZ Decision Procedures.

The application of TRIZ in design has been established for many years [12]. In designing a product that meets specific requirements, there will be design problems that need to be solved prior to the creation of a successful product. Most of the applications of TRIZ is based on the contradiction matrix; how it was applied in the 1980s and 1990s was mostly by translating a design problem into improving and worsening features before determining the inventive principles to solve the design problem [4]. Engineers then have to use their knowledge and experience to interpret the recommended inventive principles to the specific solutions to solve the design problem Although the approach of using the contradiction matrix directly will help to solve a design problem, there is a high risk of missing the actual root cause of the design problem and hence the notion of trying to solve a design problem based on a wrong root cause using the TRIZ contradiction matrix is highly likely to be unsuccessful. However, recent applications of TRIZ have stressed on the importance of identifying and determining the root cause of a design problem before the TRIZ contradiction matrix is used. Therefore, the entire process of solving design problems to create a novel torque wrench will be done based on a set of TRIZ decision procedures. These decision procedures are important to help decision makers determine the root cause of a design problem, apply the appropriate TRIZ tools and to make the right decisions when there is complexity in the decision problems, such as multiple objectives, uncertainties, risk and trade-offs [13].

There are many TRIZ tools being developed to date by TRIZ experts [11]. There is a need to help users grasp basic understanding of TRIZ decision procedures and appropriately apply these decision procedures to support their decisions. The TRIZ decision procedures taken to tackle issues and obstacles as well as the procedures taken to generate solutions are uniquely effective and consist of several steps. Most of the steps in the decision procedures are following the techniques for TRIZ inventive problem solving as shown in Fig. 2. As shown in Fig. 2, the basic TRIZ decision procedures consist of three main stages. The sub-stages of each are critical to ensure the success of the design of any new product which in this case is a new torque wrench. Only three TRIZ models (Physical Contradiction, Engineering Contradiction and Substance Field) with their corresponding tools (Separation, Satisfaction and Bypass, Contradiction Matrix and System of Standard Inventive Solutions) are shown in Fig. 2 although TRIZ has 25 tools to date.

5.1 Understanding the Problem

Understanding and defining the problem is the most important factor but is often ignored in problem solving. As explained by Gadd [11], most engineers start exploring solutions to a problem before they fully understand and define the problem. To fully understand

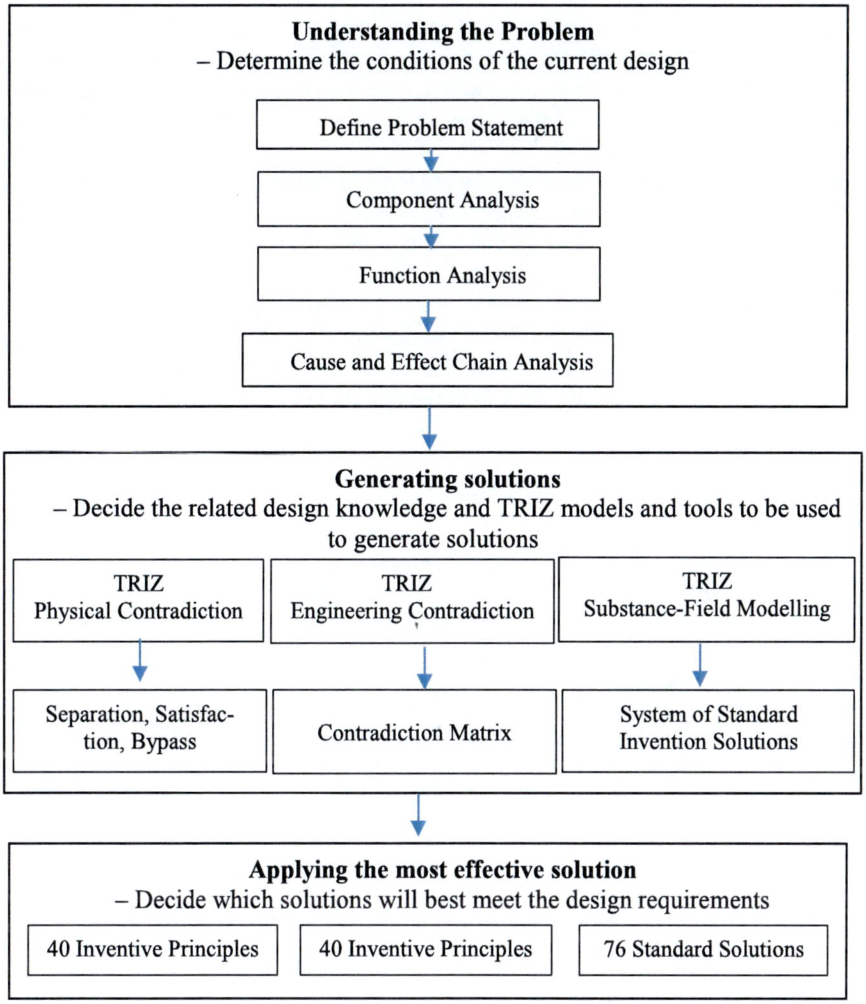

Fig. 2. Basic structure of TRIZ decision procedures

and define a design problem, engineers must determine the root cause of the design problem. This means the engineers must investigate and look for the actual cause of the design problem. The problem of over and under-tightening of fasteners emerged in the late 1800s when engineers like Bahr and Sharp were frustrated by the differences in tightening of different fasteners by different people, in other words the torque being applied to the fasteners were inconsistent. The component and function modelling were created as shown in Fig. 3 to illustrate the main components and functions between them in this design problem.

Working in companies like the water mains or an automotive assembly line, determining the torque applied to the fastener was crucial. The problem was that there was no direct way to measure the torque applied to a fastener. However, with the derivation of a way to measure force and using the Eq. (1), many solutions were developed and introduced. In this paper, a study was also conducted to understand the past and present solutions that were engineered. Many torque wrench models from the beginning to current times have been studied, while concepts and mechanisms have been thoroughly analysed and studied. Patents from the earlier wrenches to the present ones have been reviewed. The analysis and study on the past and present torque wrench can be summarised thusly, past wrenches were poorly designed with complicated, unreliable, and bulky features. The present wrenches are designed much better with more reliable and accurate measurement of torque and they are also much lighter, however, the cost of the wrenches is higher, and the complexity of the present wrenches are higher.

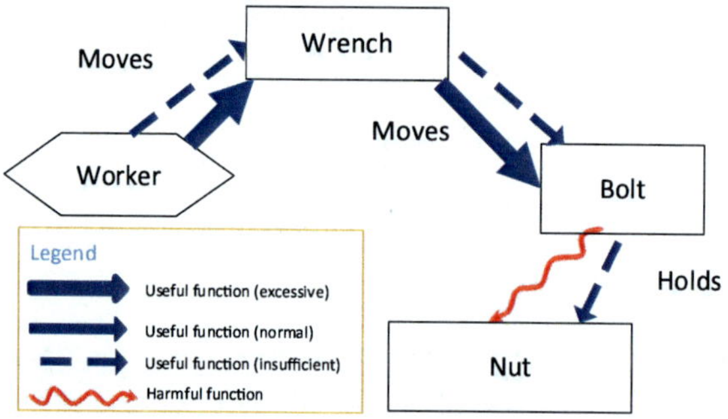

Fig. 3. Component and functioning modelling

5.2 Generating Solutions

In applying TRIZ contradiction matrix, it is important to identify the design problems related to the torque wrench. From the literature review and Cause and Effect Chain analysis (CEC) as shown in Fig. 4. Three main problems can be identified with the current torque wrench and they are as listed below:

1. Unable to ascertain the exact torque during tightening
2. If a torque wrench has a torque measurement feature, the mechanism of torque measurement is too sophisticated and hence would be costly and likely to be delicate.
3. The torque wrench is only designed for one size and is not adjustable and if it is adjustable, it can only be adjusted in a slow and cumbersome manner.

To apply TRIZ method to solve Problem 1, the problem for being unable to determine the exact torque during tightening can be translated into an engineering contradiction such as if the tightening is too much, the parts will be joined strongly but the thread may

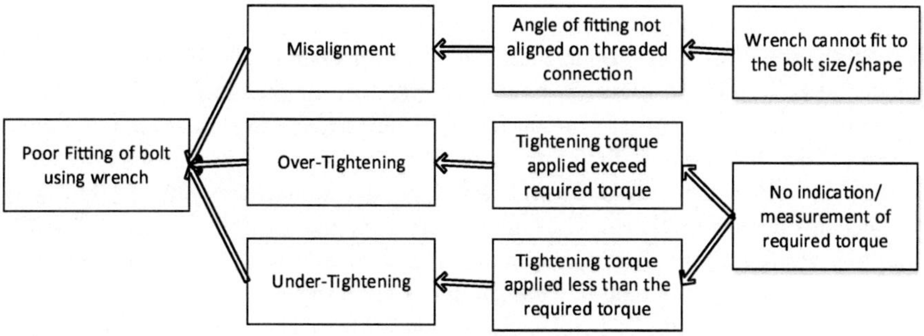

Fig. 4. Cause and Effect Chain analysis (CEC)

sustain damage. Therefore, the improving feature "strength" (Feature 14) can be selected to represent tightening too much and the worsening feature is "other harmful effects generated by system" (Feature 31), the inventive principle recommended are inventive principles: 15 (Dynamisation), 35 (Parameter Changes), 22 (Blessing in Disguise), and 2 (Taking out/Extraction).

Based on the inventive principle 15, the design will focus on applying the Dynamisation principle to enable the torque wrench to "dynamise" or move when the precise torque is reached. This means the mechanism can designed based on movement or "dynamisation" when the precise torque is achieved during the tightening of fasteners, bolts, and nuts. Therefore, based on the Dynamisation principle, the rigid or inflexible handle of the torque was redesigned to be adaptable or movable when the specified torque is attained during tightening. In the new torque wrench, when the precise torque is achieved, the cam mechanism will release the locking spring to allow the handle to move and prevent further tightening.

To solve problem 2, if the detection of torque is available then the torque can be determined precisely but the complexity of the wrench design increases. Based on the improving feature, Feature 37 (Ability to detect/Measure/Difficulty of detecting),will improve the detection feature but this will lead to the worsening feature, Feature 36 (Device Complexity). The recommended inventive principles are 15 (Dynamisation), 10 (Preliminary Action), 37 (Thermal Expansion), and 28 (Mechanics Substitution).

The Dynamisation solution is again suggested and this provides an idea on how to ascertain a precise torque can be achieved. Hence, the torque wrench was designed in a way that allowed adjustment of the torque setting by changing the length of the spring to attain the required torque by means of turning a spring cap (via rotation movement).

Problem 3 refers to the problem of catering to multiple nut sizes. A contradiction can be formed: If the catering of different nut sizes is available then the torque is adjustable to various sizes but the complexity of the wrench increases. The improving feature is Feature 35 (Adaptability/versatility) and the worsening feature is Feature 36 (device complexity). Based on the matrix, improving feature (Feature 35) against worsening feature (Feature 36) will produce recommended inventive principles: 35 (Parameter Changes), 28 (Mechanical Substitution), 6 (Universality), and 37 (Thermal expansion).

With these recommendations, Parameter Change and Universality were adopted to solve problem 3. Based on Parameter Change and Universality, the gripping mechanism design was developed to focus on how to change the parameters from axial gripping or clamping parameters (it is common for all torque wrench to grip or clamp onto two surfaces of the hex nut) to radial gripping or clamping parameters to enable the uniform gripping of various nut sizes on several surfaces of a hex nut or hex bolt. Therefore, the iris mechanism was introduced. The Universality principle implicates an idea of making a part to perform multiple functions and this idea when enhanced by the idea implicated by Parameter Change can facilitate the detail design of the iris mechanism.

Based on the inventive principles recommended by TRIZ decision procedures, the conceptual design of a new torque wrench was derived.

5.3 Applying the Most Effective Solution

The design solution involved the movement of a cam mechanism (dynamisation) to release the locking spring when the torque is reached. The special clamp is designed to be able to close on different sizes of hex nuts and bolts on all 6 sides. The new torque wrench is shown in Fig. 5.

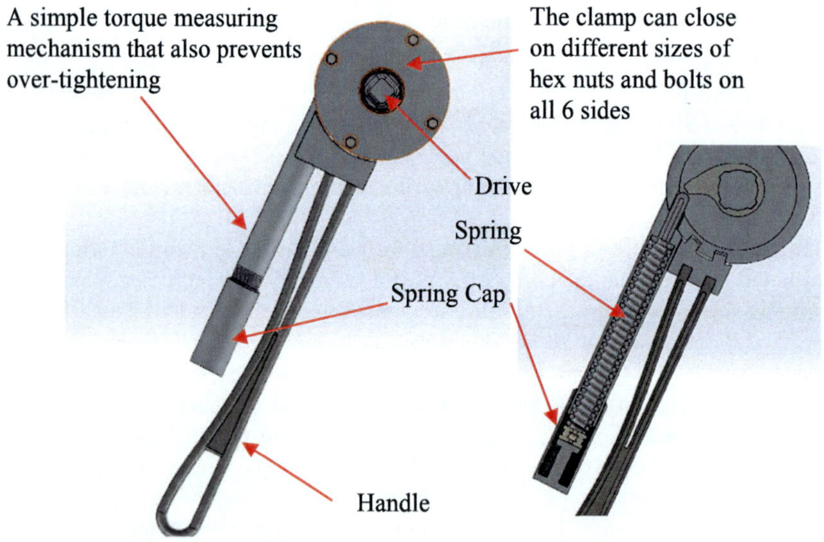

Fig. 5. The new torque wrench (left) and its cross-section view (right)

6 Results and Discussions

The entire design is made of Chrome-Vanadium steel AISI 6150. The entire design weighs approximately 6 kg. The wrench alone weighs approximately 2.5 kg. Chromium

Vanadium (Cr-V) steel is the selected material (choice of most hand tools (AISI 6150)) and is also the selected material for high performance springs (ASTM A231 AKA music wire). Cr-V was selected as it has a very high tensile strength.

6.1 Handle

Applying design study feature in Solidworks (a CAD system), the optimal dimensions of the handle were obtained for the handle (design study feature allows the user to define dimensions of the handle and using numerical analysis, the dimensions of the handle can be optimised to achieve the lightest weight with the strongest properties) and with all unwanted weight removed using topology optimisation feature of Solidworks. The weight of the entire handle is about 200 g. The handle is rigidly mounted from the 2 bars, which are attached to the middle part. The handle is designed to take a load of 200N from the furthest point of the handle as a worst-case scenario (maximum bending occurs at the furthest point away from the pivot). The load of 200 N was selected since it would create double the maximum torque set for the wrench.Using Eq. (1), the torque is calculated to be of 60Nm by multiplying the force of 200N with the perpendicular distance of 0.3m.

The design study results performed by Solidworks showed that the handle has a minimum factor of safety above 2 which is an acceptable margin, the fatigue study on the other hand shows that the handle can be loaded for 1011 times with a result of zero damage meaning it has an infinite life and would not fail.

6.2 Cam

The cam is designed to save as much weight as possible while maintaining enough stiffness and strength. The cam weighs approximately 100 g. The cam was designed to withstand high stresses to accommodate a stronger spring depending on the user's preferences.

6.3 Clamp Assembly

For the clamp assembly, a lot of weight reduction was also required to minimise the weight of the torque wrench. A design topology optimisation was carried out to shave off the unnecessary weight to achieve the minimum weight possible without compromising on the mechanical properties of the clamp.

A simulation study for the whole assembly was possible due to the perfect design fit and mates and the study showed that the safety margin and fatigue resistance were well above requirements. Although some concepts are not completely original as a lot of concept wrenches were designed in a similar way, none have been manufactured or patented based on the literature reviewed. The design of this novel torque wrench is reliable and easy to manufacture and use. When a force is applied from the wrench, the bore hole at the first disc would rotate in the tightening direction. This direction is the same direction that tightens the clamp and therefore, this same force is also applied to keep the clamp locked on the nut or bolt (Table 1).

Table 1. Comparison between the new torque wrench, Bahr's and Sharp's wrenches.

Characteristics	New torque wrench	Bahr's wrench	Sharp's wrench
Clamping adjustability	adjustable	Not adjustable	Not adjustable
Torque indication or alarm	audible	audible	dial
Over-tightening prevention	yes	yes	no

7 Conclusions

Using TRIZ decision procedures, the design was achieved while meeting the set of aims and the requirements, in which the torque wrench has a simple torque measuring mechanism and is able to prevent over-tightening. Moreover, the clamp can clamp and grip on different sizes of hex nuts and bolts on all 6 sides, preventing problems like: wearing out or rounding of edges and any movement of pivot during turning. Furthermore, the application of the inventive problem solving (TRIZ) has provided ideas on how to design a novel torque wrench to achieve the set of aims and requirements. The current design also has the flexibility to tighten a wider range of fasteners as the spring in the novel torque wrench can be readily changed.

Acknowledgment. The authors would like to thank the Ministry of Higher Education, Malaysia and Universiti Kebangsaan Malaysia for supporting the work through research grants FRGS/1/2018/TK03/UKM/02/6 and GUP-2018–124.

References

1. Sharp, J.: Torque Measuring Wrench US2007880A (1935).
2. Bahr, C., Pfefferle, G.: Torque Measuring Wrench US2074079A (1937).
3. Altshuller, G.: Innovation Algorithm:TRIZ, Systematic Innovation and Technical Creativity. Technical Innovation Center, Worcester (1999)
4. Altshuller, G.: 40 Principles: TRIZ Keys to Innovation. Technical Innovation Center, Worcester (2002)
5. Ang, M.C., Ng, K.W., Ahmad, S.A., Wahab, A.N.A.: An Engineering Design Support Tool Based on TRIZ. In: Zaman, H.B., Robinson, P., Olivier, P., Shih, T.K., Velastin, S. (eds.) IVIC 2013. LNCS, vol. 8237, pp. 115–127. Springer, Heidelberg (2013)
6. Ang, M.C., Ng, K.W., Cher, D.T.: Verifying Innovative solutions of TRIZ engineering contradiction matrix using substance-field analysis. Appl. Mech. Mater. **761**, 196–201 (2015)
7. Savransky, S.D.: Engineering of Creativity: Introduction to TRIZ Methodology of Inventive Problem Solving. CRC Press, USA (2000)
8. Terninko, J., Zusman, A., Zlotin, B.: Systematic Innovation: An Introduction to TRIZ (Theory of Inventive Problem Solving). CRC Press, USA (1998)
9. Yeoh, T.S., Yeoh, T.J., Song, C.L.: TRIZ - Systematic Innovation in Manufacturing Firstfruits Sdn Bhd, Malaysia (2009)
10. Ahmad, S.A., Ng, K.W., Airdzaman, S.H., Ang, M.C., Suliano, S.B.: Improving queuing system with limited resources using TRIZ. Int. J. Innovative Technol. Explor. Eng. **9**, 911–919 (2020)

11. Gadd, K.: TRIZ for Engineers: Enabling Inventive Problem Solving. Wiley, Singapore (2011)
12. Ng, K.W., Ang, M.C., Cher, D.T., Ahmad, S.A., Wahab, A.N.A.: Combining ARIZ with Shape Grammar to Support Designers. In: Badioze Zaman H. et al. (eds.) Advances in Visual Informatics. IVIC 2019. Lecture Notes in Computer Science 11870, pp. 305–317. Springer, Cham (2019).
13. Keeney, R.: Decision-analysis - an overview. Oper. Res. **30**, 803–838 (1982)

Using Best Worst Method for Ranking Criteria for Selection of the Best Transportation Mode During Pandemic

Yagmur Unal(✉) and Gül T. Temur

Bahcesehir University, Yildiz Mah. Ciragan Cad, Besiktas, 34349 Istanbul, Turkey
yagmur.unal@bahcesehir.edu.tr, gul.temur@eng.bau.edu.tr

Abstract. As an alternative novel model to analytical hierarchy process (AHP), the researchers have recently been attached high importance to apply Best Worst Method (BWM) in multi criteria decision making (MCDM) problems including conflicting criteria. BWM is a new methodology that decreases the consistency and makes it easier to conduct surveys among experts. During pandemic, it has been clearly seen that decision making procedures which are strong enough for highly uncertain environment becomes a very significant issue for surviving. Managers have been noticed that new factors should be taken into consideration during many conventional problems. Therefore, in this study, we goal to design a BWM based decision making methodology that helps to prioritize the criteria affecting transportation mode selection under pandemic. For this reason, (1) firstly, we aim to propose a comprehensive criteria set including not only traditional criteria but also pandemic based criteria, (2) apply surveys with experts to obtain their ideas shaped from pairwise comparisons, and (3) figure out the importance of each criterion by using BWM method. This study can be used as a guide by managers who want to have a strategic viewpoint for transportation mode selection under pandemic.

Keywords: Best Worst Method · Multi criteria decision making · Transportation mode selection

1 Introduction

Globalizing and growing economies have intensified transport activities within and between borders, generating a need for more integrated and dynamic transport systems. Creating these integrated logistics systems involves a range of complex and multi-functional decisions to support consumers from cost, quality, flexibility, and reliability perspectives (Petraška., 2018).

Moreover, choice of mode of transport is one of the most critical decisions of an effective transport system, as it has a direct effect on the transport of the right goods to the right location at the right time. In fact, the capacity of businesses to ship their goods without delay and in good condition provides a competitive advantage for them.

T. Allahviranloo et al. (Eds.): IDS 2020, AISC 1301, pp. 560–571, 2021.
https://doi.org/10.1007/978-3-030-66501-2_44

Even before the pandemic and its challenges were not in the equation, the choice of transportation mode had already demanded the incorporation of different aspects of choice, which makes it an inherently multi-criteria decision-making problem.

AHP and ANP have appeared to be the most common MCDM techniques for transport mode selection problems, so it is important to note that the use of Best Worst Method (BWM) is fairly recent and has only been introduced by Arslanhan and Tosun (2020) to determine the importance weight of the criteria.

Today, with the pandemic, there are certain responsibilities faced by transport firms to protect health and safety, which have brought new attributes to the fore. An effective logistics system has also role in mitigating the adverse social outcomes of the pandemic by supplying medical equipment and food and easing the stay-home procedure for the public (Hendrickson and Rilett, 2020) and therefore, the criteria in the selection of appropriate transport mode should prioritize immediate needs over other conventional attributes.

Four major modes of transport, namely rail, road, water, and air, have their own advantages and disadvantages in terms of different criteria (Tuzkaya and Önüt, 2008) and are thus affected by pandemic differently. Decide among these modes or combining them in the time of pandemic, therefore, will need a strategic examination which comprises all of the factors effecting logistic decisions (Stoilova and Kunchev, 2017) such as cost, time, flexibility, reliability, and now, COVID-19.

Although numerous studies have been conducted on the selection of transport modes and the prioritization of relevant criteria, there is a scarcity in the studies which have dealt with transport mode decisions during times of crisis and exceptions. Identifying and prioritizing transport mode selection criteria during unforeseen occurrences, such as pandemics, will shed light on steps that can be taken in other potential situations that may be faced in the future. Therefore, this particular research seeks to remedy this deficiency by providing criteria set not just considering frequent criteria in the literature for transport mode selection but also including pandemic perspective.

To this end, both literature in the transport mode selection criteria and the latest updates of pandemic impacts on different modes have been examined and discussed in this paper. Additionally, the importance weights of criteria affecting transportation mode selection have been calculated by Best Worst Method (BWM) after the pairwise comparisons done by experts. The importance of this study arises from the fact that it is a model specially designed for pandemic processes.

The remainder of this article structured as follows: Sect. 2 describes the importance of transport mode selection in the face of pandemic by also including a literature review to provide a base of knowledge; Sect. 3 briefly discusses the Best Worst Method. Section 4 presents the main and sub criteria by providing hierarchy, definitions, and importance weighting of the criteria. This paper is concluded by the presented results in the Sect. 5.

2 Transportation Mode Selection Problem and Pandemic Effect

2.1 Transportation Mode Selection in the Face of Pandemic

As global economic activities intensify and international trade gathered speed, the formation of integrated transportation and logistics systems has gained importance and appeared as the key of sustainable growth. Bearing transportation systems consist of numerous of subsystems in mind, planning and examining these subsystems is theoretically huge area for achieving efficiency (Stević et al., 2017) and also brings complex problems along with them. In this respect, any selection/priority/decision issue regarding sub elements of transportation requires specific research questions and criteria on their own merit.

Thus, overarching criteria sets should be constructed under the title of these elements namely, convenient supplier selections, route selections or transport mode selections. These decisions require multi criteria decision making (MCDM) processes intrinsically since using single criterion will be relatively inoperative as there are numerous perspectives to take into consideration (Rezaei et al., 2016).

One of these decisions, which is the focus point of this research, is transport mode selection. Mode choice is a vital problem for transportation managers since customers seek reliable and flexible solutions in the transportation of their goods, as they also seek for cost reduction (Černá et al., 2017). Different transportation modes convey different strategical advantages (Tuzkaya and Önüt, 2008), for instance railroad appears as a safer option in longer distances whereas roadway enables easiness in intracity transport (Černá et al., 2017).

Even though customers have a range of available routes under normal circumstances, there was still numerous parameters to conceive in the determination of optimal route to transport goods. Similar to other decisional fields in logistics, transport mode selection also requires multiple criteria to be determined, defined, and tested (Ding and Liang, 2005).

Given the importance and complexity of this problem, there is growing endeavor among researchers to address multifarious dimensions of transport mode selection and thus, concentrates these dimensions of quality, cost, flexibility, and reliability into one inclusionary criteria set. A considerable amount of literature has reported that firms and decision makers feel constrained by the trade-off in transportations such as cost versus transport time (Vannieuwenhuyse et al., 2003) and therefore these criteria sets appear as vital for both scholars and practitioners.

While there are many successful implementations in current literature of defined and weighted criteria sets, not much of the work covers the latest trends, and what we know about selection criteria for transport mode is mainly focused on empirical studies that rely on more traditional attributes. We reviewed some of these works in the following section and revisited their contribution to existing literature to strengthen our base of knowledge on the subject.

These existing criteria in literature, however, should be extended to incorporate the perspectives brought by emerging developments or downward trends, lately COVID-19. COVID-19 pandemic has got substantial impact on transportation and logistics industry by causing change in numerous parameters of trade. As countries shut down their borders and limit their logistics activities one by one, firms start to seek alternative solutions to reduce the harm of pandemic and its concomitants on their operations. Severe approaches have been implemented by logistics companies that have contributed to a shift in some of the traditional rules for freight services, but it is necessary to remember that each mode of transport has been affected at a different level and some of them have encountered more difficulties than others. Different transport modes are distinguished by numerous attributes such as cost, reliability, speed and now with their level of affection by COVID-19.

In the case of airway, cancellation of flights, some airports closing down and suspending their operations and decrement in air cargo capacities caused global decline of 19% percent in air freight volume (Mazareanu, 2020), which made aviation industry one of the most detrimentally affected transportation mode of this period. Likewise, road transport has also been affected relatively in a larger scale because of the regional or national lockdowns and quarantine procedures. Shipping operations has also suffered as port authorities have forced some ships to undergo 14 days of quarantine periods before reaching ports (Wright, 2020).

In the consideration of differently affected transportation modes, the mode selection process has become even more essential as it can provide strategic advantages during pandemic too, from the perspectives of speed, cost, flexibility, and reliability.

Additional attributes should be considered as COVID-19 and its impacts completely stopped or at least decelerated some of the vital functions of transportation services and therefore criteria sets should be reexamined.

2.2 Literature Review

Mode selection and the criteria influencing this decision have been the subject of research recently (Moschovou and Giannopoulos, 2012) and have attracted interest in existing literature. As our literature review demonstrates, transportation mode selection problem has drawn a good deal of interest from the research community (Meixell and Norbis, 2008).

We argue in our paper, however, that mode selection studies have employed a limited set of attributes that focus mainly on more conventional dimensions such as cost, quality, flexibility, and reliability, and that need to be elaborated. Therefore, in order to establish our extended criteria based on existing ones and to identify gaps due to the non-inclusion of the latest trends affecting logistic activities such as pandemics, we have generally reviewed articles that have been written in the last two decades. The majority of the present work focuses on the ranking of the relevant criteria for the selection of the mode of transport according to their level of importance and sets out consecutive criteria for the decision of the appropriate mode.

Cullinane and Toy (2000) presented systematic analysis of applications in freight mode choice literature by analyzing more than 80 articles. In their work, they have also identified most often considered attributes in the mode choice decision as cost, speed, and reliability. Following this, Meixell and Norbis (2008) has also reviewed the literature over the last 20 years to present a set of current issues in logistics management and claims the most common encountered attributes in present works. In their analysis, they have pointed out that some of the important trends are under-mentioned in the criteria sets of mode choice literature such as environmental and energy challenges, security and integration in supply chain, the role of emerging information technologies.

It has conclusively been shown that transportation of certain cargoes such as heavy and hazardous ones requires additional attributes in their criteria sets and has drawn attention in present literature. Leal Junior and D'Agosto (2011) demonstrated a modal choice model for the transportation problem of hazardous materials, by giving a field and country specific study from bioethanol in Brazil. They have also reviewed a range of indicators which are commonly applied to transport service and stated what can be additionally considered for different type of cargoes.

Petraška and Palšaitis (2012) also conducted a similar work and studied on transport mode selection problem for oversize and heavyweight cargoes by stating some of the constraints caused by the size and weight. It is important to state that studies who focuses on attributes thought to be influencing rare cargo types are also vital for the pandemic, since there are different types of mass cargoes including health supplies, vaccines, or medication.

Multi-criteria decision making has wide applications in every sub concern of transportation and logistics. Transportation mode choice, in particular, should consider different aspects of choice because of its nature impacted by numerous parameters and parties immensely. In their study based on fuzzy analytic network process approach, Tuzkaya and Önüt (2008) identifies a very detailed criteria for the selection of appropriate transport mode between Germany and Turkey by also validating the results with a case study. Similarly, Pereira et al. (2013) presented a criteria set including timeliness of delivery, transport fee, shipping time, service level and they have evaluated weightings of each attribute with AHP. Kumru and Kumru (2014) has also employed AHP in order to select most suitable transportation mode between two locations in Turkey and also claimed a comparable criteria set including cost, speed, safety, reliability, and flexibility.

A more field specified study has been presented by Wang et al., (2016) which has also applied Fuzzy AHP in the prioritization of appropriate transport mode for the case of military logistics and brought diversity in the existing literature from the perspective of fields. Another field specified study conducted by Černá et al., (2017) in the case of Slovak Transport Market which has included price, safety, time, information, flexibility, and reliability to their criteria set with the intent of evaluating the performance of 4 major transport modes.

Other multi-criteria decision-making methods than AHP and ANP have also appeared in the current literature. Stoliava and Kunchev (2017) applied DEMATEL method to their criteria set including environmental, economic, technological, and social perspectives with the difference of providing a modal choice study for intermodal transportation. Their main criteria set can be used for other preliminary studies in transportation, whether or not it is intermodal.

In their very recent study, Arslanhan and Tosun (2020) also provided a weighting of a criteria set and similar to our work, appeared as one of the few studies which have applied Best Worst Method (BWM) for weighting of criteria set. Although they have included environmental perspective to their criteria set, there is still a dominance of conventional attributes and under-mentioning of late trends.

Our paper incorporates these findings of literature as it suggests a weighting and prioritization of more comprehensive criteria set by including the pandemic perspective.

3 Best Worst Method (BWM)

Best Worst Method (BWM) which is one of the comparison based MCDM methods was firstly introduced by Rezaei (2015). As an alternative novel model to analytical hierarchy process (AHP), the researchers have recently been attached high importance to apply BWM in multi criteria decision making (MCDM) problems including conflicting criteria. BWM proposes a new methodology that decreases the consistency and make it easier to conduct surveys among experts.

In this approach, the best criterion and the worst criterion are selected by decision makers. Pairwise comparisons are conducted between the best criteria and other criteria and then the worst criteria and other criteria. The main objective is to reveal out the optimal weights and consistency ratio by utilizing a basic optimization model constructed using the comparison system. The number of comparisons is lower than AHP, therefore the potential to have inconsistency has been reduced.

The stages of BWM are given as follows (Rezaei, 2015):

Step 1) Determination of decision criteria $\left(C_1, C_2, \ldots, C_n\right)$ is completed.

Step 2) The best (the most important one) criteria (Best-B) and the worst (the least important one) criteria (Worst-W) are designed by decision makers.

Step 3) The superiority of the best decision criteria (B) over remaining criteria is figured out. Scale numbers 1 to 9 are used and 1: B is equally important for *j*; 9: B is much more important than *j*. As a result of the process the best-to-others (BO) vector is obtained like following:

$$A_B = (a_{B1}, a_{B2}, \ldots, a_{Bn})$$

(a_{Bj}, shows the superiority of B on *j* and a_{BB}, is equal to 1)

Step 4) The superiority of the all decision criteria over the worst decision (W) criteria is obtained. Scale numbers 1 to 9 are used and 1: W is equally important for *j*; 9: W is much more important than *j*. As a result of the process the others-to-worst (OW) vector is obtained. As follows;

$$A_W = \left(a_{1W}, a_{2W}, \ldots, a_{nW}\right)^T$$

(a_{jW} shows the superiority of *j* on W and a_{ww}, is equal to 1)

Step 5) Finding the optimal weights $(w_1^*, w_2^*, \ldots, w_n^*)$. In order to obtain the optimal weights, there should be maximum absolute diameters.

$max_j\left\{\left|w_B - a_{Bj}w_j\right|, \left|w_j - a_{jW}w_W\right|\right\}$ for all j is minimized as shown in Eq. (1)

$$\begin{aligned} & minmax_j\left\{\left|w_B - a_{Bj}w_j\right|, \left|w_j - a_{jW}w_W\right|\right\} \\ & \quad \text{s.t.} \\ & \sum_j w_j = 1 \\ & w_j \geq 0, \ \text{for all} j. \end{aligned} \tag{1}$$

The equation expressed above is equal to the following linear problem (as seen in Eq. (2)).

$$\begin{aligned}
&\min \xi^L \\
&\text{s.t.} \\
&\left|w_B - a_{Bj}w_j\right| \le \xi^L \text{ for all } j. \\
&\left|w_j - a_{jW}w_W\right| \le \xi^L \text{ for all } j. \\
&\sum_j w_j = 1 \\
&w_j \ge 0, \text{ for all } j.
\end{aligned} \tag{2}$$

By solving this problem, the optimal objective function value ξ^{L*} and optimal weights $(w_1^*, w_2^*, \ldots, w_n^*)$ are obtained. ξ^{L*} indicates the consistency ratio. If this value is close to 0, that means decision makers are consistent.

4 The Proposed Methodology

Determination of Criteria: As mentioned earlier, the problem of selecting the correct mode of transportation is fundamentally the MCDM problem. In this paper, Best Worst Method as a MCDM model is presented and applied for the calculation of the weights of the criteria.

The first step of the proposed methodology is to identify and define the main and sub-criteria for prioritization that have an impact on the selection of the best mode of transportation. Thirteen sub-criteria were defined for the selection of the best mode of transportation under the four main criteria clusters.

The main criteria and sub-criteria with their definitions are set out in Table 1 below.

Weighting Criteria: Belongs to steps given at Sect. 3, the surveys are designed and sent to three decision makers who have valuable experiences on this field. Their consistencies are checked and found as satisfactory. Evaluations of decision makers are aggregated by using arithmetic mean. Then, by using BWM Solver on aggregated matrices, the weights of each criterion are set. To reveal the global weights of the sub-criteria, the local weights are multiplied with the weights of corresponding parent criteria. As it is given at Table 2, the most important main criterion is Transportation Service Quality (C2). It indicates that regardless of the pandemic, service quality matters. The long-term perspective and corporate perspective are focused on long-term solutions, not short-term solutions. Short and medium-term changes do not require compromising the core strategy. Because these strategies have been created by considering the actions taken against any unexpected event or situation.

But the most important sub criterion is not included in Transportation Service Quality (C2). It is Freight Cost (C11). That means, cost keeps its popularity under uncertain environment also. Flexibility (C11), Safety (C12), Traceability (C13), Speed and Time (C14), which are the sub-criteria of Transportation Service Quality (C1), are important

Table 1. Determination of criteria

Main criteria	Definition of the main criteria	Sub criteria	Definition of the sub criteria
Transportation cost (C1)	This attribute includes all the items of cost that can be realized during the entire transportation process	Freight cost (C11)	This attribute refers to transportation cost incurred in moving goods from supply to demand point
		Damage cost (C12)	This attribute refers to cost of possible damages may be incurred at the transportation or transhipment points
		Transhipping cost (C13)	This attribute refers to cost incurred in transhipment of goods from one mode to another at transhipment points
Transportation service quality (C2)	This attribute includes all the factors affecting the efficiency and utility of transportation mode realized throughout the transportation process	Flexibility (C21)	This attribute refers to the ability to manage unexpected changes, capacity, schedule and route flexibility while accommodating the needs of the customer
		Safety (C22)	This attribute refers to protection of goods from damage or loss which can be incurred at transportation, transhipment, or warehousing stages
		Traceability (C23)	This attribute refers to common designation for place and physical condition of freight, container, vehicle, driver or any other factor associated with transportation process
		Speed and Time (C24)	This attribute refers to the perceived speed of each mode of transport by measuring the time spent during loading, unloading, transit or main trip

(continued)

Table 1. (*continued*)

Main criteria	Definition of the main criteria	Sub criteria	Definition of the sub criteria
Transportation reliability (C3)	This attribute includes the fulfilment of time, place, and contractual conditions in accordance with requirements	Travel time reliability (C31)	This attribute refers to delivery of goods to the destination within a specified and predetermined time interval
		Terminal reliability (C32)	This attribute refers to delivery of goods from predetermined origin to a predetermined destination
		Condition reliability (C33)	This attribute refers to delivery of goods in an undisturbed condition
Pandemic criteria (C4)	This attribute includes the ability to resolve the adverse pandemic impacts on transportation system	Reliability during pandemic (C41)	This attribute refers to accountability level for delivery of needed goods to the right destination at the right time during Pandemic
		Flexibility during pandemic (C42)	This attribute refers to ability of meeting unexpected changes in demand, route, capacity, or any predetermined factor due to Pandemic
		Restrictions (C43)	This attribute refers to governmental or other institutional restrictions affecting transportation modes, vehicles, or drivers such as custom regulations, health and safety regulations, and quarantine periods for drivers, waiting period on country borders

if the relative benefit is higher in cost-benefit analysis. What matters is to what extent each freight cost will reflect on quality.

Flexibility during Pandemic (C42) becomes as second sub criterion. This criterion addresses to the ability of meeting unexpected changes in demand, route, capacity, or any predetermined factor due to Pandemic. If the companies are strong enough to adapt themselves to changing conditions, they can gain competitive advantages and the threats originated from pandemic can provide good opportunities to them. On the other hand Speed and Time (C24) is the third sub-criterion. Even everyday needs started to be made not by face-to-face shopping, but by online shopping, and it became important to present such needs (for instance, fast moving consumer goods) to the consumer in a short

Table 2. Weights of main and sub criteria

Main criteria	Weights of main criteria	Sub criteria	Global weights of sub criteria
Transportation cost (C1)	0.192	Freight cost (C11)	0.568
		Damage cost (C12)	0.125
		Transhipping cost (C13)	0.306
Transportation service quality (C2)	0.373	Flexibility (C21)	0.141
		Safety (C22)	0.238
		Traceability (C23)	0.124
		Speed and time (C24)	0.498
Transportation reliability (C3)	0.225	Travel time reliability (C31)	0.302
		Terminal reliability (C32)	0.242
		Condition reliability (C33)	0.456
Pandemic criteria (C4)	0.209	Reliability during pandemic (C41)	0.162
		Flexibility during pandemic (C42)	0.500
		Restrictions (C43)	0.338

time. Traceability (C23) and Damage Cost (C12), in reverse, have the least importance. Traceability refers to common designation for place and physical condition of freight, container, vehicle, driver or any other factor associated with transportation process. Under pandemic, door-to-door delivery becomes highly important and that causes the ignorance of traceability. On the other hand, it may have been thought that the cost of damage will be dealt with by offering the option of change at all times for damaged products, or by transporting with sheltered containers that prevent damage.

5 Conclusion

Pandemic has significant impact on logistics activities by changing many micro and macro dynamics of the industry. Almost all countries close the borders for commercialization and reduce their logistics activities. Companies have begun to create flexible alternative decisions to prevent the negative effect of pandemic. The selection of the best transportation mode becomes a very significant decision for firms which aim to be flexible and offer reliable and safe solutions to the customers. This study develops a methodology based on a novel MCDM titled Best Worst Method (BWM) for transportation mode selection under pandemic. It contributes to the relevant literature by proposing an extended criteria set for the defined problem. As a result of the conducted case, it is recognized that pandemic is not strong enough to change fundamental considerations of corporates. Because, although there is a set of criteria depending on pandemic, Transportation Service Quality becomes first among four main criteria. Besides that, consideration of Freight Cost keeps its popularity under pandemic environment. The findings may highlight that for short and medium term changings, the strategic transformations are not needed.

In the future, the researchers who are interested in this topic can design a fuzzy set integrated BWM in order to increase the strength of the proposed methodology against hesitancies and linguistic evaluations. Besides that, a real case can be conducted for transportation mode selection, and the results can be discussed.

References

Arslanhan, H., Tosun, Ö.: Solving transport mode selection problem with integrated best – worst and WASPAS methods. Pamukkale Univ. J. Eng. Sci. **1000**, (2020)

Černá, L., Zitrický, V., Daniš, J.: The Methodology of selecting the transport mode for companies on the slovak transport market. Open Eng. **7**, 6–13 (2017)

Cullinane, K., Toy, N.: Identifying influential attributes in freight route/mode choice decisions: a content analysis. Transp. Res. Part E Logistics Transp. Rev. **36**(1), 41–53 (2000)

Ding, J., Liang, G.: Using fuzzy MCDM to select partners of strategic alliances for liner shipping. Inf. Sci. **173**, 197–225 (2005)

Hendrickson, C., Rilett, L.: The COVID-19 Pandemic and Transportation Engineering. J. Transp. Eng. Part A Syst. **146**(7), 01820001 (2020)

Kumru, M., Kumru, P.: Analytic hierarchy process application in selecting the mode of transport for a logistics company. J. Adv. Transp. **48**(8), 974–999 (2014)

Leal Junior, I., D'Agosto, M.: Modal choice for transportation of hazardous materials: the case of land modes of transport of bio-ethanol in Brazil. J. Cleaner Prod. **19**(2–3), 229–240 (2011)

Mazareanu, E.: Topic: Coronavirus: impact on the transportation and logistics industry worldwide. https://www.statista.com/topics/6350/coronavirus-impact-on-the-transportation-and-logistics-industry-worldwide/. Accessed 20 July 2020

Meixell, M., Norbis, M.: A review of the transportation mode choice and carrier selection literature. Int. J. Logistics Manage. **19**(2), 183–211 (2008)

Moschovou, T., Giannopoulos, G.: Modeling Freight Mode Choice in Greece. Procedia Soc. Behav. Sci. **48**, 597–611 (2012)

Pereira, M., Adelaide, M., Resgate, L., Telhada, J.: Multicriteria methodology to mode of transport choosing—the Portuguese case. In: Proceedings of the XI Congreso Galego de Estatística e Investigación de Operacións, pp. 1–15. A Coruña, Galicia, Spain (2013)

Petraška, A., Čižiūnienė, K., Prentkovskis, O., Jarašūnienė, A.: Methodology of selection of heavy and oversized freight transportation system. Transp. Telecommun. J. **19**, 45–58 (2018)

Petraška, A., Palšaitis, R.: Evaluation criteria and a route selection system For transporting oversize and heavyweight cargoes. Transport. **27**(3), 327–334 (2012)

Rezaei, J., Nispeling, T., Sarkis, J., Tavasszy, L.: A supplier selection life cycle approach integrating traditional and environmental criteria using the best worst method. J. Cleaner Prod. **135**, 577–588 (2016)

Rezaei, J.: Best-worst multi-criteria decision-making method. Omega **53**, 49–57 (2015)

Stević, Ž., Pamučar, D., Kazimieras Zavadskas, E., Ćirović, G., Prentkovskis, O.: The selection of wagons for the internal transport of a logistics company: a novel approach based on rough BWM and rough SAW methods. Symmetry **9**, 264 (2017)

Stoilova, S., Kunchev, L.: Study of criteria for evaluation of transportation with intermodal transport. Eng. Rural Dev. Proc. **16**, 349–357 (2017)

Tuzkaya, U., Önüt, S.: A fuzzy analytic network process based approach to transportation-mode selection between Turkey and Germany: a case study. Inf. Sci. **178**, 3133–3146 (2008)
Vannieuwenhuyse, B., Gelders, L., Pintelon, L.: An online decision support system for transportation mode choice. Logistics Inf. Manage. **16**, 125–133 (2003)
Wang, Y., Han, T., Chou, M.: Applying fuzzy AHP in selection of transport modes for Kinmen military logistics. J. Mar. Sci. Technol. **24**, 222–232 (2016)
Wright, R.: Pandemic strains shipping, air and rail freight operators, https://www.ft.com/content/8403ddbc-9363-11ea-899a-f62a20d54625. Accessed 20 Jul 2020

Estimation of Returns to Scale of Units Under Evaluation with Integer Data Using Non-radial Models in Data Envelopment Analysis

Z. Tavassoli(✉), M. Rostamy-Malkhalifeh, F. Hosseinzadeh lotfi, and T. Allahvirnloo

Science and Research Branch, Islamic Azad University, Tehran, Iran
tavassoli.zeinab1393@gmail.com

Abstract. In the real world, integer data plays a very important role in units; number of people working in one unit, number of industrial machines, number of teachers and students. However, many of the models in different DEA indices have been proposed for real data, and few articles have reviewed the integer data. In this paper, the efficiency of scale in units with integer data in the nature of input and output is investigated. An example of the specific behavior of units is shown in this new model compared to what was expected

Keywords: Data envelopment analysis · Left returns to scale · Right returns to scale · Integer data

1 Introduction

Data envelopment analysis was first proposed by Bunker and Charans to evaluate and calculate the performance of the units under evaluation (CHARNS et al. 1978 - BANKER et al. 1984). Gradually, it was able to determine and evaluate other indicators for the efficiency and type of performance of the units under evaluation, such as returns to scale. The output-to-input ratio has been discussed in many articles from different perspectives, including the article by Golany et al., which uses BCC to present scale returns in two modes (Boaz Golany et al. -1997). Allahyar et al., have reviewed the estimation of Returns to Scale (RTS) in Data Envelopment Analysis (DEA), presented by Golany, and Yu and proposed a new method to do this estimation. We show that the new method does not have the shortcoming compared to the previous one. Furthermore, it is able to evaluate Returns to Scale (RTS) to the right and left of the given unit in all conditions. Omidi et al. (2017) presented a method for evaluating L-RTS and R-RTS whose models are linear and there is no problem of being parametric about them. In addition, the proposed methods have the advantage of being used as a tool to assess the continuity of RTS sustainability of decision-making units by making minor changes to them and by dividing the boundary of the production capability set.

But what is problematic is the fact that the data is integer in the real world, such as the number of employees in a unit, the number of machines in a factory, etc. But modeling

T. Allahviranloo et al. (Eds.): IDS 2020, AISC 1301, pp. 572–580, 2021.
https://doi.org/10.1007/978-3-030-66501-2_45

for indicators is due to the authenticity of the data, and the assumption that the data is integer is not considered. Lozano et al. (2006) introduced new concepts and models that explicitly look at the integer input and output. Kuosmanon et al. (2009) modified the defects of the model and defined new concepts in the integer-valued DEA, after which a model was extracted for MILP. In the same year, Kazemimatin et al. (2009) introduced the concept of natural convexity and proposed a new model for variable-scale efficiency technology. Multi-objective integer models were then introduced (Wu et al. - 2011). Khezrimotlagh et al. (2013) and Kuosman (2015) subsequently refined the models. But so far, the model for scaling has not been discussed with accurate data.

Although returns to the scale of negative numbers have been discussed in the article by Allah Yaro et al. (2015), in which it first determines the efficiency of the DMU and then provides a way to determine the status of returns to scale, no model for returns to the correct data scale has been proposed so far. For example, in articles such as Peykani et al. (2018), there is a need for models for integer data.

This paper examines the efficiency of DMU-scale single-input, single-output DMUs and new models have been introduced for review while an example shows the difference between the results of changing the nature of the input and output. In the second part, the basic concepts are stated and in the third part, the nearest input is determined with one model and then with another model, which calculates the maximum amount of output that can be obtained for this input, and finally with the obtained values, the type of return to scale is determined. These models are presented in two natures, visit and output. In the fourth part, with an example, the difference between the type of return and the scale, when all the data are integer, is determined by the fact that the data is real, and in the final part, a conclusion and suggestion is provided.

2 Introductory Concepts

2.1 Overview of the Integer Data Technology

For every $DMU_j,\ j = 1, 2, \ldots, n$ where m is the input used to generate the output s. Suppose that the input vector of the outputs of each DMU_j

$$x = (x_1, x_2, \ldots, x_m)\ ,\ y = (y_1, y_2, \ldots, y_s)$$

is the input vector of each DMU output where all outputs and all inputs are integer. The technology of production as follows: (All inputs are integer - All outputs are integer).

$$T_V' = \left\{ \begin{array}{c} (\tilde{x}, \tilde{y})|\exists(\lambda_1, \lambda_2, \ldots, \lambda_{n)}, \lambda_j \geq 0, \forall_j \tilde{x}_i \geq \sum_{j=1}^n \lambda_j x_j, \tilde{y}_j \leq \sum_{j=1}^n \lambda_j y_j, \\ \sum_{j=1}^n \lambda_j = 1, i \in integer, r \in integer \end{array} \right\}$$

And using it, the following model is suggested to calculate efficiency:

$$\min\theta - \varepsilon\left(\sum_{i=1}^{p'} s_i^- + \sum_{i=1}^{q} \tilde{s}_i^- + \sum_{i=1}^{s} s_r^+\right)$$

$$\sum_{j=1}^{n} \lambda_j y_{rj} - s_r^+ = y_{ro} \qquad o \in \mathbb{R}$$

$$\sum_{j=1}^{n} \lambda_j x_{ij} + s_i^- = \tilde{x}_i \qquad i \in I$$

$$\theta x_{ip} - \tilde{s}_i^- = \tilde{x}_i \qquad i \in I$$

$$\sum_{j=1}^{n} \lambda_j = 1$$

$$\lambda_j \geq 0, \tilde{x}_i \in \mathbb{Z}, s_i^-, \tilde{s}_i^-, s_r^+ \geq 0$$

(Symbol ε denotes a non-Archimedean infinitesimal and $\left(\tilde{X}, Y\right)$ is reference point).

2.2 Return to Scale

the set of BCC output possibilities will be as follows

$$\mathrm{T}_v = \left\{ \begin{array}{c} (\mathrm{x}, \mathrm{y}) | \exists (\lambda_1, \lambda_2, \ldots, \lambda_{n)}, \lambda_j \geq 0, \forall_j x \geq \sum_{j=1}^{n} \lambda_j x_j, y \leq \sum_{j=1}^{n} \lambda_j y_j, \\ \sum_{j=1}^{n} \lambda_j = 1 \end{array} \right\}$$

$\left(x_p, y_p\right) \in \partial T_V$ is considered.
Suppose $\beta > 0$ is a constant value: we define

$$\alpha(\beta) = max\left\{\alpha | \left(\beta x_p, \alpha y_p\right) \in T_v\right\}$$

Given the definition of $\alpha(\beta)$, if $\lim_{\beta \to 1} \frac{\alpha(\beta) - 1}{\beta - 1} = \gamma$ If so:
$\gamma > 1$ then DMUP is the increasing returns to scale.
$\gamma < 1$ Then DMUP is the decreasing returns to scale.
$\gamma = 1$ Then DMUP is the constant returns to scale.

3 New Definitions and Suggested Models

3.1 The Nature of the State Input

To calculate the return to scale in the integer data, we measure the amount of output changes relative to the amount of input changes. If the input is changed to the nearest (most definitely) integer input, what is the percentage of output changes?

We accept the previous definitions, except that

Which converts x_I to βx which is an integer data and y_R to $\alpha(\beta)$ which is an integer number.

all inputs are integer and the output are also integer, which means that $x_i^I, i = 1, 2, \ldots, my_r^R, r = 1, 2, \ldots, s$ we show the magnitude of changes x_i^I with β and obtain it with the model of 2–8.

And we define $\alpha(\beta)$ as follows:

$$\alpha(\beta) = max\left\{\alpha|\begin{pmatrix}\beta x_{ip}^{I} \\ \alpha y_{rp}^{I}\end{pmatrix} \in T_v, i = 1, 2, \ldots, m, r = 1, 2, \ldots, s\right\}$$

We calculate the percentage of output-to-input changes using Formula 2–9:

And again, by comparing γ to 1, we accept the preceding rules of 2–3 for return to scale.

We determine the change value of each DMU from the following models to obtain the integer after the change:

$$\begin{aligned} & \max\ \beta \\ & \beta x_{ip} = x'_{ip} \qquad i \in Z \\ & \beta < 1 \\ & x'_p \in Z \end{aligned}$$

$$\begin{aligned} & \min\ \beta \\ & \beta x_{ip} = x'_{ip} \qquad i \in Z \\ & \beta < 1 \\ & x'_p \in Z \end{aligned}$$

It is determined and evaluates the nearest input to the unit input under evaluation.

The condition $\beta x_p = x'_p$ specifies the amount of input change so that it changes position to the nearest integer unit.

For right-handed returns to scale (the constraint $\beta > 1$ represents the input increase; the strictness of the constraint is to change the desired input state of the DMU).

Determining β helps to determine the closest input relative to the input of the unit under evaluation.

Theorem 1: In the model 2, β is between 1 and ∞ and the value of β is numerically between 1 and 2.

For Left-Returns to scale (constraint $\beta < 1$ represents input reduction; Strictness of the constraint is to change the desired input state of the DMU).

Theorem 2: In model 3, β is between 0 and 1 and the numerical value of β * is between 0 and 1.

The model for estimating the returns to scale in the integer input-integer output mode is as follows:

$$\begin{aligned} & \max\ \alpha \\ & \sum_{j=1}^{n} \lambda_j x_{ij} \leq \beta x_{ip} \qquad i \in Z \end{aligned}$$

$$\begin{aligned}
&\sum_{j=1}^{n} \lambda_j y_{ij} \geq \alpha y_{rp} \quad r \in Z \\
&\alpha y_{rp} = y'_{rp} \quad r \in Z \\
&\sum_{j=1}^{n} \lambda_j = 1 \\
&\lambda_j \geq 0, \alpha \geq 0, y'_{rp} \in Z
\end{aligned} \tag{1}$$

It has a further constraint of $\left(y'_{rp} \in Z\right), \alpha y_{rp} = y'_{rp}\ r \in Z$ compared to the previous one, which makes it possible to adjust the output changes to obtain an integer output.

3.2 The Nature of the State Output

We determine the change value of each DMU from the following models to obtain the integer after the change:

$$\begin{aligned}
&max\ \alpha \\
&\alpha y_p = y'_p \\
&\alpha < 1 \\
&y'_p \in Z
\end{aligned}$$

$$\begin{aligned}
&\min \alpha \\
&\alpha y_p = y'_p \\
&\alpha > 1 \\
&y'_p \in Z
\end{aligned} \tag{2}$$

It is determined and evaluates the nearest output to the unit input under consideration.

The model for estimating returns to scale in the integer input-integer output mode is as follows:

$$\begin{aligned}
&\min \beta \\
&\sum_{j=1}^{n} \lambda_j x_{ij} \leq \beta x_{ip} \quad i \in Z \\
&\sum_{j=1}^{n} \lambda_j y_{ij} \geq \alpha y_{rp} \quad r \in Z \\
&\beta x_{ip} = x'_{rp} \quad r \in Z \\
&\beta > 1 \\
&\sum_{j=1}^{n} \lambda_j = 1
\end{aligned}$$

$$\lambda_j \geq 0, \beta \geq 0 \tag{3}$$

The constraint $\beta > 1$ is due to the inevitability of returns to scale, and when constrained by calculating returns to a decreasing scale, it is substituted for return to increasing scale $\beta < 1$.

4 Example

4.1 An Example of the Integer Input - Integer Output Mode

A company with 13 units is considered to have an integer input-integer output. The following table shows the input-output value (Table 1).

Table 1. Information of the units under evaluation (input-output)

DMU	A	B	C	D	E	F	G	H	I	J
x	4	5	6	7	8	9	10	12	13	18
y	5	7	9	10	12	13	15	16	17	18

And in Fig. 1, the PPS associated with these units is drawn. As before, the horizontal axis is the input and the vertical axis is the output. The input and output are both integers, so the PPS is discrete (forming the intersection of the PPS gridlines). The performance value of these units is as follows (Table 2):

Table 2. Performance Dimensions of DMUs

DMU	A	B	C	D	E	F	G	H	I	J
Reference	A	B	C	D	E	F	G	H	I	J
Efficiency of DMUs	1	1	1	1	1	1	1	1	1	1
Slack value	All slacks are zero	All slacks are zero	All slacks are zero	All slacks are zero	All slacks are zero	All slacks are zero	All slacks are zero	All slacks are zero	All slacks are zero	All slacks are zero

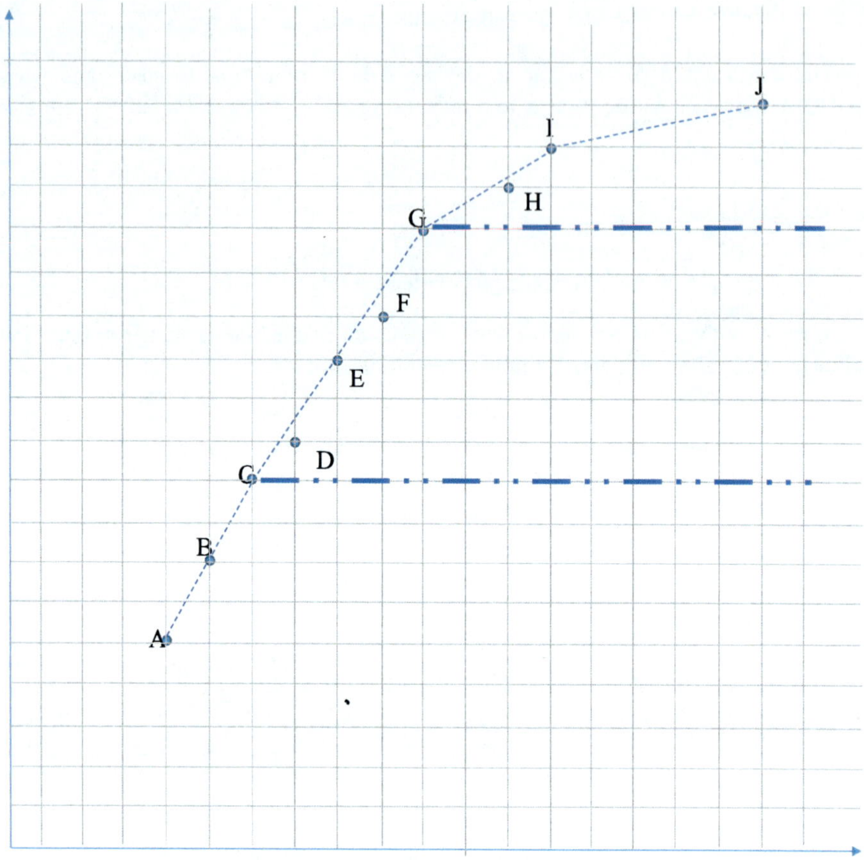

Fig. 1. PPS units under evaluation

All units are strongly efficient, and are on the PPS border (strong units are selected because if the unit is inefficient, we perform a check on the boundary to determine the type of return to scale with the unit image).

The input change value of each DMU using models 3-10 and 3-11 is shown in the table below (Table 3).

Table 3. The amount of change per DMU

DMU	A	B	C	D	E	F	G	H	I	J
β^{+}	1.250	1.200	1.6667	1.1429	1.250	1.2222	1.200	1.2500	1.2308	1.222
β^{+}	0.750	0.800	0.833	0.8571	0.75	0.778	0.800	0.75	0.7692	0.778

The scale of returns to scale and DMU to which the desired DMU is compared is given in the following Table 4.

Table 4. Unit near the DMU to evaluate the returns to scale

DMU	A	B	C	D	*E*	F	G	*H*	I	*J*
$\beta^{+}x$	5	6	7	8	9	10	11	13	14	19
$\alpha^{+}y$	7	9	10	12	13	15	15	17	17	18
γ^{+}	1.6	1.4	0.7	1.4	0.7	1.4	0	0.78	0	0
RTS type	IRS	IRS	DRS	IRS	DRS	IRS	DRS	DRS	DRS	DRS
$\beta^{-}x$	3	4	5	6	7	8	9	11	12	14
$\alpha^{-}y$	0	5	6	9	10	12	13	15	16	17
γ^{-}	4	1.4	1.3	0.7	1.3	0.7	1.3	0.75	0.8	0.3
RTS type	IRS	IRS	IRS	DRS	IRS	DRS	IRS	DRS	DRS	DRS

As can be seen from the comparison with Fig. 1, DMUs do not show the expected returns to scale. For C, D, E units the constant right returns to scale, and for units D, E, F the constant left returns to scale are expected (since they are on the assumed MPSSs), but since the boundary is no longer a continuous convex boundary (like real data), the units that are measured against them are changed and go out of their current state after which no constant ratio or ratio change occur according to a regular pattern. This pattern change occurs in the area where the expected returns on a constant scale take place.

5 Conclusion

The purpose of this paper is to propose a model to determine the type of returns to scale in the integer data. The purpose of this paper is to propose a model to determine the type of return to scale in the correct data. A separate model was defined to determine the amount of movement to the integer value. The unit to which the DMU should be measured to determine the type of return on a scale

In evaluation of the return to scale in integer input- integer output which was out of the routine procedure of return to scale, it seems that finding the links between this particular case and other ways of measuring returns to scale in real research data will be interesting with promising results.

References

Banker, R.D.: Estimating most productive scale size using data envelopment analysis. Euro. J. Oper. Res. **17**, 35–44 (1984)

Banker, R.D., Charnes, A., Cooper, W.W.: Some models for estimating technical and scale inefficiencies in data envelopment analysis. Manag. Sci. **30**, 1078–1092 (1984)

Boaz, Golany: Yu Gang Theory and Methodology Estimating returns to scale in DEA. Euro. J. Oper. Res. **103**, 28–37 (1997)

Allahyar, M., Rostamy-Malkhalifeh, M.: An improved approach for estimating returns to scale in DEA. Bull. Malaysian Math. Sci. Soc. **37**(4), 1185–1194 (2014)

Omidi, M., Rostamy-Malkhalifeh, M., Payan, A., Hosseinzadeh, F.: Determining left and Raghit Return toscale (RTS) and RTS sustainability by using linear programing problems based on simultaneous changes in inputs and outputs. J. New Res. Math. **3**, 59–79 (2017)

Lozano, S., Villa, G.: Data envelopment analysis of integer-valued inputs and outputs. Comput. Oper. Res. **33**, 3004–3014 (2006)

Kazemi, Matin R., Kuosmanen, T.: Theory of integer-valued data envelopment analysis under alternative returns to scale axioms. Omega **37**, 988–995 (2009)

Wu, J., Zhoue, Z.: Springer Science + Business Media, LLC (2011)

Khezrimotlagh, D., Salleh, S.H., Mohsenpour, Z.: A Comment on theory of integer-valued data envelopment analysis. Appl. Math. Model. **6**(116), 5769–5774 (2012)

Khezrimotlagh, D.: Different between real and integer production possibility set in data envelopment (2015)

Allahyar, M., Rostamy-Malkhalifeh, M.: Negative data in data envelopment analysis: Efficiency analysis and estimating returns to scale. Comput. Ind. Eng. **82**, 78–81 (2015)

Peykani, P., Mohammadi, E., Seyed Esmaeili, F.s.: Measuring performance, estimating most productive scale size, and benchmarking of hospitals using DEA approach: a case study in Iran. Int. J. Hospital Res. **7**(2), 21–41 (2018)

Applications in Science

A Fast Discrete Homotopy Solution Method for Two Problems in Image Deconvolution

Bryan M. Williams[1(✉)], Behzad Ghanbari[2], Ke Chen[3], and Lavdie Rada[4]

[1] Biometrics and Human Identification, Data Science Institute, Lancaster University, Lancaster, UK
b.williams6@lancaster.ac.uk
[2] Department of Basic Science, Kermanshah University of Technology, Kermanshah, Iran
[3] Centre for Mathematical Imaging Techniques and Department of Mathematical Sciences, The University of Liverpool, Liverpool, UK
[4] Bahcesehir University, Istanbul, Turkey

Abstract. Blur and noise are commonly found in real-life images. Blur occurs in the acquisition whereas noise mostly occurs due to the signal converted into a digital representation due to the sampling process. Severe blur and noise in a given image results in resolution loss and restricts accurate post-processing assessments. To improve image restoration and on the same time build a fast variational framework we propose the incorporation of restarted homotopy analysis method as an alternative compared with costly direct methods. Through experiments, we demonstrate the effectiveness of our new technique.

Keywords: Image restoration · Total variation · Image processing

1 Introduction

Image restoration has been widely studied in the last decades. The goal of image restoration is to recover the true image $u(\mathbf{x})$ from the observed noisy image

$$z(\mathbf{x}) = h * u(\mathbf{x}) + \eta(\mathbf{x})$$

where $z(\mathbf{x})$ is the observed noisy image, h is a bounded linear operator representing the convolution (named as Point Spread Function (PSF)), and $\eta(\mathbf{x})$ denotes the additive noise. The most commonly found noises in image acquisition applications are Gaussian and Poisson noise. The bounded linear operator h, which causes the blur effect in images, commonly occurs in the acquisition due to various factors such as media opacity, refractive errors and movement. Severe blur resulting loss of resolution hampers accurate image assessment and post-processing. This problem reflects a major challenge for further image outcome monitoring, object detection, accurate annotation, etc. This leads to the development of effective and accurate image restoration techniques. Modeling the inverse of such problems depends on the type of blur and many techniques

T. Allahviranloo et al. (Eds.): IDS 2020, AISC 1301, pp. 583–594, 2021.
https://doi.org/10.1007/978-3-030-66501-2_46

has been presented so far. Considering image restoration as an inverse problem, aiming to recover the true image $u(\mathbf{x})$ from the given blurred image $z(\mathbf{x})$, a classical way of solving such problems is using minimization of a variational energy which is consists of fitting and some regularization terms to the energy. Such idea was first introduced for image restoration with additive Gaussian noise by Tikhonov and Arsenin [13] and was further improved by Rudin-Osher-Fatemi [12] with the minimization of the following energy:

$$min_u \int_{\Omega} |h * u(\mathbf{x}) - z(\mathbf{x})|^2 + \lambda \int |\nabla u(\mathbf{x})|$$

where $\lambda > 0$ is a regularization parameter which balances the first and second terms. The first term is the fitting term whereas the second term represents a total variation (TV) regularizer term with edge preserving ability. Although this model improves the old Tikhonov and Arsenin [13] model, the model gives rise to some undesired effects the so-called staircase effects. In order to reduce the staircase effect, some high-order variational models were introduced [2,5,8–11,14], which contain second or higher order TV regularization terms. However, those high-order variational models lead to more complex boundary conditions.

In a recent work of Ghanbari et al. [1] a discrete restarted homotopy analysis method (DHAM) is introduced to improve the TV based denoising and segmentation models. This technique adapted the homotopy analysis method (HAM) through a sequence of three linear equations and its restarting process avoiding the using high order terms. Below we present a DHAM model for the problem of debluring evolution approaches of active contours TV based model. Through solving a typical TV inverse problem, the DHAM may be used for other un-explored nonlinear applications and does not require working out with high order nonlinear equations.

In the following, we give a short review of HAM and DHAM model for the evolution approaches of active contours TV based model for the denoising problem.

1.1 Review of the Homotopy Analysis Method

The HAM was first introduced by Liao et al. [6] as an analytic method for nonlinear problems, providing a simple way to control and adjust the convergence region and rate of solution series of nonlinear problems. The general idea of Homotopy is to transfer a nonlinear problem to a high-order linear approximation. Let $\mathcal{N}[u(\mathbf{x},\mathbf{t})] = \mathbf{0}$ be a non-linear equation with initial condition $u(\mathbf{x},\mathbf{0}) = \mathbf{u_0}(\mathbf{x})$ where $\mathcal{N}$ is a non-linear operator, $\mathbf{x} \in \mathbb{R}^{\mathbf{n}}$ and t are independent variables and $u(\mathbf{x})$ is an unknown function. We construct the zero-order deformation equation [7] from $\mathcal{N} = 0$ as follows:

$$(1-q)\mathcal{L}\left[\varphi(\mathbf{x},\mathbf{t};\mathbf{q}) - \mathbf{u_0}(\mathbf{x},\mathbf{t})\right] = q\hbar\mathcal{H}(\mathbf{x},\mathbf{t})\mathcal{N}\left[\varphi(\mathbf{x},\mathbf{t};\mathbf{q})\right]$$

where $u_0(\mathbf{x})$ is an initial estimate; $\hbar$ is a non-zero auxiliary parameter; $q \in [0,1]$ is an embedding parameter; $\varphi(\mathbf{x},\mathbf{t};\mathbf{q})$ is a function of t and q; $\mathcal{H}$ is a non-zero auxiliary function; $\mathcal{L}$ is an auxiliary operator with the property

$$\mathcal{L}[\varphi(\mathbf{x},\mathbf{t})] = \mathbf{0} \text{ where } \varphi(\mathbf{x},\mathbf{t}) = \mathbf{0}.$$

We emphasize that we have the freedom to choose the initial approximation, the auxiliary linear operator $\mathcal{L}$, the auxiliary parameter $\hbar$, and the auxiliary function $\mathcal{H}$.

By increasing q one gets

$$\begin{gathered}
\mathcal{L}\left[\varphi(\mathbf{x},\mathbf{t};\mathbf{0}) - \mathbf{u_0}(\mathbf{x},\mathbf{t})\right] = \mathbf{0}q = 0\varphi(\mathbf{x};\mathbf{0}) = \mathbf{u_0}(\mathbf{x}) \\
\vdots \\
(1-q)\mathcal{L}\left[\varphi(\mathbf{x},\mathbf{t};\mathbf{q}) - \mathbf{u_0}(\mathbf{x},\mathbf{t})\right] = q\hbar\mathcal{H}(\mathbf{x},\mathbf{t})\mathcal{N}\left[\varphi(\mathbf{x},\mathbf{t};\mathbf{q})\right]\mathbf{q} = \mathbf{q}\varphi(\mathbf{x};\mathbf{q}) \\
\vdots \\
\mathbf{0} = \hbar\mathcal{H}(\mathbf{x},\mathbf{t})\mathcal{N}\left[\varphi(\mathbf{x},\mathbf{t};\mathbf{1})\right]\mathbf{q} = \mathbf{1}\varphi(\mathbf{x};\mathbf{1}) = \mathbf{u}(\mathbf{x})
\end{gathered}$$

Since both $\hbar$ and $\mathcal{H}$ are non-zero, when $q = 0$ and $q = 1$, it holds that $\varphi(\mathbf{x};\mathbf{0}) = \mathbf{u_0}(\mathbf{x})$ and $\varphi(\mathbf{x};\mathbf{1}) = \mathbf{u}(\mathbf{x})$. Therefore, as q increases from 0 to 1, the the solution $\varphi(\mathbf{x};\mathbf{q})$ deforms from the initial estimate to the solution.

Expanding $\varphi(\mathbf{x},\mathbf{t};\mathbf{q})$ in the Taylor series with respect to q, we have

$$\varphi(\mathbf{x},\mathbf{t};\mathbf{q}) = \sum_{\mathbf{m}=\mathbf{0}}^{\infty} \mathbf{u_m}(\mathbf{x},\mathbf{t})\mathbf{q}^{\mathbf{m}} = \mathbf{u_0}(\mathbf{x},\mathbf{t}) + \sum_{\mathbf{m}=\mathbf{1}}^{\infty} \mathbf{u_m}(\mathbf{x},\mathbf{t})\mathbf{q}^{\mathbf{m}}$$

where

$$u_m(\mathbf{x},\mathbf{t}) = \frac{\mathbf{1}}{\mathbf{m}!}\frac{\partial\varphi^{\mathbf{m}}(\mathbf{x},\mathbf{t};\mathbf{q})}{\partial\mathbf{q}^{\mathbf{m}}}\bigg|_{q=0}$$

If the parameters and functions are chosen such that the approximation converges at $q = 1$, then we have

$$\varphi(\mathbf{x},\mathbf{t};\mathbf{1}) = \sum_{\mathbf{m}=\mathbf{0}}^{\infty} \mathbf{u_m}(\mathbf{x},\mathbf{t})\mathbf{q}^{\mathbf{m}}\bigg|_{\mathbf{q}=\mathbf{1}} = \mathbf{u_0}(\mathbf{x},\mathbf{t}) + \sum_{\mathbf{m}=\mathbf{1}}^{\infty} \mathbf{u_m}(\mathbf{x},\mathbf{t})$$

which must be one of the solutions to the original equation.

1.2 Review of the DHAM

As the HAM is not suited to solving partial differential equations (PDEs) because of analytical difficulties to work out the high order approximations, Ghanbari-Rada-Chen [4] modified the HAM and introduced DHAM which does not require working out high order terms. In the following we shortly review DHAM.

Many problems in image processing lead to the following minimization equation:

$$\min_u\{f(u(x))\}$$

In most of those problems Euler-Lagrange equation has been derived in the form $E(u(x)) = 0$. This leads to the following time marching scheme

$$u_t(x,t) = -E(u(x,t)).$$

For HAM, we have

$$\varphi(x,t;q)=u_0(x,t)+\sum_{m=1}^{\infty}u_m(x,t)q^m$$

and so

$$\varphi(x,t;q)|_{q=0}=u_0(x,t),\qquad \left.\frac{\partial\varphi(x,t;q)}{\partial q}\right|_{q=0}=u_1(x,t),$$

$$\left.\frac{\partial\varphi(x,t;q)}{\partial t}\right|_{q=0}=\frac{\partial u_0(x,t)}{\partial t},\qquad \left.\frac{\partial^2\varphi(x,t;q)}{\partial t\partial q}\right|_{q=0}=\frac{\partial u_1(x,t)}{\partial t}.$$

Define the above non-linear equation as

$$\mathcal{N}[\varphi(x,t;q)]=\varphi_t(x,t;q)+E(\varphi(x,t;q)) \tag{1}$$

and choosing the linear operator

$$\mathcal{L}(\varphi(x,t;q))=\varphi_t(x,t;q)+\theta\varphi(x,t;q)\qquad \text{s.t.}\qquad \mathcal{L}^{-1}=e^{-\theta t}\int_0^t e^{\theta\tau}\varphi(x,\tau;q)d\tau$$

and recall that

$$\mathcal{L}(u_m(x,t)-\chi_m u_{m-1}(x,t))=\hbar\mathcal{H}R_m(\boldsymbol{u}_{m-1}(x,t))$$

$$\left.\frac{\partial^\ell}{\partial q^\ell}(\varphi(x,t,q))\right|_{q=0}=\ell!u_\ell(x,t)$$

Solving for u_1: Letting $\varpi(x,t,0)=(\partial\varphi(x,t,q)/\partial q)|_{q=0}=u_1(x,t)$

$$\overbrace{\mathcal{L}(\varpi(x,t;0))}^{\mathcal{L}(u_1(x,t))}=\overbrace{\varpi_t(x,t;0)}^{(u_1(x,t))_t}+\theta\,\overbrace{\varpi(x,t;0)}^{u_1(x,t)}$$

and hence

$$\mathcal{L}(u_1(x,t))=(u_1(x,t))_t+\theta u_1(x,t)=\hbar\mathcal{H}R_1(\boldsymbol{u}_0(x,t))$$

which gives

$$\frac{\partial u_1(x,t)}{\partial t}+\theta u_1(x,t)=\hbar\mathcal{H}R_1(u_0(x,t))=\hbar e^{-2t}E(u_0(x,t)) \tag{2}$$

Solving for u_2: Letting

$$\varpi(x,t,0)=(\partial^2\varphi(x,t,q)/\partial q^2)/2|_{q=0}-(\partial\varphi(x,t,q)/\partial q)|_{q=0}=u_2(x,t)-u_1(x,t)$$

$$\overbrace{\mathcal{L}(\varpi(x,t;0))}^{\mathcal{L}(u_2(x,t)-u_1(x,t))}=\overbrace{\varpi_t(x,t;0)}^{(u_2(x,t)-u_1(x,t))_t}+\theta\,\overbrace{\varpi(x,t;0)}^{u_2(x,t)-u_1(x,t)}$$

and hence

$$\mathcal{L}\left(u_2(x,t)-u_1(x,t)\right)=(u_2(x,t)-u_1(x,t))_t+\theta(u_2(x,t)-u_1(x,t)) \quad (3)$$

$$=(u_2(x,t))_t-(u_1(x,t))_t+\theta u_2(x,t)-\theta u_1(x,t)=\hbar\mathcal{H}R_2\left(\boldsymbol{u}_1(x,t)\right) \quad (4)$$

which gives

$$\frac{\partial u_2(x,t)}{\partial t}+\theta u_2(x,t)=\frac{\partial u_1(x,t)}{\partial t}+\theta u_1(x,t)+\hbar\mathcal{H}R_2(\boldsymbol{u}_1(x,t)) \quad (5)$$

Solution procedure

1. Given the Euler Lagrange equation $E(u)$, write

$$\mathcal{N}[\varphi]=\varphi_t+E(\varphi).$$

2. Evaluate

$$R_m=\frac{\partial^{m-1}}{\partial q^{m-1}}\mathcal{N}[\varphi]$$

for $m=0,1,...$ using

$$\frac{\partial^m\varphi}{\partial q^m}=m!u_m(x,t).$$

3. Find u_m where u_0 is given (and not regarded as a function of time) and u_m are given by the solutions of the ODE

$$\frac{\partial u_m}{\partial t}+\theta u_m=\hbar\mathcal{H}(R_m+\chi R_{m-1})$$

whose solution may be given by

$$u_m=\chi_{m-1}u_{m-1}+\hbar e^{-\theta t}\int_0^t e^{v(\theta-2)}R_m dv$$

4. Find

$$\varphi(x,t;1)=\sum_{m=1}^{\infty}u_m(x,t)q^m$$

.

2 A Restarted Iterative Homotopy Analysis Method for Image Deblurring with Poisson Noise

The restoration of blurred images corrupted by Poisson noise is an important topic in imaging processing. Deblurring Poisson noisy images has various applications in fluorescent confocal microscopy imaging, single particle emission computed tomography (SPECT), positron emission tomography (PET), astronomical imaging, etc. One of the most successful and widely used restoration formula is obtained by optimizing the following minimisation of the variational energy functional:

$$f^P u(\mathbf{x}) = \int_\Omega \left(\Phi \left([h * u](\mathbf{x}) - z(\mathbf{x}) - z(\mathbf{x}) \ln \left(\frac{[h * u](\mathbf{x})}{z(\mathbf{x})} \right) \right) + \alpha \Psi \left(|\nabla u(\mathbf{x})|^2 \right) \right) d\mathbf{x}.$$

A particular choice for the additional function which is known to produce good results is $\Phi(s) = 2\sqrt{s + \beta}$ for small, non-negative $\beta \in \mathbb{R}_{>0}$. Note that choosing the function $\Phi(s) = s$ reduces the functional to the regularized model. The Euler Lagrange equation can be derived as

$$h^\dagger * \left(\Phi'(f_s(u)) \left(1 - \left(\frac{z}{h * u} \right) \right) \right) - \alpha \nabla \cdot \left(\Psi' \left(|\nabla u|^2 \right) \nabla u \right) = 0$$

where $f_s(u) = h * u - z - z \ln \left(\frac{h * u}{z} \right)$. The above equation can be shortly denoted as $E(u) = h^\dagger * (\Phi') = 0$.

Using DHAM one writes

$$\mathcal{N}[\varphi] = \varphi_t + f_{0,1} * [E(\varphi) f_{0,2}(\varphi)]$$

$$E(\varphi) = h^\dagger * \left(\Phi'(f_{0,3}(\varphi)) \left(1 - \frac{z}{h * \varphi} \right) \right) - \alpha \nabla \cdot (\Psi'(f_{0,4}(\varphi)) \nabla \varphi)$$

where

$$f_{0,1} = \delta \text{ or } \delta - \nabla, \quad f_{0,2}(\varphi) = 1 \text{ or } u, \quad f_{0,3}(\varphi) = h * \varphi - z - z \ln \left(\frac{h * \varphi}{z} \right), \quad f_{0,4}(\varphi) = |\nabla \varphi|^2.$$

Furthermore we write:

$$E(\varphi)|_{q=0} = h^\dagger * \left(\Phi'(f_{0,3}(u_0)) \left(1 - \frac{z}{h * u_0} \right) \right) - \alpha \nabla \cdot (\Psi'(f_{0,4}(u_0)) \nabla u_0)$$

and conduct the first derivative:

$$\begin{aligned} \frac{\partial E(\varphi)}{\partial q} &= h^\dagger * \left(\Phi''(f_{0,3}(\varphi)) \frac{\partial f_{0,3}(\varphi)}{\partial q} \left(1 - \frac{z}{h * \varphi} \right) + \Phi'(f_{0,3}(\varphi)) \left(\frac{z (h * \varphi_q)}{(h * \varphi)^2} \right) \right) \\ &- \alpha \frac{\partial \nabla \cdot (\Psi'(f_{0,4}(\varphi)) \nabla \varphi)}{\partial q} \\ &= h^\dagger * (f_{1,1}(\varphi)(h * \varphi_q)) - \alpha \frac{\partial \nabla \cdot (\Psi'(f_{0,4}(\varphi)) \nabla \varphi)}{\partial q} \end{aligned} \tag{6}$$

where

$$\frac{\partial f_{0,3}(\varphi)}{\partial q} = \left(1 - \frac{z}{h * \varphi} \right) (h * \varphi_q)$$

$$f_{1,1}(\varphi) = \Phi''(f_{0,3}(\varphi)) \left(1 - \frac{z}{h * \varphi} \right)^2 + \frac{\Phi'(f_{0,3}(\varphi)) z}{(h * \varphi)^2}$$

$$\frac{\partial \nabla \cdot (\Psi'(f_{0,4}(\varphi)) \nabla \varphi)}{\partial q} = \nabla \cdot (f_{1,2}(\varphi) \nabla \varphi_q) \tag{7}$$

$$f_{1,2}(\varphi) = 2\Psi''(f_{0,4}(\varphi))(\nabla\varphi)^2 + \Psi'(f_{0,4}(\varphi))$$

From the above we can write:

$$\left.\frac{\partial E(\varphi)}{\partial q}\right|_{q=0} = h^{\dagger} * (f_{1,1}(u_0)(h * u_1)) - \alpha\nabla\cdot(f_{1,2}(u_0)\nabla u_1)$$
$$= \sum_{i_1} \pi_{1,i_1}\left(h^{\dagger} * (f_{1,1}(u_0)(h * \xi_{1,i_1})) - \alpha\nabla\cdot(f_{1,2}(u_0)\nabla\xi_{1,i_1})\right) \quad (8)$$

$$\frac{\partial^2 E(\varphi)}{\partial q^2} = h^{\dagger} * \left(\frac{\partial f_{1,1}(\varphi)}{\partial q}(h * \varphi_q) + f_{1,1}(\varphi)\frac{\partial(h * \varphi_q)}{\partial q}\right) - \alpha\frac{\partial^2\nabla\cdot(\Psi'(f_{0,4}(\varphi))\nabla\varphi)}{\partial q^2}$$
$$= h^{\dagger} * \left(f_{2,1}(\varphi)(h * \varphi_q)^2 + f_{1,1}(\varphi)(h * \varphi_{qq})\right) - \alpha\frac{\partial^2\nabla\cdot(\Psi'(f_{0,4}(\varphi))\nabla\varphi)}{\partial q^2} \quad (9)$$

$$\frac{\partial f_{1,1}(\varphi)}{\partial q} = \frac{\partial\Phi''(f_{0,3}(\varphi))\left(1 - \frac{z}{h*\varphi}\right)^2}{\partial q} + \frac{\partial\frac{\Phi'(f_{0,3}(\varphi))z}{(h*\varphi)^2}}{\partial q}$$
$$= f_{2,1}(\varphi)(h * \varphi_q) \quad (10)$$

$$f_{2,1}(\varphi) = \Phi'''(f_{0,3}(\varphi))\left(1 - \frac{z}{h * \varphi}\right)^3 + 3\Phi''(f_{0,3}(\varphi))\left(1 - \frac{z}{h * \varphi}\right)\left(\frac{z}{(h * \varphi)^2}\right) - 2\frac{z\Phi'(f_{0,3}(\varphi))}{(h * \varphi)^3} \quad (11)$$

$$\frac{\partial^2\nabla\cdot(\Psi'(f_{0,4}(\varphi))\nabla\varphi)}{\partial q^2} = \frac{\partial\nabla\cdot(f_{1,2}(\varphi)\nabla\varphi_q)}{\partial q}$$
$$= \nabla\cdot\left(f_{2,2}(\varphi)(\nabla\varphi_q)^2 + f_{1,2}(\varphi)\nabla\varphi_{qq}\right) \quad (12)$$

$$f_{2,2}(\varphi) = 2(\nabla\varphi)\left(2\Psi'''(f_{0,4}(\varphi))(\nabla\varphi)^2 + 3\Psi''(f_{0,4}(\varphi))\right)$$

or

$$\left.\frac{\partial^2 E(\varphi)}{\partial q^2}\right|_{q=0} = h^{\dagger} * \left(f_{2,1}(u_0)(h * u_1)^2 + 2f_{1,1}(u_0)(h * u_2)\right) - \alpha\left.\frac{\partial^2\nabla\cdot(\Psi'(f_{0,4}(\varphi))\nabla\varphi)}{\partial q^2}\right|_{q=0}$$
$$= \sum_{i_1}\sum_{j_1}\pi_{1,i_1}\pi_{1,j_1}\left(h^{\dagger} * [f_{2,1}(u_0)(h * \xi_{1,i_1})(h * \xi_{1,j_1})] - \alpha\nabla\cdot\left(f_{2,2}(u_0)(\nabla\xi_{1,i_1})(\nabla\xi_{1,j_1})\right)\right)$$
$$+\sum_{i_2}\pi_{2,i_2}\left(h^{\dagger} * [2f_{1,1}(u_0)(h * \xi_{2,i_2})] - 2\alpha\nabla\cdot(f_{1,2}(\varphi)\nabla(\xi_{2,i_2}))\right) \quad (13)$$

High order derivatives can be derived in similar way. Once the derivatives are obtained we have:

$$R_1 = \mathcal{N}[\varphi]|_{q=0} = f_{0,1} * [E(u_0)f_{0,2}(u_0)]$$

$$R_2 = \left.\frac{\partial\mathcal{N}[\varphi]}{\partial q}\right|_{q=0} = (\varphi_q)_t + f_{0,1} * \left.\left[\frac{\partial E(\varphi)}{\partial q}f_{0,2}(u_0) + E(u_0)\frac{\partial f_{0,2}(\varphi)}{\partial q}\right]\right|_{q=0}$$

Experimental results of the proposed method are shown the experimental section.

3 Deblurring with Gaussian Noise

In similar way we consider Gaussian debluring problem with minimisation variational energy functional

$$f(u(\mathbf{x})) = \int_\Omega ([h * u](\mathbf{x}) - z(\mathbf{x}))^2 d\Omega + \alpha \int_\Omega |\nabla u(\mathbf{x})|_\beta d\Omega.$$

The associated Euler Lagrange equation is given by

$$E_C(u) = h^\dagger * (h * u - z) - \alpha \nabla \cdot \left(\frac{\nabla u}{|\nabla u|_\beta} \right).$$

We evaluate

$$E_C(\varphi) = h^\dagger * (h * \varphi - z) - \alpha \nabla \cdot \left(\frac{\nabla \varphi}{|\nabla \varphi|_\beta} \right) = h^\dagger * (h * u_0 - z) - \alpha \nabla \cdot \left(\frac{\nabla u_0}{|\nabla u_0|_\beta} \right) \tag{14}$$

$$\frac{\partial E_C(\varphi)}{\partial q} = h^\dagger * h * \varphi_q - \alpha\beta \nabla \cdot \left(\frac{\nabla \varphi_q}{|\nabla \varphi|_\beta^3} \right) = h^\dagger * h * u_1 - \alpha\beta \nabla \cdot \left(\frac{\nabla u_1}{|\nabla u_0|_\beta^3} \right) \tag{15}$$

$$\begin{aligned} \frac{\partial^2 E_C(\varphi)}{\partial q^2} &= h^\dagger * h * \varphi_{qq} - \alpha\beta \nabla \cdot \left(\frac{\nabla \varphi_{qq}}{|\nabla \varphi|_\beta^3} - 3 \frac{(\nabla \varphi_q)^2 \nabla \varphi}{|\nabla \varphi|_\beta^5} \right) \\ &= 2h^\dagger * h * u_2 - \alpha\beta \nabla \cdot \left(\frac{\nabla u_2}{|\nabla u_0|_\beta^3} - 3 \frac{(\nabla u_1)^2 \nabla u_0}{|\nabla u_0|_\beta^5} \right) \end{aligned} \tag{16}$$

The equations for R_m are then given by

$$R_m = \frac{\partial^{m-1}}{\partial q^{m-1}} \left(\frac{\partial \varphi}{\partial t} + E_C(\varphi) \right).$$

We now begin to evaluate the terms u_m. The initial term u_0 is given by the received data z. Moving onto the term u_1, we can note that the function R_1 may be written as the composition of functions involving space σ_m^i and those involving time π_m^i. Indeed, we have

$$R_1 = \varphi_t + E_C(\varphi) = (u_0)_t + E_C(\varphi) = \sigma_1^1, \qquad \sigma_1^1 = h^\dagger * (h * u_0 - z) - \alpha \nabla \cdot \left(\frac{\nabla u_0}{|\nabla u_0|_\beta} \right)$$

and so we can calculate

$$u_1 = \hbar e^{-\theta t} \int_0^t e^{v(\theta - 2)} R_1 dv = \hbar e^{-\theta t} \left(\int_0^t e^{v(\theta-2)} dv \right) \sigma_1^1 = \hbar e^{-\theta t} \frac{e^{t(\theta-2)} - 1}{\theta - 2} \sigma_1^1 = \pi_1^1 \sigma_1^1$$

where

$$\pi_1^1 = \hbar \frac{e^{-2t} - e^{-\theta t}}{\theta - 2}.$$

Now, we consider the case $m = 2$. It is important to note that R_2 will be a function of space and of time. We have

$$R_2 = \frac{\partial}{\partial q}(\varphi_t + E_C(\varphi)) = (u_1)_t + h^\dagger * h * u_1 - \alpha\beta\nabla \cdot \left(\frac{\nabla u_1}{|\nabla u_0|^3_\beta}\right)$$

$$= \sigma_1^1(\pi_1^1)_t + \pi_1^1\sigma_2^1, \qquad \sigma_2^1 = h^\dagger * h * (\sigma_1^1) - \alpha\beta\nabla \cdot \left(\frac{\nabla \sigma_1^1}{|\nabla u_0|^3_\beta}\right)$$

and so we can calculate

$$u_2 = u_1 + \hbar e^{-\theta t}\int_0^t e^{v(\theta-2)}R_2 dv = \left(\pi_1^1 + \hbar e^{-\theta t}\Big(\int_0^t e^{v(\theta-2)}(\pi_1^1)_v dv\Big)\right)\sigma_1^1 + \hbar e^{-\theta t}\Big(\int_0^t e^{v(\theta-2)}\pi_1^1 dv\Big)\sigma_2^1$$

$$= \pi_2^1\sigma_1^1 + \pi_2^2\sigma_2^1$$

where

$$\pi_2^1 = \pi_1^1 + \hbar e^{-\theta t}\Big(\int_0^t e^{v(\theta-2)}(\pi_1^1)_v dv\Big) = \hbar\frac{e^{-2t} - e^{-\theta t}}{\theta - 2} + \hbar^2\Big(\frac{(\theta-2)}{2(\theta-4)} - \frac{2e^{t(\theta-4)}}{(\theta-2)(\theta-4)} - \frac{\theta e^{-2t}}{2(\theta-2)}\Big)e^{-t\theta}$$

$$\pi_2^2 = \hbar e^{-\theta t}\Big(\int_0^t e^{v(\theta-2)}\pi_1^1 dv\Big) = \hbar^2\Big(\frac{-1}{2(\theta-4)} + \frac{e^{t(\theta-4)}}{(\theta-2)(\theta-4)} + \frac{e^{-2t}}{2(\theta-2)}\Big)e^{-t\theta}.$$

4 Experimental Results

4.1 Results for the Poisson Case

In order to illustrate the speed and the accuracy of the proposed DHAM restoration algorithms, we evaluate the performance of the proposed method on real images data sets in terms of restoration quality and computational speed. The PSNR has been used as an indicator of image restoration quality. In the following experiments the parameters $\alpha = .1$, maxiteration= 10^4, $\beta = 10^{-6}$, $dt = 0.01$, and $h = 1$ (step size). In each experiment, the kernel size k has been varying from 3×3 to 11×11 and precision of 10^{-1} and 10^{-2} has been used to stop the program. Figures 1, 2, 3 and 4 not only show the validity for the proposed algorithm but shows outperforming with state of art method of Dey et al. [3]not sure. In those images, we noticed that the proposed discrete DHAM has better restoration results for $n = 2$. The method shows better performance for $n = 3$ but the results are not shown due to page limitation. We noticed through experiment that increasing of $n > 3$ does not bring benefits as the rounding of error increases. To fix such a problem parameters optimization is required and might be considered as future work.

Table 1 shows the CPU and the PSNR results of a set of 10 other random images of the proposed method in comparison with TM and Dey et al. [3] model. The results show that DHAM1 competes with Dey et al. [3] by having almost same PSNR with a much higher speed, whereas DHAM2 not only improves DHAM1 speed but improves significantly the PSNR compared to Dey et al. [3].

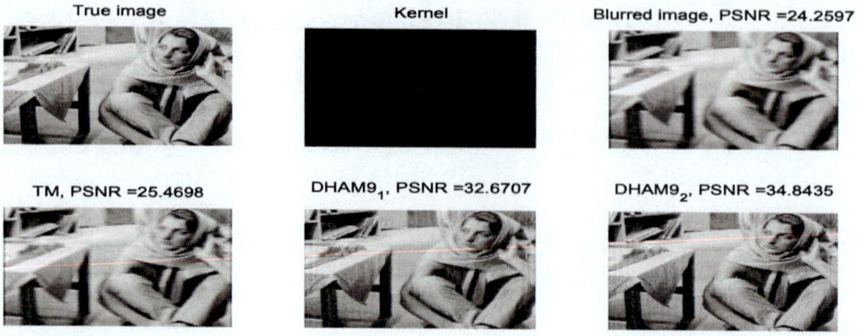

Fig. 1. Ex 1. Comparison of discrete HAM with TM for the image of Barbara.

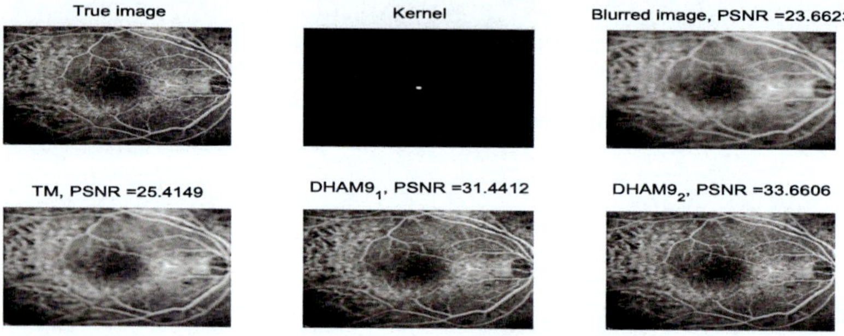

Fig. 2. Ex 2. Comparison of discrete HAM with TM for eye images.

4.2 Unconstrained Gaussian Noise Reconstruction

Figure 5 shows the validity of the proposed method for reconstruction of Gaussian unconstrained noise for two real life images. From the experiments we noticed that the DHAM evolves and finds better reconstructed data in comparison with TM, and more importantly a great CPU improvement is achieved. The parameters used are: maxiteration= 10^4, $beta = 10^{-2}$, $alpha = 10^{-3}, and$

Fig. 3. Ex 3. Comparison of discrete HAM with TM for home images

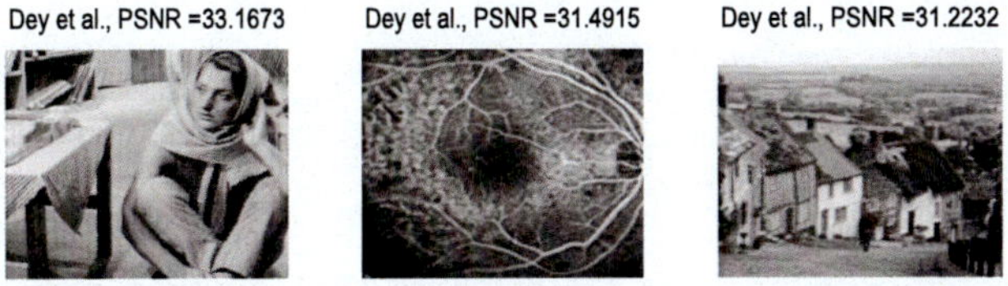

Fig. 4. Comparison of discrete HAM Model with Dey et al. [3].

Table 1. Quality and speed comparison of the proposed methods with TM and Dey et al. [3].

Image	PSNR before	Dey et al.		TM		DHAM1		DHAM2	
		PSNR	CPU	PSNR	CPU	PSNR	CPU	PSNR	CPU
$no:1$	22.45	24.75	16.35	30.03	18.55	29.76	13.35	31.21	10.90
$no:2$	23.19	24.99	16.05	27.90	17.67	27.75	14.34	28.20	11.78
$no:3$	24.26	24.80	18.00	33.10	18.00	32.67	17.65	34.84	11.65
$no:4$	23.66	24.54	14.53	31.49	18.43	31.44	14.32	33.66	8.43
$no:5$	23.93	24.67	15.26	28.60	18.57	28.57	14.71	29.39	9.30
$no:6$	25.74	26.69	15.54	35.73	19.45	35.53	14.15	37.74	9.76
$no:7$	23.65	25.86	15.04	31.22	19.30	30.86	14.65	32.72	10.98
$no:8$	23.97	25.34	16.36	28.07	19.78	28.45	14.30	29.03	13.65
$no:9$	26.17	27.00	18.00	35.22	18.00	34.63	15.45	36.70	11.53
$no:10$	22.52	25.16	15.98	31.47	18.37	31.08	14.52	33.93	10.34

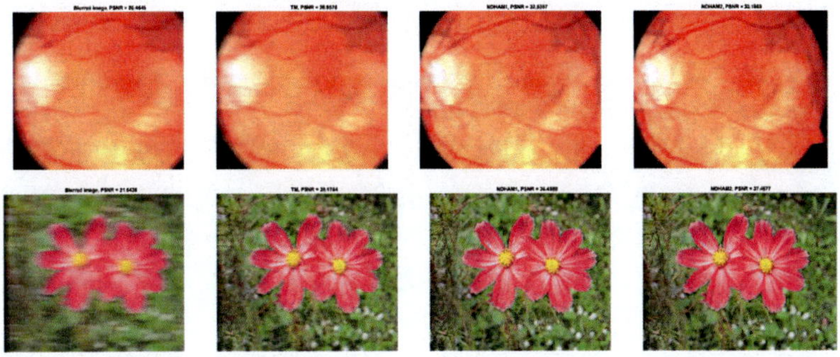

(a) Given image (b) TM (c) HAM1 (d) HAM2

Fig. 5. Unconstrained Gaussian Noise reconstruction.

$dt = 0.01$. The experimental results show similar behavior in terms of quality and speed improvement to Poison noise similar to Table 1. To avoid redundancy the results obtained are not shown in a table but can be provided to researcher.

References

1. Rada, L., Ghanbari, B., Chen, K.: A restarted iterative homotopy analysis method for two nonlinear models from image processing. Int. J. Comput. Math. **9**(3), 661–687 (2014)
2. Chan, T., Marquina, A., Mulet, P.: High-order total variation-based image restoration. SIAM J. Sci. Comput. **22**(2), 503–516 (2000)
3. Dey, N., Blanc-Feraud, L., Zimmer, C., Roux, P., Kam, Z., Olivo-Marin, J.-C., Zerubia, J.: Richardson-lucy algorithm with total variation regularization for 3d confocal microscope deconvolution. Microsc. Res. Tech. **69**(4), 260–266 (2006)
4. Ghanbari, B., Rada, L., Chen, K.: A restarted iterative homotopy analysis method for two nonlinear models from image processing. Int. J. Comput. Math. **91**(3), 661–687 (2014)
5. Fan, J., Li, F., Shen, C.: Image restoration combining a total variational filter and a fourth-order filter. J. Vis. Commun. Image Res. **18**, 322–330 (2007)
6. Liao, S.J.: The proposed homotopy analysis technique for the solution of nonlinear problems. Ph.D. Thesis, Shanghai Jiao Tong University (1992)
7. Liao, S.J.: Beyond Perturbation: Introduction to the Homotopy Analysis Method. CRC Press, Chapman Hall (2003)
8. Liu, J., Liu, G., Huang, T.: High-order tvl1-based images restoration and spatially adapted regularization parameter selection. Comput. Math. Appl. **67**, 2015–26 (2014)
9. Wang, S., Lv, X., Song, Y.: Image restoration with a high-order total variation minimization method. Appl. Math. Model **37**, 8210–8224 (2013)
10. Lysaker, O.M., Tai, X.-C.: Iterative image restoration combining total variation minimization and a second-order functional. Int. J. Comput. Vis. **66**, 5–18 (2006)
11. Tam, C., Nikolova, M., Ng, M.: Fast nonconvex nonsmooth minimization methods for image restoration and reconstruction. IEEE Trans. Image Process. **19**(12), 3073–3088 (2010)
12. Rudin, L., Osher, S., Fatemi, E.: Nonlinear total variation based noise removal algorithms. Phys. D **60**, 259–268 (1992)
13. Tikhonov, A., Arsenin, V.: Solution of Ill-posed Problems. Winston, Washington, DC (1977)
14. Liu, J., Xu, Y., Huang, T.: An augmented lagrangian algorithm for total bounded variation regularization based image deblurring. J. Franklin Inst. **351**, 3053–67 (2014)

A New Approach for Macroscopic Analysis to Improve the Technical and Economic Impacts of Urban Interchanges on Traffic Networks

Seyed Hassan Hosseini[1(✉)], Ahmad Mehrabian[2], Zhila Dehdari Ebrahimi[3], Mohsen Momenitabar[3], and Mohammad Arani[4]

[1] Department of Civil and Industrial Engineering, Sapienza University of Roma, Roma, Italy
hosseini.1851467@studenti.uniroma1.it

[2] Department of Industrial Engineering, Aliabad Katoul Branch, Islamic Azad University, Aliabad-e-Katul, Iran
ahmad.mehrabian@outlook.com

[3] Department of Transportation, Logistics and Finance, College of Business, North Dakota State University, Fargo, ND 58105, USA
{Zhila.dehdari,mohsen.momenitabar}@ndsu.edu

[4] University of Arkansas at Little Rock, 2801 S. University Ave., Little Rock, AR 72204, USA
mxarani@ualr.edu

Abstract. Pursuing three important elements including economic, safety, and traffic are the overall objective of decision evaluation across all transport projects. In this study, we investigate the feasibility of the development of city interchanges and road connections for network users. To achieve this goal, a series of minor goals are required to be met in advance including determining benefits, costs of implementing new highway interchanges, quantifying the effective parameters, the increase in fuel consumption, the reduction in travel time, and finally influence on travel speed. In this study, geometric advancement of Hakim highway, and Yadegar-e-Emam Highway were investigated in the Macro view from the cloverleaf intersection with a low capacity to a three-level directional intersection of the enhanced cloverleaf. For this purpose, the simulation was done by EMME software of INRO Company. The results of the method were evaluated by the objective of net present value (NPV), and the benefit and cost of each one was stated precisely in different years. At the end, some suggestion has been provided.

Keywords: Urban interchanges · Macroscopic analysis · Net present value · EMME software

1 Introduction

Annually, billions of dollars are spent on traffic performance improvement, smoothing traffic flow, and reducing travel time in metropolises (Bigdeli Rad and Bigdeli Rad 2018). Construction of a grade-separated interchange is a high-cost project, which takes place for physical separation of traffic routes, continuous traffic flow, decreasing delay

T. Allahviranloo et al. (Eds.): IDS 2020, AISC 1301, pp. 595–610, 2021.
https://doi.org/10.1007/978-3-030-66501-2_47

time, and increasing capacity of intersections. However, evidence has shown that the construction of a grade-separated interchange can increase travel time and delay time in the adjacent intersection (Khan and Anderson 2016). So, choosing the best type of interchange under different traffic conditions can greatly mitigate this adverse effect (Arani et al. 2020a; Arani et al. 2020d; Arani et al. 2020b; Arani et al. 2020c; Arani et al. 2019; Chan et al. 2020; Momeni Tabar et al. 2014; Momeni Tabar et al. 2020; Sutherland et al. 2018; Zhila Dehdari Ebrahimi, Raj Bridgelall 2020).

One of the most important processes of planning is evaluating several different options to achieve a common goal to ensure that the best option is chosen, and other options are excluded. Increasing the financial value of the project leads to the increasing importance of economic evaluation (Mansourkhaki and Ghanad 2014). The purpose of constructing interchanges is to creating levels through which vehicles can move in different directions without collisions with other vehicles. Traffic capacity per line decreases in at-grade interchanges (Lucietti et al. 2016). Constructing interchanges with different levels, are the best, safest, and fastest and at the same time the most expensive solution to passing cross directions. One of the ways to increase the capacity of interchanges is by promoting the geometric characteristics of interchanges (Afandi Zade et al. 2019; Chowdhury 2016). An interchange can be an adaptable solution to improve many intersection conditions either by reducing existing traffic bottlenecks or by reducing crash frequency (Mollu et al. 2017). However, the high cost of constructing an interchange limits its use to those cases where an additional expenditure can be justified. Enumeration of the specific conditions or warrants justifying an interchange at a given intersection is difficult, and in some instances, cannot be conclusively stated. Because of the wide variety of site conditions, traffic volumes, highway types, and interchange layouts, the warrants that justify an interchange may differ at each location. The following six conditions, or warrants, should be considered when determining if an interchange is justifiable at a particular site (of State Highway and Officials 2011): Design designation, Reduction of bottlenecks or congested spot, Reduction of collisions frequency and severity, Site topography, Road-user benefits, and Traffic volume warrant.

The first cloverleaf was opened in Woodbridge Township, New Jersey in 1929, at the junction of Routes 4 and 25. This is now the intersection of US 1/9 and NJ 35, and since about 2004, the interchange is no longer a cloverleaf. The classic cloverleaf allows "non-stop" full access between two busy roads. Traffic merges and weaves but does not cross at-grade; unless the interchange is heavily congested, no stop is required. The colloquial "cloverleaf" is the same as the technical "full cloverleaf", as one can omit ramps to get a partial one. The cloverleaf, which is studies here, is the simplest way to connecting two freeways. Bridges are only required to separate two roadways. If the land is expensive, which becomes a choice between tight turning radii (and lower design speed). Note that most loop ramps are banked to counteract centrifugal forces (of State Highway and Officials 2011). Typically, a cloverleaf is used where a freeway intersects a busy surface street, though many older freeway interchanges are also considered cloverleaves (Chan et al. 2020). In several places, cloverleaves have been replaced with either signalized interchanges or higher-capacity directional interchanges with flyovers.

2 Method

Implementation and exploitation of each transportation plan in road networks can have a positive effect on transport and traffic flow of a city, region, or area, resulting in improved traffic conditions. But what is of considerable importance is comparing the benefits and costs of implementing the plan to assess its economic value. In this study, a common method is presented for the economic evaluation of the implementation of a traffic plan as a grade-separated interchange. According to Fig. 1, in this study, there is a new way of looking at the highway interchanges, which is a macroscopic view. Thus, it is expressed as an element of the urban transportation network (Zhila Dehdari Ebrahimi 2017) with a very broad sphere of influence and effects of trans-regional and macroscopic view. Therefore, two logical scenarios were defined and after the simulation in macroscopic software EMME, network information was obtained. To compare the parameters of travel time, safety, environment, and fuel, these parameters are converted to economic values (dollars) to determine to what extent the plan is justifiable.

In this study, using the statistical models based on Tehran Transportation and Traffic Studies and employing EMME, the macroeconomic software, a wide variety of effects regarding the changes in the highway network was investigated. In this case, for 15 years, these effects on the highway network, due to the increase in the population and ownership of the car, and public transportation were reviewed.

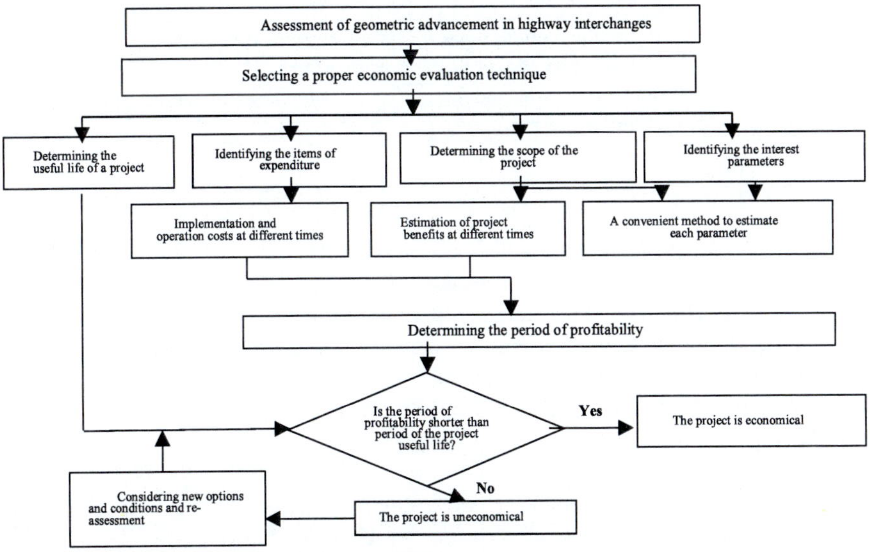

Fig. 1. The methodology of the study

Previous studies have shown that these long-term changes could be beneficial or possibly harmful to the community. On the other hand, the point that is often neglected in this assessment (despite the great importance) of highway exchange projects is the costs incurred by users of the urban transportation network during the execution of such projects. These costs include costs of inaccessibility, increasing the length of the route, changing certain routes that were used in the past, land use, etc. Typically, traffic peak-hours at 2 stages in the morning and night are the basis for designing the traffic load. But the very important point is that in analyzing and evaluating projects, it should not be evaluated economically merely according to the peak hours of the city. Due to the low traffic volume of vehicles, the time spent, and the increase in fuel consumption, there is no increase in environmental pollution. So, as the peak-based design is an overestimated design, it can be useful, but the economic assessment based on peak hours of the city has a high error rate and is somewhat irrefutable. But in the current research methodology, the evaluation is based on 24 hours a day, which indicates high accuracy. The contribution of this paper can be seen in Table 1.

Table 1. Summary of contribution (Baldauf et al. 2013)

Benefits	Aim	Method	Software
Travel time reduction	Estimating travel time in both the existing and the proposed scenarios	Macro simulation in the transportation system	EMME
Fuel consumption reduction	Estimating fuel consumption in both the existing and the proposed scenarios	Hicks and Clarkson Model	EMME
Air pollution reduction	Estimating various types of air pollutants emissions in both the existing and the proposed scenarios	Calibrated models in the laboratories of fuel consumption optimization organization	EMME
Collision reduction	Estimating the number of collisions	A linear relationship between speed and collisions	–

In this study, the method of "net present value" is recommended to assess the economic value of intersections. Net Present Value (NPV) is a formula used to determine the present value of an investment by the sum of all cash flows received from the project over time (Gonzalez-Ruiz et al. 2017). The formula for the sum of all cash flows can be rewritten as:

$$NPV = -C_0 + \sum_{i=1}^{T} \frac{C_i}{(1+r)^i} \quad (1)$$

When a company or investor takes on a project or investment, it is important to calculate an estimate of how profitable the project or investment would be (Gonzalez-Ruiz et al. 2017). NPV is equal to the equivalent value of the profits from the investment regarding the start time of the project, where the value of operational costs, maintenance cost, and cost of final equipment exploitation is considered by the monetary value of currency over time (Pfeiffer 2004). Based on economic principles, exploitation of an at-grade separated intersection would be economically justified when the desired profitability is reached before the end of its lifetime (Madanu et al. 2010). Costs and benefits of the project in the years after the start of exploitation will be the basis for determining the NPV index, in which it will be determined with software EMME. To evaluate the effectiveness of the proposed method, two different designs of at-grade separated intersections were considered at the current intersection. The proposed plan of alternative intersection is outlined in Fig. 2, which is a three-level directional intersection, where all the turnings are grade-separated and independent, which will be done with high speed. The movements are the two left-turn movements and the major routes of Mohammad Ali Jinnah in the underpass of Hakim Highway Bridge, the turn-right movement on the ground floor, and the two left-turn movements on the upper floor.

According to the proposed plan, the intersection is combined with six bridges where two of them are the main link between the eastern and western parts of the city and the other four bridges are for the ramp. Including improved areas in the main highways, this plan has about 8 KM circulation paths and ramp, 25000 m^2 bridges and more than 1000 m^2 walls.

The current plan of Hakim and Yadegar-e Emam highway is a cloverleaf intersection (Fig. 2), but the northwest ramp has occupied a very large area, increasing the length of the route and reduces driver sight distance in parts of the route. Further, at this intersection due to the low width at the main bridge, ramps, and loops, current demand cannot be satisfied. There will be heavy congestion during peak hours. In this plan, the current cloverleaf intersection will expand, and the width of ramps, loops, and bridges will increase. The geometric design of the northwest loop will become a standard loop with appropriate sight distance and the enhanced cloverleaf intersection will develop with more capacity. This plan has been shown in Fig. 3.

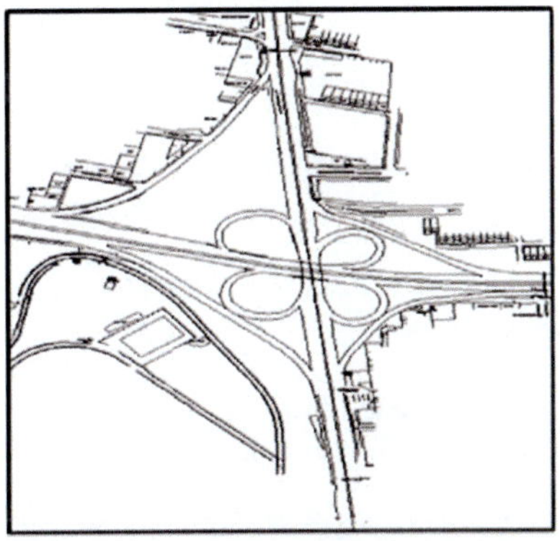

Fig. 2. The proposed plan for cloverleaf interchange versus status quo

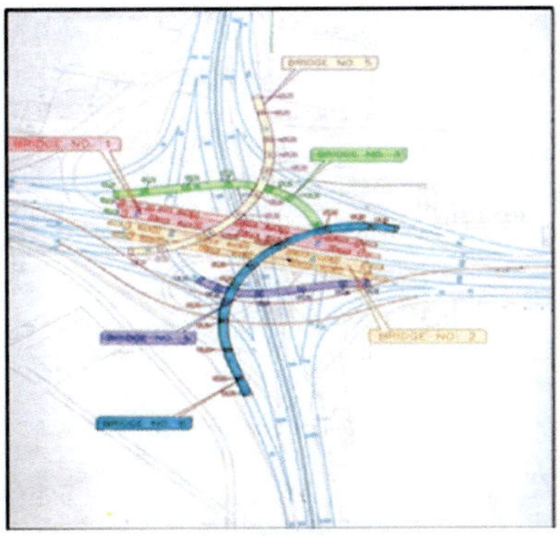

Fig. 3. Proposed plan for directional interchange versus status quo

As the simulation process in the current situation, in the construction phase and years after employment is the most important step in the estimation of costs imposed on citizens during the construction as well as the benefits arising from the utilization of the intersection. This section deals with the simulation process of transportation systems and traffic conditions in Tehran, implemented in EMME software.

Figure 4 demonstrates the general trend of demand estimated for internal trips of residents in Tehran. The demand estimation process of other trips of residents and non-residents travel is shown in Fig. 5. According to Fig. 4, at first, using calibrated models of generation and attraction, the daily trips of 560 traffic zones in Tehran are predicted. The prediction is according to non-home-based trips and home base trips (7).

Table 2. Trip generation and attraction models of Tehran for each trip purpose

Trip attraction models	Trip purpose
Work	$TA_i^w = 1/620EMPE_i + 2/420SHOP_i + 62694DB_i$
Education	$TA_i^s = 3/833VP_i * ST_i + 0/500STU_i + 26789DT_i + 9299D_{128i}$
Buy	$TA_i^{sh} = 15/760VP_i * SHOP_i + 0/195EMPE_i + 0/825HOSB_i + 15456DB_i + 3469DQ_i + 7474DF_i + 4607D_{444i} - 0/889SHOP_i * DRA_i$
Entertainment and other	$TA_i^{sr} = 122/140PARK_i + 0/040P_i + 7/364VP_i * SHOP_i + 0/304EMPE_i + 4098DR_i + 1937DF_i + 1532DQ_i - 0/279SHOP_i * DRA_i - 0/208EMPE_i * DRA_i$
Non-home base	$TA_i^{nhb} = 0/458EMPE_i + 11/526VP_i * SHOP_i + 11706DB_i + 1173DQ_i$
P_i = Per capita ownership of private car in traffic zone i ER_i = Resident employment in traffic zone i STR_i = The number of resident school students in traffic zone i $STUR_i$ = The number of resident university students in traffic zone i $EMPE_i$ = The number of staff in place of employment in traffic zone i $SHOP_i$ = The number of shops in traffic zone i ST_i = The number of students at school in traffic zone i STU_i = The number of students at university in traffic zone i $HOSPB_i$ = The number of hospital beds in traffic zone i	$PARK_i$ = The number of parks in traffic zone i DRA_i = Traffic zones covariate in traffic plan zone DB_i = Tehran Bazaar covariate (Traffic zone 1) DT_i = Tehran university covariate (Traffic zone 150) DQ_i = Main squares covariate (Imam Hossein, Enghelab, valiasr, Khorasan, Tajrish, and Imam Khomeini) in traffic zones 16, 174, 151, 121, 401, 537. DF_i = Covariate of Ghezel Ghale square, second Square of Sadeghieh mall, second mall of Nazi Abad in traffic zones 197, 232, 139. DR_i = Covariate of Mellat park, laleh park, Shahbdalzym and Behesht Zahra in traffic zones 452, 274, 148, and 444. D_{128} = Covariate of Amirkabir university, Art university and Alborz High School (Traffic zone 128) D_{444} = Covariate of Ray city (Traffic zone 444)

Due to trip purpose, some important vehicles are imported in the modeling. Note that the important vehicles are those vehicles with a major contribution to travel (9). Table 2 shows the calibration result of the trip generation and attraction model, and Table 3, Table 4, Table 5, Table 6, Table 7, and Table 8 provide the result of vehicle choice model calibration in Tehran. Vehicle choice models are Logit Models in this study.

Table 3. Trip purpose

Trip purpose	Trip generation models
Work	$T_i^w = 0/569VP_i * ER_i + 1/107ER_i$
Education	$T_i^s = 3/070VP_i * STR_i + 0/903STUR_i + 0/020DIST_i * P_i$
Buy	$T_i^{sh} = 0/061P_i + 0/414VP_i * P_i$
Entertainment	$T_i^{sr} = 0/073P_i + 0/373VP_i * P_i$
Non-home base	$T_i^{nhb} = 0/490EMPE_i + 10/213VP_i * SHOP_i + 14485DB_i + 951DQ_i$

Table 4. Vehicle choice models of Tehran for each trip purpose

Work trip		
First level	Car	$U_{CAR}^{ij} = 0.697568 - 0.034575 \times TIMCAR^{ij} + 9.008179 \times owncar^i - 0.588848 \times DESFLAG^j$
	Motorcycle	$U_{MOT}^{ij} = -0.480759 - 0.047222 \times TIMMOT^{ij} + 18.345253 \times OWNMOT^i$
	Public transportation	$U_{BUS,TAX}^{ij} = \theta\ ln\left[exp\left(U_{BUS}^{ij}\right) + EXP\left(U_{TAX}^{ij}\right)\right]$
Second level	Bus	$U_{BUS}^{ij} = 0.330393 - 0.020389 \times TIMIN^{BUS} - 0.026496 \times TIMBOT^{ij}$
	Taxi	$U_{TAXI}^{ij} = -0.048415 \times TIMTAX^{ij} + 3.100169 \times OWNCAR^i$

The results of the deviation of demand models from various vehicles (except bus) for public transportation due to new technology are provided in Table 9. Note that these models are the same for all types of intercity trips of residents (all trip purposes).

What is important in urban transportation planning is travel demand based on vehicles. Therefore, the demand matrix TP_{ij}^{mt} must be converted to travel demand matrix based on vehicle, for all vehicles except bus. As shown in Table 9, this is done using information about the average number of passengers. By adding freighter trips to the mentioned matrix, the final travel demand based on the vehicle in any period t can be achieved (similarly, except bus). According to the considered Passenger Car Equivalent

Table 5. Educational trip

Educational trip	
Bus	$U_{BUS}^{ij} = 0.8811690 - 0.012004 \times \left(TIMBIN^{ij} + TIMBOT^{ij} \right)$
Taxi	$U_{TAX}^{ij} = -0.0365572 - 0.030786 \times TIMTAX^{ij} + 8.253327 \times OWNCAR^{i}$
Car	$U_{CAR}^{ij} =$ $-1.1044833 - 0.041592 \times TIMCAR^{ij} + 11.324764 \times owncar^{i} - 0.582493 \times DESFLAG^{i}$
Minibus	$U_{MIB}^{ij} = -1.104768 \times DIST^{ij} + 6.648515 \times owncar^{i}$

Table 6. Shopping trip

Shopping trip	
Bus	$U_{BUS}^{ij} = 2.794484 - 0.013595 \times TIMBIN^{ij} - 0.015329 \times TIMBOT^{ij}$
Taxi	$U_{TAX}^{ij} = 1.967395 - 0.037180 \times TIMTAX^{ij} + 6.596312 \times OWNCAR^{i}$
Car	$U_{CAR}^{ij} = -0.015029 \times TIMCAR^{ij} + 12.443686 \times OWNCAR^{i} - 0.689367 \times DESFLAG^{j}$

Table 7. Recreational trip

Recreational trip	
Bus	$U_{BUS}^{ij} = 2.725886 - 0.009414 \times \left(TIMBIN^{ij} + TIMBOT^{ij} \right)$
Taxi	$U_{TAX}^{ij} = 2.393202 - 0.033543 \times TIMTAX^{ij} + 5.379732 \times OWNCAR^{i}$
Car	$U_{CAR}^{ij} = -0.015111 \times TIMCAR^{ij} + 13.957626 \times OWNCAR^{i} - 0.374195 \times DESFLAG^{j}$

Table 8. Non-home base trip

Non-home base trip	
Bus	$U_{BUS}^{ij} = 0.039002 - 0.008689 \times TIMBIN^{ij} - 0.041852 \times TIMBOT^{ij}$
Taxi	$U_{TAX}^{ij} = 0.334293 - 0.020176 \times TIMTAX^{ij}$
Car	$U_{CAR}^{ij} = -0.012662 \times TIMCAR^{ij} - 0.705396 \times DESFLAG^{j}$

Table 9. Results of calibration of demand deviation models

Vehicle type	b^m	a^m
Car	19/198	1/559
Taxi	2/668	4/794
Minibus	19/198	1/652
Bicycle	19/198	1/652

Table 10. Passenger car equivalent

Vehicle type	Car	Pickups	Taxi	Minibus	Bus	Motorcycle	Lorry
PCE	1	1	2	2.5	2.5	0.5	2.5

for each vehicle, these matrixes are converted as demand matrix in terms of Passenger Car Equivalent. Passenger Car Equivalent for each vehicle is provided in Table 10.

An assignment is the last step of transportation planning (UTPS) which consists of two parts. The first part is known as the car assignment. It is related to vehicles which have no fixed and certain route. Given the network that users can use, they try to minimize their travel time. The second part is the public transportation assignment, which is related to vehicles with a fixed and predetermined route, such as bus and subway. Users, according to a certain plan, prefer to minimize the expected travel time (Bigdeli Rad and Bigdeli Rad 2018). Based on the evidence in the Technical Organization of Tehran Municipality, the approximate costs of the items listed in each of the two proposals of the standard cloverleaf and directional interchanges are provided in Table 11.

Table 11. Direct costs of construction and operation

Cost type	Monetary equivalent (million $)	
	Standard cloverleaf interchange	Directional interchange
Construction	31.4	40.3
Acquisition and release	-	16.4
Annual Maintenance	0.55	0.55

3 Result

Estimation of the costs imposed on citizens during the operation and construction of interchanges will be computed according to the methods proposed in the previous chapter.

Simulation of the street network of Tehran in both the status quo and network status during the operation and construction is the basis of estimating these costs. For this purpose, the transport and traffic model in Tehran will be simulated in software EMME. The changes in network performance indicators because of constraints caused by interchanges construction are demonstrated in Table 12.

Table 12. The changes in network performance

Index		Distance Increasing (%)	Volume reduction (%)	Pollutants emission increasing (%)	Fuel consumption increasing (litter)	Wase-time increasing (hour)
Rate of Change	Directional	4.32	0.04	4.33	124325265	28620924
	Cloverleaf	35.8	0.028	3.51	98710254	22912723

Based on the method proposed in this study, the average value of travel time per person is required. For this purpose, the results related to the Tehran transportation master plan will be used, which is calculated by dividing GDP by the number of annual production hours of employed people. This index was calculated to be 2.004 dollars per hour in 2013. With increasing 28,620,924 hours in network travel time, due to construction of the directional interchange, 57.6 million dollars will be imposed on citizens. Moreover, the costs resulting from cloverleaf construction operation is estimated at 32.9 million dollars.

According to the Statistics of fuel consumption optimization organizations, fuel consumption in Tehran is 14% gasoline and 88% petrol. Due to the low percentage of gasoline and the low price, petrol is considered for all fuel consumption. The price of 1 litter petrol was equal to 0.7 dollars. Given 124,325,265 litters increase in fuel consumption caused by limitations resulting from the construction of the directional interchange, costs imposed on society are estimated at 86.6 million dollars. Also, such costs are 49.5 million dollars for constructing cloverleaf interchange.

The total direct costs resulting from accidents in Tehran are around 110 million dollars (based on the statistics presented by the Division of Insurance). Regarding value, about 4% are fatal accidents, 24% are injury accidents, and 72% are property damage accidents. With a growth of 4.32% in the traveled distances and increasing accidents, their costs are estimated to be 4.7 million dollars, which will further increase the finished cost of the project. On the other hand, the changes in traffic conditions through the black spots in the area of interest are very influential in the damage caused by accidents due to the geometric promotion of the interchange.

According to information from Tehran Traffic Police and Tehran Central Insurance Office, 3 important hot spots were determined in the plan area as demonstrated in Fig. 7. There is a direct relationship between the number of vehicles passing through the hot spots and the number of accidents and consequently the cost of accidents. A major change in the network leads to an increase or decrease in the capacity through the link. With estimating the traffic volume passing through these points before and after the promotion of the intended interchange, one can determine what percentage was the former traffic volume crossing the black spots and what percentage of the alternative will pass through routes or Bypass.

In 2006, the World Bank published a report stating that pollutants cost in Tehran traffic network is around 700 million dollars per year. These costs include deaths from air pollution, treatment including hospital stays, outpatient treatments, and the number of working days lost.

Table 13. The changes in network performance

Iran	NO_X	SO_2	CO	$NMVOC_S$
Value ($/tonne)	600	1.825	188	0.50.5

Accordingly, concerning the 4.33% increase in pollutant emission of HC, CO, and NOx based on Table 13, the additional costs imposed on the society caused by construction operation during construction of the directional interchange are estimated at 36.6 million dollars. Under similar circumstances, 21.5 million dollars is estimated for cloverleaf interchange (Table 14).

Table 14. The changes in network performance (exhaust emissions)

Pollutant	During construction (Kg)	Status quo (Kg)
CO	898952107.3	862434217.6
HC	113958772.6	108149728.5
NO_X	20669540.7	19821223.1

Thus, according to the description mentioned above, the total cost of the proposal of directional interchange and standard cloverleaf interchange at the beginning of the exploitation 2013 is presented to Table 15. Cost of the proposed plan of interchanges.

The basis of estimating these benefits is the estimation of the values of each cost factor and extent of reduction as the benefits of the plan. Thus, implementation of the three options will be discussed, including the "status quo", "directional interchange" and " standard cloverleaf interchange". These options were implemented in the Tehran transportation system simulation model with EMME software in the years after the operation. Based on the results presented in the above tables and comparing the two options of the standard cloverleaf and directional interchanges, the advantages of using these interchanges in the network in each of the studied years are provided in Fig. 4, Fig. 5, Fig. 6, and Fig. 7.

In Fig. 4, Fig. 5, Fig. 6, and Fig. 7, the comparison between benefits of using directional and cloverleaf interchanges is illustrated in different operation years for each of the benefits: reducing travel time, reducing fuel consumption, reducing accidents, and reducing pollutants.

Table 15. Cost of the proposed plan of interchanges

Cost type	Cost components	Cost (billion IRR)	
		Directional interchange	Cloverleaf interchange
Direct	Construction	403	314
	Acquisition and release	164	–
As a result of limitations in the network caused by the construction process	Travel time increasing	576	329
	Fuel consumption increasing	866	495
	Pollutants emission increasing	366	215
	Accidents increasing	47	23
Total		2422	1376

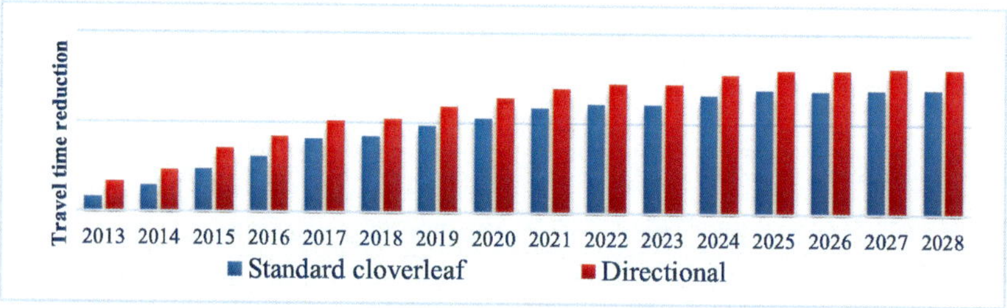

Fig. 4. Travel time reduction in different years

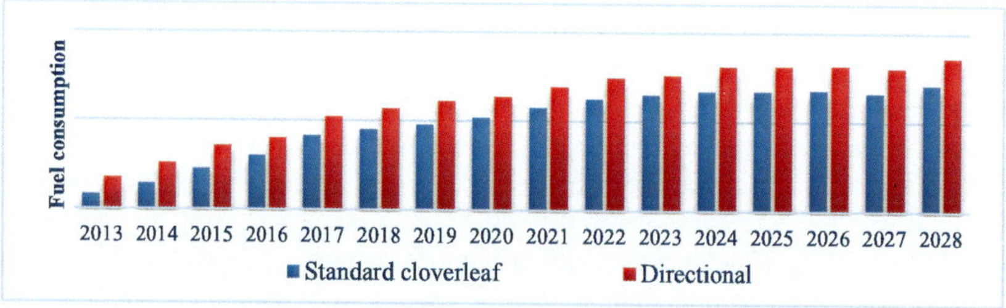

Fig. 5. Fuel consumption reduction in different years

In the comparison between the two proposed options for geometric promotion of the interchange, the directional interchange is 28% better than standard cloverleaf interchange. Also, during the construction of interchanges, this superiority is maintained. This shows that despite the initial higher cost, the directional option has higher interests during the operation. Comparison of costs and interests of directional and cloverleaf interchange in different years can be seen in Fig. 8 and Fig. 9, based on the Net Present Value index.

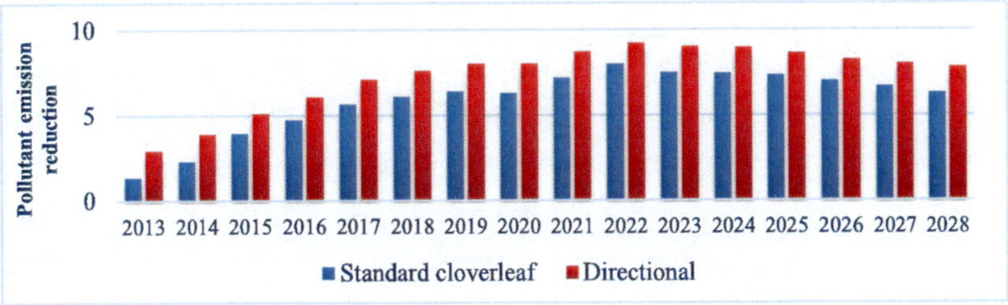

Fig. 6. Pollutant emission reduction in different years

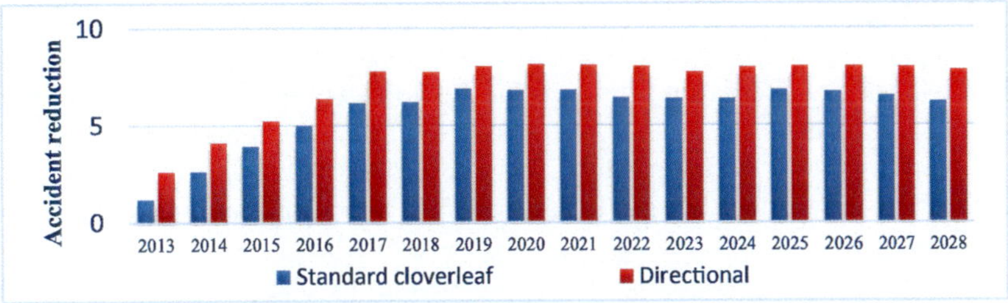

Fig. 7. Accident reduction in different years

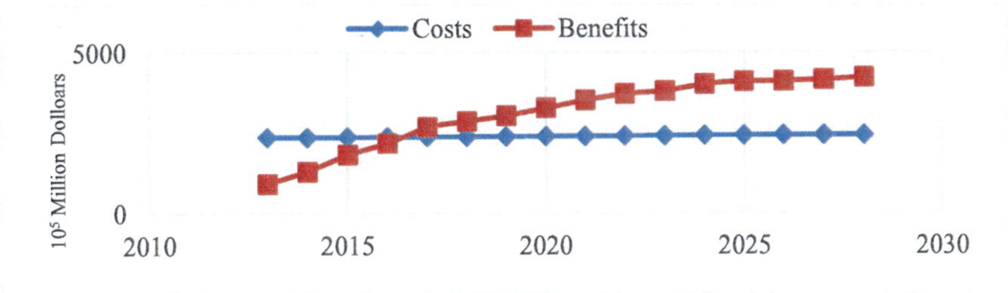

Fig. 8. Costs and benefits of directional interchange plan

According to Fig. 8 and Fig. 9, both plans have a non-negative "net present value" in the third year. Since the profitability of the project is before the end of the plan's useful life, therefore, the construction and exploitation of the studied interchanges are economically justified. It can be proven that geometric promotion in highway interchanges has a strong justification from an economic standpoint, safety, and traffic.

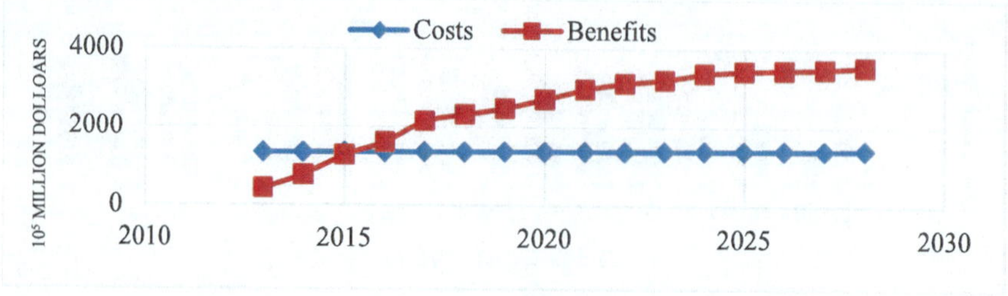

Fig. 9. Costs and benefits of the cloverleaf interchange plan

4 Conclusion

There have been two scenarios for improving the geometry characteristics of the interchange. The first one has been for increasing the capacity and transforming the status quo to improve cloverleaf interchange. The second one has been directional exchange with the three-story bridge design. Table 16 displays the process and analysis results of EMME software.

There are two methods for geometric improving the interchanges including line capacity increasing and changing the interchange to another interchange with a higher capacity. With geometric improving the cloverleaf interchange, it is concluded that method 2 has 27% economic superiority. The directional interchange plan has a $B/C = 1$ between the second and third years. After that, with much higher benefits than costs, it will have a very favorable impact on the network. The directional interchange plan has a $B/C = 1$ between the third and fourth years. Benefits increase with a lower slope than the directional option. A sensitivity analysis indicated that the fluctuations in the fuel price are most effective in improving geometric interchanges (coefficient: %55(. Also, travel costs, accidents costs, and environmental costs are 38%, 4%, and 3% respectively.

Table 16. The process and analysis results of EMME/2 software

Type	Reduction of travelled distance (%)	Reduction of fuel consumption (million litter)	Reduction of air pollution (%)	Reduction of travel time (million hours)
Directional	2.67	74	2.88	16
Cloverleaf	1.25	33	1.25	7

References

Afandi Zade, S., Bigdeli Rad, H., Shaker, H.: Using optimization and metaheuristic method to reduce the bus headway (Case study: Qazvin Bus Routes). J. Transp. Eng. **10**, 833–849 (2019)

Arani, M., Abdolmaleki, S., Liu, X.: Scenario-based simulation approach for an integrated inventory blood supply chain system. In: Winter Simulation Conference. IEEE (2020a)

Arani, M., Abdolmaleki, S., Maleki, M., Momenitabar, M., Liu, X.: A simulation-optimization technique for service level analysis in conjunction with reorder point estimation and lead-time consideration: a case study in sea port. In: Advances in Parallel & Distributed Processing, and Applications. Springer (2020b)

Arani, M., Chan, Y., Liu, X., Momenitabar, M.: A lateral resupply blood supply chain network design under uncertainties. Appl. Math. Model. (2020). https://doi.org/10.1016/j.apm.2020.12.010

Arani, M., Dastmard, M., Dehdari Ebrahimi, Z., Momenitabar, M., Liu, X.: Optimizing the total production and maintenance cost of an integrated multi-product process and maintenance planning (IPPMP) model. In: 6th IEEE International Symposium on Systems Engineering (2020). https://doi.org/10.1109/ISSE49799.2020.9272236

Arani, M., Rezvani, M.M., Davarikia, H., Chan, Y.: Routing of electric vehicles in a stochastic network with non-recurrent incidents (2019)

Baldauf, R.W., Heist, D., Isakov, V., Perry, S., Hagler, G.S.W., Kimbrough, S., Shores, R., Black, K., Brixey, L.: Air quality variability near a highway in a complex urban environment. Atmos. Environ. **64**, 169–178 (2013). https://doi.org/10.1016/j.atmosenv.2012.09.054

Bigdeli Rad, H., Bigdeli Rad, V.: A survey on the rate of public satisfaction about subway facilities in the City of Tehran using servqual model. Space Ontol. Int. J. **7**, 9–15 (2018)

Chan, Y., Fowe, J.A., Arani, M.: Routing in a stochastic network with nonrecurrent incidents: behavioral interpretation of dynamic traffic assignment. ASCE-ASME J. Risk Uncertain. Eng. Syst. Part A Civ. Eng. **6**, 4020002 (2020). https://doi.org/10.1061/AJRUA6.0001033

Chowdhury, M.S.: A simulation-based operations evaluation of major interchange configurations. In: 2016 International Conference on Transportation and Development. pp. 696–705. American Society of Civil Engineers, Reston (2016)

Gonzalez-Ruiz, J.D., Arboleda, C.A., Botero, S., Duque, E., Rojo, J.: A conceptual framework for the financing of rural highways in Colombia through shadow toll schemes. Adv. Transp. Stud. (2017)

Khan, T., Anderson, M.: Evaluating the application of diverging diamond interchange in Athens, Alabama. Int. J. Traffic Transp. Eng. **6**, 38–50 (2016). https://doi.org/10.7708/ijtte.2016.6(1).04

Lucietti, L., Hoogendoorn, C., Cré, I.: New tools and strategies for design and operation of urban transport interchanges. Transp. Res. Procedia **14**, 1240–1249 (2016). https://doi.org/10.1016/j.trpro.2016.05.195

Madanu, S., Li, Z., Abbas, M.: Life-cycle cost analysis of highway intersection safety hardware improvements. J. Transp. Eng. **136**, 129–140 (2010). https://doi.org/10.1061/(ASCE)TE.1943-5436.0000082

Mansourkhaki, A., Ghanad, H.: Optimisation of loop ramp design for cloverleaf interchanges. Proc. Inst. Civ. Eng. Transp. **167**, 248–258 (2014). https://doi.org/10.1680/tran.11.00037

Mollu, K., Cornu, J., Declercq, K., Brijs, K., Brijs, T.: The use of gantries and cantilevers at a redesigned intersection: a simulator study on lane choice and visual search behavior. Adv. Transp. Stud. **42**, 55–68 (2017)

Momenitabar, M., Akar, N., Zaghi, D., Feili, H.R., Ghaderi, M.: Fuzzy mathematical modeling of distribution network through location allocation model in a three-level supply chain design. J. Math. Comput. Sci. **9**(3), 165–174 (2014). https://doi.org/10.22436/jmcs.09.03.02

Momenitabar, M., Dehdari Ebrahimi, Z., Arani, M.: A Proposed Lean Distribution System for Solar Power Plants Using Mathematical Modeling and Simulation Technique. IEEE (2020)

American Association of State Highway, and Transportation Officials: A Policy on Geometric Design of Highways and Streets. American Association of State Highway and Transportation Officials (2011)

Pfeiffer, T.: Net present value-consistent investment criteria based on accruals: a generalisation of the residual income-identity. J. Bus. Finance Account. **31**, 905–926 (2004). https://doi.org/10.1111/j.0306-686X.2004.00561.x

Sutherland, L.F., Cook, D.J., Dixon, K.K.: Operational effects of the displaced partial cloverleaf interchange. Transp. Res. Rec. J. Transp. Res. Board **2672**, 108–119 (2018). https://doi.org/10.1177/0361198118792755

Dehdari Ebrahimi, Z., Bridgelall, R., Momenitabar, M.: Extending micro-mobility deployments: a concept and local case study. In: Transactions on Computational Science & Computational Intelligence. Springer (2020)

Dehdari Ebrahimi, Z., Momenitabar, M.: Design of mathematical modeling in a green supply chain network by collection centers in the environment. Environ. Energy Econ. Res. **1**, 153–162 (2017). https://doi.org/10.22097/eeer.2017.47242

Algebras of Concepts and Their Networks

Mohamed Khaled[1](✉) and Gergely Székely[2,3]

[1] Faculty of Engineering and Natural Sciences, Bahçeşehir University, Istanbul, Turkey
mohamed.khalifa@eng.bau.edu.tr
[2] Alfréd Rényi Institute of Mathematics, Budapest, Hungary
szekely.gergely@renyi.hu
[3] University of Public Service, Budapest, Hungary

Abstract. Concept algebras are algebras that 'talk' about concepts that one can define on a given mathematical structure. This paper is a first step in a research direction that is devoted to investigating concept algebras of specific examples of mathematical structures. To shed light on the appealing of this study, a number of conjectures and open problems is introduced. Applications in some theories of physics are discussed.

Keywords: Concept algebras · Cylindric algebras · Networks of concept algebras

1 Introduction

Concept algebras can be introduced as the algebraic counterparts to logical systems. They form an interface between logic, algebra and geometry, with numerous connections with other branches of mathematics, e.g., set theory, topology and analysis. That makes these algebras understood and appreciated by a wide range of mathematicians. They have been immensely successful in applications to a diversity of fields such as computer science, artificial intelligence, linguistics, etc.

The intrinsic merit of concept algebras is being algebras of relations. The fundamental operations are essentially the most basic operations one can consider on relations. The simplest concept algebras are Boolean algebras, which are algebras of unary relations. The theory of these algebras is directly related to the development of one of the most elementary logical systems; the sentential logics.

There are a number of algebras that are intended to correspond to relations of higher ranks. These structures are confronted with a simple extension of the Boolean parallelism between logic and algebra. For relations of rank $\alpha \geq 2$, where α can be finite or infinite, there are alternative possible algebras: cylindric algebras, diagonal-free cylindric algebras, polyadic algebras and so on. The ones which have received most attention is cylindric algebras. These algebras can be

T. Allahviranloo et al. (Eds.): IDS 2020, AISC 1301, pp. 611–622, 2021.
https://doi.org/10.1007/978-3-030-66501-2_48

regarded as multi-dimensional Boolean algebras; algebras of subsets of Cartesian spaces U^α.

Cylindric algebras were invented by Alfred Tarski around 1947 as the algebraic counterpart of first-order logic. There is an intensive literature on cylindric algebras, e.g., [7,21,22] and [3]. The last two books contains special chapters devoted to the applications of cylindric algebras and their related structures in computer science and in philosophy.

Database theory is one of the main application areas of cylindric algebras. The main concern in database design is to avoid redundancy and ensure the integrity of the database. The relational model for databases [16,17] provides a valuable formal foundation for understanding the issues of database design. A relational database consists of a set of tables (relations), where each table contains a set of records (tuples). Semantic information about particular applications is captured in the relational model by means of integrity constraints (dependencies). An important question about dependencies is the implication problem: Given a set $\Sigma \cup \{\sigma\}$ of dependencies in the class, is it the case that Σ implies σ?

It was shown that these relational data tables can be embedded into cylindric algebras. This provided an algebraic approach for data dependencies, known as cylindric dependencies [18]. These are extension of project-join dependencies to the positive reduct of first-order logic with equality. Using the theory of cylindric algebras, an equational complete axiomatization, in a similar spirit to the system of [36], was founded. With such systems, we can generate all dependencies implied by a given set Σ. Complete axiom systems have often led to efficient algorithms for dependency implication, e.g., [9,14,15,33]. More on this can be found in [19].

This paper is a contribution to the theory of cylindric algebras, but from a different perspective. We introduce these algebras as concept algebras of mathematical structures, i.e., first-order structures, and hence we use the name concept algebras. We are more interested in examples of these algebras; what are the concept algebras of specific structures? It turns out that the intensive literature of these algebras cannot answer this question even for some simple structures. We shed light on this research direction; with few seminal results and a number of open problems and conjectures.

2 Concept Algebras: Definition

A *mathematical structure* $\mathcal{M}$ is a non-empty set M together with some constants, relations and operations on M. This notion covers a wide range of structures in mathematics from graphs, through various algebraic structures such as groups, to geometries. By *formulas* we understand the usual mathematical phrases that are used to describe a given structure, these formulas are recursively built up from primitive formulas using individual variables, propositional connectives and quantifiers. (Technically, these phrases are called first-order formulas.)

Intuitively, a *concept* is a relation (of any finite arity) that can be defined on the structure at hand, using its own language. For example, given a group G, the notion of "the center $Z(G)$ of G" is a concept of groups, it is defined by the formula $\forall y\ (xy = yx)$. However, the set of "all elements of finite order in G" is not a concept of groups, because $\exists n\ (x^n = e)$ is not a formula of the language of groups (as the quantifier $\exists n$ ranges over natural numbers, not the elements of the group). From now on, when we say that φ is a formula, we understand that φ uses the (first-order) language of the structure in question. Thus, a concept is defined by a formula, and any formula defines a concept. These concepts define an algebra in a natural way. We briefly define this algebra in technical terms.

We assume a fixed enumeration $v_0, v_1, \ldots$ of the individual variables. An evaluation is an infinite sequence $\bar{a} = (a_0, a_1, \ldots) \in M^\omega$ of elements of M. We write $\mathcal{M} \models \varphi[\bar{a}]$ to denote that formula φ is true in structure $\mathcal{M}$ when each free variable v_i appears in φ is replaced by the component a_i of the evaluation $\bar{a}$. The *meaning* of φ in $\mathcal{M}$ is defined as

$$[\varphi]^{\mathcal{M}} = \{\bar{a} \in M^\omega : \mathcal{M} \models \varphi[\bar{a}]\},$$

where ω is the smallest infinite ordinal number. Thus, we correlate with each formula φ a subset $[\varphi]^{\mathcal{M}}$ of the Cartesian space M^ω. Given any other formula ψ, the set $[\psi]^{\mathcal{M}}$ coincides with $[\varphi]^{\mathcal{M}}$ if and only if ψ and φ are equivalent from the point of view of $\mathcal{M}$. In our settings, a *concept of* $\mathcal{M}$ is just the meaning of a formula in the language of $\mathcal{M}$. The *concept algebra of* $\mathcal{M}$ is the structure

$$\mathfrak{Cs}(\mathcal{M}) \stackrel{\text{def}}{=} \left(\mathsf{Cs}(\mathcal{M}), \cup, \setminus, C_i, D_{ij}\right)_{i,j<\omega},$$

where $\mathsf{Cs}(\mathcal{M})$ is the collection of all concepts of $\mathcal{M}$. It is not hard to see that $\mathsf{Cs}(\mathcal{M})$ is closed under the Boolean set-theoretic operations $\cup, \setminus$ and it contains $\emptyset$ and M^ω. The non-Boolean operations are defined as follows: for formula φ and $i, j < \omega$,

$$D_{ij} \stackrel{\text{def}}{=} [v_i = v_j]^{\mathcal{M}} = \{\bar{a} \in M^\omega : a_i = a_j\},$$

$$C_i[\varphi]^{\mathcal{M}} \stackrel{\text{def}}{=} [\exists v_i\ \varphi]^{\mathcal{M}} = \left\{\bar{a} \in M^\omega : \left(\exists \bar{a}' \in [\varphi]^{\mathcal{M}}\right)\ \bar{a} \equiv_i \bar{a}'\right\},$$

where $\bar{a} \equiv_i \bar{a}'$ means that $\bar{a}'$ agrees with $\bar{a}$ everywhere except (maybe) at the ith component. The set-theoretic operations $\cup, \setminus$ in the concept algebra represent the meaning of the propositional connectives $\vee$ (or) and $\neg$ (not).

One can think of a concept algebra as an algebra of subsets of a Cartesian space M^ω. For $i \neq j$, the elements D_{ij} are called diagonal sets, and they can be seen as the hyperplanes that consists of all points in the Cartesian space whose ith coordinate is equal to its jth coordinate. The operation C_i is called the ith cylindrification, and when it is applied to a subset X of the space M^ω, it produces the cylinder C_iX swept out by all translations of X parallel to the ith coordinate axis. Figure 1 illustrates this geometric meaning of equalities and the quantifiers, respectively.

Concept algebras, or cylindric algebras in another name, form an interface between algebra, geometry and logic; they were invented by Alfred Tarski around 1947 [21]. Properties of a mathematical structure can be read off from its concept algebra. For example, there is an equation $e(x)$ such that $e(x)$ holds for an

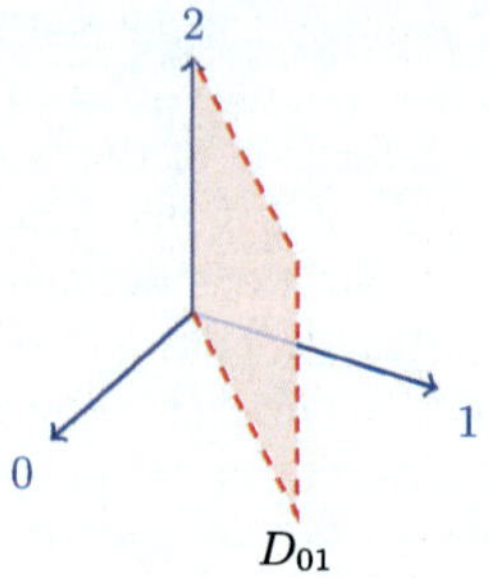

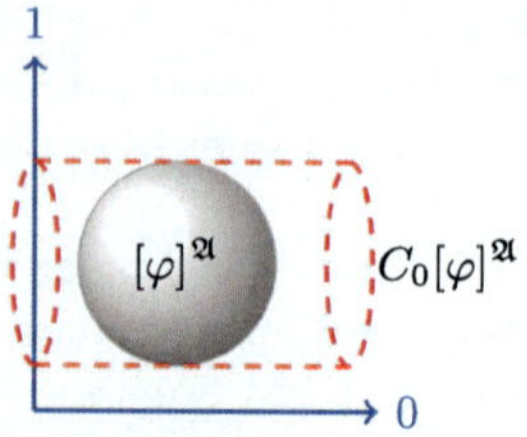

Fig. 1. The geometric twist of concept algebras

element x of the concept algebra if and only if "x is an equivalence (binary) relation," and there is an equation that holds in the concept algebra of a structure if and only if no linear order can be defined in the structure. More on this can be found in the recent article [8] and the conference presentation [5].

By correlating each mathematical structure to its concept algebra, we establish a correspondence between structures and their concept algebras. This correspondence turns out to be one-to-one in the following sense: Two structures are definitionally equivalent if and only if their concept algebras are equal. Two mathematical structures are said to be definitionally equivalent if and only if their universes coincide and the fundamental operations of either of the structures are definable in terms of the fundamental operations of the other. The significance of concept algebras for providing new insights and better understanding of mathematical structures is rather obvious.

3 Concept Algebras: Examples

The theory of concept algebras has been extensively investigated by numerous mathematicians, see, e.g., [21], but with a limited focus on concrete examples. The desire of a research direction investigating concrete algebras of concepts was voiced, e.g., by Monk in [31, p.455]. We quote:

> *"The corresponding facit of the theory of cylindric algebras is to describe the cylindric set algebras CsM for important models M. This amounts to looking at complete theories only, which is customary in model theory. It is somewhat surprising that this aspect of the theory of cylindric algebras has been almost entirely neglected. A complete description of CsM is known only in the case in which M has only one-place relations. There are many other simple structures where the description of CsM should not be difficult; for example, for M the rationals under their natural ordering."*

The concept algebras of some structures can be rather involved, see, e.g., [32]. This makes the subject the more interesting. This motivates us to do some preliminary investigation of the project suggested by Monk. Let us start with the concrete problem articulated in the quoted paragraph:

Problem 1 (Monk 2000). Give a complete description of the concept algebra of $(\mathbb{Q}, <)$, where $\mathbb{Q}$ is the set of all rational numbers and $<$ is its usual ordering.

Instead of focusing only on the ordering of rational numbers, we state a proposition about the concept algebra of any dense linear ordering without endpoints. A concept is said to be *n-dimensional concept* if it can be defined by a formula using variables only from $v_0, \ldots, v_{n-1}$ (note that there are infinitely many such formulas).

Proposition 1. *Let $\mathcal{A} = (A, <)$ be any dense linear ordering without endpoints. Every n-dimensional concept of $\mathcal{A}$ is the union of some minimal nontrivial n-dimensional concepts (called n-dimensional atoms). These atoms are the non-zero concepts corresponding to formulas of the following form: for every $i < j < n$, take one of $v_i = v_j$, $v_i < v_j$ or $v_j < v_i$ and put conjunction between these choices.*

Proof. The theory of dense linear ordering without endpoints satisfies the quantifier elimination property, i.e., for every formula φ, there is another formula ψ, such that φ is equivalent to ψ modulo the theory, but ψ is does not contain any quantifiers. Therefore, every definable concept of $\mathsf{Cs}(\mathcal{A})$ can be formulated as a Boolean combination of basic formulas $x = y$ and $x < y$, where x and y denote arbitrary individual variables.

Since every such Boolean combination can be expressed as a disjunctive normal form built up from basic formulas and negations of basic formulas, we conclude that the every concept of $\mathsf{Cs}(\mathcal{A})$ can be given as a disjunction of one or more conjuctions of formulas $x = y$, $\neg(x = y)$, $x < y$, and $\neg(x < y)$.

Since $\neg(x = y)$ is equivalent to $x < y \vee y < x$ and $\neg(x < y)$ is equivalent to $x = y \vee y < x$, by distributivity, we have that every concept of $\mathsf{Cs}(\mathcal{A})$ can be given as disjunction of conjunction of some basic formulas $x = y$ and $x < y$. Hence, every n-dimensional concept of $\mathcal{A}$ is the union of some n-dimensional atoms. Considering full disjunctive normal forms gives the complete list of possible atoms as described in the proposition. □

In principle, it is not difficult to weed out the zeros from the list of atom candidates given by Proposition 1. For example, in case of 3-dimensional concepts, we already know (via a geometric description) that there are exactly 13 atoms from the 27 candidates, see Fig. 2. Let us note, however, that Proposition 1 does not give a full description of $\mathfrak{Cs}(\mathbb{Q}, <)$ because it only lists atom candidates and does not tell which of these candidates are indeed atoms.

Here is another example of a more involved structure $(\mathbb{R}, +, \cdot, <)$. By Tarski–Seidenberg theorem, we have that the elements of $\mathfrak{Cs}(\mathbb{R}, +, \cdot, <)$ are exactly the so called *semialgebraic sets*, i.e., finite union of sets defined by a finite sequence of polynomial equations of the form $p(v_0, \ldots, v_{n-1}) = 0$ and polynomial inequalities of the form $q(v_0, \ldots, v_{n-1}) > 0$. Hence, the set of semialgebraic sets of $\mathbb{R}$ together with operations $\cup$, $\setminus$, C_i, D_{ij}, is the concept algebra $\mathfrak{Cs}(\mathbb{R}, +, \cdot, <)$.

A number of similar questions, concerning other mathematical structures, can be considered.

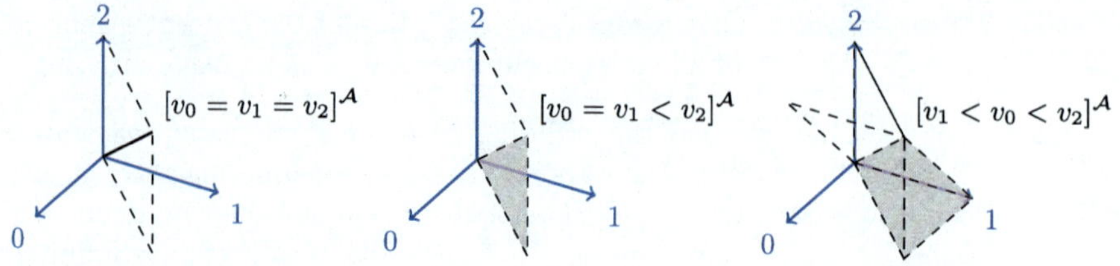

Fig. 2. The 3-dimensional atoms of $\mathfrak{Cs}(\mathcal{A})$ up to symmetries of indices

Problem 2. Describe the concept algebras of simple graphs (undirected and without loops).

Let $G = (V, E)$ be a simple graph. Define the following equivalence relation on V:

$$a \sim b \iff \text{there is an automorphism of } G \text{ that sends } a \text{ to } b.$$

We think of the equivalence classes of $\sim$ as colors. For instance, if we have three equivalence classes, then we shall call them red, blue, yellow. For example, see the $\sim$-colors of the vertices of the graph H in Fig. 3. Let Colors denote the set of all $\sim$-colors of V. Let $\alpha : n \to \mathsf{Colors}$, $\beta : n \times n \to \{-1, 1\}$ and $\gamma : n \times n \to \{-1, 1\}$. A set $X \subseteq V^\omega$ is called a $\alpha - \beta - \gamma$ choice if it consists of all infinite sequences $(a_0, a_1, \ldots)$ satisfying, for every $i, j < \omega$,

$$a_i \text{ has } \sim\text{-color } c \iff \alpha(i) = c, \quad a_i = a_j \iff \beta(i,j) = 1, \quad a_i a_j \in E \iff \gamma(i,j) = 1.$$

Conjecture 1. Each $\alpha - \beta - \gamma$ choice is definable by a formula of graph theory. Moreover, the nonempty $\alpha - \beta - \gamma$ choices are exactly the n-dimensional atoms of the concept algebra of G.

Every concept is in fact n-dimensional, for some n. So, a complete analysis of the n-dimensional concepts, for every n, yields a complete description of the full concept algebra $\mathfrak{Cs}(\mathcal{M})$. A different approach that can also lead to a complete description of $\mathfrak{Cs}(\mathcal{M})$ is studying the *concepts of arity n*. A concept is said to be *of arity n* if it can be defined by a formula whose free variables are among $v_0, \ldots, v_{n-1}$. So, a binary concept is the meaning of a formula that contains at

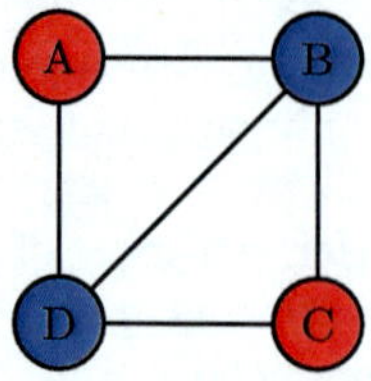

Fig. 3. The $\sim$-colors of the vertices of H

most v_0 and v_1 as free variables, regardless of how many other bounded variables it may contain.

Structure $\mathcal{N}$ is called a *reduct of* $\mathcal{M}$ if $\mathcal{N}$ is achieved from $\mathcal{M}$ by forgetting some basic relations or operations. In case $\mathcal{N}$ is a reduct of $\mathcal{M}$, the definable concepts $\mathsf{Cs}(\mathcal{N})$ of $\mathcal{N}$ is clearly a subset of $\mathsf{Cs}(\mathcal{M})$, and hence concept algebra $\mathfrak{Cs}(\mathcal{N})$ is a subalgebra of $\mathfrak{Cs}(\mathcal{M})$. The set of all concepts of arity n forms a subalgebra of a reduct of the concept algebra, we call this algebra *n-ary concept algebra*. See [1] for a precise definition of these subalgebras.

3.1 Concept Algebras in Physics

Some other preliminary results in investigating concepts of arity n in mathematical structures representing relativistic and non-relativistic spacetimes of physics were presented at a workshop in Munich [5], where they were received with great interest. Here, we summarize these results. To do this, we need some definitions.

By *relativistic spacetime* $\mathcal{S}r$, we understand the structure of timelike straight lines on $\mathbb{R}^4$. At first glance $\mathcal{S}r$ may appear as an oversimplified version of Minkowski spacetime. However, it can be shown that from timelike collinearity one can define the full-fledged scale-invariant Minkowski spacetime, i.e., lightlike separability, Minkowski-equidistance, Minkowski-orthogonality, etc. are all definable. By *classical spacetime* $\mathcal{N}t$, we understand the structure of non-horizontal straight lines on $\mathbb{R}^4$, capturing possible inertial motions in classical kinematics.

The following properties of $\mathfrak{Cs}(\mathcal{S}r)$ and $\mathfrak{Cs}(\mathcal{N}t)$ have been shown [5,6]:

Theorem 1. *The unary concepts of $\mathcal{S}r$ are trivial ($\emptyset$ and $\mathbb{R}^4$). There are exactly 16 binary concepts of $\mathcal{S}r$; moreover, any binary concept can be defined from any other nontrivial binary concept (e.g., lightlike separability can be defined from timelike separability). There are infinitely many nontrivial ternary concepts of $\mathcal{S}r$ which cannot be defined from each other. Furthermore, the corresponding ternary concept algebra is atomic.*

Theorem 2. *The unary concepts of $\mathcal{N}t$ are trivial ($\emptyset$ and $\mathbb{R}^4$). There are exactly 8 binary concepts of $\mathcal{N}t$. There are infinitely many nontrivial ternary concepts of $\mathcal{N}t$ which cannot be defined from each other. Furthermore, the corresponding ternary concept algebra is atomic.*

Using insights gathered by investigating these concept algebras, one can show interesting connections between structures $\mathcal{N}t$ and $\mathcal{S}r$. For example, it can be shown that neither $\mathcal{N}t$ nor $\mathcal{S}r$ can be interpreted into the other, see [5] and [6].

4 Networks of Concept Algebras

In this section, we will define a distance between concept algebras. This distance is intended to measure the minimum number of concept separating two concept algebras. We start by introducing a special network that can be defined for any class of similar algebras. These networks are worth investigating not only for concept algebras but for other algebras as well.

Let K be a class of similar algebras. We define the *network of* K to be the (possibly infinite) graph whose vertices are the members of K, and which has two types of edges: red edges connecting the different isomorphic members of K, and blue edges connecting any two algebras of K if they are not isomorphic, but one of them is a subalgebra of the other such that the small algebra together with one extra element generates the big algebra. For example, let S be the class of all subalgebras of the 8 element Boolean algebra having atoms a, b and $c = \neg(a \vee b)$. Then the network of S is illustrated in Fig. 4.

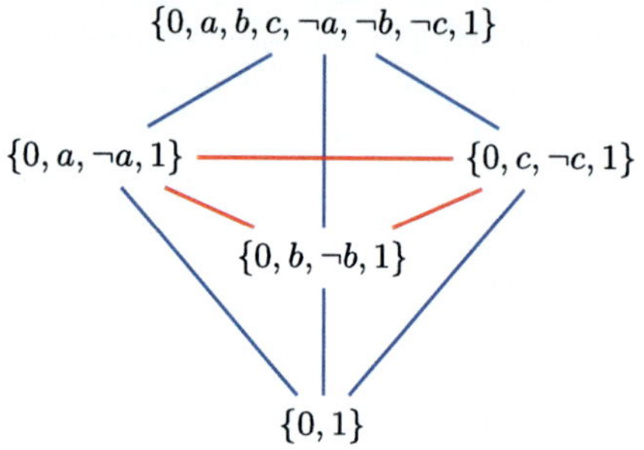

Fig. 4. A small network of Boolean algebras

Definition 1. *Let $\mathfrak{A}$ and $\mathfrak{B}$ be two algebras in K. We say that $\mathfrak{A}$ and $\mathfrak{B}$ are* connected *in K if there is a finite path connecting them in the network of K. We define the* **generator distance** *$d_{\mathsf{K}}(\mathfrak{A}, \mathfrak{B})$ between $\mathfrak{A}$ and $\mathfrak{B}$ in K as follows: If $\mathfrak{A}$ and $\mathfrak{B}$ are not connected in K, then $d_{\mathsf{K}}(\mathfrak{A}, \mathfrak{B}) \stackrel{def}{=} \infty$. Otherwise,*

$$d_{\mathsf{K}}(\mathfrak{A}, \mathfrak{B}) \stackrel{def}{=} \textit{min. number of blue edges among all the finite paths connecting } \mathfrak{A} \textit{ and } \mathfrak{B} \textit{ in } \mathsf{K}.$$

Example 1. Suppose that K is the class of all finite dimensional vector spaces over a given field, then the generator distance $d_{\mathsf{K}}(V, W) = |dim(V) - dim(W)|$ for all $V, W \in \mathsf{K}$.

It is not hard to see that d_{K} satisfies the following metric properties for each $\mathfrak{A}, \mathfrak{B}, \mathfrak{C} \in \mathsf{K}$:

- Non-negativity and identity of indiscernibles: $d_{\mathsf{K}}(\mathfrak{A}, \mathfrak{B}) \geq 0$, and $d_{\mathsf{K}}(\mathfrak{A}, \mathfrak{B}) = 0 \iff \mathfrak{A} \cong \mathfrak{B}$.
- Symmetry: $d_{\mathsf{K}}(\mathfrak{A}, \mathfrak{B}) = d_{\mathsf{K}}(\mathfrak{B}, \mathfrak{A})$.
- Triangle inequality: $d_{\mathsf{K}}(\mathfrak{A}, \mathfrak{B}) \leq d_{\mathsf{K}}(\mathfrak{A}, \mathfrak{C}) + d_{\mathsf{K}}(\mathfrak{C}, \mathfrak{B})$.

Intuitively, this distance measures the minimum number of generators needed to be added or removed from one algebra to reach the other one, through a sequence of intermediate algebras. Investigating the generator distance between algebras provides a new quantitative (and still qualitative) study that may develop our insights about the class in hand. Calculating the generator distance on a given class usually requires a profound knowledge of the class because tricky embeddings may give unexpected shortcuts in the network. As illustration, we include below our first result in this direction. More results in that concern are under preparation [10].

Free algebras play a significant role in our understanding of concept algebras, see [21, §5.6] [4,11,24–26,30]. Boolean algebras having 2^{2^k}-many elements for some natural number k are exactly the finitely generated free algebras of the variety BA of Boolean algebras. One of the facts we use about these free algebras is that they are unique up to isomorphism. For more details about Boolean algebras, we refer the reader to [20].

Proposition 2. *Let* BA *be the class of all Boolean algebras. Let* $\mathfrak{A}$ *and* $\mathfrak{B}$ *be finite Boolean algebras having* 2^{2^n} *and* 2^{2^m} *elements, respectively. Then* $d_{\mathsf{BA}}(\mathfrak{A},\mathfrak{B}) = |m-n|$.

Proof. Assume that $n \leq m$. Let k be a natural number, we know the following from the literature of Boolean algebras. The Boolean algebras of size 2^{2^k} and $2^{2^{k+1}}$ are not isomorphic, the small one is embeddable to the big one and adding one element to the small one can generate the big algebra. This shows that they are of generator distance 1 from each other.

Using this fact, we are going to show that $d_{\mathsf{BA}}(\mathfrak{A},\mathfrak{B}) \leq m-n$. To do so, let $\mathfrak{A}_i$ be a Boolean algebra having $2^{2^{n+i}}$-many elements for every $0 \leq i \leq m-n$. Then, by the triangle inequality property of the generator distance,

$$d_{\mathsf{BA}}(\mathfrak{A},\mathfrak{B}) \leq d_{\mathsf{BA}}(\mathfrak{A},\mathfrak{A}_0) + \sum_{i=0}^{m-n-1} d_{\mathsf{BA}}(\mathfrak{A}_i,\mathfrak{A}_{i+1}) + d_{\mathsf{BA}}(\mathfrak{A}_{m-n},\mathfrak{B}) = 0 + \sum_{i=0}^{m-n-1} 1 + 0 = m-n.$$

To show that $d_{\mathsf{BA}}(\mathfrak{A},\mathfrak{B}) \geq m-n$, let us first observer that, if a Boolean algebra $\mathfrak{D}$ can be generated by its subalgebra $\mathfrak{C}$ plus one element $d \in D$, then $|D| \leq |C|^2$. This is so because every element of $\mathfrak{D}$ is of form $(v \wedge d) \vee (u \wedge -d)$ for some elements v and u of $\mathfrak{C}$. Hence, in any path from $\mathfrak{A}$ to $\mathfrak{B}$, in each step, we can at most square the number of elements, and thus it takes at least $m-n$-many squarings to reach 2^{2^m} from 2^{2^n}. Consequently, $d_{\mathsf{BA}}(\mathfrak{A},\mathfrak{B}) \geq m-n$. □

In mathematics, it is important to be able to tell when two theories are the same to some extent, and if they are different, then how different they are. There are several notions of equivalence (logical equivalence, definitional equivalence, categorical equivalence, etc.) for telling when two theories are the same in a certain sense. These notions recently became of central interest in the methodology of science, see, e.g., [12,13,23,34,35].

In [27], a new quantitative method was introduced to measure the degree of non-equivalence of theories. Various distances between theories have been introduced there, which opens new opportunities to get a better understanding of the relation between nonequivalent theories. One of the central notions of that paper is *conceptual distance*, in symbols $\mathsf{Cd}(T_1, T_2)$, which refines the notion of definitional equivalence and measures the minimal number of concept adding/removing steps that are needed to be taken to reach theory T_2 from T_1.

In general, it is rather difficult to determine the concrete conceptual distance between two theories. For example, the main PhD result of K. Lefever, cf. [28,29] states, in terms of conceptual distance, that the conceptual distance between certain axiomatic theories of classical and relativistic kinematics is 1, in other words these two theories are distinguished from each other by only one concept, namely the concept of "being stationary relative to the ether." [27]

Concept algebras for theories, also known as Lindenbaum–Tarski algebras, can also be defined in a natural way. It can be shown that the generator distance between concept algebras of any two structures is the same as the conceptual distance between the theories of these two structures.[10]

Theorem 3. *Let* K *be the class of concept algebras corresponding to theories of first-order logic. Let* $\mathcal{M}_1$ *and* $\mathcal{M}_2$ *be two arbitrary mathematical structures. Then*

$$d_{\mathsf{K}}\big(\mathfrak{Cs}(\mathcal{M}_1), \mathfrak{Cs}(\mathcal{M}_2)\big) = \mathsf{Cd}\big(\mathsf{Th}(\mathcal{M}_1), \mathsf{Th}(\mathcal{M}_2)\big),$$

where $\mathsf{Th}(\mathcal{M}_1)$ *and* $\mathsf{Th}(\mathcal{M}_2)$ *denote the theories of* $\mathcal{M}_1$ *and* $\mathcal{M}_2$*, respectively.*

The connection between these two distances provided by Theorem 3 could give a powerful algebraic machinery to determine the conceptual distance between theories. This is plausible because usually many deep and complicated results in logic can be obtained using the tools of algebra [21].

Let RCA denote the variety generated by concept algebras. An equational class $\mathsf{CA} \supseteq \mathsf{RCA}$, defined by a finite schema of equations, was proposed by A. Tarski to capture the axiomatization of the class RCA. However, it was shown by Monk that RCA cannot be axiomatized by finitely many schemas. Thus, $\mathsf{RCA} \subsetneq \mathsf{CA}$, and moreover RCA seems to be "far" from its axiomatic approximation CA. The network of RCA is precisely what we call the network of concept algebras. It also reflects the network of theories. Here are some interesting questions:

Problem 3. Suppose that $\mathfrak{A}, \mathfrak{B} \in \mathsf{RCA} \subsetneq \mathsf{CA}$. Is it true that $d_{\mathsf{RCA}}(\mathfrak{A}, \mathfrak{B}) = d_{\mathsf{CA}}(\mathfrak{A}, \mathfrak{B})$?

Let $\mathcal{N}t^+$ denote classical spacetime extended with the concept of lightlike conectedness. This structure captures "later" classical kinematics, more specifically, kinematics of the nineteenth century at the time of J. C. Maxwell and the search for the "luminiferous ether." Since lightlike connectedness is definable in $\mathcal{S}r$, $\mathfrak{Cs}(\mathcal{S}r)$ is (isomorphic to) a subalgebra of $\mathfrak{Cs}(\mathcal{N}t^+)$. In fact, the bigger algebra $\mathfrak{Cs}(\mathcal{N}t^+)$ is generated by $\mathfrak{Cs}(\mathcal{S}r)$ together with the concept of simultaneity. The question of the special status of the concept of simultaneity arises here.

Problem 4 (Andréka 2017). Is it true that any concept from $\mathfrak{Cs}(\mathcal{N}t^+)$ which is not an element of the subalgebra $\mathfrak{Cs}(\mathcal{S}r)$ generates together with $\mathfrak{Cs}(\mathcal{S}r)$ the whole $\mathfrak{Cs}(\mathcal{N}t^+)$? [2]

If the answer to Andréka's question is positive, then no matter which classical concept we add to special relativity we will get later classical kinematics. Independently of the answer, solving this problem would provide an interesting insight giving a better understanding of the connection between these two key theories of physics.

Acknowledgment. We would like to express our sincere gratitude to Hajnal Andréka and István Németi for putting us together and for guiding us to initiate this project.

References

1. Ahmed, T.S., Németi, I.: On neat reducts of algebras of logic. Studia Logica **68**(2), 229–262 (2001). https://doi.org/10.1023/A:1012447223176
2. Andréka, H.: Personal communication with M. Khaled and G. Székely at Logic, Relativity and Beyond 2017 conference. Rényi Institute, Budapest (2017)
3. Andréka, H., Ferenczi, M., Németi, I.: Cylindric-like algebras and algebraic logic, vol. 22. Springer, Heidelberg (2013)
4. Andréka, H., Jónsson, B., Németi, I.: Free algebras in discriminator varieties. Algebra Universalis **28**, 401–447 (1991)
5. Andréka, H., Madarász, J.X., Németi, I., Székely, G.: Conceptual structure of spacetimes, and category of concept algebras, invited plenary prezentation at Foundations of Categorical Philosophy of Science, Munich Germany, 26–27 April 2019, Ludwig–Maximilians–Universität (2019)
6. Andréka, H., Madarász, J.X., Németi, I., Székely, G.: On the concept algebra of Minkowski spacetime (2020), in preparation
7. Andréka, H., Monk, J.D., I. Németi (editors): Algebraic Logic (Proc. Conf. Budapest 1988). Colloq. Math. Soc. J. Bolyai. **54**, North–Holland, Amsterdam (1991)
8. Andréka, H., Németi., I.: How many varieties of cylindric algebras are there. Trans. Amer. Math. Soc. **369**(12), 8903–8937 (2017)
9. Armstrong, W.W.: Dependency structures of data base relationships. In: IFIP Congress, pp. 580–583. North Holland, Amsterdam (1974)
10. Aslan, T., Khaled, M., Székely, G.: On the networks of large embeddings (2020), in preparation
11. Banerjee, A., Khaled, M.: First order logic without equality on relativized semantics. Ann. Pure Appl. Logic **169**(11), 1227–1242 (2018)
12. Barrett, T.W., Halvorson, H.: Glymour and Quine on theoretical equivalence. J. Philos. Logic **45**(5), 467–483 (2016)
13. Barrett, T.W., Halvorson, H.: Morita equivalence. Rev. Symbolic Logic **9**(3), 556–582 (2016)
14. Beeri, C., Fagin, R., Howard, J.H.: A complete axiomatization for functional and multivalued dependencies in database relations. In: Proceedings of the 1977 ACM SIGMOD International Conference on Management of Data, pp. 47–61. SIGMOD 1977, Association for Computing Machinery, New York, NY, USA (1977). https://doi.org/10.1145/509404.509414
15. Beeri, C., Vardi, M.Y.: Formal systems for tuple and equality generating dependencies. SIAM J. Comput. **13**(1), 76–98 (1984)
16. Codd, E.F.: A relational model for large shared data banks. Commun. ACM **13**(6), 377–387 (1970)

17. Codd, E.F.: Further normalization of the data base relational model. In: Rustin, R. (ed.) Data Base Systems, pp. 33–64. Prentice-Hall, Englewood Cliffs, New Jersey (1972)
18. Cosmadakis, S.S.: Database theory and cylindric lattices. In: 28th Annual Symposium on Foundations of Computer Science (SFCS 1987), pp. 411–420 (1987)
19. Düntsch, I.: Cylindric algebras and relational databases. In: [3], pp. 321–331. Springer (2012)
20. Givant, S., Halmos, P.: Introduction to Boolean Algebras. Undergraduate Texts in Mathematics, Springer (2009)
21. Henkin, L., Monk, J.D., Tarski, A.: Cylindric Algebras Parts I and II. North–Holland (1971 and 1985)
22. Henkin, L., Monk, J., Tarski, A., Andréka, H., Németi, I.: Cylindric Set Algebras. Lecture Notes in Mathematics, vol. 883. Springer-Verlag, Berlin, Heidelberg, New York (1981)
23. Hudetz, L.: Definable categorical equivalence. Philos. Sci. **86**(1), 47–75 (2019). https://doi.org/10.1086/701047
24. Khaled, M.: The free non-commutative cylindric algebras are not atomic. Logic J. IGPL **25**(5), 673–685 (2017)
25. Khaled., M.: The finitely axiomatizable complete theories of non-associative arrow frames. Adv. Math. **346**, 194–218 (2019)
26. Khaled, M., Németi, I.: Atoms in infinite dimensional free sequence-set algebras. Algebra Universalis, **80**(4), 41 (2019)
27. Khaled, M., Székely, G., Lefever, K., Friend, M.: Distances between formal theories. Rev. Symbolic Logic **13**(3), 633–654 (2020)
28. Lefever, K.: Using Logical Interpretation and Definitional Equivalence to Compare Classical Kinematics and Special Relativity Theory. Ph.D. thesis, Vrije Universiteit Brussel (2017)
29. Lefever, K., Székely, G.: Comparing classical and relativistic kinematics in first-order logic. Logique et Analyse **61**(241), 57–117 (2018)
30. Madarász, J.X., Németi, I.: Free Boolean algebras with closure operators and a conjecture of Henkin, Monk, and Tarski. Studia Sci. Math. Hungar. (Special issue on the 50th anniversary of the Alfréd Rényi Institute of Mathematics), **38**, 273–278 (2001)
31. Monk, J.D.: An introduction to cylindric set algebras. Logic J. IGPL **8**, 451–496 (2000)
32. Németi, I.: On varieties of cylindric algebras with applications to logic. Ann. Pure Appl. Logic **36**, 235–277 (1987)
33. Ullman, J.D.: Principles of databases. Computer Science, Baltimore, MD (1982)
34. Weatherall, J.O.: Are newtonian gravitation and geometrized newtonian gravitation theoretically equivalent? Erkenntnis **81**(5), 1073–1091 (2016)
35. Weatherall, J.O.: Part 1: Theoretical equivalence in physics. Philosophy Compass, **14**(5), e12592 (2019). https://onlinelibrary.wiley.com/doi/abs/10.1111/phc3.12592
36. Yannakakis, M., Papadimitriou, C.H.: Algebraic dependencies. J. Comput. Syst. Sci. **25**(1), 2–41 (1982)

Application of the CESTAC Method to Find the Optimal Iteration of the Homotopy Analysis Method for Solving Fuzzy Integral Equations

Samad Noeiaghdam[1,2(✉)] and Mohammad Ali Fariborzi Araghi[3]

[1] Baikal School of BRICS, Irkutsk National Research Technical University, Irkutsk, Russian Federation

[2] South Ural State University, Lenin Prospect 76, Chelyabinsk 454080, Russian Federation

noiagdams@susu.ru, snoei@istu.edu

[3] Department of Mathematics, Central Tehran Branch, Islamic Azad University, Tehran, Iran

fariborzi.araghi@gmail.com, m_fariborzi@iauctb.ac.ir

Abstract. The goal of this work is to apply the stochastic arithmetic (SA) in discrete form and the CESTAC (Controle et Estimation Stochastique des Arrondis de Calculs) method to validate the numerical results of solving fuzzy Fredholm and Volterra integral equations (IEs) by homotopy analysis method (HAM). The CADNA (Control of Accuracy and Debugging for Numerical Applications) library is applied for mathematical computations. Using the mentioned method, not only the optimal step and approximation of the HAM but also some of numerical instabilities can be found. Furthermore, the optimal convergence control of the HAM for solving fuzzy IEs is obtained. Main theorems are proved to show the accuracy of the HAM by applying the CESTAC method to evaluate the obtained results.

Keywords: Stochastic arithmetic · Homotopy analysis method · CADNA library · CESTAC method · Fuzzy integral equation

1 Introduction

In recent years, fuzzy mathematics have been applied in different fields of science and engineering. Specially, solving fuzzy IEs is among of so interesting topics which has been investigated by many researchers. There are many numerical methods for solving fuzzy IEs. Nouriani and Ezzati in [1] applied the Simpson quadrature rule and iterative method, Yang and Gong used the regularization method in [2] and Ziari in [3] studied the iterative methods for solving fuzzy IEs. Also, fuzzy problems were solved using the fuzzy neural network [4], the Bernoulli wavelet method [5], fuzzy differential transform method [6], hybrid of block-pulse functions and Taylor series [7], Sinc-collocation method [8] and many other methods [9–14].

T. Allahviranloo et al. (Eds.): IDS 2020, AISC 1301, pp. 623–637, 2021.
https://doi.org/10.1007/978-3-030-66501-2_49

In many studies the fuzzy mathematical problems were solved using the floating point arithmetic (FPA) without having information about disadvantages of the FPA. There are some questions about this kind of arithmetic. What is the optimal approximation for solving the problems? What is the optimal iteration of the presented method? What is the stopping condition to stop the numerical algorithm? Also, in several cases in order to show the accuracy of method the traditional absolute error was applied that depends on to have the exact solution. It can make the main question now. If we have the exact solution why we need to find the solution approximately? Moreover, in some cases the termination criterion to stop the numerical process depends on to the small value ϵ. Next important question is, if we do not have the optimal value of ϵ how can we identify the optimal step to stop the numerical algorithm? Because for large or small values of ϵ we will not have the appropriate number of iteration without having suitable accuracy.

In this research, the stochastic arithmetic is introduced generally and the CESTAC method and the CADNA library are applied to validate the HAM [15–21] for solving Fredholm and Volterra fuzzy IEs [22–27]. Using this method, a new termination criterion is presented which it depends on to two successive approximations. This condition will be stopped when the number of common significant digits (NCSDs) of Hausdorff distance between two fuzzy successive approximations to be zero [8,9,12].

The CESTAC method was applied to validate many numerical methods for solving mathematical problems. In recent years, this method was applied to validate the numerical quadrature methods [9,12,28], solve ill-posed IEs [30,31], solve linear and non-linear IEs [29–31] and others [32,33]. In this approach, the validation of numerical results can be done and based on the SA, the optimal approximation and the optimal error of the HAM can be obtained. Also, the CADNA library should be applied to calculate the results. The main difference of this library with other softwares like Mathematica and Matlab is to apply the CESTAC method. All programs in this library should be written in C/C++, Fortran or ADA. Moreover this library should be run on LINUX operating system instead of Windows.

2 Preliminaries

In this section, we present some definitions related to the fuzzy sets theory.

Definition 1 [34]. *Let $\mathcal{T}:\mathbb{R}\rightarrow[0,1]$ be the membership function of a fuzzy set, then the fuzzy number should be satisfied in the following conditions:*

- *$\mathcal{T}$ is upper semi continuous,*
- *$\mathcal{T}(z)=0$ outside some interval $[a_3,a_4]$.*

For real numbers $a_1,a_2: a_3\leq a_1\leq a_2\leq a_4$ we have

- *$\underline{\mathcal{T}}(z)$ will be monotonic increasing on $[a_3,a_1]$,*
- *$\overline{\mathcal{T}}(z)$ will be monotonic decreasing on $[a_2,a_4]$,*
- *$\mathcal{T}(z)=1, a_1\leq z\leq a_2$.*

Definition 2 [35]. *Let* $\Psi = (\psi_1, \psi_2, \psi_3)$ *be the triangular fuzzy number such that the center, left width and right width are shown by* $\psi_1, \psi_2 > 0$ *and* $\psi_3 > 0$. *Then the triangular fuzzy number can be defined as*

$$\Psi(z) = \begin{cases} 1 - \dfrac{(\psi_1 - z)}{\psi_2}, & if\ \psi_1 - \psi_2 \leq z \leq \psi_1, \\ 1 - \dfrac{(z - \psi_1)}{\psi_3}, & if\ \psi_1 \leq z \leq \psi_1 + \psi_3, \\ 0, & otherwise. \end{cases}$$

Definition 3 [35]. *For triangular fuzzy numbers* $S_1 = (c_1, c_2, c_3)$ *and* $S_2 = (d_1, d_2, d_3)$, *the sum and minus operations can be defined in the following form:*

$$S_1 + S_2 = (c_1 + d_1, c_2 + d_2, c_3 + d_3),$$

$$S_1 - S_2 = (c_1 - d_1, c_2 + d_3, c_3 + d_2).$$

Definition 4 [36]. *For two fuzzy numbers* $\widetilde{\mathcal{T}} = (\underline{\mathcal{T}}(\eta), \overline{\mathcal{T}}(\eta))$, $\widetilde{\mathcal{G}} = (\underline{\mathcal{G}}(\eta), \overline{\mathcal{G}}(\eta))$ *and* $\vartheta \in \mathbb{R}$ *the following operations can be defined*

- $\widetilde{\mathcal{T}} = \widetilde{\mathcal{G}} \Leftrightarrow \underline{\mathcal{T}}(\eta) = \underline{\mathcal{G}}(\eta)\ \&\ \overline{\mathcal{T}}(\eta) = \overline{\mathcal{G}}(\eta)$,
- $\widetilde{\mathcal{T}} \oplus \widetilde{\mathcal{G}} = (\underline{\mathcal{T}}(\eta) + \underline{\mathcal{G}}(\eta), \overline{\mathcal{T}}(\eta) + \overline{\mathcal{G}}(\eta))$,
- $(\vartheta \odot \widetilde{\mathcal{T}}) = \begin{cases} (\vartheta\underline{\mathcal{T}}(\eta), \vartheta\overline{\mathcal{T}}(\eta)), & \vartheta > 0, \\ (\vartheta\overline{\mathcal{T}}(\eta), \vartheta\underline{\mathcal{T}}(\eta)), & \vartheta < 0. \end{cases}$

Definition 5 [37,38]. *Let us* $\widetilde{\mathcal{T}} = (\underline{\mathcal{T}}(\eta), \overline{\mathcal{T}}(\eta))$ *and* $\widetilde{G} = (\underline{G}(\eta), \overline{G}(\eta))$ *be the fuzzy numbers, then the Hausdorff distance between these numbers is defined as*

$$\mathcal{D}_H(\tilde{\mathcal{T}}, \tilde{\mathrm{G}}) = \sup_{0 \leq \eta \leq 1} \left\{ \max \left\{ |\underline{\mathcal{T}}(\eta) - \underline{\mathcal{G}}(\eta)|, |\overline{\mathcal{T}}(\eta) - \overline{\mathcal{G}}(\eta)| \right\} \right\}.$$

3 Fuzzy CESTAC Method-CADNA Library

Assume that $\{\widetilde{v}_J\}_{J \geq 1}$ is a fuzzy sequence obtained from iterative or numerical methods which can be applied to find the exact fuzzy solution $\widetilde{v}$. In order to show the accuracy of method, in the computer arithmetic, researchers have been applied the following condition

$$If\ \mathcal{D}_H(\widetilde{v}_J, \widetilde{v}) \leq \epsilon\ then\ stop, \tag{1}$$

which depends on to the small positive real number ϵ and existence the exact solution. But how can we apply the Condition (1), when we do not have the exact solution? Also, if we have the exact solution, why we want to solve the problems numerically? And finally, we do not know about the optimal value of ϵ such that for large and small values of ϵ either the extra iterations can be produced or the numerical algorithm can be stopped soon without getting to a suitable approximation.

Because of the mentioned problems, the following condition based on the SA is presented

$$If\ \mathcal{D}_H(\widetilde{v}_J, \widetilde{v}_{J+1}) = @.0\ then\ stop, \tag{2}$$

where @.0 shows the informatical zero and only can be produced in the SA. It means the NCSDs of the distance between two sequential results is zero. Applying Condition (2) to find the optimal approximation, optimal error and the optimal step of the HAM are the main novelties of this study that all of them will be obtained based on the SA, the CESTAC method and the CADNA library.

As we know, when we apply the FPA we need using mathematical softwares such as Mathematica, Maple, Matlab and others. In the SA, instead of applying the mentioned softwares, we apply the CADNA library [22–25] which is based on C, C++ or FORTRAN codes and should be done on the LINUX operating system.

Consider the following rounding error with M mantissa bits

$$\mathcal{S}^{(p)} = s^{(p)} - \zeta^{(p)} 2^{E^{(p)} - \mathcal{M}} \nu^{(p)}, \quad p = 1, 2, 3, \tag{3}$$

obtained from the binary computer computations where $\tilde{s} = (s^{(1)}, s^{(2)}, s^{(3)})$ is a fuzzy number in triangular form which is reproduced by the computer precision in the form $\tilde{\mathcal{S}} = (\mathcal{S}^{(1)}, \mathcal{S}^{(2)}, \mathcal{S}^{(3)})$, $\zeta^{(p)}$ is the sign of $s^{(p)}$, $E^{(p)}$ is the binary exponents of the results and $2^{-\mathcal{M}}\nu^{(p)}$ is the omitted bit of mantissa. We know that for $\mathcal{M} = 24, 53$ the results can be obtained in single and double precisions. If we choose $\nu^{(p)}$ as chance variable of uniformly distributed on interval $[-1, 1]$, we can make perturbation on the last mantissa bit of $\widetilde{s}$. Therefore, we can produce the random variables of $\widetilde{\mathcal{S}}$ such that, the average $\mu^{(p)}$ and the standard deviation $\sigma^{(p)}$ have direct role in their precision for $p = 1, 2, 3$. If we repeat this process for l times, the exact values of $s^{(p)}$ can be approximated by $\mathcal{S}_\kappa^{(p)}, \kappa = 1, ..., l$ and we have $E(\mathcal{S}^{(p)}) = s^{(p)}$. But all of mentioned process will be done by applying the CADNA library. By the following algorithm, we can describe clearly where τ_υ is the value of $t-$distribution with $l-1$ degree of freedom and confidence interval $1-\upsilon$.

Algorithm 1:

`Step 1- Produce` l `samples of` $\mathcal{S}^{(p)}$ `as`

$$\mathcal{S}^{(p)} = \left\{\mathcal{S}_1^{(p)}, \mathcal{S}_2^{(p)}, \cdots, \mathcal{S}_l^{(p)}\right\}, \quad p = 1, 2, 3,$$

`by means of the perturbation of the last bit of mantissa.`

`Step 2- Find`

$$\mathcal{S}_{ave}^{(p)} = \frac{\sum_{\kappa=1}^{l} \mathcal{S}_\kappa^{(p)}}{l}, \quad p = 1, 2, 3.$$

`Step 3- Calculate`

$$\left(\sigma^{(p)}\right)^2 = \frac{\sum_{\kappa=1}^{l}\left(\mathcal{S}_{\kappa}^{(p)} - \mathcal{S}_{ave}^{(p)}\right)^2}{l-1}, \quad p = 1,2,3.$$

`Step 4- Calculate the NCSDs of` $\mathcal{S}^{(p)}$ `and` $\mathcal{S}_{ave}^{(p)}$ `by` [26, 27]

$$C_{\mathcal{S}_{ave}^{(p)},\mathcal{S}^{(p)}} = \log_{10}\frac{\sqrt{l}\left|\mathcal{S}_{ave}^{(p)}\right|}{\tau_{\upsilon}\sigma^{(p)}}, \quad p = 1,2,3.$$

`Step 5- If`

$$C_{\mathcal{S}_{ave}^{(p)},\mathcal{S}^{(p)}} \leq 0 \qquad or \qquad \mathcal{S}_{ave}^{(p)} = 0,$$

`then write` $\widetilde{\mathcal{S}} = @.0$.

4 Main Idea

Consider the following second kind fuzzy Fredholm IE

$$\tilde{\omega}(s) = \tilde{f}(s) \oplus \alpha \int_{g_1(s)}^{g_2(s)} K(s,r)\tilde{\omega}(r)dr, \quad g_1(s) \leq s, r \leq g_2(s), \tag{4}$$

where $\tilde{\omega}(s) = (\omega^{(1)}(s), \omega^{(2)}(s), \omega^{(3)}(s))$ and $\tilde{f}(s) = (f^{(1)}(s), f^{(2)}(s), f^{(3)}(s))$ are triangular fuzzy functions on $[a, b]$, $\alpha > 0$ and the kernel $K(s, r)$ is a positive crisp function. Now, we can write

$$\omega^{(z)}(s) = f^{(z)}(s) \oplus \alpha \int_{g_1(s)}^{g_2(s)} K(s,r)\omega^{(z)}(r)dr, \quad g_1(s) \leq s, r \leq g_2(s), \ z = 1,2,3. \tag{5}$$

Operator N can be defined as follows

$$N[\omega] = \omega^{(z)}(s) - f^{(z)}(s) - \alpha \int_{g_1(s)}^{g_2(s)} K(s,r)\omega^{(z)}(r)dr = 0.$$

According to the traditional HAM [39–42], we get

$$\Re_m(\omega_{m-1}^{\hbar,(z)}(s)) = \omega_{m-1}^{\hbar,(z)}(s) - \alpha \int_{g_1(s)}^{g_2(s)} K(s,r)\omega_{m-1}^{\hbar,(z)}(r)dr - (1-\chi_m)f^{(z)}(s).$$

And we can define the m-th order deformation equation in the following form

$$\mathcal{L}\left[\omega_m^{\hbar,(z)}(s) - \chi_m\omega_{m-1}^{\hbar,(z)}(s)\right] = \hbar\mathcal{H}(s)\left[\omega_{m-1}^{\hbar,(z)}(s) - \alpha\int_{g_1(s)}^{g_2(s)} K(s,r)\omega_{m-1}^{\hbar,(z)}(r)dr - (1-\chi_m)f^{(z)}(s)\right]. \tag{6}$$

According to [39–42], in the HAM the mentioned operators and auxiliary functions can be chosen freely. Putting $\mathcal{L} = I, \mathcal{H}(s) = 1$ and $\omega_0^{\hbar}(s) = f(s)$, the approximate solution of Eq. (5) can be obtained applying the following successive relation:

$$\begin{cases} \omega_0^{\hbar,(z)}(s) = f^{(z)}(s), \\ \omega_1^{\hbar,(z)}(s) = \hbar\Big[-\alpha \displaystyle\int_{g_1(s)}^{g_2(s)} K(s,r)\omega_0^{\hbar,(z)}(r)dr\Big], \\ \omega_m^{\hbar,(z)}(s) = \omega_{m-1}^{\hbar,(z)}(s) + \hbar\Big[\omega_{m-1}^{\hbar,(z)}(s) - \alpha \displaystyle\int_{g_1(s)}^{g_2(s)} K(s,r)\omega_{m-1}^{\hbar,(z)}(r)dr\Big], \qquad m \geq 1. \end{cases} \tag{7}$$

Finally, the approximate solution of Eq. (5) can be obtained using following formula

$$\mathcal{W}_j^{\hbar,(z)}(s) = \sum_{i=0}^{j} \omega_i^{\hbar,(z)}(s).$$

Definition 6 [26]. *For two real numbers ξ_1 and ξ_2 the NCSDs is defined in the following form*

$$\begin{cases} C_{\xi_1,\xi_2} = \log_{10}\left|\dfrac{\xi_1+\xi_2}{2(\xi_1-\xi_2)}\right| = \log_{10}\left|\dfrac{\xi_1}{\xi_1-\xi_2} - \dfrac{1}{2}\right|, & \xi_1 \neq \xi_2, \\ C_{\xi_1,\xi_1} = +\infty. \end{cases} \tag{8}$$

Theorem 1. *Assume that the approximate solution of IE (5) is obtained from $\mathcal{W}_m^{\hbar,(z)}(r) = \sum_{i=0}^{m} \omega_i^{\hbar,(z)}(r)$ and the convergence control parameter $\hbar$, then*

$$C_{\mathcal{W}_m^{\hbar,(z)},\mathcal{W}_{m+1}^{\hbar,(z)}} = C_{\mathcal{W}_m^{\hbar,(z)},\omega^{(z)}} + O(\frac{1}{m+1}). \tag{9}$$

Proof: Applying Definition 1, we can write

$$\begin{aligned} C_{\mathcal{W}_m^{\hbar,(z)},\mathcal{W}_{m+1}^{\hbar,(z)}} - C_{\mathcal{W}_m^{\hbar,(z)},\omega^{(z)}} &= \log_{10}\left|\frac{\mathcal{W}_m^{\hbar,(z)} + \mathcal{W}_{m+1}^{\hbar,(z)}}{2(\mathcal{W}_m^{\hbar,(z)} - \mathcal{W}_{m+1}^{\hbar,(z)})}\right| - \log_{10}\left|\frac{\mathcal{W}_m^{\hbar,(z)} + \omega^{(z)}}{2(\mathcal{W}_m^{\hbar,(z)} - \omega^{(z)})}\right| \\ &= \log_{10}\left|\frac{\mathcal{W}_m^{\hbar,(z)} + \mathcal{W}_{m+1}^{\hbar,(z)}}{2(\omega_{m+1}^{\hbar,(z)})}\right| - \log_{10}\left|\frac{\mathcal{W}_m^{\hbar,(z)} + \omega^{(z)}}{2(\mathcal{W}_m^{\hbar,(z)} - \omega^{(z)})}\right| \\ &= \log_{10}\left|\frac{\mathcal{W}_m^{\hbar,(z)} + \mathcal{W}_{m+1}^{\hbar,(z)}}{\mathcal{W}_m^{\hbar,(z)} + \omega^{(z)}}\right| + \log_{10}\left|\frac{\mathcal{W}_m^{\hbar,(z)} - \omega^{(z)}}{\omega_{m+1}^{\hbar,(z)}}\right|. \end{aligned} \tag{10}$$

It is clear that for m enough large, the first term of Eq. (10) can be neglected. Thus for Eq. (10) we have

$$\log_{10}\left|\frac{\mathcal{W}_m^{\hbar,(z)} - \omega^{(z)}}{\omega_{m+1}^{\hbar,(z)}}\right| = \log_{10}\left|\frac{\sum_{i=0}^{m}\omega_i^{\hbar,(z)} - \sum_{i=0}^{\infty}\omega_i^{\hbar,(z)}}{\omega_{m+1}^{\hbar,(z)}}\right|$$

$$= \log_{10}\left|\frac{\sum_{i=m+1}^{\infty}\omega_i^{\hbar,(z)}}{\omega_{m+1}^{\hbar,(z)}}\right|$$

$$= \log_{10}\left|1 + \frac{\sum_{i=m+2}^{\infty}\hbar\mathcal{L}^{-1}\left[\mathcal{H}(r)\Re_i(\omega_{i-1}^{\hbar,(z)})\right]}{\hbar\mathcal{L}^{-1}\left[\mathcal{H}(r)\Re_{m+1}(\omega_m^{\hbar,(z)})\right]}\right|.$$

Based on the HAM, we have:

$$\frac{\sum_{i=m+2}^{\infty}\hbar\mathcal{L}^{-1}\left[\mathcal{H}(r)\Re_i(\omega_{i-1}^{\hbar,(z)})\right]}{\hbar\mathcal{L}^{-1}\left[\mathcal{H}(r)\Re_{m+1}(\omega_m^{\hbar,(z)})\right]} = \frac{\sum_{i=m+2}^{\infty}\hbar\mathcal{L}^{-1}\left[\mathcal{H}(r)\frac{1}{(i-1)!}\frac{\partial^{i-1}\mathcal{N}[\Phi^{(z)}(r,v)]}{\partial v^{i-1}}\Big|_{v=0}\right]}{\hbar\mathcal{L}^{-1}\left[\mathcal{H}(r)\frac{1}{m!}\frac{\partial^{m}\mathcal{N}[\Phi^{(z)}(r,v)]}{\partial v^{m}}\Big|_{v=0}\right]}$$

$$= \frac{\hbar\mathcal{L}^{-1}\left[\mathcal{H}(r)\frac{1}{(m+1)!}\frac{\partial^{m+1}\mathcal{N}[\Phi^{(z)}(r,v)]}{\partial v^{m+1}}\Big|_{v=0}\right]}{\hbar\mathcal{L}^{-1}\left[\mathcal{H}(r)\frac{1}{m!}\frac{\partial^{m}\mathcal{N}[\Phi^{(z)}(r,v)]}{\partial v^{m}}\Big|_{v=0}\right]} + \frac{\hbar\mathcal{L}^{-1}\left[\mathcal{H}(r)\frac{1}{(m+2)!}\frac{\partial^{m+2}\mathcal{N}[\Phi^{(z)}(r,v)]}{\partial v^{m+2}}\Big|_{v=0}\right]}{\hbar\mathcal{L}^{-1}\left[\mathcal{H}(r)\frac{1}{m!}\frac{\partial^{m}\mathcal{N}[\Phi^{(z)}(r,v)]}{\partial v^{m}}\Big|_{v=0}\right]} + \cdots$$

$$= O(\frac{1}{m+1}).$$

Therefore, we can write

$$\log_{10}\left|\frac{\mathcal{W}_m^{\hbar,(z)} - \omega}{\omega_{m+1}^{\hbar,(z)}}\right| = \log_{10}\left|1 + O(\frac{1}{m+1})\right|.$$

We know that $O(\frac{1}{m+1}) << 1$, therefore $\log_{10}\left|1 + O(\frac{1}{m+1})\right|$ leads to zero when m increases. Thus, the proof is complete. □

5 Numerical Examples

In this section, two examples of the Fredholm and Volterra IEs are presented and the CESTAC method is applied to validate the numerical results. For each example, the optimal step of the HAM, the optimal error and the optimal approximation are obtained. Also, the termination criterion (2) is applied to stop the numerical process. Moreover, several $\hbar$-curves are demonstrated to show the convergence region for various η-cuts. We note that the Definition 4 is applied to transform the triangular fuzzy function $\widetilde{\omega}(s) = (\omega^{(1)}(s), \omega^{(2)}(s), \omega^{(3)}(s))$ to the parametric form $\widetilde{\omega}(z) = (\underline{\omega}(s;\eta), \overline{\omega}(s;\eta))$.

Example 1: Let us consider the fuzzy Fredholm IE [8] with $K(s,r) = \frac{s^2+r^2-2}{13}, 0 \leq s,r \leq 2$ and

$$\underline{f}(s;\eta) = s\eta + \frac{3}{26} - \frac{3}{26}\eta - \frac{1}{13}s^2 - \frac{1}{13}s^2\eta,$$

$$\overline{f}(s;\eta) = 2\,s - \eta s + \frac{3}{26}\eta + \frac{1}{13}s^2\eta - \frac{3}{26} - \frac{3}{13}s^2, \qquad 0 \leq \eta \leq 1.$$

According to Figs. 1 and 2 the convergence region of $\underline{W}(s;\eta)$ and $\overline{W}(s;\eta)$ for various η-cuts $\eta = 0;\ 0.1;\ 1$, $s = 1$ and $m = 5, 10, 15$ is $-1.7 \lesssim \hbar \lesssim -0.3$. The numerical results of the CESTAC method are presented in Table 1. Using this table, the optimal step of fuzzy HAM, optimal distance and optimal approximations can be obtained. The optimal step for solving this example is $m_{opt} = 12$. Hausdorff distance is applied to find the mentioned optimal values.

Example 2: Consider the following fuzzy Volterra IE with $K(s,r) = s - r, 0 \leq s,r \leq 1$ and

$$\underline{f}(s;\eta) = (1 - s - \frac{s^2}{2})\eta,$$

$$\overline{f}(s;\eta) = (1 - s - \frac{s^2}{2})(2 - \eta), \qquad 0 \leq \eta \leq 1.$$

The $\hbar$-curves of this example are demonstrated in Figs. 3 and 4 as $-1.5 \lesssim \hbar \lesssim -0.5$ that they are the parts of curves parallel with axis x for $m = 5, 10, 15$, $\eta = 0;\ 0.1;\ 1$, $s = 0.5$. According to Table 2, the optimal iteration, optimal step of the HAM and optimal distance are obtained for $\hbar = -1$. Optimal step of mentioned method for solving this example is $m_{opt} = 4$.

Table 1. The numerical results of Example 1 for $s = 1, \hbar = -1$.

m	η	$\underline{\mathcal{W}}_m(s;\eta)$	$\lvert\underline{\mathcal{W}}_{m+1}(s;\eta) - \underline{\mathcal{W}}_m(s;\eta)\rvert$	$\overline{\mathcal{W}}_m(s;\eta)$	$\lvert\overline{\mathcal{W}}_{m+1}(s;\eta) - \overline{\mathcal{W}}_m(s;\eta)\rvert$	Max
0	0.0	$0.384618E-001$	$0.384619E-001$	$0.1653845E+001$	$0.1653846E+001$	$0.1653846E+001$
	0.1	$0.1192310E+000$	$0.1192310E+000$	$0.1573076E+001$	$0.1573077E+001$	$0.1573077E+001$
	0.2	$0.2000002E+000$	$0.2000002E+000$	$0.149230E+001$	$0.149230E+001$	$0.149230E+001$
	0.3	$0.2807694E+000$	$0.2807694E+000$	$0.141153E+001$	$0.141153E+001$	$0.141153E+001$
	0.4	$0.3615386E+000$	$0.3615387E+000$	$0.1330769E+001$	$0.1330769E+001$	$0.1330769E+001$
	0.5	$0.4423079E+000$	$0.4423079E+000$	$0.1250000E+001$	$0.1250000E+001$	$0.1250000E+001$
	0.6	$0.5230770E+000$	$0.5230771E+000$	$0.116923E+001$	$0.116923E+001$	$0.116923E+001$
	0.7	$0.6038463E+000$	$0.6038463E+000$	$0.1088461E+001$	$0.1088461E+001$	$0.1088461E+001$
	0.8	$0.6846155E+000$	$0.6846156E+000$	$0.100769E+001$	$0.100769E+001$	$0.100769E+001$
	0.9	$0.7653847E+000$	$0.7653847E+000$	$0.9269232E+000$	$0.9269232E+000$	$0.9269232E+000$
	1.0	$0.8461539E+000$	$0.8461540E+000$	$0.8461539E+000$	$0.8461539E+000$	$0.8461539E+000$
0	-	-	-	-	**Distance =**	**0.1653846E+001**

(continued)

Table 1. (*continued*)

m	η	$\underline{W}_m(s;\eta)$	$\lvert\underline{W}_{m+1}(s;\eta) - \underline{W}_m(s;\eta)\rvert$	$\overline{W}_m(s;\eta)$	$\lvert\overline{W}_{m+1}(s;\eta) - \overline{W}_m(s;\eta)\rvert$	Max
1	0.0	0.222880E − 001	0.161739E − 001	0.1889349E + 001	0.2355033E + 000	0.2355033E + 000
	0.1	0.1156410E + 000	0.35900E − 002	0.1795996E + 001	0.222919E + 000	0.222919E + 000
	0.2	0.2089941E + 000	0.899381E − 002	0.1702643E + 001	0.210335E + 000	0.210335E + 000
	0.3	0.3023471E + 000	0.21577E − 001	0.160929E + 001	0.19775E + 000	0.19775E + 000
	0.4	0.3957002E + 000	0.341615E − 001	0.1515937E + 001	0.185167E + 000	0.185167E + 000
	0.5	0.4890533E + 000	0.467454E − 001	0.142258E + 001	0.172583E + 000	0.172583E + 000
	0.6	0.5824063E + 000	0.593292E − 001	0.1329230E + 001	0.160000E + 000	0.160000E + 000
	0.7	0.6757594E + 000	0.719131E − 001	0.1235878E + 001	0.147416E + 000	0.147416E + 000
	0.8	0.7691124E + 000	0.844969E − 001	0.1142524E + 001	0.134832E + 000	0.134832E + 000
	0.9	0.8624656E + 000	0.970808E − 001	0.1049171E + 001	0.122248E + 000	0.122248E + 000
	1.0	0.9558187E + 000	0.109664E + 000	0.9558185E + 000	0.10966E + 000	0.10966E + 000
1	-	-	-	-	**Distance =**	**0.2355033E+000**
⋮	⋮	⋮	⋮	⋮	⋮	⋮
10	0.0	0.212360E − 001	@.0	0.1978763E + 001	0.7152557E − 006	0.7152557E − 006
	0.1	0.1191124E + 000	0.4E − 007	0.1880887E + 001	@.0	@.0
	0.2	0.2169888E + 000	@.0	0.1783010E + 001	@.0	@.0
	0.3	0.3148652E + 000	@.0	0.1685134E + 001	@.0	@.0
	0.4	0.4127416E + 000	0.1490116E − 006	0.1587258E + 001	@.0	@.0
	0.5	0.5106180E + 000	@.0	0.148938E + 001	@.0	@.0
	0.6	0.6084945E + 000	@.0	0.1391505E + 001	0.4768371E − 006	0.4768371E − 006
	0.7	0.7063708E + 000	0.2980232E − 006	0.1293628E + 001	0.4768371E − 006	0.4768371E − 006
	0.8	0.8042473E + 000	0.3E − 006	0.1195752E + 001	@.0	@.0
	0.9	0.9021236E + 000	0.3576278E − 006	0.109787E + 001	@.0	@.0
	1.0	0.1000000E + 001	0.4E − 006	0.9999997E + 000	@.0	@.0
10	-	-	-	-	**Distance =**	**0.7152557E-006**
11	0.0	0.212360E − 001	@.0	0.1978763E + 001	@.0	@.0
	0.1	0.1191124E + 000	@.0	0.1880887E + 001	@.0	@.0
	0.2	0.2169888E + 000	@.0	0.1783010E + 001	0.0000000E + 000	0.0000000E + 000
	0.3	0.3148652E + 000	@.0	0.1685134E + 001	@.0	@.0
	0.4	0.4127416E + 000	@.0	0.1587257E + 001	@.0	@.0
	0.5	0.5106181E + 000	@.0	0.1489381E + 001	@.0	@.0
	0.6	0.6084945E + 000	@.0	0.139150E + 001	@.0	@.0
	0.7	0.7063708E + 000	@.0	0.129362E + 001	@.0	@.0
	0.8	0.8042473E + 000	0.5960464E − 007	0.1195752E + 001	@.0	@.0
	0.9	0.9021237E + 000	0.1192092E − 006	0.109787E + 001	0.0000000E + 000	0.1192092E − 006
	1.0	0.1000000E + 001	@.0	0.9999997E + 000	@.0	@.0
11	-	-	-	-	**Distance =**	**0.1192092E-006**
12	0.0	0.212361E − 001	@.0	0.1978763E + 001	@.0	@.0
	0.1	0.1191124E + 000	@.0	0.1880887E + 001	@.0	@.0
	0.2	0.2169888E + 000	0.0000000E + 000	0.1783010E + 001	@.0	@.0
	0.3	0.3148652E + 000	@.0	0.1685134E + 001	@.0	@.0
	0.4	0.4127417E + 000	@.0	0.1587258E + 001	@.0	@.0
	0.5	0.5106181E + 000	@.0	0.148938E + 001	@.0	@.0
	0.6	0.6084945E + 000	@.0	0.1391505E + 001	@.0	@.0
	0.7	0.7063710E + 000	@.0	0.1293629E + 001	@.0	@.0
	0.8	0.8042473E + 000	@.0	0.1195752E + 001	@.0	@.0
	0.9	0.9021237E + 000	@.0	0.1097876E + 001	@.0	@.0
	1.0	0.1000000E + 001	@.0	0.9999997E + 000	@.0	@.0
12	-	-	-	-	**Distance =**	**@.0**

Table 2. The numerical results of Example 2 for $s = 0.5, \hbar = -1$.

m	η	$\underline{\mathcal{W}}_m(s;\eta)$	$\lvert\underline{\mathcal{W}}_{m+1}(s;\eta) - \underline{\mathcal{W}}_m(s;\eta)\rvert$	$\overline{\mathcal{W}}_m(s;\eta)$	$\lvert\overline{\mathcal{W}}_{m+1}(s;\eta) - \overline{\mathcal{W}}_m(s;\eta)\rvert$	*Max*
0	0.0	0.0000000E + 000	0.0000000E + 000	0.7500000E + 000	0.7500000E + 000	0.7500000E + 000
	0.1	0.3750000E − 001	0.3750000E − 001	0.7124999E + 000	0.7125000E + 000	0.7125000E + 000
	0.2	0.7499999E − 001	0.7500000E − 001	0.6749999E + 000	0.6750000E + 000	0.6750000E + 000
	0.3	0.1125000E + 000	0.1125000E + 000	0.6375000E + 000	0.6375000E + 000	0.6375000E + 000
	0.4	0.1500000E + 000	0.150000E + 000	0.5999999E + 000	0.6000000E + 000	0.6000000E + 000
	0.5	0.1875000E + 000	0.1875000E + 000	0.5625000E + 000	0.5625000E + 000	0.5625000E + 000
	0.6	0.2250000E + 000	0.2250000E + 000	0.5249999E + 000	0.5249999E + 000	0.5249999E + 000
	0.7	0.2625000E + 000	0.2625000E + 000	0.4874999E + 000	0.4875000E + 000	0.4875000E + 000
	0.8	0.3000000E + 000	0.3000000E + 000	0.4499999E + 000	0.4499999E + 000	0.4499999E + 000
	0.9	0.3375000E + 000	0.3375000E + 000	0.4124999E + 000	0.4124999E + 000	0.4124999E + 000
	1.0	0.3750000E + 000	0.3750001E + 000	0.3749999E + 000	0.3749999E + 000	0.3750001E + 000
0	-	-	-	-	**Distance =**	**0.7500000E+000**
1	0.0	0.0000000E + 000	0.0000000E + 000	0.953125E + 000	0.203125E + 000	0.203125E + 000
	0.1	0.4765624E − 001	0.101562E − 001	0.905468E + 000	0.192968E + 000	0.192968E + 000
	0.2	0.9531248E − 001	0.203124E − 001	0.857812E + 000	0.182812E + 000	0.182812E + 000
	0.3	0.142968E + 000	0.304687E − 001	0.8101561E + 000	0.172656E + 000	0.172656E + 000
	0.4	0.1906250E + 000	0.406249E − 001	0.762500E + 000	0.162500E + 000	0.162500E + 000
	0.5	0.2382812E + 000	0.507812E − 001	0.714843E + 000	0.152343E + 000	0.152343E + 000
	0.6	0.2859374E + 000	0.609374E − 001	0.667187E + 000	0.14218E + 000	0.14218E + 000
	0.7	0.3335937E + 000	0.710937E − 001	0.619531E + 000	0.132031E + 000	0.132031E + 000
	0.8	0.3812499E + 000	0.812499E − 001	0.571875E + 000	0.12187E + 000	0.12187E + 000
	0.9	0.4289062E + 000	0.914062E − 001	0.524218E + 000	0.11171E + 000	0.11171E + 000
	1.0	0.4765625E + 000	0.101562E + 000	0.4765625E + 000	0.1015625E + 000	0.1015625E + 000
1	-	-	-	-	**Distance =**	**0.203125E+000**
2	0.0	0.0000000E + 000	0.0000000E + 000	0.957769E + 000	0.4644E − 002	0.4644E − 002
	0.1	0.4788844E − 001	0.23220E − 003	0.909880E + 000	0.4412E − 002	0.4412E − 002
	0.2	0.9577689E − 001	0.46440E − 003	0.861992E + 000	0.4179E − 002	0.4179E − 002
	0.3	0.1436653E + 000	0.6966E − 003	0.814103E + 000	0.3947E − 002	0.3947E − 002
	0.4	0.1915538E + 000	0.9288E − 003	0.766215E + 000	0.3715E − 002	0.3715E − 002
	0.5	0.2394422E + 000	0.1161E − 002	0.718327E + 000	0.3483E − 002	0.3483E − 002
	0.6	0.2873307E + 000	0.13932E − 002	0.670438E + 000	0.3251E − 002	0.3251E − 002
	0.7	0.3352191E + 000	0.1625E − 002	0.622550E + 000	0.3018E − 002	0.3018E − 002
	0.8	0.3831076E + 000	0.18576E − 002	0.574661E + 000	0.2786E − 002	0.2786E − 002
	0.9	0.4309960E + 000	0.20898E − 002	0.526773E + 000	0.2554E − 002	0.2554E − 002
	1.0	0.4788845E + 000	0.2322E − 002	0.478884E + 000	0.2322E − 002	0.2322E − 002
2	-	-	-	-	**Distance =**	**0.4644E-002**
3	0.0	0.0000000E + 000	0.0000000E + 000	0.9578092E + 000	0.39E − 004	0.39E − 004
	0.1	0.4789045E − 001	0.20E − 005	0.909918E + 000	0.37E − 004	0.37E − 004
	0.2	0.9578090E − 001	0.40E − 005	0.8620284E + 000	0.36E − 004	0.36E − 004
	0.3	0.1436713E + 000	0.60E − 005	0.814137E + 000	0.34E − 004	0.34E − 004
	0.4	0.1915618E + 000	0.80E − 005	0.766247E + 000	0.32E − 004	0.32E − 004
	0.5	0.2394522E + 000	0.10E − 004	0.718357E + 000	0.29E − 004	0.29E − 004
	0.6	0.2873427E + 000	0.12E − 004	0.670466E + 000	0.2E − 004	0.2E − 004
	0.7	0.335233E + 000	0.14E − 004	0.622576E + 000	0.26E − 004	0.26E − 004
	0.8	0.383123E + 000	0.16E − 004	0.574685E + 000	0.2E − 004	0.2E − 004
	0.9	0.4310141E + 000	0.180E − 004	0.526795E + 000	0.2E − 004	0.2E − 004
	1.0	0.478904E + 000	0.20E − 004	0.478904E + 000	0.20E − 004	0.20E − 004
3	-	-	-	-	**Distance =**	**0.39E-004**

(*continued*)

Table 2. (*continued*)

m	η	$\underline{\mathcal{W}}_m(s;\eta)$	$\lvert\underline{\mathcal{W}}_{m+1}(s;\eta) - \underline{\mathcal{W}}_m(s;\eta)\rvert$	$\overline{\mathcal{W}}_m(s;\eta)$	$\lvert\overline{\mathcal{W}}_{m+1}(s;\eta) - \overline{\mathcal{W}}_m(s;\eta)\rvert$	Max
4	0.0	$0.0000000E+000$	$0.0000000E+000$	$0.957809E+000$	@.0	@.0
	0.1	$0.4789045E-001$	@.0	$0.9099189E+000$	@.0	@.0
	0.2	$0.9578091E-001$	@.0	$0.862028E+000$	@.0	@.0
	0.3	$0.1436713E+000$	@.0	$0.8141378E+000$	@.0	@.0
	0.4	$0.1915618E+000$	@.0	$0.766247E+000$	@.0	@.0
	0.5	$0.239452E+000$	@.0	$0.7183570E+000$	@.0	@.0
	0.6	$0.2873427E+000$	@.0	$0.670466E+000$	@.0	@.0
	0.7	$0.335233E+000$	@.0	$0.622575E+000$	@.0	@.0
	0.8	$0.3831237E+000$	@.0	$0.574685E+000$	@.0	@.0
	0.9	$0.4310142E+000$	@.0	$0.526795E+000$	@.0	@.0
	1.0	$0.4789047E+000$	@.0	$0.478904E+000$	@.0	@.0
4	-	-	-	-	**Distance =**	**@.0**

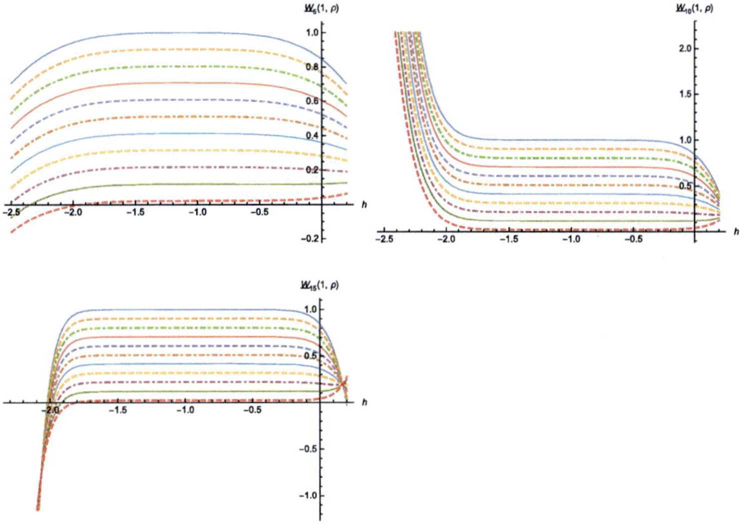

Fig. 1. The $\hbar$-curves of Example 1 for $\underline{\mathcal{W}}_m(1;\eta)$ with $\eta = 0;\ 0.1;\ 1$ and $m = 5, 10, 15$.

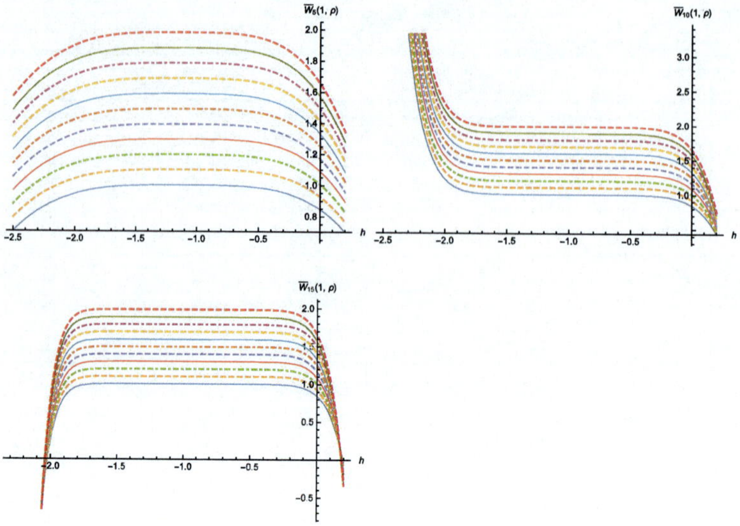

Fig. 2. The $\hbar$-curves of Example 1 for $\overline{\mathcal{W}}_m(1;\eta)$ with $\eta = 0;\ 0.1;\ 1$ and $m = 5, 10, 15$.

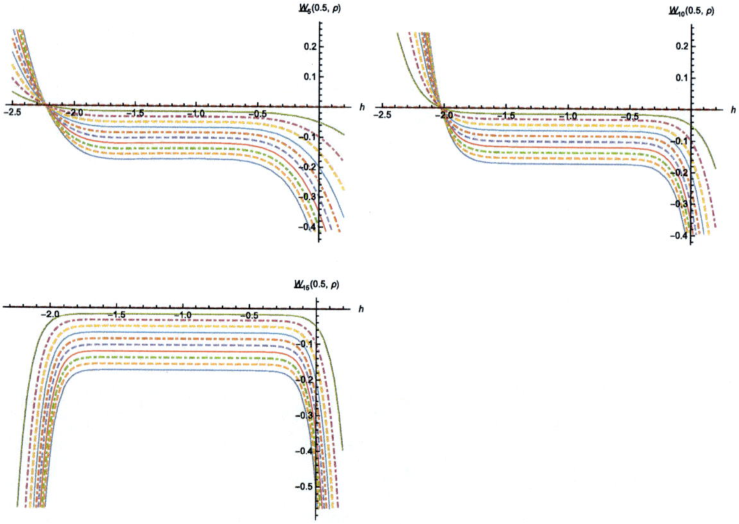

Fig. 3. The $\hbar$-curves of Example 2 for $\underline{\mathcal{W}}_m(0.5;\eta)$ with $\eta = 0;\ 0.1;\ 1$ and $m = 5, 10, 15$.

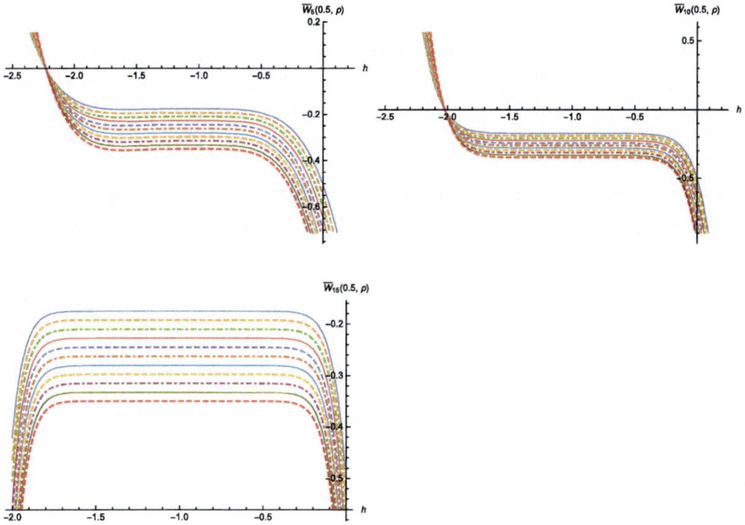

Fig. 4. The $\hbar$-curves of Example 2 for $\overline{\mathcal{W}}_m(0.5;\eta)$ with $\eta = 0;\ 0.1;\ 1$ and $m = 5, 10, 15$.

6 Conclusions

Solving the fuzzy IEs is one of interesting topics in fuzzy mathematics. There are many studies based on the FPA to solve these problems. But using the FPA we have some disadvantages for solving these problems. So we need to new arithmetic that we have several advantages than the FPA. In this study, the SA was applied to validate the numerical results of the HAM for solving second kind fuzzy IEs. Based on the CESTAC method in fuzzy case and the CADNA library, not only we do not have the disadvantages of the FPA but also the optimal step of iteration, optimal approximation and optimal distance can be found. Instead of applying the traditional absolute error to show the accuracy of method, the novel condition was presented based on Hausdorff distance. Also, the proposed theorem can illustrate the accuracy of the HAM to be applied for solving fuzzy IEs. The examples were presented to show the applications of the CESTAC method to solve the second kind fuzzy IEs.

References

1. Nouriani, H., Ezzati, R.: Application of Simpson quadrature rule and iterative method for solving nonlinear fuzzy delay integral equations. Fuzzy Sets and Systems (2020, In press)
2. Yang, H., Gong, Z.: Ill-posedness for fuzzy Fredholm integral equations of the first kind and regularization methods. Fuzzy Sets Syst. **3581**, 132–149 (2019)
3. Ziari, S.: Towards the accuracy of iterative numerical methods for fuzzy Hammerstein-Fredholm integral equations. Fuzzy Sets Syst. **37515**, 161–178 (2019)
4. Otadi, M., Mosleh, M.: Simulation and evaluation of interval-valued fuzzy linear Fredholm integral equations with interval-valued fuzzy neural network. Neurocomputing **20512**, 519–528 (2016)

5. Sahu, P.K., Saha Ray, S.: A new Bernoulli wavelet method for accurate solutions of nonlinear fuzzy Hammerstein-Volterra delay integral equations. Fuzzy Sets Syst. **30915**, 131–144 (2017)
6. Salahshour, S., Allahviranloo, T.: Application of fuzzy differential transform method for solving fuzzy Volterra integral equations. Appl. Math. Model. **37**(31), 1016–1027 (2013)
7. Mirzaee, F.: Numerical solution of Fredholm fuzzy integral equations of the second kind using hybrid of block-pulse functions and Taylor series. Ain Shams Eng. J. **5**(2), 631–636 (2014)
8. Noeiaghdam, S., Fariborzi Araghi, M.A., Abbasbandy, S.: Valid implementation of Sinc-collocation method to solve the fuzzy Fredholm integral equation. J. Comput. Appl. Math. (2020). https://doi.org/10.1016/j.cam.2019.112632
9. Fariborzi Araghi, M.A., Noeiaghdam, S.: A valid scheme to evaluate fuzzy definite integrals by applying the CADNA library. Int. J. Fuzzy Syst. Appl. **6**(4), 1–20 (2017). https://doi.org/10.4018/IJFSA.2017100101
10. Fariborzi Araghi, M.A., Noeiaghdam, S.: Homotopy analysis transform method for solving generalized Abel's fuzzy integral equations of the first kind. In: Proceeding of the 4-th Iranian Joint Congress on Fuzzy and Intelligent Systems (CFIS), IEEE (2014). https://doi.org/10.1109/CFIS.2015.7391645
11. Noeiaghdam, Z., Allahviranloo, T., Nieto, J.J.: *q*-fractional differential equations with uncertainty. Soft. Comput. **23**(19), 9507–9524 (2019)
12. Noeiaghdam, S., Fariborzi Araghi, M.A.: Finding optimal step of fuzzy Newton-Cotes integration rules by using the CESTAC method. J. Fuzzy Set Valued Anal. **2017**(2), 62–85 (2017). https://doi.org/10.5899/2017/jfsva-00383
13. Mikaeilvand, N., Noeiaghdam, Z., Noeiaghdam, S., Nieto, J.J.: A novel technique to solve the fuzzy system of equations. Mathematics **8**, 850 (2020). https://doi.org/10.3390/math8050850
14. Allahviranloo, T., Noeiaghdam, Z., Noeiaghdam, S., Nieto, J.J.: A fuzzy method for solving fuzzy fractional differential equations based on the generalized fuzzy taylor expansion. Mathematics **8**, 2166 (2020). https://doi.org/10.3390/math8122166
15. Fariborzi Araghi, M.A., Noeiaghdam, S.: Homotopy regularization method to solve the singular Volterra integral equations of the first kind. Jordan J. Math. Statist. **10**(4), 1–12 (2018)
16. Fariborzi Araghi, M.A., Noeiaghdam, S.: A novel technique based on the homotopy analysis method to solve the first kind Cauchy integral equations arising in the theory of airfoils. J. Interpol. Approximation Sci. Comput. **2016**(1), 1–13 (2016)
17. Noeiaghdam, S., Zarei, E., Barzegar Kelishami, H.: Homotopy analysis transform method for solving Abel's integral equations of the first kind. Ain Shams Eng. J. **7**, 483–495 (2016)
18. Noeiaghdam, S., Suleman, M., Budak, H.: Solving a modified nonlinear epidemiological model of computer viruses by homotopy analysis method. Math. Sci. **12**, 211–222 (2018)
19. Noeiaghdam, S.: A novel technique to solve the modified epidemiological model of computer viruses. SeMA J. (2018). https://doi.org/10.1007/s40324-018-0163-3
20. Noeiaghdam, S., Khoshrouye Ghiasi, E.: Solving a non-linear model of HIV infection for $CD4^{+}T$ cells by combining Laplace transformation and Homotopy analysis method. arXiv:1809.06232
21. Noeiaghdam, S., Khoshrouye Ghiasi, E.: An efficient method to solve the mathematical model of HIV infection for $CD8^{+}$ T-cells. arXiv:1907.01106

22. Abbasbandy, S., Fariborzi Araghi, M.A.: The use of the stochastic arithmetic to estimate the value of interpolation polynomial with optimal degree. Appl. Numer. Math. **50**, 279–290 (2004)
23. Abbasbandy, S., Fariborzi Araghi, M.A.: A reliable method to determine the ill-condition functions using stochastic arithmetic. Southwest J. Pure Appl. Math. **1**, 33–38 (2002)
24. Abbasbandy, S., Fariborzi Araghi, M.A.: Numerical solution of improper integrals with valid implementation. Math. Comput. Appl. **7**, 83–91 (2002)
25. Chesneaux, J.M.: CADNA, an ADA tool for round-off error analysis and for numerical debugging, In: Proceedings Congress on ADA in Aerospace, Barcelona (1990)
26. Vignes, J.: Discrete stochastic arithmetic for validating results of numerical software. Special Iss. Numer. Algorithms **37**, 377–390 (2004)
27. Vignes, J.: A stochastic arithmetic for reliable scientific computation. Math. Comput. Simulation **35**, 233–261 (1993)
28. Fariborzi Araghi, M.A., Noeiaghdam, S.: Dynamical control of computations using the Gauss-Laguerre integration rule by applying the CADNA library. Adv. Appl. Math. Sci. **16**, 1–18 (2016)
29. Fariborzi Araghi, M.A., Noeiaghdam, S.: Valid implementation of the Sinc-collocation method to solve linear integral equations by the CADNA library. J. Math. Model. **7**(1), 63–84 (2019)
30. Noeiaghdam, S., Fariborzi Araghi, M.A., Abbasbandy, S.: Finding optimal convergence control parameter in the homotopy analysis method to solve integral equations based on the stochastic arithmetic. Numer. Algor. **81**(1), 237–267 (2019). https://doi.org/10.1007/s11075-018-0546-7
31. Noeiaghdam, S., Fariborzi Araghi, M.A.: A novel approach to find optimal parameter in the Homotopy-regularization method for solving integral equations. Appl. Math. Inf. Sci. **14**(1), 1–8 (2020)
32. Noeiaghdam, S., Sidorov, D., Muftahov, I., Zhukov, A.V.: Control of accuracy on taylor-collocation method for load leveling problem. Bull. Irkutsk State Univ. Ser. Math. 30, 59-72 (2019). https://doi.org/10.26516/1997-7670.2019.30.59
33. Noeiaghdam, S., Sidorov, D., Sizikov, V., Sidorov, N.: Control of accuracy on Taylor-collocation method to solve the weakly regular Volterra integral equations of the first kind by using the CESTAC method. Appl. Comput. Math. **19**(1), 87–105 (2020)
34. Kaleva, O.: Fuzzy differential equations. Fuzzy Sets Syst. **24**, 301–317 (1987)
35. Fullér, R.: Neural Fuzzy Systems, Abo Akademi University (1995)
36. Ma, M., Friedman, M., Kandel, A.: Duality in fuzzy linear systems. Fuzzy Sets Syst. **109**, 55–58 (2000)
37. Anastassiou, G.A.: Fuzzy Mathematics: Approximation Theory. Springer, Heidelberg (2010)
38. Friedman, M., Ming, M., Kandel, A.: Fuzzy linear systems. Fuzzy Sets Syst. **96**, 201–209 (1998)
39. Liao, S.J.: The proposed homotopy analysis techniques for the solution of nonlinear problems. Shanghai Jiao Tong University, Shanghai (1992). (in English)
40. Liao, S.J.: Beyond Perturbation: Introduction to Homotopy Analysis Method, Chapman & Hall/CRC Press, Boca Raton (2003)
41. Liao, S.J.: On the homotopy analysis method for nonlinear problems. Appl. Math. Comput. **147**, 499–513 (2004)
42. Liao, S.J.: Homotopy Analysis Method in Nonlinear Differential Equations. Higher Education Press, Beijing and Springer-Verlag, Heidelberg (2012)

Check for updates

Boubaker Function for Solving Fractional Diffusion-Wave Equation

Amirahmad Khajehnasiri[1], Reza Ezzati[2](✉), and Mozhdeh Afshar Kermani[1]

[1] Department of Mathematics, North Tehran Branch, Islamic Azad University, Tehran, Iran
a.khajehnasiri@gmail.com

[2] Department of Mathematics, Karaj Branch, Islamic Azad University, Karaj, Iran
ezati@kiau.ac.ir

Abstract. In this paper, we developed an efficient computational method based on operational matrix of Boubaker function to find an approximate solution of fractional diffusion equations. After introducing the operational matrices of fractional integration as well as product of Boubaker function, we apply this method to reduce the original problem into a system of algebraic equations. One of the advantage of this method is the low-cost of setting up the equations without applying any projection method such as collocation, Galerkin, etc. The results of the proposed method show that this methods is very accurate and efficient.

Keywords: Fractional diffusion equations · Operational matrix · Boubaker functions · Fractional calculus

1 Introduction

Fractional diffusion equations naturally appear in history-dependent problems such as fluid flow, damped vibrations, viscoelasticity, population dynamics, heat conduction and seismology [15]. On the other hand, the multi-term time-fractional operator has been found a useful in describing complex physical and physiological systems [10], several numerical methods have been proposed in the last few years for solving fractional order diffusion-wave equations. Fractional order diffusion equations were shown to provide an adequate and accurate description of these transport processes, which exhibit anomalous diffusion [7]. However, fractional diffusion operator is a non-local operator, which generates computational and numerical difficulties that have not been encountered in the context of second-order diffusion equations. Jiang et al. [20] used some analytical techniques to solve three types of multi-term time-space Caputo Riesz fractional advection diffusion equations with non homogeneous Dirichlet boundary conditions. Li and Xu [22] improved their previous results and proposed a space-time spectral method for the equation. Reutskiy [21] have applied a combined method of separating variables and Fourier expansion with backward substitution to solve a wide class of fractional partial differential equations including diffusion-wave equations.

T. Allahviranloo et al. (Eds.): IDS 2020, AISC 1301, pp. 638–649, 2021.
https://doi.org/10.1007/978-3-030-66501-2_50

In recent years operational matrix has been extensively used in different context and emerged as a potential alternative in the field of numerical solution of partial differential equations. The operational matrix of integration has been determined for several types of orthogonal polynomials, such as Chebyshev polynomials [25], Laguerre polynomials [24], Jacobi operational matrix [16]. Ebadian have applied fractional operational matrix for solving non-linear Volterra integro-differential equations [17]. Imran Aziz have extended the Haar wavelet for the numerical solution of two-dimensional nonlinear integral equations [1]. Y. Yang in [3] also developed the Jacobi collocation method to solve the time-fractional diffusion wave equation and convergence analysis. Khajehnasiri in [2] have applied a triangular functions method for the solution of 2D nonlinear Volterra-Fredholm integro differential equations. In [6,17] the researcher employed triangular operational matrix for solving diffusion-wave equation and state-space of financial system. The authors of both [8] and [9] utilized the block-pulse operational for solving mixed Volterra-Fredholm integro-differential equation and two-dimensional integro-differential equation.

Boubaker functions (BFs) very recently have been proposed as powerful mathematical tool to evaluations wide variety of integral equations. The authors of [12] constructed BFs for solving delay fractional optimal control problems. In [11], the researchers employed BFs for solving fractional optimal control problems. In [13] Davaeifar have applied Boubaker function for solving systems of nonlinear Volterra-Fredholm integral equations. Ordokhani [18] have considered the Boubaker function for solving pantograph delay differential equation. Rabiei et al. [14] have considered the use of BFs method for solving fractional optimal control problems.

The time fractional diffusion and wave-diffusion equations can be written in the following form:

$$\frac{\partial^{\alpha} F(x,t)}{\partial t^{\alpha}} + \frac{\partial F(x,t)}{\partial t} = \frac{\partial^2 F(x,t)}{\partial x^2} + r(x,t) \quad (x,t) \in [0,1] \times [0,1], \tag{1}$$

subject to the initial conditions:

$$F(x,0) = u_0(x), \quad \frac{\partial F(x,0)}{\partial t} = u_1(x), \quad x \in [0,1], \tag{2}$$

and boundary conditions:

$$F(0,t) = k_0(t), \quad F(1,t) = k_1(t), \quad t \in [0,1], \tag{3}$$

where x and t are the space, and time variables respectively, r is a given function in $\Omega^2([0,1] \times [0,1])$ and u_0, u_1, k_0 and k_1 are given functions in $\Omega^2[0,1]$, $F(x,t)$ is a sufficiently smooth function, when $\alpha = 1$, it represents a traditional diffusion equation; and when $\alpha = 2$, it represents a traditional wave equation [4], when $0 < \alpha < 1$, Eq. (1) is a fractional diffusion equation while if $1 < \alpha < 2$, Eq. (1) is the time fractional diffusion-wave equation. However, they only considered the multi-orders lying in (0, 2) in these papers.

In the next section we will define some basic definitions and properties of the fractional integral and derivative. In the Sect. 3 we will review Boubaker and fractional Boubaker functions, and the operational matrix of product based on fractional Boubaker function. After which in "Applying the method" section we will solve fractional order diffusion-wave equation by using Boubaker functions. In Sect. 5, we apply the proposed method for some examples of fractional order diffusion-wave equation. Finally, some concluding remarks are given in "Conclusion" section.

2 Preliminaries and Basic Definitions

In this section, first we recall essential basic definitions and properties of the fractional integral and derivative [5,26].

Definition 1. *The Riemann-Liouville fractional integration, for a function $f(x)$, of order $\alpha > 0$ is given by*

$$I_{x_0}^{\alpha} f(x) = \frac{1}{\Gamma(\alpha)} \int_{x_0}^{x} (x-t)^{\alpha-1} f(t)dt, \quad \alpha > 0, \quad x > 0. \tag{4}$$

The following equations define Riemann-Liouville and Caputo fractional derivatives of order α, respectively:

$$D_{x_0}^{\alpha} f(x) = \frac{d^m}{dx^m}[I_{x_0}^{m-\alpha} f(x)], \tag{5}$$

$$D_{*x_0}^{\alpha} f(x) = I_{x_0}^{m-\alpha}[\frac{d^m}{dx^m} f(x)], \tag{6}$$

where $m-1 \leq \alpha < m$ and $n \in N$. From (4) and (5), we have

$$D_{x_0}^{\alpha} f(x) = \frac{1}{\Gamma(m-\alpha)} \frac{d^m}{dx^m} \int_{x_0}^{x} (x-t)^{m-\alpha-1} f(t)dt, \quad x > x_0 \tag{7}$$

Lemma 1. *If $m-1 < \alpha \leq m$, $m \in N$, then $D_*^{\alpha} I^{\alpha} f(x) = f(x)$, and:*

$$I^{\alpha} D_*^{\alpha} f(x) = f(x) - \sum_{k=0}^{m-1} f^{(k)}(0^+)\frac{x^k}{k!}, \quad x > 0. \tag{8}$$

3 Boubaker Polynomials

The definition of the Boubaker polynomials in the interval $[0, 1]$ is

$$B_0(x) = 1,$$
$$B_n(x) = \sum_{p=0}^{\lfloor \frac{n}{2} \rfloor} (-1)^p \binom{n-p}{p} \frac{n-4p}{m-p} x^{n-2p}$$

where $\lfloor . \rfloor$ is the floor function. The Boubaker polynomials have also a recursive relation:

$$B_n(x) = xB_{n-1}(x) - B_{n-2}(x), \quad n > 2. \tag{9}$$

The Boubaker polynomials basis can be considered as

$$\Phi(x) = \Upsilon T_m(x), \tag{10}$$

where $T_m(x) = [1, x, \cdots, x^m]^T$ and $\Upsilon = (\Lambda_{i,j})_{i,j=0}^m$ is an $(m+1) \times (m+1)$ matrix. Considering the following terms [12]

$$B_n(x) = \sum_{p=0}^{\lfloor \frac{n}{2} \rfloor} (-1)^p \binom{n-p}{p} \frac{n-4p}{m-p} x^{n-2p} = \sum_{j=0}^{k} \Lambda_{i,j} x^j,$$

or

$$B_n(x) = \sum_{j=n-\lfloor \frac{n}{2} \rfloor}^{n} (-1)^{\frac{n-j}{2}} \binom{\frac{n+j}{2}}{\frac{n-j}{2}} \frac{2j-n}{\frac{n+j}{2}} x^j = \sum_{j=0}^{k} \Lambda_{i,j} x^j,$$

we can obtain the entries of the matrix Υ for $n \geq 2$, $j = n - 2\lfloor \frac{n}{2} \rfloor, \cdots, n$, as:

$$\Lambda_{i,j} = \begin{cases} 0, & \text{if (n − j) is odd,} \\ (-1)^{\frac{n-j}{2}} \binom{\frac{n+j}{2}}{\frac{n-j}{2}} \frac{2j-n}{\frac{n+j}{2}}, & \text{if (n − j) is even,} \end{cases}$$

And based on the definition of $B_0(x)$ and $B_1(x)$, this formula is valid for $i = 1$, and also for $i = 0$, we have

$$\Lambda_{0,0} = 1, \quad \Lambda_{0,j} = 1, \quad j = 1, \cdots, k. \tag{11}$$

3.1 Function Approximation

It is clear that

$$Y = span\{B_0(x), B_1(x), \cdots, B_k(x)\} \tag{12}$$

represented a finite dimensional and closed subspace of the Hilbert space $H = L^2[0,1]$, and therefore, Y is a complete subspace, so there is a unique (best) approximation out of Y. We represent this as $\tilde{f} \in Y$ for each $f \in H$, such that

$$\forall y \in Y, \quad \|f - \tilde{f}\| \leq \|f - y\|. \tag{13}$$

Since $\tilde{f} \in Y$, there exist unique coefficients $a_0, a_1, \cdots, a_k$ such that

$$f(x) \simeq \tilde{f}(x) = \sum_{j=0}^{k} a_j B_j(x) = A^T \Phi(x), \tag{14}$$

in which T is transposition, and we can represent the vectors A and $\Phi(x)$ such as:

$$A = [a_0, a_1, \cdots, a_k]^T, \tag{15}$$

$$\Phi(x) = [B_0(x), B_1(x), \cdots, B_k(x)]^T. \tag{16}$$

We Let

$$f_j = <f, B_j> = \int_0^1 f(x)B_j(x)dt, \tag{17}$$

where $<.,.>$ denotes inner product, and then, we have

$$f_j \simeq \sum_{i=0}^{k} a_i \int_0^1 B_i(x)B_j(x)dt = \sum_{i=0}^{m} a_i r_{ij}, \quad j = 0, 1, \cdots, k, \tag{18}$$

with

$$r_{ij} = \int_0^1 B_i(x)B_j(x)dt \quad i, j = 0, 1, \cdots, k. \tag{19}$$

Now we suppose that

$$F = <f, \phi> = \begin{pmatrix} f_0 \\ f_1 \\ \vdots \\ f_k \end{pmatrix}, \quad R = [r_{ij}], \tag{20}$$

where

$$f_j = A^T[r_{0j}, r_{1j}, \cdots, r_{kj}]^T, \tag{21}$$

so we can write $F^T = A^T R$, and A can be calculated as

$$A = D^{-1}<f, \phi>, \tag{22}$$

where R is the following $(m+1) \times (m+1)$ matrix

$$R = <\Phi(x), \Phi(x)> = \int_0^1 \Phi(x)(\Phi(x))^T dt. \tag{23}$$

3.2 Operational Matrix of Fractional Integration

In this section, we derive the $(m+1) \times (m+1)$ Riemann-Liouville fractional operational matrix of integration P^α expressed by:

$$I^\alpha \Phi(x) = P^\alpha \Phi(x). \tag{24}$$

By using Eq. (22) and linearity of Riemann-Liouville fractional integral, for $n \geq 2$, we have

$$\begin{aligned} I^\alpha B_n(x) &= \sum_{p=0}^{\lfloor \frac{n}{2} \rfloor} (-1)^p \frac{n-4p}{n-p} \binom{n-p}{p} I^\alpha x^{(n-2p)} \\ &= \sum_{p=0}^{\lfloor \frac{n}{2} \rfloor} b_{n,p} x^{n-2p+\alpha} \end{aligned} \tag{25}$$

where

$$b_{n,p} = (-1)^p \frac{(n-p-1)!(n-4p)}{p!\Gamma(n-2p+\alpha+1)} \tag{26}$$

Now we can expand $x^{n-2p+\alpha}$ in terms of Boubaker polynomials as

$$x^{n-2p+\alpha} \simeq \sum_{i=0}^{k} c_{p,i} B_i(x) \tag{27}$$

with

$$c_{p,i} = \frac{\langle x^{n-2p+\alpha}, B_i(x)\rangle}{\langle B_i(x), B_i(x)\rangle}, \tag{28}$$

in which, we have calculated $c_{p,i}$ by using Eq. (22), and replaced in Eq. (25) we obtain

$$I^\alpha B_n(x) \simeq \sum_{p=0}^{\lfloor \frac{n}{2} \rfloor} b_{n,p} \sum_{i=0}^{k} c_{p,i} B_i(x) = \sum_{i=0}^{k} \left(\sum_{p=0}^{\lfloor \frac{n}{2} \rfloor} b_{n,p} c_{p,i} \right) B_i(x)$$

Now for simpler representation, we regard

$$\theta_{n,i,p} = b_{n,p} c_{p,i}, \tag{29}$$

and rewrite Eq. (29), for $i, j = 1, \cdots, k$, as

$$I^\alpha B_n(x) \simeq \left[\sum_{p=0}^{\lfloor \frac{n}{2} \rfloor} \theta_{n,0,p}, \sum_{p=0}^{\lfloor \frac{n}{2} \rfloor} \theta_{n,1,p}, \cdots, \sum_{p=0}^{\lfloor \frac{n}{2} \rfloor} \theta_{n,k,p} \right] \Phi(x) \tag{30}$$

Hence,

$$P^\alpha = \begin{pmatrix} \frac{v_{0,0}}{\Gamma(v+1)} & \frac{v_{0,1}}{\Gamma(v+1)} & \cdots & \frac{v_{0,k}}{\Gamma(v+1)} \\ \frac{v_{1,0}}{\Gamma(v+1)} & \frac{v_{1,1}}{\Gamma(v+1)} & \cdots & \frac{v_{1,k}}{\Gamma(v+1)} \\ \sum_{p=0}^{1} \theta_{2,0,p} & \sum_{p=0}^{1} \theta_{2,1,p} & \cdots & \sum_{p=0}^{1} \theta_{2,k,p} \\ \vdots & \vdots & \vdots & \vdots \\ \sum_{p=0}^{\lfloor \frac{k}{2} \rfloor} \theta_{k,0,p} & \sum_{p=0}^{\lfloor \frac{k}{2} \rfloor} \theta_{k,1,p} & \cdots & \sum_{p=0}^{\lfloor \frac{k}{2} \rfloor} \theta_{k,k,p} \end{pmatrix} \tag{31}$$

3.3 The Operational Matrix of Product

The product of two vectors of Boubaker functions satisfies the following relation

$$C^T \Phi(x) \Phi^T(x) \simeq \Phi^T(x) \tilde{C}^T \tag{32}$$

where C is any arbitrary vector and $\tilde{C}$ is an $(m+1) \times (m+1)$ matrix. We notice that $C^T \Phi(x) \Phi^T(x)$ can be approximated by $\Phi(x)$ as follows

$$C^T \Phi(x) \Phi^T(x) = [c_0(x), \cdots, c_k(x)],$$

$$c_i(x) \simeq \sum_{j=0}^{k} \tilde{c}_{i,j} B_j(x) = \tilde{C}_i^T \Phi(x) \tag{33}$$

where

$$\tilde{C}_i = [\tilde{c}_{i,0}, \tilde{c}_{i,1}, \cdots, \tilde{c}_{i,k},]^T \tag{34}$$

Using Eq. (33), we have

$$c_i^m = \langle \sum_{j=0}^{k} \tilde{c}_{i,j} B_j, B_m \rangle = \sum_{j=0}^{k} \tilde{c}_{i,j} r_{j,m}, \quad j, m = 0, 1, \cdots, k. \tag{35}$$

where $c_i^m = \langle a_i, B_m \rangle$ and $r_{j,m} = \langle B_{i,j}, B_{j,m} \rangle$. So by considering

$$C_i = [c_i^0, a_i^1, \cdots, a_i^k]^T \tag{36}$$

we have

$$C_i^T = \tilde{C}_i^T R \tag{37}$$

and

$$\tilde{C}_i^T = C_i^T R^{-1} \tag{38}$$

therefore the operational matrix of product is obtained.

4 Applying the Method

In this section we solve time fractional diffusion-wave equation with damping by using operational matrix of fractional integration of Boubaker functions. To do this, we consider the following equation:

$$\frac{\partial^\alpha f(x,t)}{\partial t^\alpha} + \frac{\partial f(x,t)}{\partial t} = \frac{\partial^2 f(x,t)}{\partial x^2} + r(x,t), \tag{39}$$

$$(x,t) \in [0,1] \times [0,1], \quad 1 < \alpha \leq 2, \tag{40}$$

with initial conditions

$$f(x,0) = u_0(x), \quad \frac{\partial f(x,0)}{\partial t} = u_1(x), \quad x \in [0,1], \tag{41}$$

and boundary conditions:

$$f(0,t) = k_0(t), \quad f(1,t) = k_1(t), \quad t \in [0,1]. \tag{42}$$

By applying the Riemann-Liouville fractional integration of order α with respect to t on both sides of (39), and using the initial conditions (41), we obtain:

$$f(x,t) - w(x,t) + (I_t^{\alpha-1} f)(x,t) = \left(I_t^\alpha \frac{\partial^2 f}{\partial x^2} \right)(x,t) + (I_t^\alpha r)(x,t), \tag{43}$$

where $w(x,t) = u_0(x) + t u_1(x) - \frac{t^{\alpha-1}}{\Gamma(\alpha)} u_0(x)$. Now we approximate $\frac{\partial^2 f(x,t)}{\partial x^2}$ by the BFs as follows:

$$\frac{\partial^2 f(x,t)}{\partial x^2} \simeq \Phi^T(x) F \Phi(t), \tag{44}$$

where $F = [u_{ij}]_{\hat{m}\times\hat{m}}$ is an unknown matrix which should be found and $\Phi(.)$ is the BFs vector that was defined in (16). Moreover, by integrating (44) two times with respect to x, we have

$$f(x,t) \simeq f(0,t) + x\left(\frac{\partial f(x,t)}{\partial x}\bigg|_{x=0}\right) + \Phi^T(x)(P^T)^2 F\Phi(t), \tag{45}$$

and by putting $x = 1$ in (45) and considering the boundary conditions (41). We obtain:

$$\frac{\partial f(x,t)}{\partial x}\bigg|_{x=0} \simeq k_1(t) - k_0(t) - \Phi^T(1)(P^T)^2 F\Phi(t). \tag{46}$$

We also expand $k_0(t)$ and $k_1(t)$ by the BFs as follows:

$$k_0(t) \simeq K_0^T\Phi(t), \quad k_1(t) \simeq K_1^T\Phi(t), \tag{47}$$

where K_0 and K_1 are the BFs coefficient vectors. By substituting (47) into (46), we obtain:

$$\begin{aligned}\frac{\partial f(x,t)}{\partial x}\bigg|_{x=0} &\simeq (K_1^T - K_0^T - \Phi^T(1)(P^T)^2 F)\Phi(t),\\ &\simeq \tilde{F}^T\Phi(t).\end{aligned} \tag{48}$$

Now by substituting (48) into (45), we have:

$$\begin{aligned}f(x,t) &\simeq \Phi^T(x)[LK_0^T + X\tilde{F}^T + (P^T)^2 F]\Phi(t),\\ &= \Phi^T(x)\Theta\Phi(t),\end{aligned} \tag{49}$$

where X and L are the BFs coefficient vectors for x and the unit step function, respectively. Furthermore, we expand $w(x,t)$ and $r(x,t)$ by the BFs as follows:

$$w(x,t) \simeq \Phi^T(x)W\Phi(t), \quad r(x,t) \simeq \Phi^T(x)R\Phi(t), \tag{50}$$

where W and R are the known BFs coefficient matrices for $w(x,t)$ and $r(x,t)$, respectively. Then by substituting (44), (49) and (50) into (43), and using fractional operational matrix of integration of BFs, we have:

$$\Phi(x)^T[\Theta + \Theta P^{(\alpha-1)} - FP^{\alpha}]\Phi(t) \simeq \Phi(x)^T[W + RP^{\alpha}]\Phi(t), \tag{51}$$

so, by replacing $\simeq$ by, $=$, we have

$$\Theta + \Theta P^{(\alpha-1)} - FP^{\alpha} = W + QP^{\alpha}. \tag{52}$$

Finally, by solving this system, we obtain an approximate solution for the problem using (49).

5 Numerical Examples

To show the effectiveness of proposed method, two fractional diffusion equations are solved. Boubaker function is utilized to approximate the unknown functions computations. As up to now there is a few alternative numerical method for these types of fractional diffusion equations, Also we report the absolute errors of the proposed computational method in some points $(x_j, t_j) \in [0,1] \times [0,1]$ as follows $\mid e(x_j, t_j) \mid = \mid \Phi(x_j)^T \Lambda \Phi(t_{t_j}) - f(x_j, t_j) \mid$.

Example 1. The following fractional diffusion-wave equation with damping is considered [19]

$$\frac{\partial^\alpha f(x,t)}{\partial t^\alpha} + x\frac{\partial f(x,t)}{\partial x} + \frac{\partial^2 f(x,t)}{\partial x^2} = 2t^\alpha + 2x^2 + 2., \tag{53}$$

$$0 \le x \le 1, \quad 0 < t \le 1,$$

the initial and boundary conditions,

$$f(0,t) = 2\frac{\Gamma(\alpha+1)}{\Gamma(2\alpha+1)}t^{2\alpha}, \quad f(x,0) = x^2 \tag{54}$$

When $\alpha = 0.5$ the exact solution of this problem is $f(0,t) = 2\frac{\Gamma(\alpha+1)}{\Gamma(2\alpha+1)}t^{2\alpha}$. To solve this example, we implement the BFs method for $\alpha = 0.5$, $t = 0.5$. Numerical results are presented in Table 1.

Example 2. Consider the following fractional diffusion-wave equation given in [8,23]

$$\frac{\partial^\alpha f(x,t)}{\partial t^\alpha} + \frac{\partial f(x,t)}{\partial t} = \frac{\partial^2 f(x,t)}{\partial x^2} + \frac{6t^{3-\alpha}}{\Gamma(4-\alpha)}e^x + 3t^2e^x - t^3e^x \tag{55}$$

$$(x,t) \in [0,1] \times [0,1], \quad 1 < \alpha \le 2, \tag{56}$$

subject to the initial conditions

$$f(x,0) = 0, \quad \frac{\partial f(x,0)}{\partial t} = 0, \tag{57}$$

and boundary conditions:

$$f(0,t) = t^3, \quad f(1,t) = et^3. \tag{58}$$

The exact solution is known and given by $f(x,t) = e^x t^3$. To solve this example, we implement the BFs method for $\alpha = 1.2$, $\alpha = 1.4$, $\alpha = 1.6$ and $\alpha = 1.8$. Numerical results are presented in Table 2.

Table 1. Absolute error for $\alpha = 0.5$, $t = 0.5s$, for Example 1

x	Exact solution	$m = 16$ [19]	$m = 16$	$m = 32$ [19]	$m = 32$
0.1	0.0094	6.093E-3	6.741E-4	6.093E-3	7.579E-4
0.2	0.0333	4.843E-3	7.001E-3	4.843E-3	6.650E-3
0.3	0.0655	1.089E-2	4.741E-2	2.750E-2	5.148E-2
0.4	0.0998	1.937E-2	4.213E-3	1.937E-2	6.254E-3
0.5	0.1300	1.000E-6	6.025E-5	1.000E-6	6.956E-6
0.6	0.1498	7.460E-3	2.742E-4	4.359E-2	4.451E-4
0.7	0.1529	1.734E-2	4.254E-2	1.734E-2	1.659E-2
0.8	0.1331	2.964E-2	6.025E-3	7.750E-2	3.041E-3
0.9	0.0842	4.443E-2	3.268E-2	4.443E-2	3.240E-2

Table 2. Numerical results for Example 2

$m = 20$				
(x, t)	$\alpha = 1.2$	$\alpha = 1.4$	$\alpha = 1.6$	$\alpha = 1.8$
$(0.0, 0.0)$	5.502E-7	1.208E-5	6.357E-4	4.0018E-4
$(0.1, 0.1)$	9.504E-7	5.053E-5	4.157E-4	3.115E-5
$(0.2, 0.2)$	5.159E-7	4.009E-5	1.241E-4	9.554E-3
$(0.3, 0.3)$	7.478E-7	6.291E-5	2.859E-6	5.458E-6
$(0.4, 0.4)$	6.321E-7	6.504E-5	2.367E-5	2.590E-7
$(0.5, 0.5)$	7.312E-7	4.312E-5	2.047E-5	6.040E-7
$(0.6, 0.6)$	6.954E-7	4.125E-6	2.108E-5	3.008E-7
$(0.7, 0.7)$	2.874E-7	7.219E-6	6.441E-6	7.428E-7
$(0.8, 0.8)$	5.257E-8	4.985E-6	3.417E-6	8.782E-8
$(0.9, 0.9)$	6.210E-8	3.457E-6	7.858E-6	6.217E-8

6 Conclusion

The purpose of this article is to develop a numerical method based on Boubaker functions for obtaining the approximate solution of fractional diffusion equations. Frits, the operational matrices of fractional integration and product of BFs has been presented. Then by using these matrices, the original problem has been reduced to an algebraic system. The benefits of this method are the low cost of setting up the equations without applying any projection method such as collocation, Galerkin, etc. For more investigation, two examples have been presented. As the numerical results show, the proposed method can be a suitable method for solving these problems.

References

1. Aziza, I., ul-Islam, S., Khan, F.: A new method based on Haar wavelet for the numerical solution of two-dimensional nonlinear integral equations. J. Comput. Appl. Math. **272**, 70–80 (2014)
2. Khajehnasiri, A.A.: Numerical solution of nonlinear 2D Volterra-Fredholm Integro-differential equations by two-dimensional triangular function. Int. J. Appl. Comput. Math. **2**, 575–591 (2016)
3. Yang, Y., Chen, Y., Huang, Y., Wei, H.: Spectral collocation method for the time-fractional diffusion-wave equation and convergence analysis. Comput. Math. Appl. **73**, 1218–1232 (2017)
4. Deng, W., Li, C., Guo, Q.: Analysis of fractional differential equations with multi-orders. Fractals **15**, 173–182 (2007)
5. Ebadian, A., Fazli, H.R., Khajehnasiri, A.A.: Solution of nonlinear fractional diffusion-wave equation by traingular functions. SeMA J. **72**, 37–46 (2015)
6. Fazli, H.R., Hassani, F., Ebadian, A., Khajehnasiri, A.A.: National economies in state-space of fractional-order financial system. Afrika Matematika **10**, 1–12 (2015)
7. Metler, R., Klafter, J.: The restaurant at the end of random walk: recent developments in the description of anomalous transport by fractional dynamics. J. Phys. A **37**, 161–208 (2004)
8. Safavi, M., Khajehnasiri, A.A.: Numerical solution of nonlinear mixed Volterra-Fredholm integro-differential equations by two-dimensional block-pulse functions. Cogent Math. Stat. **5**, 1–12 (2018)
9. Aghazadeh, N., Khajehnasiri, A.A.: Solving nonlinear two-dimensional Volterra integro-differential equations by block-pulse functions. Math. Sci. **7**, 1–6 (2013)
10. Liu, F., Anh, V., Turner, I.: Numerical solution of the space fractional Fokker-Planck equation. J. Comput. Appl. Math. **166**, 209–219 (2004)
11. Rabiei, K., Ordokhani, Y.: Boubaker hybrid functions and their application to solve fractional optimal control and fractional variational problems. Appl. Math. **5**, 541–567 (2018)
12. Rabiei, K., Ordokhani, Y., Babolian, E.: Fractional-order Boubaker functions and their applications in solving delay fractional optimal control problems. J. Vib. Control **15**, 1–14 (2017)
13. Davaeifar, S., Rashidinia, J.: Boubaker polynomials collocation approach for solving systems of nonlinear Volterra-Fredholm integral equations. J. Taibah Univ. Sci. **6**, 1182–1199 (2017)
14. Rabiei, K., Ordokhani, Y., Babolian, E.: The Boubaker polynomials and their applications to solve fractional optimal control problems. Nonlinear Dyn. **4**, 1–11 (2018)
15. Raberto, M., Scalas, E., Mainardi, F.: Waiting-times and return in high-frequency financial data: an empirical study. Phys. A **314**, 749–755 (2002)
16. Bhrawy, A.H., Doha, E.H., Baleanu, D., Ezz-Eldien, S.A.: Spectral tau algorithm based on Jacobi operational matrix for numerical solution of time fractional diffusion-wave equations. J. Comput. Phys. **293**, 142–156 (2015)
17. Ebadian, A., Khajehnasiri, A.A.: Block-pulse functions and their applications to solving systems of higher-order nonlinear Volterra integro-differential equations. Electron. J. Differ. Equ. **54**, 1–9 (2014)
18. Rabiei, K., Ordokhani, Y.: Solving fractional pantograph delay differential equations via fractional-order Boubaker polynomials. Eng. Comput. **2**, 1013–1026 (2017)

19. Chena, Y., Wua, Y., Cui, Y., Wanga, Z., Jina, D.: Wavelet method for a class of fractional convection-diffusion equation with variable coefficients. J. Comput. Sci. **1**, 146–149 (2010)
20. Liu, F., Meerschaert, M.M., McGough, R.J., Zhuang, P., Liu, Q.: Numerical methods for solving the multi-term time-fractional wave-diffusion equation. Fract. Calc. Appl. Anal. **16**, 9–25 (2013)
21. Reutskiy, S.Y.: A new semi-analytical collocation method for solving multi-term fractional partial differential equations with time variable coefficients. Appl. Math. Model. **45**, 238–254 (2017)
22. Li, X., Xu, C.: A space-time spectral method for the time fractional diffusion equation. SIAM J. Numer. Anal. **47**, 2108–2131 (2009)
23. Chen, J., Liu, F., Anh, V., Shen, S., Liu, Q., Liao, C.: The analytical solution and numerical solution of the fractional diffusion-wave equation with damping. Appl. Math. Comput. **219**, 1737–1748 (2012)
24. Saadatmandi, A., Dehghan, M.: A new operational matrix for solving fractional-order differential equations. Comput. Math. Appl. **59**, 1326–1336 (2010)
25. Doha, E.H., Bhrawy, A.H., Ezz-Eldien, S.S.: A Chebyshev spectral method based on operational matrix for initial and boundary value problems of fractional order. Comput. Math. Appl. **62**, 2364–2373 (2011)
26. Podlubny, I.: Fractional Differential Equations. Academic Press, New York (1999)

Collocation Method for Second Order Fuzzy Differential Equation Under Generalized Differentiability

E. Ahmady[1(✉)] and N. Ahmady[2]

[1] Department of Mathematics, Shahr-E-Qods Branch, Islamic Azad University, Tehran, Iran
e.ahmadi@qodsiau.ac.ir

[2] Department of Mathematics, Varamin-Pishva Branch, Islamic Azad University, Varamin, Iran

Abstract. In this paper we consider fuzzy differential equation of the second order under generalized differentiability. We give four cases by selecting different types of derivatives, and we obtain four systems for second order fuzzy differential equations. For numerical solution we propose the collocation method with positive basis functions in each cases. The proposed method is illustrated by numerical example.

Keywords: Fuzzy differential equation · Generalized differentiability · Numerical method · Collocation method

1 Introduction

In the real world, the uncertainty of nonlinear systems can be modeled by using fuzzy differential equations (FDEs). Fuzzy derivatives is very important notation in FDEs and it was first initiated by Chang and Zadeh [8], and after that Seikkala derivative, Dubois and Prade derivative, Goetschel and Voxman derivative and Hukuhara derivative were discussed by many researches [9, 11, 12, 16]. In the most of the papers fuzzy differential equation is solved under Hukuhala differentiability. Some numerical methods for FDEs under the Hukuhara differentiability concept such as the fuzzy Euler method, predictor-corrector method, Improved predictor-corrector method, Taylor method are presented in [1–3, 14].

It is well-known that the solution of FDEs under Hukuhara derivatives has the property that the diameter is non-decreasing as t increase, in order to overcome this shortcoming, Bede and Gal introduced the strongly generalized differential of set value fuzzy function [4] and studied more properties in [5–7].

Nieto et al. [15] obtained solution of first order fuzzy differential equations using generalized differentiability by interpreting the original fuzzy differential equations with two crisp ordinary differential equations. Also Khastan et al. [13], investigated the multiple solution of Nth-order fuzzy differential equations under generalized differentiability. They introduced new definitions of solution to fuzzy differential equations and by selecting different types of derivatives, they got several solutions to fuzzy initial value

T. Allahviranloo et al. (Eds.): IDS 2020, AISC 1301, pp. 650–662, 2021.
https://doi.org/10.1007/978-3-030-66501-2_51

problem for second-order fuzzy differential equations. The khastan et al. method is analytical method, in this paper we proposed the numerical method for obtaining multiple solution of second order fuzzy differential equations under generalized differentiability.

This paper is organized as follows: After a preliminary section, we introduce method in Sect. 3 and numerical example to illustrate in Sect. 4.

2 Preliminaries

In this section the most basic notations and definitions used are introduced. See for example, [10].

The set of fuzzy numbers, that is, normal, fuzzy convex, upper semi-continuous and compactly supported fuzzy sets which are defined over the real line and denoted by $\mathbb{R}_{\mathcal{F}}$. For $0 < r \leq 1$, set $[u]^r = \{t \in \mathbb{R} | u(t) \geq r\}$, and $[u]^0 = cl\{t \in \mathbb{R} | u(t) \rangle 0\}$. We represent $[u]^r = [u^-(r), u^+(r)]$, so if $u \in \mathbb{R}_{\mathcal{F}}$, the r-level set $[u]^r$ is a closed interval for all $r \in [0, 1]$. For arbitrary $u, v \in \mathbb{R}_{\mathcal{F}}$ and $k \in \mathbb{R}$, the addition and scalar multiplication are defined by $[u + v]^r = [u]^r + [v]^r, [ku]^r = k[u]^r$ respectively.

Definition 2.1. The Hausdorff distance between fuzzy numbers is given by $\mathrm{D} : \mathbb{R}_{\mathcal{F}} \times \mathbb{R}_{\mathcal{F}} \to \mathbb{R}^+ \cup \{0\}$ as

$$\mathrm{D(u, v)} = \sup_{\mathrm{r} \in [0,1]} \max\{|\underline{u}(\mathrm{r}) - \bar{\mathrm{u}}(\mathrm{r})|, |\underline{u}(\mathrm{r}) - \bar{\mathrm{u}}(\mathrm{r})|\}.$$

Consider $\mathrm{u, v, w, z} \in \mathbb{R}_{\mathcal{F}}$ and $\lambda \in \mathbb{R}$, then the following properties are well-known for metric D: (see [11])

- $D(u \oplus w, v \oplus w) = D(u, v)$,
- $D(\lambda u, \lambda v) = |\lambda| D(u, v)$,
- $D(u \oplus v, w \oplus z) \leq D(u, w) + D(v, z)$,
- $D(u \ominus v, w \ominus z) \leq D(u, w) + D(v, z)$,

as long as $u \ominus v$ and $w \ominus z$ exist, where $u, v, w, z \in \mathbb{R}_{\mathcal{F}}$.
where, $\ominus$ is the Hukuhara difference (H-difference), it means that $w \ominus v = u$ if and only if $u \oplus v = w$.

Definition 2.2. The generalized Hukuhara difference of two fuzzy numbers $\mathrm{u, v} \in \mathbb{R}_{\mathcal{F}}$ is defined as follows

$$\mathrm{u} \ominus \mathrm{v} = \mathrm{w} \Leftrightarrow \begin{cases} \mathrm{(a).u = v + w}, \\ \text{or } \mathrm{(b).v = u + (-1)w}. \end{cases}$$

Proposition 2.1. Let $\mathrm{A, B} \in \mathrm{K_n^C}$ be two compact convex set; then

- if the gH-difference exists, it is unique and it is a generalization of the usual Hukuhara difference since $A \ominus B = A \sim_h B$, whenever $A \sim_h B$ exists,
- $A \ominus A = 0$,
- if $A \ominus B$ exists in the sence (a), then $B \ominus A$ exists in sense (b) and vice versa,

- $(A \oplus B) \ominus B = A$,
- $\{0\} \ominus (A \ominus B) = (-B) \ominus (-A)$,
- we have $(A \ominus B) = (B \ominus A) = C$ if and only if $C = \{0\}$ and $A = B$.

Definition 2.3. Afuzzy valued function $f : [a, b] \to \mathbb{R}_{\mathcal{F}}$ is said to be continuous at $t_0 \in [a, b]$ if for each $\varepsilon > 0$ there is $\delta > 0$ such that $D(f(t), f(t_0)) < \varepsilon$, whenever $t \in [a, b]$ and $|t - t_0| < \delta$. We say that f is fuzzy continuous on [a, b] if f is continuous at each $t_0 \in [a, b]$.

Definition 2.4. (see [4]) Let $F : I \to \mathbb{R}$. Fix $t_0 \in I$. We say F is strongly generalized differentiable at t_0, if there exists an element $F'(t_0) \in \mathbb{R}'$ such that either

- for all $h > 0$ sufficiently closed to 0, the H-differences $F(t_0 + h) \ominus F(t_0)$, $F(t_0) \ominus F(t_0 - h)$ exist and limits (in the metric D)

$$\lim_{h \to 0+} \frac{F(t_0 + h) \ominus F(t_0)}{h} = \lim_{h \to 0+} \frac{F(t_0) \ominus F(t_0 - h)}{h} = F'(t_0),$$

or

- for all $h > 0$ sufficiently closed to 0, the H-differences $F(t_0) \ominus F(t_0 + h)$, $F(t_0 - h) \ominus F(t_0)$ exist and limits (in the metric D)

$$\lim_{h \to 0+} \frac{F(t_0) \ominus F(t_0 + h)}{-h} = \lim_{h \to 0+} \frac{F(t_0 - h) \ominus F(t_0)}{-h} = F'(t_0),$$

or

- for all $h > 0$ sufficiently closed to 0, the H-differences $F(t_0 + h) \ominus F(t_0)$, $F(t_0 - h) \ominus F(t_0)$ exist and limits (in the metric D)

$$\lim_{h \to 0+} \frac{F(t_0 + h) \ominus F(t_0)}{h} = \lim_{h \to 0+} \frac{F(t_0 - h) \ominus F(t_0)}{-h} = F'(t_0),$$

or

- for all $h > 0$ sufficiently closed to 0, the H-differences $F(t_0) \ominus F(t_0 + h)$, $F(t_0) \ominus F(t_0 - h)$ exist and limits (in the metric D)

$$\lim_{h \to 0+} \frac{F(t_0) \ominus F(t_0 + h)}{-h} = \lim_{h \to 0+} \frac{F(t_0) \ominus F(t_0 - h)}{h} = F'(t_0).$$

Remark 2.1. Four cases consider for generalized fuzzy derivatives, in this paper we only consider two first cases because in the other cases the derivatives are trivial and reduced to a crisp element.

Definition 2.4. (see [17]) Let $F : I \to \mathbb{R}_F$. we say F is (1)− differentiable on I if F is differentiable in the sense (1) and its derivative is denoted D_1F and similarly for (2)− differentiable we have D_2F.

Theorem 2.1. (see [17]) Let $F : I \to \mathbb{R}_F$ and put $[F(t)]^r = \left[\underline{f}(t, r), \bar{f}(t, r)\right]$ for each $r \in [0, 1]$.

- If F is (1)− differentiable the $\underline{f}(t, r)$ and $\bar{f}(t, r)$ are differentiable functions and

$$[\mathrm{D_1F(t)}]^{\mathrm{r}} = [\underline{f}'(t, r), \overline{f'}(t, r)]. \quad (1)$$

- If F is (2)− differentiable the $\underline{f}(t, r)$ and $\bar{f}(t, r)$ are differentiable functions and

$$[\mathrm{D_2F(t)}]^{\alpha}] = \left[\overline{f'}(t, r), \underline{f}'(t, r)\right]. \quad (2)$$

Theorem 2.2. (see [13]). Let $\mathrm{D_1^{(1)}F} : \mathrm{I} \to \mathbb{R}_\mathrm{F}$ or $\mathrm{D_2^{(1)}F} : \mathrm{I} \to \mathbb{R}_\mathrm{F}$ be fuzzy functions, where $[\mathrm{F(t)}]^{\mathrm{r}} = \left[\underline{f}(t, r), \bar{f}(t, r)\right]$

- If $\mathrm{D_1^{(1)}F}$ is (1)− differentiable, then $\underline{f}(t, r)$ and $\bar{f}(t, r)$ are differentiable functions and

$$[\mathrm{D_{1,1}^{(2)}F(t)}]^{\mathrm{r}} = \left[\underline{f}''(t, r), \overline{f''}(t, r)\right]. \quad (3)$$

- If $\mathrm{D_1^{(1)}F}$ is (2)− differentiable, then $\underline{f}(t, r)$ and $\bar{f}(t, r)$ are differentiable functions and

$$[\mathrm{D_{1,2}^{(2)}F(t)}]^{\mathrm{r}} = \left[\overline{f''}(t, r), \underline{f}''(t, r)\right]. \quad (4)$$

- If $\mathrm{D_2^{(1)}F}$ is (1)− differentiable, then $\underline{f}(t, r)$ and $\bar{f}(t, r)$ are differentiable functions and

$$[\mathrm{D_{2,1}^{(2)}F(t)}]^{\alpha} = \left[\overline{f''}(t, r), \underline{f}''(t, r)\right]. \quad (5)$$

- If $\mathrm{D_2^{(1)}F}$ is (2)− differentiable, then $\underline{f}(t, r)$ and $\bar{f}(t, r)$ are differentiable functions and

$$[\mathrm{D_{2,2}^{(2)}F(t)}]^{\mathrm{r}} = \left[\underline{f}''(t, r), \overline{f''}(t, r)\right]. \quad (6)$$

Definition 2.5. (see [13]). Let $\mathrm{F} : \mathrm{I} \to \mathbb{R}_\mathrm{F}$ ad $\mathrm{n, m} = 1, 2$. We say F is (n, m)− differentiable at $\mathrm{t_0} \in \mathrm{I}$, if $\mathrm{D_n^{(1)}F}$ exists on a neighborhood at $\mathrm{t_0}$ as fuzzy function and it is (m)− differentiable at $\mathrm{t_0}$. The second derivatives of F are denoted by $\mathrm{D_{n,m}^{(2)}F(t_0)}$ for $\mathrm{n, m} = 1, 2$.

In all of the paper the generalized Hukuhara difference of two fuzzy numbers is exists.

$\mathcal{C}_\mathcal{F}([a, b], \mathbb{R}_\mathcal{F})$ is the set of fuzzy valued function f which are defined on $[a, b]$ and be fuzzy continuous from the interior points of $[a, b]$ such that the continuity is one-sided at endpoints a, b.

$\mathcal{C}_{gH}^{k}([a, b], \mathbb{R}_\mathcal{F})$, is the space of functions f such that f and it's first k, gH-derivatives are all in $\mathcal{C}_\mathcal{F}([a, b], \mathbb{R}_\mathcal{F})$.

3 Proposed Method

In this section, we are going to solve the following second order fuzzy differential equation

$$\begin{cases} y'' + ay' + by = f(t), a > 0, b > 0, \\ \tilde{y}(0) = \tilde{y}_0, \tilde{y}'(0) = \tilde{y}_1, \end{cases} \tag{7}$$

where $\tilde{\gamma}_0, \tilde{\gamma}_1 \in \mathbb{R}_{\mathcal{F}}$ and $f : I \rightarrow \mathbb{R}_{\mathcal{F}}$ is continues fuzzy function on some interval I.

Our method for solving (7) is based on the approximation the solution by collocation method. This method is to seek an approximate solution as

$$\tilde{y}(t) = \sum_{k=0}^{2} \tilde{\alpha}_k.\emptyset_k(t). \tag{8}$$

Where $\emptyset_k(t)$ are positive basic functions whose all differentiations are positive. Now, the aim is to compute the fuzzy coefficients $\tilde{\alpha}_0, \tilde{\alpha}_1, \tilde{\alpha}_2$.

For this propose four cases are considered.

Case 1: Let $\tilde{y}(t)$ be the fuzzy $(1, 1)-$ solution, then by using Theorem 2.1 and Theorem 2.2, we have

$$\underline{y}'' + a\underline{y}' + b\underline{y} = \underline{f(t)}, \underline{y(0)} = \underline{\gamma_0}, \underline{y'(0)} = \underline{\gamma_1}, \tag{9}$$

$$\overline{y''} + a\overline{y'} + b\overline{y} = \overline{f(t)}, \overline{y(0)} = \overline{\gamma_0}, \overline{y'(0)} = \overline{\gamma_1}, \tag{10}$$

By setting Eq. (8) in (9), we have

$$\sum_{k=0}^{2} \underline{\alpha_k(r)}\emptyset_k''(t) + \mathrm{a}\sum_{k=0}^{2} \underline{\alpha_k(r)}\emptyset_k'(t) + b\sum_{k=0}^{2} \underline{\alpha_k(r)}\emptyset_k(t) = \underline{f(t, r)},$$

Therefore

$$\begin{aligned} &\underline{\alpha_0(r)}\left(\emptyset_0''(t) + a\emptyset_0'(t) + b\emptyset_0(t)\right) \\ &+ \underline{\alpha_1(r)}\left(\emptyset_1''(t) + a\emptyset_1'(t) + b\emptyset_1(t)\right) \\ &+ \underline{\alpha_2(r)}\left(\emptyset_2''(t) + a\emptyset_2'(t) + b\emptyset_2(t)\right) = \underline{f(t, r)}, \end{aligned}$$

Also for initial conditions we obtain

$$\underline{\alpha_0(r)}\emptyset_0^{(0)}(0) + \underline{\alpha_1(r)}\emptyset_1^{(0)}(0) + \underline{\alpha_2(r)}\emptyset_2^{(0)}(0) = \underline{\gamma_0(r)},$$

$$\underline{\alpha_0(r)}\emptyset_0^{(1)}(0) + \underline{\alpha_1(r)}\emptyset_1^{(1)}(0) + \underline{\alpha_2(r)}\emptyset_2^{(1)}(0) = \underline{\gamma_1(r)}.$$

Similarly we have by setting Eq. (8) in (10), we have

$$\sum_{k=0}^{2} \overline{\alpha_k(r)}\emptyset_k''(t) + \mathrm{a}\sum_{k=0}^{2} \overline{\alpha_k(r)}\emptyset_k'(t) + b\sum_{k=0}^{2} \overline{\alpha_k(r)}\emptyset_k(t) = \overline{f(t, r)},$$

Therefore

$$\overline{\alpha_0(r)}\Big(\emptyset_0''(t) + a\emptyset_0'(t) + b\emptyset_0(t)\Big) + \overline{\alpha_1(r)}\Big(\emptyset_1''(t) + a\emptyset_1'(t) + b\emptyset_1(t)\Big)$$
$$+ \overline{\alpha_2(r)}\Big(\emptyset_2''(t) + a\emptyset_2'(t) + b\emptyset_2(t)\Big) = \overline{f(t,r)}.$$

And

$$\overline{\alpha_0(r)}\emptyset_0^{(0)}(0) + \overline{\alpha_1(r)}\emptyset_1^{(0)}(0) + \overline{\alpha_2(r)}\emptyset_2^{(0)}(0) = \overline{\gamma_0(r)},$$
$$\overline{\alpha_0(r)}\emptyset_0^{(1)}(0) + \overline{\alpha_1(r)}\emptyset_1^{(1)}(0) + \overline{\alpha_2(r)}\emptyset_2^{(1)}(0) = \overline{\gamma_1(r)}.$$

By setting

$$\beta_k = \emptyset_k''(t) + a\emptyset_k'(t) + b\emptyset_k(t), \qquad \emptyset_k^{(j)}(0) = \delta_{jk}, j = 0, 1, k = 0, 1, 2$$

we have a system of linear equation $SX = Y$ such that

$$S_1 = \begin{bmatrix} \beta_0 & \beta_1 & \beta_2 \\ \delta_{00} & \delta_{01} & \delta_{02} \\ \delta_{10} & \delta_{11} & \delta_{12} \end{bmatrix}, S_2 = \begin{bmatrix} 0 & 0 & 0 \\ 0 & 0 & 0 \\ 0 & 0 & 0 \end{bmatrix}, S = \begin{bmatrix} S_1 & S_2 \\ S_2 & S_1 \end{bmatrix}$$

Where

$$X = \left[\underline{\alpha_0}(r), \underline{\alpha_1}(r), \underline{\alpha_2}(r), \overline{\alpha_0(r)}, \overline{\alpha_1(r)}, \overline{\alpha_2(r)}\right]^T,$$

$$Y = \left[\underline{f(t,r)}, \underline{\gamma_0}, \underline{\gamma_1}, \overline{f(t,r)}, \overline{\gamma_0}, \overline{\gamma_1}\right]^T.$$

By solving system $SX = Y$, the unkown coefficients are obtained.

Case 2: If $\tilde{y}(t)$ be the fuzzy $(1, 2)-$ solution therefore by using Theorem 2.1 and Theorem 2.2 we have

$$\overline{y''} + a\underline{y'} + b\underline{y} = \underline{f(t)}, \underline{y(0)} = \underline{\gamma_0}, \underline{y'(0)} = \underline{\gamma_1}, \tag{11}$$

$$\underline{y''} + a\overline{y'} + b\overline{y} = \overline{f(t)}, \overline{y(0)} = \overline{\gamma_0}, \overline{y'(0)} = \overline{\gamma_1}. \tag{12}$$

By setting (8) to (11) we have

$$\sum_{k=0}^{2}\overline{\alpha_k(r)}\emptyset_k''(t) + \mathrm{a}\sum_{k=0}^{2}\underline{\alpha_k(r)}\emptyset_k'(t) + b\sum_{k=0}^{2}\underline{\alpha_k(r)}\emptyset_k(t) = \underline{f(t,r)},$$

Therefore

$$\underline{\alpha_0(r)}\Big(a\emptyset_0'(t) + b\emptyset_0(t)\Big) + \underline{\alpha_1(r)}\Big(a\emptyset_1'(t) + b\emptyset_1(t)\Big)$$
$$+ \underline{\alpha_2(r)}\Big(a\emptyset_2'(t) + b\emptyset_2(t)\Big) + \overline{\alpha_0(r)}\emptyset_0''(t)$$
$$+ \overline{\alpha_1(r)}\emptyset_1''(t) + \overline{\alpha_2(r)}\emptyset_2''(t) = \underline{f(t,r)},$$

And

$$\underline{\alpha_0(r)}\emptyset_0^{(0)}(0)+\underline{\alpha_1(r)}\emptyset_1^{(0)}(0)+\underline{\alpha_2(r)}\emptyset_2^{(0)}(0)=\underline{\gamma_0(r)},$$

$$\underline{\alpha_0(r)}\emptyset_0^{(1)}(0)+\underline{\alpha_1(r)}\emptyset_1^{(1)}(0)+\underline{\alpha_2(r)}\emptyset_2^{(1)}(0)=\underline{\gamma_1(r)}.$$

By similarly by setting (8) to (12) we have

$$\sum_{k=0}^{2}\underline{\alpha_k(r)}\emptyset_k''(t)+\mathrm{a}\sum_{k=0}^{2}\overline{\alpha_k(r)}\emptyset_k'(t)+b\sum_{k=0}^{2}\overline{\alpha_k(r)}\emptyset_k(t)=\overline{f(t,r)},$$

Therefore

$$\overline{\alpha_0(r)}\Big(a\emptyset_0'(t)+b\emptyset_0(t)\Big)+\overline{\alpha_1(r)}\Big(a\emptyset_1'(t)+b\emptyset_1(t)\Big)$$
$$+\overline{\alpha_2(r)}\Big(a\emptyset_2'(t)+b\emptyset_2(t)\Big)+\underline{\alpha_0(r)}\emptyset_0''(t)$$
$$+\underline{\alpha_1(r)}\emptyset_1''(t)+\underline{\alpha_2(r)}\emptyset_2''(t)=\overline{f(t,r)}.$$

Also

$$\overline{\alpha_0(r)}\emptyset_0^{(0)}(0)+\overline{\alpha_1(r)}\emptyset_1^{(0)}(0)+\overline{\alpha_2(r)}\emptyset_2^{(0)}(0)=\overline{\gamma_0(r)},$$

$$\overline{\alpha_0(r)}\emptyset_0^{(1)}(0)+\overline{\alpha_1(r)}\emptyset_1^{(1)}(0)+\overline{\alpha_2(r)}\emptyset_2^{(1)}(0)=\overline{\gamma_1(r)},$$

By setting

$$\beta_k=a\emptyset_k'(t)+b\emptyset_k(t),\qquad \emptyset_k^{(j)}(0)=\delta_{jk}, j=0,1,k=0,1,2.$$

The following system of linear equation $SX=Y$ is obtained

$$S_1=\begin{bmatrix}\beta_0 & \beta_1 & \beta_2\\ \delta_{00} & \delta_{01} & \delta_{02}\\ \delta_{10} & \delta_{11} & \delta_{12}\end{bmatrix}, S_2=\begin{bmatrix}\emptyset_0'' & \emptyset_1'' & \emptyset_2''\\ 0 & 0 & 0\\ 0 & 0 & 0\end{bmatrix} S=\begin{bmatrix}S_1 & S_2\\ S_2 & S_1\end{bmatrix},$$

$$X=\left[\underline{\alpha_0}(r),\underline{\alpha_1}(r),\underline{\alpha_2}(r),\overline{\alpha_0(r)},\overline{\alpha_1(r)},\overline{\alpha_2(r)}\right]^T,$$

$$Y=\left[\underline{f(t,r)},\underline{\gamma_0},\underline{\gamma_1},\overline{f(t,r)},\overline{\gamma_0},\overline{\gamma_1}\right]^T.$$

And in this cases unknown coefficients are obtained.

Case 3: If $\tilde{y}(t)$ be the fuzzy (2, 1)– solution, by using Theorem 2.1 and Theorem 2.2 we have

$$\overline{y''}+a\overline{y'}+b\underline{y}=\underline{f(t)},\underline{y(0)}=\underline{\gamma_0},\overline{y'(0)}=\underline{\gamma_1},$$

$$\underline{y''}+a\underline{y'}+b\bar{y}=\overline{f(t)},\overline{y(0)}=\overline{\gamma_0},\underline{y'(0)}=\overline{\gamma_1},$$

Similar to previous cases and by setting

$$\beta_k = a\emptyset_k^{'}(t) + \emptyset_k^{''}(t), \emptyset_k^{(j)}(0) = \delta_{jk}, j = 0, 1, k = 0, 1, 2$$

We have the following linear equation SX = Y,

$$S_1 = \begin{bmatrix} b\emptyset_0 & b\emptyset_1 & b\emptyset_2 \\ \delta_{00} & \delta_{01} & \delta_{02} \\ 0 & 0 & 0 \end{bmatrix}, S_2 = \begin{bmatrix} \beta_0 & \beta_1 & \beta_2 \\ 0 & 0 & 0 \\ \delta_{10} & \delta_{11} & \delta_{12} \end{bmatrix}, S = \begin{bmatrix} S_1 & S_2 \\ S_2 & S_1 \end{bmatrix},$$

Where

$$X = \left[\underline{\alpha_0}(r), \underline{\alpha_1}(r), \underline{\alpha_2}(r), \overline{\alpha_0(r)}, \overline{\alpha_1(r)}, \overline{\alpha_2(r)}\right]^T,$$

$$Y = \left[\underline{f(t,r)}, \underline{\gamma_0}, \underline{\gamma_1}, \overline{f(t,r)}, \overline{\gamma_0}, \overline{\gamma_1}\right]^T.$$

And by solving the above system the unknown coefficients are obtained.

Case4: In this case we suppose that $\tilde{y}(t)$ be the fuzzy (2, 2)− solution, therefore by using theorem 2.1 and theorem 2.2. we have

$$\underline{y''} + a\overline{y'} + b\underline{y} = \underline{f(t)}, \underline{y(0)} = \underline{\gamma_0}, \overline{y'(0)} = \underline{\gamma_1},$$

$$\overline{y''} + a\underline{y'} + b\overline{y} = \overline{f(t)}, \overline{y(0)} = \overline{\gamma_0}, \underline{y'(0)} = \overline{\gamma_1},$$

By similar way and setting

$$\beta_k = b\emptyset_k(t) + \emptyset_k^{''}(t), \emptyset_k^{(j)}(0) = \delta_{jk}, j = 0, 1, k = 0, 1, 2.$$

We have the following linear equation SX = Y,

$$S_1 = \begin{bmatrix} \beta_0 & \beta_1 & \beta_2 \\ \delta_{00} & \delta_{01} & \delta_{02} \\ 0 & 0 & 0 \end{bmatrix}, S_2 = \begin{bmatrix} a\emptyset_0^{'} & a\emptyset_1^{'} & a\emptyset_2^{'} \\ 0 & 0 & 0 \\ \delta_{10} & \delta_{11} & \delta_{12} \end{bmatrix}, S = \begin{bmatrix} S_1 & S_2 \\ S_2 & S_1 \end{bmatrix},$$

$$X = \left[\underline{\alpha_0}(r), \underline{\alpha_1}(r), \underline{\alpha_2}(r), \overline{\alpha_0(r)}, \overline{\alpha_1(r)}, \overline{\alpha_2(r)}\right]^T,$$

$$Y = \left[\underline{f(t,r)}, \underline{\gamma_0}, \underline{\gamma_1}, \overline{f(t,r)}, \overline{\gamma_0}, \overline{\gamma_1}\right]^T.$$

And finally the unknown coefficients are obtained.

4 Numerical Example

Example [13]. We consider the following second order fuzzy differential equation
$y'' = \delta_0, y(0) = \gamma_0, y^{'}(0) = \gamma_1, \gamma_0 = \gamma_1 = \delta_0 = [r-1, 1-r],$

Let $\emptyset_0(t) = 1, \emptyset_1(t) = t, \emptyset_2(t) = t^2$.
If y is $(1, 1)-$ solution for the problem, then

$$\beta_k = \emptyset_k''(t) + a\emptyset_k'(t) + b\emptyset_k(t), \qquad \emptyset_k^{(j)}(0) = \delta_{jk}, j = 0, 1, k = 0, 1, 2.$$

Since $a = b = 0$ then $\beta_k = \emptyset_k''(t), \emptyset_k^{(j)}(0) = \delta_{jk}, j = 0, 1, k = 0, 1, 2$ thus

$$\beta_0 = \emptyset_0''(t) = 0, \beta_1 = \emptyset_1''(t) = 0, \beta_2 = \emptyset_2''(t) = 2,$$

$$\delta_{00} = 1, \delta_{01} = 0, \delta_{02} = 0,$$

$$\delta_{10} = 0, \delta_{11} = 1, \delta_{12} = 0,$$

$$S_1 = \begin{bmatrix} 0\,0\,2 \\ 1\,0\,0 \\ 0\,1\,0 \end{bmatrix}, S_2 = \begin{bmatrix} 0\,0\,0 \\ 0\,0\,0 \\ 0\,0\,0 \end{bmatrix}, S = \begin{bmatrix} 0\,0\,2 & 0\,0\,0 \\ 1\,0\,0 & 0\,0\,0 \\ 0\,1\,0 & 0\,0\,0 \\ 0\,0\,0 & 0\,0\,2 \\ 0\,0\,0 & 1\,0\,0 \\ 0\,0\,0 & 0\,1\,0 \end{bmatrix},$$

$$X = \left[\underline{\alpha_0(r)}, \underline{\alpha_1(r)}, \underline{\alpha_2(r)}, \overline{\alpha_0(r)}, \overline{\alpha_1(r)}, \overline{\alpha_2(r)}\right]^T,$$

$$Y = [r - 1, r - 1, r - 1, 1 - r, 1 - r, 1 - r]^T.$$

By solving SX = Y we have

$$\underline{\alpha_0(r)} = r - 1, \underline{\alpha_1(r)} = r - 1, \underline{\alpha_2(r)} = \frac{1}{2}r - \frac{1}{2},$$

$$\overline{\alpha_0(r)} = 1 - r, \overline{\alpha_1(r)} = 1 - r, \overline{\alpha_2(r)} = \frac{1}{2} - \frac{1}{2}r.$$

Thus $(1, 1)-$ solution for $t \geq 0$, is

$$\underline{\mathrm{y(t, r)}} = (r - 1) + (r - 1).t + \left(\frac{1}{2}r - \frac{1}{2}\right).t^2 \Rightarrow \underline{\mathrm{y(t, r)}} = (\mathrm{r} - 1).\left(\frac{t^2}{2} + t + 1\right),$$

$$\overline{y(t, r)} = (1 - r) + (1 - r).t + \left(\frac{1}{2} - \frac{1}{2}r\right).t^2 \Rightarrow \overline{y(t, r)} = (1 - \mathrm{r}).\left(\frac{t^2}{2} + t + 1\right).$$

If y is $(1, 2)-$ solution for the problem, then

$$\beta_k = a\emptyset_k'(t) + b\emptyset_k(t), \emptyset_k^{(j)}(0) = \delta_{jk}, j = 0, 1, k = 0, 1, 2.$$

Since $a = b = 0$ then $\beta_k = 0, \emptyset_k^{(j)}(0) = \delta_{jk}, j = 0, 1, k = 0, 1, 2$ thus

$$\beta_0 = 0, \beta_1 = 0, \beta_2 = 0, \emptyset_0''(t) = 0, \emptyset_1''(t) = 0, \emptyset_2''(t) = 2,$$

$$\delta_{00} = 1, \delta_{01} = 0, \delta_{02} = 0,$$

$$\delta_{10} = 0, \delta_{11} = 1, \delta_{12} = 0,$$

$$S_1 = \begin{bmatrix} \beta_0 \ \beta_1 \ \beta_2 \\ \delta_{00} \ \delta_{01} \ \delta_{02} \\ \delta_{10} \ \delta_{11} \ \delta_{12} \end{bmatrix} = \begin{bmatrix} 0\,0\,0 \\ 1\,0\,0 \\ 0\,1\,0 \end{bmatrix}, S_2 = \begin{bmatrix} \varnothing_0'' \ \varnothing_1'' \ \varnothing_2'' \\ 0 \ \ 0 \ \ 0 \\ 0 \ \ 0 \ \ 0 \end{bmatrix} = \begin{bmatrix} 0\,0\,2 \\ 0\,0\,0 \\ 0\,0\,0 \end{bmatrix},$$

$$S = \begin{bmatrix} 0\,0\,0 \ \ 0\,0\,2 \\ 1\,0\,0 \ \ 0\,0\,0 \\ 0\,1\,0 \ \ 0\,0\,0 \\ 0\,0\,2 \ \ 0\,0\,0 \\ 0\,0\,0 \ \ 1\,0\,0 \\ 0\,0\,0 \ \ 0\,1\,0 \end{bmatrix},$$

$$X = \left[\underline{\alpha_0}(r), \underline{\alpha_1}(r), \underline{\alpha_2}(r), \overline{\alpha_0(r)}, \overline{\alpha_1(r)}, \overline{\alpha_2(r)}\right]^T,$$

$$Y = [r-1, r-1, r-1, 1-r, 1-r, 1-r]^T.$$

By solving SX = Y we have

$$\underline{\alpha_0(r)} = r - 1, \underline{\alpha_1(r)} = r - 1, \underline{\alpha_2(r)} = \frac{1}{2} - \frac{1}{2}r,$$

$$\overline{\alpha_0(r)} = 1 - r, \overline{\alpha_1(r)} = 1 - r, \overline{\alpha_2(r)} = \frac{1}{2}r - \frac{1}{2}.$$

Thus $(1, 2)-$ solution on $(0, 1)$, is

$$\underline{\text{y(t, r)}} = (r-1) + (r-1).t + \left(\frac{1}{2} - \frac{1}{2}r\right).t^2 \Rightarrow \underline{\text{y(t, r)}} = (\text{r} - 1).\left(-\frac{t^2}{2} + t + 1\right),$$

$$\overline{y(t,r)} = (1-r) + (1-r).t + \left(\frac{1}{2}r - \frac{1}{2}\right).t^2 \Rightarrow \overline{y(t,r)} = (1-\text{r}).\left(-\frac{t^2}{2} + t + 1\right).$$

If y is $(2, 1)-$ solution for the problem, then

$$\beta_k = a\varnothing_k'(t) + \varnothing_k''(t), \qquad \varnothing_k^{(j)}(0) = \delta_{jk}, j = 0, 1, k = 0, 1, 2.$$

Since $a = 0$ then $\beta_k = \varnothing_k''(t)$, $\quad \varnothing_k^{(j)}(0) = \delta_{jk}, j = 0, 1, k = 0, 1, 2$ thus

$$\beta_0 = \varnothing_0''(t) = 0, \beta_1 = \varnothing_1''(t) = 0, \beta_2 = \varnothing_2''(t) = 2,$$

$$\delta_{00} = 1, \delta_{01} = 0, \delta_{02} = 0,$$

$\delta_{10} = 0, \delta_{11} = 1, \delta_{12} = 0,$

$$S_1 = \begin{bmatrix} b\emptyset_0 & b\emptyset_1 & b\emptyset_2 \\ \delta_{00} & \delta_{01} & \delta_{02} \\ 0 & 0 & 0 \end{bmatrix} = \begin{bmatrix} 0\,0\,0 \\ 1\,0\,0 \\ 0\,0\,0 \end{bmatrix}, S_2 = \begin{bmatrix} \beta_0 & \beta_1 & \beta_2 \\ 0 & 0 & 0 \\ \delta_{10} & \delta_{11} & \delta_{12} \end{bmatrix} = \begin{bmatrix} 0\,0\,2 \\ 0\,0\,0 \\ 0\,1\,0 \end{bmatrix},$$

$$S = \begin{bmatrix} 0\,0\,0 & 0\,0\,2 \\ 1\,0\,0 & 0\,0\,0 \\ 0\,0\,0 & 0\,1\,0 \\ 0\,0\,2 & 0\,0\,0 \\ 0\,0\,0 & 1\,0\,0 \\ 0\,1\,0 & 0\,0\,0 \end{bmatrix},$$

$$X = \left[\underline{\alpha_0(r)}, \underline{\alpha_1(r)}, \underline{\alpha_2(r)}, \overline{\alpha_0(r)}, \overline{\alpha_1(r)}, \overline{\alpha_2(r)}\right]^T,$$

$Y = [r-1, r-1, r-1, 1-r, 1-r, 1-r]^T.$
By solving SX = Y we have

$$\underline{\alpha_0(r)} = r-1, \underline{\alpha_1(r)} = 1-r, \underline{\alpha_2(r)} = \frac{1}{2} - \frac{1}{2}r,$$

$$\overline{\alpha_0(r)} = 1-r, \overline{\alpha_1(r)} = r-1, \overline{\alpha_2(r)} = \frac{1}{2}r - \frac{1}{2}.$$

Thus

$$\underline{\mathrm{y(t, r)}} = (r-1) + (1-r).t + \left(\frac{1}{2} - \frac{1}{2}r\right).t^2 \Rightarrow \underline{\mathrm{y(t, r)}} = (\mathrm{r}-1).\left(-\frac{t^2}{2} - t + 1\right),$$

$$\overline{y(t,r)} = (1-r) + (r-1).t + \left(\frac{1}{2}r - \frac{1}{2}\right).t^2 \Rightarrow \overline{y(t,r)} = (1-\mathrm{r}).\left(-\frac{t^2}{2} - t + 1\right).$$

Therefore y is (2, 1)− solution on $\left(0, \sqrt{3}-1\right)$.
If y is (2, 2)− solution for the problem, then

$$\beta_k = b\emptyset_k(t) + \emptyset_k''(t), \emptyset_k^{(j)}(0) = \delta_{jk}, j = 0, 1, k = 0, 1, 2.$$

Since $a = b = 0$ then $\beta_k = \emptyset_k''(t), \emptyset_k^{(j)}(0) = \delta_{jk}, j = 0, 1, k = 0, 1, 2$ thus

$$\beta_0 = \emptyset_0''(t) = 0, \beta_1 = \emptyset_1''(t) = 0, \beta_2 = \emptyset_2''(t) = 2,$$

$$\delta_{00} = 1, \delta_{01} = 0, \delta_{02} = 0,$$

$$\delta_{10} = 0, \delta_{11} = 1, \delta_{12} = 0,$$

$$S_1 = \begin{bmatrix} \beta_0 & \beta_1 & \beta_2 \\ \delta_{00} & \delta_{01} & \delta_{02} \\ 0 & 0 & 0 \end{bmatrix} = \begin{bmatrix} 0\,0\,2 \\ 1\,0\,0 \\ 0\,0\,0 \end{bmatrix}, S_2 = \begin{bmatrix} a\emptyset_0' & a\emptyset_1' & a\emptyset_2' \\ 0 & 0 & 0 \\ \delta_{10} & \delta_{11} & \delta_{12} \end{bmatrix} = \begin{bmatrix} 0\,0\,0 \\ 0\,0\,0 \\ 0\,1\,0 \end{bmatrix},$$

$$S = \begin{bmatrix} 0\,0\,2 & 0\,0\,0 \\ 1\,0\,0 & 0\,0\,0 \\ 0\,0\,0 & 0\,1\,0 \\ 0\,0\,0 & 0\,0\,2 \\ 0\,0\,0 & 1\,0\,0 \\ 0\,1\,0 & 0\,0\,0 \end{bmatrix},$$

$$X = \left[\underline{\alpha_0}(r), \underline{\alpha_1}(r), \underline{\alpha_2}(r), \overline{\alpha_0(r)}, \overline{\alpha_1(r)}, \overline{\alpha_2(r)}\right]^T,$$

$$Y = [r-1, r-1, r-1, 1-r, 1-r, 1-r]^T.$$

By solving SX = Y we have

$$\underline{\alpha_0(r)} = r-1, \underline{\alpha_1(r)} = 1-r, \underline{\alpha_2(r)} = \frac{1}{2}r - \frac{1}{2},$$

$$\overline{\alpha_0(r)} = 1-r, \overline{\alpha_1(r)} = r-1, \overline{\alpha_2(r)} = \frac{1}{2} - \frac{1}{2}r.$$

Thus

$$\underline{\mathrm{y(t, r)}} = (r-1) + (1-r).t + \left(\frac{1}{2}r - \frac{1}{2}\right).t^2 \Rightarrow \underline{\mathrm{y(t, r)}} = (\mathrm{r}-1).\left(\frac{t^2}{2} - t + 1\right),$$

$$\overline{y(t,r)} = (1-r) + (r-1).t + \left(\frac{1}{2} - \frac{1}{2}r\right).t^2 \Rightarrow \overline{y(t,r)} = (1-\mathrm{r}).\left(\frac{t^2}{2} - t + 1\right).$$

And y is (2, 2)− solution on (0, 1).

5 Conclusion

Collocation method with polynomial basic function introduced for solving second order fuzzy differential equation. We can be extended this method by using another basic function for obtaining higher accuracy solution for FDEs.

References

1. Abbasbandy, S., Allahviranloo, T.: Numerical solution of fuzzy differential equation by Taylor method. J. Comput. Method Appl. Math. **2**, 113–124 (2002)
2. Allahviranloo, T., Ahmady, N., Ahmady, E.: Numerical solution of fuzzy differential equations by predictor-corrector method. Inf. Sci. **177**(7), 1633–1647 (2007)
3. Allahviranloo, T., Abbasbandy, S., Ahmady, N., Ahmady, E.: Improved predictor-corrector method for solving fuzzy initial value problems. Inf. Sci. **179**, 945–955 (2009)
4. Bede, B., Gal, S.G.: Generalizations of the differentiability of fuzzy-number-valued functions with applications to fuzzy differential equations. Fuzzy Set Syst. **151**, 581–599 (2005)
5. Bede, B., Gal, S.G.: Remark on the new solutions of fuzzy differential equations. Chaos Solitons Fractals (2006)

6. Bede, B., Stefanini, L.: Solution of Fuzzy Differential Equations with generalized differentiability using LU-parametric representation. EUSFLAT **1**, 785–790 (2011)
7. Bede, B., Stefanini, L.: Generalized differentiability of fuzzy-valued functions. Fuzzy Sets Syst. **230**, 119–141 (2013)
8. Chang, S., Zadeh, L.: On fuzzy mapping and control. IEEE Trans. Syst. Cybern. **2**, 30–34 (1972)
9. Dubois, D., Prade, H.: Toward fuzzy differential calculus: Part 3. Differ. Fuzzy Sets Syst. 225–233 (1982)
10. Epperson, J.F.: An Introduction to Numerical Methods and Analysis. Wiley (2007)
11. Goetschel, R., Voxman, W.: Elementary fuzzy calculus. Fuzzy Sets Syst. **24**, 31–43 (1987)
12. Hukuhara, M.: Integration des applications mesurables dont la valeur est un compact convex. Funkcial. Ekvac. **10**, 205–229 (1967)
13. Khastan, A., Bahrami, F., Ivaz, K.: New results on multiple solutions for Nth-order fuzzy differential equation under generalized differentiability. Boundary Value Problem (Hindawi Publishing Corporation) (2009). http://dx.doi.org/10.1155/2009/395714
14. Ma, M., Friedman, M., Kandel, A.: Numerical solutions of fuzzy differential equations. Fuzzy Sets Syst. **105**, 133–138 (1999)
15. Nieto, J.J., Khastan, A., Ivaz, K.: "Numerical solution of fuzzy differential equation under generalized differentiability. Nonlinear Anal. Hybrid Syst. **3**, 700–707 (2009)
16. Seikkala, S.: On the fuzzy initial value problem. Fuzzy Sets Syst. **24**, 319–330 (1987)
17. Stefanini, L., Bede, B.: Eneralized Hukuhara differentiability of interval-valued functions and interval differential equations. Nonlinear Anal. **71**, 1311–1328 (2009)

Finding Most Preferred Solution in Imprecise DESA

Marzieh Moradi[1](✉), Elham Poudineh[2], and Mohammad Hossin Bayat[1]

[1] Department of Mathematics, Neyriz Branch, Islamic Azad University, Neyriz, Iran
marziehmoadi99@yahoo.com

[2] Department of Mathematics, Iranshahr Branch, Islamic Azad University, Iranshahr, Iran

Abstract. The basic data envelopment analysis (DEA) determines the relative efficiency of a set of decision making units (DMUs) in exact input-output environments. For imprecise data, some methods have been developed to calculate the efficiency scores. The aim of this paper is showing an equivalence relation between one of the models of data envelopment analysis with imprecise data and multi-objective linear programming (MOLP). The Relation between DEA and MOLP conduced to use interactive multiple objective models for solving the DEA problem in exact situation and find the most preferred solution. This paper will establish an equivalence model between imprecise DEA (IDEA) and MOLP and show how an IDEA problem can be solved interactively without any prior judgements by transforming it into a MOLP formulation. Various interactive multi-objective models would be used to solve IDEA problems such that PROJECT method. This PROJECT method is able to generate any efficient solution (including nonsupported efficient solutions) and also to identify the most preferred solution (MPS), even in nonconvex cases. In addition to, we will use the data envelopment scenario analysis (DESA) model since this model focuses on decreasing total input consumption and increasing total output production which results in solving one mathematical model instead of n models.

Keywords: Data envelopment analysis · Imprecise data · Interactive multiple objective linear programming · PROJECT method

1 Introduction

In many real decision and planning problems, the problem consists in optimizing (maximizing or minimizing) several conflicting objective functions. In these cases, it is impossible to find a solution which optimized all the objective functions simultaneously. In recent decades, many methods have been developed to solve multi objective linear programming (MOLP). Data envelopment analysis(DEA) and MOLP are useful for management to draw a scheming for future. DEA is used to evaluate the past preferences while MOLP is used for future preferences. Recently, the relation between DEA and MOLP, have been studied from several viewpoints by many researchers. Ebrahimnejad and Tavana (2014) proposed an interactive MOLP method for identifying target unit in DEA. Tavana et al. (2017) considered a hybrid DEA-MOLP for public school assessment

T. Allahviranloo et al. (Eds.): IDS 2020, AISC 1301, pp. 663–674, 2021.
https://doi.org/10.1007/978-3-030-66501-2_52

It would be worth mentioning that important result obtained by this equivalence relation is using interactive MOLP to solve DEA model and obtain most preferred solution (MPS).

Also, Cooper et al. (1999) introduced application of DEA whose data was imprecise. In imprecise DEA(IDEA), the data can be ordinal, interval or fuzzy, which results in a non-linear DEA model and we can transform it into a linear problem. Also, we obtain a new pair of interval DEA models will be developed for interval input and output data rather than for crisp input and output data. The final efficiency score for each DMU will be characterized by an interval bounded by the best lower bound efficiency and the best upper bound efficiency of each DMU, which we refer to as interval efficiency or efficiency interval.

In this paper we want to use the equivalence relation between MOLP and DEA models in exact environment, and expansion this relation between MOLP and DEA models with imprecise data. Following that, we use the interactive MOLP method with name PROJECT method (Luque et al. 2009) to solve the IDEA. The proposed PROJECT method belongs to the class of interactive local tradeoff methods. It is based on the projection of utility function gradients onto the tangent hyperplane of an efficient set and on a new local search procedure that inherits the advantages of the reference-point method to search for the best compromise solution within a local region. The use of a reference-point procedure makes it possible to generate any efficient solutions, even the nonsupported solutions.

On the other hand, Yang and Xu (2014) demonstrated that the minimax approach for reference point-based approaches is equivalent to input-oriented dual DEA models under certain conditions. After then Luque et al. (2015) proposed a set of new reference points called "equivalent reference points", which generates the same solution in the reference point-based approaches (minimax approach). This new set of "equivalent reference points" is the convex linear combination of two straight lines: one containing the original reference point and the other the nondominated objective solution. In addition, we would find the MPS in the IDEA models with use interactive algorithm that uses reference points, like PROJECT method.

2 Data Envelopment Scenario Analysis (DESA)

In the original DEA models, we can find the efficiency of an inefficient DMU by increasing outputs or decreasing inputs. In some cases, there is no need to reduce or increase all inputs or outputs with the same proportion. On this basis, Thanassoulis and Dyson (1992) discussed the uselessness of radial projection in target setting in some cases. In addition, they introduced the data envelopment scenario analysis model (DESA). After that, we will explain how the DEA model can deal with the target setting process. In the DESA model proposed, consider the index sets of inputs $I = \{1 \dots m\}$ and outputs and $O = \{1, \dots, s\}$ their subsets $I \equiv I_g \cup \bar{I}_g$ and $O \equiv O_g \cup \bar{O}_g$ where I_g and O_g are used to display inputs and outputs where bounds of target achievements are imposed in the construction of the following target model:

$$\underset{\lambda_j, Z_k, \theta_i}{Max} \sum_{k \in O} P_k^+ Z_k - \sum_{i \in I} P_i^- \theta_i$$

$$
\begin{aligned}
st \ & \sum_{j=1}^{n} \lambda_j x_{ij} = \theta_i x_{io} \ \ i \in I \\
& \sum_{j=1}^{n} \lambda_j y_{kj} = Z_k y_{ko} \ \ k \in O \\
& \theta_i x_{io} \geq G_i \ \ i \in I_g \\
& Z_k y_{ko} \leq G'_k \ \ k \in O_g \\
& A_i \leq \theta_i \leq 1/B_i, \ \ A_i, B_i \in [0, 1], \ \ \forall i \in I \\
& \Gamma_k \leq 1/Z_k \leq 1/\Delta_k, \ \ \Gamma_k, \Delta_k \in [0, 1], \ \ \forall k \in O \\
& \lambda_j \geq 0 \ \forall j
\end{aligned}
\tag{1}
$$

Where x_{ij} and y_{kj} are the ith input and kth output, respectively. P_i^-, P_k^+ the user-specified constants reflecting the decision makers 'preferences over the improvement of input/output components, θ_i is contraction rate of input i and Z_k is development rate of output k. G_i and G'_k are bounds for assessment targets of input i and output k, A_i, B_i are remarked as the lower and upper bounds of θ_i also Γ_k, Δ_k are regarded as the lower and upper bounds of Z_k.

3 DESA Model with Imprecise Data

In this section, we will consider the DESA with imprecise data, such that in this model we adapt model (1) to deal with computing one rather than n mathematical programming problems and reducing total input and increasing total output, this model can be formulated as follows:

$$
\max_{\lambda_j, Z_k, \theta_i} \sum_{k \in O} P_k^+ Z_k - \sum_{i \in I} P_i^- \theta_i - M \sum_{i \in I_f} s'_i - M \sum_{k \in O_f} s''_k
$$

$$
\begin{aligned}
st \ & \sum_{r=1}^{n} \sum_{j=1}^{n} \lambda_{jr} x_{ij} = \theta_i \sum_{j=1}^{n} x_{ij} \ \ i \in I \\
& \sum_{r=1}^{n} \sum_{j=1}^{n} \lambda_{jr} y_{kj} = Z_k \sum_{r=1}^{n} y_{kr} \ \ k \in O \\
& \sum_{j=1}^{n} \theta_i x_{ij} \geq G_i^L - s'_i \ \ i \in I_f \\
& \sum_{r=1}^{n} Z_k y_{kr} \leq G_k^{'U} + s''_k \ \ k \in O_f \\
& \sum_{j=1}^{n} \lambda_{jr} = 1 \ \forall r \\
& \lambda_j \geq 0 \ \forall j, \ \ \theta_i \, free, \ \ Z_k free, \ \ k \in O
\end{aligned}
$$

$$s'_i \geq 0 \ \forall i \in I_f, \ \ s''_k \ \geq 0 \ \forall k \in O_f \tag{2}$$

Where $\left[G_i^L, G_i^U\right]$ and $\left[G_k'^L, G_k'^U\right]$ indicate the interval of existing resource and bounds for total input i consumption and total output k production, respectively. s'_i, s''_k are intended for the legal of total input diminution and total output generation, respectively and M in the objective function is a penalty factor that has to be intended by decision maker.

4 Imprecise DESA Model Based on Interval Arithmetic

In this section, the authors use Wang et al. (2005) for transformed above model to two new models with name undesirable model and desirable model. The following models result in the best lower bound efficiency and the best upper bound efficiency for each DMU. These interval DEA models are suitable for interval input and output data rather than for crisp input and output data. Where $x_{ij} \in \left[x_{ij}^L, x_{ij}^U\right]$,$y_{kj} \in \left[y_{kj}^L, y_{kj}^U\right]$, $\theta_i \in \left[\theta_i^L, \theta_i^U\right]$, $Z_k \in \left[Z_k^L, Z_k^U\right]$.

$$\begin{aligned}
&\underset{\lambda_j, Z_k, \theta_i}{Max} \sum_{k \in O} P_k^+ Z_k^L - \sum_{i \in I} P_i^- \theta_i^L - M \sum_{i \in I_f} s'_i - M \sum_{k \in O_f} s''_k \\
&st \ \sum_{r=1}^{n} \sum_{j=1}^{n} \lambda_{jr} x_{ij}^L = \theta_i^L \sum_{j=1}^{n} x_{ij}^U \ \ i \in I \\
&\sum_{r=1}^{n} \sum_{j=1}^{n} \lambda_{jr} y_{kj}^U = Z_k^L \sum_{r=1}^{n} y_{kr}^L \ \ k \in O \\
&\sum_{j=1}^{n} \theta_i^L x_{ij}^U \geq G_i^U - s'_i \ \ i \in I_f \\
&\sum_{r=1}^{n} Z_k^L y_{kr}^L \leq G_k'^L + s''_k \ \ k \in O_f \\
&\sum_{j=1}^{n} \lambda_{jr} = 1 \ \forall r \\
&\lambda_{jr} \geq 0 \ \forall j, \ \ \theta_i^L \ free, \ \ Z_k^L free, \ \ k \in O \\
&s'_i \geq 0 \ \forall i \in I_f, \ \ s''_k \ \geq 0 \ \forall k \in O_f
\end{aligned} \tag{3}$$

Where G_i^U is limit to total consumption of input x_{ij}^U and $G_K'^L$ is limit to total production of output y_{kr}^L

$$\underset{\lambda_j, Z_k, \theta_i}{Max} \sum_{k \in O} P_k^+ Z_k^U - \sum_{i \in I} P_i^- \theta_i^U - M \sum_{i \in I_f} s'_i - M \sum_{k \in O_f} s''_k$$

$$
\begin{aligned}
st\ & \sum_{r=1}^{n}\sum_{j=1}^{n}\gamma_{jr}\,x_{ij}^{L} = \theta_i^{U}\sum_{j=1}^{n}x_{ij}^{L}\ \ i\in I\\
& \sum_{r=1}^{n}\sum_{j=1}^{n}\gamma_{jr}y_{kj}^{U} = Z_k^{U}\sum_{r=1}^{n}y_{kr}^{U}\ \ k\in O\\
& \sum_{j=1}^{n}\theta_i^{U}x_{ij}^{L} \ge G_i^{L} - s_i'\ \ i\in I_f\\
& \sum_{r=1}^{n}Z_k^{U}y_{kr}^{U} \le G_k^{\prime U} + s_k''\ \ k\in O_f\\
& \sum_{j=1}^{n}\lambda_{jr} = 1\ \forall r\\
& \gamma_{jr}\ge 0\ \forall j,\ \theta_i^{U}\ free,\ Z_k^{U} free,\ k\in O\\
& s_i'\ge 0\ \forall i\in I_f,\ s_k''\ \ge 0\ \forall k\in O_f
\end{aligned}
\tag{4}
$$

Where G_i^L is limit to total consumption of input x_{ij}^L and $G_K^{\prime U}$ is limit to total production of output y_{kr}^U.

According to the above definition, model (3) is the lower bound of the best possible relative efficiency and model (4) is the upper bound of the best possible relative efficiency.

5 Establish an Equivalence Relationship Between DESA with Imprecise Data and MOLP

In this section, we want to introduce a multi-objective linear program (MOLP) and show this model and imprecise model are equivalent. In desirable model, suppose that

$$
\bar{f}_k(\lambda) = \sum_{r=1}^{n}\sum_{j=1}^{n}\lambda_{jr}y_{kj}^{U} \qquad k = 1\ldots s \tag{5}
$$

$$
\tilde{f}_i(\lambda) = \sum_{r=1}^{n}\sum_{j=1}^{n}\lambda_{jr}x_{ij}^{L} \qquad i = 1\ldots m, \tag{6}
$$

$$
\lambda = (\lambda_{11}, \lambda_{jr}, \ldots, \lambda_{nn})^{T}
$$

Assume that the maximum feasible value of the kth total output and the minimum feasible value for the ith total input is are denoted by $f_i'' = \tilde{f}_i(\lambda^*), f_k' = \bar{f}_k(\lambda^*)$ respectively, where can be computing λ^* by solving the following optimization problems:

$$
\begin{aligned}
Max\ & \sum_{r=1}^{n}\sum_{j=1}^{n}\lambda_{jr}y_{kj}^{U}\\
& s.t\ \lambda\in\Omega
\end{aligned}
\tag{7}
$$

$$
\begin{aligned}
& Min \sum_{r=1}^{n}\sum_{j=1}^{n}\lambda_{jr}x_{ij}^{L} \\
& s.t\ \lambda \in \Omega
\end{aligned}
\tag{8}
$$

Such that $f' = [f_1', \ldots, f_s']$ and $f'' = [f_1'', \ldots, f_m'']$ are ideal points in object space in (5–6).

Suppose that:

$$
\sum_{j=1}^{n} x_{ij} > 0\ i = 1, \ldots, m \qquad \sum_{r=1}^{n} y_{kr} > 0\ k = 1, \ldots, s
$$

$$
w_k = \frac{1}{\sum_{r=1}^{n} y_{kr}^{U}}\quad \forall k \qquad v_i = \frac{1}{\sum_{j=1}^{n} x_{ij}^{L}}\ \forall i \tag{9}
$$

And

$$
\bar{f}_k^{*} = \frac{F^{\max}}{w_k} = F^{\max}\sum_{r=1}^{n} y_{kr}^{U},\ \tilde{f}_i^{*} = \frac{H^{\min}}{v_i} = H^{\min}\sum_{j=1}^{n} x_{ij}^{L} \tag{10}
$$

$$
\tau_i = \theta_i^{U} - H^{\min},\ \gamma_k = F^{\max} - Z_k^{U}
$$

Such that

$$
F^{\max} = \max_{1 \le k \le s}\{w_k f_k'\} = \max_{1 \le k \le s}\left\{\frac{f_k'}{\sum_{r=1}^{n} y_{kr}^{U}}\right\} \tag{11}
$$

$$
H^{\min} = \min_{1 \le i \le m}\{v_i f_i''\} = \min_{1 \le i \le m}\left\{\frac{f_i''}{\sum_{j=1}^{n} x_{ij}^{L}}\right\} \tag{12}
$$

We can write equivalence model as follow model:

$$
\begin{aligned}
& \max_{\lambda_j, Z_k, \theta_i} \sum_{k \in O} P_k^{+} Z_k^{U} - \sum_{i \in I} P_i^{-}\theta_i^{U}\frac{n!}{r!(n-r)!} \\
& st \quad \tilde{f}_i(\lambda) - \theta_i^{U}\frac{1}{v_i} = 0\ i = 1, \ldots, m \\
& Z_k^{U}\frac{1}{w_k} - \bar{f}_k(\lambda) = 0\ k = 1, \ldots, s \\
& g_i v_i \le \theta_i^{U} \qquad i \in I_f \\
& g_k' w_k \ge Z_k^{U}\ k \in O_f,\ \lambda \in \Omega
\end{aligned}
\tag{13}
$$

The first m constraint in (13) can be equivalently transformed as follows:

$$\tilde{f}_i(\lambda) - \theta_i^U \frac{1}{v_i} = 0 \Leftrightarrow v_i\tilde{f}_i(\lambda) = \theta_i^U \Leftrightarrow v_i\tilde{f}_i(\lambda) - H^{\min}$$
$$= \theta_i^U - H^{\min} \Leftrightarrow v_i(\tilde{f}_i(\lambda) - \tilde{f}_i^* = \theta_i^U - H^{\min} \tag{14}$$

Similar way the second s constraints in (13) can be equivalently transformed as follow

$$Z_k^U \frac{1}{w_k}\bar{f}_k(\lambda) = 0 \Leftrightarrow -w_k\bar{f}_k(\lambda) = -Z_k^U \Leftrightarrow F^{\max} - w_k\bar{f}_k(\lambda)$$
$$= F^{\max} - Z_k^U \Leftrightarrow w_k(\bar{f}_k^* - \bar{f}_k(\lambda) = F^{\max} - Z_k^U \tag{15}$$

Such that $P_k^+ = P_i^- = 1, k \in O, i \in I$.

Moreover, the objective function of (13) becomes:

$$\underset{\lambda_j, Z_k, \theta_i}{Max} \left(\sum_{k\in O} Z_k^U - \sum_{i\in I} \theta_i^U\right) = \underset{\lambda_j Z_k, \theta_i}{Min} \left(\sum_{i\in I} \theta_i^U + \sum_{k\in O} (-Z_k^U)\right) \Rightarrow Min\left(\sum_{i\in I} (\theta_i^U - H^{\min}\right)$$
$$+ \sum_{k\in O} (F^{\max} - Z_k^U) = Min\left(\sum_{i\in I} \tau_i + \sum_{k\in O} \gamma_k\right)$$

Also, for any $\lambda \in \Omega$, we have:

$$F^{\max} - Z_k^U \geq w_k\bar{f}_k^* - Z_k^U \geq w_k\bar{f}_k(\lambda) - Z_k^U = 0 \;\; k = 1, \ldots, s$$

$$\theta_i^U - H^{\min} \geq \theta_i^U - v_i\tilde{f}_i^* \geq \theta_i^U - v_i(\tilde{f}_i(\lambda) = 0 \;\; i = 1, \ldots, m$$

$$\bar{f}_k^* = \frac{F^{\max}}{w_k} \geq \frac{w_k f_k'}{w_k} = f_k' = \max_{\lambda\in\Omega} \bar{f}_k(\lambda) \;\; k = 1, \ldots, s$$
$$f_i^* = \frac{H^{\min}}{v_i} \leq \frac{v_i f_i''}{v_i} = f_i'' = \min_{\lambda\in\Omega} \tilde{f}_i(\lambda) \;\; i = 1, \ldots, m$$

In the end, we can rewrite the imprecise DESA model as a minimax formulation as follows:

$$\begin{aligned} & Min \sum_{i\in I} \tau_i - \sum_{k\in O} \gamma_k \\ & st \;\; v_i(\tilde{f}_i(\lambda) - \tilde{f}_i^*) = \tau_i \quad i = I \\ & \quad w_k(\bar{f}_k^* - \bar{f}_k(\lambda)) = \gamma_k \quad k = O \\ & \quad g_i v_i - H^{\min} \leq \tau_i \quad i \in I_f \\ & \quad g_k' w_k - M \leq \gamma_k \quad k \in O_f \end{aligned} \tag{16}$$

$$\sum_{j=1}^{n} \lambda_{jr} = 1, \;\; \forall r \;\; \lambda_{jr} \geq 0, \;\; M \geq 0 \; j, r = 1, \ldots, n$$

Therefore, imprecise DESA model and minimax MOLP formulation are equivalence, this result lead to using interactive methods for solving imprecise DESA models and obtain most preferred solution.

6 PROJECT Algorithm for Solving DESA Model with Imprecise Data

The aim of an interactive method in MOLP is to improve DM's utility in an iterative way, especially if the utility function is not expressed explicitly. PROJECT method is a comprehensive interactive method, this method proposed (Luque et al. 2009) based on the GRIST method and together with the methodology of reference point techniques. The GRIST method can be used to identify a normal vector at a given efficient solution on the efficient frontier. The PROJECT method, offers a better way to search for a new solution that the GRIST method as local tradeoffs provided by the DM is maintained through the use of reference point approaches. also the PROJECT method to showing the decision maker (DM) preferences in the process of assessing efficiency in the DESA model with any prior judgement. Steps of the algorithm:

Step1: the ideal values of outputs and inputs are calculated.
Optimize each of the upper bounds of total outputs and the lower bounds of total inputs, and denoted by, $\bar{f}_k = f_k'(\lambda^*), \tilde{f}_k = f_k''(\lambda^*)$.
Such that $\bar{f}(\lambda^t) = \left[\bar{f}_1(\lambda^t), \ldots, \bar{f}_k(\lambda^t), \ldots, \bar{f}_s(\lambda^t)\right]^T$ and $\tilde{f}(\lambda^t) = \left[\tilde{f}_1(\lambda^t), \ldots, \tilde{f}_i(\lambda^t), \ldots, \tilde{f}_m(\lambda^t)\right]^T$
Step2: Set the initial weighting parameters for all DMUs, and reach the initial solution of the decision variables.
Suppose for given positive weight vectors $w^t = \left[w_1^t, \ldots, \lambda_k^t, \ldots, w_s^t\right]$ and $v^t = \left[v_1^t, \ldots, \lambda_i^t, \ldots, w_m^t\right]$.
The optimal solution of the minimax model is given by $\lambda^t = \left[\lambda_{11}^t, \ldots, \lambda_{nn}^t\right]$, which must be an efficient solution. The optimal values of the dual variable of the kth and ith objective constraint $w_k(\bar{f}_k^* - \bar{f}_k(\lambda)) = \gamma_k \quad v_i(\tilde{f}_i(\lambda) - \tilde{f}_i^*) = \tau_i$ are given by β_k^t and α_i^t
Iteration $t = 0$. Select an initial point $\lambda^0 = \left[\lambda_{11}^0, \ldots, \lambda_{nn}^0\right]$. An initial solution can be obtained by solving minimax model by setting $v_i^t = \left(\frac{1}{X_I^* - f_i}\right)$ $\forall i = 1, \ldots, m$ and $w_k^t = \left(\frac{1}{f_k' - Y_k^*}\right)$ $\forall k = 1, \ldots, s$
Also, the initial dual variable values $\beta^t = \left[\beta_1^t, \ldots, \beta_s^t\right]^T$ $\alpha^t = \left[\alpha_1^t, \ldots, \alpha_s^t\right]^T$, for the first s and i constraint on the upper bounds of outputs and lower bounds of inputs.
Step3: First, calculate the normal vectors and check optimality condition.
At interaction t, estimate the normal vector. The normal vectors N^t at $\bar{f}(\lambda^t)$ and M^t at $\tilde{f}(\lambda^t)$ on the efficient frontier are given by $N^t = \left[w_1^t\beta_1^t, \ldots, w_s^t\beta_s^t\right]^T$ and $M^t = \left[v_1^t\alpha_1^t, \ldots, v_m^t\alpha_m^t\right]^T$
Next, generate reference total output and input, after then, decision maker (DM) select the most preferred solution and the interactive process will be finished. otherwise, the DM is asked to obtain local indifference tradeoffs on objectives.
Step4: determine the trade-off direction. The DM is asked to obtain local indifference tradeoffs on objectives, without loss of generality, set the first objectives $\bar{f}_1$ and $\tilde{f}_1$ as the reference objectives, then the indifference tradeoffs G_{1k}^t and R_{1i}^t between the first and kth and ith objectives, respectively, and the marginal rates of substitution G^t at $\bar{f}(\lambda^t)$ and R^t at $\tilde{f}(\lambda^t)$ are given by:

$$G_{1k}^t = -\frac{\bar{d}f_1^t}{\bar{d}f_k^t} \ and \ G^t = [1, G_{12}^t, \ldots, G_{1k}^t, \ldots, G_{1s}^t]^T, \ R_{1i}^t = -\frac{\tilde{d}f_1^t}{\tilde{d}f_i^t} \ and \ R^t = [1, R_{12}^t, \ldots, R_{1i}^t, \ldots, R_{1m}^t]^T$$

obtain the projection Δu^t and ΔS^t can be calculated using G^t and R^t as follows, denoted by $\Delta u'^t$ and $\Delta S'^t$, which provides a new tradeoff direction to improve the DM's utility:

$$\Delta \bar{u}^t = [\Delta \bar{f}_1^t, \ldots, \Delta \bar{f}_k^t, \ldots, \Delta \bar{f}_s^t] = -G^t + \left[\frac{(G^t)^T N^t}{(N^t)^T N^t}\right] N^t$$

$$\Delta \bar{s}^t = \left[\Delta \bar{f}_1''^t, \ldots, \Delta \bar{f}_k''^t, \ldots, \Delta \bar{f}_s^t\right] = R^t - \left[\frac{(R^t)^T M^t}{(M^t)^T M^t}\right] M^t$$

Step5: Calculate the desired step sizes.
The step sizes are to be determined by the largest and smallest desired step. The desired step size S^t could be determined by $\bar{\gamma} = \bar{\gamma}_{\max}^t \gamma$ *and* $\bar{\tau} = \bar{\tau}_{\min}^t \tau$ and where $\bar{\gamma}_{\max}^t$, $\bar{\tau}_{\min}^t$ are the largest permissible step sizes and regulating factors with $0 \le \gamma \le 1$ *and* $0 \le \gamma \le 1$. Suppose I^1 *and* J^1 are the sets of the subscripts of objectives for increase and decrease, respectively. The maximum step size can be determined as follows:

$$S_{\max}^t = \bar{\gamma}_{\max}^t = \max \frac{f_k' - \bar{f}_k^t}{\left|\Delta f_k'^t\right|} \ \text{and} \ S_{\min}^t = \bar{\tau}_{\min}^t = \min_{i \in J} \frac{\tilde{f}_i^t - f_i''}{\left|\Delta f_i''^t\right|}$$

The step size S^t is takin interval $[0.S_{\max}^t]$ and given that $S_l^t = \frac{1}{C_s}.S_{\max}^t \ l = 1, 2, \ldots, C_s$

The step size S^t is takin interval $\left[S_{\min}^t.0\right]$ and given that $S_l^t = \frac{1}{C_s}.S_{\min}^t \ l = 1, 2, \ldots, C_s$

Where C_s is an integer which could be regulated to change accuracy of, S^t then, the following vectors of value on the objectives are shown to the DM:

$$(f_1^t + s_1^t \cdot \Delta f_1^t, \ldots, f_s^t + s_l^t \cdot \Delta f_s^t) \qquad l = 1 \ldots C_s$$

Step6: The new reference point is defined by $\bar{q}^t = (\bar{q}_1^t, \ldots, \bar{q}_s^t)$ *and* $\tilde{q}^t = (\tilde{q}_1^t, \ldots, \tilde{q}_m^t)$
Where $\tilde{q}_i^t = \tilde{f}_i^t + \tilde{\tau}^t.\Delta f_i^t \quad \forall i = 1 \ldots m \qquad \bar{q}_k^t = \bar{f}_k^t + \bar{\gamma}^t.\Delta f_k'^t \qquad \forall k = 1 \ldots s$
And $\tilde{\mu}_i^{tt} = \frac{1}{\left|\tilde{q}_i^t - \tilde{f}_i^t\right|} \quad \forall i - 1 \ldots m \qquad \bar{\mu}_k^t = \frac{1}{\left|\bar{q}_k^t - \bar{f}_k^t\right|} \qquad \forall k = 1 \ldots s$
Let (λ^*, f^*) be a solution.
Step7: select the most preferred solutions by the decision maker.
Show the solution (λ^*, f^*) to the DM.
If $u(f^*) \ge u(f^t)$, then let $\lambda^{t+1} = \lambda^*$ and $f^{t+1} = f^*$. Set $t = t + 1$ and go to step 2.
Otherwise, let

$$\gamma^t = \gamma_{old}^t.\min\left\{\left|1 - \frac{\left\|\bar{q}_k^t - f'^*\right\|_2}{\left\|q_k^t - \bar{f}^t\right\|_2}\right|, \frac{1}{2}\right\} \le \frac{1}{2} s_{old}^t$$

$$\tau^t = \tau^t_{old}.\min\left\{\left|1 - \frac{\left\|\tilde{q}^t_k - f''^{+*}\right\|_2}{\left\|\tilde{q}^t_k - \tilde{f}^t\right\|_2}\right|, \frac{1}{2}\right\} \le \frac{1}{2} s^t_{old}$$

Decision maker select the best solutions, otherwise we go to step 6.

7 Numerical Illustration

In this section, we apply the result investigated in the previous sections to conduct a case study. Since the DEA problem can be solved interactively without any prior judgments by transforming it into multiple objective linear programming (MOLP) formulation, then, the proposed interactive PROJECT algorithm is applied to search for MPS on the efficient frontier. The case study is to examine the performance measurement of UK retail banks using the data collected Wong et al. (2004), as shown in Table 1. This table contains seven DMUs (retail banks), for each of which four inputs and two outputs are considered. The four inputs are namely number of branches, number of ATMs, number of staff and asset size. The two outputs are customer satisfaction and total revenue. On the other hand, for this study, it is assumed that the data has been changed from certain to imprecise. As shown in Table 2.

Table 1. Original data set

DMU	Bank	No. of Branch	No. of ATMs	No. of Staff	Asset size	Customer Satisfaction	Total revenue
1	Abby	0.77	2.18	2.35	2.96	6.79	10.57
2	Barclay	1.95	3.19	8.43	3.53	2.55	13.35
3	Halifax	0.80	2.10	3.21	2.41	9.17	8.14
4	HSBC	1.75	4.00	13.30	4.85	5.82	23.67
5	Lloyds TSB	2.50	4.30	9.27	2.40	6.57	14.01
6	Nat West	1.73	3.30	7.7	3.09	4.86	12.04
7	RBS	0.65	1.73	2.65	1.34	7.28	7.36

We run target model (1), (solved by LINGO, a powerful professional software package) for obtaining the efficiency score. We also solved minimax model (16) and got similar result to those obtained by target model. By the PROJECT method we are able to reach the MPS, also the more decrease of total input and the more increase of total output, thereby, yields in saving time in calculation of the method. The main difference of the suggested method in this paper was the imprecise data. The PROJECT method is coded in Microsoft Visual C++ and incorporated into the software PROMIN (Interactive MOP).

Table 2. Interval data set

DMU	Bank	No. of Branch	No.of.ATMs	No. of Staff	Asset size	Customer Satisfaction	Total revenue
1	Abby	0.77	[2.18,2.2]	[2.35,2.50]	[2.96,3]	6.79	10.57
2	Barclay	1.95	[3.19,3.20]	[8.43,8.50]	[3.53,4]	2.55	13.35
3	Halifax	0.80	[2.10, 2.2]	[3.21,3.50]	[2.41,2.5]	9.17	8.14
4	HSBC	1.75	[4.00,4.1]	[13.3,13.5]	[4.85,5]	5.82	23.67
5	Lloyds TSB	2.50	[4.30,4.7]	[9.27,9.3]	[2.40,2.43]	6.57	14.01
6	Nat West	1.73	[3.30,3.40]	[7.7,7.8]	[3.09,3.2]	4.86	12.04
7	RBS	0.65	[1.73,1.80]	[2.65,2.7]	[1.34,1.4]	7.28	7.36

8 Conclusions

In this paper, we obtained an equivalence relation between the imprecise data envelopment scenario analysis and the minimax MOLP formulation. following that, with using the interactive PROJECT method for estimating efficiency the imprecise DESA model and obtain most preferred solution. it must be pointed that any efficient solution can be generated with this interactive method, even including nonsupport efficient solutions for nonconvex problems. Also, this approach results in decreasing total input and increasing total output at the same time. Instead of solving n independent linear programming models.

References

Charnes, A., Cooper, W.W., Rhodes, E.: Measuring the efficiency of decision making units. Eur. J. Oper. Res. **2**, 429–444 (1978)

Cooper, W.W., Park, K.S., Yu, G.: IDEA and ARIDEA: models for dealing with imprecise data in DEA. Manage. Sci. **45**, 597–607 (1999)

Ebrahimnejad, A., Tavana, M.: An interactive MOLP method for identifying target units in output-oriented DEA models: The NATO enlargement problem. Measurement **52**, 124–134 (2014)

Luque, M., Yang, J., Wong, B.Y.: PROJECT method for multiobjective optimization based on gradient projection and reference points. IEEE Trans. Syst. Man Cybern.-Part A: Syst. Hum. **39**(4), 864–879 (2009)

Luque, M., Lopez-Agudo, L.A., Marcenaro-Gutierrez, O.D.: Equivalent reference points in multiobjective programming. Expert Syst. Appl. **42**, 2205–2212 (2015)

Thanassoulis, A.D., Dyson, R.G.: Estimation preferred target input-output levels using data envelopment analysis. Eur. J. Oper. Res. **56**, 80–97 (1992)

Tavana, M., Ebrahimnejad, A., Santos-Arteaga, F.J., Mansourzadeh.S.M., Matin, R.K.: A hybrid DEA-MOLP model for public school assessment and closure decision in the City of Philadelphia, Socio-Economic Planning Sciences (2017, in press)

Yang, J.B., Xu, D.L.: Interactive minimax optimization for integrated performance analysis and resource planning. Comput. Oper. Res. **46**, 78–90 (2014)

Wang, Y.M., Greatbanksa, R., Yang, J.B.: Interval efficiency assessment using data envelopment analysis fuzzy sets **153**, 347–370 (2005)
Wong, B.Y., Yang, J.B., Greatbank, R.: Using DEA and ER approach for performance measurement of UK retail banks. MCDM 2004. Whistler, Canada (2004)

Grading Homeworks, Verifying Code: How Thorough Should the Feedback Be?

Francisco Zapata, Olga Kosheleva, and Vladik Kreinovich(✉)

University of Texas at El Paso, El Paso, TX 79968, USA
fcozpt@outlook.com, {olgak,vladik}@utep.edu

Abstract. In the ideal world, we should assign many homeworks and give a thorough feedback for each homework. However, in reality, the instructor's time is limited, so we can either assign few homeworks and give a detailed feed back for all of them, or we can assign many homeworks, but give a less thorough feedback. What is the optimal thoroughness? A similar question can be raised for code verification: what is the optimal amount of feedback that should be provided to each programmer? In this paper, we provide answers to these questions.

Keywords: Grading homework · Verifying code · Optimal feedback

1 Formulation of the Problem

How thoroughly should we grade the homeworks? In the ideal world, we should assign many homeworks, and grade each of them very thoroughly. In reality, our time – and if we have a Teaching Assistant for the class, the time of the Teaching Assistant – is limited. So, we have a choice:

– we can assign few homeworks and grade all of them very thoroughly,
– or we can assign many homeworks and grade all of them less thoroughly.

What is the optimal level of thoroughness?

Comment. Doing homeworks helps students, even if these homeworks are not graded at all – since this provides the students with additional practice. There are many ways to entice the students to do homeworks even when not all homeworks will be graded:

– we can assign several problems, but grade only some of them; the selection of graded problems can be decided by a random choice after the homeworks are submitted, so the students do not know beforehand which problems will be graded;
– another idea is, after the homework is submitted, to randomly select students whose homeworks will be graded.

T. Allahviranloo et al. (Eds.): IDS 2020, AISC 1301, pp. 675–680, 2021.
https://doi.org/10.1007/978-3-030-66501-2_53

It is also possible to have a combination of these two ideas.

But no matter what we do, the same question remains: how thoroughly should we grade the homeworks?

How Thoroughly Should We Verify the Code. Usually, when a company designs software, the resulting code is not only tested to make sure it works – it is also verified by specialists who make sure that this code is well-written. Again, in the ideal world, every line of code should be thoroughly analyzed and verified, and the detailed feedback should be provided to the programmer. In practice, the verifier's time is limited, so we have a choice:

- we can verify fewer lines of code and provide more thorough feedback,
- or we can verify more lines of code and provide less thorough feedback.

What is the optimal level of thoroughness?

What We Do in this Paper. In this paper, we use the general decision techniques to come up with the optimal levels of thoroughness.

At first, we consider a simplified model in which we ignore the difference between different students (or programmers). Then, we consider a more realistic model, in which we take the difference between students (or programmers) into account.

Comment. In this paper, for clarity, all the formulas are explained on the example of teaching, when we decide on the amount of feedback provided to each homework. However, the exact same formulas and conclusions are also valid for code verification, when we have to decide on the optimal amount of feedback provided to each programmer.

2 Simplified Case, When We Ignore the Difference Between Students

Description of the Simplified Case. In this section, we consider the simplified case, when we ignore the difference between students, and assume that the effect of feedback is the same for all students.

Analysis of the Problem. The more time we spend on feedback, the more thorough this feedback, the larger the effect it has on the person who receives this feedback – be it a student or a programmer.

Usually, very little feedback is practically not helpful at all, then we have an increase, but after that, too much details will be mostly wasted – this is like micromanagement.

Notations. To describe the resulting problem in precise terms, let us introduce the following notations:

- Let us denote the effect of a piece of feedback that took t hours by $f(t)$.
- Let T denote the overall time allocated to producing this feedback.

Let us Formulate the Problem in Precise Terms. What we want is find the value n and the values $t_1, \ldots, t_n$ for which

$$t_1 + \ldots + t_n = T, \tag{1}$$

and for which, among all such tuples, the overall effect

$$E = f(t_1) + \ldots + f(t_n) \tag{2}$$

is the largest possible.

Towards Solving the Problem. Let us use the usual Lagrange multiplier technique to reduce the constrained optimization problem (1)–(2) to the unconstrained optimization problem of maximizing the following functional

$$f(t_1) + \ldots + f(t_n) + \lambda \cdot (t_1 + \ldots + t_n - T), \tag{3}$$

where λ is the Lagrange multiplier.

Differentiating the expression (3) with respect to each unknown t_i and equating the derivative to 0, we conclude that

$$f'(t_i) + \lambda = 0, \tag{4}$$

where $f'(t)$ denotes the derivative of the function $f'(t)$. Thus, for all i, we have the same value of the derivative $f'(t_i) = -\lambda$. So, in general, the corresponding values of t_i should be the same:

$$t_1 = \ldots = t_n. \tag{5}$$

Hence,

$$E = f(t_1) + \ldots + f(t_n) = n \cdot f(t_1). \tag{6}$$

From the formula (1), we can similarly conclude that $n \cdot t_1 = T$. Therefore,

$$n = \frac{T}{t_1}. \tag{7}$$

Substituting the expression (7) into the formula (6) for the overall effect, we conclude that

$$E = \frac{T}{t_1} \cdot f(t_1). \tag{8}$$

Maximizing this expression is equivalent to maximizing the expression

$$\frac{E}{T} = \frac{f(t_1)}{t_1}. \tag{9}$$

Thus, we arrive at the following conclusion.

Conclusion. The optimal time t_1 to spend on each feedback is the time that maximizes the ratio (11).

Comment. It is worth mentioning that the optimal time t_1 that needs to be spent on each feedback does not depend on the overall time T. So, if we have more time T for grading (e.g., if we get an additional teaching assistant), we should assign more homeworks – but keep the same level of thoroughness as before.

Let us Make This Conclusion More Detailed. When we have an analytical expression for $f(t)$, then the largest value of the expression (9) can be obtained if we differentiate the expression (9) with respect to t_1 and equate the derivative to 0. As a result, we get the formula

$$\frac{f'(t_1)\cdot t_1 - f(t_1)}{t_1^2} = 0, \tag{10}$$

i.e., equivalently,

$$f'(t_1)\cdot t_1 = f(t_1). \tag{11}$$

Conclusion in More Detail. The optimal time t_1 to spend on each feedback is the time that satisfies the Eq. (11).

What is the Empirical Dependence $f(t)$. According to [2] (see also [1]), the effect of each stimulus – including the feedback – is proportional to

$$f(t) = \frac{t^q}{t^q + c}, \tag{12}$$

for some empirical constants q and c.

Let us show how to solve the Eq. (11) for this function $f(t)$.

Case of Empirical Dependence: Analysis of the Problem. The derivative $f'(t)$ of the expression (12) is easy to find when we realize that

$$f(t) = 1 - \frac{c}{t^q + c}. \tag{13}$$

Differentiating the right-hand side of the formula (13), we conclude that

$$f'(t) = \frac{c\cdot q\cdot t^{q-1}}{(t^q + c)^2}. \tag{14}$$

Thus, the Eq. (11) takes the form

$$\frac{c\cdot q\cdot t_1^q}{(t_1^q + c)^2} = \frac{t_1^q}{t_1^q + c}. \tag{15}$$

If we divide both side of this equality by t_1^q and multiply both sides by $(t_1^q + c)^2$, we get the following equality:

$$c\cdot q = t_1^q + c, \tag{16}$$

hence

$$t_1^q = c \cdot (q - 1). \tag{17}$$

So, we arrive at the following conclusion.

Case of Empirical Dependence: Conclusion. The optimal time t_1 to spend on each feedback is equal to

$$t_1 = (c \cdot (q - 1))^{1/q}, \tag{18}$$

where c and q are the parameters of the dependence (12).

3 General Case, When We Take into Account that Students Are Different

Description of the General Case. Let us now consider the general case, when we take the difference between the students into account.

To be fair, we allocate equal time to each student.

Notations.

- Let s denote the number of students.
- Let $f_i(t)$ denote the effect, on the i-th students, of the feedback that the instructor produced during time t.

Analysis of the Problem: General Dependence. In the fair approach, out of the total time T, the instructor allocates time T/s to each student. Thus, what we want is find, for each student i, is the number of assignments n_i and the values $t_{i,1}, \ldots, t_{i,n_i}$ for which

$$t_{i,1} + \ldots + t_{i,n_i} = \frac{T}{s}, \tag{19}$$

and for which, among all such tuples, the overall effect

$$E_i = f_i(t_{i,1}) + \ldots + f_i(t_{i,n_i}) \tag{20}$$

is the largest possible.

Similarly to the previous section, we conclude that all the times $t_{i,1}, \ldots, t_{i,n_i}$ should be equal:

$$t_{i,1} = \ldots = t_{i,n_i}, \tag{21}$$

and the optimal value of t_{i1} is the one that maximizes the expression

$$\frac{f_i(t_{i,1})}{t_{i,1}}, \tag{22}$$

i.e., equivalently, for which the following equality holds:

$$f_i'(t_{i,1}) \cdot t_{i,1} = f_i(t_{i,1}). \tag{23}$$

General Dependence: Conclusion. The optimal time $t_{i,1}$ allocated to provide feedback for each homework of i-th students should maximize the expression (22), or, equivalently, should satisfy the Eq. (23).

Analysis of the Problem: Case of Empirical Dependence. In particular, for the case when

$$f_i(t) = \frac{t^{q_i}}{t^{q_i} + c_i} \tag{24}$$

for some coefficients q_i and c_i, the optimal time $t_{i,1}$ has the form

$$t_{i,1} = (c_i \cdot (q_i - 1))^{1/q_i}. \tag{25}$$

Empirical Dependence: Conclusion. For the empirical dependence (24), the optimal time $t_{i,1}$ allocated to provide feedback for each homework of the i-th student is determined by the formula (25).

Comment. Since we allocated the same time T/s to each student, and the optimal times $t_{i,1}$ are, in general, different for different students, this means that the number of homeworks

$$n_i = \frac{T/s}{t_{i,1}} \tag{26}$$

assigned to different students should be, in general, different:

- to some students, we assign fewer homeworks, and provide more thorough feedback, while
- to other students, we assign more homeworks, and provide less thorough feedback.

Acknowledgments. This work was supported in part by the National Science Foundation grants 1623190 (A Model of Change for Preparing a New Generation for Professional Practice in Computer Science) and HRD-1242122 (Cyber-ShARE Center of Excellence).

References

1. Bokati, L., Kreinovich, V., Ha, D.T.: How the Proportion of People Who Agree to Perform a Task Depends on the Stimulus: A Theoretical Explanation of the Empirical Formula. University of Texas at El Paso, Department of Computer Science, Technical Report UTEP-CS-20-41 (2020). http://www.cs.utep.edu/vladik/2020/tr20-41.pdf
2. Khani, M., Ahmadi, A., Hajary, H.: Distributed task allocation in multi-agent environments using cellular learning automata. Soft Comput. **23**, 1199–1218 (2019)

Image Processing for Workability of Concrete with Super Absorbent Polymer

Asghar Aryanfar[1](✉) and Irem Şanal[2]

[1] American University of Beirut, RiadEl-Solh 1107 2020, Lebanon
aryanfar@caltech.edu
[2] Bahçeşehir University, 4 ÇıRağan Cad, 34353 Istanbul, Beşiktaş, Turkey

Abstract. Super absorbent polymers (SAP) are the recent promising chemical admixtures with the potential for reducing the shrinkage, cracking, freeze/thaw and increasing the durability of the concrete [1–4]. These polymers are classified as hydrogels when cross linked and can retain exceptionally high amount of liquid solutions of their own weight [5]. In this paper, the flowability of the SAP is addressed by exploring the transient behavior in the viscoelastic properties of the SAP-contained concrete. Controlling flow within such relatively short period of time is essential for concrete functionality in construction.

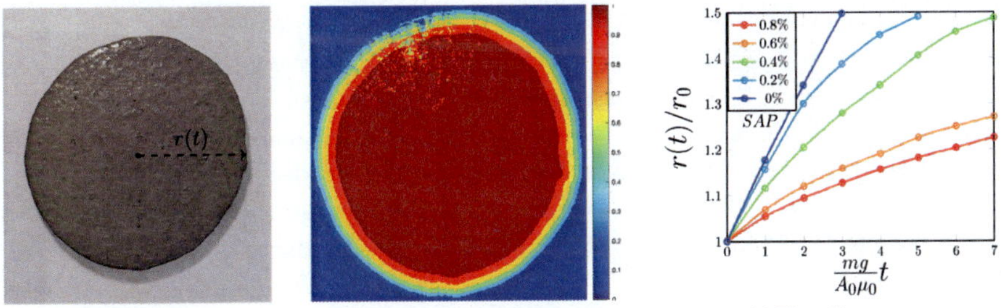

(a) Slump test. (b) Image processing. (c) Flow diagrams.

Fig. 1. Analysis.

1 Introduction

Super absorbent polymers have opened up new possibilities for controlling the control of free water in the concrete mixture, which controls both fresh and hardened state concrete properties through internal curing [6]. Such utilization as a self-curing agent saves water as the concrete dries to a large extent. Typical chemical composition is sodium salt of poly-acrylate acid $[-CH_2-CH(CO_2Na)-]_n$, a crystal-like structure classified hydrogel when cross-linked. They have capacity to take up and retain up to $\times 1000$ times their own weight. SAPs are commonly used in diapers, sanitary napkins, biomedical purposes, agriculture etc.

T. Allahviranloo et al. (Eds.): IDS 2020, AISC 1301, pp. 681–685, 2021.
https://doi.org/10.1007/978-3-030-66501-2_54

The recent introduction of SAPs into concrete has lead to promising results in terms of physical and mechanical properties of the mortar as well as reliability in both fresh and hardened states. During absorption, water is entrained and cured into the concrete leading to hydration. Subsequently, SAP particles can slowly release their inner water as humidity supply when decreased in cement paste. This process fills the pores and reduces the micro-cracks, relieving the autogenous and drying shrinkages. Consequently the strength and durability of the concrete is enhanced [7]. Therefore, the release process of water as well as its magnitude is a critical parameter for investigation of internal curing.

Predicting and controlling the workability of SAP modified concretes during the mixing process is a limiting factor for the practical uses of these promising materials in concrete industry. Therefore, the main objective in this study is to propose an image based analysis for predicting SAP concrete's workability before casting process. For this purpose we characterize the transient behavior of cementitious composite modified with SAP in fresh state where we explore the radial propagation of slump and correlate the growth to the transient viscosity $\mu(t)$. Such depiction allows to predict and control the workability during concrete casting.

2 Methodology

2.1 Materials and Design

A simple test on mixing the water with the super absorbent powder shows that it can absorb up to $100g$ of water per gram of SAP as shown in the Fig. 2.

During mixing, the SAP particles increase in volume, which is similar to air entrainment technique, utilized for improving the resistance of concrete to freeze-thaw. The entrainment is homogenous and improves hardening of concrete [8].

The control mixture with the water-to-cement ratio of $w/c = 0.4$ was prepared as reference. The next mixtures contained varying amounts of $\{0.2\%, 0.4\%, 0.8\%\}$ in SAP of the cement mass. Additional water was added

(a) Dry SAP powder. (b) SAP after hydration.

Fig. 2. Water absorption by SAP.

Table 1. Properties of SAP samples used.

$\rho(kg/m^3)$	$d_{particle}(\mu m)$	Absorption (g/g)	Desorption (%)	SAP content	SP content
1310	$150-200$	100	5	$\{0.2\%, 0.4\%, 0.8\%\}$	$\{1\%, 2\%\}$

to compensate for the absorption of mixing water by the polymers and to create mixtures showing a similar workability to the reference with a water-to-cement ratio of 0.40. Consequently, polycarboxilate based superplasticizer was used to attain a self-compacting behaviour of the concrete, with the amounts of $\{1\%, 2\%\}$. A maximum aggregate size of $2mm$ for the fine aggregates of ordinary Portland cement with the grade $CEM|42.5$. The mortar samples were prepared in the room with temperature of $20 \pm 2°C$, by mixing the solid powder with liquid for $2\,min$ at low speed of $140 \pm 5\,rev/min$ and a further $2\,min$ at high speed of $285 \pm 10\ rev/min$. The SAPs was treated by sieves, and the SAP with particle size of $150-200\,\mu m$ was selected. The detailed properties of SAP used in this study is given in Table 1.

The mini slump flow was carried out by placing the cone in the middle of a flat piece of square glass and filled with mortar. The cone is then quickly lifted, allow the mortar to flow. After certain enlargement the sample turns to steady shape.

High resolution digital images ($15\,frames/second$) were also taken at regular time intervals during the mini slump flow test to evaluate the flow behavior of materials in the subsequent computations.

2.2 Image Processing

The images from the bare concrete samples, given in Fig. 1a can be processed to get the area of the concrete in the given time $A(t)$ and the corresponding radius $r(t)$ respectively. The information from the red, green and blue values (i.e. $\{R, G, B\} \in [0, 255]$) can be transferred to a normalized gray-scale intensity image by averaging the values as:

$$I_{i,j} = \frac{R_{i,j} + G_{i,j} + B_{i,j}}{3 \times 255} \tag{1}$$

where the $I_{i,j} \in [0, 1]$ is the intensity value of the obtained grayscale image. The concrete regions can be distinguished by establishing a grayness threshold, $I_c \in [0, 1]$, which classifies the elements into black and white classes $\{B, W\} \in (0, 1)$. This value is determined iteratively from Otsu's method by minimizing the intra-class variance σ^2 as follows [9,10]:

$$\text{minimize } \sigma^2 \text{ such that:}$$

$$\begin{cases} \sigma^2 = \omega_1\sigma_0^2 + \omega_2\sigma_1^2 \\ \omega_1 + \omega_2 = 1 \end{cases} \tag{2}$$

where ω_1 and ω_2 are the fractions of black/white portions and σ_2^2 are the corresponding variances for each classified zone. The obtained binarized image provides the best approximation of the original gray-scale image since the minimization of intra-class variance σ^2 ensures the closest proximity in the values of each chosen group $(B\&W)$ is chosen. Starting from the center, the cleared concrete area (ΔA_k) in each time t_k can be obtained by propagating through the first-order neighbors until no further progress is made [11]. Figure 1b represents the computed effective incremental area in the time-span of the experiment. Consequently, we define the measure of flowability F as the normalized radius $(F := \frac{r(t)}{r_0})$ which is tracked versus time. Figure 1c represents the computed measure for 4 concrete samples varying concentration in SAP.

The obtained values for the radius $r(t)$ can be could be differentiated to get the values of the radial velocity $(u_{k+1} = \frac{r_{k+1} - r_k}{\delta t})$. The flow is gravity-driven and the shear stress $\tau(t)$ versus shear rate $(\dot{\gamma} = {}^{du}/_{dy})$ would be related with viscosity $\mu(t)$ via Eq. 3:

$$\tau(t) = \mu(t)\frac{du}{dy} \tag{3}$$

On the other hand, we can simplify the Navier-Stockes relationship to describe the flow as [12]:

$$\frac{du}{dt} + u\frac{du}{dr} = \mu(t)\frac{d^2u}{dr^2} \tag{4}$$

The Eq. 4 can be solved using the finite-difference method. Using the scheme of forward in time and space $(FTFS)$ for the experimental samples in the Fig. 1c, we get the value of the viscosity $\mu_k(t)$ as a function of SAP and SP contents, in the given time t_k as:

$$\mu_k = \left(\frac{u_{k+1} - u_k}{\delta t} + u_k\frac{u_{k+1} - u_k}{\delta r}\right)\left(\frac{u_{k+1} - 2u_k + u_{k-1}}{\delta r^2}\right)^{-1}$$

3 Results and Conclusion

The experimental and analytical results indicate that the proposed image based analysis technique has a potential to be used for automatic and practical estimate of SAP included concrete's workability before casting process. The computations reveal that the viscosity $\mu(t)$ is significantly higher in the presence of augmented SAP mixture $(SAP \uparrow \sim \mu \uparrow)$. This correlation however is opposite for the content of SP $(SP \uparrow \sim \mu\downarrow)$. Such trade-off shows that combined utilization of SAP and SP contents can be used as determining parameters to tune the flow parameters to tune and control the workability properties of the concrete effectively.

References

1. Lee, H.X.D., Wong, H.S., Buenfeld, N.R.: Self-sealing of cracks in concrete using superabsorbent polymers. Cem. Concr. Res. **79**, 194–208 (2016)
2. Schröfl, C., Mechtcherine, V., Gorges, M.: Relation between the molecular structure and the efficiency of superabsorbent polymers (SAP) as concrete admixture to mitigate autogenous shrinkage. Cem. Concr. Res. **42**(6), 865–873 (2012)
3. Justs, J., Wyrzykowski, M., Bajare, D., Lura, P.: Internal curing by superabsorbent polymers in ultra-high performance concrete. Cem. Concr. Res. **76**, 82–90 (2015)
4. Mechtcherine, V., Reinhardt, H.-W.: Application of Super Absorbent Polymers (SAP) in Concrete Construction: State-of-the-art Report Prepared by Technical Committee 225-SAP, vol. 2. Springer Science & Business Media, New York (2012)
5. Esteves, L.P.: Superabsorbent polymers: on their interaction with water and pore fluid. Cem. Concr. Compos. **33**(7), 717–724 (2011)
6. Jensen, O.M.: Use of superabsorbent polymers in concrete. Concr. Int. **35**(1), 48–52 (2013)
7. Jensen, O.M., Hansen, P.F.: Water-entrained cement-based materials: I. principles and theoretical background. Cem. Concr. Res. **31**(4), 647–654 (2001)
8. Jensen, O.M., Hansen, P.F.: Water-entrained cement-based materials: II. experimental observations. Cem. Concr. Res. **32**(6), 973–978 (2002)
9. Otsu, N.: A threshold selection method from gray-level histograms. Automatica **11**(285–296), 23–27 (1975)
10. Aryanfar, A., Hoffmann, M.R., Goddard III, W.A.: Finite-pulse waves for efficient suppression of evolving mesoscale dendrites in rechargeable batteries. Phys. Rev. E **100**(4), 042801 (2019)
11. Aryanfar, A., Goddard III, W., Marian, J.: Constriction percolation model for coupled diffusion-reaction corrosion of zirconium in pwr. Corros. Sci. **158**, 108058 (2019)
12. Munson, B.R., Okiishi, T.H., Huebsch, W.W.: Fluid Mechanics. Wiley, Singapore (2013)

Influence of Uncertainties on Reliability of Large Space Structures

Masoud Negin(✉)

Department of Civil Engineering, Bahcesehir University, Istanbul, Turkey
masoud.negin@eng.bau.edu.tr

Abstract. It is impossible to estimate and ensure the safety of a large space structure having hundreds of members without considering the influence of uncertainties. In fact, precise manufacturing and assembling of such structure is almost impossible and the existence of several imperfections such as non-rectilinearity of geometric axis of the members, material properties, connection defects, initial lack of fit, residual stresses, etc., which inherently have random nature, is inevitable. Consequently, the response of the real structure itself has also random nature. This study investigates the effects of these types of random imperfections on the reliability of large space structures using non-linear analysis and Monte Carlo simulation method. OpenSees finite element software package is used to perform the numerical analysis. It is demonstrated that these random imperfections have significant influences on the reliability assessment of space structures.

Keywords: Space structures · Random imperfections · Uncertainty · Reliability · Monte Carlo simulation

1 Introduction

Space structures are widely and increasingly used in construction of various projects to cover large open areas. Although, these structures usually have a high degree of static indeterminacy, and therefore it may be expected that they have sufficient safety against structural failures, some experiments in the progressive collapse of such structures have confirmed that the collapse of these truss-type structures can make really serious problems. In fact, existence of several imperfections such as non-rectilinearity of the geometric axis of members, material properties, connection defects, initial lack of fit and residual stresses etc. decreases the load carrying capacity of compression members and plays an important role in failure mechanism of structure. This will be crucial, especially in large space structures with hundreds or thousands of members and joints these structures typically contain. In other words, precise manufacturing and assembling of such structures is almost impossible and basically the response of these structures itself has random behavior. Therefore, in order to estimate and insure the safety of the structure, it is necessary to consider the effects of uncertainties in system parameters. In this context, numerous studies have been conducted so for by many researchers, considering the effects of some of these random variables in their investigations. Some recent

T. Allahviranloo et al. (Eds.): IDS 2020, AISC 1301, pp. 686–695, 2021.
https://doi.org/10.1007/978-3-030-66501-2_55

investigations are: Zhou *et al* (2014) studied the effect of member geometric imperfection on nonlinear geometrically buckling and seismic performance of the space steel structure composed of a reticulated shell and cable-strut system. Mousavi *et al* (2015) investigated the effects of applying different buckling modes obtained by linearized eigenvalue buckling analysis as the initial imperfection on the sensitivity analysis of double domes space structures. Chen *et al* (2016) proposed a random field model for the initial geometric imperfection of reticulated shell structures and examined the role of initial geometric imperfections on structural ultimate strength using numerical examples. Liu *et al* (2017) studied the impact of initial imperfections, such as joint position deviation, eccentricity of members, variation in cable pretension and elastic modulus of materials, on the stability of nonlinear geometries and materials of suspended domes. Kang and Liu (2018) developed a new reliability-based topology optimization framework considering spatially varying geometric uncertainties and verified the accuracy of the proposed method by Monte Carlo simulations. Zhao *et al* (2019) investigated the influence of random geometrical imperfection on the stability of two types of latticed shell structures. They considered the randomness of the initial curvature direction, initial curvature magnitude, and joint stiffness simultaneously. Cui *et al* (2020) modelled the initial geometric imperfection using the simultaneous autoregressive model and showed that this model is flexible in modelling spatially varying correlated imperfection and can cope with irregular grid system such as spherical latticed shells.

This study investigates the effect of various types of random imperfections on the reliability of large space structures using Monte Carlo simulation method. However, since performing such reliability analysis with the presence of hundreds of random variables is a complex and time consuming process, especially if we consider the nonlinear behavior of these structural systems, in this study a new simple methodology based on the classical definition of structural reliability is presented to evaluate the reliability of large space structures. In fact, we can calculate the reliability of these complex systems by modeling the buckling load of the compression members as a random variable and then use it in the forthcoming structural analysis of the structure. It should be noted that all structural analysis in the paper were carried out using the object-oriented finite elements software framework OpenSees (McKenna 2011).

2 Imperfections as Random Variables

Behavior of compression members has significant effect on failure mechanism of truss type structures. Almost all members of such a structure carry mostly axial force. When a tension member reaches its yield point, its rigidity reduces to zero and the state continues until strain hardening occurs. But when a compression member buckles under compression force, it may not be able to carry additional load. When the member keeps getting shorter, it is necessary that axial force also be reduced to maintain the equilibrium. In other words, a compression member represents ductile behavior after buckling. Finally if the applied load exceeds the elastic limit point of the structure, buckling of some compressional members leads to a sudden reduction in the load carrying capacity of the whole structure and redistribution of internal forces. If the structure can tolerate this redistribution, it may carry some additional loads; otherwise, some other members will fail, causing the progressive collapse of the system.

Behavior of a simple pin-ended compression member is a function of three basic factors namely: the slenderness ratio, the yield stress of the material and the initial imperfection of the member. This member may have a brittle or a ductile behavior based on the variation in these three factors. To model random characteristics of the buckling behavior of a compression member one pin-ended truss type member with initial geometric imperfection is considered (Fig. 1a). It is assumed that the geometric imperfection is in the shape of a half-sine wave along the length. The maximum initial deviation of the member in its mid-point is denoted by ε. This member is created in OpenSees with twenty Elastic-Perfectly Plastic non-linear displacement-based beam-column elements with equal length, integrated at 4 points along the element. The integration is based on the Gauss-Legendre quadrature rule which enforces Bernoulli beam assumptions. It is assumed that all members are made of the same circular steel tube, with an outer diameter of 48 mm, a thickness of 5.6 mm and the length of 1200 mm. This section is captured with fiber section element class in OpenSees, so that the whole section was divided into 16 subdivisions or fibers in circumferential direction and 4 subdivisions in radial direction as shown in Fig. 1a. The axial force-displacement relationship of the model is obtained through displacement control analysis using Newton-Raphson algorithm with considering both geometric and material non-linearity (McKenna 2011).

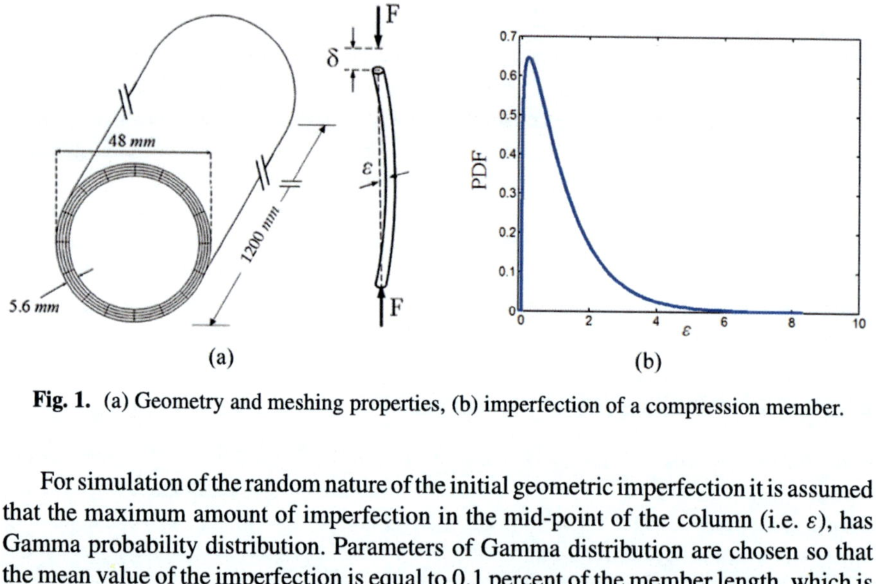

Fig. 1. (a) Geometry and meshing properties, (b) imperfection of a compression member.

For simulation of the random nature of the initial geometric imperfection it is assumed that the maximum amount of imperfection in the mid-point of the column (i.e. ε), has Gamma probability distribution. Parameters of Gamma distribution are chosen so that the mean value of the imperfection is equal to 0.1 percent of the member length, which is a typical value of the imperfection in space grid structures (i.e. $\mu_\varepsilon = 0.001L = 1.2$ mm). Furthermore, the maximum imperfection values are limited to 1 percent of the member length, because application of members with rather large imperfections can easily be avoided (i.e. $\varepsilon_{max} = 0.01L = 12$ mm). In this manner, the probability density function of the imperfection is obtained by generating random numbers in Gamma distribution with above characteristics (Fig. 1b). Then analysis is performed for each of these imperfections and the maximum load that the column carry is calculated for each case. Figure 2a shows the corresponding axial force-displacement relationship of each imperfection. Note that the graphs are drawn according to the corresponding stress values. Thus, the axial stress-strain relationship of the member and its characteristics in tension and compression are modeled as Fig. 2b in the coming analysis. It should be noted that the yield stress and module of elasticity of the used material are assumed to be 240 MPa and 210GPa, respectively. The probability distribution of the buckling stress and the corresponding statistical parameters are determined using statistical calculations. In this case the probability density function and cumulative distribution function of bucking load of the column in terms of the stresses are shown in Fig. 3. It can be seen that the post-bucking negative rigidity of the compression member has severe sensitivity to initial imperfections. The mean and the standard deviation of the buckling stress of the member were obtained as μ_s= 186 N/mm^2 and σ_s= 19 N/mm^2, respectively.

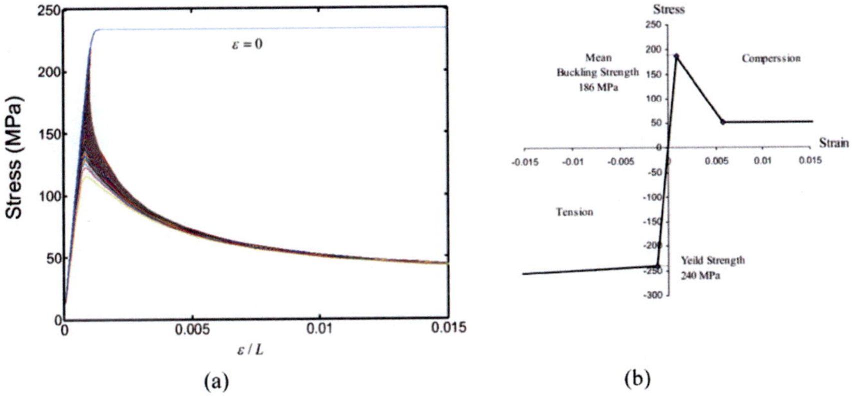

Fig. 2. (a) Force-displacement behavior of the compression member, (b) Material Properties.

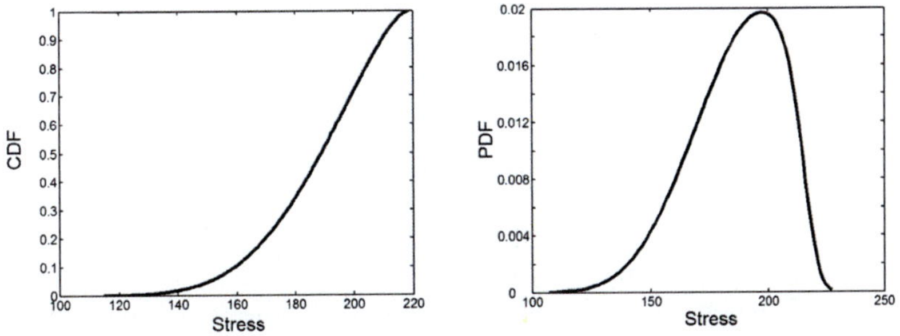

Fig. 3. CDF and PDF of the buckling stress in compression member.

3 Analytical Model

As analytical model we chose a flat double-layer space structures as shown in Fig. 4a. The geometry and dimensions of the 8 × 8 square off set double-layer grid structure, simply supported on roller bearings at the four lower corner nodes is shown in the figure. All members are steel tube sections with specifications discussed in preceding section. This model has been designed for dead loads including covering and connections equal to 50 kg/m^2 and live load equal to 200 kg/m^2. As an exception, the four diagonal members over the support nodes are considered to be solid bars with same material and geometric properties as the other members. This strengthening prevents the occurrence of local failure under high shear load over the supports. The structure has been designed using AISC allowable stress design method and it was observed that compression is the dominant response of the system.

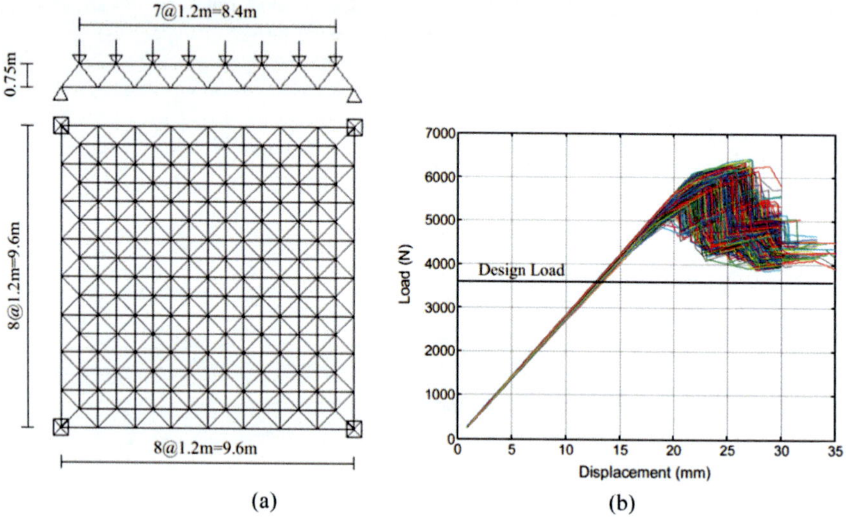

Fig. 4. (a) Double-layer grid configuration, (b) Force-displacement diagrams of the structure.

4 Reliability and Monte Carlo Simulation Method

In many situations it is impossible to mathematically describe the response of some structural systems because of random nature and complex formulations of the problem. Even when we can find a mathematical model for predicting behavior of the system, there is no closed form solution for that equation. In such situations the simulation is one of the most common techniques to gain information about complicated problems. In fact, simulation is a special technique to approximate quantities that are difficult to obtain analytically. Among many of simulation procedures the Monte Carlo simulation method is one of the most famous and common procedures in solving complicated engineering problems.

The basis of all simulation procedures is the generation of random numbers that are uniformly distributed between 0 and 1 (Nowak 2012). These random numbers usually are generated using computer programs and some common algorithms of random number generation. Many procedures and subroutines for creating random numbers with arbitrary distributions are available. Details of such procedures are beyond the scope of this paper, but it should be mentioned that currently a simple algorithm based on a standard C++ library function is implemented in OpenSees software (Haukaas 2003).

The reliability of a structure is defined as the probability that it performs its designed function without failing and is usually formulated using a failure function, $g(X_1, X_2 \ldots X_n)$, where $X_1, X_2, \ldots, X_n$ are random variables and g is called the limit state function. The violation of the limit state function is defined by the condition $g(X_1, X_2 \ldots X_n) \leq 0$, therefore the probability of failure P_f is expressed using the following expression:

$$P_f = P\left[g(X_1, X_2 \ldots X_n) \leq 0\right] = \iiint\limits_{g} f(X_1, X_2 \ldots X_n) dX_1 dX_2 \ldots dX_n, \quad (1)$$

where $(X_1, X_2 \ldots X_n)$ are the random variables and $f(X_1, X_2 \ldots X_n)$ is their joint probability density function. In practice, the probability of failure is obtained by generating a limited number of random numbers, so the calculated failure probability is only an estimate of the probability of the real failure. As a result, the probability of failure applying the Monte Carlo simulation method is calculated from the following equation (Lemaire 2013):

$$\overline{P_f} \approx E\left[P_f\right] = \frac{1}{N} \sum_{i=0}^{N} I\left[g(X_1, X_2 \ldots X_n) \leq 0\right], \quad (2)$$

where $I(X_1, X_2 \ldots X_n)$ is a function defined as:

$$I(X_1, X_2 \ldots X_n) = \begin{cases} 1, & \textit{if} \quad g(X_1, X_2 \ldots X_n) \leq 0 \\ 0, & \textit{if} \quad g(X_1, X_2 \ldots X_n) > 0 \end{cases}. \quad (3)$$

By increasing the number of simulations, this estimate will be closer to the actual value.

5 Numerical Results and Discussion

As mentioned earlier, in this study only the buckling stress of the column is considered as a random variable to model all types of initial imperfections. For this purpose, the probability distribution of the buckling stresses of the column is used as a basis for the following analysis of the double-layer space structure. In other words, in the framework of the Monte Carlo simulation method, a non-linear finite element analysis is performed for each of the buckling stresses and for each case the maximum capacity of the system is calculated. It should be mentioned that, effects of both material and geometric non-linearity is considered in the analysis. Note that to model the random nature of distribution of the imperfections through the double-layer structure, since the structure totally contains 512 pin-ended members, in each analysis 512 buckling stress values are chosen among the above stress values and then uniformly distributed among the all members. This procedure should be repeated for many times so the results would be reliable. In this study 1000 simulations were performed and the results are given as force-displacement relationship for the node in the middle bottom layer of the double-layer grid as shown in

Fig. 4b. Note that according to this figure the first maximum point is considered as failure point of the system, then the probability distribution of the failure load of the system and its corresponding statistical parameters are determined by statistical calculations. In this case, the probability density function and cumulative distribution function of the failure load of the system are obtained in Fig. 5. The characteristics of the response quantities as the mean and the standard deviation of the failure load of the system are obtained as μ_P= 5842 N and σ_P= 220 N, respectively.

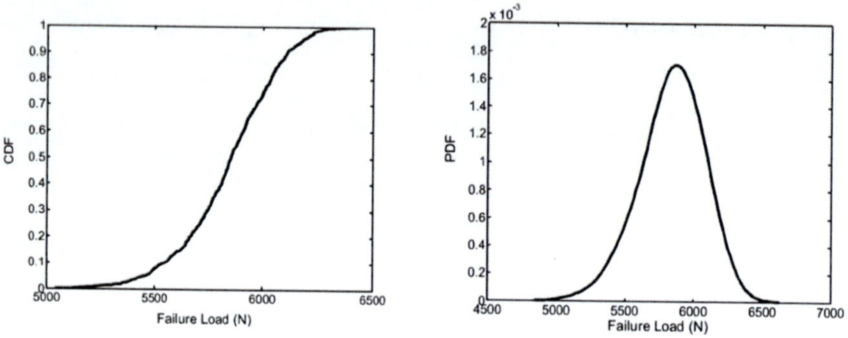

Fig. 5. PDF and CDF of the failure load for the double-layer grid structure.

About the number of simulations it should be mentioned that at we successively did 100, 200, 300 … and 1000 analyses and determined the average failure load of the structural system. Then in the next step more analyses are done and average failure loads are calculated again. After 500 analyzes, though, we observed that there was no significant change in the average failure load of the structure. Therefore, 1000 analyzes are conducted to obtain more precise and valid results.

Reliability of a system is usually shown by R and is defined in terms of probability of failure as $R = 1 - P_f$ for convenience, where P_f is the probability of failure of the system. Since the only random variable of the problem is the imperfection or the corresponding buckling stress of the members and the probability distribution of the failure load of the system is obtained (Fig. 5), thus the reliability of the system R at a specific load F_s is the probability that the failure load of the system F is equal or greater than that specific load. This can be expressed mathematically as:

$$R(F_S) = P(F > F_S) \tag{4}$$

Using this definition the reliability of double-layer space structure is calculated and the results are given in Fig. 6 and Table 1. Note that the second and third columns in the table show the amount of strength reduction in each case because of the imperfections. The last column in Table 1, gives the load factor for each case, i.e. if the designer wants to become sure about the safety of an imperfect structure, he/she must multiply the design load of the system by corresponding load factor in the last column the Table. It should be mentioned that the perfect structure theoretically is the structure without any imperfection. But in practice, construction of such an ideal structure is impossible, so in the present context, the structure that indicated maximum load carrying capacity among other structures with different imperfections is considered as the perfect or most nearest structure to the perfect structure.

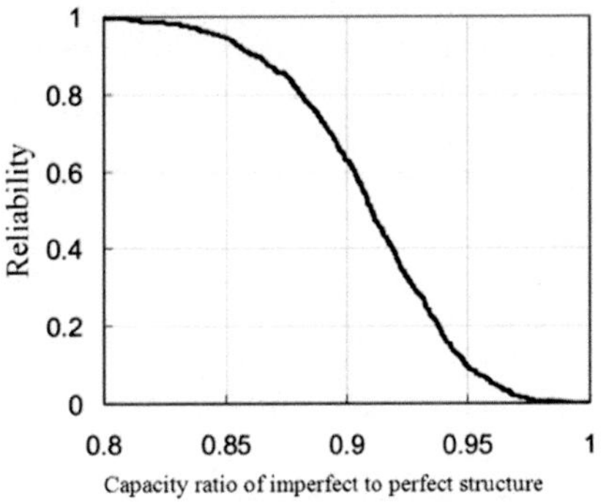

Fig. 6. Reliability of' the double-layer grid structure.

Table 1. Reliability of the double-layer grid structure.

Reliability	Capacity ratio of imperfect to perfect structure		Load Factor
	Value	%	
1	0.784	78.4	1.22
0.99	0.813	81.3	1.19
0.98	0.831	83.1	1.17
0.97	0.838	83.8	1.16
0.96	0.843	84.3	1.16
0.95	0.848	84.8	1.15
0.94	0.852	85.2	1.15
0.93	0.854	85.4	1.15
0.92	0.857	85.7	1.14
0.91	0.858	85.8	1.14
0.90	0.862	86.2	1.14
0.85	0.875	87.5	1.13
0.80	0.881	88.1	1.13
0.75	0.888	88.8	1.12
0.70	0.893	89.3	1.11

This is how in practice we can use these results. Suppose that we want to design a system with reliability equal to for example 0.99. It can be seen from Fig. 6 and also Table 1 that the capacity of the imperfect structure is 81.3% of the capacity of the perfect system, which is 18.7% less than the capacity of the perfect one. In other words, to design a system with reliability equal to 0.99, the design load of the structure must be considered 18.7% greater than the capacity of the perfect structure. Or to design a system with reliability equal to 0.95 for example, as the capacity of the imperfect structure is 84.8 percent of the capacity of the perfect one, which is 15.2% less than the perfect system, the design load of the structure must be 15.2% greater than the capacity of the perfect structure, and so on.

On the other hand, reliability analysis of complex structures such as space structures with hundreds of degrees of indeterminacy requires a lot of calculations. This can be computationally expensive, especially in the case of nonlinear analysis. Thus the above results can also be interpreted in the following way, which is the main purpose of this study. That is to design space structures with a certain level of reliability without performing complicated reliability analysis. For example, assume the goal is to design a double-layer space structure with consistent level of reliability without conducting detailed reliability analysis, in this case, if the target reliability of the system was for example 0.98, the design load of the system must be multiplied by corresponding load factor in the last column of the Table 1, in this case 1.17. Or if the target reliability of the system was 0.96 for example, the design load of the system must be multiplied by corresponding load factor in the last column of the Table 1, in this case 1.16, and so on.

6 Conclusion

The existence of some imperfections in space structures with hundreds of members and joints is almost inevitable. Yet, reliability analysis of such complex structural systems requires a lot of calculations and especially in the case of nonlinear analysis this can be computationally expensive. This study investigates and examines the effect of these random imperfections on reliability and overall strength of large space structures using simple reliability concept. Numerical analysis of a selected off set double-layer grid structure applying the Monte Carlo simulation method, indicates that these structures are highly sensitive to random imperfections. In fact, initial imperfections in selected double-layer space structure reduced the capacity of the system from 2% up to 22%. Hence, this stochasticity should be suitably considered in estimating realistic response of these structures. Furthermore, it is also shown that by applying the proposed method and drawing reliability diagrams like Fig. 6 (or Table 1) and using it in the structural design processes one can easily determine the design load of the structure with the desired safety. These diagrams help the designers to design with consistent reliability levels without conducting detailed reliability analysis for each case.

References

Chen, G., Zhang, H., Rasmussen, K.J., Fan, F.: Modeling geometric imperfections for reticulated shell structures using random field theory. Eng. Struct. **126**, 481–489 (2016)

Cui, X.Z., Li, Y.G., Hong, H.P.: Effect of spatially correlated initial geometric imperfection on reliability of spherical latticed shell considering global instability. Struct. Saf. **82**, 101895 (2020)
Haukaas, T.: Finite element reliability and sensitivity methods for performance-based engineering, Ph.D. Thesis, University of California, Berkeley (2003)
Kang, Z., Liu, P.: Reliability based topology optimization against geometric imperfections with random threshold model. Int. J. Numer. Meth. Eng. **115**(1), 99–116 (2018)
Lemaire, M.: Structural Reliability. John Wiley & Sons, Hoboken (2013)
Liu, X., Zhan, X., Zhang, A., Zhang, X., Tian, C.: Random imperfection method for stability analysis of a suspended dome. Int. J. Steel Struct. **17**(1), 91–103 (2017)
McKenna, F.: OpenSees: a framework for earthquake engineering simulation. Comput. Sci. Eng. **13**(4), 58–66 (2011)
Mousavi, M.A., Abedi, K., Chenaghlou, M.: Imperfection sensitivity analysis of double domes free form space structures. Int. J. Struct. Stab. Dyn. **15**(04), 1450067 (2015)
Andrzej, N.S., Collins, K.R.: Reliability of Structures. CRC Press, Boca Raton (2012)
Zhao, Z.W., Liu, H.Q., Liang, B., Yan, R.Z.: Influence of random geometrical imperfection on the stability of single-layer reticulated domes with semi-rigid connection. Adv. Steel Constr. **15**(1), 93–99 (2019)
Zhou, Z., Wu, J., Meng, S.: Influence of member geometric imperfection on geometrically non-linear buckling and seismic performance of suspen-dome structures. Int. J. Struct. Stab. Dyn. **14**(03), 1350070 (2014)

Investigating the Role of Renewable Energies in Integrated Energy-Water Nexus Planning under Uncertainty Using Fuzzy Logic

Afshin Ghassemi(✉) and Michael J. Scott

University of Illinois at Chicago, Chicago, IL 60607, USA
afshinghassemi@gmail.com, mjscott@uic.edu

Abstract. Energy and water systems are highly interconnected. Energy is required to extract, transmit, and treat water and wastewater, and water is needed for cooling energy systems. There is a rapid increase in demand for energy and water due to factors such as population and economic growth. In less than 30 years, the need for energy and water will nearly double globally. As the energy and water resources are limited, it is critical to have a sustainable energy-water nexus framework to meet these growing demands. Renewable energies provide substantial opportunities in energy-water nexuses by boosting energy and water reliability and sustainability and can be less water-intensive than conventional technologies. These resources, such as wind and solar power, do not need water inputs. As a result, they can be used as a supplement to the energy-water nexus portfolio. In this paper, renewable energies in energy-water nexus have been investigated for a range of possible scenarios. As renewable energy resources are not deterministic, fuzzy logic is used to model the uncertainty. The results show that renewable energies can significantly improve the energy-water nexus planning; however, the power grid reliability on renewable energy should be aligned with the level of systems uncertainty. The gap between the decisions extracted from the Fuzzy model and the deterministic model amplifies the importance of considering uncertainty to generate reliable decisions.

Keywords: Energy-water nexus · Renewable energies · Optimization under uncertainty · Fuzzy logic

1 Introduction

Energy and water are highly interconnected. Water is used to cool down power systems, and energy is used for water extraction, transmission, and treatments. Neglecting to consider the coupling between energy and water systems planning might cause disastrous consequences [1]. The meta planning system that includes energy and water planning systems at the same time is called energy-water nexus

T. Allahviranloo et al. (Eds.): IDS 2020, AISC 1301, pp. 696–702, 2021.
https://doi.org/10.1007/978-3-030-66501-2_56

[2]. Despite the growing demand for energy and water in upcoming years [3], generally, these resources are limited. The growing demand of common resources is predicted to be one of the main topics of conflicts in the upcoming years not only on interconnected resources but, in many other critical area (see [4,5]). Renewable energies such as wind and solar are considered to have unlimited capacity [6]. They also do not rely on water for power generations. However, due to these power sources' uncertain nature, building a model that generates reliable decisions under uncertainty to maximize their energy-water nexus systems efficiency is essential [7]. To the best of our knowledge, there are a few studies that offer general mathematical energy-water nexus planning models [6,8–11]. Ghassemi [12] proposed a framework for energy-water nexus planning that minimizes the total systems cost while reducing the unwanted environmental effects.

This research aims to utilize the Fuzzy logic to provide optimal solutions that work under different uncertainty levels. Fuzzy logic has been used in numerous applications in different range of fields, to deal with uncertainty [13,14]. The application of the Robust optimization [7,12], Fuzzy logic [10,15–19] and stochastic programming [20–22] Have been used on a limited scale in energy-water nexus planning and in a broader capacity in the classic scientific fields. Using the Fuzzy logic makes it possible to involve the policymakers, shareholders, and managers' opinions and also scenarios-based planning into consideration. Using the Fuzzy optimization method makes it possible to find a reliable and efficient role for renewable energies in energy-water nexus systems.

The rest of this study is organized as follows. In the next section, the energy-water systems' profile is explained. In Sect. 3, the application of the proposed Fuzzy optimization is provided. The experiments' results are provided in Sect. 4. Finally, the findings, results, and future work are discussed in Sect. 5.

2 Fuzzy Optimization Method

The Possibility, Necessity, and Credibility measures represent the chances of occurrence of Fuzzy events with an optimistic, pessimistic, and moderate view towards planning. Here, the conversion of Fuzzy chance constraints to their equivalent crisp for a specific confidence level is provided. In this paper, $\tilde{\mu}$ and γ are Fuzzy and crisp numbers. It is assumed that the uncertain parameter is a Fuzzy number with trapezoidal distribution that is showed by $\tilde{\mu}\,(\mu_1, \mu_2, \mu_3, \mu_4)$ while $\mu_1 < \mu_2 < \mu_3 < \mu_4$.

2.1 Possibility Measure

Let the triple (Θ, P(Θ), Poss) represent a possibility space that a universe set Θ is a non-empty set, covering all possible events and P(Θ) be the power set of Θ. For every $A, B \in P(\Theta)$, there are non-negative numbers, $Poss\{A\}$ and $Poss\{B\}$, the possibility measure has these features:

- $Poss\{\varnothing\} = 0$

- $Poss\{\Theta\} = 1$
- $if\ A \in P(\Theta) \Rightarrow 0 \leq Poss\{A\} \leq 1$
- $Poss\{\cup_i A_i\} = Sup_i(Poss\{A_i\})$
- $if\ A, B \in P(\Theta)\ and\ A \subseteq B \Rightarrow Poss\{A\} \leq Poss\{B\}\ (Monotonicity)$
- $if\ A, B \in P(\Theta) \Rightarrow Poss\{A \cup B\} + Poss\{A \cap B\} \leq Poss\{A\} + Poss\{B\}\ (Subadditivity)$

Let $\tilde{\mu}$ be a trapezoidal Fuzzy variable on the possibility space $(\Theta, P(\Theta), Poss)$. The possibility of fuzzy events $\{\tilde{\mu} \leq \gamma\}$ and $\{\tilde{\mu} \geq \gamma\}$ are as Eqs. 1 and 2:

$$Poss\{\tilde{\mu} \leq \gamma\} = \begin{cases} 1, & if\ \ \mu_2 \leq \gamma; \\ \dfrac{\gamma - \mu_1}{\mu_2 - \mu_1}, & if\ \ \mu_1 \leq \gamma \leq \mu_2; \\ 0, & if\ \ \mu_1 \geq \gamma. \end{cases} \tag{1}$$

$$Poss\{\tilde{\mu} \geq \gamma\} = \begin{cases} 1, & if\ \ \mu_3 \geq \gamma; \\ \dfrac{\mu_4 - \gamma}{\mu_4 - \mu_3}, & if\ \ \mu_3 \leq \gamma \leq \mu_4; \\ 0, & if\ \ \mu_4 \leq \gamma. \end{cases} \tag{2}$$

Based on the possibility measure, the Fuzzy chance constraints for the confidence level of (α) are equal to Eqs. 3 and 4:

$$Poss\{\tilde{\mu} \leq \gamma\} \geq \alpha \quad \Leftrightarrow \quad (1-\alpha)\mu_1 + \alpha\mu_2 \leq \gamma \tag{3}$$

$$Poss\{\tilde{\mu} \geq \gamma\} \geq \alpha \quad \Leftrightarrow \quad \alpha\mu_3 + (1-\alpha)\mu_4 \geq \gamma \tag{4}$$

2.2 Necessity Measure

The Necessity measure works as the dual model of the possibility measure. For every $A, B \in P(\Theta)$, the profile of the Necessity measure are presented as follows:

- $Nece\{\varnothing\} = 0$
- $Nece\{\Theta\} = 1$
- $if\ A \in P(\Theta) \Rightarrow 0 \leq Nece\{A\} \leq 1$
- $Nece\{\cap_i A_i\} = inf_i(Nece\{A_i\})$
- $if\ A, B \in P(\Theta)\ and\ A \subseteq B \Rightarrow Nece\{A\} \leq Nece\{B\}\ (Monotonicity)$
- $if\ A, B \in P(\Theta) \Rightarrow Nece\{A \cup B\} + Nece\{A \cap B\} \geq Nece\{A\} + Nece\{B\}\ (Subadditivity)$
- $if\ Poss\{A\} < 1 \Rightarrow Nece\{A\} = 0$
- $if\ Nece\{A\} > 0 \Rightarrow Poss\{A\} = 1$

The Necessity (Nece) of Fuzzy events is as Eqs. 5 and 6 and the transformed versions are as Eqs. 7 and 8:

$$Nece\{\tilde{\mu} \leq \gamma\} = \begin{cases} 1, & if\ \ \mu_4 \leq \gamma; \\ \dfrac{\gamma - \mu_3}{\mu_4 - \mu_3}, & if\ \ \mu_3 \leq \gamma \leq \mu_4; \\ 0, & if\ \ \mu_3 \geq \gamma. \end{cases} \tag{5}$$

$$Nece\{\tilde{\mu} \geq \gamma\} = \begin{cases} 1, & if \quad \mu_1 \geq \gamma; \\ \dfrac{\mu_2 - \gamma}{\mu_2 - \mu_1}, & if \quad \mu_1 \leq \gamma \leq \mu_2; \\ 0, & if \quad \mu_2 \leq \gamma. \end{cases} \tag{6}$$

$$Nece\{\tilde{\mu} \leq \gamma\} \geq \alpha \quad \Leftrightarrow \quad (1-\alpha)\mu_3 + \alpha\mu_4 \leq \gamma \tag{7}$$

$$Nece\{\tilde{\mu} \geq \gamma\} \geq \alpha \quad \Leftrightarrow \quad \alpha\mu_1 + (1-\alpha)\mu_2 \geq \gamma \tag{8}$$

2.3 Credibility Measure

Finally, the Credibility measure (Cred) of event {A} is defined as the average of its Possibility (Poss) and Necessity (Nece) measures as:

$$Cr\{A\} = \frac{1}{2}(Poss\{A\} + Nece\{A\}).$$

The profile of the Credibility measure are presented as follows:

- $Cred\{\varnothing\} = 0$
- $Nece\{\Theta\} = 1$
- $if \quad A \in P(\Theta\ ) \ \Rightarrow \ 0 \leq Cred\{A\} \leq 1$
- $if\, A_i \in P(\Theta\ )\, and\, Sup_i(Cred\{A_i\}) < 0.5 \Rightarrow Cred\{\cup_i A_i\} = Sup_i(Cred\{A_i\})$
- $if \quad A \in P(\Theta\ ) \Rightarrow Cred\{A\} + Cred\{A^C\} = 1 \quad (Self-Duality)$
- $if \quad A, B \in P(\Theta\ ) \quad and \quad A \subseteq B \Rightarrow Cred\{A\} \leq Cred\{B\} \quad (Monotonicity)$
- $if \quad A, B \in P(\Theta\ ) \Rightarrow Cred\{A \cup B\} \leq Cred\{A\} + Cred\{B\} \quad (Subadditivity)$
- $Poss\{A\} \geq Cred\{A\} \geq Nece\{A\}$

Based on the Credibility measure, transforming the Fuzzy chance constraints for the confidence level of (α) is as Eqs. 9 and 10:

$$Cred\{\tilde{\mu} \leq \gamma\} = \begin{cases} 0, & if \quad \mu_1 \geq \gamma; \\ \dfrac{\gamma - \mu_1}{2(\mu_2 - \mu_1)}, & if \quad \mu_1 \leq \gamma \leq \mu_2; \\ \dfrac{1}{2}, & if \quad \mu_2 \leq \gamma \leq \mu_3; \\ \dfrac{\gamma - 2\mu_3 + \mu_4}{2(\mu_4 - \mu_3)}, & if \quad \mu_3 \leq \gamma \leq \mu_4; \\ 1, & if \quad \mu_4 \leq \gamma. \end{cases} \tag{9}$$

$$Cred\{\tilde{\mu} \geq \gamma\} = \begin{cases} 1, & if \quad \mu_1 \geq \gamma; \\ \dfrac{2\mu_2 - \mu_1 - \gamma}{2(\mu_2 - \mu_1)}, & if \quad \mu_1 \leq \gamma \leq \mu_2; \\ \dfrac{1}{2}, & if \quad \mu_2 \leq \gamma \leq \mu_3; \\ \dfrac{\mu_4 - \gamma}{2(\mu_4 - \mu_3)}, & if \quad \mu_3 \leq \gamma \leq \mu_4; \\ 0, & if \quad \mu_4 \leq \gamma. \end{cases} \tag{10}$$

The briefed versions are presented as Eqs. 11 and 12:

$$Cred\{\tilde{\mu} \leq \gamma\} \geq \alpha \quad \Leftrightarrow \quad \begin{cases} (2-2\alpha)\mu_3 + (2\alpha-1)\mu_4 \leq \gamma & if \quad \alpha > 0.5; \\ (1-2\alpha)\mu_1 + 2\alpha\mu_2 \leq \gamma & if \quad \alpha \leq 0.5. \end{cases} \tag{11}$$

$$Cred\{\tilde{\mu} \geq \gamma\} \geq \alpha \quad \Leftrightarrow \quad \begin{cases} (2\alpha-1)\mu_1 + (2-2\alpha)\mu_2 \geq \gamma & if \quad \alpha > 0.5; \\ 2\alpha\mu_3 + (1-2\alpha)\mu_4 \geq \gamma & if \quad \alpha \leq 0.5. \end{cases} \tag{12}$$

3 Energy-Water Nexus Profile

The deterministic energy-water nexus model and the parameters such as wind generations are the same as the author's previous work [12]. The planning horizon is 7×24, and the model generates online hourly decisions. The final model is a Fuzzy, multi-period, linear mixed-integer optimization problem. The energy-water nexus model on the water side includes water extraction, transmission, treatment, and all the associated costs. On the power side, it covers thermal power systems, such as wind power and batteries. The model also has two penalty costs; (1) Unmet demand and (2) Loss of the polluted waste water before treatment.

4 Scenario Analysis

The energy-water nexus model is analyzed using three Fuzzy approaches of Possibility (optimistic planning), Necessity (pessimistic planning), and Credibility (moderate planning) to model the uncertainty for wind generation and efficiency. Trapezoidal Fuzzy numbers are used to represent the uncertain parameters. A sensitivity analysis for the confidence intervals (α) is performed to verify the performances of Fuzzy models. The results are presented in Table 1. It should be noted that:

$$PI_{i,\alpha} = \frac{\text{Total systems cost of method } (i, \alpha)}{\text{Total systems cost of the Possibility method } (\alpha = 0)} \times 10{,}000$$

where $i = Possibility, Credibility, Necessity$ and $\alpha = 0, 0.25, 0.50, 0.75, 1$.

Table 1. Total system cost comparisons of the Possibility, Necessity, and Credibility approaches

Approach	Total systems cost ($)									
	$\alpha = 0$	PI	$\alpha = 0.25$	PI	$\alpha = 0.5$	PI	$\alpha = 0.75$	PI	$\alpha = 1$	PI
Possibility	3151620000	1	3151972981	1.12	3152013953	1.25	3152051772	1.37	3152121108	1.59
Credibility	3152221959	1.91	3152253476	2.01	3152316508	2.21	3152373237	2.39	3152442573	2.61
Necessity	3152404753	2.49	3152445724	2.62	3152474089	2.71	3152518212	2.85	3152556031	2.97

As expected, the Possibility approach with $\alpha = 0$ has the least total systems cost. With an increase of α, the cost to generate reliable decisions increases as well. The Necessity approach yields the most costly choices, and the Credibility approach provides more practical decisions. The *PI* index is provided to represent the gaps between solutions. These three methods give the decision-maker flexibility to find reliable choices while respecting the additional measures (e.g., budget constraint).

5 Discussion

Increased demand for energy in the future could lead to raises in the energy system's water intensity, whereas renewable energies such as wind and solar could lower the water dependence. Renewable energy sources, such as wind power, need a low amount of water. However, they randomness around them could be more than traditional power generation such as hydro-power. To have a reliable system and maximize the efficiency of renewable energies in energy-water nexus planning, the model should be resilient against a reasonable amount of uncertainty for the key parameters. To this aim, three Fuzzy approaches with optimistic, pessimistic, and moderate conservative levels are introduced. A numerical sensitivity analysis for a range of confidence intervals is provided. The results show that utilizing Fuzzy logic, a more reliable system can be achieved, and a flexible range of solutions can be generated. For future studies, the goal is to focus on developing customized robust-Fuzzy approaches that can generate reliable decisions with partial data.

References

1. Gleick, P.H.: Water and conflict: fresh water resources and international security. Int. Secur. **18**(1), 79 (1993). http://www.jstor.org/stable/2539033?origin=crossref
2. Greenwade, G.D.: The comprehensive tex archive network (CTAN). TUGBoat **14**(3), 342–351 (1993)
3. Stillwell, A.S., King, C.W., Webber, M.E., Duncan, I.J., Hardberger, A.: The energy-water nexus in Texas. Ecol. Soc. **16**(1), 2 (2011). http://www.ecologyandsociety.org/vol16/iss1/art2/
4. Esmaeili, M., Bahrini, A., Shayanrad, S.: Using game theory approach to interpret stable policies for Iran's oil and gas common resources conflicts with Iraq and Qatar. J. Ind. Eng. Int. **11**, 543–554 (2015)
5. Mohamadi, M., Bahrini, A.: A Nash-Stackelberg equilibrium model for internet and network service providers in the demand market—a scenario-based approach. Wireless Netw. **26**(1), 449–461 (2020)
6. Muhanji, S.O., Farid, A.M.: An enterprise control methodology for the techno-economic assessment of the energy water nexus. Appl. Energy **260**, 114274 (2020). http://www.sciencedirect.com/science/article/pii/S0306261919319610
7. Ghassemi, A., Hu, M., Zhou, Z.: Robust planning decision model for an integrated water system. J. Water Resour. Plan. Manag. **143**, 1–12 (2017)

8. Tsolas, S.D., Karim, M.N., Hasan, M.F.: Optimization of water-energy nexus: a network representation-based graphical approach. Appl. Energy **224**, 230–250 (2018). http://www.sciencedirect.com/science/article/pii/S0306261918306524
9. Zhang, X., Vesselinov, V.V.: Energy-water nexus: balancing the trade-offs between two-level decision makers. Appl. Energy **183**, 77–87 (2016). http://www.sciencedirect.com/science/article/pii/S030626191631265X
10. Liu, J., Li, Y., Huang, G., Suo, C., Yin, S.: An interval fuzzy-stochastic chance-constrained programming based energy-water nexus model for planning electric power systems. Energies **10**(11), 1914 (2017). http://www.mdpi.com/1996-1073/10/11/1914
11. Wang, S., Liu, Y., Chen, B.: Multiregional input-output and ecological network analyses for regional energy-water nexus within China. Appl. Energy **227**, 353–364 (2018). http://www.sciencedirect.com/science/article/pii/S030626191731694X. Transformative Innovations for a Sustainable Future – Part III
12. Ghassemi, A.: System of systems approach to develop an energy-water nexus model under uncertainty. Ph.D. thesis (2019)
13. Peykani, P., Mohammadi, E., Pishvaee, M., Rostamy-Malkhalifeh, M., Jabbarzadeh, A.: A novel fuzzy data envelopment analysis based on robust possibilistic programming: possibility, necessity and credibility-based approaches. RAIRO Oper. Res. **52**, 1445–1463 (2018)
14. Peykani, P., Mohammadi, E., Emrouznejad, A., Pishvaee, M.S., Rostamy-Malkhalifeh, M.: Fuzzy data envelopment analysis: an adjustable approach. Expert Syst. Appl. **136**, 439–452 (2019). http://www.sciencedirect.com/science/article/pii/S0957417419304397
15. Chen, F., Huang, G., Fan, Y.: Inexact multistage fuzzy-stochastic programming model for water resources management. J. Water Resour. Plann. Manag. **141**(11), 4015027 (2015). http://dx.doi.org/10.1061/(ASCE)WR.1943-5452.0000547
16. Guo, P., Huang, G., Zhu, H., Wang, X.: A two-stage programming approach for water resources management under randomness and fuzziness. Environ. Model Softw. **25**(12), 1573–1581 (2010). http://www.sciencedirect.com/science/article/pii/S1364815210001246
17. Coelho, V.N., Coelho, I.M., Coelho, B.N., Reis, A.J., Enayatifar, R., Souza, M.J., Guimarães, F.G.: A self-adaptive evolutionary fuzzy model for load forecasting problems on smart grid environment. Appl. Energy **169**, 567–584 (2016). http://www.sciencedirect.com/science/article/pii/S0306261916301684
18. Suganthi, L., Iniyan, S., Samuel, A.A.: Applications of fuzzy logic in renewable energy systems - a review. Renew. Sustain. Energy Rev. **48**, 585–607 (2015). http://www.sciencedirect.com/science/article/pii/S136403211500307X
19. Peykani, P., Mohammadi, E., Saen, R.F., Sadjadi, S.J., Rostamy-Malkhalifeh, M.: Data envelopment analysis and robust optimization: a review. Expert Syst. **37**(4), e12534 (2020). https://onlinelibrary.wiley.com/doi/abs/10.1111/exsy.12534
20. Babayan, A., Kapelan, Z., Savic, D., Walters, G.: Least-cost design of water distribution networks under demand uncertainty. J. Water Resour. Plan. Manag. **131**(5), 375–382 (2005)
21. Weini, Z.: Water network design and management via stochastic programming. Ph.D. thesis, University of Arizona (2013)
22. Xie, Y.L., Huang, G.H., Li, W., Li, J.B., Li, Y.F.: An inexact two-stage stochastic programming model for water resources management in Nansihu Lake Basin. China. J. Environ. Manag. **127**, 188–205 (2013). http://www.sciencedirect.com/science/article/pii/S0301479713002715

IoT Based Smart Irrigation System with LoRa

Muammer Catak(✉)

College of Engineering and Technology, American University of the Middle East, Egaila, Kuwait
muammer.catak@aum.edu.kw

Abstract. Both using the water resources adequately and increasing the quantity and the quality of the agricultural products are significantly vital considering the humankind population on the Earth. Smart systems offer promising solutions for afore-mentioned issue. In literature, smart irrigation systems supplied by GSM, or Wi-Fi modules have been released. However, while smart irrigation systems with GSM are facing to power consumption problem, smart irrigation systems with Wi-Fi are having trouble about the limited coverage range. LoRa has low power consumption, so it may operate around ten years with a single battery, in addition to its long range of coverage. Therefore, a smart irrigation system equipped with LoRa has been proposed in this paper in order to increase the quality of the irrigation systems.

Keywords: Intelligent control · LoRa · LoRaWAN · Sensors · Smart systems

1 Introduction

Irrigation of agricultural land is very important issue, not only addressed by areas with limited water supplies but also water-rich countries. The right amount of water must be provided to the plants in order to increase the productivity. Issaka et al. [1] presented a brief paper smart water irrigation taking into consideration of climate change and its effect on agriculture production. According to the reports, only 50% of the used water can reach to the plants by means of the traditional irrigation systems [2,3]. Talavera et al. [4] discussed the role of IoT applications in industrial agriculture and environmental fields. They reported that humidity, temperature of the air, soil moisture, and solar radiation are recognized as universal variables in a smart agricultural application. Moreover, wind conditions in a plantation have an important role on irrigation systems considering evaporation losses [5]. According to Haijun et al. [6], appropriate irrigation systems may contribute water efficiency (from 20% to 56%) and crop yield from 3% to 50% depending on the plant type. In literature, it is reported that smart irrigation systems can achieve from 37% to 75% water saving [7,8].

T. Allahviranloo et al. (Eds.): IDS 2020, AISC 1301, pp. 703–709, 2021.
https://doi.org/10.1007/978-3-030-66501-2_57

Smart systems may propose an adequate solution for afore mentioned problems accompanying with agricultural technologies. Furthermore, a smart irrigation system can control the humidity level of the soil, and that can get information about the weather forecast aiming to enhance the harvesting. In this work, a smart irrigation system equipped with LoRa has been proposed due to its low power consumption, and its long range of coverage comparing to the other state-of-art technologies.

Long Range Wide-area Network (LoRaWAN) has an increasing popularity in engineering applications as a less-complex wireless communication protocol using the unlicensed spectrum associated with Long Range (LoRa) modulation [9].

IoT sensors mostly need to send small amount of data (uplink) and less frequently receive any data (downlink). Hence, LoRaWAN technology has been employed to transmit data of metering, smart city applications, agricultural area. The key point is to combine two phenomena such that low data rate and communication range in km [10]. LoRaWAN operates in the Industrial, Science, and Medical (ISM) radio bands, and then it must obey to some certain regulations. The maximum transmission time is bounded by duty cycle percentage such that an individual node may only switch its transmitter on for up to 36 s in an hour with duty cycle of 1%.

In this paper, design and implementation of a smart irrigation system with LoRa has been proposed. A smart irrigation system may have a crucial responsibility in not only to avoid waste of water resources but also to improve the productivity.

The structure of the paper is as follows: In Sect. 2, design and implementation of the proposed smart irrigation system has been discussed. The obtained results and discussion were presented in Sect. 3. Finally, the paper was concluded in Sect. 4.

2 Materials and Methods

The block diagram of the proposed system is illustrated in Fig. 1. It has four main parts, namely i) measurement unit (sensors), ii) microcontroller unit (MCU), iii) communication unit (LoRa), and iv) water pump unit. Watering process is heavily dependent on the geographical location, current wind conditions, moisture level of the soil, air temperature, and humidity. Therefore, each irrigation system should be calibrated in conformity with its own conditions. As shown in Fig. 1, five various sensors; rain, wind, moisture, temperature, humidity; are used in order to collect data related to environmental conditions. The collected data is sent to MCU through LoRa. MCU acts as a brain of the system, it analyses the data send by the sensors, and then it decides the action of DC pump for watering due to the predefined threshold values of each variable. MCU and the pumping part communicate each other by means of LoRa module.

LoRa is a communication technique stand on a chirp spread spectrum (CSS) modulation. Afore the frequency modulation, the corresponding data signal is

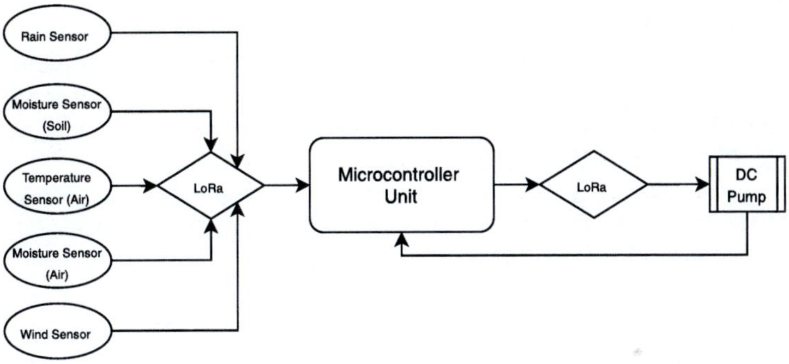

Fig. 1. Block diagram representation of the proposed system

chirped at a higher data rate. The chirp signal is reckoned as a carrier signal referring to the fundamental modulation theory. In this project, 433 MHz, which is the center frequency allowed frequency band (433.05–434.79 MHz) for ISM applications, has been used [11].

The nominal bit rate of the data signal is given [12];

$$R_b = SF \frac{\frac{4}{4+CR}}{\frac{2^{SF}}{BW}} \quad \text{bits/s} \tag{1}$$

where CR is the code rate, can be an integer from 1 to 4; SF is spreading factor from 7 to 12, BW is the modulation bandwidth in Hz (125 kHz in the study). If $(4/4 + CR)$ is defined as *Rate Code*, then the nominal bit rate is;

$$R_b = SF \frac{\text{Rate Code}}{\frac{2^{SF}}{BW}} \quad \text{bits/s} \tag{2}$$

The spreading factor (SF) of a sensor is dynamically reassigned by the gateway regarding to the power of the corresponding received signal. There are 6 levels of SF. While the SF increases the RF sensitivity also increases, which allows to signal propagates longer ranges; on the other hand, the nominal bit rate decreases, which results in increment in time on air (ToA). The LoRa specifications used in this study are given in Table 1.

Table 1. Specifications of LoRa employed for the proposed method

#	SF	Bandwidth (kHz)	Code rate	Rate code	R_b (bps)	RF sensitivity (dBm)
0	12	125	1	4/5	293	−137
1	11	125	1	4/5	537	−134.5
2	10	125	1	4/5	976	−132
3	9	125	1	4/5	1757	−129
4	8	125	1	4/5	3125	−126
5	7	125	1	4/5	5468	−123

LoRa allows the user to achieve the optimum network performance without changing bandwidth by means of the spreading factor. If the communication range is the most significant parameter in a system, then the gateway assigns the SF as 12, if the data rate is important considering the reliability of the system, the SF is most likely assigned as 7.

In this study, the following channels are employed in order to operate the system: 433.175, 433.375, 433.575, 433.825, 434.075, 434.325, 434.575, and 434.825 MHz.

An uplink transmission is followed by two downlinks receive windows, during which the motes can capture the data including information from gateway. The channels are selected randomly, and it is assumed that acting as a uniformly distributed random variable.

3 Results and Discussions

The proposed smart irrigation system was tested and validated in Kuwait. Hence, the local environmental conditions were taken into consideration while assigning threshold values of the corresponding sensors. The threshold of the air humidity is fixed at 70%. If the sensor reads more than 70% of humidity, then the system will be OFF, otherwise the system will be ON or OFF depending on the other sensors' readings. If the wind speed is greater than 5 m/s, the system will be OFF in order to avoid evaporation loss. The lower threshold of the soil moisture sensor is fixed at 10%, while the upper threshold is assigned as 40%. If the sensor reading is less than 10%, then the irrigation system should work; if the sensor reads more than 40% the irrigation process must be stopped. If the sensor reading is in between 10% and 40%, the system will continue as its current mode. The temperature sensor has also two threshold values 0 °C and 40 °C as lower and upper limits, respectively. If the temperature sensor reading is in the range between 0 °C and 40 °C, the system may start referring to the readings of the other sensors. Finally, if it is raining, the system must be at OFF mode in order to save the water.

The flowchart representations of sender and receiver sides are depicted in Fig. 2 and Fig. 3, respectively. The communication range of the system in real time has been experienced, and it works up 16 km in Kuwait urban area. In fact, the coverage distance of a LoRa system is highly dependent on the geographical conditions of the location, and the urbanization plans. The possible solution in order to increase the coverage distance and to improve the quality of the communication level could be increase the number of LoRa gateways, which may increase the application cost.

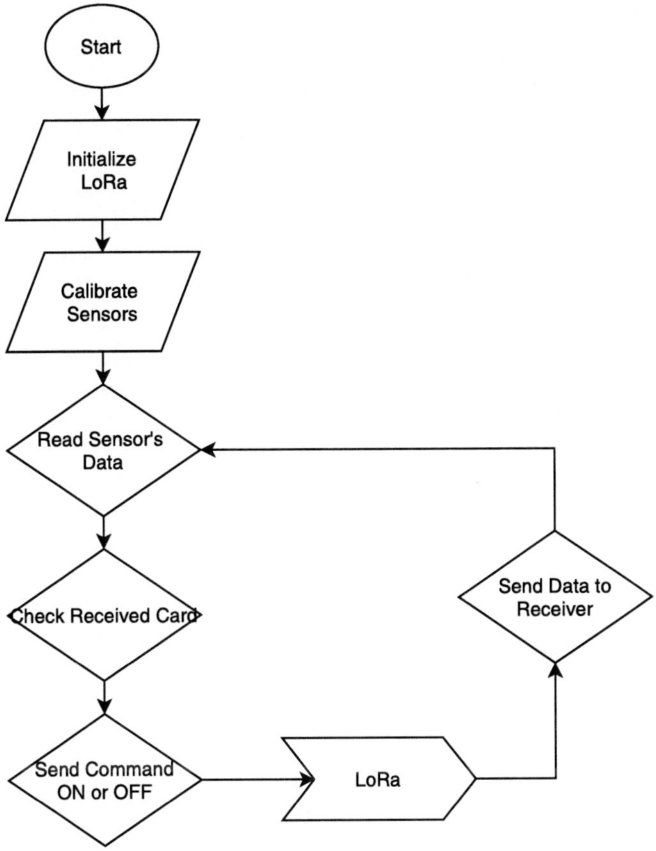

Fig. 2. Flowchart representation of sender side

By means of smart systems and automation, the agricultural industry will be improved in the direction of both the quality and the quantity of the yields. In addition, smart irrigation systems can offer watering with the appropriate amount at the proper time. Consequently, smart systems may help to avoid the soil pollution due to the over-watering, and reduce the cost including power consumption. In addition, the proposed system is convenient for small size home applications as well as industrial scales plantation applications. Please note that, such systems need periodic maintenance and calibration in order to keep the high efficiency level of the system.

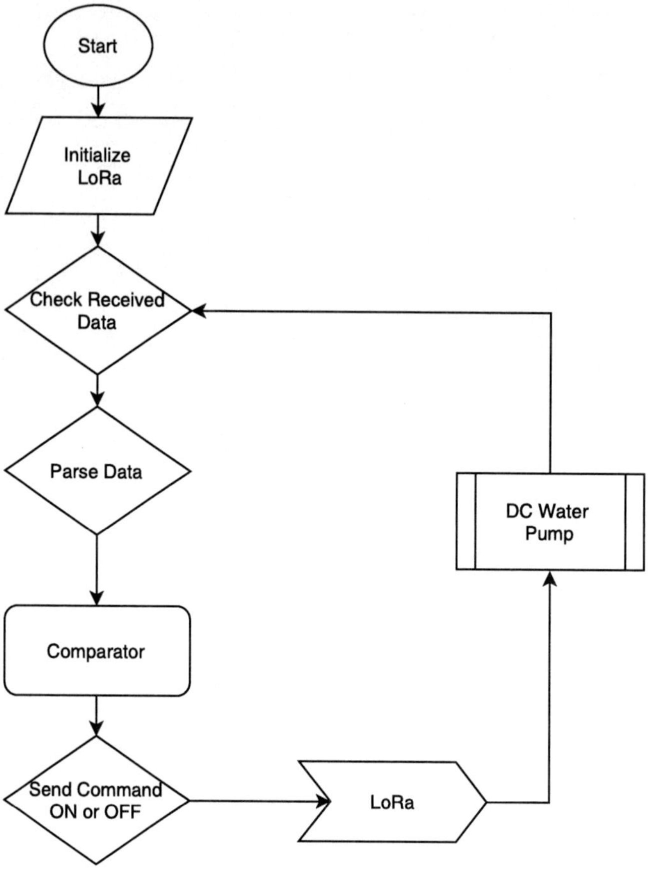

Fig. 3. Flowchart representation of receiver side

4 Conclusion

In this study, a smart irrigation system with LoRa has been presented aiming to increase the quality and the quantity of the agricultural products in Kuwait. The proposed system has advantages especially on the low power consumption and its wide coverage range comparing to the other systems. According to the results obtained from this project, smart irrigation systems can offer promising solutions for many subjects, such as soil pollution, wasting of water resources in addition to affirmative effects on the cost. For future work, a more comprehensive smart irrigation system accompanying with smart city applications can be developed.

References

1. Issaka, Z., Li, H., Yue, J., Tang, P., Darko, R.O.: Water-smart sprinkler irrigation, prerequisite to climate change adaptation: a review. J. Water Clim. Change **9**(2), 383–398 (2018)
2. Meguid, M.A.: Key features of the Egypt's water and agricultural resources. In: Conventional Water Resources and Agriculture in Egypt, pp. 39–99. Springer, Cham (2017)
3. Fao, F.: The future of food and agriculture–trends and challenges. Annual Report (2017)
4. Talavera, J.M., Tobon, L.E., Gomez, J.A., Culman, M.A., Aranda, J.M., Parra, D.T., Garreta, L.E.: Review of IoT applications in agro-industrial and environmental fields. Comput. Electron. Agric. **142**, 283–297 (2017)
5. Sinha, A., Ravikrishna, R.V.: LES of spray in crossflow-effect of droplet distortion. Int. J. Spray Combust. Dyn. **9**(1), 55–70 (2017)
6. Haijun, L., Yaohu, K., Shiping, L.: Regulation of field environmental condition by sprinkler irrigation and its effect on water use efficiency of winter wheat. Trans. Chin. Soc. Agric. Eng. **6** (2003)
7. Miller, L., Vellidis, G., Coolong, T.: Comparing a smartphone irrigation scheduling application with water balance and soil moisture-based irrigation methods: Part II–Plasticulture-grown Watermelon. HortTechnology **28**(3), 362–369 (2018)
8. Migliaccio, K.W., Morgan, K.T., Vellidis, G., Zotarelli, L., Fraisse, C., Rowland, D.L., Zurweller, B.A.: Smartphone apps for irrigation scheduling. In: 2015 ASABE/IA Irrigation Symposium: Emerging Technologies for Sustainable Irrigation-A Tribute to the Career of Terry Howell, Sr. Conference Proceedings, pp. 1–16. American Society of Agricultural and Biological Engineers (2015)
9. Mikhaylov, K., Pyhäjärvi, J., Janhunen, J.: On LoRaWAN scalability: empirical evaluation of susceptibility to inter-network interference. In: 2017 European Conference on Networks and Communications (EuCNC), pp. 1–6. IEEE, June 2017
10. Adelantado, F., Vilajosana, X., Tuset-Peiro, P., Martinez, B., Melia-Segui, J., Watteyne, T.: Understanding the limits of LoRaWAN. IEEE Commun. Mag. **55**(9), 34–40 (2017)
11. Nolan, K.E., Guibene, W., Kelly, M.Y.: An evaluation of low power wide area network technologies for the internet of things. In: 2016 International Wireless Communications and Mobile Computing Conference (IWCMC), pp. 439–444. IEEE, September 2016
12. Semtech: Lora modulation basics an1200.22. Technical report (2015)

Numerical Investigations of Physical Processes for Regularized Long Wave Equation

Ozlem Ersoy Hepson[1(✉)] and Gülsemay Yiğit[2]

[1] Department of Mathematics and Computer Science, Faculty of Science and Letters, Eskisehir Osmangazi University, Eskisehir, Turkey
ozersoy@ogu.edu.tr

[2] Department of Mathematics, Faculty of Engineering and Natural Sciences, Bahcesehir University, Istanbul, Turkey
http://www.ozlemersoy.com

Abstract. In this study, a numerical approach is used to investigate the solution of the regularized long wave (RLW) equation. A newly proposed quartic trigonometric-tension (QTT) B–spline based collocation scheme is constructed for spatial discretization. Then, the equation is transformed into a time-dependent system of differential equations, which is discretized by the Crank-Nicolson scheme. Approximate solutions of the RLW equation have successfully attained by the fully discretized system. The motions of the conservation laws of the RLW equation have also been computed numerically. Solitary wave propagation, interaction of two solitary waves, wave undulation and wave generations are simulated and results are compared to the existing literature.

Keywords: The RLW equation · Quartic trigonometric tension B-spline · Collocation method · Wave undulation · Wave generation

1 Introduction

The regularized long wave (RLW) equation attracts a remarkable interest in view of the favorable modelling of the many natural processes. This equation has been presented as forms of many physical processes such as the long–wave behavior of water waves as well as in the context of electromagnetic waves [1,2]. The model was introduced by Pregrine [1] in the study of the propagation of unidirectional water waves. Benjamin et al. [2] have contributed to the study of this equation to propose an alternative physical forms of KdV equation with the existence and uniqueness of the solutions. Olver [3] has focused on the three conservation invariants of the model equation as mass, momentum and energy characteristics.

The most popular simulations the RLW equation are available in the literature which are commonly investigated in the concept of single solitary waves, wave generation, wave undulation and interaction of two solitary waves. The study of single

T. Allahviranloo et al. (Eds.): IDS 2020, AISC 1301, pp. 710–724, 2021.
https://doi.org/10.1007/978-3-030-66501-2_58

solitary wave is conducted by means of various numerical approximations since it serves a closed analytical solution. Numerical treatment of this equation, which helps to visualize the simulations, is also at the center of interest with the relative efficiencies of the computations especially when the analytical solutions do not serve a practical approach. A few of these computational techniques consist of finite elements based on different classes of B-splines [4–17], finite differences [18–21], Haar wavelet [22], the differential quadrature [23, 24], the least squares [25–28] and pseudo spectral method [29, 30] and so forth. A closed analytical solution of the RLW equation have recently been achieved in [31].

The B–splines are smooth piecewise functions which exhibit stability preserving numerical schemes. The B–splines also present efficient formulations for the higher order derivatives that are used in the required algorithms. These functions can be defined in various degrees and aspects and can be successfully adopted as basis for the numerical methods. Alinia and Zerebnia [32] have employed three formulations to construct a new kind of B-spline form based on trigonometric and hyperbolic functions that serves efficient numerical results.

More recently, various studies have been proposed in the sense of numerical computational methods. To call a few, Hong et al. [33], conducted Fourier spectral algorithm along with the convergence of the numerical schemes. They also explored momentum conservation behavior of the model. Maharana et al. [34], employed a collocation scheme based on cubic B-spline for the RLW equation. Mittal and Rohila [35], applied a fourth order collocation scheme for the RLW and MRLW equation. Comparing present contribution with these mentioned approaches, we aimed to employ a B-spline based collocation method concerning quartic trigonometric tension proposed in [32] for the numerical solutions of the RLW equation.

The RLW equation, containing a mixed space–time derivative, is given by,

$$\frac{\partial u(x,t)}{\partial t}+\frac{\partial u(x,t)}{\partial x}+\varepsilon u(x,t)\frac{\partial u(x,t)}{\partial x}-\mu\frac{\partial^3 u(x,t)}{\partial x^2\partial t}=0,\ x\in[x_L,x_R],\ t\in(0,T] \tag{1}$$

along with the initial condition (IC) and boundary conditions (BCs):

$$u(x,0)=f(x), \tag{2}$$

$$u(x_L,t)=\sigma_1,\quad u(x_R,t)=\sigma_1,\quad u_x(x_L,t)=0,\quad u_x(x_R,t)=0, \tag{3}$$

where ε and μ denote non-dimensional positive parameters.

This article is designed as follows. In Sect. 2, the proposed B–spline is investigated with its derivatives. We also established the methodology to construct the fully discretized scheme by exploring the combination of the B–spline space discretization and Crank-Nicolson time integration. A linearization of the nonlinear term is obtained by a Taylor expansion. Section 3 serves the main results of this work. Numerical examples are attained to show the efficiency of the method and proved to be satisfied in terms of conservation laws. The natural behaviors of the model equation are presented in figures. Section 4, therefore states the conclusive remarks and future extensions of the present study.

2 Application of the QTT B-spline Collocation Method

The equally spaced distribution is conventional in the proposed B-splines. Let $[x_L, x_R]$ is uniformly partitioned as $x_j = x_L + jh$, $h = (x_R - x_L)/N$. The grids including the fictitious knots $x_{-4}, x_{-3}, x_{-2}, x_{-1}, x_{n+1}, x_{n+2}, x_{n+3}, x_{n+4}$ outside the interval constitutes the basis of QTT B-spline over the problem domain. QTT B-spline function is described as:

$$T_{j,5}(x) = r\begin{cases} \frac{\tau^2(\hat{\varphi}_{j-2})^2+2\varphi_{j-2}-2}{2\tau^2}, & [x_{j-2}, x_{j-1}],\\ -\frac{\tau^2(3h^2+6h\hat{\varphi}_{j-2}+2(\hat{\varphi}_{j-2})^2)+2K_1(\tau^2(\hat{\varphi}_{j-1})^2-2)-(6\varphi_{j-1}+2\varphi_j-4)}{2\tau^2}, & [x_{j-1}, x_j],\\ \frac{\tau^2\left(13h^2+10h\hat{\varphi}_{j-2}+2(\hat{\varphi}_{j-2})^2\right)+2K_1(11\tau^2h^2+10\tau^2h\hat{\varphi}(j-2)+4K_1\tau^2(\hat{\varphi}_{j-2})^2-8K_1+6\varphi_{j+1}-4)}{2\tau^2}, & [x_j, x_{j+1}],\\ -\frac{\tau^2(23h^2+14h\hat{\varphi}_{j-2}+2(\hat{\varphi}_{j-2})^2)+2K_1(\tau^2(\hat{\varphi}_{j+2})^2-2)+(2\varphi_{j+1}+6\varphi_{j+2}-4)}{2\tau^2}, & [x_{j+1}, x_{j+2}],\\ \frac{\tau^2(\hat{\varphi}_{j+3})^2+2\varphi_{j+3}-2}{2\tau^2}, & [x_{j+2}, x_{j+3}],\\ 0, & \text{otherwise}. \end{cases} \tag{4}$$

$r = 2h^2(1-K_1)^{-1}$, $\varphi_{j+m} = \cos(\tau(x_{j+m}-x))$, $\hat{\varphi}_{j+m} = (x_{j+m}-x)$, $K_1 = \cos(\tau h)$, $K_2 = \sin(\tau h)$ which was proposed as an alternative to unified and extended splines in [36]. The suggested B–spline function is piecewise bell–shaped function and values of the function at the connective points are presented in Table 1 using:

$$\begin{aligned}
\alpha_1 &= r(h^2\tau^2 + 2\cos(\tau h) - 2)/2\tau^2,\\
\alpha_2 &= r(h^2\tau^2 - 2\cos(\tau h)(h^2\tau^2 + 1) + 2)/2\tau^2,\\
\alpha_3 &= r(-h\tau + \sin(\tau h))/\tau,\\
\alpha_4 &= r(h\tau - 3\sin(\tau h) + 2\cos(\tau h)h\tau)/\tau,\\
\alpha_5 &= r(1 - \cos(\tau h)),\\
\alpha_6 &= r\tau\sin(\tau h),\\
\alpha_7 &= -3r\tau\sin(\tau h).
\end{aligned}$$

Table 1. $T_{j,5}(x)$ and its derivatives

	x_{j-2}	x_{j-1}	x_j	x_{j+1}	x_{j+2}	x_{j+3}
$T_{j,5}$	0	α_1	α_2	α_2	α_1	0
$T'_{j,5}$	0	α_3	α_4	$-\alpha_4$	$-\alpha_3$	0
$T''_{j,5}$	0	α_5	$-\alpha_5$	$-\alpha_5$	α_5	0
$T'''_{j,5}$	0	α_6	α_7	$-\alpha_7$	$-\alpha_6$	0

Since $T_{j,5}(x)$ are the basis function for space $\{\cos(\tau x), \sin(\tau x), x^2, x, 1\}$, approximate the unknown function u and its derivatives consisting of combination of QTT B-splines becomes as:

$$U(x,t) = \sum_{j=-2}^{N+1} \gamma_j(t)T_{j,5}(x), \tag{5}$$

where $\gamma_j(t)$ is the time-dependent variable which will be calculated by adapting the collocation method.

The unknown function u and its two derivatives at the spatial grids are obtained using QTT B-splines [32]. Thus the following set of equations is described using Eq. (5):

$$\begin{aligned} u(x_j,t_n) &\approx U_j^n = \alpha_1\gamma_{j-2}+\alpha_2\gamma_{j-1}+\alpha_2\gamma_j+\alpha_1\gamma_{j+1}, \\ u_x(x_j,t_n) &\approx (U')_j^n = \alpha_3\gamma_{j-2}+\alpha_4\gamma_{j-1}-\alpha_4\gamma_j-\alpha_3\gamma_{j+1}, \\ u_{xx}(x_j,t_n) &\approx (U'')_j^n = \alpha_5\gamma_{j-2}-\alpha_5\gamma_{j-1}-\alpha_5\gamma_j+\alpha_5\gamma_{j+1}. \end{aligned} \tag{6}$$

Time integration of the space discretized RLW equation is performed by Crank-Nicolson method. Thus, the following semi-integrated scheme is obtained as:

$$\frac{U^{n+1}-U^n}{\Delta t}+\frac{U_x^{n+1}+U_x^n}{2}+\varepsilon\frac{(UU_x)^{n+1}+(UU_x)^n}{2}-\mu\frac{U_{xx}^{n+1}+U_{xx}^n}{\Delta t}=0 \tag{7}$$

where $t^n = t^{n-1}+\Delta t$ and Δt is the time step, for the approximate solution $U^n = U(x,t^n)$ at the nth temporal stage. The nonlinear term which is linearized based on the Taylor expansion is given:

$$(UU_x)^{n+1} = U^nU_x^{n+1}+U_x^nU^{n+1}-U^nU_x^n.$$

Thus, the Eq. (7) becomes:

$$\left(1+\varepsilon\frac{\Delta t}{2}U_x^n\right)U^{n+1}+\frac{\Delta t}{2}\left(1+\varepsilon U^n\right)U_x^{n+1}-\mu U_{xx}^{n+1}=U^n-\frac{\Delta t}{2}U_x^n-\mu U_{xx}^n. \tag{8}$$

Now, using approximations in (6) into the Eq. (8), the fully-integrated scheme with the unknown coefficients γ_j, $j=-2,\cdots,N+1$ is constructed as:

$$b_1\gamma_{j-2}^{n+1}+b_2\gamma_{j-1}^{n+1}+b_3\gamma_j^{n+1}+b_4\gamma_{j+1}^{n+1}=\mathbf{C}_j^n,\quad j=0,1,\cdots,N \tag{9}$$

where

$$\mathbf{C}_j^n = U_j^n-\frac{\Delta t}{2}(U')_j^n-\mu(U'')_j^n, \tag{10}$$

$$b_1 = d_1\alpha_1+d_2\alpha_3+d_3\alpha_5,\quad b_2 = d_1\alpha_2+d_2\alpha_4-d_3\alpha_5 \tag{11}$$

$$b_3 = d_1\alpha_2-d_2\alpha_4-d_3\alpha_5,\quad b_4 = d_1\alpha_1-d_2\alpha_3+d_3\alpha_5 \tag{12}$$

$$d_1 = 1+\varepsilon\frac{\Delta t}{2}U_x^n,\quad d_2=\frac{\Delta t}{2}\left(1+\varepsilon U^n\right),\quad d_3=-\mu. \tag{13}$$

Subsequently, $(N+1)$ linear algebraic equations in $(N+4)$ unknowns are accomplished. Replacing the unknown parameters $\gamma_{-2}^{n+1}, \gamma_{-1}^{n+1}$ and γ_{n+1}^{n+1}, using the BCs $u(x_L, t) = u_x(x_L, t) = 0$ and $u(x_R, t) = 0$ three additional equations are obtained as:

$$\begin{aligned} u(x_L, t^{n+1}) &\approx U_0^{n+1} = \alpha_1\gamma_{-2}^{n+1} + \alpha_2\gamma_{-1}^{n+1} + \alpha_2\gamma_0^{n+1} + \alpha_1\gamma_1^{n+1} = 0, \\ u_x(x_L, t^{n+1}) &\approx (U')_0^{n+1} = \alpha_3\gamma_{-2}^{n+1} + \alpha_4\gamma_{-1}^{n+1} - \alpha_4\gamma_0^{n+1} - \alpha_3\gamma_1^{n+1} = 0, \\ u(x_R, t^{n+1}) &\approx U_N^{n+1} = \alpha_1\gamma_{N-2}^{n+1} + \alpha_2\gamma_{N-1}^{n+1} + \alpha_2\gamma_N^{n+1} + \alpha_1\gamma_{N+1}^{n+1} = 0, \end{aligned} \tag{14}$$

which leads to $(N+1)\times(N+1)$ system. The finalized system is solved by Gauss elimination method and the desired approximate solutions are attained.

3 Numerical Results

In this section, we propose illustrative examples to see the validity of the present method. The maximum error norm is,

$$L_\infty = \max_{0\le m\le N} |u_m - U_m| \tag{15}$$

where u and U represent the analytical and the numerical solutions, respectively. Since conservation laws must remain constants for the RLW equation during the process of the algorithm, approximate conservation invariants, corresponding to mass, momentum and energy [3], is calculated by rectangular integration formula:

$$I_1 = \int_{x_L}^{x_R} U dx, \tag{16}$$

$$I_2 = \int_{x_L}^{x_R} (U^2 + \mu(U_x)^2) dx, \tag{17}$$

$$I_3 = \int_{x_L}^{x_R} (U^3 + 3U^2) dx. \tag{18}$$

Absolute relative changes $C(I_p^t)$ are computed to see variation of the conservation laws:

$$C(I_p^t) = \left| \frac{I_p^t - I_p^0}{I_p^0} \right|, p = 1, 2, 3 \tag{19}$$

where I_p^0 are initial, I_p^t are the quantities at the time t.

3.1 Simulations of the Single Solitary Wave

The propagation of the solitary wave is resembled to show the efficiency of the QTT B-spline collocation method. The analytical solution of the RLW equation is [11]:

$$U(x,t) = 3c\text{sech}^2(k[x - x_0 - vt]), \tag{20}$$

where $3c$ and $v = 1 + \varepsilon c$ and $k = \frac{1}{2}\sqrt{(\frac{\varepsilon c}{\mu(1+\varepsilon c)})}$ represent solitary wave amplitude and wave velocity, respectively. Domain is considered as $-40 \leq x \leq 60$. The IC is taken from the exact solution and homogenous BCs are employed. We consider the parameters as $\varepsilon = \mu = 1$, $x_0 = 0$ and the amplitude $c = 0.1$ over the problem domain and process the algorithm until the time stage $t = 20$ with space split $h = 0.125$ and time split $\Delta t = 0.1$. The approximated values of the conservation constants are in accordance with the analytical characters of such variants [14]:

$$I_1 = \frac{6c}{k}, \quad I_2 = \frac{12c^2}{k} + \frac{48kc^2\mu}{5}, \quad I_3 = \frac{36c^2}{k}\left(1 + \frac{4c}{5}\right). \tag{21}$$

The solitary wave propagations at time levels $t = 0, 5, 10, 15, 20$ and L_∞ error norm graphs are presented in Fig. 1. The computational results of the model equation is presented in Table 2. It is concluded that the present method serves better results when compared to the finite element methods based on linear and cubic B-spline and gives similar accuracy when compared to the quintic and quartic B-spline oriented techniques. Absolute relative changes are documented in Table 3.

Conservation invariants and maximum error norms for $c = 0.03$ are presented in Table 4. Note that, current method gives similar solutions with [7] whereas the method serves better results when compared to the remaining studies presented in Table 4. Table 5 shows the absolute relative changes using $c = 0.03$ for the single solitary wave.

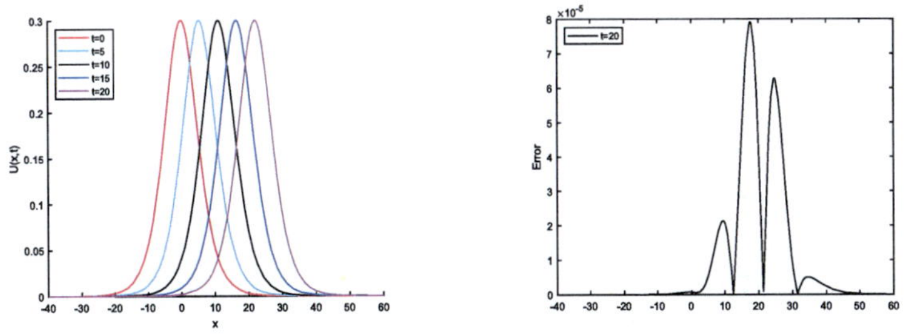

(a) (b)

Fig. 1. Single solitary wave (a) Numerical simulations (b) Maximum error norm

Table 2. Conservation invariants and maximum error norms for $c = 0.1$

	I_1^{20}	I_2^{20}	I_3^{20}	$L_\infty \times 10^5$
Present	3.9799271	0.81046247	2.5790074	7.908
[4]	3.98206	0.811164	2.58133	19.8
[13] M1	3.9798832	0.8104625	2.5790074	8.752
[13] M2	3.9798832	0.8104625	2.5790074	1.268
[14]	3.97989	0.80925	2.57501	25.398
[16]	3.9798879	0.8104622	2.5790063	7.344
Exact	3.9799497	0.8104625	2.5790074	

Table 3. Absolute relative changes for $c = 0.1$

t	$C(I_1^t)$	$C(I_2^t)$	$C(I_3^t)$
5	4.52374×10^{-6}	1.13058×10^{-9}	8.23993×10^{-10}
10	5.06426×10^{-6}	3.03338×10^{-9}	3.32620×10^{-9}
15	3.54906×10^{-6}	5.77964×10^{-9}	6.95219×10^{-9}
20	5.16445×10^{-6}	9.34539×10^{-9}	1.15241×10^{-8}

Table 4. Conservation invariants and maximum error norms for $c = 0.03$

	I_1^{20}	I_2^{20}	I_3^{20}	$L_\infty \times 10^5$
Present	2.1070702	0.1273017	0.3888046	0.41
[4]	2.10906	0.127305	0.388815	19.8
[5]	2.10460	0.127302	0.388803	43.2
[6]	2.10908	0.127318	0.388854	31.6
[7]	2.10703	0.127301	0.388804	0.41
[8]	2.10458	0.127293	0.388877	43.2
[9]	2.10902	0.12730	0.38880	43.9
[10]	2.1047075	0.1273006	0.3888025	19.8
[11]	2.10467	0.12730	0.38880	19.9
[12]	2.10467	0.127301	0.38880	43.1
[25]	2.10769	0.127260	0.388677	23.9
[28]	2.12886	0.127228	0.388571	155.0
Exact	2.104589	0.12730171	0.3888059	

3.2 Interaction of Two Solitary Wave

This subsection deals with the numerical experiments, which model the interaction of the two solitary waves. The IC is considered as [11]:

$$u(x,0) = 3c_1 sech^2(k_1(x - x_1)) + 3c_2 sech^2(k_2(x - x_2)) \tag{22}$$

Table 5. Absolute relative changes for $c = 0.03$

t	$C(I_1^t)$	$C(I_2^t)$	$C(I_3^t)$
5	5.75536×10^{-4}	5.17883×10^{-7}	2.75773×10^{-6}
10	7.12045×10^{-4}	1.53757×10^{-7}	3.12642×10^{-6}
15	5.23293×10^{-4}	7.26893×10^{-7}	2.59063×10^{-6}
20	1.43617×10^{-4}	4.45137×10^{-6}	9.60086×10^{-7}

$$c_i = \frac{4k_i^2}{1 - 4k_i^2}, \quad i = 1, 2.$$

which represent the motion of the two solitary waves moving towards one another, with the initial and terminal positions as x_1 and x_2 respectively.

Spatial domain of the problem is considered as $[0, 120]$ with $N = 400$ space partition and $\Delta t = 0.1$ time splits. Experiments are conducted for the parameters $k_1 = 0.4, k_2 = 0.3, x_1 = 15$ and $x_2 = 35$. Initial physical position depiction of the waves of heights 5.3334 and 1.686 positioned at $x = 15$ and $x = 35$ is presented in Fig. 2(a). The motion of the waves start to evolve at $t = 10$. Figure 2(b) shows that the position of the higher wave is positioned at $x = 42.6$ with the height of 5.1795 when the interaction begins. When $t = 15$, the two waves collide with each other and seem to be a newly single wave as shown in Fig. 2(c). Plot of Fig. 2(d) indicates that the two waves then begin to separate when $t = 20$ at $x = 62.1$ and $x = 72.9$, respectively. The height of the higher wave reaches 5.2905 as the height of the lower one 1.6853. Figure 2(e) indicates the continuation of the separation of the two waves at $t = 25$. The top wave on the right is positioned at $x = 87$ with the height of 5.3058 whereas the lower left is placed at $x = 69.9$ with the height of 1.6748. In the end of this physical process both waves fall apart toward the opposite directions and reappear in the initial forms in Fig. 2(f). When $t = 30$, the final height of the waves are figured out as 1.6764 and 5.3054 and they are located $x = 77.7$ and $x = 100.5$, respectively.

Conservation invariants, which are found by the current method are in good agreement with the results obtained analytically, namely given by [14]:

$$\begin{aligned} I_1 &= \frac{6c_1}{k_1} + \frac{6c_2}{k_2}, \\ I_2 &= \frac{12c_1^2}{k_1} + \frac{48k_1c_1^2\mu}{5} + \frac{12c_2^2}{k_2} + \frac{48k_2c_2^2\mu}{5}, \\ I_3 &= \frac{36c_1^2}{k_1}\left(1 + \frac{4c_1}{5}\right) + \frac{36c_2^2}{k_2}\left(1 + \frac{4c_2}{5}\right). \end{aligned}$$

Table 6 and Table 7 show the conservation invariants and the absolute relative changes of the interaction of two solitary waves, until the time stage $t = 30$, respectively.

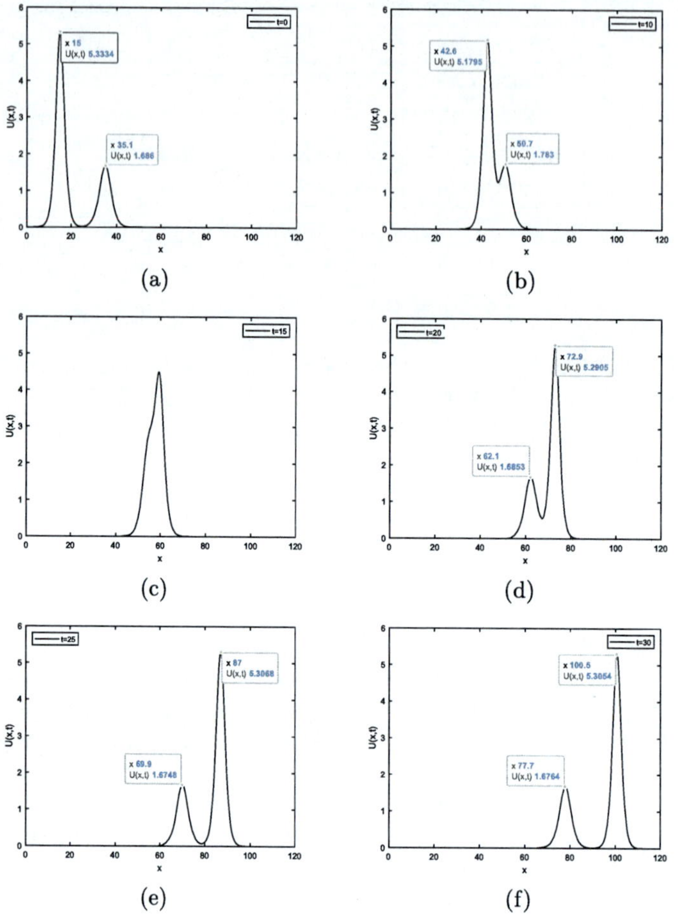

Fig. 2. Interaction of two solitary wave (a) $t = 0$ (b) $t = 10$ (c) $t = 15$ (d) $t = 20$ (e) $t = 25$ (f) $t = 30$

Table 6. Conservation invariants for interaction of two solitary waves

t	I_1^t	I_2^t	I_3^t
0	37.916502	120.521923	744.081208
5	37.916611	120.519328	744.055199
10	37.916602	120.462182	743.448341
15	37.916599	120.316487	741.873273
20	37.916592	120.506995	743.927527
25	37.916588	120.520662	744.069709
30	37.916574	120.521132	744.074296

Table 7. Absolute relative changes for interaction of two solitary waves

t	$C(I_1^t)$	$C(I_2^t)$	$C(I_3^t)$
5	2.8971×10^{-6}	2.1531×10^{-5}	3.4955×10^{-5}
10	2.6559×10^{-6}	4.9568×10^{-4}	8.5053×10^{-4}
15	2.5734×10^{-6}	1.7045×10^{-3}	2.9673×10^{-3}
20	2.3853×10^{-6}	1.2386×10^{-4}	2.0653×10^{-4}
25	2.2701×10^{-6}	1.0464×10^{-5}	1.5454×10^{-5}
30	1.8985×10^{-6}	6.5592×10^{-6}	9.2902×10^{-6}

3.3 Wave Undulation

Consider the following IC and BCs:

$$U(x,0) = 0.5U_0 \left(1 - \tanh\left(\frac{x - x_c}{d}\right)\right), \tag{23}$$

$$U(x_L,t) = \sigma_1 = U_0,\ U(x_R,t) = \sigma_2 = 0,\ U_x(x_L,t) = 0 \tag{24}$$

which produce the development of the undular bore solution for the RLW equation [11]. Here $U(x,0)$ denotes the elevation of the water above the equilibrium surface at time $t = 0$ and d represents the slope between the still water and deeper water. Still water lies to the right of this stream. The parameters $\varepsilon = 1.5$, $\mu = 1/6$, $U_0 = 0.1, x_c = 0$ are used with space step $h = 0.24$ and time step $\Delta t = 0.1$. The experiment was conducted over the space interval $-36 \leq x \leq 300$ up time $t = 250$ for $d = 2$ and $d = 5$, in Table 8 and Table 9, respectively. Amplitudes and spatial positions of peaks of the first four undulations are illustrated in Table 10 for the fixed time level $t = 250$ both $d = 2$ and $d = 5$. Physical nature depictions of the wave undulations are given by Figs. 3 and 4, for $d = 2$ and $d = 5$, respectively. Undulations are growing in time with respect to the parameter d and show similar behavior as in [11].

Table 8. Conservation invariants for wave undulation for $d = 2$

t	I_1^t	I_2^t	I_3^t	I_1^t [10]	I_2^t [10]	I_3^t [10]
0	3.61200	0.35148	1.08822	3.61200	0.35148	1.08822
50	8.98700	0.90148	2.79329	8.98700	0.90148	2.79326
100	14.36200	1.45150	4.49814	14.36200	1.45148	4.49806
150	19.73700	2.00152	6.20289	19.73700	2.00148	6.20275
200	25.11200	2.55154	7.90760	25.11200	2.55148	7.90739
250	30.48700	3.10156	9.61229	30.48700	3.10148	9.61203

Table 9. Conservation invariants for wave undulation for $d = 5$

t	I_1^t	I_2^t	I_3^t	I_1^t [10]	I_2^t [10]	I_3^t [10]
0	3.61200	0.33631	1.04097	3.61200	0.33631	1.04097
50	8.98700	0.88631	2.74646	8.98700	0.88631	2.74646
100	14.36200	1.43632	4.45170	14.36200	1.43631	4.45167
150	19.73700	1.98633	6.15664	19.73700	1.98631	6.15656
200	25.11200	2.53635	7.86142	25.11200	2.53631	7.86128
250	30.48700	3.08637	9.56614	30.48700	3.08631	9.56594

Table 10. Undulations at $t = 250$

	$d = 2$		$d = 5$	
	Position	Amplitude	Position	Amplitude
1st undulation	265.92	0.1820	264.96	0.1779
2nd undulation	254.16	0.1621	253.92	0.1534
3rd undulation	244.08	0.1446	244.08	0.1324
4th undulation	234.72	0.1305	234.96	0.1178

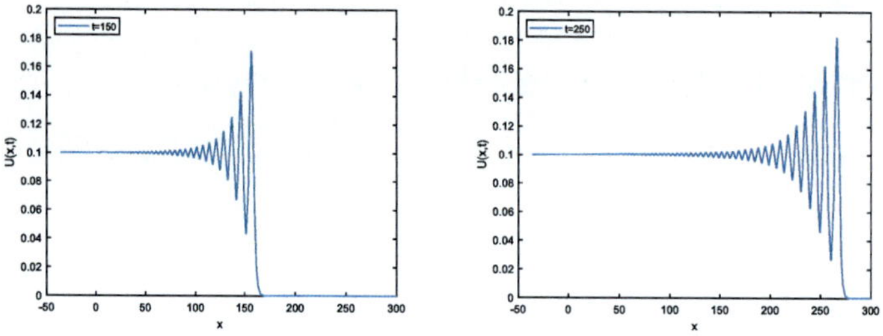

(a) (b)

Fig. 3. Wave undulation for $d = 2$ (a) $t = 150$ (b) $t = 250$

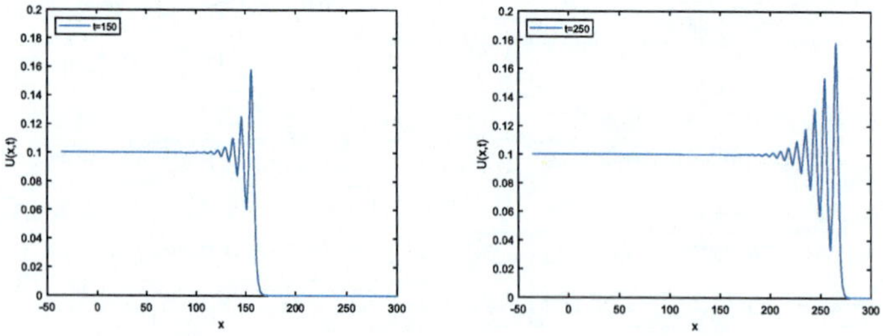

(a) (b)

Fig. 4. Wave undulation for $d = 5$ (a) $t = 150$ (b) $t = 250$

3.4 Wave Generation

The last problem is the wave generation produced by the left BC having the following the form [10,11]:

$$U(x_L,t) = \sigma_1 = \begin{cases} U_0\dfrac{t}{\tau}, & 0 \le t \le \tau \\ U_0, & \tau \le t \le t_0 - \tau \\ U_0\dfrac{t_0 - t}{\tau}, & t_0 - \tau \le t \le t_0 \\ 0 & \text{otherwise} \end{cases} \tag{25}$$

and $U(x_R,t) = \sigma_2 = 0$, $U_x(x_L,t) = 0$. The left BC on the RLW equation produces the solitary waves.

In this problem, the parameters $U_0 = 2$, $h = 0.4$, $\Delta t = 0.1$, $\tau = 0.3$, $t_0 = 20$, $\varepsilon = \mu = 1$ over the space interval $0 \le x \le 260$ are chosen. Table 11 presents the wave amplitude under the forcing period $0 \le t_0 \le 20$ on fixed time levels. Results are also compared to the works accomplished by Galerkin schemes [10]

Table 11. Amplitudes of the wave generations for $0 \le t_0 \le 20$.

t	5	4	3	2	1	I_1^t	I_2^t	I_3^t
2.5					2.30	9.767	18.100	84.573
5					2.80	20.173	43.051	215.211
7.5				2.25	3.08	30.142	66.280	334.892
10			2.02	2.53	3.25	40.018	89.139	452.204
15		2.06	2.44	2.91	3.44	60.051	135.977	693.448
20	1.94	2.38	2.74	3.14	3.54	77.538	178.308	895.464
40	1.15	2.39	3.12	3.46	3.70	79.410	177.614	901.334
60	1.15	2.36	3.09	3.53	3.74	79.410	177.625	901.439
80	1.15	2.37	3.08	3.52	3.72	79.406	177.627	901.457
100	1.15	2.36	3.07	3.51	3.76	79.409	177.627	901.461
100 [10]	1.16	2.37	3.08	3.51	3.76	79.501	177.590	900.986
100 [11]	1.18	2.38	3.08	3.52	3.77	79.567	178.497	904.993
130	1.15	2.37	3.07	3.51	3.76	79.408	177.628	901.462
130 [10]	1.16	2.37	3.08	3.51	3.76	79.499	177.593	901.003
130 [11]	1.18	2.38	3.09	3.53	3.77	79.609	178.904	907.665
150	1.15	2.37	3.08	3.75	3.52	79.409	177.628	901.462
150 [10]	1.16	2.37	3.08	3.52	3.76	79.501	177.595	901.015
150 [11]	1.18	2.38	3.09	3.54	3.78	79.178	178.904	909.466
170	1.15	2.37	3.08	3.51	3.76	79.408	177.628	901.463
170 [10]	1.16	2.37	3.07	3.52	3.76	79.500	177.596	901.026
170 [11]	1.18	2.39	3.09	3.54	3.79	79.669	179.453	911.283

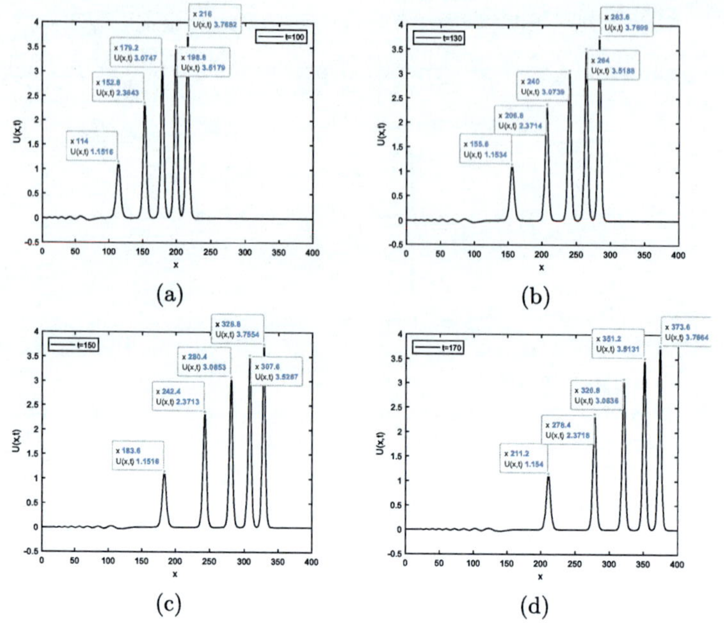

Fig. 5. Wave generation (a) $t = 100$ (b) $t = 130$ (c) $t = 150$ (d) $t = 170$

and [11] at $t = 100$, $t = 130$, $t = 150$, $t = 170$ and proved to provide similar behavior. Numerical experiments of the wave generation depictions are given in Fig. 5 at $t = 100, 130, 150, 170$. Note that, the patterns of the each five waves are conserved in time. The detailed positions and heights of each wave are presented in Fig. 5 as well. Computations show that the present method generates accurate and efficient wave generations.

4 Conclusions

In this study, numerical solutions of the RLW equation are presented with the treatment of a relatively new based B–splines. Solitary wave, interaction of two solitary waves, wave undulation and wave generation processes of the model are simulated and observed to be in accordance with the physical natures of each experiment. The numerical solutions using the collocation scheme prove to provide the efficiency through tables and figures. The computational results are found to be quite reliable which makes the algorithm applicable to the similar models. The current method can be adapted to the system of PDEs, or the fractional order of differential equations.

References

1. Peregrine, D.H.: Calculations of the development of an undular bore. J. Fluid Mech. **25**(2), 321–330 (1966)
2. Benjamin, T.B., Bona, J.L., Mahony, J.J.: Model equations for long waves in non-linear dispersive systems. Philos. Trans. R. Soc. Lond. Ser. A Math. Phys. Sci. **272**(1220), 47–78 (1972)
3. Olver, P.J.: Euler operators and conservation laws of the BBM equation. Math. Proc. Camb. Phil. Soc. **85**(143–160), 4 (1979)
4. Dogan, A.: Numerical solution of RLW equation using linear finite elements within Galerkin's method. Appl. Math. Model. **26**(7), 771–783 (2002)
5. Gardner, L.R.T., Gardner, G.A., Dag, I.: A B-spline finite element method for the RLW equation. Commun. Numer. Methods Eng. **11**, 59–98 (1995)
6. Dogan, A.: Numerical solution of regularized long wave equation using Petrov-Galerkin method. Commun. Numer. Methods Eng. **17**(7), 485–494 (2001)
7. Mohammadi, R.: Exponential B-spline collocation method for numerical solution of the generalized regularized long wave equation. Chin. Phys. B **24**(5), 050206 (2015)
8. Dag, I., Saka, B., Irk, D.: Application of cubic B-splines for numerical solution of the RLW equation. Appl. Math. Comput. **159**(2), 373–389 (2004)
9. Mei, L., Cehn, Y.: Numerical solutions of RLW equation using Galerkin method with extrapolation techniques. Comput. Phys. Commun. **183**(8), 1609–1616 (2012)
10. Saka, B., Dag, I.: A numerical solution of the RLW equation by Galerkin method using quartic B-splines. Commun. Numer. Methods Eng. **24**, 1339–1361 (2008)
11. Saka, B., Dag, I., Dogan, A.: Galerkin method for the numerical solution of the RLW equation using quadratic B-splines. Int. J. Comput. Math. **81**(6), 727–739 (2004)
12. Gorgulu, M.Z., Dag, I., Irk, D.: Simulations of solitary waves of RLW equation by exponential B-spline Galerkin method. Chin. Phys. B **26**(8), 080202 (2017)
13. Irk, D., Yildiz, P.K., Gorgulu, M.Z.: Quartic trigonometric B-spline algorithm for numerical solution of the regularized long wave equation. Turk. J. Math. **43**(1), 112–125 (2019)
14. Zaki, S.I.: Solitary waves of the splitted RLW equation. Comput. Phys. Commun. **138**(1), 80–91 (2001)
15. Esen, A., Kutluay, S.: Application of a lumped Galerkin method to the regularized long wave equation. Appl. Math. Comput. **174**(2), 833–845 (2006)
16. Dag, I., Saka, B., Irk, D.: Galerkin method for the numerical solution of the RLW equation using quintic B-splines. J. Comput. Appl. Math. **190**(1–2), 532–547 (2006)
17. Lu, C., Huang, W., Qiu, J.: An adaptive moving mesh finite element solution of the regularized long wave equation. J. Sci. Comput. **74**(1), 122–144 (2018)
18. Jain, P.C., Shankar, R., Singh, T.V.: Numerical solution of regularized long-wave equation. Commun. Numer. Methods Eng. **9**(7), 579–586 (1993)
19. Bhardwaj, D., Shankar, R.: A computational method for regularized long wave equation. Comput. Math. Appl. **40**(12), 1397–1404 (2000)
20. Kutluay, S., Esen, A.: A finite difference solution of the regularized long-wave equation. Math. Probl. Eng. **2006**, 1–14 (2006)
21. Inan, B., Bahadır, A.R.: A fully implicit finite difference scheme for the regularized long wave equation. Gen. Math. Notes **33**(2), 40 (2016)

22. Oruc, O., Bulut, F., Esen, A.: Numerical solutions of regularized long wave equation by Haar wavelet method. Mediterr. J. Math. **13**(5), 3235–3253 (2016)
23. Dağ, İ., Korkmaz, A., Saka, B.: Cosine expansion-based differential quadrature algorithm for numerical solution of the RLW equation. Numer. Methods Partial Diff. Equat. Int. J. **26**(3), 544–560 (2010)
24. Korkmaz, A., Dağ, İ.: Numerical simulations of boundary-forced RLW equation with cubic B-spline-based differential quadrature methods. Arab. J. Sci. Eng. **38**, 1151–1160 (2013)
25. Gardner, L.R.T., Gardner, G.A., Dogan, A.: A least-squares finite element scheme for the RLW equation. Commun. Numer. Methods Eng. **12**(11), 795–804 (1996)
26. Dağ, İ.: Least-squares quadratic B-spline finite element method for the regularised long wave equation. Comput. Methods Appl. Mech. Eng. **182**(1–2), 205–215 (2000)
27. Gu, H., Chen, N.: Least-squares mixed finite element methods for the RLW equations. Numer. Methods Partial Diff. Equat. Int. J. **24**(3), 749–758 (2008)
28. Dag, I., Ozer, N.: Approximation of the RLW equation by the least square cubic B spline finite element method. Appl. Math. Model. **25**(3), 221–231 (2001)
29. Guo, B.Y., Cao, W.M.: The Fourier pseudospectral method with a restrain operator for the RLW equation. J. Comput. Phys. **74**(1), 110–126 (1988)
30. Djidjeli, K., Price, W.G., Twizell, E.H., Cao, Q.: A linearized implicit pseudo-spectral method for some model equations: the regularized long wave equations. Commun. Numer. Methods Eng. **19**(11), 847–863 (2003)
31. Nuruddeen, R.I., Aboodh, K.S., Ali, K.K.: Investigating the tangent dispersive solitary wave solutions to the equal width and regularized long wave equations. J. King Saud Univ.-Sci. **32**(1), 677–681 (2020)
32. Alinia, N., Zarebnia, M.: A numerical algorithm based on a new kind of tension B-spline function for solving Burgers-Huxley equation. Numer. Algorithms **82**, 1121–1142 (2019)
33. Hong, Q., Wang, Y., Gong, Y.: Optimal error estimate of two linear and momentum-preserving Fourier pseudo-spectral schemes for the RLW equation. Numer. Methods Partial Diff. Equat. **36**(2), 394–417 (2020)
34. Maharana, N., Nayak, A.K., Jena, P.: A comparative study of regularized long wave equations (RLW) using collocation method with cubic B-spline. In: Decision Science in Action, pp. 203–216. Springer, Singapore (2019)
35. Mittal, R.C., Rohila, R.: A fourth order cubic B-spline collocation method for the numerical study of the RLW and MRLW equations. Wave Motion **80**, 47–68 (2018)
36. Wang, G., Fang, M.: Unified and extended form of three types of splines. J. Comput. Appl. Math. **216**(2), 498–508 (2008)

Numerical Solution of Fractional Cauchy Reaction-Diffusion Equation by Differential Quadrature

Gülsemay Yiğit(✉) and Soheil Salahshour

Department of Mathematics, Faculty of Engineering and Natural Sciences, Bahcesehir University, Istanbul, Turkey
gulsemay.yigit@eng.bau.edu.tr

Abstract. This study explores the numerical solution of fractional Cauchy reaction-diffusion equation by utilizing a discrete derivative technique. The present algorithm uses polynomial based differential quadrature for space discretization, which results in system of time-dependent differential equations. These systems are integrated by the Runge-Kutta method. Numerical stability is also established based on eigenvalues of space-discretized matrix. Illustrative examples are examined to support the idea. Numerical solutions accomplished by the present scheme are compared to the analytical or well-known existing methods.

Keywords: Cauchy reaction-diffusion equation · The differential quadrature · Conformable fractional derivative

1 Introduction

Numerical discretization techniques play an essential role to solve fractional partial differential equations, which model various real life problems, due to their fast and powerful results. Fractional order partial differential equations arise in many interdisciplinary fields such as biophysics, biochemistry, financial processes, [1–3] and so on. There are several descriptions that take place in the concept of fractional derivative [1–5]. Among these, conformable fractional derivative is taken into consideration in this work.

Differential quadrature (DQ) is a discrete derivative method which provides numerical solutions to differential equations with relative ease to code the employment for the necessary algorithms [6]. The differential quadrature method (DQM) was proposed by Bellman [7,8] and his associates to suggest an alternative form for the previous numerical methods such as finite difference and finite element methods. The scheme has been extensively implemented to various equations in engineering and physical disciplines [9]. The principle of the discretization scheme is to build up the formulations to construct weighting coefficients. Bellman et al [7,8] introduced two procedures to the weighting coefficients for the derivative description. First approach of Bellman does not give

T. Allahviranloo et al. (Eds.): IDS 2020, AISC 1301, pp. 725–737, 2021.
https://doi.org/10.1007/978-3-030-66501-2_59

efficient solutions, as the number of grids is increasing. This has been tackled by Quan and Chang [10,11] who proposed the application of Lagrange polynomials to construct new weighting coefficients. Shu [12] generalized the previous structures of these approaches to determine weighting coefficients using polynomial approximation theory in association with the features of linear vector space and accomplished that weighting coefficients can be figured out by appropriate preference of the base functions. This opportunity attracts many researchers to develop various formulations for establishing weighting coefficients. A few of these, incorporates with radial basis functions [13] B-splines [14–18], sinc functions [19] and so forth.

DQ method is a widely studied topic in material science [20–22], as well as physical and biological processes [23]. In [24,25] nonlinear reaction-diffusion equation in one and two dimensions are investigated through polynomial based differential quadrature with different time integration techniques. Tomasiello [26,27] explored the stability analysis of the iterative differential quadrature method.

In this work, we aim to establish an algorithm for the discretization of reaction-diffusion equation given by,

$$\frac{\partial^\alpha U(x,t)}{\partial t^\alpha} = D\frac{\partial^2 U(x,t)}{\partial x^2} + p(x,t)U(x,t) \tag{1}$$

where U is the concentration, p is the reaction parameter and $D > 0$ is the diffusion coefficient with subject to proper initial and boundary conditions,

$$U(x,0) = F(x) \tag{2}$$

$$U(0,t) = f_0(t), \quad U_x(0,t) = f_1(t) \tag{3}$$

If $\alpha = 1$, the model gets reduced to the classical Cauchy reaction-diffusion equation [28]. Numerous studies have been accomplished analytically or numerically to the fractional order form of this equation under various motions [29–31]. For further studies, we refer the reader to [32–35] for the theory and application of the fractional systems.

The object of the current work is to obtain numerical solutions by adopting polynomial based differential quadrature to the model equation. First order weighting coefficients are determined to evaluate higher order coefficients, which are used in procedures of solving the model problem. Time integration of the fractional equation is performed by Runge-Kutta method of fourth order. Matrix stability analysis is implemented related to the eigenvalues of space-discretized matrix for different space splits as well as fractional parameter.

This study is organized as follows. In Sect. 2, conformable fractional derivative is explained. In Sect. 3, the formulations of weighting coefficients matrix is proposed which will be used in the description of the method. In Sect. 4, the method is applied to model equation for space and time discretization with some stability results. Section 5 presents the numerical experiments for various reaction parameters. Concluding remarks and further extension of the present study is summarized in Sect. 6.

2 Preliminaries

In this section, we revisited conformable fractional derivative proposed by Khalil et al [4].

Definition: Suppose $f : [0, \infty) \to \Re$ be a function. The α-th order conformable fractional derivative of f is defined as

$$(T_\alpha f)(t) = \lim_{\epsilon \to 0} \frac{f(t + \epsilon t^{1-\alpha}) - f(t)}{\epsilon} \tag{4}$$

Theorem: Let $\alpha \in (0, 1]$ and f, g be α-differentiable for $t > 0$, then,

(a) $T_\alpha(af + bg) = aT_\alpha(f) + bT_\alpha(g)$ for all $a, b \in \Re$,
(b) $T_\alpha(t^\beta) = \beta t^{\beta-\alpha}$, for all $\beta \in \Re$
(c) $T_\alpha(fg) = fT_\alpha(g) + gT_\alpha(f)$,
(d) $T_\alpha(\frac{f}{g}) = \frac{gT_\alpha(f) - fT_\alpha(g)}{g^2}$,
(e) $T_\alpha(\lambda) = 0$, for all constant function as $f(t) = \lambda$,
(f) If f is differentiable, $T_\alpha(f)(t) = t^{1-\alpha}\frac{df}{dt}(t)$.

3 Description of the Differential Quadrature Method (DQM)

Differential quadrature method (DQM) uses higher order polynomial approximation to produce numerical solutions. The approximation is necessary to compute the weighting coefficients matrix, which is essential to construct numerical scheme. Let $f(x)$ be a continuous and differentiable function of x defined in $[a, b] \subset \Re$. Numerical derivative of f at each grid point x_i is expressed as the weighted sum of function values on the problem domain. Differential quadrature description of first derivative at each knots is given as [12],

$$f_x(x_i) = \sum_{j=1}^{N} w_{ij} f(x_j) \quad i = 1, 2, \cdots, N. \tag{5}$$

Here, w_{ij} is weighting coefficients of first order, N is the number of space mesh on the entire domain. The n-th order derivative formulation is defined in a similar approach as,

$$f_x^{(n)}(x_i) = \sum_{j=1}^{N} w_{ij}^{(n)} f(x_j) \quad i = 1, 2, \cdots, N, \tag{6}$$

where $w_{ij}^{(n)}$ is n-th order weighting coefficients. The main aspect of the strategy is to figure out weighting coefficients $w_{ij}^{(n)}$. The coefficients matrix for first order derivative is evaluated, then higher order derivatives are accomplished by recurrent formulas [12]. The name of the method is mentioned as polynomial based differential quadrature (PDQ) when the weighting coefficients are computed by choosing test functions as polynomials. In this work we consider Lagrange interpolation polynomials as base polynomial, which is given by,

$$e_k(x) = \frac{M(x)}{(x - x_k)M^{(1)}(x_k)}, \qquad k = 1, 2, \cdots, N \tag{7}$$

where

$$M(x) = (x - x_1)(x - x_2)\cdots(x - x_N), \quad M^{(1)}(x) = \prod_{k=1, k\neq j}^{N} (x_j - x_k). \tag{8}$$

Using polynomial approach theory, first order weighting coefficients are determined as,

$$a_{ij}^{(1)}(x) = \frac{M^{(1)}(x_i)}{(x_i - x_j)M^{(1)}(x_j)}, \quad j \neq i. \tag{9}$$

$$a_{ii}^{(1)} = - \sum_{j=1, j\neq i}^{N} a_{ij}^{(1)}, i = 1, 2, \cdots, N. \tag{10}$$

In this work, higher order derivatives in DQ formulations are attained based on matrix multiplication method.

4 Application of PDQ to the Fractional Cauchy Reaction-Diffusion Equation and Stability Analysis

Fractional Cauchy reaction-diffusion equation can be expresses as,

$$U_t^{\alpha}(x,t) = DU_{xx}(x,t) + p(x,t)U(x,t), \quad 0 < \alpha \leq 1. \tag{11}$$

Applying knots $x_i, i = 1, 2, ..., N$ to Eq. (11) gives,

$$U_t^{\alpha}(x_i,t) = DU_{xx}(x_i,t) + p(x_i,t)U(x_i,t) \tag{12}$$

Then, spatial derivatives are replaced by DQ formulations as,

$$\frac{\partial^{\alpha} U(x_i,t)}{\partial t^{\alpha}} = D\sum_{j=1}^{N} w_{ij}^{(2)} U(x_j,t) + p(x_i,t)U(x_i,t) \tag{13}$$

Now, using the properties of the conformable fractional derivative, time derivative is rewritten as,

$$\frac{\partial^{\alpha} U(x_i,t)}{\partial t^{\alpha}} = t^{1-\alpha}\frac{\partial U(x_i,t)}{\partial t} \tag{14}$$

Then, substituting the expression (14) into Eq. (13) gives the following,

$$\frac{\partial U(x_i,t)}{\partial t} = t^{\alpha-1} D \sum_{j=1}^{N} w_{ij}^{(2)} U(x_j,t) + t^{\alpha-1} p(x_i,t) U(x_i,t) \tag{15}$$

Equation (15) is expressed in the following system of ODEs,

$$\{\frac{dU}{dt}\} = [A]\{U\} + \{B\} \tag{16}$$

where $\{U\}$ is the unknown vector, {B} represent the vector containing boundary conditions, and $[A]$ is the weighting coefficient matrix. The resultant system is integrated by Runge-Kutta fourth order method.

The DQ method serves the flexibility of the selection of equally or unequally spaced knots but, it is pointed out that solutions are more stable when using Chebyshev-Gauss Lobatto (unequally spaced) points, given as [12],

$$x_i = \cos(\theta_i), \quad \theta_i = \frac{i\pi}{N}, \quad i = 0, 1, \cdots, N \tag{17}$$

which is given for the interval $[1, -1]$ and for a general problem domain [a, b] the transformation is used to get x_i as,

$$x_i = \frac{b-a}{2}(1 - \xi_i) + a. \tag{18}$$

A time dependent problem with initial and boundary conditions is asserted as,

$$\frac{\partial U}{\partial t} = l(U), \quad U(t_0) = 0, \quad t \in [t_0, T], \tag{19}$$

where l represents spatial differential operator. After PDQ discretization, Eq. (19) is transformed into ordinary differential equation. Stability of the constructed scheme for the numerical integration depends on the stability of the differential equation [12]. Let λ_i, $i = 1, 2, ..., N$ be the eigenvalues of the matrix A. Stability conditions are determined based on real and imaginary parts of the eigenvalue distribution. The Runge-Kutta methods are absolutely stable if, all the real parts of the eigenvalues are negative [16].

The eigenvalue distribution for different number of grid points for $\alpha = 0.9$ are presented in Fig. 1 for the Example 1. The effect of the fractional parameter to the stability of the method is figured out in Table 1 as showing the absolute maximum eigenvalues. As the fraction parameter α, away from 1 we can accomplish larger eigenvalues. This allows us to use larger time steps when the fraction parameter is 1, which corresponds to the integer case of the problem. Note that, as shown in Fig. 1 all the eigenvalues are negative and real-valued, which generate stable solutions. Number of space partition also affects the stability of the problem. This can be tackled by using smaller time steps as the number of spatial partitions increase.

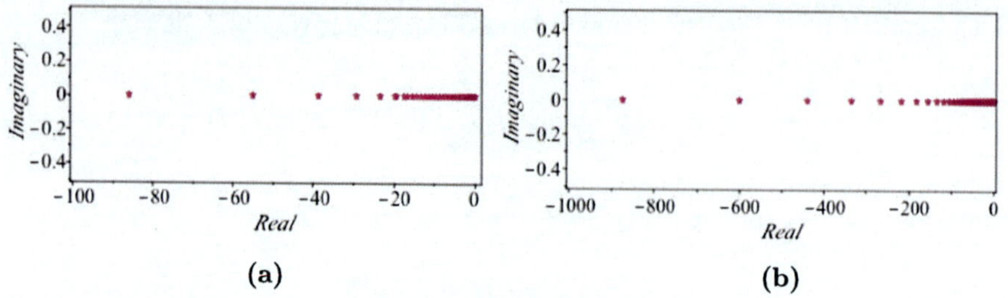

Fig. 1. Eigenvalue distribution when **(a)** $N = 50$ and **(b)** $N = 100$

Table 1. Maximum absolute eigenvalues for different fractional parameter

N	α	$\lvert \Delta t Re(\lambda_{max}) \rvert$
50	1.0	1093.4696
100	1.0	18204.460
50	0.9	2181.758
100	0.9	363222.673
50	0.8	4353.1809
100	0.8	72473.261

5 Numerical Illustrations

In this section, in order to see the efficiency of the method, we present three numerical experiments. Solutions assembled by DQM are provided in tables and figures for various α as well as different temporal stages. Concentration levels are discussed in all three cases.

Example 1. Consider the following reaction-diffusion equation with $D = 1$ and $p = -1$ [29], namely,

$$\frac{\partial^\alpha U(x,t)}{\partial t^\alpha} = \frac{\partial^2 U(x,t)}{\partial x^2} - U(x,t), \quad 0 < \alpha \leq 1, \tag{20}$$

with following initial and boundary conditions

$$U(x,0) = e^{-x} + x, \tag{21}$$

$$U(0,t) = 1, \quad U_x(0,t) = e^{-t} - 1 \tag{22}$$

In Table 2 we compared results with different methods, as the fractional parameter is sufficiently close to $\alpha \to 1$. Figure 2 describes the concentration levels for $\alpha = 0.7$, $\alpha = 0.8$, $\alpha = 0.9$, and the typical motion $\alpha = 1.0$. Numerical experiment shows that solutions decrease as fractional parameter decrease for the fixed time level $t = 1$. Graph of numerical solution is given by Fig. 3 using parameters $\Delta t = 0.001$, $N = 50$, on $0 \leq t \leq 1$, for $\alpha = 0.9$ and observed to be in accordance with the exact solution of classical Cauchy reaction-diffusion equation.

Table 2. Numerical solutions of Example 1 at $x = 1$ for different methods

t	Exact ($\alpha = 0.9$) [30]	OHAM ($\alpha = 0.9$) [30]	DQM ($\alpha = 0.9$)	Exact ($\alpha = 1$) [30]	OHAM ($\alpha = 1$) [30]	DQM ($\alpha = 1$)
0.01	1.351549394	1.35154933	1.35043616	1.357929275	1.357929275	1.357929275
0.02	1.337644269	1.33764413	1.33556352	1.348078114	1.348078112	1.348078114
0.03	1.324650319	1.324650025	1.32166032	1.338324975	1.338324957	1.338324975
0.04	1.312273662	1.312273027	1.30841187	1.328668880	1.328668812	1.328668880
0.05	1.300376341	1.300375051	1.29567132	1.319108866	1.319108682	1.319108866
0.06	1.288877082	1.288874669	1.28335228	1.309643975	1.309643571	1.309643975
0.07	1.277721838	1.277717661	1.27139735	1.300273261	1.300272481	1.300273261
0.08	1.266871806	1.266865032	1.25976540	1.290995788	1.290994417	1.290995788
0.09	1.256297601	1.256287187	1.24842531	1.281810626	1.281808382	1.281810626
0.1	1.245976083	1.245960761	1.23735259	1.272716859	1.27271338	1.272716859

Table 3. Numerical solutions of Example 2 at $x = 1$ for different methods

t	Exact ($\alpha = 1.0$)	DQM ($\alpha = 1.0$)	DQM ($\alpha = 0.9$)	DQM ($\alpha = 0.8$)
0.01	2.74587558	2.74587558	2.76589207	2.80107255
0.02	2.77430426	2.77430426	2.80952780	2.86853069
0.03	2.80358792	2.80358792	2.85245634	2.93205225
0.04	2.83374738	2.83374738	2.89540679	2.99390462
0.05	2.86480418	2.86480418	2.93872240	3.05508921
0.06	2.89678065	2.89678065	2.98260856	3.11617085
0.07	2.92969999	2.92969999	3.02720640	3.17751491
0.08	2.96358593	2.96358593	3.07262242	3.23938047
0.09	2.99846352	2.99846352	3.11894273	3.30196413
0.1	3.03435839	3.03435839	3.16624061	3.36542316

Example 2. Consider the following reaction-diffusion equation with $D = 1$ and $p = p(t)$ [29],

$$\frac{\partial^\alpha U(x,t)}{\partial t^\alpha} = \frac{\partial^2 U(x,t)}{\partial x^2} + 2tU(x,t), \quad 0 < \alpha \leq 1 \tag{23}$$

with the following initial and boundary conditions,

$$U(x,0) = e^x, \tag{24}$$

$$U(0,t) = e^{t+t^2}, \quad U_x(0,t) = e^{t+t^2} \tag{25}$$

Table 3 explains the quadrature results for $\alpha = 0.8$, $\alpha = 0.9$ and $\alpha = 1.0$ with the exact solution. Concentration levels depictions for $\alpha = 0.7$, $\alpha = 0.8$, $\alpha = 0.9$, and the typical motion $\alpha = 1.0$ are given in Fig. 4. Numerical experiment shows that solutions rapidly increase as fractional parameter decrease for the fixed time level $t = 0.1$. Figure 5 gives graph of numerical solution using parameters $\Delta t = 0.001$, $N = 50$, on $0 \leq t \leq 0.1$, for $\alpha = 0.9$ with the graph of exact solution of standard equation consisting of integer order time derivative.

Table 4. Numerical solutions of Example 3 at $x = 1$ for different methods

t	Exact ($\alpha = 0.9$) [30]	OHAM ($\alpha = 0.9$) [30]	DQM ($\alpha = 0.9$)	Exact ($\alpha = 1$) [30]	OHAM ($\alpha = 1$) [30]	DQM ($\alpha = 1$)
0.01	2.763486	2.763486	2.765434	2.745601	2.745601	2.745601
0.02	2.803306	2.803307	2.807787	2.773195	2.773195	2.773195
0.03	2.841677	2.841677	2.848634	2.801066	2.801066	2.801066
0.04	2.879315	2.879314	2.888701	2.829217	2.829217	2.829217
0.05	2.916542	2.916539	2.911225	2.857651	2.857650	2.857651
0.06	2.953541	2.953534	2.950313	2.886371	2.886370	2.886371
0.07	2.99043	2.990416	2.989114	2.915379	2.915377	2.915379
0.08	3.027289	3.027267	3.027431	2.944679	2.944675	2.944679
0.09	3.064179	3.064145	3.064741	2.974274	2.974267	2.974274
0.1	3.101147	3.101096	3.100081	3.004166	2.718282	3.004166

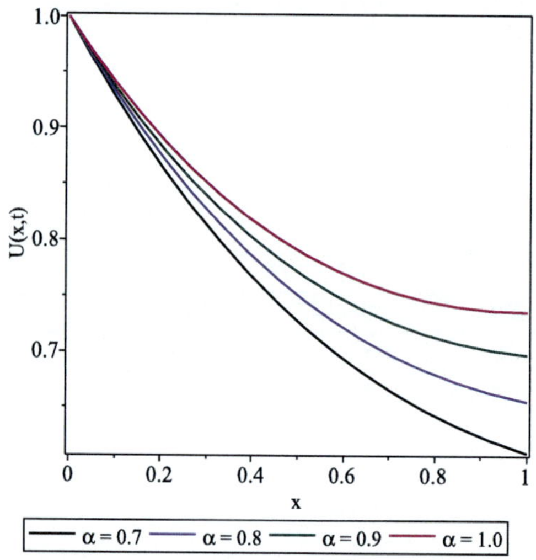

Fig. 2. Numerical solutions of Example 1 at $t = 1$

Example 3. Consider the following reaction-diffusion equation with $D = 1$ and $p = p(x)$ [29],

$$\frac{\partial^\alpha U(x,t)}{\partial t^\alpha} = \frac{\partial^2 U(x,t)}{\partial x^2} - (1 + 4x^2)U(x,t), \quad 0 < \alpha \leq 1, \tag{26}$$

with the following initial and boundary conditions,

$$U(x,0) = e^{x^2}, \tag{27}$$

$$U(0,t) = e^t, \quad U_x(0,t) = 0 \tag{28}$$

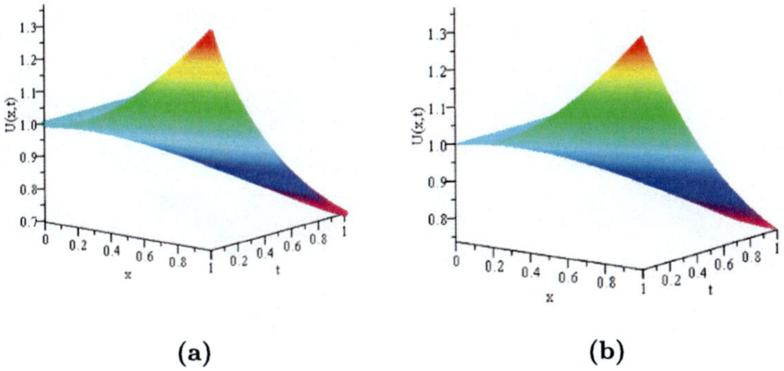

Fig. 3. Numerical (**a**) and exact (**b**) solutions of Example 1

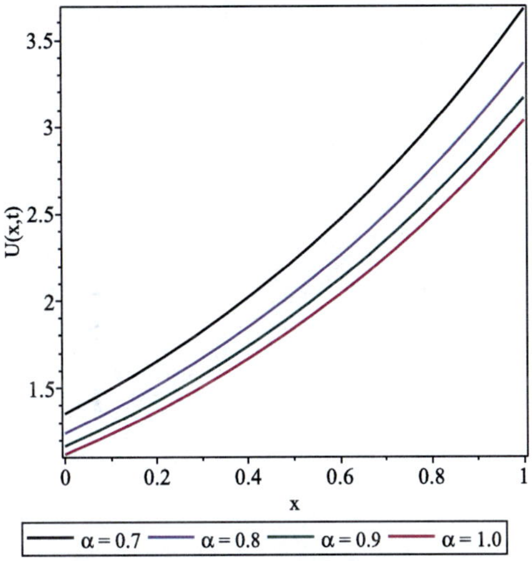

Fig. 4. Numerical solutions of Example 2 at $t = 0.1$

In Table 4 we compared results with different methods, as the fractional parameter is sufficiently close to $\alpha \to 1$. Figure 6 exhibits the concentration levels for $\alpha = 0.7$, $\alpha = 0.8$, $\alpha = 0.9$, and $\alpha = 1.0$. Numerical experiment shows that solutions increase as fractional parameter decrease for the fixed time level $t = 0.1$. Graph of numerical solution is given by Fig. 7 using parameters $\Delta t = 0.001$, $N = 50$, on $0 \leq t \leq 0.1$, for $\alpha = 0.9$ together with the exact solution of classical Cauchy reaction-diffusion equation.

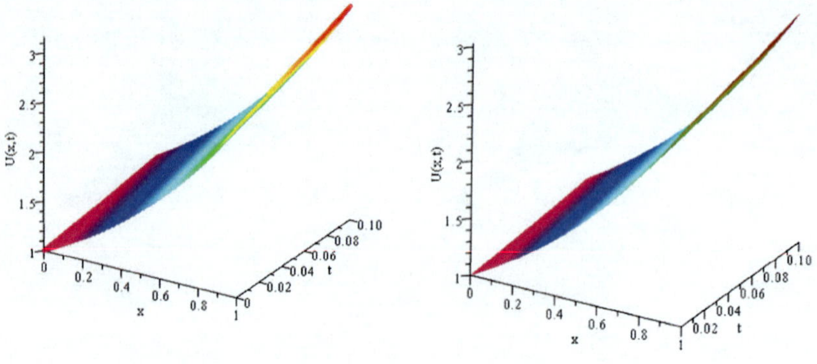

(a)

(b)

Fig. 5. Numerical **(a)** and exact **(b)** solutions of Example 2

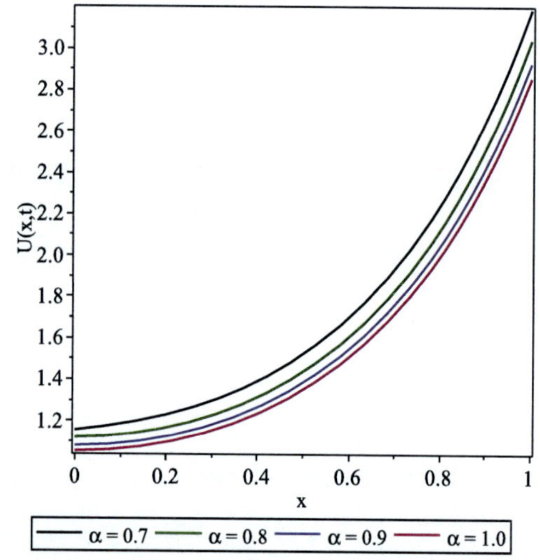

Fig. 6. Numerical solutions of Example 3 at $t = 0.1$

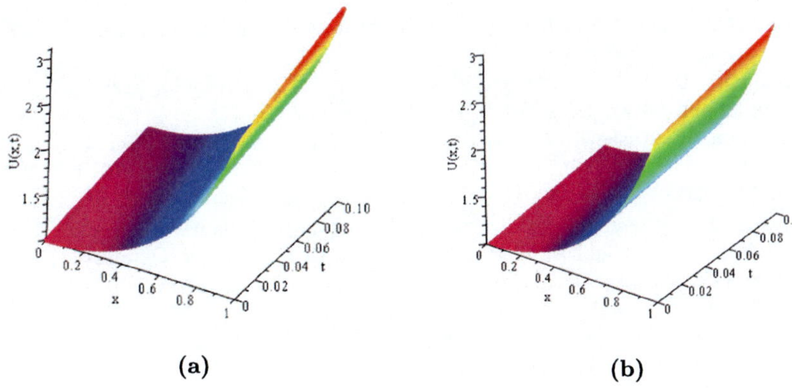

Fig. 7. Numerical (a) and exact (b) solutions of Example 3

6 Conclusion

In this study, polynomial based differential quadrature method, which is constructed based on Lagrange interpolation polynomials, is used for numerical solution of fractional Cauchy reaction-diffusion equation. The method is analyzed in terms of stability analysis, which is established by eigenvalues of the space-discretized matrix. Stability regions satisfy the criteria of Runge-Kutta fourth order method. Graphs of analytical solutions and quadrature solutions are depicted to see the effectiveness of the numerical method. The numerical solutions are presented in tables with time step $\Delta t = 0.001$. Concentration motion is analyzed with respect to the different reaction parameters as well as different fractional orders. Differential quadrature results found in the study are in good agreement with fractional reaction-diffusion equation that could be affirmed from the literature. The current method can be employed to the fractional order differential equations with the several fractional derivative descriptions.

References

1. Kilbas, A.A., Srivastava, H.M., Trujillo, J.J.: Theory and Applications of Fractional Differential Equations, vol. 204. Elsevier, Amsterdam (2006)
2. Podlubny, I.: Fractional Differential Equations: An Introduction to Fractional Derivatives, Fractional Differential Equations, to Methods of Their Solution and Some of Their Applications. Elsevier, Amsterdam (1998)
3. Atangana, A.: On the new fractional derivative and application to nonlinear Fisher's reaction-diffusion equation. Appl. Math. Comput. **273**, 948–956 (2016)
4. KKhalil, R., Al Horani, M., Yousef, A., Sababheh, M.: A new definition of fractional derivative. J. Comput. Appl. Math. **264**, 65–70 (2014)
5. Atangana, A., Baleanu, D.: New fractional derivatives with nonlocal and non-singular kernel: theory and application to heat transfer model (2016). arXiv preprint arXiv:1602.03408

6. Bert, C.W., Malik, M.: Differential quadrature method in computational mechanics: a review (1996)
7. Bellman, R., Casti, J.: Differential quadrature and long-term integration. J. Math. Anal. Appl. **34**(2), 235–238 (1971)
8. Bellman, R., Kashef, B.G., Casti, J.: Differential quadrature: a technique for the rapid solution of nonlinear partial differential equations. J. Comput. Phys. **10**(1), 40–52 (1972)
9. Wang, X.: Differential Quadrature and Differential Quadrature Based Element Methods: Theory and Applications. Butterworth-Heinemann, Oxford (2015)
10. Quan, J.R., Chang, C.T.: New insights in solving distributed system equations by the quadrature method–I. Analysis. Comput. Chem. Eng. **13**(7), 779–788 (1989)
11. Quan, J.R., Chang, C.T.: New insights in solving distributed system equations by the quadrature method–II. Numerical experiments. Comput. Chem. Eng. **13**(9), 1017–1024 (1989)
12. Shu, C.: Differential Quadrature and Its Applications in Engineering. Springer, London (2000)
13. Wu, Y.L., Shu, C.: Development of RBF-DQ method for derivative approximation and its application to simulate natural convection in concentric annuli. Comput. Mech. **29**(6), 477–485 (2002)
14. Korkmaz, A., Dağ, İ.: Cubic B-spline differential quadrature methods for the advection-diffusion equation. Int. J. Numer. Methods Heat Fluid Flow **22**(8), 1021–1036 (2012)
15. Korkmaz, A., Dağ, İ.: Cubic B-spline differential quadrature methods and stability for Burgers' equation. Eng. Comput. **30**(3), 320–344 (2013)
16. Korkmaz, A., Dağ, I.: Quartic and quintic B-spline methods for advection–diffusion equation. Appl. Math. Comput. **274**, 208–219 (2016)
17. Mittal, R.C., Dahiya, S.: A quintic B-spline based differential quadrature method for numerical solution of Kuramoto-Sivashinsky equation. Int. J. Nonlinear Sci. Numer. Simul. **18**(2), 103–114 (2017)
18. Başhan, A.: An effective application of differential quadrature method based on modified cubic B-splines to numerical solutions of the KdV equation. Turk. J. Math. **42**(1), 373–394 (2018)
19. Korkmaz, A., Dağ, İ.: Shock wave simulations using sinc differential quadrature method. Eng. Comput. **28**(6), 654–674 (2011)
20. Tomasiello, S.: A new least-squares DQ-based method for the buckling of structures with elastic supports. Appl. Math. Model. **39**(9), 2809–2814 (2015)
21. Eftekhari, S.A.: Numerical simulation of sloshing motion in a rectangular tank using differential quadrature method. KSCE J. Civil Eng. **22**(11), 4657–4667 (2018)
22. Khaniki, H.B., Hashemi, S.H.: Comparison between using generalized differential quadrature method and analytical solution in analyzing vibration behavior of nonuniform nanobeam systems. Charact. Appl. Nanomaterials **1**(2) (2018)
23. Mittal, R.C., Jiwari, R.: A higher order numerical scheme for some nonlinear differential equations: models in biology. Int. J. Comput. Methods Eng. Sci. Mech. **12**(3), 134–140 (2011)
24. Meral, G., Tezer-Sezgin, M.: Differential quadrature solution of nonlinear reaction–diffusion equation with relaxation-type time integration. Int. J. Comput. Math. **86**(3), 451–463 (2009)
25. Meral, G., Tezer-Sezgin, M.: The differential quadrature solution of nonlinear reaction-diffusion and wave equations using several time-integration schemes. Int. J. Numer. Methods Biomed. Eng. **27**(4), 485–497 (2011)

26. Tomasiello, S.: Stability and accuracy of the iterative differential quadrature method. Int. J. Numer. Meth. Eng. **58**(9), 1277–1296 (2003)
27. Tomasiello, S.: Numerical stability of DQ solutions of wave problems. Numer. Algorithms **57**(3), 289–312 (2011)
28. Lesnic, D.: The decomposition method for Cauchy reaction–diffusion problems. Appl. Math. Lett. **20**(4), 412–418 (2007)
29. Kumar, S.: A new fractional modeling arising in engineering sciences and its analytical approximate solution. Alexandria Eng. J. **52**(4), 813–819 (2013)
30. Ali, S., Bushnaq, S., Shah, K., Arif, M.: Numerical treatment of fractional order Cauchy reaction diffusion equations. Chaos Solitons Fractals **103**, 578–587 (2017)
31. Gul, H., Alrabaiah, H., Ali, S., Shah, K., Muhammad, S.: Computation of solution to fractional order partial reaction diffusion equations. J. Adv. Res. **25**, 31–38 (2020)
32. Asjad, M.I., Aleem, M., Ahmadian, A., Salahshour, S., Ferrara, M.: New trends of fractional modeling and heat and mass transfer investigation of (SWCNTs and MWCNTs)-CMC based nanofluids flow over inclined plate with generalized boundary conditions. Chin. J. Phys. **66**, 497–516 (2020)
33. Dizicheh, A.K., Salahshour, S., Ahmadian, A., Baleanu, D.: A novel algorithm based on the Legendre wavelets spectral technique for solving the Lane-Emden equations. Appl. Numer. Math. **153**, 443–456 (2020)
34. Salahshour, S., Ahmadian, A., Salimi, M., Pansera, B.A., Ferrara, M.: A new Lyapunov stability analysis of fractional-order systems with nonsingular kernel derivative. Alexandria Eng. J. **59**(5), 2985–2990 (2020)
35. Korpinar, Z., Tchier, F., İnç, M., Ragoub, L., Bayram, M.: New soliton solutions of the fractional regularized long wave burger equation by means of conformable derivative. Results Phys. **14**, 102395 (2019)

Check for updates

Numerical Solution of Linear Volterra Integral Equations Based on Operational Matrix of Integration by Using Non-uniform Haar Wavelets

Meisam Montazer, Reza Ezzati(✉), and Mohsen Fallahpour

Department of Mathematics, Karaj Branch, Islamic Azad University, Karaj, Iran
{m.montazer,ezati,m.fallahpour}@kiau.ac.ir

Abstract. In this paper, we present a numerical method for solving linear Volterra integral equations which is based on the non-uniform Haar wavelets and operational matrix of integration. Also, we present the error analysis of the proposed method. To confirm the efficiency and the accuracy of the method, we present some numerical examples.

Keywords: Non-uniform Haar wavelets · Block pulse functions · Volterra integral equations · Operational matrix · Grid points

1 Introduction

The Volterra integral equations appear in many problems such as finance, mathematics, etc. [7–14]. In the recent years, a lot of numerical methods have been developed for solving several types of integral equations [1–6]. Since, obtaining the analytical solution of these equations is generally not possible, it is important to develop the numerical method to solve these equations. The aim of this work is to present a numerical method for approximating the solution of linear Volterra integral equation

$$Y(t) = K(t) + \lambda \int_0^t P(s,t)Y(s)ds, \tag{1}$$

where $Y(t)$ is the unknown function, $P(s,t)$ is the known function, and the parameter λ is known. In the purposed method, fist we transform Volterra integral Eq. (1) to a system of linear algebraic equation with the unknown non-uniform Haar coefficients. Then, we obtain the unknown non-uniform Haar coefficients by solving this system by the known method. This article is organized as follows: In Sect. 2, we review the Block pulse functions (BPFs). In Sect. 3, we describe the non-uniform Haar wavelets and their property. The Heaviside step function is reviewed in Sect. 4. In the next section, we describe function approximation. In Sects. 6–8, a numerical method based on the operational matrix of integration of non-uniform haar wavelet is proposed. In Sect. 9, we present the error analysis. In Sect. 10, some numerical examples are given. Finally in Sect. 11, conclusion is given.

T. Allahviranloo et al. (Eds.): IDS 2020, AISC 1301, pp. 738–753, 2021.
https://doi.org/10.1007/978-3-030-66501-2_60

2 Block Pulse Functions (BPFs)

In this section, we recall the BPFs and also we describe some basic properties of these functions, see [9]. For this aim we first describe the BPFs.

2.1 Definition of BPFs

The m-set of BPFs are defined as:

$$\psi_i = \begin{cases} 1, & (i-1)\hbar \leq t \leq i\hbar, \\ 0, & elsewhere, \end{cases} \tag{2}$$

where $t \in [0, \Gamma)$, $i = 1, 2, 3, ..., m$ and $\hbar = \frac{\Gamma}{m}$. The some properties of the BPFs are as follows:

(1) Disjointness: The BPFs are disjoint with each other in $t \in [0, \Gamma)$ and

$$\psi_i(t)\psi_l(t) = \delta_{i,l}\psi_i(t),$$

where $i, l = 1, 2, 3, ..., m$ and $\delta_{i,l}$ is the Kronecker delta.

(2) Orthogonality: The set of BPFs are orthogonal with each other in $t \in [0, \Gamma)$, that is

$$\int_0^{\Gamma} \psi_i(t)\psi_l(t)dt = \hbar\delta_{i,l}, \quad i, l = 1, 2, 3, ..., m.$$

(3) Completeness: If $m \longrightarrow \infty$, then the BPF set is complete. For every $f \in L^2[0, \Gamma)$, parseval's identify holds,

$$\int_0^{\Gamma} f^2(t)dt = \sum_{i=1}^{\infty} f_i^2 \|\psi_i(t)\|^2,$$

for any real bounded function $f(t)$ which is square integrable in the interval $[0, \Gamma)$. Here the phrase:

$$\|\psi_i(t)\| = \left(\int_0^{\Gamma} \psi_i^2(t)dt\right)^{\frac{1}{2}},$$

is the norm of $\psi_i(t)$.

2.2 Vector Form

Consider the first m-terms of BPFs and write them concisely as m-vector:

$$\Psi(t) = [\psi_1(t), \psi_2(t), \psi_3(t), ..., \psi_m(t)], \quad t \in [0, 1). \tag{3}$$

By the help of disjointness property of BPFs, we conclude that:

$$\Psi(t)\Psi^T(t) = \begin{pmatrix} \psi_1(t) & 0 & 0 \dots & 0 \\ 0 & \psi_2(t) & 0 \dots & 0 \\ \vdots & \vdots & \vdots \ddots & \vdots \\ 0 & 0 & 0 \dots & \psi_m(t) \end{pmatrix}.$$

Furthermore, we have

$$\Psi^T(t)\Psi(t) = 1,$$

and

$$\Psi(t)\Psi^T(t)F = \tilde{F}\Psi(t), \tag{4}$$

where $\tilde{F}$ is an $m \times m$ diagonal matrix with the elements of the vector F on the main diagonal. Also, it is easy to show that for an $m \times m$ matrix Z

$$\Psi^T(t)Z\Psi(t) = \hat{Z}^T\Psi(t), \tag{5}$$

where $\hat{Z} = diag(Z)$ is a m-vector.

2.3 Function Approximation

An arbitrary real function $f(t)$, which is square integrable in $[0, \Gamma)$ can be expanded by the BPFs as:

$$f(t) \simeq \sum_{i=1}^{n} f_i\psi_i(t) \tag{6}$$

where f_i is the Block pulse coefficient with respect to the i-th Block pulse function $\psi_i(t)$ defined in Eq. (2). In the vector form, we have

$$f(t) \simeq F^T\Psi(t) = \Psi^T(t)F,$$

in which

$$\Psi(t) = [\psi_1(t), \psi_2(t), \psi_3(t), ..., \psi_m(t)]^T, \quad F = [f_1, f_2, f_3, ..., f_m]^T,$$

where

$$f_i = \frac{1}{\hbar}\int_0^{\Gamma} \psi_i(t)f(t)dt, \quad i = 1, 2, ..., m.$$

3 Non-uniform Haar Wavelets

Non-uniform Haar wavelets are characterized by two characteristics: the dilation parameter $j = 0, 1, ..., J$ (J is maximal level of resolution) and the translation parameter $k = 0, 1, ..., n-1$, where the integer $n = 2^j$. The number of wavelet is identified as $i = n + k + 1$. Also the maximal value is $i = 2N$, where $N = 2^J$. We choose the division points on $[a, b]$. Let us denote the length of the η-th subinterval by $\Delta x_\eta = x_\eta - x_{\eta-1}, \eta = 1, 2, ..., 2N$. We assume that $\Delta x_{\eta+1} = r\Delta x_\eta$, where $r > 1$ is constant. If we sum the lengths of these subintervals, we have

$$\Delta x_1(1 + r + r^2 + ... + r^{2N-1}) = L, \quad L = b - a,$$

or

$$\Delta x_1 = L\frac{r-1}{r^{2N}-1}.$$

Since

$$\tilde{x}(c) = \Delta x_1 \left(1 + r + ... + r^{c-1}\right) = \Delta x_1 \frac{r^c - 1}{r - 1}, \quad c = 1, 2, ..., 2N, \tag{7}$$

we have obtained the grid points as

$$\tilde{x}(c) = L \frac{r^c - 1}{r^{2N} - 1}, \quad c = 1, 2, ..., 2N. \tag{8}$$

We divide the interval $[a, b]$ into $2N$ subinterval of unequal lengths with the division points $a = \tilde{x}(0) < \tilde{x}(1) < ... < \tilde{x}(2N) = b$. Consider the following non-uniform Haar wavelet family $\{H_i\}_{i \in \mathbb{N}}$ based on [15]

$$H_i(t) = \begin{cases} 1, & \Upsilon_1(i) \le t \le \Upsilon_2(i), \\ -u_i, & \Upsilon_2(i) \le t \le \Upsilon_3(i), \\ 0, & elsewhere, \end{cases}$$

where

$$\Upsilon_1(i) = \tilde{x}(2k\theta), \quad \Upsilon_2(i) = \tilde{x}\left((2k+1)\theta\right),$$
$$\Upsilon_3(i) = \tilde{x}\left((2k+2)\theta\right), \quad \theta = N/n,$$

moreover, the wavelet coefficient u_i is calculated as follow:

$$u_i = \frac{\Upsilon_2(i) - \Upsilon_1(i)}{\Upsilon_3(i) - \Upsilon_2(i)}.$$

Clearly, these equations hold when $i > 2$. For the case $i = 1$ and $i = 2$, we have $\Upsilon_1(1) = a$, $\Upsilon_2(1) = \Upsilon_3(1) = b$, $\Upsilon_1(2) = a$, $\Upsilon_2(2) = \frac{\tilde{x}(2N)}{2}$, $\Upsilon_3(2) = b$.

3.1 Orthogonality Property

The Haar wavelets are piecewise orthogonal,

$$\int_a^b H_r(x) H_s(x) dx = \begin{cases} (b-a)2^{-j}, & r = s, \\ 0, & r \ne s. \end{cases}$$

Similarly, the non-uniform Haar wavelet defined on $[a, b]$ are piecewise orthogonal,

$$\int_a^b H_i(x) H_l(x) dx = \begin{cases} (b-a)T_i, & i = l, \\ 0, & i \ne l, \end{cases} \tag{9}$$

with

$$T_i = u_i(\Upsilon_3(i) - \Upsilon_1(i)). \tag{10}$$

4 Definition of Heaviside Step Function

The Heaviside step function hsf is a discontinuous function such that defined as follows:

$$hsf(x) = \begin{cases} 1, \, x \geq 0, \\ 0, \, x < 0. \end{cases}$$

With a useful property

$$hsf(t - \Upsilon_1(i))hsf(t - \Upsilon_2(i)) = hsf(t - \max\{\Upsilon_1(i), \Upsilon_2(i)\}), \qquad \Upsilon_1(i), \Upsilon_2(i) \in \mathbb{R}.$$

By using hsf, we can write

$$\begin{aligned} H_0(t) &= hsf(t) - hsf(t-1), \\ H_\tau(t) &= hsf(t - \Upsilon_1(i)) + (-u_i - 1)hsf(t - \Upsilon_2(i)) + (u_i)hsf(t - \Upsilon_3(i)), \end{aligned} \tag{11}$$

where

$$\tau = 2^j + k, \quad j, k \in \mathbb{N} \cup \{0\}, \quad 0 \leq k < 2^j.$$

5 Function Approximation

For $Y \in L^2[0,1]$, we can expand it in terms of non-uniform Haar wavelets as the following form

$$Y(t) = q_1 H_1(t) + \sum_{j=0}^{\infty} \sum_{k=0}^{2^j - 1} q_{2^j+k+1} H_{2^j+k+1}(t), \qquad t \in [0,1],$$

which coefficients q_i is given by

$$q_1 = \frac{1}{T_1} \int_0^1 Y(t) H_1(t) dt, \quad q_i = \frac{1}{T_i} \int_0^1 Y(t) H_i(t) dt,$$

where $i = 2^j + k + 1$, $j \geq 0$, $0 \leq k < 2^j$. By using Eq. (11), the non-uniform Haar coefficients q_i can be rewritten as:

$$q_i = \frac{1}{T_i} \left(\int_{\Upsilon_1(i)}^{\Upsilon_2(i)} Y(t) dt - u_i \int_{\Upsilon_2(i)}^{\Upsilon_3(i)} Y(t) dt \right),$$

where $j, k = 0, 1, 2, 3, \ldots$ and $0 \leq k < 2^j$. If $Y(t)$ is piecewise constant or may become approximated by piecewise constant on any subinterval, we can terminate $Y(t)$ at $2N$ finite terms as follows:

$$Y(t) \simeq q_1 H_1(t) + \sum_{j=0}^{J} \sum_{k=0}^{2^j - 1} q_{2^j+k+1} H_{2^j+k+1}(t),$$

rewriting above equation in the vector form, we have

$$Y(t) \simeq q^T H(t) = H^T(t) q, \qquad t \in [0,1],$$

where

$$q^T = [q_1, q_2, q_3, ..., q_{2N}],$$
$$H(t) = [H_1(t), H_2(t), H_3(t), ..., H_{2N}(t)]^T. \tag{12}$$

Therefore, a 2-variable function $p(s,t) \in L^2\left([0,1)^2\right)$ can be approximated in terms of the non-uniform Haar wavelets as:

$$p(s,t) = H^T(s)PH(t) = H^T(t)P^TH(s),$$

where $H(s)$ and $H(t)$ are the non-uniform Haar wavelet vectors, and $P = p_{i,l}$ is the $2N \times 2N$ non-uniform Haar wavelet coefficients matrix with (i,l)-th element of this matrix is given by

$$p_{i,l} = \frac{1}{T_i^2}\int_0^1\int_0^1 p(s,t)H_i(s)H_l(t)dtds, \qquad i,l = 1,2,...,2N,$$

also, we obtain

$$p_{i,l} = \frac{1}{T_i^2}\left(\int_{\Upsilon_1(i)}^{\Upsilon_2(i)}\int_{\Upsilon_1(l)}^{\Upsilon_2(l)} p(s,t)dtds - u_l\int_{\Upsilon_1(i)}^{\Upsilon_2(i)}\int_{\Upsilon_2(l)}^{\Upsilon_3(l)} p(s,t)dtds\right.$$
$$\left.-u_i\int_{\Upsilon_2(i)}^{\Upsilon_3(i)}\int_{\Upsilon_1(l)}^{\Upsilon_2(l)} p(s,t)dtds + u_iu_l\int_{\Upsilon_2(i)}^{\Upsilon_3(i)}\int_{\Upsilon_2(l)}^{\Upsilon_3(l)} p(s,t)dtds\right).$$

6 Relation Between Non-uniform Haar Wavelet and BPFs

In this section we will drive relation between the non-uniform Haar wavelets and BPFs. To do this, we set $\Gamma = 1$ in definition of BPFs.

Theorem 1. *Let $H(t)$ and $\Psi(t)$ are Haar wavelets vector and BPFs vector respectively. Then, we have:*

$$H(t) = Q\Psi(t), \quad m = 2^{J+1}, \quad m = 2N, \tag{13}$$

where $Q_{i,l}$ is an $m \times m$ matrix, and

$$Q_{i,l} = H_i\left(\frac{2l-1}{2m}\right), \quad i,l = 1,2,3,...,m.$$

Proof. Let $H_i(t)$, $i = 1,2,3,...,m$ be the i-th element of non-uniform Haar wavelets vector $H(t)$. By expanding $H_i(t)$ by BPFs, we can write

$$H_i(t) = \sum_{l=1}^{2N} Q_{i,l}\psi_l(t) = Q_i^T\Psi(t), \quad i = 1,2,3,...,m,$$

where Q_i is the i-th row and $Q_{i,l}$ is the (i,l)-th element of matrix Q. By using the orthogonality of BPFs, we have

$$Q_{i,l} = \frac{1}{\hbar}\int_0^1 H_i(t)\psi_l(t)dt = \frac{1}{\hbar}\int_{\frac{l-1}{m}}^{\frac{l}{m}} H_i(t)dt = m\int_{\frac{l-1}{m}}^{\frac{l}{m}} H_i(t)dt,$$

by using mean value theorem for integrals in the above equation, we can write

$$Q_{i,l} = m\left(\frac{l}{m} - \frac{l-1}{m}\right)H_i(\sigma) = H_i(\sigma), \quad \sigma \in \left(\frac{l-1}{m}, \frac{l}{m}\right).$$

Since $H_i(t)$ is a constant function on the interval $(\frac{l-1}{m}, \frac{l}{m})$, we can choose $\sigma = \frac{2l-1}{2m}$, so we have

$$Q_{i,l} = H_i\left(\frac{2l-1}{2m}\right), \quad i,l = 1,2,3,...,m.$$

Remark 1. According to definition of matrix Q in Eq. (13) it is easy to see that

$$Q^{-1} = \frac{1}{m}Q^T.$$

Remark 2. With consequence of Eqs. (4), (5) and Theorem 1 for an m-vector Y, we have

$$H(t)H^T(t)Y = \tilde{Y}H(t),$$

in which $\tilde{Y}$ is an $m \times m$ matrix as

$$\tilde{Y} = Q\bar{Y}Q^{-1},$$

where $\bar{Y} = diag(Q^{-1}Y)$. Moreover, it can be easy to show that for an $m \times m$ matrix Z

$$H^T(t)ZH(t) = \tilde{Z}^T H(t),$$

where $\tilde{Z}^T = AQ^{-1}$ and $A = diag(Q^T ZQ)$ is a m-vector.

7 Operational Matrix of Integration of the Non-uniform Haar Wavelets

In this section, we obtain the operational matrix of integration of the non-uniform Haar wavelets. For this aim we remind some useful results of the BPFs.

Lemma 1. *Let $\Psi(t)$ be the BPFs vector defined in Eq. (3). Then integration of this vector can be derived as*

$$\int_0^t \Psi(s)ds \simeq D\Psi(t),$$

where D is called the operational matrix of integration for BPFs and is given by

$$D = \frac{\hbar}{2}\begin{pmatrix} 1\,2\,2\,\ldots\,2 \\ 0\,1\,2\,\ldots\,2 \\ 0\,0\,1\,\ldots\,\vdots \\ \vdots\,\vdots\,\vdots\,\ddots\,2 \\ 0\,0\,0\,\ldots\,1 \end{pmatrix}.$$

Theorem 2. *Suppose that $H(t)$ be the non-uniform Haar wavelets vector defined in Eq. (12), the integral of this vector can be drived as*

$$\int_0^t H(s)ds \simeq \frac{1}{m}QDQ^T H(t) = \Phi H(t), \tag{14}$$

where Q is introduced in Eq. (13) and D is the operational matrix of integration of BPFs.

Proof. Let $H(t)$ be the non-uniform Haar wavelets vector. By using Theorem 1 and Lemma 1, we have

$$\int_0^t H(s)ds \simeq \int_0^t Q\Psi(s)ds = Q\int_0^t \Psi(s)ds = QD\Psi(t),$$

now, by using Theorem 1 and Remark 1, we can write

$$\int_0^t H(s)ds \simeq QD\Psi(t) = \frac{1}{m}QDQ^T H(t) = \Phi H(t).$$

8 Description of the Proposed Method

In this section, we apply the operational matrix of integration of non-uniform Harr wavelets for solving Volterra integral Eq. (1). For this purpose we approximate $Y(t)$, $K(t)$, $P(s,t)$ in Eq. (1) as follows:

$$Y(t) \simeq Y^T H(t) = H^T(t)Y, \tag{15}$$

$$K(t) \simeq K^T H(t) = H^T(t)K,$$

and

$$P(s,t) \simeq H^T(s)PH(t) = H^T(t)P^T H(s),$$

where Y and K are the non-uniform Haar wavelets coefficients vectors, and P is the non-uniform coefficient matrix. By substituting above approximations in Eq. (1), we have

$$Y^T H(t) \simeq K^T H(t) + \int_0^t \left(H^T(t)PH(s)\right)\left(H^T(s)Y\right) ds$$

$$= K^T H(t) + H^T(t)P\left(\int_0^t H(s)H^T(s)Y ds,\right)$$

by using Remark 2, we get

$$Y^T H(t) \simeq K^T H(t) + H^T(t)P\tilde{Y}\left(\int_0^t H(s)ds\right).$$

Now by applying Eq. (14), we have

$$Y^T H(t) \simeq K^T H(t) + H^T(t)P\tilde{Y}\Phi H(t)$$

by setting $\hat{Y} = P\tilde{Y}\Phi$ and using Remark 2, we get

$$Y^T H(t) \simeq K^T H(t) + \hat{Y} H(t).$$

This equation is hold for all $t \in [0, 1)$, so we can write

$$Y^T - \hat{Y} \simeq K^T. \tag{16}$$

Clearly $\hat{Y}$ is linear function of Y, Eq. (16) is a linear system of equations for unknown vector Y. After solving this linear system and determining Y, we can approximate the solution of Eq. (1) by substituting the obtained vector Y in Eq. (15).

9 Error Analysis

In this section, we peruse the convergence and error analysis of the proposed method for solving Volterra integral Eq. (1). To do this, we need the following theorems.

Theorem 3. *Suppose that $q(t) \in L^2[0, 1)$ is an arbitrary function with bounded first derivative, $|q'(t)| \leq M$, and also consider the error function*

$$e_m(t) = q(t) - \sum_{i=0}^{m-1} q_i H_i(t),$$

where $i = 2^j + k + 1$, $m = 2^{J+1}$, $J > 0$ and

$$q_i = \frac{1}{T_i}\int_0^1 H_i(t)q(t)dt = \frac{1}{T_i}\left(\int_{\Upsilon_1(i)}^{\Upsilon_2(i)} q(t)dt - u_i\int_{\Upsilon_2(i)}^{\Upsilon_3(i)} q(t)dt\right).$$

Then for $t \in [0, 1)$, we have

$$\|e_m\|_2 \simeq O(\frac{1}{m})$$

Proof. By definition of $e_m(t)$, we can write

$$\begin{aligned}\|e_m\|_2^2 &= \int_0^1 \left(q(t) - \sum_{i=0}^{m-1} q_i H_i(t)\right)^2 dt \\ &= \int_0^1 \left(\sum_{i=m}^{\infty} q_i H_i(t)\right)^2 dt \\ &= \sum_{i=m}^{\infty} q_i^2 \int_0^1 H_i^2(t)dt.\end{aligned}$$

By the mean value theorem for integrals, there are $\alpha_1 \in (\Upsilon_1(i), \Upsilon_2(i))$, $\alpha_2 \in (\Upsilon_2(i), \Upsilon_3(i))$, such that

$$\begin{aligned}q_i &= \frac{1}{T_i}\left(q(\alpha_1)\left(\Upsilon_2(i) - \Upsilon_1(i)\right) - u_i q(\alpha_2)\left(\Upsilon_3(i) - \Upsilon_2(i)\right)\right) \\ &= \frac{1}{T_i}\left(q(\alpha_1)(\Upsilon_2(i) - \Upsilon_1(i)) - \frac{\Upsilon_2(i) - \Upsilon_1(i)}{\Upsilon_3(i) - \Upsilon_2(i)}(\Upsilon_3(i) - \Upsilon_2(i))\right) \\ &= \frac{1}{T_i}\left((\Upsilon_2(i) - \Upsilon_1(i))(q(\alpha_1) - q(\alpha_2)\right) = \frac{1}{T_i}\left((\Upsilon_2(i) - \Upsilon_1(i))(\alpha_1 - \alpha_2)q'(\alpha)\right).\end{aligned}$$

From Eq. (9), it follows that

$$\begin{aligned}\|e_m\|_2^2 &= \sum_{i=m}^{\infty} \frac{1}{T_i^2}\left((\Upsilon_2(i) - \Upsilon_1(i))^2(\alpha_1 - \alpha_2)^2(q'(\alpha))^2\right)T_i \\ &\le \sum_{i=m}^{\infty} \frac{1}{T_i}\left((\Upsilon_2(i) - \Upsilon_1(i))^2(\Upsilon_3(i) - \Upsilon_1(i))^2 M^2\right).\end{aligned}$$

Now, by definitions of $\Upsilon_1(i)$, $\Upsilon_2(i)$, $\Upsilon_3(i)$, and Eq. (7) we have

$$\Upsilon_1(i) = \frac{2kN}{n}\Delta x_1, \quad \Upsilon_2(i) = \frac{(2k+1)N}{n}\Delta x_1, \quad \Upsilon_3(i) = \frac{(2k+2)N}{n}\Delta x_1,$$

therefore, we obtain

$$\begin{aligned}\Upsilon_2(i) - \Upsilon_1(i) &= \frac{N\Delta x_1}{n}, \\ \Upsilon_3(i) - \Upsilon_1(i) &= \frac{2N\Delta x_1}{n},\end{aligned}$$

and

$$\Upsilon_3(i) - \Upsilon_2(i) = \frac{N\Delta x_1}{n}.$$

Since

$$\Upsilon_2(i) - \Upsilon_1(i) \le \Upsilon_3(i) - \Upsilon_1(i),$$

we get

$$\|e_m\|_2^2 \leq M^2 \sum_{i=m}^{\infty} \frac{1}{T_i} \Big((\Upsilon_3(i) - \Upsilon_1(i))^4 \Big). \tag{17}$$

On the other hand, by applying Eq. (10), we get

$$\frac{1}{T_i} = \frac{\Upsilon_3(i) - \Upsilon_2(i)}{(\Upsilon_2(i) - \Upsilon_1(i))\,(\Upsilon_3(i) - \Upsilon_1(i))}.$$

Given that

$$\Upsilon_3(i) - \Upsilon_2(i) \leq \Upsilon_3(i) - \Upsilon_1(i),$$

we conclude that

$$\frac{1}{T_i} \leq \frac{1}{\Upsilon_2(i) - \Upsilon_1(i)} = \frac{n}{N\Delta x_1} = \frac{2^j}{N\Delta x_1}. \tag{18}$$

So, applying Eq. (18) in Eq. (17) implies

$$\begin{aligned}\|e_m\|_2^2 &\leq M^2 \sum_{j=J+1}^{\infty} \sum_{k=0}^{2^j-1} \frac{2N^3(\Delta x_1)^3}{(2^j)^3} \\ &= 2N^3(\Delta x_1)^3 \sum_{j=J+1}^{\infty} \frac{1}{(2^j)^2},\end{aligned}$$

that is,

$$\|e_m\|_2^2 \leq \frac{8M^2N^3(\Delta x_1)^3}{3} \left(\frac{1}{2^{J+1}}\right)^2 = A\left(\frac{1}{2^{J+1}}\right)^2, \tag{19}$$

where

$$A = \frac{8M^2N^3(\Delta x_1)^3}{3}.$$

Since $m = 2^{J+1}$, we can write

$$\|e_m\|_2 \simeq O(\frac{1}{m}).$$

Theorem 4. *Suppose that* $q(x,y) \in L^2\left([0,1)^2\right)$ *satisfies in* $\left|\frac{\partial^2 q}{\partial x \partial y}\right| < E$*, also*

$$e_m(x,y) = q(x,y) - \sum_{i=0}^{m-1}\sum_{l=0}^{m-1} q_{i,l} H_i(x) H_l(y),$$

where $i = 2^{j_1} + k + 1$, $l = 2^{j_2} + k + 1$, $m = 2^{J+1}$, $J > 0$ *and*

$$q_{i,l} = \frac{1}{T_i^2} \int_0^1 \int_0^1 (H_i(x) H_l(y) q(x,y))\, dx dy.$$

Then for $(x,y) \in L^2\left([0,1)^2\right)$*, we have*

$$\|e_m\|_2 \simeq O(\frac{1}{m^2}).$$

Proof. By definition of $e_m(x,y)$, we can write

$$\|e_m\|_2^2 = \int_0^1\int_0^1 \left(q(x,y) - \sum_{i=0}^{m-1}\sum_{l=0}^{m-1} q_{i,j}H_i(x)H_l(y)\right)^2 dxdy$$
$$= \int_0^1\int_0^1 \left(\sum_{i=m}^{\infty}\sum_{l=m}^{\infty} q_{i,l}H_i(x)H_l(y)\right)^2 dxdy$$
$$= \sum_{i=m}^{\infty}\sum_{l=m}^{\infty}\int_0^1\int_0^1 (q_{i,l}^2 H_i^2(x){H_l}^2(y))dxdy.$$

From the non-uniform Haar wavelet definition, mean value theorem and Theorem 3, there exist $\alpha, \alpha_1, \alpha_2$, also there exist β, β_1, β_1 such that

$$q_{i,l} = \frac{1}{{T_i}^2}\int_0^1 H_i(x)\left(\int_0^1 H_l(y)q(x,y)dy\right)dx$$
$$= \frac{1}{{T_i}^2}\int_0^1 H_i(x)\left((\Upsilon_2(l)-\Upsilon_1(l))(\beta_1-\beta_2)\frac{\partial q(x,\beta)}{\partial y}\right)dx$$
$$= \frac{1}{{T_i}^2}(\Upsilon_2(l)-\Upsilon_1(l))(\beta_1-\beta_2)\int_0^1\left(H_i(x)\frac{\partial q(x,\beta)}{\partial y}\right)dx$$
$$= \frac{1}{{T_i}^2}(\Upsilon_2(l)-\Upsilon_1(l))(\beta_1-\beta_2)(\Upsilon_2(i)-\Upsilon_1(i))(\alpha_1-\alpha_2)\frac{\partial^2 q(\alpha,\beta)}{\partial y\partial x}.$$

So, we have

$$\|e_m\|_2^2 = \sum_{i=m}^{\infty}\sum_{l=m}^{\infty}\frac{1}{{T_i}^4}(\Upsilon_2(l)-\Upsilon_1(l))^2(\beta_1-\beta_2)^2(\Upsilon_2(i)-\Upsilon_1(i))^2(\alpha_1-\alpha_2)^2\left|\frac{\partial^2 q(\alpha,\beta)}{\partial y\partial x}\right|^2 {T_i}^2$$
$$\le E^2\sum_{i=m}^{\infty}\sum_{l=m}^{\infty}\frac{1}{{T_i}^2}(\Upsilon_2(i)-\Upsilon_1(i))^2(\alpha_1-\alpha_2)^2(\Upsilon_2(l)-\Upsilon_1(l))^2(\beta_1-\beta_2)^2.$$

By using Eq. (19), we get

$$\|e_m\|_2^2 \le E^2A^2 \times \frac{1}{m^4} = \frac{W}{m^4}, \quad \textit{where } W = E^2A^2.$$

So, we conclude that

$$\|e_m\|_2 \simeq O(\frac{1}{m^2}).$$

10 Numerical Examples

Example 1. We consider the following Volterra integral equation

$$Y(t) = K(t) + \int_0^t (s-t)Y(s)ds,$$

where $K(t) = \frac{3}{4}e^{2t} + t^2 + \frac{1}{2}t - \frac{7}{4}$. The exact solution of this equation is $Y(t) = e^{2t} - 2$. Comparison of maximum absolute errors (MAE) for different values of J is shown in Table 1.

Table 1. MAE for different value of J.

J	MAE
0	0.261456
1	0.130437
2	0.094186

In Table 2, $\bar{X}_A$ is the exact solution, $\bar{X}_N$ is the numerical solution and the $\bar{X}_E$ is the absolute error (Fig. 1).

Table 2. The results of Example 1 for $J = 2$.

t	$\bar{X}_A$	$\bar{X}_N$	$\bar{X}_E$
0.1	−0.866901	−0.864632	0.002268
0.2	−0.545181	−0.523653	0.021527
0.3	−0.132069	−0.087051	0.045017
0.4	0.398410	0.470719	0.072308
0.5	1.079620	1.173806	0.094186
0.6	1.954409	2.045937	0.091528
0.7	3.077815	3.142301	0.064485
0.8	4.520533	4.497753	0.022779

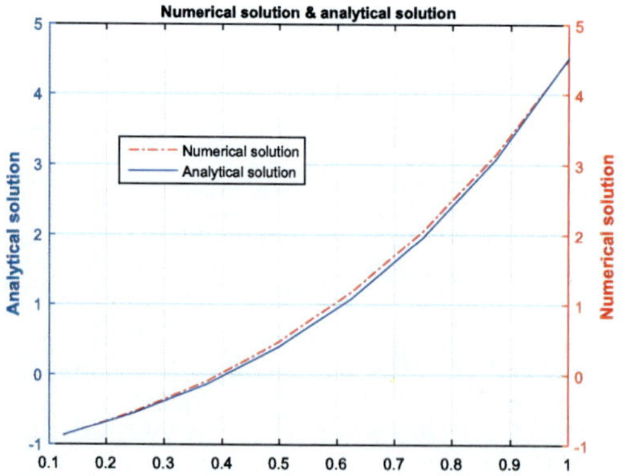

Fig. 1. Graph of the exact and numerical solutions of Example 1 for ($J = 2$).

Example 2. We consider the following Volterra integral equation

$$Y(t) = K(t) + \int_0^t (s^2 - t^2)Y(s)ds,$$

where $K(t) = 2\cosh(t) - \sinh(t) - 2t\sinh(t) + t^2 + \frac{1}{3}t^3 - \frac{3}{2}$. The exact solution of this equation is $Y(t) = \frac{1}{2} - \sinh(t)$. Comparison of MAE for different values of J is given in Table 3.

Table 3. MAE for different value of J.

J	MAE
0	0.078252
1	0.081259
2	0.075955

Table 4. The results of Example 2 for $J = 2$.

t	$\bar{X}_A$	$\bar{X}_N$	$\bar{X}_E$
0.1	0.437481	0.436462	0.001018
0.2	0.311459	0.313769	0.002310
0.3	0.182477	0.190479	0.008002
0.4	0.048515	0.070290	0.021775
0.5	−0.092523	−0.051941	0.040581
0.6	−0.242847	−0.186223	0.056623
0.7	−0.404813	−0.328858	0.075955
0.8	−0.580959	−0.506828	0.074130

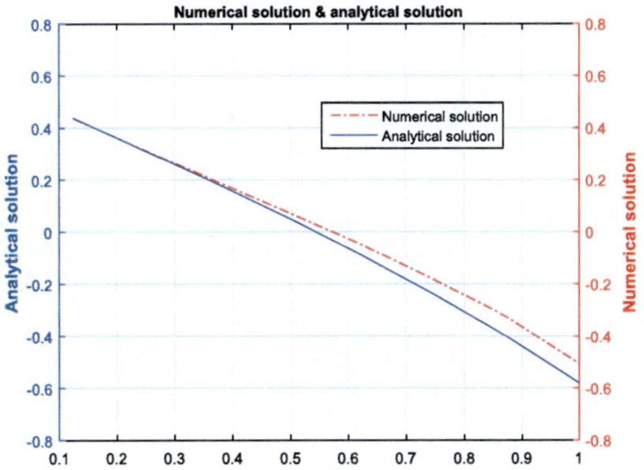

Fig. 2. Graph of the exact and numerical solutions of Example 2 for ($J = 2$).

In Table 4, $\bar{X}_A$ is the exact solution, $\bar{X}_N$ is the numerical solution and $\bar{X}_E$ is the absolute error (Fig. 2).

11 Conclusion

A computational method based on non-uniform Haar wavelets and their operational matrices is proposed for solving Volterra integral equations. The main purpose of this method is to reduce the integral equations to a linear system of algebraic equations. The convergence of the proposed method is analyzed. As a future work, this method can be used to solve the Fredholm and Volterra-Fredholm integral equations as well as integro-differential equations.

References

1. Kasumo, C.: On the approximate solutions of linear Volterra integral equations of the first kind. Appl. Math. Sci. **14**(3), 141–153 (2020)
2. Shoukralla, E.S., Ahmed, B.M.: Numerical solutions of volterra integral equations of the second kind using lagrange interpolation via the vandermonde matrix. J. Phys: Conf. Ser. **1447**(1), 012003 (2020)
3. Zheng, W., Chen, Y.: Numerical analysis for Volterra integral equation with two kinds of delay. Acta Math. Sci. **9**(2), 607–617 (2019)
4. Dawood, H.K.: Computational block-pulse functions method for solving Volterra integral equations with delay. J. Univ. Babylon **27**, 32–43 (2019)
5. Keshavarz, E., Ordokhani, Y., Razzaghi, M.: The Bernoulli wavelets operational matrix of integration and its applications for the solution of linear and nonlinear problems in calculus of variations. Appl. Math. Comput. **351**, 83–98 (2019)
6. Talaei, Y.: Chelyshkov collocation approach for solving linear weakly singular Volterra integral equations. J. Appl. Math. Comput. **60**, 201–222 (2019)
7. Gutleb, T.S., Olver, S.: A sparse spectral method for Volterra integral equations using orthogonal polynomials on the triangle. SIAM J. Numer. Anal. **58**(3), 1993–2018 (2019)
8. Talaei, Y.: Chelyshkov collocation approach for solving linear weakly singular Volterra integral equations. J. Appl. Math. Comput. **60**(1–2), 201–222 (2019)
9. Muhammad, A.M., Ayal, A.M.: Numerical solution of linear Volterra integral equation with delay using Bernstein polynomial. Int. Electron. J. Math. Educ. **14**(3), 735–740 (2019)
10. Ansari, H., Mokhtary, P.: Computational Legendre Tau method for Volterra Hammerstein pantograph integral equations. Bull. Iran. Math. Soc. **45**(2), 475–493 (2019)
11. Maleknejad, K., Torabi, P., Sauter, S.: Numerical solution of a non-linear Volterra integral equation. Vietnam J. Math. **14**(1), 5–28 (2016)

12. Mohsenyzadeh, M.: Bernoulli operational matrix method for system of linear Volterra integral equations. Int. J. Ind. Math. **8**(3), 201–207 (2016)
13. Singh, I., Kumar, S.: Haar wavelet method for some nonlinear Volterra integral equations of the first kind. J. Comput. Appl. Math. **292**, 541–552 (2016)
14. Babolian, E., Masouri, Z.: Direct method to solve Volterra integral equation of the first kind using operational matrix with block-pulse functions. J. Comput. Appl. Math. **220**(1–2), 51–57 (2008)
15. Lepik, $\ddot{U}$.: Solving integral and differential equations by the aid of non-uniform Haar wavelets. Appl. Math. Comput. **198**(1), 326–332 (2008)
16. Montazer, M., Khodabin, M., Ezzati, R., Fallahpour, M.: Non-uniform haar wavelets method for solving linear stochastic ITO - volterra integral equations. U.P.B. Sci. Bull., Ser. A **82**(3), 105–116 (2020)

On Shortest Path Problem via a Novel Neurodynamic Model: A Case Study

Amin Mansoori[1,2(✉)], Sohrab Effati[1,2], and Mohammad Eshaghnezhad[1]

[1] Department of Applied Mathematics, Ferdowsi University of Mashhad, Mashhad, Iran
am.ma7676@yahoo.com, s-effati@um.ac.ir, m_shaghnezhad@yahoo.com
[2] Center of Excellence of Soft Computing and Intelligent Information Processing (SCIIP), Ferdowsi University of Mashhad, Mashhad, Iran

Abstract. Our aim is investigating the Shortest Path Problem (SPP) by employing the Recurrent Neural Networks (RNNs). The Karush–Kuhn–Tucker (KKT) optimality conditions play an essential role to present the RNN model. In fact, the KKT conditions are reformulated as a Nonlinear Complementarity Problem (NCP). Indeed, the stability theorem of the RNN is provided. By testing some examples, we show the performance of the model. Also, we apply the approach to solve a real-world problem as a case study.

Keywords: Shortest path problem · Linear optimization problem · Recurrent neural network · Lyapunov stability

1 Introduction

One of the important problems in network programming is the shortest path problem (SPP). The problem is to find the shortest path (SP) from an origin to a specified destination in a determined network, and the aim is to minimize the total cost. The problem has many scientific and engineering applications such as road networks, routing terminal airspace air traffic, disaster decision support system, and vehicle routing problem (see [2–4,9,10,12,27,29–32,35,36]). Let $G = (V, E)$ be a directed graph, where E and V are the set of edges and vertices, respectively. The cost of a path is the sum of the weights of the edges on the path. Also, it is possible that it can be more than one path between any two vertices. Therefore, the problem is to find the best path with minimum cost. This is the so-called SPP. There are some algorithms for solving the SPP such as Dijkstra's algorithm, Bellman–Ford's algorithm, A^* search algorithm, Floyd–Warshall's algorithm, Johnson's algorithm, and Viterbi's algorithm [2,23]. Besides the classical methods, there exist some new and modified approaches. However, these algorithms may not be efficient and effective to solve the problem. The reason is for the high computational time for large scale problems. Hence, parallel approaches are effective in this way.

One of the parallel approaches for solving the optimization problems is recurrent neural networks (RNNs). They reformulate the optimization problem into a

T. Allahviranloo et al. (Eds.): IDS 2020, AISC 1301, pp. 754–770, 2021.
https://doi.org/10.1007/978-3-030-66501-2_61

dynamic system. The obtained dynamic system is in the form of the first-order differential equations. In other words, a mathematical programming problem is reformulated as a dynamic system that is called an RNN model. The main advantage of RNN approach for optimization is the inherently parallel and distributed nature. The first article on RNN model for optimization is [11]. In this case, in recent years, many papers have been published [1,5–8,16,18,19,22,26,34].

By the foregoing discussion, despite many RNNs for optimization problems, there exist some limitations for solving SPP; therefore, the current study begins by reviewing the existing literature on SPP. Following the analysis of the shortcomings and definitions of the current patterns and methods, an RNN model will be developed for solving the SPP. The presented RNN is structured with respect to the Karush–Kuhn–Tucker (KKT) conditions, and also it utilizes to solve the SPP. The construction of the given model is based on the nonlinear complementarity problem (NCP). Moreover, the stability property of the model is proved.

According to research purpose, the paper is organized as follows. Section 2 introduces the SPP. Section 3 proposes the dynamic model. In Sect. 4, the stability theorem and the global convergence property of the model are studied. Simulation results are discussed in Sect. 5. Also, we give a case study for this problem in Sect. 6. Finally, Sect. 7 states the conclusions.

2 SPP Formulation

In theory, SPP is given as a linear optimization as follows:

$$\begin{aligned} \min \quad & \sum_{(i,j)\in E} c_{ij}x_{ij} \\ s.t. \quad & \sum_{j}^{r} x_{ij} - \sum_{j}^{r} x_{ji} = \begin{cases} 1 & if\ i=1, \\ 0 & if\ i \neq 1, r \quad (i=1,\dots,r), \\ -1 & if\ i=r, \end{cases} \\ & x_{ij} = 0\ or\ 1 \qquad for\ (i,j)\in E, \end{aligned} \tag{1}$$

where c_{ij} denotes a number respecting to the edge (i, j), corresponded with the length necessary to node (i, j) from i to j and r is the number of nodes. Also, x_{ij} denotes the decision variable defined as

$$x_{ij} = \begin{cases} 1 & \text{if the edge from nodes i to j is in the path;} \\ 0 & \text{otherwise.} \end{cases}$$

If the SP is unique, then we can replace the integrality constraint in the SPP with the nonnegativity constraint. This is because the constraint coefficient matrix in

(1) has the total unimodality property; see [3]. Thus, (1) can be reformulated as

$$
\begin{aligned}
\min \quad & \sum_{i=1}^{r}\sum_{\substack{j=1\\ j\neq i}}^{r} c_{ij}x_{ij} \\
s.t. \quad & \sum_{\substack{k=1\\ k\neq i}}^{r} x_{ik} - \sum_{\substack{l=2\\ l\neq i}}^{r} x_{li} = \delta_{i1} - \delta_{ir}, \quad i = 1,\ldots,r, \\
& x_{ij} \geq 0, \quad i \neq j, \quad i = 1,\ldots,r-1, \quad j = 1,\ldots,r,
\end{aligned} \tag{2}
$$

where δ_{ij} equals to 1 when $i = j$ and equals to 0 when $i \neq j$. In accordance with the former analysis, it can be seen that the SPP is a linear programming problem.

3 RNN Model

As in the former section states, we can model the SPP as a linear programming problem. Hence the general form of the linear optimization is considered as follows:

$$
\begin{aligned}
\min \quad & c^T x \\
s.t. \quad & Ax = b, \\
& x \geq 0,
\end{aligned} \tag{3}
$$

or it can be reformulated as

$$
\begin{aligned}
\min \quad & c^T x \\
s.t. \quad & Ax \leq b, \\
& -Ax \leq -b \\
& x \geq 0.
\end{aligned} \tag{4}
$$

Also, the problem (4) can be summarized as

$$
\begin{aligned}
\min \quad & c^T x \\
s.t. \quad & Dx \leq d,
\end{aligned} \tag{5}
$$

where $D = \begin{bmatrix} A \\ -A \\ -I \end{bmatrix}$ and $d = \begin{bmatrix} b \\ -b \\ 0 \end{bmatrix}$. The KKT optimality conditions of (5) can be constructed as follows:

$$
\text{KKT Optimality Conditions}: \begin{cases} y \geq 0, \quad Dx \leq d, \\ y^T(Dx - d) = 0, \\ c + y^T D = 0, \end{cases} \tag{6}
$$

where y is the Lagrange multiplier. For introducing the RNN, we need some preliminaries. The NCP is to find a vector $x \in \mathbb{R}^n$ such that

$$x \geq 0, \quad F(x) \geq 0, \quad \langle x, F(x) \rangle = 0,$$

where $F : \mathbb{R}^n \to \mathbb{R}^n$ is a continuously differentiable function; see [25]. By reformulating the NCP to a system of nonlinear equations, some published papers have solved the problem (see [14,17]). One of the important NCP function is "min" [28], that is,

$$\phi_{\min}(x, y) = \min\{x, y\}.$$

Also, the $\phi_{\min}(x, y)$ can be differently represented as follows:

$$\phi_{\min}(x, y) = \min\{x, y\} = \frac{x + y - |x - y|}{2}.$$

The perturbed min function has the following formulation:

$$\phi^{\varepsilon}_{\min}(x, y) = \frac{x + y - |x - y + \varepsilon|}{2}, \qquad \varepsilon \to 0.$$

The following result gives the important property of $\phi^{\varepsilon}_{\min}(\cdot, \cdot)$.

Proposition 31. *For every $\varepsilon \in \mathbb{R}$, it follows that*

$$\phi^{\varepsilon}_{\min}(x, y) = 0 \Longleftrightarrow x > 0, \quad y > 0, \quad xy = f(\varepsilon),$$

where $f(\varepsilon) = \frac{\varepsilon^2}{4} + \frac{\varepsilon}{2}(x - y)$.

The proof of the above proposition is clear, so we omit the proof. Note that, $\phi^{\varepsilon}_{\min}(a, b)$ is smooth based on a and b for $\varepsilon > 0$. Here, employing the definition of NCP function, we can easily verify that the KKT optimality conditions (6) are equivalent to the following unconstrained minimization problem:

$$\min \quad \Psi(z) = \frac{1}{2}\|\Phi(z)\|^2, \tag{7}$$

where $z = (x, y)^T$ and

$$\Phi(z) = \begin{bmatrix} c + y^T D \\ \phi^{\varepsilon}_{\min}(y, -Dx + d) \end{bmatrix}.$$

It is obvious that $\Psi(z)$ is a merit function for the KKT optimality conditions (6). The merit function $\Psi(\cdot)$ in (7) is continuously differentiable for all z; see [28].

Theorem 32. *It follows that $z^* = (x^*, y^*)^T$ satisfies the equation $\Phi(z^*) = \mathbf{0}$ for every $\varepsilon \to 0$ if and only if x^* is the KKT point of* (5).

Proof. Let x^* and y^* satisfy the KKT optimality conditions (6). Therefore, from the KKT optimality conditions (6), we have $y^* \geq 0$, $-Dx^* + d \geq 0$, and ${y^*}^T (Dx^* - d) = 0$, which equal to the NCP function $\phi^{\varepsilon}_{\min}(y, -Dx + d)$. Thus, the matrix $\Phi(z^*) = \mathbf{0}$ for every $\varepsilon \to 0$. Conversely, let $z^* = (x^*, y^*)^T$ satisfy the equation $\Phi(z) = \mathbf{0}$. Therefore,

$$\begin{bmatrix} c + y^T D \\ \phi^{\varepsilon}_{\min}(y, -Dx + d) \end{bmatrix} = \mathbf{0},$$

or

$$c + y^T D = 0,$$
$$\phi^{\varepsilon}_{\min}(y, -Dx + d) = 0.$$

According to the definition of $\phi^{\varepsilon}_{\min}(\cdot, \cdot)$, the result follows. □

The RNN model associated with (3) can be described as follows:

$$\frac{dz}{dt} = -\gamma \nabla \Psi(z), \qquad z(t_0) = z_0, \tag{8}$$

where $\gamma > 0$ is the convergence rate.

4 Convergence Analysis

Here, we study the stability analysis of the proposed model (8).

Lemma 41. *The Jacobian matrix $\nabla \Phi(z)$ of the mapping $\Phi(z)$ is nonsingular.*

Proof. With a simple calculation, we have

$$\nabla \Phi(z) = \begin{bmatrix} 0 & D^T \\ -\mathrm{diag}\{\nabla_{Dx}\phi^{\varepsilon}_{\min}(y, -Dx + d)\}D & \mathrm{diag}\{\nabla_y \phi^{\varepsilon}_{\min}(y, -Dx + d)\} \end{bmatrix}.$$

Theorem 3.1 in [33] shows that $\nabla \Phi(z)$ is nonsingular. □

Lemma 42. *If A is an $n \times n$ nonsingular matrix, then the homogeneous system $AX = 0$ has only the trivial solution $X = 0$.*

Theorem 43. *It follows that $z^* = (x^*, y^*)^T$ satisfies conditions (6) if and only if it is equilibrium point of (8).*

Proof. Let z^* be the KKT point;, then according to Theorem 32, we have $\Phi(z^*) = 0$. Also, $\nabla \Psi(z) = \Phi(z)^T \nabla \Phi(z)$. Hence, we have $\nabla \Psi(z^*) = 0$. In fact, z^* is an equilibrium point of (8). For the converse, we use Lemmas 41 and 42. □

Lemma 44. *The system of (8) has the Lipschitz property.*

Proof. We are going to show that $\phi^{\varepsilon}_{\min}(\cdot,\cdot)$ is Lipschitzian. Suppose that, $\bar{x}, \bar{y}, \hat{x}, \hat{y} \in \mathbb{R}$. According to the definition of $\phi^{\varepsilon}_{\min}(\cdot,\cdot)$, we have

$$\begin{aligned}
\left\|\phi^{\varepsilon}_{\min}(\bar{x},\bar{y}) - \phi^{\varepsilon}_{\min}(\hat{x},\hat{y})\right\| &\leq \left\|\frac{\bar{x}+\bar{y}-|\bar{x}-\bar{y}+\varepsilon|}{2} - \frac{\hat{x}+\hat{y}-|\hat{x}-\hat{y}+\varepsilon|}{2}\right\| \\
&= \frac{1}{2}\left\|(\bar{x}-\hat{x})+(\bar{y}-\hat{y})-|\bar{x}-\bar{y}+\varepsilon|+|\hat{x}-\hat{y}+\varepsilon|\right\| \\
&\leq \frac{1}{2}\left(\|\bar{x}-\hat{x}\|+\|\bar{y}-\hat{y}\|+\left\||\hat{x}-\hat{y}+\varepsilon|-|\bar{x}-\bar{y}+\varepsilon|\right\|\right) \\
&\leq \frac{1}{2}\left(\|\bar{x}-\hat{x}\|+\|\bar{y}-\hat{y}\|+\|\hat{x}-\hat{y}+\varepsilon-\bar{x}+\bar{y}-\varepsilon\|\right) \\
&\leq \frac{1}{2}\left(\|\bar{x}-\hat{x}\|+\|\bar{y}-\hat{y}\|+\|\bar{x}-\hat{x}\|+\|\bar{y}-\hat{y}\|\right) \\
&= \|\bar{x}-\hat{x}\|+\|\bar{y}-\hat{y}\|.
\end{aligned}$$

Thus, $\phi^{\varepsilon}_{\min}(\cdot,\cdot)$ is a Lipschitz continuous function. □

Lemma 45. *RNN* (8) *has a unique solution* $z(t) = (x(t), y(t))^T$.

Proof. Based on Lemma 44, the system of (8) is Lipschitzian. Thus, Theorem 3.1 in [15] completes the proof for some $\tau > t_0$ over $[t_0, \tau)$ as $\tau \to \infty$. □

Lemma 46 ([15,24])**.** *An equilibrium point* x^* *is*

1. *Lyapunov stable if there exists a Lyapunov function over some neighborhood* Ω *of* x^*.
2. *asymptotically stable if [1.]exists a Lyapunov function over some neighbourhood* Ω *of* x^* *such that* $\frac{dE}{dt} < 0$ *for all* $x \neq x^* \in \Omega$.

Next result gives the main result of this section, which states the stability theorem of model (8).

Theorem 47. *The proposed model* (8) *is globally stable in the sense of Lyapunov.*

Proof. Theorem 43 and Lemma 45 give the uniqueness and the existence of the solution of (8) over $[t_0, \infty)$. Assume $\Psi(\cdot)$ as a Lyapunov function. Then, $\frac{d\Psi}{dt} = \nabla\Psi(z)^T\frac{dz}{dt}$, where $\nabla\Psi(z)$ is the Jacobian matrix. In addition, $\Psi(z^*) = 0$ and for $z \neq z^*$, $\Psi(z) > 0$. Also,

$$\begin{aligned}
\frac{d}{dt}\Psi(z) &= \nabla\Psi(z)^T\frac{dz}{dt} \\
&= -\gamma\|\nabla\Psi(z)\|^2 \\
&\leq 0.
\end{aligned}$$

Thus, $\frac{d}{dt}\Psi(z) < 0$ except for $z \neq z^*$. Therefore, employing Lemma 46 shows that (8) has the stability property in the sense of Lyapunov and it also is globally asymptotically stable. Furthermore, since the system of the model is Lipschitzian (based on Lemma 44), the proposed model is globally convergent to the optimal solution. □

5 Simulation Results

Here, several numerical simulations to display the performance of the presented model are given. The codes are done with MATLAB (ode45), and the calculations are implemented on a machine with Intel core 7 Duo processor 2 GHz and 8 GB RAM.

Example 51. *Consider the following linear programming problem obtained from an SPP in Fig. 1:*

$$\begin{aligned} \min \quad & 3x_{12} + 5x_{13} + 3x_{14} + 7x_{25} + x_{23} + x_{34} + 3x_{35} + 2x_{42} + 5x_{45} \\ s.t. \quad & x_{12} + x_{13} + x_{15} = 1, \\ & x_{23} + x_{25} - x_{42} - x_{12} = 0, \\ & x_{35} + x_{34} - x_{13} - x_{23} = 0, \\ & x_{42} + x_{45} - x_{34} - x_{14} = 0, \\ & -x_{45} - x_{25} - x_{35} = -1. \end{aligned}$$

The optimal solution for this problem is

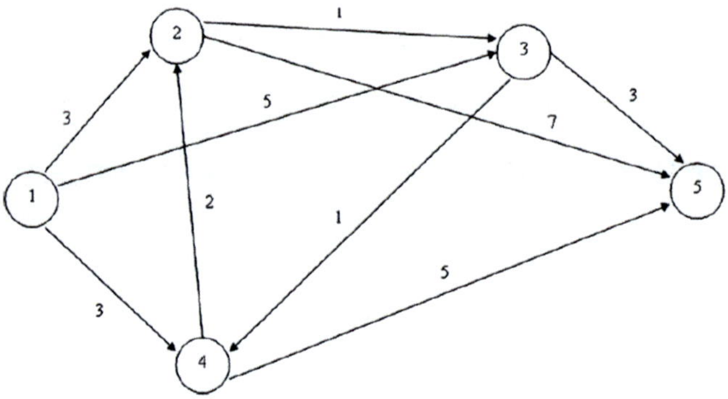

Fig. 1. The network in Example 51

$$x_{12} = x_{23} = x_{35} = 1, \quad x_{13} = x_{14} = x_{25} = x_{34} = x_{42} = x_{45} = 0.$$

We use the proposed model in (8) for solving this program. Figure 2 demonstrates the trajectories of $x(t)$ according to (8) with a random initial point and $\gamma = 3$. The shortest path according to the model (8) is (1) → (2) → (3) → (5). For comparison, we test the problem with some other models (models in [20, 21]). The results are given in Table 1.

Example 52 (Transportation network application). *Consider the transportation network depicted in Fig. 3, with 40 arcs and 23 nodes. The arc lengths are shown in Table 2. We use the presented model (8) for solving this program. Figure 4 demonstrates the behavior of $x(t)$ according to (8) with random initial point and $\gamma = 4$. The SP according to the proposed RNN (8) is* (1) → (5) → (11) → (17) → (21) → (23).

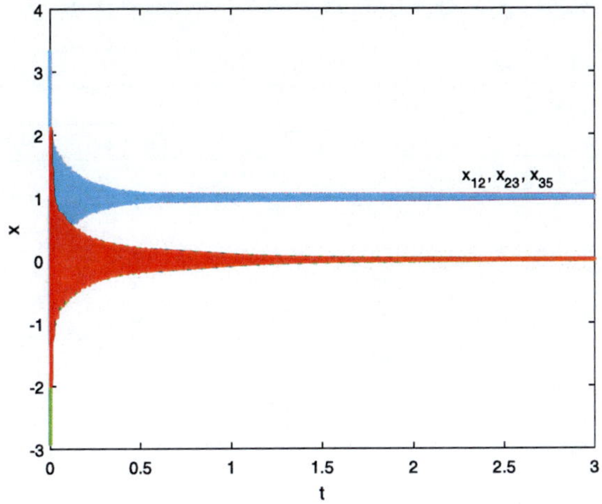

Fig. 2. Trajectories of RNN (8) with random initial points and $\gamma = 3$ in Example 51

Table 1. CPU-Time comparison about RNN (8) and RNNs in [20,21] for Example 51.

Models	RNN in (8)	RNN in [20]	RNN in [21]
CPU-Time (sec.)	0.452	0.470	0.654

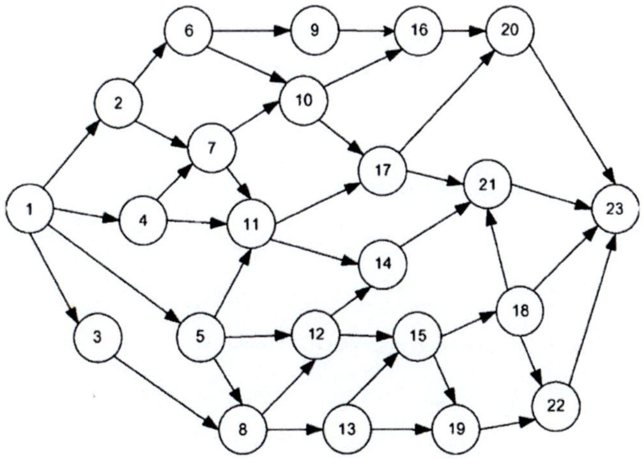

Fig. 3. A transportation network in Example 52

Table 2. Arc lengths of the transportation network in Example 52

Source node	Destination node	Arc cost	Source node	Destination node	Arc cost
1	2	14	10	17	19.5
1	3	12	11	14	10
1	4	11	11	17	10
1	5	8.5	12	14	15
2	6	12.5	12	15	14.5
2	7	11	13	15	13
3	8	13.5	13	19	18.5
4	7	21	14	21	12.5
4	11	11.5	15	18	10
5	8	10	15	19	8.5
5	11	11.5	16	20	13
5	12	14	17	20	10.5
6	9	9	17	21	7.5
6	10	12.5	18	21	17.5
7	10	11	18	22	6
7	11	7.5	18	23	8
8	12	8.5	19	22	16.5
8	13	6.5	20	23	15
9	16	8	21	23	16
10	16	14.5	22	23	5.5

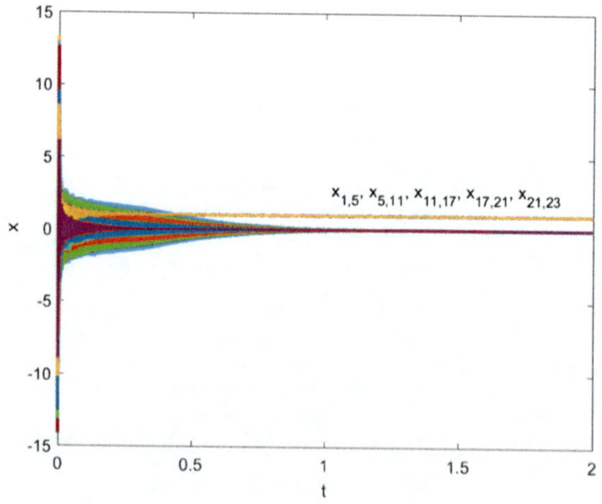

Fig. 4. Transient behaviors of RNN (8) with random initial points and $\gamma = 4$ in Example 52

Example 53. *Assume the linear optimization obtained from an SPP in Fig. 5 as follows:*

$$\begin{aligned}\min\quad & 3x_{12} + 9x_{13} + 12x_{15} + 2x_{24} + 9x_{27} + 8x_{25} + 4x_{35} + 13x_{38} \\ & + 12x_{36} + 5x_{47} + x_{57} + 16x_{59} + 6x_{58} + 2x_{68} + 15x_{79} + 4x_{89} \\ s.t.\quad & x_{12} + x_{13} + x_{15} = 1, \\ & x_{24} + x_{27} + x_{25} - x_{12} = 0, \\ & x_{35} + x_{38} + x_{36} - x_{13} = 0, \\ & x_{47} - x_{24} = 0, \\ & x_{57} + x_{59} + x_{58} - x_{15} - x_{25} - x_{35} = 0, \\ & x_{68} - x_{36} = 0, \\ & x_{79} - x_{57} - x_{27} - x_{47} = 0, \\ & x_{89} - x_{38} - x_{58} - x_{38} = 0, \\ & - x_{89} - x_{59} - x_{79} = -1.\end{aligned}$$

The optimal solution for this problem is

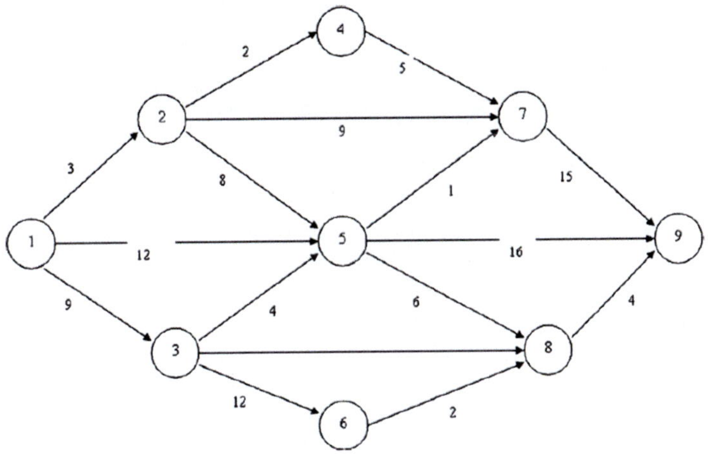

Fig. 5. The network in Example 53

$$\begin{aligned}& x_{12} = x_{25} = x_{58} = x_{89} = 1, \\ & x_{13} = x_{15} = x_{24} = x_{27} = x_{35} = x_{38} = x_{36} = x_{47} = x_{57} = x_{59} = x_{68} = x_{79} = 0.\end{aligned}$$

We use the proposed model in (8) for solving this program. Figure 6 demonstrates the behavior of $x(t)$ according to (8) with random initial point and $\gamma = 8$. The shortest path according to the proposed model (8) is (1) → (2) → (5) → (8) → (9).

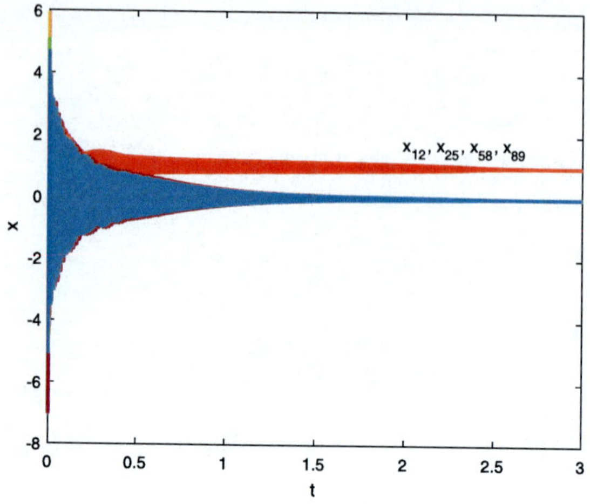

Fig. 6. Trajectories of RNN (8) with random initial points and $\gamma = 8$ in Example 53

Example 54 *(Transportation network application in gas company). A gas company owns a pipeline network to pump natural gas from its main field to its distribution center. Figure 7 depicts the network. The system has six pipeline links. The company nets* 1200 *mcf (million cubic feet) of gas per month* (c_m) *from its main field and must transport that entire amount to the distribution center. The Table 3 includes the maximum usage rates and costs corresponded to each link. The purpose of the company is to maximize the pumped gas in pipeline and minimize the cost of this performance based on the network that depicted in Fig. 7 with the costs in Table 3. Thus, the problem is formulated to minimize the ratio* $\frac{c_d}{c_m}$. *We add a real-world simulation as to show the effectiveness of*

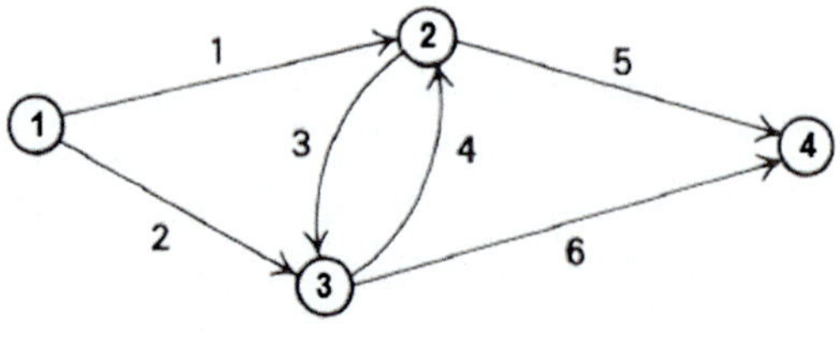

Fig. 7. The network in Example 54

Table 3. The maximum usage rates and costs in Example 54

Pipeline links	Unit	1	2	3	4	5	6
c_m: Maximum usage	$\frac{mcf}{month}$	500	900	700	400	600	1000
c_d: Tariff	$\frac{dollar}{mcf}$	20	25	10	15	20	40

the presented approach. The solution of the developed approach is the stations $(1) \to (3) \to (4)$. *Furthermore, Fig. 8 depicts the trajectories of* $x(t)$ *in (7) with random initial point and* $\gamma = 5$. *For comparison, we test the example with Dijkstra's algorithm. According to Dijkstra's algorithm, the steps of algorithm for this example can be shown as Table 4. In accordance with Dijkstra's algorithm, the optimal path for this example is* $(1) \to (3) \to (4)$, *which is the same as the optimal path obtained from RNN (8).*

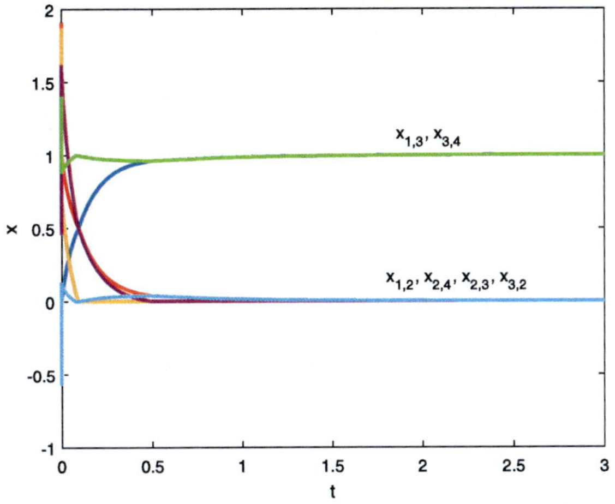

Fig. 8. Transient behaviors of RNN (8) with random initial points and $\gamma = 5$ in Example 54

6 A Case Study

Kota Kinabalu traffic network is discussed in this subsection. The ever-increasing of traffic congestion in Kota Kinabalu is caused by the abundance of vehicles on the road. Statistics showed that car ownership has been increasing from 3–4% each year and had mentioned that the number of registered vehicles had increased by 18.38% from 2012 to 2016. The increase of vehicles ownership is becoming a reflection of improvement of the nation's affordability. Therefore, nations would like to take their own vehicle, instead of public transportation to obtain a more comfortable, convenient, and efficient way to reach their destinations. If this situation remains consistent, then the traffic congestion problem will become more serious until all the road users would take a longer time to reach their destinations, especially during peak hour periods. Even though the distance is short, the congestion will cause the travel time to be longer, and consequently, burning more unnecessary fuel on the road. The aim of this part is the network

Table 4. The steps of Dijkstra's algorithm for Example 54.

Steps	Nodes	Status
1	1	Permanent
2	1	Permanent
	2	Temporary
	3	Temporary
3	1	Permanent
	2	Temporary
	3	Permanent
4	1	Permanent
	2	Temporary
	3	Permanent
	4	Temporary
5	1	Permanent
	2	Temporary
	3	Permanent
	4	Permanent
6	1	Permanent
	2	Permanent
	3	Permanent
	4	Permanent

from Inanam (source node, V1) to Centre Point (sink node, V14), where all the routes between source and sink node are established. In Fig. 9, the red nodes are part of the selected major intersections in Kota Kinabalu, which are assigned as the nodes of the network graph; see [13].

Table 5 shows the data collected in the scope of study. "Path" in Table 5 stands for the paths from a node to a node. Moreover $V1$ to $V14$ are the nodes or so-called intersection in traffic network. "Distance" stands for the length of the path from a node to a node.

The shortest path according to the developed RNN (8) is $V1 \to V2 \to V4 \to V7 \to V10 \to V14$. This case study is also solved by Dijkstra's algorithm in [13] and the solution of both techniques is the same. Figure 10 depicts the transient behavior of $x(t)$ based on RNN (8) with random initial point.

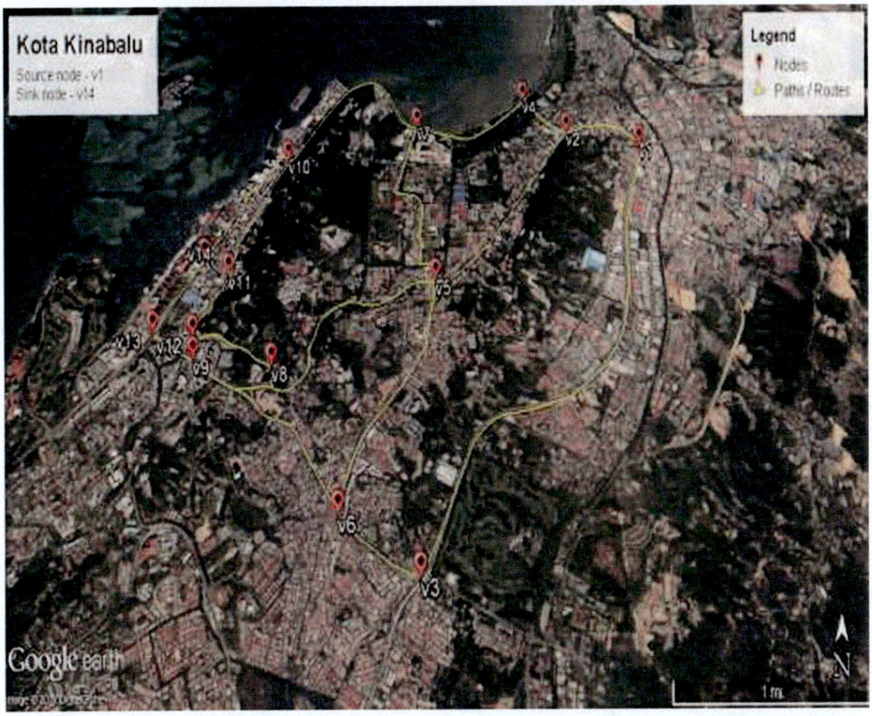

Fig. 9. Part of traffic network in kota kinabalu

Table 5. Collected data

Path	From	To	Distance	Path	From	To	Distance	Path	From	To	Distance
1	V1	V2	1.00	10	V8	V5	2.40	19	V11	V10	1.30
2	V1	V3	5.20	11	V5	V8	2.40	20	V11	V12	0.65
3	V2	V4	0.60	12	V6	V5	2.30	21	V12	V11	0.65
4	V2	V5	2.40	13	V5	V6	2.30	22	V12	V9	0.23
5	V5	V2	2.40	14	V6	V9	2.10	23	V9	V12	0.23
6	V3	V6	0.85	15	V8	V12	0.90	24	V9	V13	0.50
7	V4	V7	1.50	16	V12	V8	0.90	25	V13	V14	0.85
8	V7	V5	2.10	17	V7	V10	2.30	26	V11	V14	0.30
9	V5	V7	2.10	18	V10	V11	1.30	27	V10	V14	1.50

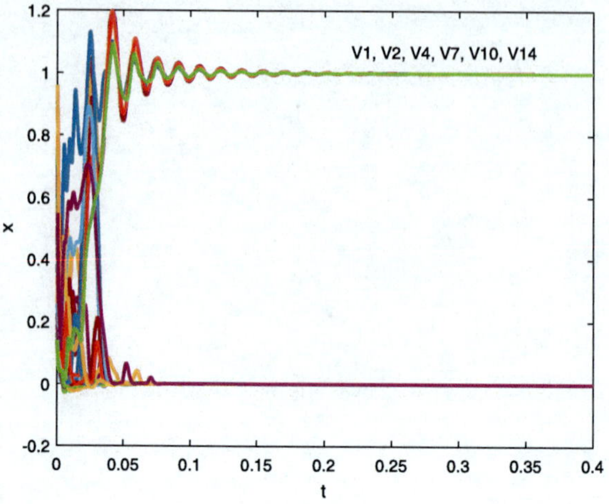

Fig. 10. Transient behaviors of RNN (8) with random initial points in case study

7 Conclusion and Discussion

Here, a new dynamic system to solve the SPP was presented. The RNN was modeled by an ODE. We should note that, the developed RNN is a one-layer structure. Simulation results showed that the presented RNN is suitable for solving the SPP in high dimension. In addition, the RNN methods do not depend on the initial point. We test all examples with random initial points, which is one of the advantages of the approach. Finally, the work is in progress to extend the other RNNs for other problems.

In accordance with theoretical and simulation results, we can summarize the advantages of the paper as follows:

- The RNN approach has accurate solutions compared with the other approaches. Also, the presented RNN is suitable to solve an SPP in high dimension.
- The RNN approaches do not depend on the initial points. This is because the model has the global convergence property.
- The RNN is utilized to solve real-world problem of transportation network application in a gas company and transportation network (see Examples 52 and 54).
- One can see the approach is useful to apply in a real-world problem (see Sect. 6: A Case Study).

Acknowledgement. This research is supported by a grant from INSF [No: 97018882].

References

1. Ahmad, I., Raja, M.A.Z., Bilal, M., Ashraf, F.: Neural network methods to solve the Lane-Emden type equations arising in thermodynamic studies of the spherical gas cloud model. Neural Comput. Appl. **28**, 929–944 (2017)
2. Ahuja, R.K., Magnanti, T.L., Orlin, J.B.: Network Flows: Theory, Algorithms, and Applications. Prentice Hall, Englewood Cliffs, New Jersey (1993)
3. Bazaraa, M.S., Shetty, C., Sherali, H.D.: Nonlinear Programming, Theory and Algorithms. Wiley, New York (1979)
4. Breugem, T., Dollevoet, T., den Heuvel, W.: Analysis of FPTASes for the multi-objective shortest path problem. Comput. Oper. Res. **78**, 44–58 (2017)
5. Cheng, L., Hou, Z.-G., Tan, M.: A neutral-type delayed projection neural network for solving nonlinear variational inequalities. IEEE Trans. Circuits Syst. II Express Briefs **55**(8), 806–810 (2008)
6. Effati, S., Mansoori, A., Eshaghnezhad, M.: An efficient projection neural network for solving bilinear programming problems. Neurocomputing **168**, 1188–1197 (2015)
7. Eshaghnezhad, M., Effati, S., Mansoori, A.: A neurodynamic model to solve nonlinear pseudo-monotone projection equation and its applications. IEEE Trans. Cybern. **47**(10), 3050–3062 (2017)
8. Eshaghnezhad, M., Rahbarnia, F., Effati, S., Mansoori, A.: An artificial neural network model to solve the fuzzy shortest path problem. Neural Process. Lett. **50**(2), 1527–1548 (2019)
9. Fischer, F., Helmberg, C.: Dynamic graph generation for the shortest path problem in time expanded networks. Math. Program. **143**(1), 257–297 (2014)
10. Golovin, D., Goyal, V., Polishchuk, V., Ravi, R., Sysikaski, M.: Improved approximations for two-stage min-cut and shortest path problems under uncertainty. Math. Program. **149**(1), 167–194 (2015)
11. Hopfield, J.J., Tank, D.W.: Neural computation of decisions in optimization problems. Biol. Cybern. **52**, 141–152 (1985)
12. Hu, X.-B., et al.: Finding the k shortest paths by ripple-spreading algorithms. Eng. Appl. Artif. Intell. **87**, 103–229 (2020)
13. Hua, T.K., Abdullah, N.: Least congested and shortest distance path in Kota Kinabalu traffic network. Sabah, Malaysia, Int. J. Eng. Sci. Res. Technol. **6**(12), 104–114 (2017)
14. Kanzow, C.: Nonlinear complementarity as unconstrained optimization. J. Optim. Theory Appl. **88**, 139–155 (1996)
15. Khalil, H.K.: Nonlinear Systems. Prentice-Hall, Michigan (1996)
16. Majeed, K., Masood, Z., Samar, R., Raja, M.A.Z.: A genetic algorithm optimized Morlet wavelet artificial neural network to study the dynamics of nonlinear Troesch's system. Appl. Soft. Comput. **56**, 420–435 (2017)
17. Mangasarian, O.L.: Equivalence of the complementarity problem to a system of nonlinear equations. SIAM J. Appl. Math. **31**, 89–92 (1976)
18. Mansoori, A., Effati, S.: An efficient neurodynamic model to solve nonlinear programming problems with fuzzy parameters. Neurocomputing **334**, 125–133 (2019)
19. Mansoori, A., Effati, S.: Parametric NCP-based recurrent neural network model: a new strategy to solve fuzzy nonconvex optimization problems. IEEE Trans. Syst. Man Cybern. Syst. (2020). https://doi.org/10.1109/TSMC.2019.2916750.
20. Mansoori, A., Erfanian, M.: A dynamic model to solve the absolute value equations. J. Comput. Appl. Math. **333**, 28–35 (2018)

21. Mansoori, A., Eshaghnezhad, M., Effati, S.: An efficient neural network model for solving the absolute value equations. IEEE Trans. Circuits Syst. II Express Briefs **65**(3), 391–395 (2018)
22. Mansoori, A., Eshaghnezhad, M., Effati, S.: Recurrent neural network model: a new strategy to solve fuzzy matrix games. IEEE Trans. Neural Netw. Learn. Syst. **30**(8), 2538–2547 (2019)
23. Mensah, D.N.A., Gao, H., Yang, L.W.: Approximation algorithm for shortest path in large social networks. Algorithms **13**, 36–48 (2020)
24. Miller, R.K., Michel, A.N.: Ordinary Differential Equations. Academic Press, Berkeley (1982)
25. Miri, S.M., Effati, S.: On generalized convexity of nonlinear complementarity functions. J. Optim. Theory Appl. **164**(2), 723–730 (2015)
26. Mohammadi, M., Mansoori, A.: A projection neural network for identifying copy number variants. IEEE J. Biomed. Health Inform. **23**(5), 2182–2188 (2018)
27. Noga, A., Yuval, E., Michal, F., Moshe, T.: Economical graph discovery. Oper. Res. **62**(6), 1236–1246 (2014)
28. Pang, J.S.: A posteriori error bounds for the linearly-constrained variational inequality problem. Math. Oper. Res. **12**, 474–484 (1987)
29. Pugliese, L.D.P., Guerriero, F., Poss, M.: The resource constrained shortest path problem with uncertain data: a robust formulation and optimal solution approach. Comput. Oper. Res. (2019). https://doi.org/10.1016/j.cor.2019.03.010
30. Sever, D., Zhao, L., Dellaert, N., Demir, E., Van Woensel, T., De Kok, T.: The dynamic shortest path problem with time-dependent stochastic disruptions. Transp. Res. Part C Emerg. Technol. **92**, 42–57 (2018)
31. Smith, O.J., Boland, N., Waterer, H.: Solving shortest path problems with a weight constraint and replenishment arcs. Comput. Oper. Res. **39**, 964–984 (2012)
32. Soueres, P., Laumond, J.-P.: Shortest paths synthesis for a car-like robot. IEEE Trans. Autom. Control **41**(5), 672–688 (1996)
33. Sun, J., Zhang, L.: A globally convergent method based on Fischer-Burmeister operators for solving second-order cone constrained variational inequality problems. Comput. Math. Appl. **58**, 1936–1946 (2009)
34. Xia, Y., Wang, J.: A bi-projection neural network for solving constrained quadratic optimization problems. IEEE Trans. Neural Netw. Learn. Syst. **27**(2), 214–224 (2015)
35. Xu, K., Zeng, Y., Zhang, Q., Yin, Q., Sun, L., Xiao, K.: Online probabilistic goal recognition and its application in dynamic shortest-path local network interdiction. Eng. Appl. Artif. Intell. **85**, 57–71 (2019)
36. Yu, H., Bertsekas, D.P.: On boundedness of Q-learning iterates for stochastic shortest path problems. Math. Oper. Res. **38**(2), 209–227 (2013)

Reliability and Sensitivity Analysis of Steel Frames with Initial Random Imperfections

Masoud Negin(✉)

Department of Civil Engineering, Bahcesehir University, Istanbul, Turkey
masoud.negin@eng.bau.edu.tr

Abstract. In this study, the effects of the different initial random imperfections on the reliability of steel frames are studied. Different types of random imperfections are considered, including imperfections in the mechanical properties of the material, imperfections in external forces and imperfections in the geometry of the structural system. Two forms of geometric imperfections, namely geometric imperfections in the coordinates of the joints and geometric imperfections of the members due to the presence of initial curvature in the theoretical direction of different structural members, which is less studied in previous investigations are taken into account. Static nonlinear pushover analysis is performed, and the reliability of the desired system is calculated using the first-order reliability and Monte Carlo simulation methods. In addition, applying the direct differentiation sensitivity analysis the importance of each of these random variables are evaluated in terms of the impact they have on the behavior of the structural system. The results show that some of the mentioned geometric imperfections are of special importance in the reliability of steel frames, which indicates how even small variations in those parameters affect the probability of failure of the structural systems.

Keywords: Steel frames · Random imperfections · Uncertainty · Reliability · Sensitivity analysis

1 Introduction

In the last two decades, the theory of structural reliability has been the subject of many studies and researches. In fact, the behavior of real structural systems is inevitably dependent on some sources of uncertainties or random parameters. Especially in the case of civil engineering structures, these uncertainties may appear in the form of variations in external forces, environmental conditions, boundary conditions, geometric parameters, or material properties. As a result, the behavior of the structures is essentially random and it is necessary to consider these uncertainties in system parameters in estimating the actual behavior of the structures.

On the other hand, in recent years, with the rapid development and advancement of digital computers, various numerical methods have been used in engineering problems to achieve the optimal design of structures. Among these, the finite element method has been a successful tool in the analysis of engineering problems in various fields.

T. Allahviranloo et al. (Eds.): IDS 2020, AISC 1301, pp. 771–784, 2021.
https://doi.org/10.1007/978-3-030-66501-2_62

The combination of finite element method with new and advanced structural reliability analysis techniques has led to the development of the reliability-based finite element method. In this method, the desired safety level is tried to be obtained in terms of the corresponding performance function of the structure. Two very common and well-known methods of structural reliability analysis are the first-order reliability method (FORM) and more accurate Monte Carlo (MC) simulation method.

In this context, numerous studies have been conducted so for by many researchers, considering the effects of some of these random variables in their investigations, some recent investigations can be summarized as [1–20]. In the present study though some mechanical properties of the material, lateral forces applied to the steel frame structures and initial geometric imperfections are considered as random variables. It should be noted that these geometric imperfections come in two forms, namely geometric imperfections in the coordinates of the nodes connecting the structural members to each other and geometric imperfections due to the presence of curvature in the initial and theoretical directions of different structural members, which is less studied in previous investigations. To calculate the reliability of the system, static and nonlinear pushover analysis in the form of first-order reliability analysis and Monte Carlo simulation is used. In addition, the sensitivity of these systems to the variations in each of the random variables of the problem and its effect on the structural behavior are evaluated.

2 Analytical Model

Numerical studies are carried out on a two-story one bay steel frame, the geometry of which is shown in Fig. 1. Stress-strain behavior of the steel is specified with three parameters: Modulus of elasticity E, yield strength F_y and strain-hardening ratio α (Fig. 2).

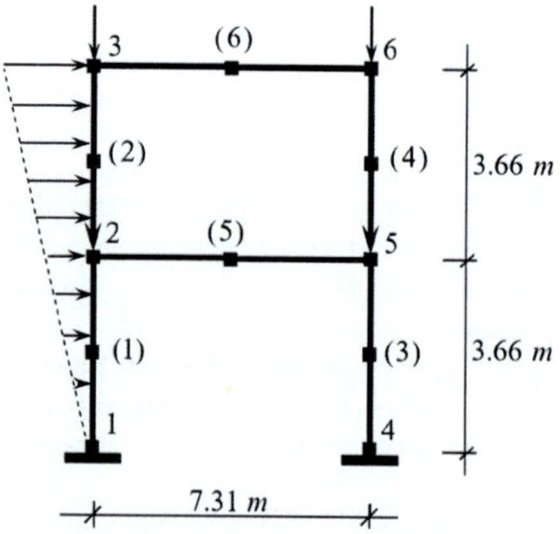

Fig. 1. Geometry of the steel frame.

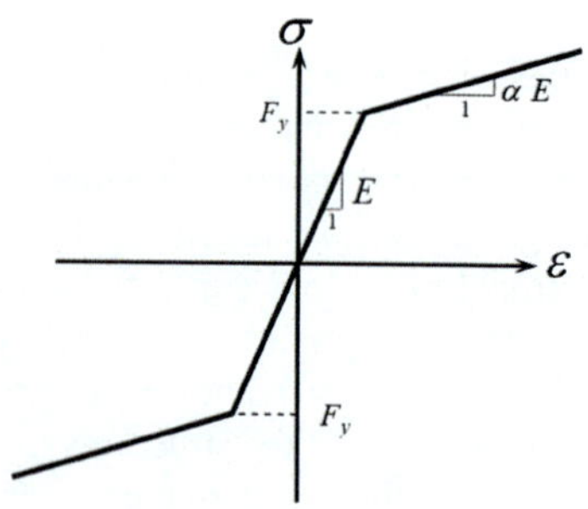

Fig. 2. Stress-strain diagram of the steel.

The node numbers of the frame used in the structural analysis, as well as the beam and column numbers are indicated in the figure. Each member of the frame is created using a non-linear beam-column element that is integrated at four points in OpenSees. Integration is done according to the Gauss-Lagrange law, where Bernoulli beam assumptions are satisfied [21]. The first and second floor columns are chosen as W14 × 53 and W14 × 43, while the first and second floor beams are chosen as W21 × 50 and W18 × 40 wide flange steel profiles, respectively. Vertical loads on the first and second floors are 100 kN and 50 kN, respectively, however, the lateral force starting from zero at the base varies linearly at the height of the frame and takes its maximum value of 200 kN at the roof level.

3 Reliability-Based Finite Element Method

Among the modern structural analysis methods, finite element method is undoubtedly one of the most powerful and successful tools in solving various engineering problems with required accuracy. Nevertheless, there are still some limitations related to the application of random parameters, such as variations in the mechanical properties of the material or in the geometric shapes and dimensions of the structural members, presence of defects, random nature of the external forces, etc. into the problem. Therefore, newer ways known as stochastic finite element methods SFEM, have become widespread to overcome these limitations. By applying the probability theory and considering the probabilistic nature of the problem variables in the analysis, SFEM has opened new horizons for engineers and researchers that can make more realistic predictions about the behavior of structures. Meanwhile, combination of finite element methods with advanced reliability analysis algorithms has led to the emergence of reliability-based finite element methods. In other words, reliability-based finite element methods refer to those methods that try to predict the random behavior of real structural systems by establishing a mutual relationship between finite element method and advanced reliability analysis methods such as simulation method or first order reliability method. It should be noted that since the failure of structures occurs mostly in the range of their nonlinear behavior, it is necessary to consider the inelastic behavior of materials as well as the geometric nonlinearity effects of the problem in these analyzes. The first link between the finite element analysis method and the first-order reliability analysis method was established in 1983 by Der Kiureghian and Taylor [22]. Since then, new, and more advanced methods have been continuously developed by various researchers and more by Der Kiureghian and his colleagues [23].

In general, the main purpose of different types of reliability analysis is to predict the probability of failure of structural systems by solving the following integral:

$$P_f = \iint \underset{g(X) \leq 0}{\cdots} \int f_X(x)dx \tag{1}$$

where X is the vector of random variables and $g(X)$ is the failure function or limit state function of the system in question. Therefore, $g(X) \leq 0$ is the failure region and $f_X(x)$ is the combined density function of the random variables [24]. A very important point to be noted here is that in general case the failure function can be defined not only in terms of the main variables of the problem, but also in terms of the response quantities of the system under study. These quantities include quantities such as strains, stresses, nodal displacements, or structural reactions that are derived from finite element analysis. This means that a finite element analysis is required at each stage to evaluate the failure function of the system, at the same time, the updated values of the random variables should be used in each step of the finite elements analysis. In fact, in this method, there is always a two-way interaction between finite element analysis and reliability evaluation and is therefore called the reliability-based finite element method.

3.1 First Order Reliability Method

In practice failure functions are usually non-linear. In these cases, the problem is not as simple as linear problems, and researchers are often forced to use approximate methods. For this purpose, usually the Taylor series expansion is used, to expand the nonlinear failure function, for example, at the mean point of random variables. In reliability contexts, this method is commonly referred to as the first order second moment FOSM method. For this purpose, assuming that X is the vector of random and main variables of the problem and $g(X)$ is the limit state function or the failure function of the system under study, using the Taylor series expansion at the mean point of the random variables, the nonlinear failure function is approximated as follows:

$$g(X) \approx g(\mu) + \nabla g\,(X - \mu) \tag{2}$$

where μ and ∇g are the mean and the gradient vectors of the failure function of the system with respect to the main random variables of the problem, respectively. As a result, the expectation and the variance of the failure function can be calculated using the following equations:

$$E[g] = \mu_g \approx g(\mu) \tag{3}$$

$$Var[g] = \sigma_g^2 \approx \nabla g \cdot \sum \cdot \nabla g^T \tag{4}$$

where $\sum$ is the covariance matrix of random variables. As a result, the reliability index β for this case is obtained from the following equation:

$$\beta = \frac{g(\mu)}{\sqrt{\nabla g \cdot \sum \cdot \nabla g^T}}. \tag{5}$$

Therefore, the corresponding probability of failure can always be calculated from the following equation [25]:

$$P_f = \Phi(-\beta). \tag{6}$$

This method, despite its simplicity, has some important disadvantages. For example, when the failure function is linearized around the mean point of the random variables, depending on the choice of failure functions this method can produce different values for the reliability index for the same problem [26]. To solve this problem, Hasofer and Lind [27] suggested to perform this linearization in the design point of the failure surface represented in normalized space. They also suggested to use the shortest distance from the origin of the coordinates to the failure surface as the reliability index β as illustrated in the two dimensional space in Fig. 3. The design point denoted by y^* is located on the failure surface with the shortest distance from the origin and indicates the most probable failure point of the system in question.

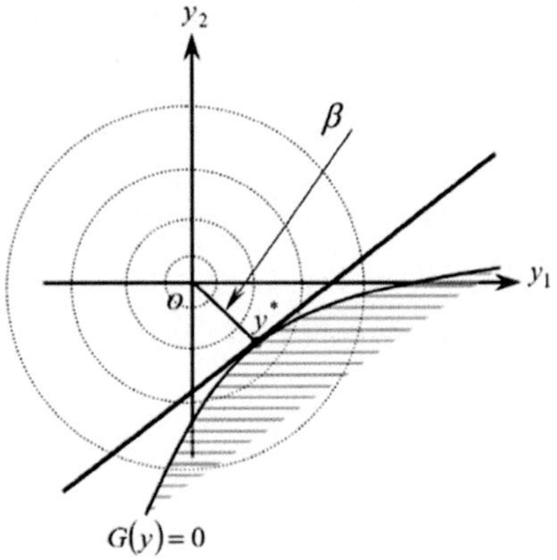

Fig. 3. Reliability index (FORM).

In this method, it is common to calculated vector α from the normalization of the gradient vector of the failure function at the design point as follows:

$$\alpha = -\frac{\nabla G}{\|\nabla G\|}. \tag{7}$$

In this case, the reliability index can be obtained from the following equation:

$$\beta = \alpha \cdot y^*. \tag{8}$$

As a result, this method is reduced to a kind of mathematical minimization problem as:

$$y^* = \min\{\ \|y\| \ \mid\ G(y) \le 0\}. \tag{9}$$

In this case, if necessary, the probability of system failure can be obtained from the approximate relationship $P_f \approx \Phi(-\beta)$. In the context of reliability analysis, this method is known as the first-order reliability method (FORM).

3.2 Monte Carlo Simulation Method

In the process of predicting the behavior of physical phenomena or structural systems, there are situations that due to the random nature and complexity of the problem formulation, the description of the behavior of the system cannot be provided mathematically; or if a mathematical model can describe the behavior of the system in question, it is not possible to obtain a direct analytical solution for the corresponding mathematical model. In such cases, due to the difficulty of formulating the problem mathematically, simulation is the only tool used to obtain the desired answers. In other words, simulation is actually a special method that can be used to obtain some numerical results related to the physical phenomena under study without performing the necessary experiments. Among the methods used in simulation, the Monte Carlo simulation method is one of the most common and widely used methods in solving complex engineering problems in various fields.

In Monte Carlo method, in general, the probability of the failure of the system is calculated by generating random numbers with the desired distribution and then performing the required number of analyzes for the case where the failure function satisfies $g(X) \leq 0$. Mathematically, this article is written as follows:

$$P_f = \int_{-\infty}^{+\infty} I\left[g(X) \leq 0\right] f_X(x) dx, \tag{10}$$

where $I\left[g(X) \leq 0\right]$ is the characteristic function of the system whose value is assumed to be equal to one when the corresponding argument is true and otherwise equal to zero. In practice, the probability of failure is obtained by creating a limited number of random numbers, so the calculated failure probability is only an estimate of the actual failure probability. As a result, the probability of failure in practice is calculated from the following equation [28]:

$$\overline{P_f} \approx E\left[P_f\right] = \frac{1}{N} \sum_{i=1}^{n} I\left[g(X_i) \leq 0\right]. \tag{11}$$

However, as the number of simulations increases, this estimate gets closer to the actual value. Although this method is generally costly and time consuming, due to the rapid increase in digital computing capacities and the improvement of analytical algorithms in simulation methods it is still widely used in solving engineering problems because of its special features and capabilities.

4 Basic Concepts

4.1 Failure Function

In reliability analyzes, the concept of limit state is used to define system failure. In other words, the limit state determines the boundary between the desired and undesirable performance of the structure. This boundary is often represented mathematically by the limit state function or the failure function. It is emphasized that the term failure may not mean the total failure of the structure or system in question. In fact, if the structure does not have the desired function the term failure is used.

In general, in reliability analysis, two types of limit state functions, namely serviceability limit state functions and strength limit state functions are common. Each of these limit states must be considered separately, but since in this paper the final capacity of steel frames is considered, only the second group, i.e. strength limit state functions are used.

4.2 Importance Measures

Most reliability algorithms can also calculate system response gradients to random parameters or variables as part of first-order reliability method FORM analysis. A very important consequence of these calculations is the introduction of criteria to measure the importance of random variables of the system under study, as a result of which it is possible to arrange these parameters in terms of their impact on system behavior or response of the structure and ultimately the reliability of the system. For this purpose, OpenSees software introduces four importance vectors of random variables [29]. The first vector is called α and is defined as follows:

$$\alpha = -\frac{\nabla G}{\|\nabla G\|}. \tag{12}$$

It can be shown that the elements of this vector show the effect of standardized normal random variables on the variance of the failure function of the system, which is linearized at the design point. It should be noted that the vector α is valid only for non-correlated random variables and has a more limited application; Therefore, this vector is not used in this study in the reliability assessment of the steel frame in question in which the random variables are defined in a correlated manner.

The second vector or importance measures is the vector γ, which is defined as follows:

$$\gamma = \frac{\alpha J_{y,\,x}\hat{D}}{\left\| \alpha J_{y,\,x}\hat{D} \right\|} \tag{13}$$

where $\hat{D}$ represents the diameter matrix of the standard deviations and $J_{y,\,x}$ is the Jacobian of transformation at the design point. The elements of the vector γ, like the vector α, represent the effect or the degree of participation of the main random variables in the variance of the linearized failure function in the original space of the mentioned variables. Also, if the random variables of the problem are independent, it is proved that the vector γ will be equal to the vector α [23].

The third vector, called δ, is defined as follows:

$$\delta = \nabla_{\mu}\beta\hat{D} \tag{14}$$

where $\nabla_{\mu}\beta$ is the gradient vector of the reliability index β with respect to the mean of the random variables of the problem with constant standard deviations. The dimensionless elements of this vector actually show the sensitivity of the reliability index β to the mean value of the random variables.

The fourth and last vector is η, which is defined by the following equation:

$$\eta = \nabla_{\sigma}\beta\hat{D} \tag{15}$$

where $\nabla_{\sigma}\beta$ is the gradient vector of the reliability index β with respect to the standard deviation the random variables of the problem with constant mean values. As can be seen, while the elements of the vector δ show the relative importance of the random variables of the problem relative to the variations in their mean values, the elements of the vector η reveal their relative importance to the variations in the standard deviation of the main random variables of the problem.

Therefore, the three vectors γ, δ, and η offer valuable perspectives and important physical interpretations on the reliability of structural systems, and later in the study, these measures will be used to evaluate the importance of random variables in the problem.

5 Numerical Results

To investigate the reliability of the steel frame, static nonlinear pushover analysis is performed, and all parameters related to the material and geometry of the structure as well as lateral forces are considered as random variables.

The modulus of elasticity E for each member is modeled as a random variable with a log-normal distribution with a mean value of 210000 MPa and a standard deviation of 5% so that the correlation coefficient of different members is assumed to be 0.3. The yield stress F_y of steel for each member is assumed to be a random variable with a log-normal distribution with a mean value of 240 MPa and a standard deviation of 10% with a correlation coefficient of 0.3. The strain-hardening ratio α for each member is considered as a random variable with log-normal distribution with mean, standard deviation, and correlation coefficient between different members as 0.02, 10% and 0.3, respectively. The initial geometric imperfections in the steel frame are modeled as random variables at the nodal coordinates in both horizontal and vertical directions. It should be noted that the initial geometric imperfections in general may have occurred in two different ways. The first type is the geometric imperfections occurred in the coordinates of the joints, which are usually caused by errors in the construction or installation stages of structure. The second type of geometric imperfections are imperfections because of the presence of initial curvature in the original and the theoretical directions of different structural members. For the geometric imperfections of the first type, each of the x and y coordinates of the connecting nodes of the members are considered as uncorrelated random variables with normal distribution. The standard deviation of the vertical coordinates of the nodes is assumed to be 10 mm. While the standard deviation of the horizontal coordinates of each node, due to the possibility of more errors in the horizontal direction in the upper floors because of lateral movement of the structure, starts from 10 mm at the base level and varies up to 20 mm at the roof level. On the other hand, for modeling the imperfections of the second type, i.e. geometric imperfections due to the presence of curvature in the initial direction of the members, as shown in Fig. 1, extra nodal points are created at the middle of all columns and beams of the frame. Each of the x coordinates of these nodes in the columns and each of the y coordinates of these nodes in the beams are considered as non-correlated random variables with normal distribution and a standard deviation of 10 mm. In addition, to study the importance and effect of geometric imperfections and material properties on the behavior of steel frames in relation to other random parameters or uncertainties, the vertical forces are considered as constant forces, while the lateral forces are considered as random variables with log-normal distribution and standard deviation equal to 20%, so that the correlation coefficient between the lateral forces in the first and second floors of the steel frame is assumed to be 0.6. Finally, considering the above assumptions, the reliability of the system is evaluated based on the following limit state or the failure functions:

$$\begin{cases} g_1 = 2\% \times h_0 - u_3 \\ g_2 = 2\% \times h - (u_3 - u_2) \\ g_3 = 2\% \times h - u_2 \end{cases} \tag{16}$$

In the first function h_0 is the total height of the frame here is 7.31 m and u_3 is the horizontal displacement of the frame at node 3 or at roof level. In the second and third functions, h is the height of the floors here 3.66 m and inside the parentheses, are the drift values from top to bottom, respectively. The reliability index and its corresponding failure probability for each of the above failure functions have been calculated using first-order reliability analysis FORM and performing 10,000 Monte Carlo MC simulations. Table 1 summarizes the results obtained for all the above cases. As mentioned, in FORM analysis, the design point, i.e. the most probable failure point, is used instead of the mean point of random variables.

Table 1. The reliability indices and corresponding failure probabilities.

	FORM		MC	
	β	P_f	β	P_f
g_1	2.0359	0.02088	2.0163	0.0219
g_2	2.0396	0.02069	2.0237	0.0215
g_3	2.0389	0.02073	2.0179	0.0218

As can be seen, the first failure function in this structure has a lower reliability index and has a more critical situation than other failure functions. In fact, this function provides the probability of system failure for the case where the horizontal displacement of the structure at the roof level is more than 2% of the total height of the structure. Therefore, in this study, we will consider this function as a system failure function. The statistical parameters of the response quantities of the steel frame, including the mean and standard deviation of the response quantities of the system defined as the horizontal displacement of node 3, i.e. the lateral displacement of the steel frame are obtained as $\mu = 7.56$ mm and $\sigma = 12.5$ mm, respectively.

As mentioned, one of the most important capabilities of reliability analysis implemented in OpenSees software is to obtain criteria for evaluating the importance of each random variable in the behavior of the system. These three vectors i.e. γ, δ, and η are the most important of these measures, which are generated together with other results and output information of FORM analysis in OpenSees. The values of these parameters are calculated for each of the random variables of the steel frame and Tables 2 shows the first 20 important random parameters of the problem in order of their importance in the behavior of the system in terms of the values of γ, δ, and η, respectively.

As can be seen in Table 2, the random parameters related to the variations in the lateral forces applied to the system shown by P in the table, as well as the variations in

Table 2. First 20 important random parameters of the steel frame.

FORM						
	γ		δ		η	
1	Node 3	P	Node 3	P	Node 3	P
2	Node 2	P	Node 2	P	Node 2	P
3	Member 5	F_y	Member 5	F_y	Member 5	F_y
4	Member 1	F_y	Member 1	F_y	Member 1	F_y
5	Member 3	F_y	Member 3	F_y	Member 3	F_y
6	Member 2	F_y	Member 2	F_y	Member 2	F_y
7	Member 4	F_y	Member 4	F_y	Member 4	F_y
8	Node 5	x	Node 5	x	Member 6	F_y
9	Member 5	E	Member 5	E	Member 5	E
10	Node 4	x	Node 4	x	Member 3	E
11	Node 6	x	Node 6	x	Member 1	E
12	Member 3	E	Member 3	E	Member 4	E
13	Node 4	y	Node 4	y	Member 5	α
14	Node 1	y	Node 1	y	Member 1	α
15	Member 1	E	Member 1	E	Member 3	α
16	Member 4	E	Member 4	E	Member 2	E
17	Member 6	F_y	Member 6	F_y	Member 6	E
18	Node 2	x	Node 2	x	Node 5	x
19	Node 3	y	Node 3	y	Node 4	x
20	Node 5	y	Node 5	y	Member 3	α

the random parameters related to the material properties of the system and mostly the yield strength of steel are among the most important parameters of the system. Random parameters related to the initial geometric imperfections in the steel frame, which are modeled as variations in the coordinates of the connecting nodes of the members and the initial curvature of the members are of considerable importance in the reliability of this system and many of these important parameters are considered as the primary parameters of the system. Among these, the horizontal displacement of nodes 5, 4, 6 and 2 and also the vertical displacement of nodes 4, 1, 3 and 6 are among the 20 important parameters of the system. Another noteworthy point is that the random variations related to the support nodes, i.e. the variations of horizontal and vertical coordinates of nodes 1 and 4 are also among these most important parameters. In other words, the geometric imperfections in the supports of the system are also of great importance in its reliability. For further investigation, the sensitivity of the steel frame to the variations of each of the random parameters of the problem in terms of their effect on the reliability of the system are obtained. This is done through reliability analyzes for different values of the standard deviations of random parameters or different values of initial geometric imperfections and the results are shown in Figs. 4 and 5. In fact, for each point shown on these curves one FORM analysis is performed in terms of different values of the standard deviations of the random variables or the initial geometric imperfections.

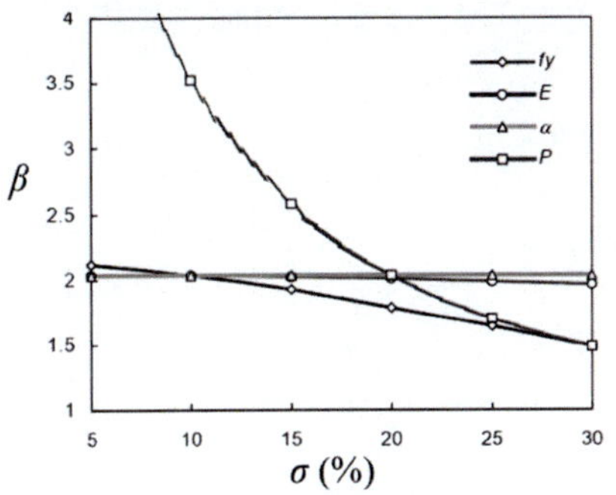

Fig. 4. Sensitivity to random variables related to lateral forces and material properties.

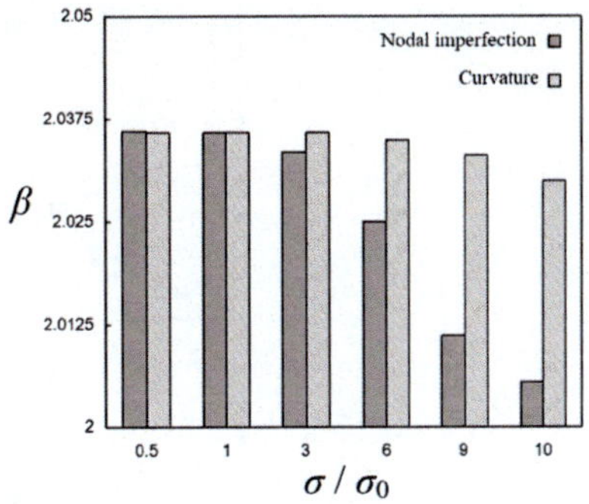

Fig. 5. Sensitivity to random variables related to geometric imperfections.

These figures show that the steel frame is very sensitive to random variations in lateral forces and yield strength of material and reliability of the structure decreases nonlinearly with increasing variations in these parameters. However, reliability of the frame is less sensitive to random variations in the modulus of elasticity as well as the strain-hardening ratio of the steel and decreases almost linearly. Furthermore, random geometric imperfections of the type of variations in the nodal coordinates of the structure have a greater effect on the capacity and ultimately on the reliability of the structure than random geometric imperfections of the type of initial curvature in the direction of structural members.

In addition, to complete the system response statistics, the lateral displacement probability distribution of the steel frame at the roof level, which is related to the horizontal displacement of node 3, can be obtained through sequential first-order reliability analysis FORM or by performing parametric studies. In this regard, to evaluate the effect of random variables related to geometric imperfections, the reliability of the system is calculated once with geometric imperfections and then without these imperfections. For this purpose, the following two conditions are considered: a) All system variables are considered as random. b) All variables except variables related to geometric imperfections are considered as random. The results are given in Fig. 6 as PDF probability density function and the corresponding complementary cumulative distribution function CCDF for the lateral displacement of the structure at the roof level. In fact, for each point on the CCDF diagram, one reliability analysis was performed. These diagrams provide some useful information about the probability distribution of lateral displacement of the frame at roof level which is of particular importance in the design processes. For example, it can be seen that the lateral displacement is always greater than 50 mm with a 74% probability; or between 40 mm and 60 mm with a 61% probability.

In addition, the mean and standard deviation of the lateral displacement of the structure at different levels of the frame are given in Fig. 7. These graphs show variations of

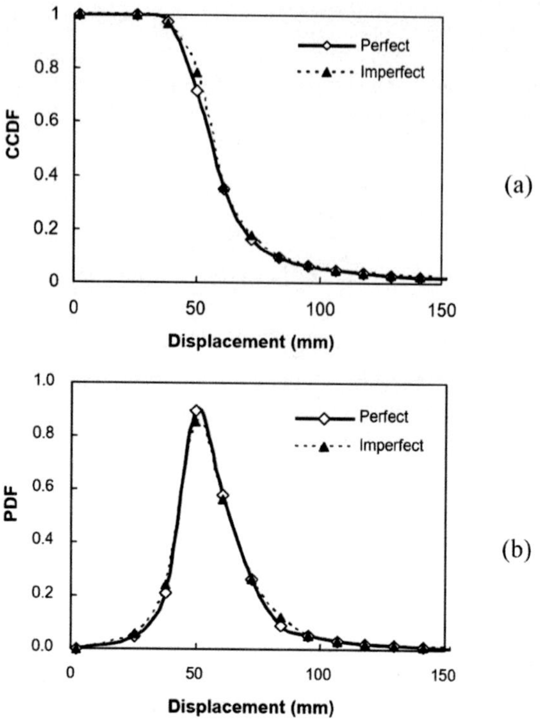

Fig. 6. (a) Complementary cumulative distribution function and (b) probability density function of lateral displacement of the structure at roof level.

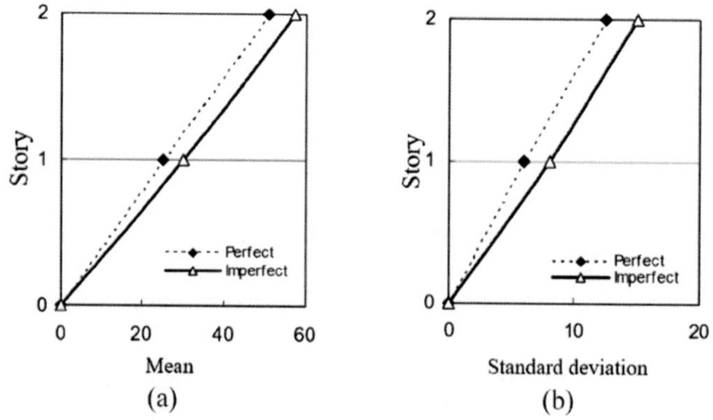

Fig. 7. (a) Mean and (b) standard deviation of the lateral displacement at roof level.

the mean and standard deviation of the lateral displacement at first and second floors of the frame, obtained by 10,000 Monte Carlo simulations analysis in two different modes, i.e. the case where the structure is considered as geometrically perfect and the case in the presence of primary geometric imperfections.

6 Conclusion

In this study, the effects of the different initial random imperfections including imperfections in the mechanical properties of the material, imperfections in external forces and imperfections in the geometry of the structural system on the reliability of steel frames are studied. The main conclusions are:

- Random parameters related to the variations of lateral forces are among the first important parameters of the steel frame and therefore due to the inherent random nature, extensive and unpredictable variations in lateral forces such as seismic forces or wind forces, they play an important role in the reliability assessment of steel frames.
- Random geometric imperfections in steel frame are of considerable importance in the reliability of these systems and many of these parameters are among the important and primary parameters. Specially, random variations in the coordinates of structural supports have important effects.
- Random geometric imperfections of the type of variations in the nodal coordinates of the structure have a greater impact than random geometric imperfections of the type of initial curvature in the direction of structural members.
- Steel frames are very sensitive to variations in random variables of yield strength of the steel, and their reliability immediately decreases nonlinearly with increasing in variations of these parameters. However, the effect of random parameters related to modulus of elasticity and strain-hardening ratio of steel is almost linear and has relatively less sensitivity.

References

1. Ellingwood, B.R.: Earthquake risk assessment of building structures. Reliab. Eng. Syst. Saf. **74**(3), 251–262 (2001)
2. Neves, L.A.C., Cruz, P.J.S., Henriques, A.A.R.: Reliability analysis of steel connection components based on FEM. Eng. Fail. Anal. **8**(1), 29–48 (2001)
3. Hadianfard, M.A., Razani, R.: Effects of semi-rigid behavior of connections in the reliability of steel frames. Struct. Saf. **25**(2), 123–138 (2003)
4. Papadrakakis, M., Tsompanakis, Y., Lagaros, N.D., Friagiadakis, M.: Reliability based optimization of steel frames under seismic loading conditions using evolutionary computation. J. Theor. Appl. Mech. **42**, 585–608 (2004)
5. Nie, J., Ellingwood, B.R.: Finite element-based structural reliability assessment using efficient directional simulation. J. Eng. Mech. **131**(3), 259–267 (2005)
6. Haukaas, T., Scott, M.H.: Shape sensitivities in the reliability analysis of nonlinear frame structures. Comput. Struct. **84**(15–16), 964–977 (2006)
7. Cardoso, J.B., de Almeida, J.R., Dias, J.M., Coelho, P.G.: Structural reliability analysis using monte carlo simulation and neural networks. Adv. Eng. Soft. **39**(6), 505–513 (2008)
8. Bojorquez, E., Ruiz, S.E., Teran-Gilmore, A.: Reliability-based evaluation of steel structures using energy concepts. Eng. Struct. **30**(6), 1745–1759 (2008)
9. Kala, Z.: Geometrically non-linear finite element reliability analysis of steel plane frames with initial imperfections. J. Civ. Eng. Manage. **18**(1), 81–90 (2012)

10. Guo, Q., Jeffers, A.E.: Finite-element reliability analysis of structures subjected to fire. J. Struct. Eng. **141**(4), 04014129 (2015)
11. Chojaczyk, A.A., Teixeira, A.P., Neves, L.C., Cardoso, J.B., Soares, C.G.: Review and application of artificial neural networks models in reliability analysis of steel structures. Struct. Saf. **52**, 78–89 (2015)
12. Thai, H.T., Uy, B., Kang, W.H., Hicks, S.: System reliability evaluation of steel frames with semi-rigid connections. J. Constr. Steel Res. **121**, 29–39 (2016)
13. Truong, V.H., Kim, S.E.: An efficient method for reliability-based design optimization of nonlinear inelastic steel space frames. Struct. Multi. Optim. **56**(2), 331–351 (2017)
14. Hazay, M., Vigh, L.G.: 11.19: seismic fragility assessment of steel frames: application of advanced reliability analysis. Ce/papers **1**(2–3), 2993-3002 (2017)
15. Agostini, B.M., Freitas, M.S.D.R., Silveira, R.A.D.M., Silva, A.R.D.D.: Structural reliability analysis of steel plane frames with semi-rigid connections. REM-Int. Eng. J. **71**(3), 333–339 (2018)
16. De Risi, R., Goda, K., Tesfamariam, S.: Multi-dimensional damage measure for seismic reliability analysis. Struct. Saf. **78**, 1–11 (2019)
17. Liu, W., Zhang, H., Rasmussen, K.J., Yan, S.: System-based limit state design criterion for 3D steel frames under wind loads. J. Constr. Steel Res. **157**, 440–449 (2019)
18. Ha, M.H., Vu, Q.V., Truong, V.H.: A deep learning-based procedure for safety evaluation of steel frames using advanced analysis. In: CIGOS 2019, Innovation for Sustainable Infrastructure, pp. 1137–1142 (2020)
19. Mohsenian, V., Padashpour, S., Hajirasouliha, I.: Seismic reliability analysis and estimation of multilevel response modification factor for steel diagrid structural systems. J. Build. Eng. **29**, 101168 (2020)
20. Ni, P., Li, J., Hao, H., Yan, W., Du, X., Zhou, H.: Reliability analysis and design optimization of nonlinear structures. Reliab. Eng. Syst. Saf. **198**, 106860 (2020)
21. McKenna, F., Fenves, G.L.: The OpenSees command language manual. University of California, Berkeley (2001)
22. Der Kiureghian, A., Taylor, R.L.: Numerical methods in structural reliability. In: Proceedings of the Fourth International Conference on Applications of Statistics and Probability in Civil Engineering, ICASP4, Florence, Italy (1983)
23. Haukaas T.: Finite Element Reliability and Sensitivity Methods for Performance-Based Engineering. Ph.D. thesis, University of California, Berkeley (2003)
24. Haukaas T., Der Kiureghian A.: A Computer Program for Nonlinear Finite Element Reliability Analysis. In: 8th International Conference on Structural Safety and Reliability, California, USA (2001)
25. Sudret, B., Der Kiureghian, A.: Stochastic Finite Element Methods and reliability: a state-of-the-art report. Department of Civil and Environmental Engineering, University of California, Berkeley (2000)
26. Nowak A.S., Collins K.R.: Reliability of Structures, McGrawHill, Boston (2000)
27. Hasofer, A.M., Lind, N.C.: Basic Notes on Load Models, The Joint Committee on Structural Safety, (2000)
28. Schueremans, L.: State of the Art Report: Structural Reliability in Design and Analysis (2003)
29. Haukaas, T., Der Kiureghian, A.: Parameter sensitivity and importance measures in nonlinear finite element reliability analysis. J. Eng. Mech. **131**(10), 1013–1026 (2005)

Solving Nonlinear Volterra-Fredholm Integral Equations via New Basis Functions

A. A. Cheraghi Tofigh and Reza Ezzati(✉)

Department of Mathematics, Karaj Branch, Islamic Azad University, Karaj, Iran
ali_chera5@yahoo.com, ezati@kiau.ac.ir

Abstract. In this paper, by introducing a class of new orthogonal basis functions (NFs), we propose a numerical method to solve the nonlinear Volterra-Fredholm integral equations of the second kind (NVFIE2). To do this, first, we present the operational matrix of integration of NFs. Then, by applying this matrix, NVFIE2 is reduced to a nonlinear system of algebraic equations. Finally, by solving some numerical examples, we show the efficiency of the proposed method.

Keywords: Volterra-Fredholm integral equations · Operational matrix · New orthogonal basis

1 Introduction

Volterra-Fredholm integral equations (VFIE) play an important role in engineering such as the population dynamics, spread of epidemics, and semi-conductor devices. Many methods have been introduced for solving these equations [2–5]. Haar wavelets and collocation method to solve NVFIE2 have applied by Babolian et.al. [1]. Maleknejad and Fadaei Yami [7] solved the system of NVFIE. The authors of [9] proposed new basis functions for solving nonlinear Volterra-Fredholm integral equations. The authors of [8] have solved nonlinear Volterra-Fredholm integral equations by introducing a new approach. In [6], the authors proposed a new computational method to obtain numerical solution of NVFIE. In [11], Ezzati et.al. applied Chebyshev polynomials for solving NVFIE. The authors of [10] used Legendre wavelets to solve VFIE.

In this article, we introduce a new orthogonal basis to solve the following NVFIE2

$$f(x) = g(x) + \int_0^x H_1(x,t)[f(t)]^{n_1} dt + \int_0^1 H_2(x,t)[f(t)]^{n_2} dt, \qquad (1)$$

where the function $g \in L^2[0,1]$, the kernels $H_1(.,.), H_2(.,.) \in L^2([0,1] \times [0,1])$ are known and f is the unknown function to be determined. To do this, we construct the operational matrix of integration on NFs, Q, as follows:

$$\int_0^t \Phi(s)ds \simeq Q\Phi(t),$$

T. Allahviranloo et al. (Eds.): IDS 2020, AISC 1301, pp. 785–795, 2021.
https://doi.org/10.1007/978-3-030-66501-2_63

where $\Phi(t) = [\phi_1(t), \phi_2(t), ..., \phi_m(t)]$ and the matrix $Q_{m\times m}$ can be uniquely determined on the basis of new orthogonal basis functions. The elements $\phi_1(t), \phi_2(t), ..., \phi_m(t)$ are the basis functions, on interval [0,1]. The important advantage of this method is to reduce the integral Eq. (1) to a system of algebraic equations.

The rest of the paper is organized as follows: Sect. 2, has devoted to introduce a new class of orthogonal basis functions. Integration operational matrix of the new basis functions is obtained in Sect. 3. we presented function approximation by using these orthogonal basis functions in Sect. 4. The efficiency an accuracy of this method, is shown by some examples in Sect. 5. Finally, we prepared concluding remarks in Sect. 6.

2 New Basis Functions (NFs)

In this section, we introduce a class of new basis functions (NFs) and some related properties.

Definition 3.1. The ith left hand and right hand functions of an m-set of new basis functions (NFs) over interval [0,1) are defined as:

$$NF1_i(x) = \begin{cases} \frac{1+cos(\frac{\pi(x-ih)}{h})}{2}, & ih \le x < (i+1)h, \\ 0, & otherwise, \end{cases} \tag{2}$$

$$NF2_i(x) = \begin{cases} \frac{1-cos(\frac{\pi(x-ih)}{h})}{2}, & ih \le x < (i+1)h, \\ 0, & otherwise, \end{cases} \tag{3}$$

where $i = 0, 2, 3, ..., m-1$ are arbitrary positive integer number and $h = \frac{1}{m}$, and $NF1_i(x)$ and $NF2_i(x)$ are the ith term of $NF1(x)$ and $NF2(x)$, respectively. By defining NFs over [0,1), we have:

$$NF1(x) = [NF1_0(x), NF1_1(x), ..., NF1_{m-1}(x)]^T$$
$$NF2(x) = [NF2_0(x), NF2_1(x), ..., NF2_{m-1}(x)]^T, \tag{4}$$

and

$$NF(x) = [NF^T1(x), NF^T2(x)]^T. \tag{5}$$

According to definition of NFs in Eqs. (4–5), we have:

$$NF1_i(x) + NF2_i(x) = \phi_i(x),$$

where $\phi_i(x)$ is the ith component of Block-Pulse functions [12].

It is obvious that $NF1_i(x)$ and $NF2_j(x)$, $i, j = 0, 1, 2, ..., m$ are disjoint. The orthogonality of NFs is derived from

$$\int_0^1 NF1_i(x)NF1_j(x)dx = \frac{3}{8}h\delta_{i,j}, \quad \int_0^1 NF1_i(x)NF2_j(x)dx = 0, \tag{6}$$

$$\int_0^1 NF2_i(x).NF2_j(x)dx = \frac{3}{8}\delta_{i,j}, \quad \int_0^1 NF2_i(x).NF1_j(x)dx = 0, \tag{7}$$

where $i, j = 0, 1, 2, \Delta\Delta\Delta, m-1$ and $\delta_{i,j}$ is the Kronecker delta function.
Also

$$NF1_i(jh) = \begin{cases} 1, & j = i, \\ 0, & j \neq i, \end{cases}$$

$$NF2_i(jh) = \begin{cases} 1, & j = i+1, \\ 0, & j \neq i+1, \end{cases}$$

where $i, j = 0, 1, 2, \Delta\Delta\Delta, m-1$.
By using disjointness property of NFs, we have:

$$NF1(x)NF1^T(x) \simeq$$

$$\begin{bmatrix} NF1_0(x) & 0 & 0 & \dots & 0 \\ 0 & NF1_1(x) & 0 & \dots & 0 \\ 0 & 0 & NF1_2(x) & \dots & 0 \\ . & . & . & & . \\ . & . & . & & . \\ . & . & . & & . \\ 0 & 0 & 0 & \dots & NF1_{m-1}(x) \end{bmatrix}_{m\times m} = diag(NF1(x)),$$

$$NF1(x)NF2^T(x) = 0_{m\times m}.$$

Similarly, we have:

$$NF2(x)NF2^T(x) \simeq diag(NF2(x)),$$

$$NF2(x)NF1^T(x) = 0_{m\times m}.$$

Therefore

$$NF(x)NF^T(x) \simeq \begin{bmatrix} diag(NF1(x)) & 0 \\ 0 & diag(NF2(x)) \end{bmatrix}_{2m\times 2m}.$$

Also

$$NF(x)NF^T(x)B \simeq \tilde{B}NF(x), \tag{8}$$

where B is a $2m$-vector and $\tilde{B} = diag(B)$. Moreover, it can be clearly concluded that for an $2m \times 2m$ matrix A:

$$NF^T(x).A.NF(x) \simeq \hat{A}^T.NF(x), \tag{9}$$

where $\hat{A}$ is an $2m$-vector with elements equal to diagonal entries of matrix A.

3 The Operational Matrix of Integration

By using the orthogonality of $NF1(x)$ and $NF2(x)$ in Eqs. (6–7), respectively, we obtain

$$\int_0^1 NF1_i(x)NF2_j(x)dx = \int_0^1 NF2_i(x)NF1_j(x)dx = \frac{1}{8}h\delta_{i,j}, \tag{10}$$

$$D = \int_0^1 NF(x)NF^T(x)dx = \frac{h}{8}\begin{bmatrix} 3I_{m\times m} & I_{m\times m} \\ I_{m\times m} & 3I_{m\times m} \end{bmatrix}_{2m\times 2m}.$$

Also we have:

$$\int_{ih}^{(i+1)h} NF1_i(x)dx = \frac{1}{2}h,$$

$$\int_{ih}^{(i+1)h} NF2_i(x)dx = \frac{1}{2}h,$$

and

$$\int_0^x NF1_i(s)ds \simeq \begin{cases} 0, & 0 \le x < ih, \\ \frac{1}{2}(\frac{h}{2}), & ih \le x < (i+1)h, \\ \frac{h}{2}, & (i+1)h \le x < 1, \end{cases}$$

therefore

$$\int_0^x NF1_i(s)ds \simeq \frac{h}{4}[0, ..., 0, 1, 2, ..., 2, 0, ..., 0, 2, ..., 2].NF(x),$$

and similarly

$$\int_0^x NF2_i(s)ds \simeq \frac{h}{4}[0, ..., 0, 1, 2, ..., 2, 0, ..., 0, 2, ..., 2].NF(x).$$

So, we have:

$$\int_0^x NF(s)ds \simeq Q.NF(x) \tag{11}$$

where $Q_{2m\times 2m}$ is called the operational matrix of integration of NFs and given by

$$Q = \frac{h}{4}\begin{bmatrix} 1 & 2 & ... & 2 & 0 & 2 & ... & 2 \\ 0 & 0 & ... & 2 & 0 & . & ... & 2 \\ . & 0 & ... & 2 & 0 & . & ... & 0 \\ 1 & 2 & ... & 2 & 0 & 2 & ... & 2 \\ 0 & 1 & ... & 2 & 0 & & ... & 2 \\ 0 & 0 & ... & 2 & 0 & 0 & ... & 2 \\ 0 & 0 & ... & 1 & 0 & 0 & ... & 0 \end{bmatrix}_{2m\times 2m}.$$

4 Function Approximation

An arbitrary real bounded function $f(x) \in L^2[0,1)$ can be expanded as:

$$f(x) \simeq \sum_{i=0}^{m-1} K1_i NF1_i(x) + \sum_{i=0}^{m-1} K2_i NF2_i(x)$$

$$= K1^T NF1(x) + K2^T NF2(x) = K^T NF(x), \tag{12}$$

where K is a $2m$-vector given by $K = [K1^T K2^T]$, and $NF1(x)$ and $NF2(x)$ are defined in Eq. (6).

The components of coefficients $K1$ and $K2$ is computed by function $f(x)$ at grid points ih and $(i+1)h$ for different h and i. Therefore,

$$K1_i = K(ih),$$

$$K2_i = K((i+1)h).$$

Also, the positive integer powers of a function $f(x)$ may be approximated by using NFs as

$$[f(x)]^r \simeq K_r^T.NF(x),$$

where K_r is a column vector whose elements are rth powers of the elements of the vector K, see ([9]).

Now, let $H(x;t)$ be an arbitrary function of two variables defined on $L^2([0;1) \times [0;1))$. Clearly, it can be expanded by NFs as the following form:

$$H(x;t) \simeq NF^T(x).H.NF(t),$$

where $NF(x)$ and $NF(t)$ are NFs vectors with $h = \frac{1}{m}$. Also H is the $2m \times 2m$ coefficients matrix as follows:

$$H = \begin{bmatrix} H11 & H12 \\ H21 & H22 \end{bmatrix},$$

where

$$[H11]_{ij} = H(ih, jh),$$

$$[H12]_{ij} = H((i+1)h, jh),$$

$$[H21]_{ij} = H(ih, (j+1)h),$$

$$[H22]_{ij} = H((i+1)h, (j+1)h).$$

Now, consider the nonlinear Volterra-Fredholm integral equations:

$$f(x) = g(x) + \int_0^x H_1(x,t)[f(t)]^{n_1} dt + \int_0^1 H_2(x,t)[f(t)]^{n_2} dt, \quad 0 \le x < 1, \tag{13}$$

where $g(x) \in L^2[0,1)$, $k(x,t) \in L^2([0,1) \times [0,1))$ are known functions, $f(x)$ is the unknown function and n_1, n_2 are positive integers. By approximating functions $f(x)$, $H(x,t)$, $[f(x)]^{n_1}$ and $[f(x)]^{n_2}$ in the matrix form, we have:

$$f(x) \simeq F^T.NF(x), \tag{14}$$

$$H_1(x,t) \simeq NF^T(x).H_1.NF(t), \tag{15}$$

$$H_2(x,t) \simeq NF^T(x).H_2.NF(t), \tag{16}$$

$$g(x) \simeq G^T.NF(x), \tag{17}$$

$$[f(x)]^{n_1} \simeq F_{n_1}^T.NF(x), \tag{18}$$

$$[f(x)]^{n_2} \simeq F_{n_2}^T.NF(x). \tag{19}$$

By substituting Eqs. (14)–(19) into Eq. (13), we obtain:

$$F^T.N(x) \simeq G^T.NF(x) + \int_0^x NF^T(x).H_1.NF(t).NF^T(t)F_{n_1}dt$$

$$+ \int_0^1 NF^T(x).H_2.NF(t).NF^T(t)F_{n_2}dt,$$

and

$$F^T.NF(x) \simeq F^T.NF(x) + NF^T(x).H_1 \int_0^x NF(t).NF^T(t).F_{n_1}dt$$

$$+NF^T(x).H_2 \int_0^1 NF(t)NF^T(t)F_{n_2}dt,$$

by using Eqs. (8) and (10) we have:

$$F^T.NF(x) \simeq G^T.NF(x) + NF^T(x).H_1 \int_0^x \tilde{F_{n_1}}.NF(t)dt + NF^T(x).H_2D.F_{n_2}.$$

Also, by using Eq. (11), we have:

$$F^T.NF(x) = G^T.NF(x) + NF^T(x).H_1\tilde{F_{n_1}}.Q.N(x) + NF^T(x).H_2.D.F_{n_2}.$$

By assuming $V = H_1\tilde{F_{n_1}}.Q$, and using Eq. (9), we will get:

$$F^T.NF(x) = G^T.NF(x) + \tilde{V}^T.NF(x) + (H_2.D.F_{n_2})^T.NF(x),$$

so

$$F = G + \tilde{V} + H_2.D.F_{n_2}.$$

The above nonlinear system of algebraic equations can be solved by using known numerical methods such as Newtons iteration method.

5 Numerical Examples

In this section, we illustrate the efficiency of the proposed method by solving some numerical examples. Also, we compare the numerical results with the exact solution.

Example 1. Consider the following nonlinear Volterra integral equation of the second kind [9]:

$$f(x) = sin(\pi x) + \frac{1}{5}\int_0^x cos(\pi x)sin(\pi t)f^3(t)dt, \tag{20}$$

with the exact solution $f(x) = sin(\pi x) + \frac{20-\sqrt{391}}{3}cos(\pi x), \quad 0 \leq x < 1.$ Table 1 and Fig. 1 show the numerical results and the exact solution.

Table 1. Comparing the absolute errors of the proposed method and the method of [9] for Example 1, for m = 32.

x_i	*Exact solution*	*Aproximate solution*	*Absolute error*	*Absolute errors in* [9]
0	0.07542668	0.07542668	0	0.01458
0.1	0.38075203	0.37104869	0.00970334	0.0741
0.2	0.64880672	0.64399452	0.00481220	0.06241
0.3	0.85335168	0.85523133	0.00187967	0.04306
0.4	0.97436464	0.97634403	0.00197938	0.0196
0.5	1.00000000	1.00000000	0	0
0.6	0.92774838	0.93122258	0.00347419	0.00537
0.7	0.76468229	0.76731357	0.00263127	0.00444
0.8	0.52676377	0.52161133	0.00515244	0.00293
0.9	0.23728195	0.22716725	0.01011469	0.00279

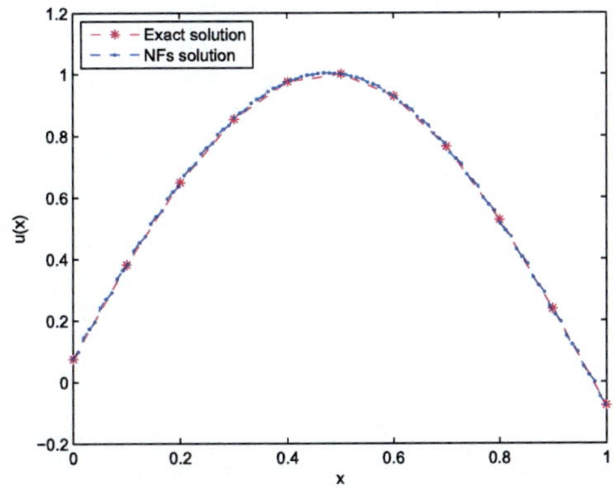

Fig. 1. Comparing the exact and approximate solutions for Example 1.

Example 2. Consider the following nonlinear Fredholm integral equations of the second kind [9]:

$$f(x) = e^{x+1} - \int_0^1 e^{x-2t} f^3(t)dt, \tag{21}$$

with the exact solution $f(x) = e^x, 0 \leq x < 1$. Table 2 and Fig. 2 compare the numerical results and the exact solution.

Table 2. The numerical results of Example 2 for m = 32.

x_i	*Exact solurion*	*Aproximate solurion*	*Absolute error*
0	1.000000	1.000000	0
0.1	1.10517091	1.10161429	0.00355662
0.2	1.2214027	1.21945907	0.02464838
0.3	1.34985880	1.35230892	–0.00245011
0.4	1.49182469	1.49676739	–0.00494269
0.5	1.64872127	1.64872127	0
0.6	1.8221188	1.81625491	0.00586388
0.7	2.01375270	2.01054810	0.00320459
0.8	2.22554092	2.22958048	– 0.00403955
0.9	2.45960311	2.46775223	–0.00814912

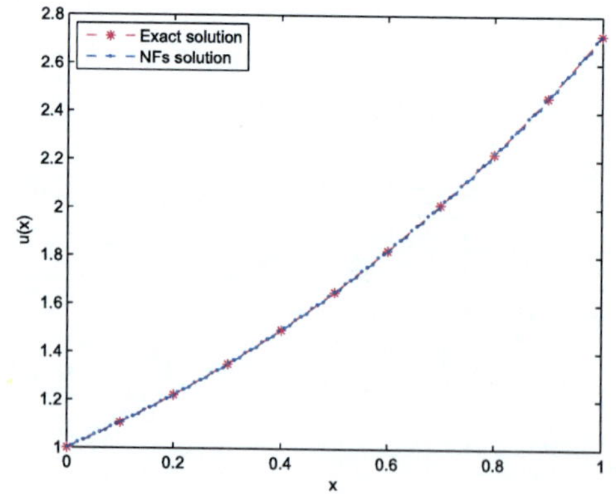

Fig. 2. Comparing the exact and approximate solutions for Example 2.

Example 3. Consider the following nonlinear Fredholm integral equations of the second kind:

$$f(x) = e^x - \frac{x}{192}(e^2 + 1) + \int_0^1 xtf^2(t)dt, \tag{22}$$

with the exact solution $f(x) = e^x, 0 \leq x < 1$. The numerical results are shown in Table 3 and Fig. 3.

Table 3. The numerical results of Example 3 for m = 32

x_i	*Exact solution*	*Aproximate solution*	*Absolute errors*
0	1.00	1	0
0.1	1.10517091	1.10161429	0.00355662
0.2	1.2214027	1.21945907	0.00194368
0.3	1.34985880	1.35230892	0.00245011
0.4	1.49182469	1.49676739	0.00494269
0.5	1.64872127	1.64872127	0
0.6	1.8221188	1.81625491	0.00586388
0.7	2.01375270	2.01054810	0.003204599
0.8	2.22554092	2.22958048	0.00403955
0.9	2.45960311	2.46775223	0.00814912

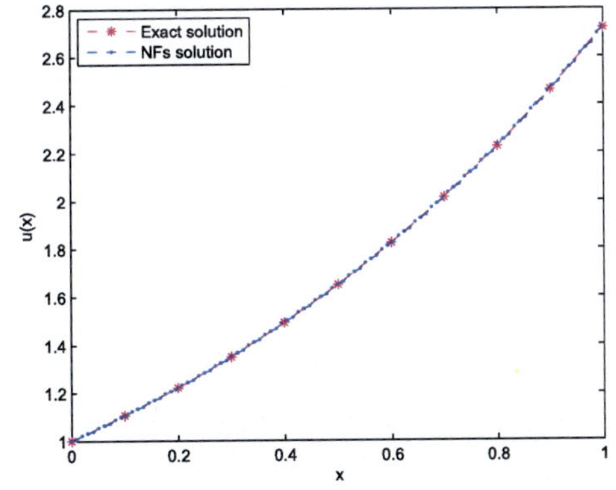

Fig. 3. Comparing the exact and approximate solutions for Example 3.

Example 4. Consider the following nonlinear Fredholm integral equations of the second kind [9]:

$$f(x) = -\frac{1}{30}x^6 + \frac{1}{3}x^4 - x^2 + \frac{5}{3}x^4 - \frac{5}{4} + \int_0^x (x-t)f^2(t)dt + \int_0^1 (x+t)f(t)dt, \quad (23)$$

with the exact solution $f(x) = x^2 - 2, 0 \leq x < 1$. Table 4 and Fig. 4 compare the numerical results with the exact solution.

Table 4. Comparing the numerical results of the proposed method and the method of [9] for Example 4 for m = 32

x_i	*Exact solution*	*Aproximate solution*	*Absolute errors*	*Absolute errors of* [9]
0.1	-1.99000000	-1.9905581	0.00055816	0.0046
0.2	-1.96000000	-1.96045762	0.00045762	0.0126
0.3	-1.91000000	-1.90875423	0.00124576	0.0303
0.4	-1.84000000	-1.83729227	0.0027077	0.0488
0.5	-1.75000000	-1.75000000	0	0.0244
0.6	-1.64000000	-1.64382405	0.00382405	0.0227
0.7	-1.51000000	-1.51216101	0.00216101	0.0212
0.8	-1.36000000	-1.35705084	0.00294915	0.0192
0.9	-1.19000000	-1.18402638	0.00597361	0.0167

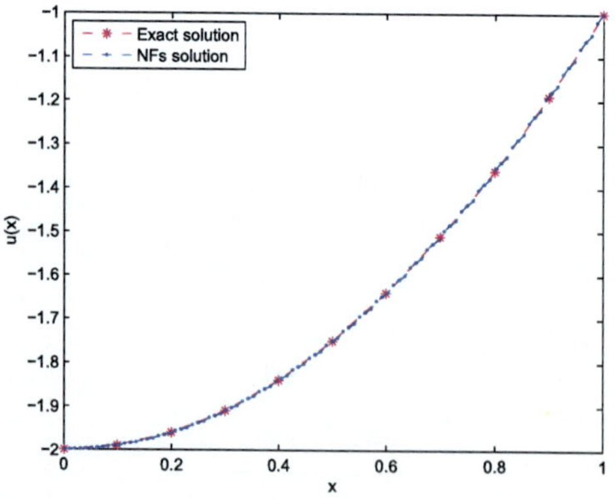

Fig. 4. Comparing the exact and approximate solutions for Example 4.

Also, comparing the numerical results of the proposed method and the method of [9] in Table 4, shows the accuracy of the proposed method.

6 Conclusion

In this paper, first, we introduced a class of new orthogonal basis functions. Then, we used these functions to construct a numerical method for solving nonlinear VFIE of the second kind. By using the proposed method, the main problem becomes a nonlinear system of algebraic equations. Finally, by solving some numerical example, we show that the proposed method can be suitable method for solving nonlinear VFIE of the second kind.

References

1. Babolian, E., Shahsavaran, A.: Numerical solution of nonlinear Fredholm and Volterra integral equations of the second kind using Haar wavelets and collocation method. Autumn **7**(3), 213-222 (2007)
2. Mirzaee, F., Hoseini, A.A.: Numerical solution of nonlinear Volterra-Fredholm integral equations using hybrid of block-pulse functions and taylor series. Alexandria Eng. J. **52**(3), 551–555 (2013)
3. Shali, J.A., Akbarfam, A.A.J., Ebadi, G.: Approximate solutions of nonlinear Volterra Fredholm integral equations. Int. J. Nonlin. Sci. **14**(4), 425–433 (2012)
4. Zarebnia, M.: Anumerical solution of nonlinear Volterra-Fredholm integral equations. J. Appl. Anal. Comput. **3**(1), 95–104 (2013)
5. Mashayekhi, S., Razzaghi, M., Tripak, O.: Solution of the nonlinear mixed Volterra-Fredholm integral equations by hybrid of block-pulse functions and bernoulli polynomials. Sci. World J. **2014**, 1–8 (2014)
6. Maleknejad, K., Hadizadeh, M.: A new computational method for Volterra Fredholm integral equation. J. Comput. Math. Appl. **37**, 37–48 (1999)
7. Maleknejad, K., Yami, M.R.F.: A computational method for system o Volterra-Fredholm integral equations. Appl. Math. Comput. **188**, 589–595 (2006)
8. Cheraghi, A.A., Tofigh, Ezzati, R.: Introducing a new approach to solve nonlinear Volterra-Fredholm integral equations. J. Pure Appl. Math. **10**(2), 175-187 (2019)
9. Paripour, M., Kamyar, M.: Numerical solution of nonlinear Volterra-Fredholm integral equations by using new basis functions. Commun. Numer. Anal. **2013**, 1–11 (2013)
10. Yousefi, S., Razzaghi, S.M.: Legendre wavelets method for the nonlinear Volterra-Fredholm integral equations. Math. Comput. Simul. **70**, 1–8 (2005)
11. Ezzati, R., Najafalizadeh, S.: Numerical solution of nonlinear Voltra-Fredholm integral equation by using Chebyshev polynomials. Math. Sci. Q. J. **5**, 14–22 (2011)
12. Jung, Z.H., Schanfelberger, W.: Block-Pulse Functions and Their Applications in Control Systems. Springer, Berlin (1992)

Time of Flights Estimations of Ultrasonic Signals Based on Numerical Methods

Nursin Bas Catak[1(✉)] and Ali R. Ansari[2]

[1] Department of Mathematics, Ege University, Izmir, Turkey
nursin.catak@ege.edu.tr
[2] Department of Mathematics and Natural Sciences, GUST, Mubarak Al-Abdullah, Kuwait
ansari.a@gust.edu.kw

Abstract. Flow measurement is of great importance in many industrial applications. Ultrasonic flow measurement using time-of-flight technique can equip adequate solutions for such processes. The estimation of time of flights of the ultrasonic signals has a key role in the measurement process under the case of ultrasonic signal technology. In this work, a robust technique is proposed based on numerical methods in order to estimate the time-of-flights of the ultrasonic signals. Due to the discrete nature of the ultrasonic signals, the proposed method shows efficient results in order to estimate the time of flights. The proposed method is validated through experimental works. According to the results, Lagrange-Taylor, Taylor (forward) numerical schemes supplied better performance over correlation based method at overall volumetric flow rates, respectively.

Keywords: Flow measurement · Time of flight · Taylor expansion · Numerical solutions · Ultrasonic signals

1 Introduction

Volumetric flow measurement is of great importance in many applications among industrial processes containing liquid components, public watering services, oil and gas pipeline systems [2,3,7,9,15]. Ultrasound-based technology has been applied in science and engineering area such as biomedical image processing, target detection, measurement of fluid flow [14]. Ultrasonic flow meters based on time-of-flight (ToF)method can supply appropriate solutions for flow measurements. The main advantages of ultrasonic flow meters can be listed as; i) having no moving mechanical apparatus, ii) suitable for corrosive and dirty fluids, iii) measurement is available for both directions [4].

Motion of ultrasound wave through the media; similarities and discrepancies of the transmitted and the received ultrasonic signals; and their some characteristics have commonly been observed in order to analyze and understand behaviors of the system under investigation [18]. A general scheme of ultrasonic flow measurements using two sensors is shown in Fig. 1. In this configuration,

T. Allahviranloo et al. (Eds.): IDS 2020, AISC 1301, pp. 796–807, 2021.
https://doi.org/10.1007/978-3-030-66501-2_64

the ultrasonic signal send by sensor-1 passes through the medium along with the same direction of the fluid flow, then it is received by sensor-2. In the following cycle, the ultrasonic signal send by sensor-2 passes through the medium moving against to the fluid flow, then it is captured by sensor-1. In theory, since sensor-1 and sensor-2 are identically the same, the received signals must be identical except a ceratin amount of time-shift due to the directions of the flow.

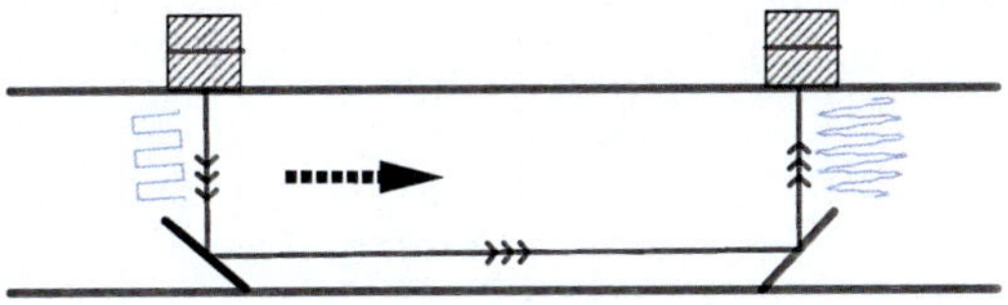

Fig. 1. The illustration of the experimental setup

A stochastic model properties of the turbulence aiming to predict the fluid flow was presented by [11]. Angrisani et al. employed discrete extended Kalman filter in order to determine the variations of ToF. The best measurement conditions to asses the performance were reported perfect orthogonality between the source and target and measurement environment free of acoustic inferences [1]. Thorn et al. discussed a three-phase flow-meter to measure precisely the flow-rate of oil-gas-water mixtures using various types of techniques including tomography, ultrasonic, nuclear magnetic resonance methods [17] (Fig. 2).

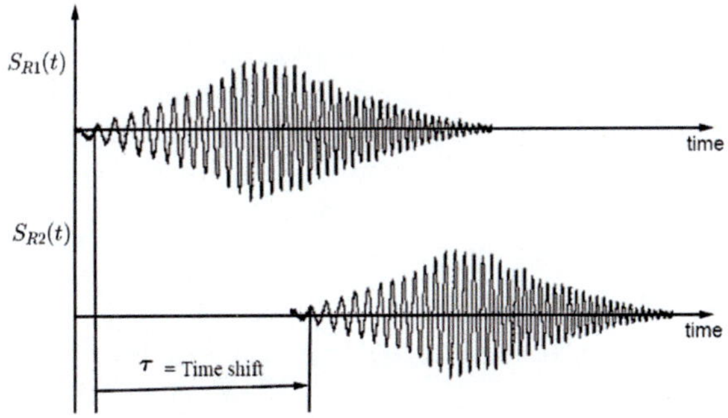

Fig. 2. The illustration of the time shift between the signals received by the sensors

Svilainis et al. discussed about the estimation of ToF in a discrete-time system. They reported that there is no need to select very high sampling rate in order to get narrow variance band of ToF [16]. A flow measurement and calibration approach based on thermal cross correlation was developed by [13]. The authors reported that the calibration factor is dependent on the system size. Interpolation techniques to minimize the measurement error of ultrasound signals has been discussed in [5]. A combination of ultrasound time shift and Doppler shift in frequency methods have been employed to measure flow velocity under the case of three-phase pile flows [6]. The effect of the various configurations of sensors for ultrasound correlation flow meters briefly discussed in [19]. A self-monitoring ultrasound gas flow meter using vortex and correlation technique was given by [10].

The cross correlation between two signals is easily calculated for a general continuous time case as;

$$R_{XY}(\tau) = \int_{-\infty}^{\infty} X(t)Y(t+\tau)\mathrm{d}\tau \tag{1}$$

The core idea of the cross correlation technique is to find the maximum $R_{XY}(\tau)$ at a specific time shift of τ. For the continuous systems, the ToF exactly equals to τ. However, for the discrete case, since τ is the number of shifting steps rather than the time the ToF is estimated as;

$$\mathrm{ToF} = \tau T_S \tag{2}$$

where T_S is the sampling period.

According to the properties of the autocorrelation function, $|R_{S_T,S_T}(\tau)| \leq R_{S_T,S_T}(0)$. It means that, the correlation between the received signals has its maximum where $\tau = \tau_1 - \tau_2$, which exactly equals to the ToF. It is also reported that the method based on cross correlation supplies more accurate results depending on the characteristics of the physical system [12].

In this paper, a robust technique based on numerical methods is presented in order to estimate the time-of-flights of the ultrasonic signals. The rest of the paper is organized as follows: In Sect. 2, theoretical background of numerical methods based on Taylor expansion have briefly been discussed. Section 3 is dedicated to results obtained from the proposed method and discussion. Finally, the paper is concluded in Sect. 4.

2 Numerical Methods Based on Taylor Series

Taylor series expansion as a type of power series can be used for the purpose of representing the analytical functions with various centers. Therefore, it is one of the fundamental topics both in complex analysis and in calculus.

2.1 Taylor Series Expansion

Taylor series expansion of a function is given as;

$$f(x+\tau) = f(x) + f'(x)\tau + \frac{f''(x)}{2!}\tau^2 + \dots + \frac{f^{(n)}(x)}{n!} + R_n \tag{3}$$

where the remainder term is

$$R_n = \frac{f^{(n+1)}(\xi)}{(n+1)!}\tau^{n+1}$$

The signals received by the transducers are denoted by $V_1[n]$ and $V_2[n]$ such that;

$$\begin{aligned} V_1[n] &= A[n]S[n] + \eta_1[n] \\ V_2[n] &= A[n+\tau_{12}]S[n+\tau_{12}] + \eta_2[n] \end{aligned} \tag{4}$$

where $S[n]$ is the transmitted signal, $\eta(t)$ is a Gaussian white noise (GWN), and $A[n]$ is the ultrasonic echo envelop.

Forward Solution. For the sake of the simplicity and better understanding the first order Taylor series expansion, that is linear approximation of the entire series, is employed in order to estimate the time-shift between two signals. Suppose that, the time-shift between $V_1[n]$ and $V_2[n]$ is τ_{12}. Hence;

$$\begin{aligned} V_2[n] &= V_1[n+\tau_{12}] \\ V_1[n+\tau_{12}] &= V_1[n] + V_1'[n]\tau_{12} \end{aligned}$$

The cost function of the forward Taylor series expansion considering the least squares error;

$$h_{12}[\tau_{12}] = \sum_{n=1}^{N} \left(V_2[n] - V_1[n] - V_1'[n]\tau_{12}\right)^2 \tag{5}$$

where N is the window length. The aim is to minimize the error (i.e. the cost function) subject to;

$$\frac{\mathrm{d}h_{12}[\tau_{12}]}{\mathrm{d}\tau_{12}} = 0$$

$$\frac{\mathrm{d}h_{12}[\tau_{12}]}{\mathrm{d}\tau_{12}} = 2\sum_{n=1}^{N} \left(V_1'[n]\right)^2 \tau_{12} \ + \ 2\sum_{n=1}^{N} V_1'[n]\left(V_1[n] - V_2[n]\right)$$

It can be written that $a_{12}\tau_{12} + b_{12} = 0$ where;

$$a_{12} = 2\sum_{n=1}^{N} (V_1'[n])^2$$
$$b_{12} = 2\sum_{n=1}^{N} V_1'[n]\,(V_1[n] - V_2[n])$$

Thereafter, the time-shift is expressed as;

$$\tau_{12} = -\frac{b_{12}}{a_{12}} \tag{6}$$

Please note that, it is important to normalize the amplitudes of the signals before time-shift estimation.

Backward Solution. The manner with the forward solution apart from the reverse sign of the time-shift, where;

$$V_1[n] = V_2[n + \tau_{21}]$$
$$V_2[n + \tau_{21}] = V_2[n] + V_2'[n]\tau_{21}$$

Thereby,the backward solution yields to the time-shift as;

$$\tau_{21} = -\frac{b_{21}}{a_{21}} \tag{7}$$

where

$$a_{21} = 2\sum_{n=1}^{N} (V_2'[n])^2$$
$$b_{21} = 2\sum_{n=1}^{N} V_2'[n]\,(V_2[n] - V_1[n])$$

In theory, $\tau_{12} = -\tau_{21}$, but in the practical cases there is no such a guarantee. Therefore, the forward and the backward schematics are combined and the results are optimized in the next section.

Optimized Solution: Lagrange-Taylor Method. To estimate the time-shift using symmetrical approach of both the forward and the backward Taylor series solutions, the cost function consists of the combination of the $h_{12}(\tau_{12})$ and $h_{21}(\tau_{21})$;

$$f(\tau_{12}, \tau_{21}) = h_{12}(\tau_{12}) + h_{21}(\tau_{21})$$

where f is the global cost function to be minimized subject to

$$g(\tau_{12}, \tau_{21}) = \tau_{12} + \tau_{21}$$

By introducing Lagrangian function [20];

$$\mathcal{L}(\tau_{12}, \tau_{21}, \lambda) = f(\tau_{12}, \tau_{21}) - \lambda g(\tau_{12}, \tau_{21})$$

and noting that

$$\nabla\mathcal{L}(\tau_{12}, \tau_{21}, \lambda) = \nabla f(\tau_{12}, \tau_{21}) - \lambda\nabla g(\tau_{12}, \tau_{21})$$

where $\nabla := \left(\frac{\partial}{\partial\tau_{12}}, \frac{\partial}{\partial\tau_{21}}\right)$, the scalar quantity λ is called a Lagrange multiplier.

At the solution for (τ_{12}, τ_{21});

$$\nabla\mathcal{L}(\tau_{12}, \tau_{21}, \lambda) = 0$$

The corresponding algebraic equation is given in matrix form;

$$\begin{bmatrix} a_{12} & 0 & 1 \\ 0 & a_{21} & 1 \\ -1 & -1 & 0 \end{bmatrix} \begin{bmatrix} \tau_{12} \\ \tau_{21} \\ \lambda \end{bmatrix} = \begin{bmatrix} -b_{12} \\ -b_{21} \\ 0 \end{bmatrix} \tag{8}$$

Lagrange-Taylor solution of the system is obtained from (8) as;

$$\tau_{12}^{L} = \frac{b_{21} - b_{12}}{a_{21} + a_{12}} \quad \text{and} \quad \tau_{21}^{L} = \frac{b_{12} - b_{21}}{a_{21} + a_{12}} \tag{9}$$

Note that, the following central difference scheme is employed in order to calculate the derivatives needed;

$$f'(x_0) = \frac{-f_2 + 8f_1 - 8f_{-1} + f_{-2}}{12h}$$

2.2 Quadratic Lagrange-Taylor Solution

In this section, quadratic form of Lagrange-Taylor solution is discussed. The aim is to improve the accuracy of the time-shift estimation.

Forward Solution. The quadratic form of Taylor expansion based on forward scheme can be written as;

$$V_2[n] = V_1[n + \tau_{12}]$$
$$V_1[n + \tau_{12}] = V_1[n] + V_1'[n]\tau_{12} + \frac{1}{2}V_1''[n]\tau_{12}^2$$

The corresponding cost function considering the least squares error is;

$$h_{12}[\tau_{12}] = \sum_{n=1}^{N}\Big(V_2[n] - V_1[n] - V_1'[n]\tau_{12} - \frac{1}{2}V_1''[n]\tau_{12}^2\Big)^2 \tag{10}$$

which is subject to;

$$\frac{\mathrm{d}h_{12}[\tau_{12}]}{\mathrm{d}\tau_{12}} = 0 \quad \text{and} \quad \frac{\mathrm{d}^2h_{12}[\tau_{12}]}{\mathrm{d}\tau_{12}^2} > 0 \tag{11}$$

Taking the derivative of $h_{12}[\tau_{12}]$ with respect to τ_{12} results in;

$$\begin{aligned}\frac{\mathrm{d}h_{12}[\tau_{12}]}{\mathrm{d}\tau_{12}} &= \sum_{n=1}^{N}\Big(V_1''[n]\Big)^2\tau_{12}^3 + 3\sum_{n=1}^{N}V_1'[n]V_1''[n]\tau_{12}^2 \\ &+ 2\sum_{n=1}^{N}\Big(\Big(V_1'[n]\Big)^2 - V_1''[n]\Big(V_1[n] + V_2[n]\Big)\Big)\tau_{12} \\ &- 2\sum_{n=1}^{N}V_1[n]\Big(V_2[n] - V_1[n]\Big)\end{aligned}$$

It can be written as;

$$a_{12}\tau_{12}^3 + b_{12}\tau_{12}^2 + c_{12}\tau_{12} + d_{12} = 0$$

where

$$\begin{aligned}a_{12} &= \sum_{n=1}^{N}\Big(V_1''[n]\Big)^2 \\ b_{12} &= 3\sum_{n=1}^{N}V_1'[n]V_1''[n] \\ c_{12} &= 2\sum_{n=1}^{N}\Big(\Big(V_1'[n]\Big)^2 - V_1''[n]\Big(V_1[n] + V_2[n]\Big)\Big) \\ d_{12} &= -2\sum_{n=1}^{N}V_1[n]\Big(V_2[n] - V_1[n]\Big)\end{aligned}$$

Backward Solution. Backward solution scheme has exactly the same manner with the forward solution. Hence, the quadratic form of Taylor expansion for the backward solution can be written as;

$$
\begin{aligned}
V_1[n] &= V_2[n+\tau_{21}] \\
V_2[n+\tau_{21}] &= V_2[n] + V_2'[n]\tau_{21} + \frac{1}{2}V_2''[n]\tau_{21}^2
\end{aligned}
$$

The corresponding cost function taking into account the least squares error is;

$$
h_{21}[\tau_{21}] = \sum_{n=1}^{N}\left(V_1[n] - V_2[n] - V_2'[n]\tau_{21} - \frac{1}{2}V_2''[n]\tau_{21}^2\right)^2 \tag{12}
$$

subject to;

$$
\frac{\mathrm{d}h_{21}[\tau_{21}]}{\mathrm{d}\tau_{21}} = 0 \quad \text{and} \quad \frac{\mathrm{d}^2 h_{21}[\tau_{21}]}{\mathrm{d}\tau_{21}^2} > 0 \tag{13}
$$

The derivative of $h_{21}[\tau_{21}]$ with respect to τ_{21} yields;

$$
\begin{aligned}
\frac{\mathrm{d}h_{21}[\tau_{21}]}{\mathrm{d}\tau_{21}} = &\sum_{n=1}^{N}\left(V_2''[n]\right)^2\tau_{21}^3 + 3\sum_{n=1}^{N}V_2'[n]V_2''[n]\tau_{12}^2 \\
&+ 2\sum_{n=1}^{N}\left(\left(V_2'[n]\right)^2 - V_2''[n]\left(V_2[n] + V_1[n]\right)\right)\tau_{21} \\
&- 2\sum_{n=1}^{N}V_2[n]\left(V_1[n] - V_2[n]\right)
\end{aligned}
$$

Note that, the following central difference scheme is employed in order to calculate the derivatives needed;

$$
\begin{aligned}
f'(x_0) &= \frac{-f_2 + 8f_1 - 8f_{-1} + f_{-2}}{12h} \\
f''(x_0) &= \frac{-f_2 + 16f_1 - 30f_0 + 16f_{-1} - f_{-2}}{12h^2}
\end{aligned}
$$

Note that, the following central difference scheme is employed in order to calculate the derivatives needed;

$$
\begin{aligned}
f'(x_0) &= \frac{-f_2 + 8f_1 - 8f_{-1} + f_{-2}}{12h} \\
f''(x_0) &= \frac{-f_2 + 16f_1 - 30f_0 + 16f_{-1} - f_{-2}}{12h^2}
\end{aligned}
$$

3 Results and Discussions

The illustration of the experimental setup was shown in Fig. 1. DN-20, i.e. the diameter of the pipeline is 20 mm, pipeline was used in order to carry out the experiments. Two ultrasonic transducers were placed onto the pipeline having the distance was 20 cm. The ultrasonic signal transmitted by a transducer with a frequency of 1 MHz, accompanying the amplitude of 3.6 V. While the original signal was a pulse train; mainly due to the echo effect the signal at the receiver side was distorted. The time shift between receiving signals measured by sensors is shown in Fig. 3.

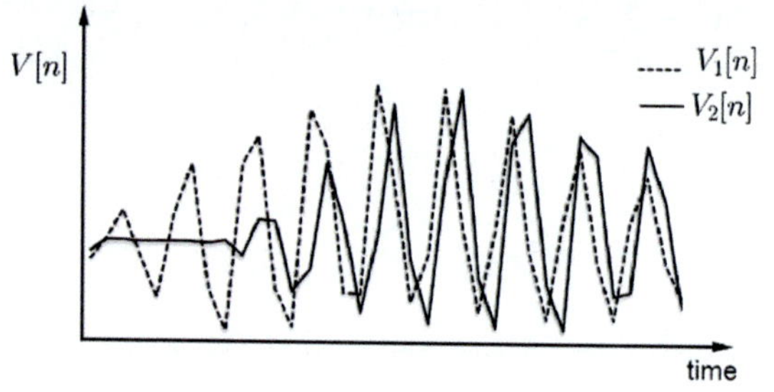

Fig. 3. The time shift between receiving signals obtained from the experiments

ToF, i.e. time delay, is related to the average velocity of the fluid passing through the pipeline, thereafter the volumetric amount of the passed fluid is calculated.

The three-point interpolation method, which is illustrated in Fig. 4, based on cosine interpolation has been employed in order to minimize the measurement error in estimating the time shift due to its discrete nature [8].

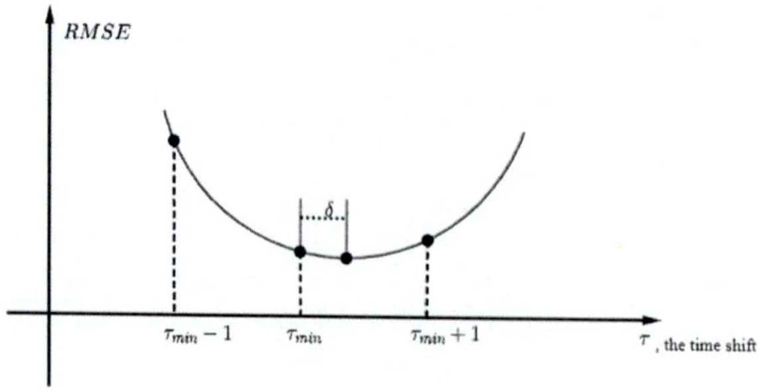

Fig. 4. Graphical explanation of the three-point interpolation methods of the local minimums

The following equations are used to calculate the fined estimation of ToF considering cosine interpolation technique;

$$
\begin{aligned}
R_{-1} &= RMSE(\tau_{min} - 1)\\
R_0 &= RMSE(\tau_{min})\\
R_1 &= RMSE(\tau_{min} + 1)\\
\omega &= \cos^{-1}(\frac{R_{-1} + R_1}{2R_0})\\
\phi &= \tan^{-1}(\frac{R_{-1} - R_1}{2R_0 \sin\omega})\\
\delta &= \frac{\phi}{\omega}\\
\text{ToF} &= (\tau + \delta)T_S
\end{aligned}
$$

In this study, the experimental data is collected from a DN-20 type ultrasonic water flow meter. Various numbers of volumetric flow rates, starting from 16 l/h to 5000 l/h, have been employed. Taking into account the calibration process of the flow meter, the most significant parameters of these experiments are stability and reliability. Therefore, standard deviation of the measured data was assigned to the main parameter in order to validate the robustness level of the proposed numerical schemes. The results obtained from the proposed methods in addition to the commonly applied correlation method are depicted in Table 1. According to the results, Lagrange-Taylor based method supplies the most accurate estimations. Moreover, Taylor approximation based methods predict the flow velocity more accurate than the commonly used correlation technique.

Table 1. Experimental results of the corresponding methods

Methods	Max τ	Min τ	Std of τ
Correlation	17.6047	13.6899	0.5704
Taylor (Forward)	0.5929	0.4535	**0.0166**
Taylor (Backward)	-0.4780	-0.6336	0.0230
Lagrange-Taylor	0.5987	0.4712	**0.0184**
Quadratic Taylor (Forward)	5.1546	4.4609	0.1336
Quadratic Taylor (Backward)	4.9745	4.4965	0.1060
Quadratic Lagrange-Taylor	4.3721	4.0908	0.0518

4 Conclusion

Sensor technologies based on ultrasonic signals are efficiently used as solutions of many problems raised in engineering area. Time-of-Flight (ToF), which is the discrepancy between the sent and the received ultrasonic signals in time, has a significant role for industrial applications. ToF can be employed in order to analyze fluid level sensing, liquid flow-metering, distance sensing, identification of fluid phenomena. Due to its nature, measurement is a discrete-time process. The proposed method has consistency with this discrete nature; thereby it will supply more accurate and reliable results of ToF measurements. Consequently, the suggested method has a great potential to improve the measurement accuracy and reliability in watering systems.

References

1. Angrisani, L., Baccigalupi, A., Moriello, R.S.L.: Ultrasonic-Based Distance Measurement Through Discrete Extended Kalman Filter. INTECH Open Access Publisher (2009)
2. Chen, X., Liu, C., Yang, D., Liu, X., Hu, L., Xie, J.: Highly accurate airflow volumetric flowmeters via pmuts arrays based on transit time. J. Microelectromech. Syst. **28**(4), 707–716 (2019)
3. Chhantyal, K., Jondahl, M.H., Viumdal, H., Mylvaganam, S.: Upstream ultrasonic level based soft sensing of volumetric flow of non-newtonian fluids in open venturi channels. IEEE Sens. J. **18**(12), 5002–5013 (2018)
4. Fernandez, J., Heller, P., Reinalter, W., Roger, M., Guillot, E.: Mass flow measurement and calibration procedures. Technical report (2014). https://doi.org/10.13140/RG.2.2.35021.23523
5. Gerasimov, S., Zhelbakov, I.: Correlation analysis of probing signals as a special case of the digital signal processing methods in an ultrasonic liquid flow measurement applications. In: 2018 International Conference on Industrial Engineering, Applications and Manufacturing (ICIEAM), pp. 1–9. IEEE (2018)
6. Hitomi, J., Murai, Y., Park, H.J., Tasaka, Y.: Ultrasound flow-monitoring and flow-metering of air-oil-water three-layer pipe flows. IEEE access **5**, 15021–15029 (2017)
7. Huhn, F., Schanz, D., Gesemann, S., Dierksheide, U., van de Meerendonk, R., Schroder, A.: Large-scale volumetric flow measurement in a pure thermal plume by dense tracking of helium-filled soap bubbles. Exp. Fluids **58**(9), 116 (2017)
8. Jacovitti, G., Scarano, G.: Discrete time techniques for time delay estimation. IEEE Trans. Signal Process. **41**(2), 525–533 (1993)
9. Jia, J., Wang, H., Millington, D.: Electrical resistance tomography sensor for highly conductive oil-water two-phase flow measurement. IEEE Sens. J. **17**(24), 8224–8233 (2017)
10. Lin, Y., Hans, V.: Self-monitoring ultrasonic gas flow meter based on vortex and correlation method. IEEE Trans. Instrum. Meas. **56**(6), 2420–2424 (2007)
11. Lysak, P.D., Jenkins, D.M., Capone, D.E., Brown, W.L.: Analytical model of an ultrasonic cross-correlation flow meter, part 2: application. Flow Meas. Instrum. **19**(1), 41–46 (2008)
12. Marioli, D., Narduzzi, C., Offelli, C., Petri, D., Sardini, E., Taroni, A.: Digital time-of-flight measurement for ultrasonic sensors. IEEE Trans. Instrum. Meas. **41**(1), 93–97 (1992)
13. Moazzeni, T., Ma, J., Jiang, Y.: Measurement and calibration of thermal cross-correlation-based flowmeter operating in harsh environment. IEEE Trans. Instrum. Meas. **62**(8), 2280–2285 (2013)

14. Rajita, G., Mandal, N.: Review on transit time ultrasonic flowmeter. In: 2016 2nd International Conference on Control, Instrumentation, Energy & Communication (CIEC), pp. 88–92. IEEE (2016)
15. Sirazitdinova, E., Pesic, I., Schwehn, P., Song, H., Satzger, M., Weingartner, D., Sattler, M., Deserno, T.M.: Stereo vision for fully automatic volumetric flow measurement in urban drainage structures. In: Videometrics, Range Imaging, and Applications XIV, vol. 10332, p. 103320M. International Society for Optics and Photonics (2017)
16. Svilainis, L., Lukoseviciute, K., Dumbrava, V., Chaziachmetovas, A.: Subsample interpolation bias error in time of flight estimation by direct correlation in digital domain. Measurement **46**(10), 3950–3958 (2013)
17. Thorn, R., Johansen, G., Hjertaker, B.: Three-phase flow measurement in the petroleum industry. Meas. Sci. Technol. **24**(1), 012003 (2012)
18. Tummers, J.S., Kerr, J.R., O'Brien, P., Kemp, P., Lucas, M.C.: Enhancing the upstream passage of river lamprey at a microhydropower installation using horizontally-mounted studded tiles. Ecol. Eng. **125**, 87–97 (2018)
19. Vogt, M., Gevers, M., Musch, T.: Evaluation of transducer configurations for ultrasound cross-correlation flowmeters. In: 2014 IEEE International Instrumentation and Measurement Technology Conference (I2MTC) Proceedings, pp. 40–44. IEEE (2014)
20. Wright, S., Nocedal, J.: Numerical optimization. Science **35**, 67–68 (1999)

Volterra Model of Energy Storage with Nonlinear Efficiency in Integrated Power Systems

Ildar Muftahov[1,2(✉)], Denis Sidorov[1,3], Aleksei Zhukov[1,3], and Dmitriy Karamov[1]

[1] Energy Systems Institute, Siberian Branch of Russian Academy of Sciences, 664033 Irkutsk, Russia
ildar_sm@mail.ru
[2] Main Computing Center, Joint Stock Company "Russian Railways", 664005 Irkutsk, Russia
[3] Baikal School of BRICS, Irkutsk National Research Technical University, 664074 Irkutsk, Russia

Abstract. The article deals with integrated power systems with renewable sources of generation in the settlements of Lake Baikal. The analysis of natural and climatic features in relation to the wind potential of the specified area is carried out. The construction of wind forecast using machine learning algorithms is considered. Global NWP forecast and machine learning model to achieve acceptable accuracy are combined. This forecast is used in the Volterra model to determine the operating parameters of energy storage units, in which the efficiency is a nonlinear function of the charge level, age and external operating conditions, i.e. the process of charging and discharging energy storage systems is described using Volterra integral equations. Such formulation of the problem will allow for the design of power systems to use a number of valuable recommendations regarding the optimal management of energy storage systems with several integrated power systems. The numerical solution of the proposed Volterra integral model was obtained using the spline collocation method.

Keywords: Volterra model · Integrated power systems · Wind forecast · Renewable sources · Energy systems

1 Volterra Model

Recently, a large number of studies have been carried out related to the use of renewable sources and energy storage in both centralized and autonomous power systems. The main factor driving the growth of installed capacity of renewable energy sources as one of the main vectors of global energy development is the environmental component caused by concerns about the increase in the average temperature of the Earth's surface [1]. At the same time, large-scale integration

T. Allahviranloo et al. (Eds.): IDS 2020, AISC 1301, pp. 808–815, 2021.
https://doi.org/10.1007/978-3-030-66501-2_65

of wind energy is challenging from a power system management perspective. This increases the total cost of energy produced and significantly limits the benefits of using renewable energy sources. One way to reduce the uncertainty associated with wind power generation is to use forecasting tools. Load and generation forecasts up to 48 h are useful for energy market operators to create daily unit commitments and economic schedules. For example, according to [2,3], an increase of only 1% of the forecast error resulted in £10 million increase in operating costs per year for one electric company in the UK. Paper [4] evaluates the performance of wind turbines, energy storage and demand response programs under deterministic and stochastic conditions, as well as the impact of wind, price, and demand uncertainties on energy hub planning. Also, one of the urgent tasks related to distributed renewable generation is to take into account the non-linearity of aging processes of storage devices that depend on the operating parameters associated with the alternating power function of the system [5–7]. In order to efficiently model the storages operation, the following Volterra integral model from [8] with constraints is employed

$$\begin{cases} \int\limits_0^t \begin{pmatrix} h_1(t,\tau,x_1(\tau),x_2(\tau),\dots,x_n(\tau))\,d\tau \\ h_2(t,\tau,x_1(\tau),x_2(\tau),\dots,x_n(\tau))\,d\tau \\ \dots \\ h_m(t,\tau,x_1(\tau),x_2(\tau),\dots,x_n(\tau))\,d\tau \end{pmatrix} = \begin{pmatrix} f_1(t) \\ f_2(t) \\ \dots \\ f_m(t) \end{pmatrix} \\ v_i(t) = \int\limits_0^t x_i(\tau)\,d\tau, \ \max\limits_{t\in[0,T]} v_i(t) \le v_{i\,max} \\ E_{i\,min}(t) \le \int\limits_0^t v_i(\tau)\,d\tau \le E_{i\,max}(t) \\ 0 < \alpha_1(t) < \alpha_2(t) < \dots < \alpha_{n-1}(t) < t, \end{cases} \tag{1}$$

where $t \in [0,T]$ and kernels $h_i(t,\tau,x_1(\tau),x_2(\tau),...,x_n(\tau))$ are jump discontinuous on the curves $a_j(t), j = 1,2,...,n-1$ and can be defined as

$$h_i(t,\tau,x_1(\tau),x_2(\tau),\dots,x_n(\tau))\,d\tau = \begin{cases} K_{i,1}(t,\tau)G_{i,1}(t,x_1(\tau)), t,\tau \in p_1 \\ K_{i,2}(t,\tau,G_{i,2}(t,x_2(\tau)), t,\tau \in p_2 \\ \dots \\ K_{i,n}(t,\tau,G_{i,n}(t,x_n(\tau)), t,\tau \in p_n \end{cases}. \tag{2}$$

Here $p_j = \{(t,\tau)|\, \alpha_{j-1}(t) < \tau \le \alpha_j(t)\}$, $\alpha_0(t) = 0$, $\alpha_n(t) = t$, $i = 1,2,...,m$, $j = 1,2,...,n$. Functions $K_{i,j}(t,\tau)$, $f_i(t)$, $\alpha_j(t)$ enjoy continuous derivatives with respect to time t for $(t,\tau) \in \bar{p_j}$, $K_{i,n}(t,t) \neq 0$, $\bar{p_j}$ is closure of the set p_j. Functions $\alpha_1(t),...,\alpha_{n1}(t)$ are nondecreasing for $t \in [0,T]$.

Here, m stands for number of grids; functions $a_j(t)$ show the proportions in which units in storage system are used in each subgrid; n is number of units in storage system for i-th grid; the elements of the matrix $K_{i,i}(t,\tau)G_{i,i}(t,x_j(\tau))$ shows efficiency of storage system of each grid, the remaining elements of the matrix show at the microgrid level the coefficients of power flow from storage systems of other grids.

Source functions $f_i(t)$ are the energy imbalance for i-th grid defined as follows

$$f_i(t) = f_{iRES}(t) + f_{iAC/DC}(t) - f_{iLoad}(t), \quad (3)$$

where functions $f_{iRES}(t)$ are the generation from renewable energy sources (RES), $f_{iAC/DC}(t)$ are AC/DC interconnection power flow between i-th grid in community and centralized system and $f_{load}(t)$ are the electric load for i-th grid. This imbalance is supposed to be covered by battery storage operated in parallel with diesel generator backup system included in $f_{iLoad}(t)$. Functions v_{imax} are maximum speed of the charge for i-th storage; E_{imin} and E_{imax} are constraints on the storage levels. The alternating power function (APF) based on $x_i(t)$ is possible to find for each storage using this model.

It should be noted that with the help of this model 1 we can determine the operating parameters of the energy storage units, such as the charge level. The 1 model is a generalization of the well-known ampere-hour integral model, but in our case we are dealing with a typical inverse problem, since we have known load unbalance functions $f_i(t)$ and must define the alternating power functions $x_i(t)$ by solving the integral equation. We also note that in our article in Sect. 3 in the description of the numerical experiment we use only wind energy as generation from renewable energy sources. To do this, in the next section, we describe in detail wind speed forecasting, since the accuracy of the Volterra model is highly dependent on the quality of the forecast.

2 Wind Speed Forecast

In order to produce efficient wind speed forecast hybrid approach was applied. We combine global NWP forecast and machine learning model to achieve acceptable accuracy. In this way we eager to get benefits of both physical and statistical approaches.

One of the largest problem in this field is an absence on the clear and representative data therefore our experiments were build on the reanalysis data. Input data [9] contain wind speed (horizontal and vertical components), atmospheric pressure and temperature obtained from ERA5 reanalysis model [10] with 0.1° x 0.1° resolution. As an additional features global NWP model with 1.5° x 1.5° output for area from (56°, 98°) to (50°, 109°) were exploited. Target forecasting point is located near from the Lake Baikal coast and Buguldeika village (52.54401°, 106.04047°). As an another additional features we use atmospheric pressure and wind speed at the following locations (Fig. 1) Bayanday (53.054821°, 105.508066°), Babushkin (51.720233°, 105.861541°), Gremyachinsk (52.799969°, 107.967595°). The distance between this locations and target point is from 67 to 135 km. Our dataset was divided into parts by time in the following way: train (2000–2013), validation (2014) and test (2015).

In order to get high-accuracy machine learning model some feature engineering approaches were introduced. First, well-known principal components analysis (PCA) [11] was applied to each variables map to reduce feature space. In this way we add 5 principal components for pressure, horizontal and vertical wind speed components. Second approach is to add periodicity features. These features include day of year (DOY) and time of a day (TOD) and their derivatives like sin(2 * pi * DOY/365), cos(2 * pi * DOY/365) or sin(2 * pi * TOD/24), cos(2 * pi * TOD/24). Another applied technique is a target mean encoding. Such a feature represents mean target value which is specific for a current categorical variable value. So we use mean target specific for combination of the month, wind direction and hour and month and hour only.

Concerning machine learning gradient boosting under decision trees was applied. We use XGBoost [12] with grid hyperparameter search. Model errors were estimated with 5-fold cross-validation.

Thus at the time of prediction, we should have the prediction made by the global NWP model, current measurements at several surrounding points, and a retrospective target location data. In the result we achieve RMSE 0.5 m per second. This model helps us to produce efficient solution for battery modeling problems considered in this paper.

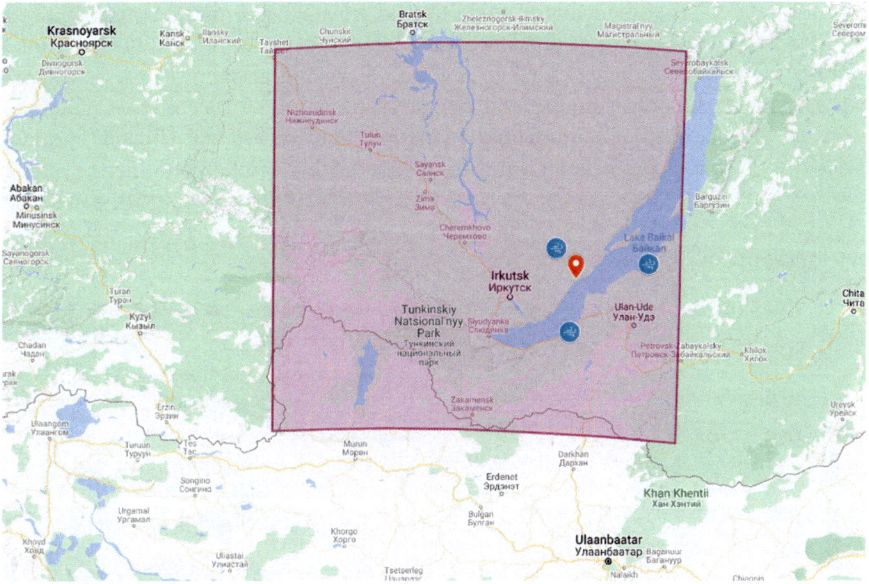

Fig. 1. Locations which are used in order to get additional pressure and wind speed features

3 Numerical Experiment

The proposed approach of battery modeling was tested on real load datasets from one resort village located on the coast of Lake Baikal (Eastern Siberia,

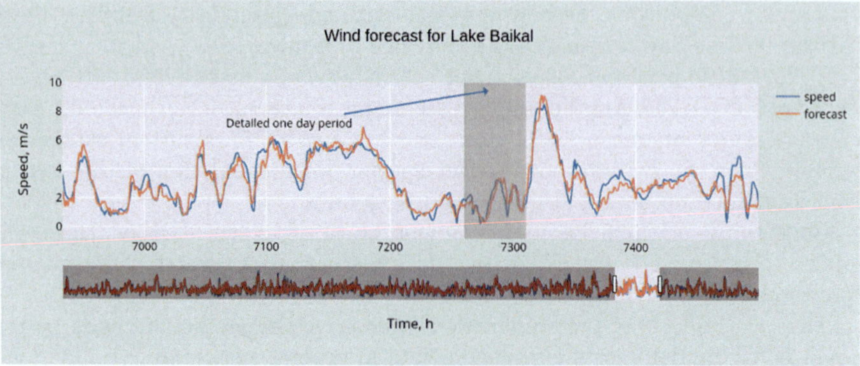

Fig. 2. Selected 24-h period for numerical experiment

Russia). Earlier in [13], for this location, the two-level optimization technique for operational and emergency control of a hybrid AC/DC community was proposed.

Alternating power functions were calculated by the system of Volterra integral Eq. (1) with efficiency described in Fig. 3. For comparison, we also performed calculations of the power function of energy storage units for a constant efficiency equal to 92% (Fig. 4). To find unknown functions we use the efficient numerical methods of collocations for solution of the weakly regular nonlinear systems

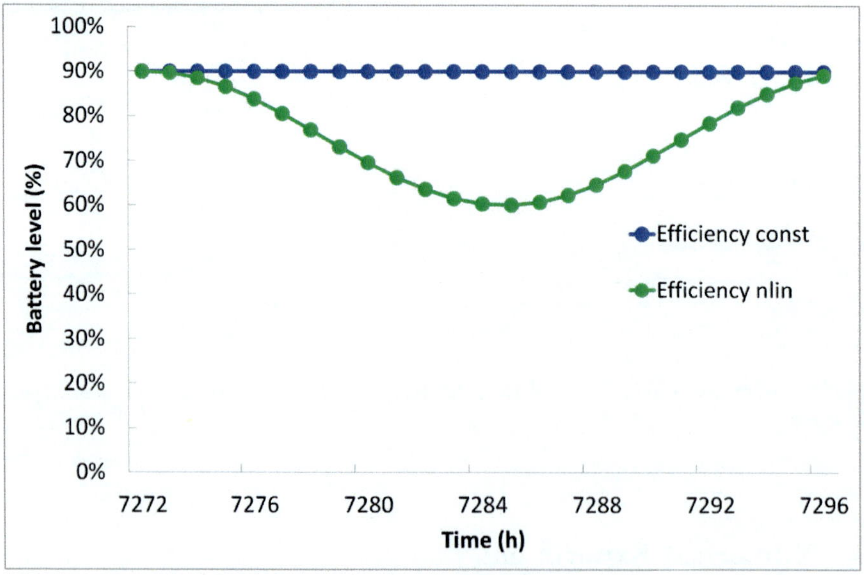

Fig. 3. Change in efficiency during the day

of Volterra integral equations of the first kind [8,14,15]. The difference in the graphs shows that the nonlinear model makes it possible to more accurately determine the operating parameters of energy storage units.

In this numerical experiment $v_{imax} = 3.5$ MW means the limitation on the maximum power with which the storage system in i-th grid can be charged and discharged, $E_{imin} = 0\%$, $E_{imax} = 100\%$. In this work, all numerical calculations were performed for one calendar year, but for clarity, in this article we took a 24-hour interval (Fig. 2).

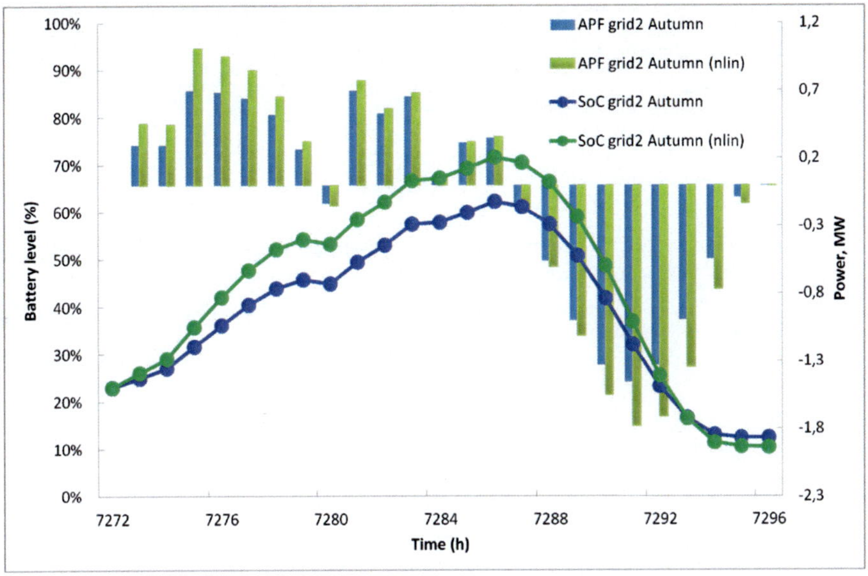

Fig. 4. Alternating power functions (APF in MW) with state of charge information (SOC in %) computed by the system of Volterra integral equation for the autumn period.

4 Conclusions

This article discusses a method for modeling energy storage systems in integrated power systems with renewable sources of generation for some locations in the nature protection zone of Lake Baikal. Such models are described in terms of an inverse problem using Volterra integral equations of the first kind with piecewise continuous kernels. It is shown that nonlinear models provide more accurate information as compared to models based on constant efficiency. These models use real information about electrical load and generation from renewable energy sources as input. Also, the article discusses in detail the problem of constructing a high-quality forecast of wind speed. As a criterion for the accuracy of the

XGBoost machine learning algorithm, the RMSE parameter was chosen, which reached approximately 0.5 m/s.

Funding

The reported study in Sects. 1 and 3 was funded by RFBR and NSFC according to the research project No. 19-58-53011/61911530132. The reported study in Sect. 2 was funded by RFBR and the Government of the Irkutsk Region, project number 20-48-383004.

Conflicts of Interest

The authors declare no conflict of interest.

References

1. Climate change 2013. The physical science basis. Working Group I Contribution to the Fifth Assessment Report of the Intergovernmental Panel on Climate Change. http://www.climatechange2013.org/images/report/WG1AR5_ALL_FINAL.pdf. Accessed 25 May 2017
2. Bunn, D., Farmer, E.: Comparative models for electrical load forecasting. Int. J. Forecast. **2**, 501–505 (1985)
3. Soares, L., Medeiros, M.: Modeling and forecasting short-term electricity load: a comparison of methods with a application to Brazilian data. Int. J. Forecast. **24**, 630–644 (2008)
4. Pazouki, S., Haghifam, M.-R.: Optimal planning and scheduling of energy hub in presence of wind, storage and demand response under uncertainty. Int. J. Electr. Power Energy Syst. **80**, 219–239 (2016)
5. Cristóbal-Monreal, I.R., Dufo-López, R.: Optimisation of photovoltaic-diesel-battery stand-alone systems minimising system weight. Energy Conversion Manag. **29**, 151–157 (2014)
6. Tsuanyo, D., Azoumah, Y., Aussel, D., Neveu, P.: Modeling and optimization of battery less hybrid PdV (photovoltaic)/Diesel systems for off-grid applications. Energy **86**, 152–163 (2015)
7. Dursun, E., Kilic, O.: Comparative evaluation of different power management strategies of a stand-alone PV/Wind/PEMFC hybrid power system. Int. J. Electr. Power Energy Syst. **34**, 81–89 (2012)
8. Sidorov, D., Aleksandr, T., Ildar, M., Aliona, D., Fang, L.: Nonlinear systems of volterra equations with piecewise smooth kernels: numerical solution and application for power systems operation. Mathematics **8**, 1257 (2020)
9. Copernicus Climate Change Service (C3S) (2017): ERA5: Fifth generation of ECMWF atmospheric reanalyses of the global climate. Copernicus Climate Change Service Climate Data Store (CDS), 07.05.2020. https://cds.climate.copernicus.eu/cdsapp#!/home
10. Hersbach, H., Bell, B., Berrisford, P., Hirahara, S., Horányi, A., Muñoz-Sabater, J., Nicolas, J., Peubey, C., Radu, R., Schepers, D., et al.: The ERA5 global reanalysis. Q. J. Roy. Meteorol. Soc. (2020)

11. Abdi, H., Williams, L.J.: Principal component analysis. Wiley Interdiscip. Rev. Comput. Statist. **2**(4), 433–459 (2010)
12. Chen, T., Guestrin, C.: IXgboost: a scalable tree boosting system. In: Proceedings of the 22nd ACM SIGKDD International Conference on Knowledge Discovery and Data Mining, pp. 785–794 (2016)
13. Denis, S., Daniil, P., Nikita, T., Dmitriy, K., Aleksei, Z., Ildar, M., Aliona, D., Fang, L., Li, Y.: Toward zero-emission hybrid AC/DC power systems with renewable energy sources and storages: a case study from lake baikal region. Energies **13**, 1226 (2020)
14. Muftahov, I.R., Sidorov, D.N.: Solvability and numerical solutions of systems of nonlinear volterra integral equations of the first kind with piecewise continuous kernels. Bull. South Ural State Univ. Ser. Math. Model. Program. Comput. Softw. (Bull. SUSU MMCS), 9(1), 130–136 (2016). https://doi.org/10.14529/mmp160111
15. Muftahov, I., Tynda, A., Sidorov, D.: Numeric solution of Volterra integral equations of the first kind with discontinuous kernels. J. Comput. Appl. Math. **313**, 119–128 (2017). https://doi.org/10.1016/j.cam.2016.09.003

Soft and Linguistic Computations

Finite Difference Approximation to Fuzzy Heat Equation in Cylindrical and Spherical Polar Co-ordinates

F. Shahmansoory and M. Afshar Kermani(✉)

Department of Mathematics, Tehran North Branch, Islamic Azad University, Tehran, Iran
Mog_afshar@yahoo.com

Abstract. In this paper, we investigate the stability of the heat equation in the polar Coordinate fuzzy parametric system, first we write the heat equation in a polar system, then express it as a parametric fuzzy, and then drive it using definitions Bede and et al., we have written the equation and finally, tested its stability using the matrices method.

These method has higher accuracy and less error, than non-fuzzy methods.

Keywords: Polar co-ordinates · Heat equation · Sustainability · Fuzzy representation

1 Introduction

With the development of numerical methods of solving equations, differential equations, some equations, encountered problems that could be solved in polar coordinates.

In general, the concept of angle and radius was used by the ancient people in the first millennium BC.

Archimedes also explained that in the bends the radius depends on the angle.

With all the work the ancient Greeks did, the polar coordinate system was not fully developed. George de Saint Vincent and Cavalieri independently introduced the concepts of polar coordinates in the mid-19th century [1–7].

Fuzzy numbers, parametric form, can be used to increase accuracy and radios error.

In this paper, we investigate the stability of the fuzzy heat equation in polar coordinates using matrix method.

2 Basie Definitions

2.1 Definition 1: Stability

We know that in the differential equation the total error equals:

$$e_{ij} = E_{ij} + R_{ij}$$

T. Allahviranloo et al. (Eds.): IDS 2020, AISC 1301, pp. 819–831, 2021.
https://doi.org/10.1007/978-3-030-66501-2_66

Where:

E_{ij}: local truncation error and
R_{ij}: Rounding error

Suppose for arbitrary column ∀ i whenever $j \to \infty$ goes, if $|E_{ij}| < M$
Is then the difference method is stable.

2.2 Definition 2: Lex's Theorem

If a linear finite difference equation is consistent, together with the problem of linear initial value, then stability ensures

That u → U converges when k and h tend to zero.

2.3 Definition 3: Theorem

When: $\rho(A) = max|\lambda_s| \leq 1$ then finite difference method is stable

Where:

λ_s: Spectral radius

2.4 Definition 4: Parametric Form of Fuzzy Number

We represent an arbitrary fuzzy number by an ordered pair of functions

$$\left(\underline{u}(r), \bar{u}(r)\right), 0 \leq r \leq 1$$

Which satisfies the following requirements:

(1) $\underline{u}(r)$ is a bounded left continuous non
Decreasing function over [0, 1]
(2) $\bar{u}(r)$: is a bounded left continuous non
Increasing function over [0, 1]
(3) $\underline{u}(r) \leq \bar{u}(r), 0 \leq r \leq 1$

2.5 Definition 5

Let F: (a, b) → F^n and $t. \in (a, b)$.

We say that F is differentiable at t. if:

1) There exist element $F'(t_0) \in F^n$ such that, for all h > 0 sufficiently near to zero
There are $F(t_0 + h) - F(t_0), F(t_0) - F(t_0 - h)$ and

$$\lim_{h \to 0+} \frac{F(t_0 + h) - F(t_0)}{h} = \lim_{h \to 0+} \frac{F(t_0) - F(t_0 - h)}{h} = \mathrm{F}'(t_0) \tag{1}$$

Or

2) There exists and element $F'(t_0) \in F^n$ such that, for all h < 0 sufficiently near to zero
There are $F(t_0 + h) - F(t_0)$, $F(t_0) - F(t_0 - h)$ and

$$\lim_{h\to 0^-} \frac{F(t_0 + h) - F(t_0)}{h} = \lim_{h\to 0^-} \frac{F(t_0) - F(t_0 - h)}{h} = \mathrm{F}'(t_0) \tag{2}$$

3 Investigation of Stability of Fuzzy Form of Heat Equation in Polar Coordinates

The fuzzy form of the finite difference approximation for the heat equation in polar coordinates polar form of heat equation:

$$\frac{\partial U}{\partial t} = \frac{\partial^2 U}{\partial r^2} + \frac{1}{r}\frac{\partial U}{\partial r} + \frac{1}{r^2}\frac{\partial^2 U}{\partial \theta^2} + \frac{\partial U}{\partial z}$$

If we consider two-dimensional we have it:

$$\frac{\partial U}{\partial z} = 0$$

So

$$\frac{\partial U}{\partial t} = \frac{\partial^2 U}{\partial r^2} + \frac{1}{r}\frac{\partial U}{\partial r} + \frac{1}{r^2}\frac{\partial^2 U}{\partial \theta^2}$$

Given the cylindrical states, because of the symmetry, the changes in it will not depend on θ, so we have:

$$\frac{\partial U}{\partial t} = \frac{\partial^2 U}{\partial r^2} + \frac{1}{r}\frac{\partial U}{\partial r}$$

Its fuzzy form is as follows:

$$\left(D_t - D_r D_r - \frac{1}{r} D_r\right)\tilde{u}(r, t) = \tilde{0}$$

So:

$$\begin{cases} D_t \underline{U}(r, t, \alpha) - (D_r D_r)\bar{u}(r, t, \alpha) - \frac{1}{r} D_r \bar{u}(r, t, \alpha) = \underline{0} \\ D_t \bar{U}(r, t, \alpha) - (D_r D_r)\underline{u}(r, t, \alpha) - \frac{1}{r} D_r \underline{u}(r, t, \alpha) = \bar{0} \end{cases} \tag{3}$$

We now use the generalized derivative definition by Bede and colleagues:
Say F is a fuzzy derivative whenever we have:
First case:

$$\lim_{h\to 0+} \frac{f(t_0 + h) - f(t_0)}{h} = \lim_{h\to 0+} \frac{f(t_0) - f(t_0 - h)}{h} = \mathrm{F}'(t_0)$$

The latter:

$$\lim_{h\to 0^-}\frac{f(t_0+h)-f(t_0)}{h}=\lim_{h\to 0^-}\frac{f(t_0)-f(t_0-h)}{h}=\mathrm{F}'(t_0)$$

Heat Equation Terms:

$$\begin{cases}\tilde{u}(0,t)=\tilde{K}_1\\ \tilde{u}(L,t)=\tilde{K}_2 \quad t>0\\ \tilde{u}(r,0)=\tilde{f}(r)\ 0<r<L\end{cases}$$

Which is fuzzy as follows:

$$\begin{cases}\underline{U}(0,t,\alpha)=\underline{K}_1(\alpha);\quad \bar{u}(0,t,\alpha)=\bar{K}_1(\alpha)\\ \underline{U}(L,t,\alpha)=\underline{K}_2(\alpha);\quad \bar{u}(L,t,\alpha)=\bar{K}_2(\alpha)\\ \underline{U}(r,0,\alpha)=\underline{f}(r,\alpha);\quad \bar{u}(r,0,\alpha)=\bar{f}(r,\alpha)\end{cases}$$

Different states of the fuzzy heat equation with respect to its derivatives
First Case:
If $\frac{\partial U}{\partial t}$ and $\frac{\partial^2 U}{\partial r^2}$ are both derived from the first case the equations will be in form (3)
The second state if $\frac{\partial U}{\partial t}$ and $\frac{\partial^2 U}{\partial r^2}$ are both fuzzy derivatives of the second type:

$$\begin{cases}D_t\bar{U}(r,t,\alpha)-(D_rD_r)\underline{u}(r,t,\alpha)-\left(\frac{1}{r}D_r\right)\underline{u}(r,t,\alpha)=\underline{0}\\ D_t\underline{U}(r,t,\alpha)-(D_rD_r)\bar{u}(r,t,\alpha)-\left(\frac{1}{r}D_r\right)\bar{u}(r,t,\alpha)=\bar{0}\end{cases}$$

Case three: if $\frac{\partial U}{\partial t}$ is the first type and $\frac{\partial^2 U}{\partial r^2}$ is the second

$$\begin{cases}D_t\underline{u}(r,t,\alpha)-(D_rD_r)\underline{u}(r,t,\alpha)-\left(\frac{1}{r}D_r\right)\bar{u}(r,t,\alpha)=\underline{0}\\ D_t\bar{u}(r,t,\alpha)-(D_rD_r)\bar{u}(r,t,\alpha)-\left(\frac{1}{r}D_r\right)\underline{u}(r,t,\alpha)=\bar{0}\end{cases}$$

Mode Four: if $\frac{\partial U}{\partial t}$ is the second type and $\frac{\partial^2 U}{\partial r^2}$ is the first:

$$\begin{cases}D_t\bar{u}(r,t,\alpha)-(D_rD_r)\bar{u}(r,t,\alpha)-\left(\frac{1}{r}D_r\right)\underline{u}(r,t,\alpha)=\underline{0}\\ D_t\underline{u}(r,t,\alpha)-(D_rD_r)\underline{u}(r,t,\alpha)-\left(\frac{1}{r}D_r\right)\bar{u}(r,t,\alpha)=\bar{0}\end{cases}$$

The central derivative in fuzzy form is defined as follows

$$\begin{cases}\underline{D_t\tilde{u}(r,t)}=\frac{\underline{u}_{i,j+1}-\bar{u}_{i,j}}{k}\\ \overline{D_t\tilde{u}(r,t)}=\frac{\bar{u}_{i,j+1}-\underline{u}_{i,j}}{k}\end{cases} \tag{4}$$

$$\begin{cases}\underline{D_rD_r(\tilde{u}(r,t))}=\frac{\underline{u}_{i+1,j}-2\bar{u}_{i,j}+\underline{u}_{i-1,j}}{h^2}\\ \overline{D_rD_r\tilde{u}(r,t)}=\frac{\bar{u}_{i+1,j}-2\underline{u}_{i,j}+\bar{u}_{i-1,j}}{h^2}\end{cases} \tag{5}$$

By replacing Eq. 4 and 5 we have the first

$$\begin{cases} \frac{\underline{u}_{i,j+1}-\bar{u}_{i,j}}{k} = \frac{\bar{u}_{i+1,j}-2\underline{u}_{i,j}+\bar{u}_{i-1,j}}{h^2} + \frac{1}{r}\frac{\bar{u}_{i,j+1}-\underline{u}_{i,j}}{k} \\ \frac{\bar{u}_{i,j+1}-\underline{u}_{i,j}}{k} = \frac{\underline{u}_{i+1,j}-2\bar{u}_{i,j}+\underline{u}_{i-1,j}}{h^2} + \frac{1}{r}\frac{\underline{u}_{i,j+1}-\bar{u}_{i,j}}{k} \end{cases}$$

If we consider $\beta = \frac{1}{r}$ and $r = \frac{K}{h^2}$
We give K as the equations:

$$\begin{cases} \underline{u}_{i,j+1} - \bar{u}_{i,j} = r\left(\bar{u}_{i+1,j} - 2\underline{u}_{i,j} + \bar{u}_{i-1,j}\right) + \beta\left(\bar{u}_{i,j+1} - \underline{u}_{i,j}\right) \\ \bar{u}_{i,j+1} - \underline{u}_{i,j} = r\left(\underline{u}_{i+1,j} - 2\bar{u}_{i,j} + \underline{u}_{i-1,j}\right) + \beta\left(\underline{u}_{i,j+1} - \bar{u}_{i,j}\right) \end{cases}$$

So:

$$\begin{cases} \underline{u}_{i,j+1} = \bar{u}_{i,j} + r\bar{u}_{i+1,j} + (-2r - \beta)\underline{u}_{i,j} + r\bar{u}_{i-1,j} + \beta\bar{u}_{i,j+1} \\ \bar{u}_{i,j+1} = \underline{u}_{i,j} + r\underline{u}_{i+1,j} + (-2r - \beta)\bar{u}_{i,j} + r\underline{u}_{i-1,j} + \beta\underline{u}_{i,j+1} \end{cases}$$

If we put Eq. 4 and 5 in the second state:

$$\begin{cases} \frac{\bar{u}_{i,j+1}-\underline{u}_{i,j}}{k} = \frac{\underline{u}_{i+1,j}-2\bar{u}_{i,j}+\underline{u}_{i-1,j}}{h^2} + \frac{1}{r}\left(\frac{\underline{u}_{i,j+1}-\bar{u}_{i,j}}{k}\right) \\ \frac{\underline{u}_{i,j+1}-\bar{u}_{i,j}}{k} = \frac{\bar{u}_{i+1,j}-2\underline{u}_{i,j}+\bar{u}_{i-1,j}}{h^2} + \frac{1}{r}\left(\frac{\bar{u}_{i,j+1}-\underline{u}_{i,j}}{k}\right) \end{cases}$$

By multiplying the equations above we have K, $r = \frac{K}{h^2}$ and $\beta = \frac{1}{r}$

$$\begin{cases} \bar{u}_{i,j+1} - \underline{u}_{i,j} = r\left(\underline{u}_{i+1,j} - 2\bar{u}_{i,j} + \underline{u}_{i-1,j}\right) + \beta\left(\underline{u}_{i,j+1} - \bar{u}_{i,j}\right) \\ \underline{u}_{i,j+1} - \bar{u}_{i,j} = r\left(\bar{u}_{i+1,j} - 2\underline{u}_{i,j} + \bar{u}_{i-1,j}\right) + \beta\left(\bar{u}_{i,j+1} - \underline{u}_{i,j}\right) \end{cases}$$

As a result:

$$\begin{cases} \bar{u}_{i,j+1} = \underline{u}_{i,j} + r\underline{u}_{i+1,j} + (-2r - \beta)\bar{u}_{i,j} + r\underline{u}_{i-1,j} + \beta\underline{u}_{i,j+1} \\ \underline{u}_{i,j+1} = \bar{u}_{i,j} + r\bar{u}_{i+1,j} + (-2r - \beta)\underline{u}_{i,j} + r\bar{u}_{i-1,j} + \beta\bar{u}_{i,j+1} \end{cases}$$

We will conclude that the second case is the same as the first one, so just to prove one is one or two states.

Proof of stability of the first case:

$$\begin{bmatrix} \underline{u}_{i,j+1} \\ \vdots \\ \vdots \\ \underline{u}_{N-1,j+1} \\ \overline{u}_{1,j+1} \\ \vdots \\ \vdots \\ \overline{u}_{N-1,j+1} \end{bmatrix} = \underbrace{\begin{bmatrix} -2r-\beta & \text{A} \;\; 0 & 1 & r \;\; \text{B} & 0 \\ \ddots & & r & \ddots & \\ & & & \ddots & r \\ 0 & -2r-\beta & 0 & r & 1 \\ 1 & r & 0 & -2r-\beta & 0 \\ r & \ddots & & \ddots & \\ \ddots & & r & & \\ 0 & r & 1 & 0 & -2r-\beta \end{bmatrix}}_{\text{p}} \begin{bmatrix} \underline{u}_{1,j} \\ \vdots \\ \vdots \\ \underline{u}_{N-1,j} \\ \overline{u}_{1,j} \\ \vdots \\ \vdots \\ \overline{u}_{N-1,j} \end{bmatrix}$$

Suffice to prove $P(p) \leq 1$
or:

$$P(A+B) \leq 1 \quad \text{And} \quad P(A-B) \leq 1 \Rightarrow P(p) \leq 1$$

Let:

$$A = (2r-\beta)I \quad \text{And} \quad B = I + r\begin{bmatrix} 0 & 1 & & 0 \\ 1 & \ddots & \ddots & \\ & \ddots & \ddots & 1 \\ 0 & & 1 & 0 \end{bmatrix}$$

$$A+B = \begin{bmatrix} 1-2r-\beta & r & & 0 \\ r & \ddots & & \ddots \\ & \ddots & \ddots & r \\ 0 & & r & 1-2r-\beta \end{bmatrix} \quad \text{and}$$

$$A-B = \begin{bmatrix} -1-2r-\beta & -r & & 0 \\ -r & \ddots & & \ddots \\ & \ddots & \ddots & -r \\ 0 & & -r & 1-2r-\beta \end{bmatrix}$$

So:

$$A+B=\begin{bmatrix}-1 & & 0\\ & \ddots & \\ 0 & & -1\end{bmatrix}+\begin{bmatrix}-2r-\beta & r & & 0\\ r & \ddots & \ddots & \\ & \ddots & \ddots & r\\ 0 & & r & -2r-\beta\end{bmatrix}$$

$$A-B=\begin{bmatrix}-1 & & 0\\ & \ddots & \\ 0 & & -1\end{bmatrix}+\begin{bmatrix}-2r-\beta & -r & & 0\\ -r & \ddots & \ddots & \\ & \ddots & \ddots & -r\\ 0 & & -r & -2r-\beta\end{bmatrix}$$

So:

$$T=\begin{bmatrix}-2 & -1 & & 0\\ -1 & \ddots & \ddots & \\ & \ddots & \ddots & -1\\ 0 & & -1 & -2\end{bmatrix} \text{ and } T'=\begin{bmatrix}-2 & 1 & & 0\\ 1 & \ddots & \ddots & \\ & \ddots & \ddots & 1\\ 0 & & 1 & -2\end{bmatrix}$$

We have:

$$\lambda(T')=-2+2\cos\frac{k\pi}{N}=-2\left(1+\cos\frac{k\pi}{N}\right)=-4\cos^2\frac{k\pi}{2N}$$

$$\lambda(T)=-2+2\cos\frac{k\pi}{N}=-4\cos^2\frac{k\pi}{2N}$$

So:

$$\begin{cases}\mathrm{A+B=I+rT'-\beta I}\\ \mathrm{A-B=-I+rT-\beta I}\end{cases}$$

$$\lambda(A+B)=1+\lambda(T')r-\beta$$

$$=1+r\left(-4\cos^2\frac{k\pi}{2N}\right)-\beta$$

$$=1+4r\cos^2\frac{k\pi}{2N}-\beta$$

$$\left|1+4r\cos^2\frac{k\pi}{2N}-\beta\right|\le 1$$

$$-1\le 1-4r-\beta\le 1$$

$$-\frac{\beta}{4}\le r\le\frac{2-\beta}{4} \tag{6}$$

In conclusion $r\le\frac{1}{2}$

$$\mathrm{A-B=rT-I-\beta I}$$

$$
\begin{aligned}
\lambda(A-B) &= r(\lambda(T)) - 1 - \beta \\
&= r\left(-4\cos^2\frac{k\pi}{2N}\right) - 1 - \beta \\
&= -4r\cos^2\frac{k\pi}{2N} - 1 - \beta \\
&= \left|-4r\cos^2\frac{k\pi}{2N} - 1 - \beta\right| \le 1 \\
&= -1 < -4r - 1 - \beta \le 1 \\
&= \frac{-1-\beta}{4} \le r \le \frac{\beta}{4}
\end{aligned} \tag{7}
$$

In conclusion $r \le \frac{1}{4}$
We Conclusion from Eqs. 6 and 7

$$
r \le \frac{1}{4}
$$

So: $P(p) \le 1$
So the first and second states are stable. □
Third mode:

$$
\begin{cases}
\dfrac{\underline{U}_{i,j+1} - \bar{U}_{i,j}}{k} - \dfrac{\underline{U}_{i+1,j} - 2\bar{U}_{i,j} + \underline{U}_{i-1,j}}{h^2} - \dfrac{1}{r}\dfrac{\bar{U}_{i,j+1} - \underline{U}_{i,j}}{k} = \underline{0} \\
\dfrac{\bar{U}_{i,j+1} - \underline{U}_{i,j}}{k} - \dfrac{\bar{U}_{i+1,j} - 2\underline{U}_{i,j} + \bar{U}_{i-1,j}}{h^2} - \dfrac{1}{r}\dfrac{\underline{U}_{i,j+1} - \bar{U}_{i,j}}{k} = \bar{0}
\end{cases}
$$

By multiplying the sides of the equations we have $\beta = \frac{1}{r}$ and $r = \frac{K}{h^2}$

$$
\begin{cases}
\underline{U}_{i,j+1} - \bar{U}_{i,j} - r\left(\underline{U}_{i+1,j} - 2\bar{U}_{i,j} + \underline{U}_{i-1,j}\right) - \beta\left(\bar{U}_{i,j+1} - \underline{U}_{i,j}\right) = \underline{0} \\
\bar{U}_{i,j+1} - \underline{U}_{i,j} - r\left(\bar{U}_{i+1,j} - 2\underline{U}_{i,j} + \bar{U}_{i-1,j}\right) - \beta\left(\underline{U}_{i,j+1} - \bar{U}_{i,j}\right) = \underline{0}
\end{cases}
$$

So:

$$
\begin{cases}
\underline{U}_{i,j+1} = \bar{U}_{i,j} + r\left(\underline{U}_{i+1,j} - 2\bar{U}_{i,j} + \underline{U}_{i-1,j}\right) + \beta\left(\bar{U}_{i,j+1} - \underline{U}_{i,j}\right) \\
\bar{U}_{i,j+1} = \underline{U}_{i,j} + r\left(\bar{U}_{i+1,j} - 2\underline{U}_{i,j} + \bar{U}_{i-1,j}\right) + \beta\left(\underline{U}_{i,j+1} - \bar{U}_{i,j}\right)
\end{cases}
$$

Thus:

$$
\begin{cases}
\underline{U}_{i,j+1} = r\left(\underline{U}_{i+1,j} + \underline{U}_{i-1,j}\right) + (1-2r)\bar{U}_{i,j} + \beta\left(\bar{U}_{i,j+1} - \underline{U}_{i,j}\right) \\
\bar{U}_{i,j+1} = r\left(\bar{U}_{i+1,j} + \bar{U}_{i-1,j}\right) + (1-2r)\underline{U}_{i,j} + \beta\left(\underline{U}_{i,j+1} - \bar{U}_{i,j}\right)
\end{cases}
$$

Fourth mode:

$$\begin{cases} \frac{\bar{U}_{i,j+1}-\underline{U}_{i,j}}{k} = \frac{\bar{U}_{i+1,j}-2\underline{U}_{i,j}+\bar{U}_{i-1,j}}{h^2} + \frac{1}{r}\frac{\underline{U}_{i,j+1}-\bar{U}_{i,j}}{k} \\ \frac{\underline{U}_{i,j+1}-\bar{U}_{i,j}}{k} = \frac{\underline{U}_{i+1,j}-2\bar{U}_{i,j}+\underline{U}_{i-1,j}}{h^2} + \frac{1}{r}\frac{\bar{U}_{i,j+1}-\underline{U}_{i,j}}{k} \end{cases}$$

By multiplying the sides of the equations we have $\beta = \frac{1}{r}$ and $r = \frac{K}{h^2}$

$$\begin{cases} \bar{U}_{i,j+1} - \underline{U}_{i,j} = r\left(\bar{U}_{i+1,j} - 2\underline{U}_{i,j} + \bar{U}_{i-1,j}\right) + \beta\left(\underline{U}_{i,j+1} - \bar{U}_{i,j}\right) \\ \underline{U}_{i,j+1} - \bar{U}_{i,j} = r\left(\underline{U}_{i+1,j} - 2\bar{U}_{i,j} + \underline{U}_{i-1,j}\right) + \beta\left(\bar{U}_{i,j+1} - \underline{U}_{i,j}\right) \end{cases}$$

So:

$$\begin{cases} \bar{U}_{i,j+1} = \underline{U}_{i,j} + r\left(\bar{U}_{i+1,j} - 2\underline{U}_{i,j} + \bar{U}_{i-1,j}\right) + \beta\left(\underline{U}_{i,j+1} - \bar{U}_{i,j}\right) \\ \underline{U}_{i,j+1} = \bar{U}_{i,j} + r\left(\underline{U}_{i+1,j} - 2\bar{U}_{i,j} + \underline{U}_{i-1,j}\right) + \beta\left(\bar{U}_{i,j+1} - \underline{U}_{i,j}\right) \end{cases}$$

Thus:

$$\begin{cases} \bar{U}_{i,j+1} = r\left(\bar{U}_{i+1,j} + \bar{U}_{i-1,j}\right) + (1-2r)\underline{U}_{i,j} + \beta\left(\underline{U}_{i,j+1} - \bar{U}_{i,j}\right) \\ \underline{U}_{i,j+1} = r\left(\underline{U}_{i+1,j} + \underline{U}_{i-1,j}\right) + (1-2r)\bar{U}_{i,j} + \beta\left(\bar{U}_{i,j+1} - \underline{U}_{i,j}\right) \end{cases}$$

Which is the same as the third case, so to prove sustainability, it is sufficient to examine one of them.

Proof of stability of the third state:

$$\begin{bmatrix} \underline{U}_{1,j+1} \\ \vdots \\ \vdots \\ \underline{U}_{N-1,j+1} \\ \overline{U}_{1,j+1} \\ \vdots \\ \vdots \\ \overline{U}_{N-1,j+1} \end{bmatrix} = \begin{bmatrix} -2r-\beta & 0 & 1 & r & 0 \\ \ddots & & r & \ddots & \\ & & & \ddots & r \\ 0 & -2r-\beta & 0 & r & 1 \\ 1 & r & 0 & -2r-\beta & 0 \\ r & \ddots & & \ddots & \\ \ddots & & r & & \\ 0 & r & 1 & 0 & -2r-\beta \end{bmatrix} \begin{bmatrix} \underline{U}_{1,j} \\ \vdots \\ \vdots \\ \underline{U}_{N-1,j} \\ \overline{U}_{1,j} \\ \vdots \\ \vdots \\ \overline{U}_{N-1,j} \end{bmatrix} +$$

$$\begin{bmatrix} 0 & & 0 & 0 & & 0 \\ & \ddots & & & \ddots & \\ & & 0 & 0 & & 0 \\ 0 & & 0 & \beta & & 0 \\ & \ddots & & & \ddots & \\ 0 & & 0 & 0 & & \beta \end{bmatrix} \begin{bmatrix} \underline{U}_{1,j+1} \\ \vdots \\ \vdots \\ \underline{U}_{N-1,j+1} \\ \overline{U}_{1,j+1} \\ \vdots \\ \vdots \\ \overline{U}_{N-1,j+1} \end{bmatrix}$$

$$(\mathrm{I}-\beta) \begin{bmatrix} \underline{U}_{1,j+1} \\ \vdots \\ \vdots \\ \underline{U}_{N-1,j+1} \\ \overline{U}_{1,j+1} \\ \vdots \\ \vdots \\ \overline{U}_{N-1,j+1} \end{bmatrix} = \begin{bmatrix} -\beta & r & 0 & 1-2r & 0 \\ & \ddots & & \ddots & \\ r & & & & \\ & & r & & \\ 0 & r & -\beta & 0 & 1-2r \\ 1-2r & 0 & -\beta & r & 0 \\ \ddots & & r & \ddots & \\ & & & \ddots & r \\ 0 & 1-2r & 0 & r & -\beta \end{bmatrix} \begin{bmatrix} \underline{U}_{1,j} \\ \vdots \\ \vdots \\ \underline{U}_{N-1,j} \\ \overline{U}_{1,j} \\ \vdots \\ \vdots \\ \overline{U}_{N-1,j} \end{bmatrix}$$

So: $U_{1,j+1} = \underbrace{(I-\beta)^{-1}(p)U_j}_{p}$

Simply prove that $P(p) \leq 1$

We have to show:

$$\lambda\Big((I-\beta)^{-1}.p\Big) \leq 1$$

$$\frac{\lambda_p}{\lambda_{I-\beta}} \leq 1$$

$$\lambda_{I-\beta} = 1-\beta \tag{8}$$

$$T = \begin{bmatrix} -2 & -1 & & 0 \\ -1 & \ddots & \ddots & \\ & \ddots & \ddots & -1 \\ 0 & & -1 & -2 \end{bmatrix} \text{ and } T' = \begin{bmatrix} -2 & 1 & & 0 \\ 1 & \ddots & \ddots & \\ & \ddots & \ddots & 1 \\ 0 & & 1 & -2 \end{bmatrix}$$

$$\begin{cases} \lambda(T) = 2 + 2\sqrt{1}\cos\frac{k\pi}{N} = 2\Big(1+\cos\frac{k\pi}{N}\Big) = 4\cos^2\frac{k\pi}{2N} \\ \lambda(T') = -2 + 2\sqrt{1}\cos\frac{k\pi}{N} = -2\Big(1-\cos\frac{k\pi}{N}\Big) = -4\sin^2\frac{k\pi}{2N} \end{cases}$$

So:

$$\lambda(A+B) =? \quad \text{and} \quad \lambda(A-B) =?$$

$$A+B=\begin{bmatrix} 1-2r-\beta & r & & 0 \\ r & \ddots & & \ddots \\ & \ddots & \ddots & r \\ 0 & & r & 1-2r-\beta \end{bmatrix} \text{ and}$$

$$A-B=\begin{bmatrix} -1-2r-\beta & -r & & 0 \\ -r & \ddots & & \ddots \\ & \ddots & \ddots & -r \\ 0 & & -r & 1-2r-\beta \end{bmatrix}$$

So we have:

$$\mathrm{A}+\mathrm{B}=\mathrm{I}+\mathrm{r}\mathrm{T}'-\beta\mathrm{I}$$

$$\mathrm{A}-\mathrm{B}=\mathrm{r}\mathrm{T}-\beta\mathrm{I}-\mathrm{I}$$

So:

$$\begin{aligned} \lambda(A+B) &= 1+\left(r\lambda\left(T'\right)\right)-\beta \\ &= 1+r\left(-4\cos^2\frac{k\pi}{2N}\right)-\beta \\ &= 1-4r\cos^2\frac{k\pi}{2N}-\beta \end{aligned}$$

Thus: $\left|1-4r\cos^2\frac{k\pi}{2N}-\beta\right|\leq 1$

$$\begin{aligned} -1 \leq 1-4r-\beta \leq 1 \\ -2 \leq -4r-\beta \leq 0 \\ -\frac{\beta}{4} \leq r \leq \frac{2-\beta}{-4} \end{aligned}$$

Given that $\beta=\frac{1}{r}$
Then:

$$\begin{array}{ll} r \geq \frac{-1}{4r} & \text{When } r \in (0,+\infty) \\ r \leq \frac{-2+\frac{1}{r}}{-4} & \text{When } r \in (-\infty,0) \end{array} \tag{9}$$

As a result: $r \in (0,+\infty)$
Also:

$$\begin{aligned} \lambda(A-B) &= r\lambda(T)-\beta-1 \\ &= r\left(4\cos^2\frac{k\pi}{2N}\right)-\beta-1 \end{aligned}$$

$$= 4r\cos^2\frac{k\pi}{2N} - \beta - 1$$
$$\left|4r\cos^2\frac{k\pi}{2N} - \beta - 1\right| \leq 1$$
$$|4r - \beta - 1| \leq 1$$
$$-1 \leq 4r - \beta - 1 \leq 1$$
$$\beta \leq 4r \leq 2 + \beta$$
$$\frac{\beta}{4} \leq r \leq \frac{2+\beta}{4}$$

Given that $\beta = \frac{1}{r}$
Then:

$$\begin{array}{l} r \geq \frac{1}{4r} \text{ When } r \in \left[\frac{1}{2}, +\infty\right) \\ r \leq \frac{2+r}{4r} \text{ When } r \in (0, 0.8] \end{array} \tag{10}$$

As a result: $r \in \left[\frac{1}{2}, 0.8\right]$
We conclusion from Eqs. 9 and 10.

$$r \in \left[\frac{1}{2}, 0.8\right]$$

So:

$$\lambda_p \leq 1$$

By pasting in relation $\in$ we have:

$$\frac{\lambda_p}{\lambda_{I-\beta}} \leq 1$$

$$\frac{\lambda_p}{1-\beta} \leq 1$$

Given that $\lambda_p \leq 1$ and $\beta = \frac{1}{r}$
Then:
$\frac{\lambda_p}{1-\frac{1}{r}} - 1 \leq 0$ When $r \in (0, 1) \cup \left[-\frac{1}{\lambda_p - 1}, +\infty\right)$
As a result $P(p) \leq 1$
So the third and fourth states are stable. □

4 Conclusions

In this paper we investigate the Fuzzy equilibrium stability in polar coordinates using the derivative of the definition of Bede et al., derivative so that this method can give stability to many finite differential equations by numerical solution that we can't prove it.

References

1. Yager, R.R.: Measures of specificity. In: Computational Intelligence: Soft Computing and Fuzzy-Neuro Integration with Applications, pp. 94–113. Springer, Heidelberg (1998)
2. Zadeh, L.A.: Fuzzy sets. Inf. Control **8**(3), 338–353 (1965)
3. Zadeh, L.A.: Probability measures of fuzzy events. J. Math. Anal. Appl. **23**(2), 421–427 (1968)
4. Zadeh, L.A.: Calculus of fuzzy restrictions. In: Zadeh, L.A., Tanaka, K.S., Fu, K.S.M. (eds.) Fuzzy sets and their applications to cognitive and decision processes, Academic press, New York and its application to approximate reasoning, part. I: information Sciences 8 199–249 (1975): part II: information sciences 8 301–357 (1975): part III: information Sciences 9 43–80 (1975)
5. Zadeh, L.A.: Fuzzy sets and information granularity. In: Gupta, M., Ragade, R., Yager, R. (eds.) Advances in Fuzzy set Theory and Applications, pp. 3–18. North Holland publishing Co., Amsterdam (1979)
6. Zadeh, L.A.: Possibility theory, soft data analysis. In: Cobb, L., Thrall, R.M. (eds.) Mathematical Frontiers of the Social and Policy Sciences, pp. 69–129. Westview Press, Boulder (1981)
7. Zadeh, L.A.: Fuzzy logic and the calcul of fuzzy rules and fuzzy graphs, multiple-Valued logic I 1–38 (1995)

Fuzzy Multi Quadric Radial Basis Functions for Solving Fractional Boundary Value Problems

M. Dirbaz[1(✉)] and F. Dirbaz[2]

[1] Department of Mathematics, Science and Research Branch, Islamic Azad University, Tehran, Iran
mohadese.dirbaz@gmail.com
[2] Department of Computer Science, University of Tabriz, Tabriz, Iran

Abstract. This paper deals with the solution of fuzzy fractional boundary value problems (FFBVP) under the fuzzy radial basis functions interpolation.

To achieve this aim, the Caputo generalized Hukuhara differentiability form is derived from the fuzzy radial basis functions interpolation.

Finally, the algorithm of solving fuzzy fractional boundary value problems is presented.

Moreover, a numerical example is solved to illustrate the ability of the fuzzy introduced method and the results presented by the error table and figures.

Keywords: Fuzzy radial basis functions · Fuzzy interpolation · Fuzzy caputo derivative · Fuzzy fractional boundary value problems

1 Introduction

Fractional calculus deals with derivative and integrals of arbitrary order. Fractional calculus has played a notable role in science and engineering. Recently, many paper on fractional calculus appeared [9, 11].

Also, fuzzy set theory is a powerful tool for modeling uncertain problems.

The first who worked on the subject of fuzzy fractional differential equations is Agarwal et al. [4]. Recently, the basic concept as a Rimann-Liouville fractional integral, Riemann-Liouville H-differentiability, Caputo type fractional derivative based on Hukuhara and generalized Hukuhara difference and strongly generalized differentiability are defined in fuzzy fractional calculus [2, 3].

Lately, fuzzy radial basic function interpolation is introduced [7]. Radial basis function is one of the accurate for construction of meshless methods [10]. Due to the good features of FRBF in accuracy and ease of operation, this function is selected to solve the problem.

The aim of the present paper is to obtain the solution of fractional boundary value problems. We introduced the fuzzy fractional multi quadric radial basis function interpolation method. To find this solution, the paper is organized as follows.

Section 2 is on readers and referred to previous studies will be used throughout the paper. Section 3 gives a solution of fractional boundary value problems. Section 4 leads with some numerical examples. Section 5 gives the conclusions.

T. Allahviranloo et al. (Eds.): IDS 2020, AISC 1301, pp. 832–840, 2021.
https://doi.org/10.1007/978-3-030-66501-2_67

2 Basic Concepts

In this section, we present the definitions and lemmas, which will be used in the paper.

Denote $\mathbb{R}_{\mathcal{F}} = \{u\colon \mathbb{R} \to [0, 1] | \text{u satisfies on the } 1-4 \text{ following properties}\}$

1. u is fuzzy convex.
2. u is normal.
3. u is upper semi-continuous.
4. Closure of $\{x \in \mathbb{R} | u(x) > 0\}$ is compact.

Then $\mathbb{R}_{\mathcal{F}}$ is called the space of fuzzy numbers.

For $0 < \alpha \leq 1$ denote $[u]^{\alpha} = \{x \in \mathbb{R} | u(x) \geq \alpha\} = \left[u^{-}(\alpha), u^{+}(\alpha)\right]$. Then from 1 to 4, it follows that the α–level set $[u]^{\alpha}$ is closed interval for all $\alpha \in [0, 1]$.

The notation $C_{\mathcal{F}}^{m}[a, b]$ is the space of fuzzy valued function that, together with their gH-drivetives of order less than or equal to m, are continues on $[0, T]$.

The notation $L_{\mathcal{F}}[a, b]$ denote the set of Lebesgue integrable fuzzy-valued function on $[a, b]$.

The notation $A_{\mathcal{F}}^{m}[a, b]$ denote the space of fuzzy-valued functions from $[a, b]$ into $\mathbb{R}_{\mathcal{F}}$ with $m-1$ gH-derivative absolutely continuous function on $[a, b]$.

The notation J_{a}^{α} is the Rremann-Liouville integral of fuzzy valued function f.

Definition 2.1 [6]. Given two fuzzy numbers $u, v \in \mathbb{R}_{\mathcal{F}}$, the generalized Hukuhara difference (gH-difference for short) is the fuzzy number w, if it exists, such that

$$u \ominus_{gH} v = w \Leftrightarrow \begin{cases} & (i).\ u = v \oplus w; \\ or & (ii).\ v = u \oplus (-1)w. \end{cases}$$

It is easy to show that (i) and (ii) are both valid if and only if w is a crisp number.

Definition 2.2 [6]. The Hausdorff distance on $\mathbb{R}_{\mathcal{F}}$ is defined by

$$D(u, v) = \sup_{\alpha \in [0,1]} \left\{[u]_{\alpha} \ominus_{gH} [v]_{\alpha *}\right\}.$$

Where for interval $[a, b]$, the norm is $\|[a, b]\|_{*} = max\{|a|, |b|\}$.

Definition 2.3 [6]. The generalized Hukuhara derivative of fuzzy valued function $f\colon (a, b) \to \mathbb{R}_{\mathcal{F}}$ at t_0 is defined as

$$f_{gH}'(t_0) = \lim_{h \to 0} \frac{1}{h}\left[f(t_0 + h) \ominus_{gH} f(t_0)\right]$$

If $f_{gH}'(t_0) \in \mathbb{R}_{\mathcal{F}}$, we say that f is generalized Hukuhara differentiable (gH-differentiable for short) at t_0.

Also we say that f is $\left[(i) - gH\right]$–differentiable at t_0 if

$$\left[f_{gH}'(t_0)\right]^{r} = \left[(f_r^{-})'(t_0), (f_r^{+})'(t_0)\right]$$

And that f is $\left[(ii) - gH\right]$– differentiable at t_0 if

$$\left[f_{gH}'(t_0)\right]^{r} = \left[(f_r^{+})'(t_0), (f_r^{-})'(t_0)\right]$$

Definition 2.4 [2]. Let f: $(a, b) \rightarrow \mathbb{R}_{\mathcal{F}}$. We say that $f(t)$ is m-th order gH-differentiable at t_0 whenever the function $f(t)$ is gH-differentiable of order $i, i = 0, 1, \ldots, m-1$ at t_0, and if there exist $f_{gH}^{(m)}(t_0) \in \mathbb{R}_{\mathcal{F}}$ such that

$$f_{gH}^{(m)}(t_0) = \lim_{h \to 0} \frac{1}{h}\left[f_{gH}^{(m-1)}(t_0 + h) \ominus_{gH} f_{gH}^{(m-1)}(t_0)\right]$$

Definition 2.5 [9]. Consider $f : (a, b) \rightarrow \mathbb{R}$, fractional derivative of $f(t)$ in Caputo sense is defined as

$$D_*^{\alpha} f(t) = \frac{1}{\Gamma(m-\alpha)} \int_a^t (t-s)^{(m-\alpha-1)} f^{(m)}(s) ds, m-1 < \alpha \leq m, m \in \mathbb{N}, t > a.$$

Definition 2.6 [1]. Let $f \in A_{\mathcal{F}}^m[a, b]$. The Caputo generalized Hukuhara differentiability of fuzzy valued function $f\left([gH]^{cf} - \textit{differentiability for short}\right)$ is defined as following:

$$\left({}_{gH}^{c}D_*^{\alpha}\right)(t) = J_a^{m-\alpha} f_{gH}^{(m)}(t) = \frac{1}{\Gamma(m-\alpha)} \int_a^t (t-s)^{(m-\alpha-1)} f_{gH}^{(m)}(s) ds$$

Where $m - 1 < \alpha \leq m, m \in \mathbb{N}, t > a$. Also we say that f is $[(i) - gH]^{cf}$-differentiable of order α if

$$\left[{}_{gH}^{c}D_*^{\alpha} f(t)\right]^r = \left[\left(D_*^{\alpha} f_r^-\right)(t), \left(D_*^{\alpha} f_r^+\right)(t)\right], \quad 0 \leq r \leq 1$$

And f is $[(ii) - gH]^{cf}$-differentiable of order α if

$$\left[{}_{gH}^{c}D_*^{\alpha} f(t)\right]^r = \left[\left(D_*^{\alpha} f_r^+\right)(t), \left(D_*^{\alpha} f_r^-\right)(t)\right], \quad 0 \leq r \leq 1$$

Where J_a^{α} is the Rremann-Liouville integral of fuzzy valued function f and $D_*^{\alpha} f$ is defined in Definition 2.5.

Definition 2.7 [8]. Let $(x_0, t_0) \in \mathbb{D}$, where $\mathbb{D} \subseteq \mathbb{R} \times \mathbb{R}^+$, then first generalized Hukuhara partial derivative ([gH-p]-derivative for short) of fuzzy-valued function $f(x, t)$: $\mathbb{D} \rightarrow \mathbb{R}_{\mathcal{F}}$ at (x_0, t_0) with respect to variable x, t are the functions $\partial_{x_{gH}} f(x_0, t_0)$ and $\partial_{t_{gH}} f(x_0, t_0)$ given by

$$\partial_{x_{gH}} f(x_0, t_0) = \lim_{h \to 0} \frac{1}{h}\left[f(x_0 + h, t_0) \ominus_{gH} f(x_0, t_0)\right]$$

$$\partial_{t_{gH}} f(x_0, t_0) = \lim_{k \to 0} \frac{1}{k}\left[f(x_0, t_0 + k) \ominus_{gH} f(x_0, t_0)\right]$$

Provided that $\partial_{x_{gH}} f(x_0, t_0)$ and $\partial_{t_{gH}} f(x_0, t_0) \in \mathbb{R}_{\mathcal{F}}$.

Definition 2.8 [8]. Let $f(x, t)$: $\mathbb{D} \rightarrow \mathbb{R}_{\mathcal{F}}$, where $\mathbb{D} \subseteq \mathbb{R} \times \mathbb{R}^+$, (x_0, t_0) and $f^-(x, t; \alpha)$ and $f^+(x, t; \alpha)$ both partial differentiable w.r.t. t at (x_0, t_0). We say that

- $f(x, t)$ is $[(i) - p]-$ differentiable w.r.t. t at (x_0, t_0) if

$$\partial_{t_{i.gH}} f(x_0, t_0; \alpha) = \left[\partial_t f^-(x_0, t_0; \alpha), \partial_t f^+(x_0, t_0; \alpha)\right]$$

- $f(x, t)$ is $[(ii) - p]-$ differentiable w.r.t. t at (x_0, t_0) if

$$\partial_{t_{ii.gH}} f(x_0, t_0; \alpha) = \left[\partial_t f^+(x_0, t_0; \alpha), \partial_t f^-(x_0, t_0; \alpha)\right]$$

Lemma 2.1 [1]. Let $m - 1 < \alpha \leq m, m \in \mathbb{N}$, and the fuzzy valued-function $f \in C^m_{\mathcal{F}}[0, T]$. Then ${}^c_{gH}D^\alpha_* \in C^m_{\mathcal{F}}[0, T]$ and ${}^c_{gH}D^\alpha_* f(0) = 0$.

Lemma 2.2 [5]. Let $\alpha, \beta > 0$ and $f\colon [0, T] \to \mathbb{R}_{\mathcal{F}}$ be a fuzzy-valued function such that $f \in L_{\mathcal{F}}[0, T]$, Then

$$J_0^\alpha J_0^\beta f = J_0^{\alpha+\beta} f$$

Lemma 2.3 [2]. Let $0 < \alpha \leq 1$ and $f\colon [0, T] \to \mathbb{R}_{\mathcal{F}}$ be a fuzzy-valued function such that $f \in A^1_{\mathcal{F}}[0, T]$, Then

$$J_0^\alpha \left({}^c_{gH}D^\alpha_* f\right)(t) = f(t) \ominus_{gH} f(0)$$

Lemma 2.4 [1]. Let $0 < \alpha \leq 1$ and $f\colon [0, T] \to \mathbb{R}_{\mathcal{F}}$ be a fuzzy-valued function such that $f \in C_{\mathcal{F}}[0, T]$, Then

$${}^c_{gH}D^\alpha_* J^\alpha_* f = f$$

Lemma 2.5 [1]. Let $f \in C^m_{\mathcal{F}}[0, T]$ and let $0 < \alpha \leq 1, \beta > 0$ be such that $k - 1 \leq \alpha + \beta \leq k$ where $k \in \mathbb{N},\ k \leq m - 1$. Then

$${}^c_{gH}D^\alpha_* {}^c_{gH}D^\beta_* f = {}^c_{gH}D^{\alpha+\beta}_* f$$

Definition 2.9 [7]. The function $\varphi = \mathbb{R} \to \mathbb{R}_{\mathcal{F}}$, is called fuzzy multi quadric radial basis functions, if

$$\varphi(x) = \sqrt{\left|x - x_j\right|^2 \oplus c^2}, \qquad j = 1, \ldots, N.$$

Where $\left|x - x_j\right|, j = 1, \ldots, N$ are fuzzy singleton numbers and $c \in \mathbb{R}_{\mathcal{F}}$ is the fuzzy shape parameter.

3 Solving Fuzzy Fractional BVP by FFRBF

In this section, in order to apply the fuzzy multi quadric radial basis function interpolation for solving fractional boundary value problems, let us consider a fuzzy fractional BVP in the form:

$$\begin{aligned} &{}^c_{gH}D^\alpha \tilde{u} + {}^m_{gH}D\tilde{u} = \tilde{f} \quad in \ \ \Omega \subset \mathbb{R} \\ &B\tilde{u} = \tilde{g} \quad on \quad \partial\Omega \end{aligned} \tag{3.1}$$

Where ${}^{c}_{gH}D^{\alpha}$ is the fuzzy fractional differential operator of α order that operates, D is differential operator. Both of the $\tilde{f}, \tilde{g}\colon \mathbb{R} \to \mathbb{R}_{\mathcal{F}}$ are known functions. $\partial\Omega$ denote the boundary of the domain Ω. Consider $\{x_j\}_{j=1}^{N}$ be the N collocation points in $\Omega \cup \partial\Omega$. (The first N points in Ω and the other M points is $\partial\Omega$). Now, from substituting the fuzzy multi quadric radial basis function $u(x) \cong \varphi(x) = \sum_{j=1}^{n} \lambda_j \sqrt{|x - x_j|^2 \oplus c^2}$ into (3.1) result is

$$\sum_{j=1}^{n} \lambda_j \varphi_j \left({}^{c}_{gH}D^{\alpha} + {}^{m}_{gH}D\right)\sqrt{|x - x_j|^2 \oplus c^2} = \tilde{f}(x)$$

$$\sum_{j=1}^{n} \lambda_j \varphi_j B \sqrt{|x - x_j|^2 \oplus c^2} = \tilde{g}(x) \tag{3.2}$$

We must chose the set of N centers, $x_1^c, \ldots, x_N^c$, such that $x_i^c \in \mathbb{R}$, we consider the centers coincident on the fuzzy interpolation nodes, it means $x_j^c = x_i$, for $i, j = 1, \ldots, N$. The approximation of unknown function $u(x)$ by fuzzy radial basis function interpolation in (3.2) obtained as following form

$$\begin{array}{ll} \sum_{j=1}^{n} \lambda_j \varphi_j \left({}^{c}_{gH}D^{\alpha} + {}^{m}_{gH}D\right)\sqrt{|x_i - x_j|^2 \oplus c^2} = \tilde{f}(x_i), & i = 1, \ldots, N - M. \\ \sum_{j=1}^{n} \lambda_j \varphi_j B \sqrt{|x_i - x_j|^2 \oplus c^2} = \tilde{g}(x_i), & i = N - M + 1, \ldots, N. \end{array} \tag{3.3}$$

The matrix form is as following

$$\begin{bmatrix} ({}^{c}_{gH}D^{\alpha} + {}^{m}_{gH}D)\sqrt{(|x_1 - x_1|)^2 \oplus c^2} & ({}^{c}_{gH}D^{\alpha} + {}^{m}_{gH}D)\sqrt{(|x_1 - x_2|)^2 \oplus c^2} & \cdots & ({}^{c}_{gH}D^{\alpha} + {}^{m}_{gH}D)\sqrt{(|x_1 - x_N|)^2 \oplus c^2} \\ \vdots & \vdots & & \vdots \\ ({}^{c}_{gH}D^{\alpha} + {}^{m}_{gH}D)\sqrt{(|x_{N-M} - x_1|)^2 \oplus c^2} & ({}^{c}_{gH}D^{\alpha} + {}^{m}_{gH}D)\sqrt{(|x_{N-M} - x_2|)^2 \oplus c^2} & & ({}^{c}_{gH}D^{\alpha} + {}^{m}_{gH}D)\sqrt{(|x_{N-M} - x_N|)^2 \oplus c^2} \\ ({}^{c}_{gH}D^{\alpha} + {}^{m}_{gH}D)\sqrt{(|x_{N-M+1} - x_1|)^2 \oplus c^2} & ({}^{c}_{gH}D^{\alpha} + {}^{m}_{gH}D)\sqrt{(|x_{N-M+1} - x_2|)^2 \oplus c^2} & \cdots & ({}^{c}_{gH}D^{\alpha} + {}^{m}_{gH}D)\sqrt{(|x_{N-M+1} - x_N|)^2 \oplus c^2} \\ \vdots & \vdots & & \vdots \\ ({}^{c}_{gH}D^{\alpha} + {}^{m}_{gH}D)\sqrt{(|x_N - x_1|)^2 \oplus c^2} & ({}^{c}_{gH}D^{\alpha} + {}^{m}_{gH}D)\sqrt{(|x_N - x_2|)^2 \oplus c^2} & \cdots & ({}^{c}_{gH}D^{\alpha} + {}^{m}_{gH}D)\sqrt{(|x_N - x_N|)^2 \oplus c^2} \end{bmatrix} \odot \begin{bmatrix} \lambda_1 \\ \vdots \\ \lambda_{N-M} \\ \lambda_{N-M+1} \\ \vdots \\ \lambda_N \end{bmatrix} = \begin{bmatrix} f(x_1) \\ \vdots \\ f(x_{N-M}) \\ f(x_{N-M+1}) \\ \vdots \\ f(x_N) \end{bmatrix} \tag{3.4}$$

Therefore obtained the $N \times N$ fully fuzzy linear system, in impact form as following form

$$\phi \odot \lambda = f \tag{3.5}$$

The fuzzy coefficients λ are obtained by solving the fully fuzzy linear system (3.5), to achieve this aim used the maximal and minimal symmetric solutions of fully fuzzy linear systems [12].

First, solved the 1-cut of the FFLS (3.5)

$$f(1) = \sum_{j=1}^{N} \phi_j(1) \odot \lambda_j \tag{3.6}$$

Where $f(1), \phi_j(1) \in \mathbb{R}$, λ_j are unknown crisp variable which will be determined by solving system (3.5).

By assigning the unknown symmetric spreads to all equation of all equation of the crisp system (3.6), obtained the following system

$$\begin{cases} \left[\underline{f}_1(r),\overline{f}_1(r)\right] = [\underline{\emptyset}_{11}(r),\overline{\emptyset}_{11}(r)][\lambda_1-\alpha_1(r),\lambda_1+\alpha_1(r)]+\cdots+[\underline{\emptyset}_{1N}(r),\overline{\emptyset}_{1N}(r)][\lambda_N-\alpha_1(r),\lambda_N+\alpha_1(r)] \\ \left[\underline{f}_2(r),\overline{f}_2(r)\right] = [\underline{\emptyset}_{21}(r),\overline{\emptyset}_{21}(r)][\lambda_1-\alpha_2(r),\lambda_1+\alpha_2(r)]+\cdots+[\underline{\emptyset}_{1N}(r),\overline{\emptyset}_{1N}(r)][\lambda_N-\alpha_2(r),\lambda_N+\alpha_2(r)] \\ \vdots \\ \left[\underline{f}_N(r),\overline{f}_N(r)\right] = [\underline{\emptyset}_{N1}(r),\overline{\emptyset}_{N1}(r)][\lambda_1-\alpha_N(r),\lambda_1+\alpha_N(r)]+\cdots+[\underline{\emptyset}_{1N}(r),\overline{\emptyset}_{1N}(r)][\lambda_N-\alpha_N(r),\lambda_N+\alpha_N(r)] \end{cases}$$

Where $\alpha_i > 0, i = 1, 2, \ldots, N$ are unknown spreads which will be determined by solving 2N above equations and $\lambda_i, i = 1, 2, \ldots, N$ are elements of the obtained vector solution of system.

Notice that, for sake of simplicity, considered that $\lambda_i - \alpha_{i1}(r), \lambda_i + \alpha_{i2}(r)$ are positive, and describe theory by such an assumption. By according to this point that, for $j = 1, \ldots, N$ always $\emptyset_j(r)$ is positive, all elements of our fuzzy interpolation matrix is positive, considered two total procedures to determine the symmetric spreads of solutions of full fuzzy linear systems, which are denoted by

$$\alpha_s^-(r) = \min_{0\le r\le 1}\{|\alpha_{i1}(r)|, |\alpha_{i2}(r)|\}, i = 1, \ldots, N$$

$$\alpha_s^+(r) = \min_{0\le r\le 1}\{|\alpha_{i1}(r)|, |\alpha_{i2}(r)|\}, i = 1, \ldots, N$$

So $\tilde{\lambda}(r) = \left(\tilde{\lambda}_1(r), \ldots, \tilde{\lambda}_N(r)\right)^t$ is following form

$$\tilde{\lambda}_i(r) = \left[\lambda_i - \alpha_s^-(r), \lambda_i + \alpha_s^-(r)\right]$$

$$\tilde{\lambda}_i(r) = \left[\lambda_i - \alpha_s^+(r), \lambda_i + \alpha_s^+(r)\right]$$

4 Numerical Analysis

In this section, to illustrate the technique, an example is solved by fuzzy fractional multi quadric radial basis functions interpolation method and the result is shown in the error table and figures.

Example 4.1. Consider the following fuzzy fractional boundary value problem:

$${}_{gH}^{x}D^{0.5}u(x) + u(x) = \frac{2x^{1.5}}{\Gamma\left({}^{2}/_{5}\right)} + x^2, x \in [0, 1]$$

$$u(0) = (0 - 0.125r, 0 + 0.125r)$$

$$u(1) = (1 - 0.125r, 1 + 0.125r)$$

Table 1. Error of fuzzy fractional multi quadric radial basis functions interpolation method

x_i	Error of method
0.1	0.00438982376
0.2	0.00440188484
0.3	0.00442002828
0.4	0.00534406622
0.5	0.00557381642
0.6	0.00560910186
0.7	0.00614975080
0.8	0.00659559648
0.9	0.00681464770

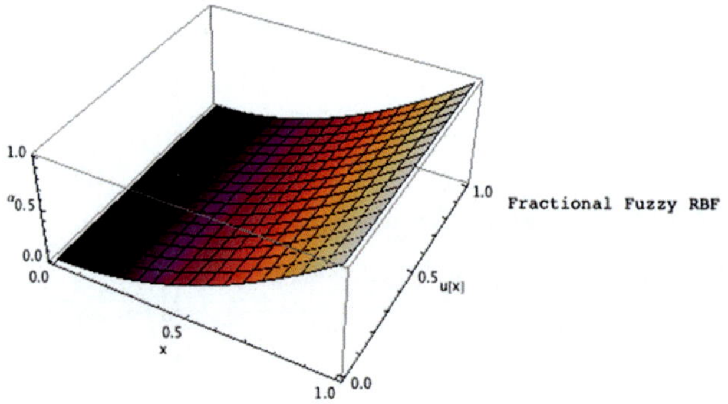

Fig. 1. Approximation solution of example 4.1 by fuzzy fractional multi quadric radial basis functions interpolation method for level $\alpha = 1$.

We apply the fuzzy fractional multi quadric radial basis functions interpolation method introduced in Sect. 3. The Error of fuzzy fractional multi quadric radial basis functions interpolation method by Hausdorff distance is given in Table 1. The Approximation solution of example by fuzzy fractional multi quadric radial basis functions interpolation method for level $\alpha = 1$ and for α level are shown in Fig. 1, and Fig. 2.

The approximated solution for level $\alpha = 1$ is plotted as following Fig. 1.

The approximated solutions for α level are plotted as following Fig. 2.

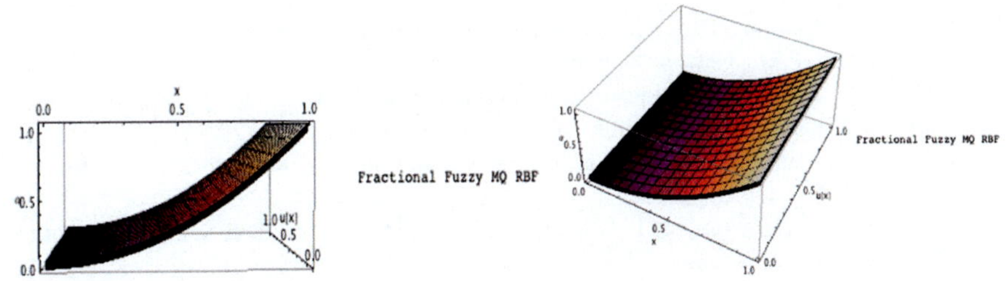

Fig. 2. Approximated solution of example 4.1 with fuzzy fractional multi quadric radial basis functions interpolation method for α level.

The plotted graphs and obtained error table show the presented fuzzy numerical method for solving fuzzy boundary value problem by using fuzzy fractional radial basis function interpolation method is stable, and presentations are acceptable solution for this problems.

5 Conclusion

In this paper, a new method was applied to solve fuzzy fractional boundary value problems. We introduced the fuzzy radial basis functions interpolation method. It is necessary derived the fuzzy fractional Caputo form the fuzzy radial basis functions. The numerical representations indicate the complete validity and efficiency of the presented method with a great potential in scientific applications. This algorithm that introduced in this paper, is a new idea for solving large category of fractional fuzzy boundary value problem such as applying another fuzzy radial basis function instead of multi quadric radial basis function.

References

1. Armand, A., Allahviranloo, T., Gouyandeh, Z.: General solution of basset equation with Caputo generalized Hukuhara derivatives. J. Appl. Anal. Comput. **6**(1) (2016)
2. Allahviranloo, T., Armand, A., Gouyandeh, Z.: Fuzzy fractional differential equations under generalized fuzzy Caputo derivative. J. Intell. Fuzzy Syst. **26**, 1481–1490 (2014)
3. Allahviranloo, T., Salahshour, S., Abbasbandy, S.: Explicit solutions of fractional differential equations with uncertainty. Soft Comput. **16**, 297–302 (2012)
4. Agarwal, R.P., Lakshmikanthama, V., Nieto, J.J.: On the concept of solution for fractional differential equations with uncertainty. Nonlinear Anal. **72**, 2859–2862 (2010)
5. Arshad, S., Lupulescu, V.: On the fractional differential equations with uncertainty. Nonlinear Anal. **74**, 3685–3693 (2011)
6. Bede, B., Stefanini, L.: Generalized differentiability of fuzzy-valued functions. Fuzzy Sets Syst. **230**, 119–141 (2013)
7. Dirbaz, M., Allahviranloo, T.: Fuzzy multiquadric radial basis functions for solving fuzzy partial differential equations. Comput. Appl. Math. **38**(4), 1–19 (2019)
8. Gouyandeh, Z., Allahviranloo, T., Abbasbandy, S., Armand, A.: A fuzzy solution of heat equation under generalized Hukuhara differentiability by fuzzy Fourier transform. Fuzzy Sets Syst. **309**, 81–97 (2017)

9. Podlubny, I.: Fractional Differential equations, Mathematics in Science and Engineering, vol. 198. Academic Press, New york (1999)
10. Sarra, S.A., Kansa, E.J.: Multiquadric Radial Basis Function Approximation Methods for the Numerical Solution of Partial Differential Equations. Science Press (2010)
11. Samko, S.G., Kilbas, A.A., Marichev, O.I.: Fractional Integral and Derivatives. Theory and Applications. Gordon and Breach, Switzerland (1993)
12. Salahshour, S., Allahviranloo, T., Khezerloo, M.: Maximal and minimal symmetric solutions of fully fuzzy linear systems. J. Comput. Appl. Math. **235** (2011)

A New Method for Ranking of Z-Numbers Based on Magnitude Value

M. Farzam[1], M. Afshar kermani[1(✉)], T. Allahviranloo[2], and Mahmoud J. S. Belaghi[2]

[1] Department of Mathematics, Tehran North Branch Islamic Azad University, Tehran, Iran
M_afshar@iau-tnb.ac.ir

[2] Faculty of Engineering and Natural Sciences, Bahcesehir University, Istanbul, Turkey

Abstract. In this paper we will introduced a new method for ranking z-numbers. This method is based on magnitude value.

We compare the z-numbers by convex compound Mag(A) and Mag(B).

Finally, we illustrate the correctness of this method by solving some examples.

Keywords: Z-numbers · Fuzzy numbers · Ranking of z-numbers · Magnitude value

1 Introduction

Real world decision issues in system analysis, decision analysis, economics, ecology and other fields are characterized by fuzziness and partial reliability of relevant information. Fuzzy theory can somewhat express the uncertainty, but it can't alone make this uncertainty in the calculation. In order to deal with such information, Zade introduced a new concept in 2011 named z-numbers [1].

Z-number shows by an ordered pair of fuzzy numbers as $z = (A, B)$ which A, the first component is a linguistic value of a variable of interest, and B, the second component, is a linguistic value of probability measure of the first one, playing a role of reliability of information.

One of the most important topics in this discussion for data analyzing and improve the decision process is ranking of z-numbers.

Different methods are presented in different papers for comparison of z-numbers, that each of them has its own special applications. For example Bakars and gegov [5] proposed a method that initially turned the z-number into a fuzzy number then by using the CPS method, they ranked the generalized fuzzy number. So the method for ranking z-numbers depends on a method for ranking generalized fuzzy numbers.

According to the previous method, it can be seen that theses methods depend on fuzzy number ranking methods, so before ranking z-numbers, a new method

T. Allahviranloo et al. (Eds.): IDS 2020, AISC 1301, pp. 841–850, 2021.
https://doi.org/10.1007/978-3-030-66501-2_68

for ranking generalized fuzzy numbers must be introduced. Chen and chen [7] introduced a method based on defuzzified values, the highest and the spreads of generalized fuzzy numbers.

As can be seen Converting z-numbers to generalized fuzzy numbers will lead to loss of original information [8].

In this paper we suggest a method without converting the z-number to a fuzzy number.

This paper is organized as follows:

Section 2 contains some definitions and concepts of fuzzy numbers and z-numbers. In Sect. 3 the proposed method for ranking z-numbers is discussed. Section 4 contains some examples to illustrate the proposed method, and the conclusion is presented in Sect. 5.

2 Preliminaries

In this section we define some definition that we use in this paper.

Definition 2.1. A fuzzy subset $\bar{A} = (a, b, c, d; w)$ of the real line R with membership function $\mu_A : R \to [0, w]$, which $0 < w < 1$, is a fuzzy number if satisfying in the following properties:

1. $\mu_{\bar{A}}$ is a continuous function from R to $[0, w]$.
2. $\mu_{\bar{A}}(x) = 0$ for all $x \in (-\infty, a]$ and $x \in [d, \infty)$.
3. $\mu_{\bar{A}}$ is strictly increasing on $[a, b]$.
4. $\mu_{\bar{A}}(x) = w$ for $x \in [b, c]$.
5. $\mu_{\bar{A}}$ is strictly decreasing on $[c, d]$.

If $w = 1$ then we say $\bar{A}$ is a normal fuzzy number.

Definition 2.2. A fuzzy number $\bar{A} = (a, b, c, d; w)$ is said a trapezoidal fuzzy number if it's membership function is as follows:

$$\mu_{\bar{A}}(x) = \begin{cases} 0 & x < a \\ \frac{x-a}{b-a} & b \leq x \leq a \\ w & b \leq x \leq c \\ \frac{d-x}{d-c} & c \leq x \leq d \\ 0 & x > d \end{cases}$$

If we have $b = c$ then $\bar{A}$ is said triangular fuzzy number.

Definition 2.3. A z-number is an ordered pair of fuzzy numbers that is shown as $Z = (\bar{A}, \bar{B})$. The first component $\bar{A}$ is a restriction on the values which x can take and the second component $\bar{B}$ is a measure of reliability of the first component.

In Fig. 2 we illustrate a z-number as $Z = (\bar{A}, \bar{B})$ by following membership functions:

$$\mu_{\bar{A}}(x) = \begin{cases} 0 & x < a_1 \\ \frac{x-a_1}{a_2-a_1} & a_2 \leq x \leq a_1 \\ w & a_2 \leq x \leq a_3 \\ \frac{a_4-x}{a_4-a_3} & a_3 \leq x \leq a_4 \\ 0 & x > a_4 \end{cases}$$

and

$$\mu_{\bar{B}}(x) = \begin{cases} 0 & x < b_1 \\ \frac{x-b_1}{b_2-b_1} & b_2 \leq x \leq b_1 \\ w & b_2 \leq x \leq b_3 \\ \frac{b_4-x}{b_4-b_3} & b_3 \leq x \leq b_4 \\ 0 & x > b_4 \end{cases}$$

where w_1 and w_2 are the maximum degree of membership functions. For $Z = (\bar{A}, \bar{B})$,

(1) Convert the $\bar{B}$ (reliability) into crisp number by using

$$\alpha = \frac{\int_X x\mu_{\bar{B}} dx}{\int_X \mu_{\bar{B}} dx}$$

where $\int$ denotes an algebraic integration.

(2) Add the weight of the $\bar{B}$ to the $\bar{A}$ (restriction). The weighted Z-number is denoted as

$$\bar{Z}^{\alpha} = \{(x, \mu_{A^{\alpha}}(x)) \mid \mu_{A^{\alpha}}(x) = \alpha\mu_A(x), x \in [0, 1]\}$$

3 The Proposed Method for Ranking Z-Numbers

Before introducing the algorithm we must note that if the z-numbers are displayed with $Z_i = (\bar{A}_i, \bar{B}_i)$ which $i = 1, 2, ..., n$, then it is easily understood that $\bar{A}_i$ is the most important part of Z_i , because the final opportunity of z-number is to describe the uncertain variable, while $\bar{b}_i$, as only a measure of reliability of $\bar{A}_i$, can influence the ranking result but can not decide the ranking result, so $\bar{A}_i$ is more important and is the main part of z-number. This point should be considered when we want to rank z-numbers.

Consider a z-number as $Z = (\bar{A}, \bar{B})$ so that $\bar{A} = (a_1, b_1, c_1, d_1; w_1)$ and $\bar{B} = (a_2, b_2, c_2, d_2; w_2)$ and their membership functions are as follows:

$$\mu_{\bar{A}}(x) = \begin{cases} \frac{w_1(x-a_1)}{b_1-a_1} & b_1 \leq x \leq a_1 \\ w_1 & b_1 \leq x \leq c_1 \\ \frac{w_1(d_1-x)}{d_1-c_1} & c_1 \leq x \leq d_1 \\ 0 & o.w \end{cases}$$

and

$$\mu_{\bar{B}}(x) = \begin{cases} \frac{w_2(x-a_2)}{b_2-a_2} & b_2 \le x \le a_2 \\ w_2 & b_2 \le x \le c_2 \\ \frac{w_2(d_2-x)}{d_2-c_2} & c_2 \le x \le d_2 \\ 0 & o.w \end{cases}$$

where w_1 and w_2 are the maximum degree of membership functions.
and

$$A_L(\alpha) = \frac{\alpha(b_1 - a_1)}{w_1} + a_1$$

$$A_U(\alpha) = \frac{\alpha(c_1 - d_1)}{w_1} + d_1 \tag{1}$$

and

$$B_L(\alpha) = \frac{\alpha(b_2 - a_2)}{w_2} + a_2$$

$$B_U(\alpha) = \frac{\alpha(c_2 - d_2)}{w_2} + d_2 \tag{2}$$

If $S : [0, 1] \to [0, 1]$ is a reduction function such that we have $S(1) = 1$ and $S(0) = 0$. The function S is said regularly whenever:

$$\int_0^1 S(\alpha)d\alpha = \frac{1}{2}$$

The magnitude value for A and B is defined as follows:

$$Mag(A) = \frac{1}{2}\int_0^{w_1} (A_L(\alpha) + A_U(\alpha) + (A_L(\alpha) + A_U(\alpha))S(\alpha))d\alpha \tag{3}$$

and

$$Mag(B) = \frac{1}{2}\int_0^{w_2} (B_L(\alpha) + B_U(\alpha) + (B_L(\alpha) + B_U(\alpha))S(\alpha))d\alpha \tag{4}$$

By using (1) and (3) and assuming that $S(\alpha) = \alpha$ we have:

$$Mag(A) = \frac{(3w_1^2 + 2)(b_1 + c_1)}{12w_1} + \frac{(3w_1 - 2)(a_1 + d_1)}{12w_1}$$

And by using (2) and (4) and assuming that $S(\alpha) = \alpha$ we have:

$$Mag(B) = \frac{(3w_2^2 + 2)(b_2 + c_2)}{12w_2} + \frac{(3w_2 - 2)(a_2 + d_2)}{12w_2}$$

If we suppose that A and B are normal numbers or w_1 and w_2 be equal by one then we have:

$$Mag(A) = \frac{5}{12}(b_1 + c_1) + \frac{1}{12}(a_1 + d_1) \tag{5}$$

and

$$Mag(B) = \frac{5}{12}(b_2 + c_2) + \frac{1}{12}(a_2 + d_2) \tag{6}$$

and finally we compare $\gamma(z_i)$ that obtain as follows:

$$\gamma(z_i) = \lambda Mag(A_i) + (1 - \lambda)Mag(B_i) \tag{7}$$

such that $\lambda \in [\frac{1}{2}, 1]$, because as mentioned before effect of A must be greater than B.

Then we rank the z-numbers by compare $\gamma(z_i)$'s as follows:

1. We say that $z_i \sim z_j$ if $R(z_i) = R(z_j)$, in the other words if we have:

$$\lambda(Mag(A_i) - Mag(A_j)) = (1 - \lambda)(Mag(B_i) - Mag(B_j)) \tag{8}$$

2. We say that $z_i \prec z_j$ if $R(z_i) < R(z_j)$, in the other words if we have:

$$\lambda(Mag(A_i) - Mag(A_j)) < (1 - \lambda)(Mag(B_i) - Mag(B_j)) \tag{9}$$

3. We say that $z_i \succ z_j$ if $R(z_i) > R(z_j)$, in the other words if we have:

$$\lambda(Mag(A_i) - Mag(A_j)) > (1 - \lambda)(Mag(B_i) - Mag(B_j)) \tag{10}$$

Theorem 3.1. This algorithm satisfy in following axioms of ranking:

a_1. For an arbitrary z-number as z_i we have $z_i \sim z_i$
a_2. For arbitrary z-numbers as z_i and z_j if we have $z_i \prec z_j$ and $z_i \succ z_j$ then we should have $z_i \sim z_j$.
a_3. If z_i, z_j and z_k be arbitrary z-numbers and we have $z_i \prec z_j$ and $z_j \prec z_k$ then we have $z_i \prec z_j$.
a_4. For three arbitrary z-numbers z_i, z_j and z_k, if we have $z_i \prec z_j$ then we should have $z_i + z_k \prec z_j + z_k$.
a_5. For arbitrary z-numbers as z_i and z_j, if $z_i \prec z_j$ and k be a crisp number such that $k > 0$ then we have $kz_i \prec kz_j$.

Proof. a_1. If we put $i = j$ in Eq. (7), then for an arbitrary z-number as z_i we have:

$$\lambda Mag(A_i) + (1 - \lambda)Mag(B_i) = \lambda Mag(A_i) + (1 - \lambda)Mag(B_i)$$

and then we have $z_i \sim z_i$.

a_2. Suppose that z_i and z_j are two z-numbers such that $z_i \prec z_j$ and $z_i \succ z_j$ so by (9) and (10) we have:

$$\lambda(Mag(A_i) - Mag(A_j)) < (1 - \lambda)(Mag(B_i) - Mag(B_j))$$

and

$$\lambda(Mag(A_i) - Mag(A_j)) > (1 - \lambda)(Mag(B_i) - Mag(B_j))$$

hence we should have:

$$\lambda(Mag(A_i) - Mag(A_j)) = (1-\lambda)(Mag(B_i) - Mag(B_j))$$

so $z_i \sim z_j$.

a_3. Suppose that z_i, z_j and z_k be three z-numbers and we have $z_i \prec z_j$ and $z_j \prec z_k$ so by (7) we should have:

$$\lambda Mag(A_i) + (1-\lambda)Mag(B_i) < \lambda Mag(A_j) + (1-\lambda)Mag(B_j)$$

and

$$\lambda Mag(A_j) + (1-\lambda)Mag(B_j) < \lambda Mag(A_k) + (1-\lambda)Mag(B_k)$$

therefore:

$$\lambda Mag(A_i) + (1-\lambda)Mag(B_i) < \lambda Mag(A_k) + (1-\lambda)Mag(B_k)$$

so we have $z_i \prec z_k$.

a_4. For three z-numbers z_i, z_j and z_k, suppose that $z_i \prec z_j$, then by (7) we have:

$$\lambda Mag(A_i) + (1-\lambda)Mag(B_i) < \lambda Mag(A_j) + (1-\lambda)Mag(B_j)$$

so we can write:

$$(\lambda Mag(A_i) + (1-\lambda)Mag(B_i)) + (\lambda Mag(A_k) + (1-\lambda)Mag(B_k))$$
$$< (\lambda Mag(A_j) + (1-\lambda)Mag(B_j)) + (\lambda Mag(A_k) + (1-\lambda)Mag(B_k))$$

hence $z_i + z_k \prec z_j + z_k$.

a_5. For arbitrary z-numbers as z_i and z_j, if $z_i \prec z_j$ and k be a crisp number such that $k > 0$ then we have:

$$k(\lambda Mag(A_i) + (1-\lambda)Mag(B_i)) < k(\lambda Mag(A_j) + (1-\lambda)Mag(B_j))$$

so $kz_i \prec kz_j$.

4 Numerical Examples

In all of the examples λ is considered equal by $\frac{1}{2}$.

Example 4.1. Suppose that $Z_i = (A_i, B_i)$ for $i = 1, 2, 3$ such that A_i and B_i are defined as follows:

$$A_1 = (1, 3, 5, 6) \qquad B_1 = (2, 4, 7, 8)$$

$$A_2 = (-2, 0, 1, 2) \qquad B_2 = (1, 2, 3, 4)$$

and

$$A_3 = (4, 6, 7, 8) \qquad B_3 = (6, 8, 9, 10)$$

by (5) and (6) we obtain that:

$$Mag(A_1) = \frac{47}{12} \quad Mag(B_1) = \frac{65}{12}$$

$$Mag(A_2) = \frac{5}{12} \quad Mag(B_2) = \frac{5}{2}$$

and

$$Mag(A_3) = \frac{77}{12} \quad Mag(B_3) = \frac{101}{12}$$

so by (7) we have:

$$\gamma(z_1) = \frac{112}{24}$$

$$\gamma(z_2) = \frac{35}{24}$$

and

$$\gamma(z_3) = \frac{178}{24}$$

so

$$z_2 \prec z_1 \prec z_3$$

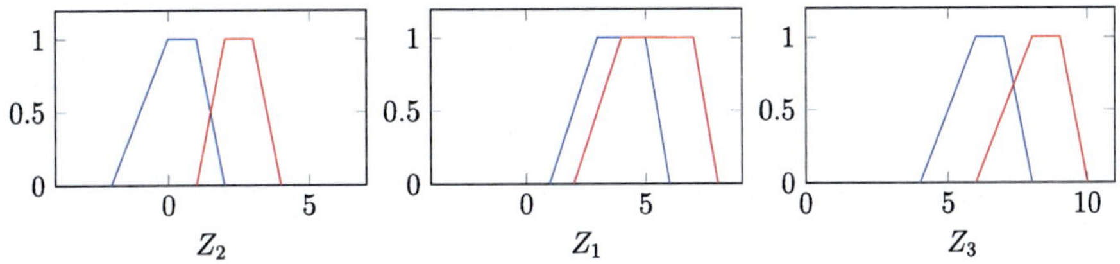

Example 4.2. Suppose that $Z_i = (A_i, B_i)$ for $i = 1, 2, 3$ such that A_i and B_i are defined as follows:

$$A_1 = (2, 4, 5, 7) \quad B_1 = (3, 5, 6, 8)$$

$$A_2 = (-1, 2, 4, 7) \quad B_2 = (1, 4, 6, 9)$$

and

$$A_3 = (1, 4, 5, 6) \quad B_3 = (2, 5, 6, 7)$$

by (5) and (6) we obtain that:

$$Mag(A_1) = \frac{54}{12} \quad Mag(B_1) = \frac{66}{12}$$

$$Mag(A_2) = \frac{36}{12} \quad Mag(B_2) = \frac{60}{12}$$

and

$$Mag(A_3) = \frac{16}{12} \quad Mag(B_3) = \frac{20}{12}$$

so by (7) we have:

$$\gamma(z_1) = \frac{120}{24} \qquad \gamma(z_2) = \frac{96}{24} \qquad \gamma(z_3) = \frac{36}{24}$$

so

$$z_3 \prec z_2 \prec z_1$$

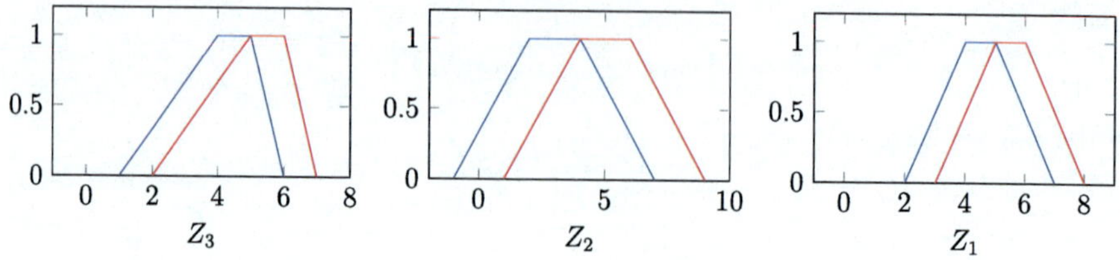

Example 4.3. Consider z-numbers $Z_i = (A_i, B_i)$ for $i = 1, 2, 3$ such that:

$$A_1 = (-3, -1, 4, 6; 1) \qquad B_1 = (1, 3, 4, 7; 1)$$

$$A_2 = (2, 4, 7, 8; 1) \qquad B_2 = (1, 4, 6, 7; 1)$$

and

$$A_3 = (-2, 1, 2, 3; 1) \qquad B_3 = (1, 2, 3, 4; 1)$$

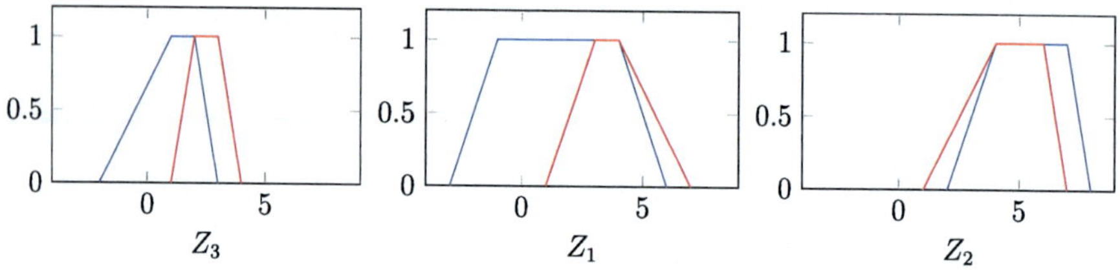

By using above algorithm we have:

$$Mag(A_1) = \frac{18}{12} \qquad Mag(B_1) = \frac{43}{12}$$

$$Mag(A_2) = \frac{65}{12} \qquad Mag(B_2) = \frac{58}{12}$$

and

$$Mag(A_3) = \frac{16}{12} \qquad Mag(B_3) = \frac{30}{12}$$

so by (7) we have:

$$\gamma(z_1) = \frac{61}{24} \qquad \gamma(z_2) = \frac{123}{24} \qquad \gamma(z_3) = 2$$

so

$$z_3 \prec z_1 \prec z_2$$

Example 4.4. Consider two z-numbers $Z_i = (A_i, B_i)$ for $i = 1, 2$ such that:

$$A_1 = (-4, -3, -2, -1; 1) \qquad B_1 = (1, 2, 3, 4; 1)$$

and

$$A_2 = (-3, -2, -1, 0; 1) \qquad B_2 = (0, 1, 2, 3; 1)$$

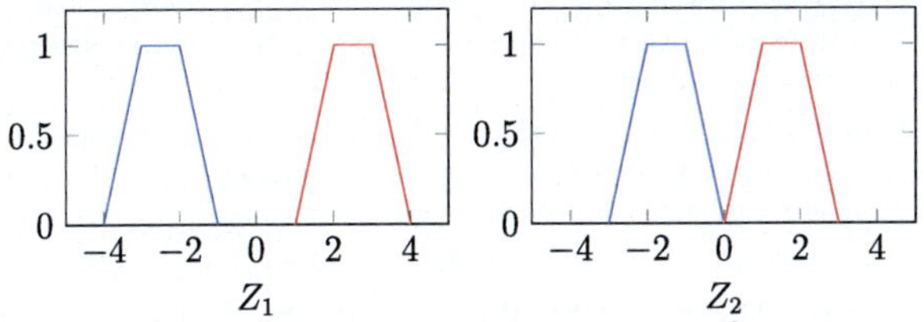

By using above algorithm and supposing that $\lambda = \frac{2}{3}$ we have:

$$Mag(A_1) = -\frac{30}{12} \qquad Mag(B_1) = \frac{30}{12}$$

and

$$Mag(A_2) = -\frac{18}{12} \qquad Mag(B_2) = \frac{18}{12}$$

so by (7) we have:

$$\gamma(z_1) = \frac{-30}{36} \qquad \gamma(z_2) = -\frac{-18}{36}$$

hence

$$z_2 \succ z_1$$

5 Conclusion

In this paper we presented a new method for ranking z-numbers. In compare with the existing methods for ranking z-numbers, the proposed method has this advantage that it does not convert z_i to fuzzy number. We considered a z-number as $z = (\bar{A}, \bar{B})$, then we compare the z-numbers by convex compound Mag(A) and Mag(B). And finally with a new ranking algorithm we ranked z-numbers.

References

1. Allahviranloo, T., Saneifard, R.: Defuzzification method for ranking fuzzy numbers based on center of gravity. Iranian J. Fuzzy Syst. **9**(6), 57–67 (2012)
2. Aliev, R.A., Alizadeh, A.V., Huseynov, O.H.: The arithmetic of discrete z-numbers. Inf. Sci. **290**, 134–155 (2015)
3. Bakar, A.S.A., Gegov, A.: Multi-layer decision methodology for ranking z-numbers. Int. J. Comput. Intell. Syst. **8**, 395–406 (2015)
4. Chen, S.M., Munif, A., Chen, G.-S., Liu, H.-C., Kuo, B.C.: Fuzzy risk analysis based on ranking generalized fuzzy numbers with different left heights and right heights. Expert Syst. Appl. **39**, 6320–6334 (2012)

5. Chu, T.C., Tsao, C.T.: Ranking fuzzy numbers with an area between the centroid point and original point. Comput. Math. Appl. **43**(1), 111–117 (2002)
6. Das, S., Guha, D.: A centroid-based ranking method of trapezoidal intuitionistic fuzzy numbers and its application to MCDM problems. Fuzzy Inf. Eng. **8**, 41–74 (2016)
7. Dat, L.Q., Yu, V.F., Chou, S.Y.: An improved ranking method for fuzzy numbers based on the centroid index. Int. J. Fuzzy Syst. **14**(3), 413–419 (2012)
8. Deng, H.: Comparing and ranking fuzzy numbers using ideal solutions. Appl. Math. Model. **38**(6), 1638–1646 (2014)
9. Ezadi, S., Allahviranloo, T., Mohammadi, S.: Two new methods for ranking of Z-numbers based on sigmoid function and sign method. Int. J. Intell. Syst. **33**(7), 1476–1487 (2018)
10. Jiang, W., Xie, C., Luo, Y., Tang, Y.: Ranking Z-numbers with an improved ranking method for generalized fuzzy numbers. J. Intell. Fuzzy Syst. **32**(3), 1931–1943 (2016)
11. Kang, B., Wei, D., Li, Y., Deng, Y.: A method of converting Z-number to classical fuzzy number. J. Inf. Comput. Sci. **3**, 703–709 (2012)
12. Mohamad, D., Shaharani, S.A., Kamis, N.H.: A z-number based decision making procedure with ranking fuzzy numbers method. In: AIP Conference Proceedings, vol. 1653, pp. 160–166 (2014)
13. Salari, M., Bagherpour, M., Wang, J.: A novel earned value management model using z-number. Int. J. Appl. Deci. Sci. **7**, 97–119 (2014)
14. Wang, X., Kerre, E.: Reasonable properties for the ordering of fuzzy quantities (I). Fuzzy Sets Syst. **118**, 375–385 (2001)
15. Wang, Y.M., Yang, J.B., Xu, D.L., Chin, K.S.: On the centroids of fuzzy numbers. Fuzzy Sets Syst. **157**, 919–926 (2006)
16. Yager, R.R.: On a View of Zadeh's Z-Numbers, pp. 90–101. Springer, Heidelberg (2012a)
17. Yager, R.R.: On Z-valuations using Zadeh's Z-numbers. Int. J. Intell. Syst. **27**(3), 259–278 (2012b)
18. Zadeh, L.A.: A note on Z-numbers. Inf. Sci. **181**, 2923–2932 (2011)

A Novel Derivations and Application of LR Fuzzy Linear System on Forces Acting on the Blocks

Ghassan Malkawi[1(✉)] and Mahdi S. Lataifeh[2]

[1] Faculty of Engineering, Math and Natural Science Division, Higher Colleges of Technology (HCT), Al Ain, UAE
gmalkawi@hct.ac.ae
[2] Physics Department, Yarmouk University, Irbid 21163, Jordan

Abstract. In this paper, a novel application of the LR-Fuzzy linear system is presented on forces acting on the blocks based on Newton's second law of motion. The definition of fuzzy forces acting on the blocks is proposed. As the tension is not obtainable or mysterious in some real-life problems, the fuzzy system provides a triangular fuzzy number reflects the oscillation and flexibility of acceleration, the approach also verifies the accuracy of gravity is not stable accordingly. The model fully transfers to the linear system for solving by the classical method in linear algebra, and tension achieves by a non-fuzzy operation. Which entitles the method to apply directly by non-fuzzy research in the application.

Crisp and fuzzy examples are presented to justify the applicability of the proposed method.

Keywords: Forces acting on the blocks · Fuzzy linear system · Triangular fuzzy number

1 Introduction

In problems concerning some applied mathematics, the linear equations system is deemed the most expressible model to use. Commonly, some coefficient values precision of these systems is untaken and often imprecise. Accordingly, the fuzzy theory is considered to overcome the shortcoming, [1] proposed the fuzzy theory to deal with the terms not constituting sets or classes in the usual mathematical context.

The pioneering achievable of modulating linear system through fuzzy theory was brought forward by Friedman and his colleagues in [2] to the fuzzy linear system (FLS), where they introduced a generic model to solve an $n \times n$ FLS through the embedding approach. In this model, the authors replaced the original FLS by a $2n \times 2n$ crisp linear system and the solution was obtained. Significant studies have been published recently concerning FLS [3–9]. Moreover, a comprehensive study of the uncertain linear system (FLS) is studied in [10].

As fuzzy systems play a major role in all filed. The main obstacle to simulating the fuzzy systems in real wide problems is fuzzy operations. In order to facilitate and

T. Allahviranloo et al. (Eds.): IDS 2020, AISC 1301, pp. 851–863, 2021.
https://doi.org/10.1007/978-3-030-66501-2_69

expend using fuzzy systems for non-fuzzy searchers, Malkawi and his colleagues in [11–23] set up new models based on the block matrix method to obtain most scenarios of fuzzy systems, additionally, the sufficient and necessary conditions were examined. An associated linear system was established to detect all feasible fuzzy solutions with no fuzzy operation used, their methods transfer the fuzzy system to the classical linear system. The ultimate solution obtains by the classical method in linear algebra like inversion matrix method, which entitles method to use with no fuzzy operation.

As an application for (Left-Right fuzzy system) LR-FLS in physics, practically, Newton's second law of motion, we assume the gravity, tension, masses, acceleration are triangular fuzzy numbers (TFN).

A Sample Problem of Block on the table connected by a rope to a hanging block [24] is studied. Three models of LR-FLS simulate the problem, first, we solve the main problem by assuming gravity as TFN to obtain acceleration and tension as TFN. Then the masses are obtained as TFNs where the tension is TFN. Finally, the vales gravity is verified as TFS by assuming acceleration and tension are TFNs. Moreover, three solver linear systems are provided, which represent a quick solver for the normal (crisp) problem, then we simulate the linear systems by three solver fuzzy linear systems to entitle the oscillation and flexibility of choosing the gravity, tension, acceleration and masses.

Based on Newton's second law of motion, we assume that $|\vec{g}|$ is nearly constant does not vary with an altitude over short vertical distance. The gravitational field g near the earth's surface decreases with increasing altitude also, slight variations in g occur with changes attitudes [24, 25].

At the earth surface $g = 9.8\frac{m}{s^2}$, note that $g = \frac{\overrightarrow{F_G}}{m}$, and $\overrightarrow{F_G} = -\frac{G_m M_E}{r^2}\hat{r}$, where $\overrightarrow{F_G}$ is the gravitational force between two masses objects, here m is the mass of the object on the earth surface, and M_E is the earth mass and r is the distance (earth radius). $\hat{r}$ is the unit vector in the direction of increase for the radius $|\vec{r}|$.

The minus sign means that the force is attractive. That means the gravitational field vector $\vec{g}$ is directed toward the centre of the earth. Note that $\overrightarrow{F_G}$ is action at a distance force. So $|\vec{g}|$ (value) is nearly constant with small variations allowed, while the acceleration of the conducted object is our examples can vary with acceptable value no restriction. Note that the acceleration is determined by the weight of the object and frictional force $\overrightarrow{F_K}$.

The rest of the paper is organized as follows. In Sect. 2, some basic definitions of fuzzy numbers are reviewed. In Sect. 3. The Forces acting on the blocks are studied in terms of fuzzy theory, then the required solver of linear and fuzzy linear systems are provided. In Sect. 4. The fuzziness of gravity value is verified. Section 5 ends the paper with a conclusion.

2 Preliminaries

The following section reviews basic definitions and remarks of the fuzzy theory, which will be needed in the sequel [14, 26].

Definition 2.1. *A fuzzy number $\tilde{u}$ is represented in parametric form by an ordered pair of functions $\tilde{u} = \left(\underline{u}(r), \overline{u}(r)\right)$, for $0 \leq r \leq 1$ which will satisfy the following conditions:*

1. $\underline{u}(r) \leq \overline{u}(r)$, *for* $0 \leq r \leq 1$.
2. $\overline{u}(r)$ *is a bounded left continuous non-increasing function on* $[0, 1]$.
3. $\underline{u}(r)$ *is a bounded left continuous non-decreasing function on* $[0, 1]$.

Definition 2.2. *(LR fuzzy number) A fuzzy number* $\tilde{m}$ *is called LR-Fuzzy number if its membership function is defined as follows:*

$$\mu_{\tilde{m}}(x) = \begin{cases} L\left(\frac{m-x}{\alpha}\right), & \text{for } x \leq m, \alpha > 0, \\ R\left(\frac{x-m}{\beta}\right), & \text{for } m \leq x, \beta > 0, \end{cases} \tag{2.1}$$

where $m, \alpha, \beta \in \mathbb{R}$.

And the function $L(.)$ *is called a left shape function if the following hold:*

1. $L(x) = L(-x)$,
2. $L(0) = 1$, $L(1) = 0$,
3. L *is non-increasing on* $[0, \infty)$.

Also, the definition of function $R(.)$, *called the right shape is similar to that of* $L(.)$, *and* m *symbolizes the mean value of* $\tilde{m}$, *while* α *and* β *are left and right spreads, respectively i.e.* α *and* β *are the coefficients of fuzziness, as the spreads increase,* $\tilde{m}$ *become fuzzier and fuzzier. It is symbolically written as* $\tilde{m} = (m, \alpha, \beta)_{LR}$.

Therefore, from between LR fuzzy number to parametric form given as:

$$\underline{u}(r) = r\alpha + m - \alpha \text{ and } \overline{u}(r) = m + \beta - \beta r. \tag{2.2}$$

Definition 2.3. *An LR fuzzy number* $\tilde{m} = (m, \alpha, \beta)_{LR}$ *is called a triangular fuzzy number where* $L = R = \max(0, 1 - x)$.

The set of triangular fuzzy numbers is denoted by $F(\mathfrak{R})$.

Definition 2.4. *(Arithmetic operations on LR fuzzy numbers) We will represent arithmetic operations for two LR Fuzzy numbers* $= (m, \alpha, \beta)_{LR}$ *and* $\tilde{n} = (n, \gamma, \delta)_{LR}$ *as follows:*

Addition:

$$(m, \alpha, \beta)_{LR} \oplus (n, \gamma, \delta)_{LR} = (m + n, \alpha + \gamma, \beta + \delta)_{LR} \tag{2.3a}$$

Opposite

$$(-1)\tilde{m} = (-m, \alpha, \beta)_{LR} = (-m, \beta, \alpha)_{LR} \tag{2.3b}$$

Subtraction:

$$(m, \alpha, \beta)_{LR} \ominus (n, \gamma, \delta)_{RL} = (m - n, \alpha + \delta, \beta + \gamma)_{LR} \tag{2.3c}$$

Scalar multiplication:

Let $\lambda \in R$ then,

$$\lambda \otimes (m, \alpha, \beta)_{LR} = \begin{cases} (\lambda m, \lambda\alpha, \lambda\beta)_{LR} & \lambda \geq 0 \\ (\lambda m, -\lambda\beta, -\lambda\alpha)_{LR} & \lambda < 0 \end{cases} \tag{2.3d}$$

Definition 2.5. *A vector* $\tilde{X} = (\tilde{x}_1, \tilde{x}_2, \ldots, \tilde{x}_n)^T$ *is called a fuzzy vector, if* $\tilde{x}_i$, $\forall i = 1, .., n$. *are fuzzy numbers.*

Definition 2.6. *A matrix* $\tilde{A} = (\tilde{a}_{ij})_{n\times n}$ *is called a left-right fuzzy matrix (shortly fuzzy matrix) if* $\tilde{a}_{ij} \in \forall i, j = 1, .., n$ *are fuzzy numbers.*

Definition 2.7. *Let* $A = (a_{ij})$ *be* $m \times n$ *crisp matrix and* $\tilde{B} = (\tilde{b}_{ij})$ *be* $n \times p$ *fuzzy matrix. We define* $A \otimes \tilde{B} = \tilde{C} = (\tilde{c}_{ij})$ *which is the* $m \times p$ *matrix where.*

$$\tilde{c}_{ij} = \sum_{k=1,\ldots,n}^{\oplus} a_{ik} \otimes \tilde{b}_{kj}. \tag{2.4}$$

Definition 2.8. *(Left-Right fuzzy linear system) Consider the* $n \times n$ *linear system of equations:*

$$\begin{cases} a_{11}\tilde{x}_1 + a_{12}\tilde{x}_2 + \ldots\ldots + a_{1n}\tilde{x}_n = \tilde{b}_1, \\ a_{21}\tilde{x}_1 + a_{22}\tilde{x}_2 + \ldots\ldots + a_{2n}\tilde{x}_n = \tilde{b}_2 \\ \vdots \\ \vdots \\ a_{n1}\tilde{x}_1 + a_{n2}\tilde{x}_2 + \ldots\ldots + a_{nn}\tilde{x}_n = \tilde{b}_n, \end{cases} \tag{2.5}$$

The matrix form of the above equations is

$$\tilde{A} \otimes \tilde{x} = \tilde{b} \tag{2.6}$$

$$\begin{pmatrix} a_{11} & a_{12} & \cdots & a_{1n} \\ a_{21} & a_{22} & \cdots & a_{2n} \\ \vdots & \vdots & \ddots & \vdots \\ a_{n1} & a_{n2} & \cdots & a_{nn} \end{pmatrix} \otimes \begin{pmatrix} \tilde{x}_1 \\ \tilde{x}_2 \\ \vdots \\ \tilde{x}_n \end{pmatrix} = \begin{pmatrix} \tilde{b}_1 \\ \tilde{b}_2 \\ \vdots \\ \tilde{b}_n \end{pmatrix}, \tag{2.7}$$

This system is called a Left-Right fuzzy linear system (LR-FLS) where the coefficient matrix $A = (a_{ij})1 \leq i, j \leq n$ *is a crisp matrix, the entries of fuzzy vectors* $\tilde{x}$, $\tilde{b}$ *are LR fuzzy numbers, and* $\tilde{x}$ *is unknown to be found.*

The next section provides the solver of linear and fuzzy linear system tension, acceleration and masses as triangular fuzzy numbers.

3 Forces Acting on the Blocks (Frictionless)

In this section, the crisp solution of forces acting on the blocks is reviewed for a frictionless surface, as a special case of Newton's second law, see [24]. Then, We assume the tension is triangular fuzzy numbers to obtain the blocks as triangular fuzzy numbers.

Finally, we assume the gravity is the triangular fuzzy number to obtain tension and acceleration as triangular fuzzy numbers. In addition, We assume the tension and acceleration are triangular fuzzy numbers to verify the value of gravity as a triangular fuzzy number. The solver crisp and fuzzy linear systems are provided for all the above cases.

The next section provides the solver linear system of tension and acceleration.

3.1 Crisp Solver of the Tension and Acceleration of a Frictionless Surface

Consider the forces acting on the blocks mass S and mass H in Fig. 1, see [24]. The sliding block with mass S is free to move along a horizontal frictionless surface and connected, by a cord that wraps over a frictionless pulley, to the hanging block with mass H. The cord and pulley have negligible masses compared to the blocks (they are massless). The hanging block H falls as the sliding block S accelerates to the right,

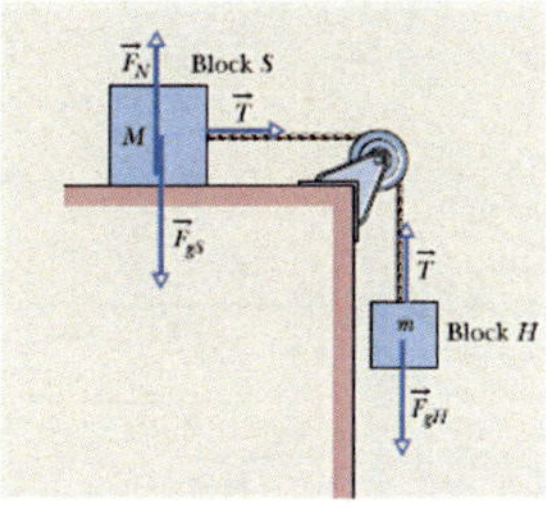

Fig. 1. The forces acting on the blocks [24].

Assume the acceleration, gravity and tension are a, g and T, respectively. Force F acting on M (Block S) and m (Bock H). According to Newton's second law

$$\sum F_x = Ma_x.$$

If we apply it on the sliding block S,

$$\sum F_x = Sa_x \Rightarrow T - 0 = Sa,$$

then

$$T = Sa. \tag{3.1}$$

The net forces in the y-direction are zero since the block S is in equilibrium, we have

$$\sum F_y = Ma_y \text{ but } a_y = 0,$$

so

$$F_N - Sg = 0. \tag{3.2}$$

Where F_N is the normal force acting on S, and Sg is the weight of the sliding block.

If we apply Newton's second law to the hanging block, forces act only along the $y-axis$ direction.

$$\sum F_y = -Ha,$$

$$Hg - T = Ha,$$

$$Hg - Ha = T. \tag{3.3}$$

Eq.'s (3.1) *and* (3.2) are correct if we assume that the rope is massless and does not stretch, also the pulley is massless and frictionless. From Eq.'s (3.1) *and* (3.3) the T *and* a are obtained as follows.

$$a = \frac{H}{S+H}g,$$

$$\frac{(H+S)a}{H} = g \tag{3.4}$$

Substitute a from Eq. (3.4) in Eq. (3.3) gives T.

$$T = S\frac{H}{S+H}g,$$

$$g = \frac{(H+S)T}{HS} \tag{3.5}$$

Hence, Eq.'s (3.4) and (3.5) may combine in a linear system and matrix form as follows:

$$\begin{cases} \frac{(H+S)T}{HS} & 0 & = g, \\ 0 & \frac{(H+S)a}{H} & = g. \end{cases} \tag{3.6}$$

$$\begin{pmatrix} \frac{(H+S)}{HS} & 0 \\ 0 & \frac{(H+S)}{H} \end{pmatrix} \begin{pmatrix} T \\ a \end{pmatrix} = \begin{pmatrix} g \\ g \end{pmatrix}. \tag{3.7}$$

The next example illustrates the Sample Problem 5.03 in [24].

Example 3.1. *The figure shows a block S(the sliding block) with mass* $M = 3.3\,kg$*. The block is free to move along a horizontal frictionless surface and connected by a cord that wraps over a frictionless pulley, to a second block H(the hanging block), with mass* $m = 2.1\,kg$*. The cord and pulley have negligible masses compared to the blocks (they are "massless"). The hanging block Hfalls as the sliding block Saccelerates to the right. Find (a) the acceleration of blocks S, H, and the tension in the cord. See Fig. 1.*

Using the solver matrix form in (3.7),

$$\begin{pmatrix} 0.78 & 0 \\ 0 & 2.57 \end{pmatrix} \begin{pmatrix} T \\ a \end{pmatrix} = \begin{pmatrix} 9.8 \\ 9.8 \end{pmatrix}, \tag{3.8}$$

$$\begin{pmatrix} T \\ a \end{pmatrix} = \begin{pmatrix} 12.56\,N \\ 3.81\,\frac{m}{s^2} \end{pmatrix}. \tag{3.9}$$

The next section obtains the solver of the tension and acceleration as triangular fuzzy numbers.

3.2 Fuzzy Solver of the Tension and Acceleration

The fuzziness of tension acceleration is discussed where the surface is frictionless. Assume the acceleration, gravity and tension are triangular fuzzy numbers $\widetilde{a}= (m^a, \alpha^a, \beta^a)$, $\widetilde{g}= (m^g, \alpha^g, \beta^g)$ and $\widetilde{T}= \left(m^T, \alpha^T, \beta^T\right)$ respectively.

Let,

$$a_{11} = \frac{(H+S)}{HS}, a_{12} = 0, a_{21} = 0, a_{22} = \frac{(H+S)}{H},$$

$$\tilde{x}_1 = \tilde{T} = \left(m^T, \alpha^T, \beta^T\right) \text{ and } \tilde{x}_2 = \tilde{a} = (m^a, \alpha^a, \beta^a),$$

$$\tilde{b}_2 = \tilde{b}_1 = \tilde{g} = \left(m^g, \alpha^g, \beta^g\right).$$

Then the LR-FLS in (2.7) with its matrix form may be written as follows,

$$\begin{cases} \frac{(H+S)}{HS} \otimes \left(m^T, \alpha^T, \beta^T\right) \oplus 0 \otimes (m^a, \alpha^a, \beta^a) = (m^g, \alpha^g, \beta^g), \\ 0 \otimes \left(m^T, \alpha^T, \beta^T\right) \oplus \frac{(H+S)}{H} \otimes (m^a, \alpha^a, \beta^a) = (m^g, \alpha^g, \beta^g), \end{cases} \tag{3.10}$$

$$\begin{pmatrix} \frac{(H+S)}{HS} & 0 \\ 0 & \frac{(H+S)}{H} \end{pmatrix} \begin{pmatrix} \tilde{T} \\ \tilde{a} \end{pmatrix} = \begin{pmatrix} \tilde{g} \\ \tilde{g} \end{pmatrix} \tag{3.11}$$

Then the solver of (3.10) and (3.11) is

$$\begin{pmatrix} \begin{pmatrix} \frac{(H+S)}{HS} & 0 \\ 0 & \frac{(H+S)}{HS} \end{pmatrix} & \begin{pmatrix} 0 & 0 \\ 0 & 0 \end{pmatrix} & \begin{pmatrix} 0 & 0 \\ 0 & 0 \end{pmatrix} \\ \begin{pmatrix} 0 & 0 \\ 0 & 0 \end{pmatrix} & \begin{pmatrix} \frac{(H+S)}{HS} & 0 \\ 0 & \frac{(H+S)}{HS} \end{pmatrix} & \begin{pmatrix} 0 & 0 \\ 0 & 0 \end{pmatrix} \\ \begin{pmatrix} 0 & 0 \\ 0 & 0 \end{pmatrix} & \begin{pmatrix} 0 & 0 \\ 0 & 0 \end{pmatrix} & \begin{pmatrix} \frac{(H+S)}{HS} & 0 \\ 0 & \frac{(H+S)}{HS} \end{pmatrix} \end{pmatrix} \begin{pmatrix} \begin{pmatrix} m^T \\ m^a \end{pmatrix} \\ \begin{pmatrix} \alpha^T \\ \alpha^a \end{pmatrix} \\ \begin{pmatrix} \beta^T \\ \beta^a \end{pmatrix} \end{pmatrix} = \begin{pmatrix} \begin{pmatrix} m^g \\ m^g \end{pmatrix} \\ \begin{pmatrix} \alpha^g \\ \alpha^g \end{pmatrix} \\ \begin{pmatrix} \beta^g \\ \beta^g \end{pmatrix} \end{pmatrix}. \tag{3.12}$$

The next example simulates the main Example 3.1 in terms of fuzziness of gravity, tension and acceleration.

Example 3.2. *If the gravity in the cord is the triangular fuzzy number* $\tilde{g} = \left(9.8, 0.8, 1.1\right)$. *in Example 3.1. Find the triangular fuzzy numbers tension and acceleration* $\tilde{a} = (m^a, \alpha^a, \beta^a)$ *and* $\tilde{T} = \left(m^T, \alpha^T, \beta^T\right)$, *respectively.*

The LR-FLS and its fuzzy matrix form in (3.10) and (3.11), respectively, may be written as follows,

$$\begin{cases} 0.78 \otimes \left(m^T, \alpha^T, \beta^T\right) \oplus 0 \otimes (m^a, \alpha^a, \beta^a) = (9.8, 0.8, 1.1), \\ 0 \otimes \left(m^T, \alpha^T, \beta^T\right) \oplus 2.57 \otimes (m^a, \alpha^a, \beta^a) = (9.8, 0.8, 1.1), \end{cases} \tag{3.13}$$

$$\begin{pmatrix} 0.78 & 0 \\ 0 & 2.57 \end{pmatrix}\begin{pmatrix} \tilde{g} \\ \tilde{a} \end{pmatrix} = \begin{pmatrix} (9.8, 0.8, 1.1) \\ (9.8, 0.8, 1.1) \end{pmatrix}, \tag{3.14}$$

using the solver in (3.12),

$$\begin{pmatrix} 0.78 & 0 & 0 & 0 & 0 & 0 \\ 0 & 2.57 & 0 & 0 & 0 & 0 \\ 0 & 0 & 0.78 & 0 & 0 & 0 \\ 0 & 0 & 0 & 2.57 & 0 & 0 \\ 0 & 0 & 0 & 0 & 0.78 & 0 \\ 0 & 0 & 0 & 0 & 0 & 2.57 \end{pmatrix}\begin{pmatrix} \begin{pmatrix} m^T \\ m^a \end{pmatrix} \\ \begin{pmatrix} \alpha^T \\ \alpha^a \end{pmatrix} \\ \begin{pmatrix} \beta^T \\ \beta^a \end{pmatrix} \end{pmatrix} = \begin{pmatrix} 9.8 \\ 9.8 \\ 0.8 \\ 0.8 \\ 1.1 \\ 1.1 \end{pmatrix} \tag{3.15}$$

Then the solution of LR-FLS in (3.13) is

$$\begin{pmatrix} \tilde{T} \\ \tilde{a} \end{pmatrix} = \begin{pmatrix} m^T, \alpha^T, \beta^T \\ m^a, \alpha^a, \beta^a \end{pmatrix} = \begin{pmatrix} (12.56, 1.03, 1.41)N \\ '' \ '' \\ (3.81, 0.31, 0.43)\frac{m}{s^2} \end{pmatrix}. \tag{3.16}$$

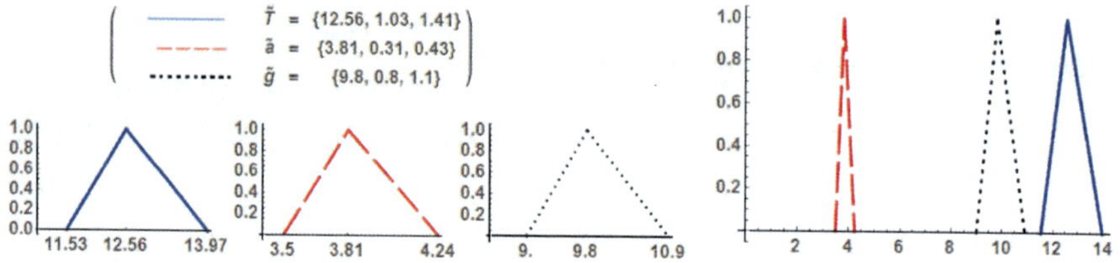

Fig. 2. Shows the fuzzy solution cancelation and tension.

Figure 2 shows the oscillation and flexibility on gravity effects for testation more than the acceleration.

The next section provides the solver linear and fuzzy linear system of masses as triangular fuzzy numbers.

3.3 Solver of Fuzzy Masses

Assume the acceleration and gravity are crisp real numbers, tension is triangular fuzzy numbers $\tilde{a} = (m^a, \alpha^a, \beta^a)$. We will find the triangular fuzzy numbers masses $\tilde{H} = \left(m^H, \alpha^H, \beta^H\right)$ and $\tilde{S} = \left(m^S, \alpha^S, \beta^S\right)$. If the surface is frictionless. From Eq.'s (3.1) and (3.3) we have a solver linear system and matrix form of gravity as follows

$$\begin{cases} gS \quad 0 \quad = T, \\ (g - a)H = T, \end{cases} \tag{3.17}$$

$$\begin{pmatrix} g & \\ & g-a \end{pmatrix}\begin{pmatrix} S \\ H \end{pmatrix} = \begin{pmatrix} T \\ T \end{pmatrix}. \tag{3.18}$$

Let $a_{11} = g, a_{12} = 0, a_{21} = H, a_{22} = g - a$.
$\tilde{x}_1 = \tilde{S} = \left(m^S, \alpha^S, \beta^S\right)$ and $\tilde{x}_2 = \tilde{H} = \left(m^H, \alpha^H, \beta^H\right)$.
$\tilde{b}_2 = \tilde{b}_1 = \tilde{T} = \left(m^T, \alpha^T, \beta^T\right)$ in

Then the LR-FLS in (2.7) with its matrix form may be written as follows, where the surface is frictionless

$$\begin{cases} g \otimes \left(m^S, \alpha^S, \beta^S\right) \oplus 0 \otimes \left(m^H, \alpha^H, \beta^H\right) = \left(m^T, \alpha^T, \beta^T\right), \\ 0 \otimes \left(m^S, \alpha^S, \beta^S\right) \oplus (g-a) \otimes \left(m^H, \alpha^H, \beta^H\right) = \left(m^T, \alpha^T, \beta^T\right), \end{cases} \tag{3.19}$$

$$\begin{pmatrix} g & 0 \\ 0 & g-a \end{pmatrix}\begin{pmatrix} \tilde{S} \\ \tilde{H} \end{pmatrix} = \begin{pmatrix} \tilde{T} \\ \tilde{T} \end{pmatrix}. \tag{3.20}$$

Then the solver of (3.19) and (3.20) depend on the sign of $g - a$, assume $g \geq a$,

$$\begin{pmatrix} \begin{pmatrix} g & 0 \\ 0 & g-a \end{pmatrix} & \begin{pmatrix} 0 & 0 \\ 0 & 0 \end{pmatrix} & \begin{pmatrix} 0 & 0 \\ 0 & 0 \end{pmatrix} \\ \begin{pmatrix} 0 & 0 \\ 0 & 0 \end{pmatrix} & \begin{pmatrix} g & 0 \\ 0 & g-a \end{pmatrix} & \begin{pmatrix} 0 & 0 \\ 0 & H \end{pmatrix} \\ \begin{pmatrix} 0 & 0 \\ 0 & 0 \end{pmatrix} & \begin{pmatrix} 0 & 0 \\ 0 & H \end{pmatrix} & \begin{pmatrix} g & 0 \\ 0 & g-a \end{pmatrix} \end{pmatrix}\begin{pmatrix} \begin{pmatrix} m^S \\ m^H \end{pmatrix} \\ \begin{pmatrix} \alpha^S \\ \alpha^H \end{pmatrix} \\ \begin{pmatrix} \beta^S \\ \beta^H \end{pmatrix} \end{pmatrix} = \begin{pmatrix} \begin{pmatrix} m^T \\ m^T \end{pmatrix} \\ \begin{pmatrix} \alpha^T \\ \alpha^T \end{pmatrix} \\ \begin{pmatrix} \beta^T \\ \beta^T \end{pmatrix} \end{pmatrix} \tag{3.21}$$

Example 3.3 *If the tension in the cord is the fuzzy number $\tilde{T} = (12.57, 0.8, 1.1)N$. The acceleration is $3.8\frac{m}{s^2}$ find the masses $\tilde{S} = \left(m^S, \alpha^S, \beta^S\right)$ and $\tilde{x}_2 = \tilde{H} = \left(m^H, \alpha^H, \beta^H\right)$.*

The LR- FLS and fuzzy matrix form in Eq's (2.7) may be written as follows,

$$\begin{cases} g \otimes \left(m^S, \alpha^S, \beta^S\right) \oplus 0 \otimes \left(m^H, \alpha^H, \beta^H\right) = \left(m^T, \alpha^T, \beta^T\right), \\ H \otimes \left(m^S, \alpha^S, \beta^S\right) \oplus (g-a) \otimes \left(m^H, \alpha^H, \beta^H\right) = \left(m^T, \alpha^T, \beta^T\right), \end{cases} \tag{3.22}$$

$$\begin{pmatrix} 0 & 3.3 \\ 2.1 & -2.1 \end{pmatrix} \otimes \begin{pmatrix} \tilde{g} \\ \tilde{a} \end{pmatrix} = \begin{pmatrix} (30.82, 2, 2.4) \\ (30.82, 2, 2.4) \end{pmatrix}, \tag{3.23}$$

As $g \geq a$ then we will use the solver of in Eq. (3.20),

$$\begin{pmatrix} \begin{pmatrix} 3.8 & 0 \\ 0 & 6 \end{pmatrix} & \begin{pmatrix} 0 & 0 \\ 0 & 0 \end{pmatrix} & \begin{pmatrix} 0 & 0 \\ 0 & 0 \end{pmatrix} \\ \begin{pmatrix} 0 & 0 \\ 0 & 0 \end{pmatrix} & \begin{pmatrix} 3.8 & 0 \\ 0 & 6 \end{pmatrix} & \begin{pmatrix} 0 & 0 \\ 0 & 0 \end{pmatrix} \\ \begin{pmatrix} 0 & 0 \\ 0 & 0 \end{pmatrix} & \begin{pmatrix} 0 & 0 \\ 0 & 0 \end{pmatrix} & \begin{pmatrix} 3.8 & 0 \\ 0 & 6 \end{pmatrix} \end{pmatrix} \begin{pmatrix} \begin{pmatrix} m^S \\ m^H \end{pmatrix} \\ \begin{pmatrix} \alpha^S \\ \alpha^H \end{pmatrix} \\ \begin{pmatrix} \beta^S \\ \beta^H \end{pmatrix} \end{pmatrix} = \begin{pmatrix} \begin{pmatrix} 12.57 \\ 12.57 \end{pmatrix} \\ \begin{pmatrix} 0.8 \\ 0.8 \end{pmatrix} \\ \begin{pmatrix} 1.1 \\ 1.1 \end{pmatrix} \end{pmatrix}, \tag{3.24}$$

Then the solution of LR-FLS in (3.22) is

$$\begin{pmatrix} \tilde{S} \\ \tilde{H} \end{pmatrix} = \begin{pmatrix} (m^S, \alpha^S, \beta^S) \\ (m^H, \alpha^H, \beta^H) \end{pmatrix} = \begin{pmatrix} (3.31, 0.21, 0.29)kg \\ '' '' \\ (2.09, 0.13, 0.18)kg \end{pmatrix}. \tag{3.25}$$

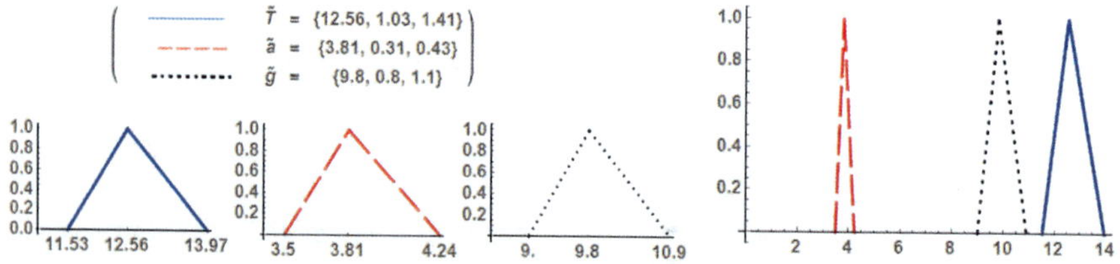

Fig. 3. Shows the fuzzy solution masses

Figure 3 shows the oscillation and flexibility on tension effects for the hanging block (S) more than the sliding block (S).

The next section verifies the fuzzy gravity value solver of the tension and acceleration as triangular fuzzy numbers.

4 Verify of Fuzzy Gravity

In this section, we assume the tension and acceleration are triangular fuzzy numbers to verify the value of gravity as a triangular fuzzy number. The solver crisp and fuzzy linear systems are provided. Assume the acceleration, gravity and tension are triangular fuzzy numbers $\tilde{a} = (m^a, \alpha^a, \beta^a)$, $\tilde{g} = (m^g, \alpha^g, \beta^g)$ and $\tilde{T} = \left(m^T, \alpha^T, \beta^T\right)$ respectively. We will verify that the value of gravity $\tilde{g}$. If the surface is frictionless. From Eq.'s (3.1) and (3.3) we have a solver linear system and matrix form of gravity as follows

$$\begin{cases} 0 & +Sg = T, \\ Hg & -Ha = T, \end{cases} \tag{4.1}$$

$$\begin{pmatrix} 0 & S \\ H & -H \end{pmatrix} \begin{pmatrix} g \\ a \end{pmatrix} = \begin{pmatrix} T \\ T \end{pmatrix}. \tag{4.2}$$

Let $a_{11} = 0, a_{12} = S, a_{21} = H, a_{22} = -H$.
$\tilde{x}_1 = \tilde{g} = (m^g, \alpha^g, \beta^g)$ and $\tilde{x}_2 = \tilde{a} = (m^a, \alpha^a, \beta^a)$.
$\tilde{b}_2 = \tilde{b}_1 = \tilde{T} = (m^T, \alpha^T, \beta^T)$,
Then the LR-FLS in (2.7) with its matrix form may be written as follows,

$$\begin{cases} 0 \otimes (m^g, \alpha^g, \beta^g) \oplus S \otimes (m^a, \alpha^a, \beta^a) = (m^T, \alpha^T, \beta^T), \\ H \otimes (m^g, \alpha^g, \beta^g) \oplus -H \otimes (m^a, \alpha^a, \beta^a) = (m^T, \alpha^T, \beta^T), \end{cases} \tag{4.3}$$

$$\begin{pmatrix} 0 & S \\ H & -H \end{pmatrix} \begin{pmatrix} \tilde{g} \\ \tilde{a} \end{pmatrix} = \begin{pmatrix} \tilde{T} \\ \tilde{T} \end{pmatrix} \tag{4.4}$$

Then the solver of (4.3) and (4.4) is

$$\begin{pmatrix} \begin{pmatrix} 0 & S \\ H & -H \end{pmatrix} & \begin{pmatrix} 0 & 0 \\ 0 & 0 \end{pmatrix} & \begin{pmatrix} 0 & 0 \\ 0 & 0 \end{pmatrix} \\ \begin{pmatrix} 0 & 0 \\ 0 & 0 \end{pmatrix} & \begin{pmatrix} 0 & S \\ H & 0 \end{pmatrix} & \begin{pmatrix} 0 & 0 \\ 0 & H \end{pmatrix} \\ \begin{pmatrix} 0 & 0 \\ 0 & 0 \end{pmatrix} & \begin{pmatrix} 0 & 0 \\ 0 & H \end{pmatrix} & \begin{pmatrix} 0 & S \\ H & 0 \end{pmatrix} \end{pmatrix} \begin{pmatrix} \begin{pmatrix} m^g \\ m^a \end{pmatrix} \\ \begin{pmatrix} \alpha^g \\ \alpha^a \end{pmatrix} \\ \begin{pmatrix} \beta^g \\ \beta^a \end{pmatrix} \end{pmatrix} = \begin{pmatrix} \begin{pmatrix} m^T \\ m^T \end{pmatrix} \\ \begin{pmatrix} \alpha^T \\ \alpha^T \end{pmatrix} \\ \begin{pmatrix} \beta^T \\ \beta^T \end{pmatrix} \end{pmatrix}. \tag{4.5}$$

The next example verifies the fuzziness of gravity and solves the acceleration as a triangular fuzzy number.

Example 4.1. *If the tension in the cord is the fuzzy number* $\tilde{T} = (12.58, 2, 2.2)N$. *Verify that the value of gravity* $\tilde{g} = (m^g, \alpha^g, \beta^g)$ *as fuzzy number value and find the acceleration fuzzy numbers* $\tilde{a} = (m^a, \alpha^a, \beta^a)$.

The LR- FLS and its fuzzy matrix form in (3.10) and (3.11), respectively, may be written as follows,

$$\begin{cases} 3.3 \otimes (m^a, \alpha^a, \beta^a) = (12.58, 2, 2.2), \\ 2.1 \otimes (m^g, \alpha^g, \beta^g) \oplus -2.1 \otimes (m^a, \alpha^a, \beta^a) = (12.58, 2, 2.2). \end{cases} \tag{4.6}$$

$$\begin{pmatrix} 0 & 3.3 \\ 2.1 & -2.1 \end{pmatrix} \otimes \begin{pmatrix} \tilde{g} \\ \tilde{a} \end{pmatrix} = \begin{pmatrix} (12.58, 2, 2.2) \\ (12.58, 2, 2.2) \end{pmatrix}, \tag{4.7}$$

using the solver in (4.5)

$$\begin{pmatrix} \begin{pmatrix} 0. & 3.3 \\ 2.1 & -2.1 \end{pmatrix} & \begin{pmatrix} 0 & 0 \\ 0 & 0 \end{pmatrix} & \begin{pmatrix} 0 & 0 \\ 0 & 0 \end{pmatrix} \\ \begin{pmatrix} 0 & 0 \\ 0 & 0 \end{pmatrix} & \begin{pmatrix} 0. & 3.3 \\ 2.1 & 0 \end{pmatrix} & \begin{pmatrix} 0 & 0 \\ 0 & 2.1 \end{pmatrix} \\ \begin{pmatrix} 0 & 0 \\ 0 & 0 \end{pmatrix} & \begin{pmatrix} 0 & 0 \\ 0 & 2.1 \end{pmatrix} & \begin{pmatrix} 0 & 3.3 \\ 2.1 & 0 \end{pmatrix} \end{pmatrix} \begin{pmatrix} \begin{pmatrix} m^g \\ m^a \end{pmatrix} \\ \begin{pmatrix} \alpha^T \\ \alpha^T \end{pmatrix} \\ \begin{pmatrix} \beta^T \\ \beta^T \end{pmatrix} \end{pmatrix} = \begin{pmatrix} \begin{pmatrix} 12.58 \\ 12.58 \end{pmatrix} \\ \begin{pmatrix} 2 \\ 2 \end{pmatrix} \\ \begin{pmatrix} 2.2 \\ 2.2 \end{pmatrix} \end{pmatrix}. \tag{4.8}$$

Then, the solution of LR-FLS in (4.6) is

$$\begin{pmatrix} \tilde{g} \\ \tilde{a} \end{pmatrix} = \begin{pmatrix} (m^g, \alpha^g, \beta^g) \\ (m^a, \alpha^a, \beta^a) \end{pmatrix} = \begin{pmatrix} (9.8, 0.29, 0.44)\frac{m}{s^2} \\ '' \; '' \\ (3.81, 0.61, 0.67)\frac{m}{s^2} \end{pmatrix}. \tag{4.9}$$

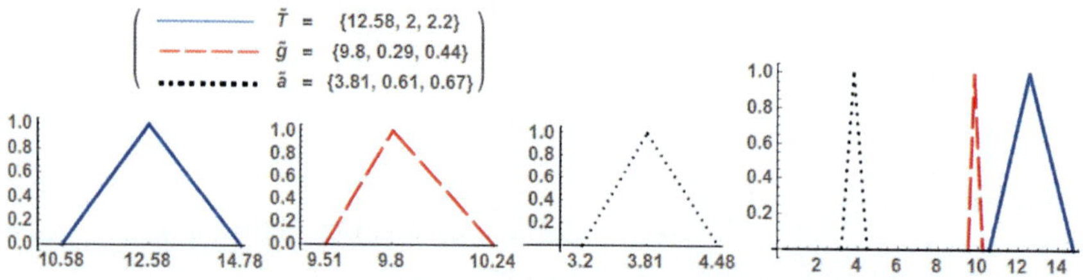

Fig. 4. Verifies the fuzzy gravity value.

Figure 4 shows the oscillation and flexibility on tension effects for gravity more than the acceleration.

5 Conclusion

The definition of fuzzy forces acting on the blocks is proposed blocks based on Newton's second law of motion. The formulated fuzzy linear system shows that oscillation and flexibility are not affected Equally for all the quantities. A fuzziness of gravity effects for testation more than the acceleration. While the fuzziness on tension effects for the hanging block (S) more than the sliding block (H). Finally, we show that the value of gravity is subject to oscillation and flexibility based on the fuzziness of testation and acceleration.

References

1. Zadeh, L.A.: Fuzzy sets. In: Fuzzy Sets, Fuzzy Logic, And Fuzzy Systems: Selected Papers by Lotfi A Zadeh, pp. 394–432. World Scientific (1996)
2. Friedman, M., Ming, M., Kandel, A.: Fuzzy linear systems. Fuzzy Sets Syst. **96**(2), 201–209 (1998)
3. Allahviranloo, T.: Numerical methods for fuzzy system of linear equations. Appl. Math. Comput. **155**(2), 493–502 (2004)

4. Dehghan, M., Hashemi, B.: Iterative solution of fuzzy linear systems. Appl. Math. Comput. **175**(1), 645–674 (2006)
5. Allahviranloo, T.: The Adomian decomposition method for fuzzy system of linear equations. Appl. Math. Comput. **163**(2), 553–563 (2005)
6. Otadi, M., Mosleh, M.: Minimal solution of fuzzy linear systems. Iran. J. Fuzzy Syst. **12**(1), 89–99 (2015)
7. Mansouri, P., Asady, B.: Iterative methods for solving fuzzy linear systems. Aust. J. Basic Appl. Sci. **5**(7), 1036–1049 (2011)
8. Allahviranloo, T., Ghanbari, M., Hosseinzadeh, A., Haghi, E., Nuraei, R.: A note on "Fuzzy linear systems." Fuzzy Sets Syst. **177**(1), 87–92 (2011)
9. Salahshour, S., Ahmadian, A., Ismail, F., Baleanu, D.: A novel weak fuzzy solution for fuzzy linear system. Entropy **18**(3), 68 (2016)
10. Allahviranloo, T.: Uncertain linear systems. In: Uncertain Information and Linear Systems, pp. 61–209. Springer, Cham (2020)
11. Daud, W.S.W., Ahmad, N., Malkawi, G.: An algorithm for solving an arbitrary triangular fully fuzzy Sylvester matrix equations. In: AIP Conference Proceedings, vol. 1905, no. 1, p. 030012. AIP Publishing (2017)
12. Daud, W.S.W., Ahmad, N., Malkawi, G.: A preliminary study on the solution of fully fuzzy Sylvester matrix equation. In: AIP Conference Proceedings, vol. 1775, no. 1, p. 030006. AIP Publishing (2016)
13. Malkawi, G., Ahmad, N., Ibrahim, H.: Finite solutions of fully fuzzy linear system. In: AIP Conference Proceedings, vol. 1635, no. 1, pp. 447–454. American Institute of Physics (2014)
14. Malkawi, G., Ahmad, N., Ibrahim, H.: Solving fully fuzzy linear system with the necessary and sufficient condition to have a positive solution. Appl. Math. Inf. Sci. **8**(3), 1003 (2014)
15. Malkawi, G., Ahmad, N., Ibrahim, H., Albayari, D.J.: A note on solving fully fuzzy linear systems by using implicit Gauss—Cholesky algorithm. Comput. Math. Model. **26**(4), 585–592 (2015)
16. Malkawi, G., Rida, I., Ahmad, N.: An associated linear system approach for solving fully fuzzy linear system with hexagonal fuzzy number. In: 2018 Advances in Science and Engineering Technology International Conferences (ASET), pp. 1–7. IEEE (2018).
17. Malkawi, G.O., Alfifi, H.Y.: The consistency of positive fully fuzzy linear system. In: AIP Conference Proceedings, vol. 1905, no. 1, p. 020004. AIP Publishing LLC (2017)
18. Ahmad, N., Malkawi, G., Ibrahim, H.: Solution of LR-fuzzy linear system with trapezoidal fuzzy number using matrix theory (2015)
19. Malkaw, G., Ahmad, N., Ibrahim, H.: An algorithm for a positive solution of arbitrary fully fuzzy linear system. Comput. Math. Model. **26**(3), 436–465 (2015)
20. Daud, W.S.W., Ahmad, N., Malkawi, G.: Positive fuzzy minimal solution for positive singular fully fuzzy Sylvester matrix equation. In: AIP Conference Proceedings, vol. 1974, no. 1, p. 020084. AIP Publishing LLC (2018)
21. Daud, W., Ahmad, N., Malkawi, G.: Solving arbitrary fully fuzzy Sylvester matrix equations and its theoretical foundation. In: AIP Conference Proceedings, vol. 2013, no. 1, p. 020026: AIP Publishing LLC (2018)
22. Daud, W.S.W., Ahmad, N., Malkawi, G.: A new solution of pair matrix equations with arbitrary triangular fuzzy numbers. Turk. J. Math. **43**(3), 1195–1217 (2019)
23. Malkawi11, G., Adnan, E., Albarghouthi, H., Elsayed, A.: The Neither Unique Nor Infinite Solution Of Fully Fuzzy Linear System for triangular fuzzy numbers (m, α, β)
24. Halliday, D., Resnick, R., Walker, J.: Fundamentals of Physics. Wiley, Hoboken (2013)
25. Serway, R.A., Jewett, J.W.: Physics for Scientists and Engineers with Modern Physics. Cengage Learning, Boston (2018)
26. Dubois, D.J.: Fuzzy Sets and Systems: Theory and Applications. Academic Press, Cambridge (1980)

Hesitant Fuzzy Linear Fractional Programming Problem

Madineh Farnam[1] and Majid Darehmiraki[2](✉)

[1] Department of Electrical Engineering, Shohadaye Hoveizeh University of Technology, Dasht-e Azadegan, Khuzestan, Iran
m.farnam@shhut.ac.ir

[2] Department of Mathematics, Behbahan Khatam Alanbia University of Technology, Behbahan, Khuzestan, Iran
darehmiraki@bkatu.ac.ir

Abstract. In last decade hesitant fuzzy theory introduced as an extension of fuzzy theory which is a powerful tool in situations that we have hesitations in recognizing the imprecise membership degree of the elements of a set. There are not many studies on hesitant fuzzy linear fractional programming (HFLFP) problems; thus in this study we investigate this kind of problems. For this aim, firstly we define all kind of symmetric hesitant fuzzy linear fractional programming problems. Secondly, we generalize the Bellman and Zadeh theory for solving this kind of problem. Thirdly we propose a new approach to solve one kind of HFLFP. Finally, a numerical example is given to show the effectiveness of our approach.

Keywords: Hesitant fuzzy set · Linear fractional programming problem · Hesitant linear fractional programming problem

1 Introduction

Linear fractional programming problem (LFPP) is accepted as one of more applicant generalization of linear programming. Due to the widespread use of the ratio of two quantities in sciences such as physics, economics, engineering and etc. linear fractional programming problems have become particularly important. LFPP can be modeled as follows:

$$(LFPP)\begin{cases} \max \frac{c^T x+\alpha}{d^T x+\beta} \\ s.t \quad Ax \le b \\ \qquad x \ge 0 \end{cases} \tag{1}$$

where $x, c, d \epsilon R^n$, $A \epsilon R^{m \times n}$ and $\alpha, \beta \epsilon R$.

In the model (1) the type of decision variables is certain and deterministic. Charnes and Cooper [1], defined transformation of variables techniques, and then solved equivalent linear program with that. For more conformity with real models, the idea of fuzzy

T. Allahviranloo et al. (Eds.): IDS 2020, AISC 1301, pp. 864–872, 2021.
https://doi.org/10.1007/978-3-030-66501-2_70

set theory that stated by Bellman and Zadeh [2] used by many researchers [3] in LFPP. The general FLFP format can be presented as follows:

$$(FLFPP)\begin{cases} \max \frac{\tilde{c}^T \tilde{x} + \tilde{\alpha}}{\tilde{d}^T \tilde{x} + \tilde{\beta}}, \\ s.t \;\; \tilde{A}\tilde{x} \lesssim \tilde{b}, \\ \tilde{x} \geq 0, \end{cases} \quad (2)$$

where $\tilde{c}, \tilde{d}, \tilde{\alpha}, \tilde{\beta}, \tilde{A}$ and $\tilde{b}$ are fuzzy sets and $\tilde{x}$ presents the vector of fuzzy decision variables. Operator $\lesssim$ is related to fuzzy constraints. Many authors considered various kind of FLFP. In some researches they did not consider fuzzy sets for all data, for example only the values of right hand sides or decision variables are fuzzy [4–7]. Of course, there are several researches which in them all numbers are fuzzy [8–10].

In recent years, as a useful extension of fuzzy sets, HFS stated by Tora and Narukawa [11, 12] to explain that kind of situations which decision makers have hesitation among some values to evaluate the various parameters.

HFSs can provide the proper and perfect structure for modeling problems in decision environments. So general model of the hesitant fuzzy linear fractional programming problem can be stated as follows:

$$(HFLFPP)\begin{cases} \max \frac{\approx{c}^T \approx{x} + \approx{\alpha}}{\approx{d}^T \approx{x} + \approx{\beta}} \\ s.t \;\; \approx{A}\approx{x} \lessapprox \approx{b}, \\ \approx{x} \geq 0, \end{cases} \quad (3)$$

where $\approx{c}, \approx{d}, \approx{\alpha}, \approx{\beta}, \approx{A}$ and $\approx{b}$ are hesitant fuzzy sets, $\approx{x}$ indicates the vector of decision variables that are hesitant fuzzy sets and the symbol $\lessapprox$ shows the hesitant fuzzy constraint. In this study we propose a solution for HFLFPP models. So the rest of this study has been constructed as follows: In Sect. 2, some related definitions and conceptions are reviewed. In Sect. 3, we review concepts of LFPP, then we proposed symmetric hesitant fuzzy linear fractional programming. In Sect. 4, we state our approach for solving one case of the symmetric hesitant fuzzy linear fractional programming models which proposed in previous section. In Sect. 5, a numerical example in HFLFPP is solved to show efficiency of the approach. Finally, in Sect. 6, some important conclusions of generalization method are presented.

2 Preliminaries and Notations

This section involves some basic notions of hesitant fuzzy set and its operations.

Definition 2.1. Xia and Xu in [13] defined the HFS as follows:

$$H = \{\langle x, h_H(x)\rangle | x \in X\},$$

where $h_H(x)$ is a set of values in [0, 1] indicating the possibe membership degree S of element like x belong to X to the set H. Most of the time, $h_H(x)$ is named an HFE. They defined some of operations on HFE as follows:

a) $h_1(x) \cup h_2(x) = \cup_{\gamma_1 \epsilon h_1(x), \gamma_2 \epsilon h_2(x)} max\{\gamma_1 , \gamma_2\}$,
b) $h_1(x) \cap h_2(x) = \cap_{\gamma_1 \epsilon h_1(x), \gamma_2 \epsilon h_2(x)} min\{\gamma_1 , \gamma_2\}$,
c) $(h_1(x))^\lambda = \cup_{\gamma_1 \epsilon h_1(x)}\{\gamma_1^\lambda\}$,
d) $\lambda(h_1(x)) = \cup_{\gamma_1 \epsilon h_1(x)}\{1 - (1 - \gamma_1)^\lambda\}$.

In next definition we generalize Bellman and Zadeh theory which need to state a decision in hesitant fuzzy spaces [2].

Definition 2.2 [14]. Let we have a hesitant fuzzy objective function $\overset{\approx}{G}$ and a hesitant constraint $\overset{\approx}{C}$ in X, which means the combintion of $\overset{\approx}{G}$ and $\overset{\approx}{C}$ make a decition like $\overset{\approx}{D}$, as a hesitant fuzzy decision caused by the intersection of $\overset{\approx}{G}$ and $\overset{\approx}{C}$. In this case, $\overset{\approx}{D} = \overset{\approx}{G} \cap \overset{\approx}{C}$ is corresponded to $\tilde{h}_{\underset{D}{\approx}} = \tau\left(\tilde{h}_{\underset{G}{\approx}}, \tilde{h}_{\underset{C}{\approx}}\right)$, such that:

$$\tilde{h}_{\underset{C}{\approx}} = \left\{\tilde{h}^1_{\underset{C}{\approx}}, \tilde{h}^2_{\underset{C}{\approx}}, \ldots, \tilde{h}^{q_C}_{\underset{C}{\approx}}\right\} \quad , \quad \tilde{h}_{\underset{G}{\approx}} = \left\{\tilde{h}^1_{\underset{G}{\approx}}, \tilde{h}^2_{\underset{G}{\approx}}, \ldots, \tilde{h}^{q_G}_{\underset{G}{\approx}}\right\},$$

where the numbers of the decision makers that establish different aspiration level for the goal and constraints are shown by q_G and q_C, respectively. Furthermore τ as a T-norm is applied to detemine the membership degree values of intersectin the hesitant fuzzy elements. This idea can extend for multi objective programming.

Definition 2.3 [15]. Let $\tau : \mathrm{H}^{(\mathrm{m})} \times \mathrm{H}^{(\mathrm{m})} \rightarrow \mathrm{H}^{(\mathrm{m})}$, that $\mathrm{H}^{(\mathrm{m})}$ is a hesitant fuzzy set with m elements and τ is a hesitant triangular norm such that, if $\mathrm{h}_1, \mathrm{h}_2, \mathrm{h}_3 \,\epsilon \mathrm{H}^{(\mathrm{m})}$then the following axioms hold:

a) *commutative property*: $\tau(h_1, h_2) = \tau(h_2, h_1)$,
b) *associative property*: $\tau(h_1, \tau(h_2, h_3)) = \tau(\tau(h_1, h_2), h_3)$,
c) *monotone property*: *if* $h_2 \leq_{H^{(m)}} h_3$ *then* $\tau(h_1, h_2) \leq_{H^{(m)}} \tau(h_1, h_3)$,
d) *neutral element*: $\tau\left(h_1, 1_{H^{(m)}}\right) = h_1$,

where $1_{H^{(m)}} = \{1, 1, \ldots, 1\}$ is a full HFE with m elements.

In this study, for the hesitant triangular norm on HFE we use the minimum operator as follows:

$$\tau\left(\tilde{h}_1, \tilde{h}_2\right) = \bigcup_{\tilde{\gamma}_1 \epsilon \tilde{h}_1, \tilde{\gamma}_2 \epsilon \tilde{h}_2} min\{\tilde{\gamma}_1, \tilde{\gamma}_2\}$$

3 LFPP and Symmetric HFLFPP Model

In the present section, at first accordance to classical LFPP, some necessary concepts and theorems reviewed, then we formulate the symmetric hesitant fuzzy linear fractional programming problems.

3.1 LFPP

The typical format of LFPP is defined in (1). For solving the classic format of LFPP that defined in (1) Charnes and Cooper [1] used the variable transformation technique. By defining $y = tx$ where $t \geq 0$, when the feasible solution set is nonempty and bounded, the problem (1) becomes to equivalent linear programming as follows:

$$\begin{cases} \max c^T y + \alpha t \\ s.t \quad Ay - bt \leq 0 \\ d^T y + \beta t = \gamma \\ y, t \geq 0, \gamma \neq 0 \end{cases} \tag{4}$$

Lemma [1]: For every (y, t) where satisfied in the constraints of model (4), we have, $t > 0$.

Theorem [1]: If x^* is an optimum solution of the problem (1), such that $sgn(\gamma) = sgn\left(d^T x^* + \beta t\right) > 0$ and (y^*, t^*) is an optimum solution of the problem (4), then $\frac{y^*}{t^*}$ is an optimum solution of the problem (1).

Theorem [16]: Assume that there is not any solution point such $(y, 0)$ with $y \geq 0$ for model (4). Also, let denominator is positive in x, then problems (1) and (4) are equivalent.

Remark: If the sign of denominator in optimum point for model (1) is negative, then we have similar problem such as model (4) with some partial variations [1].

Notice: In this study we assume that the sign of denominator in optimum point is positive.

Corollary [1]: For every nonempty and bounded set of feasible solutions of the problem (1), we can solve the following equivalent problem:

$$\begin{cases} \max c^T y + \alpha t \\ s.t \quad Ay - bt \leq 0 \\ d^T y + \beta t = 1 \\ y, t \geq 0 \end{cases} \tag{5}$$

4 Symmetric Hesitant Fuzzy Linear Fractional Programming Problem (SHFLFPP)

A typical structure of the HFLFP problem can be expressed in the model (3) which is included various cases. In this study we want to explain these kind of problems that are defining such a symmetric HFLFP problem. If several decision makers evaluate the various aspiration level for the objective function which they expect to achieve. Also the value of the right hand side of constraints consists of HFS. So, we introduce the following SHFLFPP:

$$(SHLFPP)\begin{cases} \widetilde{\widetilde{\max}} \frac{c^T x + \alpha}{d^T x + \beta} \\ s.t \quad Ax \leq \overset{\approx}{b}, \\ x \geq 0. \end{cases} \tag{6}$$

We assume that the quantity of the objective function is more than or equal to expected aspiration level $\overset{\approx}{b}_0$.

Hesitant fuzzy cases of the objective function can be included as follows:

i) The numerator is hesitant fuzzy, while the denominator is deterministic
ii) The numerator and the denominator both are hesitant fuzzy set
iii) The numerator is deterministic, while the denominator is hesitant fuzzy set

5 An Approach for Solving Symmetric Hesitant Fuzzy Linear Fractional Programming Problem (SHFLFPP)

Due to, the differences among cases (i), (ii) and (iii) the method for solving them will be different in each cases. We explain a new approach for solving case (i), by transformation $y = tx$ where $t \geq 0$, the symmetric HFLFP problem (6), when the sign of denominator in optimally is positive, can be converted to following HFLP problem:

$$(SHFLPP)\begin{cases} \widetilde{\widetilde{\max}} c^T y + \alpha t \\ s.t \quad Ay - \overset{\approx}{b} t \leq 0, \\ d^T y + \beta t = 1, \\ y, t \geq 0. \end{cases} \tag{7}$$

whenever decision variables have the following properties:

$$\begin{cases} a_i^T y - \overset{\approx}{b}_i t \leq t_i,\ t_i \leq p_i^{q_i} \, for\, i = 1, 2, \ldots, m \\ d^T y + \beta t = 1. \end{cases} \tag{8}$$

It should be noted that $p_i^{q_i}$ is the maximal admissible value of violation for t_i in the i-th constraint that is determined by q_i –th decision maker.

Then all decision variables are feasible in model (6).

Assume that r_{num} is the number of decision makers with various assessments such as $p_{num}^{q_{num}}$ $(q_{num} = 1, 2, \ldots, r_{num})$ that can establish the different aspiration levels for the goal with expected value $\overset{\approx}{b}_0$. Also $\overset{\approx}{b}_i$ denotes the i-th row of the $\overset{\approx}{b}$ and $p_i^{q_i}$ is the admissible violation respected to i-th row for model (6) that is determined by q_i –th decision maker which r_i is the number of decision makers that can record the different violation for i-th constraint.

The membership functions $\tilde{h}_{num}^{q_{num}}$ $(q_{num} = 1, 2, \ldots, r_{num})$ and $\tilde{h}_i^{q_i}$ $(q_i = 1, 2, \ldots, r_i)$, for $i = 1, 2, \ldots, m$ are defined respectively as follows:

$$\tilde{h}_{num}^{q_{num}}(y) = \begin{cases} 1 - \frac{t_{num}}{p_{num}^{q_{num}}} & c^T y + \alpha t = b_0 - t_{num}, \quad t_{num} \geq 0 \\ 1 & c^T y + \alpha t \geq b_0 \end{cases} \tag{9}$$

$$\tilde{h}_i^{q_i}(y) = \begin{cases} 1 - \frac{t_i}{p_i^{q_i}} & a_i^T y - b_i t = t_i, t_i \geq 0 \\ 1 & a_i^T y - b_i t \leq 0 \end{cases} \tag{10}$$

Now by using the extension of Bellman and Zadeh theory for decision space we have:

$$\overset{\approx}{D} = \overset{\approx}{G} \cap \overset{\approx}{C}.$$

Where $\overset{\approx}{G} = \left\{\tilde{h}^1_{num}, \tilde{h}^2_{num}, \ldots, \tilde{h}^{r_{num}}_{num}\right\}, \overset{\approx}{C} = \underset{1}{\overset{\approx}{C}} \cap \underset{2}{\overset{\approx}{C}} \cap \ldots \cap \underset{m}{\overset{\approx}{C}} = \left\{\tilde{h}^1_1, \tilde{h}^2_1, \ldots, \tilde{h}^{r_1}_1\right\} \cap$ $\left\{\tilde{h}^1_2, \tilde{h}^2_2, \ldots, \tilde{h}^{r_2}_2\right\} \cap \ldots \cap \left\{\tilde{h}^1_m, \tilde{h}^2_m, \ldots, \tilde{h}^{r_m}_m\right\}$.

According to Definition 2.2 we can obtain the hesitant fuzzy decision of case (i):

$$\tilde{h}_{\overset{\approx}{D}} = \left\{h^1_{\overset{\approx}{D}}, h^2_{\overset{\approx}{D}}, \ldots, h^{r_{num}r_1r_2\ldots r_m}_{\overset{\approx}{D}}\right\}$$

This set consists of all decision spaces in $\overset{\approx}{D}$. So we should solve $r_{num}r_1r_2\ldots r_m$ sub problems by defining $\lambda^s(s = 1, 2, \ldots, r_{num}r_1r_2\ldots r_m)$, that related to $h^s_{\overset{\approx}{D}}$ of $\tilde{h}_{\overset{\approx}{D}}$.

Thus for $s = 1, 2, \ldots, r_{num}r_1r_2\ldots r_m$ we have:

$$LP_s = \begin{cases} \max \lambda^s & \\ \lambda^s p_{num}{}^{q_{num}} + t_{num}{}^s \leq p_{num}{}^{q_{num}} & \\ \lambda^s p_i{}^{q_i} + t_i{}^s \leq p_i{}^{q_i} & i = 1, 2, \ldots, m \\ c^T y^s + \alpha t^s + t_{num}{}^s \geq b_0 & \\ a_i{}^T y^s - \overset{\approx}{b}_i t^s - t_i{}^s \leq 0 & i = 1, 2, \ldots, m \\ d^T y^s + \beta t^s = 1 & \\ t_{num}{}^s \leq p_{num}{}^{q_{num}} & \\ t_i{}^s \leq p_i{}^{q_i} & i = 1, 2, \ldots, m \\ \lambda^s, y^s, t^s, t_{num}{}^s, t_i{}^s \geq 0,\ i = 1, 2, \ldots, m & \end{cases} \tag{11}$$

Solutions of (11) and (7) are equivalent. If $t \neq o$, then $x^{*s} = \frac{y^{*s}}{t^{*s}}$ is the optimal point for (6). λ^{*s} in (11) denotes the maximal membership degree of expected aspiration level which goal and constraint can be satisfied in s-ih sub problem with $x^{*s} = \left(x^{*s}_1, x^{*s}_2, \ldots, x^{*s}_n\right)$. Therefore, by solving $r_{num}r_1r_2\ldots r_m$ LP problems the optimal point for (6) can be proposed with $\left\{x^{*1}, x^{*2}, \ldots, x^{*r_{num}r_1r_2\ldots r_m}\right\}$ in hesitant fuzzy environment. The comparison among LFP, FLFP and SHFLFP is provided in Table 1.

Table 1. The comparison among LFP, FLFP and SHFLFP.

Type of problem	Number of variables	Number of constraints	Number of LP problems
LFP	n + 1	m + 1	1
FLFP (case-*ii*)	n + m+3	3 m + 4	1
SHFLFP (case-*ii*)	n + m+3	3 m + 4	$r_{num}r_1r_2\ldots r_m$

6 Numerical Example

The efficiency of the proposed approach demonstrates in a numerical example. The results for every sub problem is calculated by using the MATLAB software.

Example 1. Find the optimum solution of following SHFLFPP:

$$\begin{cases} \widetilde{\widetilde{\max}} \frac{6x_1+3x_2}{6x_1+2x_2+1} \\ s.t \quad 3x_1 + 6x_2 \leq \tilde{\tilde{18}}, \\ 6x_1 + 2x_2 \leq \tilde{\tilde{12}}, \\ x_1, x_2 \geq 0, \end{cases} \tag{12}$$

where $b_0 = 2, p^1_{num} = 4, p^2_{num} = 3, p^1_1 = 5, p^2_1 = 6, p^1_2 = 3$ and $p^2_2 = 2$.
By transformation $y = tx$ where $t \geq 0$ model (12) becomes to:

$$\begin{cases} \widetilde{\widetilde{\max}} 6y_1 + 3y_2 \\ s.t \quad 3x_1 + 6x_2 - \tilde{\tilde{18}}t \leq 0, \\ 6y_1 + 2y_2 - \tilde{\tilde{12}}t \leq 0, \\ 6y_1 + 2y_2 + t = 1, \\ y_1, y_2, t \geq 0. \end{cases} \tag{13}$$

After calculating membership functions for goal and constraints, we have:
$\tilde{\tilde{G}} = \left\{\tilde{h}^1_{num}, \tilde{h}^2_{num}\right\}$ and $\tilde{\tilde{C}} = \tilde{\tilde{C}}_1 \cap \tilde{\tilde{C}}_2 = \left\{\tilde{h}^1_1, \tilde{h}^2_1\right\} \cap \left\{\tilde{h}^1_2, \tilde{h}^2_2\right\}$.
So we achieve 8 fuzzy subspaces:

$$h^1_{\underset{D}{\approx}} = \left\{\tilde{h}^1_{num}, \tilde{h}^1_1, \tilde{h}^1_2\right\}, h^2_{\underset{D}{\approx}} = \left\{\tilde{h}^1_{num}, \tilde{h}^1_1, \tilde{h}^2_2\right\},$$

$$h^3_{\underset{D}{\approx}} = \left\{\tilde{h}^1_{num}, \tilde{h}^2_1, \tilde{h}^1_2\right\}, h^4_{\underset{D}{\approx}} = \left\{\tilde{h}^1_{num}, \tilde{h}^2_1, \tilde{h}^2_2\right\},$$

$$h^5_{\underset{D}{\approx}} = \left\{\tilde{h}^2_{num}, \tilde{h}^1_1, \tilde{h}^1_2\right\}, h^6_{\underset{D}{\approx}} = \left\{\tilde{h}^2_{num}, \tilde{h}^1_1, \tilde{h}^2_2\right\},$$

$$h^7_{\underset{D}{\approx}} = \left\{\tilde{h}^2_{num}, \tilde{h}^2_1, \tilde{h}^1_2\right\}, h^8_{\underset{D}{\approx}} = \left\{\tilde{h}^2_{num}, \tilde{h}^2_1, \tilde{h}^2_2\right\}.$$

According to decision space we have 8 sub problem that the result of solving them summarized in Table 2.

Table 2. The result for the Example 1.

LP_s	λ^{*s}	$y^{*s} = \left(y_1^{s*}, y_2^{s*}, t^{s*}, t_{num}^{s*}, t_1^{s*}, t_2^{s*}\right)$	$x^{*s} = \left(x_1^{s*}, x_2^{s*}\right)$
LP_1	0.8361	(0, 0.4481, 0.1038, 0.6557, 0.8197, 0)	(0, 4.3170)
LP_2	0.8361	(0, 0.4481, 0.1038, 0.6557, 0.8197, 0)	(0, 4.3170)
LP_3	0.8387	(0, 0.4516, 0.0968, 0.6452, 0.9677, 0)	(0, 4.6613)
LP_4	0.8387	(0, 0.4516, 0.0968, 0.6452, 0.9677, 0)	(0, 4.6613)
LP_5	0.7872	(0, 0.4539, 0.0922, 0.6383, 1.0638, 0)	(0, 4.9230)
LP_6	0.7872	(0, 0.4539, 0.0922, 0.6383, 1.0638, 0)	(0, 4.9230)
LP_7	0.7917	(0, 0.4583, 0.0833, 0.6250, 1.2500, 0)	(0, 5.5018)
LP_8	0.7917	(0, 0.4583, 0.0833, 0.6250, 1.2500, 0)	(0, 5.5018)

7 Conclusion

In this study, an approach was presented for solving one cases of SHFLFPP. This method can extend to other cases and apply to multi objective hesitant linear fractional programming problem. Also this approach can use to practical researches such as the production programming and the business management. Simplicity of calculation and proper process are the most advantages of the proposed method.

References

1. Charnes, A., Cooper, W.W.: Programming with linear fractional functions. Nav. Res. Log. Q. **9**, 181–186 (1962)
2. Bellman, R.E., Zadeh, L.A.: Decision-making in a fuzzy environment. Manage. Sci. **17**, 141–164 (1970)
3. Dutta, D., Rao, J.R., Tiwari, R.N..: Sensitivity analysis in fuzzy linear fractional programming problem. Fuzzy Sets Syst. **48**, 211–216 (1992)
4. Chakraborty, M., Gupta, S.: Fuzzy mathematical programming for multi objective linear fractional programming problem. Fuzzy Sets Syst. **125**, 335–342 (2002)
5. Dehghan, M., Hashemi, B., Ghatee, M.: Computational methods for solving fully fuzzy linear systems. Appl. Math. Comput. **179**, 328–343 (2006)
6. Mahdavi- Amiri, N., Nasseri, S.H..: Duality in fuzzy number linear programming by use of a certain linear ranking function. Appl. Math. Comput. **180**, 206–216 (2006)
7. Safaei, N.: A new method for solving fully fuzzy linear fractional programming with a triangular fuzzy numbers. Int. J. Appl. Math. Comput. **3**, 273–281 (2014)
8. Pop, B., Stancu-Minasian, L.M.: A method of solving fully fuzzified linear fractional programming problems. J. Appl. Matematcs Comput. **27**, 227–242 (2008)
9. Lotfi, F.H., Allahviranloo, T., Jondabeh, M.A., Alizadeh, L.: Solving a fully fuzzy linear programming using lexicography method and fuzzy approximate solution. Appl. Math. Model. **33**, 3151–3156 (2009)
10. Maleki, H.R., Tata, M., Mashinchi, M.: Linear programming with fuzzy variables. Fuzzy Sets Syst. **109**, 21–33 (2000)
11. Torra, V., Narukawa, Y.: On hesitant fuzzy sets and decision. In: The 18-th IEEE International Conference on Fuzzy Systems, Jeju Island, Korea, pp. 1378–1382 (2009)

12. Torra, V.: Hesitant Fuzzy Sets. Int. J. Intell. Syst. **25**, 529–539 (2010)
13. Xia, M., Xu, Z.: Hesitant fuzzy information aggregation in decision making. Int. J. Approximation Reasoning **52**(3), 395–407 (2011)
14. Ranjbar, M., Effati, S.: Symmetric and right-hand-side hesitant fuzzy linear programming. IEEE Trans. Fuzzy Syst. **28**(2), 215–227 (2020)
15. Santos, H.S., Bedregal, B.R., Santiago, R.H., Bustince, H., Tartas, E.B.: Construction of typical hesitant triangular norms regarding Xu-Xia-partial order, In: IFSA-EUSFLAT, pp. 953–959 (2015)
16. Chakraborty, M., Gupta, S.: Fuzzy mathematical programming for multi objective linear fractional problem. Fuzzy Sets Syst. **125**, 33–342 (2002)

Improved Predictor Corrector Scheme for Solving Vasicek Interest Rate Model in Uncertain Environment

S. S. Mansouri[1], M. Gachpazan[2], and N. Ahmady[1(✉)]

[1] Department of Mathematics, Varamin-Pishva Branch, Islamic Azad University, Varamin, Iran
n.ahmadi@iauvaramin.ac.ir
[2] Department of Applied Mathematics, School of Mathematical Sciences, Ferdowsi University of Mashhad, Mashhad, Iran

Abstract. Uncertain differential equation (UDE) is a type of differential equations driven by canonical Liu's process.In this paper, we study a new numerical method for solving UDEs the improved predictor corrector. The convergence and stability of the proposed method are presented in details. Furthermore, the Vasicek interest rate model is given for numerical result.

Keywords: Fuzzy canonical liu's process · Improved predictor corrector method · Vasicek

1 Introduction

Most of phenomena and events in the real world occur unexpectedly among which are the changes in economic and political systems, collapse of governments, conflicts between tribes, wars, terrorist attacks. Thus, it is not possible to anticipate or estimate, the price of stocks, valuable papers, monetary units and precious metals accurately. Therefore, the only way find out how this factor can affect the growth or drop in the value of companies is focusing on the price of stocks.

Investigation on effects of the factors along with uncertainty theory can help better understanding and more exact modeling of these phenomena. The uncertainty theory was first introduced by Liu who then presented the concept of uncertainty measure which is powerful tool for dealing with uncertain phenomena, to facilitate measuring of uncertain events that are based on normality, monotonicity, self-duality, and maximality axioms.

Then the concept of uncertain process was proposed by Liu, [7], that introduces a particular uncertain process with stationary and independent increment named canonical Liu's process which is just like a stochastic process described by Brownian motion.

Since then some literatures have been published on the canonical Liu's process and its applications in other sciences, such as economics and optimal control have been published [28]. Then Liu was inspired by stochastic notions and ito process to introduce uncertain differential equations [10] which were driven by canonical Liu's process for better understanding the uncertain phenomena.

T. Allahviranloo et al. (Eds.): IDS 2020, AISC 1301, pp. 873–888, 2021.
https://doi.org/10.1007/978-3-030-66501-2_71

Regarding to the importance of existence and uniqueness of a solution to uncertain differential equations driven by canonical Liu's process, Liu investigated the existence and uniqueness of solution to the uncertain differential equations by employing Lipschitz and Linear growth conditions [17], and stability analysis of uncertain differential equations was given by Yao et al. [23, 24]. Many researchers managed to find analytic solutions for some special types of uncertain differential equations such as Chen and Liu [1].

From many aspects, it is a difficult task to find analytic solutions of uncertain differential equations. Therefore, it is of much importance to design some numerical methods for solving uncertain differential equations.

In the last few years, many works have been performed by several authors in numerical solutions of uncertain differential equations. As some examples to mention, Yao and Shen [19, 22] have investigated the numerical solution of uncertain differential equations by using Euler and Runge-Kutta methods respectively. Also, the numerical solutions of uncertain differential equations by Adams method [21] and Adams-Simpson method [20] have been studied.

In this paper, we aim to obtain a new numerical method for UDEs, which is more accurate than the one in [20, 21]. First, we investigate concept of α-path to UDE, which is a type of certain function that solves an associate ordinary differential equation. Then, an improved predictor corrector (IPC) method is designed to solve UDEs, which essentially solves each α-path and produces an inverse uncertainty distribution of the solution. The IPC method is generated by combining an explicit three-step method and an implicit two-step method.

The remainder of this paper is organized as follows: Sect. 2 is intended to introduce some basic concepts and theorems in uncertainty theory and uncertain differential equation. An explicit three-step method, an implicit two-step method for solving UDEs and, the IPC three-step algorithm are proposed. we introduced a brief survey of fundamental concepts from financial mathematics and some numerical examples are presented in Sect. 5. The last section contains a brief summary.

2 Preliminaries

In this section, we introduce some basic concepts and theorems about uncertainty theory and uncertain differential equation driven by canonical Liu's process, which are used throughout this paper.

2.1 Credibility Theory

Suppose that Θ is a non-empty set and $\mathcal{P}$ is the power set of Θ. Each element of $\mathbf{A}$ in $\mathcal{P}$is called an event. In order to present an axiomatic definition of uncertainty, it is necessary to assign a number $\mathbf{M}\{\mathbf{A}\}$ to each event $\mathbf{A}$ which indicates the uncertainty that $\mathbf{A}$ will occur. To ensure that the number $\mathbf{M}\{\mathbf{A}\}$ has certain mathematical properties which we intuitively expect to have an uncertainty, we accept the following four axioms [7]:

1. Axiom (Normality) $\mathbf{M}\{\Theta\} = 1$.
2. Axiom (Monotonicity) $\mathbf{M}\{\mathbf{A}\} \leq \mathbf{M}\{\mathbf{B}\}$ whenever $\mathbf{A} \subset \mathbf{B}$.

3. Axiom (Self-Duality) $\mathbf{M}\{\mathbf{A}\} + \mathbf{M}\{\mathbf{A}^c\} = 1$ for any event $\mathbf{A}$.
4. Axiom (Maximality) $\mathbf{M}\{\mathbf{U}_i\mathbf{A}_i\} = \sup_i\mathbf{M}\{\mathbf{A}_i\}$ for any events $\{\mathbf{A}_i\}$ with

$$\sup_i\mathbf{M}\{\mathbf{A}_i\} < 0.5.$$

Definition 2.1 [7]. The set function $\mathbf{M}$ is called an uncertainty measure if it satisfies the normality, monotonicity, self-duality, and maximality axioms.

A family $\mathcal{P}$ with these four properties is called a σ-algebra. The pair $(\Theta, \mathcal{P})$ is called a measurable space, and the elements of $\mathcal{P}$ is afterwards called $\mathcal{P}$-measurable sets instead of events.

Definition 2.2 [7]. Let Θ be a nonempty set, $\mathcal{P}$ the power set of Θ, and $\mathbf{M}$ a uncertainty measure. The triple $(\Theta, \mathcal{P}, \mathbf{M})$ is called an uncertainty space.

Let $(\Theta, \mathcal{P}, \mathbf{M})$ be an uncertainty space. A filtration is a family $\{\mathcal{P}_t\}_{t\,\geq\,0}$ of increasing sub-σ-algebras of $\mathcal{P}$ (i.e. $\mathcal{P}_t \subset \mathcal{P}_s \subset \mathcal{P}$ for all $0 \leq t < s < \infty$). The filtration is said to be right continuous if $\mathcal{P}_t = \bigcap_{s>t} \mathcal{P}_s$for all $t \leq 0$. When the uncertainty space is complete, the filtration is said to satisfy the usual conditions if it is right continuous and $\mathcal{P}_0$ contains all $\mathbf{M}$-null sets.

We also define $\mathcal{P}_\infty = \sigma\left(U_{t\,\geq\,0}\mathcal{P}_t\right)$ (i.e. σ-algebra generated by $U_{t\geq 0}\mathcal{P}_t$.)

$\mathcal{P}$-measurable uncertain variable is denoted by $\mathbf{L}^p\left(\Theta, \mathbf{R^d}\right)$ that will be defined later.

A process is called $\mathcal{P}$-adapted, if for all $t \in [0, t]$ the uncertain variable $x(t)$ is $\mathcal{P}$-measurable.

Definition 2.3 [7]. An uncertain variable is defined as a (measurable) function

$$\xi : (\Theta, \mathcal{P}, \mathbf{M}) \rightarrow \mathbf{R}.$$

Definition 2.4 [7]. Let ξ be an uncertain variable. Then the expected value of ξ is defined by $\mathbf{E}[\xi] = \int_0^{+\infty} \mathbf{M}\{\xi \geq r\}\mathrm{dr} - \int_{-\infty}^0 \mathbf{M}\{\xi \leq r\}\mathrm{dr}$

provided that at least one of the two integrals is finite. Furthermore, the variance is defined by $\mathbf{E}[\left(\xi - e\right)^2]$.

Let ξ and η be independent uncertain variables with finite expected values. Then for any numbers a and b, we have

$$\mathbf{E}[a\xi + b\eta] = a\mathbf{E}[\xi] + b\mathbf{E}[\eta].$$

Definition 2.5 [20]. The uncertainty distribution $\mu(x)$ of an uncertainty variable ξ is defined by

$$\mu(x) = \max\{1, 2\mathbf{M}(\xi = x)\}, x \in \mathbf{R}.$$

Definition 2.6 [20]. An uncertainty distribution $\mu(x)$ is said to be regular if it is a continuous and strictly increasing function with respect to x at which $0 < \mu(x) < 1$, and

$$\lim_{x\to-\infty}\mu(x) = 0, \lim_{x\to+\infty}\mu(x) = 1.$$

In addition, the inverse function $\mu^{-1}(\alpha)$ is called the inverse uncertainty distribution of ξ. In the following expression, we use the notation $\mathbf{x}(t)$ instead of $\mathbf{x}_t(\theta)$.

An uncertain process is essentially a sequence of uncertain variables indexed by time or space. As one of the most important types of uncertain processes, the canonical Liu's process is defined as follows.

Definition 2.7 [11]. An uncertain process $\mathbf{C}_t$ is said to be a canonical Liu's process if

1. $\mathbf{C}_0 = 0$,
2. $\mathbf{C}_t$ has stationary and independent increments,
3. every increment $\mathbf{C}_{t+s} - \mathbf{C}_s$ is a normally distributed uncertain variable with expected value **et** and variance $\sigma^2 t^2$ whose membership function is

$$\mu(x) = (1 + \exp\left(\frac{\pi|e - x|}{\sqrt{3}\sigma}\right))^{-1}, -\infty < x < +\infty.$$

The parameters e and σ are called the drift and diffusion coefficients, respectively, in addition, canonical Liu's process is said to be standard if $e = 0$ and $\sigma = 1$.

Based on canonical Liu's process, Liu integral is defined as an uncertain counterpart of Ito integral as follows.

Definition 2.8 [21]. Suppose that $x(t)$ is an uncertain process and $\mathbf{C}_t$ is a canonical Liu's process. For any partition of closed interval $[a, b]$ with $a = t_1 < t_2 < \ldots < t_{k+1} = b$, the mesh is written as
$\Delta = \max_{1 \leq i \leq k} |t_{i+1} - t_i|$.

Then the Liu integral of $x(t)$ with respect to $\mathbf{C}_t$ is
$\int_a^b x(t) d\mathbf{C}_t = \lim_{\Delta \to 0} \sum_{i=1}^{k} x(t_i).(\mathbf{C}_{t_{i+1}} - \mathbf{C}_{t_i})$
provided that the limit exists almost surely and is an uncertain variable.

2.2 Uncertain Differential Equation

In 2008, Liu [15] presented a type of uncertain differential equations driven by the canonical Liu's process.

Definition 2.9 [15]. Suppose $\mathbf{C}_t$ is a canonical Liu's process and f, g are some given functions,

$$dx(t) = f(x(t), t)dt + g(x(t), t)d\mathbf{C}_t \tag{1}$$

is called an uncertain differential equation with an initial value $x(0)$. The solution is uncertain process $x(t)$ that satisfies (1) identically in t.

Gao [4] proved the existence and uniqueness theorem of solution of uncertain differential equation under linear growth condition and Lipschitz continuous condition.

Theorem 2.1 [1]. The uncertain differential equation

$$dx(t) = f(x(t), t)dt + g(x(t), t)d\mathbf{C}_t$$

has a unique solution if the coefficients $f(t, x)$ and $g(t, x)$ satisfy the linear growth condition

$$|f(t, x)| + |g(t, x)| \leq \mathbf{L}(1 + |x|)\forall x \in \mathbf{R}, t \geq 0$$

and Lipschitz condition

$$|f(t, x) - f(t, y)| + |g(t, x) - g(t, y)| \leq \mathbf{L}|x - y|, \forall x, y \in \mathbf{R}, t \geq 0$$

for some constant $\mathbf{L}$. Moreover, the solutionis sample continuous.

Definition 2.10 [22]. The α-path $(0 < \alpha < 1)$ of an uncertain differential equation

$$dx(t) = f(x(t), t)dt + g(x(t), t)d\mathbf{C}_t$$

with initial value $x(0)$ is a deterministic function $x^{\alpha}(0)$ with respect to t that solves the corresponding ordinary differential equation

$$dx^{\alpha}(t) = f\left(x^{\alpha}(t), t\right)dt + \left|g\left(x^{\alpha}(t), t\right)\right|\mu^{-1}(\alpha)dt \tag{2}$$

where

$$\mu^{-1}(\alpha) = \frac{\sqrt{3}}{\pi}\ln\left(\frac{\alpha}{1-\alpha}\right).$$

Theorem 2.2 [22]. Let $x(t)$ and $x^{\alpha}(t)$ be the solution and α-path of uncertain differential Eq. (1), respectively. Then

$$\mathbf{M}\left\{x(t) \leq x^{\alpha}(t), \forall t\right\} = \alpha,$$

$$\mathbf{M}\{x(t) > x^{\alpha}(t), \forall t\} = 1 - \alpha.$$

3 Improved Predictor Corrector Method for Uncertain Differential Equation

Theorem 2.2 can be used to numerically solve the uncertain differential equation Eq. (1) by applying any suitable numerical method for ODEs to Eq. (2). The Improved Predictor-Corrector (IPC) method is generated by combining an explicit three-step method and an implicit two-step method. This is also an improvement of [20, 21] which is considered in the Adams and Adams-Simpson methods, respectively for Eq. (2). We first introduce IPC method for ordinary differential equations and then apply this method for UDEs.

3.1 Improved Predictor Corrector Method

For an ordinary differential equation with initial value $x(0)$

$$dx(t) = f(t, x(t))dt,$$

by linear spline interpolation for $f(t_{i-1}, x(t_{i-1})), f(t_i, x(t_i)), f(t_{i+1}, x(t_{i+1}))$ the explicit three-step method is obtained as follows:

$$\begin{cases} x(t_{i+2}) = x(t_{i-1}) + \frac{h}{2}\left[f(t_{i-1}, x(t_{i-1})) + f(t_i, x(t_i)) + 4f(t_{i+1}, x(t_{i+1}))\right] \\ x(t_{i-1}) = \alpha_0, x(t_i) = \alpha_1, x(t_{i+1}) = \alpha_2. \end{cases} \tag{3}$$

Also, the implicit two-step method can be written as:

$$\begin{cases} x(t_{i+1}) = x(t_{i-1}) + \frac{h}{2}\left[f(t_{i-1}, x(t_{i-1})) + 2f(t_i, x(t_i)) + f(t_{i+1}, x(t_{i+1}))\right], \\ x(t_{i-1}) = \alpha_0, x(t_i) = \alpha_1. \end{cases} \tag{4}$$

Using the relation between the explicit three-step method as a predictor and the implicit two-step method as a corrector.

Now, we consider the uncertain differential equation

$$dx(t) = f(x(t), t)dt + g(x(t), t)d\mathbf{C}_t$$

with initial value $x(0)$. According to Theorem 2.1, the uncertain differential equation Eq. (1) has an unique solution, and its α-path $(0 < \alpha < 1)$ is

$$\begin{cases} dx^\alpha(t) = f(x^\alpha(t), t)dt + |g(x^\alpha(t), t)|\mu^{-1}(\alpha)dt, \\ x^\alpha(0) = x(0). \end{cases} \tag{5}$$

For the sake of simplicity, we write

$$F\big(t, x^\alpha(t)\big) = f\big(t, x^\alpha(t)\big) + \big|g\big(t, x^\alpha(t)\big)\big|\mu^{-1}(\alpha).$$

Then the ordinary differential Eq. (5) is reduced to

$$dx^\alpha(t) = F\big(t, x^\alpha(t)\big)dt$$

$$x^\alpha(0) = x(0)$$

We will present IPC method for solving uncertain differential Eq. (1) as follows.

Algorithm (*Improved Predictor Corrector (IPC) three-step method for uncertain differential equations*)

To approximate the solution of the following initial value problem

$$\begin{cases} x'(t) = F(t, x(t)), & t_0 \leq t \leq T, \\ x^\alpha(t_0) = \alpha_0, x^\alpha(t_1) = \alpha_1, x^\alpha(t_2) = \alpha_2, \end{cases}$$

an arbitrary positive integer N is chosen.

Step 1. Let $h = \frac{T-t_0}{N}$,

$$w^\alpha(t_0) = \alpha_0, w^\alpha(t_1) = \alpha_1, w^\alpha(t_2) = \alpha_2,$$

Step 2. Let $i = 1$.

Step 3. Let

$$\left\{ w^{(0)\alpha}(t_{i+2}) = w^\alpha(t_{i-1}) + \frac{h}{2}\left[F^\alpha(t_{i-1}, w(t_{i-1})) + F^\alpha(t_i, w(t_i)) + 4F^\alpha(t_{i+1}, w(t_{i+1}))\right],\right.$$

Step 4. Let $t_{i+2} = t_0 + (i+2)h$,

Step 5. Let

$$\left\{ w^\alpha(t_{i+2}) = w^\alpha(t_i) + \left(\frac{h}{2}\right)F^\alpha(t_i, w(t_i)) + hF^\alpha(t_{i+1}, w(t_{i+1})) + \left(\frac{h}{2}\right)F^\alpha\left(t_{i+2}, w^{(0)}(t_{i+2})\right),\right.$$

Step 6. $i = i + 1$.

Step 7. If $i \leq N - 2$, go to step 3.

Step 8. The algorithm will end and $w^\alpha(T)$ approximates the real value of $X^\alpha(T)$.

4 Convergence

To integrate the system given in Eq. (1) from t_0 to a prefixed $T > t_0$, the interval $[t_0, T]$ will be replaced by a set of discrete equally spaced grid points $t_0 < t_1 < t_2 < \ldots < t_N = T$, and the exact solution $X(t, \alpha)$ is approximated by some $x(t, \alpha)$. The exact and approximate solutions at t_n, $0 \leq n \leq N$ are denoted by $X_n(\alpha)$, and $x_n(\alpha)$, respectively. The grid points at which the solution is calculated are $t_n = t_0 + nh$, $h = (T - t_0)/N$, $1 \leq n \leq N$.

From Eq. (4), the polygon curves

$$x(t, h, \alpha) = \{[t_0, x_0(\alpha)], [t_1, x_1(\alpha)], \ldots, [t_N, x_N(\alpha)]\},$$

are the implicit three-step approximates to $X(t, \alpha)$ over the interval $t_0 \leq t \leq t_N$. The following lemma will be applied to show the convergence of these approximates, i.e.,

$$lim_{h\to 0} x(t, h, \alpha) = X(t, \alpha)$$

Lemma 4.1. Let a sequence of numbers $\{w_n\}_{n=0}^N$ satisfy:

$$|w_{n+1}| \leq \mathbf{A}|w_n| + \mathbf{B}|w_{n-1}| + \mathbf{C}, 0 \leq n \leq N - 1$$

for some given positive constants **A**, **B**, and **C**. Then

$$|w_n| \le \left(\mathbf{A}^{n-1} + \beta_1\mathbf{A}^{n-3}B + \beta_2\mathbf{A}^{n-5}\mathbf{B}^2 + \ldots + \beta_s\mathbf{B}^{[\frac{n}{2}]}\right)|w_1|$$
$$+ \left(\mathbf{A}^{n-2}\mathbf{B} + \gamma_1\mathbf{A}^{n-4}\mathbf{B}^2 + \ldots + \gamma_t AB^{[\frac{n}{2}]}\right)|w_0| + \left(\mathbf{A}^{n-2} + \mathbf{A}^{n-3} + \ldots + 1\right)\mathbf{C}$$
$$+ \left(\delta_1\mathbf{A}^{n-4} + \delta_2\mathbf{A}^{n-5} + \ldots + \delta_m\mathbf{A} + 1\right)BC + \left(\zeta_1\mathbf{A}^{n-6} + \zeta_2\mathbf{A}^{n-7} + \ldots + \zeta_l\mathbf{A} + 1\right)\mathbf{B}^2\mathbf{C}$$
$$+ \left(\lambda_1\mathbf{A}^{n-8} + \lambda_2\mathbf{A}^{n-9} + \ldots + \lambda_p\mathbf{A} + 1\right)\mathbf{B}^3\mathbf{C} + \ldots, \textit{ when n is odd}$$

and

$$|w_n| \le \left(\mathbf{A}^{n-1} + \beta_1\mathbf{A}^{n-3}\mathbf{B}\beta_2\mathbf{A}^{n-5}\mathbf{B}^2 + \ldots + \beta_s AB^{[\frac{n}{2}]-1}\right)|w_1|$$
$$+ \left(\mathbf{A}^{n-2}\mathbf{B} + \gamma_1\mathbf{A}^{n-4}\mathbf{B}^2 + \ldots + \gamma_t\mathbf{B}^{[\frac{n}{2}]}\right)|w_0| + \left(\mathbf{A}^{n-2} + \mathbf{A}^{n-3} + \ldots + 1\right)\mathbf{C}$$
$$+ \left(\delta_1\mathbf{A}^{n-4} + \delta_2\mathbf{A}^{n-5} + \ldots + \delta_m\mathbf{A} + 1\right)BC + \left(\zeta_1\mathbf{A}^{n-6} + \zeta_2 A^{n-7} + \ldots + \zeta_l\mathbf{A} + 1\right)\mathbf{B}^2\mathbf{C}$$
$$+ \left(\lambda_1\mathbf{A}^{n-8} + \lambda_2\mathbf{A}^{n-9} + \ldots + \lambda_p\mathbf{A} + 1\right)\mathbf{B}^3\mathbf{C} + \ldots, \textit{ when n is even}$$

where $\beta_s, \gamma_t, \delta_m, \zeta_l, \lambda_p$, are constants for all s, t, m, l and p.

Theorem 4.1. For any arbitrary fixed $\alpha : \quad 0 \le \alpha \le 1$, the implicit two-step approximates of Eq. (4) converge to the exact solutions $X(t, \alpha)$ for $X \in \mathbf{C}^3[t_0, T]$.

Proof: It is sufficient to show

$$lim_{h\to 0} x_N(\alpha) = X(T, \alpha).$$

By using Taylor's theorem we have:

$$X_{n+1}(\alpha) = X_n(\alpha) + \frac{h}{2}f\left(t_{n-1}, X_{n-1}(\alpha)\right) + hf\left(t_n, X_n(\alpha)\right) + \frac{h}{2}f\left(t_{n+1}, X_{n+1}(\alpha)\right) - \frac{1}{6}h^3X'''(\xi_n),$$

where $t_n < \underline{\xi}_n$. Consequently

$$X_{n+1}(\alpha) - x_{n+1}(\alpha) = X_{n-1}(\alpha) - x_{n-1}(\alpha) + \frac{h}{2}\{f\left(t_{n-1}, X_{n-1}(\alpha)\right) - f\left(t_{n-1}, x_{n-1}(\alpha)\right)\}$$
$$+ h\{f\left(t_n, X_n(\alpha)\right) - f\left(t_n, x_n(\alpha)\right)\} + \frac{h}{2}\{f\left(t_{n+1}, X_{n+1}(\alpha)\right) - f\left(t_{n+1}, x_{n+1}(\alpha)\right)\} - \frac{1}{6}h^3X'''(\xi_n).$$

Denote $w_n = X_n(\alpha) - x_n(\alpha)$, Then

$$\left|w_{n+1}\right| \le hL_1|w_n| + \left(1 + \frac{hL_2}{2}\right)\left|w_{n-1}\right| + \left(\frac{hL_3}{2}\right)\left|w_{n+1}\right| + \frac{1}{6}h^3M,$$

where $M = \max_{t_0 \le t \le T}\left|X'''(t, \alpha)\right|$.
Set

$$L = \max\{L_1, L_2, L_3\} < \frac{1}{h},$$

then

$$\left|w_{n+1}\right| \le \left(\frac{2hL}{1 - hL}\right)\left|w_n\right| + \left(\frac{1 + hL}{1 - hL}\right)\left|w_{n-1}\right| + \left(\frac{1}{3 - 3hL}\right)h^3M,$$

are obtained, if $h \to 0$ then $w_n \to 0$, which concludes the proof.

Theorem 4.2. For any arbitrary fixed $\alpha : 0 \leq \alpha \leq 1$, the explicit three-step approximates of Eq. (3) converge to the exact solution $X(t, \alpha)$ for $X \in \mathbf{C}^3[t_0, \mathbf{T}]$.

Proof: Similar to Theorem 4.1.

Theorem 4.3. The explicit three-step method is stable.

Proof: For the explicit three-step method, there exists only one characteristic polynomial $p(\lambda) = \lambda^3 - \lambda$, then it satisfies the root condition and, therefore, it is a stable method.

Theorem 4.4. The implicit two-step method is stable.

Proof: Similar to Theorem 4.3.
Regarding the above-mentioned theorems, it is obvious that the IPC Three-step method is convergent and stable.

5 Numerical Experiments

Uncertain differential equations (UDEs) have become standard models for financial quantities such as asset prices, stock prices, interest rates, and their derivatives. We will begin with a quick survey of the most fundamental concepts from financial mathematics that are needed.

Interest Rate

A rate is charged or paid for the use of money. Also, an interest rate is often expressed as an annual percentage of the principal. It is calculated by dividing the amount of interest by the amount of principal. Interest rate often changes as a result of inflation and Federal Reserve policies.

Volatility

The relative rate at which the price of a security moves up and down.
Volatility is found by calculating the annualized standard deviation of daily change in price. If the price of a stock moves up and down rapidly over short time periods, it has high volatility. If the price almost never changes, it has low volatility.

Mean Reversion

Mean reversion is a tendency for uncertain process to remain near, or tend to return over time to a long-run average value. For example, interest rate and implied volatilities tend to exhibit mean reversion. Exchange rate and stock prices tend not to. Stock marks returns, however, do tend to exhibit the mean reversion.
In the Vasicek model, the short term interest rate is as follows

$$dx(t) = (\theta - kx(t))dt + \sigma dC_t \tag{6}$$

where we interpreted the above terms to mean

$dx(t)$: the change in the short term interest rate.

k: the speed of mean reversion. It approximately equals $1 - b$ where b is the first order autocorrelation.

θ: the average interest rate.
σ: the volatility of short rate.
dC_t: a random term so that $E(dC_t) = 0$.
Expected changes in interest rates are given by

$$E(dx(t)) = (\theta - kx(t))dt$$

Example 5.1. Consider the following Vasicek for $\theta = 4, k = 5, \sigma = 0.3$ that represents reasonable values describing the year-to-year behavior of interest rate. with given initial value $x(0) = 0$. The exact solution of (6) is
$x(t) = \frac{\theta}{k} + \exp(-kt)(x(0) - \frac{\theta}{k}) + \sigma\exp(-kt) \int_0^t \exp(ks)dC_s$
and its α-path is
$dx^\alpha(t) = (\theta - kx^\alpha(t) + \sigma\mu^{-1}(\alpha))dt.$
The inverse uncertainty distribution of $x(t)$ is

$$\psi_t^{-1}(\alpha) = \left(\frac{\theta}{k} + \frac{\sigma}{k}\mu^{-1}(\alpha)\right)(1 - \exp(-kt)).$$

Numerical solution of these equations is considered in Fig. 1, for step size = 0.01, iteration number = 1000, and T = 50 years. From Fig. 1, we can see that the three uncertainty distributions produced respectively by Runge-Kutta, Adams and Improved Predictor Corrector (IPC) methods are very close to the uncertainty distribution of the analytic solution (6). Hence, Adams, Runge-Kutta and IPC methods can solve the uncertain differential equation. We further observe Fig. 1 and find that the uncertainty distribution produced by IPC method is closer to the analytical solution compared to the other two uncertainty distributions produced by Adams method and Runge-Kutta method, respectively. Therefore, it is clear that IPC method has more higher accuracy than Adams and Runge-Kutta methods. From the above analysis, we can see the IPC method is a better way to solve the uncertain differential equation compared to Adams method and Runge-Kutta method.

In Table 1, a comparison of the numerical results of the absolute errors obtained by using the Runge-Kutta method, Adams method and IPC method with exact solutions (6) are given for the same step sizes h = 0.01 (Table 4).

Example 5.2. Let a, b be real numbers. Consider a linear uncertain differential equation

$$dx(t) = ax(t)dt + bx(t)d\mathbf{C}_t \tag{7}$$

with given initial value $x(0) = 1$. The explicit solution of Eq. (7) is

$$x(t) = x(0)\exp(at + b\mathbf{C}(t))$$

and its α-path is

$$dx^\alpha(t) = \left(ax^\alpha(t) + |bx^\alpha(t)|\mu^{-1}(\alpha)\right)dt.$$

The inverse uncertainty distribution of $x(t)$ is

$$\psi_t^{-1}(\alpha) = x(0)\exp\left(at + bt\mu^{-1}(\alpha)\right).$$

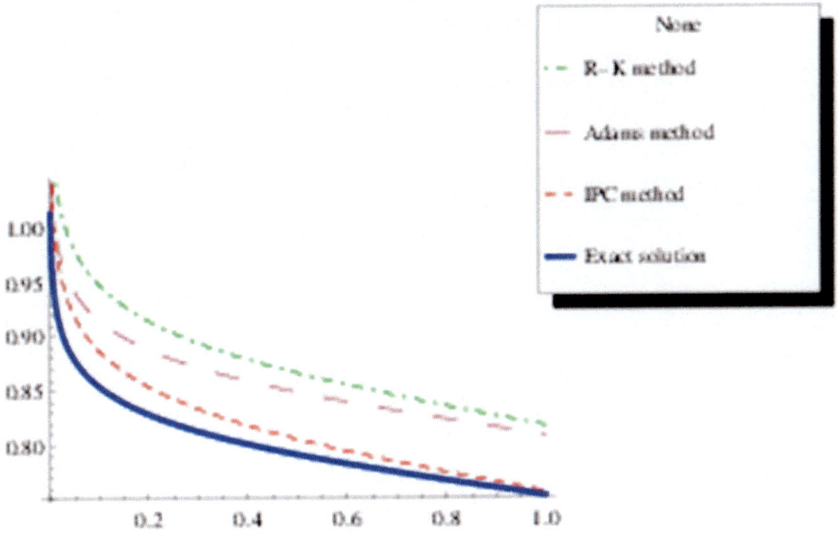

Fig. 1. The errors for for $r = 0.5$ in Example 1

Table 1. Compared Error to Adams method and IPM method.

Heading level	Error(Adams)	Error(IPM)
.1	0.0938220508540224′	0.06076827506′
.2	0.08624662223500101′	0.05930719158′
.3	0.0815564403281559′	0.05840258964′
.4	0.07802522827778158′	0.05772151989′
.5	0.07510864235613368′	0.

Table 2. Compared Error to Adams method and IPM method.

Heading level	Error(Adams)	Error(IPM)
.1	0.9281769985563846′	0.01954965844′
.2	0.6608581705328824′	0.01689519752′
.3	0.5178605149950599′	0.01530675297′
.4	0.4200838786965364′	0.01413678285′
.5	0.3451302959188256′	0.01318649342′

Table 3. Compared Error to Adams method and IPM method.

Heading level	Error(Adams)	Error(IPM)
.1	1.3512022723438073′	0.10496126228′
.2	1.2487065721012682′	0.08565417096′
.3	1.1852626006642006′	0.07384965261′
.4	1.1375031666224922′	0.06503636944′
.5	1.0980611545522911′	0.05780475874′

Table 4. Compared Error to Adams method and IPM method.

Heading level	Error(Adams)	Error(IPM)
.1	4.6266621769041905′	1.05879869257′
.2	3.592506507578089′	0.75601916594′
.3	2.9522287816477544′	0.56855900658′
.4	2.470167218514831′	0.42742128792′
.5	2.0720110193559584′	0.31084933554′

We choose the parameters as follows, $a = 0.5, b = 1.5, t = 10, N = 1000, h = \frac{t}{N} = 0.01$.

The exact, numerical solution obtained by Rung-Kutta method, Adams method and present solution using IPM are tabulated in Table 2 for h = 0.01. Also, the obtained errors are incorporated in this table. By looking at the results, one may conclude that the solution obtained by IPC method is the same as that of the exact solution. The solution plot is given in Fig. 2. By using the proposed method in Sect. 3, we can present the approximate solution for this example. A comparison of the numerical results with the exact solution is shown in Fig. 4.

Example 5.3. Let a, b be real numbers. Consider a linear uncertain differential equation

$$dx(t) = atx(t)dt + btx(t)d\mathbf{C}_t \tag{8}$$

with given initial value $x(0) = 1$. The explicit solution of Eq. (8) is
$x(t) = \exp(\frac{1}{2}at^2 + b\int_0^t dC(t))$
and its α-path is

$$dx^\alpha(t) = \left(atx^\alpha(t) + btx^\alpha(t)\mu^{-1}(\alpha)\right)dt.$$

The inverse uncertainty distribution of $x(t)$ is

$$\psi_t^{-1}(\alpha) = \exp\left(\frac{1}{2}at^2 + \frac{1}{2}|b|t^2\mu^{-1}(\alpha)\right).$$

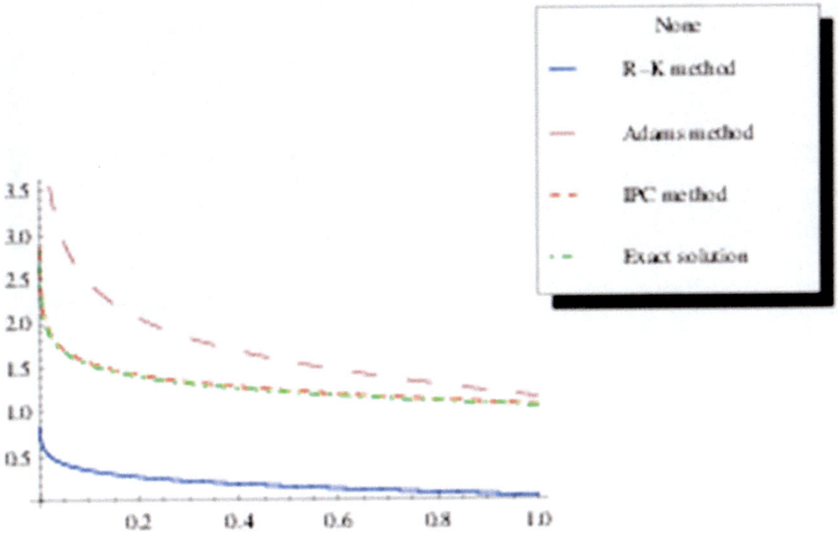

Fig. 2. The errors for for $r = 0.5$ in Example 2

We choose the parameters as follows, $a = 0.5, b = 1.5, t = 10, N = 1000, h = \frac{t}{N} = 0.01$.

We consider the test equation given in Eq. (8). In Table 3 we give the numerical solution by the above listed three numerical methods (Rung-Kutta method, Adams method, IPC method). These results show that on some fixed mesh the Rung-Kutta method and the Adams method give approximately the same accuracy, while the IPC method is more accurate (Fig. 3).

Example 5.4. Let a, b be real numbers. Consider a linear uncertain differential equation

$$dx(t) = ax(t)dt + b\exp(at)d\mathbf{C}_t \tag{9}$$

with given initial value $x(0) = 1$. The explicit solution of Eq. (9) is

$$x(t) = \exp(at)(1 + b\mathbf{C}(t))$$

and its α-path is

$$dx^{\alpha}(t) = \Big(ax^{\alpha}(t) + b\exp(at)\mu^{-1}(\alpha)\Big)dt.$$

The inverse uncertainty distribution of $x(t)$ is

$$\psi_t^{-1}(\alpha) = \Big(1 + t|b|\mu^{-1}(\alpha)\Big)\exp(at).$$

We choose the parameters as follows, $a = 0.5, b = 1.5, t = 10, N = 1000, h = \frac{t}{N} = 0.01$.

By using the proposed method in Sect. 3, we can present the approximate solution for this example. To compare the numerical results with the exact solution in Fig. 4.

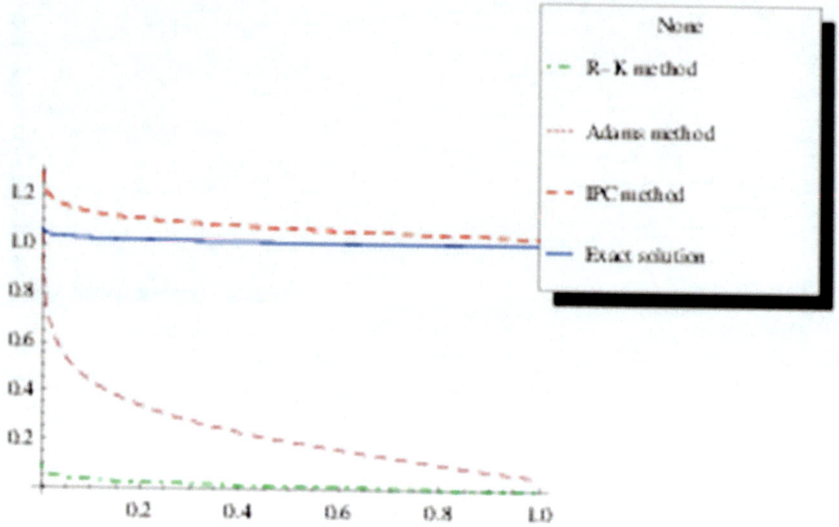

Fig. 3. The errors for for $r = 0.5$ in Example 3

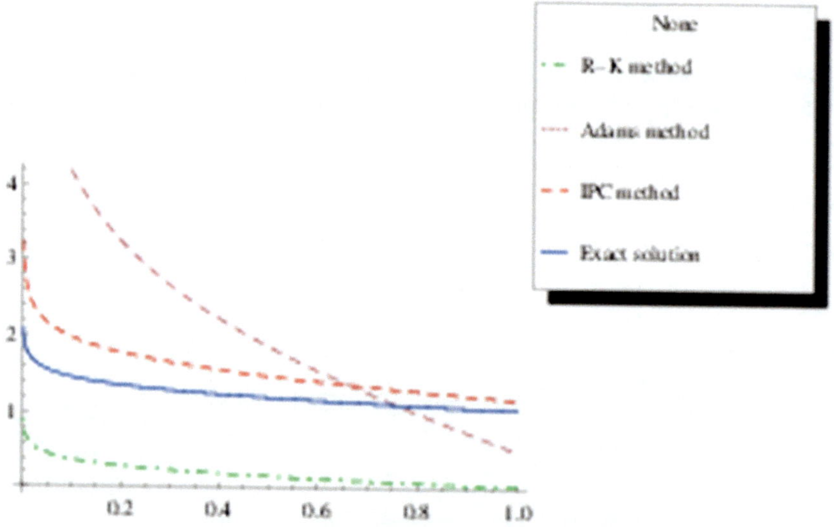

Fig. 4. The errors for for $r = 0.5$ in Example 4

6 Conclusion

Uncertain differential equation is an important tool to deal with dynamic systems in uncertain environment. In this paper, an improved predictor-corrector method is adapted and modified for solving the uncertain differential equations. Numerical examples and comparisons with exact solutions reveal that the proposed algorithm is capable of generating accurate results.

References

1. Chen, X., Qin, X.: A new existence and uniqueness theorem for fuzzy differential equations. Int. J. Fuzzy Syst. **13**(2), 148–151 (2013)
2. Chen, X., Ralescu, D.A.: Liu process and uncertain calculus. J. Uncertain. Anal. Appl.**1**, p.3 (2013)
3. Fei, W.: Uniqueness of solutions to fuzzy differential equations driven by Liu's process with non-Lipschitz coefficients. In: Internatinal Confrance On Fuzzy and Knowledge Discovery, pp. 565–569 (2009)
4. Gao, Y.: Existence and uniqueness theorem on uncertain differential equation with local Lipschitz codition. Uncertain Syst. **6**(3), 223–232 (2012)
5. Gao, J.: Credibilistic Option Pricing. J. Uncertain Syst. **2**(4), 243–247 (2008)
6. Li, X., Liu, B.: A suffcient and necessary condition for credibility measures. Int. J. Uncertainty Fuzziness Knowl. Based Syst. **14**(5), 527–535 (2004)
7. Liu, B.: Uncertainty Theory. Springer-Verlag, Berlin (2004)
8. Liu, B.: A survey of credibility theory. J. Fuzzy Optim. Decis. Mak. **5**(4), 387–408 (2006)
9. Liu, B.: Uncertainty Theory, 2nd ed. Springer-Verlag, Berlin (2007)
10. Liu, B.: Fuzzy process, hybrid process and uncertain process. J. Uncertain Syst. **2**(1), 3–16 (2008)
11. Liu, B.: Some research problems in uncertainty theory. J. Uncertain Syst. **3**(1), 3–10 (2009)
12. Liu, B.: Uncertainty Theory. A Branch of Mathematics for Modeling Human Uncertainty. Springer-Verlag, Berlin (2010)
13. Liu, B.: Why is there a need for uncertainty theory? J. Uncertain Syst. **6**(1), 3–10 (2012)
14. Liu, B.: Toward uncertain finance theory. J. Uncertain. Anal. Appl. **1,** p. 1 (2013)
15. Liu, B.: Uncertainty Theory, 4th edn. Springer-Verlag, Berlin (2015)
16. Liu, H., Ke, H., Fei, W.: Almost sure stability for uncertain differential equation. Fuzzy Optim. Decis. Mak. **13**(4), 463–473 (2014)
17. Liu, Y.: An analytic method for solving uncertain differential equations. J. Uncertain Syst. **6**(4), 244–249 (2012)
18. Peng, Z., Iwamura, K.: A sufficient and necessary condition of uncertainty distribution. J. Interdiscip. Math. **13**(3), 277–285 (2010)
19. Shen, Y., Yao, K.: Runge-Kutta method for solving uncertain differential equations. J. Uncertainty Anal. Appl. **2**(1), 3–17 (2015)
20. Wanga, X., Tauqir, Y., Moughal, A., Chena, X.: Adams-Simpson method for solving uncertain differential equation. Appl. Math. Comput., **271,** 209–219 (2015)
21. Yang, X., Ralescu, D.A.: Adams method for solving uncertain differential equations. Appl. Math. Comput. **270**, 993–1003 (2015)
22. Yao, K., Chen, X.: A numerical method for solving uncertain differential equations. J. Intell. Fuzzy Syst. **25**(3), 825–832 (2013)
23. Yao, K., Gao, J., Gao, J.: Some stability theorems of uncertain differential equation. Fuzzy Optim. **12**(1), 3–13 (2013)
24. You, C.: Multi-dimensional Liu process, differential and integral.In: Proceedings of the First Intelligent Computing Conference, China, pp. 153–158 (2007)
25. Yao,K.: A type of uncertain differential equations with an alytic solution. J. Uncertain. Anal. Appl. **1**, p. 8 (2013)

26. Yao, K., Ke, H., Sheng, Y.: Stability in mean for uncertain differential equation. Fuzzy Optim. Decis. Mak. **14**(3), 365–379 (2015)
27. You, C., Wang, W., Huo, H.: Existence and uniqueness theorems for fuzzy differential equations. J. Uncertain Syst. **7**(4), 303–315 (2013)
28. Zhu, Y.: Uncertain optimal control with application to a portfolio selection model. Cybernet. Syst. **41**(7), 535–547 (2010)

Ranking Z-Numbers Using Confidence Intervals

Mohammad Hadaegh[1], Mohammad Ali Fariborzi Araghi[1(✉)], and Somayeh Ezadi[2]

[1] Department of Mathematics, Central Tehran Branch, Islamic Azad University, Tehran, Iran
fariborzi.araghi@gmail.com

[2] Department of Mathematics, Science and Research Branch, Islamic Azad University, Tehran, Iran

Abstract. In this study, a new method for ranking Z-numbers is proposed. The suggested scheme is based on calculation of center of gravity, normal distribution function and confidence interval. In the first step, the value of the center of gravity of the confidence interval of $Z = (\tilde{A}, \tilde{B})$-numbers is calculated and considered as the confidence level. In the second step, the membership degree function of Part $\tilde{A}$ is determined using the normal distribution function. In the third step, the statistical confidence interval formula is obtained. The last step is to calculate the Z-number rank bandwidth. The results shows that the proposed method has the properties of ranking. Also, in some examples the results are compared with the existing methods and it is shown that our proposed method has a simpler algorithm.

Keywords: Ranking Z-number · Center of gravity · Normal distribution function · Confidence interval

1 Introduction

In 2011, Zade introduced a new concept in decision analysis called Z-number [12]. This concept is effectively capable of dealing with uncertainty in information about a decision that is represented as $Z = (\tilde{A}, \tilde{B})$ which consists of two parts. The first part is the constraint, $(\tilde{A})$, and the second part, the reliability of the constraint evaluation, $(\tilde{B})$. Z-number is more applicable to fuzzy numbers in decision making, risk assessment, etc. But, how to deal with the property limitations, plus the reliability is still an issue that has been discussed in the literature available to a limited and inadequate extent [1–11]. In most methods of ranking Z-numbers, they are first converted to Z^{α} numbers (generalized fuzzy numbers) and then the rank of Z^{α}-numbers is calculated. In the most of available methods for ranking Z-numbers, they are in the form of $Z = (\tilde{A}, \tilde{B})$, where $\tilde{A} = (a_1, a_2, a_3, a_4, \omega_{\tilde{A}})$ and $\tilde{B} = (b_1, b_2, b_3, b_4, \omega_{\tilde{B}})$.

While in the real world, in many cases, it is possible to model ambiguous information using Z-numbers as follows:

$$\boldsymbol{Z} = \left(\mu_{\tilde{A}}, \tilde{\boldsymbol{B}}\right).$$

The membership function can be arbitrarily selected. There is no way to rank this since Z-numbers have been rated. Therefore, decision-making by the decision maker is

T. Allahviranloo et al. (Eds.): IDS 2020, AISC 1301, pp. 889–898, 2021.
https://doi.org/10.1007/978-3-030-66501-2_72

somewhat challenging. In this paper, we propose a method for ranking Z-numbers to form $Z = (\mu_{\tilde{A}}, \tilde{B})$. Suppose fuzzy set is $\tilde{B}$ trapezoidal fuzzy number and fuzzy set $\tilde{A}$ can be any kind of fuzzy number. The rest of paper contains the following sections: In Sect. 2, some basic concepts are introduced. Section 3, describes the proposed algorithm for ranking Z-numbers. The numerical results and figures are demonstrated in Sect. 4. The conclusion remarks are mentioned in Sect. 5.

2 The Basic Concepts

In this section, some definitions and notations are brought.

Definition 2.1. A fuzzy number $\tilde{A} = (a, b, c, d; \omega)$ is described as any fuzzy subset of the real line $\mathbb{R}$ with membership function $\mu_{\tilde{A}}$ which processes the following properties [12]:

(a) $\mu_{\tilde{A}}$ is a continuous mapping from $\mathbb{R}$ to the closed interval $[0, \omega]$, $0 \leq \omega \leq 1$.
(b) $\mu_{\tilde{A}} = 0$, for all $x \in (-\infty, a]$.
(c) $\mu_{\tilde{A}}$, is strictly increasing on $[a, b]$.
(d) $\mu_{\tilde{A}} = \omega$, for all $x \in [b, c]$, where ω is constant and $\omega \in (0, 1]$.
(e) $\mu_{\tilde{A}}$, is strictly decreasing on $[c, d]$.
(f) $\mu_{\tilde{A}} = 0$, for all $x \in [d, \infty)$, where a, b, c, d are real numbers. We may let $a = -\infty, ora = b, orc = d, ord = +\infty$.

If $\omega = 1$, in (d) $\tilde{A}$ is a normal fuzzy number, and if $0 < \omega < 1$, in (d) $\tilde{A}$ is a non-normal fuzzy number. The image (opposite) of $\tilde{A} = (a, b, c, d; \omega)$ can be given by $-\tilde{A} = (-d, -c, -b, -a; \omega)$.

Definition 2.2 Z-Number. Zadeh introduced the idea of Z-numbers with an unknown variable x. A Z-number is a pair of fuzzy numbers $(\tilde{A}, \tilde{B})$ [12]. Here, $\tilde{A}$ is a fuzzy subset of constraints that X can have and $\tilde{B}$ is a fuzzy subset of the reliability of the $\tilde{A}$ component. Zadeh also defined $(x, \tilde{A}, \tilde{B})$ as a Z-valuation and indicated that this value means that x equals $(\tilde{A}, \tilde{B})$. According to the suggestion of Zadeh, this valuation for Z was observed as a constraint in x and interpreted as follows:

$$\boldsymbol{Prob}\left(\boldsymbol{xis\tilde{A}}\right)\boldsymbol{is\tilde{B}}. \tag{2.11}$$

Actually, this means that [12]

$$\boldsymbol{R(x)} : \boldsymbol{xis\tilde{A}} \rightarrow \boldsymbol{Poss(x = u)} = \boldsymbol{\mu_{\tilde{A}}(u)}, \tag{2.12}$$

$$\boldsymbol{P}\left(\boldsymbol{xis\tilde{A}}\right) = \int_{\boldsymbol{R}} \boldsymbol{\mu_{\tilde{A}}(u)P_x(u)du} \ \boldsymbol{is} \ \boldsymbol{\tilde{B}}, \tag{2.13}$$

in which $\boldsymbol{\mu_{\tilde{A}}}$ is the membership function of the fuzzy set $\tilde{A}$ and u is a value of x. $P_x(u)$ is the probability density function of x and $P(x = u)$ is the probability function of x. Since we do not know the basic probability distribution, it is clear from these data that the probability distribution is itself a fuzzy number.

Definition 2.3. The Normal Distribution Function. In probability theory, a normal (or Gaussian or Gauss or Laplace–Gauss) distribution is a type of continuous probability distribution for a real-valued random variable. The general form of its probability density function is

$$f(x) = \frac{1}{\sqrt{2\pi}\sigma} e^{-1/2(\frac{c-x}{\sigma})^2}.$$

The parameter c is the mean or expectation of the distribution (and also its median and mode); and σ is its standard deviation. The variance of the distribution is σ^2. A random variable with a Gaussian distribution is said to be normally distributed and is called a normal deviate. Normal distributions are important in statistics and are often used in the natural and social sciences to represent real-valued random variables whose distributions are not known.

3 Ranking Z-Numbers

In this section, a method for ranking Z-numbers using the center of gravity and the confidence interval is suggested.

The algorithm of the method is as follows:

Step 1: For$Z = (\mu_{\tilde{A}}, \tilde{B})$, where $\tilde{B} = (b_1, b_2, b_3, b_4, \omega_{\tilde{B}})$. Convert $\tilde{B}$ (reliability) into a crisp number by using.

$$\begin{aligned} \beta &= \frac{\int_X x\mu_{\tilde{B}}dx}{\int_X \mu_{\tilde{B}}dx}. \\ &= \frac{\int_{b_1}^{b_2} x(x-b_1)dx + \int_{b_2}^{b_3} xdx + \int_{b_3}^{b_4} x(b_4-x)dx}{\int_{b_1}^{b_2}(x-b_1)dx + \int_{b_2}^{b_3} dx + \int_{b_3}^{b_4}(b_4-x)dx}, \end{aligned} \tag{3.1}$$

where $\int$ denotes an algebraic integration; and consider the value of β as the level of confidence defined in statistics.

Step 2: For $Z_i = (\mu_{\tilde{A}_i}, \tilde{B}_i)$, If $\tilde{A}_i = \left\{a_{i1}, \ldots, a_{in}; \omega_{\tilde{A}_i}\right\}$ is converted into a standardized fuzzy number, $\tilde{A}_i^* = \left\{a_{i1}^*, \ldots, a_{in}^*; \omega_{\tilde{A}_i}\right\}$, where $a_{ij}^* = \frac{a_{ij}}{p}, \forall j = 1, 2, 3, \ldots, n., \forall i = 1, \ldots, m, \omega_{\tilde{A}_i} \in [0, 1]$ and $p = \max_{i,j}\left(a_{ij}, 1\right)$ represents the maximum value of the universe of discourse.

Step 3: Calculate c by:

$$c = \frac{1}{n}\sum_{i=1}^{n} a_i{}^*, \tag{3.2}$$

where $a_i^* = \frac{a_i}{p}, \forall i = 1, \ldots, m$ and $p = \max_i(a_i, 1)$.

Step 4: Calculate σ by:

$$\sigma = \sqrt{\frac{\sum_{i=1}^{n}(a_i{}^* - c)^2}{n-1}}, \tag{3.3}$$

Step 5: Calculate Confidence Interval

$$\left(L_{\tilde{A}_i^*}, U_{\tilde{A}_i^*}\right) = \left(c_i - \mathbf{Z}_{\frac{\alpha_i}{2}} * \frac{\sigma}{n}, c_i + \mathbf{Z}_{\frac{\alpha_i}{2}} * \frac{\sigma}{n}\right), \tag{3.4}$$

where $\alpha = 1 - \beta$. The symbol $\mathbf{Z}$ represents the normal distribution function. The value of $\mathbf{Z}_{\frac{\alpha}{2}}$ is easily determined using the $\mathbf{Z}$ table (normal distribution table in statistics and probability).

Step 6: Calculate $Rank(Z_i)$.

Let $\left(L_{\tilde{A}_1^*}, U_{\tilde{A}_1^*}\right)$ is confidence interval for Z_1, also Let $\left(L_{\tilde{A}_2^*}, U_{\tilde{A}_2^*}\right)$ is confidence interval for Z_2. We define the concept of ranking as follows:

(1) If $L_{\tilde{A}_1^*} > L_{\tilde{A}_2^*} \Rightarrow Rank(Z_1) > Rank(Z_2)$,
(2) If $L_{\tilde{A}_1^*} < L_{\tilde{A}_2^*} \Rightarrow Rank(Z_1) < Rank(Z_2)$,
(3) If $L_{\tilde{A}_1^*} = L_{\tilde{A}_2^*}$ then:

   (a) If $U_{\tilde{A}_1^*} > U_{\tilde{A}_2^*} \Rightarrow Rank(Z_1) > Rank(Z_2)$,
   (b) If $U_{\tilde{A}_1^*} < U_{\tilde{A}_2^*} \Rightarrow Rank(Z_1) < Rank(Z_2)$,
   (c) If $U_{\tilde{A}_1^*} = U_{\tilde{A}_2^*} \Rightarrow Rank(Z_1) = Rank(Z_2)$.

Actually if $L_{\tilde{A}_1^*} = L_{\tilde{A}_2^*}$ and $U_{\tilde{A}_1^*} = U_{\tilde{A}_2^*}$ then $Rank(Z_1) = Rank(Z_2)$.

Property 1. Let $Z_i = (\mu_{\tilde{A}_i^*}, \tilde{B}_i)$, if $\mu_{\tilde{A}_i^*} = 0$ and $\tilde{B}_i = (1, 1, 1, 1; 1)$, then $Rank(Z) = 0$.

Proof: Obviously, if $\mu_{\tilde{A}_i^*} = 0$ then $a_i{}^*\tilde{A}_i^*$, so$Rank(Z) = 0$.

Property 2. Let $Z_i = (\mu_{\tilde{A}_i^*}, \tilde{B}_i)$, if $\tilde{A}_i^* = \{1, 1, \ldots, 1; \omega_{\tilde{A}_i} = 1\}$ and $\tilde{B}_i = (1, 1, 1, 1; 1)$, then $Rank(Z) = 1$.

Proof: Obviously, if $\tilde{A}_i^* = \{1, 1, \ldots, 1; \omega_{\tilde{A}_i} = 1\}$ and $\tilde{B}_i = (1, 1, 1, 1; 1)$ then $c = 1$ and $\alpha = 0 \Rightarrow \mathbf{Z}_{\frac{\alpha}{2}} = 0$, so $Rank(Z) = 1$.

Property 3. Let $Z_i = (\mu_{\tilde{A}_i^*}, \tilde{B}_i)$, if $\tilde{A}_i^* = \{-1, -1, \ldots, -1; \omega_{\tilde{A}_i} = 1\}$ and $\tilde{B}_i = (1, 1, 1, 1; 1)$, then $Rank(Z) = -1$.

Proof: Obviously, if $\tilde{A}_i^* = \{-1, -1, \ldots, -1; \omega_{\tilde{A}_i} = 1\}$ and $\tilde{B}_i = (1, 1, 1, 1; 1)$ then $c = -1$ and $\alpha = 0 \Rightarrow \mathbf{Z}_{\frac{\alpha}{2}} = 0$, so $Rank(Z) = 1$.

Let K be a set of fuzzy numbers, then $f : K \rightarrow \mathbb{R}$ is defined as the rank function of fuzzy numbers, such that if $A; B \in K$ then.

- $A > B \rightarrow Rank(A) > Rank(B)$,
- $A = B \rightarrow Rank(A) = Rank(B)$.

We define $R(Z_i) \in \mathbb{R}$ as the function that ranks a Z_i-number , such that $\forall i, j \geq 1$.

- $R(Z_i) > R(Z_j) \leftrightarrow Z_i \succ Z_j$,
- $R(Z_i) = R(Z_j) \leftrightarrow Z_i \sim Z_j$.

4 The Numerical Results

Example 1. In this example, we have four sets of Z-numbers. There are two Z-numbers in each set, Z_1 and Z_2. The diagram of these Z-numbers can be seen in Fig. 1.

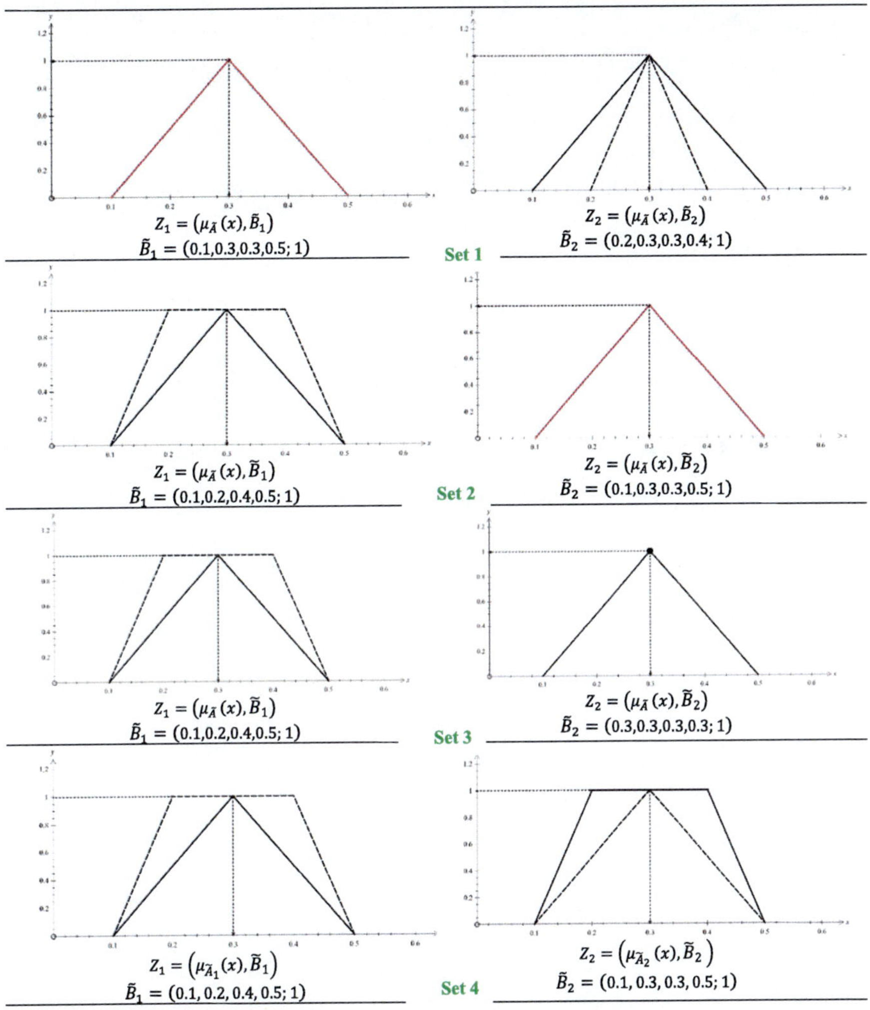

Fig. 1. Four sets of Z-numbers for Example 1.

Where $\mu_{\tilde{A}}(x)$ for Set 1, Set 2, and Set 3 are defined as follows:

$$\mu_{\tilde{A}}(x) = \begin{cases} \frac{x-0.1}{0.2} & if\ 0.1 \le x \le 0.3 \\ \frac{0.5-x}{0.2} & if\ 0.3 \le x \le 0.5 \\ 0 & Otherwise \end{cases}$$

And $\mu_{\tilde{A}_1}(x)$ of Z_1 and $\mu_{\tilde{A}_2}(x)$ of Z_2 for Set 4 are defined as follows:

$$\mu_{\tilde{A}_1}(x) = \begin{cases} \frac{x-0.1}{0.2} & if\ 0.1 \le x \le 0.3 \\ \frac{0.5-x}{0.2} & if\ 0.3 \le x \le 0.5 \\ 0 & Otherwise \end{cases},$$

$$\mu_{\tilde{A}_2}(x) = \begin{cases} \frac{x-0.1}{0.1} & if\ 0.1 \le x \le 0.2 \\ 1 & 0.2 \le x \le 0.4 \\ \frac{0.5-x}{0.1} & if\ 0.4 \le x \le 0.5 \\ 0 & Otherwise \end{cases}$$

The Ranking for $Z_1 = \left(\mu_{\tilde{A}}(x), \tilde{B}_1\right)$ and $Z_2 = \left(\mu_{\tilde{A}}(x), \tilde{B}_2\right)$ can be seen in Table 1. The results of the proposed method compared with existing methods and can be seen in Table 2.

Table 1. The calculated results by the proposed method for Example 1.

Set i	Z_i	β_i	α_i	$Z_{\frac{\alpha_i}{2}}$	c_i	$\left(L_{\tilde{A}_i^*}, U_{\tilde{A}_i^*}\right)$
Set 1	Z_1	0.3	0.7	0.6368	0.3	(0.2740, 0.3260)
	Z_2	0.3	0.7	0.6368	0.3	(0.2740, 0.3260)
	Result	$Z_1 \approx Z_2$				
Set 2	Z_1	0.3	0.7	0.6368	0.3	(0.2740, 0.3260)
	Z_2	0.3	0.7	0.6368	0.3	(0.2740, 0.3260)
	Result	$Z_1 \approx Z_2$				
Set 3	Z_1	0.3	0.7	0.6368	0.3	(0.2740, 0.3260)
	Z_2	0.3	0.7	0.6368	0.3	(0.2740, 0.3260)
	Result	$Z_1 \approx Z_2$				
Set 4	Z_1	0.3	0.7	0.6368	0.3	(0.2740, 0.3260)
	Z_2	0.3	0.7	0.6368	0.3	(0.27093, 0.32906)
	Result	$Z_1 \succ Z_2$				

Table 2. A comparison of the proposed method with the existing methods.

Methods	Set 1	Set 2	Set 3	Set 4
Mohamad's method (2014)	$Z_1 \approx Z_2$	$Z_1 \approx Z_2$	$Z_1 \approx Z_2$	$Z_1 \succ Z_2$
Bakar's method (2015)	$Z_1 \approx Z_2$	....	....	....
Kang's method (2012)	$Z_1 \approx Z_2$	$Z_1 \approx Z_2$	$Z_1 \approx Z_2$	$Z_1 \succ Z_2$
The proposed method	$Z_1 \approx Z_2$	$Z_1 \approx Z_2$	$Z_1 \approx Z_2$	$Z_1 \succ Z_2$

Example 2. Let for $Z_1 = \left(\mu_{\tilde{A}_1}(x), \tilde{B}\right)$ and $Z_2 = \left(\mu_{\tilde{A}_2}(x), \tilde{B}\right)$, where

$$\mu_{\tilde{A}_1}(x) = \begin{cases} \frac{x+5/2}{2} & if\ -\frac{5}{2} \le x \le -1/2 \\ 1 & if\ -1/2 \le x \le 0 \\ \frac{1-x}{1} & if\ 0 \le x \le 1 \\ 0 & Otherwise \end{cases},$$

$$\mu_{\tilde{A}_2}(x) = \begin{cases} \frac{x+9/4}{2} & if\ -\frac{9}{4} \le x \le -1/4 \\ \frac{3/4-x}{1} & if\ -1/4 \le x \le 3/4 \\ 0 & Otherwise \end{cases},$$

and $\tilde{B} = (0.1, 0.3, 0.3, 0.5)$, (see Fig. 2). The Ranking for Z_1 and Z_2 can be seen in Table 3.

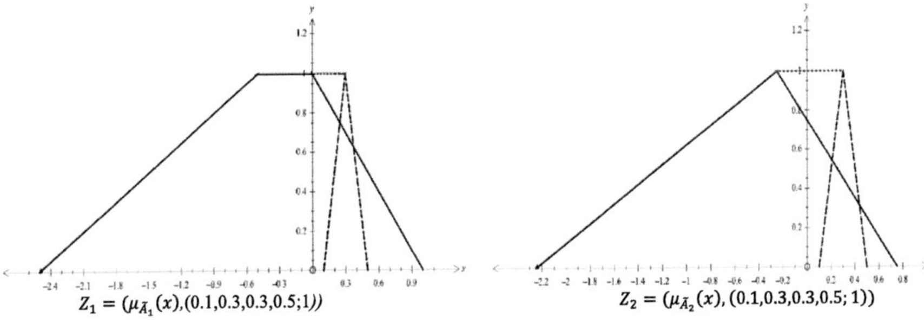

Fig. 2. Two sets of Z-numbers for Example 2.

Table 3. The calculated results by the proposed method for Example 2.

Z_i	β_i	α_i	$Z_{\frac{\alpha_i}{2}}$	c_i	$\left(L_{\tilde{A}_i^*}, U_{\tilde{A}_i^*}\right)$
Z_1	0.3	0.7	0.6368	−0.5	(−0.7343, −0.2657)
Z_2	0.3	0.7	0.6368	−0.5	(−0.70032, −0.29968)
Result					$Z_1 \prec Z_2$

Example 3. Let $\mu_{\tilde{A}_1}(x) = \begin{cases} \frac{x+0.3}{0.3} & \text{if } -0.3 \le x \le 0 \\ \frac{0.3-x}{0.3} & \text{if } 0 \le x \le 0.3 \\ 0 & \text{Otherwise} \end{cases}$,

$$\mu_{\tilde{A}_2}(x) = \begin{cases} \frac{x}{0.0484} & \text{if } 0 \le x \le 0.0484 \\ 1 & \text{if } 0.0484 \le x \le 0.0968 \\ \frac{0.9-x}{0.8032} & \text{if } 0.0968 \le x \le 0.9 \\ 0 & \text{Otherwise} \end{cases}$$

$\tilde{B} = (0.1, 0.3, 0.3, 0.5)$, (See Fig. 3). The Ranking for $Z_1 = \left(\mu_{\tilde{A}_1}(x), \tilde{B}\right)$ and $Z_2 = \left(\mu_{\tilde{A}_2}(x), \tilde{B}\right)$ in convex combination can be seen in Table 4.

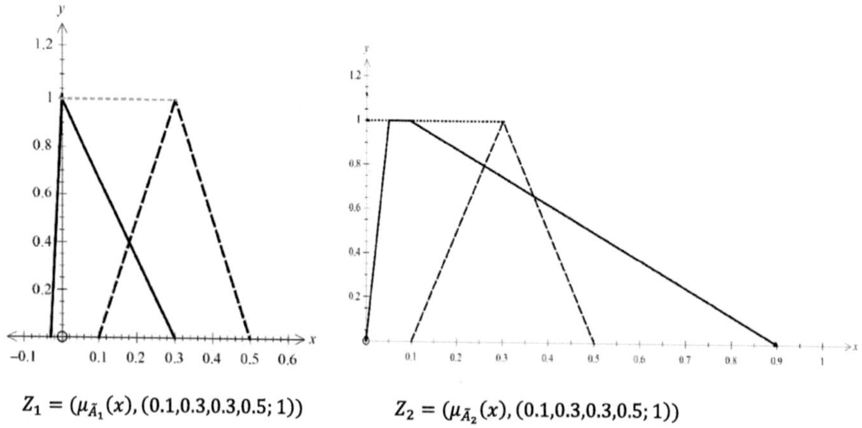

Fig. 3. Two sets of Z-numbers for Example 3.

Table 4. The calculated results by the proposed method for Example 3.

Z_i	β_i	α_i	$Z_{\frac{\alpha_i}{2}}$	c_i	$\left(L_{\tilde{A}_i^*}, U_{\tilde{A}_i^*}\right)$
Z_1	0.3	0.7	0.6368	0.0675	(0.042722, 0.092278)
Z_2	0.3	0.7	0.6368	0.2613	(0.19322, 0.32938)
Result	$Z_1 \prec Z_2$				

5 Conclusions

Decision making is an important process that can determine the productivity and development of organizations. Also in the field of medical science, Human health has received such a serious threat that the medical diagnosis becomes very worthy of study. Different methods and tools have been introduced for solving decision making problems. Each of these methods tries to prioritize effective factors in decision making so that the decision maker can make the right decision. When a person's life is in the midst of a person or when an organization's decision can have a significant impact on the community's situation, it is necessary to prioritize the factors affecting a decision. This paper attempts to use a technique based on the center of gravity, mean and standard deviation and confidence interval for ranking our Z- numbers. As a result, the method is divided into three stages. The ranking properties of the proposed method were evaluated and compared with other available methods. In addition, the results of the examples show that our proposed method has similar results to other available methods. With a simpler ranking algorithm.

References

1. Allahviranloo, T., Ezadi, S.: Z-Advanced numbers processes. Inf. Sci. **480**, 130–143 (2019)
2. Bakar, A.S.A., Gegov, A.: Multi-layer decision methodology for ranking Z-numbers. Int. J. Comput. Intell. Syst. **8**, 395–406 (2015)
3. Ezadi, S., Allahviranloo, T.: New multi-layer method for Z-number ranking using Hyperbolic Tangent function and convex combination. Intell. Autom. Soft Comput. 1–7 (2017)
4. Ezadi, S., Allahviranloo, T., Two new methods for ranking of Z-numbers based on sigmoid function and sign method. Int. J. Intell. Syst. 1–12. (2018)
5. Ezadi, S., Allahviranloo, T.: Numerical solution of linear regression based on Z-numbers by improved neural network. Intell. Autom AtIon Soft Comput. 1–11 (2017)
6. Jiang, W., Xie, Ch., Luo, Yu. and Tang, Y.: Ranking Z-numbers with an improved ranking method for generalized fuzzy numbers. J. Intell. Fuzzy Syst. (2016)
7. Kang, B., Wei, D., Li, Y., Deng, Y.: Decision making using Z-numbers under uncertain environment. J. Comput. Inf. Syst. **8**(7), 2807–2814 (2012)
8. Kang, B., Wei, D., Li, Y., Deng, Y.: A method of converting Z-number to classical fuzzy number. J. Inf. Comput. Sci. **9**(3), 703–709 (2012)
9. Mohamad, D., Shaharani, S.A., Kamis, N.H.: A Z-number based decision making procedure with ranking fuzzy numbers method. AIP Conf. Proc. **1635**, 160–166 (2014)
10. Yager, R.R.: A procedure for ordering fuzzy subsets of the unit interval. Inf. Sci. **24**, 143–161 (1981)

11. Yager, R.R.: On Z-valuations using Zadeh's Z-Numbers. Int. J. Intell. Syst. **27**, 259–278 (2012)
12. Zadeh, L.A.: A note on Z-numbers. Inf. Sci. **181**, 2923–2932 (2011)

Solution of Complex Bipolar Fuzzy Linear System

Muhammad Akram[1(✉)], Muhammad Ali[1], and Tofigh Allahviranloo[2]

[1] Department of Mathematics, University of the Punjab, New Campus, Lahore, Pakistan
m.akram@pucit.edu.pk, muhammadali6481@gmail.com
[2] Faculty of Engineering and Natural Sciences, Bahcesehir University, Istanbul, Turkey
tofigh.allahviranloo@eng.bau.edu.tr

Abstract. Decision analysis plays a vital role in various fields including operational research, information technology, engineering and social sciences. YinYang bipolar fuzzy sets (or bipolar fuzzy sets), as a generalization of fuzzy sets that extend the membership domain from $[0, 1]$ to $[-1, 0] \times [0, 1]$, were introduced by Zhang to handle double-sided information about the decision data. In this paper, we introduce two linear system of equations called general complex bipolar fuzzy linear (CBFL) system and CBFL system with complex coefficients. Further, we propose a general technique to solve the CBFL system of equations by pair of positive ($\star$) and negative ($\bullet$) of two $n \times n$ complex crisp linear systems. The necessary and sufficient condition is provided for the solution of CBFL system of equations. We describe these concepts by using some numerical examples and obtain the strong and weak solutions of the systems. Moreover, we propose a general technique to solve the CBFL system of equations with complex coefficients by pair of positive ($\star$) and negative ($\bullet$) of two $n \times n$ complex crisp linear systems with complex coefficients. Finally, we use this technique to solve the current flow circuit which is represented by CBFL system with complex coefficients and obtain the unknown current in term of bipolar fuzzy complex number.

Keywords: Complex bipolar fuzzy linear system · Complex bipolar fuzzy linear system with complex coefficients · Bipolar fuzzy complex number

1 Introduction

Initially fuzzy set was introduced by Zadeh [1] which is a powerful mathematical tool for modeling and controlling uncertain systems. Fuzzy set theory is also used for measuring the uncertainty and vagueness. Dubois and Prade [2] discussed the basic arithmetic operations of fuzzy numbers. Zhang [3,4] used terms, namely, Yin and Yang which were based on double-sided or bipolar judgemental thinking on a negative and positive side, and introduced the concept of bipolar fuzzy set in 1994. In 2011, Akram [5] applied the concept of bipolar fuzzy sets to graph theory. Alghamdi *et al.* [6] considered the multi-criteria decision-making

T. Allahviranloo et al. (Eds.): IDS 2020, AISC 1301, pp. 899–927, 2021.
https://doi.org/10.1007/978-3-030-66501-2_73

techniques in bipolar fuzzy environment. The term simultaneous linear equations plays vital role in different kind of fields like Mathematics, Physics, Networking, Circuit analysis, Economic model and attribute decision making etc. Fuzzy linear system (FLS) is a powerful tool to measure the fuzziness and uncertainty. In many places, a bipolar fuzzy linear system (BFLS) is used for linear opimization of system when system is represented in bipolar form. complex fuzzy linear system is specially used in circuit analysis to measure unknown current in complex form. First the solution of FLS was examined by Friedman *et al.* [7] and they used the embedding method to solve the system. They used the technique in which $n \times n$ FLS of equations whose coefficient matrix is a crisp number matrix and right side column vector is fuzzy number vector which is replaced by $2n \times 2n$ crisp linear system. Abbasbandy and Alavi [8] also used another method to solve $n \times n$ FLS of equations for which coefficient matrix is a crisp number matrix and right side column vector is fuzzy number vector. They used a procedure in which $n \times n$ FLS of equations is replaced by two $n \times n$ crisp linear systems. Allahviranloo [9] proposed different methods to solve uncertain linear system of equations. Some numerical techniques are proposed by different authors to solve FLS of equations for example, Jacobi, Gauss-Seidel, SOR iterative and steepest descent method in [10–13]. Some other numerical techniques are also proposed by Akram *et al.* [14–16,18] and Saqib *et al.* [17] to solve BFLS of equations, fully bipolar fuzzy linear system FBFLS of equations and Linear system of equations in m-polar fuzzy environment. Suparna and Chakraverty [19] proposed a method to solve FLS and fully fuzzy linear system (FFLS) by using fuzzy interval. The concept of fuzzy complex set and fuzzy complex number was introduced by Buckley [20] in 1989. Rahgooy [21] studied the $n \times n$ complex fuzzy linear systems with complex coefficients and proposed the method to solve complex fuzzy linear system, it was applied on problems of circuit analysis (CA) to find the current in term of complex fuzzy number. Jahantigh [22] studied the $n \times n$ complex fuzzy linear system and discussed the general technique to solve the system.

In this article, we present two different techniques to solve $n \times n$ CBFL and CBFL system with complex coefficients for which coefficient matrix is a crisp and complex crisp number respectively and right-hand side column vector is a parametric form of bipolar fuzzy complex number (BFCN). In the first technique, we propose a method to solve CBFL system we replace $n \times n$ CBFL system of equations by pair of positive ($\star$) and negative ($\bullet$) of $2n \times 2n$ complex crisp linear system. The necessary and sufficient condition is also given for the solution. Some numerical examples are also discuss to describe the efficiency of the techniques. In the second technique, we propose a method to solve CBFL system with complex coefficients in which coefficient matrix is regular complex number and right side column vector is a BFCN. We replace $n \times n$ CBFL system with complex coefficients by pair of positive ($\star$) and negative ($\bullet$) of two $n \times n$ complex linear system with complex coefficients. We utilize the current flow circuit system given in [21] which is represented by CBFL system with complex coefficients to find unknown current in the circuit in term of BFCN.

2 Preliminaries

First we revive some basic definitions like bipolar fuzzy number (BFN), fuzzy complex number (FCN), parametric form of BFN, bipolar fuzzy complex number (BFCN) and complex bipolar fuzzy linear (CBFL) system of equations.

Definition 1 [23]. A bipolar fuzzy number (BFN)

$$\xi = \prec M, N \succ = \prec [m_1, m_2, m_3, m_4], [n_1, n_2, n_3, n_4] \succ$$

is a bipolar fuzzy set of the mapping $\sigma : R \longrightarrow [0,1] \times [-1,0]$, with satisfaction degree σ_M and dissatisfaction degree σ_N such that:

$$\sigma_M(\kappa) = \begin{cases} \sigma_M^L(\kappa), \text{ if } \kappa \in [\mathrm{m}_1, \mathrm{m}_2], \\ 1, \quad \text{ if } \kappa \in [\mathrm{m}_2, \mathrm{m}_3], \\ \sigma_M^R(\kappa), \text{ if } \kappa \in [\mathrm{m}_3, \mathrm{m}_4], \\ 0, \quad \text{otherwise}, \end{cases}$$

$$\sigma_N(\kappa) = \begin{cases} \sigma_N^L(\kappa), \text{ if } \kappa \in [\mathrm{n}_1, \mathrm{n}_2], \\ -1, \quad \text{ if } \kappa \in [\mathrm{n}_2, \mathrm{n}_3], \\ \sigma_N^R(\kappa), \text{ if } \kappa \in [\mathrm{n}_3, \mathrm{n}_4], \\ 0, \quad \text{otherwise}, \end{cases}$$

and

$$\sigma_M^L(\kappa) : [m_1, m_2] \to [0,1], \quad \sigma_M^R(\kappa) : [m_3, m_4] \to [0,1],$$
$$\sigma_N^L(\kappa) : [n_1, n_2] \to [-1,0], \quad \sigma_N^R(\kappa) : [n_3, n_4] \to [-1,0],$$

where $\sigma_M^L(\kappa)$ and $\sigma_N^L(\kappa)$ represent the left membership functions for $\sigma_M(\kappa)$ and $\sigma_N(\kappa)$, respectively. Similarly, $\sigma_M^R(\kappa)$ and $\sigma_N^R(\kappa)$ represent right membership functions for $\sigma_M(\kappa)$ and $\sigma_N(\kappa)$, respectively.

Definition 2 [24]. A BFN h in a parametric form is a quadruple $\prec [\underline{h}^\star(r), \overline{h}^\star(r)], [\underline{h}^\bullet(s), \overline{h}^\bullet(s)] \succ$ of the mappings $\underline{h}^\star(r), \overline{h}^\star(r), \underline{h}^\bullet(s), \overline{h}^\bullet(s)$; $0 \leq r \leq 1$, $-1 \leq s \leq 0$ and fulfill the following restrictions:

(i) $\underline{h}^\star(r)$ is bounded left-continuous and monotonically non-decreasing mapping on a set $(0,1]$ and right-continuous at point 0,
(ii) $\overline{h}^\star(r)$ is bounded left-continuous and monotonically non-increasing mapping on a set $(0,1]$ and right-continuous at point 0,
(iii) $\underline{h}^\bullet(r)$ is bounded left-continuous monotonically non-increasing mapping on $(-1,0]$ an right-continuous at point -1,
(iv) $\overline{h}^\bullet(s)$ is bounded left-continuous monotonically non decreasing mapping on a set $(-1,0]$ and right-continuous at point -1,

(v) $\underline{h}^{\star}(r) \leq \overline{h}^{\star}(r)$,
(vi) $\underline{h}^{\bullet}(s) \leq \overline{h}^{\bullet}(s)$.

On the based of [25], we use BFN $h = \prec [\underline{h}^{\star}(r), \overline{h}^{\star}(r)], [\underline{h}^{\bullet}(s), \overline{h}^{\bullet}(s)] \succ$ in symmetric form if

$$h^c(r) = \frac{\underline{h}^{\star}(r) + \overline{h}^{\star}(r)}{2}, \quad h^c(s) = \frac{\underline{h}^{\bullet}(s) + \overline{h}^{\bullet}(s)}{2},$$

is a real number for all $0 \leq r \leq 1, -1 \leq s \leq 0$. For example, $h = \prec [\underline{h}^{\star}(r), \overline{h}^{\star}(r)], [\underline{h}^{\bullet}(s), \overline{h}^{\bullet}(s)] \succ = \prec [3+2r, 6-r], [1-s, 5+3s] \succ$ is parametric form BFN and $v = \prec [\underline{v}^{\star}(r), \overline{v}^{\star}(r)], [\underline{v}^{\bullet}(s), \overline{v}^{\bullet}(s)] \succ = \prec [-4+2r, -2-2r], [-6-s, -3+s] \succ$ is a parametric form of symmetric BFN.

Definition 3 [24]. For arbitrary BFNs $h = \prec [\underline{h}^{\star}, \overline{h}^{\star}], [\underline{h}^{\bullet}, \overline{h}^{\bullet}] \succ, g = \prec [\underline{g}^{\star}, \overline{g}^{\star}], [\underline{g}^{\bullet}, \overline{g}^{\bullet}] \succ$ and c is arbitrary scalar number, we define $h + g = m = \prec [\underline{m}^{\star}, \overline{m}^{\star}], [\underline{m}^{\bullet}, \overline{m}^{\bullet}] \succ$, $h.g = n = \prec [\underline{n}^{\star}, \overline{n}^{\star}], [\underline{n}^{\bullet}, \overline{n}^{\bullet}] \succ$ and scalar multiplication by c as follows:

(i) $\underline{m}^{\star}(r) = \underline{h}^{\star}(r) + \underline{g}^{\star}(r), \quad \overline{m}^{\star}(r) = \overline{h}^{\star}(r) + \overline{g}^{\star}(r)$,
(ii) $\underline{m}^{\bullet}(s) = \underline{h}^{\bullet}(s) + \underline{g}^{\bullet}(s), \quad \overline{m}^{\bullet}(s) = \overline{h}^{\bullet}(s) + \overline{g}^{\bullet}(s)$,
(iii) $\underline{n}^{\star}(r) = \min\{\underline{h}^{\star}.\underline{g}^{\star}, \underline{h}^{\star}.\overline{g}^{\star}, \overline{h}^{\star}.\underline{g}^{\star}, \overline{h}^{\star}.\overline{g}^{\star}\}$,
(iv) $\overline{n}^{\star}(r) = \max\{\underline{h}^{\star}.\underline{g}^{\star}, \underline{h}^{\star}.\overline{g}^{\star}, \overline{h}^{\star}.\underline{g}^{\star}, \overline{h}^{\star}.\overline{g}^{\star}\}$,
(v) $\underline{n}^{\bullet}(s) = \max\{\underline{h}^{\bullet}.\underline{g}^{\bullet}, \underline{h}^{\bullet}.\overline{g}^{\bullet}, \overline{h}^{\bullet}.\underline{g}^{\bullet}, \overline{h}^{\bullet}.\overline{g}^{\bullet}\}$,
(vi) $\overline{n}^{\bullet}(s) = \min\{\underline{h}^{\bullet}.\underline{g}^{\bullet}, \underline{h}^{\bullet}.\overline{g}^{\bullet}, \overline{h}^{\bullet}.\underline{g}^{\bullet}, \overline{h}^{\bullet}.\overline{g}^{\bullet}\}$,
(vii) If c is nonnegative then, $ch = \prec [c(\underline{h}^{\star}), c(\overline{h}^{\star})], [c(\underline{h}^{\bullet}), c(\overline{h}^{\bullet})] \succ$, But if c is negative then, $ch = \prec [c(\overline{h}^{\star}), c(\underline{h}^{\star})], [c(\overline{h}^{\bullet}), c(\underline{h}^{\bullet})] \succ$.

The collection of all BFNs under the operation addition and scaler product, represented by Ω are concave and convex cones.

Definition 4 [26]. Let ζ be a complex set. Then the fuzzy subset $\widetilde{Z}$ is called fuzzy complex set of ζ is defined by the membership function $\mu_{\widetilde{Z}} : \zeta \to [0,1]$.

Definition 5 [20]. A fuzzy complex number(FCN) $\widetilde{H}$ is a fuzzy complex set with its membership function $\pounds(h|\widetilde{H})$ is a mapping $\pounds : C \to [0,1]$, Which may be represented as $\widetilde{H} = u + iv$, where $u = (\underline{u}(r), \overline{u}(r))$ and $v = (\underline{v}(r), \overline{v}(r))$, for all $0 \leq r \leq 1$.

Definition 6. Let ζ be a complex set. Then the bipolar fuzzy subset $\widetilde{Z}$ is called bipolar fuzzy complex set of ζ is defined by the membership function $\mu_{\widetilde{Z}} : \zeta \to [-1,0] \times [0,1]$.

Definition 7. A bipolar fuzzy complex number (BFCN) $\widetilde{H}$ is bipolar fuzzy complex set with its membership function $\pounds(h|\widetilde{H})$ is a mapping $\pounds : C \to [-1,0] \times [0,1]$, which may be represented as $\widetilde{H} = u + iv$, where $u = [\underline{u}^{\star}(r), \overline{u}^{\star}(r)], [\underline{u}^{\bullet}(s), \overline{u}^{\bullet}(s)]$ and $v = [\underline{v}^{\star}(r), \overline{v}^{\star}(r)], [\underline{v}^{\bullet}(s), \overline{v}^{\bullet}(s)]$, for all $0 \leq r \leq 1$ and $-1 \leq s \leq 0$.

Definition 8 [27]. For any two arbitrary BFCN $\widetilde{h}_1 = u_1 + iv_1$ and $\widetilde{h}_2 = u_2 + iv_2$ (where u_1, u_2, v_1 and v_2 are BFN), the bipolar fuzzy complex arithmetic for addition and multiplication is given as,

(i) $\widetilde{h}_1 + \widetilde{h}_2 = (u_1 + u_2) + i(v_1 + v_2)$,
(ii) $\widetilde{h}_1 \times \widetilde{h}_2 = \{(u_1 \times u_2) - (v_1 \times v_2)\} + i\{(u_1 \times v_2) + (v_1 \times u_2)\}$.

Definition 9 [22]. Consider the $n \times n$ system of linear equations,

$$\begin{cases} l_{11}h_1 + l_{12}h_2 + ... + l_{1n}h_n & = z_1, \\ l_{21}h_1 + l_{22}h_2 + ... + l_{2n}h_n & = z_2, \\ . \\ . \\ . \\ l_{n1}h_1 + l_{n2}h_2 + ... + l_{nn}h_n & = z_n, \end{cases} \tag{1}$$

is the complex bipolar fuzzy linear (CBFL) system of equation, if all the coefficient elements l_{pq}, $1 \le p, q \le n$ is a crisp matrix of order $n \times n$ and each Z_q, $1 \le q \le n$ is a BFCN, and each unknown parameter h_p, $1 \le p \le n$ is also BFCN. The CBFL system in (1) can be written in matrix form as,

$$LH = Z, \tag{2}$$

so, we can rewrite the above CBFL system (2) as follows:

$$L(U + iV) = E + iF. \tag{3}$$

If we compare the real and imaginary part of the given equation, then we obtained two bipolar fuzzy linear system given as,

$$LU = E \quad , \quad LV = F. \tag{4}$$

Definition 10. A BFCNs vector $H = (h_1, h_2, ...h_n)^T$ is the form of $(H_q)_{(r,s)} = [\underline{u}_q^\star(r), \overline{u}_q^\star(r)], [\underline{u}_q^\bullet(s), \overline{u}_q^\bullet(s)] + i[\underline{v}_q^\star(r), \overline{v}_q^\star(r)], [\underline{v}_q^\bullet(s), \overline{v}_q^\bullet(s)] = u_q + iv_q$, where $1 \le q \le n$, $0 \le r \le 1$ and $-1 \le s \le 0$ is called the bipolar fuzzy complex solution of CBFL System of equations (3) and (4) if,

$$U' = (u_1, u_2, ..., u_n), \qquad V' = (v_1, v_2, ..., v_n),$$

are the bipolar fuzzy solution of bipolar fuzzy linear system in (4) respectively if,

$$LU' = E \quad , \quad LV' = F, \tag{5}$$

and this solution vector also satisfied the system (1) given as,

$$\begin{cases} \underline{\sum_{q=1}^{n} l_{pq} u_q^\star(r)} &= \sum_{q=1}^{n} l_{pq} \underline{u_q^\star(r)} = \underline{e}_p^\star(r), \\ \overline{\sum_{q=1}^{n} l_{pq} u_q^\star(r)} &= \sum_{q=1}^{n} l_{pq} \overline{u_q^\star(r)} = \overline{e}_p^\star(r), \\ \underline{\sum_{q=1}^{n} l_{pq} u_q^\bullet(s)} &= \sum_{q=1}^{n} l_{pq} \underline{u_q^\bullet(s)} = \underline{e}_p^\bullet(s), \\ \overline{\sum_{q=1}^{n} l_{pq} u_q^\bullet(s)} &= \sum_{q=1}^{n} l_{pq} \overline{u_q^\bullet(s)} = \overline{e}_p^\bullet(s). \end{cases} \quad (6)$$

$$\begin{cases} \underline{\sum_{q=1}^{n} l_{pq} v_q^\star(r)} &= \sum_{q=1}^{n} l_{pq} \underline{v_q^\star(r)} = \underline{f}_p^\star(r), \\ \overline{\sum_{q=1}^{n} l_{pq} v_q^\star(r)} &= \sum_{q=1}^{n} l_{pq} \overline{v_q^\star(r)} = \overline{f}_p^\star(r), \\ \underline{\sum_{q=1}^{n} l_{pq} v_q^\bullet(s)} &= \sum_{q=1}^{n} l_{pq} \underline{v_q^\bullet(s)} = \underline{f}_p^\bullet(s), \\ \overline{\sum_{q=1}^{n} l_{pq} v_q^\bullet(s)} &= \sum_{q=1}^{n} l_{pq} \overline{v_q^\bullet(s)} = \overline{f}_p^\bullet(s). \end{cases} \quad (7)$$

In parametric form solution vector h_q is given as,

$$h_q = [\underline{h}_q^\star(r), \overline{h}_q^\star(r)], [\underline{h}_q^\bullet(s), \overline{h}_q^\bullet(s)] = \prec [\underline{u}_q^\star(r) + i\underline{v}_q^\star(r), \overline{u}_q^\star(r) + i\overline{v}_q^\star(r)], [\underline{u}_q^\bullet(s) + i\underline{v}_q^\bullet(s), \overline{u}_q^\bullet(r) + i\overline{v}_q^\bullet(s)] \succ,$$

$1 \leq q \leq n$, $0 \leq r \leq 1$ and $-1 \leq s \leq 0$.

3 Solution Procedure for n × n CBFL System

To solve the CBFL system of equation in (1), first we should compare the real and imaginary part of system of equations to get two $n \times n$ bipolar fuzzy linear (BFL) systems $LU = E$ and $LV = F$.

To solve bipolar fuzzy linear (BFL) system $LU = E$, we solve pair of positive and negative $n \times n$ crisp linear system (CLS) and the right side column is function vector

$$(\underline{e}_1^\star(r), \underline{e}_2^\star(r), ..., \underline{e}_n^\star(r), \overline{e}_1^\star(r), \overline{e}_2^\star(r), ..., \overline{e}_n^\star(r))^T \, and \, (\underline{e}_1^\bullet(s), \underline{e}_2^\bullet(s), ..., \underline{e}_n^\bullet(s), \overline{e}_1^\bullet(s), \overline{e}_2^\bullet(s), ..., \overline{e}_n^\bullet(s))^T.$$

Now we rearrange the bipolar linear system , so the unknown parameters are $\underline{u}_q^\star(-\overline{u}_q^\star), \underline{u}_q^\bullet(-\overline{u}_q^\bullet)$, $1 \leq q \leq n$ and column of right side vector are

$$(\underline{e}_1^\star(r), \underline{e}_2^\star(r), ..., \underline{e}_n^\star(r), \overline{e}_1^\star(r), \overline{e}_2^\star(r), ..., \overline{e}_n^\star(r))^T \, and \, (\underline{e}_1^\bullet(s), \underline{e}_2^\bullet(s), ..., \underline{e}_n^\bullet(s), \overline{e}_1^\bullet(s), \overline{e}_2^\bullet(s), ..., \overline{e}_n^\bullet(s))^T.$$

We get first positive $2n \times 2n$ CLS for $LU = E$,
$l_{11}\underline{u}_1^\star(r) + l_{12}\underline{u}_2^\star(r) + ... + l_{1n}\underline{u}_n^\star(r) + l_{1,n+1}(-\overline{u}_1^\star(r)) + l_{1,n+2}(-\overline{u}_2^\star(r)) + ... + l_{1,2n}(-\overline{u}_n^\star(r)) = \underline{e}_1^\star(r)$,

.

.

.

$l_{n1}\underline{u}_1^\star(r) + l_{n2}\underline{u}_2^\star(r) + ... + l_{nn}\underline{u}_n^\star(r) + l_{n,n+1}(-\overline{u}_1^\star(r)) + l_{n,n+2}(-\overline{u}_2^\star(r)) + ... + l_{n,2n}(-\overline{u}_n^\star(r)) = \underline{e}_n^\star(r)$,
$l_{n+1,1}\underline{u}_1^\star(r) + l_{n+1,2}\underline{u}_2^\star(r) + ... + l_{n+1,n}\underline{u}_n^\star(r) + l_{n+1,n+1}(-\overline{u}_1^\star(r)) + l_{n+1,n+2}(-\overline{u}_2^\star(r)) + ... + l_{n+1,2n}(-\overline{u}_n^\star(r)) = \overline{e}_1^\star(r)$,

.

.

.

$l_{2n,1}\underline{u}_1^\star(r) + l_{2n,2}\underline{u}_2^\star(r) + ... + l_{2n,n}\underline{u}_n^\star(r) + l_{2n,n+1}(-\overline{u}_1^\star(r)) + l_{2n,n+2}(-\overline{u}_2^\star(r)) + ... + l_{2n,2n}(-\overline{u}_n^\star(r)) = \overline{e}_n^\star(r)$,
where l_{pq} are determined as follows:

$$c_{pq} \geq 0 \Longrightarrow u_{pq} = c_{pq}, u_{p+n,q} = c_{pq}, \qquad (8)$$
$$c_{pq} < 0 \Longrightarrow u_{p,q+n} = -c_{pq}, u_{p+n,q} = -c_{pq},$$

where $1 \leq p, q \leq n$ and any u_{pq} which is not determined from Eq. (8) is zero. Using the matrix notation, we obtained

$$MU^\star(r) = E^\star(r), \qquad (9)$$

where $M = l_{pq}$, $1 \leq p, q \leq 2n$

$(\underline{u}_1^\star(r), \underline{u}_2^\star(r), ..., \underline{u}_n^\star(r), -\overline{u}_1^\star(r), -\overline{u}_2^\star(r), ..., -\overline{u}_n^\star(r))^T and$

$(\underline{e}_1^\star(r), \underline{e}_2^\star(r), ..., \underline{e}_n^\star(r), -\overline{e}_1^\star(r), -\overline{e}_2^\star(r), ..., -\overline{e}_n^\star(r))^T$.
Now we can write negative $2n \times 2n$ CLS for $LU = E$,

$l_{11}\underline{u}_1^\bullet(s) + l_{12}\underline{u}_2^\bullet(s) + ... + l_{1n}\underline{u}_n^\bullet(s) + l_{1,n+1}(-\overline{u}_1^\bullet(s)) + l_{1,n+2}(-\overline{u}_2^\bullet(s)) + ... + l_{1,2n}(-\overline{u}_n^\bullet(s)) = \underline{e}_1^\bullet(s)$,

.

.

.

$l_{n1}\underline{u}_1^\bullet(s) + l_{n2}\underline{u}_2^\bullet(s) + ... + l_{nn}\underline{u}_n^\bullet(s) + l_{n,n+1}(-\overline{u}_1^\bullet(s)) + l_{n,n+2}(-\overline{u}_2^\bullet(s)) + ... + l_{n,2n}(-\overline{u}_n^\bullet(s)) = \underline{e}_n^\bullet(s)$,
$l_{n+1,1}\underline{u}_1^\bullet(s) + l_{n+1,2}\underline{u}_2^\bullet(s) + ... + l_{n+1,n}\underline{u}_n^\bullet(s) + l_{n+1,n+1}(-\overline{u}_1^\bullet(s)) + l_{n+1,n+2}(-\overline{u}_2^\bullet(s)) + ... + l_{n+1,2n}(-\overline{u}_n^\bullet(s)) = \overline{e}_1^\bullet(s)$,

.

.

.

$l_{2n,1}\underline{u}_1^\bullet(s) + l_{2n,2}\underline{u}_2^\bullet(s) + ... + l_{2n,n}\underline{u}_n^\bullet(s) + l_{2n,n+1}(-\overline{u}_1^\bullet(s)) + l_{2n,n+2}(-\overline{u}_2^\bullet(s)) + ... + l_{2n,2n}(-\overline{u}_n^\bullet(s)) = \overline{e}_n^\bullet(s)$,
where l_{pq} are determined as follows:

$$c_{pq} \geq 0 \Longrightarrow u_{pq} = c_{pq}, u_{p+n,q} = c_{pq}, \qquad (10)$$
$$c_{pq} < 0 \Longrightarrow u_{p,q+n} = -c_{pq}, u_{p+n,q} = -c_{pq},$$

where $1 \leq p, q \leq n$ and any u_{pq} which is not determined from Eq. (9) is zero. Using the matrix notation, we obtained

$$MU^{\bullet}(s) = E^{\bullet}(s), \tag{11}$$

where $M = l_{pq}$, $1 \leq p, q \leq 2n$

$(\underline{u}_1^{\bullet}(s), \underline{u}_2^{\bullet}(s), ..., \underline{u}_n^{\bullet}(s), -\overline{u}_1^{\bullet}(s), -\overline{u}_2^{\bullet}(s), ..., -\overline{u}_n^{\bullet}(s))^T and$

$(\underline{e}_1^{\bullet}(s), \underline{e}_2^{\bullet}(s), ..., \underline{e}_n^{\bullet}(s), -\overline{e}_1^{\bullet}(s), -\overline{e}_2^{\bullet}(s), ..., -\overline{e}_n^{\bullet}(s))^T$.

We can also write the matrix form of Eqs. (9) and (11) as,

$$\begin{pmatrix} A & B \\ B & A \end{pmatrix} \begin{pmatrix} \underline{u}^{\star}(r) \\ -\overline{u}^{\star}(r) \end{pmatrix} = \begin{pmatrix} \underline{e}^{\star}(r) \\ -\overline{e}^{\star}(r) \end{pmatrix}, \tag{12}$$

and,

$$\begin{pmatrix} A & B \\ B & A \end{pmatrix} \begin{pmatrix} \underline{u}^{\bullet}(s) \\ -\overline{u}^{\bullet}(s) \end{pmatrix} = \begin{pmatrix} \underline{e}^{\bullet}(s) \\ -\overline{e}^{\bullet}(s) \end{pmatrix}. \tag{13}$$

In the similar way we can write the solution procedure for imaginary part $LV = F$ of CBFL system (1).

First we write the matrix notation of positive part of $LV = F$ and get $2n \times 2n$ CLS,

$$MV^{\star}(r) = F^{\star}(r), \tag{14}$$

where $M = l_{pq}$, $1 \leq p, q \leq 2n$

$(\underline{v}_1^{\star}(r), \underline{v}_2^{\star}(r), ..., \underline{v}_n^{\star}(r), -\overline{v}_1^{\star}(r), -\overline{v}_2^{\star}(r), ..., -\overline{v}_n^{\star}(r))^T and$

$(\underline{f}_1^{\star}(r), \underline{f}_2^{\star}(r), ..., \underline{f}_n^{\star}(r), -\overline{f}_1^{\star}(r), -\overline{f}_2^{\star}(r), ..., -\overline{f}_n^{\star}(r))^T$.

The matrix notation for Eq. (14) is given as,

$$\begin{pmatrix} A & B \\ B & A \end{pmatrix} \begin{pmatrix} \underline{v}^{\star}(r) \\ -\overline{v}^{\star}(r) \end{pmatrix} = \begin{pmatrix} \underline{f}^{\star}(r) \\ -\overline{f}^{\star}(r) \end{pmatrix}. \tag{15}$$

Similarly, we can write the matrix notation of negative part of $LV = F$ and get $2n \times 2n$ CLS,

$$MV^{\bullet}(s) = F^{\bullet}(s), \tag{16}$$

where $M = l_{pq}$, $1 \leq p, q \leq 2n$

$(\underline{v}_1^{\bullet}(s), \underline{v}_2^{\bullet}(s), ..., \underline{v}_n^{\bullet}(s), -\overline{v}_1^{\bullet}(s), -\overline{v}_2^{\bullet}(s), ..., -\overline{v}_n^{\bullet}(s))^T and$

$(\underline{f}_1^{\bullet}(s), \underline{f}_2^{\bullet}(s), ..., \underline{f}_n^{\bullet}(s), -\overline{f}_1^{\bullet}(s), -\overline{f}_2^{\bullet}(s), ..., -\overline{f}_n^{\bullet}(s))^T$.

The matrix notation for Eq. (16) is given as,

$$\begin{pmatrix} A & B \\ B & A \end{pmatrix} \begin{pmatrix} \underline{v}^{\bullet}(s) \\ -\overline{v}^{\bullet}(s) \end{pmatrix} = \begin{pmatrix} \underline{f}^{\bullet}(s) \\ -\overline{f}^{\bullet}(s) \end{pmatrix}. \tag{17}$$

To find the solution of CBFL system of equation (1), we can find the solution of equations (12), (13), (15) and (17) by using matrix inverse method.

Theorem 1 [22]. The bipolar fuzzy complex vector solution of CBFL system (2) and (3) are equivalent.

Proof. It is sufficient to show that the solution $LH^{'} = Z^{'}$ obtained from (2) and $LH^{'} = Y^{'}$ obtained from (3) are the same where L is defined in (8) and,
$H^{'} = [(\underline{h}_1^{\star}, \overline{h}_1^{\star}], [\underline{h}_1^{\bullet}, -\overline{h}_1^{\bullet}], ..., [\underline{h}_n^{\star}, \overline{h}_n^{\star}], [\underline{h}_n^{\bullet}, \overline{h}_n^{\bullet}])^T$,
$Y^{'} = (\underline{e}_1^{\star}, ..., \underline{e}_n^{\star}, -\overline{e}_1^{\star}, ..., -\overline{e}_n^{\star})^T, (\underline{e}_1^{\bullet}, ..., \underline{e}_n^{\bullet}, -\overline{e}_1^{\bullet}, ..., -\overline{e}_n^{\bullet})^T + i(\underline{f}_1^{\star}, ..., \underline{f}_n^{\star}, -\overline{f}_1^{\star},$
$..., -\overline{f}_n^{\star})^T,$
$(\underline{f}_1^{\bullet}, ..., \underline{f}_n^{\bullet}, -\overline{f}_1^{\bullet}, ..., -\overline{f}_n^{\bullet})^T,$
$Z^{'} = (\underline{e}_1^{\star} + i\underline{f}_1^{\star}, ..., \underline{e}_n^{\star} + i\underline{f}_n^{\star}, -(\overline{e}_1^{\star} + i\overline{f}_1^{\star}), ..., -(\overline{e}_n^{\star} + i\overline{f}_n^{\star}))^T, (\underline{e}_1^{\bullet} + i\underline{f}_1^{\bullet}, ..., \underline{e}_n^{\bullet} +$
$i\underline{f}_n^{\bullet}, -(\overline{e}_1^{\bullet} + i\overline{f}_1^{\bullet}), ..., -(\overline{e}_n^{\bullet} + i\overline{f}_n^{\bullet}))^T,$
it is clear that,
$Y^{'} = (\underline{e}_1^{\star}, ..., \underline{e}_n^{\star}, -\overline{e}_1^{\star}, ..., -\overline{e}_n^{\star})^T, (\underline{e}_1^{\bullet}, ..., \underline{e}_n^{\bullet}, -\overline{e}_1^{\bullet}, ..., -\overline{e}_n^{\bullet})^T + i(\underline{f}_1^{\star}, ..., \underline{f}_n^{\star}, -\overline{f}_1^{\star}$
$, ..., -\overline{f}_n^{\star})^T,$
$(\underline{f}_1^{\bullet}, ..., \underline{f}_n^{\bullet}, -\overline{f}_1^{\bullet}, ..., -\overline{f}_n^{\bullet})^T,$
$= (\underline{e}_1^{\star} + i\underline{f}_1^{\star}, ..., \underline{e}_n^{\star} + i\underline{f}_n^{\star}, -\overline{e}_1^{\star} - i\overline{f}_1^{\star}, ..., -\overline{e}_n^{\star} - i\overline{f}_n^{\star})^T, (\underline{e}_1^{\bullet} + i\underline{f}_1^{\bullet}, ..., \underline{e}_n^{\bullet} + i\underline{f}_n^{\bullet}, -\overline{e}_1^{\bullet} - i\overline{f}_1^{\bullet},$
$..., -\overline{e}_n^{\bullet} - i\overline{f}_n^{\bullet})^T,$
$= (\underline{e}_1^{\star} + i\underline{f}_1^{\star}, ..., \underline{e}_n^{\star} + i\underline{f}_n^{\star}, -(\overline{e}_1^{\star} + i\overline{f}_1^{\star}), ..., -(\overline{e}_n^{\star} + i\overline{f}_n^{\star}))^T, (\underline{e}_1^{\bullet} + i\underline{f}_1^{\bullet}, ..., \underline{e}_n^{\bullet} + i\underline{f}_n^{\bullet}, -(\overline{e}_1^{\bullet} +$
$i\overline{f}_1^{\bullet}, ..., -(\overline{e}_n^{\bullet} + i\overline{f}_n^{\bullet}))^T,$
$= Z^{'},$

hence the solution is same, which completes the proof.

Theorem 2 [24]. For any arbitrary bipolar fuzzy complex vector Z the unique solution $\widetilde{H}$ is a bipolar fuzzy complex vector of CBFL system of equations in (1) if and only if the matrix M= $\begin{pmatrix} A & B \\ B & A \end{pmatrix}$ is non-singular, the necessary and sufficient condition that the matrices $(A+B)^{-1}$ and $(A-B)^{-1}$ exist.

Example 1. Consider the 2×2 CBFL system

$$\widetilde{h}_1 - \widetilde{h}_2 = [-6+3r, -1-2r], [-6-3s, 3s] + i[-8+4r, -4r], [-7-2s, -2+3s],$$

$$\widetilde{h}_1 + 2\widetilde{h}_2 = [17+4r, 24-3r], [9-6s, 18+3s] + i[25+7r, 36-4r], [22-3s, 29+4s].$$

To solve this CBFL system first we put $\widetilde{h}_1 = u_1 + iv_1$ and $\widetilde{h}_2 = u_2 + iv_2$ in original system and get,

$$(u_1 + iv_1) - (u_2 + iv_2) = [-6+3r, -1-2r], [-6-3s, 3s] + i[-8+4r, -4r], [-7-2s, -2+3s],$$

$$(u_1 + iv_1) + 2(u_2 + iv_2) = [17+4r, 24-3r], [9-6s, 18+3s] + i[25+7r, 36-4r], [22-3s, 29+4s],$$

compare the real and imaginary parts to both side of the equations and we obtained two BFL system of equations given as,

$$\begin{cases} u_1 - u_2 = [-6+3r, -1-2r], [-6-3s, 3s], \\ u_1 + 2u_2 = [17+4r, 24-3r], [9-6s, 18+3s], \end{cases} \tag{18}$$

and,

$$\begin{cases} v_1 - v_2 = [-8+4r, -4r], [-7-2s, -2+3s], \\ v_1 + 2v_2 = [25+7r, 36-4r], [22-3s, 29+4s], \end{cases} \tag{19}$$

to solve the BFL system in (18), first we take the positive ($\star$) part of the system (18). The extended 4×4 form of the positive ($\star$) part of system (18) is given as,

$$\begin{aligned} 1\underline{u}_1^\star + 0\underline{u}_2^\star + 0(-\overline{u}_1^\star) + 1(-\overline{u}_2^\star) &= 3r - 6, \\ 1\underline{u}_1^\star + 2\underline{u}_2^\star + 0(-\overline{u}_1^\star) + 0(-\overline{u}_2^\star) &= 4r + 17, \\ 0\underline{u}_1^\star + 1\underline{u}_2^\star + 1(-\overline{u}_1^\star) + 0(-\overline{u}_2^\star) &= 2r + 1, \\ 0\underline{u}_1^\star + 0\underline{u}_2^\star + 1(-\overline{u}_1^\star) + 2(-\overline{u}_2^\star) &= 3r - 24. \end{aligned}$$

The matrix form of the above system is

$$\begin{pmatrix} 1\,0\,0\,1 \\ 1\,2\,0\,0 \\ 0\,1\,1\,0 \\ 0\,0\,1\,2 \end{pmatrix} \begin{pmatrix} \underline{u}_1^\star(r) \\ \underline{u}_2^\star(r) \\ -\overline{u}_1^\star(r) \\ -\overline{u}_2^\star(r) \end{pmatrix} = \begin{pmatrix} 3r-6 \\ 4r+17 \\ 2r+1 \\ 3r-24 \end{pmatrix},$$

this can also be written as,

$$MU^\star = E^\star,$$

and the solution of above system is,

$$U^\star = \begin{pmatrix} \underline{u}_1^\star(r) \\ \underline{u}_2^\star(r) \\ -\overline{u}_1^\star(r) \\ -\overline{u}_2^\star(r) \end{pmatrix} = M^{-1}E^\star = \begin{pmatrix} 4/3 & -1/3 & 2/3 & -2/3 \\ -2/3 & 2/3 & -1/3 & 1/3 \\ 2/3 & -2/3 & 4/3 & -1/3 \\ -1/3 & 1/3 & -2/3 & 2/3 \end{pmatrix} \begin{pmatrix} 3r-6 \\ 4r+17 \\ 2r+1 \\ 3r-24 \end{pmatrix}$$

$$\begin{aligned} \underline{u}_1^\star(r) &= 3 + 2r \quad , \quad \underline{u}_2^\star(r) = 7 + r, \\ \overline{u}_1^\star(r) &= 6 - r \quad , \quad \overline{u}_2^\star(r) = 9 - r. \end{aligned}$$

The graphical representation for positive ($\star$) membership function of real part of complex bipolar solution number is shown below in Fig. 1.

Next we take the negative ($\bullet$) part of the system (18). The extended 4×4 form of the negative ($\bullet$) part of system (18) is given as,

$$\begin{aligned} 1\underline{u}_1^\bullet + 0\underline{u}_2^\bullet + 0(-\overline{u}_1^\bullet) + 1(-\overline{u}_2^\bullet) &= -6 - 3s, \\ 1\underline{u}_1^\bullet + 2\underline{u}_2^\bullet + 0(-\overline{u}_1^\bullet) + 0(-\overline{u}_2^\bullet) &= 9 - 6s, \\ 0\underline{u}_1^\bullet + 1\underline{u}_2^\bullet + 1(-\overline{u}_1^\bullet) + 0(-\overline{u}_2^\bullet) &= -3s, \\ 0\underline{u}_1^\bullet + 0\underline{u}_2^\bullet + 1(-\overline{u}_1^\bullet) + 2(-\overline{u}_2^\bullet) &= -18 - 3s. \end{aligned}$$

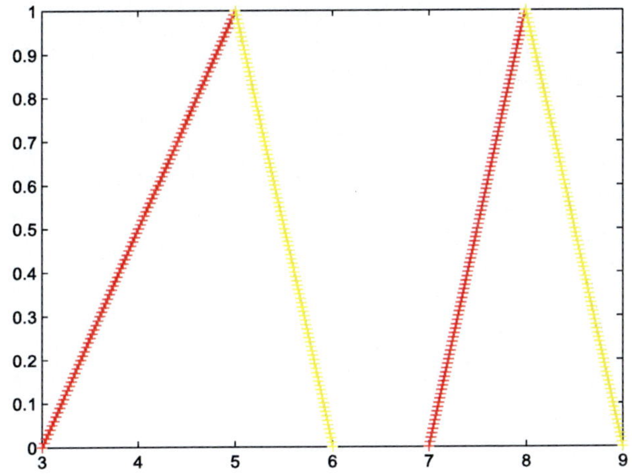

Fig. 1. The graphical representation of positive ($\star$) membership functions $\underline{u}^{\star}(r), \overline{u}^{\star}(r)$ of complex bipolar fuzzy numbers.

The matrix form of the above system is

$$\begin{pmatrix} 1\,0\,0\,1 \\ 1\,2\,0\,0 \\ 0\,1\,1\,0 \\ 0\,0\,1\,2 \end{pmatrix} \begin{pmatrix} \underline{u}_1^{\bullet}(s) \\ \underline{u}_2^{\bullet}(s) \\ -\overline{u}_1^{\bullet}(s) \\ -\overline{u}_2^{\bullet}(s) \end{pmatrix} = \begin{pmatrix} -6-3s \\ 9-6s \\ -3s \\ -18-3s \end{pmatrix},$$

this can also be written as,

$$MU^{\bullet} = E^{\bullet},$$

and the solution of above system is,

$$U^{\bullet} = \begin{pmatrix} \underline{u}_1^{\bullet}(s) \\ \underline{u}_2^{\bullet}(s) \\ -\overline{u}_1^{\bullet}(s) \\ -\overline{u}_2^{\bullet}(s) \end{pmatrix} = M^{-1}E^{\bullet} = \begin{pmatrix} 4/3 & -1/3 & 2/3 & -2/3 \\ -2/3 & 2/3 & -1/3 & 1/3 \\ 2/3 & -2/3 & 4/3 & -1/3 \\ -1/3 & 1/3 & -2/3 & 2/3 \end{pmatrix} \begin{pmatrix} -6-3s \\ 9-6s \\ -3s \\ -18-3s \end{pmatrix}$$

$$\underline{u}_1^{\bullet}(s) = 1-2s \quad , \quad \underline{u}_2^{\bullet}(s) = 4-2s,$$
$$\overline{u}_1^{\bullet}(s) = 4+s \quad , \quad \overline{u}_2^{\bullet}(s) = 7+s.$$

The graphical representation for negative ($\bullet$) membership functions of real part of complex bipolar solution number is shown below in Fig. 2.

In the similar way, to solve the BFL system in (19), first we take the positive ($\star$) part of the system (19). The extended 4×4 form of the positive ($\star$) part of

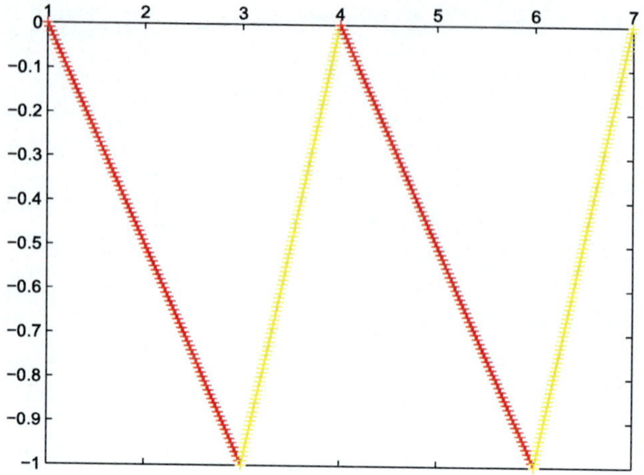

Fig. 2. The graphical representation of negative ($\bullet$) membership functions $\underline{u}^{\bullet}(s), \overline{u}^{\bullet}(s)$ of complex bipolar fuzzy numbers.

system (19) is given as,

$$\begin{aligned}
1\underline{v}_1^{\star} + 0\underline{v}_2^{\star} + 0(-\overline{v}_1^{\star}) + 1(-\overline{v}_2^{\star}) &= -8 + 4r,\\
1\underline{v}_1^{\star} + 2\underline{v}_2^{\star} + 0(-\overline{v}_1^{\star}) + 0(-\overline{v}_2^{\star}) &= 25 + 7r,\\
0\underline{v}_1^{\star} + 1\underline{v}_2^{\star} + 1(-\overline{v}_1^{\star}) + 0(-\overline{v}_2^{\star}) &= -4r,\\
0\underline{v}_1^{\star} + 0\underline{v}_2^{\star} + 1(-\overline{v}_1^{\star}) + 2(-\overline{v}_2^{\star}) &= 36 - 4r.
\end{aligned}$$

The matrix form of the above system is

$$\begin{pmatrix} 1\,0\,0\,1 \\ 1\,2\,0\,0 \\ 0\,1\,1\,0 \\ 0\,0\,1\,2 \end{pmatrix} \begin{pmatrix} \underline{v}_1^{\star}(r) \\ \underline{v}_2^{\star}(r) \\ -\overline{v}_1^{\star}(r) \\ -\overline{v}_2^{\star}(r) \end{pmatrix} = \begin{pmatrix} -8+4r \\ 25+7r \\ 4r \\ 4r-36 \end{pmatrix},$$

this can also be written as,

$$MV^{\star} = F^{\star},$$

and the solution of above system is,

$$V^{\star} = \begin{pmatrix} \underline{v}_1^{\star}(r) \\ \underline{v}_2^{\star}(r) \\ -\overline{v}_1^{\star}(r) \\ -\overline{v}_2^{\star}(r) \end{pmatrix} = M^{-1}F^{\star} = \begin{pmatrix} 4/3 & -1/3 & 2/3 & -2/3 \\ -2/3 & 2/3 & -1/3 & 1/3 \\ 2/3 & -2/3 & 4/3 & -1/3 \\ -1/3 & 1/3 & -2/3 & 2/3 \end{pmatrix} \begin{pmatrix} -8+4r \\ 25+7r \\ 4r \\ 4r-36 \end{pmatrix}$$

$$\underline{v}_1^{\star}(r) = 3r + 5 \quad , \quad \underline{v}_2^{\star}(r) = 2r + 10,$$

$$\overline{v}_1^{\star}(r) = 10 - 2r \quad , \quad \overline{v}_2^{\star}(r) = 13 - r.$$

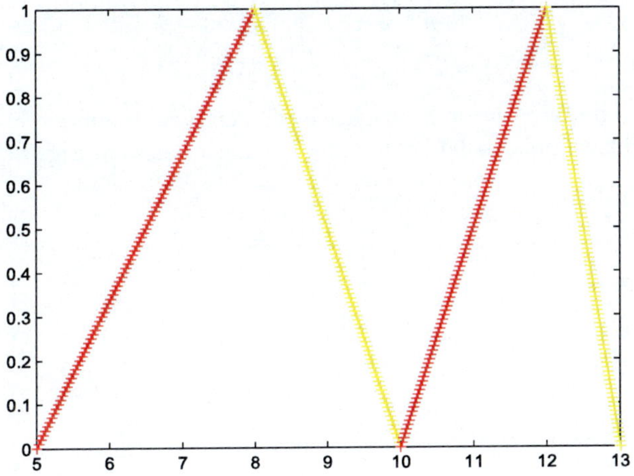

Fig. 3. The graphical representation of positive ($\star$) membership functions $\underline{v}^\star(r), \overline{v}^\star(r)$ of complex bipolar fuzzy numbers.

The graphical representation for positive ($\star$) membership functions of imaginary part of complex bipolar solution number is shown below in Fig. 3.

Next we take the negative ($\bullet$) part of the system (19). The extended 4×4 form of the negative ($\bullet$) part of system (19) is given as,

$$\begin{aligned}
1\underline{v}_1^\bullet + 0\underline{v}_2^\bullet + 0(-\overline{v}_1^\bullet) + 1(-\overline{v}_2^\bullet) &= -7 - 2s,\\
1\underline{v}_1^\bullet + 2\underline{v}_2^\bullet + 0(-\overline{v}_1^\bullet) + 0(-\overline{v}_2^\bullet) &= 22 - 3s,\\
0\underline{v}_1^\bullet + 1\underline{v}_2^\bullet + 1(-\overline{v}_1^\bullet) + 0(-\overline{v}_2^\bullet) &= 2 - 3s,\\
0\underline{v}_1^\bullet + 0\underline{v}_2^\bullet + 1(-\overline{v}_1^\bullet) + 2(-\overline{v}_2^\bullet) &= -29 - 4s.
\end{aligned}$$

The matrix form of the above system is

$$\begin{pmatrix} 1\,0\,0\,1 \\ 1\,2\,0\,0 \\ 0\,1\,1\,0 \\ 0\,0\,1\,2 \end{pmatrix} \begin{pmatrix} \underline{v}_1^\bullet(s) \\ \underline{v}_2^\bullet(s) \\ -\overline{v}_1^\bullet(s) \\ -\overline{v}_2^\bullet(s) \end{pmatrix} = \begin{pmatrix} -7-2s \\ 22-3s \\ 2-3s \\ -29-4s \end{pmatrix},$$

this can also be written as,

$$MV^\bullet = F^\bullet,$$

and the solution of above system is,

$$V^\bullet = \begin{pmatrix} \underline{v}_1^\bullet(s) \\ \underline{v}_2^\bullet(s) \\ -\overline{v}_1^\bullet(s) \\ -\overline{v}_2^\bullet(s) \end{pmatrix} = M^{-1}F^\bullet = \begin{pmatrix} 4/3 & -1/3 & 2/3 & -2/3 \\ -2/3 & 2/3 & -1/3 & 1/3 \\ 2/3 & -2/3 & 4/3 & -1/3 \\ -1/3 & 1/3 & -2/3 & 2/3 \end{pmatrix} \begin{pmatrix} -7-2s \\ 22-3s \\ 2-3s \\ -29-4s \end{pmatrix}$$

$$\underline{v}_1^{\bullet}(s) = 4 - s \quad , \quad \underline{v}_2^{\bullet}(s) = 9 - s,$$
$$\overline{v}_1^{\bullet}(s) = 7 + 2s \quad , \quad \overline{v}_2^{\bullet}(s) = 11 + s.$$

The graphical representation for negative ($\bullet$) membership functions of imaginary part of complex bipolar solution number is shown below in Fig. 4.

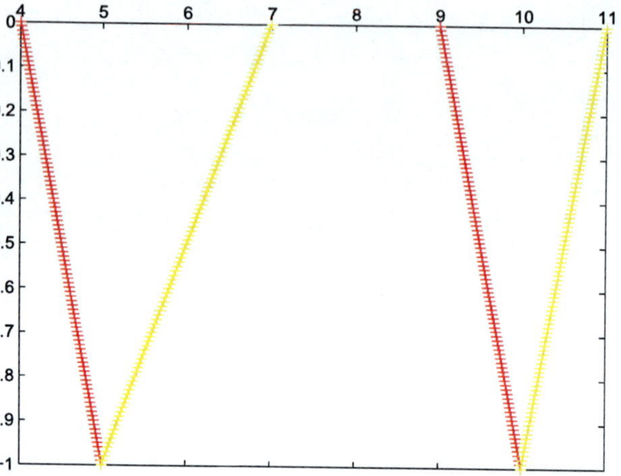

Fig. 4. The graphical representation of negative ($\bullet$) membership functions $\underline{v}^{\bullet}(s), \overline{v}^{\bullet}(s)$ of complex bipolar fuzzy numbers.

Hence the solution of CBFL system in (1) is given as,

$$\widetilde{h}_1 = u_1 + iv_1 = [3 + 2r, 6 - r], [1 - 2s, 4 + s] + i[5 + 3r, 10 - 2r], [4 - s, 7 + 2s],$$

$$\widetilde{h}_2 = u_2 + iv_2 = [7 + r, 9 - r], [4 - 2s, 7 + s] + i[10 + 2r, 13 - r], [9 - s, 11 + s].$$

3.1 Weak Complex Bipolar Fuzzy Solution

We restrict our discussion to bipolar fuzzy complex triangular (BFCT) number, i.e., $\underline{e}_q^{\star}(r)$, $\overline{e}_q^{\star}(r)$, $\underline{e}_q^{\bullet}(s)$, $\overline{e}_q^{\bullet}(s)$, $\underline{f}_q^{\star}(r)$, $\overline{f}_q^{\star}(r)$, $\underline{f}_q^{\bullet}(s)$, $\overline{f}_q^{\bullet}(s)$ and $\underline{u}_q^{\star}(r)$, $\overline{u}_q^{\star}(r)$, $\underline{u}_q^{\bullet}(s)$, $\overline{u}_q^{\bullet}(s)$, $\underline{v}_q^{\star}(r)$, $\overline{v}_q^{\star}(r)$, $\underline{v}_q^{\bullet}(s)$, $\overline{v}_q^{\bullet}(s)$ they all are linear functions of r and s, $\underline{e}_q^{\star}(1) = \overline{e}_q^{\star}(1)$, $\underline{e}_q^{\bullet}(-1) = \overline{e}_q^{\bullet}(-1)$, $\underline{f}_q^{\star}(1) = \overline{f}_q^{\star}(1)$, $\underline{f}_q^{\bullet}(-1) = \overline{f}_q^{\bullet}(-1)$ and $\underline{u}_q^{\star}(1) = \overline{u}_q^{\star}(1)$, $\underline{u}_q^{\bullet}(-1) = \overline{u}_q^{\bullet}(-1)$, $\underline{v}_q^{\star}(1) = \overline{v}_q^{\star}(1)$, $\underline{v}_q^{\bullet}(-1) = \overline{v}_q^{\bullet}(-1)$ for all $1 \leq q \leq n$. Since $(A + B)$ in non-negative but $(A + B)^{-1}$ may be negative so by this fact may be $\overline{u}_q^{\star}(r) - \underline{u}_q^{\star}(r) < 0$, $\overline{v}_q^{\star}(r) - \underline{v}_q^{\star}(r) < 0$ and $\overline{u}_q^{\bullet}(s) - \underline{u}_q^{\bullet}(s) < 0$, $\overline{v}_q^{\bullet}(s) - \underline{v}_q^{\bullet}(s) < 0$. So, by this fact $\widetilde{h}_q$ is not a BFCN and we define BFCN $G = x + iy$ as a BFCN vector as,

$$G = ([\underline{x}_1^{\star}, \overline{x}_1^{\star}], [\underline{x}_1^{\bullet}, \overline{x}_1^{\bullet}] + i[\underline{y}_1^{\star}, \overline{y}_1^{\star}], [\underline{y}_1^{\bullet}, \overline{y}_1^{\bullet}], \ldots, [\underline{x}_n^{\star}, \overline{x}_n^{\star}], [\underline{x}_n^{\bullet}, \overline{x}_n^{\bullet}] + i[\underline{y}_n^{\star}, \overline{y}_n^{\star}], [\underline{y}_n^{\bullet}, \overline{y}_n^{\bullet}])^T,$$

where,

$$\underline{x}_q^\star(r) = \min\{\underline{u}_q^\star(r), \overline{u}_q^\star(r), \underline{u}_q^\star(1)\},$$
$$\overline{x}_q^\star(r) = \max\{\underline{u}_q^\star(r), \overline{u}_q^\star(r), \underline{u}_q^\star(1)\},$$
$$\underline{x}_q^\bullet(s) = \min\{\underline{u}_q^\bullet(s), \overline{u}_q^\bullet(s), \underline{u}_q^\bullet(-1)\},$$
$$\overline{x}_q^\bullet(s) = \max\{\underline{u}_q^\bullet(s), \overline{u}_q^\bullet(s), \underline{u}_q^\bullet(-1)\},$$

and,

$$\underline{y}_q^\star(r) = \min\{\underline{v}_q^\star(r), \overline{v}_q^\star(r), \underline{v}_q^\star(1)\},$$
$$\overline{y}_q^\star(r) = \max\{\underline{v}_q^\star(r), \overline{v}_q^\star(r), \underline{v}_q^\star(1)\},$$
$$\underline{y}_q^\bullet(s) = \min\{\underline{v}_q^\bullet(s), \overline{v}_q^\bullet(s), \underline{v}_q^\bullet(-1)\},$$
$$\overline{y}_q^\bullet(s) = \max\{\underline{v}_q^\bullet(s), \overline{v}_q^\bullet(s), \underline{v}_q^\bullet(-1)\}.$$

If $\widetilde{h}_q = u_q + iv_q$, $1 \leq q \leq n$, are all bipolar fuzzy number then $\underline{u}_q^\star(r) = \underline{x}_q^\star(r)$, $\overline{u}_q^\star(r) = \overline{x}_q^\star(r)$, $\underline{u}_q^\bullet(s) = \underline{x}_q^\bullet(s)$, $\overline{u}_q^\bullet(s) = \overline{x}_q^\bullet(s)$ and $\underline{v}_q^\star(r) = \underline{y}_q^\star(r)$, $\overline{v}_q^\star(r) = \overline{y}_q^\star(r)$, $\underline{v}_q^\bullet(s) = \underline{y}_q^\bullet(s)$, $\overline{v}_q^\bullet(s) = \overline{y}_q^\bullet(s)$, $1 \leq q \leq n$ and then G is called the strong complex bipolar fuzzy solution. Otherwise, G is called the weak complex bipolar fuzzy solution. In example (1) the complex bipolar fuzzy solution was strong solution.

Example 2. Consider the 2×2 CBFL system of equations

$$\widetilde{h}_1 + \widetilde{h}_2 = [-6+r, 10-15r], [-12-6s, -5+s] + i[-5+2r, -1-2r], [-6-s, -4+s],$$
$$\widetilde{h}_1 - 2\widetilde{h}_2 = [-3+r, 4-6r], [-10-5s, -4+s] + i[2+5r, 12-5r], [7-5s, 17+5s].$$

To solve this CBFL system first we put $\widetilde{h}_1 = u_1 + iv_1$ and $\widetilde{h}_2 = u_2 + iv_2$ in original system to obtain,

$$(u_1+iv_1)+(u_2+iv_2) = [-6+r, 10-15r], [-12-6s, -5+s]+i[-5+2r, -1-2r], [-6-s, -4+s],$$

$$(u_1+iv_1)-2(u_2+iv_2) = [-3+r, 4-6r], [-10-5s, -4+s]+i[2+5r, 12-5r], [7-5s, 17+5s],$$

compare the real and imaginary parts to both side of the equations and we obtained two BFL system of equations given as,

$$\begin{cases} u_1 + u_2 = [-6+r, 10-15r], [-12-6s, -5+s], \\ u_1 - 2u_2 = [-3+r, 4-6r], [-10-5s, -4+s], \end{cases} \tag{20}$$

and,

$$\begin{cases} v_1 - v_2 = [-5+2r, -1-2r], [-6-s, -4+s], \\ v_1 + 2v_2 = [2+5r, 12-5r], [7-5s, 17+5s]. \end{cases} \tag{21}$$

To solve the BFL system in (20), first we take the positive ($\star$) part of the system (20). The extended 4×4 form of the positive ($\star$) part of system (20) is given as,

$$\begin{aligned}
1\underline{u}_1^\star + 1\underline{u}_2^\star + 0(-\overline{u}_1^\star) + 0(-\overline{u}_2^\star) &= -6 - r,\\
1\underline{u}_1^\star + 0\underline{u}_2^\star + 0(-\overline{u}_1^\star) + 2(-\overline{u}_2^\star) &= -3 + r,\\
0\underline{u}_1^\star + 0\underline{u}_2^\star + 1(-\overline{u}_1^\star) + 1(-\overline{u}_2^\star) &= 15r - 10,\\
0\underline{u}_1^\star + 2\underline{u}_2^\star + 1(-\overline{u}_1^\star) + 0(-\overline{u}_2^\star) &= 6r - 4.
\end{aligned}$$

The matrix form of the above system is

$$\begin{pmatrix} 1\,1\,0\,0 \\ 1\,0\,0\,2 \\ 0\,0\,1\,1 \\ 0\,2\,1\,0 \end{pmatrix} \begin{pmatrix} \underline{u}_1^\star(r) \\ \underline{u}_2^\star(r) \\ -\overline{u}_1^\star(r) \\ -\overline{u}_2^\star(r) \end{pmatrix} = \begin{pmatrix} -6 + r \\ -3 + r \\ 15r - 10 \\ 6r - 4 \end{pmatrix},$$

this can also be written as,

$$MU^\star = E^\star,$$

and the solution of above system is,

$$U^\star = \begin{pmatrix} \underline{u}_1^\star(r) \\ \underline{u}_2^\star(r) \\ -\overline{u}_1^\star(r) \\ -\overline{u}_2^\star(r) \end{pmatrix} = M^{-1}E^\star = \begin{pmatrix} 4/3 & -1/3 & 2/3 & -2/3 \\ -1/3 & 1/3 & -2/3 & 2/3 \\ 2/3 & -2/3 & 4/3 & -1/3 \\ -2/3 & 2/3 & -1/3 & 1/3 \end{pmatrix} \begin{pmatrix} -6 - r \\ -3 + r \\ 15r - 10 \\ 6r - 4 \end{pmatrix}$$

$$\begin{aligned}
\underline{u}_1^\star(r) &= 7r - 11 \quad , \quad \underline{u}_2^\star(r) = 5 - 6r,\\
\overline{u}_1^\star(r) &= 14 - 18r \quad , \quad \overline{u}_2^\star(r) = 3r - 4.
\end{aligned}$$

The graphical representation for positive ($\star$) membership function of real part of complex bipolar solution number is shown below in Fig. 5.

Next we take the negative ($\bullet$) part of the system (20). The extended 4×4 form of the negative ($\bullet$) part of system (20) is given as,

$$\begin{aligned}
1\underline{u}_1^\bullet + 1\underline{u}_2^\bullet + 0(-\overline{u}_1^\bullet) + 0(-\overline{u}_2^\bullet) &= -12 - 6s,\\
1\underline{u}_1^\bullet + 0\underline{u}_2^\bullet + 0(-\overline{u}_1^\bullet) + 2(-\overline{u}_2^\bullet) &= -10 - 5s,\\
0\underline{u}_1^\bullet + 0\underline{u}_2^\bullet + 1(-\overline{u}_1^\bullet) + 1(-\overline{u}_2^\bullet) &= 5 - s,\\
0\underline{u}_1^\bullet + 2\underline{u}_2^\bullet + 1(-\overline{u}_1^\bullet) + 0(-\overline{u}_2^\bullet) &= 4 - s.
\end{aligned}$$

The matrix form of the above system is

$$\begin{pmatrix} 1\,1\,0\,0 \\ 1\,0\,0\,2 \\ 0\,0\,1\,1 \\ 0\,2\,1\,0 \end{pmatrix} \begin{pmatrix} \underline{u}_1^\bullet(s) \\ \underline{u}_2^\bullet(s) \\ -\overline{u}_1^\bullet(s) \\ -\overline{u}_2^\bullet(s) \end{pmatrix} = \begin{pmatrix} -12 - 6s \\ -10 - 5s \\ 5 - s \\ 4 - s \end{pmatrix},$$

this can also be written as,

$$MU^\bullet = E^\bullet,$$

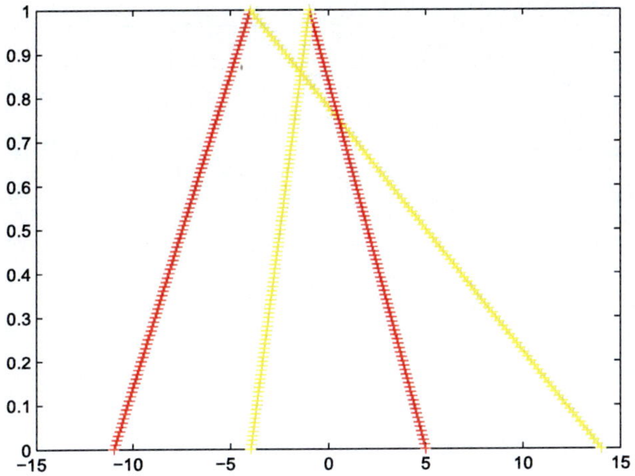

Fig. 5. The graphical representation of positive ($\star$) membership functions $\underline{u}^{\star}(r), \overline{u}^{\star}(r)$ of complex bipolar fuzzy numbers.

and the solution of above system is,

$$U^{\bullet} = \begin{pmatrix} \underline{u}_1^{\bullet}(s) \\ \underline{u}_2^{\bullet}(s) \\ -\overline{u}_1^{\bullet}(s) \\ -\overline{u}_2^{\bullet}(s) \end{pmatrix} = M^{-1}E^{\bullet} = \begin{pmatrix} 4/3 & -1/3 & 2/3 & -2/3 \\ -1/3 & 1/3 & -2/3 & 2/3 \\ 2/3 & -2/3 & 4/3 & -1/3 \\ -2/3 & 2/3 & -1/3 & 1/3 \end{pmatrix} \begin{pmatrix} -12-6s \\ -10-5s \\ 5-s \\ 4-s \end{pmatrix}$$

$$\underline{u}_1^{\bullet}(s) = -\frac{19}{3}s - 12 \quad , \quad \underline{u}_2^{\bullet}(s) = \frac{s}{3},$$
$$\overline{u}_1^{\bullet}(s) = \frac{5}{3}s - 4 \quad , \quad \overline{u}_2^{\bullet}(s) = -1 - \frac{2}{3}s.$$

The graphical representation for negative ($\bullet$) membership functions of real part of complex bipolar solution number is shown below in Fig. 6.

In the similar way, to solve the BFL system in (21), first we take the positive ($\star$) part of the system (21). The extended 4×4 form of the positive ($\star$) part of system (21) is given as,

$$1\underline{v}_1^{\star} + 1\underline{v}_2^{\star} + 0(-\overline{v}_1^{\star}) + 0(-\overline{v}_2^{\star}) = -5 + 2r,$$
$$1\underline{v}_1^{\star} + 0\underline{v}_2^{\star} + 0(-\overline{v}_1^{\star}) + 2(-\overline{v}_2^{\star}) = 2 + 5r,$$
$$0\underline{v}_1^{\star} + 0\underline{v}_2^{\star} + 1(-\overline{v}_1^{\star}) + 1(-\overline{v}_2^{\star}) = 2r + 1,$$
$$0\underline{v}_1^{\star} + 2\underline{v}_2^{\star} + 1(-\overline{v}_1^{\star}) + 0(-\overline{v}_2^{\star}) = 5r - 12.$$

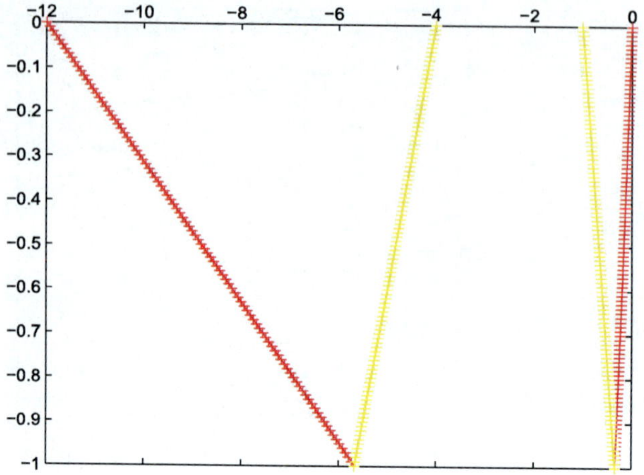

Fig. 6. The graphical representation of negative (•) membership functions $\underline{u}^{\bullet}(s), \overline{u}^{\bullet}(s)$ of complex bipolar fuzzy numbers.

The matrix form of the above system is

$$\begin{pmatrix} 1\,1\,0\,0 \\ 1\,0\,0\,2 \\ 0\,0\,1\,1 \\ 0\,2\,1\,0 \end{pmatrix} \begin{pmatrix} \underline{v}_1^{\star}(r) \\ \underline{v}_2^{\star}(r) \\ -\overline{v}_1^{\star}(r) \\ -\overline{v}_2^{\star}(r) \end{pmatrix} = \begin{pmatrix} -5+2r \\ 2+5r \\ 2r+1 \\ 5r-12 \end{pmatrix},$$

this can also be written as,

$$MV^{\star} = F^{\star},$$

and the solution of above system is,

$$V^{\star} = \begin{pmatrix} \underline{v}_1^{\star}(r) \\ \underline{v}_2^{\star}(r) \\ -\overline{v}_1^{\star}(r) \\ -\overline{v}_2^{\star}(r) \end{pmatrix} = M^{-1}F^{\star} = \begin{pmatrix} 4/3 & -1/3 & 2/3 & -2/3 \\ -1/3 & 1/3 & -2/3 & 2/3 \\ 2/3 & -2/3 & 4/3 & -1/3 \\ -2/3 & 2/3 & -1/3 & 1/3 \end{pmatrix} \begin{pmatrix} -5+2r \\ 2+5r \\ 2r+1 \\ 5r-12 \end{pmatrix}$$

$$\underline{v}_1^{\star}(r) = -r + \frac{4}{3} \quad , \quad \underline{v}_2^{\star}(r) = 3r - \frac{19}{3},$$

$$\overline{v}_1^{\star}(r) = -\frac{2}{3} + r \quad , \quad \overline{v}_2^{\star}(r) = -3r - \frac{1}{3}.$$

The graphical representation for positive (⋆) membership functions of imaginary part of complex bipolar solution number is shown below in Fig. 7.

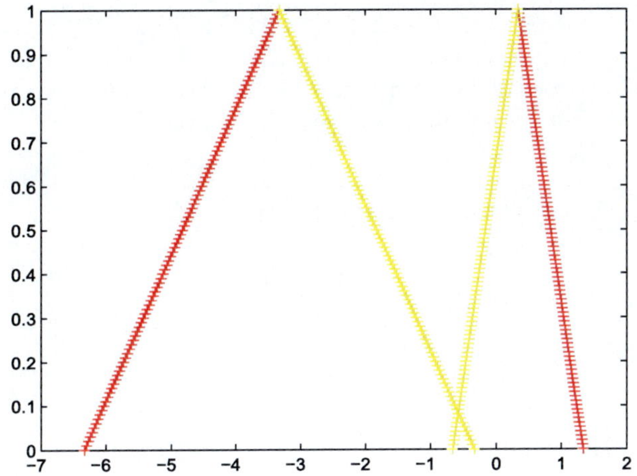

Fig. 7. The graphical representation of positive ($\star$) membership functions $\underline{v}^{\star}(r), \overline{v}^{\star}(r)$ of complex bipolar fuzzy numbers.

Next we take the negative ($\bullet$) part of the system (21). The extended 4×4 form of the negative ($\bullet$) part of system (21) is given as,

$$
\begin{aligned}
1\underline{v}_1^{\bullet} + 1\underline{v}_2^{\bullet} + 0(-\overline{v}_1^{\bullet}) + 0(-\overline{v}_2^{\bullet}) &= -6 - s,\\
1\underline{v}_1^{\bullet} + 0\underline{v}_2^{\bullet} + 0(-\overline{v}_1^{\bullet}) + 2(-\overline{v}_2^{\bullet}) &= 7 - 5s,\\
0\underline{v}_1^{\bullet} + 0\underline{v}_2^{\bullet} + 1(-\overline{v}_1^{\bullet}) + 1(-\overline{v}_2^{\bullet}) &= 4 - s,\\
0\underline{v}_1^{\bullet} + 2\underline{v}_2^{\bullet} + 1(-\overline{v}_1^{\bullet}) + 0(-\overline{v}_2^{\bullet}) &= -17 - 5s.
\end{aligned}
$$

The matrix form of the above system is

$$
\begin{pmatrix} 1\,1\,0\,0 \\ 1\,0\,0\,2 \\ 0\,0\,1\,1 \\ 0\,2\,1\,0 \end{pmatrix}
\begin{pmatrix} \underline{v}_1^{\bullet}(s) \\ \underline{v}_2^{\bullet}(s) \\ -\overline{v}_1^{\bullet}(s) \\ -\overline{v}_2^{\bullet}(s) \end{pmatrix}
=
\begin{pmatrix} -6 - s \\ 7 - 5s \\ 4 - s \\ -17 - 5s \end{pmatrix},
$$

this can also be written as,

$$MV^{\bullet} = F^{\bullet},$$

and the solution of above system is,

$$
V^{\bullet} = \begin{pmatrix} \underline{v}_1^{\bullet}(s) \\ \underline{v}_2^{\bullet}(s) \\ -\overline{v}_1^{\bullet}(s) \\ -\overline{v}_2^{\bullet}(s) \end{pmatrix}
= M^{-1}F^{\bullet} =
\begin{pmatrix} 4/3 & -1/3 & 2/3 & -2/3 \\ -1/3 & 1/3 & -2/3 & 2/3 \\ 2/3 & -2/3 & 4/3 & -1/3 \\ -2/3 & 2/3 & -1/3 & 1/3 \end{pmatrix}
\begin{pmatrix} -6 - s \\ 7 - 5s \\ 4 - s \\ -17 - 5s \end{pmatrix}
$$

$$
\underline{v}_1^{\bullet}(s) = 3s + \frac{11}{3} \quad , \quad \underline{v}_2^{\bullet}(s) = -4s - \frac{29}{3},
$$

$$\overline{v}_1^{\bullet}(s) = -3s - \frac{7}{3} \quad , \quad \overline{v}_2^{\bullet}(s) = 4s - \frac{5}{3}.$$

The graphical representation for negative (•) membership functions of imaginary part of complex bipolar solution number is shown below in Fig. 8.

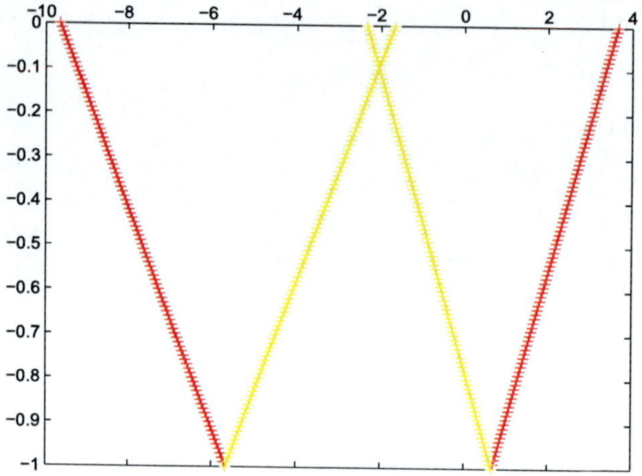

Fig. 8. The graphical representation of negative (•) membership functions $\underline{v}^{\bullet}(s), \overline{v}^{\bullet}(s)$ of complex bipolar fuzzy numbers.

Hence the solution of CBFL system in (2) is given as,

$$\widetilde{h}_1 = u_1 + iv_1 = [7r-11, 14-18r], [-\frac{19}{3}s-12, \frac{5}{3}s-4] + i[-r+\frac{4}{3}, r-\frac{2}{3}], [3s+\frac{11}{3}, -\frac{7}{3}-3s],$$

$$\widetilde{h}_2 = u_2 + iv_2 = [5-6r, 3r-4], [\frac{s}{3}, -1-\frac{2}{3}s] + i[3r-\frac{19}{3}, -\frac{1}{3}-3r], [-4s-\frac{29}{3}, 4s-\frac{5}{3}].$$

It is clear to see the complex bipolar fuzzy solution that $\overline{u}_2^{\star}(r) \leq \underline{u}_2^{\star}(r)$, $\overline{u}_2^{\bullet}(s) \leq \underline{u}_2^{\bullet}(s)$, $\overline{v}_1^{\star}(r) \leq \underline{v}_1^{\star}(r)$ and $\overline{v}_1^{\bullet}(s) \leq \underline{v}_1^{\bullet}(s)$, so that complex bipolar fuzzy solution is a weak solution and now solution is given as,

$$\widetilde{g}_1 = x_1 + iy_1 = [7r-11, 14-18r], [-\frac{19}{3}s-12, \frac{5}{3}s-4] + i[r-\frac{2}{3}, -r+\frac{4}{3}], [-\frac{7}{3}-3s, 3s+\frac{11}{3}],$$

$$\widetilde{g}_2 = x_2 + iy_2 = [3r-4, 5-6r], [-1-\frac{2}{3}s, \frac{s}{3}] + i[3r-\frac{19}{3}, -\frac{1}{3}-3r], [-4s-\frac{29}{3}, 4s-\frac{5}{3}].$$

4 Solution Procedure for Complex Bipolar Fuzzy Linear System with Complex Coefficients

In this section, we develop a method to solve CBFL system with complex coefficients in which unknown parameter, right side vector are BFCN and coefficients of unknown are complex crisp number. Rahgooy [21] suggested a method to solve $n \times n$ complex fuzzy linear system with complex coefficients in which coefficient matrix is a complex crisp matrix and unknown and right side vector are complex fuzzy number. We extent the Rahgooy technique to solve complex fuzzy linear system with complex coefficients in bipolar fuzzy environment. First we will convert BFCN in the form of interval using α-cut and then solve the CBFL system with complex coefficients using the propose method in this section. We use this technique to solve the example of circuit flow and find the current which is calculated in term of BFCN.

Definition 11. Consider the $n \times n$ complex linear system of equations given as,

$$\begin{cases} l_{11}h_1 + l_{12}h_2 + ... + l_{1n}h_n & = z_1, \\ l_{21}h_1 + l_{22}h_2 + ... + l_{2n}h_n & = z_2, \\ . \\ . \\ . \\ l_{n1}h_1 + l_{n2}h_2 + ... + l_{nn}h_n & = z_n, \end{cases} \tag{22}$$

is the complex bipolar fuzzy linear (CBFL) system of equations with complex coefficients, if all the coefficient elements l_{pq}, $1 \leq p, q \leq n$ is a complex crisp matrix of order $n \times n$ and each Z_q, $1 \leq q \leq n$ is a BFCN, and each unknown parameter h_p, $1 \leq p \leq n$ is also BFCN.
First we can write the system in (22) in positive ($\star$) and negative ($\bullet$) part separately given as,

$$\sum_{q=1}^{n} l_{pq} h_q^{\star} = z_p^{\star}, \tag{23}$$

$$\sum_{q=1}^{n} l_{pq} h_q^{\bullet} = z_p^{\bullet}, \qquad for \quad p = 1, 2, ..., n. \tag{24}$$

To solve the system we write the BFCN and crisp complex number in form of
$l_{pq} = a_{pq} + ib_{pq}$,

$h_q^{\star} = u_q^{\star} + iv_q^{\star}$,

$h_q^{\bullet} = u_q^{\bullet} + iv_q^{\bullet}$,

$z_p^\star = e_p^\star + if_p^\star,$

$z_q^\bullet = e_p^\bullet + if_p^\bullet.$

So, the above system can be written as,

$$\sum_{q=1}^{n}(a_{pq} + ib_{pq})(u_q^\star + iv_q^\star) = e_p^\star + if_p^\star,$$
$$\sum_{q=1}^{n}(a_{pq} + ib_{pq})(u_q^\bullet + iv_q^\bullet) = e_p^\bullet + if_p^\bullet, \qquad for\ p = 1, 2, ..., n.$$

Use arithmetic operations and write the above system in form,

$$\sum_{q=1}^{n} a_{pq}u_q^\star - b_{pq}v_q^\star + i\sum_{q=1}^{n} a_{pq}v_q^\star + b_{pq}p_q^\star = e_p^\star + if_p^\star,$$
$$\sum_{q=1}^{n} a_{pq}u_q^\bullet - b_{pq}v_q^\bullet + i\sum_{q=1}^{n} a_{pq}v_q^\bullet + b_{pq}p_q^\bullet = e_p^\bullet + if_p^\bullet, \qquad for\ p = 1, 2, ..., n.$$

Now we can separate the real and imaginary parts of the above system:

$$\sum_{q=1}^{n} a_{pq}u_q^\star - b_{pq}v_q^\star = e_p^\star,$$
$$\sum_{q=1}^{n} a_{pq}v_q^\star + b_{pq}p_q^\star = f_p^\star,$$
$$\sum_{q=1}^{n} a_{pq}u_q^\bullet - b_{pq}v_q^\bullet = e_p^\bullet,$$
$$\sum_{q=1}^{n} a_{pq}v_q^\bullet + b_{pq}p_q^\bullet = f_p^\bullet, \qquad for\ p = 1, 2, ..., n.$$

For solving the system we can write:
$L_1 = [a_{pq}], \qquad L_2 = [b_{pq}],$

$U^\star = [u_q^\star], \qquad V^\star = [v_q^\star],$

$U^\bullet = [u_q^\bullet], \qquad V^\bullet = [v_q^\bullet],$

$E^\star = [e_p^\star], \qquad F^\star = [f_p^\star],$

$E^\bullet = [e_p^\bullet], \qquad F^\bullet = [f_p^\bullet],$

for $p, q = 1, 2, ..., n$
and,

$$\begin{pmatrix} L_1 & -L_2 \\ L_2 & L_1 \end{pmatrix}\begin{pmatrix} U^\star \\ V^\star \end{pmatrix} = \begin{pmatrix} E^\star \\ F^\star \end{pmatrix}, \quad \begin{pmatrix} L_1 & -L_2 \\ L_2 & L_1 \end{pmatrix}\begin{pmatrix} U^\bullet \\ V^\bullet \end{pmatrix} = \begin{pmatrix} E^\bullet \\ F^\bullet \end{pmatrix}. \tag{25}$$

Example 3. We consider a simple RLC circuit [21] in which current and source all are in term of BFCN. Related circuit is shown in Fig. 9.

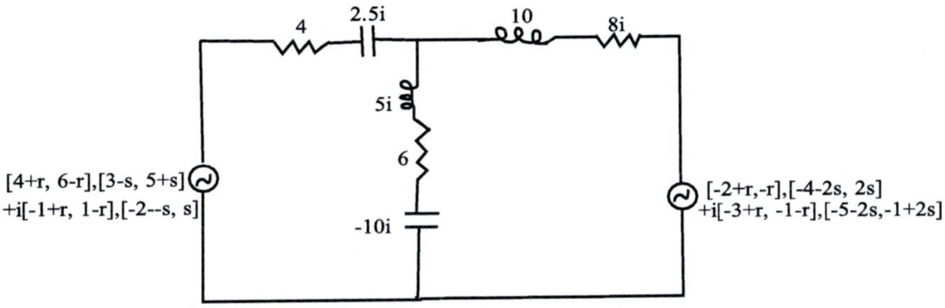

Fig. 9. A circuit with bipolar fuzzy current, source and resistance.

We use, Kirchhoff's Second law for the above circuit and write the linear equations for first and second loop is given as,

$$\begin{cases} (10-7.5i)I_1 - (6-5i)I_2 = [4+r, 6-r], [3-s, 5+s] + i[-1+r, 1-r], [-2-s, s], \\ -(6-5i)I_1 + (16+3i)I_2 = [-2+r, -r], [-4-2s, 2s] + i[-3+r, -1-r], [-5-2s, -1+2s]. \end{cases}$$

We can put $I_1 = [u_1^\star(r)], [u_1^\bullet(s)] + i[v_1^\star(r)], [v_1^\bullet(s)]$ and $I_2 = [u_2^\star(r)], [u_2^\bullet(s)] + i[v_2^\star(r)], [v_2^\bullet(s)]$ in above system then system become,
$(10 - 7.5i)\{[u_1^\star(r)], [u_1^\bullet(s)] + i[v_1^\star(r)], [v_1^\bullet(s)]\} - (6 - 5i)\{[u_2^\star(r)], [u_2^\bullet(s)] + i[v_2^\star(r)], [v_2^\bullet(s)]\} = [4+r, 6-r], [3-s, 5+s] + i[-1+r, 1-r], [-2-s, s]$,
$-(6 - 5i)\{[u_1^\star(r)], [u_1^\bullet(s)] + i[v_1^\star(r)], [v_1^\bullet(s)]\} + (16 + 3i)\{[u_2^\star(r)], [u_2^\bullet(s)] + i[v_2^\star(r)], [v_2^\bullet(s)]\} = [-2+r, -r], [-4-2s, 2s] + i[-3+r, -1-r], [-5-2s, -1+2s]$.
First, we take the positive ($\star$) part of the above system of equations,

$$\begin{cases} (10-7.5i)(u_1^\star(r) + iv_1^\star(r)) - (6-5i)(u_2^\star(r) + iv_2^\star(r)) = [4+r, 6-r] + i[-1+r, 1-r], \\ -(6-5i)(u_1^\star(r) + iv_1^\star(r)) + (16+3i)(u_2^\star(r) + iv_2^\star(r)) = [-2+r, -r] + i[-3+r, -1-r]. \end{cases} \tag{26}$$

Above system can be written in matrix form as illustrate in proposed method (25) is given as,

$$\begin{pmatrix} 10-7.5i & -(6-5i) \\ -(6-5i) & 16+3i \end{pmatrix}\begin{pmatrix} u_1^\star(r)+iv_1^\star(r) \\ u_2^\star(r)+iv_2^\star(r) \end{pmatrix} = \begin{pmatrix} (4+r,6-r)+i(-1+r,1-r) \\ (-2+r,-r)+i(-3+r,-1-r) \end{pmatrix},$$

and,

$$L_1 = \begin{pmatrix} 10 & -6 \\ -6 & 16 \end{pmatrix}, \qquad L_2 = \begin{pmatrix} -7.5 & 5 \\ 5 & 3 \end{pmatrix},$$

$$E^\star = \begin{pmatrix} (4+r,6-r) \\ (-2+r,-r) \end{pmatrix}, \qquad F^\star = \begin{pmatrix} (-1+r,1-r) \\ (-3+r,-1-r) \end{pmatrix},$$

$$U^\star = \begin{pmatrix} u_1^\star(r) \\ u_2^\star(r) \end{pmatrix}, \qquad V^\star = \begin{pmatrix} v_1^\star(r) \\ v_2^\star(r) \end{pmatrix}.$$

Now, we compare the real and imaginary parts of the system (26) and we obtained the system,

$$10u_1^\star(r) - 6u_2^\star(r) + 7.5v_1^\star(r) - 5v_2^\star(r) = (4+r,6-r),$$

$$-6u_1^\star(r) + 16u_2^\star(r) - 5v_1^\star(r) - 3v_2^\star(r) = (-2+r,-r),$$

$$-7.5u_1^\star(r) + 5u_2^\star(r) + 10v_1^\star(r) - 6v_2^\star(r) = (-1+r,1-r),$$

$$5u_1^\star(r) + 3u_2^\star(r) - 6v_1^\star(r) + 16v_2^\star(r) = (-3+r,-1-r).$$

The matrix form is given as,

$$\begin{pmatrix} 10 & -6 & 7.5 & -5 \\ -6 & 16 & -5 & -3 \\ -7.5 & 5 & 10 & -6 \\ 5 & 3 & -6 & 16 \end{pmatrix}\begin{pmatrix} u_1^\star(r) \\ u_2^\star(r) \\ v_1^\star(r) \\ v_2^\star(r) \end{pmatrix} = \begin{pmatrix} (4+r,6-r) \\ (-2+r,-r) \\ (-1+r,1-r) \\ (-3+r,-1-r) \end{pmatrix}.$$

In order to (12),

$$M_1 = \begin{pmatrix} 10 & 0 & 7.5 & 0 \\ 0 & 16 & 0 & 0 \\ 0 & 5 & 10 & 0 \\ 5 & 3 & 0 & 16 \end{pmatrix},$$

$$M_2 = \begin{pmatrix} 0 & -6 & 0 & -5 \\ -6 & 0 & -5 & -3 \\ -7.5 & 0 & 0 & -6 \\ 0 & 0 & -6 & 0 \end{pmatrix},$$

$$\begin{pmatrix} 10 & 0 & 7.5 & 0 & 0 & -6 & 0 & -5 \\ 0 & 16 & 0 & 0 & -6 & 0 & -5 & -3 \\ 0 & 5 & 10 & 0 & -7.5 & 0 & 0 & -6 \\ 5 & 3 & 0 & 16 & 0 & 0 & -6 & 0 \\ 0 & -6 & 0 & -5 & 10 & 0 & 7.5 & 0 \\ -6 & 0 & -5 & -3 & 0 & 16 & 0 & 0 \\ -7.5 & 0 & 0 & -6 & 0 & 5 & 10 & 0 \\ 0 & 0 & -6 & 0 & 5 & 3 & 0 & 16 \end{pmatrix}\begin{pmatrix} \underline{u}_1^\star(r) \\ \underline{u}_2^\star(r) \\ \underline{v}_1^\star(r) \\ \underline{v}_2^\star(r) \\ \overline{u}_1^\star(r) \\ \overline{u}_2^\star(r) \\ \overline{v}_1^\star(r) \\ \overline{v}_2^\star(r) \end{pmatrix} = \begin{pmatrix} 4+r \\ -2+r \\ -1+r \\ -3+r \\ 6-r \\ -r \\ 1-r \\ -1-r \end{pmatrix},$$

after solving the the above system we get,

$$\begin{pmatrix} \underline{u}_1^{\star}(r) \\ \underline{u}_2^{\star}(r) \\ \underline{v}_1^{\star}(r) \\ \underline{v}_2^{\star}(r) \\ \overline{u}_1^{\star}(r) \\ \overline{u}_2^{\star}(r) \\ \overline{v}_1^{\star}(r) \\ \overline{v}_2^{\star}(r) \end{pmatrix} = \begin{pmatrix} 0.3161 + 0.1379r \\ 0.0348 + 0.0307r \\ 0.0706 + 0.0379r \\ -0.2377 + 0.0307r \\ 0.3919 - 0.0379r \\ 0.0962 - 0.0307r \\ 0.1464 - 0.0379r \\ -0.1763 - 0.0307r \end{pmatrix}.$$

Solution for the positive part of the CBFL system with complex coefficients is given as,

$$\begin{cases} I_1^{\star} = u_1^{\star} + iv_1^{\star} = (0.3161 + 0.0379r, 0.3919 - 0.0379r) + i(0.0706 + 0.0379r, 0.1464 - 0.0379r), \\ I_2^{\star} = u_2^{\star} + iv_2^{\star} = (0.0348 + 0.0307r, 0.0962 - 0.0307r) + i(-0.2377 + 0.0307r, -0.1763 - 0.0307r). \end{cases}$$

The graphical representation of positive ($\star$) solution of CBFL system with complex coefficients is shown in Fig. 10.

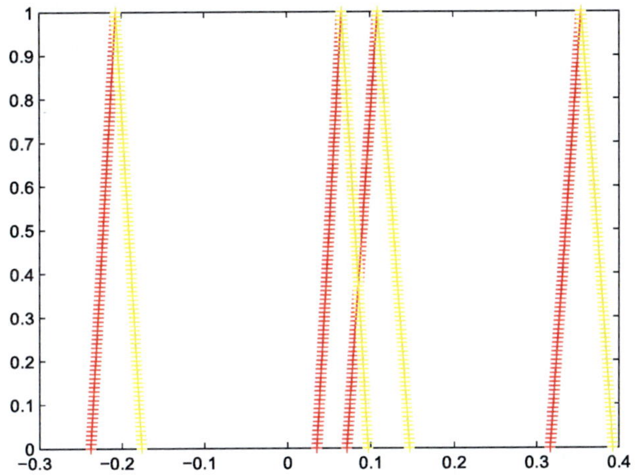

Fig. 10. The graphical representation of positive ($\star$) membership functions $I_1^{\star}(r), I_2^{\star}(r)$ of bipolar fuzzy complex numbers.

Similarly, we can take the negative ($\bullet$) part of the CBFL system with complex coefficients we get,

$$\begin{cases} (10 - 7.5i)(u_1^{\bullet}(s) + iv_1^{\bullet}(s)) - (6 - 5i)(u_2^{\bullet}(s) + iv_2^{\bullet}(s)) = [3 - s, 5 + s] + i[-2 - s, s], \\ -(6 - 5i)(u_1^{\bullet}(s) + iv_1^{\bullet}(s)) + (16 + 3i)(u_2^{\bullet}(s) + iv_2^{\bullet}(s)) = [-3 - s, -1 + s] + i[-4 - s, -2 + s]. \end{cases} \quad (27)$$

Above system can be written in matrix form as illustrate in proposed method (25) is given as,

$$\begin{pmatrix} 10-7.5i & -(6-5i) \\ -(6-5i) & 16+3i \end{pmatrix} \begin{pmatrix} u_1^{\bullet}(s)+iv_1^{\bullet}(s) \\ u_2^{\bullet}(s)+iv_2^{\bullet}(s) \end{pmatrix} = \begin{pmatrix} (3-s,5+s)+i(-2-s,s) \\ (-3-s,-1+s)+i(-4-s,-2+s) \end{pmatrix},$$

and,

$$L_1 = \begin{pmatrix} 10 & -6 \\ -6 & 16 \end{pmatrix}, \qquad L_2 = \begin{pmatrix} -7.5 & 5 \\ 5 & 3 \end{pmatrix},$$

$$E^{\bullet} = \begin{pmatrix} (3-s,5+s) \\ (-3-s,-1+s) \end{pmatrix}, \qquad F^{\bullet} = \begin{pmatrix} (-2-s,s) \\ (-4-s,-2+s) \end{pmatrix},$$

$$U^{\bullet} = \begin{pmatrix} u_1^{\bullet}(s) \\ u_2^{\bullet}(s) \end{pmatrix}, \qquad V^{\bullet} = \begin{pmatrix} v_1^{\bullet}(s) \\ v_2^{\bullet}(s) \end{pmatrix}.$$

Now, we compare the real and imaginary parts of the system (27) and we obtained the system,

$$10u_1^{\bullet}(s) - 6u_2^{\bullet}(s) + 7.5v_1^{\bullet}(s) - 5v_2^{\bullet}(s) = (3-s,5+s),$$
$$-6u_1^{\bullet}(s) + 16u_2^{\bullet}(s) - 5v_1^{\bullet}(s) - 3v_2^{\bullet}(s) = (-3-s,-1+s),$$
$$-7.5u_1^{\bullet}(s) + 5u_2^{\bullet}(s) + 10v_1^{\bullet}(s) - 6v_2^{\bullet}(s) = (-2-s,s),$$
$$5u_1^{\bullet}(s) + 3u_2^{\bullet}(s) - 6v_1^{\bullet}(s) + 16v_2^{\bullet}(s) = (-4-s,-2+s).$$

The matrix form is given as,

$$\begin{pmatrix} 10 & -6 & 7.5 & -5 \\ -6 & 16 & -5 & -3 \\ -7.5 & 5 & 10 & -6 \\ 5 & 3 & -6 & 16 \end{pmatrix} \begin{pmatrix} u_1^{\bullet}(s) \\ u_2^{\bullet}(s) \\ v_1^{\bullet}(s) \\ v_2^{\bullet}(s) \end{pmatrix} = \begin{pmatrix} (3-s,5+s) \\ (-3-s,-1-s) \\ (-2-s,s) \\ (-4-s,-2+s) \end{pmatrix}.$$

In order to (12),

$$M_1 = \begin{pmatrix} 10 & 0 & 7.5 & 0 \\ 0 & 16 & 0 & 0 \\ 0 & 5 & 10 & 0 \\ 5 & 3 & 0 & 16 \end{pmatrix},$$

$$M_2 = \begin{pmatrix} 0 & -6 & 0 & -5 \\ -6 & 0 & -5 & -3 \\ -7.5 & 0 & 0 & -6 \\ 0 & 0 & -6 & 0 \end{pmatrix},$$

$$\begin{pmatrix} 10 & 0 & 7.5 & 0 & 0 & -6 & 0 & -5 \\ 0 & 16 & 0 & 0 & -6 & 0 & -5 & -3 \\ 0 & 5 & 10 & 0 & -7.5 & 0 & 0 & -6 \\ 5 & 3 & 0 & 16 & 0 & 0 & -6 & 0 \\ 0 & -6 & 0 & -5 & 10 & 0 & 7.5 & 0 \\ -6 & 0 & -5 & -3 & 0 & 16 & 0 & 0 \\ -7.5 & 0 & 0 & -6 & 0 & 5 & 10 & 0 \\ 0 & 0 & -6 & 0 & 5 & 3 & 0 & 16 \end{pmatrix} \begin{pmatrix} \underline{u}_1^{\bullet}(s) \\ \underline{u}_2^{\bullet}(s) \\ \underline{v}_1^{\bullet}(s) \\ \underline{v}_2^{\bullet}(s) \\ \overline{u}_1^{\bullet}(s) \\ \overline{u}_2^{\bullet}(s) \\ \overline{v}_1^{\bullet}(s) \\ \overline{v}_2^{\bullet}(s) \end{pmatrix} = \begin{pmatrix} 3-s \\ -3-s \\ -2-s \\ -4-s \\ 5+s \\ -1+s \\ s \\ -2+s \end{pmatrix},$$

after solving the the above system we get,

$$\begin{pmatrix} \underline{u}_1^{\bullet}(s) \\ \underline{u}_2^{\bullet}(s) \\ \underline{v}_1^{\bullet}(s) \\ \underline{v}_2^{\bullet}(s) \\ \overline{u}_1^{\bullet}(s) \\ \overline{u}_2^{\bullet}(s) \\ \overline{v}_1^{\bullet}(s) \\ \overline{v}_2^{\bullet}(s) \end{pmatrix} = \begin{pmatrix} 0.2002 - 0.0379s \\ -0.1229 - 0.0307s \\ -0.0663 - 0.0379s \\ -0.2858 - 0.0307s \\ 0.276 + 0.0379s \\ -0.0615 + 0.0307s \\ 0.0095 + 0.0379s \\ -0.2244 + 0.0307s \end{pmatrix}.$$

Solution for the negative part of the CBFL system with complex coefficients is given as,

$$\begin{cases} I_1^{\bullet} = u_1^{\bullet} + iv_1^{\bullet} = (0.2002 - 0.0379s, 0.276 + 0.0379s) + i(-0.0663 - 0.0379s, 0.0095 + 0.0379s), \\ I_2^{\bullet} = u_2^{\bullet} + iv_2^{\bullet} = (-0.1229 - 0.0307s, -0.0615 + 0.0307s) + i(-0.2858 - 0.0307s, -0.2244 + 0.0307s). \end{cases}$$

The graphical representation of negative (•) solution of CBFL system with complex coefficients is shown in Fig. 11.

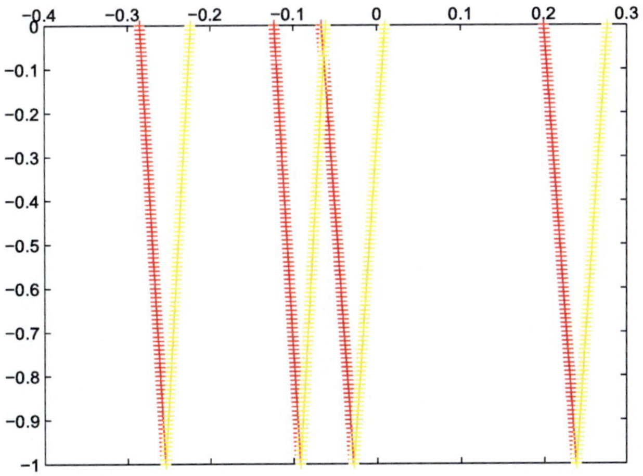

Fig. 11. The graphical representation of negative (•) membership functions $I_1^{\bullet}(s), I_2^{\bullet}(s)$ of bipolar fuzzy complex numbers.

The BFCN solution of the CBFL system with complex coefficients is given as,

$I_1 = [0.3161 + 0.0379r, 0.3919 - 0.0379r], [0.2002 - 0.0379s, 0.276 + 0.0379s] + i[0.0706 + 0.0379r, 0.1464 - 0.0379r], [-0.0663 - 0.0379s, 0.0095 + 0.0379s],$

$I_2 = [0.0348 + 0.0307r, 0.0962 - 0.0307r], [-0.1229 - 0.0307s, -0.0615 + 0.0307s] + i[-0.2377 + 0.0307r, -0.1763 - 0.0307r], [-0.2858 - 0.0307s, -0.2244 + 0.0307s].$

Conclusion. In real world problems most of the applications in different areas are solved by system of linear equations to find the unknown parameters. In different situations the unknown parameters of the systems are uncertain or vague so, to solve these parameters we represent these parameters in term of fuzzy numbers, bipolar fuzzy numbers and bipolar fuzzy complex numbers. In this paper, we have discussed two techniques to solve the CBFL and CBFL system of equations with complex coefficients. In the first method, we have extended the technique [22] to solve the CBFL system of equations in which the $n \times n$ coefficient matrix of the system is represented by crisp real numbers and right side vector is represented by BFCN. To solve the CBFL system we can replace the original system into the pair of positive ($\star$) and negative ($\bullet$) of two $n \times n$ crisp complex fuzzy linear system. In the second method, We have extended the technique [21] to solve the CBFL system of equations with complex coefficients in which the $n \times n$ coefficient matrix of the system is represented by regular complex numbers and right side vector is represented by BFCN. To solve the CBFL system with complex coefficients we can replace the original system into the pair of positive ($\star$) and negative ($\bullet$) of two $n \times n$ crisp complex fuzzy linear system with complex coefficients. Some numerical examples have also been solved to show the effectiveness of the techniques. Our proposed methods can be used to solve circuit analysis (CA) and models of artificial intelligence.

References

1. Zadeh, L.A.: Fuzzy sets. Inf. Control **8**(3), 338–353 (1965)
2. Dubois, D., Prade, H.: Operations on fuzzy numbers. Int. J. Syst. Sci. **9**, 613–626 (1978)
3. Zhang (Yin)(Yang), W.R.: Bipolar fuzzy sets. In: Fuzzy Systems Proceedings, IEEE World Congress on Computational Intelligence, pp. 835–840 (1998)
4. Zhang, W.R.: Bipolar fuzzy sets and relations: a computational framework for cognitive modeling and multiagent decision analysis. In: Proceedings of IEEE Conference, pp. 305–309 (1994)
5. Akram, M.: Bipolar fuzzy graphs. Inf. Sci. **181**, 5548–5564 (2011)
6. Alghamdi, M.A., Alshehri, N.O., Akram, M.: Multi-criteria decision-making methods in bipolar fuzzy environment. Int. J. Fuzzy Syst. **20**(6), 2057–2064 (2018)
7. Friedman, M., Ma Ming, M., Kandel, A.: Fuzzy linear systems. Fuzzy Sets Syst. **96**, 201–209 (1998)
8. Abbasbandy, S., Alavi, M.: A method for solving fuzzy linear systems. Fuzzy Syst. **2**(2), 37–43 (2005)

9. Allahviranloo, T.: Uncertain information and linear systems. In: Studies in Systems, Decision and Control, vol. 254. Springer (2019)
10. Allahviranloo, T.: Successive over relaxation iterative method for fuzzy system of linear equations. Appl. Math. Comput. **162**(1), 189–196 (2005)
11. Allahviranloo, T.: Numerical methods for fuzzy system of linear equations. Appl. Math. Comput. **155**(2), 493–502 (2004)
12. Allahviranloo, T.: The Adomian decomposition method for fuzzy system of linear equations. Appl. Math. Comput. **163**(2), 553–563 (2005)
13. Abbasbandy, S., Jafarian, A.: Steepest descent method for system of fuzzy linear equations. Appl. Math. Comput. **175**(1), 823–833 (2006)
14. Akram, M., Muhammad, G., Koam, A.N., Hussain, N.: Iterative methods for solving a system of linear equations in a bipolar fuzzy environment. Mathematics **7**(8), 728 (2019)
15. Akram, M., Muhammad, G., Hussian, N.: Bipolar fuzzy system of linear equations with polynomial parametric form. J. Intell. Fuzzy Syst. (2019). https://doi.org/10.3233/JIFS-190764
16. Akram, M., Ali, M., Allahviranloo, T.: Certain methods to solve bipolar fuzzy linear system of equations. Comput. Appl. Math. **39**, 213 (2020)
17. Saqib, M., Akram, M., Shahida, B.: Certain efficient iterative methods for bipolar fuzzy system of linear equations. J. Intell. Fuzzy Syst. (2020). https://doi.org/10.3233/JIFS-200084
18. Akram, M., Saleem, D., Allahviranloo, T.: Linear system of equations in m-polar fuzzy environment. J. Intell. Fuzzy Syst. (2019). https://doi.org/10.3233/JIFS-190744
19. Suparna, D., Chakraverty, S.: Numerical solution of interval and fuzzy system of linear equations. Doctoral dissertation (2011)
20. Buckley, J.J.: Fuzzy complex number. Fuzzy Sets Syst. **33**, 333–345 (1989)
21. Rahgooy, T., Yazdi, H.S., Monsefi, R.: Fuzzy complex system of Linear equations applied to circuit analysis. Int. J. Comput. Electr. Eng. **1**(5), 535 (2009)
22. Jahantigh, M.A., Khezerloo, S., Khezerloo, M.: Complex fuzzy linear systems. Int. J. Ind. Math. **2**(1), 21–28 (2010)
23. Akram, M., Arshad, M.: A novel trapezoidal bipolar fuzzy TOPSIS method for group decision-making. Group Decis. Negot. **28**(3), 565–584 (2019)
24. Akram, M., Muhammad, G., Allahviranloo, T.: Bipolar fuzzy linear system of equations. Comput. Appl. Math. **38**(2), 69 (2019)
25. Ming, M., Friedman, M., Kandel, A.: A new fuzzy arithmetic. Fuzzy Sets Syst. **108**(1) (1999), 83-90
26. Jha, A.: Meromorphic function of fuzzy complex variables. J. New Theory **20**, 1–12 (2018)
27. Behera, D., Chakraverty, S.: Solving fuzzy complex system of linear equations. Inf. Sci. **277**, 154–162 (2014)

Solve the Growth Model of the HIV Using Z-Number

Maryam Ardeshiri Lordejani(✉) and Mojhdeh Afshar Kermani

Department of Mathematics, Tehran North Branch Islamic Azad University, Tehran, Iran
m.ardeshiry@gmail.com, mog_afshar@yahoo.com

Abstract. In real world, most of the phenomena are based on doubt and the information which we have from various subjects such as economic, political, physics and etc., has evaluated according to verbal valuables. Every day, we see new advances in medical science and given the level of public health and information of the various strata of society compared to the past, we see that today, humans are more willing to use more accurate methods to diagnose and treat diseases. In the meantime, the use of mathematical knowledge has become an important tool for physicians. In this paper, we use differential equations to predict the spread of diseases through a population and we solve it using Z-number.

Keywords: Z-numbers · Z-number initial value problem · Growth model of the HIV

1 Basic Concepts of Z-Numbers

In this section, we try to propose a new representation for Z-numbers and some basic definitions and operations to overcome this weakness.

We denote $\mathbb{E}_z$ the space of all Z-numbers on the universe set $U_1 \times U_2$, and $\mathbb{R}_z$ the space of all Z-numbers on $\mathbb{R} \times \mathbb{R}$. If $z \in \mathbb{R}_z$, we have

$$z = (A, B), \tag{1.1}$$

where A and B are called restriction and reliability of z, respectively, [3]. In the other word, z takes the fuzziness property from part A and the probabilistic property from part B.

Remark 1.1. In this paper, according to the Continuity Correction and the Central Limit theorem in the probability and statistics theory (see [1]) and without loss of generality, we can assume that:

(i) all random variables are continuous, and
(ii) also, all of probability density functions are normal, as

$$N(t; \mu, \sigma) = \frac{1}{\sigma\sqrt{2\pi}} \exp(-\frac{(t-\mu)^2}{2\sigma^2}), \tag{1.2}$$

T. Allahviranloo et al. (Eds.): IDS 2020, AISC 1301, pp. 928–932, 2021.
https://doi.org/10.1007/978-3-030-66501-2_74

where $-\infty < t < +\infty$, the variance σ is a positive real number and the mean μ belongs to the $\mathbb{R}$. Also, if $\sigma = 1$ and $\mu = 0$, we have a standard normal probability density function.

Note that throughout this paper, we assume A is a triangular fuzzy number and B has a normal probability density function N. Hence, we represent a Z-number z as a ordered pair $z = (A, N)$.

Definition 1.2 ([2]) (Parametric Form of Z-numbers). An arbitrary Z-number $z = (A, N) \in \mathbb{R}_z$, in the parametric form is represented by an ordered pair of functions

$$(\underline{z}(r), \overline{z}(r)) = ((\underline{A}(r), \overline{A}(r)), (\underline{N}(r), \overline{N}(r))), r \in [0, 1], \tag{1.3}$$

which $\underline{z}, \overline{z} : [0, 1] \longrightarrow \mathbb{R}$, and its components satisfy the following requirements:

1. $\underline{A}(r) \leq \overline{A}(r)$ and $\underline{N}(r) \leq \overline{N}(r)$ for $r \in [0, 1]$,
2. $\underline{A}(r)$ and $\underline{N}(r)$ are bounded left-continuous non-decreasing functions over $[0, 1]$,
3. $\overline{A}(r)$ and $\overline{N}(r)$ are bounded left-continuous non-increasing functions over $[0, 1]$,
4. $\int_{-\infty}^{core(N)} \underline{N}(r)dr = \int_{core(N)}^{+\infty} \overline{N}(r)dr = \frac{1}{2}$

Definition 1.3 ([2]) (α-cut of Z-numbers). Let $z \in \mathbb{R}_z$ be a Z-number where $z = (A, N)$. For $\alpha \in [0, 1]$, the α-level set (or α-cut) of z is defined by

$$[z]^\alpha = [[A]^\alpha, [N]^\alpha] = [[\underline{A}^\alpha, \overline{A}^\alpha], [\underline{N}^\alpha, \overline{N}^\alpha]] = \begin{cases} [[\underline{A}^\alpha, \overline{A}^\alpha], [\underline{N}^\alpha, \overline{N}^\alpha]], & if \quad 0 \leq \alpha \leq \alpha_k, \\ [[\underline{A}^\alpha, \overline{A}^\alpha], [\underline{N}^{\alpha_k}, \overline{N}^{\alpha_k}]], & if \quad \alpha_k \leq \alpha \leq 1, \end{cases}$$

where $\alpha_k = h(N)$. Note that if z is normal, then this definition is valid and it is well-known that the level-cuts of Z-numbers have nesting property.

Definition 1.4 ([2]) (Z-number Initial Value Problem (ZIVP)). We consider a ZIVP as (1.4)

$$\begin{cases} x'(t) = f(t, x(t)), \\ x(t_0) = x_0 \end{cases}$$

where $t_0 \in [0, T] \subseteq \mathbb{R}_Z$ and f is a continuous mapping from $[t_0, T] \times \mathbb{R}_Z$ in to $\mathbb{R}_Z$ and x_0 is a Z-number in $\mathbb{R}_Z$.

Theorem 1.5 *(Characterization Theorem). Let $x_0 \in \mathbb{R}_Z$ and $f : [t_0, T] \times \mathbb{R}_Z \rightarrow \mathbb{R}_Z$ is a continuous Z-process such that:*

$$\begin{aligned} (i) \qquad [f(t, x(t))]^\alpha &= [f_A(t, (x_A(t), x_N(t))), f_N(t, (x_A(t), x_N(t)))]^\alpha \\ &= [[\underline{f}_A^\alpha(t, ((\underline{x}_A^\alpha(t), \overline{x}_A^\alpha(t))), \overline{f}_A^\alpha(t, ((\underline{x}_A^\alpha(t), \overline{x}_A^\alpha(t)))] \\ &\quad, [\underline{f}_N^\alpha(t, ((\underline{x}_A^\alpha(t), \overline{x}_A^\alpha(t))), \overline{f}_N^\alpha(t, ((\underline{x}_A^\alpha(t), \overline{x}_A^\alpha(t)))]] \end{aligned}$$

(ii) *Boundary function* $\underline{f}^{\alpha}_{A}, \overline{f}^{\alpha}_{A}, \underline{f}^{\alpha}_{N}$ *and* $\overline{f}^{\alpha}_{N}$ *are equicontinuous and uniformly on any bounded on any bounded set,*

(iii) *There* $L > 0$ *such that:*

$$\left|\underline{f}^{\alpha}_{A}(t,((x_{1A},x_{1N}),(y_{1A},y_{1N}))) - \underline{f}^{\alpha}_{A}(t,((x_{2A},x_{2N}),(y_{2A},y_{2N})))\right| \leq L\max\{|(x_A,x_N)-(u_A,u_N)|,|(y_A,y_N)-(v_A,v_N)|\},$$

and

$$\left|\overline{f}^{\alpha}_{A}(t,((x_{1A},x_{1N}),(y_{1A},y_{1N}))) - \overline{f}^{\alpha}_{A}(t,((x_{2A},x_{2N}),(y_{2A},y_{2N})))\right| \leq L\max\{|(x_A,x_N)-(u_A,u_N)|,|(y_A,y_N)-(v_A,v_N)|\},$$

and

$$\left|\underline{f}^{\alpha}_{N}(t,((x_{1A},x_{1N}),(y_{1A},y_{1N}))) - \underline{f}^{\alpha}_{N}(t,((x_{2A},x_{2N}),(y_{2A},y_{2N})))\right| \leq L\max\{|(x_A,x_N)-(u_A,u_N)|,|(y_A,y_N)-(v_A,v_N)|\},$$

and

$$\left|\overline{f}^{\alpha}_{N}(t,((x_{1A},x_{1N}),(y_{1A},y_{1N}))) - \overline{f}^{\alpha}_{N}(t,((x_{2A},x_{2N}),(y_{2A},y_{2N})))\right| \leq L\max\{|(x_A,x_N)-(u_A,u_N)|,|(y_A,y_N)-(v_A,v_N)|\},$$

for each $\alpha \in [0,1]$. *Then ZIVP (1.4) has a unique Z-process solution.*

Proof. [2].

2 The Growth Model of the AIDS Virus (HIV)

The growth of AIDS is an example that follows the curve of the logistic equation, derived from solving a differential equation. Populations usually grow in an exponential fashion at first, however, populations do not continue to grow forever, because food, water and other resources get used up over time. Differential equations are used to predict populations of people, animals, bacteria and viruses that are being affected by external events. The logistic equation (developed in the mid-19th century) allows for a growth term and an inhibition term. It is predicte that the AIDS epidemic will follow the pattern of the logistic equation. If

X = number of people affected by the virus at time,
P = the total population (a constant), and
c is a constant,
then the rate of growth of the virus at time is given by the differential equation:

$$\frac{dX}{dt} = cPX - cX^2 = cX(P - X) \tag{2.1}$$

The term cPX is the growth term and $-cX^2$ is the inhibition term.

Now we assume that AIDS is spreading through a city of 50000 people who take no precautions therefore $P = 50000$ and consider the Z-number initial value problem as

$$\begin{cases} X'(t) = cX(50000 - X), \\ X(t_0) = X_0 \end{cases} \tag{2.2}$$

where $t \in [0, T]$ and $X_0 :=$ (the number of people affected by the virus in some city last month, considerably close to 25 thousand, not sure) and $X_{0A} :=$ (considerably close to 25 thousand) is a triangular fuzzy number $(24, 25, 26)$ by membership function as

$$\mu_{X_{0A}}(t) = \begin{cases} 0, & t < 24 \\ t - 24, & 24 \leq t < 25 \\ 1, & t = 25 \\ 26 - t, & 25 < t \leq 26 \\ 0, & t > 26 \end{cases}$$

and $X_{0N} :=$ (not sure), presents the normal probability density function

$$N(X; \mu, \sigma) = \frac{1}{\sigma\sqrt{2\pi}} \exp(-\frac{(X-\mu)^2}{2\sigma^2}), \mu = 25 \quad and \quad \sigma = 2. \tag{2.3}$$

Let

$$X^{(k)}(t) = (X_A^{(k)}(t), X_N^{(k)}(t)), \qquad k = 0, 1, \tag{2.4}$$

where k is order of derivative and

$$[X^{(k)}(t)]^\alpha = \left([\underline{X_A}^{\alpha(k)}(t), \overline{X_A}^{\alpha(k)}(t)], [\underline{X_N}^{\alpha(k)}(t), \overline{X_N}^{\alpha(k)}(t)]\right), \tag{2.5}$$

and

$$X_0 = (X_{0A}, X_{0N}), \quad [X_0]^\alpha = \left([\underline{X_{0A}}^\alpha, \overline{X_{0A}}^\alpha], [\underline{X_{0N}}^\alpha, \overline{X_{0N}}^\alpha]\right), \tag{2.6}$$

where $\alpha \in [0, 1]$. Assume that $f(t, x(t)) = x(t)$, satisfies the recent Theorem (1.5), hence this problem has a unique solution. Now, we have:

$$\begin{aligned} \left[\underline{X_A}^\alpha(t), \overline{X_A}^\alpha(t)\right] &= \left[\underline{X_{0A}}^\alpha \frac{50000}{1 + 499e^{-50000c(t-t_0)}}, \overline{X_{0A}}^\alpha \frac{50000}{1 + 499e^{-50000c(t-t_0)}}\right] \\ &= \left[(24+\alpha)\frac{50000}{1 + 499e^{-50000ct}}, (26-\alpha)\frac{50000}{1 + 499e^{-50000ct}}\right] \end{aligned}$$

and

$$\begin{aligned} \left[\underline{X_N}^\alpha(t), \overline{X_N}^\alpha(t)\right] &= \left[\underline{X_{0N}}^\alpha \frac{50000}{1 + 499e^{-50000c(t-t_0)}}, \overline{X_{0N}}^\alpha \frac{50000}{1 + 499e^{-50000c(t-t_0)}}\right] \\ &= \left[(4+16\alpha)\frac{50000}{1 + 499e^{-50000ct}}, (36-16\alpha)\frac{50000}{1 + 499e^{-50000ct}}\right] \end{aligned}$$

hence,

$$[x(t)]^\alpha = \left[\left[(24+\alpha)\frac{50000}{1+499e^{-50000ct}}, (26-\alpha)\frac{50000}{1+499e^{-50000ct}}\right]\right.$$
$$\left., \left[(4+16\alpha)\frac{50000}{1+499e^{-50000ct}}, (36-16\alpha)\frac{50000}{1+499e^{-50000ct}}\right]\right] \quad (2.7)$$

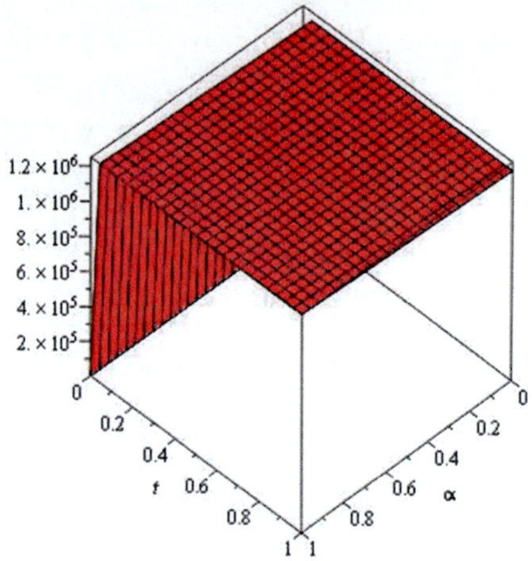

Fig. 1. Graph of a example for X_A

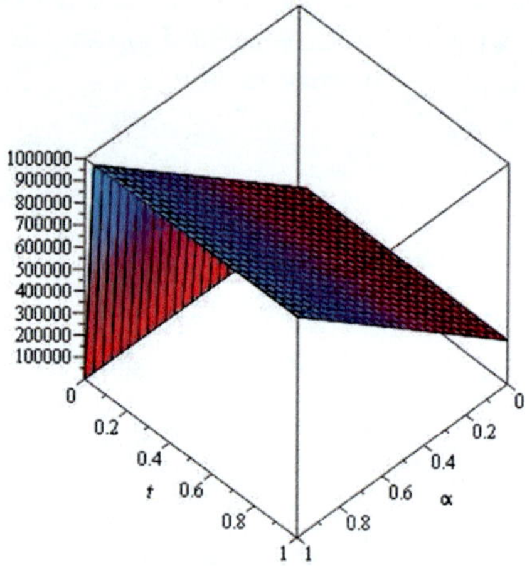

Fig. 2. Graph of a example for X_N

Figures 1 and 2 show the possibility and probability distribution of equation (2.2), respectively.

3 Conclusion

In this paper, we use differential equations to predict the spread of diseases through a population and we solve it using Z-number. According to the solutions and figures of this example, we found out that by increasing t the fuzziness (or uncertainty) tends to infinity and distribution (or probability) is fixed.

References

1. Freund, J.E.: Mathematical Statistics, 5th edn. Prentice-Hall, Upper Saddle River (1992)
2. Pirmuhammadi, S., Allahviranloo, T., Keshavarz, M.: The parametric from of Z-number and its application in Z-number in initial value problem. **32**(10) (2017). Elsevier
3. Zadeh, L.A.: A Note on Z-numbers. Inf. Sci. **181**, 2923–2932 (2011)

Solveing Fuzzy Fractional Impulsive Differential Equations by Fuzzy Fractional Adomian Decomposition Technique

Nematallah. Najafi(✉)

Department of Mathematics, Islamic Azad University,
Hamedan Branch, Hamedan, Iran
n.najafi56@yahoo.com

Abstract. In this paper, we study semi-analytical methods entitled Fuzzy Fractional Adomian Decomposition Technique (FFADT) to solve fuzzy fractional impulsive differential equations (FFIDE) using Caupto fractional derivative. At the end, first of all, FFADT in fractional case is defined and the existence and uniqueness theorem for the solution are proved and the convergence of the proposed method is considered in details. Then, we compare solutions of the FFIDE with both FFADM and Fuzzy Fractional Differential Transform Method (FFDTM). The results of the FFADT and the FFDTM show its better performance than others.

Keywords: Fractional adomian decomposition method · Fuzzy impulsive fractional differential · Generalized hukuhara differentiability · Fractional differential transform method

1 Introduction

Fuzzy set theory is the significant tool for modeling unknown problems and can be found in many branches of regional, physical, mathematical and engineering sciences. As a result many things can happen in the real world has a fuzzy meaning. The concept of the fuzzy set theory was first proposed by Zadeh, Zimmerman and Kaleva (see [20,48,49]). One of the very important branches of the fuzzy theory is fuzzy differential equations. Later many scientists, including M. Friedman, et al [16] applied numerical methods to solve this equations. Also the idea of fuzzy differential equations has been studied by scientists and engineers such as S. Abbasbandy and T. Allaviranloo [1]. They have considered new method to solve fuzzy differential equation based on fuzzy Taylor expansion as one of the branches of fuzzy differential equations . The idea of the theory of fuzzy impulsive differential equation has been emerging as an effective tool area of investigation in recent years (see [37]). Guo. M and Li. R [17] proposed fuzzy solutions for impulsive differential equations. Subsequently, the basic result for fuzzy impulsive differential equation was defined by S. Vengataasalam and et al [44].

Although the fuzzy fractional differential equations have many branches and many applications and in this research we will restrict our attention to

T. Allahviranloo et al. (Eds.): IDS 2020, AISC 1301, pp. 933–964, 2021.
https://doi.org/10.1007/978-3-030-66501-2_75

FFIDE while FFIDE are usually hard to solve analytically and the exact solution is rather difficult to be obtained. Recently, N. Najafi and T. Allaviranloo [23,24,26,28] applied FFDTM to solve FFIDE. Our aim in this paper is to study the FFADM for solving FFIDE. We will use the FFADM based on generalized Hukuhara differentiability to solve a nonlinear and linear FFIDE given by

$$ {}_cD_*^\alpha y(t) = f(t, y(t)), \quad t \in [0, T], \quad t \neq t_k \tag{1}$$

$$ \Delta y|_{t=t_k} = I_k(y(t_k^-)) \quad t = t_k \quad k = 0, 2, \ldots, m \tag{2}$$

$$ y(0) = y_0 \tag{3}$$

In this paper the set of all fuzzy numbers is denoted $\mathbb{R}_F$. Where $k = 1, 2, \ldots, m, 0 < \frac{1}{\alpha} \leq 1$, ${}_cD_*^\alpha$ denotes the Caputo fractional generalized derivative of order α, and $f : [0, T] \times \mathbb{R}_F \to \mathbb{R}_F$, is continuous fuzzy function, $I_k : \mathbb{R}_F \to \mathbb{R}_F$, is continuous function, $y_0 \in \mathbb{R}_F, 0 = t_0 < t_1 < \cdots < t_m < t_{m+1} = T$, $\Delta|_{t=t_k} = y(t_k^+) \ominus y(t_k^-)$, $y(t_k^+) = lim_{h\to 0^+} y(t_k + h)$, and $y(t_k^-) = lim_{h\to 0^-} y(t_k + h)$ represent the right and left limits of $y(t)$ at $t = t_k$. The paper is organized as follows:

We describe the basic notation and preliminaries in Sect. 2. We have stated the FFIDE such that f is generalized Hukuhara differentiable in Sect. 3. In Sect. 4, we expression fuzzy fractional Taylor series based on the concept of generalized Hukuhara differentiability. We describe solving steps solving FFIDE by FFDTM based on the concept of generalized Hukuhara differentiability in Sect. 5. In Sect. 6, we expression FFADT based on the concept of generalized Hukuhara differentiability. Some numerical examples are given to clarify the details and efficiency of the method in Sect. 7. This paper ends with conclusion in Sect. 8.

2 Preliminaries

In this portion, we review and explain several basic definitions, notations, lemmas, and propositions used all over this article.

Definition 1 *([20]). Fuzzy number y is parametric form a placenta $(\underline{y}(r), \overline{y}(r))$ for functions $\underline{y}(r)$ and $\overline{y}(r)$, which is true in the below situations:*

a_1 : $\underline{y}(r)$ be a left continuous function and monotonic bounded within $(0, 1]$ also on the right continuous in 0,
a_2 : $\overline{y}(r)$ be a left continuous function and monotonic bounded within $(0, 1]$ also on the right continuous in 0,
a_3 : $\underline{y}(r) \leq \overline{y}(r)$, $\forall r \in [10, 1]$.

Definition 2 *[7]. Suppose that $y, z \in \mathbb{R}_F$. If accessible $p \in \mathbb{R}_F$ so that $y = z \oplus p$, thus p is named the H-difference of y and z, also displayed with $p = y \ominus_H z$. The Hukuhara difference is also motivated by the problem of inverting the addition, if z, y are crisp numbers then $(z+y)-y = z$ but this is not correct if x, y are fuzzy numbers. If y and z are fuzzy numbers (and not in general fuzzy sets), so $(y+z) \ominus_H z = y$ i.e. the H-difference inverts the addition of fuzzy numbers. The gH-difference for fuzzy numbers can be defined as follows:*

Definition 3 *([24]). For $y, z, p \in \mathbb{R}_F$, the generalized Hukuhara difference for y and z,is shown by $y \ominus_{gH} z = p$ which is defined as follows*

$$y \ominus_{gH} z = p \iff \begin{cases} (i) \quad y = z \oplus p, \\ or \\ (ii) \quad z = y \oplus (-1)p. \end{cases}$$

According to r-cuts we have $[y \ominus_{gH} z]_r = [\min\{\underline{y}_r - \underline{z}_r, \overline{y}_r - \overline{z}_r\}, \max\{\underline{y}_r - \underline{z}_r, \overline{y}_r - \overline{z}_r\}]$ and the condition that $y \ominus_{gH} z = P \in \mathbb{R}_F$ is established is that

$$status(i) = \begin{cases} \underline{p}_r = \underline{y}_r - \underline{z}_r \quad and, \quad \overline{p}_r = \overline{y}_r - \overline{z}_r \\ with \quad \underline{p}_r \quad increasing, \quad \overline{p}_r \quad decreasing \quad and \quad \underline{p}_r \leq \overline{p}_r \end{cases}$$

$$status(ii) = \begin{cases} \underline{p}_r = \overline{y}_r - \overline{z}_r \quad and, \quad \overline{p}_r = \underline{y}_r - \underline{z}_r \\ with \quad \underline{p}_r \quad increasing, \quad \overline{p}_r \quad decreasing \quad and \quad \underline{p}_r \leq \overline{p}_r \end{cases}$$

We also refer the reader ([11]) for some related results in this subjec. Moreover, we assume that mapping $f : [a, b] \to \mathbb{R}_F$ is named fuzzy number valued function. The r-cuts presentation of fuzzy number valued function f shown with $[f(t)]_r = [\underline{f}(t, r), \overline{f}(t, r)]$, $a \leq t \leq b$ and $0 \leq r \leq 1$.

Definition 4 *([12]). For $0 < r \leq 1$ denoted $[y]_r = \{t \in \mathbb{R} | y(t) \geq r\} = [\underline{y}(r), \overline{y}(r)]$ also for $r = 0$ via the closure supports with $[y]_0 = cl\{t | t \in \mathbb{R}, y(t) > 0\}$ so that cl the closing subset designates. The scaller multiplication and summation are shown with $k \odot u$ and $y \oplus z$ and are defined as follows*

$$[y \oplus z]_r = [y]_r \oplus [z]_r = \{x_1 + x_2 | x_1 \in [y]_r, x_2 \in [z]_r\}, \tag{4}$$

and

$$[k \odot y]_r = k.[y]_r = \begin{cases} [k\underline{y}(r), k\overline{y}(r)], 0 \leq k, \\ [k\overline{y}(r), k\underline{y}(r)], k < 0. \end{cases}$$

The minus for fuzzy numbers $y \ominus_{gH} z$ is defined like the determined $y \oplus (-1)z$, if $y(r) = [\underline{y}(r), \overline{y}(r)]$ so $[(-1) \odot y]_r = [-\overline{y}(r), -\underline{y}(r)]$.

Definition 5 *([4]). Hausdorff distance for two fuzzy numbers is demonstrate to $d : \mathbb{R}_F \times \mathbb{R}_F \to \mathbb{R}^+ \cup \{0\}$ is defined in the form of*

$$d(y, z) = \sup_{0 < r \leq 1} \max \Big(|\underline{y}(r) - \underline{z}(r)|, |\overline{y}(r) - \overline{z}(r)| \Big).$$

Assume that $y, z, y_1, z_1 \in \mathbb{R}_F$ also $\lambda \in \mathbb{R}$, which satisfy in the under conditions

1) $d(y \oplus y_1, z \oplus z_1) = d(y, z)$,
2) $d(\lambda y, \lambda z) = |\lambda| d(y, z)$,
3) $d(y \oplus z, y_1 \oplus z_1) \leq d(y, y_1) + d(z, z_1)$,
4) $d(y \ominus_H z, y_1 \ominus_H z_1) \leq d(y, y_1) + d(z, z_1)$,

as well as $y \ominus_H z$ *also* $y_1 \ominus_H z_1$ *exists, and* $y, z, y_1, z_1 \in \mathbb{R}_F$.

Theorem 1 *([47]). By assumeing that* $f(t)$ *is the fuzzy number valued function in* $[a, \infty)$ *and it is displayed with* $(\underline{f}(t, r),\ \overline{f}(t, r))$. *Suppose* $\overline{f}(t, r)$*and* $\underline{f}(t, r)$ *are Riemann-integrable in* $[a, b]$, *also* $b \geq a$ *and* $\underline{M}(r)$, $\overline{M}(r)$ *is two positive fuzzy numbers, then*

$$\int_a^b |\underline{f}(t, r)| dt \leq \underline{M}(r),$$

and

$$\int_a^b |\overline{f}(t, r)| dt \leq \overline{M}(r),$$

$\forall b \geq a$. *Thus* $f(t)$ *is improper fuzzy Riemann-integrable in* $[a, \infty)$ *and is a fuzzy number. Moreover, the following statement is valid:*

$$[\int_a^\infty f(t) dt]_r = \Big[\int_a^\infty \underline{f}(t, r) dt, \int_a^\infty \overline{f}(t, r) dt\Big].$$

Definition 6 *[7]. let* $f : [a, b] \longrightarrow \mathbb{R}_F$ *is denominate fuzzy continuous if for discretional constant* $t_0 \in \mathbb{R}_F$ *also* $\xi > 0$, *there is a* $\delta > 0$, *so*

$$|t - t_0| < \delta, so \ \ d(f(t), f(t_0)) < \xi$$

Lemma 1. $\forall\ \alpha > 0$ *also* $\gamma > -1$

$$\int_0^t (t - s)^{k\alpha - 1} s^\gamma ds = \frac{\Gamma(k\alpha)\Gamma(\gamma + 1)}{\Gamma(k\alpha + \gamma + 1)} t^{k\alpha + \gamma}.$$

We denoted the gamma function by Γ *and we display as follows*

$$\Gamma(z) = \int_0^\infty \frac{t^{z-1}}{e^t} dt.$$

Proof: *Lemma* 2.6 [45].

Definition 7. *Suppose,* $f : (a, b) \to \mathbb{R}_F$ *is the fuzzy number valued function so Generalized Hukuhara derivative in* t_0 *as follows*

$$f'_{gH}(t_0) = \lim_{h \to 0} \frac{f(t_0 + h) \ominus_{gH} f(t_0)}{h} \tag{5}$$

if $f'_{gH}(t_0) \in \mathbb{R}_F$ *satisfying* (5) *exists, we express that* f *be generalized Hukuhara differentiable (gH-differentiable for short) in* t_0. *We express that* f *be in the form of [(i)-gH]-diferentiable in* t_0 *thus*

$$(i)\ \ [f'_{i.gH}(t_0)]_r = [\underline{f}'(t_0, r),\ \overline{f}'(t_0, r)] \tag{6}$$

Also f be in the form of [(ii)-gH]-diferentiable in t_0 thus

$$(ii)\quad [f'_{ii.gH}(t_0)]_r = [\overline{f}'(t_0, r), \underline{f}'(t_0, r)] \tag{7}$$

Definition 8 *([4]). Suppose $f : (a, b) \to \mathbb{R}_F$. We express f is gH- differentiable of the mth-order in t_0 while the function f be gH-differentiable by order $j, j = 1, 2, \ldots, n-1$, at t_0 providing gH-differentiable kind has no variation, so there is $f_{gH}^{(m)}(t_0) \in \mathbb{R}_F$ such that*

$$f_{gH}^{(m)}(t_0) = \lim_{h\to 0} \frac{f_{gH}^{(m-1)}(t_0 + h) \ominus_{gH} f_{gH}^{(m-1)}(t_0)}{h}, m \in \mathbb{N} \tag{8}$$

Definition 9 *([22]). Suppose $f : [a, b] \to \mathbb{R}_F$, then the fuzzy fractional derivative from $f(t)$ for Caputo concept explained as following*

$${}_cD_*^\alpha f(t) = \frac{1}{\Gamma(m-\alpha)} \int_a^t \frac{f^{(m)}(t)\mathrm{d}s}{(t-s)^{\alpha-m+1}}, \quad m-1 < \alpha < m, t > a \tag{9}$$

Definition 10 *([10]). Suppose that $f : [a, b] \to \mathbb{R}_F$ is fuzzy number valued function, then the fuzzy Riemann-Liouville integral for f as follows:*

$$(I_{a|t}^\alpha f)(t) = \frac{1}{\Gamma(\alpha)} \int_a^t \frac{f(t)}{(t-s)^{1-\alpha}} \mathrm{d}s,\ t > a,$$

where

$$[I_{a|t}^\alpha f(t)]_r = \left[\frac{1}{\Gamma(\alpha)} \int_a^t \frac{\underline{f}(t,r)}{(t-s)^{1-\alpha}} \mathrm{d}s,\ \frac{1}{\Gamma(\alpha)} \int_a^t \frac{\overline{f}(t,r)}{(t-s)^{1-\alpha}} \mathrm{d}s\right],$$

for $a \le s \le t$ also $0 < \alpha \le 1$.

Definition 11 *([22]). For the fuzzy number valued function f([gH]- differentiability for short) The Caputo generalized Hukuhara differentiability in the form below:*

$${}_cD_*^\alpha f_{gH}(t) = I_{a|t}^{m-\alpha} f_{gH}^{(m)}(t) = \frac{1}{\Gamma(m-\alpha)} \int_a^t \frac{f_{gH}^{(m)}(t)\mathrm{d}s}{(t-s)^{\alpha-m+1}}, \quad m-1 < \alpha < m, t > a \tag{10}$$

We assume exists the concept of gH for fuzzy function f by any order of differentiability. in addition to we express then f be in the form of $[(i) - gH]$-Caputo differentiable it t,

$$[{}_cD_{i.gH}^\alpha f(t)]_r = [{}_cD_*^\alpha \underline{f}(t, r),\ {}_cD_*^\alpha \overline{f}(t, r)], \tag{11}$$

as well as f be in the form of $[(ii) - gH]$-Caputo differentiable it t so

$$[{}_cD_{ii.gH}^\alpha f(t)]_r = [{}_cD_*^\alpha \overline{f}(t, r),\ {}_cD_*^\alpha \underline{f}(t, r)], \tag{12}$$

Here ${}_cD_*^\alpha \overline{f}(t, r)$ also $cD_*^\alpha \underline{f}(t, r)$ are defned in Defnition 9.

Definition 12 *([12]). The function f has switching point in $t_0 \in (a, b)$, if in any neighborhood v of t_0 there is points $t_1 < t_0 < t_2$ then:*

Type(I): at t_1(11) maintaining while(12) does not maintaining and at t_2(12) maintaining while (11) does not hold, or
Type(II): at t_1(12) maintaining while (11) does not maintaining and at t_2(11) maintaining while (12) does not maintaining.

Lemma 2. *We assume that $\left(I^{\alpha}_{a|t} f_{gH}\right)(t)$, for $\alpha \in (0, 1]$, $t \in [a, b]$ also $r \in [0, 1]$ is a continuous function if $f : [a, b] \to \mathbb{R}_F$ be continuous.*

Proof. $f(s)$ is the fuzzy Lebesque and integrable function by the conditions acontinuous functions. in the other words, as $\forall \alpha \in (0, 1]$, $\frac{1}{t-s)^{1-\alpha}} \geq 0$ be continuous, so $\int_a^t (t-s)^{\alpha-1} f(s) ds$ be a continuous function and as a consequence $\left(I^{\alpha}_{a|t} f_{gH}\right)(t)$ be a continuous function $\forall a \leq t \leq b$.

All over the residuum of this article, note that $f \in C^n_{gH}([a, b], \mathbb{R}_F)$ is named the collection of continuous the fuzzy number valued functions are defined in $[a, b]$. Also we show that the space for every Lebesgue integrable fuzzy number valued function in the bounded interval $[a, b] \subset \mathbb{R}$ by $L^F[a, b]$. Also we show that $C^F[a, b]$ similar the space for every continuous fuzzy number valued function in $[a, b]$.

Lemma 3. *Suppose $f \in C_{gH}(\mathbb{R}, \mathbb{R}_F), m \in \mathbb{N}$. Then the following integrals*

$$\left(I^{\alpha}_{a|t_{m-1}} f_{gH}\right)(t_{m-1}), \left(I^{\alpha}_{a|t_{m-2}}\left(I^{\alpha}_{a|t_{m-1}} f_{gH}\right)\right)(t_{m-1}),$$
$$\left(I^{\alpha}_{a|t}\left(I^{\alpha}_{a|t_1}\left(I^{\alpha}_{a|t_2} f_{gH}\right)\right)\right)(t_2), \ldots,$$
$$\left(I^{\alpha}_{a|t}(I^{\alpha}_{a|t_1} \cdots \left(I^{\alpha}_{a|t_{m-2}}\left(I^{\alpha}_{a|t_{m-1}} f_{gH}\right)\right) \cdots)\right)(t_{m-1}).$$

are continuous functions in $t_{m-1}, t_{m-2}, \ldots, t$, respectively. Here $t_{m-1}, t_{m-2}, \ldots, t > t_0 = a$ for every real numbers.

Proof. This lemma is a clearly right away generalization of Lemma 2. Proving will be accomplished by commence on $m \in \mathbb{N}$. presume the lemma maintain for (m)-times apply fuzzy fractional Riemann-Liouville integrating operator for function f, we demonstrate the details for $(m + 1)$-times apply fuzzy fractional Riemann-Liouville integrating operator for function f. By Lemma 2, so $f \in C^n_{gH}(\mathbb{R}, \mathbb{R}_F)$ then $I_{a|t_{m-1}} f(t_{m-1})$ is a continuous function at t_{m-1}. Moreover, with the help of induction theory,

$$\overbrace{\left(I^{\alpha}_{a|t_{m-2}}\left(I^{\alpha}_{a|t_{m-1}} f_{gH}\right)\right)(t_{m-1}), \ldots, \left(I^{\alpha}_{a|t}(I^{\alpha}_{a|t_1} \cdots \left(I^{\alpha}_{a|t_{m-2}}\left(I^{\alpha}_{a|t_{m-1}} f_{gH}\right)\right) \cdots)\right)(t_{m-1})}^{(m)-times} \tag{13}$$

are continuous functions in $t_{m-1}, t_{m-2}, t_{m-3}, \ldots, t$, respectively. It follows easily that

$$\overbrace{\left(I^{\alpha}_{a|t_{m+1}} I^{\alpha}_{a|t}(I^{\alpha}_{a|t_1} \cdots \left(I^{\alpha}_{a|t_{m-2}}\left(I^{\alpha}_{a|t_{m-1}} f_{gH}\right)\right) \cdots)\right)(t_{m-1})}^{(m+1)-times} \tag{14}$$

Proven the continuity of the function at the point t_{m+1}, our claim is proved.

Lemma 4 *([?]). Presume assume fuzzy number valued function f Has mapping to form $f : [a,b] \to \mathbb{R}_F$ and $f^{(m)}_{gH}$ exists, subsequent $\forall \alpha \in [0,1]$ also $m \in \mathbb{N}$,*

$$(I^{\alpha}_{a|t\,c}D^{\alpha}_{*} f)(t) = f(t) \ominus_{gH} f(a). \tag{15}$$

Lemma 5 *([26]). Let ${}_cD^{k\alpha}_{*} f : (a,b) \to \mathbb{R}_F$ be a fuzzy function continuous for $k = 1, 2, \ldots, n$ here $\alpha \in (0,1]$*
(i) $({}_cD^{(n-1)\alpha}_{i.gH} f)(t)$ and $({}_cD^{n\alpha}_{i.gH} f)(t)$ for $1 \leq n$ are [i-gH]-Caputo differentiable of order α in same type of differentiability concept, then

$$I^{\alpha}_{a|t}({}_cD^{(n+1)\alpha}_{i.gH} f)(t) = ({}_cD^{n\alpha}_{i.gH} f)(t) \ominus ({}_cD^{n\alpha}_{i.gH} f)(a) \tag{16}$$

and (ii) $({}_cD^{(n-1)\alpha}_{ii.gH} f(t)$ and $({}_cD^{n\alpha}_{ii.gH} f)(t)$ for $1 \leq n$ be in the form of [ii-gH]-Caputo differentiable of order α in same type of differentiability concept, then

$$I^{\alpha}_{a|t}({}_cD^{(n+1)\alpha}_{ii.gH} f)(t) = (-1)[({}_cD^{n\alpha}_{ii.gH} f)(a) \ominus ({}_cD^{n\alpha}_{ii.gH} f)(t)] \tag{17}$$

Lemma 6 *([8]). Let ${}_cD^{k\alpha}_{*} f_{gH} : (a,b) \to \mathbb{R}_F$ be a continuous fuzzy function for $k = 1, 2, \ldots, n$ where $\alpha \in (0,1]$*

(i) $({}_cD^{(n-1)\alpha}_{i.gH} f)(t)$ and $({}_cD^{n\alpha}_{i.gH} f_{gH})(t)$ for $1 \leq n$ are [i-gH]-Caputo differentiable of order α in same type of differentiability concept, then

$$I^{n\alpha}_{a|t}({}_cD^{n\alpha}_{i.gH} f)(t) \ominus I^{(n+1)\alpha}_{a|t}({}_cD^{(n+1)\alpha}_{i.gH} f)(t) = \frac{(t-a)^{n\alpha}}{\Gamma(n\alpha+1)}({}_cD^{n\alpha}_{i.gH} f)(a) \tag{18}$$

(ii) $({}_cD^{(n-1)\alpha}_{ii.gH} f)(t)$ and $({}_cD^{n\alpha}_{ii.gH} f)(t)$ for $1 \leq n$ be in the form of [ii-gH]-Caputo differentiable of order α in same type for differentiability concept, then

$$I^{n\alpha}_{a|t}({}_cD^{n\alpha}_{ii.gH} f)(t) \oplus (-1) I^{(n+1)\alpha}_{a|t}({}_cD^{(n+1)\alpha}_{ii.gH} f)(t) = \frac{(t-a)^{n\alpha}}{\Gamma(n\alpha+1)}({}_cD^{n\alpha}_{ii.gH} f)(a) \tag{19}$$

Theorem 2. *Fractional gH-differentiable be from $f : [a,b] \to \mathbb{R}_F$ so that kind of Caputo differentiability f in $[a,b]$ doesn't alteration. Thus,*

(i). If $f(t)$ be in the form of $[(i)-gH]$-Caputo differentiable then $\left({}_cD^{\alpha}_{i.gH} f\right)(t)$ is fuzzy Riemann-integrable in $[a,b]$ also

$$f(t) = f(a) \oplus \left(I^{\alpha}_{a|t\,c}D^{\alpha}_{i.gH} f\right)(t), \quad \forall a < t < b, \alpha \in [0,1]. \tag{20}$$

For case [(ii)-gH]-Caputo differentiable, so

$$f(t) = f(a) \ominus (-1)\left(I^{\alpha}_{a|t\,c}D^{\alpha}_{ii.gH} f\right)(t), \quad \forall a < t < b, \alpha \in [0,1]. \tag{21}$$

Proof Reason(i). We assume $f(t)$ be form the $[(i) - gH]$-Caputo differentiable, thus

$$\begin{aligned}[\left(I^{\alpha}_{a|t\,c}D^{\alpha}_{i.gH} f\right)(t)]_r &= \left[\underline{f}(t,r) - \underline{f}(a,r), \overline{f}(t,r) - \overline{f}(a,r)\right] \\ &= [f(t) \ominus_{gH} f(a)]_r, \quad r \in [0,1]\end{aligned}$$

So,

$$f(t) = f(a) \oplus \left(I^{\alpha}_{a|t\,c}D^{\alpha}_{i.gH} f\right)(t).$$

Reason(ii). We assume $f(t)$ be form the [(ii)-gH]-Caputo differentiable, thus

$$\begin{aligned}[\ominus(-1) \odot \left(I^{\alpha}_{a|t\,c}D^{\alpha}_{ii.gH} f\right)(t)]_r &= \ominus(-1) \odot \left[\overline{f}(t,r) - \overline{f}(a,r), \underline{f}(t,r) - \underline{f}(a,r)\right] \\ = \left[\underline{f}(t,r) - \underline{f}(a,r), \overline{f}(t,r) - \overline{f}(a,r)\right] &= [f(t) \ominus_{gH} f(a))]_r, \quad r \in [0,1]\end{aligned}$$

Therefore

$$f(t) = f(a) \ominus (-1)\left(I^{\alpha}_{a|t\,c}D^{\alpha}_{ii.gH} f\right)(t).$$

Lemma 7. *Consider* $f : [a,b] \to \mathbb{R}_F$ *be the fractional gH-differentiable and* $\left({}_cD^{\alpha}_{*} f_{gH}\right)(t)$ *continuous on* $[a,b]$. *Then*

$$\left(I^{\alpha}_{a|t\,c}D^{\alpha}_{i.gH} f\right)(t) := \ominus(-1) \odot \left(I^{\alpha}_{a|t\,c}D^{\alpha}_{ii.gH} f\right)(t). \tag{22}$$

Proof. Suppose, $\left({}_cD^{\alpha}_{i.gH} f\right)(t)$ is fuzzy Riemann-integrable, so

$$\begin{aligned}[\left(I^{\alpha}_{a|t\,c}D^{\alpha}_{i.gH} f\right)(t)]_r &= \left[\underline{f}(t,r) - \underline{f}(a,r), \overline{f}(t,r) - \overline{f}(a,r)\right] \\ &= \left[f(t) \ominus_{gH} f(a)\right]_r, \quad r \in [0,1]\end{aligned} \tag{23}$$

Moreover

$$\begin{aligned}&[\ominus(-1) \odot \left(I^{\alpha}_{a|t\,c}D^{\alpha}_{ii.gH} f\right)(t)]_r \\ &= \ominus(-1) \odot \left[\overline{f}(t,r) - \overline{f}(a,r), \underline{f}(t,r) - \underline{f}(a,r)\right] \\ &= \left[\underline{f}(t,r) - \underline{f}(a,r), \overline{f}(t,r) - \overline{f}(a,r)\right] = \left[f(t) \ominus_{gH} f(a))\right]_r,\end{aligned} \tag{24}$$

From Eqs. (23) and (24), the required result is obtained.

Theorem 3. *Suppose* $(({}_cD^{\alpha}_{i.gH})^j f) : [a,b] \to \mathbb{R}_F$ *and* $({}_cD^{\alpha}_{i.gH} f)(t) \in C^m_{gH}([a,b], \mathbb{R}_F)$. *then* $\forall a \le t \le b$

***(i)**. Consider $(({}_cD^{\alpha}_{i.gH})^j f)$, $j = 0,1,2,\ldots n$ be in the form of $[(i)]$-Caputo differentiable in $[a,b]$, thus*

$$\Big(({}_cD^{\alpha}_{i.gH})^j f\Big)(t) = \Big(({}_cD^{\alpha}_{i.gH})^j f\Big)(a) \oplus \Big(I^{\alpha}_{a|t}({}_cD^{\alpha}_{i.gH})^{j+1} f\Big)(t). \quad (25)$$

***(ii)**. Consider $(({}_cD^{\alpha}_{ii.gH})^j f)$, $j = 0,1,2,\ldots n$ be in form of $[(ii) - gH]$-Caputo differentiable in $[a,b]$, thus*

$$\Big(({}_cD^{\alpha}_{ii.gH})^j f\Big)(t) = \Big(({}_cD^{\alpha}_{ii.gH})^j f\Big)(a) \oplus \Big(I^{\alpha}_{a|t}({}_cD^{\alpha}_{ii.gH})^{j+1} f\Big)(t). \quad (26)$$

***(iii)**. Suppose that $(({}_cD^{\alpha}_{i.gH})^j f)$, $j = 2k-1, k \in \mathbb{N}$ be in the form of $[(i)-gH]$-Caputo differentiable also $(({}_cD^{\alpha}_{ii.gH})^j f)$, $j = 2k, k \in \mathbb{N}$ be in the form of $[(ii) - gH]$-Caputo differentiable, so*

$$\Big(({}_cD^{\alpha}_{i.gH})^j f\Big)(t) = \Big(({}_cD^{\alpha}_{i.gH})^j f\Big)(a) \ominus (-1)\Big(I^{\alpha}_{a|t}({}_cD^{\alpha}_{ii.gH})^{j+1} f\Big)(t). \quad (27)$$

***(iv)**. Suppose that $(({}_cD^{\alpha}_{ii.gH})^j f)$, $j = 2k-1$, $k \in \mathbb{N}$ be in the form of $[(ii) - gH]$-Caputo differentiable also $(({}_cD^{\alpha}_{i.gH})^j f)$, $j = 2k$, $k \in \mathbb{N}$ be in the form of $[(i) - gH]$-Caputo differentiable, so*

$$\Big(({}_cD^{\alpha}_{ii.gH})^j f\Big)(t) = \Big(({}_cD^{\alpha}_{ii.gH})^j f\Big)(a) \ominus (-1)\Big(I^{\alpha}_{a|t}({}_cD^{\alpha}_{i.gH})^{j+1} f\Big)(t). \quad (28)$$

Proof: By assuming $\Big({}_cD^{\alpha}_{*} f\Big) C^m_{gH}([a,b], \mathbb{R}_F)$, so $(({}_cD^{\alpha}_{*})^j f)$, $j = 0,1,\ldots,n$ are fuzzy Riemann-integrable. we give the proof only for parts (i) and (iii). Proving the other parts are similar. Using properties of Riemann-Liouville integral and generalized Hukuhara Caputo derivative along with Lemma 5 implies that

$$\begin{aligned}
&\Big(({}_cD^{\alpha}_{i.gH})^j f\Big)(a) \oplus \Big(I^{\alpha}_{a|t}({}_cD^{\alpha}_{i.gH})^{j+1} f\Big)(t) \\
&= \Big({}_cD^{\alpha}_{i.gH})^j f\Big)(a) \oplus ({}_cD^{\alpha}_{i.gH})^j f)(t) \ominus ({}_cD^{\alpha}_{i.gH})^j f)(a) \\
&= \big(({}_cD^{\alpha}_{i.gH})^j f\big)(t)
\end{aligned} \quad (29)$$

(iii). Let $(({}_cD^{\alpha}_{i.gH})^j f)$,$j = 2k-1$, $k \in \mathbb{N}$ be in form of $[(i) - gH]$-Caputo differentiable also for $j = 2k$, $k \in \mathbb{N}$ be form of $[(ii) - gH]$-Caputo differentiable. Using properties about Riemann-Liouville integral and generalized Hukuhara Caputo derivative along with Lemma 5 and Lemma 7 implies that

$$\begin{aligned}
&\Big(({}_cD^{\alpha}_{i.gH})^j f\Big)(a) \ominus (-1)\Big((I^{\alpha}_{a|t})({}_cD^{\alpha}_{ii.gH})^{j+1} f\Big)(t) \\
&= \Big(({}_cD^{\alpha}_{i.gH})^j f)(a) \oplus ((I^{\alpha}_{a|t})(({}_cD^{\alpha}_{i.gH})^{j+1} f)(t)\Big) \\
&= \Big(({}_cD^{\alpha}_{i.gH})^j f)(a) \oplus (({}_cD^{\alpha}_{i.gH})^j f)(t) \ominus (({}_cD^{\alpha}_{i.gH})^j f)(a)\Big) \\
&= (({}_cD^{\alpha}_{i.gH})^j f)(t)
\end{aligned}$$

Theorem 4. *Let* $(({}_cD^{\alpha}_{*})^j f) : [a,b] \to \mathbb{R}_F$ *and* $({}_cD^{\alpha}_{*} f)(t) \in C^m_{gH}([a,b],\mathbb{R}_F)$. *Then* $\forall x \in [a,b]$

(i). Consider $(({}_cD^{\alpha}_{i.gH})^j f)$, $j = 0,1,2,\dots n$ *be form of* $[(i) - gH]$*-Caputo differentiable in* $[a,b]$*, thus*

$$\Big((I^{\alpha}_{a|t})^j ({}_cD^{\alpha}_{i.gH})^j f\Big)(t) = \Big((I^{\alpha}_{a|t})^j ({}_cD^{\alpha}_{i.gH})^j f\Big)(a) \oplus \Big((I^{\alpha}_{a|t})^{j+1} ({}_cD^{\alpha}_{i.gH})^{j+1} f\Big)(t). \tag{30}$$

(ii). Consider $(({}_cD^{\alpha}_{ii.gH})^j f)$, $j = 0,1,2,\dots n$ *are* $[(ii) - gH]$*-Caputo differentiable in* $[a,b]$*, thus*

$$\Big((I^{\alpha}_{a|t})^j ({}_cD^{\alpha}_{ii.gH})^j f\Big)(t) = \Big((I^{\alpha}_{a|t})^j ({}_cD^{\alpha}_{ii.gH})^j f\Big)(a) \oplus \Big((I^{\alpha}_{a|t})^{j+1} ({}_cD^{\alpha}_{ii.gH})^{j+1} f\Big)(t). \tag{31}$$

(iii). Suppose that $(({}_cD^{\alpha}_{i.gH})^j f)$, $j = 2k-1, k \in \mathbb{N}$ *be in form of*$[(i) - gH]$*-Caputo differentiable also* $(({}_cD^{\alpha}_{ii.gH})^j f)$, $j = 2k$, $k \in \mathbb{N}$ *be in form of* $[(ii) - gH]$*-Caputo differentiable,so*

$$\Big(((I^{\alpha}_{a|t})^j ({}_cD^{\alpha}_{i.gH})^j f\Big)(t) = \Big((I^{\alpha}_{a|t})^j ({}_cD^{\alpha}_{i.gH})^j f\Big)(a) \ominus (-1)\Big((I^{\alpha}_{a|t})^{j+1} ({}_cD^{\alpha}_{ii.gH})^{j+1} f\Big)(t). \tag{32}$$

(iv). Suppose that $(({}_cD^{\alpha}_{ii.gH})^j f)$, $j = 2k-1, k \in \mathbb{N}$ *be in form of* $[(ii) - gH]$*-Caputo differentiable also* $(({}_cD^{\alpha}_{i.gH})^j f)$, $j = 2k, k \in \mathbb{N}$ *be in form of* $[(i) - gH]$*-Caputo differentiable,so*

$$\Big((I^{\alpha}_{a|t})^j ({}_cD^{\alpha}_{ii.gH})^j f\Big)(t) = \Big((I^{\alpha}_{a|t})^j ({}_cD^{\alpha}_{ii.gH})^j f\Big)(a) \ominus (-1)\Big((I^{\alpha}_{a|t})^{j+1} ({}_cD^{\alpha}_{i.gH})^{j+1} f\Big)(t). \tag{33}$$

Proof: By assuming $\Big({}_cD^{\alpha}_{*} f\Big) C^m_{gH}([a,b],\mathbb{R}_F)$, so $(({}_cD^{\alpha}_{*})^j f)$, $j = 0,1,\dots,n$ are fuzzy Riemann-integrable. we give the proof only for parts (i) and (iv). Proving the other parts are similar. Using properties about Riemann-Liouville integral and generalized Hukuhara Caputo derivative along with Lemma 5 implies that

$$\begin{aligned}
&\Big((I^{\alpha}_{a|t})^j ({}_cD^{\alpha}_{i.gH})^j f\Big)(a) \oplus \Big((I^{\alpha}_{a|t})^{j+1} ({}_cD^{\alpha}_{i.gH})^{j+1} f\Big)(t) = \\
&(I^{\alpha}_{a|t})^j \Big((({}_cD^{\alpha}_{i.gH})^j f)(a) \oplus ((I^{\alpha}_{a|t})({}_cD^{\alpha}_{i.gH})^{j+1} f)(t)\Big) \\
&= (I^{\alpha}_{a|t})^j \Big((({}_cD^{\alpha}_{i.gH})^j f)(a) \oplus (({}_cD^{\alpha}_{i.gH})^j f)(t) \ominus ({}_cD^{\alpha}_{i.gH})^j f)(a)\Big) \\
&= \Big((I^{\alpha}_{a|t})^j ({}_cD^{\alpha}_{i.gH})^j f\Big)(t)
\end{aligned} \tag{34}$$

(iv). Under the conditions stated in the part (iv), $(({}_cD^{\alpha}_{ii.gH})^j f)$, $j = 2k-1, k \in \mathbb{N}$ be in form of $[(ii) - gH]$-Caputo differentiable also $(({}_cD^{\alpha}_{i.gH})^j f)$, $j = 2k, k \in \mathbb{N}$ be in form of $[(i) - gH]$-Caputo differentiable. Using properties about Riemann-Liouville integral and generalized Hukuhara Caputo derivative along with Lemma 5 and Lemma 7 implies that

$$\Big((I^{\alpha}_{a|t})^{j}({}_{c}D^{\alpha}_{i.gH})^{j}f\Big)(a)\ominus(-1)\Big((I^{\alpha}_{a|t})^{j+1}({}_{c}D^{\alpha}_{ii.gH})^{j+1}f\Big)(t)$$
$$=\Big((I^{\alpha}_{a|t})^{j}({}_{c}D^{\alpha}_{i.gH})^{j}f\Big)(a)\oplus\Big((I^{\alpha}_{a|t})^{j+1}({}_{c}D^{\alpha}_{i.gH})^{j+1}f\Big)(t)$$
$$=(I^{\alpha}_{a|t})^{j}\Big(({}_{c}D^{\alpha}_{i.gH})^{j}f)(a)\oplus((I^{\alpha}_{a|t})(({}_{c}D^{\alpha}_{i.gH})^{j+1}f)(t)\Big)$$
$$=(I^{\alpha}_{a|t})^{j}\Big(({}_{c}D^{\alpha}_{i.gH})^{j}f)(a)\oplus(({}_{c}D^{\alpha}_{i.gH})^{j}f)(t)\ominus(({}_{c}D^{\alpha}_{i.gH})^{j}f)(a)\Big)$$
$$=(I^{\alpha}_{a|t})^{j}(({}_{c}D^{\alpha}_{i.gH})^{j}f)(t)$$

the proof is completed.

3 Fuzzy Fractional Generalized Taylor Series

IIn this section, we are going to obtain a generalized Taylor's series for fuzzy number valued functions by using the concept of Caputo generalized Hukuhara derivative. We shall derive the generalized fractional Taylor series prove in this paper to obtain an approximate solution for the fuzzy fractional boundary value problems. The proposed technique is based on generalized Taylor's formula. We define the generalized fractional Taylor series of the k-th derivative of function $f(t)$ in one variable as follows.

Definition 13 *([24]). Let $f(t)\in C^{F}[a,b]\cap L^{F}[a,b]$. And suppose that $({}_{c}D^{\alpha}_{*})^{k}f(t)\in C^{F}[a,b]$ for $k=0,1,\dots,(n\alpha-1)$ where $0<\alpha<1$, $a<\xi\leq t$ and $a<t\leq b$. Then we have*

$$[f(t)]_r=[\underline{f}(t,r),\overline{f}(t,r)],\quad r\in[0,1],$$

$$\underline{f}(t,r)=\sum_{k=0}^{n}[({}_{c}D^{\alpha}_{*})^{k}\underline{f}(a,r)]\frac{(t-a)^{k\alpha}}{\Gamma(k\alpha+1)}+\frac{({}_{c}D^{\alpha}_{*})^{(n+1)}\underline{f}(\xi,r)}{\Gamma((n+1)\alpha+1)}(t-a)^{(n+1)\alpha},\quad(35)$$

$$\overline{f}(t,r)=\sum_{k=0}^{n}[({}_{c}D^{\alpha}_{*})^{k}\overline{f}(a,r)]\frac{(t-a)^{k\alpha}}{\Gamma(k\alpha+1)}+\frac{({}_{c}D^{\alpha}_{*})^{(n+1)}\overline{f}(\xi,r)}{\Gamma((n+1)\alpha+1)}(t-a)^{(n+1)\alpha}.\quad(36)$$

Then, the following equality holds, which is
if $f(t)$ is [(i)-gH]-Caputo differentiable, then

$$[\sum_{k=0}^{\infty}\frac{(t-a)^{k\alpha}}{\Gamma(k\alpha+1)}[({}_{c}D^{\alpha}_{i.gH})^{k}f(a)]_r=$$
$$\Big[\sum_{k=0}^{\infty}\frac{(t-a)^{k\alpha}}{\Gamma(k\alpha+1)}[({}_{c}D^{\alpha}_{*})^{k}\underline{f}](a,r),\sum_{k=0}^{\infty}\frac{(t-a)^{k\alpha}}{\Gamma(k\alpha+1)}[({}_{c}D^{\alpha}_{*})^{k}\overline{f}](a,r)\Big].$$

and if $f(x)$ is [(ii)-gH]-Caputo differentiable, then

$$[\sum_{k=0}^{\infty}\frac{(t-a)^{k\alpha}}{\Gamma(k\alpha+1)}[({}_cD^{\alpha}_{ii.gH})^k f(a)]_r =$$

$$\Big[\sum_{k=0}^{\infty}\frac{(t-a)^{k\alpha}}{\Gamma(k\alpha+1)}[({}_cD^{\alpha}_{*})^k\overline{f}](a,r),\sum_{k=0}^{\infty}\frac{(t-a)^{k\alpha}}{\Gamma(k\alpha+1)}[({}_cD^{\alpha}_{*})^k\underline{f}](a,r)\Big].$$

Here, the generalized Taylor's formula is presented for Caputo [gH]-differentiable functions.

Theorem 5. *Suppose* $T=[a,a+\alpha]\subset\mathbb{R}$, *with* $\alpha>0$ *and* $f\in C^m_{gH}([a,b],\mathbb{R}_F)$. *For* $t\in T$

(i). *If* $({}_cD^{\alpha}_{i.gH}f)^j, j=0,1,\ldots,n$ *are [(i)-gH]-Caputo differentiable. Then*

$$f(t)=f(a)\oplus\tfrac{(t-a)^{\alpha}}{\Gamma(\alpha+1)}\odot[{}_cD^{\alpha}_{i.gH}f](t_0)\oplus\tfrac{(t-a)^{2\alpha}}{\Gamma(2\alpha+1)}\odot[{}_cD^{2\alpha}_{i.gH}f](a)\oplus \tfrac{(t-a)^{3\alpha}}{\Gamma(3\alpha+1)}\odot[{}_cD^{3\alpha}_{i.gH}f](t_0)\oplus\cdots\oplus\tfrac{(t-t_0)^{n\alpha}}{\Gamma(n\alpha+1)}\odot[{}_cD^{n\alpha}_{i.gH}f](t_0)\oplus R_{n+1}(a,t),$$

where

$$R_{n+1}(a,t):=\frac{[{}_cD^{(n+1)\alpha}_{i.gH}f](\xi)}{\Gamma((n+1)+1)}(t-a)^{(n\alpha)}, \tag{37}$$

and

$$[{}^c(D^{\alpha}_{i.gH}f)(t)]_r=[\frac{1}{\Gamma(m-\alpha)}\int_a^t\frac{\underline{f}^{(m)}_{gH}(s,r)ds}{(t-s)^{\alpha-m+1}},\frac{1}{\Gamma(m-\alpha)}\int_a^t\frac{\overline{f}^{(m)}_{gH}(s,r)ds}{(t-s)^{\alpha-m+1}}], \tag{38}$$

where $\alpha\le\xi\le t$, *for each* $t\in(a,b]$ *and* ${}_cD^{\alpha}_{i.gH}$ *is the Caputo fractional derivative of order* α, *and*

$$({}_cD^{\alpha}_{i.gH})^k=\overbrace{{}_cD^{\alpha}_{i.gH}\cdot{}_cD^{\alpha}_{i.gH}\cdots{}_cD^{\alpha}_{i.gH}}^{k-times}. \tag{39}$$

(ii). *If* $({}_cD^{\alpha}_{ii.gH}f)^j, j=0,1,\ldots,n$ *are [(ii)-gH]-Caputo differentiable. Then*

$$f(t)=f(a)\ominus(-1)\tfrac{(t-a)^{\alpha}}{\Gamma(\alpha+1)}\odot[{}_cD^{\alpha}_{ii.gH}f](a)\ominus(-1)\tfrac{(t-a)^{2\alpha}}{\Gamma(2\alpha+1)}\odot[{}_cD^{2\alpha}_{ii.gH}f](a) \ominus(-1)\tfrac{(t-a)^{3\alpha}}{\Gamma(3\alpha+1)}\odot[{}_cD^{3\alpha}_{ii.gH}f](a)\ominus(-1)\cdots\ominus(-1)\tfrac{(t-a)^{n\alpha}}{\Gamma(n\alpha+1)}\odot[{}_cD^{n\alpha}_{ii.gH}f](a) \ominus(-1)R_{n+1}(a,t)$$

where

$$R_{n+1}(a,t):=\frac{[{}_cD^{(n+1)\alpha}_{ii.gH}f](\xi)}{\Gamma((n+1)\alpha+1)}(t-t_0)^{(n+1)\alpha}, \tag{40}$$

and

$$[{}_c(D^{\alpha}_{ii.gH}f)(t)]_r=[\frac{1}{\Gamma(m-\alpha)}\int_a^t\frac{\overline{f}^{(m)}_{gH}(s,r)ds}{(t-s)^{\alpha-m+1}},\frac{1}{\Gamma(m-\alpha)}\int_a^t\frac{\underline{f}^{(m)}_{gH}(s,r)ds}{(t-s)^{\alpha-m+1}}]. \tag{41}$$

where $\alpha \leq \xi \leq t$, *for each* $t \in (a,b]$ *and* ${}_cD^{\alpha}_{ii.gH}$ *is the Caputo fractional derivative of order* α, *and*

$$({}_cD^{\alpha}_{ii.gH})^k = \overbrace{{}_cD^{\alpha}_{ii.gH}\cdot{}_cD^{\alpha}_{ii.gH}\cdots{}_cD^{\alpha}_{ii.gH}}^{k-times}. \tag{42}$$

***(iii)**. If* $({}_cD^{\alpha}_{i.gH}f)^j$ *are [(i)-gH]-Caputo differentiable for* $j = 2k-1$, $k \in \mathbb{N}$, *and* $({}_cD^{\alpha}_{ii.gH}f)^j$ *are* $[(ii)-gH]$*-Caputo differentiable for* $j = 2k$, $k \in \mathbb{N}\cup\{0\}$

$$\begin{aligned}&f(t) = \\ &f(a) \ominus (-1)\tfrac{(t-t_0)^{\alpha}}{\Gamma(\alpha+1)} \odot [{}_cD^{\alpha}_{ii.gH}f](a) \oplus \tfrac{(t-a)^{2\alpha}}{\Gamma(2\alpha+1)} \odot [{}_cD^{2\alpha}_{i.gH}f](a) \ominus (-1)\cdots\ominus(-1) \\ &\tfrac{(t-a)^{(2k-1)(2\alpha)}}{\Gamma((2k-1)(2\alpha+1))} \odot [{}_cD^{(2k-1)(2\alpha)}_{ii.gH}f](a) \oplus \tfrac{(t-a)^{(2k)(2\alpha)}}{\Gamma((2k)(2\alpha+1))} \odot [{}_cD^{(2k)(2\alpha)}_{i.gH}f](a)\end{aligned} \tag{43}$$

where

$$R_{n+1}(a,t) := \frac{({}_cD^{(n+1)\alpha}_{i.gH}f)(\xi)}{\Gamma((n+1)\alpha+1)}(t-a)^{(n+1)\alpha}\ominus(-1)\frac{[{}_cD^{(n+1)\alpha}_{ii.gH}f](\xi)}{\Gamma((n+1)\alpha+1)}(t-a)^{(n+1)\alpha}, \tag{44}$$

***(iv)**. If* $({}_cD^{\alpha}_{i.gH}f)^j$ *are [(i)-gH]-Caputo differentiable for* $j = 2k, k \in \mathbb{N}$, *and* $({}_cD^{\alpha}_{ii.gH}f)^j$ *are* $[(ii)-gH]$*-Caputo differentiable for* $j = 2k-1, k \in \mathbb{N}\cup\{0\}$

$$\begin{aligned}&f(t) = f(a) \oplus \tfrac{(t-a)^{\alpha}}{\Gamma(\alpha+1)} \odot [{}_cD^{\alpha}_{ii.gH}f](a) \ominus (-1)\tfrac{(t-a)^{2\alpha}}{\Gamma(2\alpha+1)} \odot [{}_cD^{2\alpha}_{i.gH}f](a) \oplus \ldots \\ &\tfrac{(t-a)^{(2k-1)(2\alpha)}}{\Gamma((2k-1)(2\alpha+1))} \odot [{}_cD^{(2k-1)(2\alpha)}_{ii.gH}f](a) \ominus (-1)\tfrac{(t-a)^{(2k)(2\alpha)}}{\Gamma((2k)(2\alpha+1))} \odot [{}_cD^{(2k)(2\alpha)}_{i.gH}f](a) \\ &\oplus\cdots\oplus R_{n+1}(a,t)\end{aligned} \tag{45}$$

where

$$R_{n+1}(t_0,t) := \frac{(D^{(n+1)\alpha}_{ii.gH}f)(\xi)}{\Gamma((n+1)\alpha+1)}(t-a)^{(n+1)(\alpha)}\ominus(-1)\frac{[{}_cD^{(n+1)\alpha}_{i.gH}f](\xi)}{\Gamma((n+1)\alpha+1)}(t-a)^{\frac{n+1}{\alpha}}, \tag{46}$$

***(v)**. Suppose that* $f \in C^n_{gH}([a,b],\mathbb{R}_f)$, $3 \leq n$. *Furthermore let* f *in* $[a,\xi]$ *are* $[(i)-gH]$*–differentiable and in* $[\xi,b]$ *are* $[(ii)-gH]$*–differentiable, in fact* ξ *is switching point type I for first order derivative of* f *and* $t_0 \in [a,\xi]$ *in neighborhood of* ξ. *Moreover suppose that second order derivative of* f *in* $\zeta_1 \in [t_0,\xi]$ *have switching point type II. So,*

$$\begin{aligned}&f(t) = f(a) \ominus (-1)[{}_cD^{\alpha}_{ii.gH}f](a) \odot \tfrac{(\xi-a)}{\Gamma(\alpha+1)} \oplus [{}_cD^{2\alpha}_{i.gH}f](a) \odot \tfrac{(a-\zeta_1)}{\Gamma(\alpha+1)} \odot \tfrac{(\xi-a)}{\Gamma(\alpha+1)} \\ &\ominus(-1)[{}_cD^{2\alpha}_{ii.gH}f](\zeta_1)\odot \\ &\left(\tfrac{(\xi-\zeta_1)^2}{\Gamma(2\alpha+1)} - \tfrac{(a-\zeta_1)^2}{\Gamma(2\alpha+1)}\right) \oplus [{}_cD^{\alpha}_{i.gH}f](\xi) \odot \tfrac{(t-\xi)}{\Gamma(\alpha+1)} \oplus [{}_cD^{2\alpha}_{i.gH}f](\xi) \odot \tfrac{(t-\xi)^{2\alpha}}{\Gamma(2\alpha+1)}\oplus \\ &I^{\alpha}_{a|\xi}\left(I^{\alpha}_{a|\zeta}\left(I^{\alpha}_{a|t_2}D^{3\alpha}_{i.gH}f(t_4)\right)\right) \ominus (-1)I^{\alpha}_{a|\xi}\left(I^{\alpha}_{\zeta_1|t_1}\left(I^{\alpha}_{\zeta_1|t_3}D^{3\alpha}_{ii.gH}f(t_5)\right)\right) \\ &\oplus I^{\alpha}_{\xi|t}\left(I^{\alpha}_{\xi|s_1}\left(I^{\alpha}_{a|s_2}D^{3\alpha}_{i.gH}f(s_3)\right)\right).\end{aligned} \tag{47}$$

Proof. Here, we will prove only the two parts (iii) and (v) and the other parts are similar. So the details are omitted.

(iii). If $\left(({}_cD^{\alpha}_{i.gH})^j f\right)$ are [(i)-gH]-Caputo differentiable for $j = 2k-1$, $k \in \mathbb{N}$, and $\left(({}_cD^{\alpha}_{ii.gH})^j f\right)$ are $[(ii)-gH]$-Caputo differentiable for $j = 2k, k \in \mathbb{N} \cup \{0\}$. By Theorem 2 we have

$$f(t) = f(a) \ominus (-1) I^{\alpha}_{a|t} (D^{\alpha}_{ii.gH} f)(t_1).$$

According to the hypothesis type of differentiability $[({}_cD^{\alpha}_{ii.gH})f](t_1)$ differentiable. So Theorem 3 gets

$$\left({}_cD^{\alpha}_{ii.gH} f\right)(t_1) = \left({}_cD^{\alpha}_{ii.gH} f\right)(a) \ominus (-1) I^{\alpha}_{a|t_1} \left(({}_cD^{\alpha}_{i.gH})^2 f\right)(t_2).$$

Now applying operator fuzzy Riemann-integral to $[{}_cD^{\alpha}_{ii.gH})f](t_1)$ gives

$$I^{\alpha}_{a|t_1}\left({}_cD^{\alpha}_{ii.gH} f\right)(t_1) = I^{\alpha}_{a|t_1}\left({}_cD^{\alpha}_{ii.gH} f\right)(a) \ominus (-1) I^{\alpha}_{a|t}\left(I^{\alpha}_{a|t_1}({}_cD^{\alpha}_{i.gH})^2 f)(t_2)\right). \tag{48}$$

The initial value problem equation (48) can be equivalent to the following integral equation

$$\begin{aligned} &I^{\alpha}_{a|t_1}\left(({}_cD^{\alpha}_{ii.gH})^1 f(t_1)\right) = \\ &\left({}_cD^{\alpha}_{ii.gH} f(a)\right)\frac{(t-a)^{\alpha}}{\Gamma(\alpha+1)} \ominus (-1) I^{\alpha}_{a|t}\left(I^{\alpha}_{a|t_1}(({}_cD^{\alpha}_{i.gH})^2 f)(t_2)\right). \end{aligned} \tag{49}$$

So Lemma 3 implies that the last double integral belongs to $\mathbb{R}_F$. Then

$$f(t) = f(a) \ominus (-1)\frac{(t-t_0)^{\alpha}}{\Gamma(\alpha+1)} \odot [({}_cD_{ii.gH})f](a) \oplus (-1) I^{\alpha}_{a|t}\left(I^{\alpha}_{a|t_1}(({}_cD^{\alpha}_{i.gH})^2 f)(t_2)\right).$$

Similarly

$$\left(({}_cD^{\alpha}_{i.gH})^2 f\right)(t_2) = \left((({}_cD^{\alpha}_{i.gH})^2 f\right)(a) \ominus (-1) I^{\alpha}_{a|t_2}\left(({}_cD^{\alpha}_{ii.gH})^3 f(t_3)\right).$$

Thus

$$\begin{aligned} &I^{\alpha}_{a|t_1}\left(({}_cD^{\alpha}_{i.gH})^2 f\right)(t_2) \\ &= \left(({}_cD^{\alpha}_{i.gH})^2 f\right)(a)\frac{(t-a)^{\alpha}}{\Gamma(\alpha+1)} \ominus (-1) I^{\alpha}_{a|t_1} I^{\alpha}_{a|t_2}\left(({}_cD^{\alpha}_{ii.gH})^3 f(t_3)\right). \end{aligned}$$

Now, applying operator $I^{\alpha}_{a|t}$ gives

$$\begin{aligned} &I^{\alpha}_{a|t} I^{\alpha}_{a|t_1}\left(({}_cD^{\alpha}_{i.gH})^2 f\right)(t_2) = \\ &\left(({}_cD^{\alpha}_{i.gH})^2 f\right)(a)\frac{(t-a)^{2\alpha}}{\Gamma(2\alpha+1)} \ominus (-1) I^{\alpha}_{a|t} I^{\alpha}_{a|t_1} I^{\alpha}_{a|t_2}\left(({}_cDf^{\alpha}_{ii.gH})^3(t_3)\right). \end{aligned}$$

Since satisfies all the other conditions for the Lemma 3, gives the last triple integral belongs to $\mathbb{R}_F$. Then

$$f(t) = f(a) \ominus (-1)\frac{(t-a)^{\alpha}}{\Gamma(\alpha+1)} \odot [({}_cD^{\alpha}_{ii.gH})^1 f](t_0) \oplus \frac{(t-a)^{2\alpha}}{\Gamma(2\alpha+1)} \odot [({}_cD^{\alpha}_{i.gH})^2 f](a) \ominus (-1)$$

$$I^{\alpha}_{a|t}\Big(I^{\alpha}_{a|t_1}\Big(I^{\alpha}_{a|t_2}((_cD^{\alpha}_{ii.gH})^3 f)\Big)\Big),$$

with the similar manner the proof for this type of differentiability will be obtained.

(iv). According to the hypothesis f is $[(ii)-gH]$−differentiable in $[t_0,\xi]$, by Theorem 2

$$f(\xi) = f(a) \ominus (-1)I^{\alpha}_{a|\xi}(((_cD^{\alpha}_{ii.gH})^1 f)(t_1)), \tag{50}$$

and f is $[(i)-gH]$−differentiable in $[\xi,b]$, so for $t\in[\xi,b]$

$$f(t) = f(\xi) \oplus I^{\alpha}_{\xi|t}\Big(((_cD^{\alpha}_{i.gH})^1 f)(s_1)\Big). \tag{51}$$

The point ξ is a switching point for differentiability so from Eq. (50) and Eq. (51), we obtain

$$f(t) = f(t_0) \ominus (-1)I^{\alpha}_{t_0|\xi}\Big(((_cD^{\alpha}_{ii.gH})^1 f)(t_1)\Big) \oplus I^{\alpha}_{\xi|t}\Big(((_cD^{\alpha}_{i.gH})^1 f)(s_1)\Big). \tag{52}$$

Consider the first integral on the right of Eq.(52). By hypothesis ζ_1 is switching point type I for gH-derivative of the function f. So ${}_cD^{\alpha}_{i.gH}f$ is $[(i)-gH]$-Caputo differentiable on $[t_0,\zeta_1]$. Using Theorem 3, we deduce that

$$\Big(((_cD^{\alpha}_{ii.gH})^1 f\Big)(\zeta_1) = \Big(((_cD^{\alpha}_{ii.gH})^1 f\Big)(t_0) \ominus (-1)I^{\alpha}_{t_0|\zeta_1}\Big(((_cD^{\alpha}_{i.gH})^2 f)(t_2)\Big). \tag{53}$$

On the other hand, we know that $(D^{\alpha}_{ii.gH}f)$ is $[(ii)-gH]$-Caputo differentiable on $[\zeta_1,\xi_1]$ and the type of differentiability does not change. Hence, from Theorem 3 for $t_1\in[\zeta_1,\xi]$, we have

$$\Big(((_cD^{\alpha}_{ii.gH})^1 f\Big)(t_1) = \Big(((_cD^{\alpha}_{ii.gH})^1 f\Big)(\zeta_1) \oplus I^{\alpha}_{\zeta|t_1}\Big((((_cD^{\alpha}_{ii.gH})^2 f)(t_3)\Big). \tag{54}$$

Substituting Eq. (53) into Eq. (54) gets

$$\begin{aligned}&\Big(((_cD^{\alpha}_{ii.gH})^1 f\Big)(t_1) =\\ &\Big(((_cD^{\alpha})^1 f_{ii.gH})(t_0) \ominus (-1)I^{\alpha}_{t_0|\zeta_1}\Big((((_cD^{\alpha}_{*})^2 f_{i.gH})(t_2)\Big) \oplus I^{\alpha}_{\zeta_1|t_1}\Big((((_cD^{\alpha}_{ii.gH})^2 f)(t_3)\Big).\end{aligned} \tag{55}$$

In this case, according to Theorem 3

$$\begin{aligned}&\Big(((_cD^{\alpha}_{i.gH})^2 f\Big)(t_2) = \Big((_cD^{\alpha}_{i.gH})^2 f_{i.gH})\Big)(a) \oplus I^{\alpha}_{a|t_2}\Big(((_cD^{\alpha}_{i.gH})^3 f)\Big)(t_4)\\ &\Longrightarrow I^{\alpha}_{a|\zeta_1}\Big((((_cD^{\alpha}_{i.gH})^2 f)\Big)(t_2) = \Big((((_cD^{\alpha}_{i.gH})^2 f)(t_0)\Big) \odot \tfrac{(\zeta_1-a)^{\alpha}}{\Gamma(\alpha+1)}\\ &\oplus I^{\alpha}_{a|\zeta_1}\Big(I^{\alpha}_{a|t_2}\Big((((_cD^{\alpha}_{i.gH})^3 f)\Big)(t_4)\Big),\end{aligned} \tag{56}$$

and

$$\begin{aligned}&\Big((((_cD^{\alpha}_{ii.gH})^2 f)\Big)(t_3) = \Big(((D^{\alpha}_{ii.gH})^2 f))\Big)(\zeta_1) \oplus I^{\alpha}_{\zeta_1|t_3}\Big(((_cD^{\alpha}_{ii.gH})^3 f)\Big)(t_5)\\ &\Longrightarrow I^{\alpha}_{\zeta_1|t_1}\Big(((_cD^{\alpha}_{ii.gH})^2 f)\Big)(t_3) = \Big((((_cD^{\alpha}_{ii.gH})^2 f)\Big)(\zeta_1) \odot \tfrac{(x_1-\zeta_1)^{\alpha}}{\Gamma(\alpha+1)}\\ &\oplus I^{\alpha}_{\zeta_1|t_1}\Big(I^{\alpha}_{\zeta_1|t_3}\Big((((_cD^{\alpha}_{ii.gH})^3 f)\Big)\Big)(t_5).\end{aligned} \tag{57}$$

The insertion of the Eqs. (57) and (56), in Eq. (55) leads to obtain

$$\Big(\big(({}_cD^{\alpha}_{ii.gH})^1 f\big)(t_1) = \Big(\big(({}_cD^{\alpha}_{ii.gH})^1 f\Big)(a) \ominus \frac{\Big(\big(({}_cD^{\alpha}_{i.gH})^2 f\big)\Big)}{\Gamma(\alpha+1)}(a) \odot (a-\zeta_1)^{\alpha}$$
$$\oplus \frac{\Big(\big(({}_cD^{\alpha}_{ii.gH})^2 f\Big)}{\Gamma(\alpha+1)}(\zeta_1) \odot (t_1-\zeta_1)^{\alpha} \ominus (-1) I^{\alpha}_{a|\zeta_1}\Big(I^{\alpha}_{a|t_2}\Big(\big(\big(({}_cD^{\alpha}_{i.gH})^3 f\big)\Big)\Big)(t_4)$$
$$\oplus I^{\alpha}_{\zeta_1|t_1}\Big(I^{\alpha}_{\zeta_1|t_3}\Big(\big(({}_cD^{\alpha}_{ii.gH})^3 f\big)\Big)\Big)(t_5). \tag{58}$$

So, the first integral on the right hand side of Eq. (52) is as follows

$$I^{\alpha}_{a|\xi}\Big(\big(\big(({}_cD^{\alpha}_{ii.gH})^1 f\big)\Big)(t_1) =$$
$$\frac{1}{\Gamma(\alpha+1)}\Big(\big(({}_cD^{\alpha}_{ii.gH})^1 f\big)\Big)(a) \odot (\xi-t_0)^{\alpha} \ominus \frac{\Big(\big(\big(({}_cD^{\alpha}_{i.gH})^2 f\big)\Big)}{\Gamma(\alpha+1)}(a)$$
$$\odot (a-\zeta_1)^{\alpha} \odot \frac{(\xi-a)^{\alpha}}{\Gamma(\alpha+1)} \oplus \frac{\Big(\big(\big(({}_cD^{\alpha}_{ii.gH})^2 f\big)\Big)}{\Gamma(2\alpha+1)}(\zeta_1)$$
$$\odot \Big((\xi-\zeta_1)^{2\alpha} - (a-\zeta_1)^{2\alpha}\Big) \ominus_{gH} (-1) I^{\alpha}_{a|\xi}\Big(I^{\alpha}_{a|\zeta_1}\Big(I^{\alpha}_{a|t_2}\Big(\big(\big(({}_cD^{\alpha}_{i.gH})^3 f\big)\Big)\Big)\Big)(t_4)$$
$$\oplus I^{\alpha}_{a|\xi}\Big(I^{\alpha}_{\xi|t_1}\Big(I^{\alpha}_{\xi|t_3}\Big(\big(\big(({}_cD^{\alpha}_{ii.gH})^3 f\big)\Big)\Big)\Big)(t_5). \tag{59}$$

Consider the second integral on the right hand side of the Eq.(52). According to the hypothesis $({}_cD^{\alpha}_{i.gH} f)^{(j)}, j=2,3$ are $[(i)-gH]-$differentiable on $[\xi, b]$, and the type of differentiability does not change . Using Theorem 3 we have

$$\Big(\big(\big(({}_cD^{\alpha}_{i.gH})^1 f\big)\Big)(s_1) = \Big(\big(\big(({}_cD^{\alpha}_{i.gH})^1 f\big)\Big)(\xi) \oplus I^{\alpha}_{\xi|s_1}\Big(\big(\big(({}_cD^{\alpha}_{i.gH})^2 f\big)\Big)(s_2), \tag{60}$$

and

$$\Big(\big(\big(({}_cD^{\alpha}_{i.gH})^2 f\big)\Big)(s_2) = \Big(\big(\big(({}_cD^{\alpha}_{i.gH})^2 f\big)\Big)(\xi) \oplus I^{\alpha}_{\xi|s_2}\Big(\big(\big(({}_cD^{\alpha}_{i.gH})^3 f\big)\Big)(s_3) \tag{61}$$

$$\Longrightarrow I^{\alpha}_{\xi|s_1}\Big(\big(\big(({}_cD^{\alpha}_{i.gH})^2 f\big)\Big)(s_2)$$
$$= \frac{\Big(\big(({}_cD^{\alpha}_{i.gH})^2 f\big)\Big)}{\Gamma(\alpha+1)}(\xi) \odot (s_1-\xi)^{\alpha} \oplus I^{\alpha}_{\xi|s_1}\Big(I^{\alpha}_{\xi|s_2}\Big(\big(\big(({}_cD^{\alpha}_{i.gH})^3 f\big)\Big)\Big)(s_3), \tag{62}$$

thus substituting Eq. (62) into Eq. (60), gives

$$\Big(\big(\big(({}_cD^{\alpha}_{i.gH})^1 f\big)\Big)(s_1) =$$
$$\Big(\big(\big(({}_cD^{\alpha}_{i.gH})^1 f\big)\Big)(\xi) \oplus \frac{\Big(\big(\big(({}_cD^{\alpha}_{i.gH})^2 f\big)\Big)}{\Gamma(\alpha+1)}(\xi) \odot (s_1-\xi)^{\alpha} \oplus I^{\alpha}_{\xi|s_1}\Big(I^{\alpha}_{\xi|s_2}\Big(\big(\big(({}_cD^{\alpha}_{i.gH})^3 f\big)\Big)\Big)(s_3). \tag{63}$$

Therefore the second integral on the right hand side of Eq. (52) obtains as follows

$$I^{\alpha}_{\xi|t}\Big((((_cD^{\alpha}_{i.gH})^1 f)\Big)(s_1) = [\frac{\Big((((_cD^{\alpha}_{i.gH})^1 f)\Big)}{\Gamma(\alpha+1)})](\xi) \odot (t-\xi)^{\alpha} \oplus \frac{\Big((((_cD^{\alpha}_{i.gH})^2 f)\Big)}{\Gamma(2\alpha+1)}(\xi)$$
$$\odot(t_1-\xi)^{2\alpha} \oplus I^{\alpha}_{\xi|t}\Big(I^{\alpha}_{\xi|s_1}\Big(I^{\alpha}_{\xi|s_2}\Big((((_cD^{\alpha}_{i.gH})^3 f)\Big)\Big)\Big)(s_3). \quad (64)$$

By substituting Eqs. (64) and (59) into Eq. (52)

$$f(t) = f(a) \ominus (-1)[_cD^{\alpha}_{ii.gH} f](a) \odot \frac{(\xi-a)}{\Gamma(\alpha+1)} \oplus [_cD^{2\alpha}_{i.gH} f](a) \odot \frac{(a-\zeta_1)}{\Gamma(\alpha+1)} \odot \frac{(\xi-a)}{\Gamma(\alpha+1)}$$
$$\ominus(-1)[_cD^{2\alpha}_{ii.gH} f](\zeta_1)\odot$$
$$\left(\frac{(\xi-\zeta_1)^2}{\Gamma(2\alpha+1)} - \frac{(a-\zeta_1)^2}{\Gamma(2\alpha+1)}\right) \oplus [_cD^{\alpha}_{i.gH} f](\xi) \odot \frac{(t-\xi)}{\Gamma(\alpha+1)} \oplus [_cD^{2\alpha}_{i.gH} f](\xi) \odot \frac{(t-\xi)^{2\alpha}}{\Gamma(2\alpha+1)} \oplus$$
$$I^{\alpha}_{a|\xi}\Big(I^{\alpha}_{a|\zeta}\Big(I^{\alpha}_{a|t_2} D^{3\alpha}_{i.gH} f(t_4)\Big)\Big) \ominus (-1) I^{\alpha}_{a|\xi}\Big(I^{\alpha}_{\zeta_1|t_1}\Big(I^{\alpha}_{\zeta_1|t_3} D^{3\alpha}_{ii.gH} f(t_5)\Big)\Big)$$
$$\oplus I^{\alpha}_{\xi|t}\Big(I^{\alpha}_{\xi|s_1}\Big(I^{\alpha}_{a|s_2} D^{3\alpha}_{i.gH} f(s_3)\Big)\Big). \quad (65)$$

This establishes the formula.

3.1 The Basic and Applied Theorems Proved by Theorem (5)

If F_{α}, G_{α}, H_{α} and W_{α} are the fractional differential transforms of functions $f(x)$, $g(x)$, $h(x)$ and $w(x)$ respectively, the following Theorems are established.

Theorem 6 *([22]). If $f(x) = g(x) \pm h(x)$, then $F_{\frac{1}{\alpha}}(k) = G_{\alpha}(K) \pm H_{\alpha}(K)$*

Theorem 7 *([22]). If $f(x) = ag(x)$, then $F_{\alpha}(k) = aG_{\alpha}(k)$*

Theorem 8 *([22]). If $f(x) = g(x)h(x)$, then $F_{\alpha}(k) = \sum_{l=0}^{k} G_{\alpha}(l)H_{\alpha}(K-l)$*

Theorem 9 *([22]). If $f(x) = g_1(x)g_2(x)g_3(x)\ldots g_{n-1}(x)g_n(x)$, then $F_{\alpha}(k) = \sum_{k_{n-1=0}}^{k} \sum_{k_{n-2=0}}^{k_{n-1}} \cdots \sum_{k_2=0}^{k_3} \sum_{k_1}^{k_2} G_1(k_1)G_2(k_2-k_1)\ldots G_{n-1}(k_{n-1}-k_{n-2})G_n(k-k_{n-1})$*

Theorem 10 *([22]). If $f(x) = D^{\beta}_{x_0}[g(x)]$, then $F_{\alpha}(k) = \frac{\Gamma(\alpha+\beta+1)}{\Gamma(k\alpha+1)} G_{\alpha}(k+\beta\alpha)$*

Theorem 11 *([22]). If $f(x) = \int_{x_0}^{x} g(t)dt$, then $F_{\alpha}(k) = \alpha\frac{G_{\alpha}(k-\alpha)}{k}$ where $\alpha \le k$*

Theorem 12 *([22]). If $f(x) = g(x)\int_{x_0}^{x} h(t)dt$, then $F_{\alpha}(k) = \sum_{k_1=\alpha}^{k} \frac{H_{k\alpha}(k_1-\alpha)}{k/\alpha} G_{k\alpha}(k-k_1)$, where, $k \geqslant \alpha$*

Theorem 13 *([22]). If $f(x) = (x-x_0)^p$, then $F_{\alpha}(k) = \delta(k-p\alpha)$ where*

$$\delta(k) = \begin{cases} 1, & ,k=0 \\ 0, & k \neq 0 \end{cases}$$

Theorem 14 *([22]). $w(x) = \int_0^x (x-t)^{k\alpha-1} g(t,y(t))dt$, then by choosing a suitable $\beta \in \mathbb{Z}^+$ such that $\frac{k}{\alpha}\beta \in \mathbb{Z}^+$ we have*

$$W_{\alpha}(k) = \sum_{l=0}^{k} B(\frac{k-l}{k}\beta + 1, k\alpha)\delta(l-\beta k\alpha)G_{\alpha}(k-l)$$

4 Describing the Fuzzy Fractional Impulsive Differential Equations

In this section, we are going to introduction the FFIDE by using concept of generalized Hukuhara differentiability.

$$ {}_cD_*^{k\alpha}y(t) = f(t, y(t)), \quad t \in [0, T], \quad t \neq t_k, \; m-1 < \frac{1}{\alpha} < m, \quad m \in \mathbb{N} \quad (66)$$

$$\Delta y|_{t=t_k} = I_k(y(t_k^-)), \quad (67)$$

$$y(0) = y_0, \quad (68)$$

where $k = 1, 2, \ldots, m$, $0 < \alpha \leq 1$, ${}_cD^\alpha$ denotes the Caputo fractional generalized derivative of order α, $y(t)$ is an unknown fuzzy function of crisp variable t and $f : [0, T] \times \mathbb{R}_F \to \mathbb{R}_F$, is continuous fuzzy function, $I_k : \mathbb{R}_F \to \mathbb{R}_F$, $y_0 \in \mathbb{R}_F$, $0 = t_0 < t_1 < \cdots < t_m < t_{m+1} = T$, $\Delta|_{t=t_k} = y(t_k^+) \ominus_{gH} y(t_k^-)$, $y(t_k^+) = lim_{h\to 0^+} y(t_k + h)$, and $y(t_k^-) = lim_{h\to 0^-} y(t_k + h)$ represent the right and left limits of $y(t)$ at $t = t_k$.

Lemma 8 *([13, 33, 34]). The initial value problem (66) under the conditions (67) and (68) is equivalent to one of the following integral equations:*

$$y(t) = y_0 \oplus \frac{1}{\Gamma(k\alpha)} \int_0^t (t-s)^{k\alpha-1} f(s, y(s))ds, \quad t \in [0, t_1], \quad (69)$$

if $y(t)$ be ${}^{cf}[(i) - gH]$-differentiable,

$$y(t) = y_0 \ominus (-1)\frac{1}{\Gamma(k\alpha)} \int_0^t (t-s)^{k\alpha-1} f(s, y(s))ds, \quad t \in [0, t_1], \quad (70)$$

if $y(t)$ be $[(ii) - gH]$-differentiable,

$$y(t) = \begin{cases} y_0 \oplus \frac{1}{\Gamma(k\alpha)} \int_0^t (t-s)^{k\alpha-1} f(s, y(s))ds, & if\, t \in [0, t_1] \\ y_0 \ominus (-1)\frac{1}{\Gamma(k\alpha)} \sum_{i=1}^k \int_{t_{i-1}}^{t_i} (t_i - s)^{k\alpha-1} f(s, y(s))ds & \\ \ominus(-1)\frac{1}{\Gamma(k\alpha)} \int_{t_k}^t (t-s)^{k\alpha-1} f(s, y(s))ds \ominus & \\ (-1)\sum_{i=1}^k I_i(y(t_i^-)), & if\, t \in (t_k, t_{k+1}] \end{cases} \quad (71)$$

if there exists a point $t_1 \in (0, t_{k+1})$ such that $y(t)$ is $[(i) - gH]-$differentiable on $[0, t_1]$ and $[(ii) - gH]-$ differentiable on (t_1, t_{k+1}).

Theorem 15 *[24]. Assume that*
$*(H_1)$ There exists a constant $0 \leq l$ such that $d(f(t, u), f(t, \overline{u})) \leq ld(u, \overline{u})$, for each $t \in [0, T]$, and each $u, \overline{u} \in \mathbb{R}_F$
(H_2) There exists a constant $0 \leq l^*$ such that $d(I_k(u), I_k(\overline{u})) \leq l^* d(u, \overline{u})$, for each $u, \overline{u} \in \mathbb{R}_F$, and $k = 1, 2, \ldots, m$.
if

$$[\frac{T^{\frac{k}{\alpha}} l(m+1)}{\Gamma(\frac{k}{\alpha}+1)} + ml^*] < 1. \quad (72)$$

Such that T is very small numbers therefore, Eqs. (66)–(68) has a unique solution on $[0, T]$.

5 Application Fractional Differential Transform Method to Solve Fuzzy Fractional Impulsive Differential Equations

Let us define $w(t) = \int_0^t (t-s)^{\frac{k}{\alpha}-1} f(s, y(s))ds$. Suppose that $Y(K)$, $W(K)$ and $F(K)$ are the fractional differential transform of the function $y(t)$, $w(t)$ and $f(t, y(t))$, respectively. Then, by choosing $\beta \in \mathbb{Z}^+$ such that $\frac{k}{\alpha}\beta \in \mathbb{Z}^+$ and using the Theorem 14, we obtain

$$Y_\alpha(k) = \begin{cases} I_k(Y_\alpha(0)) \oplus Y_\alpha(0) \oplus \frac{1}{\Gamma(k\alpha)} \\ \sum_{i=1}^{k} \left(\sum_{l=t_{i-1}}^{k} B(\frac{k-l}{k}\beta + 1, k\alpha)\delta(l - \beta k\alpha)F_\alpha(k-l) \right), \ t \in [0, t_1],\ k \geq l \\ I_0(Y_\alpha 0)) \oplus Y_\alpha(0) \ominus (-1) \frac{1}{\Gamma(\alpha)} \\ \sum_{i=1}^{k} \left(\sum_{l=t_{i-1}}^{k} B(\frac{k-l}{k}\beta + 1, k\alpha)\delta(l - \beta k\alpha)F_\alpha(k-l) \right) \ominus (-1) \\ \frac{1}{\Gamma(k\alpha)} \left(\sum_{l=1}^{k} B(\frac{k-l}{k}\beta + 1, k\alpha)\delta(l - \beta k\alpha)F_\alpha(k-l) \right) \ominus_{gH} (-1) \\ \sum_{i=1}^{k} I_i(Y_\alpha(K_i)), \qquad t \in (t_k, t_{k+1}],\ k \geq l \end{cases} \tag{73}$$

with

$$F_{\frac{1}{\alpha}}(k-l) = \begin{cases} \frac{1}{\Gamma(k\alpha+1)}[(D_{t_0}^{k\alpha})f(k-l)]_{t=t_0} & , \ \frac{k}{\alpha} \in \mathbb{Z}^+ \\ 0, \ \frac{k}{\alpha} \notin \mathbb{Z}^+ \end{cases} \tag{74}$$

Therefore, the solution of Eqs. (66)-(68) falls short of providing an explicit formula

$$y(t) = \sum_{k=0}^{\infty} Y_\alpha(K) t^{\frac{k}{\alpha}} \tag{75}$$

for $k = 0, 1, 2, \ldots, (n\alpha - 1)$, where α is the order of FFIDE

Theorem 16 *([24]). Let* $0 < \frac{1}{\alpha} \leq 1$, *then the exact solution of Eqs.* (66) – (68) *could be represented by*

$$y(t) = \sum_{k=0}^{\infty} Y_\alpha(K) t^{k\alpha} \tag{76}$$

Also, the approximate solution $y(t)$ *can be obtained by taking finitely many terms in the series representation of* $y(t)$ *so,*

$$y_N(t) = \sum_{k=0}^{N} Y_\alpha(K) t^{k\alpha} \tag{77}$$

And $y_N(t)$, $y_N^{\alpha}(t)$ convergence uniformly to the exact solution $y(t), y^{\frac{1}{\alpha}}(t)$.

Theorem 17 *([24]). Consider the FFIDE* (66)–(68) *if there exists a fixed* n *such that* $n \leq k$, $|f(t, y(t))| \leq r^k$, $0 < r \leq 1$ *and* $\sum_{i=1}^{n} I_i(Y_\alpha(n_i)) \leq r^n$, *then the numerical solution of the present method absolutely convergence.*

6 Describing the Fuzzy Fractional Adomian Decomposition Technique

The Adomian decomposition technique is applied to the following FFIDE

$$L^{\alpha}y(t) + N(y) = g_1 \tag{78}$$

$$\Delta y|_{t=t_k} = I_k(y(t_k^-)) \tag{79}$$

$$y(0) = y_0 \tag{80}$$

where $L^{\alpha} = {}_cD_*^{\alpha}{}_cD_*^{\alpha}\dots{}_cD_*^{\alpha}$ is denote the Caputo fractional generalized derivative of order α. The corresponding $L^{-\alpha}$ operator can be written in the form

$$L^{-\alpha}(.) = \frac{1}{\Gamma^n(1+\alpha)}\int_0^x\int_0^{t_n}\int_0^{t_{n-1}}\dots\int_0^{t_2}(.)(dx_1)^{\alpha}\dots(dx_{n-1})^{\alpha}(dx_n)^{\alpha}. \tag{81}$$

The nonlinear term, $N(y) = f(s, y(s))$, is expressed by an infinite series of polynomials given by:

$$f(s, y(s)) = \sum_{n=0}^{n} A_n \tag{82}$$

where the $A_n = [\underline{A_n}, \overline{A_n}], n \geq 0$ are the so-called Adomian polynomial (see [51,53]) defined by:

$$A(y_0, y_1, \dots, y_n) = \frac{1}{\Gamma(n+1)}\frac{d^n}{d\lambda^n}[N[\sum_{k=0}^{\infty}\lambda^k y_k]]|_{\lambda=0} \tag{83}$$

using the generalized Taylor's formula, then we get

$$y(x) = \sum_{k=0}^{\infty}\frac{(x-x_0)^{k\alpha}}{\Gamma(k\alpha+1)}[({}_cD_*^{\alpha})^k y](x_0) + L^{-\alpha}N(y) + L^{\frac{-1}{\alpha}}f \tag{84}$$

Here the main point is that the solution of Eqs. (78)–(80) is

$$y(x) = \sum_{n=0}^{\infty} y_n(x), \tag{85}$$

where $y_0 = \sum_{k=0}^{\infty}\frac{(x-x_0)^{k\alpha}}{\Gamma(k\alpha+1)}[({}_cD_*^{\alpha})^k y](x_0) + L^{-\alpha}f$, $y_n = +L^{-\alpha}N(y)$

Where α is he order of fractional and $N(y)$ be the Adomian polynomials. We define the fuzzy Adomian polynomials for any positive integer k and any α, $0 < \alpha < 1$, then the following equality holds, which if $y(x)$ be the from of $[(i) - gH]$- Caputo differentiability then:

$$\left[\sum_{k=0}^{\infty}\frac{(x-x_0)^{k\alpha}}{\Gamma(k\alpha+1)}[({}_cD^{\alpha}_{i.gH})^k\underline{y}](x_0)+L^{-\alpha}N(\underline{y})(x)+L^{-\alpha}\underline{f}(x)\right]_r$$
$$=\Big[\sum_{k=0}^{\infty}\frac{(x-x_0)^{k\alpha}}{\Gamma(k\alpha+1)}[({}_cD^{\alpha}_{*})^k\underline{y}](x_0,r)$$
$$+L^{-\alpha}N(\underline{y})(x,r)+L^{-\alpha}\underline{f}(x,r),\sum_{k=0}^{\infty}\frac{(x-x_0)^{k\alpha}}{\Gamma(k\alpha+1)}[({}_cD^{\alpha}_{*})^k\overline{y}](x,r)+L^{-\alpha}N(\overline{y})$$
$$+L^{-\alpha}\overline{f}(x,r)\Big]$$

if $y(x)$ be the from of $[(ii)-gH]$- Caputo differentiability then:

$$\left[\sum_{k=0}^{\infty}\frac{(x-x_0)^{k\alpha}}{\Gamma(k\alpha+1)}[({}_cD^{\alpha}_{ii.gH})^k\underline{y}](x_0)+L^{-\alpha}N(\underline{y})(x)+L^{-\alpha}\underline{f}(x)\right]_r$$
$$=\Big[\sum_{k=0}^{\infty}\frac{(x-x_0)^{k\alpha}}{\Gamma(k\alpha+1)}[({}_cD^{\alpha}_{*})^k\overline{y}](x_0,r)+L^{\frac{-1}{\alpha}}N(\overline{y})(x,r)+L^{-\alpha}\overline{f}(x,r),$$
$$\sum_{k=0}^{\infty}\frac{(x-x_0)^{k\alpha}}{\Gamma(k\alpha+1)}[({}_cD^{\alpha}_{*})^k\underline{y}](x_0,r)+L^{-\alpha}N(\underline{y})(x,r)+L^{-\alpha}\underline{f}(x,r)\Big]$$

6.1 Solving FFIDE by FFADT

The standard decomposition technique represents the solution of $y(t)$ into Eqs. (78)–(80) as the following series,

$$y(t)=\sum_{i=0}^{n}y_i(t) \tag{86}$$

substituting Eq. (82) and (83) into Eq. (71), we get

$$y_0(t)=\begin{cases}I_0(y(0))\oplus y(0), & \text{if } \ t\in[0,t_1],\\ I_0(y(0))\oplus y(0), \text{if} & t\in(t_k,t_{k+1}],\end{cases} \tag{87}$$

$$y_1(t)=\begin{cases}\frac{1}{\Gamma(k\alpha)}\int_0^t(t-s)^{\frac{k}{\alpha}-1}A_0ds, \text{if} \quad t\in[0,t_1],\\ \frac{1}{\Gamma(k\alpha)}\sum_{i=1}^{k}\int_{t_{i-1}}^{t_i}(t_i-s)^{k\alpha-1}A_0ds\ominus(-1)\\ \frac{1}{\Gamma(k\alpha)}\int_{t_k}^{t}(t-s)^{k\alpha-1}A_0ds\ominus(-1)\\ \sum_{i=1}^{k}I_i(\sum_{i=1}^{\infty}y_i(t_i)), t\in(t_k,t_{k+1}],\end{cases} \tag{88}$$

$$\vdots$$

$$y_{n+1}(t)=\begin{cases}\frac{1}{\Gamma(\frac{k}{\alpha})}\int_0^t(t-s)^{k\alpha-1}A_nds, \text{if} \quad t\in[0,t_1],\\ \frac{1}{\Gamma(k\alpha)}\sum_{i=1}^{k}\int_{t_{i-1}}^{t_i}(t_i-s)^{k\alpha-1}A_nds\ominus(-1)\\ \frac{1}{\Gamma(k\alpha)}\int_{t_k}^{t}(t-s)^{k\alpha-1}A_nds\ominus(-1)\\ \sum_{i=1}^{k}I_i(\sum_{i=1}^{\infty}y_i(t_i)), t\in(t_k,t_{k+1}],\end{cases} \tag{89}$$

substituting equations (87)–(89) into Eq (86) leads to the components of $y(t)$. We next claim prove that solving FFIDE (66)–(68) by FFADM is convergence.

Theorem 18. *The series solution $y(t,r)=\sum_{i=0}^{\infty}y_i(t,r)$ of Eqs.(87)–(89) using FFADM is convergence when $\alpha>0, |y_1(t,r)|<\infty$.*

Define the sequence of partial sums Z_n, let Z_n and Z_m be arbitrary partial sums with $n \geq m$. We are going to prove that Z_n is a Cauchy sequence in this Banach space:

$$d(Z_n(t), Z_m(t)) = max_{\forall t\in[a,b]}|\sum_{i=m+1}^{n} y_i(t,r)| = max_{\forall t\in[a,b]}|\sum_{i=m+1}^{n}$$

$$\begin{cases} \frac{1}{\Gamma(k\alpha)})\int_0^t (t-s)^{k\alpha-1}A_{i-1}ds, t\in[0,t_1] \\ \frac{1}{\Gamma(k\alpha)}\sum_{i=1}^{k}\int_{t_{i-1}}^{t_i}(t_i-s)^{k\alpha-1}A_{i-1}ds\ominus(-1) \\ \frac{1}{\Gamma(k\alpha)}\int_{t_k}^{t}(t-s)^{k\alpha-1}A_{i-1}ds\ominus(-1)\sum_{i=1}^{k}I_i(\sum_{i=1}^{\infty}y_i(t_i)), t\in[t_k,t_{k+1}] \end{cases}$$

$$= \begin{cases} \max_{\forall t\in[a,b]}|(\frac{1}{\Gamma(k\alpha)}\int_0^t (t-s)^{k\alpha-1}(\sum_{i=m}^{n-1}A_i)ds)|, t\in[0,t_1] \\ \\ max_{\forall t\in[a,b]}|(\frac{1}{\Gamma(k\alpha)}\sum_{i=1}^{k}\int_{t_{i-1}}^{t_i}(t_i-s)^{k\alpha-1}(\sum_{i=m}^{n-1}A_i)ds\ominus(-1) \\ \frac{1}{\Gamma(\alpha)}\int_{t_k}^{t}(t-s)^{k\alpha-1}(\sum_{i=m}^{n-1}A_i)ds\ominus(-1)\sum_{i=1}^{k}I_i(\sum_{i=1}^{\infty}y_i(t_i)))|, t\in[t_k,t_{k+1}] \end{cases}$$

The author in [52], deduced a new formula to Adomian polynomials A_n which can be written in the form

$$A_n = f_1(Z_n, y(Z_n)) - \sum_{i=0}^{n-1} A_i, \quad n \geq 1$$

where the partial sum $Z_n = \sum_{i=0}^{n} y_i(t)$ and $A_0 = f(Z_0, y(Z_0))$. we have

$$\sum_{i=m}^{n-1} A_i = d(f(Z_{n-1}, y(Z_{n-1})), f(Z_{m-1}, y(Z_{m-1})))$$

So

$$d(Z_n(t), Z_m(t)) \leq$$

$$\begin{cases} \max_{\forall t\in[a,b]}|(\frac{1}{\Gamma(k\alpha)}\int_0^t (t-s)^{k\alpha-1} \\ d(f(Z_{n-1}, y(Z_{n-1})), f(Z_{m-1}, y(Z_{m-1})))ds)|, t\in[0,t_1] \\ \\ \max_{\forall x\in[a,b]}|(\frac{1}{\Gamma(k\alpha)}\sum_{i=1}^{k}\int_{t_{i-1}}^{t_i}(t_i-s)^{k\alpha-1} \\ d((f(Z_{n-1}, y(Z_{n-1})), f(Z_{m-1}, y(Z_{m-1})))ds\ominus(-1)\frac{1}{\Gamma(k\alpha)} \\ \int_{t_k}^{t}(t-s)^{k\alpha-1}d(f(Z_{n-1}, y(Z_{n-1})), f(Z_{m-1}, y(Z_{m-1})))ds\ominus(-1) \\ \sum_{i=1}^{k} d(I(s, Z_n(t_k^-)), I(s, Z_m(t_k^-)))|, \\ t\in[t_k, t_{k+1}] \end{cases}$$

$$\leq \begin{cases} |\int_0^t(\frac{1}{\Gamma(k\alpha)}(t-s)^{k\alpha-1}ld(y(Z_{n-1}), y(Z_{m-1}))ds|, t\in[0,t_1] \\ |\int_{t_{i-1}}^{t_i}(\frac{1}{\Gamma(k\alpha)}\sum_{i=1}^{k}(t_i-s)^{k\alpha-1}||ld((y(Z_{n-1})), y(Z_{m-1}))|ds \\ \ominus(-1)|\int_{t_k}^{t}\frac{1}{\Gamma(k\alpha)}(t-s)^{\frac{k}{\alpha}-1}||ld((y(Z_{n-1}), y(Z_{m-1}))ds|\ominus \\ (-1)|\sum_{i=1}^{k} l^* d(Z_n(t_k^-)), Z_m(t_k^-))|, t\in[t_k, t_{k+1}] \end{cases}$$

$$\leq \begin{cases} \left(\frac{lT^{k\alpha}}{\Gamma(k\alpha+1)}\right)||d(Z_n, Z_m)||_\alpha, t \in [0, t_1] \\ \sum_{i=1}^{k}\left(\frac{lT^{k\alpha}}{\Gamma(k\alpha+1)}\right)(t_i - t_{i-1})||d(Z_n, Z_m)||_\alpha \ominus (-1)(t-t_k)\left(\frac{lT^{k\alpha}}{\Gamma(k\alpha+1)}\right)||d(Z_n, Z_m)||_\alpha \ominus \\ (-1)|\left(\frac{l^*T^{k\alpha}}{\Gamma(k\alpha+1)}\right)||d(Z_n, Z_m)||_\alpha, t \in [t_k, t_{k+1}] \end{cases}$$

$$\leq \begin{cases} \left(\frac{lT^{k\alpha}}{\Gamma(k\alpha+1)}\right)||d(Z_n, Z_m)||_\alpha, t \in [0, t_1] \\ \left(\frac{mlT^{k\alpha}}{\Gamma(k\alpha+1)}\right)||d(Z_n, Z_m)||_\alpha \ominus_{gH} (-1)\left(\frac{lT^{k\alpha}}{\Gamma(k\alpha+1)}\right)||d(Z_n, Z_m)||_\alpha \ominus_{gH} \\ (-1)|\left(\frac{l^*T^{k\alpha}}{\Gamma(k\alpha+1)}\right)||d(Z_n, Z_m)||_\alpha = [\frac{T^{k\alpha}l(m+1)}{\Gamma(k\alpha+1)} + ml^*]||d(Z_n, Z_m)||_\alpha, t \in [t_k, t_{k+1}] \end{cases}$$

Let $n = m + 1$, then

$$[\frac{T^{k\alpha}l(\rho+1)}{\Gamma(k\alpha+1)} + \rho l^*]||d(Z_n, Z_m)||_\alpha$$
$$\leq \Theta[\frac{T^{k\alpha}l(\rho+1)}{\Gamma(\frac{k}{\alpha}+1)} + \rho l^*]||d(Z_m, Z_{m-1})||_\alpha \leq \Theta^2[\frac{T^{k\alpha}l(\rho+1)}{\Gamma(k\alpha+1)} + \rho l^*]||d(Z_{m-1}, Z_{m-2})||_\alpha$$
$$\leq \cdots \leq [\frac{T^{k\alpha}l(\rho+1)}{\Gamma(k\alpha+1)} + \rho l^*]\Theta^m||d(Z_1, Z_0)||_\alpha.$$

Using the triangle inequality we have

$$||d(Z_n, Z_m)||_\alpha \leq$$
$$||d(Z_{m+1}, Z_m)||_\alpha + ||d(Z_{m+2}, Z_{m+1})||_\alpha + ||d(Z_{m+3}, Z_{m+2})||_\alpha + \cdots + ||d(Z_n, Z_{n-1})||_\alpha$$

$$\leq [\frac{T^{k\alpha}l(\rho+1)}{\Gamma(k\alpha+1)} + \rho l^*][\Theta^m + \Theta^{m+1} + \cdots + \Theta^{n-1}]||d(Z_1, Z_0)||_\alpha \leq$$

$$[T^{\frac{k}{\alpha}l(\rho+1)}\Gamma(k\alpha+1) + \rho l^*]\Theta^m[1 + \Theta + \Theta^2 + \cdots + \Theta^{n-m+1}]||d(Z_1, Z_0)||_\alpha$$
$$\leq [\frac{T^{k\alpha}l(\rho+1)}{\Gamma(k\alpha+1)} + \rho l^*]\Theta^m\left(\frac{1-\Theta^{n-m}}{1-\Theta}\right)|y_1(t,r)|.$$

Since $0 < \Theta < 1$ so,$1 - \Theta^{n-m} \leq 1$ then

$$||d(Z_n, Z_m)||_\alpha \leq [\frac{T^{k\alpha}l(\rho+1)}{\Gamma(k\alpha+1)} + \rho l^*]\frac{\Theta^m}{1-\Theta}\max|y_1(t,r)|, \tag{90}$$

but $\max|y_1(t,r)| < \infty$ then $||d(Z_n, Z_m)||_\alpha \to 0$ as $m \to \infty$ from which we conclude that $\{Z_n\}$ is a Cauchy sequence, the series $\sum_{i=0}^{\infty} y_i(t,r)$ converges and the proof is complete.

Theorem 19. *Let $\alpha > 0$, then Eq.(87)–(89)has a unique solution.*

proof: The proof is clear.

Theorem 20. *The maximum absolute truncation error of the series solution $y(t,r) = \sum_{i=0}^{\infty} y_i(t,r)$ to problem (66)–(68)is estimated to be:*

$$d(y(t), \sum_{i=0}^{m} y_i(t)) \leq [\frac{T^{k\alpha}l(\rho+1)}{\Gamma(k\alpha+1)} + \rho l^*]\frac{\Theta^m}{1-\Theta}\max|y_1(t,r)|.$$

proof: From equation (90) in Theorem 18 we have

$$||d(Z_n, Z_m)|| \leq [\frac{T^{k\alpha}l(\rho+1)}{\Gamma(k\alpha+1)} + \rho l^*]\frac{\Theta^m}{1-\Theta} \max|y_1(t,r)|.$$

As $n \to \infty$ then $Z_n \to y(t)$ so we have

$$||d(y(t), Z_m)|| \leq [\frac{T^{k\alpha}l(\rho+1)}{\Gamma(k\alpha+1)} + \rho l^*]\frac{\Theta^m}{1-\Theta} \max|y_1(t,r)|.$$

and the maximum absolute truncation error in the interval $[a, b]$ is estimated to be

$$d(y(t), \sum_{i=0}^{m} y_i(t)) \leq [\frac{T^{k\alpha}l(\rho+1)}{\Gamma(k\alpha+1)} + \rho l^*]\frac{\Theta^m}{1-\Theta} \max|y_1(t,r)|.$$

This completes the proof.

7 Numerical Examples

we demonstrate the effectiveness of the FFADT, for solving FFIDE by the following example. At last the solution obtained in this example with FFADT is compared with the solution achieved of the same example by N.Najafi and T. Allaviranloo [24] through the FFDTM.

Example 1. Let us consider the FFIDE,

$$_cD_*^{k\alpha}y(t) = \frac{ty^2(t)}{(3+t)(1+y^2(t))}, \quad t \in J := [0,1], \quad t \neq \frac{1}{2}, \quad m-1 < \frac{1}{\alpha} < m, \quad m \in \mathbb{N} \tag{91}$$

$$\Delta y|_{t=\frac{1}{2}} = \frac{|y(\frac{1^-}{2})|}{2+|y(\frac{1^-}{2})|}, \tag{92}$$

$$f(\tilde{0}, y_0(\tilde{0})) = [\tilde{0}, 1] \tag{93}$$

Set

$$I_k(t) = \frac{t}{t+2}, t \in [0, \infty)$$

$$f(t, y(t)) = \Big[\frac{t^3(r-1)}{(3+t)(1+t^2)}, \frac{t^3(1-r)}{(3+t)(1+t^2)}\Big]$$

Using the equations (87)–(89) and taking $N = 2$, $k = 2$ and $\frac{1}{\alpha} = \frac{1}{2}$, equation (91) can be transformed to:

$$y_0 = f(t, y(t)) = \Big[\frac{t^3(r-1)}{(3+t)(1+t^2)}, \frac{t^3(1-r)}{(3+t)(1+t^2)}\Big] \tag{94}$$

$$A_0 = f(\tilde{0}, y(\tilde{0})) = \Big[\tilde{0}, 1\Big] \tag{95}$$

$$y_1(t) =$$
$$\begin{cases} \frac{1}{\Gamma(k\alpha)} \int_0^t (t-s)^{\frac{k}{\alpha}-1} A_0 ds = \frac{1}{\Gamma(k\alpha)} \left[\tilde{0}, 1\right] \left(\Gamma^{\frac{k}{\alpha}} \frac{\Gamma(k\alpha)}{\Gamma(k\alpha+1)}\right) t^{k\alpha} \\ = \left[\tilde{0}, 1\right] \left(\frac{\Gamma^{k\alpha}}{\Gamma(k\alpha+1)}\right) t^{k\alpha} & t \in [0, \frac{1}{2}], \\ \frac{1}{\Gamma(k\alpha)} \sum_{i=1}^{k} \int_{t_{i-1}}^{t_i} (t_i - s)^{k\alpha-1} A_0 ds \ominus (-1) \frac{1}{\Gamma(\frac{k}{\alpha})} \int_{t_k}^{t} (t-s)^{\frac{k}{\alpha}-1} A_0 ds \ominus (-1) \\ \sum_{i=1}^{k} I_i (\sum_{i=1}^{\infty} y_i(t_i)) = \\ \left[\tilde{0}, 1\right] \sum_{i=1}^{k} \frac{\Gamma(k\alpha)}{\Gamma(k\alpha+1)} t_i^{k\alpha} \ominus (-1) \left[\tilde{0}, 1\right] \\ \frac{\Gamma(k\alpha)}{\Gamma(k\alpha+1)} t^{k\alpha} \ominus (-1) \left[\tilde{0}, 1\right] \frac{\Gamma(k\alpha}{\Gamma(k\alpha+1)} t_k^{k\alpha} \ominus (-1) [\frac{\frac{t^3(r-1)}{(3+t)(1+t^2)}}{\frac{t^3(r-1)}{(3+t)(1+t^2)}+2}, \frac{\frac{t^3(1-r)}{(3+t)(1+t^2)}}{\frac{t^3(1-r)}{(3+t)(1+t^2)}+2}] = \\ \left[\tilde{0}, \sum_{i=1}^{k} \frac{\Gamma(k\alpha)}{\Gamma(k\alpha+1)} t_i^{k\alpha} \oplus \frac{\Gamma(k\alpha)}{\Gamma(k\alpha+1)} t^{k\alpha} \ominus \frac{\Gamma(k\alpha)}{\Gamma(k\alpha+1)} t_k^{k\alpha}\right] \oplus [\frac{\frac{t^3(r-1)}{(3+t)(1+t^2)}}{\frac{t^3(r-1)}{(3+t)(1+t^2)}+2}, \frac{\frac{t^3(1-r)}{(3+t)(1+t^2)}}{\frac{t^3(1-r)}{(3+t)(1+t^2)}+2}] = \\ \left[\frac{\frac{t^3(r-1)}{(3+t)(1+t^2)}}{\frac{t^3(r-1)}{(3+t)(1+t^2)}+2}, \sum_{i=1}^{k} \frac{\Gamma(\frac{k}{\alpha})}{\Gamma(\frac{k}{\alpha}+1)} t_i^{\frac{k}{\alpha}} \oplus \frac{\Gamma(\frac{k}{\alpha})}{\Gamma(\frac{k}{\alpha}+1)} t^{\frac{k}{\alpha}} \ominus \frac{\Gamma(\frac{k}{\alpha})}{\Gamma(\frac{k}{\alpha}+1)} t_k^{\frac{k}{\alpha}} \oplus \frac{\frac{t^3(1-r)}{(3+t)(1+t^2)}}{\frac{t^3(1-r)}{(3+t)(1+t^2)}+2}\right], \\ \quad t \in (t_k, t_{k+1}], \end{cases} \tag{96}$$

Substituting Eqs.(94)–(96) into Eq.(86), hance $\sum_{i=0}^{1} y_i = y_0 + y_1$, the approximation solution for $\alpha = \frac{1}{2}$ is

$$y_{i.gH}(t) =$$
$$\begin{cases} \left[\tilde{0} \oplus \frac{t^3(r-1)}{(3+t)(1+t^2)}, \frac{\Gamma(4)t^4}{\Gamma(4+1)} \oplus \frac{t^3(1-r)}{(3+t)(1+t^2)}\right], & t \in [0, \frac{1}{2}), \\ \left[\frac{t^3(r-1)}{(3+t)(1+t^2)} \oplus \frac{\frac{t^3(r-1)}{(3+t)(1+t^2)}}{\frac{t^3(r-1)}{(3+t)(1+t^2)}+2}, \sum_{i=1}^{k} \frac{\Gamma(4)}{\Gamma(4+1)} t_i^{k\alpha} \oplus \frac{\Gamma(4)}{\Gamma(4+1)} t^{k\alpha} \ominus \frac{\Gamma(4)}{\Gamma(4+1)} t_k^{k\alpha} \right. \\ \left. \oplus \frac{\frac{t^3(1-r)}{(3+t)(1+t^2)}}{\frac{t^3(1-r)}{(3+t)(1+t^2)}+2}\right], & t \in (\frac{1}{2}, 1], \end{cases}$$

Example 2. Let us consider the FFIDE,

$${}_cD_*^{k\alpha} y(t) = \frac{|y(t)|}{10(1+|y(t)|)}, \quad t \in J := [0, 1], \quad t \neq \frac{1}{2}, \quad m-1 < \frac{1}{\alpha} < m, \quad m \in \mathbb{N}, \tag{97}$$

$$\Delta y|_{t=\frac{1}{2}} = \frac{|y(\frac{1^-}{2})|}{3 + |y(\frac{1^-}{2})|}, \tag{98}$$

$$Y_{\frac{1}{\alpha}}(0) = [\tilde{0}, \tilde{0}]. \tag{99}$$

Table 1. Numerical results of example 1 for $\underline{y_{i.gH}}(t)$

r/t	0.55	0.60	0.65	0.70	0.75	0.80	0.85	0.90	0.95	1
0.55	−0.0139	−0.0098	−0.0057	−0.0016	0.0024	0.0065	0.0106	0.0147	0.0188	0.0229
0.60	−0.0166	−0.0112	−0.0057	−0.0003	0.0051	0.0106	0.0161	0.0215	0.0271	0.0324
0.65	−0.0193	−0.0122	−0.0051	0.0020	0.0091	0.0170	0.0233	0.0304	0.0375	0.0446
0.70	−0.02199	−0.0129	−0.0038	0.0053	0.0145	0.0236	0.0327	0.0418	0.0509	0.6002
0.75	−0.0245	−0.0130	−0.0015	0.0100	0.0216	0.0331	0.0446	0.0561	0.0676	0.0791
0.80	−0.0267	−0.0128	0.0020	0.0163	0.0307	0.0450	0.0594	0.0737	0.0881	0.1024
0.85	−0.0286	−0.0109	0.0067	0.0244	0.0421	0.0598	0.0775	0.0951	0.1128	0.1305
0.90	−0.0300	−0.0085	0.01306	0.0346	0.0562	0.0778	0.0993	0.1209	0.1425	0.1640
0.95	−0.0310	−0.0049	0.0212	0.0472	0.0733	0.9936	0.1254	0.1515	0.1776	0.2036
1	−0.0312	5.55E–1	0.1313	0.0625	0.0938	0.0125	0.1563	0.1875	0.2188	0.2500

Table 2. Numerical results of example 1 for $\overline{y_{i.gH}}(t)$

r/t	0.55	0.60	0.65	0.70	0.75	0.80	0.85	0.90	0.95	1
0.55	0.0139	0.0098	0.0057	0.0016	−0.0024	−0.0065	−0.0106	−0.0147	−0.0188	−0.0229
0.60	0.0166	0.0112	0.0057	0.0003	− 0.0051	−0.0106	−0.0161	−0.0215	− 0.271	−0.0324
0.65	0.0193	0.0122	0.0051	−0.0020	−0.0091	−0.0170	−0.0233	−0.0304	-0.0375	−0.0446
0.70	0.02199	0.0129	0.0038	−0.0053	−.0145	−0.0236	−0.0327	−0.0418	−0.0509	−0.6002
0.75	0.0245	0.0130	0.0015	−0.0100	−0.0216	− 0.0331	−0.0446	−0.0561	−0.0676	−0.0791
0.80	0.0267	0.0128	−0.0020	−0.0163	−0.0307	− 0.0450	−0.0594	−0.0737	−0.0881	−0.1024
0.85	0.0286	0.0109	−0.0067	−0.0244	−0.0421	− 0.0598	−0.0775	−0.0951	−0.1128	−0.1305
0.90	0.0300	0.0085	−0.01306	−0.0346	−0.0562	−0.0778	−0.0993	−0.1209	−0.1425	−0.1640
0.95	0.0310	0.0049	−0.0212	−0.0472	−0.0733	−0.9936	−0.1254	−0.1515	−0.1776	−0.2036
1	0.0312	−5.55E–1	−0.1313	−0.0625	−0.0938	−0.0125	−0.1563	−0.1875	−0.2188	−0.2500

Set

$$I_k(t) = \frac{t}{t+3}, t \in [0, \infty) \tag{100}$$

$$f(t, y(t)) = \left[\frac{(r-1)t}{10(1+t)}, \frac{(1-r)t}{10(1+t)}\right] \tag{101}$$

Using the equations (87)–(89) and taking $N = 2, k = 2$ and $\frac{1}{\alpha} = \frac{1}{2}$, equation (91) can be transformed to:

$$y_0 = f(t, y(t)) = \left[\frac{(r-1)t}{10(1+t)}, \frac{(1-r)t}{10(1+t)}\right] \tag{102}$$

$$A_0 = f(\tilde{0}, y(\tilde{0})) = \left[\tilde{0}, 1\right] \tag{103}$$

Table 3. Numerical results of example 2 for $\underline{y_{ii.gH}}(t)$

r/t	0.55	0.60	0.65	0.70	0.75	0.80	0.85	0.90	0.95	1
0.55	−0.0140	0.01787	0.0497	0.0816	0.1135	0.1434	0.1773	0.2091	0.2410	0.2729
0.60	−0.0280	0.0055	0.0389	0.0723	0.1058	0.1392	0.1726	0.2060	0.2395	0.2729
0.65	−0.0453	−0.0099	0.0255	0.0609	0.0962	0.1316	0.1669	0.2022	0.2376	0.2729
0.70	−0.0664	−0.0287	0.0091	0.0468	0.0845	0.1222	0.1599	0.1976	0.2352	0.2729
0.75	−0.0920	−0.0514	−0.0108	0.0298	0.0703	0.1109	0.1514	01919	0.2324	0.2729
0.80	−0.1229	−0.0787	−0.0347	0.0093	0.0533	0.0973	0.1412	0.1851	0.2290	0.2729
0.85	−0.1595	−0.1113	−0.0632	−0.0151	0.03330	0.0810	0.1290	0.1770	0.2250	0.2729
0.90	−0.2029	−0.1496	−0.0969	−0.0439	0.0090	0.0619	0.1147	0.1675	0.2202	0.2729
0.95	−0.2539	−0.1951	−0.1364	−0.0778	−0.0192	0.0394	0.0978	0.1562	0.2146	0.2729
1	−0.3135	−0.2480	−0.1826	−0.1173	−0.0521	0.0131	0.0782	0.1431	0.2081	0.2729

Table 4. Numerical results of example 2 for $\overline{y_{ii.gH}}(t)$

r/t	0.55	0.60	0.65	0.70	0.75	0.80	0.85	0.90	0.95	1
0.55	0.0138	−0.0180	−0.0499	−0.0817	−0.1136	−0.1454	−0.1772	−0.20921	−0.2410	−0.2729
0.60	0.0276	−0.0057	−0.0391	−0.0725	−0.1059	−0.1393	−0.1727	−0.2060	−0.2395	−0.2729
0.65	0.0447	0.0094	−0.0258	−0.0611	−0.0964	−0.1317	−0.1670	−0.2023	−0.2376	−0.2729
0.70	0.0655	0.0280	−0 .0096	−0.0472	−0.0848	−0.1224	−0.1610	−0.1976	−0.2352	−0.2729
0.75	0.0907	0.0504	0.0100	−0.0303	−0.0707	−0.1111	−0.1515	−0.1920	−0.2324	−0.2729
0.80	0.1209	0.0773	0.0336	−0.0101	−0.0539	−0.0976	−0.1414	−0.1852	−0.2290	−0.2729
0.85	0.1569	0.1093	0.0616	0.0139	−0.0338	−0.0816	−0.1293	−0.1772	−0.2250	−0.2729
0.90	0. 1994	0.1471	0.0947	0.0439	−0.0101	−0.0626	−0.1151	−0.1676	−0.2202	−0.2729
0.95	0.2491	0.1914	0.1335	0.0756	0.0177	−0.0403	−0.0984	−0.1565	−0.2147	−0.2729
1	0.3071	0.2430	0.1788	0.1145	0.0501	−0.0143	−0.0789	−0.1435	−0.2081	− 0.2729

$$y_1(t) = \tag{104}$$

$$\begin{cases} \frac{1}{\Gamma(k\alpha)}\int_0^t (t-s)^{k\alpha-1}A_0 ds = \frac{1}{\Gamma(k\alpha)}\left[\tilde{0},1\right]\left(\Gamma^{k\alpha}\frac{\Gamma(k\alpha)}{\Gamma(k\alpha+1)}\right)t^{k\alpha} \\ = \left[\tilde{0},1\right]\left(\frac{\Gamma^{k\alpha}}{\Gamma(k\alpha+1)}\right)t^{k\alpha} & t\in[0,\frac{1}{2}], \\ \\ \frac{1}{\Gamma(k\alpha)}\sum_{i=1}^{k}\int_{t_{i-1}}^{t_i}(t_i-s)^{k\alpha-1}A_0 ds \ominus (-1)\frac{1}{\Gamma(k\alpha)}\int_{t_k}^{t}(t-s)^{k\alpha-1}A_0 ds \ominus (-1) \\ \sum_{i=1}^{k} I_i(\sum_{i=1}^{\infty} y_i(t_i)) = \\ \left[\tilde{0},1\right]\sum_{i=1}^{k}\frac{\Gamma(k\alpha)}{\Gamma(k\alpha+1)}t_i^{k\alpha}\ominus(-1)\left[\tilde{0},1\right]\frac{\Gamma(k\alpha)}{\Gamma(k\alpha+1)}t^{k\alpha}\ominus(-1) \\ \left[\tilde{0},1\right]\frac{\Gamma(k\alpha}{\Gamma(k\alpha+1)}t_k^{\frac{k}{\alpha}}\ominus(-1) \\ [\frac{\frac{t(r-1)}{10(1+t)}}{\frac{t(r-1)}{10(1+t)}+3},\frac{\frac{t(1-r)}{10(1+t)}}{\frac{t(1-r)}{10(1+t)}+3}] = \\ \left[\tilde{0},\sum_{i=1}^{k}\frac{\Gamma(\frac{k}{\alpha})}{\Gamma(k\alpha+1)}t_i^{k\alpha}\oplus\frac{\Gamma(k\alpha)}{\Gamma(k\alpha+1)}t^{k\alpha}\ominus\frac{\Gamma(k\alpha)}{\Gamma(k\alpha+1)}t_k^{k\alpha}\right]\oplus[\frac{\frac{t(r-1)}{10(1+t)}}{\frac{t(r-1)}{10(1+t)}+3},\frac{\frac{t(1-r)}{10(1+t)}}{\frac{t(1-r)}{10(1+t)}+3}] = \\ [\frac{\frac{t(r-1)}{10(1+t)}}{\frac{t(r-1)}{10(1+t)}+3},\sum_{i=1}^{k}\frac{\Gamma(\frac{k}{\alpha})}{\Gamma(\frac{k}{\alpha}+1)}t_i^{\frac{k}{\alpha}}\oplus\frac{\Gamma(\frac{k}{\alpha})}{\Gamma(\frac{k}{\alpha}+1)}t^{\frac{k}{\alpha}}\ominus\frac{\Gamma(\frac{k}{\alpha})}{\Gamma(\frac{k}{\alpha}+1)}t_k^{\frac{k}{\alpha}} & ,t\in(t_k,t_{k+1}], \\ \oplus\frac{\frac{t(1-r)}{10(1+t)}}{\frac{t(1-r)}{10(1+t)}+3}] \end{cases} \tag{105}$$

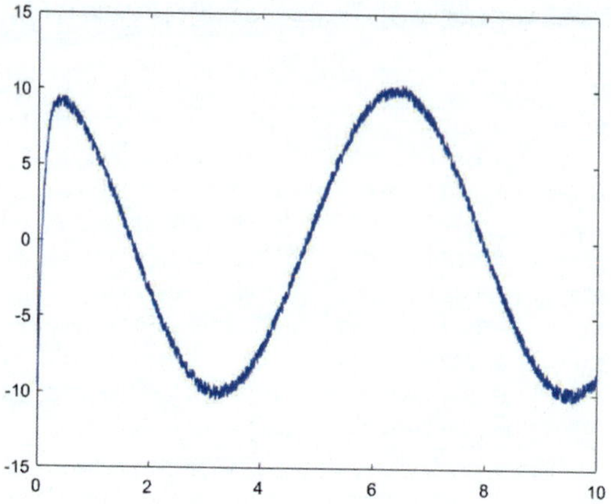

(a) Left hand level sets of the gH-derivative of solutions of example 1, for $\underline{y_{i.gH}}$,

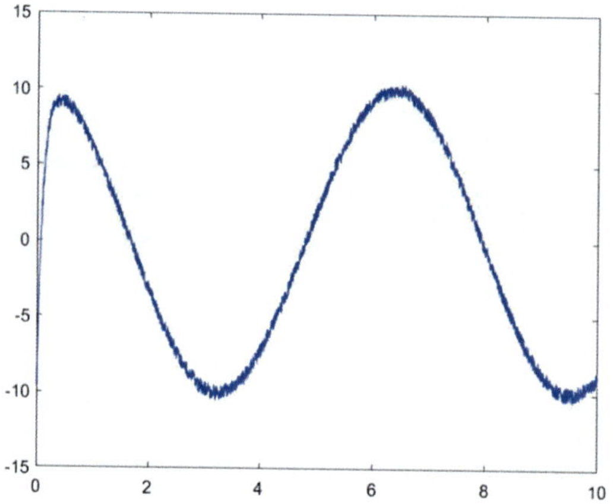

(b) Right hand level sets of the gH-derivative of solutions of example 1, for $\overline{y_{i.gH}}$.

Fig. 1.

Substituting Eqs.(102)–(104) into Eq. (86), hance $\sum_{i=0}^{1} y_i = y_0 + y_1$, the approximation solution for $\frac{1}{\alpha} = \frac{1}{2}$, is

$$y_{ii.gH}(t) = \begin{cases} [\tilde{0} \oplus \frac{t(r-1)}{10(1+t)}, \frac{\Gamma(4)t^4}{\Gamma(4+1)} \oplus \frac{t(1-r)}{10(1+t)}]\ t \in [0, \frac{1}{2}), \\ [\sum_{i=1}^{k} (\frac{\Gamma(4)}{\Gamma(4+1)} t_i^{k\alpha} \oplus \frac{\Gamma(4)}{\Gamma(4+1)} t^{k\alpha} \ominus \frac{\Gamma(4)}{\Gamma(4+1)} t_k^{k\alpha}) \oplus \frac{\frac{t(1-r)}{10(1+t)}}{\frac{t(1-r)}{10(1+t)}+3}], t \in (\frac{1}{2}, 1], \end{cases}$$

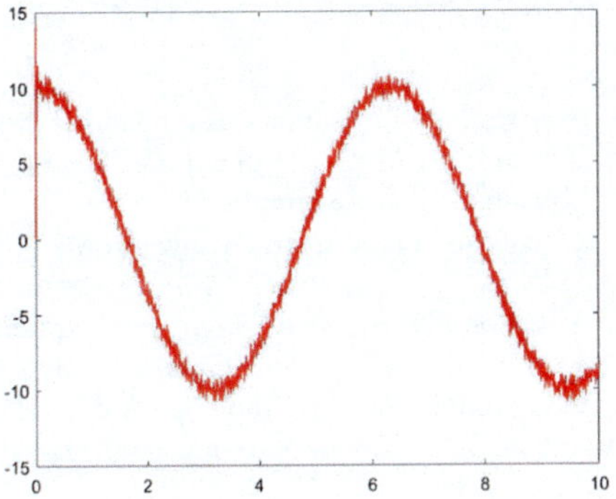

(a) Left hand level sets of the gH-derivative of solutions of example 2, for $\underline{y_{ii.gH}}$,

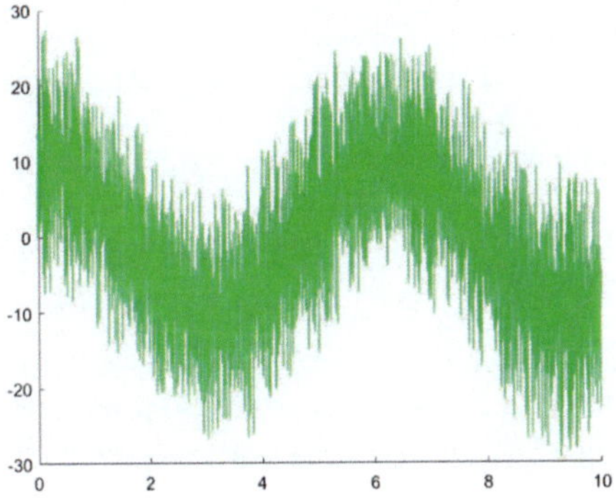

(b) Right hand level sets of the gH-derivative of solutions of example 2, for $\overline{y_{ii.gH}}$.

Fig. 2.

8 Conclusions

In this article a new technique was introduced by presenting FFADT expansion according to gH-differentiability . The conclusions of the test examples show that FFADT conclusion are equal to FFDTM and also, it is demonstrated that the reduced FFDTM solves linear and nonlinear FFIDE without using any complicated polynomials like the FFADT. Furthermore, the obtained series solution by the reduced FFDTM converges faster than those obtained by the FFADT .

References

1. Abbasbandy, S., Allahviranloo, T.: Numerical solution of fuzzy differential equation by Taylor method. J. Comput. Meth. Appl. Math. **2**, 113–124 (2002)
2. Allahviranloo, T., Abbasbandy, S., Ahmady, N., Ahmady, E.: Improved predictor-corrector method for solving fuzzy initial value problem. Inf. Sci. **179**, 945–955 (2009)
3. Allahviranloo, T., Abbasbandy, S., Salahshour, S.: Explicit solution of fractional differential equation with uncertainty. Soft Comput. **16**, 297–302 (2012)
4. Allahviranloo, T., Gouyandeh, Z., Armand, A.: A full fuzzy method for solving differential equation based on Taylor expansion. J. Intell. Fuzzy Syst. **1**, 1–16 (2015)
5. Allahviranloo, T., Gouyandeh, Z., Armand, A., Ghadiri, H.: Existence and uniqueness of solution for fuzzy fractional Volterra-Fredholm integro-differential equation. J. Fuzzy Set Valued Anal. **2013**, 1–9 (2013)
6. Abbasbandy, S., Allahviranloo, T., Balooch, S.M.R., Salahshour, S.: Fuzzy local fractional differential equation. Int. J. Ind. Math. **4**(3), 231–246 (2012)
7. Armand, A., Allahviranloo, T., Gouyandeh, Z.: General solution of Basset equation with Caputo generalized Hukuhara derivative. J. Appl. Anal. Comput. **6**, 119–130 (2016)
8. Anastassiou, G.A.: Fuzzy Mathematics: Approximation Theory, vol. 251. Springer, Heidelberg (2010)
9. Anastassiou, G.A.: Fuzzy fractional calculus and ostrowski inequality. J. Fuzzy Math. **19**(3), 577–590 (2011)
10. Armand, A., Mohammadi, S.: Existence and uniqueness for fractional differential equations with uncertainty. J. Unccertainty Math. Sci. **2014**, 1–9 (2014)
11. Bede, B.: Generalizations of the differentiability of fuzzy-number-valued functions with applications to fuzzy differential equations. Fuzzy Sets Syst. **151**, 581–599 (2005)
12. Bede, B., Stefanini, L.: Generalized differentiability of fuzzy-valued function. Fuzzy Sets Syst. **230**, 119–141 (2013)
13. Benchohra, M., Seba, D.: Impulsive fractional differential equations. Electron. J. Qual. Theor. Diff. Equat. Spec. Ed. **I**(8), 1–14 (2009)
14. Benchohra. M, Slimani. B.A.: Existence and uniqueness of solutions to impulsive fractional, differential equations. Electron. J. Diff. Equat. (2009)
15. Epperson, J.F.: An Introduction to Numerical Methods a Analysis. John Wiley and Sons, Hoboken (2007)
16. Friedman, M., Ming, M., Kandel, A.: Numerical solution of fuzzy differential and integral equations. Fuzzy Sets Syst. **106**, 35–48 (1999)
17. Guo, M., Li, R.: Impulsive functional differential inclusions and fuzzy population models. Fuzzy Set Syst. **138**, 601–615 (2003)
18. JD, M.: Fractional calculus and the Taylor-Riemann series, Undergrad J Math (2006)
19. Karthikeyan, K., Chanran, C.: Existence results for functional fractional fuzzy impulsive differential equation. Int. J. Contemp. Math. Sci **6**(39), 1941–1954 (2011)
20. Kaleva, O.: Fuzzy differential equations. Fuzzy Sets Syst. **24**, 301–317 (1987)
21. Lakshmikanthm, V., Vatsala, A.S.: Basic theory of fractional differential equation. Nonlinear Anal. Theor. Meth. Appl. **69**(8), 2677–2682 (2008)

22. Najafi, N: Computing the fuzzy fractional differential transform method based on the concept of generalized Hukuhara differentiability, Conference: https://ictm.sums.ac.ir/Dorsapax/userfiles/Sub1/etelaeiye/third-international-conference-on-decision-science-ids-2018.jpg, At iran thran
23. Najafi, N.: Solving fuzzy impulsive fractional differential equations by Homotopy pertourbation method. Int. J. Math. Model. Comput. (IJM2C), **8**(3), 2–3 (2018)
24. Najafi, N., Allahviranloo, T.: Semi-analytical methods for solving fuzzy impulsive fractional differential equations. J. Intell. Fuzzy Syst. **33**(6), 3539–3560 (2017)
25. Najafi, N., Paripour, M., Lotfi, T.: A new computational method for fuzzy Laplace transforms by the differential transformation method. Math. Inverse Prob. **2**(1), 16–31 (2015)
26. Najafi, N., Allahviranloo, T.: Combining fractional differential transform method and reproducing kernel Hilbert spacemethod to solve fuzzy impulsive fractional differential equations. Comput. Appl. Math. **39**, 122 (2020). https://doi.org/10.1007/s40314-020-01140-8
27. Najafi, N.: Method for solving nonlenar initial value problems by combining homotopy perturbation and fuzzy reprodusing kernel hilbert spac methods (2018)
28. Najafi, N., Biranvand, N.: Solving fuzzy impulsive fractional differential equations by reproducing kernel Hilbert space method. Int. J. Math. Model. Comput. **10**(01), 37–56 (2020)
29. Najafi, N., Ezadi, S., Karamikabir, N.: Application of the Fourier transform via Adomian's decomposition method. J. Interpolation Approximation Sci. Comput. **2017**(2), 30–37 (2017)
30. Paripour, M., Najafi, N.: Fuzzy integration using homotopy perturbation method. J. Fuzzy Set Valued Anal. **2013**, 1–6 (2013)
31. Dirbaz, M., Dirbaz, F.: Numerical solution of impulsive fuzzy initial value problem by modified Euler method. J. Fuzzy Set Valued Anal. **1**, 50–57 (2016)
32. Paripour, M., Komak Yari, M.: Existence and uniqueness of solutions for Fuzzy quadratic integral equation of fractional order. J. Intell. Fuzzy Syst. **32**(3), 2327–2338 (2017)
33. Guo, T.L., Jiang, W.: Impulsive problems for fractional differential equations with boundary value conditions. Comput. Math. Appl. **64**, 3281–3292 (2012)
34. Dwijendra, Alka Chaddha, Pandey, N.: Approximations of solutions for an impulsive fractional differential equation with a deviated argument. Int. J. Appl. Comput. Math. **2**, 269–289 (2016). https://doi.org/10.1007/s40819-015-0059-1
35. Odibat, Z., Momani, S.H.: An algorithm for the numerical solution of differential equation of fractional order. J. Appl. Math. Inform. **26**(1), 15–27 (2008)
36. Podlubny, I.: Fractional Differential Equations. Academic Press, San Diego, CA (1999)
37. Ramesh, R.: Existence and uniqueness theorem for a solution of fuzzy impulsive differential equations. Ital. J. Pure Appl. Math. **33**, 345–366 (2014)
38. Ruban, S., Saradha, M.: Solving hybrid fuzzy fractional differential equations by improved Euler method. Math. Theor. Model. **5**(5) (2015)
39. Salahshour, S., Allahviranloo, T., Abbasbandy, S.: Solving fuzzy fractional differential equations by fuzzy Laplace transforms. Commun. Nonlinear Sci. Numer. Simul. **17**(3), 1372–1381 (2012)
40. Song, S., Wu, C.: Existence and uniqueness of solutions to Cauchy problem of fuzzy differential equations. Fuzzy Sets Syst. **110**, 55–67 (2000)
41. Stefanini, L., Bede, B.: Generalized Hukuhara differentiability of interval-valued function and interval differential equation. Nonlinear Anal. **71**, 1311–1328 (2009)

42. Thongmoon, M., Pusjuso, S.: The numerical solutions of differential transform method and the Laplace transform method for a system of differential equations. Nonlinear Anal. Hybrid Syst. **4**(3), 425–431 (2010)
43. Usero, D.: Fractional Taylor series for Caputo Fractional derivatives. Elsevier, Construction of numerical schemes (2008)
44. Vengataasalam, S., Ramesh, R.: Existence of fuzzy solutions for impulsive semilinear differential equations with nonlocal condition. Int. J. Pure Appl. Math. **95**(2), 297–308 (2014)
45. Wei, W., Yang, Y.: On some impulsive fractional differential equations in banach spaces. Opuscula Math. **30**(4), 507–525 (2010)
46. Wang, Q., Lu, D., Fang, Y.: Stability analysis of impulsive fractional differential systems. Appl. Math. Lett. **40**, 1–6 (2015)
47. Wu, H.C.: The improper fuzzy Riemann integral and its numerical integration. Inf. Sci. **111**, 109–137 (1999)
48. Zadeh, L.A.: Fuzzy sets. Inf. Control **8**, 338–353 (1965)
49. Zimmerman, H.J.: Fuzzy Set Theory and its Applications. Kluwer Academic, NewYork (1996)
50. Zhou, J.K.: Differential Transformation and its Applications for Electrical Circuits. Huazhong University Press, Wuhan, China (1986). (in Chinese)
51. Wazwaz, A.M.: A new algorithm for calculating Adomian polynomials for nonlinear operators. Appl. Math. Comput. **111**, 53–69 (2000)
52. El-kalla, I.L.: Error analysis of Adomian series solution to a class of nonlinear differential equations. Appl. Math. E-Notes **7**, 214–221 (2007)
53. Babolian. E, Biazar. J.: Solution of a system of nonlinear Volterra integral equations by Adomian decomposition method. Far East J. Math. Sci. **2**(6), 935-945 (200)

Solving Linear Systems Based on Z-Numbers

S. M. Hashemi Moosavi[1(✉)], M. A. Fariborzi Araghi[1], and S. Ezadi[2]

[1] Department of Mathematics, Central Tehran Branch, Islamic Azad University, P.O. Box 1955847881, Tehran, Iran
smr.hashemi.moosavi@gmail.com

[2] Department of Mathematics, Science and Research Branch, Islamic Azad University, Tehran, Iran

Abstract. As we know, one of the most important issues in fuzzy mathematics is to find the solution of a linear system, which may play a key role in solving an optimal response to a linear programming problem. The Z-number, which has not been invented by Zadeh (decision matrix) for a long time, can draw fuzzy statistics with a degree of balance and ranking. This matrix creates a problem that eventually becomes a linear programming problem, so then we can find the optimal solution to the problem. With the help of one of the estimation methods and techniques on the decision matrix, which is very old, it will be easy to act. Let a general linear system as $\tilde{A}X = \tilde{B}$; so there may be three states that one of the $\tilde{A}$ or $\tilde{B}$ or both of them are fuzzy. In this case, if operated with the Z-number approach, three states $\tilde{A}_Z$, $\tilde{B}_Z$ or $\tilde{A}_Z$ and $\tilde{B}_Z$ will occur, which simply after conversion and normalization to a definite and normal linear algebraic system from the methods of solving linear systems by applying Mathematica, Maple or other numerical packages, we will reach the final solution of the system, which may be the definitive optimal solution of a linear program.

Keywords: Z-numbers · Fuzzy number · Decision matrix · Linear systems

1 Introduction

It is clear that in the real world, ambiguity is a serious and stressful event. The amount of data being generated is very large, and the methods of analysis found on this data are not decisive. We have significant potential for logical analysis of data that is unspecified and inaccurate, in which case they are incomplete; so, in general, the minimum formality of this new method can be applied very clearly in this work. As we know, in the current real situation, a new linear program repairs all the undesirable parameters involved by experts. In fact, proprietary models are not accurate, and the decision maker has to conclude the contract with ambiguity and doubt in order to describe the existing processes. Thus, here

T. Allahviranloo et al. (Eds.): IDS 2020, AISC 1301, pp. 965–974, 2021.
https://doi.org/10.1007/978-3-030-66501-2_76

a new idea that could more comprehensively and reliably model the problems of linear planning could be felt, in which case for the first time Zadeh [20] with an idea in this regard, especially the Z-number hoped to come up with the idea of "Z-number" to complete its fuzzy view, and as we know throughout history in recent decades, classical fuzzy sets perform many functions in areas such as fuzzy control. Zadeh made fuzzy decisions and so on and the results were very satisfying, but not as good as all the limitations and reliability [4–8,12,13,16,21]. Compared to conventional fuzzy numbers, the Z-number has the complementary skill to illuminate human knowledge, and this Z-number can provide the interpretation of the constraints and reliability, and is much closer to reality in terms of achieving the desired end results.

However, this flexibility is not felt in the use of classical fuzzy numbers. For more on fuzzy numbers, see [1,9–11,18,19]. In fact, the Z-number can enter the issues, especially linear programming, and in particular the linear systems, with high reliability and taking into account the desired limitations, and lead us to the real solution of linear planning and systems other issues that can be addressed in the form of a linear program or linear system.

In this paper, the Z-number is used to solve linear systems and multilateral decisions, and the decision framework is presented using a planned Z-number, such as a numerical example. The rest of sections is as follows:

In the second section, the basic concepts are stated. In the third section, the main idea for solving linear systems with Z-number approach is presented. The fourth section provides a numerical example. Concluding remarks are mentioned in Sect. 3.

2 Preliminaries

Definition 1 (Triangular fuzzy numbers (TFN)). *A fuzzy number $\tilde{A}$ is a TFN [7] defined by a triad (a_1, a_2, a_3) in which the parameters a_1, a_2 and a_3 are real values and its membership function is generally defined as follows:*

$$\mu_{\tilde{A}}(x) = \begin{cases} \frac{x-a_1}{a_2-a_1} & , \quad a_1 \leqslant x \leqslant a_2, \\ \frac{a_2-x}{a_3-a_2} & , \quad a_2 \leqslant x \leqslant a_3, \\ 0 & , \quad otherwise. \end{cases} \tag{1}$$

Definition 2 (The Z-number [14]). *The Z-number is an ordered pair of fuzzy numbers that are defined as follows:*

$$Z = (\tilde{A}, \tilde{R}),$$

$\tilde{A}$ (First component): Limitations on values with real value of definite variable,
$\tilde{R}$ (Second component): A criterion for reliability of the first component.

Definition 3 (Ranking triangular fuzzy numbers). *Suppose* $\tilde{A} = (a_1, a_2, a_3)$ *and* $\tilde{B} = (b_1, b_2, b_3)$ *are two fuzzy triangular numbers. The rank of triangular fuzzy numbers* $\tilde{A}$ *and* $\tilde{B}$ *can be obtained as follows:*

$$R(\tilde{A}) = \frac{a_1 + a_2 + a_3}{3}, \quad R(\tilde{B}) = \frac{b_1 + b_2 + b_3}{3}. \tag{2}$$

Remark 1. The standard representation of multiplication $\tilde{A}$ and $\tilde{B}$ is also defined as follows:

$$\begin{aligned} R(\tilde{A} \otimes \tilde{B}) &= R(\tilde{A}) \times R(\tilde{B}) \\ &= \Big(\frac{a_1 + a_2 + a_3}{3}\Big) \times \Big(\frac{b_1 + b_2 + b_3}{3}\Big). \end{aligned} \tag{3}$$

Remark 2. We should convert expert's reliability into crisp number according to (4):

$$\text{(Conversion Factor) } \alpha = \frac{\int \mu_{\tilde{R}}(x)dx}{\int \mu_{\tilde{R}}(x)dx}. \tag{4}$$

Theorem 1 ([15])

$$E_{\tilde{A}^\alpha}(x) = \alpha E_{\tilde{A}}(x), \qquad (x \in X, \quad \alpha \in \mathbb{R}),$$

Subject to

$$\mu_{\tilde{A}^\alpha}(x) = \alpha \mu_{\tilde{A}}(x), \qquad (x \in X).$$

Theorem 2 ([15])

$$E_{\tilde{Z}'}(x) = \alpha E_{\tilde{A}}(x), \qquad (x \in \sqrt{\alpha} X),$$

Subject to

$$\mu_{\tilde{Z}'}(x) = \mu_{\tilde{A}}\Big(\frac{x}{\sqrt{\alpha}}\Big), \qquad (x \in \sqrt{\alpha} X).$$

Theorem 3 ([15])

$$E_{\tilde{Z}'}(x) = E_{\tilde{A}^\alpha}(x).$$

Theorem 4 (HM Theorem). *Each linear system based on Z-number:*

$$(Z\text{-}number)\ \tilde{A}_Z X = \tilde{B}_Z\ (Z\text{-}number), \tag{5}$$

where X is crisp, it can be converted to a definite linear system as follows:

$$(Crisp)\ \tilde{A}''_Z X = \tilde{B}''_Z\ (Crisp) \tag{6}$$

Proof. To convert a Z-number system to a definite system, it is sufficient to perform the following process for the $\tilde{A}_Z$ and $\tilde{B}_Z$ matrices:

$$(Z\text{-number})\ \tilde{A}_Z \xrightarrow{\text{(Fuzzy)}} \tilde{A}'_Z \xrightarrow{\text{(Crisp)}} \tilde{A}''_Z\ \text{(Crisp)},$$
$$(Z\text{-number})\ \tilde{B}_Z \xrightarrow{\text{(Fuzzy)}} \tilde{B}'_Z \xrightarrow{\text{(Crisp)}} \tilde{B}''_Z\ \text{(Crisp)}.$$

After this process, we are faced with a typical (definitive) linear system:

$$\text{(Crisp)}\ \tilde{A}''_Z X = \tilde{B}''_Z\ \text{(Crisp)}.$$

Therefore, instead of solving the system based on Z-number (5), it is enough to solve the definite (ordinary) (6) system.

Example 1. Assume an expert gives his opinion as follows:

$$\tilde{A} = (0.7, 0.8, 0.9, 1; 1),$$
$$\tilde{R} = (0.8, 0.9, 1; 1).$$

The expert's knowledge can be expressed to Z-number as:

$$\tilde{Z} = (\tilde{A}, \tilde{R}) = \left[(0.7, 0.8, 0.9, 1; 1), (0.8, 0.9, 1; 1)\right].$$

We should to convert expert's reliability into crisp number according to (7):

$$\begin{aligned} \alpha &= \frac{\int x\mu_R(x)dx}{\int \mu_R(x)dx} = 0.9; \\ \tilde{Z}^{\alpha} &= \{0.7, 0.8, 0.9, 1; 0.9\} \end{aligned} \tag{7}$$

Convert the weighted Z-number to regular fuzzy number according to the proposed approach:

$$\begin{aligned} \text{(Converted to the Fuzzy)}\ \tilde{Z}' &= \{\sqrt{0.9} \times 0.7, \sqrt{0.9} \times 0.8, \sqrt{0.9} \times 0.9, \sqrt{0.9} \times 1; 1\} \\ &= \{0.9487 \times 0.7, \ldots, 0.9487 \times 1; 1\} \\ &= \{0.6641, 0.759, 0.854, 0.949; 1\}. \end{aligned}$$

Note 1. The notion Z-number introduced by Zadeh in 2011 has more capability to describe the uncertain information.

3 Main Idea

The operation method (algorithm) for solving linear programming [2,3,17,18] and linear system with Z-number approach has 6 steps as follows:

Algorithm 1

Step 1. Create a K_{ij} fuzzy decision matrix as a formal component:

$$K_{ij} = Z_{ij}(\tilde{A}, \tilde{R}), \qquad (i = 1, 2, \ldots, m, \quad j = 1, 2, \ldots, n),$$

Z_{ij} : An evaluation of jth criterion from ith selection,
$\tilde{A}$: A limit for the Z-number,
$\tilde{R}$: An assurance for limitation of $\tilde{A}$ in the Z-number.
For example, the phrase "travel time is definitely important" with the number (H; VH) can be used to estimate the description of the sentence (Very High: VH).

Step 2. We turn linguistic importance into mathematical principles based on fuzzy structures. For example, if the language evaluation (H; VH) is with the Z-number, then according to the linguistic membership function, its mathematical value is as follows:

$$Z = \big((a, b, 1); (c, 1, 1)\big); \qquad (0 < a, b, c < 1).$$

Step 3. Become three fuzzy (with $\sqrt{\alpha}$ coefficient).

Step 4. The ranking function, update the Z-number to the definite number (Crisp):

$$\text{(Ranking Function) } r_{ij} = \frac{x_{ij}}{\sqrt{\sum_{k=1}^{m} x_{kj}^2}}, \qquad (i = 1, 2, \ldots, m \quad j = 1, 2, \ldots, n). \tag{8}$$

where
x_{ij} : Data's of problem, and
r_{ij} : Normalization of problem's data.

Step 5. At this stage, the decision matrix or the coefficients of the linear system as definite (non-fuzzy) numbers (Crisp numbers) becomes a linear programming problem or a conventional algebraic linear system.

Step 6. We will use one of the common methods and conventional software of Mathematica, Maple or other numerical packages to get the solution of the optimal problem or to the unique result of the linear system, In order to find the optimal result to the problem of linear programming or the unique solution of the linear system.

Note 2. Here, by providing a practical numerical example, we will perform the last steps and determine the optimal result to the linear programming problem or the unique solution of the system. For simplicity, we assume:

$$\mu_{\tilde{R}}(x) = \mu_{\tilde{A}}(x).$$

4 Numerical Experiments

Here, with a system of three equations and three unknowns with Z-number approach and applying of Algorithm 1, we present the method of solving this type of system:

$$\tilde{A}_Z X = \tilde{B}_Z; \qquad \mu_{\tilde{R}}(x) = \mu_{\tilde{A}}(x).$$

In this system, it is assumed that the coefficients ($\tilde{A}_Z$) and the constants ($\tilde{B}_Z$) are both Z-number and $x_1, x_2, x_3 \in \mathbb{R}$, $X = [x_1, x_2, x_3]$ and we have:

$$\tilde{A}_Z = \begin{bmatrix} [(9,10,12),(0.75,1,1)] & [(70,1001,120),(0.25,0.5,0.75)] & [(4,5,6),(0.5,0.75,1)] \\ [(20,24,25),(0.5,0.75,1)] & [(60,70,100),(0.75,1,1)] & [(7,8,10),(0.5,0.75,1)] \\ [(15,15,15),(0.5,0.75,1)] & [(70,80,90),(0.5,0.75,1)] & [(1,4,7),(0.5,0.75,1)] \end{bmatrix}$$

$$\tilde{B}_Z^T = [[(0.75,1,1),(0.75,1,1)]\ [(0.5,0.75,1),(0.75,1,1)]\ [(1,4,7),(0.75,1,1)]$$

Here are the six steps of Algorithm 1 to solve the Z-number system:

Step 1. $\tilde{A}_Z$ and $\tilde{B}_Z$ matrices are defined numerically and in Z-numbers.

Step 2. Linguistic Significance based on fuzzy structures, it has become a mathematical principle. For example, linguistic evaluation (VH), H with Z-number and according to the function of linguistic membership, its mathematical value is $[(1, 1, 0.75)]$ and $[(1, 0.75, 0.5)]$.

Step 3. Simplify fuzzy information in order to avoid complexity in the accuracy of the operation during the evaluation. In this regard, we calculate the modified $\tilde{A}_Z$ matrix. It should be noted that the $\tilde{B}_Z$ matrix is acceptable and does not require modification. For this purpose, the Z-number must be converted to fuzzy numbers, which is enough to calculate the value of the following equation:

$$\text{(Convert the weighted } Z\text{-number to regular fuzzy number)}\ \sqrt{\alpha} = \sqrt{\frac{\int x\mu_{\tilde{R}}(x)dx}{\int \mu_{\tilde{R}}(x)dx}}.$$

In this part, a method of transforming Z-number to classical fuzzy number is proposed according to the fuzzy expectation of fuzzy number, and a simple example is used to illustrate the procedure of the proposed approach.

Definition 4 ([15]). *Let a fuzzy set $\tilde{A}$ be defined an a universe X may be given as $\tilde{A} = \{(x, \mu_{\tilde{A}}(x)); x \in X\}$, where $\mu_{\tilde{A}} : X \to [0, 1]$ is the membership function $\tilde{A}$, $x \in X$ in $\tilde{A}$.*

$$E_{\tilde{A}}(x) = \int_x x\mu_{\tilde{A}}(x)dx. \tag{9}$$

The fuzzy expectation of a fuzzy set is denoted as (9), which is not the same as the meaning of the expectation of probability space. It can be considered as the information. Strength supporting the fuzzy set $\tilde{A}$.

The proposed method of converting Z-number to regular fuzzy number, assume a Z-number $Z = (\tilde{A}, \tilde{R})$:

$$\tilde{A} = \{(x, \mu_{\tilde{A}}(x))) : x \in [0,1]\},$$
$$\tilde{R} = \{(x, \mu_{\tilde{R}}(x))) : x \in [0,1]\},$$

where $\mu_{\tilde{A}}(x)$ is a trapezoid membership function and $\mu_{\tilde{R}}(x)$ is a triangular membership function.

I. Convert the second part (reliability) into a crisp number:

$$\alpha = \frac{\int x\mu_{\tilde{R}}(x)dx}{\int \mu_{\tilde{R}}(x)dx}.$$

II. Add the weight of the second part (reliability) Z-number can be denoted as

$$\tilde{Z}^{\alpha} = \{(x, \mu_{\tilde{A}^{\alpha}}(x)) : \mu_{\tilde{A}^{\alpha}}(x) = \alpha\mu_{\tilde{A}}(x), x \in [0,1]\}.$$

Now to multiply the Z-numbers to fuzzy numbers, just multiply $\sqrt{\alpha}$ with multiple calculations $\sqrt{\alpha}$:

$$\alpha = \frac{\int x\mu_{R}(x)dx}{\int \mu_{R}(x)dx} \approx 0.0001323.$$

so

$$\sqrt{\alpha} = \sqrt{0.0001323} \approx 0.0115,$$

and multiplying this number by the components of the Z-number matrix, the modification of the Z-number is done to the triangle fuzzy numbers, and the following matrix is obtained. For example, we have:

$$9\sqrt{\alpha} = 9\sqrt{0.0001323} = 0.1035 \approx 0.10.$$

After the necessary calculations, the modified matrix $\tilde{A}'_Z$ is obtained:

$$\tilde{A}'_Z = \begin{bmatrix} [(0.10,0.11,0.14),(0.75,1,1)] & [(0.16,0.23,0.27),(0.25,0.5,0.75)] & [(0.13,0.16,0.19),(0.5,0.75,1)] \\ [(0.23,0.27,0.28),(0.5,0.75,1)] & [(0.14,0.16,0.23),(0.75,1,1)] & [(0.22,0.25,0.32),(0.5,0.75,1)] \\ [(0.17,0.17,0.17),(0.5,0.75.1)] & [(0.16,0.18.0.20),(0.5,0.75,1)] & [(0.03,0.13,0.22),(0.5,0.75,1)] \end{bmatrix}.$$

So we show the $\tilde{B}_Z$ matrix to $\tilde{B}'_Z$ without any change.

Step 4. Here, using the ranking function, the matrix $\tilde{A}'_Z$ is used to determine the definite number of numbers (Crisp Numbers) by performing $\otimes$ to the modified $\tilde{A}''_Z$ matrix; (With definite components) we deliver. We do the same for $\tilde{B}'_Z$.

Note 3. If $\tilde{A} = (a_1, a_2, a_3)$ and $\tilde{B} = (b_1, b_2, b_3)$, are two triangular fuzzy numbers. We determine the rank of triangular fuzzy numbers $\tilde{A}$ and $\tilde{B}$ as follows:

$$\tilde{A} = \frac{a_1 + a_2 + a_3}{3},$$
$$\tilde{B} = \frac{b_1 + b_2 + b_3}{3}$$

Then; here, you can specify the "(Crisp)$\tilde{A}''_Z$" matrix:

$$(\text{Crisp})\tilde{A}''_Z = \begin{bmatrix} [(0.10,0.11,0.14)\otimes(0.75,1,1)] & [(0.16,0.23,0.27)\otimes(0.25,0.5,0.75)] & [(0.13,0.16,0.19)\otimes(0.5,0.75,1)] \\ [(0.23,0.27,0.28)\otimes(0.5,0.75,1)] & [(0.14,0.16,0.23)\otimes(0.75,1,1)] & [(0.22,0.25,0.32)\otimes(0.5,0.75,1)] \\ [(0.17,0.17,0.17)\otimes(0.5,0.75.1)] & [(0.16,0.18.0.20)\otimes(0.5,0.75,1)] & [(0.03,0.13,0.22)\otimes(0.5,0.75,1)] \end{bmatrix}.$$

In the same way, we determine (Crisp)$\tilde{B}''_Z$:

$$(\text{Crisp})\tilde{B}''_Z = [[(0.75,1,1)\otimes(0.75,1,1)]\ [(0.5,0.75,1)\otimes(0.75,1,1)]\ [(1,4,7)\otimes(0.75,1,1))]].$$

Step 5. After determining the product of the multiplications ($R(\tilde{A} \otimes \tilde{B})$), the matrices $\tilde{A}''_Z$ and $\tilde{B}''_Z$ which all the derivatives are obtuse (real) and non-fuzzy numbers are obtained. So after the necessary calculations and normalization such as:

$$R((0.10, 0.11, 0.14) \otimes (0.75, 1, 1)) = R(0.10, 0.11, 0.14) \times R(0.75, 1, 1)$$
$$= \frac{0.10 + 0.11 + 0.14}{3} \times \frac{0.75 + 1 + 1}{3} \simeq 0.11.$$

We will reach the final matrices with the following real tactics:

$$(\text{Crisp})\tilde{A}''_Z = \begin{bmatrix} \textcircled{0.11} & 0.11 & 0.12 \\ 0.19 & 0.16 & 0.19 \\ 0.13 & 0.14 & 0.09 \end{bmatrix}, \quad (\text{Crisp})\tilde{B}''_Z = \begin{bmatrix} 0.84 \\ 0.69 \\ 0.46 \end{bmatrix}.$$

So, here we face with a simple linear equation:

$$(\text{Crisp})\tilde{A}''_Z \cdot X = (\text{Crisp})\tilde{B}''_Z. \tag{10}$$

Step 6. We solve the system of linear equations with (10), which is defined in the set of real numbers $\mathbb{R}$, by one of the methods of solving or conventional software of Mathematica:

$$X = ((\text{Crisp})\tilde{A}''_Z)^{-1} \cdot ((\text{Crisp})\tilde{B}''_Z). \tag{11}$$

With the necessary calculations from (11), we will reach the final result of the system (10):

$$X = [x_1\ x_2\ x_3]^T = \begin{bmatrix} -24.75 \\ 17.50 \\ 13.66 \end{bmatrix}.$$

Example 2 (Numerical example for linear programming [2,3,17,18]*).* If we assume that for three cars M_1, M_2 and M_3 with linguistic criteria based on the idea of the Z-number and the effect of cost parameters, travel time and comfort of the $\tilde{A}_Z$ matrix is obtained. In the same way solve the system and make the necessary conversions:

$$(Z\text{-number})\ \tilde{A}_Z \xrightarrow{(\text{Fuzzy})} \tilde{A}'_Z \xrightarrow{(\text{Crisp})} \tilde{A}''_Z$$

Finally, as a linear programming problem, we formulate the following solution and solve the L_P, which has the max Z objective function and the specified linear constraints, by the simplex method, and we will achieve the optimal solution:

$$(\text{The objective function})\ \max Z = 0.11x_1 + 0.19x_2 + 0.13x_3$$

Restrictions of linear programming are:

$$\begin{aligned} 0.11x_1 + 0.16x_2 + 0.14x_3 &\leqslant 0.69, \\ 0.12x_1 + 0.19x_2 + 0.09x_3 &\leqslant 0.46 \end{aligned} \tag{12}$$

The system (12) is a linear programming system that can be solved with one of the simplest methods such as simplex and get the optimal (definite) result:

$$(x_1 = 0, \quad x_2 = 0.19, \quad x_3 = 4.71) : \max Z = 0.65.$$

Therefore, this general method for solving any linear system of equations and linear programming can be easily applied and efficient.

5 Conclusion

As we have seen, to solve a linear system with a Z-number approach and use the proposed algorithm, we were able to solve the system in three following stages:

$$(Z\text{-number system})\tilde{A}_Z X = \tilde{B}_Z \xrightarrow{(\text{Fuzzy})} \tilde{A}'_Z X = \tilde{B}'_Z \xrightarrow{(\text{Crisp})} \tilde{A}''_Z X = \tilde{B}''_Z.$$

We turned it into a definite linear equation system (set of real numbers).

By solving the obtained crisp system via one of the usual methods or software to solve the real system, we get the final result. With the same algorithm, linear systems can be operated with more than three unknowns (even for linear programming problems) and the final result to the problems can be obtained.

References

1. Allahviranloo, T., Ezadi, S.: *Z*-advanced numbers processes. Inf. Sci. **480**, 130–143 (2019)
2. Araghi, M.A.F., Hashemi Moosavi, S.M.R.: Fundamentals of Fuzzy Mathematics. Meraje Ghalam (2020)

3. Bandyopadhyay, S., Raha, S., Nayak, P.K.: Matrix game with z-numbers. Int. J. Fuzzy Logic Intell. Syst. **15**(1), 60–71 (2015)
4. Deng, Y., Chan, F.T., Wu, Y., Wang, D.: A new linguistic MCDM method based on multiple-criterion data fusion. Expert Syst. Appl. **38**(6), 6985–6993 (2011)
5. Deng, Y., Chen, Y., Zhang, Y., Mahadevan, S.: Fuzzy Dijkstra algorithm for shortest path problem under uncertain environment. Appl. Soft Comput. **12**(3), 1231–1237 (2012)
6. Deng, Y., Wu, J., Sun, X., Xu, P., Zhang, Y.: Aggregation of interval numbers in group decision making. ICIC Express Lett. **5**(4), 1057–1062 (2011)
7. Deng, Y., Zhenfu, Z., Qi, L.: Ranking fuzzy numbers with an area method using radius of gyration. Comput. Math. Appl. **51**(6–7), 1127–1136 (2006)
8. Deng, Y.L.Q.: An improved optimal fuzzy information fusion method and its application in group decision. Proc. Russ. Acad. Sci. Theory Control Syst. RAS (4), 34–43 (2005)
9. Ezadi, S., Allahviranloo, T.: New multi-layer method for Z-number ranking using hyperbolic tangent function and convex combination. Intell. Autom. Soft Comput. 1–7 (2017)
10. Ezadi, S., Allahviranloo, T.: Numerical solution of linear regression based on Z-numbers by improved neural network. Intell. Autom. Soft Comput. 1–11 (2017)
11. Ezadi, S., Allahviranloo, T., Mohammadi, S.: Two new methods for ranking of Z-numbers based on sigmoid function and sign method. Int. J. Intell. Syst. **33**(7), 1476–1487 (2018)
12. Farooq Hasan, M., Sultana, A.: A new approach to solve fuzzy linear equation $A \cdot X + B = C$. IOSR J. Math. 22–30 (2017)
13. Friedman, M., Ming, M., Kandel, A.: Fuzzy linear systems. Fuzzy Sets Syst. **96**(2), 201–209 (1998)
14. Kang, B., Wei, D., Li, Y., Deng, Y.: Decision making using Z-numbers under uncertain environment. J. Comput. Inf. Syst. **8**(7), 2807–2814 (2012)
15. Kang, B., Wei, D., Li, Y., Deng, Y.: A method of converting Z-number to classical fuzzy number. J. Inf. Comput. Sci. **9**(3), 703–709 (2012)
16. Madi, E.N., Tap, A.O.M., Osman, A.: Fuzzy TOPSIS method in the selection of investment boards by incorporating operational risks. Proc. World Congr. Eng. **1**, 291–295 (2011)
17. Namakin, A., Najafi, S.E., Fallah, M., Javadi, M.: A new evaluation for solving the fully fuzzy data envelopment analysis with z-numbers. Symmetry **10**(9), 384 (2018)
18. Revathy, M., Sudha, A.S.: Solving linear programming problem by means of fuzzy Z-numbers. J. Phys: Conf. Ser. **1139**, 012055 (2018)
19. Yager, R.R.: On Z-valuations using Zadeh's Z-numbers. Int. J. Intell. Syst. **27**(3), 259–278 (2012)
20. Zadeh, L.A.: A note on Z-numbers. Inf. Sci. **181**(14), 2923–2932 (2011)
21. Zhang, Y., Deng, X., Wei, D., Deng, Y.: Assessment of e-commerce security using AHP and evidential reasoning. Expert Syst. Appl. **39**(3), 3611–3623 (2012)

Z-Numbers for Uncertainty Formulation

Zeynab Motamedi Pour[1](✉), Tofigh Allahviranloo[2], and Mozhdeh Afshar Kermani[3]

[1] Department of Applied Mathematics, Science and Research Branch, Islamic Azad University, Tehran, Iran
Z.motamedi1989@gmail.com
[2] Faculty of Engineering and Natural Sciences, Bahcesehir University, Istanbul, Turkey
[3] Department of Applied Mathematics, Tehran North Branch, Islamic Azad University, Tehran, Iran
mog_afshar@yahoo.com

Abstract. This paper aims to formulate the uncertainty caused by the loss of information of several problems by utilizing the Z-numbers. To accomplish this aim, this paper investigates how to convert real numbers and fuzzy numbers to Z-numbers. Moreover, a novel system of the Z-linear equations is developed.

Keywords: Z-numbers · Fuzzy numbers · Z- linear systems · Ranking

1 Introduction

Overall, the decision-making process is a mental process that everyone is dealing with throughout his or her lives. The decision-making process mainly takes place in the light of an individual's attitudes, culture, perceptions, beliefs and values, personality, knowledge, and insight, in which some of them may interact with each other. In contemporary management, decision-making is introduced as the process of solving a problem at hand. Thus, the decision-making is often called as a problem-solving technique. Here, it is worthwhile to mention that in any intelligent decision-making, a rational basis for any decision-making process, based on which the formulation and implementation of actions and programs must be provided with a focus on the above elements. Unfortunately, there is no guarantee of access to complete information about these key variables in the important decisions that planners may encounter with them. More importantly, it is very difficult to fully understand these key variables in the real world. Consequently, a deep and comprehensive understanding of each of these variables should be generated using available methods and tools. In this matter, many researchers have paid attention to the fuzzy logic to exhibit uncertainty [1–5, 9]. Nevertheless, the fuzzy sets of information reliability have not appropriately been provided. Thus, resolving this challenge can be considered as a critical problem.

For the first time, Zadeh [13] identified this problem by introducing the Z-numbers concept, and then solved this problem to some extent. Recall that Z-numbers is a pair fuzzy number in the form (A, B), where A denotes a fuzzy subset of the constraints that x values can have, and B refers to a fuzzy subset of the component A confidence scale.

T. Allahviranloo et al. (Eds.): IDS 2020, AISC 1301, pp. 975–985, 2021.
https://doi.org/10.1007/978-3-030-66501-2_77

The contract (*x*, *A*, *B*) can also be introduced as a value-Z. Semantically, with the value proposition provided in [13], the variable *x* can be interpreted in such a way that the probability that *x* takes a value of *A* is equal to *B*. That is, it means as follows:

$$Prob(x \text{ is } A) \text{ is } B$$

It should be mentioned that as these numbers are directly related to quantity, it can be observed that they have more ability to describe than fuzzy numbers, which do not have this feature. In this regard, although several researchers have conducted many studies in various fields based on Z-numbers [7, 8, 10–13], how to formulate the uncertainty into Z-numbers is challenging in turn. As such, this paper aims to facilitate this aforementioned challenge. Finally, the Z-linear equation system is introduced as well.

The structure of the paper is organized as follows: In the second part, the basic concepts are stated. In the third section, the formulation of uncertainty in the form of Z-numbers is investigated. At the end, the conclusion is provided.

2 Preliminary Definitions

In this section, some definitions and notations are defined.

Definition 2.1. Fuzzy Set
A fuzzy Set *A* is defined on a universe *X* may be introduced as follows:

$$A = \{< x, \mu_A(x) > |x \in X\} \tag{1}$$

where $\mu_A : X \to [0, 1]$ is the membership function *A*. The membership value $\mu_A(x)$ describes the degree of belongingness of $x \in X$ in *A*.

Definition 2.2. A triangular fuzzy number $\tilde{A}$ can be defined by a triplet (a_1, a_2, a_3), where the membership can be determind as follows *A* triangular fuzzy number $\tilde{A} = (a_1, a_2, a_3)$ can be considered in Eq. (2).

$$\mu_{\tilde{A}}(x)\begin{cases} 0, x \in (-\infty, a_1) \\ \frac{x-a_1}{a_2-a_1}, x \in [a_1, a_2] \\ \frac{a_3-x}{a_3-a_2}, x \in [a_2, a_3] \\ 0, x \in (a_3, +\infty,) \end{cases} \tag{2}$$

Definition 2.3. Z-numbers
AZ-number [13] is introduced as an ordered pair $Z = (A, B)$, where *A* is a fuzzy number playing a rol of a fuzzy constraint on values that a random variable x may take: *x* is *A* and *B* is a fuzzy number with a membership function $\mu_B : [0, 1] \to [0, 1]$ playing a role of a fuzzy constraint on the probability measure of *A*.*p*(*A*) is *B*.

3 Mathematical Formulation of Uncertainty Using Z-Numbers

In this section, the uncertainty modeling method is developed using Z-numbers in a few examples.

3.1 Converting a Real Data to Z-Numbers

In this section, a novel method is developed to properly evaluate the uncertainty with Z-numbers for sensor data fusion in fault diagnosis. It should be noted that sensor data is a real value. The advantages of the proposed model are as follows:

1 - As the Z-number can express more vague and uncertain information than fuzzy numbers, it is suitable to measure the uncertainty in the sensor data fusion system.
2 - The reliability of a sensor is measured using a fuzzy number, which is more reasonable and robust than a crisp value.
3 - The fuzzy reliability is achieved from the original feature information of sensors that not only can decrease the information lost, but also can avoid to generate unreasonable results following improper data processing.

The following subsections will lead to form Z-number.

3.1.1 The Component A: The Measurement of Fuzziness

In fault diagnosis, considering the interference in the sensors' work environment, a single measurement is often difficult to match the true value. The sensor data may vary with the measuring time, place, and the working conditions. Therefore, feature information has fuzziness to some degree. Here, suppose a device working stably at a period of time Δt, then the Gaussian membership function of the fault feature variable can be determined based on the range of input variables. Besides, suppose X is a sample space of the input variable. Then, the membership function of the detected variable can be described as follows:

$$\mu_A(x) : X \to [0, 1], \quad x \in X \tag{3}$$

The $\mu_A(x)$ is formulated as:

$$\mu_A(x) = e^{\left(-\frac{(x-\bar{x})^2}{2\sigma^2}\right)} \tag{4}$$

where k is the number of the detected variable. Besides, the following equations represent the sample mean and the sample variance, respectively.

$$\bar{x} = \frac{1}{k}\sum_{i=1}^{k} x_i \tag{5}$$

$$\sigma = \sqrt{\frac{\sum_{i=1}^{k}(x_i - \bar{x})^2}{k-1}} \tag{6}$$

3.1.2 The Component B: The Measurement of the Fuzzy Reliability

Figure 1 displays the flow chart of the proposed method to model the component *B*. The specifics of producing fuzzy reliability are illustrated as follows.

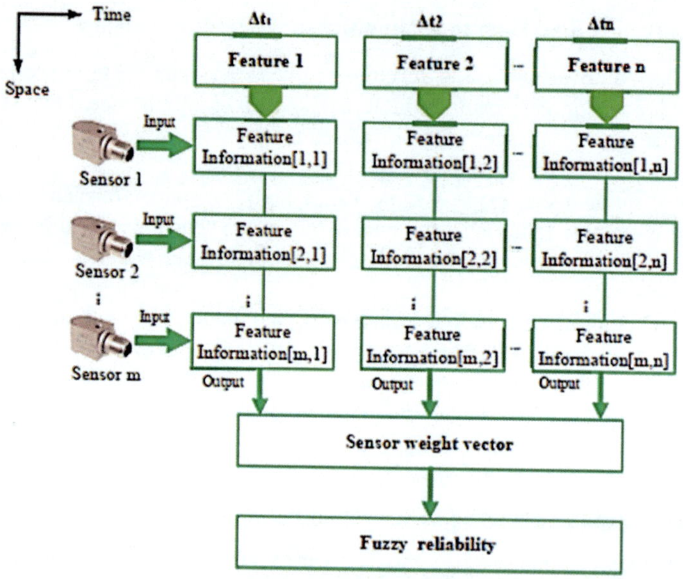

Fig. 1. The procedure of modeling component B (reliability) of sensors [6]

1. As shown in Table 1, $m \times n$ groups' feature information is determined with Eqs. (4)–(6). In this paper, the feature information refers to the membership function of the feature variable. The expression $\mu_{A_{ij}}(x)$ refers to the feature information of the *i*-th sensor *Si* with respect to the *j*-th feature variable v_j.

Table 1. The obtained $m \times n \leftarrow$ groups feature information

Sensor S_i	Feature variable v_i $(j = i \ldots n)$			
$(i = 1 \ldots m)$	v_1	v_2	…	v_n
S_1	$\mu_{A_{11}}(x)$	$\mu_{A_{12}}(x)$	…	$\mu_{A_{1n}}(x)$
S_2	$\mu_{A_{21}}(x)$	$\mu_{A_{22}}(x)$	…	$\mu_{A_{2n}}(x)$
…	…	…	…	…
S_m	$\mu_{A_{m1}}(x)$	$\mu_{A_{m2}}(x)$	…	$\mu_{A_{n1}}(x)$

2. After that, the similarities of the feature information of the m sensors are calculated. Hence, a similarity matrix of the j-th feature variable, $\mathrm{SM_j}$, can be obtained as follows:

$$\begin{matrix} & & \mu^j_{A1}(x) \ \ldots \ \mu^j_{Am(x)} \\ \mathrm{SM_j} = & \begin{matrix} \mu^j_{A1}(x) \\ \vdots \\ \mu^j_{Am(x)} \end{matrix} & \begin{bmatrix} S^j_{11} & \cdots & S^j_{11} \\ \vdots & \vdots & \vdots \\ S^j_{m1} & \ldots & S^j_{mm} \end{bmatrix} \end{matrix} \tag{7}$$

where $\mathrm{j} = 1, \ldots, \mathrm{n}$ and $S^j_{il}(i, l = 1, \ldots, m)$ represents the similarity of two pieces of feature information $\mu^j_{Ai}(x)$ and $\mu^j_{Al}(x)$ in regard to the j-th feature variable, formulated as:

$$S^j_{il} = \sup min\left\{\mu^j_{Ai}(x), \mu^j_{Al}(x)\right\} \tag{8}$$

The above equation means that the similarity can be determined by a minimum operation between two pieces of feature information. As such, it is assigned by the maximal value of the minimums, i.e., the maximum value of the intersection of the functions $\mu^j_{Ai}(x)$ and $\mu^j_{Al}(x)$.

3. Calculate the support degree and credibility degree of the sensor Si. The support degree $sup(S^j_i)$ of the sensor Si related to the j-th feature variable can be defined as:

$$Sup\left(S^j_i\right) = \sum_{l=1, l\neq i}^{m} S^j_{il} \tag{9}$$

Then, the support degree Crd^j_i of the sensor $\mathrm{S_i}$ related to the j-th feature variable is formulated as:

$$Crd^j_i = \frac{Sup\left(S^j_i\right)}{\sum_{i=1}^{m} Sup\left(S^j_i\right)} \tag{10}$$

Obviously, it can be observed that $\sum\limits_{i=1}^{m} Crd^j_i = 1$; thus, the credibility degree is actually a weight, which denotes the relative importance of the sensors.

4. Construct the sensor weight vector w_i, which is provided as follows:

$$w_i = \left(\omega^1_i, \omega^2_i, \ldots, \omega^j_i, \ldots, \omega^n_i\right) = \left(Crd^1_i, Crd^2_i, \ldots, Crd^j_i, \ldots, Crd^n_i\right) \tag{11}$$

Next, the fuzzy reliability of the sensor $\mathrm{S_i}$, $\mathrm{B_i} = (a_i, b_i, c_i)$, is defined as:

$$B_i = \left(\frac{\min\limits_j Crd^j_i}{\max\limits_{i,j}\left(Crd^j_i\right)}, \frac{\min\limits_j Crd^j_i + \max\limits_j Crd^j_i}{2\max\limits_{i,j}\left(Crd^j_i\right)}, \frac{\max\limits_j Crd^j_i}{\max\limits_{i,j}\left(Crd^j_i\right)}\right) \tag{12}$$

where (a_i, b_i, c_i) represents a triplet fuzzy number; $\max_{i,j}\left(Crd_i^j\right)$ represents the max value of the Crd_i^j. In conclusion, the Z-information of the i-th sensor with respect to the j-th feature variable can be expressed as:

$$Z_i^j = \left(A_i^j, B_i^j\right) = \left(\mu_{Ai}^j(x), \mathrm{B_i}\right) \tag{13}$$

3.2 Convert Fuzzy Data to Z-Numbers

Consider a fuzzy problem whose right side is fuzzy

$$A\tilde{X} = \tilde{b} \tag{14}$$

The generative relation (14) can also be shown as follows

$$\begin{bmatrix} a_{11} & \cdots & a_{1n} \\ \vdots & \ddots & \vdots \\ a_{n1} & \cdots & a_{nn} \end{bmatrix} \begin{bmatrix} \tilde{x}_1 \\ \vdots \\ \tilde{x}_n \end{bmatrix} = \begin{bmatrix} \tilde{b}_1 \\ \vdots \\ \tilde{b}_n \end{bmatrix} \tag{15}$$

Suppose $\tilde{b}_i = (b_{i1}, b_{i2}, b_{i3})$ is a triangular fuzzy number. Now, three methods are suggested to convert the fuzzy data of the above device.

3.2.1 Normal Distribution Method

In this section, we first convert the $\tilde{b}_i = (b_{i1}, b_{i2}, b_{i3})$ to Z - valuation. To do this, calculate the reliability of the constraint, B.

Algorithm for calculating B:

Step 1. Calculate $\bar{x}_i$;

$$\bar{x}_i = \frac{b_{i1} + b_{i2} + b_{i3}}{3} \tag{16}$$

Step 2. Calculate σ;

$$\sigma = \sqrt{\frac{\sum_{i=1}^{n} (x_i - \bar{x})^2}{n - 1}} \tag{17}$$

Step 3. Calculate $p(b_{i1} \le x \le b_{i3})$;

$$p(b_{i1} \le x \le b_{i3}) = p\left(\frac{b_{i1} - \bar{x}_i}{\sigma} \le \frac{x - \bar{x}_i}{\sigma} \le \frac{b_{i3} - \bar{x}_i}{\sigma}\right)$$

$$p(b_{i1} \le x \le b_{i3}) = p\left(z_{i1} \le z_{b_{i1}} \le z_{i2}\right) \tag{18}$$

where $z_{b_{i1}}$ is normal distribution, while both $z_{i1} = \frac{b_{i1} - \bar{x}_i}{\sigma}$ and $z_{i2} = \frac{b_{i3} - \bar{x}_i}{\sigma}$ are real numbers. Therefore, we have:

$$p\left(z_{i1} \le z_{b_{i1}} \le z_{i2}\right) = p\left(z_{b_{i1}} \le z_{i2}\right) - p\left(z_{b_{i1}} \le z_{i1}\right) \tag{19}$$

The symbol z represents the normal distribution function. The value of z_{i1} and z_{i2} are easily determined using the z table (normal distribution table in statistics and probability).

4 Z-Number Linear System

In this section, we introduce a system of linear equations based on Z-numbers (system of Z-linear equations). We define the system of Z-linear equations in the following general form.

$$AX = b \tag{20}$$

The Z-linear equation system can be divided into the following categories:

1) System of linear equations with coefficients based on Z-numbers

$$[A_{m\times n}]^Z X_{n\times 1} = [b_{m\times 1}]^Z$$

2) Linear equation device with fuzzy variables

$$A_{m\times n}\left[\tilde{X}_{n\times 1}\right]^Z = [b_{m\times 1}]^Z$$

3) Fully Z-linear equation machine

$$[A_{m\times n}]^Z\left[\tilde{X}_{n\times 1}\right]^Z = [b_{m\times 1}]^Z$$

Where $[.]^Z$ denotes the valuation symbol of Z-number.

5 Numerical Example

Example 5.1. Consider the following data (Table 2).

Now, it is desired to convert the following data into a Z-number. Therefore, with the contents of Sect. 3.1.1, we performthis conversion step by step as follows:

We calculate the mean and variance of the data. The results are reported in Table 3. Subsequently, the SM_1, SM_2 and SM_3 matrices are obtained as follows:

$$SM_1\begin{bmatrix} 1\ 0.8187\ 0.9694 \\ 0.8187\ 1\ 0.9512 \\ 0.9694\ 0.9512\ 1 \end{bmatrix}, \quad SM_2\begin{bmatrix} 1\ 0.9507\ 0.5651 \\ 0.9507\ 1\ 0.8922 \\ 0.5651\ 0.8922\ 1 \end{bmatrix}, \quad SM_2\begin{bmatrix} 1\ 0.9507\ 0.5651 \\ 0.9507\ 1\ 0.8922 \\ 0.5651\ 0.8922\ 1 \end{bmatrix}$$

In addition, we calculate the support degree and credibility degree of the sensor Si, using Eq. (9) and (10), and the obtained results are listed in Table 4.

According to Eq. (11) and (12), we have

$$Z_1^1 = \left(e^{-\frac{(x-0.1429)^2}{2\times 1.4e-7}}, (0.8225, 0.8473, 0.8721)\right)$$

$$Z_1^2 = \left(e^{-\frac{(x-0.1424)^2}{2\times 1.5e-7}}, (0.8225, 0.8473, 0.8721)\right)$$

Table 2. Experimental data of the test sample [6]

1X			2X			3X		
Sensor 1	Sensor 2	Sensor 3	Sensor 1	Sensor 2	Sensor 3	Sensor 1	Sensor 2	Sensor 3
0.1421	0.1424	0.142	0.104	0.1058	0.1068	0.163	0.161	0.1604
0.1426	0.142	0.1421	0.1046	0.117	0.1063	0.1629	0.1603	0.1609
0.1422	0.1422	0.1426	0.1052	0.1068	0.1057	0.1627	0.1605	0.1587
0.1423	0.1426	0.1421	0.1032	0.1084	0.1091	0.1626	0.1616	0.1574
0.1433	0.1431	0.1434	0.1054	0.1092	0.1094	0.1582	0.1618	0.1572
0.144	0.1428	0.1427	0.1058	0.1078	0.1067	0.1624	0.1584	0.1598
0.1439	0.1426	0.1424	0.1056	0.1076	0.1109	0.1627	0.1592	0.1597
0.1437	0.1424	0.1422	0.105	0.102	0.1111	0.1598	0.1606	0.1567
0.1436	0.1422	0.1425	0.1028	0.108	0.1112	0.1594	0.1614	0.1571
0.1432	0.1416	0.1412	0.1048	0.1076	0.1096	0.1617	0.1619	0.1566
0.1434	0.1424	0.1418	0.1078	0.106	0.1074	0.1621	0.1614	0.1578
0.1437	0.1429	0.1422	0.1056	0.1038	0.1109	0.1615	0.1609	0.1597
0.1428	0.1424	0.1436	0.106	0.105	0.1116	0.1618	0.161	0.1563
0.1424	0.1423	0.1432	0.1074	0.1048	0.111	0.162	0.1612	0.1572
0.1427	0.1421	0.1424	0.108	0.1046	0.1113	0.1615	0.1615	0.1619
0.1431	0.142	0.1434	0.1064	0.1044	0.1106	0.1611	0.1606	0.1613
0.1425	0.1423	0.1434	0.1046	0.106	0.111	0.1612	0.1604	0.1617
0.1428	0.142	0.1424	0.1054	0.106	0.1091	0.1616	0.1605	0.1604
0.1422	0.1425	0.1426	0.1036	0.1048	0.108	0.1608	0.1591	0.1608
0.1421	0.1426	0.1432	0.1032	0.1055	0.1044	0.1615	0.1611	0.1602

Table 3. Sample average and sample variance of the experimental data.

	1X			2X			3X		
Sensor 1	Sensor 2	Sensor 3	Sensor 1	Sensor 2	Sensor 3	Sensor 1	Sensor 2	Sensor 3	
Average	0.1429	0.1424	0.1426	0.1052	0.1091	0.1615	0.1607	0.1591	
Variance	4.04E − 07	1.22E − 07	3.97E − 07	2.16E − 06	4.86E − 06	1.51E − 06	8.5E − 07	3.54E − 06	

$$Z_1^3 = \left(e^{-\frac{(x-0.1426)^2}{2\times3.1e-7}}, (0.8225, 0.8473, 0.8721)\right)$$

$$Z_2^1 = \left(e^{-\frac{(x-0.1052)^2}{2\times2e-6}}, (08433, 0.9217, 1)\right)$$

$$Z_2^2 = \left(e^{-\frac{(x-0.1066)^2}{2\times8.2e-6}}, (08433, 0.9217, 1)\right)$$

Table 4. The support degree Sup and credibility degree Crd of the sensors under different features.

	1X		2X		3X	
	Sup	*Crd*	*Sup*	*Crd*	*Sup*	*Crd*
S_1	1.7881	03264	1.5158	0.3147	1.7225	0.3337
S_2	1.7699	0.3231	1.8429	0.3827	1.7996	0.3487
S_3	1.9206	0.3506	1.4573	0.3026	1.6393	0.3176

$$Z_2^3 = \left(e^{-\frac{(x-0.1091)^2}{2\times 3.7e-6}}, (08433, 0.9217, 1)\right)$$

$$Z_3^1 = \left(e^{-\frac{(x-0.1615)^2}{2\times 1.3e-6}}, (0.7911, 0.8538, 0.9164)\right)$$

$$Z_3^2 = \left(e^{-\frac{(x-0.1607)^2}{2\times 9.5e-7}}, (0.7911, 0.8538, 0.9164)\right)$$

$$Z_3^3 = \left(e^{-\frac{(x-0.1591)^2}{2\times 3.1e-6}}, (0.7911, 0.8538, 0.9164)\right)$$

Example 2. Consider the following fuzzy linear equation system

$$5\widetilde{x_1} + \widetilde{x_2} + 2\widetilde{x_3} = (8r + 2, 16 - 6r) = (2, 10, 16)$$

$$2\widetilde{x_1} + 6\widetilde{x_2} + 2\widetilde{x_3} = (10r + 2, 20 - 8r) = (2, 12, 20)$$

$$\widetilde{x_1} + 3\widetilde{x_2} + 7\widetilde{x_3} = (11r + 7, 22 - 4r) = (7, 18, 22)$$

We want to convert the above device into a system of Z-linear equations. For this purpose, we will have

$$b_1 = (2, 10, 16)$$

$$\bar{x}_1 = \frac{2 + 10 + 16}{3} = 9.33, \qquad \sigma_1 = 5.7348$$

$$p(2 \le x_1 \le 16) = p\left(\frac{2 - 9.3333}{5.7348} \le \mathcal{Z}_{b_1} \le \frac{16 - 9.3333}{5.7348}\right)$$

$$p(-1.2787 \le \mathcal{Z}_{b_1} \le 1.1624) = p(0 \le \mathcal{Z}_{b_1} \le 1.1624) + p(0 \le \mathcal{Z}_{b_1} \le 1.2787)$$
$$= 0.3770 + 0.3980 = 0.775$$

and

$$b_2 = (2, 12, 20)$$

$$\bar{x}_2 = \frac{2+12+20}{3} = 11.3333, \qquad \sigma_2 = 7.3639$$

$$p(2 \le x_2 \le 20) = p\left(\frac{2-11.3333}{7.3639} \le \mathcal{Z}_{b_2} \le \frac{20-11.3333}{7.3639}\right)$$

$$p\big(-1.2674 \le \mathcal{Z}_{b_2} \le 1.1769\big) = p(0 \le Z \le 1.1769) + p(0 \le Z \le 1.2674)$$
$$= 0.3790 + 0.3962 = 0.7752$$

$$b_3 = (7, 18, 22)$$

$$\bar{x}_3 = \frac{7+18+22}{3} = 15.6666, \qquad \sigma_3 = 6.3420$$

$$p(7 \le x_3 \le 22) = p\left(\frac{7-15.6666}{6.3420} \le \mathcal{Z}_{b_3} \le \frac{22-15.6666}{6.3420}\right)$$

$$p\big(-1.3665 \le \mathcal{Z}_{b_3} \le 0.9986\big)$$

$$p(0 \le Z \le Z_0) = p(0 \le Z \le 0.9986) + p(0 \le Z \le 1.3665) = 0.3389 + 0.4131 = 0.752$$

So,

$$5[x_1]^z + [x_2]^z + 2[x_3]^z = ((8r+2, 16-6r), 0.775) = ((2, 10, 16), 0.775)$$

$$2[x_1]^z + 6[x_2]^z + 2[x_3]^z = ((10r+2, 20-8r), 0.7752) = ((2, 12, 20), 0.7752)$$

$$[x_1]^z + 3[x_2]^z + 7[x_3]^z = ((11r+7, 22-4r), 0.752) = ((7, 18, 22), 0.752)$$

6 Concluding Remarks

In the real world, uncertainty is a common occurrence. Today's, much information on several decisions are mainly generated based on uncertainty. Some people may have a considerable ability to make rational decisions based on inconclusive information, which also have the ability to make rational decisions using inaccurate and incomplete information. In this paper, both formulating and representing this ability was carried out using the concept of Z-numbers. In this paper, the concept of Z-numbers was investigated to formulate data with uncertainty. To do so, the system of Z-linear equations was discussed.

In the following, some numerical methods were provided to solve the system of fuzzy linear equations.

References

1. Allahviranloo, T., Ghanbari, M., Hosseinzadeh, A., Haghi, E., Nuraei, R.: A note on fuzzy linear systems. Fuzzy Sets Syst. **177**, 87–92 (2011)
2. Allahviranloo, T.: Numerical methods for fuzzy system of linear equations. Appl. Math. Comput. **155**, 493–502 (2004)
3. Allahviranloo, T., Ghanbari, M.: Solving fuzzy linear systems by Homotopy perturbation method. Int. J. Comput. Cogn. **8**(2), 24–30 (2010)
4. Allahviranloo, T., Afshar Kermani, M.: Solution of a fuzzy system of linear equation. Appl. Math. Comput. **175**, 519–531 (2006)
5. Asady, B., Abbasbandy, S., Alavi, M.: Fuzzy general linear systems. Appl. Math. Comput. **169**, 34–40 (2005)
6. Wen, C., Xu, X.: Theories and Applications in Multi-Source Uncertain Information Fusion—Fault Diagnosis and Reliability Evaluation. Beijing Science Press, Beijing (2012)
7. Ezadi, S., Allahviranloo, T.: Two new methods for ranking of Z-numbers based on sigmoid function and sign method. Int. J. Intell. Syst. **33**(7), 1476–1487 (2018)
8. Ezadi, S., Allahviranloo, T.: New multi-layer method for Z-number ranking using hyperbolic tangent function and convex combination. Intell. Autom. Soft Comput. **24**(1), 218–221 (2018)
9. Friedman, M., Ming, M., Kandel, A.: Fuzzy linear systems fuzzy sets and systems **96**, 201–209 (1998)
10. Jiang, W., Xie, Ch., Luo, Yu., Tang, Y.: Ranking Z-numbers with an improved ranking method for generalized fuzzy numbers. J. Intell. Fuzzy Syst. **32**(3), 1931–1943 (2017)
11. Shalila Bhanu, M., Velammal, G.: Operation on Zadeh's Z-numbers, May–June 2015, pp. 88–94 (2015)
12. Yager, R.: On Z-valuations using Zadeh's Z-numbers. Int. J. Intell. Syst. **27**, 259–278 (2012)
13. Zadeh, L.A.: A note on Z-numbers. Inf. Sci. **181**, 2923–2932 (2011)

Author Index

T. Allahviranloo et al. (Eds.): IDS 2020, AISC 1301, pp. 987–989, 2021.
https://doi.org/10.1007/978-3-030-66501-2